FORMULAS IN GEOMETRY

Rectangle

$A = lw$

$P = 2l + 2w$

Triangle

$A = \frac{1}{2}bh$

Heron's Formula:

$A = \sqrt{s(s-a)(s-b)(s-c)}$

where $s = \frac{a+b+c}{2}$

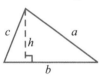

Circle

$A = \pi r^2$

$C = 2\pi r$

Trapezoid

$A = \frac{1}{2}h(b+c)$

Right Cylinder

$V = (Area\ of\ Base)h$

Rectangular Prism

$V = lwh$

$SA = 2lh + 2wh + 2lw$

Rectangular Pyramid

$V = \frac{1}{3}lwh$

Sphere

$V = \frac{4}{3}\pi r^3$

$SA = 4\pi r^2$

Right Circular Cylinder

$V = \pi r^2 h$

$SA = 2\pi r^2 + 2\pi rh$

Cone

$V = \frac{1}{3}\pi r^2 h$

PROPERTIES OF ABSOLUTE VALUE

For all real numbers a and b:

$|a| \geq 0$ \qquad $|-a| = |a|$

$a \leq |a|$ \qquad $|ab| = |a||b|$

$\left|\dfrac{a}{b}\right| = \dfrac{|a|}{|b|}, \ b \neq 0$

$|a+b| \leq |a|+|b|$ (the triangle inequality)

PROPERTIES OF EXPONENTS AND RADICALS

$a^n \cdot a^m = a^{n+m}$ \qquad $\left(a^n\right)^m = a^{nm}$

$\dfrac{a^n}{a^m} = a^{n-m}$ \qquad $(ab)^n = a^n b^n$

$a^{-n} = \dfrac{1}{a^n}$ \qquad $\left(\dfrac{a}{b}\right)^n = \dfrac{a^n}{b^n}$

$a^{\frac{1}{n}} = \sqrt[n]{a}$ \qquad $a^{\frac{m}{n}} = \sqrt[n]{a^m} = \left(\sqrt[n]{a}\right)^m$

$\sqrt[n]{ab} = \sqrt[n]{a}\sqrt[n]{b}$ \qquad $\sqrt[n]{\dfrac{a}{b}} = \dfrac{\sqrt[n]{a}}{\sqrt[n]{b}}$

$\sqrt[m]{\sqrt[n]{a}} = \sqrt[mn]{a}$

SPECIAL PRODUCT FORMULAS

$(A - B)(A + B) = A^2 - B^2$

$(A + B)^2 = A^2 + 2AB + B^2$

$(A - B)^2 = A^2 - 2AB + B^2$

FACTORING SPECIAL BINOMIALS

$A^2 - B^2 = (A - B)(A + B)$

$A^3 - B^3 = (A - B)(A^2 + AB + B^2)$

$A^3 + B^3 = (A + B)(A^2 - AB + B^2)$

THE QUADRATIC FORMULA

The solutions of the equation $ax^2 + bx + c = 0$ are:

$$x = \frac{-b \pm \sqrt{b^2 - 4ac}}{2a}$$

DISTANCE FORMULA

$$d = \sqrt{(x_2 - x_1)^2 + (y_2 - y_1)^2}$$

MIDPOINT FORMULA

$$\left(\frac{x_1 + x_2}{2}, \frac{y_1 + y_2}{2}\right)$$

SLOPE OF A LINE

$$m = \frac{y_2 - y_1}{x_2 - x_1}$$

Horizontal lines $y = c$ have slope 0.

Vertical lines $x = c$ have undefined slope.

PARALLEL AND PERPENDICULAR LINES

Given a line with slope m:

slope of parallel line $= m$

slope of perpendicular line $= -\dfrac{1}{m}$

FORMS OF LINEAR EQUATIONS

Standard form: $ax + by = c$

Slope-intercept form: $y = mx + b$

Point-slope form: $y - y_1 = m(x - x_1)$

STANDARD FORM OF A CIRCLE

The standard form of the equation for a circle of radius r and center (h, k) is:

$$(x - h)^2 + (y - k)^2 = r^2$$

COMPOUND INTEREST

An investment of P dollars, compounded n times per year at an annual interest rate of r, has a value after t years of:

$$A(t) = P\left(1 + \frac{r}{n}\right)^{nt}$$

An investment compounded continuously has an accumulated value of $A(t) = Pe^{rt}$.

PROPERTIES OF LOGARITHMS

Let a be a positive real number not equal to 1, let x and y be positive real numbers, and let r be any real number:

$\log_a x = y$ and $x = a^y$ are equivalent

$\log_a 1 = 0$ $\qquad\qquad$ $\log_a a = 1$

$\log_a(a^x) = x$ $\qquad\qquad$ $a^{\log_a x} = x$

$\log_a(xy) = \log_a x + \log_a y$

$\log_a\left(\dfrac{x}{y}\right) = \log_a x - \log_a y$

$\log_a(x^r) = r\log_a x$

CHANGE OF BASE FORMULA

For $a, b, x > 0; a, b \neq 1$:

$$\log_b x = \frac{\log_a x}{\log_a b}$$

DETERMINANT OF A 2×2 MATRIX

The determinant of a 2×2 matrix A is given by the formula:

$$|A| = a_{11}a_{22} - a_{21}a_{12}$$

PROPERTIES OF SIGMA NOTATION

For sequences $\{a_n\}$ and $\{b_n\}$ and a constant c:

$$\sum_{i=1}^{n}(a_i + b_i) = \sum_{i=1}^{n}a_i + \sum_{i=1}^{n}b_i \qquad \sum_{i=1}^{n}ca_i = c\sum_{i=1}^{n}a_i$$

$$\sum_{i=1}^{n}a_i = \sum_{i=1}^{k}a_i + \sum_{i=k+1}^{n}a_i \text{ (for any } 1 \leq k \leq n-1)$$

SUMMATION FORMULAS

$$\sum_{i=1}^{n}1 = n \qquad\qquad \sum_{i=1}^{n}i = \frac{n(n+1)}{2}$$

$$\sum_{i=1}^{n}i^2 = \frac{n(n+1)(2n+1)}{6} \qquad \sum_{i=1}^{n}i^3 = \frac{n^2(n+1)^2}{4}$$

SEQUENCES AND SERIES

Arithmetic

For an arithmetic sequence $\{a_n\}$ with common difference d:

General term:

$$a_n = a_1 + (n-1)d$$

Partial sum: $S_n = na_1 + d\left(\dfrac{(n-1)n}{2}\right) = \left(\dfrac{n}{2}\right)(a_1 + a_n)$

Geometric

For a geometric sequence $\{a_n\}$ with common ratio r:

General term:

$$a_n = a_1 r^{n-1}$$

Partial sum: $S_n = \dfrac{a_1(1 - r^n)}{1 - r}$, if $r \neq 0, 1$

Infinite sum: $S = \sum_{n=0}^{\infty}a_1 r^n = \dfrac{a_1}{1-r}$, if $|r| < 1$

PERMUTATION FORMULA

$$_nP_k = \frac{n!}{(n-k)!}$$

COMBINATION FORMULA

$$_nC_k = \binom{n}{k} = \frac{n!}{k!(n-k)!}$$

BINOMIAL THEOREM

$$(A + B)^n = \sum_{k=0}^{n}\binom{n}{k}A^{n-k}B^k$$

COMPLEX NUMBERS & DEMOIVRE'S THEOREM

$z = a + bi$ $\qquad\qquad$ $|z| = \sqrt{a^2 + b^2}$

$\tan\theta = \dfrac{b}{a}$

$z = |z|(\cos\theta + i\sin\theta)$

$z^n = |z|^n(\cos n\theta + i\sin n\theta)$ $\qquad$ $z^n = |z|^n e^{in\theta}$

$$w_k = |z|^{\frac{1}{n}}\left[\cos\left(\frac{\theta + 2k\pi}{n}\right) + i\sin\left(\frac{\theta + 2k\pi}{n}\right)\right]$$

$$w_k = |z|^{\frac{1}{n}}e^{i\left(\frac{\theta + 2k\pi}{n}\right)}$$

where $k = 0, 1, ..., n-1$

PRECALCULUS

Second Edition

PAUL SISSON

Editor: Margaret Gibbs
Assistant Editor: Danielle C. Bess
Project Manager: Kimberly Cumbie
Vice President, Research and Development: Marcel Prevuznak
Editorial Assistants: Susan Fuller, Barbara Miller, Claudia Vance, Nina Waldron
Layout Design: Tracy Carr, Nancy Derby, Rachel A. I. Link, Jennifer Moran, Tee Jay Zajac
Layout: Cenveo® Publisher Services
Art: Cenveo® Publisher Services, Kristina Feczer, Ayvin Samonte
Cover Design: Tee Jay Zajac

Photograph Credits:
BigStockPhoto.com, iStockPhoto.com, and Digital Vision with the exception of:
298: NASA

A division of Quant Systems, Inc.
546 Long Point Road, Mount Pleasant, SC 29464

Printed in the United States of America
Library of Congress Control Number: 2013940891

ISBN: 978-1-938891-30-4

Table of Contents

Preface vii

Number Systems and Equations and Inequalities of One Variable

1.1 Real Numbers and Algebraic Expressions 3
1.2 Properties of Exponents and Radicals 19
1.3 Polynomials and Factoring 39
1.4 The Complex Number System 52
1.5 Linear Equations in One Variable 60
1.6 Linear Inequalities in One Variable 71
1.7 Quadratic Equations 81
1.8 Rational and Radical Equations 98
Chapter 1 Project 116
Chapter 1 Summary 117
Chapter 1 Review 123
Chapter 1 Test 129

Introduction to Equations and Inequalities of Two Variables

2.1 The Cartesian Coordinate System 135
2.2 Linear Equations in Two Variables 149
2.3 Forms of Linear Equations 158
2.4 Parallel and Perpendicular Lines 169
2.5 Linear Inequalities in Two Variables 179
2.6 Introduction to Circles 189
Chapter 2 Project 196
Chapter 2 Summary 197
Chapter 2 Review 200
Chapter 2 Test 204

Relations, Functions, and Their Graphs

3.1 Relations and Functions 209
3.2 Linear and Quadratic Functions 225
3.3 Other Common Functions 240
3.4 Variation and Multivariable Functions 251
3.5 Transformations of Functions 260
3.6 Combining Functions 274
3.7 Inverses of Functions 287
Chapter 3 Project 298
Chapter 3 Summary 299
Chapter 3 Review 302
Chapter 3 Test 307

Polynomial Functions

4.1 Introduction to Polynomial Equations and Graphs 311
4.2 Polynomial Division and the Division Algorithm 327
4.3 Locating Real Zeros of Polynomials 340
4.4 The Fundamental Theorem of Algebra 351
4.5 Rational Functions and Rational Inequalities 363
Chapter 4 Project 380
Chapter 4 Summary 381
Chapter 4 Review 384
Chapter 4 Test 390

Exponential and Logarithmic Functions

5.1 Exponential Functions and Their Graphs 395
5.2 Applications of Exponential Functions 404
5.3 Logarithmic Functions and Their Graphs 417
5.4 Properties and Applications of Logarithms 428
5.5 Exponential and Logarithmic Equations 441
Chapter 5 Project 450
Chapter 5 Summary 451
Chapter 5 Review 453
Chapter 5 Test 457

Trigonometric Functions

6.1 Radian and Degree Measure of Angles...461
6.2 Trigonometric Functions of Acute Angles...477
6.3 Trigonometric Functions of Any Angle 492
6.4 Graphs of Trigonometric Functions 506
6.5 Inverse Trigonometric Functions 524
Chapter 6 Project 537
Chapter 6 Summary 539
Chapter 6 Review 543
Chapter 6 Test 547

Trigonometric Identities and Equations

7.1 Fundamental Identities and Their Uses 553
7.2 Sum and Difference Identities 563
7.3 Product-Sum Identities..576
7.4 Trigonometric Equations..587
Chapter 7 Project 597
Chapter 7 Summary 598
Chapter 7 Review 601
Chapter 7 Test 604

Additional Topics in Trigonometry

8.1 The Law of Sines and the Law of Cosines...609
8.2 Polar Coordinates and Polar Equations...626
8.3 Parametric Equations 640
8.4 Trigonometric Form of Complex Numbers..652
8.5 Vectors in the Cartesian Plane..665
8.6 The Dot Product and Its Uses 677
Chapter 8 Project 688
Chapter 8 Summary 690
Chapter 8 Review 694
Chapter 8 Test 699

Conic Sections

9.1 The Ellipse 705
9.2 The Parabola 720
9.3 The Hyperbola 730
9.4 Rotation of Conics 743
9.5 Polar Form of Conic Sections 755
Chapter 9 Project 768
Chapter 9 Summary 769
Chapter 9 Review 771
Chapter 9 Test 776

Systems of Equations

10.1 Solving Systems by Substitution and Elimination 781
10.2 Matrix Notation and Gaussian Elimination 795
10.3 Determinants and Cramer's Rule 810
10.4 The Algebra of Matrices 823
10.5 Inverses of Matrices 833
10.6 Partial Fraction Decomposition 844
10.7 Linear Programming 854
10.8 Nonlinear Systems of Equations and Inequalities 863
Chapter 10 Project 875
Chapter 10 Summary 876
Chapter 10 Review 879
Chapter 10 Test 886

An Introduction to Sequences, Series, Combinatorics, and Probability

11.1 Sequences and Series 891
11.2 Arithmetic Sequences and Series 906
11.3 Geometric Sequences and Series 915
11.4 Mathematical Induction 928
11.5 An Introduction to Combinatorics 939
11.6 An Introduction to Probability 955
Chapter 11 Project 966
Chapter 11 Summary 967
Chapter 11 Review 970
Chapter 11 Test 975

Answer Key 979
Index 1061

Introduction

Why do students study precalculus? Is it because of an innate love of math? For those with such a love, for those who have glorious visions of a world beyond algebra, precalculus unlocks the door to calculus and higher mathematics. For other students, those for whom calculus is a tool and not necessarily a thing of beauty, precalculus provides a solid foundation for successfully tackling the subject.

Whether you dream of integrals and derivatives or perceive yourself to be merely laying the groundwork for success in later math, this book prepares you for the next step. Its primary purpose is to provide you with the skills and concepts necessary to achieve a mastery of calculus. If you are coming from a course with a title such as "College Algebra" or "Algebra," you will encounter many familiar topics in this book. You will also discover new concepts that are not usually included in college algebra courses, such as the elements of trigonometry.

The book begins with two chapters that will ease the transition from your previous courses. While you may find a few topics that are unfamiliar, most of the material is included to help refresh your memory and for you to have as a reference throughout the course. The rest of the book is devoted to covering concepts from algebra in greater depth and introducing new topics that will prepare you for calculus.

This second edition of Paul Sisson's *Precalculus* offers a straightforward writing style, updated layout elements and graphics, a new emphasis on graphing, and notes to students about the techniques and thought processes required when approaching a problem. Whether calculus is near the beginning or end of your math experience, best wishes to you on this step in your journey.

Features

Chapter Openers

Each chapter begins with a list of sections and an engaging preview of an application appearing in the chapter.

Historical Contexts

Each chapter includes a brief introduction to the historical context of the math that follows. Mathematics is a human endeavor, and knowledge of how and why a particular idea developed is of great help in understanding it. Too often, math is presented in cold, abstract chunks completely divorced from the rest of reality. While a (very) few students may be able to master material this way, most benefit from an explanation of how math ties into the rest of what people were doing at the time it was created.

Topics

Each section begins with a list of topics. These concise objectives are a helpful guide for both reference and class preparation.

Chapter

5

Exponential and Logarithmic Functions

5.1   Exponential Functions and Their Graphs
5.2   Applications of Exponential Functions
5.3   Logarithmic Functions and Their Graphs
5.4   Properties and Applications of Logarithms
5.5   Exponential and Logarithmic Equations
      Chapter 5 Project
      Chapter 5 Summary
      Chapter 5 Review
      Chapter 5 Test

By the end of this chapter you should be able to:

What if you had to predict the amount of time it will take for the earth's population to reach 20 billion (ignoring the effect of limited resources on population growth)? How would you perform this calculation?

By the end of this chapter, you'll be able to apply the skills regarding exponential and logarithmic functions to estimate the behavior of growing populations, interest-earning accounts, and decaying elements. On page 448 you'll encounter a problem like the population problem given above. You'll master this type of problem using tools such as the Summary of Logarithmic Properties, found on page 441.

Introduction

This chapter introduces the [...] that underlies most of the [...]

Many problems that we s[...] with two variables. The ex[...] of a two-dimensional syste[...] situation graphically, but i[...] to evolve. Some of the g[...] philosopher René Descar[...]

Descartes

field of analytic geometry [...] appendix to a volume of scientific philosophy, Descartes laid out the basic principles by which algebraic problems could be construed as geometric problems, and the methods by which solutions to the geometric problems could be interpreted algebraically. As many later mathematicians expanded upon Descartes' work, analytic geometry came to be an indispensable tool in understanding and solving problems of both an algebraic and geometric nature.

In this chapter we will use the Cartesian coordinate system, named in honor of Descartes, primarily to study linear equations and inequalities in two variables. As we will see, a graph is one of the best ways to describe the solutions of such problems. Sections 2.2

The Ellipse

TOPICS

1. Overview of conic sections
2. The standard form of an ellipse
3. Planetary orbits

TOPIC

Overview of Conic Sections

The three types of conic sections—ellipses, parabolas, and hyperbolas—are so named because all three types of curves arise from intersecting a plane with a circular cone. As shown in Figure 1, an **ellipse** is a closed curve resulting from the intersection of a cone with a plane that intersects only one *nappe* of the cone (the part of the cone on one side of the vertex). A **parabola** results from intersecting a cone with a plane that is parallel to a line on the surface of the cone passing through the vertex (a parabola also intersects only one nappe). Finally, a **hyperbola** is the intersection of a cone with a plane that intersects both nappes. In each case, we

Definitions, Properties, Procedures, and Theorems

All definitions and theorems are clearly identified and set off in blue boxes that are highly visible and easily found again. Important terms appear in bold print when first defined, and other useful terms appear in italic font.

DEFINITION

Symmetry of Equations

We say that an equation in x and y is **symmetric with respect to**:

1. The **y-axis** if replacing x with $-x$ results in an equivalent equation
2. The **x-axis** if replacing y with $-y$ results in an equivalent equation
3. The **origin** if replacing x with $-x$ and y with $-y$ results in an equivalent equation

PROPERTIES

Geometric Meaning of Multiplicity

If c is a real zero of multiplicity k of a polynomial p (alternatively, if $(x-c)^k$ is a factor of p), the graph of p will touch the x-axis at $(c, 0)$ and:

PROCEDURE

Formulas of Inverse Functions

Let f be a one-to-one function, and assume that f is defined by a formula. To find a formula for f^{-1}, perform the following steps:

Step 1: Replace $f(x)$ in the definition of f with the variable y. The result is an

THEOREM

Composition of Functions and Inverses

Given a function f and its inverse f^{-1}, the following statements are true.

$$f\left(f^{-1}(x)\right) = x \text{ for all } x \in \mathrm{Dom}\left(f^{-1}\right), \text{ and}$$

$$f^{-1}\left(f(x)\right) = x \text{ for all } x \in \mathrm{Dom}(f).$$

Cautions

Many common errors are pointed out, along with how to correct them. These are set apart in bright red boxes.

CAUTION!

Errors in working with logarithms often arise from incorrect recall of the logarithmic properties. The table below highlights some common mistakes.

Incorrect Statements	Correct Statements
$\log_a(x+y) = \log_a x + \log_a y$	$\log_a(xy) = \log_a x + \log_a y$
$\log_a(xy) = (\log_a x)(\log_a y)$	$\log_a(xy) = \log_a x + \log_a y$
$\dfrac{\log_a x}{\log_a y} = \log_a x - \log_a y$	$\log_a\left(\dfrac{x}{y}\right) = \log_a x - \log_a y$
$\dfrac{\log_a x}{\log_a y} = \log_a\left(\dfrac{x}{y}\right)$	$\log_a\left(\dfrac{x}{y}\right) = \log_a x - \log_a y$
$\dfrac{\log_a(xz)}{\log_a(yz)} = \dfrac{\log_a x}{\log_a y}$	$\dfrac{\log_a(xz)}{\log_a(yz)} = \dfrac{\log_a x + \log_a z}{\log_a y + \log_a z}$

We will not always be replacing the arguments of functions with numbers. In many instances, we will have reason to replace the argument of a function with another variable or possibly a more complicated algebraic expression. Keep in mind that this just involves substituting something for the placeholder used in defining the function.

Numerous Examples

Each section contains many examples that illustrate the concepts presented and the skills to be mastered. The examples are clearly set off from the accompanying text in green boxes, and the exercises refer the student to the relevant examples to study.

EXAMPLE 5

Evaluating Functions

Given the function $f(x) = 3x^2 - 2$, evaluate:

a. $f(a)$ **b.** $f(x+h)$ **c.** $\dfrac{f(x+h)-f(x)}{h}$

Note:
The expression in part c. of this example is called the difference quotient of a function, and is used heavily in calculus.

Solutions:

a. $f(a) = 3a^2 - 2$

This is just a matter of replacing x with a.

b. $f(x+h) = 3(x+h)^2 - 2$

$$= 3(x^2 + 2xh + h^2) - 2$$

$$= 3x^2 + 6xh + 3h^2 - 2$$

Here we replace x with $x+h$ and simplify the result.

c. $\dfrac{f(x-h)-f(x)}{h} = \dfrac{(3x^2 + 6xh + 3h^2 - 2) - (3x^2 - 2)}{h}$

$$= \dfrac{6xh + 3h^2}{h}$$

$$= \dfrac{h(6x+3h)}{h}$$

$$= 6x + 3h$$

We can use the result from above in simplifying this expression.

Simplify.

Factor out h, so that we can cancel out the h in the denominator.

Applications

Many exercises and examples illustrate practical applications, keeping students engaged.

EXAMPLE 5

Calculating Distance

The distance from Shreveport, LA to Austin, TX by one route is 325 miles. If Kevin made the trip in five and a half hours, what was his average speed?

Texas

Austin

Shreveport

Louisiana

Note:
With application or mathematical modeling problems, it often helps to list the variables in the problem. As you read through the problem, fill in the values you know, so you can determine which variable you need to find.

Solution:

We know that $d = 325$ miles and $t = 5.5$ hours, and need to solve for r.

$$d = rt$$
$$325 = 5.5r$$

Chapter Project

Each project describes a plausible scenario related to the concepts of the chapter, and is suitable for an individual or group assignment.

Chapter 6 Project

Trigonometric Applications

At the Paris Observatory in 1851, Jean Foucault used a long pendulum to prove that the Earth is rotating. As it swings, the pendulum appears to change its path. However, it is not the pendulum that changes path, but the room rotating underneath it. At the North Pole the Earth revolves 360° underneath the pendulum over 24 hours. The path of a pendulum at the equator does not revolve at all; instead, the pendulum travels in a huge circle while the Earth spins. At points between the two, the pendulum cannot show how far it travels, but it can show how much the Earth is revolving underneath it. To calculate how much the Earth revolves in a particular location, use the following equation:

Earth's axis

path of pendulum

Foucault Pendulum

$$\text{degrees of revolution} = 360° \sin (\text{latitude of location})$$

Location	Latitude	Location	Latitude
United Nations, NY	40°44′58″	St. Isaac's Cathedral, Russia	59°53′02″

Technology Topics

Many sections include a Technology Topic, which demonstrates how to use a graphing calculator (with an emphasis on the TI-84 Plus) to study the concepts and problems found in the section.

TOPIC 　　　**Graphing Inequalities**

Just as we can use a calculator to graph linear equations, we can also graph linear inequalities like $y < -3x + 5$ on a calculator. Press ▬▬ and type in the right-hand side of the inequality. Pressing ▬▬▬ would display the line $y = -3x + 5$, which is the boundary line of the solution set. If we test the point $(0,0)$, we find that it is a solution to the inequality, so we know to shade the half-plane below the boundary line. On the ▬▬ screen, use the left arrow to move the cursor to the left of the Y1, where there is a small diagonal line. Press ENTER until that line appears with shading like in the screen below.

With that selection, pressing ▬▬▬ will display the boundary line with the half-plane below it shaded, which is the solution to the inequality.

Exercises

Sketch the graphs of the following hyperbolas, using asymptotes as guides. Determine the coordinates of the foci in each case. See Examples 1 and 2.

1. $\dfrac{(x+3)^2}{4} - \dfrac{(y+1)^2}{9} = 1$ **2.** $\dfrac{(y-2)^2}{25} - \dfrac{(x+2)^2}{9} = 1$

3. $4y^2 - x^2 - 24y + 2x = -19$ **4.** $x^2 - 9y^2 + 4x + 18y - 14 = 0$

5. $9x^2 - 25y^2 = 18x - 50y + 241$ **6.** $9x^2 - 16y^2 + 116 = 36x + 64y$

7. $\dfrac{x^2}{16} - \dfrac{(y-2)^2}{4} = 1$ **8.** $\dfrac{(y-1)^2}{9} - (x+3)^2 = 1$

9. $9y^2 - 25x^2 - 36y - 100x = 289$ **10.** $9x^2 + 18x = 4y^2 + 27$

11. $9x^2 - 16y^2 - 36x + 32y - 124 = 0$ **12.** $x^2 - y^2 + 6x - 6y = 4$

13. $\dfrac{(y-2)^2}{64} - \dfrac{(x+7)^2}{49} = 1$ **14.** $\dfrac{(y-4)^2}{49} - \dfrac{(x+2)^2}{16} = 1$

15. $\dfrac{(x+1)^2}{64} - \dfrac{(y+7)^2}{4} = 1$ **16.** $\dfrac{(x+10)^2}{16} - \dfrac{(y+8)^2}{25} = 1$

Find the center, foci, and vertices of each hyperbola that the equation describes.

17. $\dfrac{(x+3)^2}{4} - \dfrac{(y-2)^2}{9} = 1$ **18.** $\dfrac{(y-2)^2}{16} - \dfrac{(x+1)^2}{9} = 1$

19. $3(x-1)^2 - (y+4)^2 = 9$ **20.** $(y-2)^2 - 2(x-4)^2 = 4$

21. $(x+2)^2 - 5(y-1)^2 = 25$ **22.** $6(y+2)^2 - (x+1)^2 = 12$

23. $2x^2 + 12x - y^2 - 2y + 9 = 0$ **24.** $y^2 - 9x^2 + 6y + 72x - 144 = 0$

25. $x^2 - 4y^2 - 2x = 0$ **26.** $4y^2 - x^2 + 32y + 2x + 47 = 0$

27. $4x^2 - y^2 - 64x + 10y + 167 = 0$ **28.** $4x^2 - 9y^2 - 36y - 72 = 0$

g.

h.

Chapter Summary

Each chapter ends with a concise summary of the concepts learned and the skills acquired, arranged by section and topic.

Chapter Review

and

Chapter Test

Immediately following each Chapter Summary is a Chapter Review and Chapter Test. Each presents several more problems pertaining to the major ideas of the chapter.

Chapter Summary

A summary of concepts and skills follows each chapter. Refer to these summaries to make sure you feel comfortable with the material in the chapter. The concepts and skills are organized according to the section title and topic title in which the material is first discussed.

7.1: Fundamental Identities and Their Uses

Previously Encountered Identities
• Reciprocal identities, quotient identities, cofunction identities, period identities, even identities, odd identities, and Pythagorean Identities

---

Chapter Review

Section 8.1

Solve the following problem.

1. The base of a 25-ft ladder is positioned 7 ft away from an office building situated on a slight hill, and the ladder and ground form a 62° angle. At what angle and at what height does the ladder touch the building?

Construct a triangle, if possible, using the following information.

Chapter Test

Use trigonometric identities to simplify the expressions. There may be more than one correct answer.

1. $\sin^2(-x) - 5\sec^2 x \cot^2 x$

2. $\dfrac{\tan\beta\cos\beta}{\sin(-\beta)}$

Verify the identities.

3. $1 + \sin^2\theta = \csc^2\theta - \cot^2\theta + \sin^2\theta$

4. $(\sin^2 x)(\csc^2 x - 1) = \cos^2 x$

Use the suggested substitution to rewrite the given expression as a trigonometric expression. For all problems, assume $0 \le \theta \le \dfrac{\pi}{2}$.

5. $\sqrt{x^2 + 169}$, $\dfrac{x}{13} = \tan\theta$

6. $\sqrt{128 - 2x^2}$, $x = 8\sin\theta$

Show how the identity below follows from the first Pythagorean Identity.

7. $\tan^2 x + 1 = \sec^2 x$

Use the sum and difference identities to determine the exact value of each of the following expressions.

8. $\sin\dfrac{25\pi}{12}$

9. $\tan 165°$

Use the sum and difference identities to rewrite each of the following expressions as a trigonometric function of a single number, and then evaluate the result.

10. $\cos 131° \cos 28° + \sin 131° \sin 28°$

11. $\dfrac{\tan\dfrac{5\pi}{8} + \tan\dfrac{\pi}{4}}{1 - \tan\dfrac{5\pi}{8}\tan\dfrac{\pi}{4}}$

Formula Sheets

Three pages in the front of the text and three pages in the back of the text detail the most important formulas, theorems, graphs, and identities covered in precalculus.

SUM AND DIFFERENCE IDENTITIES

$\sin(u+v) = \sin u \cos v + \cos u \sin v$

$\sin(u-v) = \sin u \cos v - \cos u \sin v$

$\cos(u+v) = \cos u \cos v - \sin u \sin v$

$\cos(u-v) = \cos u \cos v + \sin u \sin v$

$\tan(u+v) = \dfrac{\tan u + \tan v}{1 - \tan u \tan v}$ $\tan(u-v) = \dfrac{\tan u - \tan v}{1 + \tan u \tan v}$

THE LAWS OF SINES AND COSINES

The Law of Sines
$$\frac{\sin A}{a} = \frac{\sin B}{b} = \frac{\sin C}{c}$$

The Law of Cosines
$a^2 = b^2 + c^2 - 2bc \cos A$
$b^2 = a^2 + c^2 - 2ac \cos B$
$c^2 = a^2 + b^2 - 2ab \cos C$

GRAPHS OF COMMON FUNCTIONS

$f(x) = x$ $f(x) = x^2$ $f(x) = x^3$ $f(x) = \dfrac{1}{x}$

FORMULAS IN GEOMETRY

Rectangle
$A = lw$
$P = 2l + 2w$

Triangle
$A = \dfrac{1}{2}bh$

Heron's Formula:
$A = \sqrt{s(s-a)(s-b)(s-c)}$
where $s = \dfrac{a+b+c}{2}$

Circle
$A = \pi r^2$
$C = 2\pi r$

Trapezoid
$A = \dfrac{1}{2}h(b+c)$

Right Cylinder
$V = (Area\ of\ Base)h$

Rectangular Prism
$V = lwh$
$SA = 2lh + 2wh + 2lw$

Rectangular Pyramid
$V = \dfrac{1}{3}lwh$

Sphere
$V = \dfrac{4}{3}\pi r^3$
$SA = 4\pi r^2$

Right Circular Cylinder
$V = \pi r^2 h$
$SA = 2\pi r^2 + 2\pi rh$

Cone
$V = \dfrac{1}{3}\pi r^2 h$

PROPERTIES OF ABSOLUTE VALUE
For all real numbers a and b:

$|a| \ge 0$ $|-a| = |a|$

$a \le |a|$ $|ab| = |a||b|$

$\left|\dfrac{a}{b}\right| = \dfrac{|a|}{|b|},\ b \ne 0$

$|a+b| \le |a| + |b|$ (the triangle inequality)

PROPERTIES OF EXPONENTS AND RADICALS

$a^n \cdot a^m = a^{n+m}$ $(a^n)^m = a^{nm}$

$\dfrac{a^n}{a^m} = a^{n-m}$ $(ab)^n = a^n b^n$

$a^{-n} = \dfrac{1}{a^n}$ $\left(\dfrac{a}{b}\right)^n = \dfrac{a^n}{b^n}$

$a^{\frac{1}{n}} = \sqrt[n]{a}$ $a^{\frac{m}{n}} = \sqrt[n]{a^m} = \left(\sqrt[n]{a}\right)^m$

$\sqrt[n]{ab} = \sqrt[n]{a}\sqrt[n]{b}$ $\sqrt[n]{\dfrac{a}{b}} = \dfrac{\sqrt[n]{a}}{\sqrt[n]{b}}$

$\sqrt[n]{\sqrt[m]{a}} = \sqrt[mn]{a}$

SPECIAL PRODUCT FORMULAS

$(A-B)(A+B) = A^2 - B^2$

$(A+B)^2 = A^2 + 2AB + B^2$

$(A-B)^2 = A^2 - 2AB + B^2$

FACTORING SPECIAL BINOMIALS

$A^2 - B^2 = (A-B)(A+B)$

$A^3 - B^3 = (A-B)(A^2 + AB + B^2)$

$A^3 + B^3 = (A+B)(A^2 - AB + B^2)$

THE QUADRATIC FORMULA
The solutions of the equation $ax^2 + bx + c = 0$ are:
$$x = \frac{-b \pm \sqrt{b^2 - 4ac}}{2a}$$

DISTANCE FORMULA
$$d = \sqrt{(x_2 - x_1)^2 + (y_2 - y_1)^2}$$

MIDPOINT FORMULA
$$\left(\frac{x_1 + x_2}{2}, \frac{y_1 + y_2}{2}\right)$$

SLOPE OF A LINE
$$m = \frac{y_2 - y_1}{x_2 - x_1}$$

Horizontal lines $y = c$ have slope 0.
Vertical lines $x = c$ have undefined slope.

Acknowledgements

I am very grateful to all the people at Hawkes Learning Systems for their support and dedication to this project. In particular, many thanks to the editorial team, Marcel Prevuznak (development director), Emily Cook (marketing director), and James Hawkes.

I am also grateful to the following for their many insightful comments and reviews:

Dhruba Adhikari   *Mississippi University for Women*
Froozan Afiat   *College of Southern Nevada*
Donna Ahlrich   *Holmes Community College*
Dora Ahmadi   *Morehead State University*
Eva Allen   *Indian River State College*
Anna Pat Alpert   *Navarro College*
Larry Anderson   *Louisiana State University Shreveport*
Lisa Anglin   *Holmes Community College*
Marchetta Atkins   *Alcorn State University*
Russ Baker   *Howard Community College*
Madelaine Bates   *Bronx Community College of the City University of New York*
Shari Beck   *Navarro College*
Sage Bentley   *Navarro College*
Richard Alan Blanton   *Morehead State University*
Stephanie Blue   *Holmes Community College*
Brent Bollich   *South Louisiana Community College*
Stephanie Burton   *Holmes Community College*
Candace Carter-Stevens   *Mississippi Valley State University*
Brenda Cates   *Mount Olive College*
Deanna Caveny   *College of Charleston*
Brenda Chapman   *Trident Technical College*
Michelle DeDeo   *University of North Florida*
Gilbert Eyabi   *Anderson University*
Hamidullah Farhat   *Hampton University*
Mary Ellen Foley   *Louisiana State University Shreveport*
Terry Fung   *Kean University*
Nathan Gastineau   *Arkansas State University*
Mark Goldstein   *West Virginia Northern Community College–New Martinsville*
Leslie Gomes   *University of Arkansas Community College at Morrilton*
Heidi Griffin   *Arkansas State University*
Joshua Hanes   *Mississippi University for Women*
Bobbie Jo Hill   *Coastal Bend College*
Leslie Horton   *Delta State University*
Christopher Imm   *Johnson County Community College*
Abdusamad Kabir   *Bowie State Univeristy*
Bathi Kasturiarachi   *Kent State University at Stark*
Richard LeBorne   *Tennessee Technological University*

Bill Lepowsky   *Laney College*
Alice Lou   *Columbia College*
Heidi Lyman   *South Seattle Community College*
Richard Mabry   *Louisiana State University Shreveport*
Katherine Malone   *Fort Scott Community College*
Monica Meissen   *University of Dubuque*
Virginia Metcalf   *Somerset Community College*
Angela Miles   *Holmes Community College*
Mike Miller   *Minnesota State University Moorhead*
Cailin Mistrille   *University of Arkansas Community College at Morrilton*
Charles Naffziger   *Central Oregon Community College*
Paula Norris   *Delta State University*
Carol Okigbo   *Minnesota State University Moorhead*
Bonnie Oppenheimer   *Mississippi University for Women*
Ron Palcic   *Johnson County Community College*
Nancy Parkerson   *Holmes Community College*
Jennie Pegg   *Holmes Community College*
Stan Perrine   *Charleston Southern University*
Kimberly Potters   *Eastern New Mexico University*
Brenda Reed   *Navarro College*
Harriette Roadman   *Community College of Allegheny County*
David Rule   *Holmes Community College*
Joan Sallenger   *Midlands Technical College–Beltline*
Mike Schramm   *Indian River State College*
Christopher Schroeder   *Morehead State University*
Elizabeth Schubert   *Saddleback College*
Barbara Sehr   *Indiana University Kokomo*
Mack Smith   *Delta State University*
Carlos Spaht   *Louisiana State University Shreveport*
Mary Jane Sterling   *Bradley University*
Gloria Stone   *State University of New York at Oswego*
Gail Stringer   *Somerset Community College*
Preety Tripathi   *State University of New York at Oswego*
Al Vekovius   *Louisiana State University Shreveport*
Vance Waggener   *Trident Technical College*
Danae Watson   *University of Arkansas Community College at Morrilton*
Bill Weber   *University of Wyoming*
Elizabeth White   *Trident Technical College*
Ralph L. Wildy Jr.   *Georgia Military College Augusta Campus*
Mary Beth Williams   *Eastern New Mexico University*
Raymond Williams   *Mississippi Valley State University*
Clifton Wingard   *Delta State University*
Shaochen Yang   *Mississippi University for Women*
Lixin Yu   *Alcorn State University*

I would also like to thank John Scofield, Charlie Hunter, Miles Davis, Herbie Hancock, the Dave Matthews Band, Groove Collective, Carlos Santana, and Medeski Martin & Wood, among others, for their unwitting but much-appreciated assistance.

Finally, my very great thanks to my wife Cindy for her support and understanding at all times.

To the Student

There is a saying among math instructors that you may have heard: "Math is not a spectator sport." While this may sound trite, it is undeniably true. Mathematics is not something you can learn by watching someone else do it. You have probably had the experience of watching an instructor solve a problem and marveling at how easy it seems, only to find that a nearly identical problem is much harder to solve at home or on a test.

The key point is that you have to practice mathematics in order to learn mathematics. Make sure you do the homework problems your instructor assigns you, not only because that's how mathematics is learned but also because it gives you insight into the types of problems that your instructor considers to be important.

Beyond that, there are some other key ideas to keep in mind. One is that very, very few people will fully grasp a mathematical concept or master a skill on the first try. If you find yourself lost after reading a section of this book (or any math book), don't despair. Just remember what is puzzling you, take a break, and try it again when you're fresh. Most math is learned in a cyclic process of plowing ahead until lost, backing up and rereading, and then plowing ahead a bit further. A math book is not like a novel: you shouldn't expect to read it cover to cover a single time.

Finally, make sure you take advantage of opportunities to learn from your instructor and peers. Go to class and pay attention to what your instructor emphasizes and learn from his or her unique insight into the material. Work with friends when possible, and ask others for help if they understand something you haven't gotten yet. And when you have the opportunity to explain some math to someone else, take advantage of it. Teaching mathematics to others is an amazingly effective way to improve your own understanding.

Hawkes Learning Systems: *Precalculus*

Overview

Hawkes Learning Systems specializes in interactive mathematics courseware with a unique, mastery-based approach to student learning. Hawkes Learning Systems: *Precalculus* is designed to help students develop a solid foundation and understanding of precalculus. Through its topical mastery approach, it has been proven to promote and increase student success. The courseware consists of three learning modes: Instruct, Practice, and Certify. Additional auxiliary instructional tools including videos, teaching slides, and eBook options are also available.

Exploring the Courseware

Our mastery-based courseware engages students in the learning process, so they successfully learn, understand, and demonstrate competencies for assigned topics. Just-in-time feedback and a student-centered learning environment offer a systematic approach to learning that includes the following learning modes, each with its own unique, differentiating features:

Instruct

Instruct is a multimedia presentation of each lesson that correlates with the content in the textbook.

Instruct Features:

- Definitions, rules, properties
- Example problems
- Audio narration
- Instructional video tutorials
- Interactive questions and animations

Practice

Practice presents students with an unlimited number of algorithmically generated problems and intelligent feedback on incorrect answers. Within the Practice mode, the *Interactive Tutor* encourages active learning.

Practice Features:

- Explain Error: Targeted feedback specific to the student's mistake explains not only what is wrong, but *why*.
- Step By Step: Breaks each problem down into smaller steps offering feedback and solutions at each point.
- Solution: Offers guided solutions for every problem.
- Instructor Connect: Allows for monitoring of more challenging problems.
- Performance Report: Tracks homework preparedness.

Certify

Certify is an assignment that holds students accountable for learning the material at a defined proficiency level while removing learning aids.

Certify Features:

- Motivational, non-penalty approach
- Algorithmically generated, free-response assignments
- Comparable question sets to those offered in the Practice mode
- A Mastery Progress meter that allows students to visually track their progress toward topic mastery

Video

View instructional videos anytime, anywhere at **HawkesTV.com**

Feel free to contact support for questions or technical help with Hawkes Learning Systems: *Precalculus*.

Support Center: support.hawkeslearning.com
Chat: chat.hawkeslearning.com
E-mail: support@hawkeslearning.com
Phone: 843.571.2825

Number Systems and Equations and Inequalities of One Variable

1.1   Real Numbers and Algebraic Expressions

1.2   Properties of Exponents and Radicals

1.3   Polynomials and Factoring

1.4   The Complex Number System

1.5   Linear Equations in One Variable

1.6   Linear Inequalities in One Variable

1.7   Quadratic Equations

1.8   Rational and Radical Equations

Chapter 1 Project

Chapter 1 Summary

Chapter 1 Review

Chapter 1 Test

By the end of this chapter you should be able to:

What if while you were picking strawberries and dropping them into a bucket, your brother was sneaking strawberries out of the bucket to eat? How would this affect the rate at which you can fill the bucket?

By the end of this chapter, you'll be able to work with algebraic expressions, exponents, and polynomials, solve equations of one variable, operate with rational expressions, and solve formulas involving radicals. You'll encounter the answer to the berry-picking question on page 115. You'll master this type of problem using the techniques for solving rational equations, found on page 104.

Introduction

In this chapter, we review the terminology, the notation and properties of the real number system frequently encountered in algebra, the extension of the real number system to the larger set of complex numbers, and the basic algebraic methods we use to solve equations and inequalities.

We begin with a discussion of common subsets of the set of real numbers. Certain types of numbers are important from both a historical and a mathematical perspective. There is archaeological evidence that people used the simplest sort of numbers, the counting or natural numbers, as far back as 50,000 years ago. Over time, many cultures discovered needs for various refinements to the number system, resulting in the development of such classes of numbers as the integers, the rational numbers, the irrational numbers, and ultimately, the complex numbers, a number system which contains the real numbers.

Pythagoras

Many of the ideas in this chapter have a history dating as far back as Egyptian and Babylonian civilizations of around 3000 BC, with later developments and additions due to Greek, Hindu, and Arabic mathematicians. Much of the material was also developed independently by Chinese mathematicians. It is a tribute to the necessity, utility, and objectivity of mathematics that so many civilizations adopted so much mathematics from abroad, and that different cultures operating independently developed identical mathematical concepts so frequently.

As an example of the historical development of just one concept, consider the notion of an irrational number. The very idea that a real number could be irrational (which simply means not rational, or not the ratio of two integers) is fairly sophisticated, and it took some time for mathematicians to come to this realization. The Pythagoreans, members of a school founded by the Greek philosopher Pythagoras in southern Italy around 540 BC, discovered that the square root of 2 was such a number, and there is evidence that for a long time $\sqrt{2}$ was the only known irrational number. A member of the Pythagorean school, Theodorus of Cyrene, later showed (c. 425 BC) that $\sqrt{3}, \sqrt{5}, \sqrt{6}, \sqrt{7}, \sqrt{8}, \sqrt{10}, \sqrt{11}, \sqrt{12}, \sqrt{13}, \sqrt{14}, \sqrt{15}$, and $\sqrt{17}$ also are irrational. It wasn't until 1767 that European mathematician, Johann Lambert, showed that the famous number π is irrational, and the modern rigorous mathematical description of irrational numbers is due to work by Richard Dedekind in 1872.

As you review the concepts in Chapter 1, keep the larger picture firmly in mind. All of the material presented in this chapter was developed over long periods of time by many different cultures with the aim of solving problems important to them.

Real Numbers and Algebraic Expressions

TOPICS

1. Common subsets of real numbers

2. The real number line

3. Order on the real number line

4. Set-builder notation and interval notation

5. Basic set operations and Venn diagrams

6. Absolute value and distance

7. Components and terminology of algebraic expressions

8. The field properties and their use in algebra

TOPIC

Common Subsets of Real Numbers

Some types of numbers occur so frequently in mathematics that they have been given special names and symbols. These names will be used throughout this book and in later math classes when referring to members of the following sets:

DEFINITION

Types of Real
Numbers

The Natural (or Counting) Numbers: This is the set of numbers $\mathbb{N} = \{1, 2, 3, 4, 5, ...\}$.

The Whole Numbers: This is the set of natural numbers and 0: $\{0, 1, 2, 3, 4, 5, ...\}$. Again, we can list only the first few members of this set.

The Integers: This is the set of natural numbers, their negatives, and 0. As a list, this is the set $\mathbb{Z} = \{..., -4, -3, -2, -1, 0, 1, 2, 3, 4, ...\}$.

The Rational Numbers: This is the set, with symbol $\mathbb{Q}$ (for quotient), of ratios of integers (hence the name). That is, any rational number can be written in the form $\dfrac{p}{q}$, where p and q are both integers and $q \neq 0$. When written in decimal form, rational numbers either terminate or have a repeating pattern of digits past some point.

The Irrational Numbers: Every real number that is not rational is, by definition, irrational. In decimal form, irrational numbers are nonterminating and nonrepeating.

The Real Numbers: Every set above is a subset of the set of real numbers, which is denoted $\mathbb{R}$. Every real number is either rational or irrational, and no real number is both.

The following figure shows the relationships among the subsets of $\mathbb{R}$ defined on the previous page. This figure indicates, for example, that every natural number is automatically a whole number, and also an integer, and also a rational number.

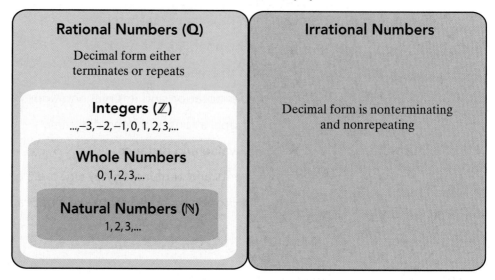

Figure 1: The Real Numbers

EXAMPLE 1

Types of Real Numbers

Consider the set $S = \left\{-15,\, -7.5,\, -\dfrac{7}{3},\, 0,\, \sqrt{2},\, 1.\overline{6},\, \sqrt{9},\, \pi,\, 10^{17}\right\}$.

a. The natural numbers in S are $\sqrt{9}$ and 10^{17}. $\sqrt{9}$ is a natural number since $\sqrt{9} = 3$.

b. The whole numbers in S are 0, $\sqrt{9}$, and 10^{17}.

c. The integers in S are $-15, 0, \sqrt{9}$, and 10^{17}.

d. The rational numbers in S are $-15,\, -7.5,\, -\dfrac{7}{3},\, 0,\, 1.\overline{6},\, \sqrt{9}$, and 10^{17}. The numbers -7.5 and $1.\overline{6}$ are both rational numbers since $-7.5 = \dfrac{-15}{2}$ and $1.\overline{6} = \dfrac{5}{3}$ (the bar over the last digit indicates that the digit repeats indefinitely). Note that any integer p is also a rational number, since it can be written as $\dfrac{p}{1}$.

e. The only irrational numbers in S are $\sqrt{2}$ and π.

TOPIC 2

The Real Number Line

Mathematicians often depict the set of real numbers as a horizontal line, with each point on the line representing a unique real number (so each real number is associated with a unique point on the line). The real number corresponding to a given point is called the **coordinate** of that point. Thus one (and only one) point on the real number line represents the number 0, and this point is called the **origin**. Points to the right of the origin represent positive real numbers, while points to the left of the origin represent negative real numbers. Figure 2 is an illustration of the real number line with several points plotted. Note that two irrational numbers are plotted, though their locations on the line are approximations.

$$-\pi \qquad -1 \quad 0 \qquad \sqrt{2} \qquad\qquad 5$$

Figure 2: The Real Number Line

TOPIC 3

Order on the Real Number Line

Representing the real numbers as a line leads naturally to the idea of *ordering* the real numbers.

DEFINITION

Inequality Symbols
(Order)

Symbol	Reading	Meaning
$a < b$	"a is **less than** b"	a lies to the left of b on the number line.
$a \leq b$	"a is **less than or equal to** b"	a lies to the left of b or is equal to b.
$b > a$	"b is **greater than** a"	b lies to the right of a on the number line.
$b \geq a$	"b is **greater than or equal to** a"	b lies to the right of a or is equal to a.

The two symbols $<$ and $>$ are called *strict* inequality signs, while the symbols $\leq$ and $\geq$ are *nonstrict* inequality signs.

CAUTION!

Remember that order is defined by the placement of real numbers on the number line, *not* by magnitude (its distance from zero). For instance, $-36 < 5$ because -36 lies to the left of 5 on the number line. Also, be aware that the negation of the statement $a \leq b$ is the statement $a > b$. Furthermore, if $a \leq b$ and $a \geq b$, then it must be the case that $a = b$.

TOPIC 4   Set-Builder Notation and Interval Notation

To describe the solutions to equations and inequalities, we need a precise, consistent way of expressing sets of real numbers. **Set-builder notation** is a general method of describing the elements that belong to a given set. **Interval notation** is a way of describing certain subsets of the real line.

DEFINITION

Set-Builder Notation

The notation $\{x\,|\,x \text{ has property } P\}$ is used to describe a set of real numbers, all of which have the property P. This can be read "the set of all real numbers x having property P."

The symbol "|" is also read as "such that," so the above notation can also be read "the set of all real numbers x, *such that* x has property P."

EXAMPLE 2

Set-Builder Notation

a. $\{x\,|\,x \text{ is an even integer}\}$ is another way of describing the set $\{\ldots,-4,-2,0,2,4,\ldots\}$. We could also describe this set as $\{2n\,|\,n \text{ is an integer}\}$, since every even integer is a multiple of 2.

b. $\{x\,|\,x \text{ is an integer such that } -3 \leq x < 2\}$ describes the set $\{-3,-2,-1,0,1\}$.

DEFINITION

The Empty Set

A set with no elements is called the **empty set** or the **null set**, and is denoted by the symbol $\varnothing$.

The empty set can arise from a set defined using set-builder notation; for example, the set $\{y\,|\,y > 1 \text{ and } y \leq -4\}$ is equivalent to the empty set, since no real number y satisfies the stated property.

Sets that consist of all real numbers bounded by two endpoints, possibly including those endpoints, are called **intervals**. Intervals can also consist of a portion of the real line extending indefinitely in either direction from just one endpoint.

We can describe such sets with set-builder notation, but intervals occur frequently enough that special notation has been devised to define them succinctly.

DEFINITION

Interval Notation

Interval Notation	Set-Builder Notation	Meaning
(a,b)	$\{x \mid a < x < b\}$	all real numbers strictly between a and b
$[a,b]$	$\{x \mid a \le x \le b\}$	all real numbers between a and b, including both a and b
$(a,b]$	$\{x \mid a < x \le b\}$	all real numbers between a and b, including b but not a
$(-\infty, b)$	$\{x \mid x < b\}$	all real numbers less than b
$[a, \infty)$	$\{x \mid x \ge a\}$	all real numbers greater than or equal to a

Intervals of the form (a,b) are called **open** intervals, while those of the form $[a,b]$ are **closed** intervals. The interval $(a,b]$ is **half-open** (or **half-closed**). Of course, a half-open interval may be open at either endpoint, as long as it is closed at the other. The symbols $-\infty$ and ∞ indicate that the interval extends indefinitely in the left and the right directions, respectively. Note that $(-\infty, b)$ excludes the endpoint b, while $[a, \infty)$ includes the endpoint a.

CAUTION!

The symbols $-\infty$ and ∞ are just that: symbols! They are not real numbers, so they cannot be solutions to a given equation. The fact that they are symbols, and not numbers, means that they can never be included in a set of real numbers. For this reason, a parenthesis always appears next to either $-\infty$ or ∞; a bracket should never appear next to either infinity symbol.

EXAMPLE 3

Intervals of Real Numbers

a. The interval $(2,8)$ represents the set $\{x \mid 2 < x < 8\}$. This interval is open at both endpoints, so neither 2 nor 8 are included in the set.

b. The interval $[-5,-1]$ is another way to write the set $\{x \mid -5 \le x \le -1\}$. This interval is closed at both endpoints, so both −5 and −1 are included in the set.

c. The interval $[-3,10)$ stands for the set $\{x \mid -3 \le x < 10\}$. This interval is closed at the left endpoint, −3, and open at the right endpoint, 10.

d. The interval $(4, \infty)$ stands for the set $\{x \mid x > 4\}$. Since the interval is open on the left endpoint, it is the set of numbers greater than (but not equal to) 4.

e. The interval $(-\infty, \infty)$ is just another way of describing the entire set of real numbers.

TOPIC 5   Basic Set Operations and Venn Diagrams

The sets that arise most frequently in algebra are sets of real numbers, and these sets are often the solutions of equations or inequalities. We will need to combine two or more such sets through the set operations of **union** and **intersection**. These operations are defined on sets in general, not just sets of real numbers, and can be illustrated by means of Venn diagrams.

A **Venn diagram** is a pictorial representation of a set or sets, and it indicates, through shading, the outcome of set operations such as union and intersection. In the following definition, these two operations are first defined with set-builder notation and then demonstrated with a Venn diagram. The symbol $\in$ is read "is an element of."

DEFINITION

Union

In this definition, A and B denote two sets, and are represented in the Venn diagram by circles. The operation of union is depicted in the diagram by shading.

The **union** of A and B, denoted $A \cup B$, is the set $\{x \mid x \in A \text{ or } x \in B\}$. That is, an element x is in $A \cup B$ if it is in the set A, the set B, or both. Note that the union of A and B contains both individual sets.

DEFINITION

Intersection

In this definition, A and B denote two sets, and are represented in the Venn diagram by circles. The operation of intersection is depicted in the diagram by shading.

The **intersection** of A and B, denoted $A \cap B$, is the set $\{x \mid x \in A \text{ and } x \in B\}$. That is, an element x is in $A \cap B$ if it is in both A and B. Note that the intersection of A and B is contained in each individual set.

EXAMPLE 4

Union and
Intersection of
Intervals

Simplify each of the following set expressions.

a. $(-2,4]\cup[0,9]$ **b.** $(-2,4]\cap[0,9]$

c. $[3,4)\cap(4,9)$ **d.** $(-\infty,4]\cup(-1,\infty)$

Solutions:

a. $(-2,4]\cup[0,9]=(-2,9]$ Since these two intervals overlap, their union is described with a single interval.

b. $(-2,4]\cap[0,9]=[0,4]$ This intersection of two intervals can also be described with a single interval.

c. $[3,4)\cap(4,9)=\varnothing$ These two intervals have no elements in common, so their intersection is the empty set.

d. $(-\infty,4]\cup(-1,\infty)=(-\infty,\infty)$ The union of these two intervals is the entire set of real numbers.

EXAMPLE 5

Union and
Intersection

Simplify each of the following set expressions.

a. $\{1,2\}\cup\{0,3\}$ **b.** $\{x,y,z\}\cap\{w,x\}$

c. $\mathbb{Z}\cup\mathbb{R}$ **d.** $\mathbb{Z}\cap\mathbb{R}$

Solutions:

a. The union of the two sets consists of all elements in either set: $\{0,1,2,3\}$.

b. The intersection consists only of elements in both sets: $\{x\}$.

c. Since the integers are all also real numbers, the union of these two sets is simply the set of real numbers $\mathbb{R}$. We say that $\mathbb{Z}$ is *contained* in $\mathbb{R}$.

d. Similarly, since all integers are also real numbers, the integers are the elements contained in both sets. Thus, the intersection is $\mathbb{Z}$.

TOPIC **6**

Absolute Value and Distance

In addition to order, the depiction of the set of real numbers as a line leads to the notion of *distance*. Physically we understand distance as a number, indicating how close two objects are to one another. The idea of **absolute value** gives us a way to define distance in a mathematical setting.

DEFINITION

Absolute Value

The **absolute value** of a real number a, denoted as $|a|$, is defined by:

$$|a| = \begin{cases} a & \text{if } a \geq 0 \\ -a & \text{if } a < 0 \end{cases}$$

The absolute value of a number is also referred to as its magnitude; it is the non-negative number corresponding to its distance from the origin. Note that 0 is the only real number whose absolute value is 0.

DEFINITION

Distance on the Real Number Line

Given two real numbers a and b, the **distance** between them is defined to be $|a-b|$. In particular, the distance between a and 0 is $|a-0|$ or just $|a|$.

The distance from a to b should be the same as the distance from b to a. The definition confirms this intuition; note that it does not state whether a or b is smaller. This is because $|a-b| = |b-a|$. No matter which number we call a, the distance is the same.

EXAMPLE 6

Absolute Value

a. $|17-3| = |3-17| = 14$. 17 and 3 are 14 units apart.

b. $|-\pi| = |\pi| = \pi$. Both $-\pi$ and π are π units from 0.

c. $-|-5| = -5$. Note that the negative sign outside the absolute value symbol is not affected by the absolute value. Compare this with the fact that $-(-5) = 5$.

d. $|\sqrt{7}-2| = \sqrt{7}-2$. Even without a calculator, we know $\sqrt{7}$ is larger than 2 (since $2 = \sqrt{4}$), so $\sqrt{7}-2$ is positive and hence $|\sqrt{7}-2| = \sqrt{7}-2$.

e. $|\sqrt{7}-19| = 19-\sqrt{7}$. In contrast to the last example, we know $\sqrt{7}-19$ is negative, so its absolute value is $-(\sqrt{7}-19) = 19-\sqrt{7}$.

The list of properties below can all be derived from the definition of absolute value.

PROPERTIES

Properties of Absolute Value

For all real numbers a and b:

1. $|a| \geq 0$ (The absolute value of a number is never negative.)

2. $|-a| = |a|$

3. $a \leq |a|$

4. $|ab| = |a| |b|$

5. $\left|\dfrac{a}{b}\right| = \dfrac{|a|}{|b|}, \; b \neq 0$

6. $|a+b| \leq |a| + |b|$ (This is called the triangle inequality, as it is a reflection of the fact that one side of a triangle is never longer than the sum of the other two sides.)

EXAMPLE 7

Using Absolute Value Properties

a. $|(-3)(5)| = |-15| = 15 = |-3||5|$

b. $1 = |-3+4| \leq |-3| + |4| = 7$

c. $7 = |-3-4| \leq |-3| + |-4| = 7$

d. $\left|\dfrac{-3}{7}\right| = \dfrac{|-3|}{|7|} = \dfrac{3}{7}$

TOPIC 7

Components and Terminology of Algebraic Expressions

Algebraic expressions are made up of constants and variables, combined by the operations of addition, subtraction, multiplication, division, exponentiation and the taking of roots. **Constants** like 6 and –3, are fixed numbers, while **variables** like x and y are usually letters that represent unspecified numbers. To **evaluate** a given expression means to replace the variables (if there are any) with specific numbers, perform the indicated mathematical operations and simplify the result.

The **terms** of an algebraic expression are those parts joined by addition (or subtraction), while the **factors** of a term are the individual parts of the term that are joined by multiplication (or division). The **coefficient** of a term is the constant factor of the term, while the remaining part of the term is the **variable factor**.

EXAMPLE 8

Terminology of Algebraic Expressions

Consider the algebraic expression $-17x\left(x^2 + 4y\right) + 5\sqrt{x} - 13$.

a. This expression contains three terms; $-17x\left(x^2 + 4y\right)$, $5\sqrt{x}$, and –13. The terms are combined by addition and subtraction to form the whole expression.

b. The factors of the term $-17x\left(x^2 + 4y\right)$ are –17, x, and $\left(x^2 + 4y\right)$. The factors are combined by multiplication to form the whole term. The coefficient of $-17x\left(x^2 + 4y\right)$ is –17, and the variable part is $x\left(x^2 + 4y\right)$.

EXAMPLE 9

Evaluate the following algebraic expressions.

a. $5x^3 - 16$ for $x = 4$

b. $-3x^2 - 2(x + y)$ for $x = -2$ and $y = 3$

Solutions:

In both cases, we simply "plug in" the given values for each variable, then simplify.

a. $5(4)^3 - 16 = 5(64) - 16$

$$= 320 - 16$$
$$= 304$$

b. $-3(-2)^2 - 2(-2 + 3) = -3(4) - 2(1)$

$$= -12 - 2$$
$$= -14$$

TOPIC 8

The Field Properties and Their Use in Algebra

The set of real numbers forms what is known mathematically as a *field*, and consequently, the properties below are called *field properties*. These properties also apply to the set of complex numbers, which is a larger field containing the real numbers. We will discuss complex numbers in Section 1.4.

PROPERTIES

Field Properties

In this table, a, b, and c represent arbitrary real numbers. The first five properties apply to addition and multiplication, while the last combines the two.

Name of Property	Additive Version	Multiplicative Version
Closure	$a + b$ is a real number	ab is a real number
Commutative	$a + b = b + a$	$ab = ba$
Associative	$a + (b + c) = (a + b) + c$	$a(bc) = (ab)c$
Identity	$a + 0 = 0 + a = a$	$a \cdot 1 = 1 \cdot a = a$
Inverse	$a + (-a) = 0$	$a \cdot \dfrac{1}{a} = 1 \ (\text{for } a \neq 0)$
Distributive	$a(b + c) = ab + ac$	

While the field properties are of fundamental importance to algebra, they imply further properties that are often of more immediate use.

PROPERTIES

Cancellation Properties

Let A, B, and C be algebraic expressions.

Additive Cancellation: Adding the same quantity to both sides of an equation results in an equivalent equation.

$$\text{If } A = B, \text{ then } A + C = B + C.$$

Multiplicative Cancellation: Multiplying both sides of an equation by the same *nonzero* constant results in an equivalent equation.

$$\text{If } A = B \text{ and } C \neq 0, \text{ then } A \cdot C = B \cdot C.$$

PROPERTIES

Zero-Factor Property

Let A and B represent algebraic expressions. If the product of A and B is 0, then at least one of A and B is itself 0.

$$AB = 0 \text{ implies that } A = 0 \text{ or } B = 0 \text{ (or both)}.$$

EXAMPLE 10

Properties of Real Numbers

a.
$$y + 12 = 18$$
$$y + 12 + (-12) = 18 + (-12)$$
$$y = 6$$

Using additive cancellation, we add -12 to both sides, then simplify.

This shows that the equation $y + 12 = 18$ is equivalent to the equation $y = 6$.

b.
$$-6x = 30$$
$$-6x\left(-\frac{1}{6}\right) = 30\left(-\frac{1}{6}\right)$$
$$x = -5$$

Using multiplicative cancellation, we multiply both sides by $-\frac{1}{6}$, then simplify.

Thus, the equation $-6x = 30$ is equivalent to $x = -5$. We can see how cancellation properties can help us *solve* equations for variables.

c. The equation $(x - y)(x + y) = 0$ means that either $x - y = 0$ or $x + y = 0$, by the Zero-Factor Property.

1. a. $19, 2^5$

b. $19, \dfrac{0}{15}, 2^5$

c. $19, \dfrac{0}{15}, 2^5, -33$

d. $19, -4.3, \dfrac{0}{15}, 2^5,$ -33

e. $-\sqrt{3}$

f. All

2. a. $\sqrt{16}, |-8|$

b. $\sqrt{16}, |-8|$

c. $\sqrt{16}, -1, |-8|$

d. $\sqrt{16}, 3.\bar{3}, -1, \dfrac{22}{7},$ $|-8|$

e. $5\sqrt{7}, 4\pi$

f. All

3.
$-4.5 \quad -1 \quad 2.5$

4.
$-24 \quad 2 \quad 15$

5.
$5.1 \; 5.2 \qquad 5.8$

6.
$0 \quad \dfrac{1}{2} \; \dfrac{5}{6}$

7. $<, \le$ **8.** $<, \le$

9. $\le, \ge$ **10.** $<, \le$

11. $>, \ge$ **12.** $>, \ge$

13. $2a + b > c$

14. $2 \le x$

15. $2c \le 3d$

16. $6 + x \ge 4x$

17. $\{n \mid n$ is an integer and $5 \le n \le 105\}$

18. $\{n \mid n$ is a prime$\}$

19. $\{2^n \mid n$ is a whole number$\}$

20. $\{3n \mid n$ is an integer and $-2 \le n \le 3\}$

21. $\left\{\left|\dfrac{1}{n}\right| \; n$ is an odd integer$\right\}$

Exercises

Which elements of the following sets are **a.** natural numbers, **b.** whole numbers, **c.** integers, **d.** rational numbers, **e.** irrational numbers, **f.** real numbers? See Example 1.

1. $\left\{19, -4.3, -\sqrt{3}, \dfrac{0}{15}, 2^5, -33\right\}$ **2.** $\left\{5\sqrt{7}, 4\pi, \sqrt{16}, 3.\bar{3}, -1, \dfrac{22}{7}, |-8|\right\}$

Plot the real numbers in the following sets on a number line. Choose the unit length appropriately for each set.

3. $\{-4.5, -1, 2.5\}$ **4.** $\{-24, 2, 15\}$ **5.** $\{5.1, 5.2, 5.8\}$ **6.** $\left\{0, \dfrac{1}{2}, \dfrac{5}{6}\right\}$

Select all of the symbols from the set $\{<, \le, >, \ge\}$ that can be placed in the blank to make each statement true.

7. $12 \underline{\quad} 14$ **8.** $-102 \underline{\quad} 9$ **9.** $3 \underline{\quad} 3$

10. $-50 \underline{\quad} -45$ **11.** $-3.4 \underline{\quad} -3.5$ **12.** $\dfrac{-1}{4} \underline{\quad} \dfrac{-1}{3}$

Write each statement as an inequality, using the appropriate inequality symbol.

13. "$2a + b$ is strictly greater than c"   **14.** "2 is less than or equal to x"

15. "$2c$ is no more than $3d$"   **16.** "$6 + x$ is greater than or equal to $4x$"

Describe each of the following sets using set-builder notation. There may be more than one correct way to do this. See Example 2.

17. $\{5, 6, 7, \ldots, 105\}$ **18.** $\{2, 3, 5, 7, 11, 13, 17, \ldots\}$

19. $\{1, 2, 4, 8, 16, 32, \ldots\}$ **20.** $\{-6, -3, 0, 3, 6, 9\}$

21. $\left\{\ldots, \dfrac{1}{3}, \dfrac{1}{5}, \dfrac{1}{7}, \dfrac{1}{9}, \ldots\right\}$ **22.** $\{0, 1, 2, 3, 4, 5, \ldots\}$

Write each set as an interval using interval notation. See Example 3.

23. $\{x \mid -3 \le x < 19\}$ **24.** $\{x \mid x < 4\}$ **25.** $x < 15$

26. $-9 \le x \le 6$ **27.** The positive real numbers

28. The nonnegative real numbers

Graph the following intervals.

29. $[5, 14)$ **30.** $[-9, -1]$ **31.** $(0, 2)$

32. $(-3, 18]$ **33.** $(-\infty, 7]$ **34.** $(25, \infty)$

22. $\{n|n$ is a whole number$\}$

23. $[-3,19)$

24. $(-\infty,4)$

25. $(-\infty,15)$

26 $[-9,6]$

27. $(0,\infty)$

28. $[0,\infty)$

29. 5   14

30. -9   -1

31. 0 2

32. -3   18

33. 7

34. 25

35. $[-7,7)$

36. $(-5,4]$

37. $\varnothing$

38. $[3,4]$

39. $(-\infty,\infty)$

40. $[-\pi,21)$

41. $\mathbb{Z}$

42. $\mathbb{R}$

43. $\mathbb{Z}$

44. $\{-4\}$

45. -11

46. 4

47. $\sqrt{5}-\sqrt{3}$

48. $\sqrt{2}$

49. 1 **50.** -5

51. -15 **52.** -12

53. 5 **54.** 0

55. 6 **56.** 11

57. $3x^2y^3,\ -2\sqrt{x+y},7z$

58. $-2,\sqrt{x+y}$

59. $1,8.5,-14$

60. $8.5,x$

61. $\dfrac{-5x}{2yz},-8x^5y^3,6.9z$

62. $\dfrac{-5}{2},-8,6.9$

Simplify the following set expressions. See Example 4.

35. $[-7,7)\cup(2,5)$

36. $(-5,2]\cup(2,4]$

37. $(-5,2]\cap(2,4]$

38. $[3,5]\cap[2,4]$

39. $(-\infty,4]\cup(0,\infty)$

40. $(-\infty,\infty)\cap[-\pi,21)$

Simplify the following set expressions. See Example 5.

41. $\mathbb{Q}\cap\mathbb{Z}$

42. $\mathbb{N}\cup\mathbb{R}$

43. $\mathbb{N}\cup\mathbb{Z}\cap\mathbb{Q}$

44. $(-4.8,-3.5)\cap\mathbb{Z}$

Evaluate the absolute value expressions. See Examples 6 and 7.

45. $-|-11|$

46. $|3-7|$

47. $\left|\sqrt{3}-\sqrt{5}\right|$

48. $\left|-\sqrt{2}\right|$

49. $\dfrac{|-x|}{|x|}(x\neq0)$

50. $-|4-9|$

51. $-|-4-|-11||$

52. $-\left|-\sqrt{|-9|}-|-9|\right|$

Find the distance on the real number line between each pair of numbers given. See Example 6.

53. $a=8,b=3$

54. $a=5,b=5$

55. $a=4,b=-2$

56. $a=-12,b=-1$

Identify the components of the algebraic expressions, as indicated. See Example 8.

57. Identify the terms in the expression $3x^2y^3-2\sqrt{x+y}+7z$.

58. Identify the factors in the term $-2\sqrt{x+y}$.

59. Identify the coefficients in the expression $x^2+8.5x-14y^3$.

60. Identify the factors in the term $8.5x$.

61. Identify the terms in the expression $\dfrac{-5x}{2yz}-8x^5y^3+6.9z$.

62. Identify the coefficients in the expression $\dfrac{-5x}{2yz}-8x^5y^3+6.9z$.

63. -98

64. 10

65. $6\pi + 8$

66. 1

67. -20

68. $\dfrac{-1}{3}$

69. Commutative

70. Commutative

71. Distributive

72. Associative

73. Associative

74. Multiplicative inverse

75. Multiplicative cancellation; $\dfrac{1}{5}$

76. Multiplicative cancellation; $-\dfrac{1}{14}$

77. Additive cancellation; x

78. Zero-Factor Property

79. Multiplicative cancellation; 6

80. Zero-Factor Property

81. Additive cancellation; $-2x + y$

82. Zero-Factor Property

Evaluate each expression for the given values of the variables. See Example 9.

63. $3x^3 + 5x - 2$ for $x = -3$.

64. $\sqrt{2x} + \dfrac{3x}{4}$ for $x = 8$.

65. $-3\pi y + 8x + y^3$ for $x = 2$ and $y = -2$.

66. $y\sqrt{x^3 - 2} + \sqrt{x - 2y} - 3y$ for $x = 3$ and $y = -\dfrac{1}{2}$.

67. $|x - 9y| - (8z - 8)$ for $x = -3$, $y = 1$, and $z = 5$.

68. $\dfrac{x^2 y^3}{8z} - \dfrac{|2xy|}{8z}$ for $x = 2$, $y = -1$, and $z = 3$.

Identify the property that justifies each of the following statements.

69. $(x - y)(z^2) = (z^2)(x - y)$

70. $3 - 7 = -7 + 3$

71. $4(y - 3) = 4y - 12$

72. $-3(4x^6 z) = (-3)(4)(x^6 z) = -12x^6 z$

73. $4 + (-3 + x) = (4 - 3) + x = 1 + x$

74. $(x + y)\left(\dfrac{1}{x + y}\right) = 1$

Identify the property that justifies each of the following statements. If one of the cancellation properties is being used to transform an equation, identify the quantity that is being added to both sides or the quantity by which both sides are being multiplied. See Example 10.

75. $25x^3 = 10y \Leftrightarrow 5x^3 = 2y$

76. $-14y = 7 \Leftrightarrow y = -\dfrac{1}{2}$

77. $14 - x = 2x \Leftrightarrow 14 = 3x$

78. $(a + b)(x) = 0 \Rightarrow a + b = 0$ or $x = 0$

79. $\dfrac{x}{6} + \dfrac{y}{3} - 2 = 0 \Leftrightarrow x + 2y - 12 = 0$

80. $x^2 z = 0 \Rightarrow x^2 = 0$ or $z = 0$

81. $5 + 3x - y = 2x - y \Leftrightarrow 5 + x = 0$

82. $(x - 3)(x + 2) = 0 \Rightarrow x - 3 = 0$ or $x + 2 = 0$

83.

84. JR > Freddie > Sarah > Aubrey > Elizabeth

85. 20 miles

86. If age = x,

$\{x \mid x < 2\} = [0,2) \rightarrow$ free;

$\{x \mid 2 \le x < 12\}$ $= [2,12) \rightarrow \$3$;

$\{x \mid 12 \le x < 65\}$ $= [12,65) \rightarrow \$7$;

$\{x \mid x \ge 65\}$ $= [65, \infty) \rightarrow \$5$

87. If sugar = s,

$\{s \mid 3 \le s \le 4\}$ $= [3,4]$

If walnuts = w,

$\left\{w \mid \frac{1}{2} \le w \le \frac{2}{3}\right\}$ $= \left[\frac{1}{2}, \frac{2}{3}\right]$

88. $66

89. 131.11° C

90. $102

Solve the following application problems.

83. Jess, Stan, Nina, and Michele are in a marathon. Twenty-five minutes after beginning, Jess has run 3.4 miles, Stan has run 4 miles, Nina has run 2.25 miles, and Michele has walked 1.6 miles. Using 0 as the beginning point, plot each competitor's location on a real number line using an appropriate interval.

84. Freddie, Sarah, Elizabeth, JR, and Aubrey are trying to line up by height for a photo shoot. JR is the tallest and Elizabeth is the shortest. Freddie is taller than Sarah, and Sarah is taller than Aubrey. Express their lineup using appropriate inequality symbols.

85. Sue boards an eastbound train in Center Station at the same time Joy boards a westbound train in Center Station. After riding the Straight Line for 20 minutes, Sue's train has traveled 13 miles east, while Joy's train (also on the Straight Line) has traveled 7 miles west. Find the distance between the two trains at this time. (Assume the Straight Line is true to its name and that the tracks lie literally along a straight line.)

86. The admission prices at the local zoo are as follows:

Admission Prices	
Children under 2	free
Children under 12	$3
Adults	$7
Seniors (65 and up)	$5
**Open 10 am–11 pm daily	

Express the age range for each of these prices in set-builder notation and interval notation.

87. A particular fudge recipe calls for at least 3 but no more than 4 cups of sugar and at least $\frac{1}{2}$ but no more than $\frac{2}{3}$ of a cup of walnuts. Express the amount of sugar and nuts needed in both set-builder and interval notation.

88. At the beginning of the month, your checking account contains $128. For your birthday, your mother deposits $50 and your grandmother deposits $25. After you write three checks for $17, $23, and $62, you make a deposit of $41. At the end of the month, your bank removes half of the balance to put in your savings account and then charges you a $5 fee for doing so. How much do you have remaining in your checking account?

89. A particular liquid boils at 268°F. Given the formula $C = \frac{5}{9}(F - 32)$ for converting temperatures from Celsius (C) to Fahrenheit (F), find the boiling point of this liquid in the Celsius scale. Round your answer to the nearest hundredth.

90. Stephen received $75 as a gift from his aunt. With this money, he decided to start saving to buy the newest gaming console, which costs $398 after tax. After working two weeks at his part-time job, he got one check for $123 and a second check for $98. How much more does Stephen need to save to buy his gaming console?

91. Derek's BMI is 23.7 and his weight status is normal.

92. 2.19 square meters

93. 187 ft

94. It is the same number you began with. Explanations will vary.

95. The union contains the citizens with brown hair or blue eyes and the intersection contains the citizens with brown hair and blue eyes.

96. No, because all natural numbers can be expressed as fractions.

97. Yes, all whole numbers are also integers, but the negative integers are not whole numbers, so all integers are not whole numbers.

98. Answers will vary.

99. Answers will vary.

100. Answers will vary.

91. Body mass index, abbreviated BMI, is one way doctors determine an adult's weight status. A BMI below 18.5 is considered underweight, the range 18.5–24.9 is normal, the range 25.0–29.9 is overweight, and a BMI above 30.0 indicates obesity. The formula used to determine BMI is

$$\text{BMI} = 703\left(\frac{\text{weight in pounds}}{(\text{height in inches})^2}\right).$$

Derek weighs 180 pounds and is 73 inches tall. Use this formula to determine Derek's BMI and weight status. Round your answer to the nearest tenth.

92. The Du Bois Method provides a formula used to estimate your body's surface area in meters squared: $\text{BSA} = 0.007184(\text{height})^{0.725}(\text{weight})^{0.425}$ where *height* is in centimeters and *weight* is in kilograms. Assume Juan is 193 cm tall and weighs 88 kg. Use the Du Bois Method to estimate his body's surface area in square meters. Round your answer to the nearest hundredth.

93. Samantha drops a tennis ball from the top of the mathematics building. If it takes the ball 3.42 seconds to hit the ground, use the formula

$$\text{distance} = \frac{1}{2}(\text{acceleration})(\text{time})^2$$

to find the height of the building, which is equivalent to the distance the ball falls. Use the value of 32 ft/s^2 for the acceleration of a falling object. Round your answer to the nearest foot.

3.42 seconds

94. Choose a number. Multiply it by 3 and then add 4. Now multiply by 2 and subtract 8. Finally divide by 6. What do you notice about your final answer? Explain why you got this as a result.

95. After taking a poll in her town, Sally began grouping the citizens into various sets. One set contained all the citizens with brown hair and another set contained all the citizens with blue eyes. What do you know about the citizens who would be listed in the union of these two sets? What do you know about the citizens who would be listed in the intersection of these two sets?

96. Can a natural number be irrational? Explain.

97. Are all whole numbers also integers? Are all integers also whole numbers? Explain your answers.

98. In your own words, define absolute value.

99. Write a short paragraph explaining the similarities and differences between > and ≥.

100. In your own words, describe the difference between a union and an intersection of two sets.

Properties of Exponents and Radicals

TOPICS

1 . Natural number and integer exponents

2 . Properties of exponents

3 . Scientific notation

4 . Working with geometric formulas

5 . Roots and radical notation

6 . Simplifying and combining radical expressions

7 . Rational number exponents

TOPIC

Natural Number and Integer Exponents

As we progress in Chapter 1, we will encounter a variety of algebraic expressions. As discussed in Section 1.1, algebraic expressions consist of constants and variables combined by the basic operations of addition, subtraction, multiplication, and division, along with exponentiation and the taking of roots. In this section, we will explore the meaning of exponentiation and the properties of exponents.

DEFINITION

Natural Number Exponents

If a is any real number and if n is any natural number, then $a^n = \underbrace{a \cdot a \cdot \ldots \cdot a}_{n \text{ factors}}$.

In the expression a^n, a is called the **base**, and n is the **exponent**. The process of multiplying n factors of a is called "raising a to the nth power," and the expression a^n may be referred to as "the nth power of a" or "a to the nth power." Note that a^1 is simply a.

Other phrases are also commonly used to denote the raising of something to a power, especially when the power is 2 or 3. For instance, a^2 is often referred to as "a squared" and a^3 is often referred to as "a cubed."

EXAMPLE 1

Natural Number Exponents

a. $4^3 = 4 \cdot 4 \cdot 4 = 64$. Thus "four cubed is sixty-four."

b. $(-3)^2 = (-3)(-3) = 9$. Thus "negative three, squared, is nine."

c. $-3^4 = -(3 \cdot 3 \cdot 3 \cdot 3) = -81$. Note that, by order of operations, the exponent 4 applies only to the number 3. After raising 3 to the 4th power, the result is multiplied by -1.

d. $-(-2)^3 \cdot 5^2 = -\big((-2) \cdot (-2) \cdot (-2)\big)(5 \cdot 5) = -(-8)(25) = 200$.

e. $x^3 \cdot x^4 = (x \cdot x \cdot x)(x \cdot x \cdot x \cdot x) = x^7$.

The ultimate goal is to give meaning to the expression a^n for any real number n and to do so in such a way that the properties of exponents hold consistently. For example, analysis of Example 1e above leads to the observation that if n and m are natural numbers, then

$$a^n \cdot a^m = \underbrace{a \cdot a \cdot \ldots \cdot a}_{n \text{ factors}} \cdot \underbrace{a \cdot a \cdot \ldots \cdot a}_{m \text{ factors}} = \underbrace{a \cdot \ldots \cdot a \cdot a \cdot \ldots \cdot a}_{n+m \text{ factors}} = a^{n+m}.$$

To extend the meaning of a^n to the case where $n = 0$, we might start by noting that the following statement should be true:

$$a^0 \cdot a^m = a^{0+m} = a^m$$

This suggests the following:

DEFINITION

0 as an Exponent

For any real number $a \neq 0$, we define $a^0 = 1$. 0^0 is **undefined**, just as division by 0 is undefined.

With this small extension of the meaning of exponents, let us continue. Consider the following table:

The exponent is decreased by one at each step.	$3^3 = 27$ $3^2 = 9$ $3^1 = 3$ $3^0 = 1$ $3^{-1} = ?$ $3^{-2} = ?$ $3^{-3} = ?$	The result is $\dfrac{1}{3}$ of the result from the previous line.

In order to maintain the pattern that has begun to emerge, we are led to complete the table with $3^{-1} = \dfrac{1}{3}$, $3^{-2} = \dfrac{1}{9}$, and $3^{-3} = \dfrac{1}{27}$. In general, negative integer exponents are defined as follows.

DEFINITION

Negative Integer
Exponents

For any real number $a \neq 0$ and for any natural number n, $a^{-n} = \dfrac{1}{a^n}$. Since any negative integer is the negative of a natural number, this defines exponentiation by negative integers.

EXAMPLE 2

Simplifying
Exponents

a. $\dfrac{y^2}{y^7} = \dfrac{\cancel{y} \cdot \cancel{y}}{y \cdot y \cdot y \cdot y \cdot y \cdot \cancel{y} \cdot \cancel{y}} = \dfrac{1}{y \cdot y \cdot y \cdot y \cdot y} = \dfrac{1}{y^5} = y^{-5}.$

b. $\dfrac{6x^2}{-3x^2} = \dfrac{6}{-3} = -2.$ Note that the variable x cancels out entirely, if $x \neq 0$.

c. $5^0 \cdot 5^{-3} = 5^{0-3} = 5^{-3} = \dfrac{1}{5^3} = \dfrac{1}{125}.$ Note that $5^0 = 1$, as does a^0 for any $a \neq 0$.

d. $\dfrac{1}{t^{-3}} = \dfrac{1}{\dfrac{1}{t^3}} = 1 \cdot \dfrac{t^3}{1} = t^3.$

e. $\left(x^2 y\right)^3 = \left(x^2 y\right)\left(x^2 y\right)\left(x^2 y\right) = x^2 \cdot x^2 \cdot x^2 \cdot y \cdot y \cdot y = x^6 y^3.$

TOPIC 2 Properties of Exponents

The table below lists the properties of exponents that are used frequently in algebra. Most of these properties have been illustrated already in Examples 1 and 2. All of them can be readily demonstrated by applying the definition of integer exponents.

PROPERTIES

Properties of
Exponents

Throughout this table, a and b may be taken to represent constants, variables, or more complicated algebraic expressions. The letters n and m represent integers.

	Property	Example
1.	$a^n \cdot a^m = a^{n+m}$	$3^3 \cdot 3^{-1} = 3^{3+(-1)} = 3^2 = 9$
2.	$\dfrac{a^n}{a^m} = a^{n-m}$	$\dfrac{7^9}{7^{10}} = 7^{9-10} = 7^{-1}$
3.	$a^{-n} = \dfrac{1}{a^n}$	$5^{-2} = \dfrac{1}{5^2} = \dfrac{1}{25}$ and $x^3 = \dfrac{1}{x^{-3}}$
4.	$\left(a^n\right)^m = a^{nm}$	$\left(2^3\right)^2 = 2^{3 \cdot 2} = 2^6 = 64$
5.	$(ab)^n = a^n b^n$	$(7x)^3 = 7^3 x^3 = 343x^3$ and $\left(-2x^5\right)^2 = (-2)^2 \left(x^5\right)^2 = 4x^{10}$
6.	$\left(\dfrac{a}{b}\right)^n = \dfrac{a^n}{b^n}$	$\left(\dfrac{3}{x}\right)^2 = \dfrac{3^2}{x^2} = \dfrac{9}{x^2}$ and $\left(\dfrac{1}{3z}\right)^2 = \dfrac{1^2}{(3z)^2} = \dfrac{1}{9z^2}$

Assume every expression is defined for the given properties.

EXAMPLE 3

Properties of Exponents

Simplify the following expressions by using the properties of exponents. Write the final answers with only positive exponents. (As in the table of properties, it is assumed that every expression is defined.)

a. $\left(17x^4 + 5x^2 + 2\right)^0$

b. $\dfrac{\left(x^2y^3\right)^{-1}z^{-2}}{x^3z^{-3}}$

c. $\dfrac{\left(-2x^3y^{-1}\right)^{-3}}{\left(18x^{-3}\right)^0 (xy)^{-2}}$

d. $\left(7xz^{-2}\right)^2\left(5x^2y\right)^{-1}$

Note:
There are often many ways to simplify an expression; the order in which you apply the properties of exponents will not change the result.

Solutions:

a. $\left(17x^4 + 5x^2 + 2\right)^0 = 1$

Any nonzero expression with an exponent of 0 is equal to 1.

b. $\dfrac{\left(x^2y^3\right)^{-1}z^{-2}}{x^3z^{-3}} = \dfrac{x^{-2}y^{-3}z^{-2}}{x^3z^{-3}}$

Apply Property 4.

$= \dfrac{z^3}{x^3x^2y^3z^2}$

Apply Property 3 several times to reach this point.

$= \dfrac{z}{x^5y^3}$

Then apply Properties 1 and 2.

c. $\dfrac{\left(-2x^3y^{-1}\right)^{-3}}{\left(18x^{-3}\right)^0 (xy)^{-2}} = \dfrac{(-2)^{-3}x^{-9}y^3}{x^{-2}y^{-2}}$

Begin by applying Property 4 in the numerator, Property 5 in the denominator.

$= (-2)^{-3}x^{-9-(-2)}y^{3-(-2)}$

Then simplify using Property 2.

$= (-2)^{-3}x^{-7}y^5$

$= \dfrac{y^5}{-8x^7}$

Unlike the previous example, Property 3 gets applied at the very end.

$= -\dfrac{y^5}{8x^7}$

d. $\left(7xz^{-2}\right)^2\left(5x^2y\right)^{-1} = \dfrac{49x^2z^{-4}}{5x^2y}$

$= \dfrac{49}{5yz^4}$

Note that the variable x no longer appears in the expression.

CAUTION! 〰〰〰〰〰〰〰〰〰〰〰〰〰〰〰〰〰〰〰〰〰〰〰〰〰〰〰〰〰〰〰

Many errors can be made in applying the properties of exponents as a result of forgetting the exact form of the properties. The first column below contains examples of some common errors. The second column contains the corrected statements.

Incorrect Statements	Correct Statements
$x^2 x^5 = x^{10}$	$x^2 x^5 = x^{2+5} = x^7$
$2^4 2^3 = 4^7$	$2^4 2^3 = 2^{4+3} = 2^7$
$(3+4)^2 = 3^2 + 4^2$	$(3+4)^2 = 7^2$
$(x^2 + 3y)^{-1} = \dfrac{1}{x^2} + \dfrac{1}{3y}$	$(x^2 + 3y)^{-1} = \dfrac{1}{x^2 + 3y}$
$(3x)^2 = 3x^2$	$(3x)^2 = 3^2 x^2 = 9x^2$
$\dfrac{x^5}{x^{-2}} = x^3$	$\dfrac{x^5}{x^{-2}} = x^{5-(-2)} = x^7$

TOPIC 3

Scientific Notation

Scientific notation is an important application of exponents. Scientific notation uses the properties of exponents to rewrite very large and very small numbers in a less clumsy form. Very large and very small numbers arise naturally in a variety of situations, and working with them without scientific notation is an unwieldy and error-prone process.

DEFINITION ═══════════════════════════════════════

Scientific Notation

A number is in **scientific notation** when it is written in the form

$$a \times 10^n$$

where $1 \le |a| < 10$ and n is an integer. If n is a positive integer, the number is large in magnitude, and if n is a negative integer, the number is small in magnitude (close to 0). The number a itself can be either positive or negative, and the sign of a determines the sign of the number as a whole.

CAUTION! 〰〰〰〰〰〰〰〰〰〰〰〰〰〰〰〰〰〰〰〰〰〰〰〰〰〰〰〰〰〰〰

The sign of the exponent n in scientific notation does *not* determine the sign of the number as a whole. The sign of n only determines if the number is large (positive n) or small (negative n) in magnitude.

EXAMPLE 4

Scientific Notation

a. The distance from Earth to the Sun is approximately 93,000,000 miles. The scientific notation 9.3×10^7 is equal to 93,000,000, as 93,000,000 is obtained from 9.3 by moving the decimal point 7 places to the right:

$$9.3 \times 10^7 = 9\underbrace{3000000}_{7\ places}$$

b. The mass of an electron, in kilograms, is approximately

$$0.00000000000000000000000000000911,$$

clearly not a convenient number to work with. We can count that the decimal point has been moved 31 places to the left, beginning with 9.11. Thus in scientific notation,

$$0.00000000000000000000000000000911 = 9.11 \times 10^{-31}.$$

c. The speed of light in a vacuum is approximately 3×10^8 meters/second. In standard (nonscientific) notation, this number is written as 300,000,000.

We can also use the properties of exponents to simplify computations involving two or more numbers that are large or small in magnitude, as illustrated by the next set of examples.

EXAMPLE 5

Simplifying Expressions with Scientific Notation

Simplify the following expressions, writing your answer either in scientific or standard notation, as appropriate.

a. $\dfrac{\left(3.6 \times 10^{-12}\right)\left(-6 \times 10^4\right)}{1.8 \times 10^{-6}}$

b. $\dfrac{\left(7 \times 10^{34}\right)\left(3 \times 10^{-12}\right)}{6 \times 10^{-7}}$

Note:
We use the associative property to multiply the powers of 10 separately from the remaining values.

Solutions:

a. $\dfrac{\left(3.6 \times 10^{-12}\right)\left(-6 \times 10^4\right)}{1.8 \times 10^{-6}} = \dfrac{(3.6)(-6)}{1.8} \times 10^{-12+4-(-6)}$

$= -12 \times 10^{-2}$

$= -0.12$

This answer can be written conveniently in standard notation.

b. $\dfrac{\left(7 \times 10^{34}\right)\left(3 \times 10^{-12}\right)}{6 \times 10^{-7}} = \dfrac{(7)(3)}{6} \times 10^{34+(-12)-(-7)}$

$= 3.5 \times 10^{29}$

This answer is best written in scientific notation.

TOPIC 4

Working with Geometric Formulas

Exponents occur frequently when geometric formulas are considered. Some problems require nothing more than using a basic geometric formula, but others will require a bit more work. Often, the exact geometric formula that you need to solve a given problem can be derived from simpler formulas.

We will look at several examples of how a new geometric formula is built up from known formulas. The general rule of thumb in each case is to break down the task at hand into smaller pieces that can be easily handled.

EXAMPLE 6

Geometric Formulas

Find formulas for each of the following:

a. The surface area of a box

b. The surface area of a soup can

c. The volume of a birdbath in the shape of half of a sphere

d. The volume of a gold bar whose shape is a right trapezoidal cylinder

Note:
For reference, many basic geometric formulas are listed on the inside front cover of this book.

Solutions:

a. A box whose six faces are all rectangular is characterized by its length l, its width w, and its height h. The formula for the surface area of a box, is just the sum of the areas of the six sides, and the formula for the area of a rectangle $A = l \cdot w$ can be applied separately to each side. If we let S stand for the total surface area, we obtain the formula $S = lw + lw + lh + lh + hw + hw$ or $S = 2lw + 2lh + 2hw$.

b. A soup can, an example of a right circular cylinder, is characterized by its height h and the radius r of the circle that makes up the base (or the top). To determine the surface area of this shape, imagine removing the top and bottom surfaces, cutting the soup can along the dotted line as shown, and flattening out the curved piece of metal making up the side. The flattened piece of metal is a rectangle with height h and width $2\pi r$. Do you see why? The width of the rectangle is the same as the circumference of the circular top and base, and the circumference of a circle is $2\pi r$. Thus the surface area of the curved side is $2\pi rh$. We also know that the area of a circle is πr^2, so if we let S stand for the surface area of the entire can, we have $S = \pi r^2 + \pi r^2 + 2\pi rh$, or $S = 2\pi r^2 + 2\pi rh$.

c. The volume of a sphere of radius r is $\frac{4}{3}\pi r^3$, and the birdbath has the shape of half a sphere. So if we let V stand for the birdbath's volume,

$$V = \left(\frac{1}{2}\right)\left(\frac{4}{3}\pi r^3\right), \text{ or } V = \frac{2}{3}\pi r^3.$$

d. A *right cylinder* is the three-dimensional object generated by extending a plane region along an axis perpendicular to itself for a certain distance. (Such objects are often called prisms when the plane region is a polygon.) The volume of any right cylinder is the product of the area of the plane region and the distance that region has been extended perpendicular to itself. The gold bar in this example is a right cylinder based on a trapezoid, as shown. The area of a trapezoid is $\frac{1}{2}(B+b)h$ and the bar has length l, so its volume is $V = \frac{1}{2}(B+b)hl$. This can also be written as $V = \frac{(B+b)hl}{2}$.

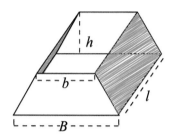

TOPIC 5   Roots and Radical Notation

Taking the nth root of an expression (where n is a natural number) is the opposite operation of exponentiation by n. For example, the equation $4^3 = 64$ (which can be read as "four cubed is sixty-four") implies that the cube root of 64 is 4 (written $\sqrt[3]{64} = 4$). Similarly, the equation $2^4 = 16$ leads us to write $\sqrt[4]{16} = 2$ (read "the fourth root of sixteen is two").

DEFINITION

Radical Notation

Case 1: n is an even natural number. If a is a nonnegative real number and n is an even natural number, $\sqrt[n]{a}$ is the nonnegative real number b with the property that $b^n = a$. That is, $\sqrt[n]{a} = b$ if and only if $a = b^n$. Note that $\left(\sqrt[n]{a}\right)^n = a$ and $\sqrt[n]{a^n} = a$.

Case 2: n is an odd natural number. If a is any real number and n is an odd natural number, $\sqrt[n]{a}$ is the real number b (whose sign will be the same as the sign of a) with the property that $b^n = a$. Again, $\sqrt[n]{a} = b$ if and only if $a = b^n$, $\left(\sqrt[n]{a}\right)^n = a$, and $\sqrt[n]{a^n} = a$.

The expression $\sqrt[n]{a}$ gives the nth root of a in **radical notation**. The natural number n is called the **index**, a is the **radicand**, and $\sqrt{}$ is called a **radical sign**. By convention, $\sqrt[2]{}$ is usually simply written as $\sqrt{}$.

The important distinction between the two cases is that when n is even, $\sqrt[n]{a}$ is defined only when a is nonnegative, whereas if n is odd, $\sqrt[n]{a}$ is defined for all real numbers a. This difference will be remedied in Section 1.4 by the introduction of *complex numbers*.

DEFINITION

Perfect Powers

A **perfect square** is an integer equal to the square of another integer. The square root of a perfect square is always an integer.

A **perfect cube** is an integer equal to the cube of another integer. The cube root of a perfect cube is always an integer.

EXAMPLE 7

Radical Notation

a. $\sqrt[5]{-32} = -2$ because $(-2)^5 = -32$.

b. $\sqrt[4]{-16}$ is not a real number, as no real number raised to the fourth power is -16.

c. $-\sqrt[4]{16} = -2$. Note that the fourth root of 16 is a real number, which is then multiplied by -1.

d. $\sqrt{0} = 0$. In fact, $\sqrt[n]{0} = 0$ for any natural number n.

e. $\sqrt[n]{1} = 1$ for any natural number n. $\sqrt[n]{-1} = -1$ for any odd natural number n.

f. $\sqrt[3]{-\dfrac{27}{64}} = -\dfrac{3}{4}$ because $\left(-\dfrac{3}{4}\right)^3 = -\dfrac{27}{64}$.

g. $\sqrt[5]{-\pi^5} = -\pi$ because $(-\pi)^5 = (-1)^5 \pi^5 = -\pi^5$.

h. $\sqrt[4]{(-3)^4} = \sqrt[4]{81} = 3$. In general, if n is an even natural number, $\sqrt[n]{a^n} = |a|$ for any real number a. Remember, though, that $\sqrt[n]{a^n} = a$ if n is an odd natural number.

EXAMPLE 8

The Pythagorean Theorem

Given a right triangle with sides of length a and b, the Pythagorean Theorem states that the length of the hypotenuse c is given by $c = \sqrt{a^2 + b^2}$. In the Pythagorean Theorem, find the following:

a. The radicand

b. The index

c. The value of c if $a = 5$ and $b = 12$

d. The value of c if $a = 1$ and $b = 2$

Solutions:

a. The radicand is the quantity beneath the radical sign, $a^2 + b^2$.

b. Because no index is indicated, the index is 2.

c. $c = \sqrt{a^2 + b^2}$

$c = \sqrt{(5)^2 + (12)^2}$ Substitute.

$c = \sqrt{169}$ Since 169 is a perfect square, the

$c = 13$ solution is an integer.

d. $c = \sqrt{a^2 + b^2}$

$c = \sqrt{(1)^2 + (2)^2}$ Substitute.

$c = \sqrt{5}$ Since 5 is not a perfect square, the solution is not an integer.

TOPIC 6

Simplifying and Combining Radical Expressions

When solving equations that contain radical expressions, it is often helpful to simplify the expressions first. The following definition establishes clear rules for simplifying radical expressions.

DEFINITION

Simplified Form of Radical Expressions

A radical expression is in **simplified form** when:

1. The radicand contains no factor with an exponent greater than or equal to the index of the radical.

2. The radicand contains no fractions.

3. The denominator, if there is one, contains no radical.

4. The greatest common factor of the index and any exponent occurring in the radicand is 1. That is, the index and any exponent in the radicand have no common factor other than 1.

Before illustrating the process of simplifying radicals, we will review a few useful properties of radicals. These properties can be proved using nothing more than the definition of roots, but their validity will be more clear once we have discussed rational exponents later in this section.

PROPERTIES

Properties of Radicals

Let a and b represent constants, variables, or more complicated algebraic expressions. The letters n and m represent natural numbers. Assume that all expressions are defined and are real numbers.

	Property	Example
1.	$\sqrt[n]{ab} = \sqrt[n]{a}\sqrt[n]{b}$	$\sqrt[3]{3x^6y^2} = \sqrt[3]{3} \cdot \sqrt[3]{x^3} \cdot \sqrt[3]{x^3} \cdot \sqrt[3]{y^2} = \sqrt[3]{3} \cdot x \cdot x \cdot \sqrt[3]{y^2} = x^2\sqrt[3]{3y^2}$
2.	$\sqrt[n]{\dfrac{a}{b}} = \dfrac{\sqrt[n]{a}}{\sqrt[n]{b}}$	$\sqrt[4]{\dfrac{x^4}{16}} = \dfrac{\sqrt[4]{x^4}}{\sqrt[4]{16}} = \dfrac{|x|}{2}$
3.	$\sqrt[m]{\sqrt[n]{a}} = \sqrt[mn]{a}$	$\sqrt[3]{\sqrt{64}} = \sqrt[3]{\sqrt[2]{64}} = \sqrt[6]{64} = 2$

EXAMPLE 9

Simplifying Radical Expressions

Simplify the following radical expressions:

a. $\sqrt[3]{-16x^8y^4}$ **b.** $\sqrt{8z^6}$ **c.** $\sqrt[3]{\dfrac{72x^2}{y^3}}$

Note:
Begin by factoring the radicand, looking for perfect powers that match the index. This makes it easier to recognize what terms can be "pulled out" of the radical.

Solutions:

a. $\sqrt[3]{-16x^8y^4} = \sqrt[3]{(-2)^3 \cdot 2 \cdot x^3 \cdot x^3 \cdot x^2 \cdot y^3 \cdot y}$ Factor out perfect cubes inside the radical.

$\qquad = -2x^2y\,\sqrt[3]{2x^2y}$

b. $\sqrt{8z^6} = \sqrt{2^2 \cdot 2 \cdot \left(z^3\right)^2}$ Factor out perfect squares.

$\qquad = |2z^3|\sqrt{2}$

$\qquad = 2|z^3|\sqrt{2}$ Since the index, 2, is even, absolute value signs are needed around the factor of z^3.

c. $\sqrt[3]{\dfrac{72x^2}{y^3}} = \dfrac{\sqrt[3]{8 \cdot 9 \cdot x^2}}{\sqrt[3]{y^3}}$

$\qquad = \dfrac{2\sqrt[3]{9x^2}}{y}$ Note that, in this case, simplifying rationalizes the denominator.

CAUTION!

As with the properties of exponents, many mistakes arise from forgetting the properties of radicals. One common error is to rewrite $\sqrt{a+b}$ as $\sqrt{a} + \sqrt{b}$. These two expressions are not equal! To convince yourself of this, evaluate the two expressions with actual constants in place of a and b. Using $a = 9$ and $b = 16$, we see that $5 = \sqrt{9+16} \neq \sqrt{9} + \sqrt{16} = 7$.

Rationalizing denominators sometimes requires more effort than in Example 3c, while sometimes it is impossible! The following methods will, however, take care of two common cases.

Case 1: Denominator is a single term containing a root.

If the denominator is a single term containing a factor of $\sqrt[n]{a^m}$, we take advantage of the fact that $\sqrt[n]{a^m} \cdot \sqrt[n]{a^{n-m}} = \sqrt[n]{a^m \cdot a^{n-m}} = \sqrt[n]{a^n}$ and that this last expression is either a or $|a|$, depending on whether n is odd or even. Now, if we multiply the denominator by a factor of $\sqrt[n]{a^{n-m}}$ we must also multiply the numerator by the same factor. Thus in this case we multiply the fraction by $\dfrac{\sqrt[n]{a^{n-m}}}{\sqrt[n]{a^{n-m}}}$.

For example, $\dfrac{1}{\sqrt{a}} = \dfrac{1}{\sqrt{a}} \cdot \dfrac{1}{1} = \dfrac{1}{\sqrt{a}} \cdot \dfrac{\sqrt{a}}{\sqrt{a}} = \dfrac{\sqrt{a}}{a}$.

Case 2: Denominator consists of two terms, one or both of which are square roots.

Let $A + B$ represent the denominator of the fraction under consideration, where at least one of A and B stands for a square root term. We will take advantage of the fact that $(A+B)(A-B) = A^2 - B^2$ and that the exponents of 2 will eliminate the square root (or roots) initially in the denominator. Just as in case 1, we can't multiply the denominator by $A - B$ unless we multiply the numerator by this same factor. The method is thus to multiply the fraction by $\dfrac{A-B}{A-B}$. The factor $A - B$ is called the **conjugate radical** expression of $A + B$.

For example, $\dfrac{1}{\sqrt{a} + \sqrt{b}} = \dfrac{1}{\sqrt{a} + \sqrt{b}} \cdot \dfrac{1}{1} = \dfrac{1}{\sqrt{a} + \sqrt{b}} \cdot \dfrac{\sqrt{a} - \sqrt{b}}{\sqrt{a} - \sqrt{b}} = \dfrac{\sqrt{a} - \sqrt{b}}{a - b}$.

EXAMPLE 10

Rationalizing the Denominator

Simplify the following radical expressions:

a. $\dfrac{1}{\sqrt{x}}$

b. $\sqrt[5]{\dfrac{-4x^6}{8y^2}}$

c. $\dfrac{4}{\sqrt{7} + \sqrt{3}}$

d. $\dfrac{-\sqrt{5x}}{5 - \sqrt{x}}$

Note:
The first two examples
follow Case 1. In
general, it still helps to
factor the radicands
before further
simplifying.
The second pair
of examples have
two terms in the
denominator, and
thus follow Case 2.
Begin by multiplying
the numerator and
denominator by the
conjugate radical of
the denominator.

Solutions:

a. $\dfrac{1}{\sqrt{x}} = \dfrac{1}{\sqrt{x}} \cdot \dfrac{\sqrt{x}}{\sqrt{x}}$

$\qquad = \dfrac{\sqrt{x}}{x}$

Multiply the numerator and denominator by $\sqrt{x}$.

b. $\sqrt[5]{\dfrac{-4x^6}{8y^2}} = \dfrac{\sqrt[5]{-4x \cdot x^5}}{\sqrt[5]{8y^2}}$

$\qquad = \dfrac{-x\sqrt[5]{4x}}{\sqrt[5]{2^3 y^2}} \cdot \dfrac{\sqrt[5]{2^2 y^3}}{\sqrt[5]{2^2 y^3}}$

$\qquad = \dfrac{-x\sqrt[5]{16xy^3}}{\sqrt[5]{2^5 y^5}}$

$\qquad = \dfrac{-x\sqrt[5]{16xy^3}}{2y}$

Since $2^3 y^2 \cdot 2^2 y^3 = 2^5 y^5$, a perfect fifth power, multiply the numerator and denominator by $\sqrt[5]{2^2 y^3}$.

c. $\dfrac{4}{\sqrt{7}+\sqrt{3}} = \left(\dfrac{4}{\sqrt{7}+\sqrt{3}}\right)\left(\dfrac{\sqrt{7}-\sqrt{3}}{\sqrt{7}-\sqrt{3}}\right)$

$\qquad = \dfrac{4\left(\sqrt{7}-\sqrt{3}\right)}{7-3}$

$\qquad = \dfrac{4\left(\sqrt{7}-\sqrt{3}\right)}{4}$

$\qquad = \sqrt{7}-\sqrt{3}$

Multiply the numerator and denominator by the conjugate of the denominator, then simplify.

d. $\dfrac{-\sqrt{5x}}{5-\sqrt{x}} = \left(\dfrac{-\sqrt{5x}}{5-\sqrt{x}}\right)\left(\dfrac{5+\sqrt{x}}{5+\sqrt{x}}\right)$

$\qquad = \dfrac{-5\sqrt{5x}-\sqrt{5x^2}}{25-x}$

$\qquad = \dfrac{-5\sqrt{5x}-x\sqrt{5}}{25-x}$

Multiply the numerator and denominator by the conjugate of the denominator.

Simplify. Note that we can write $-x\sqrt{5}$ instead of $-|x|\sqrt{5}$ since the original expression is not real if x is negative.

Frequently, a sum of two or more radical expressions can be combined into one. This can be done if the radical expressions are **like radicals**, meaning that they have the same index and the same radicand. Often, you may have to simplify the radical expressions before determining if they are like or not.

Combining Radicals

EXAMPLE 11

Combine the radical expressions, if possible.

a. $-3\sqrt{8x^5} + \sqrt{18x}$ **b.** $\sqrt[3]{54x^3} + \sqrt{50x^2}$ **c.** $\sqrt{\dfrac{1}{12}} - \sqrt{\dfrac{25}{48}}$

Solutions:

a. $-3\sqrt{8x^5} + \sqrt{18x} = -3\sqrt{2^2 \cdot 2 \cdot x^4 \cdot x} + \sqrt{2 \cdot 3^2 \cdot x}$ Simplify each radical separately.

$\qquad\qquad\qquad = -6x^2\sqrt{2x} + 3\sqrt{2x}$ Now we can see that the radicals have the same index and radicand.

$\qquad\qquad\qquad = \left(-6x^2 + 3\right)\sqrt{2x}$

b. $\sqrt[3]{54x^3} + \sqrt{50x^2} = \sqrt[3]{2 \cdot 3^3 \cdot x^3} + \sqrt{2 \cdot 5^2 \cdot x^2}$ The radicands are the same, but the indices are not, so the terms cannot be combined.

$\qquad\qquad\qquad = 3x\sqrt[3]{2} + 5|x|\sqrt{2}$

c. $\sqrt{\dfrac{1}{12}} - \sqrt{\dfrac{25}{48}} = \dfrac{1}{\sqrt{2^2 \cdot 3}} - \dfrac{\sqrt{5^2}}{\sqrt{4^2 \cdot 3}}$ Simplify the radicals.

$\qquad\qquad\qquad = \dfrac{1}{2\sqrt{3}} \cdot \dfrac{\sqrt{3}}{\sqrt{3}} - \dfrac{5}{4\sqrt{3}} \cdot \dfrac{\sqrt{3}}{\sqrt{3}}$ Rationalize denominators.

$\qquad\qquad\qquad = \dfrac{2\sqrt{3}}{4 \cdot 3} - \dfrac{5 \cdot \sqrt{3}}{4 \cdot 3}$ Multiply the first term by $\dfrac{2}{2}$ to get a common denominator.

$\qquad\qquad\qquad = -\dfrac{3\sqrt{3}}{12} = -\dfrac{\sqrt{3}}{4}$

TOPIC

Rational Number Exponents

We can now return to defining exponentiation and give meaning to a^r when r is a rational number.

DEFINITION

Rational Number
Exponents

Meaning of $a^{\frac{1}{n}}$: If n is a natural number and if $\sqrt[n]{a}$ is a real number, then $a^{\frac{1}{n}} = \sqrt[n]{a}$.

Meaning of $a^{\frac{m}{n}}$: If m and n are natural numbers with $n \neq 0$, if m and n have no common factors greater than 1, and if $\sqrt[n]{a}$ is a real number, then $a^{\frac{m}{n}} = \sqrt[n]{a^m} = \left(\sqrt[n]{a}\right)^m$. Either $\sqrt[n]{a^m}$ or $\left(\sqrt[n]{a}\right)^m$ can be used to evaluate $a^{\frac{m}{n}}$, as they are equal. $a^{-\frac{m}{n}}$ is defined to be $\dfrac{1}{a^{\frac{m}{n}}}$.

In addition to giving meaning to rational exponentiation, this definition describes how to convert between radical notation and exponential notation. Often, one notation is much more convenient than the other, so converting between the two can be a crucial step in solving problems.

Although originally stated only for integer exponents, the properties of exponents listed earlier also hold for rational exponents (and for real exponents as well).

EXAMPLE 12

Simplifying Expressions

Simplify each of the following expressions, writing your answer using the same notation as the original expression.

a. $27^{-\frac{2}{3}}$

b. $\sqrt[9]{-8x^6}$

c. $\left(5x^2+3\right)^{\frac{8}{3}}\left(5x^2+3\right)^{-\frac{2}{3}}$

d. $\sqrt[5]{\sqrt[3]{x^2}}$

e. $\dfrac{5x-y}{\left(5x-y\right)^{-\frac{1}{3}}}$

Note:
When simplifying, it is good practice to write the final answer in the same form as the original expression. However, it is often useful to convert between notations to make available a particular method of simplification. The first two examples exhibit this type of strategy.

Solutions:

a. $27^{-\frac{2}{3}}=\left(27^{\frac{1}{3}}\right)^{-2}$

$=3^{-2}$

$=\dfrac{1}{3^2}$

$=\dfrac{1}{9}$

Writing $27^{-\frac{2}{3}}=\left(27^{-2}\right)^{\frac{1}{3}}$ is also a valid first step, but it leads to a messier calculation.

b. $\sqrt[9]{-8x^6}=-\sqrt[9]{2^3 x^6}$

$=-2^{\frac{3}{9}}x^{\frac{6}{9}}$

$=-2^{\frac{1}{3}}x^{\frac{2}{3}}$

$=-\sqrt[3]{2x^2}$

Rewrite the expression using rational exponents.

Note that the exponents in the radicand and the index now have no common factors other than 1.

c. $\left(5x^2+3\right)^{\frac{8}{3}}\left(5x^2+3\right)^{-\frac{2}{3}}=\left(5x^2+3\right)^{\left(\frac{8}{3}\right)-\left(\frac{2}{3}\right)}$

$=\left(5x^2+3\right)^{\frac{6}{3}}$

$=\left(5x^2+3\right)^2$

The bases are the same, so we add the exponents.

d. $\sqrt[5]{\sqrt[3]{x^2}}=\sqrt[15]{x^2}$

Apply the property $\sqrt[m]{\sqrt[n]{a}}=\sqrt[mn]{a}$.

e. $\dfrac{5x-y}{\left(5x-y\right)^{\frac{-1}{3}}}=\left(5x-y\right)^{1-\left(\frac{-1}{3}\right)}$

$=\left(5x-y\right)^{\frac{4}{3}}$

Apply the property $\dfrac{a^n}{a^m}=a^{n-m}$ to write the expression as a single term.

The following examples are a bit more complex.

EXAMPLE 13

Simplifying Radical Expressions

a. Simplify the expression $\sqrt[4]{x^2}$.

b. Write $\sqrt[3]{2} \cdot \sqrt{3}$ as a single radical.

Solutions:

a. $\sqrt[4]{x^2} = \left(x^2\right)^{\frac{1}{4}}$

$= |x|^{\frac{1}{2}}$

$= \sqrt{|x|}$

Since the original expression is defined for all real numbers, but $\sqrt{x}$ is defined only for nonnegative real numbers, we need absolute value bars.

b. $\sqrt[3]{2} \cdot \sqrt{3} = 2^{\frac{1}{3}} \cdot 3^{\frac{1}{2}}$

$= 2^{\frac{2}{6}} \cdot 3^{\frac{3}{6}}$

$= \left(2^2\right)^{\frac{1}{6}} \left(3^3\right)^{\frac{1}{6}}$

$= 4^{\frac{1}{6}} \cdot 27^{\frac{1}{6}}$

$= 108^{\frac{1}{6}}$

$= \sqrt[6]{108}$

Rewrite the expression using rational exponents.

Write the two exponents with a common denominator.

By applying $a^{nm} = \left(a^n\right)^m$ we can give both terms the same exponent.

This allows us to combine the terms using the property $a^n b^n = \left(ab\right)^n$.

1. 16

2. -16

3. 81

4. $\dfrac{1}{7}$

5. x^3

6. n^7

7. $\dfrac{3}{t^5}$

8. x^3

9. $2n^8$

10. $\dfrac{x^8 z^3}{y^3}$

11. $\dfrac{y^2}{s^3 z^{11}}$

12. $\dfrac{1}{3y^2 z}$

13. $\dfrac{1}{9m^2 - 8n^6}$

14. $\dfrac{c^2}{9a^7 b^3}$

15. $\dfrac{25m^8}{n^4}$

Exercises

Simplify each of the following expressions, writing your answer with only positive exponents. See Examples 1 and 2.

1. $(-2)^4$

2. -2^4

3 $3^2 \cdot 3^2$

4. $\dfrac{7^4}{7^5}$

5. $\dfrac{x^5}{x^2}$

6. $n^2 \cdot n^5$

Use the properties of exponents to simplify each of the following expressions, writing your answer with only positive exponents. See Examples 1, 2, and 3.

7. $\dfrac{3t^{-2}}{t^3}$

8. $9^0 x^3 y^0$

9. $\dfrac{2n^3}{n^{-5}}$

10. $\dfrac{x^7 y^{-3} z^{12}}{x^{-1} z^9}$

11. $\dfrac{s^5 y^{-5} z^{-11}}{s^8 y^{-7}}$

12. $\dfrac{\left(3yz^{-2}\right)^0}{3y^2 z}$

13. $\left[9m^2 - \left(2n^2\right)^3\right]^{-1}$

14. $\dfrac{\left(-3a\right)^{-2} \left(bc^{-2}\right)^{-3}}{a^5 c^4}$

15. $\left[\left(5m^4 n^{-2}\right)^{-1}\right]^{-2}$

16. $\dfrac{64a^6}{b^{15}}$

17. $27x^9$

18. x^3y^{12}

19. $\dfrac{1}{5z^6-81x^{12}}$

20. $9z^2$

21. -9.12×10^8

22. 2.1×10^{-7}

23. $32,000,000$

24. 0.0001934

25. 3.1536×10^7

26. $3,000,000,000$

27. 0.0003937

28. 2.605×10^{-7}

29. 2.8×10^{-3}

30. 1.2×10^{-9}

31. -11

32. 1.5×10^8

33. 1×10^{-8}

34. 1.2×10^{13}

35.–37. Answers will vary.

38. -3

39. -3

40. Not real

41. $-\dfrac{3}{5}$

42. $\dfrac{5}{11}$

43. 2

44. 2

45. $\dfrac{2}{5}$

16. $\left[\left(4a^2b^{-5}\right)^{-1}\right]^{-3}$ **17.** $\left[\left(3^{-1}x^{-1}y\right)\left(x^2y\right)^{-1}\right]^{-3}$ **18.** $\left[\dfrac{100^0\left(x^{-1}y^3\right)^{-1}}{x^2y}\right]^{-3}$

19. $\left(5z^6-\left(3x^3\right)^4\right)^{-1}$ **20.** $\left[\dfrac{y^6\left(xy^2\right)^{-3}}{3x^{-3}z}\right]^{-2}$

Convert each number from scientific notation to standard notation, or vice versa, as indicated. See Example 4.

21. $-912,000,000$; convert to scientific   **22.** 0.00000021; convert to scientific

23. 3.2×10^7; convert to standard   **24.** 1.934×10^{-4}; convert to standard

25. There are approximately $31,536,000$ seconds in a calendar year. Express the number of seconds in scientific notation.

26. Together, the 46 human chromosomes are estimated to contain some 3.0×10^9 base pairs of DNA. Express the number of pairs of DNA in standard notation.

27. A white blood cell is approximately 3.937×10^{-4} inches in diameter. Express this diameter in standard notation.

28. The probability of winning the lottery with one dollar is approximately 0.0000002605. Express this probability in scientific notation.

Evaluate each expression using the properties of exponents. Use a calculator only to check your final answer. See Example 5.

29. $\dfrac{\left(2\times10^3\right)\left(7\times10^{-2}\right)}{\left(5\times10^4\right)}$ **30.** $\dfrac{\left(8\times10^{-3}\right)\left(3\times10^{-2}\right)}{\left(2\times10^5\right)}$

31. $\left(2\times10^{-13}\right)\left(5.5\times10^{10}\right)\left(-1\times10^3\right)$ **32.** $\left(6\times10^{21}\right)\left(5\times10^{-19}\right)\left(5\times10^4\right)$

33. $\dfrac{4\times10^{-6}}{\left(5\times10^4\right)\left(8\times10^{-3}\right)}$ **34.** $\dfrac{\left(4.6\times10^{12}\right)\left(9\times10^3\right)}{\left(1.5\times10^8\right)\left(2.3\times10^{-5}\right)}$

Apply the definition of integer exponents to demonstrate the following properties.

35. $a^n\cdot a^m=a^{n+m}$ **36.** $\left(a^n\right)^m=a^{nm}$ **37.** $(ab)^n=a^nb^n$

Evaluate the following radical expressions. See Example 7.

38. $-\sqrt{9}$ **39.** $\sqrt[3]{-27}$ **40.** $\sqrt{-25}$

41. $\sqrt[3]{-\dfrac{27}{125}}$ **42.** $\sqrt{\dfrac{25}{121}}$ **43.** $-\sqrt[3]{-8}$

44. $\sqrt[4]{\sqrt{16}-\sqrt[3]{-27}+\sqrt{81}}$ **45.** $\sqrt{\dfrac{\sqrt[3]{-64}}{-\sqrt{144}-\sqrt{169}}}$

46. $-2x^2y^3$

47. $\left|x^3\right|\sqrt{2y}$

48. $x^2y^7z^3$

49. $\dfrac{1}{2\left|xy^3\right|}$

50. $\dfrac{ab^4}{3c^2}$

51. $\dfrac{\left|x^3\right|y^2}{2}$

52. $\dfrac{y^6z^5}{2x^7}$

53. $2xy^2\sqrt[5]{x^2}$

54. $\dfrac{\sqrt[3]{36x^2y^2}}{3y^2}$

55. $-\dfrac{|a|\sqrt{2}}{2}$

56. $2\sqrt{7}+2\sqrt{2}$

57. $\sqrt{6}+\sqrt{3}$

58. $\dfrac{x+\sqrt{2x}}{x-2}$

59. $\dfrac{x+2\sqrt{xy}+y}{x-y}$

60. $3x\sqrt[3]{2x}$

61. $-|y|\sqrt{3x}$

62. Not possible

63. 0

64. $5x^2\sqrt[5]{x^3}$

65. $4z\sqrt[3]{2z}$

66. $\left|x^3\right|$

67. $\dfrac{1}{8}$

68. $\left(3x^2-4\right)^2$

69. 27

70. 4

71. $\dfrac{1}{125}$

Simplify the following radical expressions. See Example 9.

46. $\sqrt[3]{-8x^6y^9}$ **47.** $\sqrt{2x^6y}$ **48.** $\sqrt[7]{x^{14}y^{49}z^{21}}$

49. $\sqrt{\dfrac{x^2}{4x^4y^6}}$ **50.** $\sqrt[3]{\dfrac{a^3b^{12}}{27c^6}}$ **51.** $\sqrt[4]{\dfrac{x^{12}y^8}{16}}$

52. $\sqrt[5]{\dfrac{y^{30}z^{25}}{32x^{35}}}$ **53.** $\sqrt[5]{32x^7y^{10}}$

Simplify the following radicals by rationalizing the denominators. See Example 10.

54. $\sqrt[3]{\dfrac{4x^2}{3y^4}}$ **55.** $\dfrac{-\sqrt{3a^3}}{\sqrt{6a}}$ **56.** $\dfrac{10}{\sqrt{7}-\sqrt{2}}$

57. $\dfrac{3}{\sqrt{6}-\sqrt{3}}$ **58.** $\dfrac{\sqrt{x}}{\sqrt{x}-\sqrt{2}}$ **59.** $\dfrac{\sqrt{x}+\sqrt{y}}{\sqrt{x}-\sqrt{y}}$

Combine the radical expressions, if possible. See Example 11.

60. $\sqrt[3]{-16x^4}+5x\sqrt[3]{2x}$ **61.** $\sqrt{27xy^2}-4\sqrt{3xy^2}$

62. $\sqrt{7x}-\sqrt[3]{7x}$ **63.** $-x^2\sqrt[3]{54x}+3\sqrt[3]{2x^7}$

64. $\sqrt[5]{32x^{13}}+3x\sqrt[5]{x^8}$ **65.** $\sqrt[3]{-16z^4}+6z\sqrt[3]{2z}$

Simplify the following expressions, writing your answer using the same notation as the original expression. See Example 12.

66. $\sqrt[3]{\sqrt[4]{x^{36}}}$ **67.** $32^{-\frac{3}{5}}$ **68.** $\left(3x^2-4\right)^{\frac{1}{3}}\left(3x^2-4\right)^{\frac{5}{3}}$

69. $81^{\frac{3}{4}}$ **70.** $(-8)^{\frac{2}{3}}$ **71.** $625^{-\frac{3}{4}}$

72. $\sqrt[3]{\sqrt[5]{y^{25}}}$ **73.** $\dfrac{(a-b)^{-\frac{2}{3}}}{(a-b)^{-2}}$

Convert the following expressions from radical notation to exponential notation, or vice versa. Simplify each expression in the process, if possible.

74. $256^{-\frac{3}{4}}$ **75.** $\sqrt[12]{x^3}$ **76.** $\sqrt[6]{\dfrac{2}{72}}$ **77.** $\left(36n^4\right)^{\frac{5}{6}}$

72. $y\sqrt[3]{y^2}$

73. $(a-b)^{\frac{4}{3}}$

74. $\dfrac{1}{64}$

75. $x^{\frac{1}{4}}$

76. $6^{-\frac{1}{3}}$

77. $6n^3\sqrt[3]{36n}$

78. $\sqrt[4]{125}$

79. x^3

80. y^4

81. $\sqrt[16]{16,807}$

82.–84. Answers may vary.

85. $\dfrac{1}{2}bhl$

86. $\pi r^2 h$

87. $180\pi\,\text{in.}^3$

88. $585\,\text{m}^3$

89. $81\pi d\,\text{ft}^3$

90. $14(N+M)\,\text{ft}^2$

91. $2\pi r^2$

Simplify the following expressions. See Example 13.

78. $\sqrt{5}\cdot\sqrt[4]{5}$ **79.** $\sqrt[3]{x^7}\cdot\sqrt[9]{x^6}$ **80.** $\sqrt[5]{y^{16}}\cdot\sqrt[25]{y^{20}}$ **81.** $\sqrt[4]{7}\cdot\sqrt[16]{7}$

Apply the definition of rational exponents to demonstrate the following properties.

82. $\sqrt[n]{ab}=\sqrt[n]{a}\cdot\sqrt[n]{b}$ **83.** $\sqrt[n]{\dfrac{a}{b}}=\dfrac{\sqrt[n]{a}}{\sqrt[n]{b}}$ **84.** $\sqrt[m]{\sqrt[n]{a}}=\sqrt[mn]{a}$

Complete the following word problems as indicated.

85. The prism shown below is a right triangular cylinder, where the triangular base is a right triangle. Find the volume of the prism in terms of b, h, and l.

86. Determine the volume of the right circular cylinder shown, in terms of r and h.

87. Matt wants to let people in the future know what life is like today, so he goes shopping for a time capsule. Capacity, along with price and quality, is an important consideration for him. One time capsule he looks at is a right circular cylinder with a hemisphere on each end. Find the volume of the time capsule, given that the length l of the cylinder is 16 inches and the radius r is 3 inches.

l = 16 in.

88. Bill and Dee are buying a new house. The house is a right cylinder based on a trapezoid atop a rectangular prism. The bases of the trapezoid are $B=10$ m and $b=8$ m, and the length of the house is $l=15$ m. The height of the house up to the bottom of the roof is $H=3$ m, and the height of the roof is $h=1$ m. Find the volume of the house.

89. Construct the expression for the volume of water contained in an above-ground circular swimming pool that has a diameter of 18 feet, assuming the water has a uniform depth of d feet.

90. The floor of a rectangular bedroom measures N feet wide and M feet long. The height of the walls is 7 feet. Construct an expression for the number of square feet of wallpaper needed to cover all the walls. (Ignore the presence of doors and windows.)

91. The interior surface of the birdbath in Example 6c needs to be painted with a waterproof (and nontoxic) coating. Construct the expression for the interior surface area.

92. $bh + bl + hl +$
$l\sqrt{b^2 + h^2}$

93. $3d^2\sqrt{3}$; $3.326\ \text{cm}^2$

94. $1651\ \text{cm}^2$; no

95. $2.998 \times 10^8\ \text{m/s}$

96. Answers will vary.

97. Answers will vary.

98. Answers will vary.

99. Because a positive number squared is positive and a negative number squared is positive.

100. Because a root is the same as a fractional exponent.

92. The prism shown below is a triangular right cylinder, where the triangular base is a right triangle. Find the surface area of the prism in terms of $b, h,$ and l.

93. A jeweler decides to construct a pendant for a necklace by simply attaching equilateral triangles to each edge of a regular hexagon. The edge length of one of the points of the resulting star is $d = 0.8$ cm. Find the formula for the area of the star in terms of d and then evaluate for $d = 0.8$ cm (rounding to the nearest thousandth). Use the fact that the area of an equilateral triangle of side length d is $A = \dfrac{d^2\sqrt{3}}{4}$.

$d = 0.8\ \text{cm}$

94. Ilyana has made a home for her pet guinea pig (Ralph) in the shape of a right triangular cylinder. Before she can put the new home in Ralph's cage, she must paint it with a nontoxic outer coat. If the front of the home has a base of 17.5 cm and a height of 15 cm and the length of the home is 25 cm, what is the surface area of Ralph's home, rounded to the nearest square centimeter? The small bottle of nontoxic coating will cover up to 1500 cm². Will the small bottle

contain enough nontoxic coating to cover Ralph's home?

$h = 15$ cm

$b = 17.5$ cm

95. Einstein's Theory of Special Relativity tells us that $E = mc^2$, where E is energy (in joules, J), m is mass (in kilograms, kg), and c is the speed of light (in meters per second, m/s). This equation may also be written as $\sqrt{\dfrac{E}{m}} = c.$ Assume you know $E = 418{,}400$ J and $m = 4.655 \times 10^{-12}$ kg. Use this information to estimate the speed of light. Round your answer, as necessary.

96. Give a few examples of instances in which it would be more useful to use scientific notation rather than standard.

97. In February of 2006, the US national debt was approximately 8.2 trillion dollars. How is saying 8.2 trillion similar to scientific notation? How is it different?

98. In your own words, explain why $a^0 = 1$.

99. Explain, in your own words, why the square root of a negative number is not a real number.

100. Explain, in your own words, why exponents and roots are evaluated at the same time in the order of operations.

1.3

Polynomials and Factoring

TOPICS

1. The terminology of polynomial expressions

2. The algebra of polynomials

3. Common factoring methods:
   - Greatest Common Factor
   - Factoring by Grouping
   - Factoring Special Binomials
   - Factoring Trinomials
   - Factoring Expressions Containing Fractional Exponents

TOPIC The Terminology of Polynomial Expressions

Polynomials are a special class of algebraic expressions. Each term in a polynomial consists only of a number multiplied by variables (if it is multiplied by anything at all) raised to positive integer exponents.

The number in any such term is called the **coefficient** of the term, and the sum of the exponents of the variables is the **degree of the term**. If a given term consists only of a nonzero number a (known as a **constant term**), its degree is defined to be 0, while the term 0 is not assigned any degree at all. The polynomial as a whole is also given a degree: the **degree of a polynomial** is the largest of the degrees of the individual terms.

Every polynomial consists of some finite number of terms. Polynomials consisting of a single term are called **monomials**, those consisting of two terms are **binomials**, and those consisting of three terms are **trinomials**. The following example illustrates the use of the above terminology as applied to some specific polynomials.

EXAMPLE 1

Polynomial Expressions

a. The expression $14x^6$ is a monomial in the variable x. The coefficient of the term is 14 and the degree of the term is 6, which is also the degree of the polynomial.

b. The polynomial $-3x^4y^2 + 5.4x^3y^4$ is a binomial in the two variables x and y. The degree of the first term is 6, and the degree of the second term is 7, so the degree of the polynomial as a whole is 7. The coefficient of the first term is -3 and the coefficient of the second term is 5.4.

c. The single number 5 can be considered a polynomial. In particular, it is a monomial of degree 0. The rationale for assigning degree 0 to nonzero constants such as 5 is that 5 can be thought of as $5x^0$ (or 5 times *any* variable raised to the 0 power). The coefficient of this monomial is itself: 5.

d. The polynomial $\frac{2}{3}x^3y^5 - z + y^{10} + 3$ has four terms and is a polynomial in three variables. If one of the terms of a polynomial consists only of a number, it is referred to as the constant term. The degree of this polynomial is 10, and the degrees of the individual terms are, from left to right, 8, 1, 10 and 0. The coefficients are $\frac{2}{3}$, $-1, 1$, and 3.

e. The expression $x^3 + 2x + 7x^{-1}$ is **not** a polynomial because it contains a term in which a variable is raised to a *negative* exponent.

f. The expression $\sqrt{x}$ is also **not** a polynomial, because it is equivalent to an expression with a variable raised to a *fractional* exponent, $x^{\frac{1}{2}}$.

The majority of the polynomials in this book are polynomials of a single variable and can be described generically as follows.

DEFINITION

Polynomials of a Single Variable

A polynomial in the variable x of degree n can be written in the form

$$a_n x^n + a_{n-1} x^{n-1} + \ldots + a_1 x + a_0,$$

where $a_n, a_{n-1}, \ldots, a_1, a_0$ are numbers, $a_n \neq 0$ and n is a positive integer. This form is called **descending order**, because the powers descend from left to right. The **leading coefficient** of this polynomial is a_n.

TOPIC 2   The Algebra of Polynomials

The variables in polynomials, as in all algebraic expressions, represent numbers, and consequently polynomials can be added, subtracted, multiplied and divided according to the field properties discussed in Section 1.1.

We will discuss addition (and thus subtraction) and multiplication of polynomials here, while division of polynomials will be covered in Chapter 4.

To add two or more polynomials, we use the field properties to combine **like terms** (also called **similar terms**). These are terms that have the same variables raised to the same powers. When subtracting one polynomial from another, we first distribute the minus sign over the terms of the polynomial being subtracted and then add.

EXAMPLE 2

Adding and Subtracting Polynomials

Add or subtract the polynomials as indicated.

a. $\left(2x^3y - 3y + z^2x\right) + \left(3y + z^2 - 3xz^2 + 4\right)$

b. $\left(4ab^3 - b^3c\right) - \left(4 - b^3c\right)$

Note:
The first step is always to identify and group like terms. The like terms are then combined using the distributive property.

Solutions:

a. $\left(2x^3y - 3y + z^2x\right) + \left(3y + z^2 - 3xz^2 + 4\right)$

$= 2x^3y + \left(-3y + 3y\right) + \left(xz^2 - 3xz^2\right) + z^2 + 4$

$= 2x^3y + y\left(-3 + 3\right) + xz^2\left(1 - 3\right) + z^2 + 4$

$= 2x^3y - 2xz^2 + z^2 + 4$

Note that the terms z^2x and $-3xz^2$ are similar, since multiplication is commutative.

Combine like terms.

b. $\left(4ab^3 - b^3c\right) - \left(4 - b^3c\right)$

$= 4ab^3 - b^3c - 4 + b^3c$

$= 4ab^3 - 4 + \left(-b^3c + b^3c\right)$

$= 4ab^3 - 4$

Begin by distributing the minus sign over all the terms in the second polynomial.

Combine like terms.

Polynomials are multiplied using the same properties. The distributive property is of particular importance in multiplying polynomials correctly. In general, the product of two polynomials, one with n terms and one with m terms, will consist initially of nm terms. If any of the resulting terms are similar, they are then combined in the last step.

EXAMPLE 3

Multiplying Polynomials

Multiply the polynomials, as indicated.

a. $\left(2x^2y - z^3\right)\left(4 + 3z - 3xy\right)$

b. $\left(3ab - a^2\right)\left(ab + a^2\right)$

Note:
We use the distributive property first, and multiply each term of the first polynomial by each term of the second.

Solutions:

a. $\left(2x^2y - z^3\right)\left(4 + 3z - 3xy\right)$

$= 2x^2y\left(4 + 3z - 3xy\right) - z^3\left(4 + 3z - 3xy\right)$

$= 8x^2y + 6x^2yz - 6x^3y^2 - 4z^3 - 3z^4 + 3xyz^3$

Multiply the second polynomial by each term of the first.

None of the resulting terms are similar, so the final answer is a polynomial of 6 terms.

b. $\left(3ab - a^2\right)\left(ab + a^2\right)$

$= 3ab\left(ab + a^2\right) - a^2\left(ab + a^2\right)$

$= 3a^2b^2 + 3a^3b - a^3b - a^4$

$= 3a^2b^2 + 2a^3b - a^4$

Multiply the second polynomial by each term of the first.

We combine the two similar terms to obtain the final trinomial.

EXAMPLE 4

The FOIL Method

When a binomial is multiplied by another binomial, as in Example 3b, the acronym FOIL is commonly used as a reminder of the four necessary products. Again consider the product $\left(3ab - a^2\right)\left(ab + a^2\right)$. Each letter represents a pair of terms to multiply:

First $\left(3ab - a^2\right)\left(ab + a^2\right)$ $(3ab)(ab) = 3a^2b^2$

Outer $\left(3ab - a^2\right)\left(ab + a^2\right)$ $(3ab)\left(a^2\right) = 3a^3b$

Inner $\left(3ab - a^2\right)\left(ab + a^2\right)$ $\left(-a^2\right)(ab) = -a^3b$

Last $\left(3ab - a^2\right)\left(ab + a^2\right)$ $\left(-a^2\right)\left(a^2\right) = -a^4$

Once again, we sum the resulting terms, combining like terms to find the same answer as before; $\left(3ab - a^2\right)\left(ab + a^2\right) = 3a^2b^2 + 2a^3b - a^4$. While the FOIL acronym is a helpful tool, it is important to realize that it is nothing more than a special application of the distributive property.

DEFINITION

Special Product Formulas

In the following, A and B represent algebraic expressions.

1. $(A - B)(A + B) = A^2 - B^2$

2. $(A + B)^2 = A^2 + 2AB + B^2$

3. $(A - B)^2 = A^2 - 2AB + B^2$

TOPIC 3   Common Factoring Methods

Factoring, in general, means reversing the process of multiplication in order to find two or more expressions whose product is the original expression. For instance, the factored form of the polynomial $x^2 - 3x + 2$ is $(x - 2)(x - 1)$, since if the product $(x - 2)(x - 1)$ is expanded we obtain $x^2 - 3x + 2$ (as you should verify). Factoring is thus a logical topic to discuss immediately after multiplication, but the process of factoring is not as straightforward as multiplication.

We say that a polynomial with integer coefficients is **factorable** if it can be written as a product of two or more polynomials, all of which also have integer coefficients. If this cannot be done, we say the polynomial is **irreducible** (over the integers), or **prime**. The goal of factoring a polynomial is to **completely factor** it: to write it as a product of prime polynomials. This means no factors can be factored any further.

Method 1: Greatest Common Factor. Factoring out those factors common to all the terms in an expression is the simplest factoring method and should be done first if possible. The **Greatest Common Factor** (GCF) among all the terms is the product of all the factors common to each.

EXAMPLE 5

The Greatest Common Factor Method

Use the Greatest Common Factor method to factor the following polynomials.

a. $12x^5 - 4x^2 + 8x^3z^3$ **b.** $\left(a^2 - b\right) - 3\left(a^2 - b\right)$

Solutions:

Note:
Pay attention to how the distributive property works in reverse.

a.
$$12x^5 - 4x^2 + 8x^3z^3$$
$$= \left(4x^2\right)\left(3x^3\right) + \left(4x^2\right)(-1) + \left(4x^2\right)\left(2xz^3\right)$$
$$= \left(4x^2\right)\left(3x^3 - 1 + 2xz^3\right)$$

As we noted above, $4x^2$ is the greatest common factor.

b.
$$\left(a^2 - b\right) - 3\left(a^2 - b\right)$$
$$= \left(a^2 - b\right)(1) + \left(a^2 - b\right)(-3)$$
$$= \left(a^2 - b\right)(1 - 3)$$
$$= -2\left(a^2 - b\right)$$

One common source of error in factoring is to forget factors of 1.

Method 2: Factoring by Grouping. If the terms of the polynomial are grouped in a suitable way, the GCF method may apply to each group, and a common factor might be found among the groups. **Factoring by Grouping** is the name given to this process, and it is important to realize that it is a trial and error process. Your first attempt at grouping and factoring may not succeed, so you may have to try several different ways of grouping the terms.

EXAMPLE 6

Factoring by Grouping

Use the Factoring by Grouping method to factor the following polynomials.

a. $6x^2 - y + 2x - 3xy$ **b.** $ax - ay - bx + by$

Solutions:

a.
$$6x^2 - y + 2x - 3xy$$
$$= \left(6x^2 + 2x\right) + (-y - 3xy)$$
$$= 2x(3x + 1) + y(-1 - 3x)$$
$$= 2x(3x + 1) - y(3x + 1)$$
$$= (3x + 1)(2x - y)$$

The first and third terms have a GCF of $2x$, while the second and fourth have a GCF of y, so we group accordingly.

After factoring the two groups, we notice that $3x + 1$ and $-1 - 3x$ differ only by a minus sign (and the order). This means $3x + 1$ can be factored out.

b.
$$ax - ay - bx + by$$
$$= a(x - y) - b(x - y)$$
$$= (x - y)(a - b)$$

In this problem, we could also have grouped the first and third terms, and the second and fourth terms, and obtained the same result.

CAUTION!

One common error in factoring is to stop after groups within the original polynomial have been factored. For instance, while we have done some factoring to achieve the expression $2x(3x+1)+y(-1-3x)$ in Example 6a, this is *not* in factored form. An polynomial is only factored if it is written as a *product* of two or more factors, while $2x(3x+1)+y(-1-3x)$ is a *sum* of two expressions.

Method 3: Factoring Special Binomials. Three types of binomials can always be factored by following the patterns outlined below. You should verify these patterns by multiplying out the products on the right-hand side of each one.

DEFINITION

Factoring Special Binomials

In the following, A and B represent algebraic expressions.

Difference of Two Squares: $A^2 - B^2 = (A-B)(A+B)$

Difference of Two Cubes: $A^3 - B^3 = (A-B)(A^2 + AB + B^2)$

Sum of Two Cubes: $A^3 + B^3 = (A+B)(A^2 - AB + B^2)$

Note that $A^2 + B^2$ cannot be factored in the real sense.

EXAMPLE 7

Factoring Special Binomials

Factor the following binomials using the special binomial patterns.

a. $49x^2 - 9y^6$

b. $27a^6b^{12} + c^3$

c. $125y^3 - 8z^3$

d. $64 - (x+y)^3$

Solutions:

a. $49x^2 - 9y^6$ A difference of two squares.

$= (7x)^2 - (3y^3)^2$ $A = 7x, B = 3y^3$.

$= (7x - 3y^3)(7x + 3y^3)$ $A^2 - B^2 = (A-B)(A+B)$.

b. $27a^6b^{12} + c^3$ A sum of two cubes.

$= (3a^2b^4)^3 + (c)^3$ $A = 3a^2b^4, B = c$.

$= \left(\underbrace{3a^2b^4}_{A} + \underbrace{c}_{B}\right)\left(\underbrace{(3a^2b^4)^2}_{A^2} - \underbrace{(3a^2b^4)(c)}_{AB} + \underbrace{(c)^2}_{B^2}\right)$ $A^3 + B^3 = (A+B)(A^2 - AB + B^2)$.

$= (3a^2b^4 + c)(9a^4b^8 - 3a^2b^4c + c^2)$

c. $125y^3 - 8z^3$ A difference of two cubes.

$= (5y)^3 - (2z)^3$ $A = 5y, B = 2z.$

$= (5y - 2z)\left((5y)^2 + (5y)(2z) + (2z)^2\right)$ $A^3 - B^3 = (A - B)(A^2 + AB + B^2).$

$= (5y - 2z)\left(25y^2 + 10yz + 4z^2\right)$

d. $64 - (x + y)^3$ In this difference of two cubes, the

$= 4^3 - (x + y)^3$ second cube is itself a binomial. But

the factoring pattern still applies,

$= (4 - (x + y))\left(4^2 + 4(x + y) + (x + y)^2\right)$ leading to the final factored form of

the original binomial.

$= (4 - x - y)\left(16 + 4x + 4y + x^2 + 2xy + y^2\right)$

Method 4: Factoring Trinomials. In factoring a trinomial of the form $ax^2 + bx + c$ the goal is to find two binomials $px + q$ and $rx + s$ such that

$$ax^2 + bx + c = (px + q)(rx + s).$$

Since $(px + q)(rx + s) = prx^2 + (ps + qr)x + qs$, we seek $p, q, r,$ and s such that $a = pr$, $b = ps + qr$ and $c = qs$:

$$ax^2 + bx + c = \underbrace{pr}_{a}\, x^2 + \underbrace{(ps + qr)}_{b}x + \underbrace{qs}_{c}$$

In general, this may require trial and error, but the following guidelines will help.

Case 1: Leading Coefficient is 1. In this case, p and r must both be 1, so we only need q and s such that $x^2 + bx + c = x^2 + (q + s)x + qs$. That is, we need two integers whose sum is b, the coefficient of x, and whose product is c, the constant term.

EXAMPLE 8

Factoring a Trinomial

To factor $x^2 + x - 12$ we can begin by writing $x^2 + x - 12 = \left(x + \boxed{?}\right)\left(x + \boxed{?}\right)$ and then try to find two integers to replace the question marks. The two integers we seek must have a product of -12, and the fact that the product is negative means that one integer must be positive and one negative. The only possibilities are $\{1, -12\}$, $\{-1, 12\}$, $\{2, -6\}$, $\{-2, 6\}$, $\{3, -4\}$, and $\{-3, 4\}$, and when we add the requirement that the sum must be 1, we are left with $\{-3, 4\}$. Thus $x^2 + x - 12 = (x - 3)(x + 4)$.

Case 2: Leading Coefficient is not 1. In this case, trial and error may still be an effective way to factor the trinomial $ax^2 + bx + c$ especially if a, b, and c are relatively small in magnitude. If, however, trial and error seems to be taking too long, the following steps use factoring by grouping to minimize the amount of guessing required.

PROCEDURE

Factoring a Trinomial by Grouping

To factor the trinomial $ax^2 + bx + c$:

Step 1: Multiply a and c.

Step 2: Factor ac into two integers whose sum is b. If no such factors exist, the trinomial is irreducible over the integers.

Step 3: Rewrite b in the trinomial with the sum found in step 2, and distribute. The resulting polynomial of four terms may now be factored by grouping.

EXAMPLE 9

Factoring a Trinomial by Grouping

To factor the trinomial $6x^2 - x - 12$ by trial and error, we would begin by noting that if it can be factored, the factors must be of the form $\left(x + \boxed{?}\right)\left(6x + \boxed{?}\right)$ or $\left(2x + \boxed{?}\right)\left(3x + \boxed{?}\right)$. If we use the grouping method , we form the product $(6)(-12) = -72$ and then factor -72 into two integers whose sum is -1. The two numbers -9 and 8 work, so we write $6x^2 - x - 12 = 6x^2 + (-9 + 8)x - 12 = 6x^2 - 9x + 8x - 12$. Now proceed by grouping:

$$6x^2 - 9x + 8x - 12$$
$$= 3x(2x - 3) + 4(2x - 3)$$
$$= (2x - 3)(3x + 4)$$

Some trinomial expressions are known as "perfect square trinomials" because their factored form is the square of a binomial expression. For example, $x^2 - 6x + 9 = (x - 3)^2$. (Either the trial and error method or factoring by grouping can give us this answer.) In general, such trinomials will have one of the following two forms.

DEFINITION

Perfect Square Trinomials

In the following, A and B represent algebraic expressions.

$$A^2 + 2AB + B^2 = (A + B)^2$$
$$A^2 - 2AB + B^2 = (A - B)^2$$

EXAMPLE 10

Perfect Square Trinomials

Factor the algebraic expression $x^2 + 10x + 25$.

Solution:

The expression appears to be in the form of a perfect square trinomial, but we need to check that the value of the middle term follows the above pattern. Taking $A = x$

and $B = 5$, we see that $2AB = 10x$, so the expression does match the perfect square trinomial form.

Thus the factored form of $x^2 + 10x + 25 = (x+5)^2$ and $x^2 + 10x + 25$ is a perfect square trinomial.

Method 5: Factoring Expressions Containing Fractional Exponents. This last method does not apply to polynomials, as polynomials cannot have fractional exponents. It will, however, be very useful in solving problems later in this book and in other math classes. The method applies to negative fractional exponents as well as positive.

To factor an algebraic expression that has fractional exponents, identify the smallest exponent among the terms, then factor out the variable raised to that smallest exponent from each of the terms. Factor out any other common factors and simplify if possible.

EXAMPLE 11

Factoring Expressions with Fractional Exponents

Factor each of the following algebraic expressions.

a. $3x^{-\frac{2}{3}} - 6x^{\frac{1}{3}} + 3x^{\frac{4}{3}}$

b. $(x-1)^{\frac{1}{2}} - (x-1)^{-\frac{1}{2}}$

Solutions:

a. $3x^{-\frac{2}{3}} - 6x^{\frac{1}{3}} + 3x^{\frac{4}{3}}$

$= 3x^{-\frac{2}{3}}\left(1 - 2x + x^2\right)$

$= 3x^{-\frac{2}{3}}\left(x^2 - 2x + 1\right)$

$= 3x^{-\frac{2}{3}}(x-1)(x-1)$

$= 3x^{-\frac{2}{3}}(x-1)^2$

Under the guidelines above, we factor out $3x^{-\frac{2}{3}}$. Note that we use the properties of exponents to obtain the terms in the second factor.

We notice that the second factor is a second-degree trinomial, and is itself factorable. In fact, it is an example of a perfect square trinomial.

b. $(x-1)^{\frac{1}{2}} - (x-1)^{-\frac{1}{2}}$

$= (x-1)^{-\frac{1}{2}}\left((x-1) - 1\right)$

$= (x-1)^{-\frac{1}{2}}(x-2)$

In this example we factor out $(x-1)^{-\frac{1}{2}}$ again using the properties of exponents to obtain the terms in the second factor.

Exercises

Answers (left column):

1. Not a polynomial
2. Degree 7 trinomial
3. Degree 11; polynomial of four terms
4. Degree 3 monomial
5. Degree 0 monomial
6. Monomial of no degree
7. Degree 4 binomial
8. Degree 7 monomial
9. Degree 2 trinomial
10. Not a polynomial
11. Degree 5 binomial
12. Degree 6 trinomial
13. $-x^{13}+7x^{11}-4x^{10}+9$
   a. 13   b. -1
14. $-9x^{10}+9x^8$
   a. 10   b. -9
15. $2s^6-10s^5+4s^3$
   a. 6   b. 2
16. $-2x^5+x^2+4$
   a. 5   b. -2
17. $9y^6-3y^5+y-2$
   a. 6   b. 9
18. $6n^2+4n-3$
   a. 2   b. 6
19. πz^5+8z^2-2z+1
   a. 5   b. π
20. $-3y^7+12y^6-6y^5$
   a. 7   b. -3
21. $-4x^3y-6y-x^2z$
22. $2x^3-9x^2-7$
23. $x^2y+xy^2+6x-6y$
24. $8x^2-2$
25. $-3ab$
26. $2x^4+x^3+5$

Classify each of the following expressions as either a polynomial or not a polynomial. For those that are polynomials, identify the degree of the polynomial, and the number of terms (use the words monomial, binomial, and trinomial if applicable). See Example 1.

1. $3x^{\frac{3}{2}}-2x$
2. $17x^2y^5+2z^3-4$
3. $5x^{10}+3x^3-2y^3z^8+9$
4. πx^3
5. 8
6. 0
7. $7^3xy^2+4y^4$
8. abc^2d^3
9. $4x^2+7xy+5y^2$
10. $3n^4m^{-3}+n^2m$
11. $\dfrac{y^2z}{4}+2yz^4$
12. $6x^4y+3x^2y^2+xy^5$

Write each of the following polynomials in descending order, and identify **a.** the degree of the polynomial, and **b.** the leading coefficient.

13. $-4x^{10}-x^{13}+9+7x^{11}$
14. $9x^8-9x^{10}$
15. $4s^3-10s^5+2s^6$
16. $4-2x^5+x^2$
17. $9y^6-2+y-3y^5$
18. $4n+6n^2-3$
19. $8z^2+\pi z^5-2z+1$
20. $-6y^5-3y^7+12y^6$

Add or subtract the polynomials, as indicated. See Example 2.

21. $\left(-4x^3y+2xz-3y\right)-\left(2xz+3y+x^2z\right)$
22. $\left(4x^3-9x^2+1\right)+\left(-2x^3-8\right)$
23. $\left(x^2y-xy-6y\right)+\left(xy^2+xy+6x\right)$
24. $\left(5x^2-6x+2\right)-\left(4-6x-3x^2\right)$
25. $\left(a^2b+2ab+ab^2\right)-\left(ab^2+5ab+a^2b\right)$
26. $\left(x^4+2x^3-x+5\right)-\left(x^3-x-x^4\right)$
27. $\left(xy-4y+xy^2\right)+\left(3y-x^2y-xy\right)$
28. $\left(-8x^4+13-9x^2\right)-\left(8-2x^4\right)$

Multiply the polynomials, as indicated. See Examples 3 and 4.

29. $\left(3a^2b+2a-3b\right)\left(ab^2+7ab\right)$
30. $\left(x^2-2y\right)\left(x^2+y\right)$
31. $\left(3a+4b\right)\left(a-2b\right)$
32. $\left(x+xy+y\right)\left(x-y\right)$
33. $\left(6x-3y\right)\left(x+6y\right)$
34. $\left(5y+x\right)\left(4y-2x\right)$

27. $xy^2 - x^2y - y$

28. $-6x^4 - 9x^2 + 5$

29. $3a^3b^3 + 21a^3b^2$
    $+ 2a^2b^2 + 14a^2b$
    $- 3ab^3 - 21ab^2$

30. $x^4 - x^2y - 2y^2$

31. $3a^2 - 2ab - 8b^2$

32. $x^2 + x^2y - xy^2 - y^2$

33. $6x^2 + 33xy - 18y^2$

34. $20y^2 - 6xy - 2x^2$

35. $7y^4 - 34xy^2 - 5x^2$

36. $3y^4 - 4xy^2 - 7x^2$

37. $6x^3y^3 - 3x^3y$
    $+ 36x^2y^3 + 4x^2y^2$
    $- 18x^2y + 24xy^2$

38. $2x^3y^3 - 10x^2y^3$
    $+ 4x^2y^2 - 20xy^2$
    $- 6x^3y + 30x^2y$

39. $m(4mn + 16m^2 + 7)$

40. $(3a^2b)(1 + a - 3b)$

41. $6(a - b^2)$

42. $3x^3y(1 - 3x + 4y)$

43. $2x(x^5 - 7x^2 + 4)$

44. $9x^4y(3x^3 + x^2 - z)$

45. $(x^3 - y)(x^3 - y - 1)$

46. $3y^3(2x + 3 - 4xy)$

47. $4y^2(3y^4 - 2 - 4y^3)$

48. $(2x + y^2)^4 \times$
    $\left(1 - (2x + y^2)^2\right)$

49. $(a^2 + b)(a - b)$

50. $(x - 2y)(a - 2b)$

51. $z(1 + z)(1 + z^2)$

35. $(7y^2 + x)(y^2 - 5x)$

36. $(y^2 + x)(3y^2 - 7x)$

37. $(6xy^2 - 3x + 4y)(x^2y + 6xy)$

38. $(2xy^2 + 4y - 6x)(x^2y - 5xy)$

Factor each polynomial by factoring out the greatest common factor. See Example 5.

39. $4m^2n + 16m^3 + 7m$

40. $3a^2b + 3a^3b - 9a^2b^2$

41. $5(a - b^2) + (a - b^2)$

42. $3x^3y - 9x^4y + 12x^3y^2$

43. $2x^6 - 14x^3 + 8x$

44. $27x^7y + 9x^6y - 9x^4yz$

45. $(x^3 - y)^2 - (x^3 - y)$

46. $6xy^3 + 9y^3 - 12xy^4$

47. $12y^6 - 8y^2 - 16y^5$

48. $(2x + y^2)^4 - (2x + y^2)^6$

Factor each polynomial by grouping. See Example 6.

49. $a^3 + ab - a^2b - b^2$

50. $ax - 2bx - 2ay + 4by$

51. $z + z^2 + z^3 + z^4$

52. $x^2 + 3xy + 3y + x$

53. $nx^2 - 2y - 2x^2 + ny$

54. $2ac - 3bd + bc - 6ad$

55. $ax - 5bx + 5ay - 25by$

56. $3ac - 5bd + bc - 15ad$

Use the special factoring patterns to factor the following binomials. See Example 7.

57. $4x^2 - 121$

58. $64z^3 + 216$

59. $49a^2 - 144b^2$

60. $x^3 - 27y^3$

61. $25x^4y^2 - 9$

62. $27a^9 + 8b^{12}$

63. $x^3 - 1000y^3$

64. $64x^6 - 125y^3z^9$

65. $m^6 + 125n^9$

66. $49a^6 - 9b^2c^4$

67. $27x^6 - 8y^{12}z^3$

68. $(3x - 6)^2 - (y - 2x)^2$

69. $16z^2y^4 - 9x^8$

70. $512x^6 + 729y^3$

71. $343y^9 - 27x^3z^6$

72. $(2x + y^2)^2 - (y^2 - 3)^2$

52. $(x+3y)(x+1)$

53. $(n-2)(x^2+y)$

54. $(2a+b)(c-3d)$

55. $(a-5b)(x+5y)$

56. $(3a+b)(c-5d)$

57. $(2x-11)(2x+11)$

58. $8(2z+3)\times$
$(4z^2-6z+9)$

59. $(7a-12b)(7a+12b)$

60. $(x-3y)\times$
$(x^3+3xy+9x^2)$

61. $(5x^2y-3)(5x^2y+3)$

62. $(3a^3+2b^4)\times$
$(9a^6-6a^3b^4+4b^8)$

63. $(x-10y)\times$
$(x^2+10xy+100y^2)$

64. $(4x^2-5yz^3)\times$
$\left(\begin{array}{l}16x^4+20x^2yz^3\\ +25y^2z^6\end{array}\right)$

65. $(m^2+5n^3)\times$
$(m^4-5m^2n^3+25n^6)$

66. $(7a^3-3bc^2)\times$
$(7a^3+3bc^2)$

67. $(3x^2-2y^4z)\times$
$\left(\begin{array}{l}9x^4+6x^2y^4z\\ +4y^8z^2\end{array}\right)$

68. $(5x-y-6)\times$
$(x+y-6)$

69. $(4y^2z-3x^4)\times$
$(4y^2z+3x^4)$

70. $(8x^2+9y)\times$
$\left(\begin{array}{l}64x^4-72x^2y\\ +81y^2\end{array}\right)$

71. $(7y^3+3xz^2)\times$
$\left(\begin{array}{l}49y^6+21xy^3z^2\\ +9x^2z^4\end{array}\right)$

72. $(2x+3)(2x+2y^2-3)$

73. $(x+5)(x-3)$

74. $(x+3)^2$

75. $(x-1)^2$

Factor the following trinomials. See Examples 8, 9, and 10.

73. $x^2+2x-15$

74. x^2+6x+9

75. x^2-2x+1

76. x^2-5x+6

77. x^2-4x+4

78. x^2+5x+4

79. $y^2+14y+49$

80. $x^2-3x-18$

81. $x^2+13x+22$

82. y^2+y-42

83. y^2-9y+8

84. $6x^2+5x-6$

85. $5a^2-37a-24$

86. $25y^2+10y+1$

87. $5x^2+27x-18$

88. $6y^2-13y-8$

89. $16y^2-25y+9$

90. $10m^2+29m+10$

91. $8a^2-2a-3$

92. $20y^2+21y-5$

93. $12y^2-19y+5$

94. $10y^2-11y-6$

Factor the following algebraic expressions. See Example 11.

95. $(2x-1)^{-\frac{3}{2}}+(2x-1)^{-\frac{1}{2}}$

96. $2x^{-2}+3x^{-1}$

97. $7a^{-1}-2a^{-3}b$

98. $(3z+2)^{\frac{5}{3}}-(3z+2)^{\frac{2}{3}}$

99. $10y^{-2}-2y^{-5}x$

100. $4y^{-3}+12y^{-4}$

101. $(5x+7)^{\frac{7}{3}}-(5x+7)^{\frac{4}{3}}$

102. $(8x+6)^{-\frac{7}{2}}-(8x+6)^{-\frac{1}{2}}$

103. $7y^{-1}+5y^{-4}$

104. $5x^{-4}-4x^{-5}y$

76. $(x-2)(x-3)$

77. $(x-2)^2$

78. $(x+1)(x+4)$

79. $(y+7)^2$

80. $(x-6)(x+3)$

81. $(x+11)(x+2)$

82. $(y+7)(y-6)$

83. $(y-8)(y-1)$

84. $(2x+3)(3x-2)$

85. $(5a+3)(a-8)$

86. $(5y+1)^2$

87. $(x+6)(5x-3)$

88. $(2y+1)(3y-8)$

89. $(16y-9)(y-1)$

90. $(2m+5)(5m+2)$

91. $(4a-3)(2a+1)$

92. $(4y+5)(5y-1)$

93. $(4y-5)(3y-1)$

94. $(2y-3)(5y+2)$

95. $2x(2x-1)^{\frac{-3}{2}}$

96. $x^{-2}(3x+2)$

97. $a^{-3}\left(7a^2-2b\right)$

98. $(3z+1)(3z+2)^{\frac{2}{3}}$

99. $2y^{-5}\left(5y^3-x\right)$

100. $4y^{-4}(y+3)$

101. $(5x+7)^{\frac{4}{3}}(5x+6)$

102. $(8x+6)^{-\frac{7}{2}}\times$ $\left(1-(8x+6)^3\right)$

103. $y^{-4}\left(7y^3+5\right)$

104. $x^{-5}(5x-4y)$

Solve the following application problems.

105. Pneumothorax is a disease in which air or gas collects between the lung and the chest wall, causing the lung to collapse. When this disease is evident, the following formula is used to determine the degree of collapse of the lungs, represented as a percent:

$$\text{Degree} = 100\left(1 - \frac{L^3}{H^3}\right)$$

In this formula, L is the diameter of one lung and H is the diameter of one hemithorax (or half the chest cavity). Is this formula a polynomial? If so, find its degree and the number of terms. If not, explain.

106. You are trying to find a formula for the area of a certain trapezoid. You know the height of the trapezoid is x^2, the bottom base is $2x^2+4$, and the top base is $6x+2$. Insert these values into the formula for the area of a trapezoid. Is the result a polynomial? If so, find the degree of the polynomial, the leading coefficient, and the number of terms in the polynomial. If not, explain.

107. a. Given a rectangular picture frame with sides of $2x+1$ and x^3+4, find the area of the picture frame. Is the result a polynomial? If so, find the degree of the polynomial, the leading coefficient, and the number of terms in the polynomial. If not, explain.

b. Now find the perimeter of the picture frame. Is this a polynomial? If so, find the degree of the polynomial, the leading coefficient, and the number of terms in the polynomial. If not, explain.

$2x+1$

x^3+4

105. No; a variable in the denominator is equivalent to a variable with a negative exponent.

106. Yes; degree $= 4$; leading coefficient $= 1$; terms $= 3$

107. a. Yes; degree $= 4$; leading coefficient $= 2$; terms $= 4$

b. Yes; degree $= 3$; leading coefficient $= 2$; terms $= 3$

1.4 The Complex Number System

TOPICS

1. The imaginary unit *i* and its properties

2. The algebra of complex numbers

3. Roots and complex numbers

 TOPIC 1

The Imaginary Unit *i* and Its Properties

In Section 1.2 we encountered a problem with the real number system: there is a lack of symmetry in the definition of roots of real numbers. Recall that so far we have defined even roots only for nonnegative numbers, but we have defined odd roots for both positive and negative numbers (as well as 0).

This asymmetry is a reflection of the fact that the real number system is not *algebraically complete*. Roughly, this means that there are polynomial equations with real coefficients that have no real solutions! Consider the following question:

For a given nonzero real number a, how many solutions does the equation $x^2 = a$ have?

As we have noted previously, the equation has two solutions $\left(x = \sqrt{a} \text{ and } x = -\sqrt{a}\right)$ if *a* is positive, but no (real) solutions if *a* is negative. The following definition changes this situation.

DEFINITION

The Imaginary Unit *i*

The **imaginary unit *i*** is defined as $i = \sqrt{-1}$. In other words, *i* has the property that its square is –1: $i^2 = -1$.

This allows us to immediately define square roots of negative numbers in general, as follows.

DEFINITION

Square Roots of Negative Numbers

If *a* is a positive real number, $\sqrt{-a} = i\sqrt{a}$. Note that by this definition, and by a logical extension of exponentiation, $\left(i\sqrt{a}\right)^2 = i^2\left(\sqrt{a}\right)^2 = -a$.

The Number i	**EXAMPLE 1**

a. $\sqrt{-16} = i\sqrt{16} = i(4) = 4i$. As is customary, we write a constant such as 4 before letters in algebraic expressions, even if, as in this case, the letter is not a variable. Remember that i has a fixed meaning: i is the square root of -1.

b. $\sqrt{-8} = i\sqrt{8} = i(2\sqrt{2}) = 2i\sqrt{2}$. As is customary, again, we write the radical factor last. You should verify that $(2i\sqrt{2})^2$ is indeed -8.

c. $i^3 = i^2 i = (-1)(i) = -i$, and $i^4 = i^2 i^2 = (-1)(-1) = 1$. The simple fact that $i^2 = -1$ allows us, by our extension of exponentiation, to determine i^n for any natural number n.

d. $(-i)^2 = (-1)^2 i^2 = i^2 = -1$. This shows that $-i$ also has the property that its square is -1.

The **powers of i** follow a pattern that repeats with every fourth power:

$$i^1 = i \qquad\qquad\qquad i^{4n+1} = i$$
$$i^2 = -1 \qquad\qquad\qquad i^{4n+2} = -1$$
$$i^3 = -i \qquad\qquad\qquad i^{4n+3} = -i$$
$$i^4 = 1 \qquad\qquad\qquad i^{4n} = 1$$

Powers of i	**EXAMPLE 2**

Compute the following powers of i.

a. i^9 **b.** i^{28} **c.** i^{102}

Solutions:

a. When we divide 9 by 4, we have a remainder of 1. This means this power of i takes the form i^{4n+1}, so $i^9 = i$.

b. When we divide 28 by 4, the remainder is 0. This means this power of i is of the form i^{4n}, so $i^{28} = 1$.

c. When we divide 102 by 4, the remainder is 2. This means that this power of i fits the form i^{4n+2}, so $i^{102} = -1$.

The definition of the imaginary unit i leads to the following definition of complex numbers.

DEFINITION

Complex Numbers

For any two real numbers a and b, the sum $a + bi$ is a **complex number**. The collection $\mathbb{C} = \{a + bi \mid a \text{ and } b \text{ are both real}\}$ is called the set of complex numbers and is another example of a field. The number a is called the **real part** of $a + bi$, and the number b is called the **imaginary part**. If the imaginary part of a given complex number is 0, the number is simply a real number. If the real part of a given complex number is 0, the number is a **pure imaginary number**.

Note that the set of real numbers is a subset of the complex numbers: every real number *is* a complex number with 0 as the imaginary part. The set of complex numbers is the largest set of numbers that will appear in this text.

Do not be misled by the names into thinking that complex numbers, with their possible imaginary parts, are unimportant or physically meaningless. In many applications, complex numbers, even pure imaginary numbers, arise naturally and have important implications. For instance, the fields of electrical engineering and fluid dynamics both rely on complex number arithmetic.

TOPIC 2

The Algebra of Complex Numbers

The set of complex numbers is a field, so the field properties discussed in Section 1.1 apply. In particular, every complex number has an additive inverse (its negative), and every nonzero complex number has a multiplicative inverse (its reciprocal). Further, sums and products (and hence differences and quotients) of complex numbers are complex numbers, and can be written in the standard form $a + bi$. Given several complex numbers combined by the operations of addition, subtraction, multiplication, or division, the goal is to *simplify* the expression into the standard form $a + bi$.

Sums, differences, and products of complex numbers are easily simplified by remembering the definition of i and by thinking of every complex number $a + bi$ as a binomial.

PROCEDURE

Simplifying Complex Expressions

Step 1: Add, subtract, or multiply the complex numbers, as required, by treating every complex number $a + bi$ as a polynomial expression. Remember, though, that i is not actually a variable. Treating $a + bi$ as a binomial in i is just a handy device.

Step 2: Complete the simplification by using the fact that $i^2 = -1$.

EXAMPLE 3

Simplifying Complex
Expressions

Simplify the following complex number expressions.

a. $(4+3i)+(-5+7i)$

b. $(-2+3i)-(-3+3i)$

c. $(3+2i)(-2+3i)$

d. $(2-3i)^2$

Note:
Remember that a complex number is not simplified until it has the form $a + bi$.

Solutions:

a. $(4+3i)+(-5+7i) = (4-5)+(3+7)i$
$$= -1+10i$$

As if adding polynomials, we combine the real parts, then the imaginary parts.

b. $(-2+3i)-(-3+3i) = (-2+3i)+(-(-3)-3i)$
$$= (-2+3)+(3-3)i$$
$$= 1+0i$$
$$= 1$$

Begin by distributing the minus sign over the second complex number.

c. $(3+2i)(-2+3i) = -6+9i-4i+6i^2$
$$= -6+(9-4)i+6(-1)$$
$$= -6+5i-6$$
$$= -12+5i$$

After multiplying, combine the two terms containing i and rewrite i^2 as -1.

d. $(2-3i)^2 = (2-3i)(2-3i)$
$$= 4-6i-6i+9i^2$$
$$= 4-12i+9(-1)$$
$$= -5-12i$$

Squaring this complex number also leads to four terms, which we simplify as in part c.

DEFINITION

Complex Conjugates

Given any complex number $a + bi$, the complex number $a - bi$ is called its **complex conjugate**.

A very useful property of the complex conjugate is demonstrated below; the product of any complex number and its complex conjugate is a *real* number.

$$(a+bi)(a-bi) = a^2 - abi + abi - b^2i^2 = a^2 + b^2$$

This fact is critical in dividing complex numbers. In order to simplify a quotient of complex numbers, we need to rewrite it in the standard form $a + bi$. We simplify the quotient of two complex numbers by multiplying the numerator and denominator of the fraction by the complex conjugate of the denominator. This multiplication leaves a real number in the denominator so that a straightforward simplification leads to the standard form.

Note that this process is very similar to the process we used when rationalizing the denominator of a radical expression in Section 1.2.

EXAMPLE 4

Dividing Complex Numbers

Simplify the following expressions.

a. $\dfrac{2+3i}{3-i}$ **b.** $(4-3i)^{-1}$ **c.** $\dfrac{1}{i}$

Note:
Always begin by finding the complex conjugate of the denominator, then multiply the numerator and denominator by this conjugate.

Solutions:

a. $\dfrac{2+3i}{3-i} = \left(\dfrac{2+3i}{3-i}\right)\left(\dfrac{3+i}{3+i}\right)$ $3+i$ is the complex conjugate of the denominator.

$= \dfrac{(2+3i)(3+i)}{(3-i)(3+i)}$ Multiply the numerator and the denominator by the complex conjugate.

$= \dfrac{6+2i+9i+3i^2}{9+3i-3i-i^2}$

$= \dfrac{3+11i}{10} = \dfrac{3}{10}+\dfrac{11}{10}i$ We can often leave the answer in the form $\dfrac{3+11i}{10}$.

b. $(4-3i)^{-1} = \dfrac{1}{4-3i}$ Rewrite the original expression as a fraction.

$= \left(\dfrac{1}{4-3i}\right)\left(\dfrac{4+3i}{4+3i}\right)$ Then multiply the top and bottom by the complex conjugate of the denominator and proceed as in part a.

$= \dfrac{4+3i}{(4-3i)(4+3i)}$

$= \dfrac{4+3i}{16-9i^2}$

$= \dfrac{4+3i}{25} = \dfrac{4}{25}+\dfrac{3}{25}i$

c. $\dfrac{1}{i} = \left(\dfrac{1}{i}\right)\left(\dfrac{-i}{-i}\right)$ Here we write the reciprocal of the imaginary unit as a complex number. With this as a starting point, we could now calculate i^{-2}, i^{-3}, ...

$= \dfrac{-i}{-i^2}$

$= \dfrac{-i}{1} = -i$

TOPIC `3` **Roots and Complex Numbers**

We have now defined $\sqrt{a}$ without ambiguity: given a positive real number a, $\sqrt{a}$ is the positive real number whose square is a, and $\sqrt{-a}$ is defined to be $i\sqrt{a}$. These are called the **principal square roots**, to distinguish them from $-\sqrt{a}$ and $-i\sqrt{a}$ respectively. (Remember, both $\sqrt{a}$ and $-\sqrt{a}$ are square roots of a.)

CAUTION!

In simplifying radical expressions, we have made frequent use of the properties that if $\sqrt{a}$ and $\sqrt{b}$ are real numbers, then

$$\sqrt{a}\sqrt{b} = \sqrt{ab} \text{ and } \frac{\sqrt{a}}{\sqrt{b}} = \sqrt{\frac{a}{b}}.$$

There is a subtle but important condition in the above statement: $\sqrt{a}$ and $\sqrt{b}$ must both be *real* numbers. If this condition is not met, these properties of radicals do not necessarily hold. For instance,

$$\sqrt{(-9)(-4)} = \sqrt{36} = 6, \text{ but } \sqrt{-9}\sqrt{-4} = (3i)(2i) = 6i^2 = -6.$$

In order to apply either of these two properties, then, first simplify any square roots of negative numbers by rewriting them as pure imaginary numbers.

EXAMPLE 5

Roots and Complex Numbers

Simplify the following expressions.

a. $\left(2-\sqrt{-3}\right)^2$ **b.** $\dfrac{\sqrt{4}}{\sqrt{-4}}$

Solutions:

a. $\left(2-\sqrt{-3}\right)^2 = \left(2-\sqrt{-3}\right)\left(2-\sqrt{-3}\right)$

$= 4 - 4\sqrt{-3} + \sqrt{-3}\sqrt{-3}$

$= 4 - 4i\sqrt{3} + \left(i\sqrt{3}\right)^2$ Each $\sqrt{-3}$ is converted to $i\sqrt{3}$ before multiplying.

$= 4 - 4i\sqrt{3} - 3$

$= 1 - 4i\sqrt{3}$

b. $\dfrac{\sqrt{4}}{\sqrt{-4}} = \dfrac{2}{2i}$ We simplify each radical before dividing.

$= \dfrac{1}{i}$ We already simplified $\dfrac{1}{i}$ in Example 4c, so

$= -i$ we quickly obtain the correct answer of $-i$.

1. $5i$

2. $2i\sqrt{3}$

3. $-3i\sqrt{3}$

4. $-10i$

5. $4i\sqrt{2x}$

6. $|x|i$

7. $i\sqrt{29}$

8. $-8i$

9. $1-3i$

10. $9+2i$

11. $8-6i$

12. $-i$

13. $-5+6i$

14. 10

15. $16-30i$

16. $19-43i$

17. i

18. 97

19. -11

20. 1

21. $40-42i$

22. $31-17i$

23. -9

24. $1+2i$

25. $1+5i$

26. 6

27. $-1-4i$

28. $125i$

29. $7i$

30. $-\dfrac{3}{5}+\dfrac{4}{5}i$

31. $3+i$

32. $\dfrac{1}{5}+\dfrac{2}{5}i$

33. $-i$

34. $\dfrac{2}{29}-\dfrac{5}{29}i$

35. $-i$

36. i

37. $10-2i$

38. $\dfrac{2}{13}+\dfrac{3}{13}i$

39. $\dfrac{14}{37}+\dfrac{10}{37}i$

40. 1

41. $\dfrac{21}{17}-\dfrac{1}{17}i$

Exercises

Evaluate the following square root expressions. See Example 1.

1. $\sqrt{-25}$ **2.** $\sqrt{-12}$ **3.** $-\sqrt{-27}$ **4.** $-\sqrt{-100}$

5. $\sqrt{-32x}$ **6.** $\sqrt{-x^2}$ **7.** $\sqrt{-29}$ **8.** $\left(-i\right)^2\sqrt{-64}$

Simplify the following complex expressions by adding, subtracting, or multiplying as indicated. See Examples 2 and 3.

9. $(4-2i)-(3+i)$ **10.** $(4-i)(2+i)$ **11.** $(3-i)^2$

12. i^7 **13.** $(7i-2)+(3i^2-i)$ **14.** $(3+i)(3-i)$

15. $(5-3i)^2$ **16.** $(5+i)(2-9i)$ **17.** i^{13}

18. $(9-4i)(9+4i)$ **19.** $11i^{314}$ **20.** i^{132}

21. $(7-3i)^2$ **22.** $(4-3i)(7+i)$ **23.** $(3i)^2$

24. $(1+i)+i$ **25.** $i(5-i)$ **26.** $i^{11}\left(\dfrac{6}{i^3}\right)$

27. $(10i^2-9i)+(9+5i)$ **28.** $(-5i)^3$ **29.** $i^7\left(\dfrac{49}{7i^2}\right)$

Simplify the following quotients. See Example 4.

30. $\dfrac{1+2i}{1-2i}$ **31.** $\dfrac{10}{3-i}$ **32.** $\dfrac{i}{2+i}$ **33.** $\dfrac{1}{i^9}$

34. $(2+5i)^{-1}$ **35.** i^{-25} **36.** $\dfrac{1}{i^{27}}$ **37.** $\dfrac{52}{5+i}$

38. $(2-3i)^{-1}$ **39.** $\dfrac{4i}{5+7i}$ **40.** i^{-4} **41.** $\dfrac{5+i}{4+i}$

42. $7 + 6i\sqrt{2}$

43. $-5 + 2i\sqrt{6}$

44. $-3i$

45. 8

46. $-3\sqrt{2}$

47. $-\dfrac{7}{3}i$

48. -4

49. $22 + 10i\sqrt{3}$

50. $\dfrac{-6}{5}i$

51. $6 + 3j$ ohms

52. $48 - 22j$ volts

53. $11 - 2j$ ohms

54. Answers will vary.

Simplify the following expressions. See Example 5.

42. $\left(3 + \sqrt{-2}\right)^2$

43. $\left(1 + \sqrt{-6}\right)^2$

44. $\dfrac{\sqrt{18}}{\sqrt{-2}}$

45. $\left(\sqrt{-32}\right)\left(-\sqrt{-2}\right)$

46. $\left(\sqrt{-9}\right)\left(\sqrt{-2}\right)$

47. $\dfrac{\sqrt{-98}}{3i\sqrt{-2}}$

48. $\left(\sqrt{-8}\right)\left(\sqrt{-2}\right)$

49. $\left(5 + \sqrt{-3}\right)^2$

50. $\dfrac{\sqrt{-72}}{5i\sqrt{-2}}$

Solve the following application problems.

51. Electrical engineers often use j, rather than i, to represent imaginary numbers. This is to prevent confusion with their use of i, which often represents current. Under this convention, assume the impedance of a particular part of a series circuit is $4 - 3j$ ohms and the impedance of another part of the circuit is $2 + 6j$ ohms. Find the total impedance of the circuit. (Impedances in series are simply added.)

52. Consider the formula $V = IZ$, where V is voltage (in volts), I is current (in amps), and Z is impedance (in ohms). If you know the current of a circuit is $5 - 4j$ amps and the impedance is $8 + 2j$ ohms, find the voltage.

53. If you know the voltage of a circuit is $35 + 5j$ volts and the current is $3 + j$ amps, find the impedance.

54. Explain why it may be useful to be able to use imaginary numbers in real world math.

Linear Equations in One Variable

TOPICS

1. Equations and the meaning of solutions
2. Solving linear equations in one variable
3. Solving absolute value equations
4. Solving equations for one variable
5. Distance and interest problems

TOPIC 1

Equations and the Meaning of Solutions

In mathematics, an **equation** is a statement that two expressions are equal. An equation like $5 + 3 = 8$ that contains no variables is always true or false. Equations without variables are covered in arithmetic. In algebra, we are interested in equations with variables and the question of when these equations are true statements.

DEFINITION

Types of Equations

If an equation is always true for any allowable value(s) of the variable(s), then the equation is an **identity**. We will see that identities can always be transformed into true statements without variables.

For example, the equation $x = x$ is an identity; it is true for all values of x.

If an equation is never true, it is a **contradiction**. Often, contradictions can be written as false statements without variables.

The equation $y = y + 1$ is a contradiction; no matter what value we choose for y, the equation is false. The solution set of this equation is thus $\varnothing$, the empty set.

The third (and most common) possibility is that the equation is true for some values of the variable(s) and false for others. These equations are called **conditional**, and the goal in solving such equations is to discover the set of values that the variable(s) can be replaced by to make the equation true. This set is called the **solution set** of the equation and any one element of the solution set is called a **solution** of the equation.

The equation $3z + 1 = 7$ is a conditional equation. For example, when $z = 2$ the equation is true; for all other values of z, the equation is false. The solution set of the equation is thus $\{2\}$.

TOPIC 2

Solving Linear Equations in One Variable

Two equations that have the same solution set are called **equivalent equations**, and transforming an equation into an equivalent but simpler equation constitutes much of the work of algebra. The field properties and the cancellation properties introduced in Section 1.1 are the tools we use to transform equations into equivalent equations which have clear solution sets.

DEFINITION

Linear Equations in One Variable

A **linear equation in one variable**, say the variable x, is an equation that can be transformed into the form $ax + b = 0$, where a and b are real numbers and $a \neq 0$. Such equations are also called **first-degree** equations, as x is raised to the first power.

Linear equations in one variable are arguably the least complicated type of equation to solve. Once the equation has been written in the form $ax + b = 0$, the additive cancellation property implies

$$ax = -b, \qquad \text{(add } -b \text{ to both sides)}$$

and then the multiplicative cancellation property implies

$$x = -\frac{b}{a}. \qquad \text{(multiply both sides by } \frac{1}{a} \text{)}$$

EXAMPLE 1

Solving Linear Equations in One Variable

Solve the following linear equation in one variable: $4(x+2)+2 = 3x - 3(2x-1)$.

$$4(x+2)+2 = 3x - 3(2x-1)$$
$$\downarrow$$
$$4x + 8 + 2 = 3x - 6x - (-3)$$
$$\downarrow$$
$$4x + 10 = -3x + 3$$
$$\downarrow$$
$$7x = -7$$
$$\downarrow$$
$$x = -1$$

1. Clear parentheses using the distributive property.

2. Combine like terms on each side of the equation.

3. Use the additive cancellation property to arrange all variables on one side of the equation and all constants on the opposite side.

4. Use the multiplicative cancellation property to arrive at an equivalent equation of the form $x = c$.

EXAMPLE 2

Solving Linear Equations in One Variable

Solve the following linear equations in one variable.

a. $0.25(x-3)+0.08 = 0.15x$

b. $\dfrac{y}{6} + \dfrac{2y-1}{2} = \dfrac{y+1}{3}$

c. $3x-7 = 3(x-2)$

d. $5x+12 = 5(x+3)-3$

Note:
These examples all follow the four step process shown in Example 1, but each one has a quirk you might see in other problems.

Solutions:

a. $0.25(x-3)+0.08 = 0.15x$

$25(x-3)+8 = 15x$

$25x-75+8 = 15x$

$10x = 67$

$x = 6.7$

One approach to solving an equation with decimals is to multiply both sides by the power of 10 that will eliminate the decimals. In this problem, multiplying both sides by 100 results in a simpler linear equation. Another approach is to keep the decimals throughout the solution process.

b. $\dfrac{y}{6} + \dfrac{2y-1}{2} = \dfrac{y+1}{3}$

$6\left(\dfrac{y}{6} + \dfrac{2y-1}{2}\right) = 6\left(\dfrac{y+1}{3}\right)$

$6\cdot\dfrac{y}{6} + 6\cdot\dfrac{2y-1}{2} = 6\cdot\dfrac{y+1}{3}$

$y + 3(2y-1) = 2(y+1)$

$y + 6y - 3 = 2y + 2$

$y + 6y - 2y = 2 + 3$

$5y = 5$

$y = 1$

Although it is not necessary, it is often easier to work a problem if you get rid of any fractions by multiplying both sides of the equation by the least common denominator (LCD). Remember to multiply every term by the LCD.

Note the cancellation that has occurred. We have simplified $\dfrac{6}{6}$ to 1, $\dfrac{6}{2}$ to 3, and $\dfrac{6}{3}$ to 2.

c. $3x-7 = 3(x-2)$

$3x-7 = 3x-6$

$3x-3x = -6+7$

$0 = 1$

In this problem, the variable cancels out and we are left with a false statement.

Thus, the equation is a contradiction and has no solutions. The solution set is $\varnothing$.

d. $5x+12 = 5(x+3)-3$

$5x+12 = 5x+15-3$

$5x-5x = 15-3-12$

$0 = 0$

In this problem, the variable cancels out and we are left with a true statement.

Thus, the equation is an identity. The solution set is all real numbers, $\mathbb{R}$.

TOPIC 3

Solving Absolute Value Equations

Linear absolute value equations in one variable are closely related to linear equations and are solved in a similar fashion. The difference is that a linear absolute value equation contains at least one variable term inside absolute value symbols; if these symbols were removed, the equation would be linear.

These equations are solved by recognizing that the absolute value of any quantity equals either the original quantity or its negative, depending on whether the initial quantity is positive or negative. In general, every absolute value term in an equation leads to two equations with the absolute value signs removed. For example,

$$|ax+b| = c \text{ means } ax+b=c \text{ or } -(ax+b)=c.$$

Thus, to solve a linear absolute value equation, we rewrite the equation as two (or more) linear equations without absolute value signs.

EXAMPLE 3

Solving Absolute
Value Equations

Solve the absolute value equations.

a. $|3x-2|=1$ **b.** $|x-4|=|2x+1|$ **c.** $|6x-7|+5=3$

Note:
Begin by rewriting each absolute value term as two new equations, one with a positive sign and one with a negative sign.

Solutions:

a.
$$|3x-2|=1$$

$$3x-2=1 \quad \text{or} \quad -(3x-2)=1$$
$$3x=3 \qquad\qquad 3x-2=-1$$
$$x=1 \qquad\qquad 3x=1$$
$$x=\frac{1}{3}$$

Rewrite the absolute value equation without absolute value bars.

The result is two linear equations that can be solved using the method illustrated earlier in this section.

Now check each solution in the original equation: $|3(1)-2|=|3-2|=|1|=1$, and $\left|3\left(\frac{1}{3}\right)-2\right|=|1-2|=|-1|=1$, so the solution set is $x=\frac{1}{3}$, 1.

b. This equation has two absolute value terms, which leads to four linear equations when the absolute value bars are removed.

$$|x-4|=|2x+1| \rightarrow \begin{cases} +(x-4)=+(2x+1) \\ +(x-4)=-(2x+1) \\ -(x-4)=+(2x+1) \\ -(x-4)=-(2x+1) \end{cases}$$

Note that the two equations in blue are equivalent, as are the two equations in magenta. Thus, we have two linear equations to solve:

$$|x-4| = |2x+1|$$

$$x-4 = 2x+1 \quad \text{or} \quad -(x-4) = 2x+1$$
$$-x = 5 \qquad\qquad\qquad -3x = -3$$
$$x = -5 \qquad\qquad\qquad x = 1$$

We proceed as before; applying the distributive property and combining like terms.

Finally, we check the apparent solutions in the original equation.

$$|(-5)-4| = |2(-5)+1| \qquad\qquad |(1)-4| = |2(1)+1|$$
$$|-9| = |-9| \qquad\qquad\qquad |-3| = |3|$$
$$9 = 9 \qquad\qquad\qquad\qquad 3 = 3$$

Both apparent solutions are actual solutions, so the solution set is $x = -5,\ 1$.

c. $$|6x-7| + 5 = 3$$

Isolate the absolute value term

$$|6x-7| = -2$$

$$6x-7 = -2 \quad \text{or} \quad -(6x-7) = -2$$
$$6x = 5 \qquad\qquad\qquad -6x = -9$$
$$x = \frac{5}{6} \qquad\qquad\qquad x = \frac{3}{2}$$

Again, we rewrite the original equation as two linear equations.

When we check the solutions this time, we find that neither apparent solution actually solves the equation!

$$\left|6\left(\frac{5}{6}\right)-7\right| + 5 = 3 \qquad\qquad \left|6\left(\frac{3}{2}\right)-7\right| + 5 = 3$$
$$|5-7| = -2 \qquad\qquad\qquad |9-7| = -2$$
$$|-2| = -2 \qquad\qquad\qquad |2| = -2$$
$$2 \neq -2 \qquad\qquad\qquad 2 \neq -2$$

Thus, the equation is a contradiction, and the solution set is $\varnothing$.

Example 3c illustrates an important point about absolute value equations. Note that as we checked each solution, we encountered an equation with an absolute value term on one side and a negative value on the other. Since any absolute value expression is automatically nonnegative, these types of equations have no solution!

It is good practice to isolate the absolute value term on one side of the equation before rewriting as multiple linear equations. Once we rewrite $|6x-7|+5=3$ in the form $|6x-7|=-2$, we can immediately recognize the equation has no solution.

CAUTION!

Absolute value equations are one class of equations (there are others, as we shall see) in which it is very important to check your final answer in the original equation, as the apparent solutions obtained by the above method may not solve the original absolute value equation. An apparent solution that does not solve the original problem is called an **extraneous solution**.

TOPIC **Solving Equations for One Variable**

One common task in applied mathematics is to solve a given equation in two or more variables for one of the variables. **Solving for a variable** means to transform the equation into an equivalent one in which the specified variable is isolated on one side of the equation. For linear equations we accomplish this by the same methods we have used in the previous examples.

EXAMPLE 4

Solving Linear Equations for One Variable

Solve the following equations for the specified variable. All of the equations are formulas that arise in application problems, and they are linear in the specified variable.

a. $P = 2l + 2w$. Solve for w.   **b.** $A = P\left(1 + \dfrac{r}{m}\right)^{mt}$. Solve for P.

c. $S = 2\pi r^2 + 2\pi r h$. Solve for h.

Note:
Nonlinear equations may be linear when solved for a particular variable. In this case, we can still apply the same methods of cancellation used in the prior examples.

Solutions:

a.
$$P = 2l + 2w$$
$$P - 2l = 2w$$
$$\frac{P - 2l}{2} = w$$
$$w = \frac{P - 2l}{2}$$

This is the formula for the perimeter P of a rectangle of length l and width w.

The last equation is equivalent to the preceding one, but it is conventional to put the specified variable on the left side of the equation.

b.
$$A = P\left(1 + \frac{r}{m}\right)^{mt}$$
$$\frac{A}{\left(1 + \dfrac{r}{m}\right)^{mt}} = P$$
$$P = A\left(1 + \frac{r}{m}\right)^{-mt}$$

This is the formula for compound interest. If principal P is invested at an annual rate r for t years, compounded m times a year, the value of the investment at time t is A. This formula is linear in the variables P and A, but not in m, t, or r.

We use the properties of exponents to find a cleaner solution.

c. $S = 2\pi r^2 + 2\pi rh$

This is the formula for the surface area of a right circular cylinder of radius r and height h. It is linear in the variables S and h, but not in r.

$$S - 2\pi r^2 = 2\pi rh$$

$$\frac{S - 2\pi r^2}{2\pi r} = h$$

$$h = \frac{S - 2\pi r^2}{2\pi r}$$

TOPIC 5

Distance and Interest Problems

Many applications lead to equations more complicated than those that we have studied so far, but good examples of linear equations arise from distance and simple interest problems. This is because the basic distance and simple interest formulas are linear in all of their variables.

Distance: $d = rt$, where d is the distance traveled at rate r for time t.

Simple Interest: $I = Prt$, where I is the interest earned on principal P invested at rate r for time t.

EXAMPLE 5

Calculating Distance

The distance from Shreveport, LA to Austin, TX by one route is 325 miles. If Kevin made the trip in five and a half hours, what was his average speed?

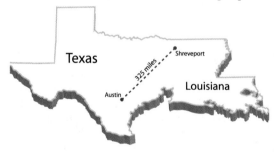

Note:
With application or mathematical modeling problems, it often helps to list the variables in the problem. As you read through the problem statement, fill in this list and determine which variable to solve for.

Solution:

We know that $d = 325$ miles and $t = 5.5$ hours, and need to solve for r.

$$d = rt$$
$$325 = 5.5r$$

$$\frac{325}{5.5} = r$$

$$r = 59.1 \text{ miles/hour (rounded to nearest tenth)}$$

EXAMPLE 6

Calculating Average Interest

Julie invested $1500 in a risky high-tech stock on January 1st. On July 1st, her stock is worth $2100. She knows that her investment does not earn interest at a constant rate, but she wants to determine her average annual rate of return at this point in the year. What is the average annual rate of return she has earned so far?

Solution:

The interest that Julie has earned in half a year is $600 (or $2100 − $1500). Replacing P with 1500, t with $\frac{1}{2}$, and I with 600 in the formula $I = Prt$, we have:

$$600 = (1500)\left(\frac{1}{2}\right)r$$

$$\frac{1200}{1500} = r$$

$$r = 0.8$$

$$r = 80\% \text{ average rate of return per year}$$

1. $\mathbb{R}$ (Identity)

2. $t = -5$

3. $x = 1$

4. $y = -1$

5. $w = -3$

6. $\varnothing$ (Contradiction)

7. $\mathbb{R}$ (Identity)

8. $x = \dfrac{15}{2}$

9. $\varnothing$ (Contradiction)

10. $w = 7$

11. $m = 7$

12. $p = 7$

13. $x = 3.7$

14. $x = 1.25$

15. $x = 1.05$

16. $\mathbb{R}$ (Identity)

17. $y = -5$

18. $w = 63$

Exercises

Solve the following linear equations. See Examples 1 and 2.

1. $3x + 5 = 3(x + 3) - 4$

2. $-3(2t - 4) = 7(1 - t)$

3. $5(2x - 1) = 3(1 - x) + 5x$

4. $\dfrac{y + 5}{4} = \dfrac{1 - 5y}{6}$

5. $3w + 5 = 2(w + 3) - 4$

6. $3x + 5 = 3(x + 3) - 5$

7. $\dfrac{4s - 3}{2} + \dfrac{7}{4} = \dfrac{8s + 1}{4}$

8. $\dfrac{4x - 3}{2} + \dfrac{3}{8} = \dfrac{7x + 3}{4}$

9. $\dfrac{4z - 3}{2} + \dfrac{3}{8} = \dfrac{8z + 3}{4}$

10. $3(2w + 13) = 5w + w\left(7 - \dfrac{3}{w}\right)$

11. $\dfrac{6}{7}(m - 4) - \dfrac{11}{7} = 1$

12. $0.08p + 0.09 = 0.65$

13. $0.6x + 0.08 = 2.3$

14. $0.9x + 0.5 = 1.3x$

15. $0.73x + 0.42(x - 2) = 0.35x$

16. $\dfrac{8y - 2}{4} + \dfrac{6}{8} = \dfrac{16y + 2}{8}$

17. $\dfrac{3}{7}(y - 2) - \dfrac{14}{7} = -5$

18. $6(5w - 5) = -31(3 - w)$

19. $\mathbb{R}$ (Identity)

20. $y = -9$

21. $\mathbb{R}$ (Identity)

22. $\varnothing$ (Contradiction)

23. $x = 3$

24. $z = 12$

25. $\varnothing$ (Contradiction)

26. $x = -1, \dfrac{7}{3}$

27. $y = -\dfrac{1}{3}, -3$

28. $\varnothing$ (Contradiction)

29. $x = \dfrac{1}{3}$

30. $x = -\dfrac{3}{2}, 2$

31. $x = -311, 420$

32. $x = -9, 11$

33. $x = -\dfrac{4}{5}, 2$

34. $x = -\dfrac{9}{2}, -3$

35. $\varnothing$ (Contradiction)

36. $\varnothing$ (Contradiction)

37. $x = -2, 2$

38. $x = 2$

39. $x = 0$

40. $x = -\dfrac{1}{2}$

41. $x = -99$

42. $x = 5$

43. $x = \dfrac{1}{4}$

44. $z = 28$

45. $x = \dfrac{1}{7}$

19. $\dfrac{7x-5}{4} + \dfrac{14}{8} = \dfrac{14x+4}{8}$

20. $\dfrac{3}{11}(y-2) - \dfrac{33}{11} = -6$

21. $3z + 3 = 3(z+4) - 9$

22. $4y + 9 = 4(y+4) - 10$

23. $2.8x + 1.2 = 3.2x$

24. $0.73z + 0.34 = 9.1$

25. $0.24x + 0.58(x-6) = 0.82x - 3.67$

Solve the following absolute value equations. See Example 3.

26. $|3x - 2| = 5$

27. $-|3y + 5| + 6 = 2$

28. $|4x + 3| + 2 = 0$

29. $|6x - 2| = 0$

30. $|-8x + 2| = 14$

31. $|2x - 109| = 731$

32. $|4x - 4| - 40 = 0$

33. $|5x - 3| = 7$

34. $|4x + 15| = 3$

35. $-|6x + 1| = 11$

36. $|-14y + 3| + 3 = 2$

37. $|3x - 2| - 1 = |5 - x|$

38. $|x + 3| = |x - 7|$

39. $|2 - x| = |2 + x|$

40. $|x| = |x + 1|$

41. $|x + 97| = |x + 101|$

42. $|x - 3| - |x - 7| = 0$

43. $\left|x + \dfrac{1}{4}\right| = \left|x - \dfrac{3}{4}\right|$

44. $|z - 51| - |z - 5| = 0$

45. $\left|x - \dfrac{5}{7}\right| = \left|x + \dfrac{3}{7}\right|$

46. $|6y - 3| = |5y + 5|$

Solve the following equations for the indicated variable. See Example 4.

47. Circumference of a Circle: $C = 2\pi r$; solve for r

48. Ideal Gas Law: $PV = nRT$; solve for T

49. Velocity: $v^2 = v_0^{\,2} + 2ax$; solve for a

50. Area of a Trapezoid: $A = \dfrac{1}{2}(B + b)h$; solve for h

51. Temperature Conversions: $C = \dfrac{5}{9}(F - 32)$; solve for F

52. Volume of a Right Circular Cone: $V = \dfrac{1}{3}\pi r^2 h$; solve for h

53. Surface Area of a Cube: $A = 2lw + 2wh + 2hl$; solve for h

54. Distance: $d = rt_1 + rt_2$; solve for r

46. $y = -\dfrac{2}{11}, 8$

47. $r = \dfrac{C}{2\pi}$

48. $T = \dfrac{PV}{nR}$

49. $a = \dfrac{v^2 - v_0^2}{2x}$

50. $h = \dfrac{2A}{(B+b)}$

51. $F = \dfrac{9}{5}C + 32$

52. $h = \dfrac{3V}{\pi r^2}$

53. $h = \dfrac{A - 2lw}{2w + 2l}$

54. $r = \dfrac{d}{t_1 + t_2}$

55. $m = \dfrac{2K}{v^2}$

56. $t = \dfrac{A}{Pr} - \dfrac{1}{r}$

57. $\dfrac{19}{3}$ hours, or 6 hours and 20 minutes

58. 4.5 hours

59. 13.5 miles

60. 3.5 hours

61. $390

62. 28%

63. 7.5%

64. 96.2%, 103.9%

65. 26 feet by 26 feet

55. Kinetic Energy of Protons: $K = \dfrac{1}{2}mv^2$; solve for m

56. Finance: $A = P(1 + rt)$; solve for t

Solve the following application problems. See Examples 5 and 6.

57. A riverboat leaves port and proceeds to travel downstream at an average speed of 15 miles per hour. How long will it take for the boat to arrive at the next port, 95 miles downstream?

58. Two trucks leave a warehouse at the same time. One travels due east at an average speed of 45 miles per hour, and the other travels due west at an average speed of 55 miles per hour. After how many hours will they be 450 miles apart?

59. Two cars leave a rest stop at the same time and proceed to travel down the highway in the same direction. One travels at an average rate of 62 miles per hour, and the other at an average rate of 59 miles per hour. How far apart are the two cars after four and a half hours?

60. Two trains are 630 miles apart, heading directly toward each other. The first train is traveling at 95 mph, and the second train is traveling at 85 mph. How long will it be before the trains pass each other?

61. Two brothers, Rick and Tom, each inherit $10,000. Rick invests his inheritance in a savings account with an annual return of 2.25%, while Tom invests his in a CD paying 6.15% annually. How much more money does Tom have than Rick after 1 year?

62. Sarah, sister to Rick and Tom in the previous problem, also inherits $10,000, but she invests her inheritance in a global technology mutual fund. At the end of 1 year, her investment is worth $12,800. What has her effective annual rate of return been?

63. Bob buys a large screen digital TV priced at $9500, but pays $10,212.50 with tax. What is the rate of tax where Bob lives?

64. Will and Matt are brothers. Will is 6 feet, 4 inches tall, and Matt is 6 feet, 7 inches tall. How tall is Will as a percentage of Matt's height? How tall is Matt as a percentage of Will's height?

65. A farmer wants to fence in three square garden plots situated along a road, as shown, and he decides not to install fencing along the edge of the road. If he has 182 feet of fencing material total, what dimensions should he make each square plot?

66. 95, 96, and 97

67. 53, 55, and 57

68. $21.50

69. 36.4%

66. Find three consecutive integers whose sum is 288. (**Hint:** If n represents the smallest of the three, then $n+1$ and $n+2$ represent the other two numbers.)

67. Find three consecutive odd integers whose sum is 165. (**Hint:** If n represents the smallest of the three, then $n+2$ and $n+4$ represent the other two numbers.)

68. Kathy buys last year's best selling novel, in hardcover, for $15.05. This is a 30% discount from the original price. What was the original price?

69. The highest point on Earth is the peak of Mount Everest. If you climbed to the top, you would be approximately 29,035 feet above sea level. Remembering that a mile is 5280 feet, what percentage of the height of the mountain would you have to climb to reach a point two miles above sea level?

Linear Inequalities in One Variable

TOPICS

1. Solving linear inequalities

2. Solving compound linear inequalities

3. Solving absolute value inequalities

4. Translating inequality phrases

TOPIC 1

Solving Linear Inequalities

If the equality symbol in a linear equation is replaced with $<, \leq, >,$ or $\geq$, the result is a **linear inequality**. One difference between linear equations and linear inequalities is the way in which the solutions are described. Typically, the solution of a linear inequality consists of some interval of real numbers; such solutions can be described graphically or with interval notation. The process of obtaining the solution, however, is much the same as the process for solving linear equations, with the one important difference discussed below.

When solving linear inequalities, the field properties outlined in Section 1.1 all still apply, and we often use the distributive and commutative properties in order to simplify one or both sides of an inequality. The additive version of the two cancellation laws is also used in the same way as in solving equations. The one difference lies in applying the multiplicative version of cancellation. When dealing with linear *equations* we can multiply both sides by a positive or negative value and obtain an equivalent equation. When dealing with linear *inequalities*, some problems arise when multiplying both sides by a negative value.

EXAMPLE 1

Multiplying
Inequalities by
Negative Numbers

Consider the following two inequalities: $-3 < 2$ and $x < 0$. Observe what happens if we multiply both sides of each inequality by -1.

1. The statement $-3 < 2$ is clearly true, but if we multiply both sides by -1, we obtain the false statement $3 < -2$.

2. Now consider the inequality $x < 0$. If we multiply both sides by -1, we have the inequality $-x < 0$. But these two statements can't both be true!

These examples show that multiplicative cancellation must behave a bit differently for linear inequalities. Note that if we reverse the inequality sign in our results, we actually get true statements. This provides a clue to how we approach multiplicative cancellation in the case of linear inequalities.

PROPERTIES

In this table, A, B, and C represent algebraic expressions and D represents a nonzero constant. Each of the properties is stated for the inequality symbol <, but they are also true for the other three symbols (when substituted below).

Property	Description
If $A < B$, then $A + C < B + C$.	Adding the same quantity to both sides of an inequality results in an equivalent inequality.
If $A < B$ and $D > 0$, then $A \cdot D < B \cdot D$.	If both sides of an inequality are multiplied by a positive constant, the sense of the inequality is unchanged.
If $A < B$ and $D < 0$, then $A \cdot D > B \cdot D$.	If both sides are multiplied by a negative constant, the sense of the inequality is reversed.

Keep in mind that multiplying (or dividing) both sides of an inequality by a negative quantity requires reversing, or "flipping" the inequality symbol. We will see this several times in the examples to follow.

EXAMPLE 2

Solve the following inequalities, using interval notation to describe the solution set.

a. $5 - 2(x - 3) \le -(1 - x)$ **b.** $\dfrac{3(a - 2)}{2} < \dfrac{5a}{4}$

Solutions:

a. $5 - 2(x - 3) \le -(1 - x)$

$\qquad 5 - 2x + 6 \le -1 + x$ Begin by using the distributive property, then combine like terms.

$\qquad -2x + 11 \le -1 + x$

$\qquad\qquad -3x \le -12$ Now, all we need to do is divide by −3.

$\qquad\qquad\qquad x \ge 4$ Note the reversal of the inequality symbol.

In interval notation, the solution is $[4, \infty)$.

b. $\dfrac{3(a-2)}{2} < \dfrac{5a}{4}$

$4\left(\dfrac{3(a-2)}{2}\right) < 4\left(\dfrac{5a}{4}\right)$

$6(a-2) < 5a$

$6a - 12 < 5a$

$a < 12$

Just as with equations, fractions in inequalities can be eliminated by multiplying both sides by the least common denominator.

Since we do not need to multiply or divide by a negative value, the sense of the inequality does not change.

Thus, in interval notation, the solution is $(-\infty, 12)$.

The solutions in Example 2 were described using interval notation, but solutions can also be described by set-builder notation or by graphing. Graphing a solution to an inequality can lead to a better understanding of which real numbers solve the inequality.

The symbols used in this text for graphing intervals are the same as the symbols in interval notation. Parentheses are used to indicate excluded endpoints of intervals and brackets are used when the endpoints are included in the interval. The portion of the number line that constitutes the interval is then shaded. (Other commonly used symbols in graphing are open circles for parentheses and filled-in circles for brackets.)

For example, the two solutions above are graphed as follows:

a.
4

b.
12

EXAMPLE 3

Graphing Intervals of Real Numbers

Graph the following intervals.

a. $[-3, 6]$ **b.** $(-\infty, 5]$ **c.** $[2, 9)$

Solutions:

a. ◄─┼─[─┼─┼─┼─┼─┼─┼─┼─┼─┼─┤─►
    −3   6

Both endpoints are included in the interval.

b. ◄─┼─┼─┼─┼─┼─┼─┤─┼─┼─┼─┼─►
                  5

The left-hand side of the graph extends to negative infinity.

c. ◄─┼─[─┼─┼─┼─┼─┼─┼─)─┼─┼─►
    2   9

The left endpoint is included in the solution, while the right endpoint is excluded.

TOPIC 2 — Solving Compound Linear Inequalities

A **compound inequality** is a statement containing two inequality symbols and can be interpreted as two distinct inequalities joined by the word "and" (this means both inequalities must be true). We'll explore how to solve such inequalities with an application example you are likely to encounter at some point.

EXAMPLE 4

Calculating Final Grades

The final grade in a class depends on the grades of 5 exams, each worth a maximum of 100 points. Suppose Janice's scores on the first four tests are 67, 82, 73, and 85. What scores can she make on the fifth test to get a B in the class?

Solution:

If the B range corresponds to an average greater than or equal to 80 and less than 90, and if x represents the fifth test score, we need to solve the following compound inequality:

$$80 \le \frac{67 + 82 + 73 + 85 + x}{5} < 90$$

This could be solved by breaking it into the two inequalities

$$80 \le \frac{67 + 82 + 73 + 85 + x}{5} \quad \text{and} \quad \frac{67 + 82 + 73 + 85 + x}{5} < 90,$$

but it is more efficient to solve both at the same time as a compound inequality.

$$80 \le \frac{67 + 82 + 73 + 85 + x}{5} < 90 \qquad \text{First, simplify the middle term.}$$

$$80 \le \frac{307 + x}{5} < 90$$

$$400 \le 307 + x < 450 \qquad \text{Just as if we were working with a single inequality, we begin by multiplying by 5,}$$

$$93 \le x < 143 \qquad \text{then subtract 307 from all three parts.}$$

Thus, the mathematical solution to the compound inequality is $[93, 143)$. However, this is not the solution to our application problem!

Why not? There is an additional restriction on the solution set based on the context of the problem; each exam is worth a maximum of 100 points. This means that any value in the calculated solution set greater than 100 does not apply, making the actual solution $[93, 100]$.

93   100

Note:
We follow the same process in solving compound inequalities as we do for standard ones. The only difference is that operations are applied to all three "sides" of the statements.

EXAMPLE 5

Compound Linear
Inequalities

Solve the following compound inequalities.

a. $-1 < 3 - 2x \leq 5$

b. $2(2x - 1) \leq 4x + 2 \leq 4(x + 1)$

Solutions:

a. $-1 < 3 - 2x \leq 5$ Begin by subtracting 3 from all three expressions.

$-4 < -2x \leq 2$ Since we divide each expression by -2, we must reverse each inequality symbol.

$2 > x \geq -1$

$-1 \leq x < 2$ The final compound inequality is identical to the one before it, but has been written so that the smaller number appears first.

In interval notation, the solution is $[-1, 2)$.

b. $2(2x - 1) \leq 4x + 2 \leq 4(x + 1)$ First, apply the distributive property to simplify.

$4x - 2 \leq 4x + 2 \leq 4x + 4$

$-2 \leq 2 \leq 4$ The variable disappears from the inequality, and we are left to assess whether the statement is true.

Since we are left with a true statement, the compound inequality is true for all values of the variable x, and the solution set is $(-\infty, \infty)$.

TOPIC   ## Solving Absolute Value Inequalities

An **absolute value inequality** is an inequality in which some variable expression appears inside absolute value symbols. In the problems that we will study, the inequality would be linear if the absolute value symbols were not there.

The geometric meaning of absolute value provides the method by which absolute value inequalities are solved. Recall that $|x|$ represents the distance between x and 0 on the real number line. If a is a positive real number, the inequality $|x| < a$ means that x is less than a units from 0, and the inequality $|x| > a$ means that x is greater than a units from 0 (similar interpretations hold for the symbols $\leq$ and $\geq$). This means that absolute value inequalities can be written without absolute values as follows:

$$|x| < a \Leftrightarrow -a < x < a$$

and

$$|x| > a \Leftrightarrow x < -a \text{ or } x > a$$

The two absolute value inequalities can serve as a template for rewriting more complicated inequalities without absolute values. Take note of the fact that the $<$ symbol leads to a set of two inequalities that must *both* be true, while the $>$ symbol leads to a solution in which *either* of two inequalities must hold.

EXAMPLE 6

**Absolute Value
Inequalities**

Solve the following absolute value inequalities.

a. $\left|4-2x\right|>6$ **b.** $2\left|3y-2\right|+3\le11$

c. $\left|5+2s\right|\le-3$ **d.** $\left|5+2s\right|\ge-3$

Solutions:

a. $$\left|4-2x\right|>6$$

$$4-2x<-6 \qquad\qquad 4-2x>6$$

$$-2x<-10 \quad\text{or}\quad -2x>2$$

$$x>5 \qquad\qquad\qquad x<-1$$

We can immediately rewrite the inequality without absolute values and begin solving the two independent inequalities.

After dividing by −2, we need to reverse the sense of the inequality.

The graph of the solution is:

The solution is $\left(-\infty,-1\right)\cup\left(5,\infty\right)$.

b. $2\left|3y-2\right|+3\le11$

$$\left|3y-2\right|\le4$$

$$-4\le3y-2\le4$$

$$-2\le3y\le6$$

$$-\frac{2}{3}\le y\le2$$

Thus, the solution is $\left[-\frac{2}{3},2\right]$.

Isolate the term containing absolute values by subtracting 3 from both sides, then dividing both sides by 2.

After rewriting the inequality as described earlier, we have a compound inequality to solve.

Graphically, the solution can be written as:

c. $\left|5+2s\right|\le-3$

The solution is $\varnothing$.

Just as in Example 3c of Section 1.5, we conclude that the solution set is the empty set, as it is impossible for the absolute value of any expression to be negative.

d. $\left|5+2s\right|\ge-3$

The solution is $\mathbb{R}$.

Since every absolute value is greater than or equal to 0, then the equation is true for all s.

TOPIC 4   Translating Inequality Phrases

Many real-world applications leading to inequalities involve notions such as "is no greater than," "at least as large as," "does not exceed," and so on. Phrases such as these all have precise mathematical translations that use one of the four inequality symbols.

Let's look at how the phrases translate when variables are added.

Translating Inequality
Phrases

EXAMPLE 7

"x is no greater than y"

This means that x is not greater than y, which is the same as saying x is less than or equal to y, so this translates to $x \leq y$.

"x is at least as large as y"

If x is at least as large as y, then it can either be as large as (equal to) y or larger (greater) than y, so this phrase translates to $x \geq y$.

"x does not exceed y"

Compare this to the first phrase; the words "is no" carry the same meaning as "does not," and "greater than" is a synonym for "exceed." The two phrases have the same meaning, and so "x does not exceed y" also translates to $x \leq y$.

While it is important to be able to reason out what inequality a particular phrase represents, it is also useful to have a reliable technique for doing these translations.

Given a statement like "x (phrase) y," one method is to ask whether the statement makes sense if x is less than y, if x is equal to y, and if x is greater than y. The answers to these three questions uniquely determine the appropriate inequality symbol.

Applying the process to the first two phrases above and one new phrase, we have:

Phrase	Can x be less than y?	Can x equal y?	Can x be greater than y?	Inequality
"x is no greater than y"	Yes	Yes	No	$x \leq y$
"x is at least as large as y"	No	Yes	Yes	$x \geq y$
"y exceeds x"	Yes	No	No	$x < y$

====== **EXAMPLE 8** ======

Applications of Inequalities

Express each of the following problems as an inequality, and then solve the inequality.

a. The average daily high temperature in Santa Fe, NM over the course of three days exceeded 75. Given that the high on the first day was 72 and the high on the third day was 77, what was the minimum high temperature on the second day?

b. As a test for quality at a plant manufacturing silicon wafers for computer chips, a random sample of 10 batches of 1000 wafers each must not detect more than 5 defective wafers per batch on average. In the first 9 batches tested, the average number of defective wafers per batch is found to be 4.78 (to the nearest hundredth). What is the maximum number of defective wafers that can be found in the 10th batch for the plant to pass the quality test?

1. $\{-9, 3.14, -2.83, 1, -3, 4\}$

2. $\{12, 8\}$

3. $\{-2.83, 1, -3\}$

4. $\{3.14, 1\}$

5. $(-\infty, -3]$

6. $(-\infty, -6]$

7. $(-\infty, 4.8)$

8. $(19, \infty)$

9. $(-\infty, 2.25)$

10. $[1, \infty)$

11. $\left(-\infty, \dfrac{3}{2}\right)$

12. $(-\infty, 2)$

13. $\left(-\infty, -\dfrac{3}{11}\right]$

14. $(-\infty, -1.2)$

Solutions:

a. We'll begin building the inequality with an expression for calculating the average. Letting x represent the high temperature on the second day, we have:

$$\frac{72 + x + 77}{3}$$

What inequality symbol do we use? The problem states the average *exceeded* 75. To exceed is to be greater than, so we use the $>$ symbol:

$$\frac{72 + x + 77}{3} > 75$$

We then proceed to solve the inequality.

$$\frac{72 + x + 77}{3} > 75$$

$$149 + x > 225$$

$$x > 76$$

Thus, the high temperature on the second day exceeded 76 degrees.

b. The phrase "must not detect more than 5 defective wafers per batch on average" means the average number must be less than or equal to 5. Let x denote the maximum number of defective wafers in the last batch.

$$\frac{(9)(4.78) + x}{10} \le 5$$

The number of defective wafers found in the first 9 batches is $(9)(4.78) = 43.02$.

$$43.02 + x \le 50$$

$$x \le 6.98$$

Since it is not possible to have a fractional number of wafers, there must have been 43 defective wafers in the first 9 batches, so the maximum allowable number of defective wafers in the final batch is 7.

15. $(7, \infty)$

16. $(-\infty, -1]$

17. $(35, \infty)$

18. $\left(\dfrac{23}{18}, \infty\right)$

19. $(-3, \infty)$

20. $[8, \infty)$

21. $(-0.11, \infty)$

22. $[1.21, \infty)$

23. $(1, 5]$

24. $[4, 8]$

25. $(-10, 6]$

26. $\left(-\dfrac{4}{3}, 5\right]$

27. $[-8, -2)$

28. $(-\infty, \infty)$

29. $(21, 69]$

30. $(-25, -15]$

31. $\left(\dfrac{23}{7}, \dfrac{25}{7}\right)$

Exercises

Determine which elements of $S = \{12, -9, 3.14, -2.83, 1, 5.24, 8, -3, 4\}$ satisfy each inequality below.

1. $7y - 33.6 < -8.6 + 2y$

2. $-2.2y - 18.8 \geq 5.2(1 - y)$

3. $-40 < 4y - 8 \leq 4$

4. $-4 < -2(z - 2) \leq 2$

Solve the following linear inequalities. Describe the solution set using interval notation and by graphing. See Examples 2 and 3.

5. $4 + 3t \leq t - 2$

6. $x - 7 \geq 5 + 3x$

7. $5y - 24 < -9.6 + 2y$

8. $-\dfrac{v + 2}{3} > \dfrac{5 - v}{2}$

9. $4.2x - 5.6 < 1.6 + x$

10. $8.5y - 3.5 \geq 2.5(3 - y)$

11. $-2(3 - x) < -2x$

12. $\dfrac{1 - x}{5} > \dfrac{-x}{10}$

13. $4w + 7 \leq -7w + 4$

14. $-5(p - 3) > 19.8 - p$

15. $\dfrac{6f - 2}{5} < \dfrac{5f - 3}{4}$

16. $\dfrac{u - 6}{7} \geq \dfrac{2u - 1}{3}$

17. $0.04n + 1.7 < 0.13n - 1.45$

18. $2k + \dfrac{3}{2} < 5k - \dfrac{7}{3}$

19. $\dfrac{4x + 4}{5} > \dfrac{3x + 2.6}{4}$

20. $-1.4z - 19.6 \geq 4.4(1 - z)$

21. $6m + \dfrac{7}{4} > \dfrac{4m + 5.8}{5}$

22. $-3.9n - 5.4 \geq 6.2(2 - 3n)$

Solve the following compound inequalities. Describe the solution set using interval notation and by graphing. See Examples 4 and 5.

23. $-4 < 3x - 7 \leq 8$

24. $5 \leq 2m - 3 \leq 13$

25. $-36 < 3x - 6 \leq 12$

26. $2 < 3(x + 2) \leq 21$

27. $-8 \leq \dfrac{z}{2} - 4 < -5$

28. $6(x - 1) < 2(3x + 5) \leq 6x + 10$

29. $3 < \dfrac{w + 3}{8} \leq 9$

30. $4 \leq \dfrac{p + 7}{-2} < 9$

31. $\dfrac{1}{3} < \dfrac{7}{6}(l - 3) < \dfrac{2}{3}$

32. $-10 < -2(4 + y) \leq 9$

33. $\dfrac{1}{4} \leq \dfrac{g}{2} - 3 < 5$

34. $-1.2 \leq \dfrac{x + 3}{-5} \leq 0.2$

35. $0.08 < 0.03c + 0.13 \leq 0.16$

32. $\left[-\dfrac{17}{2}, 1\right)$

$-\frac{17}{2}$ 1

33. $\left[\dfrac{13}{2}, 16\right)$

$\frac{13}{2}$ 16

34. $[-4, 3]$

-4 3

35. $\left(-\dfrac{5}{3}, 1\right]$

$-\frac{5}{3}$ 1

36. $(-\infty, -3] \cup [7, \infty)$

-3 7

37.
$\left(-\infty, -\dfrac{7}{2}\right) \cup \left(\dfrac{15}{2}, \infty\right)$

$-\frac{7}{2}$ $\frac{15}{2}$

38. $\left[\dfrac{1}{2}, \dfrac{5}{2}\right]$

$\frac{1}{2}$ $\frac{5}{2}$

39. $\left(-\infty, \dfrac{1}{2}\right) \cup \left(\dfrac{5}{2}, \infty\right)$

$\frac{1}{2}$ $\frac{5}{2}$

40. $(-11, 1)$

-11 1

41. $\varnothing$

42. $[-10, 4]$

-10 4

43. $(-\infty, 2) \cup (6, \infty)$

2   6

44. $(2, 6)$

2   6

45. $\varnothing$

46. $(-\infty, -4] \cup [8, \infty)$

-4 8

Solve the following absolute value inequalities. Describe the solution set using interval notation and by graphing. See Example 6.

36. $|x - 2| \geq 5$

37. $|4 - 2x| > 11$

38. $4 + |3 - 2y| \leq 6$

39. $4 + |3 - 2y| > 6$

40. $2|z + 5| < 12$

41. $7 - \left|\dfrac{q}{2} + 3\right| \geq 12$

42. $4|z + 3| \leq 28$

43. $-3|4 - t| < -6$

44. $-3|4 - t| > -6$

45. $3|4 - t| < -6$

46. $7 - |4 - 2y| \leq -5$

47. $11 - \left|\dfrac{w}{4} + 1\right| \geq 12$

48. $5.5 + |x - 7.2| \leq 3.5$

49. $6 - 5|x + 2| \geq -4$

50. $|2x - 1| < x + 4$

51. $|3t + 4| > -8$

52. $2 < |6w - 2| + 7$

Solve the following application problems. See Examples 7 and 8.

53. In a class in which the final course grade depends entirely on the average of four equally weighted 100-point tests, Cindy has scored 96, 94, and 97 on the first three. The professor has announced that there will be a fifteen-point bonus problem on the fourth test, and that anyone who finishes the semester with an average of more than 100 will receive an A+. What range of scores on the fourth test will give Cindy an A for the semester (an average between 90 and 100, inclusive), and what range will give Cindy an A+?

54. In a series of 30 racquetball games played to date, Larry has won 10, giving him a winning average so far of 33.3% (to the nearest tenth). If he continues to play, what interval describes the number of games he must now win in a row to have an overall winning average greater than 50%?

55. Assume that the national average SAT score for high school seniors is 1020 out of 1600. A group of 7 students receive their scores in the mail, and 6 of them look at their scores. Two students scored 1090, one got an 1120, two others each got a 910, and the sixth student received an 880. What range of scores can the seventh student receive to pull the group's average above the national average?

56. The United States government tries to keep the inflation rate below 5.0% on an annual basis. Assume that inflation rates for the first 3 quarters of a given year are as follows: 5.2%, 4.3%, and 4.7%. What range of inflation rates for the final quarter would satisfy the government's goal?

47. $\varnothing$

48. $\varnothing$

49. $[-4, 0]$

-4 0

50. $(-1, 5)$

-1 5

51. $(-\infty, \infty)$

52. $(-\infty, \infty)$

53. $[73, 113]$ for an A, $(113, 115]$ for an A+.

54. $(10, \infty)$ or $[11, \infty)$

55. $(1140, 1600]$

56. $[0, 5.8)$

Quadratic Equations

TOPICS

1. Solving quadratic equations by factoring
2. Solving "perfect square" quadratic equations
3. Solving quadratic equations by completing the square
4. The quadratic formula
5. Gravity problems
6. Solving quadratic-like equations
7. Solving general polynomial equations by factoring
8. Solving polynomial-like equations by factoring

TOPIC

Solving Quadratic Equations by Factoring

In Section 1.5, we studied first-degree polynomial equations in one variable. We will now learn how to solve **second-degree polynomial** equations in one variable, which we commonly call **quadratic** equations.

Recall that our method for solving linear equations is straightforward, and that the method always works for *any* such equation. By the end of this section, we will develop a method for solving one-variable quadratic equations that is also guaranteed to work.

We will begin with a formal definition of quadratic equations and then proceed to study those quadratic equations that can be solved by applying the factoring skills learned in Section 1.3.

DEFINITION

Quadratic Equations

A **quadratic equation in one variable**, say the variable x, is an equation that can be transformed into the form

$$ax^2 + bx + c = 0,$$

where a, b, and c are real numbers and $a \neq 0$. Such equations are also called **second-degree** equations, as x is raised to the second power.

The key to using factoring to solve a quadratic equation, or indeed any polynomial equation, is to rewrite the equation so that 0 appears by itself on one side. This allows us to use the Zero-Factor Property discussed in Section 1.1.

If the trinomial $ax^2 + bx + c$ can be factored, it can be written as a product of two linear factors A and B. The Zero-Factor Property then implies that the only way for $ax^2 + bx + c$ to be 0 is if one (or both) of A and B is 0. This is all we need to solve the equation.

EXAMPLE 1

Solving Quadratic Equations by Factoring

Solve the quadratic equations by factoring.

a. $5x^2 + 10x = 0$ **b.** $s^2 + 9 = 6s$ **c.** $x^2 + \dfrac{5x}{2} = \dfrac{3}{2}$

Solutions:

a. $\quad 5x^2 + 10x = 0$

$\quad 5x(x+2) = 0$

$\quad 5x = 0 \quad \text{or} \quad x + 2 = 0$

$\quad x = 0 \quad \text{or} \quad x = -2$

An alternate approach in this example is to divide both sides by 5 at the very beginning. This would lead to the equation $x(x+2) = 0$, which gives us the same solution set of $\{0, -2\}$.

b. $\quad s^2 + 9 = 6s$

$\quad s^2 - 6s + 9 = 0$

$\quad (s-3)^2 = 0$

$\quad s - 3 = 0 \quad \text{or} \quad s - 3 = 0$

$\quad s = 3$

Again, we rewrite the equation with 0 on one side, then factor the quadratic polynomial.

In this example, the two linear factors are the same. In such cases, the solution is called a *double root* or a *root of multiplicity 2*.

c. $\quad x^2 + \dfrac{5x}{2} = \dfrac{3}{2}$

$\quad 2x^2 + 5x = 3$

$\quad 2x^2 + 5x - 3 = 0$

$\quad (2x-1)(x+3) = 0$

$\quad 2x - 1 = 0 \quad \text{or} \quad x + 3 = 0$

$\quad x = \dfrac{1}{2} \quad \text{or} \quad x = -3$

To make the polynomial easier to factor, we multiply both sides by the LCD.

Although we could factor $2x^2 + 5x$, this would not do us any good. We must have 0 on one side in order to apply the Zero-Factor Property.

After factoring, we have two linear equations to solve. The solution set is $\left\{ \dfrac{1}{2}, -3 \right\}$.

TOPIC 2

Solving "Perfect Square" Quadratic Equations

The factoring method is fine when it works, but there are two potential problems with the method: (1) the second-degree polynomial in question might not factor over the integers, and (2) even if the polynomial does factor, the factored form may not be obvious.

In some cases where the factoring method is unsuitable, the solution can be obtained by using our knowledge of square roots. If A is an algebraic expression and if c is a constant, the equation $A^2 = c$ implies $A = \pm\sqrt{c}$.

If a quadratic equation can be written in the form $A^2 = c$, we can use the previous observation to obtain two linear equations that can be easily solved.

EXAMPLE 2

Perfect Square Quadratic Equations

Solve the quadratic equations by taking square roots.

a. $(2x+3)^2 = 8$ **b.** $(x-5)^2 + 4 = 0$

Note:
In the factoring method, we move all terms to one side. Here, we isolate a term that is squared, ideally with only a constant on the other side.

Solutions:

a. $(2x+3)^2 = 8$

$$2x+3 = \pm\sqrt{8}$$

$$2x+3 = \pm 2\sqrt{2}$$

$$2x = -3 \pm 2\sqrt{2}$$

$$x = \frac{-3 \pm 2\sqrt{2}}{2}$$

We begin by taking the square root of each side, keeping in mind that there are two numbers whose square is 8.

We solve the two linear equations at once by subtracting 3 from both sides and then dividing both sides by 2. The solution set is

$$\left\{ \frac{-3+2\sqrt{2}}{2}, \frac{-3-2\sqrt{2}}{2} \right\}.$$

Note that if we expand the expression to attempt the factoring method, we get $4x^2 + 12x + 1 = 0$, which we are unable to factor.

b. $(x-5)^2 + 4 = 0$

$$(x-5)^2 = -4$$

$$x-5 = \pm\sqrt{-4}$$

$$x-5 = \pm 2i$$

$$x = 5 \pm 2i$$

Before taking square roots, we isolate the perfect square algebraic expression on one side and put the constant on the other.

In this example, taking square roots leads to two complex number solutions. (See Section 1.4 for a review of complex numbers.) The solution set is $\{5+2i, 5-2i\}$.

Again, let's see what happens if we attempt the factoring method. The resulting equation is $x^2 - 10x + 29 = 0$, which is not factorable.

TOPIC 3

Solving Quadratic Equations by Completing the Square

Just like the factoring method, the square root method depends on the equation fitting a particular form. If the quadratic equation under consideration appears in the form $A^2 = c$, the method works well (even if the solutions wind up being complex, as in Example 2b). But what if the equation doesn't have the form $A^2 = c$?

The method of **completing the square** allows us to write any arbitrary quadratic equation $ax^2 + bx + c = 0$ in the desired square root form. This method is outlined in the following procedure.

PROCEDURE

Completing the
Square

Step 1: Write the equation $ax^2 + bx + c = 0$ in the form $ax^2 + bx = -c$.

Step 2: Divide by a, if $a \neq 1$, so that the coefficient of x^2 is 1: $x^2 + \dfrac{b}{a}x = -\dfrac{c}{a}$.

Step 3: Divide the coefficient of x by 2, square the result, and add this to both sides:

$$x^2 + \frac{b}{a}x + \left(\frac{b}{2a}\right)^2 = -\frac{c}{a} + \left(\frac{b}{2a}\right)^2.$$

Step 4: The trinomial on the left side will now be a perfect square trinomial. That is, it can be written as the square of a binomial.

At this point, the equation will have the form $A^2 = c$ and can be solved by taking the square root of both sides.

Don't try to memorize the formulas outlined in the steps above; instead, practice applying each of the four steps to many different quadratic equations.

EXAMPLE 3

Completing the
Square

Solve the quadratic equations by completing the square.

a. $x^2 - 2x - 6 = 0$ **b.** $9x^2 + 3x = 2$

Note:
Sometimes, a step in the process is not required. In the first example, the coefficient on the squared term is 1, so there is no need to follow Step 2.

Solutions:

a. $x^2 - 2x - 6 = 0$

$x^2 - 2x = 6$ Step 1: Move the constant to the right-hand side.

$x^2 - 2x + 1 = 6 + 1$ Step 3: Divide -2 (the coefficient of x) by 2 to get -1 and add $(-1)^2 = 1$ to both sides.

$(x-1)^2 = 7$ Step 4: Factor the perfect square trinomial.

$x - 1 = \pm\sqrt{7}$ Taking square roots leads to two easily solved linear equations.

$x = 1 \pm \sqrt{7}$

b. $9x^2 + 3x = 2$ Begin with Step 2: Divide each term by 9 (and simplify the resulting fractions).

$x^2 + \dfrac{1}{3}x = \dfrac{2}{9}$

$x^2 + \dfrac{1}{3}x + \dfrac{1}{36} = \dfrac{2}{9} + \dfrac{1}{36}$ Step 3: Half of the coefficient of x is $\dfrac{1}{6}$, and the square of $\dfrac{1}{6}$, is $\dfrac{1}{36}$. We add this to both sides.

$\left(x + \dfrac{1}{6}\right)^2 = \dfrac{1}{4}$ Step 4: Factor the resulting trinomial.

$x + \dfrac{1}{6} = \pm\dfrac{1}{2}$ We take the square root of each side, and solve the resulting linear equations.

$x = -\dfrac{1}{6} \pm \dfrac{1}{2}$ Since the answer of $-\dfrac{1}{6} \pm \dfrac{1}{2}$ can be simplified, we do so to obtain the final answer.

$x = \dfrac{1}{3}, -\dfrac{2}{3}$

If a quadratic can be factored as $(x-p)(x-q)$, then p and q solve the equation $(x-p)(x-q) = 0$. Note that this is just an extension of the Zero-Factor Property.

How does the quadratic $9x^2 + 3x - 2$ factor? This quadratic comes from Example 3b, so we might guess factors of $x - \dfrac{1}{3}$ and $x + \dfrac{2}{3}$. But,

$$\left(x - \frac{1}{3}\right)\left(x + \frac{2}{3}\right) = x^2 + \frac{1}{3}x - \frac{2}{9}.$$

It is not surprising that the product of these two factors has a leading coefficient of 1, since they each individually have a leading coefficient of 1. To get the correct leading coefficient, we multiply by 9 (this reverses the step we took in completing the square):

$$9\left(x - \frac{1}{3}\right)\left(x + \frac{2}{3}\right) = 9\left(x^2 + \frac{1}{3}x - \frac{2}{9}\right) = 9x^2 + 3x - 2$$

Alternatively, we can factor 9 into two factors of 3 and rearrange the products:

$$9\left(x - \frac{1}{3}\right)\left(x + \frac{2}{3}\right) = 3\left(x - \frac{1}{3}\right) \cdot 3\left(x + \frac{2}{3}\right) = (3x - 1)(3x + 2)$$

TOPIC 4   The Quadratic Formula

The method of completing the square will always serve to solve any equation of the form $ax^2 + bx + c = 0$. Why not just solve $ax^2 + bx + c = 0$ once and for all? Given that a, b, and c represent arbitrary constants, the ideal result would be to find a formula for the solutions of $ax^2 + bx + c = 0$ based on a, b, and c. That is exactly what the quadratic formula is: a formula that gives the solution to *any* equation of the form $ax^2 + bx + c = 0$. We will derive the formula now by completing the square.

$$ax^2 + bx + c = 0$$

$$x^2 + \frac{b}{a}x = -\frac{c}{a}$$

We begin with Steps 1 and 2, moving the constant to the right-hand side and dividing by a.

$$x^2 + \frac{b}{a}x + \frac{b^2}{4a^2} = -\frac{c}{a} + \frac{b^2}{4a^2}$$

$$\left(x + \frac{b}{2a}\right)^2 = -\frac{4ac}{4a^2} + \frac{b^2}{4a^2}$$

We next divide $\dfrac{b}{a}$ by 2 to get $\dfrac{b}{2a}$ and add $\left(\dfrac{b}{2a}\right)^2 = \dfrac{b^2}{4a^2}$ to both sides. Note that to add the fractions on the right, we need a common denominator of $4a^2$.

$$\left(x + \frac{b}{2a}\right)^2 = \frac{b^2 - 4ac}{4a^2}$$

$$x + \frac{b}{2a} = \pm\frac{\sqrt{b^2 - 4ac}}{2a}$$

Taking square roots leads to two linear equations, which we then solve for x.

$$x = \frac{-b}{2a} \pm \frac{\sqrt{b^2 - 4ac}}{2a}$$

$$x = \frac{-b \pm \sqrt{b^2 - 4ac}}{2a}$$

Since the fractions have the same denominator, they are easily added to obtain the final formula.

DEFINITION

The solutions of the general quadratic equation $ax^2 + bx + c = 0$, with $a \neq 0$, are given by the **quadratic formula**: $x = \dfrac{-b \pm \sqrt{b^2 - 4ac}}{2a}$.

The expression beneath the radical, $b^2 - 4ac$, is called the **discriminant**. Its value determines the number and type (real or complex) of solutions:

Discriminant	Number of Distinct Solutions	Type of Solutions	Notes
$b^2 - 4ac > 0$	2	Real	The solutions are always different.
$b^2 - 4ac = 0$	1	Real	This solution is a double root.
$b^2 - 4ac < 0$	2	Complex	The solutions are complex conjugates.

EXAMPLE 4

Solve the quadratic equations using the quadratic formula.

a. $8x^2 - 4x = 1$ **b.** $t^2 + 6t + 13 = 0$

Solutions:

a. $8x^2 - 4x = 1$

$8x^2 - 4x - 1 = 0$

$a = 8, \ b = -4, \ c = -1$

$x = \dfrac{-(-4) \pm \sqrt{(-4)^2 - 4(8)(-1)}}{2(8)}$

$x = \dfrac{4 \pm \sqrt{16 + 32}}{16}$

$x = \dfrac{4 \pm \sqrt{48}}{16}$

$x = \dfrac{4 \pm 4\sqrt{3}}{16}$

$x = \dfrac{1 \pm \sqrt{3}}{4}$

Before applying the quadratic formula, move all the terms to one side so a, b, and c can be identified correctly.

Apply the quadratic formula by making the appropriate replacements for a, b, and c.

The discriminant, 48, is positive.

We can cancel out the common factor of 4 in the numerator and denominator.

Thus, the solutions are $x = \dfrac{1 + \sqrt{3}}{4}$ and $x = \dfrac{1 - \sqrt{3}}{4}$; two unique, real solutions.

b. $t^2 + 6t + 13 = 0$ The equation is already in the proper form to apply the quadratic formula.

$$a = 1, \ b = 6, \ c = 13$$

$$t = \frac{-(6) \pm \sqrt{(6)^2 - 4(1)(13)}}{2(1)}$$ Substitute the values for a, b, and c.

$$t = \frac{-6 \pm \sqrt{36 - 52}}{2}$$

$$t = \frac{-6 \pm \sqrt{-16}}{2}$$ The discriminant is negative, so we know the solutions will be complex.

$$t = \frac{-6 \pm 4i}{2}$$ Again, we cancel out a common factor.

$$t = -3 \pm 2i$$

Thus, the solutions are $t = -3 + 2i$ and $t = -3 - 2i$; two complex conjugate solutions.

Calculating the discriminant can be a very useful tool; it provides a quick check of whether solutions are reasonable, and later on we will find it helpful in classifying the graphs of quadratic equations.

EXAMPLE 5

The Discriminant

For each of the following quadratic equations, calculate the discriminant and determine the number and type of solutions:

a. $-2x^2 + 12x - 18 = 0$ **b.** $5x^2 + 7x + 2 = 0$ **c.** $x^2 - 4x + 9 = 0$

Solutions:

We identify the values of a, b, and c, then calculate the discriminant $b^2 - 4ac$:

a. $-2x^2 + 12x - 18 = 0$

$$a = -2, \ b = 12, \ c = -18$$

We substitute, then calculate the discriminant:

$$b^2 - 4ac = (12)^2 - 4(-2)(-18) = 144 - 144 = 0$$

Since the discriminant is zero, we know there will be one real solution.

b. $5x^2 + 7x + 2 = 0$

$$a = 5, \ b = 7, \ c = 2$$

$$b^2 - 4ac = (7)^2 - 4(5)(2) = 49 - 40 = 9$$

This time, the discriminant is positive, so there are two distinct real solutions.

c. $x^2 - 4x + 9 = 0$

$$a = 1, \quad b = -4, \quad c = 9$$

$$b^2 - 4ac = (-4)^2 - 4(1)(9) = 16 - 36 = -20$$

The discriminant is negative, so the equation has two complex conjugate solutions.

While we can solve any quadratic equation using the quadratic formula, it is not always the easiest or most efficient method of solution. For example; if an equation is already in factored form, it is much easier to read off the solutions by applying the Zero-Factor Property than to multiply out the factors and then apply the quadratic formula. Similarly, if an equation is already in the form $A^2 = c$, it's much easier to apply the square root method than to use the quadratic formula. In the following example, try to use the easiest, most efficient method to solve the different quadratic equations.

EXAMPLE 6

Methods of Solving Quadratic Equations

Solve each of the following quadratic equations, identifying the most efficient method of solution.

a. $4x^2 - 25 = 0$

b. $(2x - 3)^2 = 7$

c. $3x^2 - 11x - 4 = 0$

d. $3x^2 - 10x - 4 = 0$

Solutions:

a. The left-hand side is a difference of squares, so factoring is the easiest method:

$$4x^2 - 25 = 0$$

$(2x - 5)(2x + 5) = 0$ Factor the difference of squares.

$2x - 5 = 0 \ \text{ or } \ 2x + 5 = 0$ We need to solve two linear equations.

$x = \dfrac{5}{2} \ \text{ or } \quad x = -\dfrac{5}{2}$ We have two unique real solutions.

b. The left-hand side is already a squared quantity, but the right-hand side is not zero, so we want to use the square root method:

$$(2x - 3)^2 = 7$$

$2x - 3 = \pm\sqrt{7}$ Take the square root of both sides.

$2x = 3 \pm \sqrt{7}$ Simplify the linear equation.

$x = \dfrac{3 \pm \sqrt{7}}{2}$ We have two unique real solutions.

c. While the quadratic formula certainly works, this quadratic equation is factorable:

$$3x^2 - 11x - 4 = 0$$

$$(3x+1)(x-4) = 0$$

Use trial and error or factoring by grouping to factor the trinomial into two binomials.

$$3x+1 = 0 \quad \text{or} \quad x-4 = 0$$

The Zero-Factor Property gives us two linear equations to solve.

$$x = -\frac{1}{3} \quad \text{or} \quad x = 4$$

d. Here, the equation is not factorable, so we use the quadratic formula:

$$a = 3, \quad b = -10, \quad c = -4$$

$$x = \frac{-(-10) \pm \sqrt{(-10)^2 - 4(3)(-4)}}{2(3)}$$

Identify the values of a, b, and c, then substitute into the quadratic formula.

$$x = \frac{10 \pm \sqrt{148}}{6}$$

All that remains is to simplify the solutions.

$$x = \frac{10 \pm 2\sqrt{37}}{6}$$

$$x = \frac{5 \pm \sqrt{37}}{3}$$

TOPIC 5 Gravity Problems

When an object near the surface of the Earth is moving under the influence of gravity alone, its height above the surface is described by a quadratic polynomial in the variable t, where t stands for time (usually measured in seconds).

The phrase "moving under the influence of gravity alone" means that all other forces that could potentially affect the object's motion, such as air resistance or mechanical lifting forces, are either negligible or absent. The phrase "near the surface of the Earth" means that we are considering objects that travel short vertical distances relative to the Earth's radius; the following formula doesn't apply, for instance, to rockets shot into orbit. As an example, think of someone throwing a baseball into the air on a windless day. After the ball is released, gravity is the only force acting on it.

If we let h represent the height at time t of such an object,

$$h = -\frac{1}{2}gt^2 + v_0 t + h_0,$$

where g, v_0, and h_0 are all constants: g is the force due to gravity, v_0 is the initial vertical velocity which the object has when $t = 0$, and h_0 is the height of the object when $t = 0$ (we normally say that ground level corresponds to a height of 0). If t is measured in seconds and h in feet, g is 32 ft/s^2. If t is measured in seconds and h in meters, g is 9.8 m/s^2.

Many applications involving the above formula will result in a quadratic equation that must be solved for t. In some cases, one of the two solutions must be discarded as meaningless in the context of the given problem.

EXAMPLE 7

Gravity Problems

Robert stands on the topmost tier of seats in a baseball stadium, and throws a ball out onto the field with a vertical upward velocity of 60 ft/s. The ball is 50 feet above the ground at the moment he releases the ball. When does the ball land?

Solution:

First, note that although the thrown ball has a horizontal velocity as well as a vertical velocity (otherwise it would go straight up and come straight back down), it is irrelevant in this question. All we are interested in is when the ball lands on the ground ($h = 0$). If we wanted to determine where in the field the ball lands, we would have to know the horizontal velocity as well.

We have the following information: $h_0 = 50$ ft and $v_0 = 60$ ft/s. Since the units in the problem are feet and seconds, we know to use $g = 32$ ft/s². What we are interested in determining is the time, t, when the height, h, of the ball is 0. Therefore we need to solve the quadratic equation $0 = -16t^2 + 60t + 50$ for t.

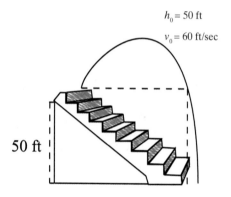

$$h_0 = 50 \text{ ft}$$
$$v_0 = 60 \text{ ft/sec}$$

50 ft

$0 = -16t^2 + 60t + 50$

$0 = 8t^2 - 30t - 25$

To simplify the calculations, we can begin by dividing both sides of the equation by -2.

$t = \dfrac{30 \pm \sqrt{900 + 800}}{16}$

The trinomial is not factorable, so we apply the quadratic formula.

$t = \dfrac{30 \pm 10\sqrt{17}}{16}$

Simplify the radical, then simplify the fraction, which contains two solutions.

$t = \dfrac{15 \pm 5\sqrt{17}}{8}$

$t \approx \cancel{-0.70}, 4.45$

The solutions are most meaningful in decimal form. The negative solution represents a time before the ball is thrown, so we discard it. The ball lands 4.45 seconds after being thrown.

TOPIC 6

Solving Quadratic-Like Equations

A polynomial equation of degree n in one variable, say x, is an equation that can be written in the form $a_n x^n + a_{n-1} x^{n-1} + \ldots + a_1 x + a_0 = 0$, where each a_i is a constant and $a_n \neq 0$. As we have seen, such equations can always be solved if $n = 1$ or $n = 2$, but in general there is no method for solving polynomial equations that is guaranteed to find all solutions. There are formulas, called the *cubic* and *quartic* formulas, that solve third- and fourth-degree polynomial equations, but there are no formulas to solve polynomial equations of degree five (or higher)! Moreover, many nonpolynomial equations have no solution method that is guaranteed to work.

Just because there isn't a formula for solving these equations doesn't mean they can never be solved! In this section, we'll use factoring, the Zero-Factor Property and our knowledge of quadratic equations to solve higher-degree polynomial equations.

The Zero-Factor Property applies whenever a product of any finite number of factors is equal to 0; if $A_1 \cdot A_2 \cdot \ldots \cdot A_n = 0$, then at least one of the A_i's must equal 0. Recall that we used the Zero-Factor Property to solve quadratic equations. This means that if we can rewrite an equation in a quadratic form, we can use the Zero-Factor Property to solve this equation as well.

DEFINITION

Quadratic-Like
Equations

An equation is **quadratic-like**, or **quadratic in form**, if it can be written in the form

$$aA^2 + bA + c = 0,$$

where a, b, and c are constants, $a \neq 0$, and A is an algebraic expression. Such equations can be solved by first solving for A and then solving for the variable in the expression A. This method of solution is called **substitution**.

EXAMPLE 8

Quadratic-Like
Equations

Solve the quadratic-like equations.

a. $\left(x^2 + 2x\right)^2 - 7\left(x^2 + 2x\right) - 8 = 0$ **b.** $y^{\frac{2}{3}} + 4y^{\frac{1}{3}} - 5 = 0$

Solutions:

a. $\left(x^2 + 2x\right)^2 - 7\left(x^2 + 2x\right) - 8 = 0$

$A^2 - 7A - 8 = 0$

$(A - 8)(A + 1) = 0$

$A = 8, -1$

$A = 8$ or $A = -1$

$x^2 + 2x = 8$ $x^2 + 2x = -1$

$x^2 + 2x - 8 = 0$ $x^2 + 2x + 1 = 0$

$(x + 4)(x - 2) = 0$ $(x + 1)^2 = 0$

$x = -4$ or $x = 2$ or $x = -1$

Making the substitution $A = x^2 + 2x$ transforms the quadratic-like equation into a quadratic equation that can be solved by factoring.

Once we have solved for A, we replace A with $x^2 + 2x$ and solve for x.

Note that -1 is a double root, while -4 and 2 are single roots.

While the substitution method does not necessarily introduce extraneous solutions, you should still check that each solution solves the original quadratic-like equation.

b. $y^{\frac{2}{3}} + 4y^{\frac{1}{3}} - 5 = 0$

$\left(y^{\frac{1}{3}}\right)^2 + 4\left(y^{\frac{1}{3}}\right) - 5 = 0$

$A^2 + 4A - 5 = 0$

$(A+5)(A-1) = 0$

$A = -5, 1$

Using properties of exponents, we can see that the substitution $A = y^{\frac{1}{3}}$ will make this equation quadratic.

We can solve this equation by factoring.

$A = -5$ or $A = 1$

$y^{\frac{1}{3}} = -5$ $y^{\frac{1}{3}} = 1$

$y = (-5)^3$ $y = (1)^3$

$y = -125$ $y = 1$

Now that we've solved for A, we reverse the substitution and solve for y in each case.

Once again, you should confirm that both values do indeed solve the original equation $y^{\frac{2}{3}} + 4y^{\frac{1}{3}} - 5 = 0$.

TOPIC 7 — Solving General Polynomial Equations by Factoring

If an equation consists of a polynomial on one side and 0 on the other, and if the polynomial can be factored completely, then the equation can be solved by using the Zero-Factor Property. If the coefficients in the polynomial are all real, the polynomial can, in principle, be factored into a product of first-degree and second-degree factors. In practice, however, this may be difficult to accomplish unless the degree of the polynomial is small or the polynomial is easily recognizable as a special product. For higher-degree polynomials, the GCF factoring method and factoring by grouping can be very effective. You may want to review some of the factoring techniques from Section 1.3 as you work through these problems.

EXAMPLE 9

Solving Equations by Factoring

Solve the equations by factoring.

a. $x^4 = 9$

b. $y^3 + y^2 - 4y - 4 = 0$

c. $8t^3 - 27 = 0$

d. $3x^4 + 18x^3 - 21x^2 = 0$

Solutions:

Note:
While checking solutions is always a good practice, solving by factoring does not produce any extraneous solutions.

a. $x^4 = 9$

$x^4 - 9 = 0$

$(x^2 - 3)(x^2 + 3) = 0$

$x^2 = 3$ or $x^2 = -3$

$x = \pm\sqrt{3}$ or $x = \pm\sqrt{-3}$

$x = \pm\sqrt{3}$ or $x = \pm i\sqrt{3}$

After isolating 0 on one side, we see the polynomial is a difference of two squares, which can always be factored.

The Zero-Factor Property gives us two equations, both of which can be solved by taking square roots.

b. $\quad y^3 + y^2 - 4y - 4 = 0$ We factor the initial equation by grouping.

$$y^2(y+1) - 4(y+1) = 0$$

$$(y+1)(y^2 - 4) = 0$$ We can factor the second term further, since it is a difference of squares.

$$(y+1)(y-2)(y+2) = 0$$

$$y = -1 \text{ or } y = 2 \text{ or } y = -2$$ There are three solutions by the Zero-Factor Property.

c. $\quad 8t^3 - 27 = 0$ The polynomial in this case is a difference of two cubes, which can always be factored.

$$(2t)^3 - 3^3 = 0$$

$$(2t - 3)(4t^2 + 6t + 9) = 0$$

$$2t - 3 = 0 \quad \text{or} \quad 4t^2 + 6t + 9 = 0$$ The Zero-Factor Property gives us two equations to solve. One of the equations is quadratic, and we can use the quadratic formula to solve it.

$$2t = 3 \qquad t = \frac{-(6) \pm \sqrt{6^2 - 4(4)(9)}}{2(4)}$$

$$t = \frac{3}{2} \qquad t = \frac{-6 \pm \sqrt{36 - 144}}{8}$$

$$t = \frac{-6 \pm \sqrt{-108}}{8}$$

$$t = \frac{-6 \pm 6i\sqrt{3}}{8}$$ Thus, we have three solutions to the original equation.

$$\text{or} \quad t = \frac{-3 \pm 3i\sqrt{3}}{4}$$

d. $3x^4 + 18x^3 - 21x^2 = 0$

$$3x^2(x^2 + 6x - 7) = 0$$ Factoring out the GCF of $3x^2$ yields a quadratic term that we can factor.

$$3x^2(x-1)(x+7) = 0$$

$$x^2(x-1)(x+7) = 0$$ The Zero-Factor Property yields three solutions to the original equation.

$$x = 0, 1, \text{or} -7$$

TOPIC **8**

Solving Polynomial-Like Equations by Factoring

The last equations that we will consider in this section are equations that are not polynomials, but which can be solved using the methods we have developed above. We have already seen one such equation in Example 8b: the equation $y^{\frac{2}{3}} + 4y^{\frac{1}{3}} - 5 = 0$ is quadratic-like, and can be solved using polynomial methods.

Like the equation in Example 8b, some polynomial-like equations can be solved by substitution, transforming them into polynomial equations. Other equations can be solved by rewriting the equation so that 0 appears on one side, factoring the equation, and then applying the Zero-Factor Property. Often, equations involving rational exponents can be solved by factoring out a common factor, as in the following examples.

EXAMPLE 10

Solving Equations by Factoring

Solve the equations by factoring.

a. $x^{\frac{7}{3}} + x^{\frac{4}{3}} - 2x^{\frac{1}{3}} = 0$

b. $(x-1)^{\frac{1}{2}} - (x-1)^{-\frac{1}{2}} = 0$

Solutions:

a.
$$x^{\frac{7}{3}} + x^{\frac{4}{3}} - 2x^{\frac{1}{3}} = 0$$

$$x^{\frac{1}{3}}\left(x^2 + x - 2\right) = 0$$

$$x^{\frac{1}{3}}(x+2)(x-1) = 0$$

$$x^{\frac{1}{3}} = 0 \quad \text{or} \quad x+2=0 \quad \text{or} \quad x-1=0$$
$$x = 0 \quad \text{or} \quad x = -2 \quad \text{or} \quad x = 1$$

Recall that in cases like this, we factor out x raised to the lowest exponent. In this case, the remaining factor is a factorable trinomial.

The Zero-Factor Property leads to three simple equations.

b.
$$(x-1)^{\frac{1}{2}} - (x-1)^{-\frac{1}{2}} = 0$$

$$(x-1)^{-\frac{1}{2}}\left((x-1)-1\right) = 0$$

$$(x-1)^{-\frac{1}{2}}(x-2) = 0$$

$$(x-1)^{-\frac{1}{2}} = 0 \quad \text{or} \quad x-2=0$$

$$\frac{1}{(x-1)^{\frac{1}{2}}} = 0 \quad \text{or} \quad x = 2$$

$$x = 2$$

Again, we factor out the common algebraic expression raised to the lowest exponent.

This equation leads to two equations, only one of which has a solution. (Note that there is no value for x which would solve the first of the two equations.)

The original equation has only one solution.

1. $\left\{-1, \dfrac{3}{2}\right\}$

2. $\{0, 3\}$

3. $\{7\}$

4. $\left\{\dfrac{-3}{2}, -3\right\}$

5. $\left\{-1, \dfrac{8}{3}\right\}$

6. $\left\{\pm\dfrac{3}{2}\right\}$

7. $\{3, 11\}$

8. $\{2\}$

9. $\{3, 11\}$

10. $\{-5, 3\}$

Exercises

Solve the following quadratic equations by factoring. See Example 1.

1. $2x^2 - x = 3$

2. $4x^2 - 12x = 0$

3. $x^2 - 14x + 49 = 0$

4. $y(2y+9) = -9$

5. $x^2 - \dfrac{8}{3} = \dfrac{5x}{3}$

6. $4x^2 - 9 = 0$

7. $3x^2 + 33 = 2x^2 + 14x$

8. $5x^2 + 2x + 3 = 4x^2 + 6x - 1$

9. $(x-7)^2 = 16$

10. $(x+1)^2 - 1 = 15$

11. $\{0, 6\}$

12. $\left\{2 \pm i\sqrt{5}\right\}$

13. $\{17, 19\}$

14. $\left\{-\dfrac{5}{3}, \dfrac{1}{3}\right\}$

15. $\left\{\dfrac{1}{2} \pm \sqrt{2}\right\}$

16. $\{-5, 9\}$

17. $\{-5, -3\}$

18. $\{-5, 2\}$

19. $\left\{-5, \dfrac{3}{2}\right\}$

20. $\left\{-\dfrac{7}{2}, \dfrac{9}{2}\right\}$

21. $\left\{\dfrac{13}{2}, \dfrac{15}{2}\right\}$

22. $\{-16, -6\}$

23. $\left\{-\dfrac{4}{3}, 1\right\}$

24. $\{0.17 \pm 0.86i\}$

25. $\{-1\}$

26. $\left\{0, \dfrac{2}{3}\right\}$

27. $\left\{-\dfrac{3}{2}, 5\right\}$

28. $\left\{\dfrac{-1 \pm \sqrt{13}}{6}\right\}$

29. $\{-4.5 \pm 4.5i\}$

30. $\{8, 14\}$

31. $\left\{\pm \dfrac{9}{8}\right\}$

32. $\left\{-3, \dfrac{3}{10}\right\}$

33. $\left\{\dfrac{-1 \pm \sqrt{7}}{2}\right\}$

Solve the following quadratic equations by the square root method. See Example 2.

11. $(x-3)^2 = 9$

12. $(a-2)^2 = -5$

13. $(y-18)^2 - 1 = 0$

14. $9 = (3s+2)^2$

15. $(2x-1)^2 = 8$

16. $x^2 - 4x + 4 = 49$

Solve the following quadratic equations by completing the square. See Example 3.

17. $x^2 + 8x + 7 = -8$

18. $2x^2 + 6x - 10 = 10$

19. $2x^2 + 7x - 15 = 0$

20. $4x^2 - 4x - 63 = 0$

21. $4x^2 - 56x + 195 = 0$

22. $y^2 + 22y + 96 = 0$

Solve the following quadratic equations by using the quadratic formula. See Example 4.

23. $3x^2 - 4 = -x$

24. $2.6z^2 - 0.9z + 2 = 0$

25. $a(a+2) = -1$

26. $3x^2 - 2x = 0$

27. $4x^2 - 14x - 27 = 3$

28. $6x^2 + 5x - 4 = 3x - 2$

Solve the following quadratic equations using any appropriate method.

29. $y^2 + 9y = -40.50$

30. $(z-11)^2 = 9$

31. $256t^2 - 324 = 0$

32. $(9y-6)^2 = 121y^2$

33. $2x^2 + 8x - 3 = 6x$

34. $4z^2 + 14z = 10z - 3$

35. $x^2 - 6x = 27$

36. $y^2 - 2y + 1 = -289$

37. $-3(b+5)^2 = -768$

38. $7x^2 - 42x = 0$

39. $5x^2 - 5x - 10 = 0$

40. $4w^2 + 10w + 5 = 3w^2 + 18w - 10$

Use the connection between solutions of quadratic equations and polynomial factoring to answer the following questions. See the discussion after Example 3.

41. Factor the quadratic $x^2 - 6x + 13$.

42. Factor the quadratic $9x^2 - 6x - 4$.

43. Factor the quadratic $4x^2 + 12x + 1$.

44. Factor the quadratic $25x^2 - 10x + 2$.

45. Determine b and c so that the equation $x^2 + bx + c = 0$ has the solution set $\{-3, 8\}$.

Solve the following quadratic-like equations. See Example 8.

46. $(x-1)^2 + (x-1) - 12 = 0$

47. $(y-5)^2 - 11(y-5) + 24 = 0$

48. $(x^2+1)^2 + (x^2+1) - 12 = 0$

49. $(x^2-13)^2 + (x^2-13) - 12 = 0$

50. $(x^2-2x+1)^2 + (x^2-2x+1) - 12 = 0$

51. $2y^{\frac{2}{3}} + y^{\frac{1}{3}} - 1 = 0$

34. $\left\{\dfrac{-1\pm i\sqrt{2}}{2}\right\}$

35. $\{-3, 9\}$

36. $\{1\pm 17i\}$

37. $\{-21, 11\}$

38. $\{0, 6\}$

39. $\{-1, 2\}$

40. $\{3, 5\}$

41. $(x-3-2i)\times$
$(x-3+2i)$

42. $\left(3x-1-\sqrt{5}\right)\times$
$\left(3x-1+\sqrt{5}\right)$

43. $\left(2x+3-2\sqrt{2}\right)\times$
$\left(2x+3+2\sqrt{2}\right)$

44. $(5x-1-i)\times$
$(5x-1+i)$

45. $b=-5$ and $c=-24$

46. $\{-3, 4\}$

47. $\{8, 13\}$

48. $\left\{\pm\sqrt{2},\ \pm i\sqrt{5}\right\}$

49. $\{\pm 4,\ \pm 3\}$

50. $\{1\pm 2i, 1\pm\sqrt{3}\}$

51. $\left\{-1, \dfrac{1}{8}\right\}$

52. $\left\{\dfrac{1}{8}, 27\right\}$

53. $\left\{1, 5, 3\pm\sqrt{10}\right\}$

54. $\{\pm 2i,\ \pm 3\}$

55. $\left\{\pm i\sqrt{13},\ \pm 3i\right\}$

56. $\{-1, \pm 2, 3\}$

57. $\left\{\dfrac{1}{16}, 16\right\}$

52. $2x^{\frac{2}{3}} - 7x^{\frac{1}{3}} + 3 = 0$

54. $\left(y^2-5\right)^2 + 5\left(y^2-5\right) - 36 = 0$

56. $\left(t^2-t\right)^2 - 8\left(t^2-t\right) + 12 = 0$

58. $3x^{\frac{2}{3}} - x^{\frac{1}{3}} - 2 = 0$

53. $\left(x^2-6x\right)^2 + 4\left(x^2-6x\right) - 5 = 0$

55. $\left(x^2+7\right)^2 + 8\left(x^2+7\right) + 12 = 0$

57. $2x^{\frac{1}{2}} - 5x^{\frac{1}{4}} + 2 = 0$

59. $5y^{\frac{2}{3}} + 33y^{\frac{1}{3}} + 18 = 0$

Solve the following polynomial equations by factoring. See Example 9.

60. $a^3 - 3a^2 = a - 3$

62. $x^4 + 5x^2 - 36 = 0$

64. $5s^3 + 6s^2 - 20s = 24$

66. $16a^4 = 81$

68. $14x^3 + 27x^2 - 20x = 0$

70. $27x^3 + 64 = 0$

61. $2x^3 + x^2 + 2x + 1 = 0$

63. $y^3 + 8 = 0$

65. $8a^3 - 27 = 0$

67. $6x^3 + 8x^2 = 14x$

69. $5z^3 + 28z^2 = 49z$

71. $x^3 - 4x^2 + x = 4$

Solve the following equations by factoring. See Example 10.

72. $3x^{\frac{11}{3}} + 2x^{\frac{8}{3}} - 5x^{\frac{5}{3}} = 0$

74. $(t+4)^{\frac{2}{3}} + 2(t+4)^{\frac{8}{3}} = 0$

76. $2x^{\frac{13}{5}} - 5x^{\frac{8}{5}} + 2x^{\frac{3}{5}} = 0$

78. $x^{\frac{11}{2}} - 6x^{\frac{9}{2}} + 9x^{\frac{7}{2}} = 0$

80. $(3x-3)^{-\frac{1}{3}} - 5(3x-3)^{-\frac{4}{3}} = 0$

73. $y^{\frac{7}{2}} - 5y^{\frac{5}{2}} + 6y^{\frac{3}{2}} = 0$

75. $(y-6)^{-\frac{5}{2}} + 7(y-6)^{-\frac{3}{2}} = 0$

77. $(2x-5)^{\frac{1}{3}} - 3(2x-5)^{-\frac{2}{3}} = 0$

79. $5y^{\frac{11}{3}} + 3y^{\frac{8}{3}} - 2y^{\frac{5}{3}} = 0$

81. $(y+3)^{\frac{2}{5}} + 4(y+3)^{\frac{7}{5}} = 0$

Use the connection between solutions of polynomial equations and polynomial factoring to answer the following questions.

82. Find $b, c,$ and d so the equation $x^3 + bx^2 + cx + d = 0$ has solutions of $-3, -1,$ and 5.

83. Find $b, c,$ and d so the equation $x^3 + bx^2 + cx + d = 0$ has solutions of $-2, 0,$ and 6.

84. Find b and c so the equation $x^3 + bx^2 + cx = 0$ has solutions of $0, 1,$ and -7.

85. Find $a, c,$ and d so the equation $ax^3 + 4x^2 + cx + d = 0$ has solutions of $-4, 6,$ and -6.

86. Find $a, b,$ and d so the equation $ax^3 + bx^2 + 3x + d = 0$ has solutions of $-3, -\dfrac{1}{2},$ and 0.

87. Find $a, b,$ and c so the equation $ax^3 + bx^2 + cx + 6 = 0$ has solutions of $-\dfrac{3}{5}, \dfrac{2}{3},$ and 1.

58. $\left\{1, -\dfrac{8}{27}\right\}$

59. $\left\{-216, -\dfrac{27}{125}\right\}$

60. $\{\pm 1, 3\}$

61. $\left\{-\dfrac{1}{2}, \pm i\right\}$

62. $\{\pm 2, \pm 3i\}$

63. $\left\{-2, 1 \pm i\sqrt{3}\right\}$

64. $\left\{\pm 2, -\dfrac{6}{5}\right\}$

65. $\left\{\dfrac{3}{2}, \dfrac{-3 \pm 3i\sqrt{3}}{4}\right\}$

66. $\left\{\pm \dfrac{3}{2}, \pm \dfrac{3i}{2}\right\}$

67. $\left\{-\dfrac{7}{3}, 0, 1\right\}$

68. $\left\{-\dfrac{5}{2}, 0, \dfrac{4}{7}\right\}$

69. $\left\{-7, 0, \dfrac{7}{5}\right\}$

70. $\left\{-\dfrac{4}{3}, \dfrac{2 \pm 2i\sqrt{3}}{3}\right\}$

71. $\{4, \pm i\}$

72. $\left\{-\dfrac{5}{3}, 0, 1\right\}$

73. $\{0, 2, 3\}$

74. $\left\{-4, \dfrac{-8 \pm i\sqrt{2}}{2}\right\}$

75. $\left\{\dfrac{41}{7}\right\}$

76. $\left\{0, \dfrac{1}{2}, 2\right\}$

77. $\{4\}$

78. $\{0, 3\}$

79. $\left\{-1, 0, \dfrac{2}{5}\right\}$

80. $\left\{\dfrac{8}{3}\right\}$

81. $\left\{-3, -\dfrac{13}{4}\right\}$

82. $b = -1$, $c = -17$, and $d = -15$

Solve the following application problems. See Example 7.

88. How long would it take for a ball dropped from the top of a 144-foot building to hit the ground?

89. Suppose that instead of being dropped, as in problem 88, a ball is thrown upward with a velocity of 40 feet per second from the top of a 144-foot building. Assuming it misses the building on the way back down, how long after being thrown will it hit the ground?

90. A slingshot is used to shoot a BB at a velocity of 96 feet per second straight up from ground level. When will the BB reach its maximum height of 144 feet?

91. A rock is thrown upward with a velocity of 20 meters per second from the top of a 24 meter high cliff, and it misses the cliff on the way back down. When will the rock be 7 meters from ground level? (Round your answer to the nearest tenth.)

$h_0 = 24$ m

$v_0 = 20$ m/sec

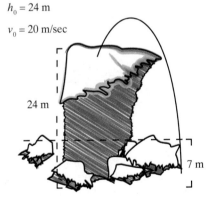

24 m

7 m

92. Luke, an experienced bungee jumper, leaps from a tall bridge and falls toward the river below. The bridge is 170 feet above the water and Luke's bungee cord is 110 feet long unstretched. When will Luke's cord begin to stretch? (Round your answer to the nearest tenth.)

170 ft

83. $b = -4$, $c = -12$, and $d = 0$

84. $b = 6$ and $c = -7$

85. $a = 1$, $c = -36$, and $d = -144$

86. $a = 2$, $b = 7$, and $d = 0$

87. $a = 15$, $b = -16$, and $c = -5$

88. 3 seconds

89. 4.5 seconds

90. 3 seconds

91. 4.8 seconds

92. 2.6 seconds

1.8

Rational and Radical Equations

TOPICS

1. Simplifying rational expressions
2. Combining rational expressions
3. Simplifying complex rational expressions
4. Solving rational equations
5. Work-rate problems
6. Solving radical equations
7. Solving equations with positive rational exponents

TOPIC   1

Simplifying Rational Expressions

Many equations contain fractions in which a variable appears in the denominator, and the presence of such fractions can make the solution process challenging. We will learn how to work with a class of fractions called *rational expressions* and develop a general method for solving equations that contain such expressions.

DEFINITION

Rational Expressions

A **rational expression** is an expression that can be written as a *ratio* of two polynomials $\dfrac{P}{Q}$. Of course, such a fraction is undefined for any value(s) of the variable(s) for which $Q = 0$. A rational expression is simplified or reduced when P and Q have no common factors (other than 1 and -1).

To simplify rational expressions, we factor the polynomials in the numerator and denominator completely and then cancel any common factors.

It is important to remember, however, that the simplified rational expression may be defined for values of the variable (or variables) that the original (unsimplified) expression is not, and the two versions are equal only where they are both defined. That is, if A, B, and C are algebraic expressions,

$$\frac{AC}{BC} = \frac{A}{B} \text{ only where } B \neq 0 \text{ and } C \neq 0.$$

This will be important when working with equations containing rational expressions, which can have extraneous solutions that are not defined in the original rational expressions.

EXAMPLE 1

Rational Expressions

Simplify the following rational expressions, and indicate values of the variable that must be excluded.

a. $\dfrac{x^3 - 8}{x^2 - 2x}$

b. $\dfrac{x^2 - x - 6}{3 - x}$

Note:
Always begin by factoring both the numerator and denominator. Note that once the denominator has been factored, we can immediately determine which values of the variable must be excluded.

Solutions:

a. $\dfrac{x^3 - 8}{x^2 - 2x} = \dfrac{\cancel{(x-2)}\left(x^2 + 2x + 4\right)}{x\cancel{(x-2)}}$

Cancel the common factor of $x - 2$.

$= \dfrac{x^2 + 2x + 4}{x}, \qquad x \neq 0, 2$

Even though the final expression is defined when $x = 2$, the first and last expressions are equal only where both are defined.

b. $\dfrac{x^2 - x - 6}{3 - x} = \dfrac{(x+2)\cancel{(x-3)}}{-\cancel{(x-3)}}$

The denominator is already factored, but we bring out a factor of -1 from the denominator in order to cancel a common factor of $x - 3$.

$= \dfrac{x + 2}{-1}$

$= -x - 2, \qquad x \neq 3$

Note that the original and simplified versions are only equal for $x \neq 3$.

CAUTION!

Remember that only common *factors* can be cancelled! A very common error is to think that common terms from the numerator and denominator can be cancelled. For instance, the statement $\dfrac{x + 4}{x^2} = \dfrac{4}{x}$ is **incorrect**. It is not possible to factor $x + 4$ at all, and the x that appears in the numerator is not a factor that can be cancelled with one of the x's in the denominator. The expression $\dfrac{x + 4}{x^2}$ is already completely simplified.

TOPIC 2

Combining Rational Expressions

Rational expressions can be combined by addition, subtraction, multiplication, and division the same way that numerical fractions are. In order to add or subtract two rational expressions, a common denominator must first be found. In order to multiply two rational expressions, the two numerators are multiplied and the two denominators are multiplied. Finally, in order to divide one rational expression by another, the first is multiplied by the reciprocal of the second. It is generally best to factor all numerators and denominators before combining rational expressions.

EXAMPLE 2

Combining Rational Expressions

Add or subtract the rational expressions, as indicated.

a. $\dfrac{2x-1}{x^2+x-2} - \dfrac{2x}{x^2-4}$

b. $\dfrac{x+1}{x+3} + \dfrac{x^2+x-2}{x^2-x-6} - \dfrac{x^2-2x+9}{x^2-9}$

Note:
Begin by finding the LCD. The LCD will be the product of all the unique factors among the denominators of the original expressions (raised to powers when needed).

Solutions:

a. $\dfrac{2x-1}{x^2+x-2} - \dfrac{2x}{x^2-4}$

$= \dfrac{2x-1}{(x+2)(x-1)} - \dfrac{2x}{(x+2)(x-2)}$

The LCD is $(x-2)(x+2)(x-1)$.

$= \dfrac{x-2}{x-2} \cdot \dfrac{2x-1}{(x+2)(x-1)} - \dfrac{x-1}{x-1} \cdot \dfrac{2x}{(x+2)(x-2)}$

Multiply each term by the appropriate fraction to obtain the LCD.

$= \dfrac{2x^2-5x+2}{(x-2)(x+2)(x-1)} - \dfrac{2x^2-2x}{(x-1)(x+2)(x-2)}$

After subtracting the second numerator from the first, we are done. Note that there are no common factors to cancel.

$= \dfrac{-3x+2}{(x-2)(x+2)(x-1)}$

b. $\dfrac{x+1}{x+3} + \dfrac{x^2+x-2}{x^2-x-6} - \dfrac{x^2-2x+9}{x^2-9}$

Again factor all the polynomials.

$= \dfrac{x+1}{x+3} + \dfrac{\cancel{(x+2)}(x-1)}{(x-3)\cancel{(x+2)}} - \dfrac{x^2-2x+9}{(x-3)(x+3)}$

Note that the second rational expression can be reduced. We do this before determining the LCD.

$= \dfrac{x-3}{x-3} \cdot \dfrac{x+1}{x+3} + \dfrac{x+3}{x+3} \cdot \dfrac{x-1}{x-3} - \dfrac{x^2-2x+9}{(x-3)(x+3)}$

Multiply each term by the appropriate fraction to get the LCD $(x-3)(x+3)$.

$= \dfrac{x^2-2x-3+x^2+2x-3-x^2+2x-9}{(x-3)(x+3)}$

Combine like terms.

$= \dfrac{x^2+2x-15}{(x-3)(x+3)}$

Simplify the resulting numerator.

$= \dfrac{(x+5)\cancel{(x-3)}}{\cancel{(x-3)}(x+3)} = \dfrac{x+5}{x+3}$

After factoring the resulting numerator, there is a common factor that can be cancelled.

====== **EXAMPLE 3** ======

Combining Rational Expressions

Multiply or divide the rational expressions, as indicated.

a. $\dfrac{x^2+3x-10}{x+3} \cdot \dfrac{x-3}{x^2-x-2}$

b. $\dfrac{x^2+5x-14}{3x} \div \dfrac{x^2-4x+4}{9x^3}$

Solutions:

a. $\dfrac{x^2+3x-10}{x+3} \cdot \dfrac{x-3}{x^2-x-2}$

We begin by factoring both numerators and denominators.

$= \dfrac{(x+5)(x-2)}{x+3} \cdot \dfrac{x-3}{(x-2)(x+1)}$

Then write the product of the two rational expressions as a single fraction.

$= \dfrac{(x+5)\,\cancel{(x-2)}\,(x-3)}{(x+3)\,\cancel{(x-2)}\,(x+1)}$

Since we already factored the polynomials, the common factors are easily identified.

$= \dfrac{(x+5)(x-3)}{(x+3)(x+1)}$

b. $\dfrac{x^2+5x-14}{3x} \div \dfrac{x^2-4x+4}{9x^3}$

We divide the first rational expression by the second by inverting the second fraction and multiplying. Note that we factor all of the polynomials and invert the second fraction in one step.

$= \dfrac{(x+7)(x-2)}{3x} \cdot \dfrac{9x^3}{(x-2)^2}$

$= \dfrac{\overset{3}{\cancel{9}}\,\overset{2}{x^{\cancel{3}}}\,(x+7)\,\cancel{(x-2)}}{\cancel{3}\,\cancel{x}\,(x-2)^{\cancel{2}}}$

Now we proceed to cancel common factors (including constant factors) to obtain the final answer.

$= \dfrac{3x^2(x+7)}{x-2}$

TOPIC 3

Simplifying Complex Rational Expressions

A **complex rational expression** is a fraction in which the numerator or denominator (or both) contains at least one rational expression. Complex rational expressions can always be rewritten as simple rational expressions. One way to do this is to simplify the numerator and denominator individually and then divide the numerator by the denominator as in Example 3b. Another way, which is frequently faster, is to multiply the numerator and denominator by the LCD of all the fractions that make up the complex rational expression. This method will be illustrated in the next two examples.

EXAMPLE 4

Complex Rational Expressions

Simplify the complex rational expressions.

a. $\dfrac{\dfrac{1}{x+h}-\dfrac{1}{x}}{h}$

b. $\dfrac{x^{-1}-y^{-1}}{x^{-2}-y^{-2}}$

Solutions:

a. $\dfrac{\dfrac{1}{x+h}-\dfrac{1}{x}}{h}=\dfrac{\dfrac{1}{x+h}-\dfrac{1}{x}}{\dfrac{h}{1}}\cdot\dfrac{(x+h)(x)}{(x+h)(x)}$

It may be helpful to write the denominator as a fraction, as we have done here, in order to determine that the LCD of all the fractions making up the overall expression is $(x+h)(x)$.

$=\dfrac{\dfrac{(x+h)(x)}{x+h}-\dfrac{(x+h)(x)}{x}}{(h)(x+h)(x)}$

We multiply the numerator and denominator by the LCD (so we are multiplying the overall expression by 1).

$=\dfrac{x-(x+h)}{(h)(x+h)(x)}$

Simplify the resulting numerator.

$=\dfrac{-h}{(h)(x+h)(x)}$

Cancel out the common factor h to arrive at the final answer.

$=\dfrac{-1}{x(x+h)}$

b. $\dfrac{x^{-1}-y^{-1}}{x^{-2}-y^{-2}}=\dfrac{\dfrac{1}{x}-\dfrac{1}{y}}{\dfrac{1}{x^2}-\dfrac{1}{y^2}}$

This expression is also a complex rational expression, which we see once we rewrite the terms that have negative exponents as fractions.

$=\dfrac{\dfrac{1}{x}-\dfrac{1}{y}}{\dfrac{1}{x^2}-\dfrac{1}{y^2}}\cdot\dfrac{x^2y^2}{x^2y^2}$

The LCD in this case is x^2y^2, so we multiply the top and bottom by this and factor the resulting polynomials.

$=\dfrac{xy^2-x^2y}{y^2-x^2}$

Simplify the numerator and denominator, then factor.

$=\dfrac{xy(y-x)}{(y-x)(y+x)}$

Cancel the common factor to obtain the final simplified expression.

$=\dfrac{xy}{y+x}$

TOPIC 4   Solving Rational Equations

A **rational equation** is an equation that contains at least one rational expression, while any nonrational expressions are polynomials. Our general approach to solving such equations is to multiply each term in the equation by the LCD of all the rational expressions; this has the effect of converting rational equations into polynomial equations, which we have already learned how to solve.

There is one important difference between rational and polynomial equations: it is quite possible that one or more rational expressions in a rational equation are not defined for some values of the variable. Of course, these values cannot possibly be solutions of the equation, and must be excluded from the solution set. However, these excluded values may appear as solutions of the polynomial equation derived from the original rational equation. These are extraneous solutions! Errors can be avoided by keeping track of what values of the variable are disallowed and/or checking all solutions in the original equation.

EXAMPLE 5

Solving Rational Equations

Solve the following rational equations.

a. $\dfrac{x^3 + 3x^2}{x^2 - 2x - 15} = \dfrac{4x + 5}{x - 5}$ **b.** $\dfrac{3x^2}{5x - 1} - 1 = 0$

Solutions:

a.
$$\frac{x^3 + 3x^2}{x^2 - 2x - 15} = \frac{4x + 5}{x - 5}$$

$$\frac{x^2 \cancel{(x+3)}}{(x-5)\cancel{(x+3)}} = \frac{4x + 5}{x - 5}$$

$$(x-5)\cdot \frac{x^2}{(x-5)} = (x-5)\cdot \frac{4x + 5}{x - 5}$$

$$x^2 = 4x + 5$$

$$x^2 - 4x - 5 = 0$$

$$(x-5)(x+1) = 0$$

$$x = \cancel{5}, -1$$

$$x = -1$$

In order to cancel factors, and in order to determine the LCD, we begin by factoring all the numerators and denominators. This also tells us the very important fact that 5 and −3 cannot be solutions of the equation.

After multiplying both sides of the equation by the LCD, we have a second-degree polynomial equation that can be solved by factoring.

Note that we already determined that 5 cannot be a solution. It must be discarded.

b.
$$\frac{3x^2}{5x-1} - 1 = 0$$

$$(5x-1)\cdot\frac{3x^2}{5x-1} - (5x-1) = 0$$

$$3x^2 - 5x + 1 = 0$$

$$x = \frac{5 \pm \sqrt{25-12}}{6}$$

$$x = \frac{5 \pm \sqrt{13}}{6}$$

There is no factoring possible in this problem, so we just note that $\frac{1}{5}$ is the one value for x that must be excluded.

After multiplying through by the LCD, we have a second-degree polynomial equation that can be solved by the quadratic formula.

Since neither of the roots is $\frac{1}{5}$, both solve the original equation.

TOPIC 5   Work-Rate Problems

Many seemingly different applications fall into a class of problems known as work-rate problems. What these applications have in common is two or more "workers" acting in unison to complete a task. The workers can be, for example, employees on a job, machines manufacturing a part, or inlet and outlet pipes filling or draining a tank. Typically, each worker is capable of doing the task alone, and works at an individual rate regardless of whether others are involved.

The goal in a work-rate problem is usually to determine how fast the task at hand can be completed, either by the workers together or by one of the workers individually. There are two keys to solving a work-rate problem:

(1)   The rate of work is the reciprocal of the time needed to complete the task.

If a given job can be done by a worker in x units of time, the worker works at a rate of $\frac{1}{x}$ jobs per unit of time. For instance, if Jane can overhaul an engine in 2 hours, her rate of work is $\frac{1}{2}$ of the job per hour. If a faucet can fill a sink in 5 minutes, its rate is $\frac{1}{5}$ of the sink per minute. Of course, this also means that the time needed to complete a task is the reciprocal of the rate of work: if a faucet fills a sink at the rate of $\frac{1}{5}$ of the sink per minute, it takes 5 minutes to fill the sink.

(2)   Rates of work are "additive."

This means that, in the ideal situation, two workers working together on the same task will have a combined rate of work that is the sum of their individual rates. For instance, if Jane's rate in overhauling an engine is $\frac{1}{2}$ and Ted's rate is $\frac{1}{3}$, their combined rate is $\frac{1}{2}+\frac{1}{3}$, or $\frac{5}{6}$. That is, together they can overhaul an engine in $\frac{6}{5}$ hours, or one hour and twelve minutes.

EXAMPLE 6

Filling a Pool

One hose can fill a swimming pool in 12 hours. The owner buys a second hose that can fill the pool at twice the rate of the first one. If both hoses are used together, how long does it take to fill the pool?

Solution:

The rate of work of the first hose is $\dfrac{1}{12}$, so the rate of the second hose is $\dfrac{1}{6}$. If we let x denote the time needed to fill the pool when both hoses are used together, the sum of the two individual rates must equal $\dfrac{1}{x}$. So we need to solve the equation $\dfrac{1}{12}+\dfrac{1}{6}=\dfrac{1}{x}$.

$$\frac{1}{12}+\frac{1}{6}=\frac{1}{x}$$

$$x+2x=12$$

$$3x=12$$

$$x=4$$

As is typical, this work-rate problem leads to a rational equation which we can solve using the methods of this section.

After multiplying by the LCD, $12x$, we are left with a polynomial equation (linear in this case).

Thus, using both hoses, it will take 4 hours to fill the pool.

EXAMPLE 7

Filling a Pool Poorly

The pool owner in Example 6 is a bit clumsy, and one day proceeds to fill his empty pool with the two hoses but accidentally turns on the pump that drains the pool also. Fortunately, the pump rate is slower than the combined rate of the two hoses, and the pool fills anyway, but it takes 10 hours to do so. At what rate can the pump empty the pool?

Solution:

First, let's translate the information in the problem into a work-rate rational equation. If we let x denote the time it takes the pump to empty the pool, we can say that the pump has a filling rate of $-\dfrac{1}{x}$ (since emptying is the opposite of filling). Since the two hoses can fill the pool in 4 hours, the combined filling rate of the two hoses is $\dfrac{1}{4}$. Finally, the total rate at which the pool is filled on this unfortunate day is $\dfrac{1}{10}$.

Again, the sum of the individual rates is equal to the combined rate, so the rational equation that reflects this situation is:

$$\frac{1}{4}-\frac{1}{x}=\frac{1}{10}$$

$$5x-20=2x$$

$$3x=20$$

$$x=\frac{20}{3}$$

Again, we multiply each term by the LCD, $20x$, to arrive at a polynomial equation to solve.

Thus, working alone, the pump can empty the pool in $\dfrac{20}{3}$ hours, or 6 hours and 40 minutes.

TOPIC 6   Solving Radical Equations

The last one-variable equations we will discuss are those that contain radical expressions. A **radical equation** is an equation that has at least one radical expression containing a variable, while any nonradical expressions are polynomial terms. As with the rational equations discussed previously, we will develop a general method of solution that converts a given radical equation into a polynomial equation. We will see that just as with rational equations, we must check our potential solutions carefully to see if they actually solve the original radical equation.

Our method of solving rational equations involved multiplying both sides of the equation by an algebraic expression (the LCD of all the rational expressions), and in some cases potential solutions had to be discarded because they led to division by 0 in one or more of the rational expressions. Something similar can happen with radical equations.

Since our goal is to convert a given radical equation into a polynomial equation, one reasonable approach is to raise both sides of the equation to whatever power is necessary to "undo" the radical (or radicals). The problem is that this does *not* result in an equivalent equation; remember that we can only transform an equation into an equivalent equation by adding the same quantity to both sides or by multiplying both sides by a nonzero quantity. We won't *lose* any solutions by raising both sides of an equation to the same power, but we may *gain* some extraneous solutions. We identify these and discard them by checking all of our eventual solutions in the original equation. A simple example will make this clear.

EXAMPLE 8

Causing Extraneous Solutions

Consider the equation $x = -3$.

Solution:

This equation is so basic that it is its own solution. But for this demonstration, suppose we square both sides, obtaining the equation

$$x^2 = 9.$$

This second-degree equation can be solved by factoring the polynomial $x^2 - 9$ or by taking the square root of both sides, and in either case we obtain the solution set $\{-3, 3\}$. That is, by squaring both sides of the original equation, we gained a second (extraneous) solution.

PROCEDURE

Solving Radical
Equations

Step 1: Begin by isolating the radical expression on one side of the equation. If there is more than one radical expression, choose one to isolate on one side.

Step 2: Raise both sides of the equation by the power necessary to "undo" the isolated radical. That is, if the radical is an nth root, raise both sides to the nth power.

Step 3: If any radical expressions remain, simplify the equation if possible and then repeat steps 1 and 2 until the result is a polynomial equation. When a polynomial equation has been obtained, solve the equation using polynomial methods.

Step 4: Check your solutions in the original equation! Any extraneous solutions must be discarded.

If the equation contains many radical expressions, and especially if they have different indices, eliminating all the radicals may be a long process! The equations that we will solve will not require more than a few repetitions of steps 1 and 2.

EXAMPLE 9

Radical Equations

Solve the radical equations.

a. $\sqrt{1-x} - 1 = x$ **b.** $\sqrt{x+1} + \sqrt{x+2} = 1$ **c.** $\sqrt[4]{x^2+8x+7} - 2 = 0$

Solutions:

a. $\sqrt{1-x} - 1 = x$

$\sqrt{1-x} = x+1$ Isolate the radical expression.

$\left(\sqrt{1-x}\right)^2 = (x+1)^2$ Since we have square root, square both sides.

$1-x = x^2 + 2x + 1$ The result is a second-degree polynomial equation that can be solved by factoring.

$0 = x^2 + 3x$

$0 = x(x+3)$

$x = 0, -3$ We have two apparent solutions to check.

Now we need to check each apparent solution in the original equation:

$$\sqrt{1-0} - 1 = 0 \qquad\qquad \sqrt{1-(-3)} - 1 = -3$$

$$\sqrt{1} - 1 = 0 \qquad\qquad\qquad \sqrt{4} - 1 = -3$$

$$0 = 0 \qquad\qquad\qquad\qquad\quad 1 \neq -3$$

Thus, -3 is an extraneous solution, so the solution set is $\{0\}$.

b. $\sqrt{x+1}+\sqrt{x+2}=1$

$$\sqrt{x+1}=1-\sqrt{x+2}$$
$$\left(\sqrt{x+1}\right)^2=\left(1-\sqrt{x+2}\right)^2$$

This equation has two radical expressions, so we isolate one of them initially.

$$x+1=1-2\sqrt{x+2}+x+2$$
$$2\sqrt{x+2}=2$$

We square both sides to eliminate the isolated radical, and then proceed to simplify and isolate the remaining radical.

$$\sqrt{x+2}=1$$
$$x+2=1$$
$$x=-1$$

Finally, we square both sides again and solve the polynomial equation.

We check the apparent solution in the original equation:

$$\sqrt{(-1)+1}+\sqrt{(-1)+2}=1$$
$$\sqrt{0}+\sqrt{1}=1$$
$$1=1$$

Thus, the solution set is $\{-1\}$.

c. $\sqrt[4]{x^2+8x+7}-2=0$

$$\sqrt[4]{x^2+8x+7}=2$$

First isolate the radical.

$$\left(\sqrt[4]{x^2+8x+7}\right)^4=2^4$$

In this case, the radical is a fourth root, so we raise both sides to the fourth power.

$$x^2+8x+7=16$$
$$x^2+8x-9=0$$
$$(x+9)(x-1)=0$$
$$x=-9,1$$

The resulting second-degree equation can again be solved by factoring.

Once again, we check the apparent solutions in the original equation:

$$\sqrt[4]{(-9)^2+8(-9)+7}-2=0 \qquad \sqrt[4]{(1)^2+8(1)+7}-2=0$$
$$\sqrt[4]{16}-2=0 \qquad \sqrt[4]{16}-2=0$$
$$2-2=0 \qquad 2-2=0$$
$$0=0 \qquad 0=0$$

Both apparent solutions are actual solutions, so the solution set is $\{-9,1\}$.

========== EXAMPLE 10 ==========

Escape Speed

The speed required for an object to escape from the gravitational pull of a planet is called the **escape speed** of the planet. The escape speed is given by the equation $v_e = \sqrt{\dfrac{2GM}{r}}$, where v_e is the escape speed, G is the universal gravitation constant, M is the mass of the planet, and r is the radius of the planet. Solve this equation for r.

Solution:

We follow the same procedure for solving radical equations.

$$v_e = \sqrt{\frac{2GM}{r}}$$ The radical expression is already isolated.

$$v_e^2 = \frac{2GM}{r}$$ Square both sides to eliminate the radical.

$$r = \frac{2GM}{v_e^2}$$ Solve for r.

TOPIC   7   Solving Equations with Positive Rational Exponents

In Section 1.7, we encountered equations with rational exponents that we solved by factoring or by quadratic methods. Now that we have learned how to solve radical equations, we have another option for solving equations with *positive* rational exponents. Recall that we defined positive rational exponents in terms of radicals in Section 1.2.

DEFINITION

Positive Rational Number Exponents

Meaning of $a^{\frac{m}{n}}$: If m and n are natural numbers with $n \neq 0$, if m and n have no common factors greater than 1, and if $\sqrt[n]{a}$ is a real number, then $a^{\frac{m}{n}} = \sqrt[n]{a^m} = \left(\sqrt[n]{a}\right)^m$.

This means that an equation with positive rational exponents can be rewritten as one with radical terms, which we can then solve using the methods for solving radical equations.

━━━━━━━━ **EXAMPLE 11** ━━━━━━━━

Rational Exponents

Solve the following equations with rational exponents.

a. $x^{\frac{2}{3}} - 9 = 0$

b. $\left(32x^2 - 32x + 17\right)^{\frac{1}{4}} = 3$

Solutions:

a. $x^{\frac{2}{3}} - 9 = 0$

The term containing the rational exponent can be rewritten as a radical expression, so we will begin by isolating that term.

$x^{\frac{2}{3}} = 9$

$\sqrt[3]{x^2} = 9$

Rewrite the left-hand side as a radical.

$x^2 = 9^3$

Cubing both sides eliminates the cube root.

$x = 9^{\frac{3}{2}}$

$x = \left(9^{\frac{1}{2}}\right)^3$

$x = (\pm 3)^3$

$x = \pm 27$

Raising both sides to the $\frac{1}{2}$ power solves the equation for x, but we can evaluate the expression on the right-hand side.

Note that both +3 and −3 must be considered.

Plugging the values in, we see that $(27)^{\frac{2}{3}} = 9$ and $(-27)^{\frac{2}{3}} = 9$, so both are solutions to the original equation.

b. $\left(32x^2 - 32x + 17\right)^{\frac{1}{4}} = 3$

The exponent of $\frac{1}{4}$ indicates we will need to raise both sides to the fourth power.

$\sqrt[4]{32x^2 - 32x + 17} = 3$

$32x^2 - 32x + 17 = 3^4$

$32x^2 - 32x + 17 = 81$

$32x^2 - 32x - 64 = 0$

$32\left(x^2 - x - 2\right) = 0$

We are left with a second-degree polynomial equation that can be solved by factoring. Note that both solutions again satisfy the original equation.

$x^2 - x - 2 = 0$

$(x - 2)(x + 1) = 0$

$x = 2, -1$

Confirm, by plugging in to the original equation, that both apparent solutions truly solve the equation with rational exponents.

1. $\dfrac{2x+1}{x-5}$

$x \neq -3, 5$

2. $\dfrac{x+6}{x(x+3)}$

$x \neq -3, 0, 1$

3. $x(x-1)$

$x \neq -3$

4. $\dfrac{x-2}{x+2}$

$x \neq -2, 2$

5. $\dfrac{2x-3}{x-2}$

$x \neq -5, 2$

6. $\dfrac{x^3}{x-2}$

$x \neq 1, 2$

7. $2x-3$

$x \neq -7$

8. $4x^2+6x+9$

$x \neq \dfrac{3}{2}$

9. $\dfrac{x^3+9x^2+11x+19}{(x-3)(x+5)}$

10. $\dfrac{x^3+2x-3}{(x-2)(x+1)}$

11. $\dfrac{13x}{(x-3)(x+5)}$

12. $\dfrac{x+1}{x-3}$

13. $\dfrac{x^3+4x^2-7x+18}{(x+3)(x-3)}$

14. $\dfrac{2y^2-2y+24}{(y-2)(y+4)}$

15. $\dfrac{x+2}{x-6}$

16. $\dfrac{y^3-4y^2+3y+6}{(y+3)(y-3)}$

17. $y-1$

Exercises

Simplify the following rational expressions, indicating which real values of the variable must be excluded. See Example 1.

1. $\dfrac{2x^2+7x+3}{x^2-2x-15}$

2. $\dfrac{x^2+5x-6}{x^3+2x^2-3x}$

3. $\dfrac{x^3+2x^2-3x}{x+3}$

4. $\dfrac{x^2-4x+4}{x^2-4}$

5. $\dfrac{2x^2+7x-15}{x^2+3x-10}$

6. $\dfrac{x^4-x^3}{x^2-3x+2}$

7. $\dfrac{2x^2+11x-21}{x+7}$

8. $\dfrac{8x^3-27}{2x-3}$

Add or subtract the rational expressions, as indicated, and simplify your answer. See Example 2.

9. $\dfrac{x-3}{x+5}+\dfrac{x^2+3x+2}{x-3}$

10. $\dfrac{x^2-1}{x-2}-\dfrac{x-1}{x+1}$

11. $\dfrac{x+2}{x-3}-\dfrac{x-3}{x+5}-\dfrac{1}{x^2+2x-15}$

12. $\dfrac{x+1}{x-3}+\dfrac{x^2+3x+2}{x^2-x-6}-\dfrac{x^2-2x-3}{x^2-6x+9}$

13. $\dfrac{x^2+1}{x-3}+\dfrac{x-5}{x+3}$

14. $\dfrac{y+2}{y-2}+\dfrac{y-6}{y+4}+\dfrac{4}{y^2+2y-8}$

15. $\dfrac{x+2}{x-6}+\dfrac{x^2+5x+6}{x^2-3x-18}-\dfrac{x^2-4x-12}{x^2-12x+36}$

16. $\dfrac{y^2+2}{y+3}-\dfrac{y-4}{y-3}$

Multiply or divide the rational expressions, as indicated, and simplify your answer. See Example 3.

17. $\dfrac{y-2}{y+1}\cdot\dfrac{y^2-1}{y-2}$

18. $\dfrac{2x^2-5x-12}{x-3}\cdot\dfrac{x^2-x-6}{x-4}$

19. $\dfrac{z^2+2z+1}{2z^2+3z+1}\cdot\dfrac{2z^2-5z-3}{z+1}$

20. $\dfrac{y^2-11y+24}{y+6}\div\dfrac{y^2+5y-24}{y+6}$

21. $\dfrac{y^2+8y+16}{5y^2+22y+8}\cdot\dfrac{5y^2-13y-6}{y+4}$

22. $\dfrac{4z^2+20z-56}{z^2-8z+12}\div\dfrac{5z^2+43z+56}{15z^2-66z-144}$

23. $\dfrac{3b^2+9b-84}{b^2-5b+4}\div\dfrac{5b^2+37b+14}{-10b^2+6b+4}$

24. $\dfrac{3x^2-x-10}{x-1}\cdot\dfrac{x^2-1}{6x^2+x-15}\div\dfrac{x^2-x-2}{2x^2+5x-12}$

18. $(x+2)(2x+3)$

19. $z-3$

20. $\dfrac{y-8}{y+8}$

21. $y-3$

22. 12

23. -6

24. $x+4$

25. $\dfrac{x^2+9}{6x-3}$

26. $\dfrac{y-x}{x+y}$

27. $\dfrac{2x^2}{x+1}$

28. $\dfrac{x+y}{x^2y^2}$

29. $\dfrac{-xy-y^2}{x}$

30. $\dfrac{m+n}{mn}$

31. $\dfrac{x}{y}$

32. $\dfrac{x}{y}$

33. x^2y^2

34. xy

35. $\dfrac{11x}{7y}$

36. $\dfrac{5z-3x}{z^2}$

37. $\left\{-2,-\dfrac{3}{2}\right\}$

38. $\left\{\dfrac{3\pm\sqrt{21}}{-2}\right\}$

39. $\left\{3\pm\sqrt{10}\right\}$

40. $\{0\}$

41. $\{\pm i\}$

42. $\{2,3\}$

Simplify the complex rational expressions. See Example 4.

25. $\dfrac{\dfrac{3}{x}+\dfrac{x}{3}}{2-\dfrac{1}{x}}$

26. $\dfrac{\dfrac{1}{x}-\dfrac{1}{y}}{\dfrac{1}{x}+\dfrac{1}{y}}$

27. $\dfrac{6x-6}{3-\dfrac{3}{x^2}}$

28. $\dfrac{x^{-2}-y^{-2}}{y-x}$

29. $\dfrac{\dfrac{1}{x^2}-\dfrac{1}{y^2}}{\dfrac{1}{y^3}-\dfrac{1}{xy^2}}$

30. $\dfrac{\dfrac{m}{n}-\dfrac{n}{m}}{m-n}$

31. $\dfrac{\dfrac{1}{y}-\dfrac{1}{x+3}}{\dfrac{1}{x}-\dfrac{y}{x^2+3x}}$

32. $\dfrac{x+y^{-1}}{x^{-1}+y}$

33. $\dfrac{x^2-y^2}{y^{-2}-x^{-2}}$

34. $\dfrac{xy^{-1}+\left(\dfrac{x}{y}\right)^{-1}}{x^{-2}+y^{-2}}$

35. $\dfrac{\dfrac{1}{7y}+\dfrac{1}{x-2}}{\dfrac{1}{11x}+\dfrac{7y}{11x^2-22x}}$

36. $\dfrac{25x^{-2}-9z^{-2}}{\dfrac{5z+3x}{x^2}}$

Solve the following rational equations. See Example 5.

37. $\dfrac{2x^3+4x^2}{x^2-4x-12}=\dfrac{-7x-6}{x-6}$

38. $\dfrac{-x^2}{x-1}-3=0$

39. $\dfrac{3}{x-2}+\dfrac{2}{x+1}=1$

40. $\dfrac{x}{x-1}+\dfrac{2}{x-3}=-\dfrac{2}{x^2-4x+3}$

41. $\dfrac{1}{t-3}+\dfrac{1}{t+2}=\dfrac{t}{t-3}$

42. $\dfrac{z}{6+z}+\dfrac{z-1}{6-z}=\dfrac{z}{6-z}$

43. $\dfrac{y}{y-1}+\dfrac{2}{y-3}=\dfrac{y^2}{y^2-4y+3}$

44. $\dfrac{2}{2x+1}-\dfrac{x}{x-4}=\dfrac{-3x^2+x-4}{2x^2-7x-4}$

45. $\dfrac{2}{2b+1}+\dfrac{2b^2-b+4}{2b^2-7b-4}=\dfrac{b}{b-4}$

46. $\dfrac{2}{n+3}+\dfrac{3}{n+2}=\dfrac{6}{n}$

43. $\{-2\}$

44. $\{-2,2\}$

45. $\varnothing$

46. $\left\{\dfrac{-17\pm\sqrt{145}}{2}\right\}$

47. $\dfrac{x-2}{x+2}$

48. $y+1$

49.
$\dfrac{\left(z^2-11z+54\right)(z-9)}{(z-2)}$

50. $\dfrac{3y^2+10y+17}{2(y+3)(y+9)}$

51. $\dfrac{2y^2+5y-4}{y+1}$

52. $\{0\}$

53. $\{-1\}$

54. $\{1\}$

55. $\{-7,-2\}$

56. $\{1\}$

57. $\varnothing$

58. $\{3\}$

59. $\{4,44\}$

60. $\{15\}$

61. $\varnothing$

62. $\left\{\dfrac{4}{5},2\right\}$

63. $\{-2,1\}$

64. $\{1\}$

65. $\{0,5\}$

66. $\{10\}$

67. $\varnothing$

68. $\{4\}$

69. $\left\{\dfrac{1}{4},1\right\}$

70. $\{2\}$

71. $\left\{\pm\dfrac{8}{27}\right\}$

72. $\left\{\pm\dfrac{125}{343}\right\}$

73. $\{-2,5\}$

74. $\{\pm 3\}$

75. $\{7,10\}$

Perform the indicated operations on the following rational expressions, and simplify your answer.

47. $\left(\dfrac{x^2-3x}{x^2+6x-27}-\dfrac{2}{x+9}\right)\cdot\dfrac{x+9}{x+2}$

48. $\dfrac{2y(y-1)}{y^2+6y-16}\div\dfrac{2}{y+8}-\dfrac{2}{y-2}$

49. $\left(\dfrac{z^2-17z+30}{z^2+2z-8}+\dfrac{6}{z-2}\right)\div\dfrac{1}{z^2-5z-36}$

50. $\dfrac{y+3}{2y+18}+\dfrac{y^2+2y+4}{y^2+3y-54}\cdot\dfrac{y-6}{y+3}$

51. $\dfrac{y^2+2y-15}{y+1}\cdot\left(\dfrac{y^2+3y+4}{y^2+3y-10}+\dfrac{y+4}{y+5}\right)\div\dfrac{y-3}{y-2}$

Solve the following radical equations. See Example 9.

52. $\sqrt{4-x}-x=2$

53. $\sqrt{3y+4}+\sqrt{5y+6}=2$

54. $\sqrt{x^2-4x+4}+2=3x$

55. $\sqrt{50+7s}-s=8$

56. $\sqrt[4]{x^2-x}=\sqrt[4]{x-1}$

57. $\sqrt[4]{2x+3}=-1$

58. $\sqrt{11x+3}+4x=18$

59. $\sqrt{5x+5}=\sqrt{4x-7}+2$

60. $\sqrt{x+1}+10=x-1$

61. $\sqrt{x^2-10}-1=x+1$

62. $\sqrt[3]{5x^2-14x}=-2$

63. $\sqrt[5]{7t^2+2t}=\sqrt[5]{5t^2+4}$

64. $\sqrt{14y^2-18y+4}+2=2y$

65. $\sqrt{9x+4}=\sqrt{7x+1}+1$

66. $\sqrt{4z+41}+3=z+2$

Solve the following equations. See Example 11.

67. $(x+3)^{\frac{1}{4}}+2=0$

68. $(2x-5)^{\frac{1}{4}}=(x-1)^{\frac{1}{4}}$

69. $(2x-1)^{\frac{2}{3}}=x^{\frac{1}{3}}$

70. $\left(3y^2+9y-5\right)^{\frac{1}{2}}=y+3$

71. $z^{\frac{4}{3}}-\dfrac{16}{81}=0$

72. $x^{\frac{2}{3}}-\dfrac{25}{49}=0$

73. $(x-2)^{\frac{2}{3}}=(14-x)^{\frac{1}{3}}$

74. $\left(x^2+7\right)^{-\frac{3}{2}}=\dfrac{1}{64}$

75. $(y-2)^{\frac{2}{3}}=(13y-66)^{\frac{1}{3}}$

76. $l = \dfrac{gT^2}{4\pi^2}$

77. $a = \pm\sqrt{c^2 - b^2}$

78. $c = \pm\sqrt{\dfrac{E}{m}}$

79. $m = \dfrac{k}{\omega^2}$

80. $r = \sqrt[3]{\dfrac{3V}{4\pi}}$

81. $v = \pm\sqrt{\dfrac{Fr}{m}}$

82. $t = \pm\sqrt{\dfrac{1.227r}{a}}$

83. $h = \pm\sqrt{\dfrac{w}{23}}$

84. $r = \sqrt[3]{\dfrac{T^2 GM}{4\pi^2}}$

85. $c = \pm\sqrt{\dfrac{2gm}{r}}$

86. $v =$

$\pm\sqrt{\dfrac{2(ME - mgh)}{m}}$

87. $b = \pm\sqrt{c^2 - a^2}$

88. $L = \dfrac{\pm\sqrt{Z^2 - R^2}}{\omega}$

$+ \dfrac{1}{\omega^2 C}$

89. $a = \sqrt[3]{\dfrac{uP^2}{4\pi^2}}$

Solve the following formulas for the indicated variable. See Example 10.

76. $T = 2\pi\sqrt{\dfrac{l}{g}}$. This is the formula for the period T of a pendulum of length l. Solve this formula for l.

77. $c = \sqrt{a^2 + b^2}$. This is the formula for the length of the hypotenuse c of a right triangle. Solve this formula for a.

78. Einstein's Theory of Relativity states that $E = mc^2$. Solve this equation for c.

79. $\omega = \sqrt{\dfrac{k}{m}}$. This is the formula for the angular frequency ω of a mass m suspended from a spring of spring constant k. Solve this formula for m.

80. $V = \dfrac{4}{3}\pi r^3$. This is the formula for the volume of a sphere with radius r. Solve the equation for r.

81. $F = \dfrac{mv^2}{r}$. This is the formula for the force on an object in circular motion. Solve the equation for v.

82. The formula for lateral acceleration, used in automotives, is $a = \dfrac{1.227r}{t^2}$. Solve this equation for t.

83. The ideal body weight for a male may be found using the formula $w = 23h^2$. Solve this equation for h.

84. Kepler's Third Law is $T^2 = \dfrac{4\pi^2 r^3}{GM}$. It relates the period T of a planet to the radius r of its orbit and the Sun's mass M. Solve this formula for r.

85. $r = \dfrac{2gm}{c^2}$. This is the Schwarzschild Radius Formula used to find the radius of a black hole in space. Solve the equation for c.

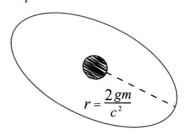

$r = \dfrac{2gm}{c^2}$

86. The total mechanical energy of an object with mass m at height h in a closed system can be written as $ME = \dfrac{1}{2}mv^2 + mgh$. Solve for v, the velocity of the object, in terms of the given quantities.

87. Recall, the Pythagorean Theorem states that $a^2 + b^2 = c^2$. Solve the Pythagorean Theorem for b.

88. In a circuit with an AC power source, the total impedance Z depends on the resistance R, the capacitance C, the inductance L, and the frequency of the current ω according to: $Z = \sqrt{R^2 + \left(\omega L - \dfrac{1}{\omega C}\right)^2}$. Solve this equation for the inductance L.

89. The formula used to find the orbital period for circular Keplerian orbits is $P = \dfrac{2\pi}{\sqrt{\dfrac{u}{a^3}}}$. Solve this equation for a.

90. $\dfrac{35}{12}$ hours, or 2 hours and 55 minutes

91. $\dfrac{4}{3}$ minutes, or 1 minute and 20 seconds

92. 7.5 hours

93. 4 hours and 12 hours

94. 20 weeks

95. 90 minutes

96. 45 minutes

97. 9.1 hours

Solve the following application problems. See Examples 6 and 7.

90. If Joanne were to paint her living room alone, it would take 5 hours. Her sister Lisa could do the job in 7 hours. How long would it take them working together?

91. The hot water tap can fill a given sink in 4 minutes. If the cold water tap is turned on as well, the sink fills in 1 minute. How long would it take for the cold water tap to fill the sink alone?

92. The hull of Jack's yacht needs to be cleaned. He can clean it by himself in 5 hours, but he asks his friend Thomas to help him. If it takes 3 hours for the two men to clean the hull of the boat, how long would it have taken Thomas alone?

93. Two hoses, one of which has a flow rate three times the other, can together fill a tank in 3 hours. How long does it take each of the hoses individually to fill the tank?

94. Officials begin to release water from a full man-made lake at a rate that would empty the lake in 12 weeks, but a river that can fill the lake in 30 weeks is replenishing the lake at the same time. How long does it take to empty the lake?

95. In order to flush deposits from a radiator, a drain that can empty the entire radiator in 45 minutes is left open at the same time it is being filled at a rate that would fill it in 30 minutes. How long does it take for the radiator to fill?

96. Jimmy and Janice are picking strawberries. Janice can fill a bucket in a half hour, but Jimmy continues to eat the strawberries that Janice has picked at a rate of one bucket per 1.5 hours. How long does it take Janice to fill her bucket?

97. A farmer can plow a given field in 2 hours less time than it takes his son. If they acquire two tractors and work together, they can plow the field in 5 hours. How long does it take the father alone? Round your answer to the nearest tenth of an hour.

Polynomials

A chemistry professor calculates final grades for her class using the polynomial

$$A = 0.3f + 0.15h + 0.4t + 0.15p,$$

where A is the final grade, f is the final exam, h is the homework average, t is the chapter test average, and p is the semester project.

The following is a table containing the grades for various students in the class:

Name	Final Exam	Homework Avg.	Test Avg.	Project
Alex	77	95	79	85
Ashley	91	95	88	90
Barron	82	85	81	75
Elizabeth	75	100	84	80
Gabe	94	90	90	85
Lynn	88	85	80	75

1. Find the course average for each student, rounded to the nearest tenth.

2. Who has the highest total score?

3. Why is the total grade raised more with a grade of 100 on the final exam than with a grade of 100 on the semester project?

4. Assume you are a student in this class. With 1 week until the final exam, you have a homework average of 85, a test average of 85, and a 95 on the semester project. What score must you make on the final exam to achieve at least a 90.0 overall? (Round to the nearest tenth.)

Chapter Summary

A summary of concepts and skills follows each chapter. Refer to these summaries to make sure you feel comfortable with the material in the chapter. The concepts and skills are organized according to the section title and topic title in which the material is first discussed.

1.1: Real Numbers and Algebraic Expressions

Common Subsets of Real Numbers
- The sets $\mathbb{N}, \mathbb{Z}, \mathbb{Q},$ and $\mathbb{R},$ as well as *whole* numbers and *irrational* numbers
- Identifying numbers as elements of one or more of the common sets

The Real Number Line
- Plotting numbers on the real number line
- The *origin* of the real line, and its relation to negative and positive numbers

Order on the Real Number Line
- The four inequality symbols $<, \leq, >,$ and $\geq$
- Distinguishing between strict and nonstrict inequalities

Set-Builder Notation and Interval Notation
- Using set-builder notation to define sets
- The empty set
- Interval notation and its relation to inequality statements

Basic Set Operations and Venn Diagrams
- The meaning and use of Venn diagrams
- The definition of the set operations *union* and *intersection* and their application to intervals

Absolute Value and Distance
- The definition of absolute value on the real line
- The relationship between absolute value and distance
- Distance between two real numbers
- Properties of absolute value

Components and Terminology of Algebraic Expressions
- Identifying *terms* of an expression, and identifying the *coefficient* of a term
- Distinguishing between *constants* and *variables* in an expression
- Identifying *factors* of a term

The Field Properties and Their Use in Algebra
- The *closure, commutative, associative, identity, inverse,* and *distributive* properties of the real numbers
- Using the *cancellation* and *zero-factor* properties

1.2: Properties of Exponents and Radicals

Natural Number and Integer Exponents
- The meaning of exponential notation, and the distinction between *base* and *exponent*
- The extension of natural number exponents to integer exponents
- The definition of exponentiation by 0
- The equivalence of a^{-n} and $\dfrac{1}{a^n}$ for $a \neq 0$

Properties of Exponents
- The properties of exponents and their use in simplifying expressions
- Recognition of common errors made in trying to apply properties of exponents

Scientific Notation
- The definition of the scientific notation form of a number
- Using scientific notation in expressing numerical values

Working with Geometric Formulas
- Using geometric formulas with exponents

Roots and Radical Notation
- The definition of nth roots and the meaning of *index* and *radicand*

Simplifying and Combining Radical Expressions
- The meaning of *simplification* as applied to radical expressions
- The properties of radicals and their use in simplifying expressions
- The use of *conjugate radical expressions* in rationalizing denominators (and numerators)
- The meaning of *like radicals*

Rational Number Exponents
- The extension of integer exponents to rational exponents
- Applying properties of exponents to simplify expressions containing rational exponents

1.3: Polynomials and Factoring

The Terminology of Polynomial Expressions
- The definition of *polynomial*
- The meaning of *coefficient, leading coefficient, descending order, degree of a term,* and *degree of a polynomial*
- The meaning of *monomial, binomial,* and *trinomial*

The Algebra of Polynomials
- The meaning of *like,* or *similar,* terms
- Polynomial addition, subtraction, and multiplication

Common Factoring Methods
- The meaning of *factorable* and *irreducible* (or *prime*) polynomials
- The mechanics of using the *greatest common factor, factoring by grouping, special binomials,* and *factoring trinomials* methods
- The extension of the methods to expressions containing fractional exponents

1.4: The Complex Number System

The Imaginary Unit *i* and Its Properties
- The definition of *i*
- The definition of *complex numbers,* and the identification of their *real* and *imaginary* parts

The Algebra of Complex Numbers
- Simplifying complex number expressions
- The use of *complex conjugates*

Roots and Complex Numbers
- The meaning of *principal square root*
- Understanding when properties of radicals apply and when they do not

1.5: Linear Equations in One Variable

Equations and the Meaning of Solutions
- The three categories of equations: *identities*, *contradictions*, and *conditionals*
- The *solution set* of an equation, and *equivalent* equations

Solving Linear Equations in One Variable
- The definition of a *linear*, or *first-degree*, *equation in one variable*
- *Equivalent equations*
- General method of solving linear equations

Solving Absolute Value Equations
- Algebraic and geometric meaning of absolute value expressions in equations
- *Extraneous solutions* to equations

Solving Equations for One Variable
- Solving an equation (or a formula) for a specified variable

Distance and Interest Problems
- Basic distance and interest problems leading to linear equations

1.6: Linear Inequalities in One Variable

Solving Linear Inequalities
- The definition of a *linear inequality*
- The use of *cancellation properties* in solving inequalities

Solving Compound Linear Inequalities
- The definition of a *compound linear inequality*

Solving Absolute Value Inequalities
- The definition of an *absolute value inequality*
- The geometric meaning of inequalities containing an absolute value expression

Translating Inequality Phrases
- The meaning of commonly encountered inequality phrases

1.7: Quadratic Equations

Solving Quadratic Equations by Factoring
- The definition of a *quadratic*, or *second-degree*, *equation in one variable*
- Using the Zero-Factor Property to solve quadratic equations

Solving "Perfect Square" Quadratic Equations
- Solving quadratic equations of the general form $A^2 = c$

Solving Quadratic Equations by Completing the Square
- Using the method of *completing the square* to solve arbitrary quadratic equations

The Quadratic Formula
- Applying the method of completing the square to a generic quadratic equation
- The *quadratic formula*

Gravity Problems
- The general equation of the motion of an object under the influence of gravity

Solving Quadratic-Like Equations
- The definition of an nth-*degree polynomial equation in one variable*
- *Quadratic-like equations*

Solving General Polynomial Equations by Factoring
- Extending the use of the Zero-Factor Property to higher-degree polynomial equations

Solving Polynomial-Like Equations by Factoring
- Extending the use of the Zero-Factor Property to selected nonpolynomial equations

1.8: Rational and Radical Equations

Simplifying Rational Expressions
- The definition of a *rational expression*
- Simplifying rational expressions

Combining Rational Expressions
- Addition, subtraction, multiplication, and division of rational expressions

Simplifying Complex Rational Expressions
- The definition of a *complex rational expression*
- Two methods for simplifying complex rational expressions

Solving Rational Equations
- The meaning of the phrase *rational equation*
- A general method of solving rational equations

Work-Rate Problems
- Types of problems that fall into the classification of *work-rate problems*
- The concept of rate of work as the reciprocal of time needed to complete the work
- The concept of the *additivity* of rates of work

Solving Radical Equations
- The definition of a *radical equation*
- The importance of checking for extraneous solutions to radical equations
- A general method of solving radical equations

Solving Equations with Positive Rational Exponents
- Using the method of solving radical equations to solve equations with positive rational exponents

1. a. 2^3

   b. $2^3,\ 0$

   c. $-\sqrt{4},\ 2^3,\ 0$

   d. All except $\sqrt{17}$

   e. $\sqrt{17}$

   f. All

2. $\left\{\dfrac{1}{2n}\middle|\ n \text{ is a natural}\right.$

   $\text{number}\}$

3. $[4,17)$

4. $[-8,-1]$

5. -7

6. -1

7. $\dfrac{x^2}{2y},\ 12.1x,\ -\sqrt{y+5}$

8. $\dfrac{1}{2},\ 12.1,\ -1$

9. $\dfrac{4\pi}{3}-36$

10. -66

11. Commutative property

12. Multiplicative cancellation; $\dfrac{1}{4}$

Chapter Review

Section 1.1

Which elements of the following sets are **a.** natural numbers, **b.** whole numbers, **c.** integers, **d.** rational numbers, **e.** irrational numbers, **f.** real numbers?

1. $\left\{\dfrac{3}{7},\ -\sqrt{4},\ 2^3,\ 5.3,\ \left|-2.1\right|,\ \sqrt{17},\ 0\right\}$

Describe the following set using set-builder notation. There may be more than one correct way to do this.

2. $\left\{\dfrac{1}{2},\dfrac{1}{4},\dfrac{1}{6},\dfrac{1}{8},\dfrac{1}{10},\dots\right\}$

Write each set as an interval using interval notation.

3. $4 \le x < 17$

4. $\left\{x\middle|-8 \le x \le -1\right\}$

Evaluate the absolute value expressions.

5. $-\left|-4-3\right|$

6. $-\dfrac{\left|x\right|}{\left|-x\right|}$

Identify the components of the algebraic expressions, as indicated.

7. Identify the terms in the expression $\dfrac{x^2}{2y}+12.1x-\sqrt{y+5}$.

8. Identify the coefficients in the expression $\dfrac{x^2}{2y}+12.1x-\sqrt{y+5}$.

Evaluate each expression for the given values of the variables.

9. $7y^2-\dfrac{1}{3}\pi xy+8x^3$ for $x=-2$ and $y=2$

10. $3\sqrt{\dfrac{xy}{3}}-2y^2$ for $x=2$ and $y=6$

Identify the property that justifies each of the following statements. If one of the cancellation properties is being used to transform an equation, identify the quantity that is being added to both sides or the quantity by which both sides are being multiplied.

11. $-4+x=x-4$

12. $12a^2=8b \Leftrightarrow 3a^2=2b$

13. $(-4, 13]$

14. $[5, 8)$

15. Melissa, Monica, Peter, Liz, James

16. $\dfrac{-t^9}{2s^7}$

17. $\dfrac{27z^3}{y^6}$

18. -0.0003005

19. 6.952×10^7

20. 4.152×10^{12}

21. 2.0×10^{-8}

22. 5

23. $\sqrt[6]{3}$

24. $5x^{10}$

25. $\dfrac{3\sqrt{x} - 3\sqrt{2}}{x - 2}$

26. $\dfrac{2y\sqrt[3]{9x^2 y}}{3}$

27. $3|x|\sqrt{2xy} - 2x\sqrt[3]{2xy}$

28. $62 - 20\sqrt{6}$

29. $\dfrac{1}{x^{\frac{7}{4}}}$

30. $56x^{11}$

31. $\dfrac{4000\pi}{3}$ in.^{3}

Simplify the following unions and intersections of intervals.

13. $(-4, 8) \cup [5, 13]$ **14.** $(-4, 8) \cap [5, 13]$

15. Liz, Monica, Peter, James, and Melissa are comparing their ages. Liz is older than Peter and Melissa is the youngest. James is the oldest and Peter is older than Monica. Order them from youngest to oldest.

Section 1.2

Use the properties of exponents to simplify each of the following expressions, writing your answer with only positive exponents.

16. $\dfrac{-4t^0 \left(s^2 t^{-2}\right)^{-3}}{2^3 s t^{-3}}$ **17.** $\left[\left(3y^{-2} z\right)^{-1}\right]^{-3}$

Convert each number from scientific notation to standard notation, or vice versa, as indicated.

18. -3.005×10^{-4}; convert to standard   **19.** $69{,}520{,}000$; convert to scientific

Evaluate each expression, using the properties of exponents. Use a calculator only to check your final answer.

20. $\left(3.46 \times 10^8\right)\left(1.2 \times 10^4\right)$ **21.** $\dfrac{2.4 \times 10^{-12}}{(1.2) \times 10^{-4}}$

Evaluate the following radical expressions.

22. $\sqrt{3^2 + 4^2}$ **23.** $\dfrac{\sqrt[3]{\sqrt{15}}}{\sqrt{\sqrt[3]{5}}}$

Simplify the following radical expressions. Rationalize all denominators and use only positive exponents.

24. $\sqrt{25x^{20}}$ **25.** $\dfrac{3}{\sqrt{x} + \sqrt{2}}$ **26.** $\sqrt[3]{\dfrac{8x^2}{3y^{-4}}}$

Simplify the following expressions.

27. $\sqrt{18x^3 y} - \sqrt[3]{16x^4 y}$ **28.** $\left(2\sqrt{3} - 5\sqrt{2}\right)^2$

Convert the following expressions from radical notation to exponential notation, or vice versa. Simplify each expression in the process, if possible.

29. $\sqrt{x^{-5}} \cdot \sqrt[4]{x^3}$ **30.** $\left(49x^4\right)^{\frac{1}{2}} \left(16x^{12}\right)^{\frac{3}{4}}$

31. Sam is making a piñata in the shape of a sphere and needs to know how much candy to buy to fill it. If the radius of the piñata is 10 inches, what is the volume of the piñata?

32. $m^4 - 5m^3 +$

$3m^2 + 2$

33. $-8x^2y + 8xy + y$

34.

$3x^3 - 4x^2y^3 + 3xy - 4y^4$

35. $5a^2 - 7a^2b +$

$27ab - 35ab^2 + 10b^2$

36. $4xy(2x^2y + x^2 - 3y)$

37. $(2x - 5y)(x + 3)$

38. $(2a + 1)(3a - 5)$

39. $(2a + 3b^2)(2a - 3b^2)$

40. $(3x - 2y)^{\frac{2}{3}}$

$\left[(3x - 2y)^{\frac{2}{3}} - 1\right]$

41. $x^{-2}(8 + 5x)$

42. $-2i\sqrt{2x}$

43. 3

44. $2 + 7i$

45. $-30 + 10i$

46. $-\dfrac{7}{25} + \dfrac{24}{25}i$

47. $-4\sqrt{3}$

48. $62 - 16i\sqrt{2}$

49. $\dfrac{-3i\sqrt{3}}{2}$

50. No solution

51. All real numbers

52. 6.25

Section 1.3

Add or subtract the polynomials, as indicated.

32. $(-4m^2 - 5m^3 + 4) + (m^4 + 7m^2 - 2)$ **33.** $(2xy + 3x) - (8x^2y - 6xy + 3x - y)$

Multiply the polynomials, as indicated.

34. $(x^2 + y)(3x - 4y^3)$ **35.** $(a + 5b)(5a - 7ab + 2b)$

Factor each of the following polynomials.

36. $8x^3y^2 + 4x^3y - 12xy^2$ **37.** $2x^2 + 6x - 5xy - 15y$

38. $6a^2 - 7a - 5$ **39.** $4a^2 - 9b^4$

Factor the following algebraic expressions.

40. $(3x - 2y)^{\frac{4}{3}} - (3x - 2y)^{\frac{2}{3}}$ **41.** $8x^{-2} + 5x^{-1}$

Section 1.4

Evaluate the following square root expressions.

42. $-\sqrt{-8x}$ **43.** $i^3\sqrt{-9}$

Simplify the following expressions.

44. $(7 - 2i) + (9i - 5)$ **45.** $(3 - i)(6i^2 - 4)$ **46.** $\dfrac{3 + 4i}{3 - 4i}$

Simplify the following square root expressions.

47. $(\sqrt{-3})(\sqrt{-16})$ **48.** $(8 - \sqrt{-2})^2$ **49.** $\dfrac{2i\sqrt{-27}}{\sqrt{-16}}$

Section 1.5

Solve the following linear equations.

50. $2y - (1 - y) = y + 2(y - 1)$ **51.** $\dfrac{x}{2} - \dfrac{1}{3} = x - \dfrac{1}{3} - \dfrac{x}{2}$

52. $-0.2x - 0.5 = -0.4x + 0.75$

53. $\{3, 4\}$

54. $\{1, 3\}$

55. $\{-3, 1\}$

56. $\{-3, 4\}$

57. $\left\{\dfrac{-3}{2}\right\}$

58. $b_2 = \dfrac{2A}{h} - b_1$

59. $C = \dfrac{5}{9}(F - 32)$

60. 246.7 miles

61. $\$85$

62. $[7, \infty)$

63. $(4, \infty)$

64. $[4, \infty)$

65. $(1, \infty)$

66. $[-7, 4)$

67. $(7, 27]$

68. $(-5, -1)$

69. No solution

70. $(-\infty, -6] \cup [8, \infty)$

71. $(-\infty, -26] \cup [-7, \infty)$

Solve the following absolute value equations.

53. $|2x - 7| = 1$ **54.** $|2y - 5| - 1 = |3 - y|$ **55.** $|w - 5| = |3w + 1|$

Solve the following absolute value equations geometrically and algebraically.

56. $|-2x + 1| = 7$ **57.** $|x + 4| - |x - 1| = 0$

Solve the following equations for the indicated variable.

58. Area of a Trapezoid: $A = \dfrac{(b_1 + b_2)h}{2}$; solve for b_2

59. Temperature Conversions: $F = \dfrac{9}{5}C + 32$; solve for C

Solve the following application problems.

60. Two trains leave the station at the same time in opposite directions. One travels at an average rate of 90 miles per hour, and the other at an average rate of 95 miles per hour. How far apart are the two trains after an hour and twenty minutes? Round your answer to the nearest tenth of a mile.

61. Two firefighters, Jake and Rose, each have $\$5000$ to invest. Jake invests his money in a money market account with an annual return of 3.25%, while Rose invests hers in a CD paying 4.95% annually. How much more money does Rose have than Jake after 1 year?

Section 1.6

Solve the following inequalities. Describe the solution set using interval notation and by graphing.

62. $-8x + 3 \geq -9x + 10$

63. $4(2x - 5) < -3(-3x + 8)$

64. $\dfrac{-2(x - 1)}{3} \leq \dfrac{-2x}{4}$

65. $3.1(2x - 1) > 7.2 - 4.1x$

66. $-14 < -2(3 + y) \leq 8$

67. $2 < \dfrac{x + 1}{4} \leq 7$

68. $-5|3 + t| > -10$

69. $3 + |2x - 1| < 1$

70. $-2|x - 1| + |3x - 3| \geq 7$

71. $6 + \dfrac{x}{5} \leq \dfrac{4}{5}$ or $5 + 2x \geq x - 2$

72. $\left\{-\dfrac{2}{5},3\right\}$

73. $\{2\pm 3i\}$

74. $\left\{4\pm\sqrt{2}\right\}$

75. $\left\{\dfrac{19\pm\sqrt{701}}{17}\right\}$

76. $\left\{\pm 1,\pm\sqrt{2}\right\}$

77. $\{-27,8\}$

78. $\{\pm 2,\pm 3\}$

79. $\left\{\pm\sqrt{2},4\right\}$

80. $\left\{0,\dfrac{1}{2},2\right\}$

81. $\{-1,0,4\}$

82. $\left\{\dfrac{3}{4}\right\}$

83. $b=-2$ and $c=-8$

84. 10

85. $\dfrac{x+3}{x-3}$, $x\neq 0,\pm 3$

86. $\dfrac{x+3}{x^2+3x+9}$, $x\neq 3$

87. $\dfrac{-2}{x}$

88. $\dfrac{1}{a+1}$

89. $\dfrac{x+3}{3}$

90. $-x-y$

Section 1.7

Solve the following quadratic equations.

72. $5x^2-13x-6=0$

73. $2(x-2)^2=-18$

74. $x^2-8x+14=0$

75. $1.7z^2-3.8z-2=0$

Solve the following quadratic-like equations.

76. $(x^2+2)^2-7(x^2+2)+12=0$

77. $y^{\frac{2}{3}}+y^{\frac{1}{3}}-6=0$

78. $x^4-13x^2+36=0$

Solve the following equations by factoring.

79. $x^3-4x^2-2x+8=0$

80. $2x^3+2x=5x^2$

81. $x^{\frac{7}{2}}-3x^{\frac{5}{2}}-4x^{\frac{3}{2}}=0$

82. $(x-1)^{-\frac{1}{2}}+4(x-1)^{\frac{1}{2}}=0$

Use the connection between solutions of polynomial equations and polynomial factoring to answer the following questions.

83. Find b and c so the equation $x^3+bx^2+cx=0$ has solutions of $-2,0$, and 4.

84. Given that the equation $x^2-6x+m-1=0$ has only one root, find m.

Section 1.8

Simplify the following rational expressions, indicating which real values of the variable must be excluded.

85. $\dfrac{x^3+6x^2+9x}{x^3-9x}$

86. $\dfrac{x^2-9}{x^3-27}$

Perform the indicated operations on the rational expressions and simplify your answer.

87. $\dfrac{1}{x}-\dfrac{3}{x+2}-\dfrac{6}{x^2+2x}$

88. $\dfrac{a^3-8}{a^2-4}\div\dfrac{a^3+2a^2+4a}{a^3+2a^2}\cdot\dfrac{1}{a^2+a}$

Simplify the complex rational expressions.

89. $\dfrac{\dfrac{x}{3}-\dfrac{3}{x}}{-\dfrac{3}{x}+1}$

90. $\dfrac{\dfrac{x}{y}-\dfrac{y}{x}}{x^{-1}-y^{-1}}$

91. $\pm i$

92. $\left\{-\dfrac{1}{3}\right\}$

93. $\dfrac{x^2}{x+1}$

94. 2

95. $\{-4\}$

96. $\{13\}$

97. $\{-1, 8\}$

98. $r = \sqrt{\dfrac{3V}{\pi h}}$

99. $\dfrac{24}{7}$

Solve the following rational equations.

91. $\dfrac{1}{x+2} + \dfrac{1}{x-3} - \dfrac{x}{x-3} = 0$

92. $\dfrac{y}{y-1} + \dfrac{1}{y-4} = \dfrac{y^2}{y^2-5y+4}$

Perform the indicated operations on the following rational expressions, and simplify your answer.

93. $\left(x-1+\dfrac{2}{x+1}\right) \div \left(1+\dfrac{1}{x^2}\right)$

94. $\dfrac{x-3}{x+2} \cdot \left(\dfrac{x+2}{x-3} + \dfrac{x-3}{x-4} - \dfrac{17-4x}{x^2-7x+12}\right)$

Solve the following equations.

95. $\sqrt{-4-x} - 4 = x$

96. $\sqrt{5x-1} = 4 + \sqrt{x+3}$

97. $\left(x^2+x-16\right)^{\frac{1}{3}} = 2(x-1)^{\frac{1}{3}}$

98. The formula for the volume of a cone with radius r and height h is $V = \dfrac{1}{3}\pi r^2 h$. Solve the equation for r.

99. Jim cleans a house in 6 hours. John cleans the same house in 8 hours. How long does it take together?

Answers (left margin):

1. a. $\sqrt{1}$, $|-21|$

b. $\sqrt{1}$, $|-21|$

c. $\sqrt{1}$, -1, $|-21|$

d. $\sqrt{1}$, $7.\overline{6}$, -1, $\dfrac{2}{9}$, $|-21|$

e. $2\sqrt{3}$, 5π

f. $2\sqrt{3}$, 5π, $\sqrt{1}$, $7.\overline{6}$, -1, $\dfrac{2}{9}$, $|-21|$

2. $(-7,9]$ **3.** $[14,\infty)$

4. -9

5. $\sqrt{11}-\sqrt{5}$

6. 6

7. -0.0002004

8. 5.224×10^{7}

9. $\dfrac{18y^{2}}{x^{4}z^{5}}$

10. $\dfrac{|a|}{2}$

11. $\dfrac{-4y}{x^{3}}$

12. $3x^{2}y+7xy-2z+4xz$

13. $20x^{3}y+10x^{2}y^{2}+8x^{2}y+4xy^{2}-12x-6y$

14. $(2a+1)(3a-5)$

15. $(6x^{3}+y)(6x^{3}-y)$

Chapter Test

Which elements of the following sets are **a.** natural numbers, **b.** whole numbers, **c.** integers, **d.** rational numbers, **e.** irrational numbers, **f.** real numbers?

1. $\left\{2\sqrt{3},\ 5\pi,\ \sqrt{1},\ 7.\overline{6},\ -1,\ \dfrac{2}{9},\ |-21|\right\}$

Write each set as an interval using interval notation.

2. $\{x\,|\,-7<x\le9\}$ **3.** $\{x\,|\,x\ge14\}$

Evaluate the absolute value expressions.

4. $-|11-2|$ **5.** $\left|\sqrt{5}-\sqrt{11}\right|$

Evaluate the expression for the given values of the variables.

6. $x^{2}z^{3}+5\sqrt{3x-2y}$ for $x=2,y=1,$ and $z=-1$.

Convert each number from scientific notation to standard notation, or vice versa, as indicated.

7. -2.004×10^{-4}; convert to standard.

8. $52{,}240{,}000$; convert to scientific.

Simplify the following expression, writing your answer with only positive exponents.

9. $\dfrac{3^{2}x^{-4}\left(y^{2}z\right)^{-2}}{\left(2z^{-3}\right)^{-1}y^{-6}}$

Simplify the following radical expressions. Rationalize all denominators and use only positive exponents.

10. $\dfrac{\sqrt{3a^{3}}}{\sqrt{12a}}$ **11.** $\sqrt[3]{-64x^{-9}y^{3}}$

Simplify the following expressions by performing the indicated operations.

12. $(5x^{2}y+7xy-z)-(2x^{2}y+z-4xz)$ **13.** $(5x^{2}y+2xy-3)(4x+2y)$

Factor each of the following polynomials.

14. $3x^{4}+3x^{2}y-x^{2}y^{3}-y^{4}$ **15.** $36x^{6}-y^{2}$

16. $5i - 5$

17. $-\dfrac{1}{2}$

18. $4+i$ **19.** $-8i$

20. $y = \dfrac{14}{9}$;

    conditional

21. $16 = 126$;

    contradiction

22. $\left\{-\dfrac{10}{7}, 0\right\}$

23. $v_0 = \dfrac{h+16t^2}{t}$

24. $(-1, 7]$

25. $(-3, 13]$

26. Under $420

27. $\left\{-4, \dfrac{5}{2}\right\}$

28. $\{-3, -2\}$

29. $\{-6, 4\}$

30. $\left\{\pm 3i, \pm\sqrt{2}\right\}$

31. $\left\{0, \dfrac{1\pm\sqrt{13}}{4}\right\}$

32. $\left\{-\dfrac{1}{3}\right\}$

33. $\dfrac{3t^2 + 8t}{3t+12}$; $t \neq -4, -5$

34. $\dfrac{3a^3 + a}{5a+1}$

35. $\dfrac{2x+1}{(2x-1)(x-8)}$

Simplify the following expressions by performing the indicated operations.

16. $\left(8i + 3i^3\right) - \left(4i^4 - i^6\right)$

17. $\left(\dfrac{i^7}{10}\right)\left(\dfrac{5}{i^9}\right)$

18. $\dfrac{17}{4-i}$

19. $\left(2 - \sqrt{-4}\right)^2$

Solve the following equations.

20. $5(3y - 2) = (4y + 4) + 2y$

21. $\dfrac{9a+4}{3} = \dfrac{6(2a+7)}{4}$

22. $|7z + 5| + 3 = 8$

Solve the following equation for the indicated variable.

23. $h = -16t^2 + v_0 t$; solve for v_0.

Solve the following inequalities. Describe the solution set using interval notation and by graphing.

24. $-8 < 3x - 5 \le 16$

25. $-3(x-1) < 12$ and $x - 4 \le 9$

26. Kim wants to keep her bills under $1800 a month. If each month her rent is $550, her utilities are $80, food is $420, entertainment is $250, and she has $80 in other expenses, how much can she afford to spend on car payments?

Solve the following equations.

27. $2x^2 + 3x - 10 = 10$

28. $2x^2 + 7x = x^2 + 2x - 6$

29. $(t+2)^2 - 2(t+2) = 24$

30. $x^4 + 7x^2 - 18 = 0$

31. $4x^{\frac{18}{7}} - 2x^{\frac{11}{7}} - 3x^{\frac{4}{7}} = 0$

32. $(2x+1)^{-\frac{1}{2}} - 3(2x+1)^{\frac{1}{2}} = 0$

Simplify the following rational expression, indicating which real values of the variable must be excluded.

33. $\dfrac{3t^3 + 23t^2 + 40t}{3t^2 + 27t + 60}$

Perform the indicated operations on the rational expressions and simplify your answer.

34. $\dfrac{3a^3 + 5}{5a+1} + \dfrac{a-5}{5a+1}$

35. $\dfrac{4x^2 - 30x - 16}{2x^3 - 19x^2 + 24x} \div \dfrac{4x^2 - 34x + 16}{2x^2 - 3x}$

Simplify the complex rational expression.

36. $\dfrac{b-a}{4a+4b}$

36. $\dfrac{\dfrac{1}{2a}-\dfrac{1}{2b}}{\dfrac{2}{a}+\dfrac{2}{b}}$

37. $\{-5\}$

38. $\{2\}$

39. $\{4,5\}$

Solve the following equations.

37. $\dfrac{2}{x+1}-\dfrac{x}{x-3}=\dfrac{3x-21}{x^2-2x-3}$

38. $\sqrt{2x^2+8x+1}-x-3=0$

39. $\left(2x^2-18x+67\right)^{\frac{1}{3}}=3$

Chapter

2

Introduction to Equations and Inequalities of Two Variables

2.1   The Cartesian Coordinate System

2.2   Linear Equations in Two Variables

2.3   Forms of Linear Equations

2.4   Parallel and Perpendicular Lines

2.5   Linear Inequalities in Two Variables

2.6   Introduction to Circles

Chapter 2 Project

Chapter 2 Summary

Chapter 2 Review

Chapter 2 Test

By the end of this chapter you should be able to:

What if, as a sales manager, you were asked to forecast next year's sales for your company? How could you use previous data to make an informed and accurate prediction?

By the end of this chapter, you'll be able to apply the skills regarding linear inequalities in two variables to applications concerning price, weight, volume, time, materials, and more. On page 168, you'll solve problems like the one given above. You'll master this type of problem using the formula for the slope of a line on page 158 and the equation of a line in two variables.

Introduction

This chapter introduces the Cartesian coordinate system, the two-dimensional framework that underlies most of the material throughout the remainder of this text.

Descartes

Many problems that we seek to solve with mathematics are most naturally described with two variables. The existence of two variables in a problem leads naturally to the use of a two-dimensional system in which to work, especially when we attempt to depict the situation graphically, but it took many centuries for the ideas presented in this chapter to evolve. Some of the greatest accomplishments of the French mathematician and philosopher René Descartes (1596–1650) were his contributions to the then fledgling field of analytic geometry, the marriage of algebra and geometry. In *La géométrie*, an appendix to a volume of scientific philosophy, Descartes laid out the basic principles by which algebraic problems could be construed as geometric problems, and the methods by which solutions to the geometric problems could be interpreted algebraically. As many later mathematicians expanded upon Descartes' work, analytic geometry came to be an indispensable tool in understanding and solving problems of both an algebraic and geometric nature.

In this chapter we will use the Cartesian coordinate system, named in honor of Descartes, primarily to study linear equations and inequalities in two variables. As we will see, a graph is one of the best ways to describe the solutions of such problems. Sections 2.2 through 2.5 will introduce the basic means by which we can construct graphs and consequently shift between the algebraic and geometric views of a given linear equation or linear inequality.

The Cartesian coordinate system will continue to play a prominent role as we proceed to other topics, such as relations and functions, in later chapters. Mastery of the foundational concepts in this chapter will be essential to understanding these related ideas.

2.1

The Cartesian Coordinate System

TOPICS

1. The Cartesian coordinate system
2. The graph of an equation
3. The distance and midpoint formulas
T. Graphing an equation

TOPIC **1**

The Cartesian Coordinate System

In Chapter 1, we studied the algebra of a single variable, we learned how to solve equations and inequalities in one variable, and how to isolate a single variable in equations with more than one variable.

While many important problems can be studied and solved using one variable, many more problems require two or more variables. In this chapter, you will learn how to write and solve equations and inequalities in two variables.

The first question we need to answer is how to express a solution to an equation in two variables. Recall that a solution to an equation in one variable (for example, x) is any value of x that when substituted in the equation results in a true statement. Consider an equation in two variables x and y. A particular solution of the equation, if there is one, must consist of a value for x and a *corresponding* value for y. The solution consists of a *pair* of numbers, called an ordered pair.

DEFINITION

Ordered Pairs

An **ordered pair** (a,b) consists of two real numbers a and b such that the order of a and b matters. That is, $(a,b) = (b,a)$ if and only if $a = b$. The number a is called the **first coordinate** and the number b is called the **second coordinate**.

We can then write a solution to an equation in two variables as an ordered pair (x, y).

The ordered pair notation is also very useful for graphing the solutions to equations and inequalities in two variables. For one variable problems, we used the real number line, a one-dimensional coordinate system, to graph solutions. Think about how difficult it would be to interpret solutions if we plotted the values of both variables on a single number line. Because the two variables are linked in each solution, a two-dimensional coordinate system is a more natural place to graph solutions of two-variable equations and inequalities. The coordinate system we use is named after René Descartes (pronounced "day-cart"), the 17th century French mathematician largely responsible for its development.

DEFINITION

The Cartesian
Coordinate System

The **Cartesian coordinate system** (also called the **Cartesian plane**) consists of two perpendicular real number lines (each called an **axis**) intersecting at the 0 point of each line. The point of intersection is called the **origin** of the system, and the four quarters defined by the two lines are called the **quadrants** of the plane, numbered as indicated below in Figure 1. Because the Cartesian plane consists of two crossed real lines, it is often given the symbol $\mathbb{R} \times \mathbb{R}$, or $\mathbb{R}^2$. Each point P in the plane is identified by an ordered pair. The first coordinate indicates the horizontal displacement of the point from the origin, and the second coordinate indicates the vertical displacement. Figure 1 is an example of a Cartesian coordinate system, and illustrates how several ordered pairs are **graphed**, or **plotted**.

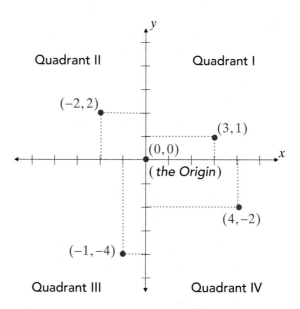

Figure 1: The Cartesian Plane

CAUTION!

Unfortunately, mathematics uses parentheses to denote ordered pairs as well as open intervals, which sometimes leads to confusion. Context is the key to interpreting notation correctly. For instance, in the context of solving a one-variable inequality, the notation $(-2, 5)$ most likely refers to the open interval with endpoints at -2 and 5, while in the context of solving an equation in two variables, $(-2, 5)$ probably refers to a point in the Cartesian plane.

EXAMPLE 1

Plotting Points in the
Cartesian Coordinate
System

Plot the following ordered pairs on the Cartesian plane, and identify which quadrant they lie in (or which axis they lie on).

a. $(2,3)$ **b.** $(-5,0)$ **c.** $(1,-3)$

d. $(-2,4)$ **e.** $(-6,-6)$ **f.** $(0,5)$

Note:
While there is no required method when plotting points, it is helpful to establish a set pattern that *you* follow; for example, you may always count the horizontal value first, then the vertical displacement.

Solutions:

a.

b.

c.

d.

e.
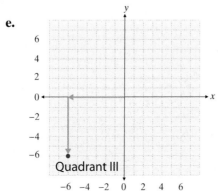

f.

TOPIC 2   The Graph of an Equation

A solution of an equation in x and y must consist of a value a for x and a corresponding value b for y. It is natural to write such a solution as an ordered pair (a, b), and equally natural to graph it as a point in the plane whose coordinates are a and b. In this context we refer to the horizontal number line as the **x-axis**, the vertical number line as the **y-axis**, and the two coordinates of the ordered pair (a, b) as the **x-coordinate** and the **y-coordinate**.

As we will see, an equation in x and y usually consists of far more than one ordered pair (a, b). The **graph of an equation** is a plot in the Cartesian plane of *all* of the ordered pairs that make up the solution set of the equation.

We can make rough sketches of the graphs of many equations just by plotting enough solutions to give us a sense of the entire solution set. We can find individual ordered pair solutions of a given equation by selecting numbers that seem appropriate for one of the variables and then solving the equation for the other variable. This changes the task of solving a two-variable equation into that of solving a one-variable equation, and we have all the methods of Chapter 1 at our disposal to accomplish this. The process is illustrated in Example 2.

EXAMPLE 2

Graphing Equations in Two Variables

Sketch graphs of the following equations by plotting points.

a. $2x - 5y = 10$ **b.** $x^2 + y^2 - 6x = 0$ **c.** $y = x^2 - 2x$

Solutions:

Note:
Rather than solving a new equation each time you substitute a value, it is more efficient to solve the equation for one variable before making substitutions (see Section 1.5). This method is shown in Example 2b.

a.

x	y
-3	?
0	?
?	0
?	5
1	?

$2x - 5y = 10$ →

x	y
-3	$-\dfrac{16}{5}$
0	-2
5	0
$\dfrac{35}{2}$	5
1	$-\dfrac{8}{5}$

In each row in the first table, we select a value for one of the two variables.

Once we substitute a value, the equation $2x - 5y = 10$ can be solved for the other variable. An example of this is shown below the table.

This gives us a list of 5 ordered pairs that can be plotted, though the ordered pair $\left(\dfrac{35}{2}, 5\right)$ is off the coordinate system we draw.

$$2(-3) - 5y = 10$$
$$-6 - 5y = 10$$
$$-5y = 16$$
$$y = -\frac{16}{5}$$

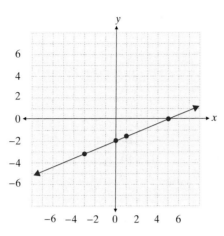

The four ordered pairs appear to lie on a straight line, and this is indeed the case. To gain more confidence in this fact, we could continue to plot more solutions of the equation, and we would find they all lie along the line that has been drawn through the four plotted ordered pairs. The infinite number of solutions of the equation are depicted by the line drawn through the plotted points.

b. For this example, we solve the original equation for y, then substitute several values for x to generate a series of y-values.

$$x^2 + y^2 - 6x = 0$$
$$y^2 = 6x - x^2$$
$$y = \pm\sqrt{6x - x^2}$$

x	y
0	?
1	?
2	?
3	?
4	?
5	?
6	?

$x^2 + y^2 - 6x = 0$

x	y
0	0
1	$\pm\sqrt{5} \approx \pm 2.2$
2	$\pm 2\sqrt{2} \approx \pm 2.8$
3	± 3
4	$\pm 2\sqrt{2} \approx \pm 2.8$
5	$\pm\sqrt{5} \approx \pm 2.2$
6	0

Again, we plot enough solutions to feel confident in sketching the entire solution set. Note that for $x < 0$ and $x > 6$, the corresponding y would be imaginary, and thus irrelevant when graphing the equation. Similarly, for $y < -3$ and for any $y > 3$, the corresponding x would be a complex number (you can use the quadratic formula to verify this).

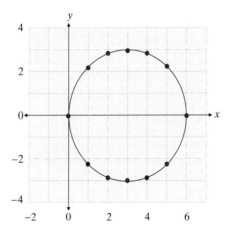

Once we plot enough solutions, the graph of the equation begins to take the shape of a circle. In fact, the graph is a circle, with center $(3, 0)$ and a radius of 3, but we will not be able to prove this claim until Section 2.6.

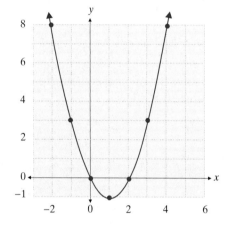

c.

x	y
0	?
2	?
1	?
−1	?
3	?
−2	?
4	?

$y = x^2 - 2x$

x	y
0	0
2	0
1	−1
−1	3
3	3
−2	8
4	8

Since this equation is already solved for y, we use a table of x-values and substitute them in the given equation. Again, enough points should be plotted to give some idea of the nature of the entire solution set of the equation.

The graph of $y = x^2 - 2x$ is a shape known as a *parabola*. We will encounter these shapes again in Section 3.2, and we will be able, at that time, to prove that our rough sketch at the left is indeed the graph of the solution set of $y = x^2 - 2x$.

Plotting points is, for the most part, easily accomplished, but it is also a rather crude and tedious method, and there are a few concerns about graphing by plotting points:

Have we really plotted enough points to accurately "fill in" the gaps and sketch the entire solution set of each equation? Is filling in the gaps justified in the first place? What proof do we have that *all* the ordered pairs along our sketches actually solve the corresponding equation? Finally, is there a faster and more sophisticated way to determine the graph of an equation?

These concerns are not trivial. Throughout this chapter, we will address them for linear equations. Much of the rest of this textbook works to answer these questions for more complicated equations and graphs. Regardless, plotting points is an important skill since it is so useful when dealing with new or unfamiliar situations.

TOPIC 3

The Distance and Midpoint Formulas

Throughout the rest of this text, we will have reasons for wanting to know, on occasion, the *distance* between two points in the Cartesian plane. We already have the tools necessary to answer this question, and we will now derive a formula that we can apply whenever necessary.

Let (x_1, y_1) and (x_2, y_2) be the coordinates of two points in the plane. By drawing the dotted lines parallel to the coordinate axes as shown in Figure 2, we can form a right triangle. Note that we are able to determine the coordinates of the vertex at the right angle from the other two vertices (x_1, y_1) and (x_2, y_2).

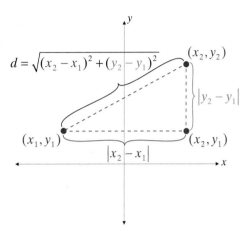

Figure 2: The Distance Formula

The lengths of the two legs of the triangle are easy to find, as they are just distances between numbers on real number lines. (The absolute value symbols are present since in general, $x_2 - x_1$ and $y_2 - y_1$ could be positive or negative.) Recall that the Pythagorean Theorem states that for a right triangle the sum of the squares of its legs is equal to the square of the hypotenuse ($a^2 + b^2 = c^2$). We can apply the Pythagorean Theorem to determine the distance labeled in Figure 2.

$$d^2 = \left(\left|x_2 - x_1\right|\right)^2 + \left(\left|y_2 - y_1\right|\right)^2 \text{, so}$$

$$d = \sqrt{\left(x_2 - x_1\right)^2 + \left(y_2 - y_1\right)^2}$$

Notice that the absolute value symbols are not necessary in the final formula, as any quantity squared is automatically nonnegative.

THEOREM

Distance Formula

The **distance** between two points (x_1, y_1) and (x_2, y_2) in the Cartesian plane is given by the following formula:

$$d = \sqrt{\left(x_2 - x_1\right)^2 + \left(y_2 - y_1\right)^2}$$

EXAMPLE 3

Using the Distance Formula

Calculate the distance between the following pairs of points.

a. $(-4, -2)$ and $(-7, 2)$ **b.** $(5, 1)$ and $(-1, 3)$

Solutions:

a. $d = \sqrt{\left((-4)-(-7)\right)^2 + \left((-2)-2\right)^2}$
     Substitute the coordinates of each point into the distance formula.

$= \sqrt{3^2 + (-4)^2}$
     Simplify.

$= \sqrt{9 + 16}$

$= \sqrt{25}$

$= 5$
     Note that we only take the positive square root, since we are calculating a distance.

b. $d = \sqrt{\left(5-(-1)\right)^2 + \left(1-3\right)^2}$
     Again, substitute the coordinates into the distance formula, then simplify.

$= \sqrt{36 + 4}$

$= \sqrt{40}$
     Simplify the radical by factoring out $2^2 = 4$.

$= 2\sqrt{10}$

We will also want to be able to determine the midpoint of a line segment in the plane. That is, given two points (x_1, y_1) and (x_2, y_2), we want to know the coordinates of the point exactly halfway between the two given points.

Consider the points plotted in Figure 3. The x-coordinate of the midpoint is the average of the two x-coordinates of the given points, and the y-coordinate of the midpoint is the average of the two y-coordinates.

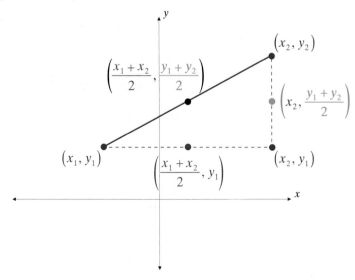

Figure 3: The Midpoint Formula

Since x_1 and x_2 are numbers on a real number line, and y_1 and y_2 are numbers on a (different) real number line, determining the averages of these two pairs of numbers is straightforward: the average of the x-coordinates is $\dfrac{x_1 + x_2}{2}$ and the average of the y-coordinates is $\dfrac{y_1 + y_2}{2}$. Putting these two coordinates together gives us the desired formula.

THEOREM

Midpoint Formula

The **midpoint** between two points (x_1, y_1) and (x_2, y_2) in the Cartesian plane has the following coordinates:

$$\left(\frac{x_1 + x_2}{2}, \frac{y_1 + y_2}{2} \right)$$

EXAMPLE 4

Using the Midpoint Formula

Calculate the midpoint of the line connecting each pair of points.

a. $(5, 1)$ and $(-1, 3)$ | **b.** $(3, 0)$ and $(-6, 11)$

Solutions:

a. $\left(\dfrac{5 + (-1)}{2}, \dfrac{1 + 3}{2} \right) = (2, 2)$

b. $\left(\dfrac{3 + (-6)}{2}, \dfrac{0 + 11}{2} \right) = \left(-\dfrac{3}{2}, \dfrac{11}{2} \right)$

In each case, we simply substitute the coordinates of each point into the midpoint formula. This has the effect of averaging both x-coordinates and both y-coordinates.

TOPIC

Graphing an Equation

If we were trying to sketch the graph of the equation $y = 0.1x^4 - 2.2x^2 + 2.4x + 4.5$, the method of plotting enough ordered pairs that solve the equation would not be very efficient, or accurate. We can use a calculator to graph this equation.

Press 🔘 Y= and type in the equation next to Y1. To type in the variable, x, press 🔘 X,T,θ,n .

Press ⬛GRAPH⬛ and the following graph should appear:

Notice that you can't see the very bottom of the curve. Oftentimes when we use a calculator to graph equations, we have to adjust the viewing window to see the whole graph. To do so, press ⬛WINDOW⬛. The default window displays the graph with x- and y-values ranging from –10 to 10. This window can be changed by changing the values for Xmin, Xmax, Ymin, and Ymax. Since the graph that appears descends below our viewing screen, we need to change the Ymin to something smaller, like –20.

Press ⬛GRAPH⬛ again and the calculator will display the graph again with the new window settings.

Keep in mind that the screen is not square: one unit on the x-axis looks longer than one unit on the y-axis, so the picture will not be an accurate representation unless the window is set to a ratio of about 3:2. One way to attain a window with this ratio is to press ⬛ZOOM⬛ and select 5:ZSquare. This will change the values in the Window screen.

Finally, notice that the equation we graphed has two variables, specifically x and y, and is solved for y. An equation must be in this form in order to graph it on a calculator. For example, in order to graph the equation $4x + 2y = 1$, we would first have to solve the equation for y: $y = -2x + \dfrac{1}{2}$.

1.

2.

3.

4.

5.

6.

7.

8.

9. III
10. Negative y-axis
11. IV
12. Negative x-axis
13. Positive x-axis
14. I
15. III

Exercises

Plot the following sets of points in the Cartesian plane. See Example 1.

1. $\{(-3,2),(5,-1),(0,-2),(3,0)\}$

2. $\{(-4,0),(0,-4),(-3,-3),(3,-3)\}$

3. $\{(3,4),(-2,-1),(-1,-3),(-3,0)\}$

4. $\{(2,2),(0,3),(4,-5),(-1,3)\}$

5. $\{(0,5),(-3,2),(2,4),(1,1)\}$

6. $\{(8,3),(-3,4),(-4,-6),(3,-4)\}$

7. $\{(-5,-4),(3,2),(4,5),(-2,-1),(-4,-4),(1,1)\}$

8. $\{(-2,5),(0,1),(1,-1),(1,-3),(0,0),(-1,2),(0,-2)\}$

Identify the quadrant in which each point lies, if possible. If a point lies on an axis, specify which part (positive or negative) of which axis (x or y). See Example 1.

9. $(-2,-4)$ **10.** $(0,-12)$ **11.** $(4,-7)$ **12.** $(-2,0)$ **13.** $(9,0)$

14. $(3,26)$ **15.** $(-4,-7)$ **16.** $(0,1)$ **17.** $(17,-2)$ **18.** $\left(-\sqrt{2},4\right)$

19. $(-1,1)$ **20.** $(-4,0)$ **21.** $(3,-9)$ **22.** $(0,0)$ **23.** $(4,3)$

24. $(-3,-11)$ **25.** $(0,-97)$ **26.** $\left(\dfrac{1}{3},0\right)$

Determine appropriate settings on a graphing calculator so that each of the given points will lie within the viewing window. Answers will vary slightly.

27. $\{(-4,1),(2,8),(5,7)\}$

28. $\{(12,3),(5,-11),(-9,6)\}$

29. $\{(3,2),(-2,4),(5,-3)\}$

30. $\{(30,55),(40,25),(-80,-10)\}$

31. $\{(3.75,-8.5),(-5.25,6.0),(7.5,-2.25)\}$

32. $\{(63,99),(-87,34),(45,-22)\}$

For each of the following equations, determine the value of the missing entries in the accompanying table of ordered pairs. Then plot the ordered pairs and sketch your guess of the complete graph of the equation. See Example 2.

16. Positive y-axis

17. IV

18. II

19. II

20. Negative x-axis

21. IV

22. Origin (x-axis and y-axis)

23. I

24. III

25. Negative y-axis

26. Positive x-axis

27. $X = [-5, 6]$; $Y = [-8, 9]$

28. $X = [-10, 13]$; $Y = [-12, 7]$

29. $X = [-3, 6]$; $Y = [-4, 5]$

30. $X = [-90, 50]$; $Y = [-20, 60]$

31. $X = [-6, 8]$; $Y = [-9, 7]$

32. $X = [-90, 70]$; $Y = [-30, 100]$

33. $\{(0, -3),\ (2, 0),$

$\left(3, \dfrac{3}{2}\right),\ (4, 3)\}$

34. $\{(-1, 0),\ (1, 4),$

$(-2, 1)\text{ or }(0, 1),$

$(2, 9),\ (-3, 4)\}$

33. $6x - 4y = 12$

x	y
0	?
?	0
3	?
?	3

34. $y = x^2 + 2x + 1$

x	y
?	0
1	?
?	1
2	?
-3	?

35. $x = y^2$

x	y
0	?
1	?
4	?
9	?
?	$-\sqrt{2}$

36. $5x - 2 = -y$

x	y
?	0
0	?
1	?
?	7
-2	?

37. $x^2 + y^2 = 9$

x	y
0	?
?	0
-1	?
1	?
?	2

38. $y = -x^2$

x	y
0	?
-1	?
1	?
-2	?
2	?

Determine **a.** the distance between the following pairs of points, and **b.** the midpoint of the line segment joining each pair of points. See Examples 3 and 4.

39. $(-2, 3)$ and $(-5, -2)$

40. $(-1, -2)$ and $(2, 2)$

41. $(0, 7)$ and $(3, 0)$

42. $\left(-\dfrac{1}{2}, 5\right)$ and $\left(\dfrac{9}{2}, -7\right)$

43. $(-2, 0)$ and $(0, -2)$

44. $(5, 6)$ and $(-3, -2)$

45. $(13, -14)$ and $(-7, -2)$

46. $(-8, 3)$ and $(2, 11)$

47. $(-3, -3)$ and $(5, -9)$

48. $(7, -7)$ and $(-7, -6)$

49. $(5, -4)$ and $(-1, 5)$

50. $(4, 6)$ and $(2, -7)$

51. $(8, 8)$ and $(-2, -2)$

52. $\left(3, \dfrac{26}{5}\right)$ and $\left(9, -\dfrac{14}{5}\right)$

53. Given $(10, 4)$ and $(x, -2)$, find x such that the distance between these two points is 10.

54. Given $(1, y)$ and $(13, -3)$, find y such that the distance between these two points is 15.

55. Given $(x, 3)$ and $(-6, y)$, find x and y such that the midpoint between these two points is $(2, 2)$.

Find the perimeter of the triangle whose vertices are the specified points in the plane.

35. $\{(0,0),\ (1,\pm1),$
$(4,\pm2),\ (9,\pm3),$
$(2,-\sqrt{2})\}$

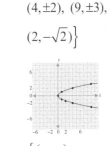

36. $\left\{\left(\dfrac{2}{5},0\right),\ (0,2),\right.$
$(1,-3),\ (-1,7),$
$\left.(-2,12)\right\}$

37. $\{(0,\pm3,\ (\pm3,0),$
$(-1,\pm2\sqrt{2}),$
$(1,\pm2\sqrt{2}),$
$(\pm\sqrt{5},2)\}$

38. $\{(0,0),\ (-1,-1),$
$(1,-1),(-2,-4),$
$(2,-4)\}$

39. $\sqrt{34}\ ,\ \left(\dfrac{-7}{2},\dfrac{1}{2}\right)$

40. $5,\ \left(\dfrac{1}{2},0\right)$

56. $(-2,3),(-2,1),$ and $(-5,-2)$

57. $(-1,-2),(2,-2),$ and $(2,2)$

58. $(6,-1),(-6,4),$ and $(9,3)$

59. $(3,-4),(-7,0),$ and $(-2,-5)$

60. $(-3,7),(5,1),$ and $(-3,-14)$

61. $(-12,-3),(-7,9),$ and $(9,-3)$

Use the distance and midpoint formulas to answer the following geometry and application problems.

62. Prove that the triangle with vertices at the points $(1,1)$, $(-2,-5)$, and $(3,0)$ is a right triangle. Then determine the area of the triangle.

63. Prove that the triangle with vertices at the points $(-2,2)$, $(1,-2)$, and $(2,5)$ is isosceles. Then determine the area of the triangle. (**Hint:** Make use of the midpoint formula.)

64. Prove that the triangle with vertices at the points $(5,1)$, $(-3,7)$, and $(8,5)$ is a right triangle. Then determine the area of the triangle.

65. Prove that the triangle with vertices at the points $(1,2)$, $(-2,0)$, and $(3,5)$ is isosceles. Then determine the area of the triangle. (**Hint:** Make use of the midpoint formula.)

66. Prove that the triangle with vertices at the points $(2,2)$, $(6,3)$, and $(4,11)$ is a right triangle. Then determine the area of the triangle.

67. Prove that the triangle with vertices at the points $(2,-1)$, $(4,3)$, and $(-2,-3)$ is isosceles. Then determine the area of the triangle. (**Hint:** Make use of the midpoint formula.)

68. Prove that the polygon with vertices at the points $(-2,-1)$, $(6,5)$, $(-2,5)$, and $(6,-1)$ is a rectangle. Then determine the area of the rectangle. (**Hint:** It may help to plot the points before you begin.)

69. Plot the points $(-3,3)$, $(-5,-2)$, $(3,-2)$ and $(1,3)$ to demonstrate they are the vertices of a trapezoid. Then determine the area of the trapezoid.

70. Two college friends are taking a weekend road trip. Friday they leave home and drive 87 miles north for a night of dinner and dancing in the city. The next morning they drive 116 miles east to spend a day at the beach. If they drive straight home from the beach the next day, how far do they have to travel on Sunday?

71. Your backpacker's guide contains a grid map of Paris, with each unit on the grid representing 0.25 kilometers. If the Eiffel Tower is located at $(-8,1)$ and the Arc de Triomphe is located at $(-8,4)$, what is the direct distance (not walking distance, which would have to account for bridges and roadways) between the two monuments in kilometers?

41. $\sqrt{58}$, $\left(\dfrac{3}{2}, \dfrac{7}{2}\right)$

42. 13, $(2, -1)$

43. $2\sqrt{2}$, $(-1, -1)$

44. $8\sqrt{2}$, $(1, 2)$

45. $4\sqrt{34}$, $(3, -8)$

46. $2\sqrt{41}$, $(-3, 7)$

47. 10, $(1, -6)$

48. $\sqrt{197}$, $\left(0, -\dfrac{13}{2}\right)$

49. $3\sqrt{13}$, $\left(2, \dfrac{1}{2}\right)$

50. $\sqrt{173}$, $\left(3, -\dfrac{1}{2}\right)$

51. $10\sqrt{2}$, $(3, 3)$

52. 10, $\left(6, \dfrac{6}{5}\right)$

53. $x = 2$ or 18

54. $y = -12$ or 6

55. $x = 10$, $y = 1$

56. $2 + 3\sqrt{2} + \sqrt{34}$

57. 12

58. $18 + \sqrt{226}$

59. $2\sqrt{29} + \sqrt{26} + 5\sqrt{2}$

60. 48

61. 54

62. Area $= \dfrac{15}{2}$

63. Area $= \dfrac{25}{2}$

64. Area $= 25$

65. Area $= \dfrac{5}{2}$

66. Area $= 17$

67. Area $= 6$

68. Area $= 48$

69. Area $= 30$

70. 145 miles

71. 1.25 kilometers

72. $(6, -3)$, 3.54 kilometers

72. Your hotel, located at $(-1, -2)$ on the map from Exercise 70, is advertised as exactly halfway between the Eiffel Tower and Notre Dame. What are the grid coordinates of Notre Dame on your map? Find the direct distance from the Eiffel Tower to Notre Dame, rounded to the nearest hundredth of a kilometer.

73. The navigator of a submarine plots the position of the submarine and surrounding objects using a rectangular coordinate system, where each block is one square meter.

   a. If his submarine is located at $(50, 231)$ and the mobile base to which he is heading is located at $(83, 478)$, how far is he from the mobile base?

   b. Suppose there is another submarine located halfway between the first submarine and the mobile base. What is the position of the second sub?

74. At the entrance to Paradise Island Theme Park you are given a map of the park that is in the form of a grid, with the park entrance located at $(-5, -5)$. After walking past three rides and the restrooms, you arrive at the Tsunami Water Ride, which is located at $(-3, -1)$ on the grid. If you have traveled halfway along a straight line to your favorite ride, Thundering Tower, where on the grid is your favorite ride located? How far is Thundering Tower from the park entrance on the map?

73. a. 249.19 meters

   b. $\left(\dfrac{133}{2}, \dfrac{709}{2}\right)$

74. $(-1, 3)$ and $4\sqrt{5}$

Linear Equations in Two Variables

TOPICS

1. Recognizing linear equations in two variables

2. Intercepts of the coordinate axes

3. Horizontal and vertical lines

T. Finding intercepts

TOPIC 1

Recognizing Linear Equations in Two Variables

Example 2a in Section 2.1 was our first encounter with a linear equation in two variables. Our first goal in this section is to recognize when an equation in two variables is linear; we want to know when the solution set of an equation is a straight line in the Cartesian plane.

DEFINITION

Linear Equations in Two Variables

A **linear equation in two variables**, say the variables x and y, is an equation that can be written in the form $ax + by = c$, where $a, b,$ and c are constants and a and b are not both zero. This form of such an equation is called the **standard form**.

Of course, an equation may be linear but not appear in standard form. Some algebraic manipulation is often necessary in order to determine if a given equation is linear. We will see in the next section that there are other forms of linear equations that are useful in different situations. For now, we will focus on the standard form when identifying linear equations.

EXAMPLE 1

Linear Equations in Two Variables

Determine if the following equations are linear equations.

a. $3x - (2 - 4y) = x - y + 1$

b. $3x + 2(x + 7) - 2y = 5x$

c. $\dfrac{x+2}{3} - y = \dfrac{y}{5}$

d. $7x - (4x - 2) + y = y + 3(x - 1)$

e. $4x^3 - 2y = 5x$

f. $x^2 - (x - 3)^2 = 3y$

Solutions:

a. $3x - (2 - 4y) = x - y + 1$

$3x - 2 + 4y = x - y + 1$ First, apply the distributive property.

$3x - x + 4y + y = 1 + 2$ Arrange the variables on one side.

$2x + 5y = 3$ Combine like terms. The equation is linear.

b. $3x + 2(x + 7) - 2y = 5x$ Begin, again, with the distributive property.

$3x + 2x + 14 - 2y = 5x$

$5x - 5x - 2y = -14$ Move the variables to one side.

$-2y = -14$ Combine like terms. The x variable disappears, indicating a coefficient of 0, but the coefficient on y is nonzero, so the equation is still linear.

$y = 7$

c. $\dfrac{x+2}{3} - y = \dfrac{y}{5}$ For this equation, we need to separate the fraction into a variable part and a constant part.

$\dfrac{1}{3}x + \dfrac{2}{3} - y = \dfrac{1}{5}y$

$\dfrac{1}{3}x - y - \dfrac{1}{5}y = -\dfrac{2}{3}$ Once again, we move all the variables to one side, then combine like terms.

$\dfrac{1}{3}x - \dfrac{6}{5}y = -\dfrac{2}{3}$ The equation is linear. Note that we could also have begun by clearing the fractions.

d. $7x - (4x - 2) + y = y + 3(x - 1)$ After simplifying this equation, we see that the coefficient on both x and y is 0. Thus, the equation is not linear.

$7x - 4x + 2 + y = y + 3x - 3$

$3x - 3x + y - y = -3 - 2$

$0 = -5$ Further, the equation simplifies to a false statement, so it actually has no solutions!

e. $4x^3 - 2y = 5x$ The presence of the cubed term in this already simplified equation makes it clearly not linear.

f. $x^2 - (x - 3)^2 = 3y$ First, expand the squared binomial term.

$x^2 - x^2 + 6x - 9 = 3y$ In contrast to the last equation, when we simplify this equation the result clearly is linear.

$6x - 3y = 9$

TOPIC 2 ## Intercepts of the Coordinate Axes

Often, the goal in working with a given linear equation is to graph its solution set. Since two points determine a line, all we need to do to graph the solution set of a linear equation is to find two different solutions.

If the equation under consideration is in the two variables x and y, it is natural to call the point where the graph crosses the x-axis the **x-intercept** and the point where it crosses the y-axis the **y-intercept**. If the line does indeed cross both axes, the two intercepts are easy to find: the y-coordinate of the x-intercept is 0, and the x-coordinate of the y-intercept is 0.

DEFINITION

The x- and y-Intercepts

Given a graph in the Cartesian plane, any point where the graph intersects the x-axis is called an **x-intercept**, and any point where the graph intersects the y-axis is called a **y-intercept**.

All x-intercepts are of the form $(c, 0)$ and all y-intercepts are of the form $(0, c)$.

EXAMPLE 2

Finding Intercepts and Graphing Linear Equations

Find the x- and y-intercepts of the following equations, and then graph each equation.

a. $3x - 4y = 12$

b. $4x - (3 - x) + 2y = 7$

Solutions:

a.
$$3x - 4y = 12$$

$$3(0) - 4y = 12 \qquad 3x - 4(0) = 12$$

$$y = -3 \qquad\qquad x = 4$$

y-intercept: $(0, -3)$ x-intercept: $(4, 0)$

To find the two intercepts, first set x equal to 0 and solve for y, then set y equal to 0 and solve for x.

This gives us the coordinates of the two intercepts, which we plot below.

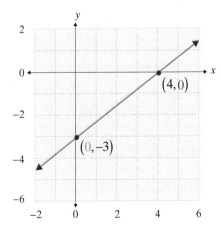

Once we have plotted the intercepts, drawing a straight line through them gives us the graph of the equation.

b.
$$4x - (3 - x) + 2y = 7$$
$$5x + 2y = 10$$

$$5(0) + 2y = 10 \qquad 5x + 2(0) = 10$$
$$y = 5 \qquad\qquad x = 2$$
y-intercept: $(0, 5)$ x-intercept: $(2, 0)$

Again, find the two intercepts by setting the appropriate variables equal to 0.

Solving the resulting equations in one variable yields the intercept solutions.

Plot the two intercepts, then draw the line passing through these two points. Note that in both graphs, the location of the origin has been chosen in order to conveniently plot the intercepts.

TOPIC 3

Horizontal and Vertical Lines

If a linear equation doesn't have two intercepts, the graphing process in Example 2 does not work. When might this happen? One case is when the line passes through the origin, $(0,0)$. Then, the x-intercept and y-intercept are the same point, instead of two distinct points. The other possibility is that the graph may be parallel to an axis (and thus not have one intercept); this happens when the equation is a horizontal or vertical line. In order to graph these equations, a second point (not an intercept) must be found in order to have two points to connect with a line.

Equations of horizontal or vertical lines are missing one of the two variables. In the absence of any other information, it is impossible to know if the solutions of equations like $x = 4$ or $y = -3$ consist of a point on the real number line or a line in the Cartesian plane. You must rely on the context of the problem to know how many variables should be considered. Throughout this chapter, all equations are assumed to be in two variables unless otherwise stated, so an equation of the form $ax = d$ or $by = d$ should be thought of as representing a line in the plane.

Consider an equation of the form $ax = d$. The variable y is absent, so *any* value for y will give a solution as long as we pair it with $x = \dfrac{d}{a}$. Thinking of the solution set as a set of ordered pairs, the solution consists of ordered pairs with a fixed first coordinate and arbitrary second coordinate. This describes, geometrically, a vertical line with an x-intercept of $\left(\dfrac{d}{a}, 0\right)$. Similarly, the equation $by = d$ represents a horizontal line with y-intercept equal to $\left(0, \dfrac{d}{b}\right)$.

EXAMPLE 3

Graphing Horizontal and Vertical Lines

Graph the following equations.

a. $5x = 0$

b. $2x - 2 = 3$

c. $3x + 2(x + 7) - 2y = 5x$

Solutions:

a. $5x = 0$

$\qquad x = 0$

The first step is to divide both sides by 5, leaving the simple equation $x = 0$.

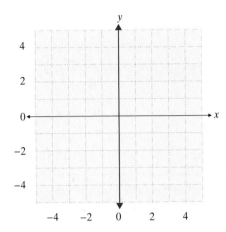

The graph of this equation is the *y*-axis, as all ordered pairs on the *y*-axis have an *x*-coordinate of 0.

This equation is unique in that it has an infinite number of *y*-intercepts (since each point on the graph is on the *y*-axis) and one *x*-intercept (the origin).

Similarly, the equation $y = 0$ has an infinite number of *x*-intercepts and one *y*-intercept.

b. $2x - 2 = 3$

$\qquad x = \dfrac{5}{2}$

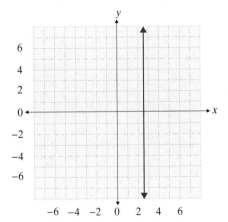

Upon simplifying, it is apparent that this equation also represents a vertical line, this time passing through $\dfrac{5}{2}$ on the *x*-axis.

c. $3x + 2(x + 7) - 2y = 5x$

$$y = 7$$

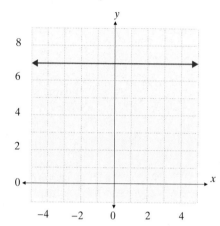

We encountered this equation in Example 1b, and have already written it in standard form as shown.

The graph of this equation is the horizontal line consisting of all those ordered pairs whose y-coordinate is 7.

TOPIC

Finding Intercepts

Since all linear equations can be solved for the variable y, we can use the calculator to graph them. Doing so enables us to use the calculator to find the x- and y-intercepts. Suppose we've graphed the equation $y = 2x - 6$.

To find the y-intercept, press TRACE . Since the y-intercept occurs where $x = 0$, use the arrows to move the cursor along the line until the x-value is zero. Alternatively, just press 0 and ENTER to place the cursor at that point. The corresponding y-value, -6, is shown at the bottom of the screen. So the point $(0, -6)$ is the y-intercept.

We will find the x-intercept using a different method. Press 2ND TRACE to access the CALC menu and select 2: zero and press ENTER. The screen should now display the graph with the words "Left Bound?" shown at the bottom. Use the arrows to move the cursor anywhere to the left of where the line crosses the x-axis and press ENTER. The screen should now say "Right Bound?" Use the right arrow to move the cursor to the right of where the line crosses the x-axis and press ENTER again. The text should now read "Guess?" Press ENTER a third time and the x- and y-values of the x-intercept will appear at the bottom of the screen.

1. Yes

2. Yes

3. No

4. No

5. No

6. Yes

7. No

8. Yes

9. Yes

10. Yes

11. Yes

12. No

13. Yes

14. Yes

15. No

16. Yes

17. No

18. Yes

19. No

20. Yes

21. No

22. No

23. Yes

24. Yes

So the x-intercept is $(3,0)$. Both of these techniques can be used to find the x- and y-intercepts of any equation graphed with a calculator, not just linear equations.

25.

26.

27.

28.

29.

30.

Exercises

Determine if the following equations are linear. See Example 1.

1. $3x + 2(x - 4y) = 2x - y$

2. $9x + 4(y - x) = 3$

3. $9x^2 - (x+1)^2 = y - 3$

4. $3x + xy = 2y$

5. $8 - 4xy = x - 2y$

6. $\dfrac{x-y}{2} + \dfrac{7y}{3} = 5$

7. $\dfrac{6}{x} - \dfrac{5}{y} = 2$

8. $3x - 3(x - 2y) = y + 1$

9. $2y - (x + y) = y + 1$

10. $(3 - y)^2 - y^2 = x + 2$

11. $x^2 - (x-1)^2 = y$

12. $(x+y)^2 - (x-y)^2 = 1$

13. $x(y+1) = 16 - y(1-x)$

14. $\dfrac{x-3}{2} = \dfrac{4+y}{5}$

15. $x - 2x^2 + 3 = \dfrac{x-7}{2}$

16. $x - 3 = \dfrac{4x+17}{5}$

17. $13x - 17y = y(7 - 2x)$

18. $y^2 - 3y = (1+y)^2 - 2x$

19. $x - 1 = \dfrac{2y}{x} - x$

20. $3x - 4 = 89(x - y) - y$

21. $x - x(1+x) = y - 3x$

22. $x^2 - 2x = 3 - x^2 + y$

23. $\dfrac{2y-5}{14} = \dfrac{x-3}{9}$

24. $16x = y(4 + (x-3)) - xy$

31.

32.

33.

34.

35.

36.

37.

38.

39.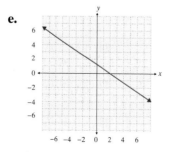

Determine the *x*- and *y*-intercepts of the following linear equations, if possible, and then graph the equations. See Examples 2 and 3.

25. $4x - 3y = 12$

26. $y - 3x = 9$

27. $5 - y = 10x$

28. $y - 2x = y - 4$

29. $3y = 9$

30. $2x - (x + y) = x + 1$

31. $x + 2y = 7$

32. $y - x = x - y$

33. $y = -x$

34. $2x - 3 = 1 - 4y$

35. $3y + 7x = 7(3 + x)$

36. $4 - 2y = -2 - 6x$

37. $x + y = 1 + 2y$

38. $3y + x = 2x + 3y + 4$

39. $3(x + y) + 1 = x - 5$

Match each equation to the correct graph.

40. $y = 2x + 3$

41. $2x + 3y = 4$

42. $2x - 1 = 5$

43. $y + 3 - x = 3$

44. $4y + 3 = 11$

45. $5y - x - 1 = 4y + 3x + 5$

a.

b.

c.

d.

e.

f.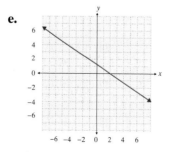

40. b

41. e

42. a

43. c

44. d

45. f

46. $y = \dfrac{c - ax}{b}$

47. $a = P - b - c$

48. $w = \dfrac{S - 2lh}{2(l + h)}$

49. $j = 24,000 + 9b;$

$b = \dfrac{j - 24,000}{9}$

Yes

50. $m = 26d + 1100;$

$d = \dfrac{m - 1100}{26}$

Yes

Solve each equation for the specified variable.

46. Standard Form of a Line: $ax + by = c.$ Solve for y.

47. Perimeter of a Triangle: $P = a + b + c.$ Solve for a.

48. Surface Area of a Rectangular Solid: $S = 2lw + 2wh + 2lh.$ Solve for w.

Solve the following application problems.

49. In your history class, you were told that the current population of Jamaica is approximately 24,000 more than 9 times the population of the Bahamas. Using j to represent the population of Jamaica and b to represent the population of the Bahamas, write this in the form of an equation. Then solve your equation for b to find an equation representing the population of the Bahamas. Are these equations linear?

50. The lowest point in the ocean, the bottom of the Mariana Trench, is about 1100 feet deeper than 26 times the depth of the lowest point on land, the Dead Sea. Find an equation to express the depth of the Mariana Trench, m, in terms of the depth of the Dead Sea, d. Then solve your equation for d to find the depth of the Dead Sea in terms of the depth of the Mariana Trench. Are these equations linear?

Mariana Trench   Dead Sea

Forms of Linear Equations

TOPICS

1. The slope of a line

2. Slope-intercept form of a line

3. Point-slope form of a line

TOPIC

The Slope of a Line

There are several ways to characterize a given line in the plane. We have already used one way repeatedly: two distinct points in the Cartesian plane determine a line. Another, often more useful, approach is to identify just one point on the line and to indicate how "steeply" the line is rising or falling as we scan the plane from left to right. It turns out that a single number is sufficient to convey this notion of "steepness."

DEFINITION

The Slope of a Line

Let L stand for a given line in the Cartesian plane, and let (x_1, y_1) and (x_2, y_2) be the coordinates of any two distinct points on L. The **slope** of the line L is the ratio $\dfrac{y_2 - y_1}{x_2 - x_1}$ which can be described in words as "change in y over change in x" or "rise over run."

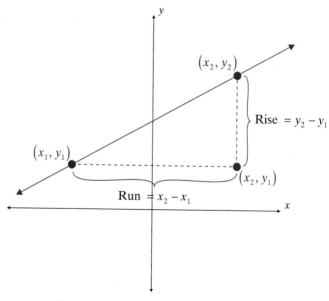

Figure 1: Rise and Run Between Two Points

In the line drawn in Figure 1, the ratio $\dfrac{y_2 - y_1}{x_2 - x_1}$ is positive, the line rises from the lower left to the upper right, and we say that the line has a positive slope. If the rise and run have opposite signs, the slope of the line is negative and the line under consideration would fall from the upper left to the lower right.

CAUTION!

It doesn't matter how you assign the labels (x_1, y_1) and (x_2, y_2) to the two points you are using to calculate slope, but it *is* important that you are consistent as you apply the formula. You cannot change the order in which you are subtracting as you determine the numerator and denominator in the slope formula.

Correct: $\dfrac{y_2 - y_1}{x_2 - x_1}$ or $\dfrac{y_1 - y_2}{x_1 - x_2}$ **Incorrect:** $\dfrac{y_1 - y_2}{x_2 - x_1}$ or $\dfrac{y_2 - y_1}{x_1 - x_2}$

EXAMPLE 1

Calculating the Slope of a Line

Determine the slopes of the lines passing through the following pairs of points in $\mathbb{R}^2$.

a. $(-4, -3)$ and $(2, -5)$ **b.** $\left(\dfrac{3}{2}, 1\right)$ and $\left(1, -\dfrac{4}{3}\right)$ **c.** $(-2, 7)$ and $(1, 7)$

Solutions:

a. $\dfrac{-3 - (-5)}{-4 - 2} = \dfrac{2}{-6} = -\dfrac{1}{3}$

We calculate the slope in two ways; first set $(x_1, y_1) = (-4, -3)$ and $(x_2, y_2) = (2, -5)$.

$\dfrac{-5 - (-3)}{2 - (-4)} = \dfrac{-2}{6} = -\dfrac{1}{3}$

We get the same result by setting $(x_1, y_1) = (2, -5)$ and $(x_2, y_2) = (-4, -3)$.

b. $\dfrac{1 - \left(-\dfrac{4}{3}\right)}{\dfrac{3}{2} - 1} = \dfrac{\dfrac{7}{3}}{\dfrac{1}{2}} = \dfrac{7}{3} \cdot \dfrac{2}{1} = \dfrac{14}{3}$

The final answer tells us that the line through the two points rises 14 units for every run of 3 units horizontally.

c. $\dfrac{7 - 7}{-2 - 1} = \dfrac{0}{-3} = 0$

These two points have the same y-coordinate and thus lie on a horizontal line.

The formula $\dfrac{y_2 - y_1}{x_2 - x_1}$ is only valid if $x_2 - x_1 \neq 0$, since division by zero is undefined. What lines have undefined slope? If $x_2 - x_1 = 0$, then the line has two points with the same x-coordinate, which defines a *vertical* line. The other extreme is that the numerator is 0 (and the denominator is nonzero), in which case the slope is 0. For what sorts of lines will this happen? If two points on a line have the same y-coordinate, that is if $y_1 = y_2$, the line must be *horizontal*.

PROPERTIES

Horizontal lines, which can be written in the form $y = c$, have a **slope of 0.**

Vertical lines, which can be written in the form $x = c$, have an **undefined slope.**

EXAMPLE 2

Calculating the Slope
of a Line

Determine the slopes of the lines defined by the following equations.

a. $4x - 3y = 12$

b. $2x + 7y = 9$

c. $x = -\dfrac{3}{4}$

d. $y = 9$

Note:
Intercepts are often good points to use in calculating the slope, since they have at least one coordinate equal to zero.

Solutions:

a. First, we find two points on the line by calculating the intercepts.

$$4x - 3y = 12$$

$$4(0) - 3y = 12 \quad \text{and} \quad 4x - 3(0) = 12$$

$$-3y = 12 \quad \text{and} \quad 4x = 12$$

$$y = -4 \quad \text{and} \quad x = 3$$

Recall that the x-intercept is found by setting y equal to 0 and solving for x, and vice versa for the y-intercept.

y-intercept: $(0, -4)$ x-intercept: $(3, 0)$

$$\text{slope} = \frac{-4 - 0}{0 - 3} = \frac{-4}{-3} = \frac{4}{3}$$

Once we have two points, we apply the slope formula.

b. x-intercept: $\left(\dfrac{9}{2}, 0\right)$

second point on the line: $(1, 1)$

$$\text{slope} = \frac{1 - 0}{1 - \dfrac{9}{2}} = \frac{1}{-\dfrac{7}{2}} = -\frac{2}{7}$$

In this example, we have found the x-intercept. We do not have to find both intercepts; the point $(1,1)$ is clearly on the line and is simple to use in calculation.

c. The equation is of the form $x = c$, and is a vertical line. Therefore, the slope is undefined.

d. This equation is of the form $y = c$, and is a horizontal line. Therefore, it has a slope of 0.

Note that the line in Example 2a has a positive slope and the line in Example 2b has a negative slope. Without graphing these lines, we know that the first line will rise from the lower left to the upper right part of the plane, while the second line will fall from the upper left to the lower right. You should practice your graphing skills and verify that these observations are indeed correct.

TOPIC [2]

Slope-Intercept Form of a Line

Example 2 illustrates the most elementary way of determining the slope of a line from an equation. With a little work, we can develop a faster method for determining not only the slope of a line, but also the y-intercept.

Consider a nonvertical line in the plane. The variable y must appear in the linear equation that describes the line (otherwise the line would be vertical), so the equation can be solved for y. The result will be an equation of the form $y = mx + b$, where m and b are constants, and it turns out that these constants provide a lot of information about the graph of the line.

Suppose that (x_1, y_1) and (x_2, y_2) are two points that lie on the line $y = mx + b$. Then, it must be the case that $y_1 = mx_1 + b$ and $y_2 = mx_2 + b$. If we use these two points to determine the slope of the line, we obtain:

$$\text{slope} = \frac{y_2 - y_1}{x_2 - x_1} = \frac{(mx_2 + b) - (mx_1 + b)}{x_2 - x_1} = \frac{m(x_2 - x_1)}{x_2 - x_1} = m$$

Now, let's calculate the y-intercept of this line. As usual, we substitute 0 for x and then solve for y:

$$y = mx + b$$
$$y = m(0) + b$$
$$y = b$$

So, the y-intercept is $(0, b)$. Thus, the two constants m and b describe the slope and y-intercept of the line. As such, we call this the *slope-intercept* form of a linear equation.

DEFINITION

Slope-Intercept Form of a Line

If the equation of a nonvertical line in x and y is solved for y, the result is an equation in **slope-intercept form**:

$$y = mx + b.$$

The constant m is the slope of the line, and the y-intercept of the line is $(0, b)$. If the variable x does not appear in the equation, the slope is 0 and the equation is simply of the form $y = b$ and is a horizontal line.

We can make use of the slope-intercept form of a line to graph the line, as illustrated in the following example.

EXAMPLE 3

Slope-Intercept Form of a Line

Use the slope-intercept form of the line to graph the equation $4x - 3y = 6$.

Solution:

$$4x - 3y = 6$$
$$-3y = -4x + 6$$
$$y = \frac{4}{3}x - 2$$

Solving the equation for y puts it in slope-intercept form. Once we have done this, we know that the line has a slope of $\frac{4}{3}$ and crosses the y-axis at -2.

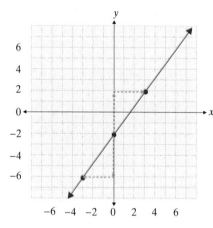

Immediately, we can plot the y-intercept.

A second point can now be found by using the fact that slope is "rise over run." This means a second point must lie 4 units up and 3 units to the right, i.e. at $(3, 2)$.

Alternatively, we could locate a second point by moving down 4 units and moving to the left 3 units.

In some cases, we can also make use of the slope-intercept form to find the equation of a line that has certain properties.

EXAMPLE 4

Slope-Intercept Form of a Line

Find the equation of the line that passes through the point $(0, 3)$ and has a slope of $-\frac{3}{5}$. Then graph the line.

Note:
If a line is already in slope-intercept form, it's usually easier to graph the line by plotting the y-intercept and then using the slope to find a second point.

Solution:

First, we write the equation of this line in slope-intercept form. We are given the y-intercept of $(0, 3)$ and the slope of $-\frac{3}{5}$.

$$y = mx + b$$

$$y = -\frac{3}{5}x + 3$$

We can immediately write down the equation since the only information we need is the y-intercept and the slope.

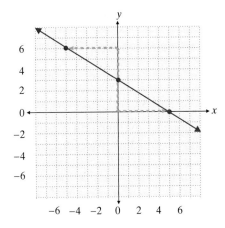

We first plot the y-intercept, then move down 3 units and to the right 5 units to find a second point.

Or, we could have plotted a second point by moving up 3 units and to the left 5 units. Both methods make use of the fact that the slope is $-\dfrac{3}{5}$.

TOPIC 3 Point-Slope Form of a Line

As in Example 4, we can easily find the slope-intercept form of a line given its slope and y-intercept. In the most general case, we would like to be able to construct an equation of a line given the slope of the line and *any* point on the line (not just the y-intercept). This would allow us to find the equation of a line given only two points on that line (since we can determine the slope from two points). The *point-slope* form of a line meets these requirements.

Suppose we know that a given line has slope m and passes through the point (x_1, y_1). If we plug any two points on the line into the slope formula, we must get a result of m. Choosing (x_1, y_1) and the generic point (x, y), the definition of slope states that:

$$\frac{y - y_1}{x - x_1} = m, \text{ so that}$$

$$y - y_1 = m(x - x_1).$$

A simple rearrangement of the slope equation leads to a linear equation defined by the slope and the coordinates of a single, arbitrary point. This is the point-slope form.

DEFINITION

Point-Slope Form of a Line

The **point-slope form** of the equation for the line passing through the point (x_1, y_1) with slope m is

$$y - y_1 = m(x - x_1).$$

Note that m, x_1, and y_1 are all constants, while x and y are variables. Note also that since the line, by definition, has slope m, vertical lines cannot be described in this form.

EXAMPLE 5

Point-Slope Form of
a Line

Find the equation, in slope-intercept form, of the line that passes through the point $(-2, 5)$ with slope 3.

Note:
When asked for an equation in either standard form or slope-intercept form, it's frequently easiest to write the point-slope form and then convert the equation to the desired form.

Solution:

Since we are given the slope of the line and a point on the line, we can substitute directly into the point-slope form, then solve for y to obtain the slope-intercept form.

$$y - y_1 = m(x - x_1)$$
$$y - 5 = 3(x - (-2)) \qquad \text{The point-slope form.}$$
$$y - 5 = 3(x + 2)$$
$$y - 5 = 3x + 6$$
$$y = 3x + 11 \qquad \text{The slope-intercept form.}$$

We know that two distinct points in the plane are sufficient to determine a line. With our knowledge of the point-slope form of a line, we can now easily deduce the equation for the line determined by two points.

EXAMPLE 6

Point-Slope Form of
a Line

Find the equation, in slope-intercept form, of the line that passes through the two points $(-3, -2)$ and $(1, 6)$.

Solution:

We already have a point (actually, two) on the line, but we still need the slope to use the point-slope form. We can calculate this using the two points and the slope formula:

$$m = \frac{-2 - 6}{-3 - 1} = \frac{-8}{-4} = 2$$

Now we can substitute into the point-slope form, then solve for y to obtain the desired slope-intercept equation.

$$y - y_1 = m(x - x_1)$$
$$y - 6 = 2(x - 1) \qquad \text{Note that no matter which point we substitute into}$$
$$y - 6 = 2x - 2 \qquad \text{the point-slope form, the resulting slope-intercept}$$
$$y = 2x + 4 \qquad \text{equation is the same.}$$

We close this section with a summary of the different forms of linear equations, what information we need to write them, and what they are each most useful for.

Standard Form: $ax + by = c$

Information Required: Typically, we arrive at the standard form when given a linear equation in another form.

Potential Uses: The standard form is most useful for easily calculating the x- and y-intercepts.

Slope-Intercept Form: $y = mx + b$

Information Required: The slope m and the y-intercept $(0,b)$.

Potential Uses: The slope-intercept form makes it very easy to find the y-intercept and slope, and therefore to graph the line.

Point-Slope Form: $y - y_1 = m(x - x_1)$

Information Required: The slope m and a point on the line (x_1, y_1) or two points on the line (x_1, y_1) and (x_2, y_2).

Potential Uses: The point-slope form allows us to find the equation for a line when the y-intercept is unknown.

1. -4

2. $-\dfrac{6}{5}$

3. 0

4. 0

5. Undefined

6. $-\dfrac{5}{2}$

7. $\dfrac{2}{3}$

8. 8

9. $\dfrac{1}{6}$

10. $\dfrac{5}{8}$

11. -7

12. 13

13. -3

14. -6

15. $-\dfrac{9}{13}$

16. 4

17. $-\dfrac{1}{4}$

18. 3

19. 0

================= **Exercises** =================

Determine the slopes of the lines passing through the specified points. See Example 1.

1. $(0, -3)$ and $(-2, 5)$ **2.** $(-3, 2)$ and $(7, -10)$ **3.** $(4, 5)$ and $(-1, 5)$

4. $(3, -1)$ and $(-7, -1)$ **5.** $(3, -5)$ and $(3, 2)$ **6.** $(0, 0)$ and $(-2, 5)$

7. $(-2, 1)$ and $(-5, -1)$ **8.** $\left(\dfrac{1}{2}, -7\right)$ and $\left(\dfrac{3}{4}, -5\right)$ **9.** $\left(10, \dfrac{1}{5}\right)$ and $\left(4, -\dfrac{4}{5}\right)$

10. $(-2, 4)$ and $(6, 9)$ **11.** $(0, -21)$ and $(-3, 0)$ **12.** $(-3, -5)$ and $(-2, 8)$

13. $\left(\dfrac{1}{3}, 9\right)$ and $(2, 4)$ **14.** $(29, -17)$ and $(31, -29)$ **15.** $(7, 4)$ and $(-6, 13)$

Determine the slopes of the lines defined by the following equations. See Example 2.

16. $8x - 2y = 11$ **17.** $2x + 8y = 11$

18. $12x - 4y = -9$ **19.** $4y = 13$

20. 1

21. Undefined

22. $\dfrac{1}{15}$

23. 2

24. $-\dfrac{5}{6}$

25. $\dfrac{7}{6}$

26. -6

27. $-\dfrac{5}{2}$

28.

29.

30.

31.

32.

33.

20. $\dfrac{x-y}{3}+2=4$

21. $7x=2$

22. $3y-2=\dfrac{x}{5}$

23. $3-y=2(5-x)$

24. $3(2y-1)=5(2-x)$

25. $\dfrac{x+2}{3}+2(1-y)=-2x$

26. $2y-7x=4y+5x$

27. $x-7=\dfrac{2y-1}{-5}$

Use the slope-intercept form of each line to graph the equations. See Example 3.

28. $6x-2y=4$

29. $3y+2x-9=0$

30. $5y-15=0$

31. $x+4y=20$

32. $\dfrac{x-y}{2}=-1$

33. $3x+7y=8y-x$

34. $-4x-4y=8$

35. $-5x+3y+16=0$

36. $3x=3y-21$

Find the equation, in slope-intercept form, of the line with the given y-intercept and slope. See Example 4.

37. point $(0,-3)$; slope of $\dfrac{3}{4}$

38. point $(0,5)$; slope of -3

39. point $(0,-7)$; slope of $-\dfrac{5}{2}$

40. point $(0,6)$; slope of 4

41. point $(0,-9)$; slope of -5

42. point $(0,2)$; slope of $\dfrac{1}{2}$

Find the equation, in standard form, of the line passing through the given point with the given slope.

43. point $(-1,-3)$; slope of $\dfrac{3}{2}$

44. point $(6,0)$; slope of $\dfrac{5}{4}$

45. point $(-3,5)$; slope of 0

46. point $(-2,-13)$; undefined slope

47. point $(3,-1)$; slope of 10

48. point $(-1,3)$; slope of $-\dfrac{2}{7}$

49. point $(5,11)$; slope of -3

50. point $(5,-9)$; slope of $-\dfrac{1}{2}$

34.

35.

36.
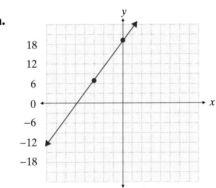

37. $y = \dfrac{3}{4}x - 3$

38. $y = -3x + 5$

39. $y = -\dfrac{5}{2}x - 7$

40. $y = 4x + 6$

41. $y = -5x - 9$

42. $y = \dfrac{1}{2}x + 2$

43. $3x - 2y = 3$

44. $5x - 4y = 30$

45. $y = 5$

46. $x = -2$

47. $10x - y = 31$

48. $2x + 7y = 19$

49. $3x + y = 26$

50. $x + 2y = -13$

51. $4x + 3y = 5$

52. $y = 3$

53. $x = 2$

54. $3x - 10y = -47$

55. $y = -1$

Find the equation, in standard form, of the line passing through the specified points. See Example 6.

51. $(-1, 3)$ and $(2, -1)$

52. $(1, 3)$ and $(-2, 3)$

53. $(2, -2)$ and $(2, 17)$

54. $(-9, 2)$ and $(1, 5)$

55. $(3, -1)$ and $(8, -1)$

56. $\left(\dfrac{4}{3}, 1\right)$ and $\left(\dfrac{2}{5}, \dfrac{3}{7}\right)$

57. $(-2, 8)$ and $(5, 6)$

58. $(8, -10)$ and $(8, 0)$

59. $(7, 5)$ and $(-9, 5)$

60. $(7, 7)$ and $(9, -8)$

61. $\left(\dfrac{2}{3}, \dfrac{5}{4}\right)$ and $\left(\dfrac{3}{5}, \dfrac{9}{8}\right)$

62. $(-5, -5)$ and $(10, -11)$

Match each equation or description to the correct graph from the following options.

63. $-3x - 2y = 17$

64. $-4y + 10 = -4x$

65. $-6y + 9 = \dfrac{x}{-2}$

66. point $(-9, 7)$; slope $\dfrac{4}{3}$

67. point $(-2, 4)$; slope -2

68. point $(0, -5)$; slope -9

a.

b.

c.

d.
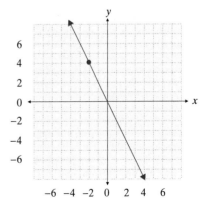

56. $30x - 49y = -9$

57. $2x + 7y = 52$

58. $x = 8$

59. $y = 5$

60. $15x + 2y = 119$

61. $15x - 8y = 0$

62. $2x + 5y = -35$

63. c

64. f

65. e

66. a

67. d

68. b

69. a. $2225

   b. $2100

   c. $0.25

70. $267,000

71. $325

72. $V = -2766.67t + 51,500$; 18.61 years

e.

f.

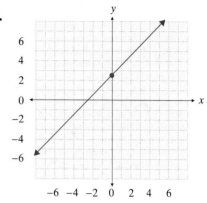

Solve the following application problems.

69. A bottle manufacturer has determined that the total cost (C) in dollars of producing x bottles is $C = 0.25x + 2100$.

   a. What is the cost of producing 500 bottles?

   b. What are the fixed costs (costs incurred even when 0 bottles are produced)?

   c. What is the increase in cost for each bottle produced?

70. Sales at Glover's Golf Emporium have been increasing linearly for the past couple of years. Last year, sales were $163,000. This year, sales were $215,000. If sales continue to increase at this linear rate, predict the sales for next year.

71. Amy owns stock in Trimetric Technologies. If the stock had a value of $2500 in 2003 when she purchased it, what has been the average change in value per year if in 2005 the stock was worth $3150?

72. For tax and accounting purposes, businesses often have to depreciate equipment values over time. One method of depreciation is the straight-line method. Three years ago Hilde Construction purchased a bulldozer for $51,500. Using the straight-line method, the bulldozer has now depreciated to a value of $43,200. If V equals the value at the end of year t, write a linear equation expressing the value of the bulldozer over time. How many years from the purchase date will the value equal 0? (Round your answer to the nearest hundredth.)

Parallel and Perpendicular Lines

TOPICS

1. Slopes of parallel lines

2. Slopes of perpendicular lines

TOPIC 1

Slopes of Parallel Lines

In this section, we will explore the relationship between slope and the geometric concepts of parallel and perpendicular lines. This will allow us to use algebra to construct lines parallel or perpendicular to a given line. We will begin with parallel lines; consider the following figure showing "rise" and "run" of two parallel lines:

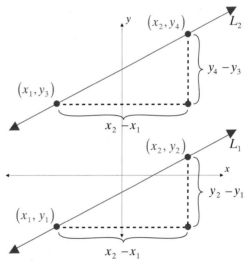

Figure 1: Slopes of Parallel Lines

In Figure 1, we can see that if we choose the right points on each line, the rise and run are equal. Further, no matter which points we choose, the ratio of rise to run (which we know as the slope) is equal for parallel lines. Thus, we have a very simple algebraic definition for parallel lines.

THEOREM

Slopes of Parallel Lines

Two nonvertical lines with slopes m_1 and m_2 are **parallel** if and only if $m_1 = m_2$. Also, two vertical lines (with undefined slopes) are always parallel to each other.

This fact gives us a straightforward way of finding the equations of lines parallel to a given line. First, we calculate the slope of the given line, and then construct new lines using that same slope. Usually, it will be easier to use the slope-intercept or point-slope form of a linear equation, since the slope appears directly.

EXAMPLE 1

Finding Equations of
Parallel Lines

Find equations for two lines parallel to each of the lines given below.

a. $y = -\dfrac{2}{3}x + 4$ **b.** $10x - 2y = 14$

Note:
When given a line in standard form, it is usually easier to find its slope by rewriting it in slope-intercept form than to find two points on the line and calculate the slope directly.

Solutions:

a. This line is already in slope-intercept form, so we immediately know the slope of the line is $-\dfrac{2}{3}$. Any line parallel to this one must also have a slope of $-\dfrac{2}{3}$. To find two parallel lines, we can simply change the value of the y-intercept.

$$y = -\dfrac{2}{3}x + 1 \text{ and } y = -\dfrac{2}{3}x - 10$$

b. This line is in standard form. Our first step is to rewrite it in slope-intercept form.

$$10x - 2y = 14$$
$$-2y = -10x + 14 \qquad \text{Subtract 10x from both sides.}$$
$$y = \dfrac{-10}{-2}x + \dfrac{14}{-2} \qquad \text{Divide each term by } -2.$$
$$y = 5x - 7 \qquad \text{Simplify.}$$

Again, once the line is in slope-intercept form, we can change the y-intercept to find two lines parallel to the original line.

$$y = 5x \text{ and } y = 5x + 8$$

EXAMPLE 2

Finding Equations of
Parallel Lines

Find the equation, in slope-intercept form, for the line which is parallel to the line $3x + 5y = 23$ and which passes through the point $(-2, 1)$.

Solution:

Again, our first step is to write the initial equation in slope-intercept form.

$$3x + 5y = 23$$
$$5y = -3x + 23$$
$$y = -\dfrac{3}{5}x + \dfrac{23}{5}$$

This tells us that the slope of the line whose equation we seek is $-\dfrac{3}{5}$. We also know that the line is to pass through $(-2, 1)$, so we can use the point-slope form to obtain the desired equation.

$$y - y_1 = m(x - x_1)$$

$$y - 1 = -\frac{3}{5}(x - (-2))$$

Begin by substituting our known information into the point-slope form: $m = -\frac{3}{5}$, $(x_1, y_1) = (-2, 1)$.

$$y - 1 = -\frac{3}{5}(x + 2)$$

$$y - 1 = -\frac{3}{5}x - \frac{6}{5}$$

$$y = -\frac{3}{5}x - \frac{1}{5}$$

The instructions asked for the equation in slope-intercept form, so we solve for y to obtain the final answer.

We can also use the knowledge that parallel lines have the same slope to answer questions that are more geometric in nature.

EXAMPLE 3

Identifying a Quadrilateral

Determine if the quadrilateral (four-sided figure) graphed below is a parallelogram (a quadrilateral in which both pairs of opposite sides are parallel).

Solution:

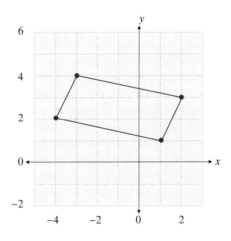

The four vertices are plotted in the picture to the left, and the sides of the quadrilateral drawn. The figure is a parallelogram if the left and right sides are parallel and the top and bottom sides are parallel. The slopes of the left and right sides are, respectively,

$$\frac{4-2}{-3-(-4)} = 2 \quad \text{and} \quad \frac{3-1}{2-1} = 2,$$

and the slopes of the top and bottom sides are, respectively,

$$\frac{4-3}{-3-2} = -\frac{1}{5} \quad \text{and} \quad \frac{2-1}{-4-1} = -\frac{1}{5}.$$

Thus the figure is indeed a parallelogram.

TOPIC 2   Slopes of Perpendicular Lines

The relationship between the slopes of perpendicular lines is a bit less obvious. Consider a nonvertical line L_1, and two points (x_1, y_1) and (x_2, y_2) on the line, as shown in Figure 2. These two points can be used, to calculate the slope m_1 of L_1, with the result that $m_1 = \dfrac{a}{b}$, where $a = y_2 - y_1$ and $b = x_2 - x_1$.

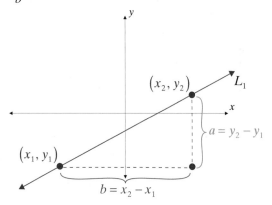

Figure 2: Definition of a and b

If we now draw a line L_2 perpendicular to L_1, we can use a and b to determine the slope m_2 of line L_2. There are an infinite number of lines that are perpendicular to L_1; one of them is drawn in Figure 3.

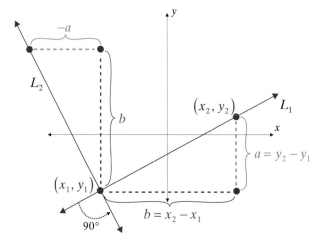

Figure 3: Perpendicular Lines

Note that in rotating the line L_1 by 90 degrees to obtain L_2, we have also rotated the right triangle drawn with dashed lines, so the sides of the triangle are the same length. But to travel along the line L_2 from the point (x_1, y_1) to the second point drawn requires a positive rise and a negative run, whereas the rise and run between (x_1, y_1) and (x_2, y_2) are both positive. In other words, $m_2 = -\dfrac{b}{a}$, the negative reciprocal of the slope m_1.

This relationship always exists between the slopes of two perpendicular lines, assuming neither one is vertical. Of course, if one line is vertical, any line perpendicular to it will be horizontal with a slope of zero, while if one line is horizontal, any line perpendicular to it will be vertical with undefined slope. This is summarized below.

THEOREM

Slopes of Perpendicular Lines

Suppose m_1 and m_2 represent the slopes of two lines, neither of which is vertical. The two lines are **perpendicular** if and only if $m_1 = -\dfrac{1}{m_2}$ (equivalently, $m_2 = -\dfrac{1}{m_1}$ and $m_1 m_2 = -1$). If one of two perpendicular lines is vertical, the other is horizontal, and the slopes are, respectively, undefined and zero.

The following examples illustrate how we can use the relationship between slopes of perpendicular lines to solve problems.

EXAMPLE 4

Finding Equations of Perpendicular Lines

For each line given below, find the equation of a perpendicular line.

a. $y = -\dfrac{4}{9}x + 2$ **b.** The line passing through the points $(-1, 3)$ and $(4, 1)$.

Solutions:

a. This line is in slope-intercept form, so we immediately identify the slope of $-\dfrac{4}{9}$. The slope of any perpendicular line must equal $\dfrac{9}{4}$, the negative reciprocal of the original slope. Thus, one solution is $y = \dfrac{9}{4}x$.

b. Since we only need the slope of the original line, there is no need to find its equation; we can calculate the slope directly from the given points.

$$m = \frac{1-3}{4-(-1)} = -\frac{2}{5}$$

Again, the slope of a line perpendicular to the line through the given points must have a slope equal to the negative reciprocal of $-\dfrac{2}{5}$, which is $\dfrac{5}{2}$. One perpendicular line is $y = \dfrac{5}{2}x + 6$.

EXAMPLE 5

Finding Equations of Perpendicular Lines

Find the equation, in standard form, of the line that passes through the point $(-3, 13)$ and that is perpendicular to the line $y = -7$.

Solution:

Note:
Remember that if you encounter a horizontal or vertical line, you cannot use the slope formulas to find a perpendicular line.

The line $y = -7$ is a horizontal line, and hence any line perpendicular to it must be a vertical line, having the form $x = c$. Since the perpendicular line must pass through the point $(-3, 13)$, the desired solution is $x = -3$.

Given a pair of lines, we can use their slopes to determine if they are parallel, perpendicular, or neither. Note that the only information we need is the slope! The equations do not have to be written in the same form, and we do not need to know anything about their intercepts. Find the most efficient way to calculate the slope of each line to avoid any unnecessary work.

EXAMPLE 6

Identifying Parallel and Perpendicular Lines

For each pair of lines, determine if the lines are parallel, perpendicular, or neither.

a. $3x - 7y = 12$ and $14x + 6y = -5$

b. $y - \dfrac{263}{4} = 9\left(x + \dfrac{77}{13}\right)$ and the line passing through the points $(0, 4)$ and $(2, 22)$

c. $y = \dfrac{3}{4}x + 1$ and $y = \dfrac{4}{3}x - 5$

Solutions:

Note:
A pair of lines cannot be *both* parallel and perpendicular.

a. Both equations are in standard form, so our first step is to rewrite them in slope-intercept form to identify the slopes.

$$3x - 7y = 12 \qquad\qquad 14x + 6y = -5$$
$$-7y = -3x + 12 \qquad\qquad 6y = -14x - 5$$
$$y = \frac{3}{7}x - \frac{12}{7} \qquad\qquad y = -\frac{7}{3}x - \frac{5}{6}$$

Are the lines parallel?   No, the slopes are not equal.

Are the lines perpendicular?   Yes, the slopes are negative reciprocals of each other.

Thus, the lines are perpendicular.

b. One line is in point-slope form, so we can see its slope is 9. We calculate the slope of the other line using the two points given.

$$m = \frac{22 - 4}{2 - 0} = \frac{18}{2} = 9$$

Are the lines parallel?   Yes, the slopes are equal.

Thus, the lines are parallel.   Note that we didn't need to find the
                                 equation of the second line.

c. Both lines are in slope-intercept form, so we can read off the slopes: $\dfrac{3}{4}$ and $\dfrac{4}{3}$.

Are the lines parallel?   No, the slopes are not equal.

Are the lines perpendicular?   No, the slopes are reciprocals, not *negative*
                                 reciprocals.

Thus, the lines are neither parallel nor perpendicular.

Exercises

Find the equation, in slope-intercept form, for the line parallel to the given line and passing through the indicated point. See Examples 1 and 2.

1. Parallel to $y - 4x = 7$ and passing through $(-1, 5)$.

2. Parallel to $6x + 2y = 19$ and passing through $(-6, -13)$.

3. Parallel to $3x + 2y = 3y - 7$ and passing through $(3, -2)$.

4. Parallel to $2 - \dfrac{y - 3x}{3} = 5$ and passing through $(0, -2)$.

5. Parallel to $y - 4x = 7 - 4x$ and passing through $(23, -9)$.

6. Parallel to $2(y - 1) + \dfrac{x + 3}{5} = -7$ and passing through $(-5, 0)$.

7. Parallel to $6y - 4 = -3(1 - 2x)$ and passing through $(-2, -2)$.

8. Parallel to $5 - \dfrac{7y + 5x}{2} = 1$ and passing through $(4, 1)$.

9. Parallel to $2(y - 1) - \dfrac{7x + 1}{3} = -3$ and passing through $(1, 10)$.

10. Parallel to $8y - 6 = -3(4 - x)$ and passing through $(11, -5)$.

Each set of four ordered pairs below defines the vertices, in counterclockwise order, of a quadrilateral. Determine if the quadrilateral is a parallelogram. See Example 3.

11. $\{(-2, 2), (-5, -2), (2, -3), (5, 1)\}$ **12.** $\{(-1, 6), (-4, 7), (-2, 3), (1, 1)\}$

13. $\{(-3, 3), (-2, -2), (3, -1), (2, 4)\}$ **14.** $\{(-2, -3), (-3, -6), (1, -2), (2, 1)\}$

1. $y = 4x + 9$

2. $y = -3x - 31$

3. $y = 3x - 11$

4. $y = 3x - 2$

5. $y = -9$

6. $y = -\dfrac{1}{10}x - \dfrac{1}{2}$

7. $y = x$

8. $y = -\dfrac{5}{7}x + \dfrac{27}{7}$

9. $y = \dfrac{7}{6}x + \dfrac{53}{6}$

10. $y = \dfrac{3}{8}x - \dfrac{73}{8}$

11. Yes

12. No

13. Yes

14. Yes

15. Yes

16. Yes

17. No

18. No

19. No

20. No

21. No

22. Yes

23. No

24. Yes

25. No

26. Yes

27. Yes

28. No

29. No

30. Yes

31. $y = -\dfrac{1}{3}x - 1$

32. $y = 3x$

33. $y = 7$

34. $y = x$

15. $\{(-6,-2), (-1,0), (-3,4), (-8,2)\}$

16. $\{(-3,-2), (3,-3), (5,2), (-1,3)\}$

17. $\{(-1,-1), (5,1), (3,5), (-2,3)\}$

18. $\{(0,1), (6,0), (7,4), (1,6)\}$

Determine if the two lines in each problem below are parallel. See Example 6.

19. $y = 8x + 7$ and $y = -8x + 7$

20. $x - 5y = 2$ and $5x - y = 2$

21. $2x - 3y = (x - 1) - (y - x)$ and $-2y - x = 9$

22. $3 - (2y + x) = 7(x - y)$ and $\dfrac{5y + 1}{4} = 3 + 2x$

23. $6 = -12(x - y) + y$ and $13y = -12x + 3$

24. $\dfrac{2x - 3y}{3} = \dfrac{x - 1}{6}$ and $2y - x = 3$

25. $\dfrac{x - y}{2} = \dfrac{x + y}{3}$ and $\dfrac{2x + 3}{5} - 4y = 1 + 2y$

26. $5 - (4y + 3x) = 5(x - y)$ and $y + 4 = 5 + 8x$

27. $7x - 2(x + 3) = 5y - x$ and $-6x = 1 - 5y$

28. $\dfrac{2y + 11x}{3} = x + 1$ and $7x - 8y = 9x + 7$

29. $\dfrac{x - y}{5} = \dfrac{x + y}{3} - 1$ and $7 = -2(x - y) + 6y$

30. $2x + 5y = 14$ and the line passing through the points $(8, -5)$ and $(3, -3)$

Find the equation, in slope-intercept form, for the line perpendicular to the given line and passing through the indicated point. See Examples 4 and 5.

31. Perpendicular to $3x + 2y = 3y - 7$ and passing through $(3, -2)$.

32. Perpendicular to $6y + 2x = 1$ and passing through $(-4, -12)$.

33. Perpendicular to $-y + 3x = 5 - y$ and passing through $(-2, 7)$.

34. Perpendicular to $x + y = 5$ and passing through the origin.

35. $y = -\dfrac{1}{4}x - \dfrac{3}{4}$

36. $y = -5x + 15$

37. $y = x + 3$

38. $y = \dfrac{x}{5} + \dfrac{17}{5}$

39. $y = -3x + 28$

40. $y = x - 10$

41. No

42. Yes

43. No

44. Yes

45. No

46. No

47. No

48. Yes

49. Yes

50. Yes

51. No

52. No

35. Perpendicular to $x = \dfrac{1}{4}y - 3$ and passing through $(1, -1)$.

36. Perpendicular to $2(y + x) - 3(x - y) = -9$ and passing through $(2, 5)$.

37. Perpendicular to $4x + 8y = 4y - 3$ and passing through $(-2, 1)$.

38. Perpendicular to $\dfrac{3x - y}{4} = \dfrac{4x - 5}{2}$ and passing through $(8, 5)$.

39. Perpendicular to $4(y + x) - 8(x - y) = -1$ and passing through $(6, 10)$.

40. Perpendicular to $\dfrac{3x + 4}{3} - 3y = 1 - 4y$ and passing through $(2, -8)$.

Determine if the two lines in each problem below are perpendicular. See Example 6.

41. $x - 5y = 2$ and $5x - y = 2$ **42.** $y = 5x + 4$ and $y = -\dfrac{1}{5}x - 9$

43. $3x + y = 2$ and $x + 3y = 2$ **44.** $\dfrac{3x - y}{3} = x + 2$ and $x = 9$

45. $5x - 6(x + 1) = 2y - x$ and $2y - (x + y) = 4y + x$

46. $-6y + 3x = 7$ and $8x - 3(x + 1) = 3y - x$

47. $-x = -\dfrac{2}{5}y + 2$ and $5y = 2x$

48. $\dfrac{7x - 5y}{4} = x + 2$ and $-3y - 3x = 2x + 4$

49. $3(4 - x) = 6y + 3$ and $-3y - 2x = 3 - 8x$

50. $\dfrac{x - 1}{2} + \dfrac{3y + 2}{3} = -9$ and $3y - 5x = x + 5$

51. $1 - \dfrac{2y - 5x}{2} = 7x + 4$ and $9x - 2y = 11$

52. $y - \dfrac{2}{3} = 4\left(x + \dfrac{7}{11}\right)$ and the line passing through the points $(-2, 4)$ and $(7, -14)$

53. No

54. Yes

55. No

56. Yes

57. Yes

58. No

59. $41\frac{2}{3}$ ft

Each set of four ordered pairs below defines the vertices, in counterclockwise order, of a quadrilateral. Use the ideas in this section to determine if the quadrilateral is a rectangle.

53. $\{(-2, 2), (-5, -2), (2, -3), (5, 1)\}$ **54.** $\{(2, -1), (-2, 1), (-3, -1), (1, -3)\}$

55. $\{(1, 2), (3, -3), (9, -1), (7, 4)\}$ **56.** $\{(5, -7), (1, -13), (28, -31), (32, -25)\}$

57. $\{(-5, -1), (0, -6), (5, -1), (0, 4)\}$ **58.** $\{(-3, -3), (3, -2), (1, 2), (-5, 1)\}$

Solve the following application problem.

59. A construction company is building a new suspension bridge that has support cables attached to a center tower at various heights. One cable is attached at a height of 30 feet and connects to the ground 50 feet from the base of the tower. If the support cables should run parallel to each other, how far from the base should the company attach a cable whose other end is connected to the tower at a height of 25 feet?

Linear Inequalities in Two Variables

TOPICS

1. Solving linear inequalities in two variables

2. Solving linear inequalities joined by "and" or "or"

T. Graphing inequalities

TOPIC 1

Solving Linear Inequalities in Two Variables

Just as in one variable, linear inequalities in two variables have much in common with linear equations in two variables. The first similarity is in the definition: if the equality symbol in a linear equation in two variables is replaced with $<, >, \leq,$ or $\geq$, the result is a **linear inequality in two variables**. In other words, a linear inequality in the two variables x and y is an inequality that can be written in the form

$$ax + by < c, ax + by > c,$$
$$ax + by \leq c, \text{ or } ax + by \geq c,$$

where $a, b,$ and c are constants and a and b are not both 0.

Another similarity lies in the solution process. The solution set of a linear inequality in two variables consists of all the ordered pairs in the Cartesian plane that lie on one side of a line in the plane, possibly including those points on the line. The first step in solving such an inequality then is to identify and graph this line. This line is simply the graph of the equation that results from replacing the inequality symbol in the original problem with an equality symbol.

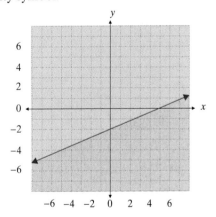

Figure 1: Dividing the Plane

Any line divides the plane into two **half-planes**; given a linear inequality, all of the points in one of the two half-planes will solve the inequality. In addition, the points on the **boundary line** also solve the inequality if the inequality symbol is $\leq$ or $\geq$, and this fact must be denoted graphically.

PROCEDURE

Solving Linear
Inequalities in Two
Variables

Step 1: Graph the line that results from replacing the inequality symbol with =.

Step 2: Make the line solid if the inequality symbol is ≤ or ≥ (not strict) and dashed if the symbol is < or > (strict). A solid line indicates that points on the line are included in the solution set while a dashed line indicates that points on the line are excluded from the solution set.

Step 3: Determine which of the half-planes defined by the boundary line solve the inequality by substituting a **test point** from one of the two half-planes into the inequality. If the resulting numerical statement is true, all the points in the same half-plane as the test point solve the inequality. Otherwise, the points in the other half-plane solve the inequality. Shade in the half-plane that solves the inequality.

EXAMPLE 1

Solving Linear
Inequalities

Solve the following linear inequalities by graphing their solution sets.

a. $3x + 2y < 12$ **b.** $x - y \leq 0$ **c.** $x > 3$

Note:
Whenever possible, use the origin as a test point, since it makes for a very simple calculation.

Solutions:

a.
$$3x + 2y = 12$$

$$3(0) + 2y = 12 \qquad 3x + 2(0) = 12$$

$$y = 6 \qquad\qquad x = 4$$

x-intercept of boundary: $(4, 0)$

y-intercept of boundary: $(0, 6)$

To graph the boundary line, replace the inequality symbol with an equal sign, then find the x- and y-intercepts.

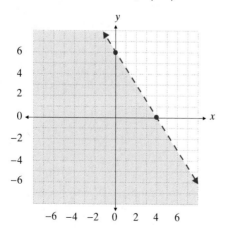

The graph of the equation is drawn with a dashed line since the inequality is strict.

The origin $(0, 0)$ clearly lies on one side of the boundary line, so we can use it as our test point.

When both x and y in the inequality are replaced with 0, we obtain the true statement $0 < 12$. This tells us to shade the half-plane that contains $(0, 0)$.

b. $x - y \le 0$

$$x - y = 0$$

$$y = x$$

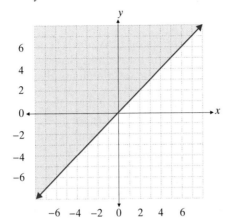

In slope-intercept form, the equation for the boundary is $y = x$, a line with a y-intercept of $(0,0)$ and a slope of 1.

Graph the boundary with a solid line since the inequality is not strict. Thus, points on the boundary do solve the inequality.

This time, the origin cannot be used as a test point, since it lies directly on the boundary line. The point $(1,-1)$ lies below the boundary, and if we substitute $x = 1$ and $y = -1$ into the inequality $x - y \le 0$, we obtain the false statement $2 \le 0$. Thus, we shade the half-plane that does not contain $(1,-1)$.

c. $x > 3$

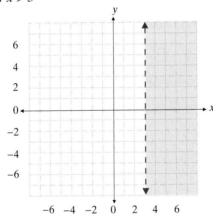

In this example, the boundary line is a vertical line with x-intercept $(3,0)$.

Since the inequality is strict, we graph the boundary line as a dashed line.

The origin can be used as a convenient test point, and results in the false statement $0 > 3$, leading us to shade the half-plane to the right of the boundary.

Another way to decide which half-plane to shade is to consider each inequality after solving for y (or for x if y is not present, as in Example 1c). The inequality in Example 1a is satisfied by all those ordered pairs for which y is less than $-\dfrac{3}{2}x + 6$. That is, if the line $y = -\dfrac{3}{2}x + 6$ is graphed (with a dashed line), the solution set of the inequality consists of all those ordered pairs whose y-coordinate is *less than* (lies below) $-\dfrac{3}{2}$ times the x-coordinate, plus 6 (those ordered pairs below the boundary line). Similarly, in Example 1b, we shade in those ordered pairs for which the y-coordinate is *greater than or equal to* the x-coordinate. Finally, in Example 1c, we shade in the ordered pairs for which the x-coordinate is *greater than* (lies to the right of) 3.

TOPIC 2   Solving Linear Inequalities Joined by "And" or "Or"

Often, we need to identify those points in the Cartesian plane that satisfy more than one inequality. In the problems that we will examine, we will identify the portion of the plane that satisfies two or more linear inequalities joined by the word "and" or the word "or."

In Section 1.1, we defined the union of two sets A and B, denoted $A \cup B$, as the set containing all elements that are in set A **or** set B, and we defined the intersection of two sets A and B, denoted $A \cap B$, as the set containing all elements that are in both A **and** B.

If we let A denote the portion of the plane that solves one inequality and B the portion of the plane that solves a second inequality, then $A \cup B$ represents the solution set of the two inequalities joined by the word "or" and $A \cap B$ represents the solution set of the two inequalities joined by the word "and."

Visually, we can think of the union as the combining of two regions, while we can think of the intersection as the overlaps between two regions. Keep in mind that $A \cup B$ *contains* both of the sets A and B (and so is at least as large as either one individually), while $A \cap B$ is *contained in* both A and B (and so is no larger than either individual set).

To find the solution sets in the following problems, we will solve each linear inequality individually and then form the union or the intersection of the individual solutions, as appropriate.

EXAMPLE 2

Solving Linear Inequalities

Graph the solution sets that satisfy the following inequalities.

a. $5x - 2y < 10$ and $y \leq x$ **b.** $x + y < 4$ or $x \geq 4$

Solutions:

a. $5x - 2y < 10$ and $y \leq x$

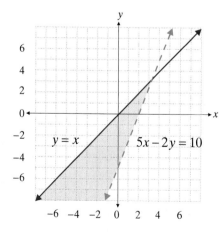

First graph the individual inequalities. We graph the line $5x - 2y = 10$ with a dashed line and the line $y = x$ with a solid line.

We then note that the half-plane lying above $5x - 2y = 10$ solves the first inequality (verify this by using the test point $(0,0)$) and that the half-plane lying below $y = x$ solves the second inequality (this can be verified using the test point $(1,-1)$).

Since the two inequalities are joined by the word "and", we shade in the intersection of the two half-planes, resulting in the graph to the left.

b. $x + y < 4$ or $x \geq 4$

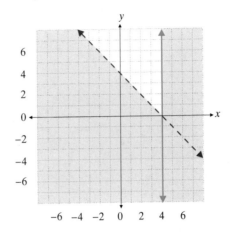

Again, begin by solving each inequality separately. We graph the line $x + y = 4$ with a dashed line, and the line $x = 4$ with a solid line, and note that the half-plane below $x + y = 4$ solves the first inequality and that the half-plane to the right of $x = 4$ solves the second.

Since the two inequalities are joined by "or," we shade the *union* of the two half-planes. Note that this results in a larger solution set than either of the two individual solutions. Any ordered pair in the shaded region will satisfy one or both of the inequalities.

Unions and intersections of regions of the plane can also arise when solving inequalities involving absolute values. In Section 1.6, we saw that an inequality of the form $|x| < a$ can be rewritten as the joint condition $x > -a$ and $x < a$, which corresponds to the intersection of two sets. Similarly, an inequality of the form $|x| > a$ can be rewritten as $x < -a$ or $x > a$, a union of two sets.

In the Cartesian plane, solutions of such inequalities will be, respectively, intersections and unions of half-planes. The next example demonstrates how such solution sets can be found.

EXAMPLE 3

Solving Absolute Value Linear Inequalities

Graph the solution set in $\mathbb{R}^2$ that satisfies the joint conditions $|x - 3| > 1$ and $|y - 2| \leq 3$.

Solution:

First, we need to find the solution to each individual inequality.

$$|x - 3| > 1$$
$$x - 3 > 1 \text{ or } x - 3 < -1$$
$$x > 4 \quad \text{or} \quad x < 2$$

Begin by rewriting the absolute value inequality as two linear inequalities.

Simplify. We have a union of solutions.

$$|y - 2| \leq 3$$
$$y - 2 \leq 3 \text{ and } y - 2 \geq -3$$
$$y \leq 5 \quad \text{and} \quad y \geq -1$$

Again, rewrite the inequality without absolute value signs.

This time, we have an intersection of solutions.

Now we can graph the solutions to the individual conditions:

$x < 2$ or $x > 4$ $y \geq -1$ and $y \leq 5$

                    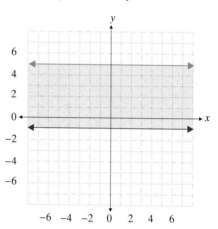

We now intersect the solution sets to obtain the final answer:

$|x - 3| > 1$ and $|y - 2| \leq 3$

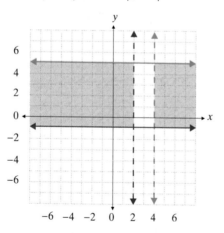

We will return to the notion of finding solutions to two or more inequalities when we study *linear programming* in Section 10.7.

TOPIC

Graphing Inequalities

Just as we can use a calculator to graph linear equations, we can also graph linear inequalities like $y < -3x + 5$ on a calculator. Press and type in the right-hand side of the inequality. Pressing ⬤GRAPH would display the line $y = -3x + 5$, which is the boundary line of the solution set. If we test the point $(0,0)$, we find that it is a solution to the inequality, so we know to shade the half-plane below the boundary line. On the ⬤Y= screen, use the left arrow to move the cursor to the left of the Y1, where there is a small diagonal line. Press ENTER until that line appears with shading like in the screen below.

With that selection, pressing ⬤GRAPH will display the boundary line with the half-plane below it shaded, which is the solution to the inequality.

In order to shade the half-plane above the boundary line, which would be the solution to $y > -3x + 5$, we would need to return to the ⬤Y= screen, again place the cursor to the left of the Y1, press ENTER until the diagonal line appeared with shading above it, and press ⬤GRAPH.

CAUTION!

Notice that the boundary line appears in the calculator as a solid line. However, because we are finding the solution to a strict inequality, the boundary line is not included in the answer.

1.

2.

3.

4.

5.

6.

7.

8.

9.

10.

Exercises

Graph the solution sets of the following linear inequalities. See Examples 1 through 3.

1. $x - 3y < 6$

2. $y < 2x - 1$

3. $x > \dfrac{3}{4}y$

4. $x - 3y \geq 6$

5. $3x - y \leq 2$

6. $\dfrac{2x - y}{4} > 1$

7. $y < -2$

8. $x + 1 \geq 0$

9. $x + y < 0$

10. $x + y > 0$

11. $-(y - x) > -\dfrac{5}{2} - y$

12. $-2y \leq -x + 4$

13. $5(y + 1) \geq -x$

14. $3x - 7y \geq 7(1 - y) + 2$

15. $x - y < 2y + 3$

16. $y > -3x - 6$ or $y \leq 2x - 7$

17. $y \geq -2$ and $y > 1$

18. $y \geq -2x - 5$ and $y \leq -6x - 9$

19. $y \leq 4x + 4$ and $y > 7x + 7$

20. $x - 3y \geq 6$ and $y > -4$

21. $x - 3y \geq 6$ or $y > -4$

22. $3x - y \leq 2$ and $x + y > 0$

23. $x > 1$ and $y > 2$

24. $x > 1$ or $y > 2$

25. $x + y > -2$ and $x + y < 2$

26. $y > -2$ and $2y > -3x - 4$

27. $3y > x + 2$ or $4y \leq -x - 2$

28. $y \leq -x$ and $2y + 3x > -4$

29. $5x + 6y < -30$ and $x \geq 2$

30. $6y - 2x > -6$ or $y > 6$

31. $x > -3$ or $y \geq 4$

32. $-2y < -3x - 6$ or $-3y \geq -6x - 18$

33. $x < 6$ and $x \geq -5$

34. $|x - 3| < 2$

35. $|x - 3| > 2$

36. $|3y - 1| \leq 2$

37. $|2x - 4| > 2$

38. $1 - |y + 3| < -1$

39. $|x + 1| < 2$ and $|y - 3| \leq 1$

40. $|x - 3| \geq 1$ or $|y - 2| \leq 1$

41. $|x - y| < 1$

42. $|x + y| \geq 1$

43. $|4x - 2y - 3| \leq 5$

44. $|2x - 3| \geq 1$ or $|2y + 3| \geq 1$

45. $|y - 3x| \leq 2$ and $|y| < 2$

11.

12.

13.

14.

15.

16.

17.

18.

19.

20.

Match the following inequalities to the appropriate graph.

46. $-8y + 5x \geq -8y + 5$

47. $x < -2$ and $x \geq -5$

48. $|-7x - 4y + 23| \leq 16$

49. $y \leq 3x - 6$ and $y > 2x - 4$

50. $|3y - 2x| > 17$ and $|y + 6| \geq 1$

51. $4(y + 2) < -x$

52. $-y < 6x + 3$ or $4y \geq 3x - 6$

53. $|7x + 4| \leq 5$ or $|7y + 4| \leq 5$

a.

b.

c.

d.

e.

f.

g.

h.

21.

22.

23.

24.

25.

26.

Solve the following application problems.

54. It costs Happy Land Toys $5.50 in variable costs per doll produced. If total costs must remain less than $200, write a linear inequality describing the relationship between cost and dolls produced.

55. Trish is having a garden party where she wants to have several arrangements of lilies and orchids for decoration. The lily arrangements cost $12 each and the orchids cost $22 each. If Trish wants to spend less than $150 on flowers, write a linear inequality describing the number of each arrangement she can purchase. Graph the inequality.

56. Rob has 300 feet of fencing he can use to enclose a small rectangular area of his yard for a garden. Assuming Rob may or may not use all the fencing, write a linear inequality describing the possible dimensions of his garden. Graph the inequality.

57. Flowertown Canoes produces two types of canoes. The two-person model costs $73 to produce and the one-person model costs $46 to produce. Write a linear inequality describing the number of each canoe the company can produce and keep costs under $1750. Graph the inequality.

30.

31.

32.

33.

34.

35.

36.

37.

38.

39.

40.

41.

42.

43.

44.

45.

46. d

47. h

48. f

49. b

50. a

51. g

52. e

53. c

54. $5.5x < 200$

55. $12x + 22y < 150$

56. $x + y \leq 150$

57. $73x + 46y < 1750$

27.

28.

29.

Introduction to Circles

TOPICS

1. Standard form of a circle

2. Graphing circles

TOPIC **1**

Standard Form of a Circle

The focus of this chapter so far has been on equations that define lines in the Cartesian plane. But, linear equations are just the beginning. Much of the rest of this book will deal with more interesting graphs in the plane and the equations that represent them. As a preview, we will close this chapter with a brief introduction to one class of nonlinear equations, circles.

Circles in the plane can be described mathematically with just two pieces of information: the circle's center and the circle's radius. To be concrete, suppose (h,k) is the location of the circle's center, and suppose the radius is given by the positive real number r. Our goal is to develop an equation in the two variables x and y so that every solution (x,y) of the equation corresponds to a point on the circle.

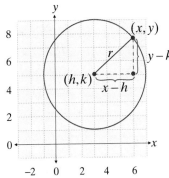

Figure 1: Graph of a Circle

The main tool that we need is the distance formula derived in Section 2.1. As we can see in Figure 1 above, every point (x,y) on the circle is at the same distance r from the circle's center (h,k), so the distance formula tells us that

$$r = \sqrt{(x-h)^2 + (y-k)^2}.$$

This form of the equation is in many ways the most natural, but the equations of circles are usually presented in the radical-free form that results from squaring both sides.

$$r^2 = (x-h)^2 + (y-k)^2$$

DEFINITION

Standard Form of a Circle

The **standard form** of the equation for a circle of radius r and center (h,k) is

$$(x-h)^2 + (y-k)^2 = r^2.$$

EXAMPLE 1

**Standard Form
of a Circle**

Find the standard form of the equation for the circle with radius 3 and center $(-2, 7)$.

Solution:

We are given $h = -2$, $k = 7$, and $r = 3$, so we plug these values into the standard form

$$\left(x - (-2)\right)^2 + \left(y - 7\right)^2 = 3^2.$$

This can be simplified as follows

$$\left(x + 2\right)^2 + \left(y - 7\right)^2 = 9.$$

EXAMPLE 2

**Standard Form
of a Circle**

Find the standard form of the equation for the circle with a diameter whose endpoints are $(-4, -1)$ and $(2, 5)$.

Note:
Sometimes you will need to find the center and radius of the circle using other geometric facts.

Solution:

The midpoint of a diameter of a circle is the circle's center, so we can use the midpoint formula to find the center (h, k).

$$(h, k) = \left(\frac{-4 + 2}{2}, \frac{-1 + 5}{2}\right) = (-1, 2)$$

The distance from either diameter endpoint to the center gives us the circle's radius. Since we ultimately will want r^2, we can use a slight variation of the distance formula to determine

$$r^2 = \left(-4 - (-1)\right)^2 + \left(-1 - 2\right)^2 = 9 + 9 = 18.$$

Thus, the equation of the circle is

$$\left(x + 1\right)^2 + \left(y - 2\right)^2 = 18.$$

EXAMPLE 3

Standard Form of a Circle

Find the standard form of the equation for the circle graphed below.

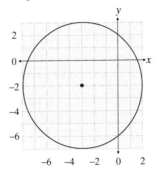

Solution:

We can identify the center by looking at the graph; it lies at $(h,k) = (-3,-2)$. To find the radius, we could find any point on the circle, then use the distance formula. A convenient choice is the point $(-3,3)$ since it is directly horizontally across from the center. We can simply count off the distance to find that the radius is 5. Plugging our information into the standard form equation, we have

$$\left(x-(-3)\right)^2 + \left(y-(-2)\right)^2 = 5^2$$
$$(x+3)^2 + (y+2)^2 = 25$$

TOPIC 2 Graphing Circles

We will often need to reverse the process illustrated in the first three examples. That is, given an equation for a circle, we will need to determine the circle's center and radius and possibly graph the circle. If the equation is given in standard form, this is straightforward since the standard form presents all the information needed to graph a circle.

EXAMPLE 4

Graphing Circles

Sketch the graph of the circle defined by $(x-2)^2 + (y+3)^2 = 4$.

Solution:

The only preliminary step is to slightly rewrite the equation in the form

$$(x-2)^2 + \left(y-(-3)\right)^2 = 2^2.$$

From this, we see that $(h,k) = (2,-3)$ and that $r = 2$. Thus, the graph of the equation is as follows:

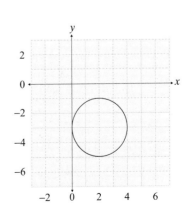

More often, the equation for a circle will not be given to us in quite so neat a fashion. We may have to apply some algebraic manipulation in order to determine that a given equation describes a circle and to determine the center and radius of that circle. Fortunately, the algebraic technique of *completing the square* (Section 1.7) is usually all that is required.

EXAMPLE 5

Graphing Circles and Completing the Square

Sketch the graph of the equation $x^2 + y^2 + 8x - 2y = -1$.

Solution:

We need to complete the square in the variable x and the variable y, and we do so as follows:

$$x^2 + y^2 + 8x - 2y = -1$$

Begin by rearranging the equation so we can complete the square for each variable.

$$\left(x^2 + 8x\right) + \left(y^2 - 2y\right) = -1$$

$$\left(x^2 + 8x + 16\right) + \left(y^2 - 2y\right) = -1 + 16$$

The coefficient on x is 8, so we add $4^2 = 16$ to both sides, then rewrite the x terms.

$$\left(x + 4\right)^2 + \left(y^2 - 2y\right) = 15$$

$$\left(x + 4\right)^2 + \left(y^2 - 2y + 1\right) = 15 + 1$$

The coefficient on y is -2, so we add $\left(-1\right)^2 = 1$ to both sides, then rewrite the y terms.

$$\left(x + 4\right)^2 + \left(y - 1\right)^2 = 16$$

We now see that the equation does indeed describe a circle, and that the center of the circle is $(-4, 1)$ and the radius is 4. The graph appears below.

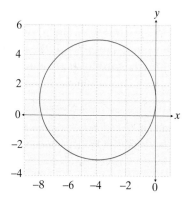

1. $(x+4)^2+(y+3)^2=25$

2. $x^2+y^2=9$

3. $(x-7)^2+(y+9)^2=9$

4. $(x+2)^2+(y-2)^2=4$

5. $x^2+y^2=6$

6. $(x-6)^2+(y-3)^2=64$

7. $\left(x-\sqrt{5}\right)^2+\left(y-\sqrt{3}\right)^2$
$=16$

8. $\left(x-\dfrac{5}{3}\right)^2+\left(y-\dfrac{8}{5}\right)^2$
$=8$

9. $(x-7)^2+(y-2)^2=4$

10. $(x-3)^2+(y-3)^2=4$

11. $(x+3)^2+(y-8)^2=2$

12. $x^2+y^2=104$

13. $(x-4)^2+(y-8)^2=10$

14. $(x-12)^2+(y+4)^2=$
522

15. $x^2+y^2=85$

16. $(x-13)^2+(y+2)^2=$
26

17. $\left(x+\dfrac{7}{2}\right)^2+\left(y-\dfrac{17}{2}\right)^2$
$=\dfrac{53}{2}$

18. $\left(x-\dfrac{13}{2}\right)^2+y^2=\dfrac{45}{4}$

19. $(x+6)^2+\left(y-\dfrac{3}{2}\right)^2$
$=\dfrac{125}{4}$

20. $\left(x-\dfrac{9}{2}\right)^2+\left(y-\dfrac{7}{2}\right)^2$
$=\dfrac{13}{2}$

21. $\left(x+\dfrac{13}{2}\right)^2+\left(y+7\right)^2$
$=\dfrac{365}{4}$

Exercises

Find the standard form of the equation for the circle. See Examples 1 and 2.

1. Center $(-4,-3)$; radius 5

2. Center at origin; radius 3

3. Center $(7,-9)$; radius 3

4. Center $(-2, 2)$; radius 2

5. Center $(0,0)$; radius $\sqrt{6}$

6. Center $(6,3)$; radius 8

7. Center $\left(\sqrt{5}, \sqrt{3}\right)$; radius 4

8. Center $\left(\dfrac{5}{3}, \dfrac{8}{5}\right)$; radius $\sqrt{8}$

9. Center $(7,2)$; passes through $(7,0)$

10. Center $(3,3)$; passes through $(1,3)$

11. Center $(-3,8)$; passes through $(-4,9)$

12. Center $(0,0)$; passes through $(2,10)$

13. Center $(4,8)$; passes through $(1,9)$

14. Center $(12,-4)$; passes through $(-9,5)$

15. Center at the origin; passes through $(6,-7)$

16. Center $(13,-2)$; passes through $(8,-3)$

17. Endpoints of a diameter are $(-8,6)$ and $(1,11)$

18. Endpoints of a diameter are $(5,3)$ and $(8,-3)$

19. Endpoints of a diameter are $(-7,-4)$ and $(-5,7)$

20. Endpoints of a diameter are $(2,3)$ and $(7,4)$

21. Endpoints of a diameter are $(0,0)$ and $(-13,-14)$

22. Endpoints of a diameter are $(4,10)$ and $(0,3)$

23. Endpoints of a diameter are $(0,6)$ and $(8,0)$

24. Endpoints of a diameter are $(6,9)$ and $(4,9)$

22. $(x-2)^2 + \left(y - \dfrac{13}{2}\right)^2$

$= \dfrac{65}{4}$

23. $(x-4)^2 + (y-3)^2 = 25$

24. $(x-5)^2 + (y-9)^2 = 1$

25. $(x-2)^2 + y^2 = 4$

26. $x^2 + (y-4)^2 = 36$

27. $(x-2)^2 + (y-4)^2 = 49$

28. $(x+2)^2 + (y+1)^2 = 16$

29. $(x+3)^2 + (y+2)^2 = 64$

30.

$(0,0), r = 5$

31.

$(0,0), r = 6$

32.

$(0,3), r = 4$

33.

$(0,8), r = 3$

34.

$(-2,0), r = 13$

35.

$(8,0), r = 2\sqrt{2}$

Find the standard form of the equation for the circle. See Example 3.

25.

26.

27.

28.

29.

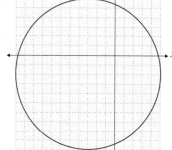

Sketch a graph of the equation and find the center and radius of each circle. See Examples 4 and 5.

30. $x^2 + y^2 = 25$

31. $x^2 + y^2 = 36$

32. $x^2 + (y-3)^2 = 16$

33. $x^2 + (y-8)^2 = 9$

34. $(x+2)^2 + y^2 = 169$

35. $(x-8)^2 + y^2 = 8$

36. $(x-9)^2 + (y-4)^2 = 49$

37. $(x+5)^2 + (y+4)^2 = 4$

36.

$(9, 4), r = 7$

37.

$(-5, -4), r = 2$

38.

$(-2, 7), r = 8$

39.

$(5, -5), r = \sqrt{5}$

40.

$(1, -5), r = 5$

41.

$(2, -2), r = 4$

42.

$(-3, 0), r = 2$

43.

$(0, -5), r = 4$

38. $(x + 2)^2 + (y - 7)^2 = 64$

40. $x^2 + y^2 - 2x + 10y + 1 = 0$

42. $x^2 + y^2 + 6x + 5 = 0$

44. $x^2 + y^2 - x - y = 2$

46. $(x - 5)^2 + y^2 = 225$

48. $(x - 3)^2 + (y + 2)^2 = 81$

50. $(x + 2)^2 + (y - 1)^2 = 16$

52. $x^2 + (y + 2)^2 = 49$

54. $x^2 + y^2 + 8x = 9$

44.

$\left(\dfrac{1}{2}, \dfrac{1}{2}\right), r = \sqrt{\dfrac{5}{2}}$

45.

$(1, -3), r = 2\sqrt{2}$

46.

$(5, 0), r = 15$

47.

$(0, 0), r = 8$

39. $(x - 5)^2 + (y + 5)^2 = 5$

41. $x^2 + y^2 - 4x + 4y - 8 = 0$

43. $x^2 + y^2 + 10y + 9 = 0$

45. $x^2 + y^2 + 6y - 2x = -2$

47. $4x^2 + 4y^2 = 256$

49. $x^2 + y^2 - 6x + 4y - 3 = 0$

51. $(x - 1)^2 + y^2 = 9$

53. $x^2 + y^2 - 4x + 8y - 16 = 0$

55. $4x^2 + 4y^2 - 24x + 24y = 28$

48.

$(3, -2), r = 9$

49.

$(3, -2), r = 4$

50.

$(-2, 1), r = 4$

51.

$(1, 0), r = 3$

52.

$(0, -2), r = 7$

53.

$(2, -4), r = 6$

54.

$(-4, 0), r = 5$

55.

$(3, -3), r = 5$

Chapter 2 Project

Using the Pythagorean Theorem

Assume that a company that builds radio towers has hired you to supervise the installation of steel support cables for several newly built structures. Your task is to find the point at which the cables should be secured to the ground. Assume that the cables reach from ground level to the top of each tower. The cables have been precut by a subcontractor and have been labeled for each tower. Finding the correct distance from the base is necessary because each cable must be grounded before being attached to a tower to avoid damaging the equipment by electric shock.

The following is your work list for this week.

Tower Name	Tower Height	Cable Length	Distance from Base
Shelbyville Tower	58 ft	75 ft	_____
Brockton Tower	100 ft	125 ft	_____
Springfield Tower	77 ft	98 ft	_____
Ogdenville Tower	130 ft	170 ft	_____

1. Use the Pythagorean formula $\left(a^2 + b^2 = c^2\right)$ to determine how far from the base of the towers to attach the cables to the ground.

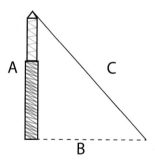

2. Could you determine the length of cable needed if you knew the height of a particular tower and the distance to the grounding point? If so, what length of cable would you need for a 150-foot tower if the grounding point should be 100 feet from the base?

Chapter Summary

A summary of concepts and skills follows each chapter. Refer to these summaries to make sure you feel comfortable with the material in the chapter. The concepts and skills are organized according to the section title and topic title in which the material is first discussed.

2.1: The Cartesian Coordinate System

The Cartesian Coordinate System
- The meaning of the *Cartesian plane*, as well as the terms *axes*, *origin*, *quadrant*, and *ordered pair*
- The relation between ordered pairs and points in the plane

The Graph of an Equation
- The correspondence between an equation in two variables and its *graph* in the Cartesian plane

The Distance and Midpoint Formulas
- The meaning of *distance* in the plane, and the use of the *distance formula* in determining the distance between two points
- The use of the *midpoint formula* in finding the point midway between two given points in the plane

2.2: Linear Equations in Two Variables

Recognizing Linear Equations in Two Variables
- The definition of a *linear equation in two variables*
- The *standard form* of a linear equation

Intercepts of the Coordinate Axes
- The geometric meaning of the *x-intercept* and the *y-intercept* of a line, and how to determine them

Horizontal and Vertical Lines
- The forms of equations that correspond to horizontal and vertical lines

2.3: Forms of Linear Equations

The Slope of a Line
- The meaning of *slope*, and how to determine the slope of the line passing through two given points
- The meaning of a slope of zero, and the meaning of undefined slope

Slope-Intercept Form of a Line
- The definition of the *slope-intercept form* of a line
- Obtaining the slope-intercept form of a given line

Point-Slope Form of a Line
- The definition of the *point-slope form* of a line
- Using the point-slope form to obtain equations of lines with prescribed properties

2.4: Parallel and Perpendicular Lines

Slopes of Parallel Lines
- The relation between the slopes of parallel lines in the plane

Slopes of Perpendicular Lines
- The relation between the slopes of perpendicular lines in the plane
- The relation between slopes of perpendicular lines when one of the lines is horizontal or vertical

2.5: Linear Inequalities in Two Variables

Solving Linear Inequalities in Two Variables
- The meaning of a *linear inequality in two variables*, and its solution set in the plane
- The difference between strict and nonstrict inequalities, and how the difference is indicated graphically
- The use of *test points* in determining the solution set of an inequality

Solving Linear Inequalities Joined by "And" or "Or"
- The meaning of "and" and "or" when used to join two or more linear inequalities
- The solution set of two or more joined inequalities

2.6: Introduction to Circles

Standard Form of a Circle
- The definition of the standard form of the equation of a circle with radius r and center (h, k)

Graphing Circles
- Obtaining the center and radius of a circle given only the equation
- Using the method of *completing the square* to convert the equation of a circle to standard form

1.

2.

3.

4. Origin (both axes)

5. Positive x-axis

6. Quadrant IV

7.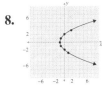

$(2,0),(0,-3),\left(-1,-\dfrac{9}{2}\right),$

$\left(\dfrac{2}{3},-2\right),(-2,-6)$

8.

$(0,\pm 2),\left(-\dfrac{4}{3},0\right),$

$(1,-\sqrt{7}),(-1,\pm 1),\left(\dfrac{5}{3},3\right)$

9. a. $\sqrt{2}$

   b. $\left(\dfrac{5}{2},-\dfrac{13}{2}\right)$

10. a. 10

   b. $(0,-6)$

11. a. $2\sqrt{13}$

   b. $(-5,3)$

12. a. $\sqrt{97}$

   b. $\left(\dfrac{1}{2},1\right)$

13. 2

14. $2+3\sqrt{2}+\sqrt{34}$

15. 24

16.

$\left(\sqrt{5}\right)^{2}+\left(\sqrt{80}\right)^{2}=\left(\sqrt{85}\right)^{2};$

$\text{Area}=\dfrac{1}{2}\sqrt{5}\sqrt{80}=10$

17. No

18. Yes

19. Yes

20. Yes

21. No

22. No

23.

24.

25.

26.

27.

28.

29. 12

30. −4

31. Undefined

32.

33.

34.

35. $x - y = 5$

36. $3x - 2y = -12$

37. $y = \dfrac{5}{9}x - 2$

Chapter Review

Section 2.1

Plot the following sets of points in the Cartesian plane.

1. $\left\{(7, 3), (-2, 4), (3, 0), (-1, -6)\right\}$ **2.** $\left\{(4, -4), (-6, 3), (-3, -1), (-4, 2)\right\}$

3. $\left\{(2, 1), (-4, 5), (3, -7), (2, 3)\right\}$

Identify the quadrant in which each point lies, if possible. If a point lies on an axis, specify which part (positive or negative) of which axis (x or y).

4. $(0, 0)$ **5.** $(1, 0)$ **6.** $(3, -2)$

For each of the following equations, determine the value of the missing entries in the accompanying table of ordered pairs. Then plot the ordered pairs and sketch your guess of the complete graph of the equation.

7. $3x - 2y = 6$

x	y
?	0
0	?
−1	?
?	−2
−2	?

8. $3x = y^2 - 4$

x	y
0	?
?	0
?	$-\sqrt{7}$
−1	?
?	3

Determine **a.** the distance between the following pairs of points, and **b.** the midpoint of the line segment joining each pair of points.

9. $(2, -6)$ and $(3, -7)$ **10.** $(-4, -3)$ and $(4, -9)$

11. $(-3, 6)$ and $(-7, 0)$ **12.** $(5, -1)$ and $(-4, 3)$

13. Given $A(-4, 2)$, $B(x, y)$, and $C(1, -1)$, find $x + y$ if C is the midpoint of the line segment $\overline{AB}$.

Find the perimeter of the triangle whose vertices are the specified points in the plane.

14. $(-3, 2)$, $(-3, 0)$, and $(-6, -3)$ **15.** $(8, -3)$, $(2, -3)$, and $(2, 5)$

16. Use the distance formula to prove that the triangle with vertices at the points $(-2, 2)$, $(0, 3)$, and $(4, -5)$ is a right triangle and determine the area of the triangle.

38. $y = -\dfrac{7}{3}x + 9$

39. $9x - 2y = 31$

40. $2x + 6y = 9$

41. $W = 0.08s + 2800$

42. Neither

43. Perpendicular

44. Neither

45. $y = 3x + 10$

46. $y = \dfrac{1}{6}x + 4$

47. $y = 2x - 3$

48. $y = -\dfrac{10}{3}x + \dfrac{5}{3}$

49. $y = -\dfrac{4}{3}x + 6$

50. $y = -3x - 19$

51. $x = 7$

52. $y = \dfrac{7}{5}x - \dfrac{7}{5}$

53. Yes

54. No

55.

56.

57.

58.

Section 2.2

Determine if the following equations are linear.

17. $3x + y(4 - 2x) = 8$

18. $y - 3(y - x) = 8x$

19. $9x^2 - (3x + 1)^2 = y - 3$

20. $8x - 3y = 4(x - 1) + y$

21. $2x(3y - 1) = 7$

22. $3x^2 + 2 = (x + 2)^2 - 1$

Determine the x- and y-intercepts of the following linear equations, if possible, and then graph the equations.

23. $4y - 12 = 8x$

24. $3(2y + 1) = 5y - 4x + 3$

25. $2x + y - 2 = 2(3 + x)$

26. $3y - 4x = -2(3x - y)$

27. $2x + 3y = 18$

28. $4x + y = 12 + y$

Section 2.3

Determine the slopes of the lines passing through the specified points.

29. $(-2, 5)$ and $(-3, -7)$

30. $(3, 6)$ and $(7, -10)$

31. $(3, 5)$ and $(3, -7)$

Use the slope-intercept form of each line to graph the equations.

32. $6x - 3y = 9$

33. $2y + 5x + 9 = 0$

34. $15y - 5x = 0$

Find the equation, in standard form, of the line passing through the given point with the given slope.

35. point $(4, -1)$; slope of 1

36. point $(-2, 3)$; slope of $\dfrac{3}{2}$

Find the equation, in slope-intercept form, of the line with the given y-intercept and slope.

37. point $(0, -2)$; slope of $\dfrac{5}{9}$

38. point $(0, 9)$; slope of $-\dfrac{7}{3}$

Find the equation, in standard form, of the line passing through the specified points.

39. $(5, 7)$ and $(3, -2)$

40. $\left(\dfrac{3}{2}, 1\right)$ and $\left(-3, \dfrac{5}{2}\right)$

41. A sales person receives a monthly salary of $2800 plus a commission of 8% of sales. Write a linear equation for the sales person's monthly wage W, in terms of monthly sales, s.

59.

60.

61.

62.

63.

64.

65.

66.

67. $3x + 4y \geq 1500$

68.
$$\left(x - \sqrt{5}\right)^2 + \left(y + \sqrt{2}\right)^2 = 16$$

Section 2.4

Determine if the two lines in each problem below are perpendicular, parallel, or neither.

42. $x - 4y = 3$ and $4x - y = 2$

43. $3x + y = 2$ and $x - 3y = 25$

44. $\dfrac{3x - y}{3} = x + 2$ and $\dfrac{y}{3} + x = 9$

Find the equation, in slope-intercept form, for the line parallel to the given line and passing through the indicated point.

45. Parallel to $y - 3x = 10$ and passing through $(-2, 4)$.

46. Parallel to $3(y + 1) = \dfrac{x - 3}{2}$ and passing through $(-6, 3)$.

47. Parallel to $y = 2x + 1$ and passing through $(1, -1)$.

48. Parallel to $3y - 2 = -5(2x - 1)$ and passing through $(2, -5)$.

Find the equation, in slope-intercept form, for the line perpendicular to the given line and passing through the indicated point.

49. Perpendicular to $y = \dfrac{3}{4}x - 1$ and passing through $(6, -2)$.

50. Perpendicular to $2(y - 3) = \dfrac{2x + 3}{3}$ and passing through $(-5, -4)$.

51. Perpendicular to $y = 8$ and passing through $(7, 1)$.

52. Perpendicular to $5x + 7y - 2 = 10$ and passing through $\left(\dfrac{2}{7}, -1\right)$.

Each set of four ordered pairs below defines the vertices, in counterclockwise order, of a quadrilateral. Determine if the quadrilateral is a rectangle.

53. $\left\{(-2, 1), (-1, -1), (3, 1), (2, 3)\right\}$

54. $\left\{(-2, 2), (-3, -1), (2, -3), (2, 1)\right\}$

Section 2.5

Graph the solution sets of the following linear inequalities.

55. $x - 2y < 4$

56. $y < 3x + 2$

57. $\dfrac{4x + y}{3} \geq 2$

58. $7x - 2y \geq 8$ and $y < 5$

59. $x - 4y \geq 6$ or $y > -2$

60. $y - x > 0$ and $x < 2$

61. $|2x + 5| < 3$

62. $|2x - 1| < 5$

69. $(x+2)^2 + y^2 = 18$

70.
$(x-2)^2 + (y+1)^2 = 20$

71.
$(x+2)^2 + (y-5)^2 = 18$

72. Center: $(-3, 1)$,
      Radius: $2\sqrt{2}$

73. ± 3

74.
$r = 4$
$(h,k) = (-5,2)$

75.
$r = \sqrt{10}$
$(h,k) = (0,3)$

76.
$r = 3$
$(h,k) = (1,-4)$

77.
$r = \sqrt{29}$
$(h,k) = (-3,5)$

63. $|x-y| < 3$

64. $-5 + |x-3| > -1$

65. $|2x+1| < 3$ or $|y+3| \geq 4$

66. $|x| > 4$ and $\left|\dfrac{2y-1}{3}\right| < 3$

67. A candle store makes a \$3 profit for every novelty candle sold and a \$4 profit for every accompanying candle holder sold. Write a linear inequality describing the number of each type of item that needs to be sold in order to make a total profit of at least \$1500.

Section 2.6

Find the standard form of the equation for each circle described below.

68. Radius 4; center $\left(\sqrt{5}, -\sqrt{2}\right)$

69. Endpoints of a diameter are $(1,-3)$ and $(-5,3)$.

70. Center at $(2,-1)$; passes through $(4,3)$

71. Endpoints of a diameter are $(1,2)$ and $(-5,8)$.

72. What is the radius and center of the circle $(x+3)^2 + (y-1)^2 = 8$?

73. Given that point $(a, 4)$ is on the circle $x^2 + y^2 = 25$, find a.

Sketch a graph of the circle defined by the given equation. Then state the radius and center of the circle.

74. $(x+5)^2 + (y-2)^2 = 16$

75. $x^2 + (y-3)^2 = 10$

76. $(x-1)^2 + (y+4)^2 = 9$

77. $x^2 + y^2 + 6x - 10y = -5$

1. Quadrant III

2. Negative y-axis

3.

$(\pm 2,0),(0,\pm 2),\left(-1,\pm\sqrt{3}\right),$

$\left(1,\pm\sqrt{3}\right),(0,2)$

4. $5, -1$

5. $\sqrt{26}+2\sqrt{5}+\sqrt{10}$

6. 1

7. No

8. Yes, $y \neq -1$

9. $(0,-2),(6,0)$

10. $-\dfrac{3}{4}$

11. Undefined

12. $-\dfrac{16}{3}$

13. $2y - x = -8$

14.

15. $55,000

16. $y = 3x - 11$

17. $4, -2$

18. $y = 2x + 11$

19. Parallel

20. Neither

21. Answers will vary.

22. Answers will vary.

23. $m = 4$;
y-intercept $= 1$

Chapter Test

Identify the quadrant in which each point lies, if possible. If a point lies on an axis, specify which part (positive or negative) of which axis (x or y).

1. $(-3,-1)$ **2.** $(0,-2)$

For the following equation, determine the value of the missing entries in the accompanying table of ordered pairs. Then plot the ordered pairs and sketch your guess of the complete graph of the equation.

3. $x^2 + y^2 = 4$

x	y
?	0
0	?
−1	?
1	?
?	2

4. Given $(3,2)$ and $(4,y)$, find y such that the distance between these two points is $\sqrt{10}$.

5. Find the perimeter of the triangle whose vertices are $(2,3)$, $(1,-2)$, and $(-1,2)$.

6. ΔKLM is isosceles and $KL = KM$. If $L(-2,0)$, $M(4,0)$, and $K(x,6)$, find x.

Determine if the following equations are linear.

7. $6x - 5y + xy = 1$ **8.** $\dfrac{x}{2} - \dfrac{y^2 - 1}{y+1} - 3 = 0$

9. Find the x- and y-intercepts of $2x - 6y = 12$, then graph the equation.

Determine the slopes of the lines passing through the specified points.

10. $(1,-2)$ and $(-3,1)$ **11.** $(-2,3)$ and $(-2,1)$

12. Given the points $A(-2,4)$ and $B(x,-1)$ find x if the line connecting the points has a slope of $\dfrac{3}{2}$.

13. Find the equation, in standard form, of the line with a slope of $\dfrac{1}{2}$ passing through point $(2,-3)$.

14. Graph the line that passes through point $(4,-3)$ and has a slope of -2.

15. Assume your salary was $30,000 in 2010 and $40,000 in 2012. If your salary follows a linear growth pattern, what will your salary be in 2015?

24. m is undefined; no y-intercept

25. a. $x + y = -1$
b. $-x + y = -2$

26. a. $x = 2$
b. $y = 5$

27.

28.

29.

30.

31.

32.

16. Find the equation, in slope-intercept form, of the line parallel to $1 - \dfrac{y - 3x}{2} = 4$ and passing through $(2, -5)$.

17. Given that the line $4x + (a - 2)y = 6$ and the line $ax + 2y = 9$ are parallel to each other, find a.

18. Find the equation, in slope-intercept form, of the line perpendicular to $3x - y = 2(x - 1) - 3y$ and passing through $(-4, 3)$.

Determine if each of the following sets of lines is parallel, perpendicular, or neither.

19. $y = 2x - 1$ and $y = 2x + 1$

20. $2(x - 3) - \dfrac{y - 1}{2} = 0$ and $-x + 3y = 6$

Given points $P(0, -4)$, $Q(8, -3)$, $R(4, 4)$, and $S(-4, 3)$ show each of the following.

21. Quadrilateral $PQRS$ is a parallelogram.

22. Parallelogram $PQRS$ is a rhombus.

Determine the slopes and y-intercepts of the following linear equations, if possible, and then graph the equations.

23. $4x - y + 1 = 0$

24. $3x + 2 = 0$

For each linear equation below, find the equation in standard form of a line **a.** parallel to the given line at the specified point, and **b.** perpendicular to the given line at the specified point.

25. $x + y = 3$ at $\left(\dfrac{1}{2}, \dfrac{-3}{2}\right)$

26. $x = 4$ at $(2, 5)$

Graph the solution sets of the following linear inequalities.

27. $\dfrac{x}{2} \geq 1 - y$

28. $x < -2$ and $y \geq -1$

29. $x > 1$ or $y < -2$

30. $2 < x + y \leq 4$

31. $|y + 2| < 3$

32. $|x + y| \geq 2$

33.
$(x+3)^2 + (y+2)^2 = 5$

34. $x^2 + y^2 = 169$

35. Center: $(-1,3)$,
Radius: 4

36.

37. $B(6,3)$, Equation:
$(x-2)^2 + (y-2)^2 = 17$

Find the standard form of the equation for each circle described below.

33. Center $(-3,-2)$, radius $\sqrt{5}$

34. Center at the origin; passes through $(5,12)$

35. Find the radius and center of the circle $x^2 + y^2 + 2x - 6y - 6 = 0$.

36. Sketch a graph of the following equation: $x^2 + y^2 + 2x + 4y = 4$

37. Given the diameter of a circle is $\overline{AB}$ and the center is at $(2,2)$, find the coordinates of point B if point A lies at $(-2,1)$. Find the equation of this circle.

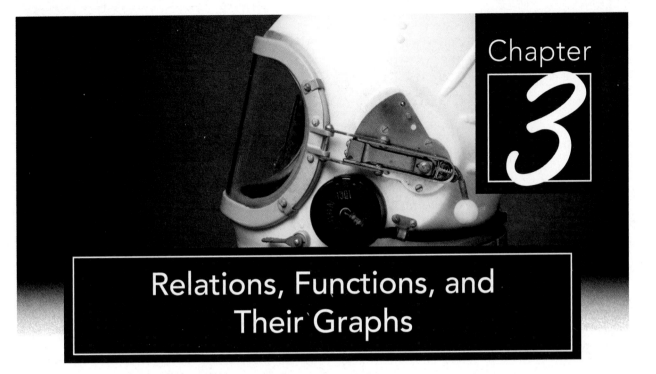

Chapter 3

Relations, Functions, and Their Graphs

3.1   Relations and Functions

3.2   Linear and Quadratic Functions

3.3   Other Common Functions

3.4   Variation and Multivariable Functions

3.5   Transformations of Functions

3.6   Combining Functions

3.7   Inverses of Functions

Chapter 3 Project

Chapter 3 Summary

Chapter 3 Review

Chapter 3 Test

By the end of this chapter you should be able to:

What if you visited outer space in a space shuttle? Knowing how much you weigh on Earth, how much would you weigh 1000 miles above Earth?

By the end of this chapter, you'll be able to describe and manipulate relations, functions, and their graphs. To calculate your weight in outer space, you'll need to solve a variation problem like the one on page 258. You'll master this type of problem using the definition of Inverse Variation, found on page 252.

Introduction

This chapter begins with a study of *relations*, which are generalizations of the equations in two variables discussed in Chapter 2, and then moves on to the more specialized topic of *functions*. As concepts, relations and functions are more abstract, but at the same time far more powerful and useful than the equations studied thus far in this text. Functions, in particular, lie at the heart of a great deal of the mathematics that you will encounter from this point on.

Leibniz

The history of the function concept serves as a good illustration of how mathematics develops. One of the first people to use the idea in a mathematical context was the German mathematician and philosopher Gottfried Leibniz (1646–1716), one of two people (along with Isaac Newton) usually credited with the development of calculus. Initially, Leibniz and other mathematicians tended to use the term to indicate that one quantity could be defined in terms of another by some sort of algebraic expression, and this (incomplete) definition of function is often encountered even today in elementary mathematics. As the problems that mathematicians were trying to solve increased in complexity, however, it became apparent that functional relations between quantities existed in situations where no algebraic expression defining the function was possible. One example came from the study of heat flow in materials, in which a description of the temperature at a given point at a given time was often given in terms of an infinite sum, not an algebraic expression.

The result of numerous refinements and revisions of the function concept is the definition that you will encounter in this chapter, and is essentially due to the German mathematician Lejeune Dirichlet (1805–1859). Dirichlet also refined our notion of what is meant by a *variable*, and gave us our modern understanding of *dependent* and *independent* variables, all of which you will soon encounter.

The proof of the power of functions lies in the multitude and diversity of their applications. As you work through Chapter 3, pay special attention to how function notation works. A solid understanding of what function notation means is essential to using functions.

3.1

Relations and Functions

TOPICS

1. Relations, domain, and range

2. Functions and the vertical line test

3. Function notation and function evaluation

4. Implied domain of a function

TOPIC 1   Relations, Domain, and Range

In Chapter 2 we saw many examples of equations in two variables. Any such equation automatically defines a relation between the two variables present, in the sense that each ordered pair on the graph of the equation relates a value for one variable (namely, the first coordinate of the ordered pair) to a value for the second variable (the second coordinate). Many applications of mathematics involve relating one variable to another, and we will spend much of the rest of this book studying this.

DEFINITION

Relations, Domain, and Range

A **relation** is a set of ordered pairs. Any set of ordered pairs automatically relates the set of first coordinates to the set of second coordinates, and these sets have special names. The **domain** of a relation is the set of all the first coordinates, and the **range** of a relation is the set of all second coordinates.

Relations can be described in many different ways. We have already noted that an equation in two variables describes a relation, as the solution set of the equation is a collection of ordered pairs. Relations can also be described with a simple list of ordered pairs (if the list is not too long), with a picture in the Cartesian plane, and by many other means.

The following example demonstrates some of the common ways of describing relations and identifies the domain and range of each relation.

EXAMPLE 1

Relations, Domains, and Ranges

a. The set $R = \left\{ (-4, 2), (6, -1), (0, 0), (-4, 0), \left(\pi, \pi^2 \right) \right\}$ is a relation consisting of five ordered pairs.

The domain of R is the set $\{-4, 6, 0, \pi\}$, as these four numbers appear as first coordinates in the relation. Note that it is not necessary to list the number -4 twice in the domain, even though it appears twice as a first coordinate in the relation.

The range of R is the set $\{2, -1, 0, \pi^2\}$, as these are the numbers that appear as second coordinates. Again, it is not necessary to list 0 twice in the range, even though it is used twice as a second coordinate in the relation.

The *graph* of this relation is simply a picture of the five ordered pairs plotted in the Cartesian plane, as shown below.

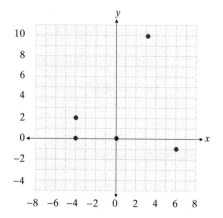

b. The equation $-3x + 7y = 13$ describes a relation. Using the skills we learned in Chapter 2, we can graph the solution set below:

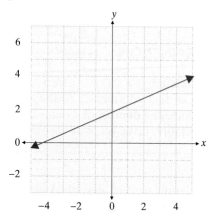

Unlike the last example, this relation consists of an infinite number of ordered pairs, so it is not possible to list them all as a set. One of the ordered pairs in the relation is $(-2, 1)$, since $-3(-2) + 7(1) = 13$. The domain and range of this relation are both the set of real numbers, since every real number appears as both a first coordinate and a second coordinate in the relation.

c. The picture below describes a relation.

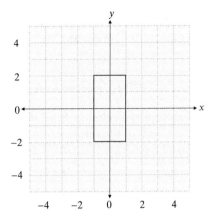

Some of the elements of the relation are $(-1, 1)$, $(-1, -2)$, $(-0.3, 2)$, $(0, -2)$, and $(1, -0.758)$, but this is another example of a relation with an infinite number of elements so we cannot list all of them. Using interval notation, the domain of this relation is the closed interval $[-1, 1]$ and the range is the closed interval $[-2, 2]$.

d. The picture below describes another relation, similar to the last but still different. The shading indicates that all ordered pairs lying inside the rectangle, as well as those actually on the rectangle, are elements of the relation.

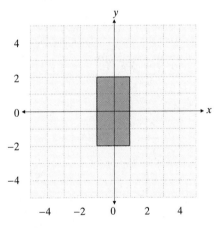

The domain is again the closed interval $[-1, 1]$ and the range is again the closed interval $[-2, 2]$, but this relation is not identical to the last example. For instance, the ordered pairs $(0, 0)$ and $(0.2, 1.5)$ are elements of this relation but are not elements of the relation in Example 1c.

e. Although we will almost never encounter relations in this text that do not consist of ordered pairs of real numbers, nothing in our definition prevents us from considering more exotic relations. For example, the set $S = \{(x, y) \mid x \text{ is the mother of } y\}$ is a relation among people. Each element of the relation consists of an ordered pair of a mother and her child. The domain of S is the set of all mothers, and the range of S is the set of all people. (Although advances in cloning are occurring rapidly, as of the writing of this text, no one has yet been born without a mother!)

TOPIC 2 — Functions and the Vertical Line Test

As important as relations are in mathematics, a special type of relation, called a function, is of even greater use.

DEFINITION

Functions

A **function** is a relation in which every element of the domain is paired with *exactly one* element of the range. Equivalently, a function is a relation in which no two distinct ordered pairs have the same first coordinate.

Note that there is a difference in the way domains and ranges are treated in the definition of a function: the definition allows for the two distinct ordered pairs to have the same second coordinate, as long as their first coordinates differ. A picture helps in understanding this distinction:

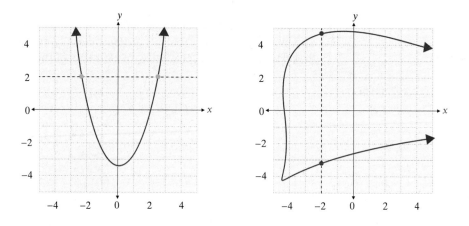

Figure 1: Definition of Functions

The relation on the left in Figure 1 has pairs of points that share the same *y*-value (one such pair is indicated in green). This means that some elements of the range are paired with more than one element of the domain. However, each element of the domain is paired with exactly one element of the range. Thus, this relation is a function.

On the other hand, the relation on the right in Figure 1 has pairs of points that share the same *x*-value (one such pair is indicated in red). This means that some elements of the domain are paired with more than one element of the range. This relation is not a function.

EXAMPLE 2

Is the Relation a Function?

For each relation in Example 1, identify whether the relation is also a function.

Solutions:

a. The relation in Example 1a is not a function because the two ordered pairs $(-4, 2)$ and $(-4, 0)$ have the same first coordinate. If either one of these ordered pairs were deleted from the relation, the relation would be a function.

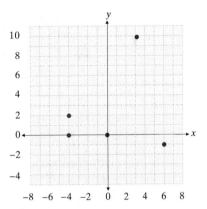

b. The relation in Example 1b is a function. Any two distinct ordered pairs that solve the equation $-3x + 7y = 13$ have different first coordinates. This can also be seen from the graph of the equation. If two ordered pairs have the same first coordinate, they must be aligned vertically, and no two ordered pairs on the graph of $-3x + 7y = 13$ have this property.

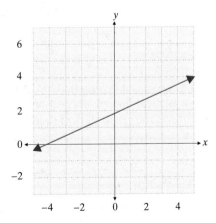

c. The relation in Example 1c is not a function. To prove that a relation is not a function, it is only necessary to find two ordered pairs with the same first coordinate, and the pairs $(0, 2)$ and $(0, -2)$ show that this relation fails to be a function.

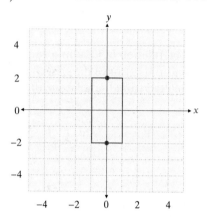

d. The relation in Example 1d is also not a function. In fact, we can use the same two ordered pairs as in the previous part to prove this fact.

e. Finally, the relation in Example 1e also fails to be a function. Think of two people who have the same mother; this gives us two ordered pairs with the same first coordinate: (mother, child 1) and (mother, child 2). Thus, the relation is not a function.

In Example 2b, we noted that two ordered pairs in the plane have the same first coordinate only if they are aligned vertically. We could also have used this criterion to determine that the relation in Example 1a is not a function, since the two ordered pairs $(-4, 2)$ and $(-4, 0)$ clearly lie on the same vertical line. This visual method of determining whether a relation is a function, called the **vertical line test**, is very useful when an accurate graph of the relation is available.

THEOREM

The Vertical Line Test

If a relation can be graphed in the Cartesian plane, the relation is a **function** if and only if no vertical line passes through the graph more than once. If even *one* vertical line intersects the graph of the relation two or more times, the relation fails to be a function.

CAUTION!

Note that vertical lines that miss the graph of a relation entirely don't prevent the relation from being a function; it is only the presence of a vertical line that hits the graph two or more times that indicates the relation isn't a function. The next example illustrates some more applications of the vertical line test.

EXAMPLE 3

Functions and the
Vertical Line Test

a. The relation $R = \{(-3, 2), (-1, 0), (0, 2), (2, -4), (4, 0)\}$, graphed below, is a function. Any given vertical line in the plane either intersects the graph once or not at all.

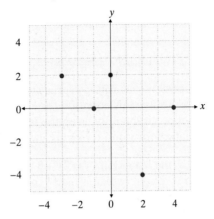

b. The relation graphed below is not a function, as there are many vertical lines that intersect the graph more than once. The dashed line is one such vertical line.

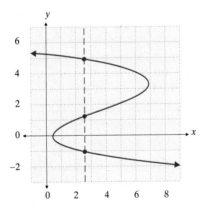

c. The relation graphed below is a function. In this case, every vertical line in the plane intersects the graph exactly once.

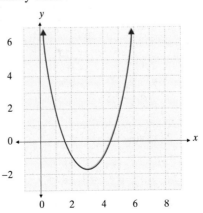

TOPIC 3

Function Notation and Function Evaluation

When a function is defined with an equation in two variables, one represents the domain (usually x) and one represents the range (usually y). Because functions assign each element of the domain exactly one element of the range, we can solve the equation for y. This leads to a special notation for functions, called **function notation**.

DEFINITION

Function Notation

Suppose a function is represented by an equation in two variables, say x and y, and we can solve this equation for y, the variable representing the range. We can name the function (frequently using the letter f), and write it in **function notation** by solving the equation for y and replacing y with $f(x)$.

With the function $y = 2x - 3$ as an example, using function notation we write: $f(x) = 2x - 3$, which is read "f of x equals two times x, minus three." Function notation can also indicate what to do with a specific value of x; $f(4)$, read "f of 4" tells us to plug the value 4 into the formula given for f. The result is $f(4) = 2(4) - 3$, $f(4) = 5$, which we read "f of 4 equals 5."

DEFINITION

Independent and Dependent Variables

Given an equation representing a function $y = f(x)$, we call x the **independent variable** and y the **dependent variable**, since the value of y depends on the value of x we input into the formula for f.

EXAMPLE 4

Function Notation

Each of the following equations in x and y represents a function. Rewrite each one using function notation, and then evaluate each function at $x = -3$.

a. $y = \dfrac{3}{x} + 2$

b. $7x + 3 = 2y - 1$

c. $y - 5 = x^2$

d. $\sqrt{1-x} - 2y = 6$

Solutions:

a. $y = \dfrac{3}{x} + 2$

The equation is already solved for y.

$f(x) = \dfrac{3}{x} + 2$

To write the function in function notation, replace y with $f(x)$.

$f(-3) = \dfrac{3}{-3} + 2 = 1$

Substitute $x = -3$ and evaluate. This means the point $(-3, 1)$ is on the graph of f.

b. $7x + 3 = 2y - 1$

$7x - 2y = -4$

$-2y = -7x - 4$

$y = \dfrac{7}{2}x + 2$

$g(x) = \dfrac{7}{2}x + 2$

$g(-3) = \dfrac{7}{2}(-3) + 2 = -\dfrac{17}{2}$

The first step is to solve the equation for the dependent variable y.

We can name the function anything at all. Typical names of functions are f, g, h, etc. We will use g to differentiate this function from the one in part a.

Now evaluate g at –3. The point $\left(-3, -\dfrac{17}{2}\right)$ is on the graph of g.

c. $y - 5 = x^2$

$y = x^2 + 5$

$h(x) = x^2 + 5$

$h(-3) = (-3)^2 + 5 = 14$

Again, begin by solving for y.

To distinguish this function, use a different name.

Substitute –3 into the function. The point $(-3, 14)$ is on the graph of h.

d. $\sqrt{1-x} - 2y = 6$

$-2y = 6 - \sqrt{1-x}$

$y = -3 + \dfrac{\sqrt{1-x}}{2}$

$j(x) = -3 + \dfrac{\sqrt{1-x}}{2}$

$j(-3) = -3 + \dfrac{\sqrt{1-(-3)}}{2}$

$= -3 + \dfrac{2}{2} = -2$

As usual, the process begins by solving for y.

Generally, we avoid using i as a function name, since i also represents the imaginary unit.

Substitute –3 and then simplify to evaluate $j(-3)$.

This tells us that $(-3, -2)$ is on the graph of j.

CAUTION!

By far the most common error made when encountering functions for the first time is to think that f(x) stands for the product of f and x. This is entirely wrong! While it is true that parentheses are often used to indicate multiplication, they are also used in defining functions.

DEFINITION

Argument of a
Function

In defining a function f, such as $f(x) = 2x - 3$, the critical idea is the formula. We can use any symbol at all as the variable in defining the formula that we have named f. For instance, $f(n) = 2n - 3$, $f(z) = 2z - 3$, and $f(\$) = 2(\$) - 3$ all define exactly the same function. The variable (or symbol) that is used in defining a given function is called its **argument**, and serves as nothing more than a placeholder.

We will not always be replacing the arguments of functions with numbers. In many instances, we will have reason to replace the argument of a function with another variable or possibly a more complicated algebraic expression. Keep in mind that this just involves substituting something for the placeholder used in defining the function.

EXAMPLE 5

Evaluating Functions

Given the function $f(x) = 3x^2 - 2$, evaluate:

a. $f(a)$ **b.** $f(x + h)$ **c.** $\dfrac{f(x+h) - f(x)}{h}$

Solutions:

Note:
The expression in part c. of this example is called the difference quotient of a function, and is used heavily in calculus.

a. $f(a) = 3a^2 - 2$

This is just a matter of replacing x with a.

b. $f(x + h) = 3(x + h)^2 - 2$

$\qquad = 3(x^2 + 2xh + h^2) - 2$

$\qquad = 3x^2 + 6xh + 3h^2 - 2$

Here we replace x with $x + h$ and simplify the result.

c. $\dfrac{f(x-h) - f(x)}{h} = \dfrac{(3x^2 + 6xh + 3h^2 - 2) - (3x^2 - 2)}{h}$

$\qquad = \dfrac{6xh + 3h^2}{h}$

$\qquad = \dfrac{h(6x + 3h)}{h}$

$\qquad = 6x + 3h$

We can use the result from above in simplifying this expression.

Simplify.

Factor out h, so that we can cancel out the h in the denominator.

There is one final piece of function notation that is often encountered, especially in later math classes such as calculus.

DEFINITION

Domain and
Codomain Notation

The notation $f: A \to B$ (read "f defined from A to B" or "f maps A to B") implies that f is a function from the set A to the set B. The symbols indicate that the domain of f is the set A, and that the range of f is a subset of the set B. In this context, the set B is often called the **codomain** of f.

Note that while the notation $f: A \rightarrow B$ implies the domain of f is the entire set A, there is no requirement for the range of f to be all of B. If it so happens that the range of f actually *is* the entire set B, f is said to be *onto* B (or, more formally, to be a *surjective* function). The next example illustrates how this notation is typically encountered, and also points out some of the subtleties inherent in these notions.

EXAMPLE 6

Domain, Codomain, and Range

Identify the domain, the codomain, and the range of each of the following functions.

a. $f: \mathbb{R} \rightarrow \mathbb{R}$ by $f(x) = x^2$

b. $g: \mathbb{R} \rightarrow [0, \infty)$ by $g(x) = x^2$

c. $h: \mathbb{Z} \rightarrow \mathbb{Z}$ by $h(x) = x^2$

d. $j: \mathbb{N} \rightarrow \mathbb{R}$ by $j(x) = x^2$

Solutions:

a. The "$f: \mathbb{R} \rightarrow \mathbb{R}$" portion of the statement tells us that a function f on the real numbers is about to be defined, and that each value of the function will also be a real number. That is, the domain and codomain of f are both $\mathbb{R}$.

The "$f(x) = x^2$" portion tells us the details of how the function acts. Namely, it returns the square of each real number it is given. Since the square of any real number is nonnegative, and since every nonnegative real number is the square of some real number, the range of f is the interval $[0, \infty)$.

b. The function g is very similar to the function f in part a. The only difference is that the notation "$g: \mathbb{R} \rightarrow [0, \infty)$" tells us in advance that the codomain of g is the nonnegative real numbers. Note that the domain of g is $\mathbb{R}$ and the range of g is the same as the range of f. But since the range of g is the same as the codomain of g, the function g is said to be onto, or surjective.

This points out that the quality of being "onto" depends entirely on how the codomain of the function is specified. If it is no larger than the range of the function, then the function is onto.

c. The function "$h: \mathbb{Z} \rightarrow \mathbb{Z}$ by $h(x) = x^2$" has a domain and codomain of $\mathbb{Z}$. But if we think about the result of squaring any given integer (positive or negative), we quickly see that the range of h is the set $\{0, 1, 4, 9, 16, 25, ...\}$. That is, the range of h consists of those integers which are squares of other integers. Since the range of h is not the same as the codomain, h is not onto.

d. The function "$j: \mathbb{N} \rightarrow \mathbb{R}$ by $j(x) = x^2$" is one final variation on the squaring function. The action of j is the same as that of the previous three functions, but this time the domain is specified to be the natural numbers (the positive integers) and the codomain is the entire set of real numbers. Since the range of j is the set $\{1, 4, 9, 16, 25, ...\}$, which is not the same as the codomain, j is not onto.

TOPIC 4   Implied Domain of a Function

Occasionally, the domain of a function is made clear by the function definition. However, it is often up to us to determine the domain, to find what numbers may be "plugged into" the function so that the output is real number. In these cases, the domain of the function is *implied* by the formula defining the function. For instance, any values for the argument of a function that result in division by zero or an even root of a negative number must be excluded from the domain of that function.

EXAMPLE 7

Implied Domain
of a Function

Determine the domain of the following functions.

a. $f(x) = 5x - \sqrt{3-x}$ **b.** $g(x) = \dfrac{x-3}{x^2-1}$

Solutions:

a. Looking at the formula, we can identify what may cause the function to be undefined.

$f(x) = 5x - \sqrt{3-x}$ We can always multiply a number by 5, but taking the square root of a negative number is undefined.

The square root term is defined as long as $3 - x \geq 0$. Solving this inequality for x, we have $x \leq 3$.

Using interval notation, the domain of the function f is the interval $(-\infty, 3]$.

b. Again, we first identify potential "dangers" in the formula for this function.

$g(x) = \dfrac{x-3}{x^2-1}$ We can safely substitute any value in the numerator, but we can't let the denominator equal zero.

The denominator will equal zero whenever $x^2 - 1 = 0$. This tells us that we must exclude $x = -1$ and $x = 1$ from the domain.

In interval notation, the domain of g is $(-\infty, -1) \cup (-1, 1) \cup (1, \infty)$.

---

1. Dom = { -2 },
   Ran = { 5, 3, 0, -9 }

2. Dom = {0, -5, 3, 5},
   Ran = { 0, 2, 3 }

3. Dom = $\{\pi, -2\pi, 3, 1\}$,
   Ran = { 2, 4, 0, 7 }

4. Dom = { 3, -4 },
   Ran = { 3, 8, -2 }

5. Dom = $\mathbb{Z}$,
   Ran = even integers

6. Dom = $\{\pi\}$,
   Ran = $\mathbb{Q}$

Exercises

For each relation below, describe the domain and range. See Example 1.

1. $R = \{(-2,5), (-2,3), (-2,0), (-2,-9)\}$ **2.** $S = \{(0,0), (-5,2), (3,3), (5,3)\}$

3. $A = \{(\pi, 2), (-2\pi, 4), (3,0), (1,7)\}$ **4.** $B = \{(3,3), (-4,3), (3,8), (3,-2)\}$

5. $T = \{(x, y) \mid x \in \mathbb{Z} \text{ and } y = 2x\}$ **6.** $U = \{(\pi, y) \mid y \in \mathbb{Q}\}$

7. Dom $= \mathbb{Z}$,

   Ran $=$
   $\{\ldots, -2, 1, 4, \ldots\}$

8. Dom $=$ multiples of 5,

   Ran $=$ multiples of 3

9. Dom $=$ Ran $= \mathbb{R}$

10. Dom $=$ Ran $= \mathbb{R}$

11. Dom $= [0, \infty)$,

   Ran $= \mathbb{R}$

12. Dom $= \mathbb{R}$,

   Ran $= [0, \infty)$

13. Dom $= \mathbb{R}$,

   Ran $= \{-1\}$

14. Dom $= \{3\}$,

   Ran $= \mathbb{R}$

15. Dom $= \{0\}$,

   Ran $= \mathbb{R}$

16. Dom $= \mathbb{R}$,

   Ran $= \{7\pi^2\}$

17. Dom $= [-3, 1]$,
   Ran $= [0, 4]$

18. Dom $= \mathbb{R}$,
   Ran $= (-\infty, 0]$

19. Dom $= [0, 3]$,
   Ran $= [1, 5]$

20. Dom $= [-1, 2]$,
   Ran $= [-1, 5]$

21. Dom $= [-1, 3]$,
   Ran $= [-4, 3]$

22. Dom $= [-1, 3]$,
   Ran $= [-4, 3]$

23. Dom $=$ All males
   with siblings,
   Ran $=$ All people
   who have brothers

24. Dom $=$ All parents
   of girls,
   Ran $=$ All females

7. $C = \left\{(x, 3x+4)\,\middle|\, x \in \mathbb{Z}\right\}$

8. $D = \left\{(5x, 3y)\,\middle|\, x \in \mathbb{Z} \text{ and } y \in \mathbb{Z}\right\}$

9. $3x - 4y = 17$

10. $x + y = 0$

11. $x = |y|$

12. $y = x^2$

13. $y = -1$

14. $x = 3$

15. $x = 4x$

16. $y = 7\pi^2$

17.

18.

19.

20.

21.

22.

23. $V = \left\{(x, y)\,\middle|\, x \text{ is the brother of } y\right\}$

24. $W = \left\{(x, y)\,\middle|\, y \text{ is the daughter of } x\right\}$

25. Not a function;
$(-2, 5)$ and $(-2, 3)$

26. Function

27. Function

28. Not a function;
$(4, 5)$ and $(4, -1)$

29. Not a function;
$(6, -1)$ and $(6, 4)$

30. Function

31. Not a function;
$(-1, 0)$ and $(-1, 4)$

32. Function

33. Function

34. Not a function;
$(1, -3)$ and $(1, 3)$

35. Function

36. Not a function;
$(2, -2)$ and $(2, 3)$

37. Function; Dom $=$
$(-\infty, 0) \cup (0, \infty)$

38. Not a function

39. Not a function

40. Function;
Dom $= \mathbb{R}$

41. Function; Dom $=$
$(-\infty, -2) \cup (-2, \infty)$

42. Not a function

43. Function;
Dom $= \mathbb{R}$

44. Function;
Dom $= \mathbb{R}$

45. Not a function

46. Function;
Dom $= [4, \infty)$

47. $f(x) = -6x^2 + 2x$
$f(-1) = -8$

Determine which of the relations below is a function. For those that are not, identify two ordered pairs with the same first coordinate. See Examples 2 and 3.

25. $R = \left\{(-2, 5), (2, 4), (-2, 3), (3, -9)\right\}$ **26.** $S = \left\{(3, -2), (4, -2)\right\}$

27. $T = \left\{(-1, 2), (1, 1), (2, -1), (-3, 1)\right\}$ **28.** $U = \left\{(4, 5), (2, -3), (-2, 1), (4, -1)\right\}$

29. $V = \left\{(6, -1), (3, 2), (6, 4), (-1, 5)\right\}$ **30.** $W = \left\{(2, -3), (-2, 4), (-3, 2), (4, -2)\right\}$

31.

32.

33.

34.

35.

36.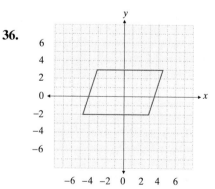

48. $f(x) =$

$\dfrac{\sqrt[3]{x} - x^2 + 3x - 1}{2}$

$f(-1) = -3$

49. $f(x) = \dfrac{-x + 10}{3}$

$f(-1) = \dfrac{11}{3}$

50. $f(x) = 5x^2 - 3$

$f(-1) = 2$

51. $f(x) = -2x - 10$

$f(-1) = -8$

52. $f(x) = x - \dfrac{5}{9}$

$f(-1) = -\dfrac{14}{9}$

53. a. 10

   b. $x^2 + x - 2$

   c. $2ax + 3a + a^2$

   d. $x^4 + 3x^2$

54. a. $\sqrt{2}$

   b. $\sqrt{x-1}$

   c. $\sqrt{x+a} - \sqrt{x}$

   d. $|x|$

55. a. 8

   b. $3x - 1$

   c. $3a$

   d. $3x^2 + 2$

56. a. -11

   b. $-x^2 + 2x - 8$

   c. $-2ax - a^2$

   d. $-x^4 - 7$

57. a. -2

   b. $-6x + 16$

   c. $-6a$

   d. $-6x^2 + 10$

Determine whether each of the following relations is a function. If it is a function, give the relation's domain.

37. $y = \dfrac{1}{x}$ **38.** $x = y^2 - 1$ **39.** $x + y^2 = 0$ **40.** $y = 2x^2 - 4$ **41.** $y = \dfrac{x-1}{x+2}$

42. $x^2 + y^2 = 1$ **43.** $y = |x - 2|$ **44.** $y = x^3$ **45.** $y^2 - x^2 = 3$ **46.** $y = \sqrt{x - 4}$

Rewrite each of the relations below as a function of x. Then evaluate the function at $x = -1$. See Example 4.

47. $6x^2 - x + 3y = x + 2y$ **48.** $2y - \sqrt[3]{x} = x - (x-1)^2$

49. $\dfrac{x + 3y}{5} = 2$ **50.** $x^2 + y = 3 - 4x^2 + 2y$

51. $y - 2x^2 = -2(x + x^2 + 5)$ **52.** $\dfrac{9y + 2}{6} = \dfrac{3x - 1}{2}$

For each function below, determine **a.** $f(2)$, **b.** $f(x-1)$, **c.** $f(x+a) - f(x)$, and **d.** $f(x^2)$. See Example 5.

53. $f(x) = x^2 + 3x$ **54.** $f(x) = \sqrt{x}$

55. $f(x) = 3x + 2$ **56.** $f(x) = -x^2 - 7$

57. $f(x) = 2(5 - 3x)$ **58.** $f(x) = 2x^2 + \sqrt[4]{x}$

59. $f(x) = \sqrt{1 - x} - 3$ **60.** $f(x) = \dfrac{-\sqrt{1 - x} + 5}{2}$

Determine the difference quotient $\dfrac{f(x+h) - f(x)}{h}$ of each of the following functions. See Example 5c.

61. $f(x) = x^2 - 5x$ **62.** $t(x) = x^3 + 2$

63. $h(x) = \dfrac{1}{x + 2}$ **64.** $g(x) = 6x^2 - 7x + 3$

65. $f(x) = 5x^2$ **66.** $f(x) = (x + 3)^2$

67. $f(x) = 2x - 7$ **68.** $f(x) = \sqrt{x}$

69. $f(x) = x^{\frac{1}{2}} - 4$ **70.** $f(x) = \dfrac{3}{x}$

58. a. $8 + \sqrt[4]{2}$

   b. $2x^2 - 4x + 2 + \sqrt[4]{x - 1}$

   c. $4ax + 2a^2 + \sqrt[4]{x + a} - \sqrt[4]{x}$

   d. $2x^4 + \sqrt{|x|}$

59. a. $i - 3$

    b. $\sqrt{2-x} - 3$

    c. $\sqrt{1-x-a}$
    $-\sqrt{1-x}$

    d. $\sqrt{1-x^2} - 3$

60. a. $\dfrac{-i+5}{2}$

    b. $\dfrac{-\sqrt{2-x}+5}{2}$

    c. $\dfrac{\sqrt{1-x}-\sqrt{1-(x+a)}}{2}$

    d. $\dfrac{-\sqrt{1-x^2}+5}{2}$

61. $2x + h - 5$

62. $3x^2 + h^2 + 3xh$

63. $\dfrac{-1}{(x+h+2)(x+2)}$

64. $12x + 6h - 7$

65. $5(2x+h)$

66. $2x + h + 6$

67. 2

68. $\dfrac{\sqrt{x+h}-\sqrt{x}}{h}$

69. $\dfrac{\sqrt{x+h}-\sqrt{x}}{h}$

70. $\dfrac{-3}{x(x+h)}$

Identify the domain, the codomain, and the range of each of the following functions. See Example 6.

71. $f : \mathbb{R} \to \mathbb{R}$ by $f(x) = 3x$

72. $g : \mathbb{Z} \to \mathbb{Z}$ by $g(x) = 3x$

73. $f : \mathbb{Z} \to \mathbb{Z}$ by $f(x) = x + 5$

74. $g : [0, \infty) \to \mathbb{R}$ by $g(x) = \sqrt{x}$

75. $h : \mathbb{N} \to \mathbb{N}$ by $h(x) = x + 5$

76. $h : \mathbb{N} \to \mathbb{R}$ by $h(x) = \dfrac{x}{2}$

Determine the implied domain of each of the following functions. See Example 7.

77. $f(x) = \sqrt{x-1}$

78. $g(x) = \sqrt[5]{x+3} - 2$

79. $h(x) = \dfrac{3x}{x^2 - x - 6}$

80. $f(x) = (2x+6)^{\frac{1}{2}}$

81. $g(x) = \sqrt[4]{2x^2 + 3}$

82. $h(x) = \dfrac{3x^2 - 6x}{x^2 - 6x + 9}$

83. $s(x) = \dfrac{2x}{1-3x}$

84. $f(x) = (x^2 - 5x + 6)^3$

85. $c(x) = \dfrac{x-1}{2-x}$

86. $g(x) = \dfrac{5}{\sqrt{3-x^2}}$

87. $f(x) = \sqrt{x+6} + 1$

88. $g(x) = -5x^2 - 4x$

89. $h(x) = \dfrac{-3(-5+5x)}{x}$

90. $h(x) = \sqrt{3-x}$

71. Dom = Cod = Ran = $\mathbb{R}$

72. Dom = Cod = $\mathbb{Z}$; Ran = $\{..., -3, 0, 3, ...\}$

73. Dom = Cod = Ran = $\mathbb{Z}$

74. Dom = $[0, \infty)$; Cod = $\mathbb{R}$; Ran = $[0, \infty)$

75. Dom = Cod = $\mathbb{N}$; Ran = $\{6, 7, 8, ...\}$

76. Dom = $\mathbb{N}$; Cod = $\mathbb{R}$; Ran = $\left\{\dfrac{1}{2}, 1, \dfrac{3}{2}, 2, ...\right\}$

77. $[1, \infty)$

78. $\mathbb{R}$

79. $(-\infty, -2) \cup (-2, 3) \cup (3, \infty)$

80. $[-3, \infty)$

81. $\mathbb{R}$

82. $(-\infty, 3) \cup (3, \infty)$

83. $\left(-\infty, \dfrac{1}{3}\right) \cup \left(\dfrac{1}{3}, \infty\right)$

84. $\mathbb{R}$

85. $(-\infty, 2) \cup (2, \infty)$

86. $(-\sqrt{3}, \sqrt{3})$

87. $[-6, \infty)$

88. $\mathbb{R}$

89. $(-\infty, 0) \cup (0, \infty)$

90. $(-\infty, 3]$

Linear and Quadratic Functions

TOPICS

1. Linear functions and their graphs
2. Quadratic functions and their graphs
3. Maximization/minimization problems
T. Maximum/minimum of graphs

TOPIC 1

Linear Functions and Their Graphs

Much of the next several sections of this chapter will be devoted to gaining familiarity with some of the types of functions that commonly arise in mathematics. We will discuss two classes of functions in this section, beginning with linear functions.

Recall that a linear equation is an equation whose graph consists of a straight line in the Cartesian plane. Similarly, a linear function is a function whose graph is a straight line. We can define such functions algebraically as follows.

DEFINITION

Linear Functions

A **linear function**, say f, of one variable, say the variable x, is any function that can be written in the form $f(x) = mx + b$, where m and b are real numbers. If $m \neq 0$, $f(x) = mx + b$ is also called a **first-degree function**.

In the last section, we learned that a function defined by an equation in x and y can be written in function form by solving the equation for y and then replacing y with $f(x)$. This process can be reversed, so the linear function $f(x) = mx + b$ appears in equation form as $y = mx + b$, a linear equation written in slope-intercept form. Thus, the graph of a linear function is a straight line with slope m and y-intercept $(0, b)$.

As we noted in Section 3.1, the graph of a function is a plot of all the ordered pairs that make up the function; that is, the graph of a function f is the plot of all the ordered pairs in the set $\{(x, y) \mid f(x) = y\}$. We have a great deal of experience in plotting such sets if the ordered pairs are defined by an equation in x and y, but we have only plotted a few functions that have been defined with function notation. Any function of x defined with function notation can be written as an equation in x and y by replacing $f(x)$ with y, so the graph of a function f consists of a plot of the ordered pairs in the set $\{(x, f(x)) \mid x \in \text{domain of } f\}$.

Consider the function $f(x) = -3x + 5$. Figure 1 contains a table of four ordered pairs defined by the function and a graph of the function with the four ordered pairs noted.

x	$f(x)$
-1	8
0	5
1	2
2	-1

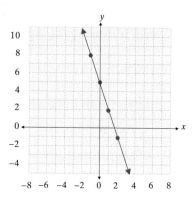

Figure 1: Graph of $f(x) = -3x + 5$

Again, note that every point on the graph of the function in Figure 1 is an ordered pair of the form $(x, f(x))$; we have simply highlighted four of them with dots.

We could have graphed the function $f(x) = -3x + 5$ by noting that it is a straight line with a slope of -3 and a y-intercept of 5. We use this approach in the following example.

EXAMPLE 1

Graphing Linear Functions

Graph the following linear functions.

a. $f(x) = 3x + 2$

b. $g(x) = 3$

Note:
A function cannot represent a vertical line (since it fails the vertical line test). Vertical lines can represent the graphs of equations, but not functions.

Solutions:

a.

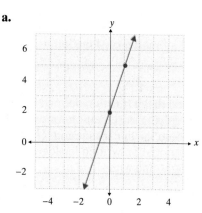

The function f is a line with a slope of 3 and a y-intercept of 2.

To graph the function, plot the ordered pair (0, 2) and locate another point on the line by moving up 3 units and over to the right 1 unit, giving the ordered pair (1, 5). Once these two points have been plotted, connecting them with a straight line completes the process.

b.

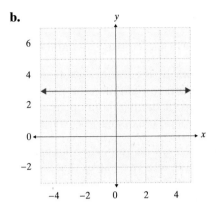

The graph of the function g is a straight line with a slope of 0 and a y-intercept of 3.

A linear function with a slope of 0 is also called a **constant** function, as it turns any input into one fixed constant—in this case the number 3. The graph of a constant function is always a horizontal line.

TOPIC 2

Quadratic Functions and Their Graphs

In Section 1.7, we learned how to solve quadratic equations in one variable. We will now study quadratic *functions* of one variable and relate this new material to what we already know.

DEFINITION

Quadratic Functions

A **quadratic function**, or **second-degree function**, of one variable is any function that can be written in the form $f(x) = ax^2 + bx + c$, where a, b, and c are real numbers and $a \neq 0$.

The graph of any quadratic function is a roughly U-shaped curve known as a **parabola**. We will study parabolas further in Chapter 9, but in this section we will learn how to graph parabolas as they arise in the context of quadratic functions.

The graph in Figure 2 is the most basic example of a parabola; it is the graph of the quadratic function $f(x) = x^2$, and the table that appears alongside the graph contains a few of the ordered pairs on the graph.

x	$f(x)$
-3	9
-1	1
0	0
2	4

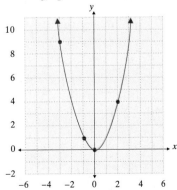

Figure 2: Graph of $f(x) = x^2$

DEFINITION

Figure 2 demonstrates two key characteristics of parabolas:

> There is one point, known as the **vertex**, where the graph "changes direction." Scanning the graph from left to right, it is the point where the graph stops going down and begins to go up (if the parabola opens upward) or stops going up and begins to go down (if the parabola opens downward).

> Every parabola is symmetric with respect to its **axis**, a line passing through the vertex dividing the parabola into two halves that are mirror images of each other. This line is also called the **axis of symmetry**.

Every parabola that represents the graph of a quadratic function has a vertical axis, but we will see parabolas later in the text that have nonvertical axes. Finally, parabolas can be relatively skinny or relatively broad, meaning that the curve of the parabola at the vertex can range from very sharp to very flat.

We will develop our graphing method by working from the answer backward. We will first see what effects various mathematical operations have on the graphs of parabolas, and then see how this knowledge lets us graph a general quadratic function.

To begin, the graph of the function $f(x) = x^2$, shown in Figure 2, is the basic parabola. We already know its characteristics: its vertex is at the origin, its axis is the y-axis, it opens upward, and the sharpness of the curve at its vertex will serve as a convenient reference when discussing other parabolas.

Now consider the function $g(x) = (x-3)^2$, obtained by replacing x in the formula for f with $x - 3$. We know x^2 is equal to 0 when $x = 0$. What value of x results in $(x-3)^2$ equaling 0? The answer is $x = 3$. In other words, the point $(0, 0)$ on the graph of f corresponds to the point $(3, 0)$ on the graph of g. With this in mind, examine the table and graph in Figure 3.

x	$g(x)$
0	9
2	1
3	0
5	4

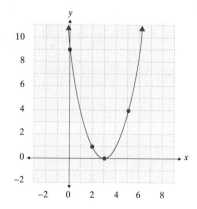

Figure 3: Graph of $g(x) = (x-3)^2$

Notice that the shape of the graph of g is identical to that of f, but it has been shifted over to the right by 3 units. This is our first example of how we can manipulate graphs of functions, a topic we will fully explore in Section 3.5.

Now consider the function h obtained by replacing the x in x^2 with $x + 7$. As with the functions f and g, $h(x) = (x+7)^2$ is nonnegative for all values of x, and only one value for x will return a value of 0: $h(-7) = 0$. Compare the table and graph in Figure 4 with those in Figures 2 and 3.

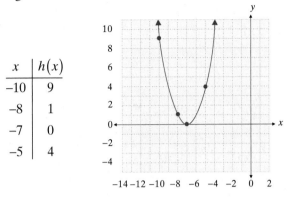

x	$h(x)$
-10	9
-8	1
-7	0
-5	4

Figure 4: Graph of $h(x) = (x+7)^2$

So we have seen how to shift the basic parabola to the left and right: the graph of $g(x) = (x-h)^2$ has the same shape as the graph of $f(x) = x^2$, but it is shifted h units to the right if h is positive and h units to the left if h is negative.

How do we shift a parabola up and down? To move the graph of $f(x) = x^2$ up by a fixed number of units, we need to add that number of units to the second coordinate of each ordered pair. Similarly, to move the graph down we subtract the desired number of units from each second coordinate. To see this, consider the table and graphs for the two functions $j(x) = x^2 + 5$ and $k(x) = x^2 - 2$ in Figure 5.

x	$j(x)$	$k(x)$
-3	14	7
-1	6	-1
0	5	-2
2	9	2

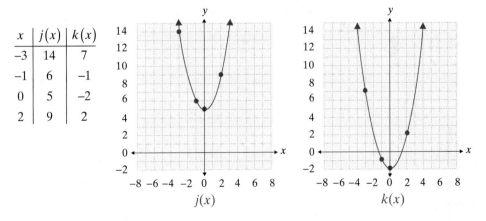

$j(x)$ $k(x)$

Figure 5: Graph of $j(x) = x^2 + 5$ and $k(x) = x^2 - 2$

Finally, how do we make a parabola skinnier or broader? To make the basic parabola skinnier (to make the curve at the vertex sharper), we need to stretch the graph vertically. We can do this by multiplying the formula x^2 by a constant a greater than 1 to obtain the formula ax^2. Multiplying the formula x^2 by a constant a that lies between 0 and 1 makes the parabola broader (it makes the curve at the vertex flatter). Finally, multiplying x^2 by a negative constant a turns all of the nonnegative outputs of f into nonpositive outputs, resulting in a parabola that opens downward instead of upward.

Compare the graphs of $l(x) = 6x^2$ and $m(x) = -\dfrac{1}{2}x^2$ in Figure 6 to the basic parabola $f(x) = x^2$, which is shown as a green curve.

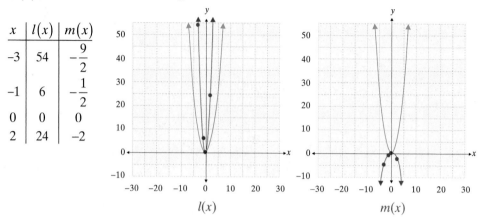

x	$l(x)$	$m(x)$
-3	54	$-\dfrac{9}{2}$
-1	6	$-\dfrac{1}{2}$
0	0	0
2	24	-2

Figure 6: Graph of $l(x) = 6x^2$ and $m(x) = -\dfrac{1}{2}x^2$

The following form of a quadratic function brings together all of the above ways of altering the basic parabola $f(x) = x^2$.

DEFINITION

Vertex Form of a Quadratic Function

The graph of the function $g(x) = a(x - h)^2 + k$, where a, h, and k are real numbers and $a \neq 0$, is a parabola whose vertex is at (h, k). The parabola is narrower than $f(x) = x^2$ if $|a| > 1$, and is broader than $f(x) = x^2$ if $0 < |a| < 1$. The parabola opens upward if a is positive and downward if a is negative.

Where does this definition come from? Consider this construction of the vertex form:

$f(x) = x^2$ Begin with the basic parabola, with vertex $(0, 0)$.

$g(x) = (x - h)^2$ This represents a horizontal shift of h.

$g(x) = (x - h)^2 + k$ This step adds a vertical shift of k, making the new vertex (h, k).

$g(x) = a(x - h)^2 + k$ Finally, apply the stretch/compress factor of a. If a is negative, the parabola opens downward.

The question now is: given a quadratic function $f(x) = ax^2 + bx + c$, how do we determine the location of its vertex, whether it opens upward or downward, and whether it is skinnier or broader than the basic parabola? All of this information is available if the equation is in vertex form, so we need to convert the formula $ax^2 + bx + c$ into the form $a(x - h)^2 + k$. It turns out that we can *always* do this by completing the square on the first two terms of the expression.

EXAMPLE 2

Graphing Quadratic Functions

Sketch the graph of the function $f(x) = -x^2 - 2x + 3$. Locate the vertex and the x-intercepts.

Note:
Finding and plotting the x-intercepts is a great way to see the shape of the function.

Solution:

First, identify the vertex of the function by completing the square as shown below.

$$f(x) = -x^2 - 2x + 3$$

First, factor out the leading coefficient of -1 from the first two terms.

$$= -\left(x^2 + 2x\right) + 3$$

Complete the square on the x^2 and $2x$ terms.

$$= -\left(x^2 + 2x + 1\right) + 1 + 3$$

$$= -\left(x + 1\right)^2 + 4$$

Because of the -1 in front of the parentheses, this amounts to adding -1 to the function, so we compensate by adding 1 as well.

Completing the square places the equation in vertex form, and we rewrite the expression $-\left(x+1\right)^2 + 4$ as $-\left(x - \left(-1\right)\right)^2 + 4$, so the vertex is $\left(-1, 4\right)$.

The instructions also ask us to identify the x-intercepts. An x-intercept of the function f is any point on the x-axis where $f(x) = 0$, so we need to solve the equation $-x^2 - 2x + 3 = 0$. This can be done by factoring:

$$-x^2 - 2x + 3 = 0$$

First, divide each term by -1.

$$x^2 + 2x - 3 = 0$$

Factor into two binomials.

$$(x + 3)(x - 1) = 0$$

The Zero-Factor Property gives us the x-intercepts.

$$x = -3, 1$$

Therefore, the x-intercepts are located at $\left(-3, 0\right)$ and $\left(1, 0\right)$.

The vertex form of the function, $f(x) = -\left(x+1\right)^2 + 4$, tells us that this quadratic opens downward, has its vertex at $\left(-1, 4\right)$, and is neither skinnier nor broader than the basic parabola. We now also know that it crosses the x-axis at -3 and 1. Putting this all together, we obtain the following graph.

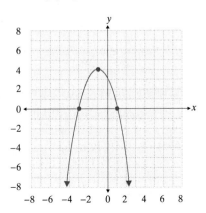

In Section 1.7, we completed the square on the generic quadratic equation to develop the quadratic formula. We can use a similar approach to transform the standard form of a quadratic function into vertex form.

$$f(x) = ax^2 + bx + c$$

$$= a\left(x^2 + \frac{b}{a}x\right) + c$$

As always, begin by factoring the leading coefficient a from the first two terms.

$$= a\left(x^2 + \frac{b}{a}x + \frac{b^2}{4a^2}\right) - a\left(\frac{b^2}{4a^2}\right) + c$$

To complete the square, add the square of half of $\frac{b}{a}$ inside the parentheses. We need to balance the equation by subtracting $a\left(\frac{b^2}{4a^2}\right)$ outside the parentheses, then simplify.

$$= a\left(x + \frac{b}{2a}\right)^2 - \frac{b^2}{4a} + c$$

$$= a\left(x + \frac{b}{2a}\right)^2 + \frac{4ac - b^2}{4a}$$

THEOREM

Vertex of a
Quadratic Function

Given a quadratic function $f(x) = ax^2 + bx + c$, the graph of f is a parabola with a vertex given by:

$$\left(-\frac{b}{2a}, f\left(\frac{-b}{2a}\right)\right) = \left(-\frac{b}{2a}, \frac{4ac - b^2}{4a}\right).$$

EXAMPLE 3

Using the Vertex
Formula

Find the vertex of the following quadratic functions using the vertex formula.

a. $f(x) = x^2 - 4x + 8$ **b.** $g(x) = 3x^2 + 5x - 1$

Note:
If the x-coordinate of the vertex is simple, use substitution to find the y-coordinate. If the x-coordinate is complicated, use the explicit formula (the right-hand form in the definition above).

Solutions:

a. Begin by using the formula to find the x-coordinate of the vertex:

$$1x^2 - 4x + 8$$ Note that the value of a is 1.

$$-\frac{b}{2a} = -\frac{(-4)}{2(1)} = 2$$ Substitute a and b into the formula and simplify.

At this point, we need to decide how to find the y-coordinate. Since the x-coordinate is an integer, substitute it directly into the original equation, finding $f\left(-\frac{b}{2a}\right)$.

$$f(2) = 2^2 - 4(2) + 8$$

$$= 4$$

Thus, the vertex of the graph of $f(x)$ is $(2, 4)$.

b. Again, begin by finding the x-coordinate of the vertex.

$3x^2 + 5x - 1$

$$-\frac{b}{2a} = -\frac{(5)}{2(3)} = -\frac{5}{6}$$ Substitute a and b into the formula and simplify.

Here, the x-coordinate is a fraction, so substituting it into the original equation leads to messy calculations. Instead, use the explicit formula to find the y-coordinate.

$$\frac{4ac - b^2}{4a} = \frac{4(3)(-1) - (5)^2}{4(3)}$$ Substitute a, b, and c into the formula and simplify.

$$= \frac{-12 - 25}{12}$$

$$= -\frac{37}{12}$$

Thus, the vertex of the graph of $g(x)$ is $\left(-\dfrac{5}{6}, -\dfrac{37}{12}\right)$.

TOPIC 3

Maximization/Minimization Problems

Many applications of mathematics involve determining the value (or values) of the variable x that return either the maximum or minimum possible value of some function $f(x)$. Such problems are called Max/Min problems for short. Examples from business include minimizing cost functions and maximizing profit functions. Examples from physics include maximizing a function that measures the height of a rocket as a function of time and minimizing a function that measures the energy required by a particle accelerator.

If we have a Max/Min problem involving a quadratic function, we can solve it by finding the vertex. Recall that the vertex is the only point where the graph of a parabola changes direction. This means it will be the minimum value of a function (if the parabola opens upward) or the maximum value (if the parabola opens downward).

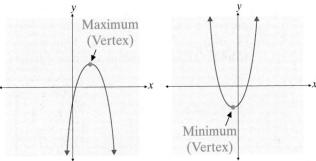

Parabola opening downward   Parabola opening upward

Figure 7: Maximum/Minimum Values of Quadratic Functions

─────────────────────── EXAMPLE 4 ───────────────────────

Fencing a Garden

A farmer plans to use 100 feet of spare fencing material to form a rectangular garden plot against the side of a long barn, using the barn as one side of the plot. How should he split up the fencing among the other three sides in order to maximize the area of the garden plot?

Solution:

If we let x represent the length of one side of the plot, as shown in the diagram below, then the dimensions of the plot are x feet by $100 - 2x$ feet. A function representing the area of the plot is $A(x) = x(100 - 2x)$.

If we multiply out the formula for A, we recognize it as a quadratic function $A(x) = -2x^2 + 100x$. This is a parabola opening downward, so the vertex will be the maximum point on the graph of A.

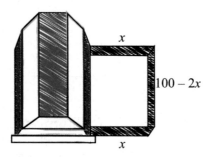

Using the vertex formula we know that the vertex of A is the ordered pair $\left(-\dfrac{100}{2(-2)}, A\left(-\dfrac{100}{2(-2)} \right) \right)$, or $(25, A(25))$. Thus, to maximize area, we should let $x = 25$, and so $100 - 2x = 50$. The resulting maximum possible area, 25×50, or 1250 square feet, is also the value $A(25)$.

TOPIC   Maximum/Minimum of Graphs

As we've seen, finding the maximum or minimum possible values of some function $f(x)$ can be extremely important, and we have a method for doing so when the function is quadratic. But what if we wanted to find the minimum of the function $f(x) = x^4 + 2x^3 - 7x^2 + 2x - 4$? One way is to graph it on a calculator, shown below with the following window settings: $\mathsf{Xmin} = -5, \mathsf{Xmax} = 5, \mathsf{Ymin} = -100, \mathsf{Ymax} = 10$.

To find the minimum, press 2ND TRACE to access the CALC menu and select 3: minimum. (If we were trying to find the maximum, we would select 4: maximum.)

1.

2.

3.

4.

The screen should now display the graph with the words "**Left Bound?**" shown at the bottom. Use the arrows to move the cursor anywhere to the left of where the minimum appears to be and press **ENTER**. The screen should now say "**Right Bound?**" Use the right arrow to move the cursor to the right of where the minimum appears to be and press **ENTER** again. The text should now read "**Guess?**" Press **ENTER** a third time and the x- and y-values of the minimum will appear at the bottom of the screen.

So the minimum is approximately $(-2.809, -46.920)$.

================== **Exercises** ==================

Graph the following linear functions. See Example 1.

1. $f(x) = -5x + 2$

2. $g(x) = \dfrac{3x - 2}{4}$

3. $h(x) = -x + 2$

4. $p(x) = -2$

5. $g(x) = 3 - 2x$

6. $r(x) = 2 - \dfrac{x}{5}$

7. $f(x) = -2(1 - x)$

8. $a(x) = 3\left(1 - \dfrac{1}{3}x\right) + x$

9. $f(x) = 2 - 4x$

10. $g(x) = \dfrac{2x - 8}{4}$

11. $h(x) = 5x - 10$

12. $k(x) = 3x - \dfrac{2 + 6x}{2}$

13. $m(x) = \dfrac{-x + 25}{10}$

14. $q(x) = 1.5x - 1$

15. $w(x) = (x - 2) - (2 + x)$

Graph the following quadratic functions, locating the vertices and x-intercepts (if any) accurately. See Example 2.

16. $f(x) = (x - 2)^2 + 3$

17. $g(x) = -(x + 2)^2 - 1$

18. $h(x) = x^2 + 6x + 7$

19. $F(x) = 3x^2 + 2$

20. $G(x) = x^2 - x - 6$

21. $p(x) = -2x^2 + 2x + 12$

22. $q(x) = 2x^2 + 4x + 3$

23. $r(x) = -3x^2 - 1$

24. $s(x) = \dfrac{(x - 1)^2}{4}$

5.

6.

7.

8.

9.

10.

11.

12.

13.

14.

15.

25. $m(x) = x^2 + 2x + 4$ **26.** $n(x) = (x+2)(2-x)$ **27.** $p(x) = -x^2 + 2x - 5$

28. $f(x) = 4x^2 - 6$ **29.** $k(x) = 2x^2 - 4x$ **30.** $q(x) = (x+10)(x-2) + 36$

Match the following functions with their graphs.

31. $f(x) = (8x - 14) - (-17 + 2x)$

16. Vertex: $(2, 3)$
No x-int.

a.   **b.**

32. $f(x) = -x^2 + 2x$

17. Vertex: $(-2, -1)$
No x-int.

33. $f(x) = x^2 + 7x + 6$

c.   **d.**

34. $f(x) = 3x - \dfrac{7 + 8x}{3}$

18. Vertex: $(-3, -2)$
x-int.:

$x = -3 - \sqrt{2},$

$x = -3 + \sqrt{2}$

35. $f(x) = \dfrac{6}{2} - \dfrac{2}{8}x$

e.   **f.**

36. $f(x) = 2\left(2 - \dfrac{8}{5}x\right) + x$

19. Vertex: $(0, 2)$
No x-int.

37. $f(x) = \dfrac{x^2 - 8x + 16}{2}$

g.   **h.**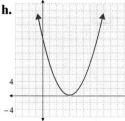

38. $f(x) = (x-5)(x+3) + 16$

20. Vertex: $\left(\dfrac{1}{2}, -\dfrac{25}{4}\right)$
x-int.: $x = -2, 3$

21. Vertex: $\left(\dfrac{1}{2}, \dfrac{25}{2}\right)$

x-int.: $x = -2, 3$

22. Vertex: $(-1, 1)$
No x-int.

23. Vertex: $(0, -1)$
No x-int.

24. Vertex: $(1, 0)$,
x-int.: $x = 1$

25. Vertex: $(-1, 3)$
No x-int.

26. Vertex: $(0, 4)$
x-int.: $x = -2, 2$

27. Vertex: $(1, -4)$
No x-int.

28. Vertex: $(0, -6)$
x-int.: $x = \pm\dfrac{\sqrt{6}}{2}$

Solve the following application problems. See Example 4.

39. Cindy wants to construct three rectangular dog-training arenas side-by-side, as shown, using a total of 400 feet of fencing. What should the overall length and width be in order to maximize the area of the three combined arenas? (Suggestion: let x represent the width, as shown, and find an expression for the overall length in terms of x.)

40. Among all the pairs of numbers with a sum of 10, find the pair whose product is maximum.

41. Among all rectangles that have a perimeter of 20, find the dimensions of the one whose area is largest.

42. Find the point on the line $2x + y = 5$ that is closest to the origin. (**Hint:** Instead of trying to minimize the distance between the origin and points on the line, minimize the square of the distance.)

43. Among all the pairs of numbers (x, y) such that $2x + y = 20$, find the pair for which the sum of the squares is minimum.

44. A rancher has a rectangular piece of sheet metal that is 20 inches wide by 10 feet long. He plans to fold the metal into a three-sided channel and weld two other sheets of metal to the ends to form a watering trough 10 feet long, as shown. How should he fold the metal in order to maximize the volume of the resulting trough?

10 feet

45. Find a pair of numbers whose product is maximum if the pair must have a sum of 16.

46. Search the Seas cruise ship has a conference room offering unlimited internet access that can hold up to 60 people. Companies can reserve the room for groups of 38 or more. If the group contains 38 people, the company pays $60 per person. The cost per person is reduced by $1 for each person in excess of 38. Find the size of the group that maximizes the income for the owners of the ship and find this income.

29. Vertex: $(1, -2)$
x-int.: $x = 0, 2$

30. Vertex: $(-4, 0)$
x-int.: $x = -4$

31. g

32. e

33. a

34. f

35. c

36. b

37. h

38. d

39. Width of 50 feet, length of 100 feet

40. 5 and 5

41. Width and length are 5

42. $(2, 1)$

43. $(8, 4)$

44. The dimensions should be 5 inches by 10 inches by 10 feet

45. 8 and 8

46. 49 people; $2401

47. 11,250 square feet

48. 12 and 24

49. 500 rooms

50. 375 units

51. 1500 cars

52. 25 sets of golf clubs

53. 6050 square feet

54. 112 feet

47. The back of George's property is a creek. George would like to enclose a rectangular area, using the creek as one side and fencing for the other three sides, to create a pasture for his two horses. If he has 300 feet of material, what is the maximum possible area of the pasture?

300 feet of fencing

48. Find a pair of numbers whose product is maximum if two times the first number plus the second number is 48.

49. The total revenue for Thompson's Studio Apartments is given as the function

$$R(x) = 100x - 0.1x^2,$$

where x is the number of rooms rented. What number of rooms rented produces the maximum revenue?

50. The total revenue of Tran's Machinery Rental is given as the function

$$R(x) = 300x - 0.4x^2,$$

where x is the number of units rented. What number of units rented produces the maximum revenue?

51. The total cost of producing a type of small car is given by

$$C(x) = 9000 - 135x + 0.045x^2,$$

where x is the number of cars produced. How many cars should be produced to incur minimum cost?

52. The total cost of manufacturing a set of golf clubs is given by

$$C(x) = 800 - 10x + 0.20x^2,$$

where x is the number of sets of golf clubs produced. How many sets of golf clubs should be manufactured to incur minimum cost?

53. The owner of a parking lot is going to enclose a rectangular area with fencing, using an existing fence as one of the sides. The owner has 220 feet of new fencing material (which is much less than the length of the existing fence). What is the maximum possible area that the owner can enclose?

For each of the following three problems, use the formula $h(t) = -16t^2 + v_0 t + h_0$ for the height at time t of an object thrown vertically with velocity v_0 (in feet per second) from an initial height of h_0 (in feet).

54. Sitting in a tree, 48 feet above ground level, Sue shoots a pebble straight up with a velocity of 64 feet per second. What is the maximum height attained by the pebble?

64 ft/sec

48 ft

55. A ball is thrown upward with a velocity of 48 feet per second from the top of a 144-foot building. What is the maximum height of the ball?

56. A rock is thrown upward with a velocity of 80 feet per second from the top of a 64-foot-high cliff. What is the maximum height of the rock?

Use a graphing calculator to graph the following quadratic functions. Then determine the vertex and x-intercepts.

57. $f(x) = 2x^2 - 16x + 31$

58. $f(x) = -x^2 - 2x + 3$

59. $f(x) = x^2 - 8x - 20$

60. $f(x) = x^2 - 4x$

61. $f(x) = 25 - x^2$

62. $f(x) = 3x^2 + 18x$

63. $f(x) = x^2 + 2x + 1$

64. $f(x) = 3x^2 - 8x + 2$

65. $f(x) = -x^2 + 10x - 4$

66. $f(x) = \dfrac{1}{2}x^2 + x - 1$

55. 180 feet

56. 164 feet

57. Vertex: $(4, -1)$,

x-int.: $x = \dfrac{8 \pm \sqrt{2}}{2}$

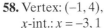

58. Vertex: $(-1, 4)$,
x-int.: $x = -3, 1$

59. Vertex: $(4, -36)$,
x-int.: $x = -2, 10$

60. Vertex: $(2, -4)$,
x-int.: $x = 0, 4$

61. Vertex: $(0, 25)$,
x-int.: $x = -5, 5$

62. Vertex: $(-3, -27)$,
x-int.: $x = -6, 0$

63. Vertex: $(-1, 0)$,
x-int.: $x = -1$

64. Vertex: $\left(\dfrac{4}{3}, \dfrac{-10}{3}\right)$,

x-int.: $x = \dfrac{4}{3} \pm \dfrac{\sqrt{10}}{3}$

65. Vertex: $(5, 21)$,
x-int.: $x = 5 \pm \sqrt{21}$

66. Vertex: $\left(-1, -\dfrac{3}{2}\right)$,

x-int.: $x = -1 \pm \sqrt{3}$

3.3 Other Common Functions

TOPICS

1. Functions of the form ax^n

2. Functions of the form $\dfrac{a}{x^n}$

3. Functions of the form $ax^{\frac{1}{n}}$

4. The absolute value function

5. The greatest integer function

6. Piecewise-defined functions

In Section 3.2, we investigated the behavior of linear and quadratic functions, but these are just two types of commonly occurring functions; there are many other functions that arise naturally in solving various problems. In this section, we will explore several other classes of functions, building up a portfolio of functions to be familiar with.

TOPIC 1

Functions of the Form ax^n

We already know what the graph of any function of the form $f(x) = ax$ or $f(x) = ax^2$ looks like, as these are, respectively, simple linear and quadratic functions. What happens to the graphs as we increase the exponent, and consider functions of the form $f(x) = ax^3$, $f(x) = ax^4$, etc.?

The behavior of a function of the form $f(x) = ax^n$, where a is a real number and n is a natural number, falls into one of two categories. Consider the graphs in Figure 1:

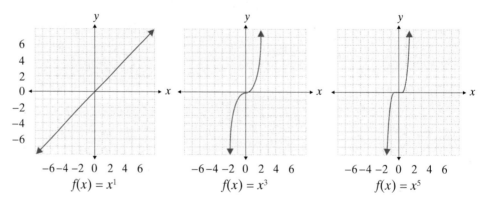

Figure 1: Odd Exponents

The three graphs in Figure 1 show the behavior of $f(x) = x^n$ for the first three odd exponents. Note that in each case, the domain and the range of the function are both the entire set of real numbers; the same is true for higher odd exponents as well. Now, consider the graphs in Figure 2:

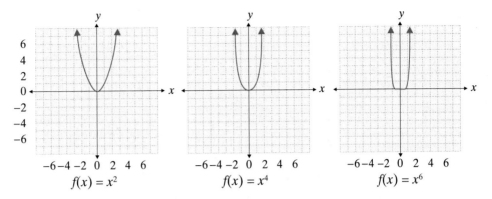

Figure 2: Even Exponents

These three functions are also similar to one another. The first one is the basic parabola we studied in Section 3.2. The other two bear some similarity to parabolas, but are flatter near the origin and rise more steeply for $|x| > 1$. For any function of the form $f(x) = x^n$ where n is an even natural number, the domain is the entire set of real numbers and the range is the interval $[0, \infty)$.

Multiplying a function of the form x^n by a constant a has the effect that we noticed in Section 3.2. If $|a| > 1$, the graph of the function is stretched vertically; if $0 < |a| < 1$, the graph is compressed vertically; and if $a < 0$, the graph is reflected with respect to the x-axis. We can use this knowledge, along with plotting a few specific points, to quickly sketch graphs of any function of the form $f(x) = ax^n$.

EXAMPLE 1

Functions of the Form ax^n

Sketch the graphs of the following functions.

a. $f(x) = \dfrac{x^4}{5}$

b. $g(x) = -x^3$

Solutions:

a.

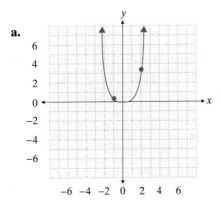

The graph of the function f will have the same basic shape as the function x^4, but compressed vertically because of the factor of $\dfrac{1}{5}$. To make the sketch more accurate, calculate the coordinates of a few points on the graph. The graph to the left illustrates that $f(-1) = \dfrac{1}{5}$ and that $f(2) = \dfrac{16}{5}$.

b.

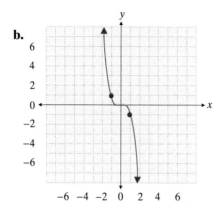

We know that the function g will have the same shape as the function x^3, but reflected over the x-axis because of the factor of -1. The graph to the left illustrates this.

We also plot a few points on the graph of g, namely $(-1, 1)$ and $(1, -1)$, as a check.

TOPIC   2

Functions of the Form $\dfrac{a}{x^n}$

We could also describe the following functions as having the form ax^{-n}, where a is a real number and n is a natural number. Once again, the graphs of these functions fall roughly into two categories, as illustrated in Figures 3 and 4.

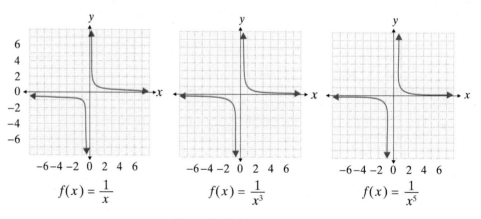

$$f(x) = \frac{1}{x} \qquad f(x) = \frac{1}{x^3} \qquad f(x) = \frac{1}{x^5}$$

Figure 3: Odd Exponents

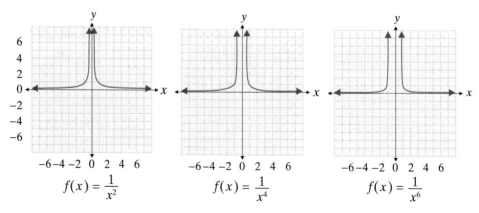

$$f(x) = \frac{1}{x^2} \qquad f(x) = \frac{1}{x^4} \qquad f(x) = \frac{1}{x^6}$$

Figure 4: Even Exponents

As with functions of the form ax^n, increasing the exponent on functions of the form $\dfrac{a}{x^n}$ sharpens the curve of the graph near the origin. Note that the domain of any function of the form $f(x) = \dfrac{a}{x^n}$ is $(-\infty, 0) \cup (0, \infty)$, but that the range depends on whether n is even or odd. When n is odd, the range is also $(-\infty, 0) \cup (0, \infty)$, and when n is even the range is $(0, \infty)$.

EXAMPLE 2

Functions of the Form $\dfrac{a}{x^n}$

Sketch the graph of the function $f(x) = -\dfrac{1}{4x}$.

Solution:

The graph of the function f is similar to that of the function $\dfrac{1}{x}$, with two differences. We obtain the formula $-\dfrac{1}{4x}$ by multiplying $\dfrac{1}{x}$ by $-\dfrac{1}{4}$, a negative number between -1 and 1. So one difference is that the graph of f is the reflection of $\dfrac{1}{x}$ with respect to the x-axis. The other difference is that the graph of f is compressed vertically.

With the above in mind, we can calculate the coordinates of a few points (such as $\left(-\dfrac{1}{4}, 1\right)$ and $\left(1, -\dfrac{1}{4}\right)$) and sketch the graph of f as shown below.

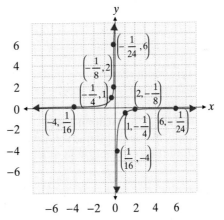

TOPIC 3

Functions of the Form $ax^{\frac{1}{n}}$

Using radical notation, these are functions of the form $a\sqrt[n]{x}$, where a is again a real number and n is a natural number. Square root and cube root functions, in particular, are commonly seen in mathematics.

Functions of this form again fall into one of two categories, depending on whether n is odd or even. To begin with, note that the domain and range are both the entire set of real numbers when n is odd, and that both are the interval $[0, \infty)$ when n is even. Figures 5 and 6 illustrate the two basic shapes of functions of this form.

Figure 5: Odd Roots

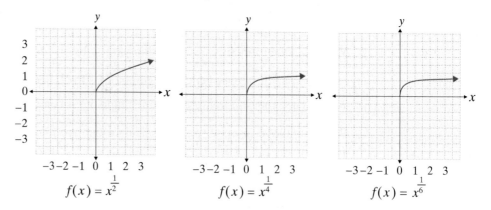

Figure 6: Even Roots

At this point, you may be thinking that the graphs in Figures 5 and 6 appear familiar. The shapes in Figure 5 are the same as those seen in Figure 1, but rotated by 90 degrees and reflected with respect to the x-axis. Similarly, the shapes in Figure 6 bear some resemblance to those in Figure 2, except that half of the graphs appear to have been erased. This resemblance is no accident, given that n^{th} roots undo n^{th} powers. We will explore this observation in much more detail in Section 3.7.

TOPIC 4

The Absolute Value Function

The basic absolute value function is $f(x) = |x|$. Note that for any value of x, $f(x)$ is nonnegative, so the graph of f should lie on or above the x-axis. One way to determine its exact shape is to review the definition of absolute value:

$$|x| = \begin{cases} x & \text{if } x \geq 0 \\ -x & \text{if } x < 0 \end{cases}$$

This means that for nonnegative values of x, $f(x)$ is a linear function with a slope of 1, and for negative values of x, $f(x)$ is a linear function with a slope of -1. Both linear functions have a y-intercept of 0, so the complete graph of f is as shown in Figure 7.

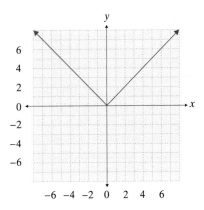

Figure 7: The Absolute Value Function

The effect of multiplying $|x|$ by a real number a is what we have come to expect: if $|a| > 1$, the graph is stretched vertically; if $0 < |a| < 1$, the graph is compressed vertically; and if a is negative, the graph is reflected with respect to the x-axis.

EXAMPLE 3

The Absolute Value Function

Sketch the graph of the function $f(x) = -2|x|$.

Solution:

The graph of f will be a vertically stretched version of $|x|$, reflected over the x-axis. As always, we can plot a few points to verify that our reasoning is correct. In the graph below, we have plotted the values of $f(-4)$ and $f(2)$.

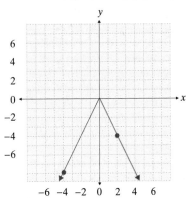

TOPIC 5

The Greatest Integer Function

DEFINITION

The Greatest Integer Function

The greatest integer function, $f(x) = [\![x]\!]$, is a function commonly encountered in computer science applications. It is defined as follows: the **greatest integer of x** is the largest integer less than or equal to x. For instance, $[\![4.3]\!] = 4$ and $[\![-2.9]\!] = -3$ (note that -3 is the largest integer to the left of -2.9 on the real number line).

Careful study of the greatest integer function reveals that its graph must consist of intervals where the function is constant, and that these portions of the graph must be separated by discrete "jumps," or breaks, in the graph. For instance, any value for x chosen from the interval $[1, 2)$ results in $f(x) = 1$, but $f(2) = 2$. Similarly, any value for x chosen from the interval $[-3, -2)$ results in $f(x) = -3$, but $f(-2) = -2$.

Our graph of the greatest integer function must somehow indicate this repeated pattern of jumps. In cases like this, it is conventional to use an open circle on the graph to indicate that the function is either undefined at that point or is defined to be another value. Closed circles are used to emphasize that a certain point really does lie on the graph of the function. With these conventions in mind, the graph of the greatest integer function appears in Figure 8.

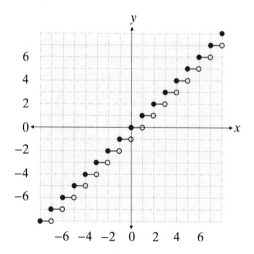

Figure 8: The Greatest Integer Function

TOPIC 6

Piecewise-Defined Functions

There is no rule stating that a function needs to be defined by a single formula. In fact, we have worked with such a function already; in evaluating the absolute value of x, we use one formula if x is greater than or equal to 0 and a different formula if x is less than 0. Obviously, we can't have two rules govern the same input, but we can have multiple formulas on separate pieces of a function's domain.

DEFINITION

Piecewise-Defined
Function

A **piecewise-defined** function is a function defined in terms of two or more formulas, each valid for its own unique portion of the real number line. In evaluating a piecewise-defined function f at a certain value for x, it is important to correctly identify which formula is valid for that particular value.

EXAMPLE 4

Piecewise-Defined
Function

Note:
Always play close attention to the boundary points of each interval. Remember that only one rule applies at each point.

Sketch the graph of the function $f(x) = \begin{cases} -2x - 2 & \text{if } x \le -1 \\ x^2 & \text{if } x > -1 \end{cases}$.

Solution:

The function f is a piecewise function with a different formula for two intervals. To graph f, graph each portion separately, making sure that each formula is applied only on the appropriate interval.

The function f is a linear function on the interval $(-\infty, -1]$ and a quadratic function on the interval $(-1, \infty)$.

The complete graph appears below, with the points $f(-4) = 6$ and $f(2) = 4$ noted in particular.

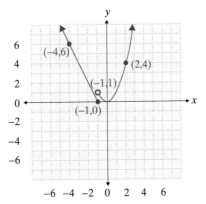

Note the use of a closed circle at $(-1, 0)$ to emphasize that this point is part of the graph, and the use of an open circle at $(-1, 1)$ to indicate that this point is not part of the graph. That is, the value of $f(-1)$ is 0, not 1.

1.

2.

3.

4.

5.

6.

7.

8.

9.

Exercises

Sketch the graphs of the following functions. Pay particular attention to intercepts, if any, and locate these accurately. See Examples 1 through 4.

1. $f(x) = -x^3$

2. $g(x) = 2x^2$

3. $F(x) = \sqrt{x}$

4. $h(x) = \dfrac{1}{x}$

5. $p(x) = -\dfrac{2}{x}$

6. $q(x) = -\sqrt[3]{x}$

7. $G(x) = -|x|$

8. $k(x) = \dfrac{1}{x^3}$

9. $G(x) = \dfrac{\sqrt{x}}{2}$

10. $H(x) = 0.5\sqrt[3]{x}$

11. $r(x) = 3|x|$

12. $p(x) = \dfrac{-1}{x^2}$

13. $W(x) = \dfrac{x^4}{16}$

14. $k(x) = \dfrac{x^3}{9}$

15. $h(x) = 2\sqrt[3]{x}$

16. $S(x) = \dfrac{4}{x^2}$

17. $d(x) = 2x^5$

18. $f(x) = -x^2$

19. $r(x) = \dfrac{\sqrt[3]{x}}{3}$

20. $s(x) = -2|x|$

21. $t(x) = \dfrac{x^6}{4}$

22. $f(x) = 2[\![x]\!]$

23. $P(x) = -[\![x]\!]$

24. $m(x) = \left[\!\!\left[\dfrac{x}{2}\right]\!\!\right]$

10.

11.

12.

13.

14.

15.

16.

17.

18.

19.

25. $f(x) = \begin{cases} 3-x & \text{if } x < -2 \\ \sqrt[3]{x} & \text{if } x \geq -2 \end{cases}$

26. $g(x) = \begin{cases} -x^2 & \text{if } x \leq 1 \\ x^2 & \text{if } x > 1 \end{cases}$

27. $r(x) = \begin{cases} \dfrac{1}{x} & \text{if } x < 1 \\ -x & \text{if } x > 1 \end{cases}$

28. $p(x) = \begin{cases} x+1 & \text{if } x < -2 \\ x^3 & \text{if } -2 \leq x < 3 \\ -1-x & \text{if } x \geq 3 \end{cases}$

29. $q(x) = \begin{cases} -1 & \text{if } x \in \mathbb{Z} \\ 1 & \text{if } x \notin \mathbb{Z} \end{cases}$

30. $s(x) = \begin{cases} \dfrac{x^2}{3} & \text{if } x < 0 \\ -\dfrac{x^2}{3} & \text{if } x \geq 0 \end{cases}$

31. $v(x) = \begin{cases} x^2 & \text{if } -1 \leq x \leq 1 \\ |x| & \text{if } x < -1 \text{ or } x > 1 \end{cases}$

32. $M(x) = \begin{cases} x & \text{if } x \in \mathbb{Z} \\ -x & \text{if } x \notin \mathbb{Z} \end{cases}$

33. $h(x) = \begin{cases} -|x| & \text{if } x < 2 \\ [\![x]\!] & \text{if } x \geq 2 \end{cases}$

34. $u(x) = \begin{cases} [\![x]\!] & \text{if } x \leq 1 \\ 2x-2 & \text{if } x > 1 \end{cases}$

20.

24.

28.

32.

21.

25.

29.

33.

22.

26.

30.

34.

23.

27.

31.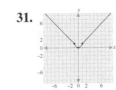

35. f

36. d

37. g

38. a

39. h

40. e

41. b

42. c

Match the following functions to their graphs.

35. $f(x) = -2x^4$

36. $f(x) = -\dfrac{7}{9x^4}$

37. $f(x) = -\dfrac{7\sqrt[3]{x}}{3}$

38. $f(x) = -\dfrac{8}{9}|x|$

39. $f(x) = -4\sqrt{x}$

40. $f(x) = \dfrac{3}{7}|x|$

41. $f(x) = \begin{cases} -4x - 12 & \text{if } x \le -3 \\ \dfrac{5}{10}x^2 & \text{if } x > -3 \end{cases}$

42. $f(x) = \begin{cases} -\dfrac{1}{3}|x| & \text{if } x < 2 \\ \dfrac{x}{2} & \text{if } x \ge 2 \end{cases}$

a.

b.

c.

d.

e.

f.

g.

h.

3.4

Variation and Multivariable Functions

TOPICS

1. Direct variation
2. Inverse variation
3. Joint variation
4. Multivariable functions

TOPIC 1

Direct Variation

A number of natural phenomena exhibit the mathematical property of variation: one quantity varies (or changes) as a result of a change in another quantity. One example is the electrostatic force of attraction between two oppositely charged particles, which varies in response to the distance between the particles. Another example is the distance traveled by a falling object, which varies as time increases. Of course, the principle underlying variation is that of functional dependence; in the first example, the force of attraction is a function of distance, and in the second example the distance traveled is a function of time.

We have now gained enough familiarity with functions that we can define the most common forms of variation.

DEFINITION

Direct Variation

We say that y **varies directly as the nth power of** x (or that y is **proportional to the nth power of** x) if there is a nonzero constant k (called the **constant of proportionality**) such that

$$y = kx^n.$$

Many variation problems involve determining what, exactly, the constant of proportionality is in a given situation. This can be easily done if enough information is given about how the various quantities in the problem vary with respect to one another, and once k is determined many other questions can be answered. The following example illustrates the solution of a typical direct variation problem.

EXAMPLE 1

Direct Variation

Hooke's Law says that the force exerted by the spring in a spring scale varies directly with the distance that the spring is stretched. If a 5-pound mass suspended on a spring scale stretches the spring 2 inches, how far will a 13-pound mass stretch it?

Solution:

The first equation tells us that $F = kx$, where F represents the force exerted by the spring and x represents the distance that the spring is stretched. When a mass is suspended on a spring scale (and is stationary), the force exerted upward by the spring must equal the force downward due to gravity, so the spring exerts a force of 5 pounds when a 5-pound mass is suspended from it. So the second sentence tells us that

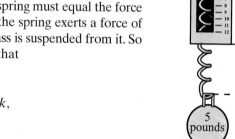

$$5 = 2k,$$

or $k = \dfrac{5}{2}$. We can now answer the question:

$$13 = \frac{5}{2}x$$

$$\frac{26}{5} = x.$$

So the spring stretches 5.2 inches when a 13-pound mass is suspended from it.

TOPIC 2

Inverse Variation

In many situations, an increase in one quantity results in a corresponding decrease in another quantity, and vice versa. Again, this is a natural illustration of a functional relationship between quantities, and an appropriate name for this type of relationship is *inverse variation*.

DEFINITION

Inverse Variation

We say that y **varies inversely as the nth power of** x (or that y is **inversely proportional to the nth power of** x) if there is a nonzero constant k such that

$$y = \frac{k}{x^n}.$$

The method of solving an inverse variation problem is identical to that seen in the first example. First, write an equation that expresses the nature of the relationship (including the as-yet-unknown constant of proportionality). Second, use the given information to determine the constant of proportionality. Third, use the knowledge gained to answer the question.

EXAMPLE 2

Inverse Variation

The weight of a person, relative to the Earth, is inversely proportional to the square of the person's distance from the center of the Earth. Using a radius for the Earth of 6370 kilometers, how much does a 180-pound man weigh when flying in a jet 9 kilometers above the Earth's surface?

Solution:

If we let W stand for the weight of a person and d the distance between the person and the Earth's center, the first sentence tells us that

$$W = \frac{k}{d^2}.$$

The second sentence gives us enough information to determine k. Namely, we know that $W = 180$ (pounds) when $d = 6370$ (kilometers). Solving the equation for k and substituting in the values that we know, we obtain

$$k = Wd^2 = (180)(6370)^2 \approx 7.3 \times 10^9.$$

When the man is 9 kilometers above the Earth's surface, we know $d = 6379$, so the man's weight while flying is

$$W = \frac{(180)(6370)^2}{(6379)^2}$$
$$= 179.49 \text{ pounds.}$$

Flying is not, therefore, a terribly effective way to lose weight.

TOPIC 3

Joint Variation

In more complicated situations, it may be necessary to identify more than two variables and to express how the variables relate to one another. And it may very well be the case that one quantity varies directly with respect to some variables and inversely with respect to others. For instance, the force of gravitational attraction F between two bodies of mass m_1 and mass m_2 varies directly as the product of the masses and inversely as the square of the distance between the masses: $F = \dfrac{km_1m_2}{d^2}$.

When one quantity varies directly as two or more other quantities, the word *jointly* is often used.

DEFINITION

Joint Variation

We say that z **varies jointly as** x and y (or that z is **jointly proportional to x and y**) if there is a nonzero constant k such that

$$z = kxy.$$

If z varies **jointly as the nth power of x and the mth power of** y, we write

$$z = kx^n y^m.$$

EXAMPLE 3

Joint Variation

The volume of a right circular cylinder varies jointly as the height and the square of the radius. Express this relationship in equation form.

Solution:

This simple problem merely asks for the form of the variation equation. If we let V stand for the volume of a right circular cylinder, r for its radius, and h for its height, we would write

$$V = kr^2h.$$

Of course, we are already familiar with this volume formula and know that the constant of proportionality is actually π, so we could provide more information and write

$$V = \pi r^2 h.$$

TOPIC 4   Multivariable Functions

The topic of variation provides an excellent opportunity to introduce functions that depend on two or more arguments. Examples abound in both pure and applied mathematics, and as you progress through calculus and later math classes you will encounter such functions frequently.

Consider again how the force of gravity F between two objects depends on the masses m_1 and m_2 of the objects and the distance d between them:

$$F = \frac{km_1m_2}{d^2}.$$

If we change any of the three quantities m_1, m_2, and d, the force F changes in response. A slight extension of our familiar function notation leads us then to express F as a function of m_1, m_2, and d and to write

$$F(m_1, m_2, d) = \frac{km_1m_2}{d^2}.$$

In fact, we can be a bit more precise and replace the constant of proportionality k with G, the Universal Gravitational Constant. Through many measurements in many different experiments, G has been determined to be approximately 6.67×10^{-11} N·m^2/kg^2 (N stands for the unit of force called the *Newton*; 1 Newton of force gives a mass of 1 kg an acceleration of 1 m/s^2). If we use this value for G, we must be sure to measure the masses of the objects in kilograms and the distance between them in meters.

The next example illustrates an application of this function to which all of us on Earth can relate.

EXAMPLE 4

Finding the Force of Gravity

Determine the approximate force of gravitational attraction between the Earth and the Moon.

Solution:

The mass of the Earth is approximately 6.0×10^{24} kg and the mass of the Moon is approximately 7.4×10^{22} kg. The distance between these two bodies varies, but on average it is 3.8×10^8 m. Using function notation, we would write

$$F\left(6.0 \times 10^{24}, 7.4 \times 10^{22}, 3.8 \times 10^8\right) = \frac{\left(6.67 \times 10^{-11}\right)\left(6.0 \times 10^{24}\right)\left(7.4 \times 10^{22}\right)}{\left(3.8 \times 10^8\right)^2}$$

$$= 2.1 \times 10^{20} \text{ N}.$$

It is this force of mutual attraction that keeps the Moon in orbit about the Earth.

Exercises

Mathematical modeling is the process of finding a function that describes how quantities or variables relate to one another. The function is called the mathematical model. Find the mathematical model for each of the following verbal statements.

1. $A = kbh$

2. $V = \dfrac{4}{3}kr^3$

3. $W = \dfrac{k}{d^2}$

4. $P = \dfrac{k}{V}$

5. $r = \dfrac{k}{t}$

6. $S = 4kr^2$

7. $x = ky^3z^2$

8. $a = \dfrac{kb^2}{c}$

9. $y = 18\sqrt{5}$

10. $y = 0.04$

11. $y = 60\sqrt[3]{2}$

12. $y = 9375$

13. $y = 0.75$

14. $y = 0.0024$

15. $z = 48$

16. $z = 432$

17. $z = 112$

18. $z = 12,544$

19. 256 feet

20. 896 CDs

1. A varies directly as the product of b and h.

2. V varies directly as the product of four-thirds and r cubed.

3. W varies inversely as d squared.

4. P varies inversely as V.

5. r varies inversely as t.

6. S varies directly as the product of four and r squared.

7. x varies jointly as the cube of y and the square of z.

8. a varies jointly as the square of b and inversely as c.

Solve the following variation problems. See Examples 1, 2, and 3.

9. Suppose that y varies directly as the square root of x, and that $y = 36$ when $x = 16$. What is y when $x = 20$?

10. Suppose that y varies inversely as the cube of x, and that $y = 0.005$ when $x = 10$. What is y when $x = 5$?

11. Suppose that y varies directly as the cube root of x, and that $y = 75$ when $x = 125$. What is y when $x = 128$?

12. Suppose that y is proportional to the 5th power of x, and that $y = 96$ when $x = 2$. What is y when $x = 5$?

13. Suppose that y varies inversely as the square of x, and that $y = 3$ when $x = 4$. What is y when $x = 8$?

14. Suppose that y is inversely proportional to the 4th power of x, and that $y = 1.5$ when $x = 4$. What is y when $x = 20$?

15. z varies directly as the square of x and inversely as y. If $z = 36$ when $x = 6$ and $y = 7$, what value does z have when $x = 12$ and $y = 21$?

16. Suppose that z varies jointly as the square of x and the cube of y, and that $z = 768$ when $x = 4$ and $y = 2$. What is z when $x = 3$ and $y = 2$?

17. Suppose that z is jointly proportional to x and y, and that $z = 90$ when $x = 1.5$ and $y = 3$. What is z when $x = 0.8$ and $y = 7$?

18. Suppose that z is jointly proportional to x and the cube of y, and that $z = 9828$ when $x = 13$ and $y = 6$. What is z when $x = 7$ and $y = 8$?

19. The distance that an object falls from rest, when air resistance is negligible, varies directly as the square of the time. A stone dropped from rest travels 144 feet in the first 3 seconds. How far does it travel in the first 4 seconds?

20. A record store manager observes that the number of CDs sold seems to vary inversely as the price per CD. If the store sells 840 CDs per week when the price per CD is $15.99, how many does he expect to sell if he lowers the price to $14.99?

21. 20.60

22. 315 pounds

23. 6.7 meters

24. 20 amps

25. 1.25 centimeters

21. A person's Body Mass Index (BMI) is used by physicians to determine if a patient's weight falls within reasonable guidelines relative to the patient's height. The BMI varies directly as a person's weight in pounds and inversely as the square of a person's height in inches. Given that a 6-foot-tall man weighing 180 pounds has a BMI of 24.41, what is the BMI of a woman weighing 120 pounds with a height of 5 feet 4 inches?

22. The force necessary to keep a car from skidding as it travels along a circular arc varies directly as the product of the weight of the car and the square of the car's speed, and inversely as the radius of the arc. If it takes 241 pounds of force to keep a 2200-pound car moving 35 miles per hour on an arc whose radius is 750 feet, how many pounds of force would be required if the car were to travel 40 miles per hour?

23. If a beam of width w, height h, and length l is supported at both ends, the maximum load that the beam can hold varies directly as the product of the width and the square of the height, and inversely as the length. A given beam 10 meters long with a width of 10 centimeters and a height of 5 centimeters can hold a load of 200 kilograms when the beam is supported at both ends. If the supports are moved inward so that the effective length of the beam is shorter, the beam can support more load. What should the distance between the supports be if the beam has to hold a load of 300 kilograms?

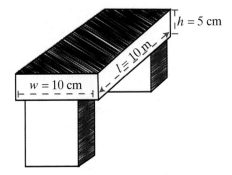

24. In a simple electric circuit connecting a battery and a light bulb, the current I varies directly as the voltage V but inversely as the resistance R. When a 1.5-volt battery is connected to a light bulb with resistance 0.3 ohms (Ω), the current that travels through the circuit is 5 amps. Find the current if the same light bulb is connected to a 6-volt battery.

25. The amount of time it takes for water to flow down a drainage pipe is inversely proportional to the square of the radius of the pipe. If a pipe of radius 1 cm can empty a sink in 25 seconds, find the radius of a pipe that would allow the sink to drain completely in 16 seconds.

26. 532 inches

27. 34.54 inches

28. 70 hot dogs

29. 164.7872 in.²

30. 123.52 pounds

31. 9 watts

32. $\mathrm{BMI}(w,h) = \dfrac{703w}{h^2}$

33. $P(\sigma, \varepsilon) = \dfrac{\sigma^2}{2\varepsilon}$

34. $V(r,h) = 3.14r^2 h$

26. The perimeter of a square varies directly as the length of the side of a square. If the perimeter of a square is 308 inches when one side is 77 inches, what is the perimeter of a square when the side is 133 inches?

27. The circumference of a circle varies directly as the diameter. A circular pizza slice has a length of 6.5 inches when the circumference of the pizza is 40.82 inches. What would the circumference of a pizza be if the pizza slice has a length of 5.5 inches?

28. A hot dog vendor has determined that the number of hot dogs she sells a day is inversely proportional to the price she charges. The vendor wants to decide if increasing her price by 50 cents will drive away too many customers. On average, she sells 80 hot dogs a day at a price of $3.50. How many hot dogs can she expect to sell if the price is increased by 50 cents?

29. The surface area of a right circular cylinder varies directly as the sum of the radius times the height and the square of the radius. With a height of 18 in. and a radius of 7 in., the surface area of a right circular cylinder is 1099 in.² What would the surface area be if the height equaled 5 in. and the radius equaled 3.2 in.?

30. The gravitational force, F, between an object and the Earth is inversely proportional to the square of the distance from the object to the center of the Earth. If an astronaut weighs 193 pounds on the surface of the Earth, what will this astronaut weigh 1000 miles above the Earth? Assume that the radius of the Earth is 4000 miles.

31. In an electrical schematic, the voltage across a load is directly proportional to the power used by the load but inversely proportional to the current through the load. If a computer is connected to a wall outlet and the computer needs 18 volts to run and absorbs 54 watts of power, the current through the computer is 3 amps. Find the power absorbed by the computer if the same 18-volt computer is attached to a circuit with a loop current of 0.5 amps.

Express the indicated quantities as functions of the other variables. See Example 3.

32. A person's Body Mass Index (BMI) varies directly as a person's weight in pounds and inversely as the square of a person's height in inches. Given that a 6-foot-tall man weighing 180 pounds has a BMI of 24.41, express BMI as a function of weight (w) and height (h).

33. The electric pressure varies directly as the square of the surface charge density (σ) and inversely as the permittivity (ε). If the surface charge density is 6 coulombs per unit area and the free space permittivity equals 3, the pressure is equal to 6 N/m². Express the electric pressure as a function of surface charge density and permittivity.

34. The volume of a right circular cylinder varies directly as the radius squared times the height of the cylinder. If the radius is 7 and the height is 4, the volume is equal to 615.44. Determine the expression of the volume of a right circular cylinder.

35. $10\sqrt{3}$

36. $a = \dfrac{b}{3}; \dfrac{7}{3}$

37. $a = \dfrac{9b^2}{4}; 36$

38. $a = \dfrac{6b^2}{\sqrt{c}}; 32$

39. $a = 3bc; 108$

40. $a = \dfrac{7b^3}{24c}; \dfrac{3}{8}$

41. $P = 2.15g; \$43$

42. $I = \dfrac{9i}{d^2}; 11.25$ fc

43. $F = \dfrac{15d}{9}; 12$ cm

44. $V = 3.14r^2h;$
904.32 in.³

45. $V = \dfrac{800}{P}; 200$ cm³

46. $F = \dfrac{2ab}{3c}; 16$

47. $R = \dfrac{0.000009l}{d^2};$
17.28 ohms

Find an equation for the relationship given and then use the equation to find the unknown value.

35. The variable a is proportional to $\sqrt{b}$. If $a = 15$ when $b = 9$, what is a when $b = 12$?

36. The variable a varies directly as b. If $a = 3$ when $b = 9$, what is a when $b = 7$?

37. The variable a varies directly as the square of b. If $a = 9$ when $b = 2$, what is a when $b = 4$?

38. The variable a is proportional to the square of b and varies inversely as the square root of c. If $a = 108$ when $b = 6$ and $c = 4$, what is a when $b = 4$ and $c = 9$?

39. The variable a varies jointly as b and c. If $a = 210$ when $b = 14$ and $c = 5$, what is the value of a when $b = 6$ and $c = 6$?

40. The variable a varies directly as the cube of b and inversely as c. If $a = 9$ when $b = 6$ and $c = 7$, what is the value of a when $b = 3$ and $c = 21$?

41. The price of gasoline purchased varies directly with the number of gallons of gas purchased. If 16 gallons of gas are purchased for $34.40, what is the price of purchasing 20 gallons?

42. The illumination, I, of a light source varies directly as the intensity, i, and inversely as the square of the distance, d. If a light source with an intensity of 500 cp (candlepower) has an illumination of 20 fc (foot-candles) at a distance of 15 feet, what is the illumination at a distance of 20 feet?

43. The force exerted by a spring varies directly with the distance that the spring is stretched. A hanging spring will stretch 9 cm if a weight of 15 grams is placed on the end of the spring. How far will the spring stretch if the weight is increased to 20 grams?

44. The volume of a cylinder varies jointly as its height and the square of its radius. If a cylinder has the measurements $V = 301.44$ cubic inches, $r = 4$ inches, and $h = 6$ inches, what is the volume of a cylinder that has a radius of 6 inches and a height of 8 inches?

45. The volume of a gas in a storage container varies inversely as the pressure on the gas. If the volume is 100 cubic centimeters under a pressure of 8 grams, what would be the volume of the gas if the pressure was decreased to 4 grams?

46. F is jointly proportional to a and b and varies inversely as c. If $F = 10$ when $a = 6$, $b = 5$, and $c = 2$, what is the value of F when $a = 12$, $b = 6$, and $c = 3$?

47. The resistance of a wire varies directly as its length and inversely as the square of the diameter. When a wire is 500 feet long and has a diameter of 0.015 in., it has a resistance of 20 ohms. What is the resistance of a wire that is 1200 feet long and has a diameter of 0.025 in.?

3.5 Transformations of Functions

TOPICS

1. Shifting, reflecting, and stretching graphs
2. Symmetry of functions and equations

TOPIC 1

Shifting, Reflecting, and Stretching Graphs

Much of the material in this section was introduced in Section 3.2, in our discussion of quadratic functions. You may want to review the ways in which the basic quadratic function $f(x) = x^2$ can be shifted, stretched, and reflected as you work through the more general ideas here.

THEOREM

Horizontal Shifting/Translation

Let $f(x)$ be a function, and let h be a fixed real number. If we replace x with $x - h$, we obtain a new function $g(x) = f(x - h)$. The graph of g has the same shape as the graph of f, but shifted to the right by h units if $h > 0$ and shifted to the left by h units if $h < 0$.

EXAMPLE 1

Horizontal Shifting/Translation

Sketch the graphs of the following functions.

a. $f(x) = (x+2)^3$ **b.** $g(x) = |x-4|$

Note:
Begin by identifying the underlying function that is being shifted.

Solutions:

a.

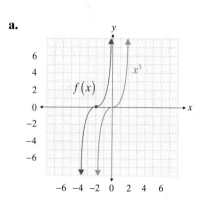

The basic function being shifted is x^3.

Begin by drawing the basic cubic shape (the shape of $y = x^3$).

Since x is replaced by $x + 2$, the graph of $f(x)$ is the graph of x^3 shifted to the left by 2 units.

Note, for example, that $(-2, 0)$ is one point on the graph.

b.

The basic function being shifted is $|x|$.

Start by graphing the basic absolute value function.

The graph of $g(x) = |x - 4|$ has the same shape, but shifted to the right by 4 units.

Note, for example, that $(4, 0)$ lies on the graph of g.

CAUTION!

The minus sign in the expression $x - h$ is critical. When you see an expression in the form $x + h$ you must think of it as $x - (-h)$.

Consider a specific example: replacing x with $x - 5$ shifts the graph 5 units to the *right*, since 5 is positive. Replacing x with $x + 5$ shifts the graph 5 units to the *left*, since we have actually replaced x with $x - (-5)$.

THEOREM

Vertical Shifting/
Translation

Let $f(x)$ be a function whose graph is known, and let k be a fixed real number. The graph of the function $g(x) = f(x) + k$ is the same shape as the graph of f, but shifted upward if $k > 0$ and downward if $k < 0$.

EXAMPLE 2

Vertical Shifting/
Translation

Sketch the graphs of the following functions.

a. $f(x) = \dfrac{1}{x} + 3$

b. $g(x) = \sqrt[3]{x} - 2$

Solutions:

a.

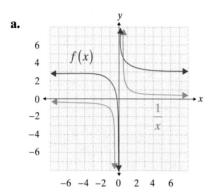

The basic function being shifted is $\dfrac{1}{x}$.

The graph of $f(x) = \dfrac{1}{x} + 3$ is the graph of $y = \dfrac{1}{x}$ shifted up 3 units.

Note that this doesn't change the domain.

However, the range is affected; the range of f is $(-\infty, 3) \cup (3, \infty)$.

b.

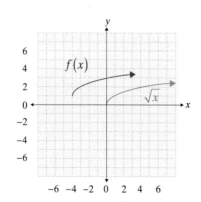

The basic function being shifted is $\sqrt[3]{x}$.

Begin by graphing the basic cube root shape.

To graph $g(x) = \sqrt[3]{x} - 2$, we shift the graph of $y = \sqrt[3]{x}$ down by 2 units.

EXAMPLE 3

Horizontal and Vertical Shifting

Sketch the graph of the function $f(x) = \sqrt{x+4} + 1$.

Solution:

The basic function being shifted is $\sqrt{x}$.

Begin by graphing the basic square root shape.

In $f(x)$ we have replaced x with $x+4$, so shift the basic function 4 units left.

Then shift the resulting function 1 unit up.

THEOREM

Reflecting with Respect to the Axes

Given a function $f(x)$:

1. The graph of the function $g(x) = -f(x)$ is the reflection of the graph of f with respect to the x-axis.

2. The graph of the function $g(x) = f(-x)$ is the reflection of the graph of f with respect to the y-axis.

In other words, a function is reflected with respect to the x-axis by multiplying the entire function by -1, and reflected with respect to the y-axis by replacing x with $-x$.

EXAMPLE 4

Reflecting with Respect to the Axes

Sketch the graphs of the following functions.

a. $f(x) = -x^2$

b. $g(x) = \sqrt{-x}$

Note:
We state that a function is reflected with respect to particular axis. Visually, this means the function is reflected over (across) that axis.

Solutions:

a.

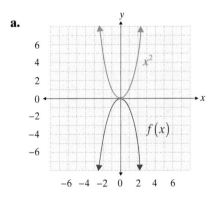

To graph $f(x) = -x^2$, begin with the graph of the basic parabola $y = x^2$.

The entire function is multiplied by -1, so reflect the graph over the x-axis, resulting in the original shape turned upside down.

Note that the domain is still the entire real line, but the range of f is the interval $(-\infty, 0]$.

b.

To graph $g(x) = \sqrt{-x}$, begin by graphing $y = \sqrt{x}$, the basic square root.

In $g(x)$, x has been replaced by $-x$, so reflect the graph with respect to the y-axis.

Note that this changes the domain but not the range. The domain of g is the interval $(-\infty, 0]$ and the range is $[0, \infty)$.

THEOREM

Vertical Stretching and Compressing

Let $f(x)$ be a function and let a be a positive real number.

1. The graph of the function $g(x) = af(x)$ is stretched vertically compared to the graph of f if $a > 1$.

2. The graph of the function $g(x) = af(x)$ is compressed vertically compared to the graph of f if $0 < a < 1$.

EXAMPLE 5

Vertical Stretching and Compressing

Sketch the graphs of the following functions.

a. $f(x) = \dfrac{\sqrt{x}}{10}$

b. $g(x) = 5|x|$

Note:
When graphing stretched or compressed functions, it may help to plot a few points of the new function.

Solutions:

a.

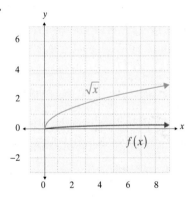

Begin with the graph of $\sqrt{x}$.

The shape of $f(x)$ is similar to the shape of $\sqrt{x}$ but all of the y-coordinates have been multiplied by the factor of $\dfrac{1}{10}$, and are consequently much smaller.

b.

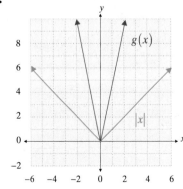

Begin with the graph of the absolute value function.

In contrast to the last example, the graph of $g(x) = 5|x|$ is stretched compared to the standard absolute value function.

Every second coordinate is multiplied by a factor of 5.

If the function g is obtained from the function f by multiplying f by a negative real number, think of the number as the product of –1 and a positive real number (namely, its absolute value). This is a simple example of a function going under multiple transformations. When dealing with more complicated functions, undergoing numerous transformations, we need a procedure for untangling the individual transformations in order to find the correct graph.

PROCEDURE

Order of
Transformations

If a function g has been obtained from a simpler function f through a number of transformations, g can be understood by looking for transformations in this order:

1. Horizontal shifts

2. Stretching and compressing

3. Reflections

4. Vertical shifts

Consider, for example, the function $g(x) = -2\sqrt{x+1} + 3$, which has been "built up" from the basic square root function through a variety of transformations.

1. First, $\sqrt{x}$ has been transformed into $\sqrt{x+1}$ by replacing x with $x + 1$, and we know that this corresponds graphically to a shift to the left of 1 unit.

2. Next, the function $\sqrt{x+1}$ has been multiplied by 2 to get the function $2\sqrt{x+1}$, and we know that this has the effect of stretching the graph of $\sqrt{x+1}$ vertically.

3. The function $2\sqrt{x+1}$ has then been multiplied by –1, giving us $-2\sqrt{x+1}$, and the graph of this is the reflection of $2\sqrt{x+1}$ with respect to the x-axis.

4. Finally, the constant 3 has been added to $-2\sqrt{x+1}$, shifting the entire graph upward by 3 units. These transformations are illustrated, in order, in Figure 1, culminating in the graph of $g(x) = -2\sqrt{x+1} + 3$.

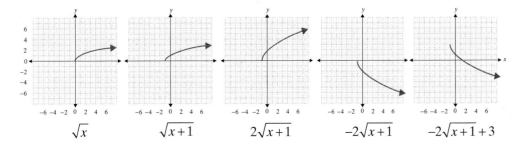

Figure 1: Building the Graph of $g(x) = -2\sqrt{x+1} + 3$

EXAMPLE 6

Order of
Transformations

Sketch the graph of the function $f(x) = \dfrac{1}{2-x}$.

Solution:

The basic function that f is similar to is $\dfrac{1}{x}$. Following the order of transformations:

1. If we replace x by $x + 2$ (shifting the graph 2 units to the left), we obtain the function $\dfrac{1}{x+2}$, which is closer to what we want.

2. There does not appear to be any stretching or compressing transformation.

3. If we replace x by $-x$, we have $\dfrac{1}{-x+2} = \dfrac{1}{2-x}$, which is equal to f. This reflects the graph of $\dfrac{1}{x+2}$ with respect to the y-axis.

4. Since we have already found f, we know there is no vertical shift.

The entire sequence of transformations is shown below, ending with the graph of f.

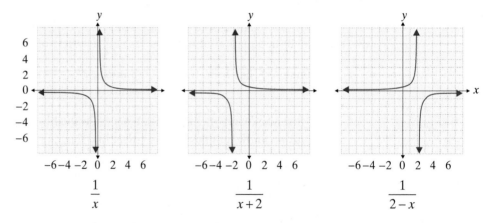

$$\frac{1}{x} \qquad\qquad \frac{1}{x+2} \qquad\qquad \frac{1}{2-x}$$

Note: An alternate approach to graphing $f(x) = \dfrac{1}{2-x}$ is to rewrite the function in the form $f(x) = -\dfrac{1}{x-2}$. In this form, the graph of f is the graph of $\dfrac{1}{x}$ shifted two units to the right, and then reflected with respect to the x-axis. The result is the same, as you should verify. Rewriting an equation in a different form never changes its graph.

TOPIC 2   Symmetry of Functions and Equations

We know that replacing x with $-x$ reflects the graph of a function with respect to the y-axis, but what if $f(-x) = f(x)$? In this case the original graph is the same as the reflection! This means the function f is symmetric with respect to the y-axis.

DEFINITION

y-Axis Symmetry

The graph of a function f has **y-axis symmetry**, or is **symmetric with respect to the y-axis**, if $f(-x) = f(x)$ for all x in the domain of f. Such functions are called **even** functions.

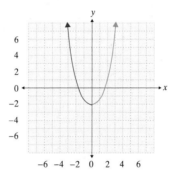

Figure 2: A Function with **y**-Axis Symmetry

Functions whose graphs have y-axis symmetry are called even functions because polynomial functions with only even exponents form one large class of functions with this property. Consider the function $f(x) = 7x^8 - 5x^4 + 2x^2 - 3$. This function is a polynomial of four terms, all of which have even degree. If we replace x with $-x$ and simplify the result, we obtain the function f again:

$$f(-x) = 7(-x)^8 - 5(-x)^4 + 2(-x)^2 - 3$$
$$= 7x^8 - 5x^4 + 2x^2 - 3$$
$$= f(x)$$

Be aware, however, that such polynomial functions are not the only even functions. We will see many more examples as we proceed.

There is another class of functions for which replacing x with $-x$ results in the exact negative of the original function. That is, $f(-x) = -f(x)$ for all x in the domain, and this means changing the sign of the x-coordinate of a point on the graph also changes the sign of the y-coordinate.

What does this mean geometrically? Suppose f is such a function, and that $(x, f(x))$ is a point on the graph of f. If we change the sign of both coordinates, we obtain a new point that is the original point reflected through the origin (we can also think of this as reflected over the y-axis, then the x-axis).

For instance, if $\left(x, f(x)\right)$ lies in the first quadrant, $\left(-x, -f(x)\right)$ lies in the third, and if $\left(x, f(x)\right)$ lies in the second quadrant, $\left(-x, -f(x)\right)$ lies in the fourth. But since $f(-x) = -f(x)$, the point $\left(-x, -f(x)\right)$ can be rewritten as $\left(-x, f(-x)\right)$.

Written in this form, we know that $\left(-x, f(-x)\right)$. is a point on the graph of f, since *any* point of the form $\left(?, f(?)\right)$ lies on the graph of f. So a function with the property $f(-x) = -f(x)$ has a graph that is symmetric with respect to the origin.

DEFINITION

Origin Symmetry

The graph of a function f has **origin symmetry**, or is **symmetric with respect to the origin**, if $f(-x) = -f(x)$ for all x in the domain of f. Such functions are called **odd** functions.

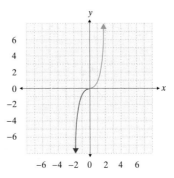

Figure 3: A Function with Origin Symmetry

As you might guess, such functions are called odd because polynomial functions with only odd exponents serve as simple examples. For instance, the function $f(x) = -2x^3 + 8x$ is odd:

$$f(-x) = -2(-x)^3 + 8(-x)$$
$$= -2(-x^3) + 8(-x)$$
$$= 2x^3 - 8x$$
$$= -f(x)$$

As far as functions are concerned, y-axis and origin symmetry are the two principal types of symmetry. What about x-axis symmetry? It is certainly possible to draw a graph that displays x-axis symmetry; but unless the graph lies entirely on the x-axis, such a graph cannot represent a function. Why not? Draw a few graphs that are symmetric with respect to the x-axis, then apply the Vertical Line Test to these graphs. In order to have x-axis symmetry, if (x, y) is a point on the graph, then $(x, -y)$ must also be on the graph, and thus the graph can not represent a function.

This brings us back to relations. Recall that any equation in x and y defines a relation between the two variables. There are three principal types of symmetry that equations can possess.

DEFINITION

Symmetry of
Equations

We say that an equation in x and y is **symmetric with respect to**:

1. The **y-axis** if replacing x with $-x$ results in an equivalent equation
2. The **x-axis** if replacing y with $-y$ results in an equivalent equation
3. The **origin** if replacing x with $-x$ and y with $-y$ results in an equivalent equation

Knowing the symmetry of a function or an equation can serve as a useful aid in graphing. For instance, when graphing an even function it is only necessary to graph the part to the right of the y-axis, as the left half of the graph is the reflection of the right half with respect to the y-axis. Similarly, if a function is odd, the left half of its graph is the reflection of the right half through the origin.

EXAMPLE 7

Symmetry of
Equations

Sketch the graphs of the following relations, making use of symmetry.

   a. $f(x) = \dfrac{1}{x^2}$ **b.** $g(x) = x^3 - x$ **c.** $x = y^2$

Note:
If you don't know where to begin when sketching a graph, plotting points often helps you understand the basic shape.

Solutions:

a.

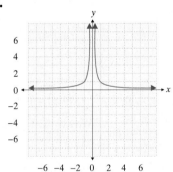

This relation is a function, one that we already graphed in Section 3.3. Note that it is indeed an even function and exhibits y-axis symmetry:

$$f(-x) = \frac{1}{(-x)^2}$$

$$= \frac{1}{x^2}$$

$$= f(x)$$

b.

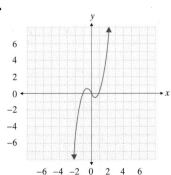

While we do not yet have the tools to graph general polynomial functions, we can obtain a good sketch of $g(x) = x^3 - x$.

First, g is odd: $g(-x) = -g(x)$ (verify this).

If we calculate a few values, such as $g(0) = 0$,

$g\left(\dfrac{1}{2}\right) = -\dfrac{3}{8}$, $g(1) = 0$, and $g(2) = 6$, and then

reflect these through the origin, we get a good idea of the shape of g.

c.

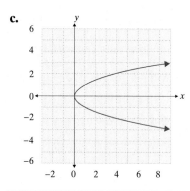

The equation $x = y^2$ is not a function, but it is a relation in x and y that has x-axis symmetry. If we replace y with $-y$ and simplify the result, we obtain the original equation:

$$x = (-y)^2$$
$$x = y^2$$

The upper half of the graph is the function $y = \sqrt{x}$, so drawing this and its reflection gives us the complete graph of $x = y^2$.

Summary of Symmetry

The first column in the table below summarizes the behavior of a graph in the Cartesian plane if it possesses any of the three types of symmetry we covered. If the graph is of an equation in x and y, the algebraic method in the second column can be used to identify the symmetry. The third column gives the algebraic method used to identify the type of symmetry if the graph is that of a function $f(x)$. Finally, the fourth column contains an example of each type of symmetry.

A graph is symmetric with respect to:	If the graph is of an equation in x and y, the equation is symmetric with respect to:	If the graph is of a function $f(x)$, the function is symmetric with respect to:	Example:
The y-axis if whenever the point (x, y) is on the graph, the point $(-x, y)$ is also on the graph.	The y-axis if replacing x with $-x$ results in an equivalent equation.	The y-axis if $f(-x) = f(x)$. We say the function is even.	
The x-axis if whenever the point (x, y) is on the graph, the point $(x, -y)$ is also on the graph.	The x-axis if replacing y with $-y$ results in an equivalent equation.	Not applicable (unless the graph consists only of points on the x-axis).	
The origin if whenever the point (x, y) is on the graph, the point $(-x, -y)$ is also on the graph.	The origin if replacing x with $-x$ and y with $-y$ results in an equivalent equation.	The origin if $f(-x) = -f(x)$. We say the function is odd.	

1. $f(x) = x^2$

2. $f(x) = \dfrac{1}{x}$

3. $f(x) = \sqrt[3]{x}$

4. $f(x) = |x|$

5. $f(x) = \sqrt{x}$

6. $f(x) = \dfrac{1}{x^2}$

7. $f(x) = \sqrt{x}$

8. $f(x) = x^3$

9. $f(x) = x^2$

10. $f(x) = \sqrt{x}$

11. $f(x) = |x|$

12. $f(x) = x^3$

13.

Dom = Ran = $\mathbb{R}$

14.

Dom = $\mathbb{R}$,
Ran = $[0, \infty)$

15.
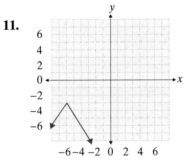
Dom = $\mathbb{R}$,
Ran = $(-\infty, 2]$

16.

Dom = $[-3, \infty)$,
Ran = $[-1, \infty)$

Exercises

For each function or graph below, determine the basic function that has been shifted, reflected, stretched, or compressed.

1. $f(x) = -(1-x)^2 + 2$

2. $f(x) = \dfrac{1}{x-4} + 5$

3. $f(x) = \sqrt[3]{x+6} - 2$

4. $f(x) = -2 + 2|x-3|$

5. $f(x) = \sqrt{x+2} - 5$

6. $f(x) = \dfrac{1}{(x+2)^2} + 1$

7. $f(x) = \dfrac{\sqrt{-x}}{2} + 4$

8. $f(x) = (x+6)^3$

9.

10.

11.

12.

17.

Dom = $\mathbb{R}$,
Ran = $[0, \infty)$

18.

Dom = Ran = $\mathbb{R}$

19.

Dom = $(-\infty, 2]$,
Ran = $[0, \infty)$

20.

Dom = $\mathbb{R}$,
Ran = $[3, \infty)$

21.

Dom = $(-\infty,3)\cup(3,\infty)$, Ran = $(0,\infty)$

22.

Dom = $(-\infty,0)\cup(0,\infty)$, Ran = $(-\infty,-2)\cup(-2,\infty)$

23.

Dom = $(-\infty,2)\cup(2,\infty)$, Ran = $(-\infty,0)\cup(0,\infty)$

24.

Dom = $(-\infty,0]$, Ran = $[2,\infty)$

Sketch the graphs of the following functions by first identifying the more basic functions that have been shifted, reflected, stretched, or compressed. Then determine the domain and range of each function. See Examples 1 through 6.

13. $f(x)=(x+2)^3$ **14.** $G(x)=|x-4|$ **15.** $p(x)=-(x+1)^2+2$

16. $g(x)=\sqrt{x+3}-1$ **17.** $q(x)=(1-x)^2$ **18.** $r(x)=-\sqrt[3]{x}$

19. $s(x)=\sqrt{2-x}$ **20.** $F(x)=\dfrac{|x+2|}{3}+3$ **21.** $w(x)=\dfrac{1}{(x-3)^2}$

22. $v(x)=\dfrac{1}{3x}-2$ **23.** $f(x)=\dfrac{1}{2-x}$ **24.** $k(x)=\sqrt{-x}+2$

25. $b(x)=\sqrt[3]{x+2}-5$ **26.** $R(x)=4-|2x|$ **27.** $S(x)=(3-x)^3$

28. $g(x)=-\dfrac{1}{x+1}$ **29.** $h(x)=\dfrac{x^2}{2}-3$ **30.** $W(x)=1-|4-x|$

31. $W(x)=-\dfrac{|x-1|}{4}$ **32.** $S(x)=\dfrac{1}{x^2}+3$ **33.** $V(x)=-3\sqrt{x-1}+2$

34. $g(x)=x^2-6x+9$ **(Hint:** Find a better way to write the function.)

Write a formula for each of the functions described below.

35. Use the function $g(x)=x^2$. Move the function 3 units to the left and 4 units down.

36. Use the function $g(x)=x^2$. Move the function 4 units to the right and 2 units up.

37. Use the function $g(x)=x^2$. Reflect the function across the x-axis and move it 6 units up.

38. Use the function $g(x)=x^2$. Move the function 2 units to the right and reflect across the y-axis.

25.

Dom = Ran = $\mathbb{R}$

26.

Dom = $\mathbb{R}$, Ran = $(-\infty,4]$

27.

Dom = Ran = $\mathbb{R}$

28.

Dom = $(-\infty,-1)\cup(-1,\infty)$, Ran = $(-\infty,0)\cup(0,\infty)$

29.

Dom = $\mathbb{R}$, Ran = $[-3,\infty)$

30.

Dom = $\mathbb{R}$, Ran = $(-\infty,1]$

31.

Dom = $\mathbb{R}$, Ran = $(-\infty,0]$

32.

Dom = $(-\infty,0)\cup(0,\infty)$, Ran = $(3,\infty)$

33.

Dom = $[1,\infty)$, Ran = $(-\infty,2]$

34.

Dom = $\mathbb{R}$, Ran = $[0,\infty)$

35. $f(x) = (x+3)^2 - 4$

36. $f(x) = (x-4)^2 + 2$

37. $f(x) = -x^2 + 6$

38. $f(x) = (-x-2)^2$
$= (x+2)^2$

39. $f(x) = (-x+1)^3$

40. $f(x) = (x-10)^3 + 4$

41. $f(x) = -\sqrt{x+5}$

42. $f(x) = \sqrt{-x} - 3$

43. $f(x) = -|-x+7|$

44. $f(x) = -|x-8| - 2$

39. Use the function $g(x) = x^3$. Move the function 1 unit to the left and reflect across the y-axis.

40. Use the function $g(x) = x^3$. Move the function 10 units to the right and 4 units up.

41. Use the function $g(x) = \sqrt{x}$. Move the function 5 units to the left and reflect across the x-axis.

42. Use the function $g(x) = \sqrt{x}$. Reflect the function across the y-axis and move it down 3 units.

43. Use the function $g(x) = |x|$. Move the function 7 units to the left, reflect across the x-axis, and reflect across the y-axis.

44. Use the function $g(x) = |x|$. Move the function 8 units to the right, 2 units up, and reflect across the x-axis.

Determine if each of the following relations is a function. If so, determine whether it is even, odd, or neither. Also determine if it has y-axis symmetry, x-axis symmetry, origin symmetry, or none of the above, and then sketch the graph of the relation. See Example 7.

45. $f(x) = |x| + 3$

46. $g(x) = x^3$

47. $h(x) = x^3 - 1$

48. $w(x) = \sqrt[3]{x}$

49. $x = -y^2$

50. $3y - 2x = 1$

51. $x + y = 1$

52. $F(x) = (x-1)^2$

53. $x = y^2 + 1$

54. $x = |2y|$

55. $g(x) = \dfrac{x^2}{5} - 5$

56. $m(x) = \sqrt[3]{x} - 1$

57. $xy = 2$

58. $x + y^2 = 3$

45.

Even function;
y-axis symmetry

46.

Odd function;
Origin symmetry

47.

Neither;
No symmetry

48.

Odd function;
Origin symmetry

49.

Not a function;
x-axis symmetry

50.

Neither;
No symmetry

51.

Neither;
No symmetry

52.

Neither;
No symmetry

53.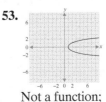

Not a function;
x-axis symmetry

54.

Not a function;
x-axis symmetry

55.

Even function;
y-axis symmetry

56.

Neither;
No symmetry

57.

Odd function;
Origin symmetry

58.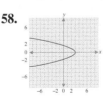

Not a function;
x-axis symmetry

3.6 Combining Functions

TOPICS

1. Combining functions arithmetically
2. Composing functions
3. Decomposing functions
4. Recursive graphics

TOPIC

Combining Functions Arithmetically

In Section 3.5, we gained experience in building new functions from old ones by shifting, reflecting and stretching the old functions. In this section, we will explore more ways of building functions.

We begin with four arithmetic ways of combining two or more functions to obtain new functions. The basic operations are very familiar to you: addition, subtraction, multiplication, and division. The difference is that we are applying these operations to functions, but as we will see, the arithmetic combination of functions is based entirely on the arithmetic combination of numbers.

DEFINITION

Addition, Subtraction, Multiplication, and Division of Functions

Let f and g be two functions. The **sum** $f + g$, **difference** $f - g$, **product** fg, and **quotient** $\dfrac{f}{g}$ are four new functions defined as follows:

$$(f + g)(x) = f(x) + g(x)$$

$$(f - g)(x) = f(x) - g(x)$$

$$(fg)(x) = f(x)g(x)$$

$$\left(\frac{f}{g}\right)(x) = \frac{f(x)}{g(x)}, \text{ provided that } g(x) \neq 0$$

The domain of each of these new functions consists of the common elements (or the intersection of elements) of the domains of f and g individually, with the added condition that in the quotient function we have to omit those elements for which $g(x) = 0$.

With the above definition, we can determine the sum, difference, product, or quotient of two functions at one particular value for x, or find a formula for these new functions based on the formulas for f and g, if they are available.

━━━━━━━━━━ **EXAMPLE 1** ━━━━━━━━━━

**Combining Functions
Arithmetically**

Given that $f(-2)=5$ and $g(-2)=-3$, find $(f-g)(-2)$ and $\left(\dfrac{f}{g}\right)(-2)$.

Solution:

By the definition of the difference and quotient of functions,

$$(f-g)(-2) = f(-2) - g(-2)$$
$$= 5 - (-3)$$
$$= 8,$$

and

$$\left(\frac{f}{g}\right)(-2) = \frac{f(-2)}{g(-2)}$$
$$= \frac{5}{-3}$$
$$= -\frac{5}{3}.$$

━━━━━━━━━━ **EXAMPLE 2** ━━━━━━━━━━

**Combining Functions
Arithmetically**

Given the two functions $f(x)=4x^2-1$ and $g(x)=\sqrt{x}$, find $(f+g)(x)$ and $(fg)(x)$.

Solution:

By the definition of the sum and product of functions,

$$(f+g)(x) = f(x) + g(x)$$
$$= 4x^2 - 1 + \sqrt{x},$$

and

$$(fg)(x) = f(x)g(x)$$
$$= \left(4x^2 - 1\right)\left(\sqrt{x}\right)$$
$$= 4x^{\frac{5}{2}} - x^{\frac{1}{2}}.$$

What are the domains of $f+g$ and fg? We first need to find the domains of the individual functions f and g.

Domain of f:　　$(-\infty,\infty)$ since f is a quadratic function

Domain of g:　　$[0,\infty)$ since square roots of negative numbers are undefined

Since the domain of two functions combined arithmetically is the intersection of the individual domains, $f+g$ and fg both have a domain of $[0,\infty)$.

EXAMPLE 3

Combining Functions Arithmetically

Given the graphs of f and g below, determine the domain of $f + g$ and $\dfrac{f}{g}$ and evaluate $(f + g)(1)$ and $\left(\dfrac{f}{g}\right)(1)$.

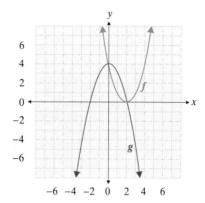

Solution:

From the graph, we can see that the domain of both f and g is the set of all real numbers $(-\infty, \infty)$. This means that the domain of $f + g$ is also $(-\infty, \infty)$. To find the domain of the quotient, we need to check where $g(x) = 0$. The graph shows us that this occurs when $x = \pm 2$, so the domain of $\dfrac{f}{g}$ is all real numbers *except* 2 and –2:

$$(-\infty, -2) \cup (-2, 2) \cup (2, \infty)$$

To evaluate the new functions, we need to find $f(1)$ and $g(1)$ using the graph:

We can see that $f(1) = 1$ and $g(1) = 3$, which means:

$$(f + g)(1) = 1 + 3 = 4 \quad \text{and} \quad \left(\frac{f}{g}\right)(1) = \frac{1}{3} = \frac{1}{3}.$$

TOPIC 2

Composing Functions

A fifth way of combining functions is to form the *composition* of one function with another. Informally speaking, this means to apply one function to the output of another function. The symbol for composition is an open circle.

DEFINITION

Composing Functions

Let f and g be two functions. The **composition** of f and g, denoted $f \circ g$, is the function defined by $(f \circ g)(x) = f\big(g(x)\big)$. The domain of $f \circ g$ consists of all x in the domain of g for which $g(x)$ is in turn in the domain of f. The function $f \circ g$ is read "f composed with g," or "f of g."

The diagram in Figure 1 is a schematic of the composition of two functions. To calculate $(f \circ g)(x)$ we first apply the function g, calculating $g(x)$, then apply the function f to the result, calculating $f(g(x)) = (f \circ g)(x)$.

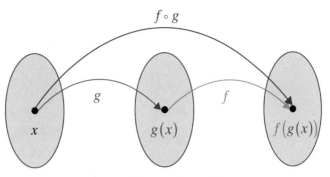

Figure 1: Composition of f and g

As with the four arithmetic ways of combining functions, we can evaluate the composition of two functions at a single point, or find a formula for the composition if we have been given formulas for the individual functions.

CAUTION!

Note that the order of f and g is important. In general, we can expect the function $f \circ g$ to be *different* from the function $g \circ f$. In formal terms, the composition of two functions, unlike the sum and product of two functions, is not commutative.

EXAMPLE 4

Composing Functions

Given $f(x) = x^2$ and $g(x) = x - 3$, find:

a. $(f \circ g)(6)$ **b.** $(g \circ f)(6)$

c. $(f \circ g)(x)$ **d.** $(g \circ f)(x)$

Solutions:

a. Since $(f \circ g)(6) = f(g(6))$, the first step is to calculate $g(6)$:

$$g(6) = 6 - 3 = 3$$

Then, apply f to the result:

$$(f \circ g)(6) = f(g(6)) = f(3) = 3^2 = 9.$$

b. This time, we begin by finding $f(6)$:

$$f(6) = 6^2 = 36$$

Now, apply g to the result:

$$g\big(f(6)\big) = g(36) = 36 - 3 = 33.$$

c. To find the formula for $f \circ g$ we apply the definition of composition, then simplify:

$(f \circ g)(x) = f\big(g(x)\big)$	Write out the definition of composition.
$= f(x - 3)$	Substitute the formula for $g(x)$.
$= (x - 3)^2$	Apply the formula for $f(x)$.
$= x^2 - 6x + 9$	Simplify.

d. To find a formula for the function $g \circ f$ we follow the same process:

$(g \circ f)(x) = g\big(f(x)\big)$	Write out the definition of composition.
$= g(x^2)$	Substitute the formula for $f(x)$.
$= x^2 - 3$	Apply the formula for $g(x)$; the result is already simplified.

Note that once we have found formulas $f \circ g$ and $g \circ f$ we can answer the first two parts by directly plugging into these formulas:

$$(f \circ g)(6) = 6^2 - 6(6) + 9 = 9$$

$$(g \circ f)(6) = 6^2 - 3 = 33$$

CAUTION!

When evaluating the composition $(f \circ g)(x)$ at a point x, there are two reasons the value might be undefined:

x is not in the domain of g. Then $g(x)$ is undefined and we can't evaluate $f\big(g(x)\big)$.

$g(x)$ is not in the domain of f. Then $f\big(g(x)\big)$ is undefined and we can't evaluate it.

In either case, $(f \circ g)(x) = f\big(g(x)\big)$ is undefined, and x is not in the domain of $(f \circ g)(x)$.

EXAMPLE 5

Domains of
Compositions of
Functions

Let $f(x) = \sqrt{x-5}$ and $g(x) = \dfrac{2}{x+1}$. Evaluate the following:

a. $(f \circ g)(-1)$ **b.** $(f \circ g)(1)$

Solutions:

a. $(f \circ g)(-1) = f(g(-1))$

But, if we try to evaluate $g(-1)$, we see that it is undefined, so $(f \circ g)(-1)$ is also undefined.

b. $(f \circ g)(1) = f(g(1))$

First, we evaluate $g(1)$.

$$g(1) = \frac{2}{1+1} = \frac{2}{2} = 1$$

We plug this result into $f(x)$ but see that $\sqrt{1-5} = \sqrt{-4}$ is undefined. Thus, $(f \circ g)(1)$ is also undefined.

EXAMPLE 6

Domains of
Compositions of
Functions

Let $f(x) = x^2 - 4$ and $g(x) = \sqrt{x}$. Find formulas and state the domains for:

a. $f \circ g$ **b.** $g \circ f$

Solutions:

a. $(f \circ g)(x) = f(g(x))$

$\qquad\qquad\quad = f(\sqrt{x})$ Substitute the formula for $g(x)$ into $f(x)$.

$\qquad\qquad\quad = (\sqrt{x})^2 - 4$ Simplify.

$\qquad\qquad\quad = x - 4$

While the domain of $x - 4$ is the set of all real numbers, the domain of $f \circ g$ is $[0, \infty)$ since only nonnegative numbers can be plugged into g.

b. $(g \circ f)(x) = g(f(x))$

$\qquad\qquad\quad = g(x^2 - 4)$ Substitute the formula for $f(x)$ into $g(x)$.

$\qquad\qquad\quad = \sqrt{x^2 - 4}$ The answer is already simplified.

The domain of $g \circ f$ consists of all x for which $x^2 - 4 \geq 0$, or $x^2 \geq 4$. We can write this in interval form as $(-\infty, -2] \cup [2, \infty)$.

TOPIC 3

Decomposing Functions

Often, functions can be best understood by recognizing them as a composition of two or more simpler functions. We have already seen an instance of this: shifting, reflecting, stretching, and compressing can all be thought of as a composition of two or more functions. For example, the function $h(x) = (x-2)^3$ is a composition of the functions $f(x) = x^3$ and $g(x) = x-2$:

$$f(g(x)) = f(x-2)$$
$$= (x-2)^3$$
$$= h(x).$$

To "decompose" a function into a composition of simpler functions, it is usually best to identify what the function does to its argument from the inside out. That is, identify the first thing that is done to the variable, then the second, and so on. Each action describes a less complex function, and can be identified as such. The composition of these functions, with the innermost function corresponding to the first action, the next innermost corresponding to the second action, and so on, is then equivalent to the original function.

Decomposition can often be done in several different ways. Consider, for example, the function $f(x) = \sqrt[3]{5x^2 - 1}$. Below we illustrate just a few of the ways f can be written as a composition of functions. Be sure you understand how each of the different compositions is equivalent to f.

1. $g(x) = \sqrt[3]{x}$ $\qquad\qquad g(h(x)) = g(5x^2 - 1)$
    $h(x) = 5x^2 - 1$ $\qquad\qquad\quad = \sqrt[3]{5x^2 - 1}$
    $\qquad\qquad\qquad\qquad\qquad\quad = f(x)$

2. $g(x) = \sqrt[3]{x-1}$ $\qquad\quad g(h(x)) = g(5x^2)$
    $h(x) = 5x^2$ $\qquad\qquad\quad\ = \sqrt[3]{5x^2 - 1}$
    $\qquad\qquad\qquad\qquad\qquad\quad = f(x)$

3. $g(x) = \sqrt[3]{x}$ $\qquad\qquad g(h(j(x))) = g(h(x^2))$
    $h(x) = 5x - 1$ $\qquad\qquad\qquad\ = g(5x^2 - 1)$
    $j(x) = x^2$ $\qquad\qquad\qquad\qquad = \sqrt[3]{5x^2 - 1}$
    $\qquad\qquad\qquad\qquad\qquad\qquad = f(x)$

EXAMPLE 7

Decomposing
Functions

Decompose the function $f(x) = |x^2 - 3| + 2$ into:

a. a composition of two functions   **b.** a composition of three functions

Solutions:

a. $g(x) = |x| + 2$
$h(x) = x^2 - 3$

$g(h(x)) = g(x^2 - 3)$
$= |x^2 - 3| + 2$
$= f(x)$

b. $g(x) = x + 2$
$h(x) = |x - 3|$
$j(x) = x^2$

$g(h(j(x))) = g(h(x^2))$
$= g(|x^2 - 3|)$
$= |x^2 - 3| + 2$
$= f(x)$

Note: These are **not** the only possible solutions for the decompositions of $f(x)$.

TOPIC **4**

Recursive Graphics

Recursion, in general, refers to using the output of a function as its input, and repeating the process a certain number of times. In other words, recursion refers to the composition of a function with itself, possibly many times. Recursion has many varied uses, one of which is a branch of mathematical art.

There is some special notation to describe recursion. If f is a function, $f^2(x)$ is used in this context to stand for $f(f(x))$, or $(f \circ f)(x)$, not $(f(x))^2$! Similarly, $f^3(x)$ stands for $f(f(f(x)))$, or $(f \circ f \circ f)(x)$, and so on. The functions f^2, f^3, ... are called **iterates** of f, with f^n being the nth **iterate** of f.

Some of the most famous recursively generated mathematical art is based on functions whose inputs and outputs are complex numbers. Recall from Section 1.4 that every complex number can be expressed in the form $a + bi$, where a and b are real numbers and i is the imaginary unit. A one-dimensional coordinate system, such as the real number line, is insufficient to graph complex numbers, but complex numbers are easily graphed in a two-dimensional coordinate system.

To graph the number $a + bi$, we treat it as the ordered pair (a, b) and plot it as a point in the Cartesian plane, where the horizontal axis represents pure real numbers and the vertical axis represents pure imaginary numbers.

Benoit Mandelbrot used the function $f(z) = z^2 + c$, where both z and c are variables representing complex numbers, to generate the image known as the Mandelbrot set in the 1970s. The basic idea is to evaluate the sequence of iterates $f(0) = 0^2 + c = c$,

$f^2(0) = f(c) = c^2 + c, \ f^3(0) = f(c^2 + c) = (c^2 + c)^2 + c, \ldots$ for various complex numbers c and determine if the sequence of complex numbers stays close to the origin or not. Those complex numbers c that result in so-called "bounded" sequences are colored black, while those that lead to unbounded sequences are colored white. The author has used similar ideas to generate his own recursive art, as described below.

1. a. 2   **b.** –8

   c. –15   **d.** $-\dfrac{3}{5}$

2. a. –1   **b.** 1
   c. 0   **d.** 0

3. a. –3   **b.** –1
   c. 2   **d.** 2

4. a. –3   **b.** 1

   c. 2   **d.** $\dfrac{1}{2}$

5. a. 12   **b.** 18
   c. –45   **d.** –5

6. a. –4   **b.** 8

   c. –12   **d.** $-\dfrac{1}{3}$

7. a. 3   **b.** 1
   c. 2   **d.** 2

8. a. 4   **b.** 3

   c. $\dfrac{7}{4}$ **d.** 7

9. a. 6   **b.** 0
   c. 9   **d.** 1

10. a. –1   **b.** 3

   c. –2   **d.** $-\dfrac{1}{2}$

11. a. 5   **b.** –1

   c. 6   **d.** $\dfrac{2}{3}$

12. a. –1   **b.** –3
   c. –2   **d.** –2

13. a. 3   **b.** 5
   c. –4   **d.** –4

14. a. –3   **b.** –5
   c. –4   **d.** –4

15. a. $|x| + \sqrt{x}$,

   Dom $= [0, \infty)$

   b. $\dfrac{|x|}{\sqrt{x}}$,

   Dom $= (0, \infty)$

The image "*i* of the storm" reproduced here is based on the function $f(z) = \dfrac{(1-i)z^4 + (7+i)z}{2z^5 + 6}$, where again z is a variable that will be replaced with complex numbers. The image is actually a picture of the complex plane, with the origin in the very center of the golden ring. The golden ring consists of those complex numbers that lie a distance between 0.9 and 1.1 units from the origin. The rules for coloring other complex numbers in the plane are as follows: given an initial complex number z not on the gold ring, $f(z)$ is calculated. If the complex number $f(z)$ lies somewhere on the gold ring, the original number z is colored the deepest shade of green. If not, the iterate $f^2(z)$ is calculated.

If this result lies in the gold ring, the original z is colored a bluish shade of green. If not, the process continues up to the 12th iterate $f^{12}(z)$, using a different color each time. If $f^{12}(z)$ lies in the gold ring, z is colored red, and if not the process halts and z is colored black.

The idea of recursion can be used to generate any number of similar images, with the end result usually striking and often surprising even to the creator.

Exercises

In each of the following problems, use the information given to determine **a.** $(f + g)(-1)$, **b.** $(f - g)(-1)$, **c.** $(fg)(-1)$, and **d.** $\left(\dfrac{f}{g}\right)(-1)$. See Examples 1, 2, and 3.

1. $f(-1) = -3$ and $g(-1) = 5$

2. $f(-1) = 0$ and $g(-1) = -1$

3. $f(x) = x^2 - 3$ and $g(x) = x$

4. $f(x) = \sqrt[3]{x}$ and $g(x) = x - 1$

5. $f(-1) = 15$ and $g(-1) = -3$

6. $f(x) = \dfrac{x+5}{2}$ and $g(x) = 6x$

7. $f(x) = x^4 + 1$ and $g(x) = x^{11} + 2$

8. $f(x) = \dfrac{6-x}{2}$ and $g(x) = \sqrt{\dfrac{x}{-4}}$

9. $f = \{(5, 2), (0, -1), (-1, 3), (-2, 4)\}$ and $g = \{(-1, 3), (0, 5)\}$

10. $f = \{(3, 15), (2, -1), (-1, 1)\}$ and $g(x) = -2$

16. a. $x^2 + \sqrt[3]{x} - 1$,
Dom = $\mathbb{R}$

b. $\dfrac{x^2 - 1}{\sqrt[3]{x}}$,
Dom = $(-\infty, 0) \cup (0, \infty)$

17. a. $x^2 + x - 2$,
Dom = $\mathbb{R}$

b. $\dfrac{1}{x+1}$,
Dom = $(-\infty, -1) \cup (-1, 1) \cup (1, \infty)$

18. a. $x^{\frac{3}{2}} + x - 3$,
Dom = $[0, \infty)$

b. $\dfrac{x^{\frac{3}{2}}}{x-3}$,
Dom = $[0, 3) \cup (3, \infty)$

19. a. $x^3 + 3x - 8$,
Dom = $\mathbb{R}$

b. $\dfrac{3x}{x^3 - 8}$,
Dom = $(-\infty, 2) \cup (2, \infty)$

20. a. $x^3 + \sqrt{x-2} + 4$,
Dom = $[2, \infty)$

b. $\dfrac{x^3 + 4}{\sqrt{x-2}}$,
Dom = $(2, \infty)$

21. a. $-2x^2 + |x+4|$,
Dom = $\mathbb{R}$

b. $\dfrac{-2x^2}{|x+4|}$,
Dom = $(-\infty, -4) \cup (-4, \infty)$

11.

12.

13.

14.
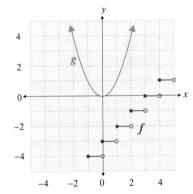

In each of the following problems, find **a.** the formula and domain for $f+g$, and **b.** the formula and domain for $\dfrac{f}{g}$. See Examples 2 and 3.

15. $f(x) = |x|$ and $g(x) = \sqrt{x}$

16. $f(x) = x^2 - 1$ and $g(x) = \sqrt[3]{x}$

17. $f(x) = x - 1$ and $g(x) = x^2 - 1$

18. $f(x) = x^{\frac{3}{2}}$ and $g(x) = x - 3$

19. $f(x) = 3x$ and $g(x) = x^3 - 8$

20. $f(x) = x^3 + 4$ and $g(x) = \sqrt{x-2}$

21. $f(x) = -2x^2$ and $g(x) = |x+4|$

22. $f(x) = 6x - 1$ and $g(x) = x^{\frac{2}{3}}$

22. a. $x^{\frac{2}{3}} + 6x - 1$, Dom = $\mathbb{R}$

b. $\dfrac{6x-1}{x^{\frac{2}{3}}}$, Dom = $(-\infty, 0) \cup (0, \infty)$

23. 2

24. π^2
25. 0
26. 4
27. 8
28. 5

29. 3
30. 3
31. 1
32. 2
33. $\dfrac{1}{3}$

34. a. $\sqrt{x^2-1}$,

Dom $= (-\infty, -1] \cup [1, \infty)$

b. $x - 1$,

Dom $= \mathbb{R}$

35. a. $\dfrac{1}{x-1}$,

Dom $= (-\infty, 1) \cup (1, \infty)$

b. $\dfrac{1}{x} - 1$,

Dom $= (-\infty, 0) \cup (0, \infty)$

36. a. $\dfrac{4-2x}{3x}$,

Dom $= (-\infty, 0) \cup (0, \infty)$

b. $\dfrac{3}{4x-2}$,

Dom $=$

$\left(-\infty, \dfrac{1}{2}\right) \cup \left(\dfrac{1}{2}, \infty\right)$

37. a. $1 - \sqrt{x}$,

Dom $= [0, \infty)$

b. $\sqrt{1-x}$,

Dom $= (-\infty, 1]$

38. a. $\left|x^3 - 2\right|$,

Dom $= \mathbb{R}$

b. $\left|x-3\right|^3 + 1$,

Dom $= \mathbb{R}$

39. a. $x^2 - 4x + 3$,

Dom $= \mathbb{R}$

b. $x^2 + 2x - 3$,

Dom $= \mathbb{R}$

40. a. $\sqrt{\dfrac{x-1}{2}}$,

Dom $= [1, \infty)$

b. $\dfrac{\sqrt{x-1}+1}{2}$,

Dom $= [1, \infty)$

In each of the following problems, use the information given to determine $(f \circ g)(3)$. See Examples 4 and 5.

23. $f(-5) = 2$ and $g(3) = -5$ **24.** $f(\pi) = \pi^2$ and $g(3) = \pi$

25. $f(x) = x^2 - 3$ and $g(x) = \sqrt{x}$ **26.** $f(x) = \sqrt{x^2 - 9}$ and $g(x) = 1 - 2x$

27. $f(x) = 2 + \sqrt{x}$ and $g(x) = x^3 + x^2$ **28.** $f(x) = x^{\frac{3}{2}} - 3$ and $g(x) = \left|\dfrac{4x}{3}\right|$

29. $f(x) = \sqrt{x+6}$ and $g(x) = \sqrt{4x-3}$ **30.** $f(x) = \sqrt{\dfrac{3x}{14}}$ and $g(x) = x^4 - x^3 - x^2 - x$

31.

32.

33.

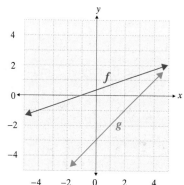

In each of the following problems, find **a.** the formula and domain for $f \circ g$, and **b.** the formula and domain for $g \circ f$. See Example 6.

34. $f(x) = \sqrt{x-1}$ and $g(x) = x^2$ **35.** $f(x) = \dfrac{1}{x}$ and $g(x) = x - 1$

36. $f(x) = \dfrac{4x-2}{3}$ and $g(x) = \dfrac{1}{x}$ **37.** $f(x) = 1 - x$ and $g(x) = \sqrt{x}$

38. $f(x) = |x-3|$ and $g(x) = x^3 + 1$ **39.** $f(x) = x^2 + 2x$ and $g(x) = x - 3$

41. a. $|x|^3 + |x|^2$
$-5|x| + 3,$
Dom = $\mathbb{R}$

b. $|x^3 + 4x^2| - 1,$
Dom = $\mathbb{R}$

42. a. $-3x^2 - 4,$
Dom = $\mathbb{R}$

b. $9x^2 - 12x + 6,$
Dom = $\mathbb{R}$

43. a. $\dfrac{x^2 + 7}{2},$
Dom = $\mathbb{R}$

b. $\dfrac{x^2 + 4x + 7}{2},$
Dom = $\mathbb{R}$

44. $g(x) = \sqrt[3]{x},$
$h(x) = 3x^2 - 1,$
$f(x) = g(h(x))$

45. $g(x) = \dfrac{2}{x},$
$h(x) = 5x - 1,$
$f(x) = g(h(x))$

46. $g(x) = |x| + 3,$
$h(x) = x - 2,$
$f(x) = g(h(x))$

47. $g(x) = x + \sqrt{x} - 5,$
$h(x) = x + 2,$
$f(x) = g(h(x))$

48. $g(x) = |x| + 7,$
$h(x) = x^3 - 5x,$
$f(x) = g(h(x))$

40. $f(x) = \sqrt{x-1}$ and $g(x) = \dfrac{x+1}{2}$

41. $f(x) = x^3 + 4x^2$ and $g(x) = |x| - 1$

42. $f(x) = -3x + 2$ and $g(x) = x^2 + 2$

43. $f(x) = x + 2$ and $g(x) = \dfrac{x^2 + 3}{2}$

Write the following functions as a composition of two functions. Answers will vary. See Example 7.

44. $f(x) = \sqrt[3]{3x^2 - 1}$

45. $f(x) = \dfrac{2}{5x - 1}$

46. $f(x) = |x - 2| + 3$

47. $f(x) = x + \sqrt{x + 2} - 5$

48. $f(x) = |x^3 - 5x| + 7$

49. $f(x) = \dfrac{\sqrt{x-3}}{x^2 - 6x + 9}$

50. $f(x) = \sqrt{2x^3 - 3} - 4$

51. $f(x) = |x^2 + 3x| - 3$

52. $f(x) = \dfrac{3}{4x - 2}$

In each of the following problems, use the information given to find $g(x)$.

53. $f(x) = |x + 3|$ and $(f + g)(x) = |x + 3| + \sqrt{x + 5}$

54. $f(x) = x$ and $(f \circ g)(x) = \dfrac{x + 12}{-3}$

55. $f(x) = x^2 - 3$ and $(f - g)(x) = x^3 + x^2 + 4$

56. $f(x) = x^2$ and $(g \circ f)(x) = \sqrt{-x^2 + 5} + 4$

Solve the following application problems.

57. The volume of a right circular cylinder is given by the formula $V = \pi r^2 h$. If the height h is three times the radius r, show the volume V as a function of r.

58. The surface area S of a wind sock is given by the formula $S = \pi r \sqrt{r^2 + h^2}$, where r is the radius of the base of the wind sock and h is the height of the wind sock. As the wind sock is being knitted by an automated knitter, the height h increases with time t according to the formula $h(t) = \dfrac{1}{4}t^2$. Find the surface area S of the wind sock as a function of time t and radius r.

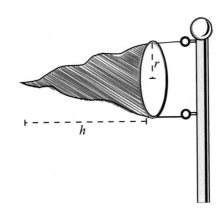

49. $g(x) = \dfrac{\sqrt{x}}{x^2}$,

$h(x) = x - 3$,

$f(x) = g(h(x))$

50. $g(x) = \sqrt{x} - 4$,

$h(x) = 2x^3 - 3$,

$f(x) = g(h(x))$

51. $g(x) = x - 3$,

$h(x) = \left| x^2 + 3x \right|$,

$f(x) = g(h(x))$

52. $g(x) = \dfrac{3}{x}$,

$h(x) = 4x - 2$,

$f(x) = g(h(x))$

53. $g(x) = \sqrt{x + 5}$

54. $g(x) = \dfrac{x + 12}{-3}$

55. $g(x) = -x^3 - 7$

56. $\sqrt{-x + 5} + 4$

57. $V = 3\pi r^3$

58. $S = \pi r \sqrt{r^2 + \dfrac{t^4}{16}}$

59. $V = \dfrac{1}{12} \pi r^2 t^2$

60. $c(t) = 2040 +$

$17{,}400t - 43.5t^2$

61. $(f \circ g)(x) = \sqrt[3]{\dfrac{-x^3}{3x^2 - 9}}$,

$(f \circ g)(-x) = \sqrt[3]{\dfrac{x^3}{3x^2 - 9}}$

$= -(f \circ g)(x)$

62. $(fg)(x) = 2x^2 - 1$,

$(fg)(-x) = 2x^2 - 1$

$= (fg)(x)$

59. The volume V of the wind sock described in the previous question is given by the formula $V = \dfrac{1}{3}\pi r^2 h$ where r is the radius of the wind sock and h is the height of the wind sock. If the height h increases with time t according to the formula $h(t) = \dfrac{1}{4}t^2$, find the volume V of the wind sock as a function of time t and radius r.

60. A widget factory produces n widgets in t hours of a single day. The number of widgets the factory produces is given by the formula $n(t) = 10{,}000t - 25t^2$, $0 \le t \le 9$. The cost c in dollars of producing n widgets is given by the formula $c(n) = 2040 + 1.74n$. Find the cost c as a function of time t.

61. Given two odd functions f and g, show that $f \circ g$ is also odd. Verify this fact with the particular functions $f(x) = \sqrt[3]{x}$ and $g(x) = \dfrac{-x^3}{3x^2 - 9}$. Recall that a function is odd if $f(-x) = -f(x)$ for all x in the domain of f.

62. Given two even functions f and g, show that the product is also even. Verify this fact with the particular functions $f(x) = 2x^4 - x^2$ and $g(x) = \dfrac{1}{x^2}$. Recall that a function is even if $f(-x) = f(x)$ for all x in the domain of f.

As mentioned in Topic 4, a given complex number c is said to be in the Mandelbrot set if, for the function $f(z) = z^2 + c$, the sequence of iterates $f(0)$, $f^2(0)$, $f^3(0)$, ... stays close to the origin (which is the complex number $0 + 0i$). It can be shown that if any single iterate falls more than 2 units in distance (magnitude) from the origin, then the remaining iterates will grow larger and larger in magnitude. In practice, computer programs that generate the Mandelbrot set calculate the iterates up to a predecided point in the sequence, such as $f^{50}(0)$, and if no iterate up to this point exceeds 2 in magnitude, the number c is admitted to the set. The magnitude of a complex number $a + bi$ is the distance between the point (a, b) and the origin, so the formula for the magnitude of $a + bi$ is $\sqrt{a^2 + b^2}$.

Use the above criterion to determine, without a calculator or computer, if the following complex numbers are in the Mandelbrot set or not.

63. $c = 0$ **64.** $c = 1$ **65.** $c = i$ **66.** $c = -1$ **67.** $c = 1 + i$

68. $c = -i$ **69.** $c = 1 - i$ **70.** $c = -1 - i$ **71.** $c = 2$ **72.** $c = -2$

63. Yes

64. No

65. Yes

66. Yes

67. No

68. Yes

69. No

70. No

71. No

72. Yes

Inverses of Functions

TOPICS

1. Inverses of relations
2. Inverse functions and the horizontal line test
3. Finding inverse function formulas

TOPIC 1

Inverses of Relations

In many problems, "undoing" one or more mathematical operations plays a critical role in the solution process. For instance, to solve the equation $3x + 2 = 8$, the first step is to "undo" the addition of 2 on the left-hand side (by subtracting 2 from both sides) and the second step is to "undo" the multiplication by 3 (by dividing both sides by 3). In the context of more complex problems, the "undoing" process is often a matter of finding and applying the inverse of a function.

We begin with the more general idea of the inverse of a relation. Recall that a relation is just a set of ordered pairs; the inverse of a given relation is the set of these ordered pairs with the first and second coordinates of each exchanged.

DEFINITION

Inverse of a Relation

Let R be a relation. The **inverse of R**, denoted R^{-1}, is the relation defined by:

$$R^{-1} = \left\{ (b, a) \middle| (a, b) \in R \right\}.$$

EXAMPLE 1

Finding the Inverse of a Relation

Determine the inverse of each of the following relations. Then graph each relation and its inverse, and determine the domain and range of both.

   a. $R = \left\{ (4, -1), (-3, 2), (0, 5) \right\}$ **b.** $y = x^2$

Solutions:

   a. $R = \left\{ (4, -1), (-3, 2), (0, 5) \right\}$ For each ordered pair, switch the first and second coordinates (x- and y-coordinates).

       $R^{-1} = \left\{ (-1, 4), (2, -3), (5, 0) \right\}$

Recall that the domain is the set of first coordinates, and the range is the set of second coordinates.

$$R: \quad \text{Domain} = \{4, -3, 0\} \quad \text{Range} = \{-1, 2, 5\}$$
$$R^{-1}: \quad \text{Domain} = \{-1, 2, 5\} \quad \text{Range} = \{4, -3, 0\}$$

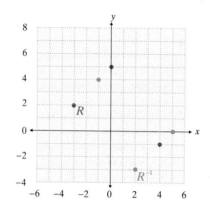

In the graph to the left, R is in purple and its inverse is in green. The relation R consists of three ordered pairs, and its inverse is simply these three ordered pairs with the coordinates exchanged. Note that the domain of R is the range of R^{-1} and vice versa.

b. $R = \{(x, y) \mid y = x^2\}$

$R^{-1} = \{(x, y) \mid x = y^2\}$

$R: \quad \text{Domain} = \mathbb{R} \quad \text{Range} = [0, \infty)$

$R^{-1}: \quad \text{Domain} = [0, \infty) \quad \text{Range} = \mathbb{R}$

In this problem, R is described by the given equation in x and y. The inverse relation is the set of ordered pairs in R with the coordinates exchanged, so we can describe the inverse relation by just exchanging x and y in the equation, as shown at left.

Note that the shape of the graph of the relation and its inverse are essentially the same.

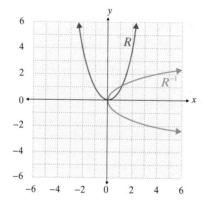

Consider the graphs of the two relations and their respective inverses in Example 1. By definition, an ordered pair (b, a) lies on the graph of a relation R^{-1} if and only if (a, b) lies on the graph of R, so it shouldn't be surprising that the graphs of a relation and its inverse bear some resemblance to one another. Specifically, they are mirror images of one another with respect to the line $y = x$. If you were to fold the Cartesian plane in half along the line $y = x$ in the two examples above, you would see that the points in R and R^{-1} coincide with one another.

The two relations in Example 1 illustrate another important point. Note that in both cases, R is a function, as its graph passes the vertical line test. By the same criterion, R^{-1} in Example 1a is also a function, but R^{-1} in Example 1b is not. The conclusion to be drawn is that even if a relation is a function, its inverse may or may not be a function.

TOPIC 2   ## Inverse Functions and the Horizontal Line Test

We have a convenient graphical test for determining when a relation is a function (the Vertical Line Test); we would like to have a similar test about determining when the inverse of a relation is a function.

In practice, we will only be concerned with the question of when the inverse of a function f, denoted f^{-1}, is itself a function.

CAUTION!

We are faced with another example of the reuse of notation. f^{-1} does *not* stand for $\dfrac{1}{f}$ when f is a function! We use an exponent of -1 to indicate the reciprocal of a number or an algebraic expression, but when applied to a function or a relation it stands for the inverse relation.

Assume that f is a function. f^{-1} will only be a function itself if its graph passes the vertical line test; that is, only if each element of the domain of f^{-1} is paired with exactly one element of the range of f^{-1}. This is identical to saying that each element of the range of f is paired with exactly one element of the domain of f. In other words, every *horizontal* line in the plane must intersect the graph of f no more than once.

THEOREM

The Horizontal
Line Test

Let f be a function. We say that the graph of f passes the **horizontal line test** if every horizontal line in the plane intersects the graph no more than once. If f passes the horizontal line test, then f^{-1} is also a function.

Of course, the horizontal line test is only useful if the graph of f is available to study. We can also phrase the above condition in a nongraphical manner. The inverse of f will only be a function if for every pair of distinct elements x_1 and x_2 in the domain of f, we have $f(x_1) \neq f(x_2)$. This criterion is important enough to merit a name.

DEFINITION

One-to-One
Functions

A function f is **one-to-one** if for every pair of distinct elements x_1 and x_2 in the domain of f, we have $f(x_1) \neq f(x_2)$. This means that every element of the range of f is paired with exactly one element of the domain of f.

To sum up: the inverse f^{-1} of a function f is also a function if and only if f is one-to-one and f is one-to-one if and only if its graph passes the horizontal line test.

EXAMPLE 2

Determine if the following functions have inverse functions.

a. $f(x) = |x|$　　　　　　　**b.** $g(x) = (x+2)^3$

Note:
Even when a function f does not have an inverse *function*, it always has an inverse *relation*.

Solutions:

a. The function f does not have an inverse function, a fact demonstrated by showing that its graph does not pass the horizontal line test. We can also prove this algebraically: although $-3 \neq 3$, we have $f(-3) = f(3)$. Note that it only takes two ordered pairs to show that f does not have an inverse function.

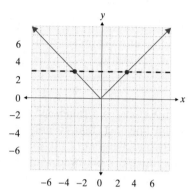

b. The graph of g is the standard cubic shape shifted horizontally two units to the left. We can see this graph passes the horizontal line test, so g has an inverse function.

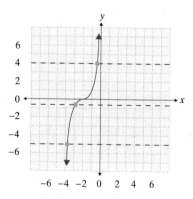

Algebraically, any two distinct elements of the domain of g lead to different values when plugged into g, so g is one-to-one and hence has an inverse function.

Consider the function in Example 2a again. As we noted, the function $f(x) = |x|$ is not one-to-one, and so cannot have an inverse function. However, if we *restrict the domain* of f by specifying that the domain is the interval $[0, \infty)$, the new function, with this restricted domain, is one-to-one and has an inverse function. Of course, this **restriction of domain** changes the function; in this case the graph of the new function is the right-hand half of the graph of the absolute value function.

TOPIC [3]

Finding Inverse Function Formulas

In applying the notion of the inverse of a function, we will often begin with a formula for f and want to find a formula for f^{-1}. This will allow us, for instance, to transform equations of the form

$$f(x) = y \text{ into the form } x = f^{-1}(y).$$

Before we discuss the general algorithm for finding a formula for f^{-1}, consider the problem with which we began this section. If we define $f(x) = 3x + 2$, the equation $3x + 2 = 8$ can be written as $f(x) = 8$. Note that f is one-to-one, so f^{-1} is a function. If we can find a formula for f^{-1}, we can transform the equation into $x = f^{-1}(8)$. This is a complicated way to solve this equation, but it illustrates how to find inverses.

What should the formula for f^{-1} be? Consider what f does to its argument. The first action is to multiply x by 3, and the second is to add 2. To "undo" f, we need to negate these two actions in reverse order: subtract 2 and then divide the result by 3. So,

$$f^{-1}(x) = \frac{x-2}{3}.$$

Applying this to the problem at hand, we obtain

$$x = f^{-1}(8) = \frac{8-2}{3} = 2.$$

This method of analyzing a function f and then finding a formula for f^{-1} by undoing the actions of f in reverse order is conceptually important and works for simple functions. For other functions, however, the following algorithm may be necessary as a standardized way to find the inverse formula.

PROCEDURE

Formulas of Inverse Functions

Let f be a one-to-one function, and assume that f is defined by a formula. To find a formula for f^{-1}, perform the following steps:

Step 1: Replace $f(x)$ in the definition of f with the variable y. The result is an equation in x and y that is solved for y at this point.

Step 2: Interchange x and y in the equation.

Step 3: Solve the new equation for y.

Step 4: Replace the y in the resulting equation with $f^{-1}(x)$.

===== EXAMPLE 3 =====

Finding Formulas of Inverse Functions

Find the inverse of each of the following functions.

a. $f(x) = (x-1)^3 + 2$

b. $g(x) = \dfrac{x-3}{2x+1}$

Solutions:

a.

$f(x) = (x-1)^3 + 2$	Following the algorithm shows us how the steps of the original function get "undone."
$y = (x-1)^3 + 2$	First, replace $f(x)$ with y.
$x = (y-1)^3 + 2$	Next, switch x and y in the equation.
$x - 2 = (y-1)^3$	To solve the resulting equation for y, first subtract 2 from both sides.
$\sqrt[3]{x-2} = y - 1$	Take the cube root of both sides.
$\sqrt[3]{x-2} + 1 = y$	Add 1 to both sides.
$f^{-1}(x) = \sqrt[3]{x-2} + 1$	Replace y with $f^{-1}(x)$.

b.

$g(x) = \dfrac{x-3}{2x+1}$	The inverse of the function g is most easily found by the algorithm.
$y = \dfrac{x-3}{2x+1}$	The first step is to replace $g(x)$ with y.
$x = \dfrac{y-3}{2y+1}$	The second step is to interchange x and y in the equation.
$x(2y+1) = y - 3$	
$2xy + x = y - 3$	We now have to solve the equation for y. Begin by clearing the equation of fractions, and then proceed to collect all the terms that contain y on one side.
$2xy - y = -x - 3$	
$y(2x-1) = -x - 3$	
$y = \dfrac{-x-3}{2x-1}$	Factoring out the y on the left-hand side and dividing by $2x - 1$ completes the process.
$g^{-1}(x) = \dfrac{-x-3}{2x-1}$	The last step is to rename the formula $g^{-1}(x)$.

Remember that the graphs of a relation and its inverse are mirror images of one another with respect to the line $y = x$; this is still true if the relations are functions. We can demonstrate this fact by graphing the function and its inverse from Example 3a above, as shown in Figure 1.

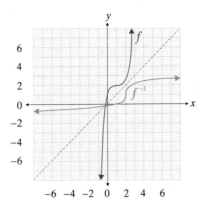

Figure 1: Graph of a Function and Its Inverse

We can use the functions and their inverses from Example 3 to illustrate one last important point. The key characteristic of the inverse of a function is that it undoes the function. This means that if a function and its inverse are composed together, in either order, the resulting function has no effect on any allowable input!

THEOREM

Composition of
Functions and
Inverses

Given a function f and its inverse f^{-1}, the following statements are true.

$$f\left(f^{-1}(x)\right) = x \text{ for all } x \in \text{Dom}\left(f^{-1}\right), \text{ and}$$

$$f^{-1}\left(f(x)\right) = x \text{ for all } x \in \text{Dom}(f).$$

For example, given $f(x) = (x-1)^3 + 2$ and $f^{-1}(x) = (x-2)^{\frac{1}{3}} + 1$:

$$f\left(f^{-1}(x)\right) = f\left((x-2)^{\frac{1}{3}} + 1\right)$$

$$= \left((x-2)^{\frac{1}{3}} + 1 - 1\right)^3 + 2$$

$$= \left((x-2)^{\frac{1}{3}}\right)^3 + 2$$

$$= x - 2 + 2$$

$$= x.$$

A similar calculation shows that $f^{-1}\left(f(x)\right) = x$, as you should verify.

As another example, consider $g(x) = \dfrac{x-3}{2x+1}$ and $g^{-1}(x) = \dfrac{-x-3}{2x-1}$:

1.

Dom = { 2, –1, –2 }

Ran = { –4, 3, 0 }

2.

Dom = {–3, –1, 1, 4}

Ran = { –3, –1, 0, 4 }

3.

Dom = Ran = $\mathbb{R}$

4.

Dom = $[2, \infty)$

Ran = $\mathbb{R}$

5.

Dom = $\mathbb{R}$

Ran = $[0, \infty)$

6.

Dom = $[0, \infty)$

Ran = $(-\infty, 0]$

7.

Dom = $\mathbb{R}$

Ran = $\mathbb{R}$

$$g^{-1}\big(g(x)\big) = g^{-1}\left(\frac{x-3}{2x+1}\right)$$

$$= \frac{-\dfrac{x-3}{2x+1} - 3}{2\left(\dfrac{x-3}{2x+1}\right) - 1}$$

$$= \left(\frac{-\dfrac{x-3}{2x+1} - 3}{2\left(\dfrac{x-3}{2x+1}\right) - 1}\right)\left(\frac{2x+1}{2x+1}\right)$$

$$= \frac{-x+3-6x-3}{2x-6-2x-1}$$

$$= \frac{-7x}{-7}$$

$$= x$$

Similarly, $g\big(g^{-1}(x)\big) = x$, as you should verify.

Exercises

Graph the inverse of each of the following relations, and state its domain and range. See Example 1.

1. $R = \{(-4, 2), (3, 2), (0, -1), (3, -2)\}$ **2.** $S = \{(-3, -3), (-1, -1), (0, 1), (4, 4)\}$

3. $y = x^3$ **4.** $y = |x| + 2$

5. $x = |y|$ **6.** $x = -\sqrt{y}$

7. $y = \dfrac{1}{2}x - 3$ **8.** $y = -x + 1$

9. $y = \sqrt{x} + 2$ **10.** $T = \{(4, 2), (3, -1), (-2, -1), (2, 4)\}$

11. $x = y^2 - 2$ **12.** $y = 2\sqrt{x}$

8.

Dom = $\mathbb{R}$

Ran = $\mathbb{R}$

9.

Dom = $[2, \infty)$

Ran = $[0, \infty)$

10.

Dom = {2, –1, 4}

Ran = {4, 3, –2, 2}

11.

Dom = $\mathbb{R}$

Ran = $[-2, \infty)$

12.

$\text{Dom} = [0, \infty)$

$\text{Ran} = [0, \infty)$

13.

$\text{Dom} = \mathbb{R}$

$\text{Ran} = \mathbb{R}$

14. Not a one-to-one
function
$f(-5) = f(1) = 5$

15. Not a one-to-one
function
$f(-1) = f(1) = 1$

16.

$\text{Dom} = \mathbb{R}$

$\text{Ran} = \mathbb{R}$

17. Restrict to $[0, \infty)$

18. Inverse exists

19. Inverse exists

20. Restrict to $(0, \infty)$

21. Inverse exists

22. Restrict to $[0, \infty)$

23. Inverse exists

24. Inverse exists

25. Restrict to $[2, \infty)$

26. Inverse exists

27. Restrict to $[12, \infty)$

28. Restrict to $[0, \infty)$

29. $f^{-1}(x) = (x+2)^3$

30. $g^{-1}(x) = \dfrac{x+3}{4}$

Determine if each of the following functions is a one-to-one function. If so, graph the inverse of the function and state its domain and range.

13.

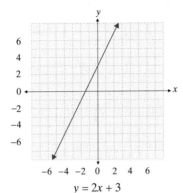

$y = 2x + 3$

14.

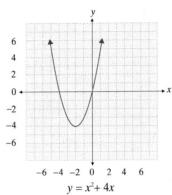

$y = x^2 + 4x$

15.

$y = \dfrac{1}{x^2}$

16.

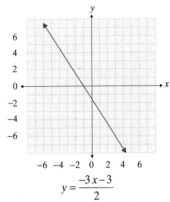

$y = \dfrac{-3x - 3}{2}$

Determine if the following functions have inverse functions. If not, suggest a domain to restrict the function to so that it would have an inverse function (answers will vary). See Example 2.

17. $f(x) = x^2 + 1$ **18.** $g(x) = (x-2)^3 - 1$ **19.** $h(x) = \sqrt{x+3}$

20. $s(x) = \dfrac{1}{x^2}$ **21.** $G(x) = 3x - 5$ **22.** $F(x) = -x^2 + 5$

23. $r(x) = -\sqrt{x^3}$ **24.** $b(x) = \dfrac{1}{x}$ **25.** $f(x) = x^2 - 4x$

26. $m(x) = \dfrac{13x - 2}{4}$ **27.** $H(x) = |x - 12|$ **28.** $p(x) = 10 - x^2$

Find a formula for the inverse of each of the following functions. See Example 3.

29. $f(x) = x^{\frac{1}{3}} - 2$ **30.** $g(x) = 4x - 3$ **31.** $r(x) = \dfrac{x-1}{3x+2}$

32. $s(x) = \dfrac{1-x}{1+x}$ **33.** $F(x) = (x-5)^3 + 2$ **34.** $G(x) = \sqrt[3]{3x-1}$

35. $V(x) = \dfrac{x+5}{2}$ **36.** $W(x) = \dfrac{1}{x}$ **37.** $h(x) = x^{\frac{3}{5}} - 2$

31. $r^{-1}(x) = \dfrac{-2x-1}{3x-1}$

32. $s^{-1}(x) = \dfrac{1-x}{1+x}$

33. $F^{-1}(x) =$

$(x-2)^{\frac{1}{3}} + 5$

34. $G^{-1}(x) = \dfrac{x^3+1}{3}$

35. $V^{-1}(x) = 2x-5$

36. $W^{-1}(x) = \dfrac{1}{x}$

37. $h^{-1}(x) = (x+2)^{\frac{5}{3}}$

38. $A^{-1}(x) = (x^5-1)^{\frac{1}{3}}$

39. $J^{-1}(x) = \dfrac{x-2}{3x}$

40. $k^{-1}(x) = \dfrac{3x-4}{x+1}$

41. $h^{-1}(x) = (x-6)^{\frac{1}{7}}$

42. $F^{-1}(x) = (9x+3)^{\frac{1}{5}}$

43. $r^{-1}(x) = \dfrac{x^5}{2}$

44. $P^{-1}(x) = \dfrac{x^{\frac{1}{3}}-2}{3}$

45. $f^{-1}(x) = \dfrac{x^3}{54}$

46. $q^{-1}(x) =$

$(x-2)^{\frac{1}{2}} + 2$

47.–56. Answers will vary.

57. b

58. f

59. e

60. c

38. $A(x) = (x^3+1)^{\frac{1}{5}}$

39. $J(x) = \dfrac{2}{1-3x}$

40. $k(x) = \dfrac{x+4}{3-x}$

41. $h(x) = x^7+6$

42. $F(x) = \dfrac{3-x^5}{-9}$

43. $r(x) = \sqrt[5]{2x}$

44. $P(x) = (2+3x)^3$

45. $f(x) = 3(2x)^{\frac{1}{3}}$

46. $q(x) = (x-2)^2 + 2, \ x \geq 2$

In each of the following problems, verify that $f\left(f^{-1}(x)\right) = x$ and that $f^{-1}\left(f(x)\right) = x$.

47. $f(x) = \dfrac{3x-1}{5}$ and $f^{-1}(x) = \dfrac{5x+1}{3}$

48. $f(x) = \sqrt[3]{x+2} - 1$ and $f^{-1}(x) = (x+1)^3 - 2$

49. $f(x) = \dfrac{2x+7}{x-1}$ and $f^{-1}(x) = \dfrac{x+7}{x-2}$

50. $f(x) = x^2, \ x \geq 0$ and $f^{-1}(x) = \sqrt{x}$

51. $f(x) = 2x-3$ and $f^{-1}(x) = \dfrac{x+3}{2}$

52. $f(x) = \sqrt[3]{x+1}$ and $f^{-1}(x) = x^3 - 1$

53. $f(x) = \dfrac{1}{x}$ and $f^{-1}(x) = \dfrac{1}{x}$

54. $f(x) = \dfrac{x-5}{2x+3}$ and $f^{-1}(x) = \dfrac{3x+5}{1-2x}$

55. $f(x) = (x-2)^2, \ x \geq 2$ and $f^{-1}(x) = \sqrt{x} + 2, \ x \geq 0$

56. $f(x) = \dfrac{1}{1+x}$ and $f^{-1}(x) = \dfrac{1-x}{x}$

Match the following functions with the graphs of the inverses of the functions. The graphs are labeled **a.** through **f.**

57. $f(x) = x^3$

58. $f(x) = x - 5$

59. $f(x) = \sqrt{x-4}$

60. $f(x) = x^2, \ x \geq 0$

61. $f(x) = \dfrac{x}{4}$

62. $f(x) = \sqrt[3]{x+1}$

a.

b.

c.

d.

e.

f.

61. a

62. d

63. 73 1 53 13 97 73
29 57 17 73

64. 184 72 96 96 72
160 144 8 72 112
160 120 32 8 200

65. FRISBEE
VOLLEYBALL
AND
HORSESHOES

66. REMEMBER
YOUR
SUNBLOCK

67. CATCH A WAVE

68. BEACH FUN IN
THE SUN

An inverse function can be used to encode and decode words and sentences by assigning each letter of the alphabet a numerical value (A = 1, B = 2, C = 3, ..., Z = 26). Example: Use the function $f(x) = x^2$ to encode the word PRECALCULUS. The encoded message would be 256 324 25 9 1 144 9 441 144 441 361. The word can then be decoded by using the inverse function $f^{-1}(x) = \sqrt{x}$. The inverse values are 16 18 5 3 1 12 3 21 12 21 19 which translates back to the word PRECALCULUS. Encode or decode the following words using the numerical values A = 1, B = 2, C = 3, ..., Z = 26.

63. Encode the message SANDY SHOES using the function $f(x) = 4x - 3$.

64. Encode the message WILL IT RAIN TODAY using the function $f(x) = 8x$.

65. The following message was encoded using the function $f(x) = 8x - 7$. Decode the message.

41 137 65 145 9 33 33 169 113 89 89 33 193 9 1 89 89 1 105 25 57 113 137 145 33 145 57 113 33 145

66. The following message was encoded using the function $f(x) = 5x + 1$. Decode the message.

91 26 66 26 66 11 26 91 126 76 106 91 96 106 71 11 61 76 16 56

67. The following message was encoded using the function $f(x) = x^3$. Decode the message.

27 1 8000 27 512 1 12167 1 10648 125

68. The following message was encoded using the function $f(x) = -3 - 5x$. Decode the message.

−13 −28 −8 −18 −43 −33 −108 −73 −48 −73 −103 −43 −28 −98 −108 −73

Chapter 3 Project

The Ozone Layer

As time goes on, there is continually increasing awareness, controversy, and legislation regarding the ozone layer and other environmental issues. The hole in the ozone layer over the south pole disappears and reappears annually, and one model for its growth assumes the hole is circular and that its radius grows at a constant rate of 2.6 kilometers per hour.

1. Write the area of the circle as a function of the radius, r.

2. Assuming that t is measured in hours, that $t = 0$ corresponds to the start of the annual growth of the hole, and that the radius of the hole is initially 0, write the radius as a function of time, t.

3. Write the area of the circle as a function of time, t.

4. What is the radius after 3 hours?

5. What is the radius after 5.5 hours?

PHOTO COURTESY OF NASA

6. What is the area of the circle after 3 hours?

7. What is the area of the circle after 5.5 hours?

8. What is the average rate of change of the area from 3 hours to 5.5 hours?

9. What is the average rate of change of the area from 5.5 hours to 8 hours?

10. Is the average rate of change of the area increasing or decreasing as time passes?

Chapter Summary

A summary of concepts and skills follows each chapter. Refer to these summaries to make sure you feel comfortable with the material in the chapter. The concepts and skills are organized according to the section title and topic title in which the material is first discussed.

3.1: Relations and Functions

Relations, Domain, and Range
- The definition of *relation* as a set of ordered pairs
- The definition of *domain* and *range* as, respectively, the set of first coordinates and the set of second coordinates for a given relation
- The correspondence between a relation and its graph in the Cartesian plane

Functions and the Vertical Line Test
- The definition of a *function* as a special type of relation
- The meaning of the *vertical line test* as applied to the graph of a relation and in identifying functions

Function Notation and Function Evaluation
- The meaning of *function notation*
- Evaluation of a function for a given *argument*
- The role of the argument as a placeholder in defining a function
- The definition of *domain* and *codomain*

Implied Domain of a Function
- Determining the domain of a function when it is not stated explicitly

3.2: Linear and Quadratic Functions

Linear Functions and Their Graphs
- The definition of a *linear*, or *first-degree*, *function*
- The graph of a linear function

Quadratic Functions and Their Graphs
- The definition of a *quadratic*, or *second-degree*, *function*
- The graph of a quadratic function, including the location of the vertex and the *x*- and *y*-intercepts
- Finding the vertex form of a quadratic function

Maximization/Minimization Problems
- The role of *completing the square* in locating the maximum or minimum value of a quadratic function

3.3: Other Common Functions

Functions of the Form ax^n
- The basic form of the graph of ax^n when n is even
- The basic form of the graph of ax^n when n is odd

Functions of the Form $\dfrac{a}{x^n}$
- The basic form of the graph of $\dfrac{a}{x^n}$ when n is even
- The basic form of the graph of $\dfrac{a}{x^n}$ when n is odd

Functions of the Form $ax^{\frac{1}{n}}$
- The basic form of the graph of $ax^{\frac{1}{n}}$ when n is even
- The basic form of the graph of $ax^{\frac{1}{n}}$ when n is odd

The Absolute Value Function
- The basic form of the graph of the absolute value function

Piecewise-Defined Functions
- The definition of *piecewise-defined function*: a function defined in terms of two or more formulas, each valid for its own unique portion of the real number line

3.4: Variation and Multivariable Functions

Direct Variation
- Concept of *direct variation*
- Applications of direct variation

Inverse Variation
- Concept of *inverse variation*
- Applications of inverse variation

Joint Variation
- Concept of *joint variation*
- Applications of joint variation

Multivariable Functions
- Evaluating multivariable functions

3.5: Transformations of Functions

Shifting, Stretching, and Reflecting Graphs
- Replacing the argument x with $x - h$ to shift a graph h units horizontally
- Adding k to a function to shift its graph k units vertically
- Multiplying a function by -1 to reflect its graph with respect to the x-axis
- Replacing the argument x with $-x$ to reflect a graph with respect to the y-axis
- Multiplying a function by an appropriate constant to stretch or compress its graph
- Determining the order in which to evaluate transformations

3.5: Transformations of Functions (cont.)

Symmetry of Functions and Equations
- The meaning of *y-axis symmetry*
- The meaning of *x-axis symmetry*
- The meaning of *origin symmetry*
- The meaning of *even* and *odd* functions

3.6: Combining Functions

Combining Functions Arithmetically
- *Sums*, *differences*, *products*, and *quotients* of functions, and how to evaluate such combinations
- Identifying the domain of an arithmetic combination of functions

Composing Functions
- The meaning of *composition* of functions
- Determining a formula for the composition of two functions, and evaluating a composition of functions for a given argument
- Identifying the domain of a composition of functions

Decomposing Functions
- The decomposition of complicated functions into simpler functions

Recursive Graphics
- *Recursion* as an application of composition, and the meaning of the *iterates* of a function

3.7: Inverses of Functions

Inverses of Relations
- The definition of the *inverse* of a relation, and the correspondence between the graph of a relation and the graph of its inverse

Inverse Functions and the Horizontal Line Test
- The concept of the *inverse* of a function, and the notation used
- Identifying when a function has an inverse function
- The correspondence between the *horizontal line test* and whether a function is *one-to-one*
- Restriction of domain

Finding Inverse Function Formulas
- Constructing the formula for the inverse of a function by "undoing" the function step by step
- The algorithmic approach to finding the inverse of a function

1. Dom $= \{-3,-1,0,3,4\}$,
Ran $= \{-1,0,3,4\}$;
Yes

2. Dom $= \mathbb{R}$,
Ran $= \{2\}$; Yes

3. Dom $= \mathbb{Z}$,
Ran $=$
$\{...,-4,0,4,8,...\}$;
Yes

4. Dom $= \mathbb{R}$,
Ran $= \mathbb{R}$; Yes

5. Dom $= [-6,\infty)$,
Ran $= \mathbb{R}$; No

6. Dom $= [0,\infty)$,
Ran $= [4,\infty)$; Yes

7. Dom $= \mathbb{R}$,
Ran $= \{-5\}$; Yes

8. Dom $= \{-2,-4\}$,
Ran $= \{-1,5\}$; Yes

9. Dom $= \mathbb{R}$,
Ran $= (-\infty,3]$; Yes

10. $f(x) =$
$3\sqrt{x+11}-4$
$f(-2) = 5$

11. $f(x) = -x^2 +5x$
$f(-2) = -14$

12. $\sqrt{x+h}$

13. $\dfrac{\sqrt{x+h}-\sqrt{x}}{x+h}$

14. $\sqrt[3]{(x+h)^2}$

15. $\dfrac{\sqrt[3]{(x+h)^2}-\sqrt[3]{x^2}}{h}$

16. Dom $= \mathbb{N}$,
Cod $= \mathbb{R}$,
Ran $= \left\{\dfrac{3}{4},\dfrac{3}{2},\dfrac{9}{4},...\right\}$

17. Dom $= \mathbb{R}$,
Cod $= \mathbb{R}$,
Ran $= \mathbb{R}$

18. $\mathbb{R}$

19. $(-\infty,1)\cup(1,\infty)$

Chapter Review

Section 3.1

For each relation below, describe the domain and range and determine whether or not the relation is a function.

1. $f = \{(-3,0),(-1,4),(0,3),(3,3),(4,-1)\}$

2. $R = \{(x,2) | x \in \mathbb{R}\}$

3. $S = \{(x,4x) | x \in \mathbb{Z}\}$

4. $3x - 4y = 17$

5. $x = y^2 - 6$

6. $x = \sqrt{y-4}$

7. $y = -5$

8.

9.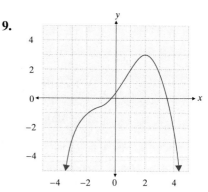

Rewrite each of the relations below as a function of x. Then evaluate the function at $x = -2$.

10. $\dfrac{y+4}{\sqrt{x+11}} - 3y = 3(1-y)$

11. $x^2 - 4x + 3y = x + 2y$

Given $f(x) = \sqrt{x}$ and $g(x) = \sqrt[3]{x^2}$, evaluate the following.

12. $f(x+h)$

13. $\dfrac{f(x+h)-f(x)}{x+h}$

14. $g(x+h)$

15. $\dfrac{g(x+h)-g(x)}{h}$

20. **21.** **22.** **23.**

24.

25.

26.

27.

28.

29.

30. 125

31.

32.

33.

34.

Identify the domain, codomain, and range of each of the following functions.

16. $g: \mathbb{N} \to \mathbb{R}$ by $g(x) = \dfrac{3x}{4}$

17. $h: \mathbb{R} \to \mathbb{R}$ by $h(x) = 5x + 1$

Determine the implied domain of each of the following functions.

18. $f(x) = \sqrt{|-x - 3|}$

19. $f(x) = \dfrac{x}{1 - x}$

Section 3.2

Graph the following linear functions.

20. $g(x) = \dfrac{2x - 6}{3}$

21. $p(x) = -2x - 3$

22. $k(x) = 2(1 - 2x) + x$

23. $f(x) = \dfrac{4x + 3}{2} - 2x$

Graph the following quadratic functions, locating the vertices and x-intercepts (if any) accurately.

24. $f(x) = (x - 1)^2 - 1$

25. $g(x) = -(x + 3)^2 - 2$

26. $p(x) = x^2 - 2$

27. $k(x) = -x^2 + 4x$

28. $h(x) = x^2 + 2x - 3$

29. $f(x) = -x^2 + 5$

Solve the following maximization/minimization problems by analyzing the appropriate quadratic function.

30. The total revenue for McDaniel's Storage Plus is given as the function
$$R(x) = -0.4x^2 + 100x - 5250,$$
where x is the number of storage units rented. What number of units rented produces the maximum revenue?

Section 3.3

Sketch the graphs of the following functions. Pay particular attention to intercepts, if any, and locate these accurately.

31. $f(x) = -4|x|$ **32.** $g(x) = 3\sqrt{x}$ **33.** $r(x) = \dfrac{1}{x^2}$

34. $p(x) = -2x^4$ **35.** $q(x) = -\dfrac{1}{x^3}$ **36.** $k(x) = \dfrac{\sqrt[3]{x}}{2}$

35.   **36.**   **37.**   **38.**

39.

40.

41.

42.

43. $V = khr^2$

44. $y = \dfrac{ka^3}{\sqrt{b}}$

45. $y = 112$

46. $y = 72$

47. About 1226 videos per month

48. 7.44×10^7 meters

49.

Dom = Ran = $\mathbb{R}$

50.

Dom = $\mathbb{R}$,

Ran = $[0, \infty)$

51.

Dom =

$(-\infty, -2) \cup (-2, \infty)$,

Ran = $(0, \infty)$

37. $f(x) = \begin{cases} x^2 & \text{if } x < 1 \\ \dfrac{1}{x} & \text{if } x \geq 1 \end{cases}$

38. $g(x) = \begin{cases} (x+1)^2 - 1 & \text{if } x \leq 0 \\ \sqrt[3]{x} & \text{if } x > 0 \end{cases}$

39. $h(x) = \begin{cases} -|x| & \text{if } x < 3 \\ (x-4)^2 + 1 & \text{if } x \geq 3 \end{cases}$

40. $f(x) = \begin{cases} x^2 & \text{if } x \leq -2 \\ \dfrac{1}{x^2} & \text{if } x > -2 \end{cases}$

41. $q(x) = \begin{cases} 3x - 1 & \text{if } x < 1 \\ x^4 & \text{if } x \geq 1 \end{cases}$

42. $g(x) = \begin{cases} 2|x| & \text{if } x < 2 \\ \sqrt{x} & \text{if } x \geq 2 \end{cases}$

Section 3.4

Find the mathematical model for each of the following verbal statements.

43. V varies directly as the product of r squared and h.

44. y varies directly as the cube of a and inversely as the square root of b.

Solve the following variation problems.

45. Suppose that y varies directly as the square of x, and that $y = 567$ when $x = 9$. What is y when $x = 4$?

46. Suppose that y is inversely proportional to the square root of x, and that $y = 45$ when $x = 64$. What is y when $x = 25$?

47. A video store manager observes that the number of videos rented seems to vary inversely as the price of a rental. If the store's customers rent 1050 videos per month when the price per rental is \$3.49, how many videos per month does he expect to rent if he lowers the price to \$2.99?

48. Determine the approximate distance between the Earth, which has a mass of approximately 6.4×10^{24} kg, and an object that has a mass of 6.42×10^{22} kg, if the gravitational force equals approximately 4.95×10^{21} N. Remember, $F = \dfrac{km_1 m_2}{d^2}$ and the Universal Gravitational Constant equals 6.67×10^{-11} N·m²/kg².

Section 3.5

Sketch the graphs of the following functions by first identifying the more basic functions that have been shifted, reflected, stretched, or compressed. Then determine the domain and range of each function.

49. $f(x) = (x-1)^3 + 2$

50. $G(x) = 4|x+3|$

51. $m(x) = \dfrac{1}{(x+2)^2}$

52. $g(x) = -\sqrt[3]{x} + 4$

53. $r(x) = \dfrac{1}{x-2} - 3$

54. $f(x) = \sqrt{x-1} + 3$

Write a formula for each of the functions described below.

55. Use the function $g(x) = x^2$. Move the function 1 unit right and 2 units down.

52.

Dom = Ran = $\mathbb{R}$

53.

Dom =
$(-\infty, 2) \cup (2, \infty)$,

Ran =
$(-\infty, -3) \cup (-3, \infty)$

54.

Dom = $[1, \infty)$,
Ran = $[3, \infty)$

55. $f(x) = (x-1)^2 - 2$

56. $f(x) = -|x - 3|$

57. $f(x) = -\sqrt{x} + 4$

58. Even function;
y-axis symmetry

59. Not a function;
x-axis symmetry

60. a. $-\dfrac{3}{2}$ **b.** $-\dfrac{5}{2}$
  c. -1 **d.** -4

61. a. 7   **b.** -3
  c. 10   **d.** $\dfrac{2}{5}$

62. a. -2 **b.** 18
  c. -80 **d.** $-\dfrac{4}{5}$

63. a. $x^2 + \sqrt{x}$,
  Dom = $[0, \infty)$
  b. $x^{\frac{3}{2}}$,
  Dom = $(0, \infty)$

56. Use the function $g(x) = |x|$. Move the functions 3 units right and reflect across the x-axis.

57. Use the function $g(x) = \sqrt{x}$. Reflect the function across the x-axis and move it 4 units up.

Determine if each of the following relations is a function. If so, determine whether it is even, odd, or neither. Also determine if it has y-axis symmetry, x-axis symmetry, origin symmetry, or none of the above.

58. $y = \dfrac{1}{x^2} + 1$ **59.** $x = -5|y|$

Section 3.6

In each of the following problems, use the information given to determine **a.** $(f + g)(2)$,

b. $(f - g)(2)$, **c.** $(fg)(2)$, and **d.** $\left(\dfrac{f}{g}\right)(2)$.

60. $f(x) = -x^2 + x$ and $g(x) = \dfrac{1}{x}$ **61.** $f(x) = \sqrt{2x}$ and $g(x) = x + 3$

62. $f = \{(0,4),(2,8)\}$ and $g = \{(-2,2),(0,3),(2,-10)\}$

In each of the following problems, find **a.** the formula and domain for $f + g$, and **b.** the formula and domain for $\dfrac{f}{g}$.

63. $f(x) = x^2$ and $g(x) = \sqrt{x}$ **64.** $f(x) = \dfrac{1}{x-2}$ and $g(x) = \sqrt[3]{x}$

65. $f(x) = 3x$ and $g(x) = (x-1)^2$ **66.** $f(x) = x^2 - 4$ and $g(x) = \sqrt[3]{x} - 1$

In each of the following problems, use the information given to determine $(f \circ g)(3)$.

67. $f(x) = -x + 1$ and $g(x) = -x - 1$ **68.** $f(x) = \dfrac{x^{-1}}{18} - 3$ and $g(x) = \dfrac{x-4}{x^3}$

69. $f(-3) = 4$ and $g(3) = -3$ **70.** $f(x) = \dfrac{x}{3}$ and $g(x) = -\sqrt{x+1}$

In each of the following problems, find **a.** the formula and domain for $f \circ g$, and **b.** the formula and domain for $g \circ f$.

71. $f(x) = 4x - 1$ and $g(x) = x^3 + 2$ **72.** $f(x) = \dfrac{1}{\sqrt{x-4}}$ and $g(x) = x + 2$

73. $f(x) = 2x^2 + 1$ and $g(x) = x - 4$ **74.** $f(x) = 3x$ and $g(x) = \sqrt{x-3}$

Write the following functions as a composition of two functions. Answers will vary.

75. $f(x) = \dfrac{3}{3x^2 + 1}$ **76.** $f(x) = \dfrac{\sqrt{x+2}}{x^2 + 4x + 4}$

64. a. $\dfrac{1}{x-2}+\sqrt[3]{x}$,

Dom $=$
$(-\infty,2)\cup(2,\infty)$

b. $\dfrac{1}{\sqrt[3]{x}\,(x-2)}$,

Dom $=$
$(-\infty,0)\cup(0,2)\cup$
$(2,\infty)$

65. a. x^2+x+1,

Dom $=\mathbb{R}$

b. $\dfrac{3x}{(x-1)^2}$,

Dom $=$
$(-\infty,1)\cup(1,\infty)$

66. a. $x^2+\sqrt[3]{x}-5$,

Dom $=\mathbb{R}$

b. $\dfrac{x^2-4}{\sqrt[3]{x}-1}$,

Dom $=$
$(-\infty,1)\cup(1,\infty)$

67. 5

68. $-\dfrac{9}{2}$

69. 4

70. $-\dfrac{2}{3}$

71. a. $4x^3+7$,

Dom $=\mathbb{R}$

b. $(4x-1)^3+2$,

Dom $=\mathbb{R}$

72. a. $\dfrac{1}{\sqrt{x-2}}$,

Dom $=(2,\infty)$

b. $\dfrac{1}{\sqrt{x-4}}+2$,

Dom $=(4,\infty)$

In each of the following problems, use the information given to find $g(x)$.

77. $f(x)=6x-1$ and $(f\circ g)(x)=x+3$

78. $f(x)=\sqrt{x}+3$ and $(g\circ f)(x)=\dfrac{2}{\sqrt{x}+3}+1$

Section 3.7

Graph the inverse of each of the following relations, and state its domain and range.

79. $R=\{(3,4),(-1,5),(0,2),(-6,-1)\}$

80. $y=3x+1$

81. $y=\dfrac{\sqrt{x}}{2}$

Find a formula for the inverse of each of the following functions.

82. $r(x)=\dfrac{2}{7x-1}$

83. $g(x)=\dfrac{4x-3}{x}$

84. $f(x)=x^{\frac{1}{5}}-6$

85. $p(x)=2\sqrt{x-1}+3$

86. $f(x)=\dfrac{2x-3}{x+1}$

87. $f(x)=\sqrt[3]{x+2}-1$

88. $f(x)=(x-1)^2-3,\ x\ge 1$

In the following problem, verify that $f\left(f^{-1}(x)\right)=x$ and that $f^{-1}\left(f(x)\right)=x$.

89. $f(x)=\dfrac{6x-7}{2-x}$ and $f^{-1}(x)=\dfrac{2x+7}{6+x}$

73. a. $2x^2-16x+33$,

Dom $=\mathbb{R}$

b. $2x^2-3$,

Dom $=\mathbb{R}$

74. a. $3\sqrt{x-3}$,

Dom $=[3,\infty)$

b. $\sqrt{3x-3}$,

Dom $=[1,\infty)$

75. $g(x)=\dfrac{3}{x}$

$h(x)=3x^2+1$

$f(x)=(g\circ h)(x)$

76. $g(x)=\dfrac{\sqrt{x}}{x^2}$

$h(x)=x+2$

$f(x)=(g\circ h)(x)$

77. $g(x)=\dfrac{x+4}{6}$

78. $g(x)=\dfrac{2}{x}+1$

79.

Dom $=\{-1,2,4,5\}$,
Ran $=\{-6,-1,0,3\}$

80.

Dom $=$ Ran $=\mathbb{R}$

81.

Dom $=[0,\infty)$,

Ran $=[0,\infty)$

82. $r^{-1}(x)=\dfrac{x+2}{7x}$

83. $g^{-1}(x)=\dfrac{3}{4-x}$

84. $f^{-1}(x)=(x+6)^5$

85. $p^{-1}(x)=\dfrac{(x-3)^2}{4}+1$

86. $f^{-1}(x)=\dfrac{-x-3}{x-2}$

87. $f^{-1}(x)=(x+1)^3-2$

88. $f^{-1}(x)=\sqrt{x+3}+1$

$x\ge -3$

89. Answers will vary.

1. $-x^2 + 2(x-h)$

$-h^2 + 2xh$

2. $-2x - h + 2$

3. $f(x) = 3x; -6$

4. $f(x) = \dfrac{8x}{4-3x}; \dfrac{-8}{5}$

5. $\left(-\infty, \dfrac{1}{2}\right]$

6. $(-\infty, -2) \cup (-2, 2)$

$\cup (2, \infty)$

7.

8.

9. $-\dfrac{25}{4}$

10.

11. $y = 48$

12. 64 feet

13. $f(x) = (x+2)^2 + 4$

14. a. -2

b. -6

c. -8

d. -2

15. $-x^4 - 2x^2 - 2$

16. $0, -2$

========

Chapter Test

Given $f(x) = -x^2 + 2x$, evaluate each of the following.

1. $f(x-h)$

2. $\dfrac{f(x+h) - f(x)}{h}$

Rewrite each of the relations below as a function of x. Then evaluate the function at $x = -2$.

3. $y + 3x - x^2 = 2y - x^2$

4. $\dfrac{1}{x} - \dfrac{2}{y} = \dfrac{3}{4}$

Determine the implied domain of each of the following functions.

5. $f(x) = \sqrt{2 - 4x}$

6. $f(x) = \dfrac{x+1}{x^2 - 4}$

Graph the following quadratic functions, locating the vertices, x-intercepts (if any), and y-intercepts.

7. $y = -(x-1)^2 + 2$

8. $y = x^2 + 2x - 8$

9. Given $f(x) = \begin{cases} -|2x| & \text{if } x > 1 \\ \dfrac{1}{x^2} & \text{if } x \le 1 \end{cases}$, find $f(3) - f(-2)$.

Sketch a graph of the following function.

10. $y = \begin{cases} -x^2 & \text{if } |x| \le 1 \\ x & \text{if } x < -1 \text{ or } x > 1 \end{cases}$

11. Suppose that y varies jointly with the cube of x and the square root of z. If $y = 270$ when $x = 3$ and $z = 25$, what is y when $x = 2$ and $z = 9$?

12. The distance that an object falls from rest, when air resistance is negligible, varies directly as the square of the time. A stone dropped from rest travels 400 feet in the first 5 seconds. How far did it travel in the first 2 seconds?

Write a formula for the function described below.

13. Use the function $f(x) = x^2$. Move the function 2 units left and 4 units up.

14. Given that $f(-2) = -4$ and $g(-2) = 2$, determine **a.** $(f+g)(-2)$, **b.** $(f-g)(-2)$, **c.** $(f \cdot g)(-2)$, and **d.** $\left(\dfrac{f}{g}\right)(-2)$.

17. $g(x) = \sqrt{x}$

$h(x) = 2x^2 + 1$

$f(x) = (g \circ h)(x)$

18. $g(x) = |x| - 2$

$h(x) = x^2 - x$

$f(x) = (g \circ h)(x)$

19. $f^{-1}(x) = (x-3)^4$,

$x \geq 3$

20. $f^{-1}(x) = \sqrt{4-x}$

21. $f^{-1}(x) = \dfrac{2}{x} + 1$

22. $f^{-1}(x) = \dfrac{x^7}{3}$

23. -3

24. -3

25. 7

26. y-axis

27. Origin

28. $\dfrac{-dx+b}{cx-a}$

29. $\dfrac{4}{5}$

30. a. -64

b. 1

---

15. Given $f(x) = -x^2 - 1$, find $(f \circ f)(x)$.

16. Given $f(x) = x^2 - 1$, $g(x) = x + 1$, and $(f \circ g)(x) = 0$, find x.

Write the following functions as a composition of two functions. Answers will vary.

17. $f(x) = \sqrt{2x^2 + 1}$

18. $f(x) = |x^2 - x| - 2$

Find a formula for the inverse of each of the following functions.

19. $f(x) = x^{\frac{1}{4}} + 3$

20. $f(x) = -x^2 + 4,\ x \geq 0$

21. $f(x) = \dfrac{2}{x-1}$

22. $f(x) = \sqrt[7]{3x}$

23. Given $f(x) = 2x - a + 4$, if $f^{-1}(3) = -2$, then find a.

24. Given $f(x) = x + a$ and $(f \circ f)^{-1}(-2) = 4$, find a.

25. If $f(x)$ is linear function such that $f(1) = 1$ and $f(2) = 3$, find $f(4)$.

For each function below, determine if it has y-axis symmetry, x-axis symmetry, origin symmetry, or none of these.

26. $y = x^4 - 2x^2 + 3$

27. $y = \dfrac{x^3 - x}{2}$

28. Find the inverse of $f(x) = \dfrac{ax+b}{cx+d}$.

29. If $f(x) = \dfrac{x+4}{3}$ and $f(x) + f^{-1}(x) = 0$, find x.

30. Given $f(x) = \sqrt[3]{x}$ and $g(x) = x - 2$, determine **a.** $(f^{-1} \circ g)(-2)$, and

b. $(f \circ g^{-1})(-1)$.

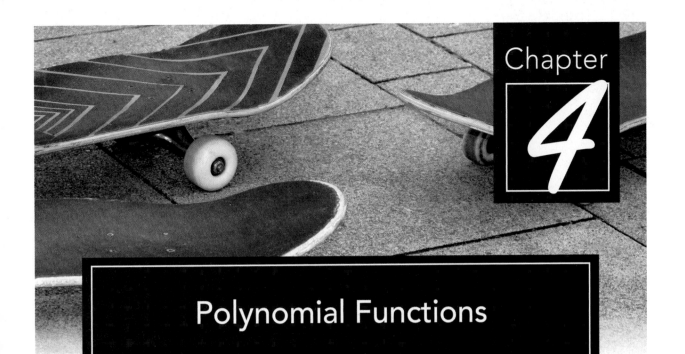

Polynomial Functions

4.1   Introduction to Polynomial Equations and Graphs

4.2   Polynomial Division and the Division Algorithm

4.3   Locating Real Zeros of Polynomials

4.4   The Fundamental Theorem of Algebra

4.5   Rational Functions and Rational Inequalities

Chapter 4 Project

Chapter 4 Summary

Chapter 4 Review

Chapter 4 Test

By the end of this chapter you should be able to:

What if you owned a skateboard manufacturing company? How would you decide how many skateboards to produce in order to maximize your profits?

By the end of this chapter, you'll be able to apply the skills regarding polynomials to solve business application problems. You'll encounter the skateboard problem on page 325. You'll master this type of problem using techniques for Solving Polynomial Inequalities, found on page 318.

Introduction

In Chapter 3, we studied properties of functions in general and learned the nomenclature and notation commonly used when working with functions and their graphs. In this chapter, we narrow our focus and concentrate on polynomial functions.

We have, of course, already seen many examples of polynomial functions but have barely scratched the surface as far as obtaining a deep understanding of polynomials is concerned. We will soon see many mathematical methods that are peculiar to polynomials and make many observations that are relevant only to polynomials. Some of these methods include polynomial division and the Rational Zero Theorem, and many of the observations point out the strong connection between factors of polynomials, solutions of polynomial equations, and (when appropriate) graphs of polynomial functions. Our deeper understanding of polynomial functions will then make solving polynomial inequalities a far easier task than it would be otherwise.

The discussion of polynomials concludes with the Fundamental Theorem of Algebra, a theorem that makes a deceptively simple claim about polynomials, but one that nevertheless manages to tie together nearly all of the methods and observations of the chapter. The German mathematician Carl Friedrich Gauss (1777–1855), one of the towering figures in the history of mathematics, first proved this theorem in 1799. In doing so, he accomplished something that many brilliant people (among them Isaac Newton and Leonhard Euler) had attempted, and it is all the more remarkable that Gauss did so in his doctoral dissertation at the age of twenty-two! The Fundamental Theorem of Algebra operates on many levels: philosophically, it can be seen as one of the most elegant and fundamental arguments for the necessity of complex numbers, while pragmatically it is of great importance in solving polynomial equations. It thus ties together observations about polynomials that originated with Italian mathematicians of the 16th century, and points the way toward later work on polynomials by such people as Niels Abel and Évariste Galois.

Gauss

The chapter ends with an opportunity to put to use many of the skills acquired thus far. An understanding of the subject of the last section, rational functions, depends not only on a knowledge of polynomials (of which rational functions are ratios), but also of x- and y-intercepts, factoring, and transformations of functions.

Introduction to Polynomial Equations and Graphs

TOPICS

1. Zeros of polynomials and solutions of polynomial equations
2. Graphing factored polynomials
3. Solving polynomial inequalities
T. Finding zeros of polynomials

TOPIC

Zeros of Polynomials and Solutions of Polynomial Equations

At this point, we have studied how linear and quadratic polynomial functions behave, and we have tools guaranteed to solve all linear and quadratic equations. We have also studied some elementary higher-degree polynomials (those of the form ax^n). In this section, we begin a more complete exploration of higher-degree polynomials.

Not surprisingly, the complexity of polynomial functions increases with the degree; higher-degree polynomials are usually more difficult to graph accurately, and we cannot necessarily expect to find exact solutions to higher-degree polynomial equations.

In order to make our work as general as possible, we will refer to a generic n^{th}-degree polynomial function $p(x) = a_n x^n + a_{n-1} x^{n-1} + \ldots + a_1 x + a_0$, where n is a nonnegative integer, a_n, a_{n-1}, ..., a_1, a_0 all represent constants (that may be real or complex), and $a_n \neq 0$. **Throughout this chapter, $p(x)$ will refer to this generic polynomial.** We begin by identifying which values of the variable make a polynomial function equal to zero.

DEFINITION

Zeros of a Polynomial

The number k (k may be a complex number) is a **zero** of the polynomial function $p(x)$ if $p(k) = 0$. This is also expressed by saying that k is a **root** of the polynomial or a **solution** of the equation $p(x) = 0$.

The task of determining the zeros of a polynomial arises in many contexts, two of which are solving polynomial equations and graphing polynomials.

DEFINITION

Polynomial Equations

A **polynomial equation in one variable**, say the variable x, is an equation that can be written in the form $a_n x^n + a_{n-1} x^{n-1} + \ldots + a_1 x + a_0 = 0$, where n is a nonnegative integer, a_n, a_{n-1}, ..., a_1, a_0 are constants. Assuming $a_n \neq 0$, we say such an equation is of **degree** n and call a_n the **leading coefficient**.

Just as with linear and quadratic equations (which are polynomial equations of degree 1 and 2, respectively), a polynomial equation may not appear in the form of the above definition. The first task is often to rewrite the equation so that one side is zero. Then, the zeros of the polynomial on the other side are the solutions of the equation.

Note that, given a polynomial equation in the form $p(x) = 0$, the zeros of p are precisely the solutions of the equation (and vice versa).

EXAMPLE 1

Solutions of Polynomial Equations

Verify that the given values of x solve the corresponding polynomial equations.

a. $6x^2 - x^3 = 12 + 5x; \ x = 4$ **b.** $x^2 = 2x - 5; \ x = 1 + 2i$

c. $\dfrac{x}{1-i} = 3x^2; \ x = 0$

Note:
When verifying a zero, there is no need to rewrite the equation in the form of the definition.

Solutions:

a. $6x^2 - x^3 = 12 + 5x$

$6(4)^2 - (4)^3 \overset{?}{=} 12 + 5(4)$ Substitute $x = 4$ throughout the equation.

$96 - 64 \overset{?}{=} 12 + 20$ Simplify both sides.

$32 = 32$ This is a true statement, so 4 is a solution.

b. $x^2 = 2x - 5$

$(1+2i)^2 \overset{?}{=} 2(1+2i) - 5$ Substitute $x = 1 + 2i$ in the equation.

$1 + 4i + 4i^2 \overset{?}{=} 2 + 4i - 5$ Expand both sides.

$1 + 4i - 4 \overset{?}{=} 2 + 4i - 5$ Simplify, using the fact that $i^2 = -1$.

$-3 + 4i = -3 + 4i$ This is a true statement, so $1+2i$ is a solution.

c. $\dfrac{x}{1-i} = 3x^2$

$\dfrac{0}{1-i} \overset{?}{=} 3(0)^2$ Substitute 0 for x in the equation.

$0 = 0$ After simplifying, we arrive at a true statement. Thus, 0 is a solution to the polynomial equation.

TOPIC 2   Graphing Factored Polynomials

Consider a generic polynomial function $p(x)$ with all real coefficients. Our goal is to be able to sketch the graph of such a function, paying particular attention to the behavior of p as $x \to -\infty$ and as $x \to \infty$, and the x- and y-intercepts of p. We will begin by looking at the behavior of a polynomial function as $x \to \pm\infty$.

The graph of $p(x)$ is similar to the graph of $a_n x^n$ for values of x that are very large in magnitude; the leading term of $p(x)$ dominates the behavior. Take the function $f(x) = x^4 - 3x^3 - 5x^2 + 8$. The graph below show that as x gets very large, $f(x)$ takes values very similar to x^4 and thus has a very similar graph.

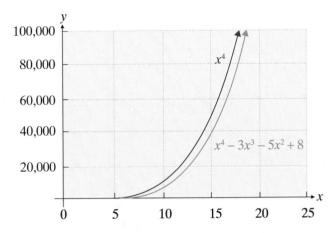

Figure 1: Similar Behavior as $x \to \infty$

This means we can understand the behavior of all polynomials as $x \to \pm\infty$ just by understanding how functions of the form $f(x) = a_n x^n$ behave. We know that if n is even, $x^n \to \infty$ as $x \to -\infty$ and as $x \to \infty$, and if n is odd, then $x^n \to -\infty$ as $x \to -\infty$ and $x^n \to \infty$ as $x \to \infty$. We also know that if a_n is negative, the graph undergoes a reflection across the x-axis; this reverses the sign of every y-value, changing the behavior as $x \to \pm\infty$. Figure 2 below shows a few examples.

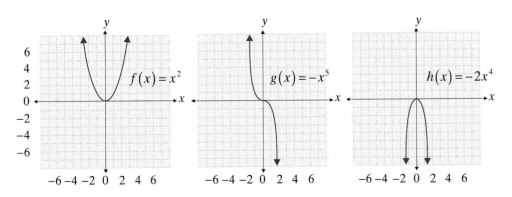

Figure 2: Examples of Behavior as $x \to \infty$

PROPERTIES

Behavior of
Polynomials as
$x \to \pm\infty$

Given a polynomial function $p(x)$ with degree n, the behavior of $p(x)$ as $x \to \pm\infty$ can be determined from the leading term $a_n x^n$ using the table below.

	n is even	n is odd
a_n is positive	as $x \to -\infty$, $p(x) \to +\infty$ as $x \to +\infty$, $p(x) \to +\infty$ The graph rises to the left and rises to the right.	as $x \to -\infty$, $p(x) \to -\infty$ as $x \to +\infty$, $p(x) \to +\infty$ The graph falls to the left and rises to the right.
a_n is negative	as $x \to -\infty$, $p(x) \to -\infty$ as $x \to +\infty$, $p(x) \to -\infty$ The graph falls to the left and falls to the right.	as $x \to -\infty$, $p(x) \to +\infty$ as $x \to +\infty$, $p(x) \to -\infty$ The graph rises to the left and falls to the right.

Near the origin, however, the graph of $p(x)$ is likely to be quite different from the graph of $a_n x^n$. Recall our example from before; x^4 and $f(x) = x^4 - 3x^3 - 5x^2 + 8$ both behave the same as $x \to \pm\infty$. Observe how differently these functions behave near the origin.

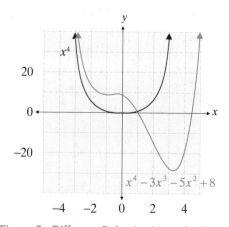

Figure 3: Different Behavior Near the Origin

Finding the x- and y-intercepts of a polynomial function will help us sketch a more complete graph. Given a polynomial in the form $p(x) = a_n x^n + a_{n-1} x^{n-1} + \ldots + a_1 x + a_0$, finding the y-intercept is not difficult, as it simply requires evaluating $p(0)$:

$$p(0) = a_n (0)^n + a_{n-1} (0)^{n-1} + \ldots + a_1 (0) + a_0$$

$$= a_0$$

Thus, the y-intercept is $(0, a_0)$. On the other hand, finding the x-intercepts requires solving the polynomial equation $0 = a_n x^n + a_{n-1} x^{n-1} + \ldots + a_1 x + a_0$, which often takes more effort.

Writing the equation in a different form can make our task much easier. If we factor a given polynomial f into a product of linear factors, each linear factor with real coefficients corresponds to an x-intercept of the graph of f. To see why this is true, consider the polynomial function

$$f(x) = (3x - 5)(x + 2)(2x - 6).$$

To determine the x-intercepts of this polynomial we need to solve the equation

$$0 = (3x - 5)(x + 2)(2x - 6).$$

The Zero-Factor Property tells us that the only solutions are those values of x for which $3x - 5 = 0$, $x + 2 = 0$, or $2x - 6 = 0$. Solving these three linear equations gives us the x-coordinates of the three x-intercepts of f: $\left\{ \dfrac{5}{3}, -2, 3 \right\}$.

Working with a polynomial in factored form almost always makes finding the x-intercepts easier, but we do have to adjust how we calculate the y-intercept and the behavior as $x \to \pm\infty$. If a polynomial is in factored form, we can not simply read off the value of the y-intercept. However, substituting $x = 0$ is not much more work:

$$f(0) = (3(0) - 5)((0) + 2)(2(0) - 6)$$

$$= (-5)(2)(-6)$$

$$= 60$$

Similarly, when a polynomial is in factored form, we can't directly see the term $a_n x^n$ to determine the end behavior. Instead of multiplying out f completely, we just determine how the leading x^3 term arises. The third-degree term comes from multiplying together the $3x$ from the first factor, the x from the second factor, and the $2x$ from the third factor. Thus, $a_n x^n = 6x^3$. Since the leading coefficient is positive and the degree is odd, we know that $f(x) \to -\infty$ as $x \to -\infty$ and $f(x) \to \infty$ as $x \to \infty$.

Putting it all together, along with a few computed values of f, we obtain the sketch in Figure 4 (note the difference in the horizontal and vertical scales).

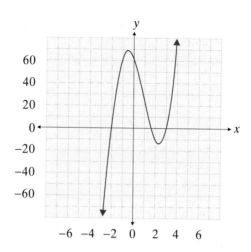

Figure 4: Graph of $f(x) = (3x-5)(x+2)(2x-6)$

This method of graphing factored polynomials raises an important question: What do we do if we are given a polynomial in nonfactored form? In other words, how do we solve the generic n^{th}-degree polynomial equation $a_n x^n + a_{n-1} x^{n-1} + \ldots + a_1 x + a_0 = 0$? We'll study these questions throughout the rest of this chapter, so for the remainder of this section all polynomials will either be given in factored form or be factorable with the tools we already possess.

EXAMPLE 2

Graphing Polynomial Functions

Sketch the graphs of the following polynomial functions, paying particular attention to the x-intercept(s), the y-intercept, and the behavior as $x \to \pm\infty$.

a. $f(x) = -x(2x+1)(x-2)$ **b.** $g(x) = x^2 + 2x - 3$

c. $h(x) = x^4 - 1$

Note:
As always, plotting additional points will help in sketching an accurate graph.

Solutions:

a. Begin with the x-intercepts. Using the Zero-Factor Property, we solve $f(x) = 0$.

$$-x(2x+1)(x-2) = 0$$

$$x = -\frac{1}{2}, 0, 2$$

Thus, the x-intercepts are $\left(-\frac{1}{2}, 0\right), (0,0),$ and $(2,0)$.

We could plug in $x = 0$ to find the y-intercept, but we already found it when calculating the x-intercepts! The y-intercept is the origin $(0,0)$.

All that remains is to determine the end behavior. Find the highest-degree term by multiplying $(-x)(2x)(x) = -2x^3$. The leading coefficient is negative, and the degree is odd, so $f(x) \to \infty$ as $x \to -\infty$ and $f(x) \to -\infty$ as $x \to \infty$.

Putting all this together, we obtain the following graph.

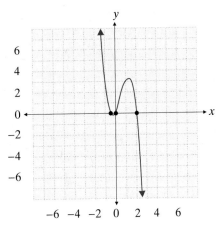

b. This polynomial is not in factored form. Before factoring, we can collect information about the y-intercept and behavior as $x \to \pm\infty$.

$g(x) = x^2 + 2x - 3$, so $a_0 = -3$ and the leading term is x^2. This means that the y-intercept is $(0, -3)$ and that $g(x) \to \infty$ as $x \to \pm\infty$.

Now, to find the x-intercepts, we factor the polynomial.

$$x^2 + 2x - 3 = 0$$
$$(x + 3)(x - 1) = 0$$
$$x = -3, 1$$

Thus, the x-intercepts are $(-3, 0)$ and $(1, 0)$.

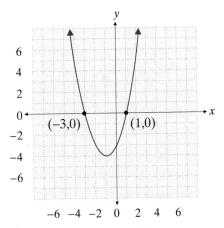

c. As with the previous example, we determine the y-intercept and end behavior before factoring. $h(x) = x^4 - 1$, so $a_0 = -1$ and the leading term is x^4. This tells us that the y-intercept is $(0, -1)$ and that $h(x) \to \infty$ as $x \to \pm\infty$.

Once again, we factor the polynomial to calculate the x-intercepts.

$$x^4 - 1 = 0$$
$$(x^2 - 1)(x^2 + 1) = 0 \qquad \text{A difference of squares.}$$
$$(x - 1)(x + 1)(x^2 + 1) = 0 \qquad \text{Another difference of squares.}$$
$$x = -1, 1 \qquad \text{Only the real solutions lead to } x\text{-intercepts. Here,}$$
$$\text{the } x\text{-intercepts are } (-1, 0) \text{ and } (1, 0).$$

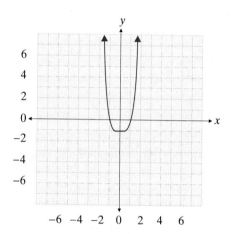

TOPIC 3 — Solving Polynomial Inequalities

DEFINITION

Polynomial Inequalities

A **polynomial inequality** is any inequality that can be written in the form:

$$p(x) < 0, \; p(x) \le 0, \; p(x) > 0, \text{ or } p(x) \ge 0,$$

where $p(x)$ is a polynomial function.

If we have an accurate graph of the function, solving a polynomial inequality is very straightforward; simply read where the function is positive and where it is negative.

EXAMPLE 3

Solving Polynomial Inequalities Using a Graph

Solve the polynomial inequalities, given the graph of $f(x) = (x + 3)(x + 1)(x - 2)$.

a. $(x + 3)(x + 1)(x - 2) < 0$

b. $(x + 3)(x + 1)(x - 2) \ge 0$

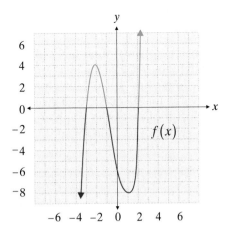

Solutions:

a. From the graph, we see that $f(x) < 0$ on the intervals $(-\infty, -3)$ and $(-1, 2)$. Since the inequality is strict, we do not include the endpoints. Thus, the solution to this inequality is $(-\infty, -3) \cup (-1, 2)$.

b. Similarly, the graph reveals that $f(x) > 0$ on the intervals $(-3, -1)$ and $(2, \infty)$. This inequality is not strict, so we include the x-values where $f(x) = 0$. Therefore, the solution to this inequality is $[-3, -1] \cup [2, \infty)$.

What if we do not have an accurate graph of the function? The graphing methods we have covered so far tell us if a polynomial $p(x)$ is positive or negative as $x \to \pm\infty$, but give no information about the sign of $p(x)$ near the origin.

As such, we need an algebraic method for solving polynomial inequalities when a detailed graph is not available. Having such a method will be a helpful tool in sketching graphs of polynomials.

Our method depends on a property of polynomials called **continuity**. While a more formal definition of continuity requires tools from calculus, intuitively, a continuous function has no "breaks" in it, in other words its graph can be drawn without lifting the pencil off the paper.

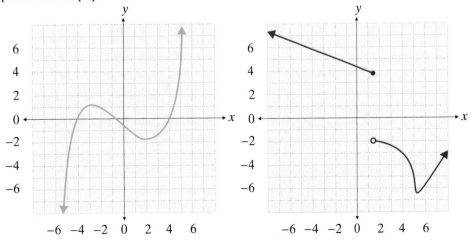

Figure 5: Continuous Function   Figure 6: Discontinuous Function

Why is continuity important? Look at the graphs in Figures 5 and 6. If a graph is not continuous, it can change sign without passing through the x-axis (thus, the value of the function "skips" zero). However, with a polynomial (or any continuous function), the only way for the graph to change sign is to pass through zero! This means that between each pair of zeros, the graph of $p(x)$ is always positive or always negative. Knowing this gives us a method for solving polynomial inequalities.

PROCEDURE

Solving Polynomial Inequalities: Sign-Test Method

To solve a polynomial inequality $p(x) < 0$, $p(x) \leq 0$, $p(x) > 0$, or $p(x) \geq 0$:

Step 1: Find the real zeros of $p(x)$. Equivalently, find the real solutions of $p(x) = 0$.

Step 2: Place the zeros on a number line, splitting it into intervals.

Step 3: Within each interval, select a **test point** and evaluate p at that number. If the result is positive, then $p(x) > 0$ for all x in the interval. If the result is negative, then $p(x) < 0$ for all x in the interval.

Step 4: Write the solution set, consisting of all of the intervals that satisfy the given inequality. If the inequality is not strict (uses $\leq$ or $\geq$), then the zeros are included in the solution set as well.

EXAMPLE 4

Solving Polynomial Inequalities: Sign-Test Method

Solve the polynomial inequality $(x - 2)(x + 5)(x - 4) \leq 0$.

Solution:

Follow the steps in the procedure for the Sign-Test Method.

Note:

When choosing test points, integers (especially 0) are usually easiest to work with.

Step 1: Find the zeros of $p(x) = (x - 2)(x + 5)(x - 4)$. By the Zero-Factor Property, we can see that $p(x) = 0$ when $x = -5$, 2, or 4.

Step 2: Place the zeros on a number line.

$$-5 \qquad 2 \quad 4$$

This splits the number line into the intervals

$(-\infty, -5)$, $(-5, 2)$, $(2, 4)$, and $(4, \infty)$.

Step 3: Evaluate $p(x)$ for a test point in each interval.

Interval	Test Point	Evaluate	Result
$(-\infty, -5)$	$x = -6$	$p(-6) = (-6 - 2)(-6 + 5)(-6 - 4)$ $= (-8)(-1)(-10)$ $= -80$	$p(x) < 0$ on $(-\infty, -5)$ **Negative**
$(-5, 2)$	$x = 0$	$p(0) = (0 - 2)(0 + 5)(0 - 4)$ $= (-2)(5)(-4)$ $= 40$	$p(x) > 0$ on $(-5, 2)$ **Positive**
$(2, 4)$	$x = 3$	$p(3) = (3 - 2)(3 + 5)(3 - 4)$ $= (1)(8)(-1)$ $= -8$	$p(x) < 0$ on $(2, 4)$ **Negative**
$(4, \infty)$	$x = 6$	$p(6) = (6 - 2)(6 + 5)(6 - 4)$ $= (4)(11)(2)$ $= 88$	$p(x) > 0$ on $(4, \infty)$ **Positive**

Step 4: Write the solution set to the original inequality, $p(x) \leq 0$. From our table, we see that $p(x)$ is negative on the intervals $(-\infty, -5)$ and $(2, 4)$. Since the inequality is not strict, we need to include the zeros in our solution set. Thus, the solution to the inequality is $(-\infty, -5] \cup [2, 4]$.

TOPIC Finding Zeros of Polynomials

In Chapter 2, we saw how to find the x-intercepts of a linear equation on a calculator. The same method can be used to find the x-intercepts, or zeros, of any function graphed on a calculator. The main difference is that with linear functions, there can be no more than one zero, but other functions might have more. Consider the graph of the function $f(x) = x^2 + 4x - 6$:

We can see that there are two zeros that appear to be located near $x = -5$ and $x = 1$. To check more accurately, press **2ND** **TRACE** to access the CALC menu, then select 2:zero. The screen should now display the graph with the words "Left Bound?" shown at the bottom. Choose which zero you want to find and use the arrows to move

1.–17. Answers will
vary.

18. Yes

19. Yes

20. No

21. Yes

22. No

23. Yes

24. $-2, 0, 3$

25. $1 \pm 2i$

26. $\pm 1, \pm i\sqrt{2}$

27. $-3, \dfrac{1}{2}$

28. $\dfrac{1}{3}$

29. $\pm\sqrt{3}, \pm\sqrt{5}$

30. $-8, 0, 9$

31. $-\dfrac{5}{2}$

32. $\dfrac{1}{2}, 5$

33. $0, 4 \pm 3i$

34. $\pm 2, \pm 3$

35. $\pm 1, \pm 2i\sqrt{2}$

36. 4ᵗʰ-degree;
lead coef = 2;
$p(x) \to \infty$ as
$x \to -\infty$ and $x \to \infty$

37. 7ᵗʰ-degree;
lead coef = 4;
$j(x) \to -\infty$ as $x \to -\infty$
$j(x) \to \infty$ as $x \to \infty$

38. 4ᵗʰ-degree;
lead coef = 24;
$r(x) \to \infty$ as
$x \to -\infty$ and $x \to \infty$

39. 5ᵗʰ-degree;
lead coef = -6;
$h(x) \to \infty$ as $x \to -\infty$
$h(x) \to -\infty$ as $x \to \infty$

the cursor anywhere to the left of that intercept and press **ENTER**. The screen should now say "Right Bound?" Use the right arrow to move the cursor to the right of that same intercept and press **ENTER** again. (Be sure there is only one x-intercept between what you select as the left bound and what you select as the right bound.) The text should now read "Guess?" Press **ENTER** a third time and the x- and y-values of that x-intercept will appear at the bottom of the screen.

The leftmost zero occurs at approximately $x \approx -5.162$. Note that sometimes, as in the example above, the display will read a very small number rather than exactly zero. To find the other zero, repeat the process, this time focusing on the rightmost zero. We find that it occurs at $x \approx 1.162$.

Exercises

Verify that the given values of x solve the corresponding polynomial equations. See Example 1.

1. $9x^2 - 4x = 2x^3 + 15$; $x = -1$ **2.** $x^2 - 4x = -13$; $x = 2 - 3i$

3. $x^2 + 13 = 4x$; $x = 2 + 3i$ **4.** $3x^3 + (5 - 3i)x^2 = (2 + 5i)x - 2i$; $x = i$

5. $9x^2 - 4x = 2x^3 + 15$; $x = 3$ **6.** $9x^2 - 4x = 2x^3 + 15$; $x = \dfrac{5}{2}$

7. $3x^3 + (5 - 3i)x^2 = (2 + 5i)x - 2i$; $x = -2$

8. $x^5 - 10x^4 - 80x^2 = 32 - 80x - 40x^3$; $x = 2$

9. $4x^5 - 8x^4 - 12x^3 = 16x^2 - 25x - 69$; $x = 3$

10. $x^2 - 4x - 12 = 0$; $x = 6$ **11.** $23x^7 - 12x^5 = 63x^4 - 3x^2$; $x = 0$

12. $x^2 + 74 = 10x$; $x = 5 + 7i$ **13.** $4x^2 + 32x + (8 + i)x^3 = -8$; $x = 2i$

14. $8x - 17 = x^2$; $x = 4 - i$ **15.** $(5 - 3i)x - 3x = 4 - 6i$; $x = 2$

16. $x^6 - x^5 + 7x^4 + x^3 - 9x = -1$; $x = 1$ **17.** $6x^7 - 3x^5 = 3x^4 - 6x^2$; $x = -1$

40. 5th-degree;
lead coef $= -2$;
$g(x) \to \infty$ as $x \to -\infty$
$g(x) \to -\infty$ as $x \to \infty$

41. 4th-degree;
lead coef $= -2$;
$f(x) \to -\infty$ as
$x \to -\infty$ and $x \to \infty$

42.

$f(x) \to -\infty$ as $x \to -\infty$;
$f(x) \to \infty$ as $x \to \infty$

x-int: $(-4, 0)$,
$(-2, 0)$, $(3, 0)$;
y-int: $(0, -24)$

43.

$g(x) \to \infty$ as $x \to -\infty$;
$g(x) \to -\infty$ as $x \to \infty$

x-int: $(-4, 0)$,
$(-2, 0)$, $(3, 0)$;
y-int: $(0, 24)$

44.

$f(x) \to -\infty$ as $x \to -\infty$;
$f(x) \to \infty$ as $x \to \infty$

x-int: $(-5, 0)$,
$(2, 0)$;
y-int: $(0, 20)$

Determine if the given values of x are zeros of the corresponding polynomial equations. See Example 1.

18. $16x = x^3 + x^2 + 20$; $x = -5$

19. $x^4 - 13x^2 + 12 = -x^3 + x$; $x = -1$

20. $x^4 - 3x^3 - 10x^2 = 0$; $x = 2$

21. $4x^5 - 216x^2 = 36x^3 - 24x^4$; $x = -6$

22. $x^3 - 8ix + 30 = 15x + 2x^2 + 16i$; $x = -i$ **23.** $x^3 - 7x^2 + 4x - 28 = 0$; $x = 2i$

Solve the following polynomial equations by factoring and/or using the quadratic formula, making sure to identify all the solutions. See Section 1.7 for review, if necessary.

24. $x^3 - x^2 - 6x = 0$

25. $x^2 - 2x + 5 = 0$

26. $x^4 + x^2 - 2 = 0$

27. $2x^2 + 5x = 3$

28. $9x^2 = 6x - 1$

29. $x^4 - 8x^2 + 15 = 0$

30. $x^3 - x^2 = 72x$

31. $x^2 + 5x = -\dfrac{25}{4}$

32. $2x^2 + 5 = 11x$

33. $x^4 - 8x^3 + 25x^2 = 0$

34. $x^4 - 13x^2 + 36 = 0$

35. $x^4 + 7x^2 = 8$

For each of the following polynomials, determine the degree and the leading coefficient; then describe the behavior of the graph as $x \to \pm\infty$.

36. $p(x) = 2x^4 - 3x^3 - 6x^2 - x - 23$

37. $j(x) = 4x^7 + 5x^5 + 12$

38. $r(x) = (3x + 5)(x - 2)(2x - 1)(4x - 7)$ **39.** $h(x) = -6x^5 + 2x^3 - 7x$

40. $g(x) = (x - 5)^3 (2x + 1)(-x - 1)$

41. $f(x) = -2(x + 4)(x - 4)(x^2)$

For each of the following polynomial functions, describe the behavior of its graph as $x \to \pm\infty$ and identify the x- and y-intercepts. Use this information to sketch the graph of each polynomial. See Example 2.

42. $f(x) = (x - 3)(x + 2)(x + 4)$

43. $g(x) = (3 - x)(x + 2)(x + 4)$

44. $f(x) = (x - 2)^2 (x + 5)$

45. $h(x) = -(x + 2)^3$

46. $r(x) = x^2 - 2x - 3$

47. $s(x) = x^3 + 3x^2 + 2x$

48. $f(x) = -(x - 2)(x + 1)^2 (x + 3)$

49. $g(x) = (x - 3)^5$

45.

$h(x) \to \infty$ as $x \to -\infty$;
$h(x) \to -\infty$ as $x \to \infty$

x-int: $(-2, 0)$;
y-int: $(0, -8)$

46.

$r(x) \to \infty$ as $x \to \pm\infty$

x-int: $(-1, 0)$,
$(3, 0)$;
y-int: $(0, -3)$

47.

$s(x) \to -\infty$ as $x \to -\infty$;
$s(x) \to \infty$ as $x \to \infty$

x-int: $(-2, 0)$,
$(-1, 0), (0, 0)$;
y-int: $(0, 0)$

48.

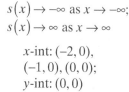

$f(x) \to -\infty$ as $x \to \pm\infty$

x-int: $(-3, 0)$,
$(-1, 0), (2, 0)$;
y-int: $(0, 6)$

In Exercises 50 through 55, use the behavior as $x \to \pm\infty$ and the intercepts to match each polynomial with one of the graphs labeled **a.** through **f.**

50. $g(x) = (x+1)^2 (x-3)^2$

51. $h(x) = 1 - (x+2)^2$

52. $f(x) = (x-1)(x+2)(3-x)$

53. $r(x) = x^2 - x - 6$

54. $s(x) = (x-1)^3 - 2$

55. $f(x) = (x-3)^2 (4x+1)(x+2)(x-2)$

a.   **b.**

c.   **d.**

e.   **f.**

Match each of the following functions to the appropriate description.

56. $z(x) = (x-1)(x+2)(4-x)$

57. $r(x) = x^2 - 6x - 7$

58. $s(x) = x^3 + 3x^2 + 2x$

59. $g(x) = (x-1)(x+4)(3-x)$

60. $s(x) = x^3 + 5x^2 + 4x$

61. $s(x) = x^2 - 5x - 6$

a. x-cubed curve increasing to infinity, has x-intercepts of 0, –1, and –2, and crosses the y-axis at 0.

b. parabola that opens up, has x-intercepts at 6 and –1, crosses the y-axis at –6.

c. x-cubed curve increasing to infinity, has x-intercepts of 0, –1, and –4, and crosses the y-axis at 0.

d. parabola that opens up, has x-intercepts at 7 and –1, crosses the y-axis at –7.

e. x-cubed curve decreasing to infinity, has x-intercepts at 1, 4, and –2, crosses the y-axis at –8.

f. x-cubed curve decreasing to infinity, has x-intercepts of 1, 3, and –4, and crosses the y-axis at –12.

49.

$g(x) \to -\infty$ as $x \to -\infty$;
$g(x) \to \infty$ as $x \to \infty$

x-int: $(3, 0)$;
y-int: $(0, -243)$

50. d

51. e

52. b

53. a

54. c

55. f

56. e

57. d

58. a

59. f

60. c

61. b

62. $[-2, 3]$

63. $(-\infty, -2) \cup (3, \infty)$

64. $(-\infty, -2) \cup$

$(-2, 1) \cup (1, \infty)$

65. $(-\infty, -2) \cup (-1, 0)$

66. $[-3, -1] \cup [2, \infty)$

67. $[-2, 1] \cup [3, \infty)$

68. $(-\infty, -6) \cup (0, 5)$

69. $[-5, -1] \cup [1, 4]$

70. $(-\infty, 0) \cup (0, \infty)$

71. $\left(-\dfrac{1}{2}, 2\right)$

72. $(-\infty, -4) \cup (3, \infty)$

73. $(-\infty, -4) \cup (2, 3)$

74. All integers
between 10 and
30, inclusive

Solve the following polynomial inequalities. See Example 4.

62. $x^2 - x - 6 \le 0$

63. $x^2 > x + 6$

64. $(x+2)^2 (x-1)^2 > 0$

65. $x^3 + 3x^2 + 2x < 0$

66. $(x-2)(x+1)(x+3) \ge 0$

67. $(x-1)(x+2)(3-x) \le 0$

68. $-x^3 - x^2 + 30x > 0$

69. $(x^2 - 1)(x-4)(x+5) \le 0$

70. $x^4 + x^2 > 0$

71. $4x^2 < 6x + 4$

72. $x^2(x+4)(x-3) > 0$

73. $(x-3)(x+4)(2-x) > 0$

Solve the following application problems.

74. A small start-up skateboard company projects that the cost per month of manufacturing x skateboards will be $C(x) = 10x + 300$, and the revenue per month from selling x skateboards will be $r(x) = -x^2 + 50x$. Given that profit is revenue minus cost, for what value(s) of x will the company break even or make a profit?

75. A manufacturer has determined that the revenue from the sale of x cordless telephones is given by $r(x) = -x^2 + 15x$. The cost of producing x telephones is $C(x) = 135 - 17x$. Given that profit is revenue minus cost, for what value(s) of x will the company break even or make a profit?

76. The revenue from the sale of x fire extinguishers is estimated to be $r(x) = 9 - x^2$. The total cost of producing x fire extinguishers is $C(x) = 209 - 33x$. Given that profit is revenue minus cost, for what value(s) of x will the company break even or make a profit?

77. A manufacturer has determined that the cost and revenue of producing and selling x telescopes are $C(x) = 253 - 7x$ and $r(x) = 27x - x^2$, respectively. Given that profit is revenue minus cost, for what value(s) of x will the company break even or make a profit?

78. A company that produces and sells compact refrigerators has found that the revenue from the sale of x compact refrigerators is $r(x) = -x^2 + 30x - 370$. The cost function is given by $C(x) = 6 - 25x$. Given that profit is revenue minus cost, for what value(s) of x will the company break even or make a profit?

79. The population of sea lions on an island is represented by the function $L(m) = 110m^2 - 0.35m^4 + 750$, where m is the number of months the sea lions have been observed on the island. Given this information, how many more months will there be sea lions on the island?

75. All integers between 5 and 27, inclusive

76. All integers between 8 and 25, inclusive

77. All integers between 11 and 23, inclusive

78. All integers between 8 and 47, inclusive

79. About 17.9 months

80. Between 3490 and 17,740 MP3 players, inclusive.

81. About 141.4 weeks

80. An electronics company is deciding whether or not to begin producing MP3 players. The company must determine if a profit can be made on the MP3 players. The profit function is modeled by the equation $P(x) = x + 0.27x^2 - 0.0015x^3 - 300$, where x is the number of MP3 players produced in hundreds. Given this equation, how many MP3 players must the company produce to make a profit?

81. The population of mosquitoes in a city in Florida is modeled by the function $M(w) = 200w^2 - 0.01w^4 + 1200$, where w is the number of weeks since the town began spraying for mosquitoes. How many weeks will it take for all the mosquitoes to die?

Polynomial Division and the Division Algorithm

TOPICS

1. The division algorithm and the remainder theorem

2. Polynomial long division

3. Synthetic division

4. Constructing polynomials with given zeros

TOPIC

The Division Algorithm and the Remainder Theorem

Recall from arithmetic that if a natural number is divided by a smaller one (the divisor), the result can be expressed as a whole number (called the quotient) plus a remainder that is less than the divisor. The polynomial version of this concept is called the *division algorithm*, and it will prove to be very useful in helping us factor, evaluate, and graph polynomials.

THEOREM

The Division Algorithm

Let $p(x)$ and $d(x)$ be polynomials such that $d(x) \neq 0$ and with the degree of d less than or equal to the degree of p. Then there are unique polynomials $q(x)$ and $r(x)$, called the **quotient** and the **remainder**, respectively, such that

$$\underbrace{p(x)}_{\text{dividend}} = \underbrace{q(x)}_{\text{quotient}} \cdot \underbrace{d(x)}_{\text{divisor}} + \underbrace{r(x)}_{\text{remainder}}.$$

Either the degree of the remainder, r, is less than the degree of the **divisor** d, or the remainder is 0, in which case we say d **divides evenly** into the polynomial p. If the remainder is 0, the two polynomials q and d are **factors** of p.

If we divide every term in the equation $p(x) = q(x) \cdot d(x) + r(x)$ by the polynomial d, we obtain another form of the division algorithm:

$$\frac{p(x)}{d(x)} = q(x) + \frac{r(x)}{d(x)}$$

This version of the division algorithm will be very useful when we study rational functions in Section 4.5.

In many cases, we may need to divide a given polynomial p by a divisor of the form $d(x) = x - k$. The division algorithm tells us that the remainder is guaranteed to be either 0, or else a polynomial of degree 0 (since the degree of d is 1). In either case, the remainder polynomial is guaranteed to be simply a number, so $p(x) = q(x)(x - k) + r$ where r is a constant.

What does this mean if k happens to be a zero of the polynomial p? If this is the case, $p(k) = 0$, so

$$0 = p(k)$$
$$= q(k)(k - k) + r$$
$$= r.$$

The remainder is 0! Thus, if k is a zero of the polynomial p, $p(x) = q(x)(x - k)$. Conversely, if $x - k$ divides a given polynomial p evenly, then we know $p(x) = q(x)(x - k)$ and hence $p(k) = q(k)(k - k) = 0$, so k is a zero of p. Together, these two observations constitute a major tool that we use in graphing polynomials and in solving polynomial equations, summarized below.

THEOREM

Zeros and Linear Factors

The number k is a **zero** of a polynomial $p(x)$ if and only if the linear polynomial $x - k$ is a factor of p. In this case, $p(x) = q(x)(x - k)$ for some quotient polynomial q. This also means that k is a solution of the polynomial equation $p(x) = 0$, and if p is a polynomial with real coefficients and if k is a real number, then k is an x-intercept of p.

This reasoning also leads to a more general conclusion called the *remainder theorem*.

THEOREM

The Remainder Theorem

If the polynomial $p(x)$ is divided by $x - k$, the remainder is $p(k)$. That is,

$$p(x) = q(x)(x - k) + p(k).$$

TOPIC 2   Polynomial Long Division

To make use of the division algorithm and the remainder theorem, we need to be able to actually divide one polynomial by another. Polynomial long division is the analog of numerical long division and provides the means for dividing any polynomial by another of equal or smaller degree.

We will begin looking at polynomial long division with an example, then write out a formal procedure. As you follow Example 1, notice the similarities between polynomial division and numerical division.

EXAMPLE 1

Polynomial Long
Division

Divide the polynomial $x^2 + 2x - 24$ by the polynomial $x + 6$.

Solution:

$$x + 6 \overline{\smash{)}x^2 + 2x - 24}$$

Set up the division by arranging the terms of each polynomial in descending order of powers of x.

$$\begin{array}{r} x \\ x + 6 \overline{\smash{)}x^2 + 2x - 24} \end{array}$$

Divide the first term in the dividend, x^2, by the first term in the divisor, x.

$$\begin{array}{r} x \\ x + 6 \overline{\smash{)}x^2 + 2x - 24} \\ -\left(x^2 + 6x\right) \end{array}$$

Multiply each term in the divisor by the result. We align like terms under those in the dividend. There is a minus sign because the next step is to subtract these terms.

$$\begin{array}{r} x \\ x + 6 \overline{\smash{)}x^2 + 2x - 24} \\ -\left(x^2 + 6x\right) \\ \hline -4x - 24 \end{array}$$

Subtract $x^2 + 6x$ from $x^2 + 2x$. Then, bring down -24 from the original dividend. This forms a new dividend to continue the process.

$$\begin{array}{r} x \;-\; 4 \\ x + 6 \overline{\smash{)}x^2 + 2x - 24} \\ -\left(x^2 + 6x\right) \\ \hline -4x - 24 \end{array}$$

We then apply the steps to the new dividend; divide, multiply, and subtract. At this point, there is nothing to bring down from the original dividend.

$$\begin{array}{r} -\left(-4x - 24\right) \\ \hline 0 \end{array}$$

After subtracting, we are left with 0, which tells us that the remainder is 0.

Thus, the quotient is $x - 4$ with a remainder of 0.

PROCEDURE

Polynomial Long
Division

Step 1: Arrange the terms of each polynomial in descending order.

Step 2: Divide the first term in the dividend by the first term in the divisor. This gives the first term of the quotient.

Step 3: Multiply the entire divisor by the result (the first term of the quotient) and write this beneath the dividend so that like terms line up.

Step 4: Subtract the product from the dividend.

Step 5: Bring down the rest of the original dividend, forming a new dividend.

Step 6: Repeat the process with the new dividend. Continue until the degree of the remainder is less than the degree of the divisor.

EXAMPLE 2

Polynomial Long
Division

Divide the polynomial $6x^5 - 5x^4 + 10x^3 - 15x^2 - 19$ by the polynomial $2x^2 - x + 3$.

Solution:

For each cycle of the long division procedure, the quotient is shown in pink, the product in blue, and the subtraction (and bringing down) in green.

$$
\begin{array}{r}
3x^3 \\
2x^2 - x + 3 \overline{\smash{\big)}\, 6x^5 - 5x^4 + 10x^3 - 15x^2 + 0x - 19} \\
-\left(6x^5 - 3x^4 + 9x^3\right) \\
\hline
-2x^4 + x^3 - 15x^2 + 0x - 19
\end{array}
$$

When arranging the terms of the dividend, we insert a placeholder of $0x$. This makes it easier to keep like terms aligned, which prevents errors.

$$
\begin{array}{r}
3x^3 - x^2 \\
2x^2 - x + 3 \overline{\smash{\big)}\, 6x^5 - 5x^4 + 10x^3 - 15x^2 + 0x - 19} \\
-\left(6x^5 - 3x^4 + 9x^3\right) \\
\hline
-2x^4 + x^3 - 15x^2 + 0x - 19 \\
-\left(-2x^4 + x^3 - 3x^2\right) \\
\hline
-12x^2 + 0x - 19
\end{array}
$$

Divide.

Multiply.

Subtract and bring down.

$$
\begin{array}{r}
3x^3 - x^2 - 6 \\
2x^2 - x + 3 \overline{\smash{\big)}\, 6x^5 - 5x^4 + 10x^3 - 15x^2 + 0x - 19} \\
-\left(6x^5 - 3x^4 + 9x^3\right) \\
\hline
-2x^4 + x^3 - 15x^2 + 0x - 19 \\
-\left(-2x^4 + x^3 - 3x^2\right) \\
\hline
-12x^2 + 0x - 19 \\
-\left(-12x^2 + 6x - 18\right) \\
\hline
-6x - 1
\end{array}
$$

To complete the division, repeat the procedure one more time.

At this point, the process halts, as the degree of $-6x - 1$ is smaller than the degree of the divisor.

Thus, the solution is $3x^3 - x^2 - 6 + \dfrac{-6x - 1}{2x^2 - x + 3}$. We can also say the quotient is $3x^3 - x^2 - 6$ with a remainder of $-6x - 1$.

CAUTION!

Although polynomial long division is a straightforward process, one common error is to forget to distribute the minus sign in each step as one polynomial is subtracted from the one above it. A good way to avoid this error is to put parentheses around the polynomial being subtracted, as shown in Examples 1 and 2.

Long division also works on polynomials with complex coefficients. When graphing polynomials, we work with those that have only real coefficients, but complex values can arise in intermediate steps of the graphing process. Further, in solving polynomial equations, we have seen (in some quadratic equations) that complex numbers may be the *only* solutions. Thus, we need to be able to handle division with complex numbers.

EXAMPLE 3

Polynomial Long Division with Complex Numbers

Divide $p(x) = x^4 + 1$ by $d(x) = x^2 + i$.

Solution:

$$x^2 + 0x + i \overline{) x^4 + 0x^3 + 0x^2 + 0x + 1}$$

Insert placeholders in both polynomials.

$$\begin{array}{r} x^2 \\ x^2 + 0x + i \overline{) x^4 + 0x^3 + 0x^2 + 0x + 1} \\ -\left(x^4 + 0x^3 + ix^2\right) \\ \hline -ix^2 + 0x + 1 \end{array}$$

The procedure is exactly the same as with all real coefficients.

Notice that we can use placeholders in the intermediate steps as well.

$$\begin{array}{r} x^2 \qquad -i \\ x^2 + 0x + i \overline{) x^4 + 0x^3 + 0x^2 + 0x + 1} \\ -\left(x^4 + 0x^3 + ix^2\right) \\ \hline -ix^2 + 0x + 1 \\ -\left(-ix^2 + 0x + 1\right) \\ \hline 0 \end{array}$$

When complex numbers are involved, we may need complex number arithmetic.

In the product step, use the fact that $(i)(-i) = -i^2 = 1$.

The remainder is zero, so we are finished.

Thus, the quotient is $x^2 - i$. There is no remainder, which tells us that the quotient is a factor of $p(x)$. In fact, we can write $x^4 + 1 = \left(x^2 + i\right)\left(x^2 - i\right)$.

TOPIC 3

Synthetic Division

Synthetic division is a shortened version of polynomial long division that can be used when the divisor is of the form $x - k$ for some constant k.

Synthetic division is more efficient because it omits the variables in the division process. Instead of various powers of the variable, synthetic division uses a tabular arrangement to keep track of the coefficients of the dividend and, ultimately, the coefficients of the quotient and the remainder. Consider the long division of $-2x^3 + 8x^2 - 9x + 7$ by $x - 2$ shown below.

$$
\begin{array}{r}
-2x^2 + 4x - 1 \\
x-2\overline{)-2x^3 + 8x^2 - 9x + 7} \\
\underline{-\left(-2x^3 + 4x^2\right)} \\
4x^2 - 9x + 7 \\
\underline{-\left(4x^2 - 8x\right)} \\
-x + 7 \\
\underline{-\left(-x + 2\right)} \\
5
\end{array}
$$

First, we will place the constant k and the coefficients of the dividend in a row:

$$ x-2\overline{)-2x^3 + 8x^2 - 9x + 7} \qquad \rightarrow \qquad 2\rfloor \ \ -2 \ \ 8 \ \ -9 \ \ 7 $$

Note that the number 2, which corresponds to k in the form $x - k$, appears without the minus sign. This is a very important and easily overlooked fact. When dividing by $x - k$ using synthetic division, the number that appears in the upper left is k.

Now, look at the key subtractions that occur in the long division:

$$
\begin{array}{r}
-2x^2 + 4x - 1 \\
x-2\overline{)-2x^3 + 8x^2 - 9x + 7} \\
\underline{-\left(-2x^3 + 4x^2\right)} \\
4x^2 - 9x + 7 \\
\underline{-\left(4x^2 - 8x\right)} \\
-1x + 7 \\
\underline{-\left(-x + 2\right)} \\
5
\end{array}
$$

Because the x term of the divisor has a coefficient of 1, the result of each subtraction is the coefficient of the next term of the quotient. The result of the final subtraction, of course, is the remainder. Further, each subtraction begins with the coefficient of the dividend in the same "column." Thus, the result of our synthetic division should look like this:

$$\underline{2\rfloor}\ \ -2\ \ 8\ \ -9\ \ 7$$

$$\overline{\qquad -2\ \ 4\ \ -1\ \ 5}$$

All that remains is to figure out how to get from the top row to the bottom row. Let's look one more time at the long division, except that we'll distribute the negative sign on each subtraction step, turning it into an addition.

$$
\begin{array}{r}
-2x^2+4x-1 \\
x-2\overline{\smash{\big)}-2x^3+8x^2-9x+7} \\
\underline{+\left(2x^3-4x^2\right)} \\
4x^2-9x+7 \\
\underline{+\left(-4x^2+8x\right)} \\
-x+7 \\
\underline{+\left(x-2\right)} \\
5
\end{array}
$$

Note the following two observations:

1. The first coefficient of the quotient matches the first coefficient of the dividend.

2. To find the next coefficient of the quotient, add the product of k and the previous coefficient of the quotient to the corresponding coefficient of the dividend.

Figure 1 illustrates these calculations and then gives the completed table.

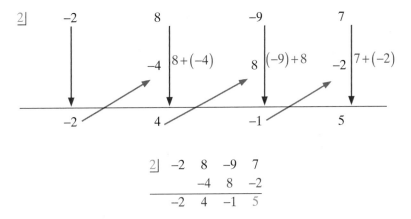

Figure 1: Synthetic Division

PROCEDURE

Synthetic Division

Step 1: If the divisor is $x - k$, write down k, followed by the coefficients of the dividend.

Step 2: Write the leading coefficient of the dividend on the bottom row.

Step 3: Multiply k by the value placed on the bottom, and place the product in the next column, in the second row.

Step 4: Add these values, giving a new value in the bottom row.

Step 5: Repeat this process until the table is complete.

Step 6: The numbers in the bottom row are the coefficients of the quotient, plus the remainder (which is the final value in this row). Note that the first term of the quotient will have degree one less than the first term of the dividend.

By the remainder theorem, synthetic division also provides a quick means of determining $p(k)$ for a given polynomial p, since $p(k)$ is the remainder when $p(x)$ is divided by $x - k$. Of course, if $p(k) = 0$ then we know that k is a zero of $p(x)$ and that $x - k$ is a factor of $p(x)$. These facts are used in Example 4.

EXAMPLE 4

Synthetic Division

For each polynomial p below, divide p by $x - k$ using synthetic division. Use the result to determine if the given k is a zero. If not, determine $p(k)$.

a. $p(x) = -2x^4 + 11x^3 - 5x^2 - 3x + 15; \quad k = 5$

b. $p(x) = 3x^8 + 9x^7 - x^3 - 3x^2 + x - 1; \quad k = -3$

Solutions:

Note:
Remember, unlike in long division, we add the vertically aligned values of synthetic division.

a.

$$\underline{5\rfloor} \quad -2 \quad 11 \quad -5 \quad -3 \quad 15$$

$$\overline{ -2}$$

Place k in the upper-left corner, then write the coefficients of p on the top line.

$$\underline{5\rfloor} \quad -2 \quad 11 \quad -5 \quad -3 \quad 15$$
$$ -10$$
$$\overline{ -2 \quad 1}$$

Multiply $-2 \cdot 5 = -10$ and write down the result. Then add $11 + (-10)$ to get 1, the next coefficient, shown in purple.

$$\underline{5\rfloor} \quad -2 \quad 11 \quad -5 \quad -3 \quad 15$$
$$ -10 \quad 5 \quad 0 \quad -15$$
$$\overline{ -2 \quad 1 \quad 0 \quad -3 \quad 0}$$

Continue the process, showing products in red and quotient coefficients in purple. The last result, in green, is the remainder of 0.

Because the remainder is 0, we know that $k = 5$ is a zero of $p(x)$. With this information, we can factor $p(x)$ as follows:

$$-2x^4 + 11x^3 - 5x^2 - 3x + 15 = \left(-2x^3 + x^2 - 3\right)(x - 5)$$

b. While placeholders for missing terms are useful in long division, they are *necessary* when doing synthetic division.

$$\underline{-3|}\ \ 3\ \ 9\ \ 0\ \ 0\ \ 0\ \ -1\ \ -3\ \ 1\ \ -1$$

$$\overline{\ \ 3}$$

Set up the synthetic division, including placeholders for x^6, x^5, and x^4.

$$\underline{-3|}\ \ 3\ \ \ \ 9\ \ 0\ \ 0\ \ 0\ \ -1\ \ -3\ \ 1\ \ -1$$
$$\ \ -9\ \ 0\ \ 0\ \ 0\ \ \ \ 0\ \ \ \ 3\ \ 0\ \ -3$$
$$\overline{\ \ 3\ \ \ \ 0\ \ 0\ \ 0\ \ 0\ \ -1\ \ \ \ 0\ \ 1\ \ -4}$$

Proceed with the synthetic division. Once again, the products are shown in red, while the coefficients of the new quotient appear in purple.

This time the remainder (shown in green) is not zero. This means that $k = -3$ is not a zero of $p(x)$. The remainder theorem tells us that $p(-3) = -4$.

Examine Example 4b again. The standard way to determine $p(-3)$ is to simplify $p(-3) = 3(-3)^8 + 9(-3)^7 - (-3)^3 - 3(-3)^2 + (-3) - 1$. This is a tedious calculation, but the result, $3(6561) + 9(-2187) - (-27) - 3(9) + (-3) - 1$, does indeed equal -4. Compare these calculations with the far simpler synthetic division used in the example to see how useful synthetic division can be when evaluating polynomials.

Just as with long division, we can perform synthetic division on polynomials with complex coefficients (as long as the divisor is still first-degree).

EXAMPLE 5

Synthetic Division (with Complex Numbers)

Compute using synthetic division: $\dfrac{-3x^3 + (5 - 2i)x^2 + (-4 + i)x + (1 - i)}{x - 1 + i}$.

Solution:

$$\underline{1-i|}\ \ -3\ \ 5-2i\ \ -4+i\ \ 1-i$$

$$\overline{\ \ -3}$$

Set up the synthetic division. The dividend has several complex coefficients.

$$\underline{1-i|}\ \ -3\ \ \ \ 5-2i\ \ \ -4+i\ \ \ 1-i$$
$$\ \ \ \ \ \ \ -3+3i\ \ \ \ 3-i\ \ \ -1+i$$
$$\overline{\ \ -3\ \ \ \ 2+i\ \ \ \ \ -1\ \ \ \ \ \ 0}$$

Throughout the division we need complex number arithmetic. In particular,
$$(1-i)(2+i) = 2 - i - i^2 = 3 - i.$$

Thus, the quotient is $-3x^2 + (2 + i)x - 1$.

TOPIC 4 — Constructing Polynomials with Given Zeros

The last topic in this section concerns reversing the division process. We now know the connection between a polynomial's zeros and factors: k is a zero of the polynomial $p(x)$ if and only if $x-k$ is a factor of $p(x)$. We can make use of this fact to construct polynomials that have certain desired properties, as illustrated in Example 6.

EXAMPLE 6

Constructing Polynomials

Construct a polynomial that has the given properties.

a. Third-degree, zeros of $-3, 2$, and 5, and goes to $-\infty$ as $x \to \infty$.

b. Fourth-degree, zeros of $-5, -2, 1$, and 3, and y-intercept at $(0, 15)$.

Note:
When constructing polynomials from a set of factors, it is easier to keep the result in factored form until the last step. Often, it is fine to leave the polynomial in factored form.

Solutions:

a. We need $p(x)$ to have zeros $-3, 2$, and 5, so it must have linear factors of $(x+3)$, $(x-2)$, and $(x-5)$. Three linear factors gives us a third-degree polynomial, so there can be no more factors. Putting these together, we have:

$$p(x) = (x+3)(x-2)(x-5)$$

But does $p(x) \to -\infty$ as $x \to \infty$? No, if we multiply out, the leading term of this polynomial would be x^3, which has a positive leading coefficient. To fix this, we multiply the entire polynomial by -1:

$$p(x) = -(x+3)(x-2)(x-5)$$
$$= -x^3 + 4x^2 + 11x - 30$$

b. Once again, $p(x)$ is a product of linear factors, identified by the required zeros:

$$p(x) = (x+5)(x+2)(x-1)(x-3)$$

Our second condition is that the y-intercept must be $(0, 15)$. If we substitute $x = 0$, we see that $p(0) = (5)(2)(-1)(-3) = 30$, so the y-intercept is $(0, 30)$. To fix this, we might try subtracting 15 from the polynomial:

$$p(x) \overset{?}{=} (x+5)(x+2)(x-1)(x-3) - 15$$

But, we can not do this because it causes $-5, -2, 1$ and 3 to no longer be zeros! Instead, we multiply $p(x)$ by $\dfrac{1}{2}$.

$$p(x) = \frac{1}{2}(x+5)(x+2)(x-1)(x-3)$$
$$= \frac{1}{2}x^4 + \frac{3}{2}x^3 - \frac{15}{2}x^2 - \frac{19}{2}x + 15$$

1. $3x^2 - x + 1 +$
$\dfrac{5x-1}{2x^2+2}$

2. $5x - 1 + \dfrac{-4}{x+2}$

3. $x - 2 +$
$\dfrac{-2}{x^2-4x+4}$

4. $7x^3 - x^2 - 5x +$
$\dfrac{5x}{x^2+1}$

5. $4x^2 - 14x + 29 -$
$\dfrac{65}{x+2}$

6. $x^2 + 5x + 11 +$
$\dfrac{25}{x-3}$

7. $x^3 + 6x^2 - 2x + 5 +$
$\dfrac{2x+5}{3x^2-1}$

8. $x^3 - x^2 + 2x - 3$

9. $2x^3 - 3x^2 + 2x - 5$

10. $7x^2 - x + 3$

11. $x^3 + 3x^2 + 10x$
$+ 10 + \dfrac{22}{x-3}$

12. $2x^3 - 2x -$
$3 + \dfrac{2x+4}{x^2+1}$

13. $3x^2 + 5x +$
$9 + \dfrac{45}{3x-5}$

14. $-2x^2 + 5 +$
$\dfrac{-5x-2}{2x^3+x}$

15. $2x - 5 + \dfrac{7}{x+3}$

16. $5x^2 + x - 3$

<table><thead><tr><th colspan="2"> Exercises </th></tr></thead></table>

Use polynomial long division to rewrite each of the following fractions in the form $q(x) + \dfrac{r(x)}{d(x)}$, where $d(x)$ is the denominator of the original fraction, $q(x)$ is the quotient, and $r(x)$ is the remainder. See Examples 1 through 3.

1. $\dfrac{6x^4 - 2x^3 + 8x^2 + 3x + 1}{2x^2 + 2}$

2. $\dfrac{5x^2 + 9x - 6}{x + 2}$

3. $\dfrac{x^3 - 6x^2 + 12x - 10}{x^2 - 4x + 4}$

4. $\dfrac{7x^5 - x^4 + 2x^3 - x^2}{x^2 + 1}$

5. $\dfrac{4x^3 - 6x^2 + x - 7}{x + 2}$

6. $\dfrac{x^3 + 2x^2 - 4x - 8}{x - 3}$

7. $\dfrac{3x^5 + 18x^4 - 7x^3 + 9x^2 + 4x}{3x^2 - 1}$

8. $\dfrac{9x^5 - 10x^4 + 18x^3 - 28x^2 + x + 3}{9x^2 - x - 1}$

9. $\dfrac{2x^5 - 5x^4 + 7x^3 - 10x^2 + 7x - 5}{x^2 - x + 1}$

10. $\dfrac{14x^5 - 2x^4 + 27x^3 - 3x^2 + 9x}{2x^3 + 3x}$

11. $\dfrac{x^4 + x^2 - 20x - 8}{x - 3}$

12. $\dfrac{2x^5 - 3x^2 + 1}{x^2 + 1}$

13. $\dfrac{9x^3 + 2x}{3x - 5}$

14. $\dfrac{-4x^5 + 8x^3 - 2}{2x^3 + x}$

15. $\dfrac{2x^2 + x - 8}{x + 3}$

16. $\dfrac{5x^5 + x^4 - 13x^3 - 2x^2 + 6x}{x^3 - 2x}$

17. $\dfrac{2x^3 - 3ix^2 + 11x + (1 - 5i)}{2x - i}$

18. $\dfrac{9x^3 - (18 + 9i)x^2 + x + (-2 - i)}{x - 2 - i}$

19. $\dfrac{3x^3 + ix^2 + 9x + 3i}{3x + i}$

20. $\dfrac{35x^4 + (14 - 10i)x^3 - (7 + 4i)x^2 + 2ix}{7x - 2i}$

Use synthetic division to determine if the given value for k is a zero of the corresponding polynomial. If not, determine $p(k)$. See Example 4.

21. $p(x) = 32x^5 - 80x^4 + 80x^3 - 40x^2 + 10x + 2; \ k = 1$

22. $p(x) = 32x^5 - 80x^4 + 80x^3 - 40x^2 + 10x + 2; \ k = \dfrac{1}{2}$

23. $p(x) = 12x^4 - 7x^3 - 32x^2 - 7x + 6; \ k = 2$

24. $p(x) = 12x^4 - 7x^3 - 32x^2 - 7x + 6; \ k = 1$

25. $p(x) = 12x^4 - 7x^3 - 32x^2 - 7x + 6; \ k = \dfrac{1}{3}$

17. $x^2 - ix + 6 + \dfrac{1+i}{2x-i}$

18. $9x^2 + 1$

19. $x^2 + 3$

20. $5x^3 + 2x^2 - x$

21. $p(1) = 4$

22. $p\left(\dfrac{1}{2}\right) = 3$

23. k is a zero

24. $p(1) = -28$

25. k is a zero

26. $p(-2) = 17 - 19i$

27. $p(1) = 12$

28. k is a zero

29. k is a zero

30. k is a zero

31. k is a zero

32. $p(2) = 5$

33. k is a zero

34. $p(-2) = -32$

35. k is a zero

36. $p(2) = 1$

37. $p(5) = -2$

38. k is a zero

39. k is a zero

40. k is a zero

41. $x^2 - 4x + 2 + \dfrac{-1}{x+5}$

42. $-2x^4 + 3x^2 - x + 1$

43. $x^7 - 3x^2 + \dfrac{3}{x+1}$

44. $x^7 - 3x^2 + \dfrac{-2}{x-5}$

26. $p(x) = 2x^2 - (3-5i)x + (3-9i);\ k = -2$

27. $p(x) = 8x^4 - 2x + 6;\ k = 1$

28. $p(x) = x^4 - 1;\ k = 1$

29. $p(x) = x^5 + 32;\ k = -2$

30. $p(x) = 3x^5 + 9x^4 + 2x^2 + 5x - 3;\ k = -3$

31. $p(x) = 2x^2 - (3-5i)x + (3-9i);\ k = -3i$

32. $p(x) = x^2 - 6x + 13;\ k = 2$

33. $p(x) = x^2 - 6x + 13;\ k = 3 - 2i$

34. $p(x) = 3x^3 - 13x^2 - 28x - 12;\ k = -2$

35. $p(x) = 3x^3 - 13x^2 - 28x - 12;\ k = 6$

36. $p(x) = 2x^3 - 8x^2 - 23x + 63;\ k = 2$

37. $p(x) = 2x^3 - 8x^2 - 23x + 63;\ k = 5$

38. $p(x) = x^4 - 3x^3 - 3x^2 + 11x - 6;\ k = 1$

39. $p(x) = x^4 - 3x^3 - 3x^2 + 11x - 6;\ k = -2$

40. $p(x) = x^4 - 3x^3 - 3x^2 + 11x - 6;\ k = 3$

Use synthetic division to rewrite each of the following fractions in the form $q(x) + \dfrac{r(x)}{d(x)}$, where $d(x)$ is the denominator of the original fraction, $q(x)$ is the quotient, and $r(x)$ is the remainder. See Example 5.

41. $\dfrac{x^3 + x^2 - 18x + 9}{x+5}$

42. $\dfrac{-2x^5 + 4x^4 + 3x^3 - 7x^2 + 3x - 2}{x-2}$

43. $\dfrac{x^8 + x^7 - 3x^3 - 3x^2 + 3}{x+1}$

44. $\dfrac{x^8 - 5x^7 - 3x^3 + 15x^2 - 2}{x-5}$

45. $\dfrac{4x^3 - (16+4i)x^2 + (14+4i)x + (-6-2i)}{x-3-i}$

46. $\dfrac{x^6 - 2x^5 + 2x^4 + 4x^2 - 8x + 8}{x-1+i}$

45. $4x^2 - 4x + 2$

46. $x^5 - (1+i)x^4 + 4x + (-4-4i)$

47. $x^4 - x^3 - x^2 - 7x - 14 - \dfrac{10}{x-2}$

48. $x^4 + x^2 - 2x - 6$

49. $x^3 - x^2 + x$

50. $x^5 + x^2 - 14$

51. $2x^2 - 4ix + 17 + \dfrac{8+48i}{x-3i}$

52. $4x^4 - 2x^3 + 8x^2 + 4x$

53. $f(x) = -x^2 - x + 12$

54. $f(x) = -2x^3 + 4x^2 + 10x - 12$

55. $f(x) = -x^2 + 4x - 13$

56. $f(x) = -2x^3 + 4x^2 + 2ix + (-6+2i)$

57. $f(x) = x^4 - 12x^3 + 54x^2 - 108x + 81$

58. $f(x) = -4x^2 + 5x + 6$

59. $f(x) = 3x^4 + 9x^3 - 9x^2 - 21x + 18$

60. $f(x) = 2x^3 - 12x^2 + 22x - 12$

61. $SA = (x+5)(x+2)$ $= x^2 + 7x + 10$

47. $\dfrac{x^5 - 3x^4 + x^3 - 5x^2 + 18}{x-2}$

48. $\dfrac{x^5 - 3x^4 + x^3 - 5x^2 + 18}{x-3}$

49. $\dfrac{x^4 + (i-1)x^3 + (1-i)x^2 + ix}{x+i}$

50. $\dfrac{x^6 + 8x^5 + x^3 + 8x^2 - 14x - 112}{x+8}$

51. $\dfrac{2x^3 - 10ix^2 + 5x + (8-3i)}{x-3i}$

52. $\dfrac{4x^5 - 6x^4 + 10x^3 - 4x^2 - 4x}{x-1}$

Construct a polynomial function with the stated properties. See Example 6.

53. Second-degree, zeros of -4 and 3, and goes to $-\infty$ as $x \to -\infty$.

54. Third-degree, zeros of -2, 1, and 3, and a y-intercept of -12.

55. Second-degree, zeros of $2-3i$ and $2+3i$, and a y-intercept of -13.

56. Third-degree, zeros of $1-i$, $2+i$, and -1, and a leading coefficient of -2.

57. Fourth-degree and a single x-intercept of 3.

58. Second-degree, zeros of $-\dfrac{3}{4}$ and 2, and a y-intercept of 6.

59. Fourth-degree, zeros of -3, -2, and 1, and a y-intercept of 18.

60. Third-degree, zeros of 1, 2, and 3, and passes through the point $(4, 12)$.

Solve the following application problem.

61. A box company makes a variety of boxes, all with volume given by the formula $x^3 + 10x^2 + 31x + 30$. If the height is given by $x + 3$, what is the formula for the surface area of the base?

4.3 Locating Real Zeros of Polynomials

TOPICS

1. The Rational Zero Theorem
2. Descartes' Rule of Signs
3. Bounds of real zeros
4. The Intermediate Value Theorem

TOPIC 1

The Rational Zero Theorem

Given a polynomial function $p(x)$, we now know that k is a zero if and only if $x - k$ is a factor of p. Furthermore, if $x - k$ is a factor of p, we can use either polynomial long division or synthetic division to actually divide p by $x - k$ and find the quotient polynomial q, allowing us to write $p(x) = (x - k)q(x)$. This leaves us with a polynomial q of smaller degree and brings us one step closer to factoring p completely.

What we are lacking is a method for finding the zeros of p when it doesn't easily factor. The techniques from the last two sections cannot be put to use until we have some way of locating a zero k as a starting point.

Unfortunately, it can be proven that there is no formula that, like the quadratic formula, identifies all the zeros of a polynomial of degree five or higher. We do, however, have tools that give us hints about where to look for zeros of a given polynomial. In this section, we will study several such tools, beginning with the Rational Zero Theorem.

THEOREM

The Rational Zero Theorem

If $f(x) = a_n x^n + a_{n-1} x^{n-1} + \ldots + a_1 x + a_0$ is a polynomial with integer coefficients with $a_n \neq 0$, then any rational zero of f must be of the form $\dfrac{p}{q}$, where p is a factor of the constant term a_0 and q is a factor of the leading coefficient a_n.

CAUTION!

Before applying the Rational Zero Theorem, we note two things it *doesn't* do:

1. The theorem doesn't necessarily find even a single zero of a polynomial; instead, it identifies a list of rational numbers that could *potentially* be zeros.

2. The theorem says nothing about irrational or complex zeros. If a polynomial has zeros that are either irrational or complex, we must resort to other means to find them.

EXAMPLE 1

The Rational Zero Theorem

For each of the polynomials that follow, list all of the potential rational zeros. Then write the polynomial in factored form and identify the actual zeros.

a. $f(x) = 2x^3 + 5x^2 - 4x - 3$ **b.** $g(x) = 27x^4 - 9x^3 - 33x^2 - x - 4$

Note:
After generating the list of possible rational zeros, it's easiest to use synthetic division to check the zeros.

Solutions:

a. To apply the Rational Zero Theorem, find the factors of a_0 and a_3.

Factors of $a_0 : \pm\{1, 3\}$ Note that we take both the positive and negative factors into consideration.

Factors of $a_3 : \pm\{1, 2\}$

Possible rational zeros: $\pm\left\{1, 3, \dfrac{1}{2}, \dfrac{3}{2}\right\}$ *If there are any rational zeros, they will come from this set of 8 numbers.*

Now perform synthetic division on a trial-and-error basis with the potential rational zeros.

$$\begin{array}{r|rrr} 1 & 2 & 5 & -4 & -3 \\ & & 2 & 7 & 3 \\ \hline & 2 & 7 & 3 & 0 \end{array}$$

Performing synthetic division with $k = 1$ gives a remainder of 0.

Thus, we can factor $f(x)$ as follows:

$f(x) = (x-1)(2x^2 + 7x + 3)$ Use the result from synthetic division.

$\qquad = (x-1)(2x+1)(x+3)$ Factor the quadratic.

Actual zeros: $\left\{1, -\dfrac{1}{2}, -3\right\}$ Apply the Zero-Factor Property to determine the actual zeros.

b. Again, begin by listing the factors of the leading coefficient and the constant term.

Factors of $a_0 : \pm\{1, 2, 4\}$

Factors of $a_4 : \pm\{1, 3, 9, 27\}$

Possible rational zeros : $\pm\left\{1, 2, 4, \dfrac{1}{3}, \dfrac{2}{3}, \dfrac{4}{3}, \dfrac{1}{9}, \dfrac{2}{9}, \dfrac{4}{9}, \dfrac{1}{27}, \dfrac{2}{27}, \dfrac{4}{27}\right\}$

While it may be daunting to consider 24 potential rational zeros, appreciate the fact that the Rational Zero Theorem has eliminated all rational numbers except these 24! Before we begin trial-and-error, consider a few tips for choosing possible zeros:

1. Begin with integer values. The synthetic division and resulting quotient will usually be simpler than when trying fractions.

2. Begin with values near 1. This will keep the numbers in calculations smaller, allowing you to test values a bit more quickly.

$$\underline{-1|}\;\; \begin{array}{rrrrr} 27 & -9 & -33 & -1 & -4 \\ & -27 & 36 & -3 & 4 \\ \hline 27 & -36 & 3 & -4 & 0 \end{array}$$

Synthetic division with $k = -1$ tells us that $(x+1)$ is a factor of $g(x)$.

The resulting quotient is still a cubic, so we proceed with synthetic division. It also leads to another consideration when selecting potential zeros:

3. If the value k is a zero, do not forget to test k with the quotient, as $(x-k)$ can appear as a factor multiple times.

$$\underline{-1|}\;\; \begin{array}{rrrr} 27 & -36 & 3 & -4 \\ & -27 & 63 & -66 \\ \hline 27 & -63 & 66 & -70 \end{array}$$

In this case, the synthetic division fails to find a zero. We simply proceed with the next candidate.

$$\dfrac{4}{3}\bigg|\;\; \begin{array}{rrrr} 27 & -36 & 3 & -4 \\ & 36 & 0 & 4 \\ \hline 27 & 0 & 3 & 0 \end{array}$$

This time, the remainder is 0, so we have uncovered another factor, leaving us with a quadratic quotient.

We can then write the factorization using the result from synthetic division. The last step in factoring comes from solving a quadratic equation.

$$g(x) = (x+1)\left(x - \frac{4}{3}\right)\left(27x^2 + 3\right)$$

$$= 27(x+1)\left(x - \frac{4}{3}\right)\left(x - \frac{i}{3}\right)\left(x + \frac{i}{3}\right)$$

By solving the equation $27x^2 + 3 = 0$, we find $27x^2 + 3 = 27\left(x - \dfrac{i}{3}\right)\left(x + \dfrac{i}{3}\right)$.

$$= (x+1)(3x-4)(3x-i)(3x+i)$$

Actual zeros: $\left\{-1, \dfrac{4}{3}, \dfrac{i}{3}, -\dfrac{i}{3}\right\}$

TOPIC 2   Descartes' Rule of Signs

Descartes' Rule of Signs is another tool to aid us in our search for zeros. Unlike the Rational Zero Theorem, Descartes' Rule doesn't identify candidates; instead, it gives us guidelines on how many positive real zeros and how many negative real zeros we can expect a given polynomial to have. Again, this tool tells us nothing about complex zeros.

THEOREM

Descartes' Rule
of Signs

Let $f(x) = a_n x^n + a_{n-1} x^{n-1} + \ldots + a_1 x + a_0$ be a polynomial with real coefficients, and assume $a_n \neq 0$. A **variation in sign** of f is a change in the sign of one coefficient of f to the next, either from positive to negative or vice versa.

1. The number of **positive real zeros** of f is either the number of variations in sign of $f(x)$ or is less than this number by a positive even integer.

2. The number of **negative real zeros** of f is either the number of variations in sign of $f(-x)$ or is less than this number by a positive even integer.

Note that in order to apply Descartes' Rule of Signs, it is critical to first write the terms of the polynomial in descending order. Also, unless the number of variations in sign is 0 or 1, the rule does not give us a definitive answer for the number of zeros to expect. For instance, if the number of variations in sign of $f(x)$ is 4, we know only that there will be 4, 2, or 0 positive real zeros.

EXAMPLE 2

Descartes' Rule
of Signs

Use Descartes' Rule of Signs to determine the possible numbers of positive and negative real zeros of each of the following polynomials.

a. $f(x) = 2x^3 + 3x^2 - 14x - 21$ b. $g(x) = 3x^3 - 10x^2 + \dfrac{51}{4}x - \dfrac{13}{4}$

Solutions:

a. $f(x) = 2x^3 + 3x^2 \underset{\text{sign change}}{-14x} - 21$ One sign change in $f(x)$.

$f(-x) = 2(-x)^3 + 3(-x)^2 - 14(-x) - 21$ Plug in $-x$, then simplify.

$= -2x^3 \underset{\text{sign change}}{+3x^2} + 14x \underset{\text{sign change}}{-21}$ Two sign changes in $f(-x)$.

By Descartes' Rule of Signs, there is exactly 1 positive real zero and either 2 or 0 negative real zeros of $f(x)$.

b. $g(x) = 3x^3 - 10x^2 + \dfrac{51}{4}x - \dfrac{13}{4}$ Three changes of sign in $g(x)$.

$g(-x) = 3(-x)^3 - 10(-x)^2 + \dfrac{51}{4}(-x) - \dfrac{13}{4}$ Plug in $-x$, then simplify.

$= -3x^3 - 10x^2 - \dfrac{51}{4}x - \dfrac{13}{4}$ There are no sign changes in $g(-x)$.

By Descartes' Rule of Signs, there are either 3 or 1 positive real zeros, but no negative real zeros of $g(x)$.

TOPIC 3   Bounds of Real Zeros

Although the Rational Zero Theorem and Descartes' Rule of Signs are useful for determining the zeros of a polynomial, Examples 1 and 2 show that more guidance would certainly be welcome, especially guidance that reduces the number of potential zeros that must be tested by trial and error. The following theorem does just that.

THEOREM

Upper and Lower Bounds of Zeros

Let $f(x)$ be a polynomial with real coefficients, a positive leading coefficient, and degree ≥ 1. Let a be a negative number and b be a positive number. Then:

1. No real zero of f is larger than b (we say b is an **upper bound** of the zeros of f) if the last row in the synthetic division of $f(x)$ by $x-b$ contains no negative numbers. That is, b is an upper bound of the zeros if the quotient and remainder have no negative coefficients when $f(x)$ is divided by $x-b$.

2. No real zero of f is smaller than a (we say a is a **lower bound** of the zeros of f) if the last row in the synthetic division of $f(x)$ by $x-a$ has entries that alternate in sign (0 can count as either positive *or* negative).

Example 3 revisits the polynomial $f(x) = 2x^3 + 3x^2 - 14x - 21$ that we studied in Example 2a and illustrates the use of the above theorem.

EXAMPLE 3

Finding Bounds of Real Zeros

Use synthetic division to identify upper and lower bounds of the real zeros of the polynomial $f(x) = 2x^3 + 3x^2 - 14x - 21$.

Note:
Finding the smallest upper bound and largest lower bound possible will help us eliminate as many potential zeros as possible.

Solution:

Begin by testing any positive number as a potential upper bound.

$$
\begin{array}{r|rrrr}
2 & 2 & 3 & -14 & -21 \\
  &   & 4 & 14 & 0 \\
\hline
  & 2 & 7 & 0   & -21
\end{array}
$$

Synthetic division shows that 2 is not necessarily an upper bound, as the last row contains a negative number.

It is best to begin with a small value, then test progressively larger ones, as this will help in finding the smallest upper bound.

$$
\begin{array}{r|rrrr}
3 & 2 & 3 & -14 & -21 \\
  &   & 6 & 27 & 39 \\
\hline
  & 2 & 9 & 13 & 18
\end{array}
$$

The number 3 is an upper bound according to the theorem, as all of the coefficients in the last row are nonnegative.

This tells us that all real zeros (including irrational zeros) of f are less than or equal to 3.

We continue by testing a value for the lower bound.

$$\begin{array}{r|rrrr} -3 & 2 & 3 & -14 & -21 \\ & & -6 & 9 & - \\ \hline & 2 & -3 & -5 & - \end{array}$$

The synthetic division has not been completed, because as soon as the signs in the last row cease to alternate, we know -3 is not a lower bound.

Move on by testing a lower number.

$$\begin{array}{r|rrrr} -4 & 2 & 3 & -14 & -21 \\ & & -8 & 20 & -24 \\ \hline & 2 & -5 & 6 & -45 \end{array}$$

We find that -4 is a lower bound, as the signs in the last row alternate. Remember that if a 0 appears, it can be counted as either positive or negative, whichever leads to a sequence of alternating signs.

Thus, we see that -4 is a lower bound. Combined with the upper bound, we now know that all real zeros of f lie in the interval $[-4, 3]$.

EXAMPLE 4

Finding the Zeros of a Polynomial

Use the results of Example 3, in conjunction with the Rational Zero Theorem, to find the actual zeros of $f(x) = 2x^3 + 3x^2 - 14x - 21$.

Solution:

Start by finding the potential rational zeros:

Factors of $a_0 : \pm\{1, 3, 7, 21\}$

Factors of $a_3 : \pm\{1, 2\}$

Possible rational zeros: $\pm\left\{1, 3, 7, 21, \dfrac{1}{2}, \dfrac{3}{2}, \dfrac{7}{2}, \dfrac{21}{2}\right\}$

Now, apply the lower and upper bounds. This allows us to eliminate any potential zeros greater than 3 or less than -4:

Possible rational zeros: $\left\{1, -1, 3, -3, \dfrac{1}{2}, -\dfrac{1}{2}, \dfrac{3}{2}, -\dfrac{3}{2}, -\dfrac{7}{2}\right\}$

Now, use synthetic division to test potential rational zeros.

$$\begin{array}{r|rrrr} -\dfrac{3}{2} & 2 & 3 & -14 & -21 \\ & & -3 & 0 & 21 \\ \hline & 2 & 0 & -14 & 0 \end{array}$$

The quotient is $2x^2 - 14$. We can find the remaining two zeros by using the square root method.

$$2x^2 - 14 = 0$$
$$2x^2 = 14$$
$$x^2 = 7$$
$$x = \pm\sqrt{7}$$

Thus, the actual zeros of $f(x)$ are $\left\{ -\dfrac{3}{2}, \sqrt{7}, -\sqrt{7} \right\}$.

CAUTION!

Don't read more into the Upper and Lower Bounds Theorem than is actually there. For instance, -3 actually *is* a lower bound of the zeros of $f(x) = 2x^3 + 3x^2 - 14x - 21$, but the theorem is not powerful enough to indicate this. The work in Example 3 shows that -4 is a lower bound, but the theorem fails to spot the fact that -3 is a better lower bound. The trade-off for this weakness in the theorem is that it is quickly and easily applied.

TOPIC

The Intermediate Value Theorem

The last technique for locating zeros that we study makes use of a property of polynomials called *continuity*, which we briefly discussed when solving polynomial inequalities. Although continuity of functions will not be discussed in this text, one consequence of continuity is that the graph of a continuous function has no "breaks" in it. That is, assuming that the function can be graphed at all, it can be drawn without lifting your pencil.

THEOREM

Intermediate Value Theorem

Assume that $f(x)$ is a polynomial with real coefficients, and that a and b are real numbers with $a < b$. **If $f(a)$ and $f(b)$ differ in sign, then there is at least one point c such that $a < c < b$ and $f(c) = 0$.** That is, at least one zero of f lies between a and b.

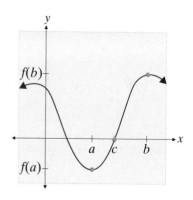

CAUTION! ∿∿∿

The Intermediate Value Theorem can only tell us that there *is* a zero between two *x*-values, it can not prove that a zero *does not* exist between two values. If $f(a)$ and $f(b)$ do not differ in sign, there may still be one or more zeros between a and b.

We can use the Intermediate Value Theorem to prove that a zero of a given polynomial must lie in a particular interval. Repeated application of this process allows us to "hone in" on the zero, generating a good approximation.

EXAMPLE 5

Intermediate Value Theorem

a. Show that $f(x) = x^3 + 3x - 7$ has a zero between 1 and 2.

b. Find an approximation of the zero to the nearest tenth.

Solutions:

a. To use Intermediate Value Theorem, we need to calculate $f(1)$ and $f(2)$.

$$f(1) = (1)^3 + 3(1) - 7 = -3 \qquad\qquad f(1) \text{ is negative.}$$
$$f(2) = (2)^3 + 3(2) - 7 = 7 \qquad\qquad f(2) \text{ is positive.}$$

Because $f(1)$ and $f(2)$ differ in sign, the Intermediate Value Theorem states that f has a zero between 1 and 2.

b. To estimate this zero to the nearest tenth, we plug in more values to shrink the interval where the zero could potentially lie. We begin with 1.5.

$$f(1.5) = (1.5)^3 + 3(1.5) - 7 = 0.875$$

We see that $f(1.5)$ is positive, but small. Since $f(1)$ is negative, we might expect that the zero is slightly less than 1.5.

$$f(1.4) = (1.4)^3 + 3(1.4) - 7 = -0.056$$

Now the Intermediate Value Theorem tells us that the zero lies between 1.4 and 1.5. We need to test one more point to determine the zero to the nearest tenth.

$$f(1.45) = (1.45)^3 + 3(1.45) - 7 = 0.398625$$

Once more, the Intermediate Value Theorem narrows the interval to $(1.4, 1.45)$, which shows that the value of the zero, to the nearest tenth, is 1.4.

1. $\pm\left\{\begin{matrix}\dfrac{1}{3},\dfrac{2}{3},1,\dfrac{4}{3}, \\ 2,\dfrac{8}{3},4,8\end{matrix}\right\}$,

$\left\{-4,\dfrac{1}{3},2\right\}$

2. $\pm\left\{\begin{matrix}\dfrac{1}{2},1,\dfrac{3}{2},2,\dfrac{5}{2}, \\ 3,5,6,\dfrac{15}{2},10, \\ 15,30\end{matrix}\right\}$,

$\left\{-\dfrac{3}{2},2,5\right\}$

3. $\pm\left\{\begin{matrix}1,2,3,4,6, \\ 8,12,24\end{matrix}\right\}$,

$\left\{\pm 2i,2,3\right\}$

4. $\pm\left\{1,13\right\}$,

$\left\{-1,2\pm 3i\right\}$

5. $\pm\left\{1,2,7,14\right\}$,

$\left\{1,2,7\right\}$

6. $\pm\left\{\begin{matrix}1,2,5,10, \\ 25,50\end{matrix}\right\}$,

$\left\{\pm 5i,-2,1\right\}$

7. $\pm\left\{\begin{matrix}\dfrac{1}{2},1,\dfrac{3}{2},\dfrac{5}{2}, \\ 3,5,\dfrac{15}{2},15\end{matrix}\right\}$,

$\left\{-1,\dfrac{5}{2},3\right\}$

8. $\pm\left\{\begin{matrix}1,2,4,5, \\ 10,20\end{matrix}\right\}$,

$\left\{4,1-2i,1+2i\right\}$

9. $\pm\left\{\dfrac{1}{3},1,3\right\}$,

$\left\{-1,1,-i,i\right\}$

10. $\pm\left\{\begin{matrix}1,2,3,4,6, \\ 8,12,24\end{matrix}\right\}$,

$\left\{-2,2,-\sqrt{6},\sqrt{6}\right\}$

Exercises

List all of the potential rational zeros of the following polynomials. Then use polynomial division and the quadratic formula, if necessary, to identify the actual zeros. See Example 1.

1. $f(x)=3x^3+5x^2-26x+8$ **2.** $g(x)=-2x^3+11x^2+x-30$

3. $p(x)=x^4-5x^3+10x^2-20x+24$ **4.** $h(x)=x^3-3x^2+9x+13$

5. $q(x)=x^3-10x^2+23x-14$ **6.** $r(x)=x^4+x^3+23x^2+25x-50$

7. $s(x)=2x^3-9x^2+4x+15$ **8.** $t(x)=x^3-6x^2+13x-20$

9. $j(x)=3x^4-3$ **10.** $k(x)=x^4-10x^2+24$

11. $m(x)=x^3+11x^2-x-11$ **12.** $g(x)=x^3-6x^2-5x+30$

Using the Rational Zero Theorem or your answers to the preceding problems, solve the following polynomial equations.

13. $x^4+x-2=-2x^4+x+1$ **14.** $x^4+10=10x^2-14$

15. $x^3-3x^2+9x+13=0$ **16.** $3x^3+5x^2=26x-8$

17. $x^4+10x^2-20x=5x^3-24$ **18.** $-2x^3+11x^2+x=30$

19. $2x^3-12x^2+26x=40$ **20.** $2x^3+9x^2+4x=15$

21. $x^4+x^3+23x^2=50-25x$ **22.** $x^3+23x=10x^2+14$

23. $x^3+11x^2=11+x$ **24.** $-6x^2+x^3=5x-30$

Use Descartes' Rule of Signs to determine the possible numbers of positive and negative real zeros of each of the following polynomials. See Example 2.

25. $f(x)=x^3+8x^2+17x+10$ **26.** $g(x)=x^3+2x^2-5x-6$

27. $f(x)=x^3-6x^2+3x+10$ **28.** $g(x)=x^3+6x^2+11x+6$

29. $f(x)=x^4-5x^3-2x^2+40x-48$ **30.** $g(x)=x^3+3x^2+3x+9$

31. $f(x)=x^4-25$ **32.** $g(x)=x^4-7x^3+5x^2+31x-30$

33. $f(x)=5x^5-x^4+2x^3+x-9$ **34.** $g(x)=-6x^7-x^5-7x^3-2x$

35. $f(x)=-5x^{11}-14x^9-10x^7-15x^5$ **36.** $g(x)=2x^4+7x^3+28x^2+112x-64$

11. $\pm\{1,11\}$,
$\{-11,-1,1\}$

12. $\pm\begin{Bmatrix}1,2,3,5,6,\\10,15,30\end{Bmatrix}$,
$\{-\sqrt{5},\sqrt{5},6\}$

13. $\{-1,1,-i,i\}$

14. $\{-2,2,-\sqrt{6},\sqrt{6}\}$

15. $\{-1,2-3i,2+3i\}$

16. $\{-4,\dfrac{1}{3},2\}$

17. $\{-2i,2i,2,3\}$

18. $\{-\dfrac{3}{2},2,5\}$

19. $\{4,1-2i,1+2i\}$

20. $\{-3,-\dfrac{5}{2},1\}$

21. $\{-5i,5i,-2,1\}$

22. $\{1,2,7\}$

23. $\{-11,-1,1\}$

24. $\{-\sqrt{5},\sqrt{5},6\}$

25. 0 pos., 3 or 1 neg.

26. 1 pos., 2 or 0 neg.

27. 2 or 0 pos., 1 neg.

28. 0 pos., 3 or 1 neg.

29. 3 or 1 pos., 1 neg.

30. 0 pos., 3 or 1 neg.

31. 1 pos., 1 neg.

32. 3 or 1 pos., 1 neg.

33. 3 or 1 pos., 0 neg.

34. 0 pos., 0 neg.

35. 0 pos., 0 neg.

36. 1 pos., 3 or 1 neg.

37. $[-5,1]$

38. $[-2,3]$

39. $[-1,6]$

40. $[-6,1]$

Use synthetic division to identify upper and lower bounds of the real zeros of the following polynomials (answers will vary). See Example 3.

37. $f(x)=x^3+4x^2+x-4$

38. $f(x)=2x^3-3x^2-8x-3$

39. $f(x)=x^3-6x^2+3x+10$

40. $g(x)=x^3+6x^2+11x+6$

41. $f(x)=x^4-5x^3-2x^2+40x-48$

42. $g(x)=x^3+3x^2+3x+9$

43. $f(x)=x^4-25$

44. $g(x)=x^4-7x^3+5x^2+31x-30$

45. $f(x)=2x^3-7x^2-28x-12$

46. $g(x)=x^5+x^4-9x^3-x^2+20x-12$

Using your answers to the preceding problems, polynomial division, and the quadratic formula, if necessary, find all of the zeros of the following polynomials.

47. $f(x)=x^3+4x^2-x-4$

48. $f(x)=2x^3-3x^2-8x-3$

49. $f(x)=x^3-6x^2+3x+10$

50. $g(x)=x^3+6x^2+11x+6$

51. $f(x)=x^4-5x^3-2x^2+40x-48$

52. $g(x)=x^3+3x^2+3x+9$

53. $f(x)=x^4-25$

54. $g(x)=x^4-7x^3+5x^2+31x-30$

55. $f(x)=2x^3-7x^2-28x-12$

56. $g(x)=x^5+x^4-9x^3-x^2+20x-12$

Use the Intermediate Value Theorem to show that each of the following polynomials has a real zero between the indicated values. See Example 5.

57. $f(x)=5x^3-4x^2-31x-6$; -3 and -1

58. $f(x)=x^4-9x^2-14$; 1 and 4

59. $f(x)=x^4+2x^3-10x^2-14x+21$; 2 and 3

60. $f(x)=-x^3+2x^2+13x-26$; -4 and -3

Show that each of the following equations must have a solution between the indicated real numbers.

61. $14x+10x^2=x^4+2x^3+21$; 2 and 3

62. $x^3-2x^2=13(x-2)$; -4 and -3

41. $[-3,6]$

42. $[-3,1]$

43. $[-3,3]$

44. $[-2,7]$

45. $[-3,6]$

46. $[-4,3]$

47. $\{-4,-1,1\}$

48. $\{-1,-\dfrac{1}{2},3\}$

49. $\{-1,2,5\}$

50. $\{-3,-2,-1\}$

51. $\{2,3,\pm2\sqrt{2}\}$

52. $\{-3,\pm i\sqrt{3}\}$

53. $\pm\{\sqrt{5},i\sqrt{5}\}$

54. $\{-2,1,3,5\}$

55. $\{-2,-\dfrac{1}{2},6\}$

56. $\{-3,1,\pm2\}$

57. $f(-3)=-84$,
$f(-1)=16$

58. $f(1)=-22$,
$f(4)=98$

59. $f(2)=-15$,
$f(3)=24$

60. $f(-4)=18$,
$f(-3)=-20$

61. $f(2) = 15,$
$f(3) = -24$

62. $f(-4) = -18,$
$f(-3) = 20$

63. Answers will vary.

64. $\{-1, 2, 5\}$

65. $\left\{-6, \dfrac{1}{4}, 1\right\}$

66. $\left\{-\dfrac{7}{3}, \pm 1\right\}$

67. $\left\{-3, \pm\sqrt{3}, \dfrac{1}{2}\right\}$

68. $\left\{-1, -\dfrac{2}{5}, 2, \dfrac{7}{3}\right\}$

69. $\left\{\dfrac{-9 \pm \sqrt{41}}{2}, \dfrac{1}{2}, 2\right\}$

70. $\left\{\pm 3, -2, -\dfrac{1}{3}\right\}$

71. $\{-4, -3, 1, 2\}$

72. $\left\{-5, \pm 1, \dfrac{5}{2}, 4\right\}$

73. $\{\pm 5, \pm 6, \pm 8\}$

74. $\{-3, -2, -1\}$

75. $\{-3, 4, 5\}$

76. $\left\{-9, \pm\sqrt{2}\right\}$

77. $\left\{-\dfrac{2}{3}, \dfrac{1}{2}, 7\right\}$

78. $\left\{-3, -\dfrac{1}{2}, 8\right\}$

79. $\{-4, -3, 2\}$

80. $\left\{\pm 2, -\dfrac{3}{4}, 1\right\}$

81. $\left\{\pm i\sqrt{7}, 0, 3\right\}$

82. $\left\{\sqrt[3]{-2}, \pm\dfrac{\sqrt{6}}{2}, \dfrac{5}{3}\right\}$

83. $\left\{\sqrt[3]{-5}, -\dfrac{3}{4}, 2\right\}$

63. Construct a proof of the Rational Zero Theorem by following the suggested steps.

a. Assuming $\dfrac{p}{q}$ is a zero of the polynomial $f(x) = a_n x^n + a_{n-1} x^{n-1} + \ldots + a_1 x + a_0$, show that the equation $a_n\left(\dfrac{p}{q}\right)^n + a_{n-1}\left(\dfrac{p}{q}\right)^{n-1} + \ldots + a_1\left(\dfrac{p}{q}\right) + a_0 = 0$ can be written in the form $a_n p^n + a_{n-1} p^{n-1} q + \ldots + a_1 p q^{n-1} = -a_0 q^n$.

b. It can be assumed that $\dfrac{p}{q}$ is written in lowest terms (that is, the greatest common divisor of p and q is 1). By examining the left-hand side of the last equation above, show that p must be a divisor of the right-hand side, and hence a factor of a_0.

c. By rearranging the equation so that all terms with a factor of q are on one side, use a similar argument to show that q must be a factor of a_n.

Using any of the methods discussed in this section as guides, find all of the real zeros of the following functions.

64. $f(x) = 3x^3 - 18x^2 + 9x + 30$

65. $f(x) = -4x^3 - 19x^2 + 29x - 6$

66. $f(x) = 3x^5 + 7x^4 + 12x^3 + 28x^2 - 15x - 35$

67. $f(x) = 2x^4 + 5x^3 - 9x^2 - 15x + 9$

68. $f(x) = -15x^4 + 44x^3 + 15x^2 - 72x - 28$

69. $f(x) = 2x^4 + 13x^3 - 23x^2 - 32x + 20$

70. $f(x) = 3x^4 + 7x^3 - 25x^2 - 63x - 18$

71. $f(x) = x^5 + 7x^4 + 5x^3 - 43x^2 - 42x + 72$

72. $f(x) = 2x^5 - 3x^4 - 47x^3 + 103x^2 + 45x - 100$

73. $f(x) = x^6 - 125x^4 + 4804x^2 - 57,600$

Using any of the methods discussed in this section as guides, solve the following equations.

74. $x^3 + 6x^2 + 11x = -6$

75. $x^3 - 7x = 6\left(x^2 - 10\right)$

76. $x^3 + 9x^2 = 2x + 18$

77. $6x^3 + 14 = 41x^2 + 9x$

78. $4x^3 = 18x^2 + 106x + 48$

79. $3x^3 + 15x^2 - 6x = 72$

80. $8x^4 + 24 + 8x = 2x^3 + 38x^2$

81. $x^4 + 7x^2 = 3x^3 + 21x$

82. $6x^6 - 10x^5 - 9x^4 + 27x^3 = 20x^2 + 18x - 30$

83. $4x^5 - 5x^4 + 20x^2 = 6x^3 + 25x + 30$

The Fundamental Theorem of Algebra

TOPICS

1. The Fundamental Theorem of Algebra

2. Multiple zeros and their geometric meaning

3. Conjugate pairs of zeros

4. Summary of polynomial methods

TOPIC

The Fundamental Theorem of Algebra

We are now ready to tie together all that we have learned about polynomials, and we begin with a powerful but deceptively simple-looking statement called the Fundamental Theorem of Algebra.

THEOREM

The Fundamental Theorem of Algebra

If p is a polynomial of degree n, with $n \geq 1$, then p has **at least one zero**. That is, the equation $p(x) = 0$ has at least one solution. It is important to note that the zero of p, and consequently the solution of $p(x) = 0$, may be a nonreal complex number.

Mathematicians began to suspect the truth of this statement in the first half of the 17th century, but a convincing proof did not appear until the German mathematician Carl Friedrich Gauss (1777–1855) provided one in his doctoral dissertation in 1799 at just 22 years of age!

Although the proof of the Fundamental Theorem of Algebra is beyond the scope of this text, we can use it to prove a consequence that summarizes much of the previous three sections. The following theorem has great implications in solving polynomial equations and in graphing real-coefficient polynomial functions. It tells us that our goal of factoring a polynomial completely is always at least theoretically possible.

THEOREM

The Linear Factors Theorem

Given the polynomial $p(x) = a_n x^n + a_{n-1} x^{n-1} + \ldots + a_1 x + a_0$, where $n \geq 1$ and $a_n \neq 0$, p can be factored as $p(x) = a_n (x - c_1)(x - c_2) \cdots (x - c_n)$, where $c_1, c_2, \ldots, c_n$ are constants (possibly nonreal complex constants and not necessarily distinct). In other words, **an nth-degree polynomial can be factored as a product of n linear factors**.

Proof:

The Fundamental Theorem of Algebra tells us that $p(x)$ has at least one zero; call it c_1. Using the division algorithm, we know $(x - c_1)$ is a factor of p, and we can write

$$p(x) = (x - c_1)q_1(x),$$

where $q_1(x)$ is a polynomial of degree $n - 1$. Note that the leading coefficient of q_1 must be a_n, since we divided p by $(x - c_1)$, a polynomial with leading coefficient of 1.

If the degree of q_1 is 0 (that is, if $n = 1$), then $q_1(x) = a_1$ and $p(x) = a_1(x - c_1)$. Otherwise, q_1 is of degree 1 or larger, and by the Fundamental Theorem of Algebra, q_1 itself has at least one zero; call it c_2. By the same reasoning, then, we can write

$$p(x) = (x - c_1)(x - c_2)q_2(x),$$

where q_2 is a polynomial of degree $n - 2$, also with leading coefficient a_n. We can perform this process a total of n times (and no more), at which point we have the desired result:

$$p(x) = a_n(x - c_1)(x - c_2)\cdots(x - c_n)$$

CAUTION!

The Linear Factors Theorem *does not* tell us the following things:

1. The theorem does not tell us that a polynomial has all real zeros. Some, or all, of the constants $c_1, c_2, \ldots, c_n$ may be nonreal complex numbers.

2. The theorem does not tell us that a polynomial has n *distinct* zeros. Some, or all, of the constants $c_1, c_2, \ldots, c_n$ may be identical.

3. The theorem does tell us that any polynomial can be written as a product of linear factors; it does not tell us *how to determine* the linear factors.

In the case where all of the coefficients of $p(x) = a_n x^n + a_{n-1}x^{n-1} + \ldots + a_1 x + a_0$ are real, the Linear Factors Theorem tells us that the graph of p has *at most n x*-intercepts (and can only have *exactly n x*-intercepts if all n zeros are real and distinct). Indirectly, the theorem tells us something more: the graph of p can have at most $n - 1$ *turning points*. A **turning point** of a graph is a point where the graph changes behavior from decreasing to increasing or vice versa. These facts are summarized below.

THEOREM

Interpreting the
Linear Factors
Theorem

The graph of an nth**-degree polynomial function has at most n x-intercepts and at most $n-1$ turning points**. This also means that an nth-degree polynomial function has at most n zeros.

Figure 1 illustrates that the degree of a polynomial only gives us an upper bound on the number of x-intercepts and turning points. Note that the graph of the 4th-degree polynomial f has just two x-intercepts and only one turning point, while the graph of the 3rd-degree polynomial g has three x-intercepts and two turning points.

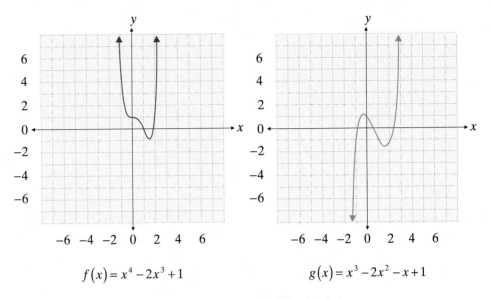

$$f(x) = x^4 - 2x^3 + 1 \qquad g(x) = x^3 - 2x^2 - x + 1$$

Figure 1: **x-Intercepts and Turning Points**

TOPIC 2

Multiple Zeros and Their Geometric Meaning

We know that, for example, the functions $(x-3)$, $(x-3)^2$, and $(x-3)^{15}$ are not the same and do not behave the same way. Yet, they have the same set of zeros: $\{3\}$. We need a way to classify functions in which a particular linear factor appears more than once.

DEFINITION

Multiplicity of Zeros

If the linear factor $(x-c)$ appears $k > 0$ times in the factorization of a polynomial (or as $(x-c)^k$), we say the number c is a **zero of multiplicity k**.

If we are graphing a polynomial p for which c is a real zero of multiplicity k, then c is certainly an x-intercept of the graph of p, but the behavior of the graph near c depends on two characteristics:

1. Whether k is equal to or greater than 1.

2. Whether k is even or odd.

Before generalizing these concepts, let's look at a few examples.

Consider the function $f(x) = (x-1)^3$. We see that 1 is the only zero of f, and it is a zero of multiplicity 3. From our work with transformations of functions, we know the graph of f is the basic cubic shape shifted to the right by 1 unit, as shown in Figure 2. Note that the graph of f appears to "flatten out" near the zero.

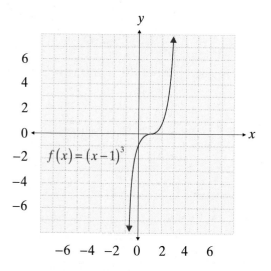

Figure 2: Zero of Multiplicity 3

Compare the behavior of $f(x) = (x-1)^3$ near its zero to the behavior of the function $g(x) = (x+2)^4$ near its own zero of -2, a zero of multiplicity 4. Figure 3 shows the graph of g. Again, the graph of g flattens out near the zero. Unlike the zero of multiplicity 3, in this case we see that the function has the same sign before and after the zero.

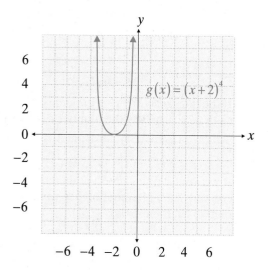

Figure 3: Zero of Multiplicity 4

PROPERTIES

Geometric Meaning
of Multiplicity

If c is a real zero of multiplicity k of a polynomial p (alternatively, if $(x-c)^k$ is a factor of p), the graph of p will touch the x-axis at $(c, 0)$ and:

- Cross through the x-axis if k is odd

- Stay on the same side of the x-axis if k is even

Further, if $k > 1$, the graph of p will "flatten out" near $(c, 0)$.

With an understanding of how a zero's multiplicity affects a polynomial, constructing a reasonably accurate sketch of the graph becomes easier.

EXAMPLE 1

Graphing Polynomial
Functions

Sketch the graph of the polynomial $f(x) = (x+2)(x+1)^2(x-3)^3$.

Solution:

We begin with the steps from before. Since f has even degree (6) and a positive leading coefficient (1), we know the end behavior: $f(x) \to \infty$ as $x \to \pm\infty$.

Then plug in $x = 0$ to find the y-intercept.

$$f(0) = (0+2)(0+1)^2(0-3)^3$$

$$= -54$$

Thus, f has its y-intercept at $(0, -54)$.

Using our knowledge of multiplicity, we can determine that f crosses the x-axis at -2 and 3, but not at -1, and that the graph of f flattens out near -1 and 3. Putting all of this together, we obtain the sketch below.

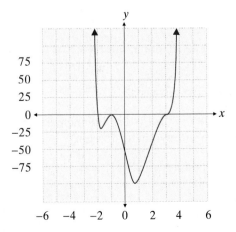

Note the extreme difference in the scales of the two axes.

To fill in portions of the graph between zeros accurately, we must still compute a few values of the function. For instance, $f(1) = -96$ and $f(2) = -36$. This also gives us a way to double-check our analysis of the behavior on either side of a zero.

TOPIC 3   Conjugate Pairs of Zeros

If all of the coefficients of a polynomial $p(x)$ are real, then p is a function that transforms real numbers into other real numbers, and consequently p can be graphed in the Cartesian plane. Nonetheless, it is very possible that some of the constants $c_1, c_2, \ldots, c_n$ in the factored form of p, $p(x) = a_n(x - c_1)(x - c_2)\cdots(x - c_n)$, might be nonreal complex numbers. For example, $x^2 + 1 = (x - i)(x + i)$. It turns out that such complex roots must occur in pairs.

THEOREM

The Conjugate Roots Theorem

Let $p(x) = a_n x^n + a_{n-1} x^{n-1} + \ldots + a_1 x + a_0$ be a polynomial with only real coefficients. If the complex number $a + bi$ is a zero of p, then so is the complex number $a - bi$. In terms of the linear factors of p, this means that if $x - (a + bi)$ is a factor of p, then so is $x - (a - bi)$.

We can make use of this fact in several ways. For instance, if we are given one nonreal zero of a real-coefficient polynomial, we automatically know a second zero. The theorem is also useful when constructing polynomials with specified properties.

Another consequence of the Conjugate Roots Theorem is that every polynomial with real coefficients can be factored into a product of linear factors and irreducible quadratic factors with real coefficients. Each irreducible quadratic factor with real coefficients corresponds to the product of two factors of the form $x - (a + bi)$ and $x - (a - bi)$.

EXAMPLE 2

Factoring Polynomials

Given that $4 - 3i$ is a zero of the polynomial $f(x) = x^4 - 8x^3 + 200x - 625$, factor f completely.

Solution:

By the Conjugate Roots Theorem, since $4 - 3i$ is a zero of f, we know that $4 + 3i$ is a zero as well.

This gives us two ways to proceed:

1. We could divide f by $x - (4 - 3i)$ and then divide the result by $x - (4 + 3i)$ (most efficiently done with synthetic division).

2. Or, we could multiply $x - (4 - 3i)$ and $x - (4 + 3i)$ and divide f by their product (using polynomial long division).

In either case, we will be left with a quadratic polynomial that we know we can factor.

If we take the second approach, the first step is as follows:

$$\left(x-(4-3i)\right)\left(x-(4+3i)\right)=\left(x-4+3i\right)\left(x-4-3i\right)$$

$$= x^2-4x-3ix-4x+16+12i+3ix-12i-9i^2$$

$$= x^2-8x+25$$

Now we divide f by this product:

$$
\begin{array}{r}
x^2 \qquad\quad -25 \\
x^2-8x+25{\overline{\smash{\big)}\,x^4-8x^3+\;0x^2+200x-625}} \\
\underline{-\left(x^4-8x^3+25x^2\right)} \\
-25x^2+200x-625 \\
\underline{-\left(-25x^2+200x-625\right)} \\
0
\end{array}
$$

The quotient, x^2-25, is a difference of two squares and is easily factored, giving us our final result:

$$f(x)=(x-4+3i)(x-4-3i)(x-5)(x+5)$$

EXAMPLE 3

Constructing Polynomials

Construct a 4th-degree real-coefficient polynomial function f with zeros of 2, –5, and $1+i$ such that $f(1)=12$.

Solution:

Since $1+i$ is one of the zeros and f is to have only real coefficients, $1-i$ must be a zero as well by the Conjugate Roots Theorem. Based on this, f must be of the form

$$f(x)=a_n\left(x-(1+i)\right)\left(x-(1-i)\right)(x-2)(x+5)$$

for some real constant a_n. Of course, we must find a_n so that $f(1)=12$. In order to do this, we begin by multiplying out $\left(x-(1+i)\right)\left(x-(1-i)\right)$:

$$\left(x-(1+i)\right)\left(x-(1-i)\right)=(x-1-i)(x-1+i)$$

$$= x^2-2x+2$$

We then plug in $x=1$ and $f(1)=12$, then solve for a_n.

$$f(1)=a_n\left(1^2-2(1)+2\right)(1-2)(1+5) \qquad \text{Substitute } x=1.$$

$$12=a_n(1)(-1)(6) \qquad \text{Substitute } f(1)=12.$$

$$12=-6a_n \qquad \text{Solve for } a_n.$$

$$-2=a_n$$

In factored form, the polynomial is $f(x)=-2(x-1-i)(x-1+i)(x-2)(x+5)$, which, if multiplied out, is $f(x)=-2x^4-2x^3+28x^2-52x+40$.

TOPIC 4   Summary of Polynomial Methods

All of the methods that you have learned in this chapter may be useful in solving a particular polynomial problem, whether it focuses on graphing a polynomial function, solving a polynomial equation, or solving a polynomial inequality. Now that all of the methods have been introduced, it makes sense to summarize them and see how they contribute to the big picture.

Recall that, in general, an nth-degree polynomial function has the form

$$p(x) = a_n x^n + a_{n-1} x^{n-1} + \ldots + a_1 x + a_0,$$

where $a_n \neq 0$ and any (or all) of the coefficients may be nonreal complex numbers. Keep in mind that it only makes sense to talk about graphing p in the Cartesian plane if all of the coefficients are real. Similarly, a polynomial inequality in which p appears on one side only makes sense if all the coefficients are real. For this reason, most of the polynomials in this text have only real coefficients.

Nevertheless, complex numbers often arise when working with polynomials, as some of the numbers $c_1, c_2, \ldots, c_n$ in the factored form of p,

$$p(x) = a_n (x - c_1)(x - c_2) \cdots (x - c_n),$$

may be nonreal even if all of $a_1, a_2, \ldots, a_n$ are real. The fact that p can, in principle, be factored is a direct consequence of the Fundamental Theorem of Algebra.

Factoring p into a product of linear factors as shown is the central point in solving a polynomial equation and (when the coefficients of p are real) in graphing a polynomial and solving a polynomial inequality. Specifically,

- The solutions of the polynomial equation $p(x) = 0$ are the numbers $c_1, c_2, \ldots, c_n$.

- When $a_1, a_2, \ldots, a_n$ are all real, the x-intercepts of the graph of p are the real numbers in the list $c_1, c_2, \ldots, c_n$. If a given c_i appears in the list k times, it is a *zero of multiplicity* k. If an x-intercept of p is of multiplicity k, the behavior of p near that x-intercept depends on whether k is even or odd. Any nonreal zeros in the list must appear in conjugate pairs.

- When $a_1, a_2, \ldots, a_n$ are all real, the solution of the polynomial inequality $p(x) > 0$ consists of all the open intervals on the x-axis where the graph of p lies strictly above the x-axis. The solution of $p(x) < 0$ consists of all the open intervals where the graph of p lies strictly below the x-axis. The solutions of $p(x) \geq 0$ and $p(x) \leq 0$ consist of closed intervals. Testing a single point in each interval suffices to determine the sign of p on that interval.

The remaining topics discussed in this chapter are observations and techniques that aid us in filling in the details of the big picture.

- The observation that the degree of a polynomial and the sign of its leading coefficient tell us how the graph of the polynomial behaves as $x \to -\infty$ and as $x \to \infty$.

- The observation that the graph of p crosses the y-axis at the easily computed point $(0, p(0))$.

- The technique of polynomial long division, useful in dividing one polynomial by another of the same or smaller degree.

- The technique of synthetic division, a shortcut that applies when dividing a polynomial by a polynomial of the form $x - k$. Recall that the remainder of this division is the value $p(k)$.

- The Rational Zero Theorem, which provides a list of potential rational zeros for polynomials with integer coefficients.

- Descartes' Rule of Signs, which provides guidance on the number of positive and negative real zeros that a real-coefficient polynomial might have.

- The Upper and Lower Bounds rule, which indicates an interval in which to search for all the zeros of a real-coefficient polynomial.

- The Intermediate Value Theorem, which can be used to "hone in" on a real zero of a given polynomial.

As you solve various polynomial problems, try to keep the big picture in mind. Often, it is useful to literally keep a picture, namely the graph of the polynomial, in mind even if the problem does not specifically involve graphing.

Exercises

Throughout these exercises, a graphing calculator or a computer algebra system may be helpful in identifying zeros and in checking your graphing, if permitted by your instructor.

Sketch the graph of each factored polynomial. See Example 1.

1. $f(x) = (x+1)^4 (x-2)^3 (x-1)$

2. $g(x) = -x^3 (x-1)(x+2)^2$

3. $f(x) = -x(x+2)(x-1)^2$

4. $g(x) = (x+2)(x-1)^3$

5. $f(x) = (x-1)^4 (x-2)(x-3)$

6. $g(x) = (x+1)^2 (x-2)^3$

7. $f(x) = (x-4)(x+2)^2 (x-3)^3$

8. $g(x) = (x+3)(x-1)^5$

1.

2.

3.

4.

5.

6.

7.

8.

Use all available methods to factor each of the following polynomials completely, and then sketch the graph of each one. See Example 1.

9.

$f(x) = (x+2)^3 (x-1)^2$

10.

$p(x) = 2\left(x - \dfrac{5}{2}\right)(x+1)^2$

11.

$s(x) = -(x-3)(x+1)$
$\times \left(x - \sqrt{5}\right)\left(x + \sqrt{5}\right)$

12.

$f(x) = -(x-2)^3$

13.

$H(x) = (x-2)(x+1) \times$
$\left(x - \sqrt{3}\right)\left(x + \sqrt{3}\right)$

14.

$h(x) = (x-1)^2 (x-3)^3$

9. $f(x) = x^5 + 4x^4 + x^3 - 10x^2 - 4x + 8$ **10.** $p(x) = 2x^3 - x^2 - 8x - 5$

11. $s(x) = -x^4 + 2x^3 + 8x^2 - 10x - 15$ **12.** $f(x) = -x^3 + 6x^2 - 12x + 8$

13. $H(x) = x^4 - x^3 - 5x^2 + 3x + 6$

14. $h(x) = x^5 - 11x^4 + 46x^3 - 90x^2 + 81x - 27$

15. $f(x) = 2x^3 + 11x^2 + 20x + 12$ **16.** $g(x) = x^4 + 3x^3 - 5x^2 - 21x - 14$

Use all available methods to solve each polynomial equation. Use the Linear Factors Theorem to make sure you find the appropriate number of solutions, counting multiplicity.

17. $x^5 + 4x^4 + x^3 = 10x^2 + 4x - 8$ **18.** $x^4 + 15 = 2x^3 + 8x^2 - 10x$

19. $x^4 + x^3 + 3x^2 + 5x - 10 = 0$ **20.** $x^3 - 9x^2 = 30 - 28x$

21. $x^5 + x^4 - x^3 + 7x^2 - 20x + 12 = 0$ **22.** $2x^4 - 5x^3 - 2x^2 + 15x = 0$

23. $x^5 + 15x^3 + 16 = x^4 + 15x^2 + 16x$ **24.** $x^3 - 5 = 5x^2 - 9x$

Use all available methods (in particular, the Conjugate Roots Theorem, if applicable) to factor each of the following polynomials completely, making use of the given zero if one is given. See Example 2.

25. $f(x) = x^4 - 9x^3 + 27x^2 - 15x - 52$; $3 - 2i$ is a zero.

26. $g(x) = x^3 - (1-i)x^2 - (8-i)x + (12-6i)$; $2-i$ is a zero.

27. $f(x) = x^3 - (2+3i)x^2 - (1-3i)x + (2+6i)$; 2 is a zero.

28. $p(x) = x^4 - 2x^3 + 14x^2 - 8x + 40$; $2i$ is a zero.

29. $n(x) = x^4 - 4x^3 + 6x^2 + 28x - 91$; $2+3i$ is a zero.

30. $G(x) = x^4 - 14x^3 + 98x^2 - 686x + 2401$; $7i$ is a zero.

31. $f(x) = x^4 - 3x^3 + 5x^2 - x - 10$

32. $g(x) = x^6 - 8x^5 + 25x^4 - 40x^3 + 40x^2 - 32x + 16$

15.

$f(x) = (x+2)^2 (2x+3)$

16.

$g(x) = (x+1)(x+2)$
$\left(x - \sqrt{7}\right)\left(x + \sqrt{7}\right)$

17. $\{-2, 1\}$

18. $\{-1, 3, \pm\sqrt{5}\}$

19. $\left\{-2, 1, \pm i\sqrt{5}\right\}$

20. $\left\{3, 3 \pm i\right\}$

21. $\left\{-3, 1, \pm 2i\right\}$

22. $\left\{0, -\dfrac{3}{2}, 2 \pm i\right\}$

23. $\left\{\pm 1, \pm 4i\right\}$

24. $\left\{1, 2 \pm i\right\}$

25. $(x-3+2i)\times$
$(x-3-2i)\times$
$(x-4)(x+1)$

26. $(x-2+i)(x+3)$
$\times(x-2)$

27. $(x-1-3i)\times$
$(x-2)(x+1)$

28. $(x-1-3i)\times$
$(x-1+3i)\times$
$(x-2i)(x+2i)$

29. $\left(x+\sqrt{7}\right)\left(x-\sqrt{7}\right)\times$
$(x-2+3i)$
$(x-2-3i)$

30. $(x+7i)(x-7i)\times$
$(x-7)^2$

31. $(x-2)(x+1)\times$
$(x-1+2i)\times$
$(x-1-2i)$

32. $(x-2)^4(x-i)(x+i)$

33. $x(x-3)(x-1)\times$
$(x+11)$

34. $(x+3)^2(x-3)^2$
$(x-1)$

35. $(x-2)(x-2+4i)\times$
$(x-2-4i)$

33. $r(x) = x^4 + 7x^3 - 41x^2 + 33x$

34. $d(x) = x^5 - x^4 - 18x^3 + 18x^2 + 81x - 81$

35. $P(x) = x^3 - 6x^2 + 28x - 40$

36. $g(x) = x^6 - x^4 - 16x^2 + 16$

Construct polynomial functions with the stated properties. See Example 3.

37. Third-degree, only real coefficients, -1 and $5+i$ are two of the zeros, y-intercept is -52.

38. Fourth-degree, only real coefficients, $\sqrt{7}$ and $i\sqrt{5}$ are two of the zeros, y-intercept is -35.

39. Fifth-degree, 1 is a zero of multiplicity 3, -2 is the only other zero, leading coefficient is 2.

40. Fifth-degree, only real coefficients, 0 is the only real zero, $1+i$ is a zero of multiplicity 1, leading coefficient is 1.

41. Fourth-degree, only real coefficients, x-intercepts are 0 and 6, $-2i$ is a zero, leading coefficient is 3.

42. Fifth-degree, -2 is a zero of multiplicity 2, another integer is a zero of multiplicity 3, y-intercept is 108, leading coefficient is 1.

43. Third-degree, only real coefficients, -4 and $3+i$ are two of the zeros, y-intercept is -40.

44. Fifth-degree, 1 is a zero of multiplicity 4, -2 is the only other zero, leading coefficient is 4.

45. Third-degree, only real coefficients, -4 and $4+i$ are two of the zeros, y-intercept is -68.

36. $(x+2i)(x-2i)$
$(x+1)(x-1)$
$(x+2)(x-2)$

37. $f(x) = -2x^3 + 18x^2$
$-32x - 52$

38. $f(x) = x^4 - 2x^2 - 35$

39. $f(x) = 2x^5 + 2x^4$
$-10x^3 - 2x^2$
$+16x - 8$

40. $f(x) = x^5 - 2x^4 + 2x^3$

41. $f(x) = 3x^4 - 18x^3$
$+12x^2 - 72x$

42. $f(x) = x^5 + 13x^4$
$+67x^3 + 171x^2 + 216x$
$+108$

43. $f(x) = -x^3 + 2x^2 + 14x - 40$

44. $f(x) = 4x^5 - 8x^4 - 8x^3$
$+32x^2 - 28x + 8$

45. $f(x) = -x^3 + 4x^2 + 15x - 68$

Solve the following application problems.

46. An open-top box is to be constructed from a 10 inch by 18 inch sheet of tin by cutting out squares from each corner as shown and then folding up the sides. Let $V(x)$ denote the volume of the resulting box.

    a. Write $V(x)$ as a product of linear factors.

    b. For which values of x is $V(x) = 0$?

    c. Which answers from part **b.** are physically possible?

47. An open-top box is to be constructed from a 10 inch by 15 inch sheet of tin by cutting out squares from each corner and then folding up the sides. Let $V(x)$ denote the volume of the resulting box.

    a. Write $V(x)$ as a product of linear factors.

    b. For which values of x is $V(x) = 0$?

    c. Which of your answers from part **b.** are physically possible?

48. An open-top box is to be constructed from a 9 inch by 17 inch sheet of tin by cutting out squares from each corner and then folding up the sides. Let $V(x)$ denote the volume of the resulting box.

    a. Write $V(x)$ as a product of linear factors.

    b. For which values of x is $V(x) = 0$?

    c. Which of your answers from part **b.** are physically possible?

49. Assume $f(x)$ is an nth-degree polynomial with real coefficients. Explain why the following statement is true: If n is even, the number of turning points is odd and if n is odd, the number of turning points is even.

46. a. $V(x) = 4x(5-x)(9-x)$
     b. $x = 0, x = 5, x = 9$
     c. $x = 0$ and $x = 5$

47. a. $V(x) = 2x(15-2x)(5-x)$
     b. $x = 0, x = 5, x = \dfrac{15}{2}$
     c. $x = 0$ and $x = 5$

48. a. $V(x) = x(17-2x)(9-2x)$
     b. $x = 0, x = \dfrac{9}{2}, x = \dfrac{17}{2}$
     c. $x = 0$ and $x = \dfrac{9}{2}$

49. Answers will vary.

4.5

Rational Functions and Rational Inequalities

TOPICS

1. Definitions and useful notation

2. Vertical asymptotes

3. Horizontal and oblique asymptotes

4. Graphing rational functions

5. Solving rational inequalities

TOPIC 1

Definitions and Useful Notation

The study of polynomials leads directly to a study of rational functions, which are ratios of polynomials. Since rational functions can have variables in the denominators of fractions, their behavior can be significantly more complex than that of polynomials.

DEFINITION

Rational Functions

A **rational function** is a function that can be written in the form

$$f(x) = \frac{p(x)}{q(x)},$$

where $p(x)$ and $q(x)$ are polynomial functions and $q(x) \neq 0$. Even though q is not allowed to be identically zero, there will often be values of x for which $q(x)$ is zero, and at these values the function is undefined. Consequently, the **domain of f** consists of all real numbers except those for which $q(x) = 0$.

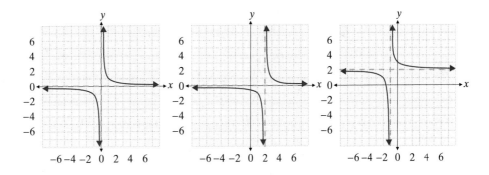

Figure 1: Graphs of Three Rational Functions

Each of the three graphs in Figure 1 has a new feature: vertical and horizontal dashed lines. These dashed lines are *not* part of the function, they are examples of *asymptotes*, and they serve as guides to understanding the function. Roughly speaking, an asymptote is a line that the graph of a function approaches, but does not touch. Three kinds of asymptotes will appear in our study of rational functions: vertical, horizontal, and oblique.

DEFINITION

Vertical Asymptotes

The vertical line $x = c$ is a **vertical asymptote** of a function f if $f(x)$ increases in magnitude without bound as x approaches c. Examples of vertical asymptotes appear in Figure 2. The graph of a rational function cannot intersect a vertical asymptote.

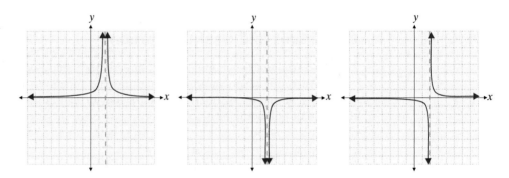

Figure 2: Vertical Asymptotes

To understand how vertical asymptotes arise, let's observe what happens to the function $f(x) = \dfrac{1}{x}$ as x gets closer to 0 from the left and from the right.

x	$f(x) = \dfrac{1}{x}$		x	$f(x) = \dfrac{1}{x}$
-1	-1		1	1
-0.1	-10		0.1	10
-0.01	-100		0.01	100
-0.001	-1000		0.001	1000
-0.0001	$-10,000$		0.0001	$10,000$
-0.00001	$-100,000$		0.00001	$100,000$

Table 1: Values of $f(x) = \dfrac{1}{x}$ as x Approaches 0

We can see that as x gets closer and closer to 0 (where the function is undefined), the value of f increases in magnitude without bound. The graph reflects this, as the curve gets steeper and steeper, never touching the line $x = 0$. This type of behavior occurs in all rational functions as x approaches a value where the function is undefined.

DEFINITION

Horizontal
Asymptotes

The horizontal line $y = c$ is a **horizontal asymptote** of a function f if $f(x)$ approaches the value c as $x \to -\infty$ or as $x \to \infty$. Examples of horizontal asymptotes appear in Figure 3. The graph of a rational function may intersect a horizontal asymptote near the origin, but will eventually approach the asymptote from one side only as x increases in magnitude.

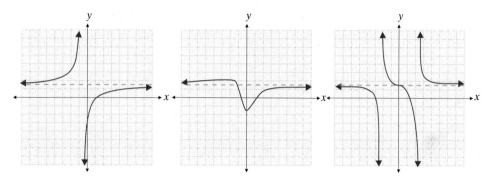

Figure 3: Horizontal Asymptotes

DEFINITION

Oblique Asymptotes

A nonvertical, nonhorizontal line may also be an asymptote of a function f. Examples of **oblique** (or **slant**) **asymptotes** appear in Figure 4. Again, the graph of a rational function may intersect an oblique asymptote near the origin, but will eventually approach the asymptote from one side only as $x \to \infty$ or $x \to -\infty$.

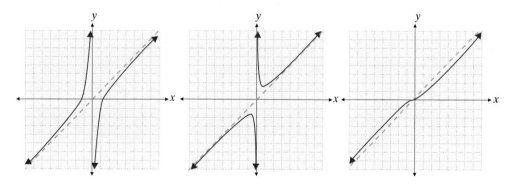

Figure 4: Oblique Asymptotes

As Figures 2, 3, and 4 illustrate, the behavior of rational functions with respect to asymptotes can vary considerably. In order to describe the behavior of a given rational function more easily, we have specific asymptote notation.

DEFINITION

Asymptote Notation

The notation $x \rightarrow c^-$ is used when describing the behavior of a graph as x approaches the value c from the left (the negative side). The notation $x \rightarrow c^+$ is used when describing behavior as x approaches c from the right (the positive side). The notation $x \rightarrow c$ is used when describing behavior that is the same on both sides of c.

Figure 5 illustrates how the above notation can be used to describe the behavior of functions.

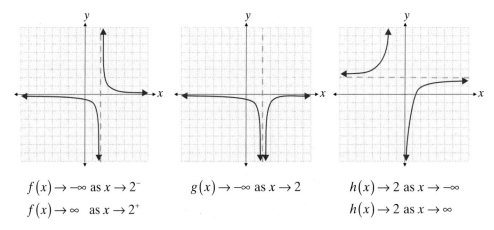

$$f(x) \rightarrow -\infty \text{ as } x \rightarrow 2^-$$
$$f(x) \rightarrow \infty \text{ as } x \rightarrow 2^+$$

$$g(x) \rightarrow -\infty \text{ as } x \rightarrow 2$$

$$h(x) \rightarrow 2 \text{ as } x \rightarrow -\infty$$
$$h(x) \rightarrow 2 \text{ as } x \rightarrow \infty$$

Figure 5: Asymptote Notation

TOPIC 2

Vertical Asymptotes

With the above notation and examples as background, we are ready to delve into the details of identifying asymptotes for rational functions.

THEOREM

Equations for Vertical Asymptotes

If the rational function $f(x) = \dfrac{p(x)}{q(x)}$ has been written in reduced form (so that p and q have no common factors), the vertical line $x = c$ is a **vertical asymptote** of f if and only if c is a zero of the polynomial q. In other words, f has vertical asymptotes at the x-intercepts of q.

Note that the numerator of a rational function is irrelevant in locating the vertical asymptotes, assuming that all common factors in the fraction have been canceled. However, if the numerator and denominator share a common factor of $(x-c)$, the value c will be out of the domain of the function, but the line $x = c$ will not be a vertical asymptote.

EXAMPLE 1

Vertical Asymptotes

Find the domains and the equations for the vertical asymptotes of the following functions.

a. $f(x) = \dfrac{32}{x+2}$ **b.** $g(x) = \dfrac{x^2+1}{x^2+2x-15}$ **c.** $h(x) = \dfrac{x^2-x}{x-1}$

Solutions:

Note:
We must always find the domain before cancelling common factors. Even if a zero is removed from the denominator when finding the reduced form, that value is not part of the domain.

a. To answer both questions, we need to calculate the zeros of the denominator (which is already in factored form).

$$x+2 = 0$$
$$x = -2$$

Since the function f is in reduced form, this zero is the only point excluded from the domain. This means the domain of f is $(-\infty,-2)\cup(-2,\infty)$.

Further, we know that f has a vertical asymptote of $x = -2$.

b. In this case, we need to factor the denominator before calculating the domain and vertical asymptotes.

$$g(x) = \frac{x^2+1}{x^2+2x-15} = \frac{x^2+1}{(x+5)(x-3)}$$

Since the values -5 and 3 both make the denominator zero, the domain of g is $(-\infty,-5)\cup(-5,3)\cup(3,\infty)$.

The numerator is a sum of two squares and cannot be factored, so the rational function is already in reduced form. This means that the equations of the two vertical asymptotes are $x = -5$ and $x = 3$.

c. The denominator is already in factored form, so we can see that its only zero is at $x = 1$. Thus, the domain of h is $(-\infty,1)\cup(1,\infty)$.

Now that we have found the domain, we can look for common factors to cancel.

$$h(x) = \frac{x^2-x}{x-1} = \frac{x(x-1)}{x-1}$$
$$= x$$

By cancelling the common factor of $(x-1)$, we have the reduced form $h(x) = x$, which applies only for values in the domain of h (all real numbers except for 1). Since the reduced form of h has no denominator, h has no vertical asymptotes.

TOPIC 3

Horizontal and Oblique Asymptotes

To determine horizontal and oblique asymptotes, we are interested in the behavior of a function $f(x)$ as $x \to -\infty$ and as $x \to \infty$. If f is a rational function, f is a ratio of two polynomials p and q, so we can begin by considering the effect p and q have on one another. Consider a rational function $f(x) = \dfrac{p(x)}{q(x)}$ in which the polynomial p has degree n and the polynomial q has degree m. We know from long division that f equals a polynomial of degree $n - m$ plus a remainder term. The key fact is that the behavior of rational functions as the magnitude of x gets very large tends to approach the behavior of the *quotient*. Thus, the horizontal and oblique asymptotes of a rational function depend on the difference in degrees between p and q.

THEOREM

Equations for
Horizontal and
Oblique Asymptotes

Let $f(x) = \dfrac{p(x)}{q(x)}$ be a rational function, where p is an nth-degree polynomial with leading coefficient a_n and q is an mth-degree polynomial with leading coefficient b_m, and $p(x)$ and $q(x)$ have no common factors other than constants. Then the asymptotes of f are found as follows:

1. If $n < m$, the horizontal line $y = 0$ (the x-axis) is the **horizontal asymptote** for f.

2. If $n = m$, the horizontal line $y = \dfrac{a_n}{b_m}$ is the **horizontal asymptote** for f.

3. If $n = m + 1$, the line $y = g(x)$ is an **oblique asymptote** for f, where g is the quotient polynomial obtained by dividing p by q. (The remainder polynomial is irrelevant.)

4. If $n > m + 1$, there is **no** straight line **horizontal** or **oblique asymptote** for f.

EXAMPLE 2

Horizontal and
Oblique Asymptotes

Find the equation for the horizontal or oblique asymptote of the following functions.

a. $f(x) = \dfrac{x^2 + 1}{x^2 + 2x - 15}$

b. $g(x) = \dfrac{x^3 + x^2 + 2x + 2}{x^2 + 9}$

c. $h(x) = \dfrac{3x^4 + 10x - 7}{x^6 + x^5 - x^2 - 1}$

d. $j(x) = \dfrac{2x^4 - 3x^2 + 8}{x^2 - 25}$

Note:
Always begin by comparing the degrees of the numerator and the denominator.

Solutions:

a. First, note that the degree of the numerator of f equals the degree of the denominator of f. This means that the line $y = \dfrac{a_n}{b_m} = 1$ is the horizontal asymptote of f.

b. Here, the degree of the numerator is one more than the degree of the denominator, so we know g has an oblique asymptote, equal to the quotient polynomial of the numerator and denominator. To find it, we need to perform polynomial division:

$$x^2 + 9 \overline{\smash{\big)}\, x^3 + x^2 + 2x + 2} \quad \overset{\displaystyle x + 1}{}$$
$$\underline{-\left(x^3 + 0x^2 + 9x\right)}$$
$$x^2 - 7x + 2$$
$$\underline{-\left(x^2 + 0x + 9\right)}$$
$$-7x - 7$$

This tells us that $g(x) = x + 1 + \dfrac{-7x - 7}{x^2 + 9}$, but we only need the quotient, $x + 1$, to find that the equation for the oblique asymptote is $y = x + 1$.

c. In this case, the degree of the numerator of h is two *less* than the degree of the denominator. This means that the line $y = 0$ is the horizontal asymptote of h.

d. For $j(x)$, the degree of the numerator is two *more* than the degree of the denominator. Thus, j has no horizontal or oblique asymptotes.

TOPIC **4**

Graphing Rational Functions

Much of our experience in graph sketching will be useful as we graph rational functions. In addition to the standard steps of identifying the x-intercepts (if any) and y-intercept (if there is one), we will make use of asymptotes when graphing rational functions. The following is a list of suggested steps.

PROCEDURE

Graphing Rational Functions

Given a rational function f,

Step 1: Factor the denominator in order to determine the domain of f. Any points excluded from the domain may appear as "holes" in the graph or as vertical asymptotes.

Step 2: Factor the numerator as well and cancel any common factors.

Step 3: Examine the remaining factors in the denominator to determine the equations for any vertical asymptotes.

Step 4: Compare the degrees of the numerator and denominator to determine if there is a horizontal or oblique asymptote. If so, find its equation.

Step 5: Determine the y-intercept, if 0 is in the domain of f.

Step 6: Determine the x-intercepts, if there are any, by setting the numerator of the reduced fraction equal to 0.

Step 7: Plot enough points to determine the behavior of f between x-intercepts and between vertical asymptotes.

EXAMPLE 3

Graphing Rational Functions

Sketch the graphs of the following rational functions.

a. $f(x) = \dfrac{x^2 - x}{x - 1}$ **b.** $g(x) = \dfrac{x^2 + 1}{x^2 + 2x - 15}$ **c.** $h(x) = \dfrac{x^3 + x^2 + 2x + 2}{x^2 + 9}$

Solutions:

a. The denominator of f is already factored, so we know that the domain of f consists of all real numbers except for $x = 1$.

As in Example 1c, we factor the numerator to see if there are any common factors.

$$f(x) = \frac{x^2 - x}{x - 1} = \frac{x(x - 1)}{x - 1}$$
$$= x$$

This means that, except for at $x = 1$, where f is undefined, we have $f(x) = x$. We already know how to graph this function, so the remaining steps are unnecessary. The graph of f is the line $y = x$, excluding the point $(1, 1)$, since $x = 1$ is not in the domain of f. The result is that a "hole" appears in the graph f at $x = 1$.

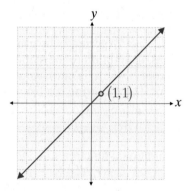

b. In Example 1b, we factored the denominator as follows.

$$g(x) = \frac{x^2 + 1}{x^2 + 2x - 15} = \frac{x^2 + 1}{(x + 5)(x - 3)}$$

This means the domain of g excludes the values $x = -5$ and $x = 3$. Since the numerator cannot be factored, we also know that the lines $x = -5$ and $x = 3$ are vertical asymptotes of g.

Next, we look at the degrees of the numerator and denominator. As we saw in Example 2a, the degrees are the same, so the line $y = 1$ is the horizontal asymptote. Plotting the asymptotes provides us a framework for graphing $g(x)$.

Setting $x = 0$, we find that the y-intercept lies at $\left(0, -\dfrac{1}{15}\right)$. Since there is no real solution to the equation $x^2 + 1 = 0$, g has no x-intercepts.

Plotting a few points in each region between asymptotes gives us an idea of the general shape of the graph, shown below.

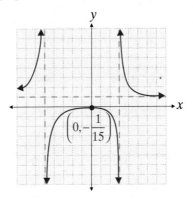

c. The denominator of this function cannot be factored, so there are no restrictions on the domain of h. Further, we saw in Example 2b that this function has an oblique asymptote of $y = x + 1$.

As usual, we calculate the y-intercept by substituting $x = 0$.

$$h(0) = \frac{0^3 + 0^2 + 2(0) + 2}{0^2 + 9} = \frac{2}{9}$$

There are different approaches to finding the x-intercepts. Looking at the numerator, we might guess that -1 is a zero of the numerator. A quick calculation confirms this: $(-1)^3 + (-1)^2 + 2(-1) + 2 = 0$. This means we can factor the numerator. Using synthetic or long division, we have $(x+1)(x^2 + 2)$. Thus, $(-1, 0)$ is the only x-intercept.

With the intercepts and a few other plotted points, we obtain the graph of h.

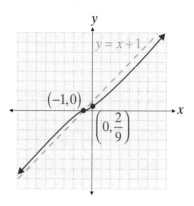

TOPIC 5 Solving Rational Inequalities

Now that we have discussed rational functions and equations, we can consider rational inequalities. We will see that solving rational inequalities involves similar work as solving polynomial inequalities.

DEFINITION

Rational Inequalities

A **rational inequality** is any inequality that can be written in the form:

$$f(x) < 0, \ f(x) \le 0, \ f(x) > 0, \ \text{or} \ f(x) \ge 0,$$

where $f(x)$ is a rational function.

Just as with polynomial inequalities, solving rational inequalities is relatively simple if we have an accurate graph of the function. All we need to do is identify the intervals on which the function is positive and negative, then determine which intervals satisfy the inequality.

However, since graphing rational functions is even more difficult than graphing polynomial functions, we should depend on an algebraic method.

Recall that to solve a polynomial inequality, we made use of the fact that polynomial functions are continuous. This let us apply the *Sign-Test Method*, in which testing a single point on each interval (between x-intercepts) described the sign behavior of the entire function.

Although rational functions are not always continuous, we know exactly where they are discontinuous: at their vertical asymptotes. This means that we can apply the Sign-Test Method, as long as we account for a possible sign change at each vertical asymptote.

PROCEDURE

Solving Rational
Inequalities:
Sign-Test Method

To solve a rational inequality $f(x) < 0$, $f(x) \leq 0$, $f(x) > 0$, or $f(x) \geq 0$, where the rational function $f(x) = \dfrac{p(x)}{q(x)}$ is in reduced form:

Step 1: Find the real zeros of the numerator $p(x)$. These values are the **zeros** of f.

Step 2: Find the real zeros of the denominator $q(x)$. These values are the locations of the **vertical asymptotes** of f.

Step 3: Place the values from steps 1 and 2 on a number line, splitting it into intervals.

Step 4: Within each interval, select a **test point** and evaluate f at that number. If the result is positive, then $f(x) > 0$ for all x in the interval. If the result is negative, then $f(x) < 0$ for all x in the interval.

Step 5: Write the **solution set**, consisting of all of the intervals that satisfy the given inequality. If the inequality is not strict (uses $\leq$ or $\geq$), then the zeros of p are included in the solution set as well. The zeros of q are never included in the solution set (as they are not in the domain of f).

EXAMPLE 4

Solving Rational
Inequalities

Solve the rational inequality $\dfrac{x^2 + 1}{x^2 + 2x - 15} > 0$.

Solution:

Begin by finding the zeros of the numerator and denominator. As we've seen in previous examples, the rational expression can be factored as follows:

$$\frac{x^2 + 1}{x^2 + 2x - 15} = \frac{x^2 + 1}{(x + 5)(x - 3)}$$

Thus, we can see that the numerator has no zeros, while the denominator has zeros at $x = -5$ and $x = 3$.

Therefore, we place the values -5 and 3 on a number line.

$$-5 \qquad\qquad\qquad 3$$

This splits the number line into the intervals $(-\infty, -5)$, $(-5, 3)$, $(3, \infty)$. We then evaluate $f(x)$ for a test point in each interval.

Interval	Test Point	Evaluate	Result
$(-\infty, -5)$	$x = -6$	$f(-6) = \dfrac{(-6)^2 + 1}{(-6)^2 + 2(-6) - 15}$ $= \dfrac{37}{9}$	$f(x) > 0$ on $(-\infty, -5)$ **Positive**
$(-5, 3)$	$x = 0$	$f(0) = \dfrac{(0)^2 + 1}{(0)^2 + 2(0) - 15}$ $= -\dfrac{1}{15}$	$f(x) < 0$ on $(-5, 3)$ **Negative**
$(3, \infty)$	$x = 4$	$f(4) = \dfrac{(4)^2 + 1}{(4)^2 + 2(4) - 15}$ $= \dfrac{17}{9}$	$f(x) > 0$ on $(3, \infty)$ **Positive**

Then, the solution to the inequality consists of those intervals where $f(x) > 0$, which is the set $(-\infty, -5) \cup (3, \infty)$.

CAUTION!

It is very important to write rational inequalities in their proper form before solving them. Attempting to simplify them may cause us to neglect asymptotes. Suppose, for example, we tried to solve the inequality

$$\frac{x}{x+2} < 3$$

by multiplying through by $x + 2$, thus clearing the inequality of fractions. The result would be
$$x < 3x + 6,$$

a simple linear inequality. We can solve this to obtain $x > -3$, or the interval $(-3, \infty)$. But does this really work?

The number $-\dfrac{5}{2}$ is in this interval, but $\dfrac{-\dfrac{5}{2}}{-\dfrac{5}{2} + 2} = 5$,

which is not less than 3. Also, note that -4 solves the inequality, but -4 is not in the interval $(-3, \infty)$.

What went wrong? By multiplying through by $x + 2$, we made the assumption that $x + 2$ was positive, since we did not worry about possibly reversing the inequality symbol. But we don't know beforehand if $x + 2$ is positive or not, since we are trying to solve for the variable x!

To solve this inequality correctly, we have to write it in the standard form of a rational inequality, as shown in Example 5.

EXAMPLE 5

Rational Inequalities

Solve the rational inequalities $\dfrac{x}{x+2} < 3$ and $\dfrac{x}{x+2} \le 3$.

Solution:

Most of the work can be done for both inequalities at the same time.

We begin by subtracting 3 from both sides and then write the left-hand side as a single rational function, giving us:

$$\frac{x}{x+2} - 3 < 0 \quad \text{and} \quad \frac{x}{x+2} - 3 \le 0$$

$$\frac{x}{x+2} - \frac{3(x+2)}{x+2} < 0 \qquad \frac{x}{x+2} - \frac{3(x+2)}{x+2} \le 0$$

$$\frac{-2x-6}{x+2} < 0 \qquad \frac{-2x-6}{x+2} \le 0$$

Then factor −2 from the numerator of the fraction and divide both sides by −2 (reversing the inequality symbol) to obtain the simpler inequalities

$$\frac{x+3}{x+2} > 0 \quad \text{and} \quad \frac{x+3}{x+2} \ge 0.$$

Now that the inequalities are in standard form, we can follow the procedure for solving rational inequalities.

The zeros of the numerator and denominator are −3 and −2, respectively. Placing these values on a number line,

$$-3 \; -2$$

we have the intervals $(-\infty, -3)$, $(-3, -2)$, and $(-2, \infty)$.

Interval	Test Point	Evaluate	Result
$(-\infty, -3)$	$x = -4$	$f(-4) = \dfrac{(-4)+3}{(-4)+2}$ $= \dfrac{1}{2}$	$f(x) > 0$ on $(-\infty, -3)$ **Positive**
$(-3, -2)$	$x = -2.5$	$f(-2.5) = \dfrac{(-2.5)+3}{(-2.5)+2}$ $= -1$	$f(x) < 0$ on $(-3, -2)$ **Negative**
$(-2, \infty)$	$x = 0$	$f(0) = \dfrac{(0)+3}{(0)+2}$ $= \dfrac{3}{2}$	$f(x) > 0$ on $(-2, \infty)$ **Positive**

1. $x = 1$

2. $x = -3$

3. No vertical asymptote

4. $x = 2$

5. $x = 2$

6. $x = -1$

7. $x = 0$

8. $x = -3$, $x = 3$

9. $x = -\dfrac{1}{2}$

10. $x = -1$

11. No vertical asymptote

12. $x = 3$

13. $x = 7$

14. $x = -5$, $x = \dfrac{3}{2}$

15. $x = -2$

16. No vertical asymptote

17. $x = -2$, $x = 2$

18. $x = 1$

19. $y = 0$

20. $y = x - 3$

21. No horizontal or oblique asymptote

22. $y = -1$

23. $y = 0$

24. $y = \dfrac{1}{2}$

25. $y = 2$

26. $y = -3$

27. $y = 3x + 6$

28. $y = x$

29. $y = 0$

30. $y = x + 1$

31. $y = 0$

32. No horizontal or oblique asymptote

33. $y = x - 11$

The final step is to evaluate which intervals satisfy each inequality. The solution to the first inequality is simply the union of the two intervals where f is positive:

$$(-\infty, -3) \cup (-2, \infty).$$

For the second inequality, we have to decide which endpoints to include. We include $x = -3$, since this is a zero of the rational function, but we do not include $x = -2$, since the value is not in the domain of f. Thus, the solution to the second inequality is:

$$(-\infty, -3] \cup (-2, \infty).$$

Exercises

Find equations for the vertical asymptotes, if any, for each of the following rational functions. See Example 1.

1. $f(x) = \dfrac{5}{x-1}$

2. $f(x) = \dfrac{x^2 + 3}{x + 3}$

3. $f(x) = \dfrac{x^2 - 4}{x + 2}$

4. $f(x) = \dfrac{-3x + 5}{x - 2}$

5. $f(x) = \dfrac{3x^2 + 1}{x - 2}$

6. $f(x) = \dfrac{x^2 + 2x}{x + 1}$

7. $f(x) = \dfrac{x^2 - 4}{2x - x^2}$

8. $f(x) = \dfrac{x + 2}{x^2 - 9}$

9. $f(x) = \dfrac{x^2 - 2x - 3}{2x^2 - 5x - 3}$

10. $f(x) = \dfrac{2x^2 + 2x - 4}{x^2 + 2x + 1}$

11. $f(x) = \dfrac{x^3 - 27}{x^2 + 5}$

12. $f(x) = \dfrac{x^2 + 5}{x^3 - 27}$

13. $f(x) = \dfrac{x^2 - 1}{x^2 - 8x + 7}$

14. $f(x) = \dfrac{2x^2 + 7x - 14}{2x^2 + 7x - 15}$

15. $f(x) = \dfrac{x^3 - 6x^2 + 11x - 6}{x^3 + 8}$

16. $f(x) = \dfrac{x^2 - 2x - 15}{x - 5}$

17. $f(x) = \dfrac{x^2 - 16}{x^2 - 4}$

18. $f(x) = \dfrac{x^2 + 4x + 4}{x^2 + x - 2}$

Find equations for the horizontal or oblique asymptotes, if any, for each of the following rational functions. See Example 2.

19. $f(x) = \dfrac{5}{x-1}$

20. $f(x) = \dfrac{x^2 + 3}{x + 3}$

21. $f(x) = \dfrac{x^4 - 4}{x^2 + 2}$

22. $f(x) = \dfrac{x^2 - 4}{2x - x^2}$

23. $f(x) = \dfrac{x + 2}{x^2 - 9}$

24. $f(x) = \dfrac{x^2 - 2x - 3}{2x^2 - 5x - 3}$

34. $y = -\dfrac{1}{5}$

35. $y = 5x + 4$

36. $y = 2x + 1$

37.

38.

39.

40.

41.

42.

43.

44.

45.

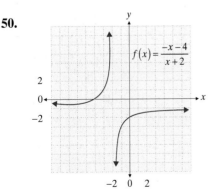

25. $f(x) = \dfrac{2x^2 + 2x - 4}{x^2 + 2x + 1}$

26. $f(x) = \dfrac{-3x + 5}{x - 2}$

27. $f(x) = \dfrac{3x^2 + 1}{x - 2}$

28. $f(x) = \dfrac{x^3 - 27}{x^2 + 5}$

29. $f(x) = \dfrac{x^2 + 5}{x^3 - 27}$

30. $f(x) = \dfrac{x^2 + 2x}{x + 1}$

31. $f(x) = \dfrac{x^2 - 81}{x^3 + 7x - 12}$

32. $f(x) = \dfrac{x^3 - 3x^2 + 2x}{x - 7}$

33. $f(x) = \dfrac{x^2 - 9x + 4}{x + 2}$

34. $f(x) = \dfrac{-x^5 + 2x^2}{5x^5 + 3x^3 - 7}$

35. $f(x) = \dfrac{5x^2 - x + 12}{x - 1}$

36. $f(x) = \dfrac{2x^2 - 5x + 6}{x - 3}$

Sketch the graphs of the following rational functions, making use of your work in the problems above and additional information about intercepts and any other points that may be useful. See Example 3.

37. $f(x) = \dfrac{5}{x - 1}$

38. $f(x) = \dfrac{x^2 + 3}{x + 3}$

39. $f(x) = \dfrac{x^2 - 4}{x + 2}$

40. $f(x) = \dfrac{x^2 - 4}{2x - x^2}$

41. $f(x) = \dfrac{x + 2}{x^2 - 9}$

42. $f(x) = \dfrac{x^2 - 2x - 3}{2x^2 - 5x - 3}$

43. $f(x) = \dfrac{2x^2 + 2x - 4}{x^2 + 2x + 1}$

44. $f(x) = \dfrac{-3x + 5}{x - 2}$

45. $f(x) = \dfrac{3x^2 + 1}{x - 2}$

46. $f(x) = \dfrac{x^3 - 27}{x^2 + 5}$

47. $f(x) = \dfrac{x^2 + 5}{x^3 - 27}$

48. $f(x) = \dfrac{x^2 + 2x}{x + 1}$

For each graph below, find any **a.** vertical asymptotes, **b.** horizontal asymptotes, **c.** oblique asymptotes, **d.** visible x-intercepts, or **e.** visible y-intercepts.

49.

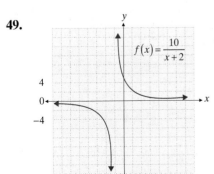

$f(x) = \dfrac{10}{x + 2}$

50.

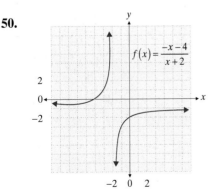

$f(x) = \dfrac{-x - 4}{x + 2}$

46.

47.

48.

49. a. $x = -2$
b. $y = 0$
c. None
d. None
e. $(0, 5)$

50. a. $x = -2$
b. $y = -1$
c. None
d. $(-4, 0)$
e. $(0, -2)$

51. a. $x = 9$
b. $y = 0$
c. None
d. None
e. $\left(0, -\dfrac{1}{3}\right)$

52. a. $x = -3$
b. $y = 3$
c. None
d. $(3, 0)$
e. $(0, -3)$

53. a. $x = -1, x = 1$
b. None
c. $y = x$
d. $\left(\sqrt[3]{3}, 0\right)$
e. $(0, 3)$

54. a. $x = 2$
b. $y = 0$
c. None
d. $\left(\pm\dfrac{\sqrt{3}}{3}, 0\right)$
e. $\left(0, \dfrac{1}{8}\right)$

51.

52.

53.

54.

55.

56.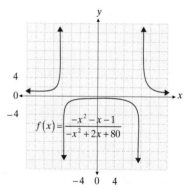

Solve the following rational inequalities. See Examples 4 and 5.

57. $2x < \dfrac{4}{x+1}$

58. $\dfrac{5}{x-2} \geq \dfrac{3x}{x-2}$

59. $\dfrac{5}{x-2} > \dfrac{3}{x+2}$

60. $\dfrac{x}{x^2-x-6} \leq \dfrac{-1}{x^2-x-6}$

61. $\dfrac{x}{x^2-x-6} \leq \dfrac{-2}{x^2-x-6}$

62. $x > \dfrac{1}{x}$

63. $\dfrac{4}{x-3} \leq \dfrac{4}{x}$

64. $\dfrac{x-7}{x-3} \geq \dfrac{x}{x-1}$

65. $\dfrac{x}{x^2+3x+2} > \dfrac{1}{x^2+3x+2}$

66. $\dfrac{1}{x-4} \geq \dfrac{1}{x+1}$

67. $\dfrac{x}{x+1} \geq \dfrac{x+1}{x}$

68. $\dfrac{x}{x^2-2x-3} > \dfrac{3}{x^2-2x-3}$

55. a. $x = 1$

b. None

c. $y = 3x$

d. None

e. $(0, -3)$

56. a. $x = -8$, $x = 10$

b. $y = 1$

c. None

d. None

e. $\left(0, -\dfrac{1}{80}\right)$

57. $(-\infty, -2) \cup (-1, 1)$

58. $\left[\dfrac{5}{3}, 2\right)$

59. $(-8, -2) \cup (2, \infty)$

60. $(-\infty, -2) \cup [-1, 3)$

61. $(-\infty, -2) \cup (-2, 3)$

62. $(-1, 0) \cup (1, \infty)$

63. $(0, 3)$

64. $(-\infty, 1) \cup \left[\dfrac{7}{5}, 3\right)$

65. $(-2, -1) \cup (1, \infty)$

66. $(-\infty, -1) \cup (4, \infty)$

67. $(-\infty, -1) \cup \left[-\dfrac{1}{2}, 0\right)$

68. $(-1, 3) \cup (3, \infty)$

69. a.

b. Joan's rabbit population reaches a maximum of 200 rabbits.

Solve the following application problems.

69. Joan raises rabbits, and the population of her rabbit colony follows the formula

$$p(t) = \frac{200t}{t+1}$$

where $t \geq 0$ represents the number of months since she began.

a. Sketch the graph of $p(t)$ for $t \geq 0$.

b. What happens to Joan's rabbit population in the long run?

70. If an object is placed a distance x from a lens with a focal length of f, the image of the object will appear a distance y on the opposite side of the lens, where x, f, and y are related by the equation $\dfrac{1}{x} + \dfrac{1}{y} = \dfrac{1}{f}$.

a. Express y as a function of x and f.

b. Graph your function for a lens with a focal length of 30 mm ($f = 30$). What happens to y as the distance x increases?

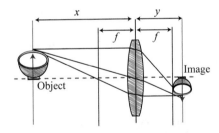

70. a. $y = \dfrac{fx}{x - f}$

b.

y decreases

71. At t minutes after injection, the concentration (in mg/L) of a certain drug in the bloodstream of a patient is given by the formula

$$c(t) = \frac{20t}{t^2 + 1}.$$

a. Sketch the graph of $c(t)$ for $t \geq 0$.

b. What happens to the concentration of the drug in the long run?

71. a.

b. The concentration of the drug disappears in the long run.

Polynomial Functions

Ace Automobiles is an international auto manufacturer that has just finished the design for a new sports car. In order to produce this car, they must lease a new assembly plant, purchase new robotic equipment, and hire new staff.

Ace has projected the monthly costs of these expenditures for budgeting purposes. According to their estimates, the plant lease and all utilities will cost $72,000; the depreciation on the new equipment will be $130,000; salaries will cost $480,000; and all other combined overhead will be $47,000. Thus the monthly total for all expenses will be $729,000.

The cost of raw materials to produce each car will be $4500.

The cost function per month for manufacturing x new cars will be
$$C(x) = 4500x + 729{,}000.$$
The revenue per month from selling x cars will be
$$r(x) = 13{,}500x.$$

1. How many cars per month must the new plant make in order to show a profit, if profit is revenue minus cost?

2. If the plant manager decides to increase staff and create two shifts, resulting in an additional $300,000 in salaries with all other expenses remaining unchanged, how many cars would the plant then have to produce in a month to be profitable?

3. Assume now that the plant at full capacity (including the second shift) can produce a maximum of 75 cars per month. What minimum price per car must Ace Automobiles set in order to maintain a profit?

Chapter Summary

A summary of concepts and skills follows each chapter. Refer to these summaries to make sure you feel comfortable with the material in the chapter. The concepts and skills are organized according to the section title and topic title in which the material is first discussed.

4.1: Introduction to Polynomial Equations and Graphs

Zeros of Polynomials and Solutions of Polynomial Equations
- The connection between *zeros* of polynomials and *roots* or *solutions* of polynomial equations
- The connection between x-intercepts and real zeros of polynomials with real coefficients

Graphing Factored Polynomials
- The geometric meaning of linear factors of a polynomial
- Behavior of a polynomial as $x \to \pm\infty$ based on its degree and the sign of its leading coefficient

Solving Polynomial Inequalities
- The use of the graph of a polynomial in solving a polynomial inequality
- The Sign-Test Method for solving polynomial inequalities

4.2: Polynomial Division and the Division Algorithm

The Division Algorithm and the Remainder Theorem
- The meaning of the terms *quotient*, *divisor*, *dividend*, and *remainder* as applied to polynomial division
- The *division algorithm* and what it implies about the degree of the remainder
- The connection between zeros of a polynomial and its linear factors
- The *remainder theorem*

Polynomial Long Division
- The method of *polynomial long division*

Synthetic Division
- The method of *synthetic division*
- Using synthetic division to evaluate polynomials for given values

Constructing Polynomials with Given Zeros
- Constructing polynomials with desired properties

4.3: Locating Real Zeros of Polynomials

The Rational Zero Theorem
- Using the *Rational Zero Theorem* to identify potential rational zeros

Descartes' Rule of Signs
- Using *Descartes' Rule of Signs* to identify the possible numbers of positive and negative real zeros

Bounds of Real Zeros
- Determining *upper* and *lower bounds* for the real zeros of a polynomial

The Intermediate Value Theorem
- Using the *Intermediate Value Theorem* to approximate real zeros of a polynomial

4.4: The Fundamental Theorem of Algebra

The Fundamental Theorem of Algebra
- The *Fundamental Theorem of Algebra* and what it says about solutions of polynomial equations
- The *Linear Factors Theorem* and what it says about the number of zeros of a polynomial based on its degree
- How to determine the maximum number of x-intercepts and *turning points* of the graph of a polynomial

Multiple Zeros and Their Geometric Meaning
- The meaning of *multiplicity* as applied to zeros of a polynomial
- The geometric implications of zeros of a given multiplicity

Conjugate Pairs of Zeros
- The *Conjugate Roots Theorem* and what it says about polynomials with real coefficients
- Using the conjugate roots theorem to construct polynomials with desired properties

Summary of Polynomial Methods
- Using all the tools of the chapter to graph polynomials, factor polynomials, and solve polynomial equations and inequalities

4.5: Rational Functions and Rational Inequalities

Definitions and Useful Notation
- The definition of a *rational function*, and how to determine domains of rational functions
- The meaning of *vertical*, *horizontal*, and *oblique asymptotes*

Vertical Asymptotes
- How to find the vertical asymptotes for a given rational function

Horizontal and Oblique Asymptotes
- How to determine if a rational function has a horizontal or oblique asymptote based on the degrees of the numerator and denominator
- How to find the horizontal asymptote of a given rational function, if there is one
- How to find the oblique asymptote of a given rational function, if there is one

Graphing Rational Functions
- How to use knowledge of a rational function's domain, intercepts, asymptotes, and symmetry to sketch its graph

Solving Rational Inequalities
- How to use graphing knowledge to solve *rational inequalities*

1.–4. Answers will vary.

5. $\pm\sqrt{2}, \pm\sqrt{5}$

6. $0, \pm\sqrt{2}, \pm i$

7. $\pm\sqrt{2}$

8. $-4, -2, 0$

9. $0, \dfrac{-1 \pm \sqrt{5}}{2}$

10. $-2 \pm i\sqrt{3}$

11. x-int: $-2, 1, 3$
   y-int: 6
   $f(x) \to -\infty$ as $x \to -\infty$;
   $f(x) \to \infty$ as $x \to \infty$

12. x-int: $2, -1$
   y-int: 4
   $f(x) \to \infty$ as $x \to \pm\infty$

13. x-int: $1, 4$
   y-int: 4
   $g(x) \to \infty$ as $x \to \pm\infty$

14. x-int: $0, -2, -5$
   y-int: 0
   $h(x) \to \infty$ as $x \to -\infty$;
   $h(x) \to -\infty$ as $x \to \infty$

Chapter Review

Section 4.1

Verify that the given values of x solve the corresponding polynomial equations.

1. $4x^3 - 5x^2 = -3x + 18;\ x = 2$

2. $x^2 - 6x = -13;\ x = 3 + 2i$

3. $x^3 + x = 6x^2 - 164;\ x = 5 - 4i$

4. $x^3 + (1+4i)x = (7-2i)x^2 - 2i + 36;\ x = -2i$

Solve the following polynomial equations by factoring and/or using the quadratic formula, making sure to identify all the solutions.

5. $x^4 - 7x^2 + 10 = 0$

6. $x^5 - x^3 - 2x = 0$

7. $x^4 + 4 = 4x^2$

8. $6x^2 + 8x = -x^3$

9. $x^4 + x^3 = x^2$

10. $x^2 + 4x + 7 = 0$

For each of the following polynomial functions, describe the behavior of its graph as $x \to \pm\infty$ and identify the x- and y-intercepts. Use this information to then sketch the graph of each polynomial.

11. $f(x) = (x+2)(x-1)(x-3)$

12. $f(x) = (x-2)^2(x+1)^2$

13. $g(x) = x^2 - 5x + 4$

14. $h(x) = -x^3 - 7x^2 - 10x$

Solve the following polynomial inequalities.

15. $2x^2 + 15 \le 11x$

16. $(x-3)^2(x+1)^2 > 0$

17. $(x-4)(x+2)(x^2-1) \le 0$

18. $x^3 - 2x^2 - 8x \ge 0$

19. $x^2(x-2)(1-x) < 0$

20. $-3x^2 + 7x - 2 > 0$

21. A manufacturer has determined that the revenue from the sale of x video games is given by $r(x) = -x^2 + 12x$. The cost of producing x video games is $C(x) = 120 - 22x$. Given that profit is revenue minus cost, what value(s) for x will give the company a nonnegative profit?

15. $\left[\dfrac{5}{2}, 3\right]$

16. $(-\infty, -1) \cup (-1, 3) \cup (3, \infty)$

17. $[-2, -1] \cup [1, 4]$

18. $[-2, 0] \cup [4, \infty)$

19. $(-\infty, 0) \cup (0, 1) \cup (2, \infty)$

20. $\left(\dfrac{1}{3}, 2\right)$

21. All integers between 4 and 30, inclusive

22. $4x^2 - 3x + 3$

$+ \dfrac{7}{2x^2 - 1}$

23. $11x + 35 + \dfrac{100}{x-3}$

24. $x^2 - 3x + 4$

$- \dfrac{5x + 16}{x^2 + 3x + 2}$

25. $2x^3 + 2x^2 - 2x$

$-3 + \dfrac{-2x - 2}{x^2 - x}$

26. $x^2 - 6 + \dfrac{-4 + 7i}{2x + i}$

27. $p(1) = 90$

28. k is a zero

29. $p\left(\dfrac{2}{3}\right) = -\dfrac{7}{3}$

30. $x^3 + x^2 + 2x + 7$

31. $-x^3 + 2x^2 - 7x + 23$

32. $x^4 + 3x^2 - 5 + \dfrac{23}{x+2}$

33.

$-x^3 + 7x^2 + x - 3 + \dfrac{-1}{x-1}$

34. $x^3 + 4x^2 - x + 3$

Section 4.2

Use polynomial long division to rewrite each of the following fractions in the form $q(x) + \dfrac{r(x)}{d(x)}$, where $d(x)$ is the denominator of the original fraction, $q(x)$ is the quotient, and $r(x)$ is the remainder.

22. $\dfrac{8x^4 - 6x^3 + 2x^2 + 3x + 4}{2x^2 - 1}$

23. $\dfrac{11x^2 + 2x - 5}{x - 3}$

24. $\dfrac{x^4 - 3x^2 + x - 8}{x^2 + 3x + 2}$

25. $\dfrac{2x^5 - 4x^3 - x^2 + x - 2}{x^2 - x}$

26. $\dfrac{2x^3 + ix^2 - 12x - 4 + i}{2x + i}$

Use synthetic division to determine if the given value for k is a zero of the corresponding polynomial. If not, determine $p(k)$.

27. $p(x) = 6x^5 - 23x^4 - 95x^3 + 70x^2 + 204x - 72; \; k = 1$

28. $p(x) = 48x^4 + 10x^3 - 51x^2 - 10x + 3; \; k = \dfrac{1}{6}$

29. $p(x) = 18x^5 - 87x^4 + 110x^3 - 28x^2 - 16x + 3; \; k = \dfrac{2}{3}$

Use synthetic division to rewrite each of the following fractions in the form $q(x) + \dfrac{r(x)}{d(x)}$, where $d(x)$ is the denominator of the original fraction, $q(x)$ is the quotient, and $r(x)$ is the remainder.

30. $\dfrac{x^4 - 2x^3 - x^2 + x - 21}{x - 3}$

31. $\dfrac{-x^4 - x^3 - x^2 + 2x + 69}{x + 3}$

32. $\dfrac{x^5 + 2x^4 + 3x^3 + 6x^2 - 5x + 13}{x + 2}$

33. $\dfrac{-x^4 + 8x^3 - 6x^2 - 4x + 2}{x - 1}$

34. $\dfrac{x^4 + (4 - 2i)x^3 - (1 + 8i)x^2 + (3 + 2i)x - 6i}{x - 2i}$

35. $f(x) = x^2 - 4x - 12$

36. $y = \dfrac{1}{2}(x+4)^4$

37. $y = 2(x^2 - 4)(x - 3)$

38. $\pm\{1, 2, 3, 6\}$,
$\{-3, -1, 2\}$

39. $\pm\left\{\dfrac{1}{2}, 1, \dfrac{3}{2}, 3, \dfrac{9}{2}, 9\right\}$,
$\left\{1, \dfrac{3}{2}, 3\right\}$

40. $\pm\left\{\dfrac{1}{2}, 1, \dfrac{3}{2}, 3, \dfrac{9}{2}, 9\right\}$,
$\left\{-3, \dfrac{2 \pm i\sqrt{2}}{2}\right\}$

41. $\pm\{1, 3, 9\}$,
$\{-3, -1\}$

42. $\{-3, -1, 2\}$

43. $\left\{1, \dfrac{3}{2}, 3\right\}$

44. $\left\{-3, \dfrac{2 \pm i\sqrt{2}}{2}\right\}$

45. $\{-3, -1\}$

46. 2 or 0 pos.,
2 or 0 neg.

47. 4, 2 or 0 pos.,
2 or 0 neg.

48. $[-2, 6]$

49. $[-3, 7]$

50. $\left\{-\dfrac{3}{2}, 3, 4\right\}$

51. $\left\{-\dfrac{5}{2}, -\dfrac{1}{2}, 7\right\}$

52. $f(-2) = 73$
$f(0) = -5$

53. $f(2) = 3$
$f(4) = -15$

Construct a polynomial function with the stated properties.

35. Second-degree, zeros of –2 and 6, and goes to ∞ as $x \to \infty$.

36. Fourth-degree and a single x-intercept of –4 and y-intercept $(0, 128)$.

37. Third-degree, zeros of ± 2 and 3 and passing through the point $(4, 24)$.

Section 4.3

List all of the potential rational zeros of the following polynomials. Then use polynomial division and the quadratic formula, if necessary, to identify the actual zeros.

38. $f(x) = x^4 + 3x^3 - 3x^2 - 11x - 6$

39. $g(x) = 2x^3 - 11x^2 + 18x - 9$

40. $h(x) = 2x^3 + 2x^2 - 9x + 9$

41. $p(x) = x^4 + 8x^3 + 22x^2 + 24x + 9$

Using the Rational Zero Theorem or your answers to the preceding problems, solve the following polynomial equations.

42. $2x^4 - 6x^2 = -6x^3 + 22x + 12$

43. $2x^3 - 9x^2 + 18x = 9 + 2x^2$

44. $2x^3 + 9 = 9x - 2x^2$

45. $x^4 - x^5 = -x^5 - 8x^3 - 22x^2 - 24x - 9$

Use Descartes' Rule of Signs to determine the possible numbers of positive and negative real zeros of each of the following polynomials.

46. $f(x) = 2x^4 - 3x^3 - x^2 + 3x + 10$

47. $g(x) = x^6 - 4x^5 - 2x^4 + x^3 - 6x^2 - 11x + 6$

Use synthetic division to identify integer upper and lower bounds of the real zeros of the following polynomials.

48. $f(x) = 2x^3 - 11x^2 + 3x + 36$

49. $g(x) = 4x^3 - 16x^2 - 79x - 35$

Using your answers to the preceding problems, polynomial division, and the quadratic formula, if necessary, find all of the zeros of the following polynomials.

50. $f(x) = 2x^3 - 11x^2 + 3x + 36$

51. $g(x) = 4x^3 - 16x^2 - 79x - 35$

Use the Intermediate Value Theorem to show that each of the following polynomials has a real zero between the indicated values.

52. $f(x) = 2x^4 - 6x^3 + x - 5$; -2 and 0

53. $f(x) = -x^3 + 3x^2 + x - 3$; 2 and 4

54. $\{\pm 1, 2, 3\}$

55. $\{\pm 3i, 4\}$

56. $\{-3, -2, -1\}$

57. $\{3, 2 \pm \sqrt{3}\}$

58. $\{1, \pm 3i\}$

59. $\left\{-\dfrac{1}{2}, 2\right\}$

60. $\left\{-\dfrac{4}{3}, \pm i\sqrt{2}\right\}$

61.

62.

63. $(x^2 + 1)(x - 3)$

64. $(x + 1)(x - 1) \times$
$(x - 2)(x^2 + x + 1)$

65. $\left\{-1, \dfrac{5}{3}, \pm 2i\right\}$

66. $\left\{\pm\sqrt{3}, -\dfrac{1}{2}\right\}$

67. $\{\pm 2i, -3, \pm 1\}$

Using any of the methods discussed in this section as guides, find all of the real zeros of the following functions.

54. $f(x) = x^4 - 5x^3 + 5x^2 + 5x - 6$

55. $g(x) = x^3 - 4x^2 + 9x - 36$

56. $f(x) = x^3 + 6x^2 + 11x + 6$

57. $f(x) = x^3 - 7x^2 + 13x - 3$

Using any of the methods discussed in this section as guides, solve the following equations.

58. $x^4 - 2x^3 + 10x^2 = 9(2x - 1)$ **59.** $2x^3 = 7x^2 - 4x - 4$

60. $-8 = 3x^3 + 4x^2 + 6x$

Section 4.4

Throughout these exercises, a graphing calculator or a computer algebra system may be helpful in identifying zeros and in checking your graphing, if permitted by your instructor.

Sketch the graph of each factored polynomial.

61. $f(x) = (x + 4)^2 (x - 1)$ **62.** $g(x) = x(x - 3)(x + 4)^3$

Use all available methods to factor each of the following polynomials completely, and then sketch the graph of each one.

63. $f(x) = x^3 - 3x^2 + x - 3$ **64.** $f(x) = x^5 - x^4 - 2x^3 - x^2 + x + 2$

Use all available methods (e.g., the Rational Zero Theorem, Descartes' Rule of Signs, polynomial division, etc.) to solve each polynomial equation. Use the Linear Factors Theorem to make sure you find the appropriate number of solutions, counting multiplicity.

65. $3x^5 + x^4 + 5x^3 = x^2 + 28x + 20$ **66.** $8x^5 + 12x^4 - 18x^3 - 35x^2 = 18x + 3$

67. $x^5 + 3x^4 + 3x^3 + 9x^2 = 4(x + 3)$

68. $(x-2+i)\times$
$(x-2-i)\times$
$(7x-2)(2x-7)$

69. $(x-5i)(x+5i)$
$\times(x-6)(x+1)$

70. $(x-1-i)(x-1+i)$
$\times(2x+1)(x+3)$

71.
$(x+3)\left(x-\dfrac{1+i\sqrt{19}}{4}\right)$

$\times\left(x-\dfrac{1-i\sqrt{19}}{4}\right)$

72. $f(x)=2x^4+7x^3$
$-18x^2+67x-30$

73. $f(x)=x^5+3x^4$
$-3x^3-17x^2$
$-18x-6$

74. $f(x)=x^5-3x^4$
$+8x^2-9x+3$

75. $x=\dfrac{5}{2}$

76. No vertical
asymptote

77. $x=0$

78. $x=3$

79. $y=2x+9$

80. $y=\dfrac{1}{3}$

81. None

82. $y=0$

83.

Use all available methods (in particular, the Conjugate Roots Theorem, if applicable) to factor each of the following polynomials completely, making use of the given zero.

68. $f(x)=14x^4-109x^3+296x^2-321x+70$; $2+i$ is a zero

69. $f(x)=x^4-5x^3+19x^2-125x-150$; $-5i$ is a zero

70. $f(x)=2x^4+3x^3-7x^2+8x+6$; $1+i$ is a zero

71. $f(x)=4x^3+10x^2-x+15$; -3 is a zero

Construct polynomial functions with the stated properties.

72. Fourth-degree, only real coefficients, $\dfrac{1}{2}$ and $1+2i$ are two of the zeros, y-intercept is -30, leading coefficient is 2.

73. Fifth-degree, only real coefficients, -1 is a zero of multiplicity 3, $\sqrt{6}$ is a zero, y-intercept is -6, leading coefficient is 1.

74. Fifth-degree, only real coefficients, 1 is a zero of multiplicity 3, $\sqrt{3}$ is a zero, y-intercept is 3, leading coefficient is 1.

Section 4.5

Find equations for the vertical asymptotes, if any, for each of the following rational functions.

75. $f(x)=\dfrac{4}{2x-5}$

76. $f(x)=\dfrac{x^2-3x+2}{x-1}$

77. $f(x)=\dfrac{x^2-1}{x-x^2}$

78. $f(x)=\dfrac{x^2-x-6}{x^2-6x+9}$

Find equations for the horizontal or oblique asymptotes, if any, for each of the following rational functions.

79. $f(x)=\dfrac{2x^3+5x^2-1}{x^2-2x}$

80. $f(x)=\dfrac{x^2-x+8}{3x^2-7}$

81. $f(x)=\dfrac{x^2-9}{x+3}$

82. $f(x)=\dfrac{x^2+2x-3}{(x+1)^3}$

84.

85.

86.

Sketch the graphs of the following rational functions.

83. $\dfrac{2x}{x+1}$

84. $\dfrac{4x^2}{x^2+3x}$

85. $\dfrac{x^2+2}{x+2}$

86. $\dfrac{x+1}{x^2-4}$

Solve the following rational inequalities.

87. $\dfrac{7}{x+3} \geq \dfrac{2x}{x+3}$

88. $\dfrac{x}{x^2-5x+6} \leq \dfrac{3}{x^2-5x+6}$

89. $\dfrac{x-4}{x+3} < \dfrac{x}{x-2}$

90. $\dfrac{x-2}{x+3} < 2$

87. $\left(-3, \dfrac{7}{2}\right]$

88. $(-\infty, 2)$

89. $\left(-3, \dfrac{8}{9}\right) \cup (2, \infty)$

90. $(-\infty, -8) \cup (-3, \infty)$

Chapter Test

1.–2. Answers will vary.

3. $0, 5, -4$

4. $\dfrac{3 \pm i\sqrt{3}}{2}$

5. x-int: $1, \pm 2$

    y-int: -4

    $f(x) \to \infty$ as $x \to -\infty$

    $f(x) \to -\infty$ as $x \to \infty$

6. x-int: 2

    y-int: -8

    $f(x) \to -\infty$ as $x \to -\infty$

    $f(x) \to \infty$ as $x \to \infty$

7. $(-6, 1)$

8. All real numbers except $-1, 3$

9. All real numbers

10. $40 \le x \le 50$

11. $x^3 + x^2 + x + 1$

12. $3x^2 i - 6x + i - 3$

13. $x^3 - 3x^2 + 8x - 16$

    $+ \dfrac{28}{x+2}$

14. k is a zero

15. $p\left(\dfrac{1}{2}\right) = \dfrac{5}{2}$

16. $p(-2) = -41$

Verify that the given values of x solve the corresponding polynomial equations.

1. $x^2 - 2x = -3$; $x = 1 + i\sqrt{2}$

2. $x^3 - 2ix^2 = 2i - x$; $x = i$

Solve the following polynomial equations by factoring and/or using the quadratic formula, making sure to identify all the solutions.

3. $x^3 - x^2 = 20x$

4. $x^2 - 3x + 3 = 0$

For each of the following polynomial functions, describe the behavior of its graph as $x \to \pm\infty$ and identify the x- and y-intercepts. Use this information to sketch the graph of each polynomial equation.

5. $f(x) = (1 - x)(x^2 - 4)$

6. $f(x) = (x - 2)^3$

Solve the following polynomial inequalities.

7. $x^2 + 5x < 6$

8. $(x + 1)^2 (x - 3)^2 > 0$

9. $x^2 + x \ge -1$

10. A manufacturer has determined that the revenue from the sale of x shoes is given by $r(x) = -x^2 + 60x$. The cost of producing x shoes is $C(x) = 2000 - 30x$. Given that profit is revenue minus cost, what value(s) for x will give the company a nonnegative profit?

Use polynomial long division to rewrite each of the following fractions in the form $q(x) + \dfrac{r(x)}{d(x)}$, where $d(x)$ is the denominator of the original fraction, $q(x)$ is the quotient, and $r(x)$ is the remainder.

11. $\dfrac{x^4 - 1}{x - 1}$

12. $\dfrac{6x^3 i - 3x^2 + (20i - 6)x + 3 + 9i}{2x - 3i}$

13. $\dfrac{x^4 - x^3 + 2x^2 - 4}{x + 2}$

Use synthetic division to determine if the given value of k is a zero of the corresponding polynomial. If not, determine $p(k)$.

14. $p(x) = 10x^4 - 20x^3 + 8x^2 - x + 3$; $k = 1$

15. $p(x) = 16x^4 - 20x^3 + 8x^2 - 2x + 3$; $k = \dfrac{1}{2}$

16. $p(x) = 4x^3 - x^2 + x - 3$; $k = -2$

17. $y = (x+2)(1-x)$

18. $y = x(x+3)(x-2)$

19. $-2, 1, 3$

20. $0, -2, \pm 2i$

21. 0 pos., 3 or 1 neg.

22. 2 or 0 pos., 1 neg.

23. $(2x-1)(x-2)$
    $\times (x+1)$

24. $(x-2)^3(x+3)$

25. $y = -2x^2 + 4x - 6$

26. $y = (x^2 - 2x + 2)(x-2)$

27. No

28. $f(x) = \left(x + \dfrac{2}{3}\right)$
    $(15x^3 - 6x + 4)$
    $+ \dfrac{34}{3}; f\left(\dfrac{-2}{3}\right) = \dfrac{34}{3}$

29. $x = -1$

30. None

31. $y = 0$

32. $y = x - 2$

Construct a polynomial function with the stated properties.

17. Second-degree, zeros of –2 and 1, and goes to $-\infty$ as $x \to -\infty$.

18. Third-degree, zeros of –3, 0, 2, and y-intercept 0.

Solve the following polynomial equations.

19. $x^3 - 2x^2 - 5x + 6 = 0$

20. $x^4 + 2x^3 + 4x^2 + 8x = 0$

Use Descartes' Rule of Signs to determine the possible numbers of positive and negative real zeros of each of the following polynomials.

21. $f(x) = 2x^3 + 3x^2 + 5x + 2$

22. $f(x) = x^3 - 7x^2 + x + 10$

Use all available methods to factor each of the following polynomials completely and then sketch the graph of each one.

23. $f(x) = 2x^3 - 3x^2 - 3x + 2$

24. $f(x) = x^4 - 3x^3 - 6x^2 + 28x - 24$

Construct polynomial functions with the stated properties.

25. Second-degree, only real coefficients, $1 + i\sqrt{2}$ is one zero, y-intercept is –6.

26. Third-degree, real coefficients, zeros 2, $1 + i$, and $1 - i$, leading coefficient is 1.

27. Use synthetic division to determine if $(x+3)$ is a factor of
    $f(x) = 3x^3 + 4x^2 - 18x - 3$.

Express the following function in the form of $f(x) = (x-k)q(x) + r$ for the given value of k, and demonstrate that $f(k) = r$.

28. $f(x) = 15x^4 + 10x^3 - 6x^2 + 14, \quad k = -\dfrac{2}{3}$

Find equations for the vertical asymptotes, if any, for each of the following rational functions.

29. $y = \dfrac{-2}{x+1}$

30. $f(x) = \dfrac{x^3 + 2x^2 - x - 2}{x^3 + 8}$

Find the equations for the horizontal or oblique asymptotes, if any, for each of the following rational functions.

31. $f(x) = \dfrac{3}{x-3}$

32. $f(x) = \dfrac{x^2 + 2}{x+2}$

33.

34.

35. $x > 2$

36. $1 < x \leq \dfrac{5}{3}$

    or $x > 2$

37. No; explanations will vary.

Sketch the graphs of the following rational functions.

33. $f(x) = \dfrac{x^2 - 1}{x - 1}$

34. $f(x) = \dfrac{2x^2 - 2x - 4}{x^2 - 2x + 1}$

Solve the following rational inequalities.

35. $3x > \dfrac{-3}{x - 2}$

36. $\dfrac{1}{x - 2} + \dfrac{2}{x - 1} \geq 0$

37. If the graph of a rational function f has a vertical asymptote at $x = -2$, is it possible to sketch the graph without lifting your pencil from the paper? Explain.

Chapter 5

Exponential and Logarithmic Functions

5.1   Exponential Functions and Their Graphs

5.2   Applications of Exponential Functions

5.3   Logarithmic Functions and Their Graphs

5.4   Properties and Applications of Logarithms

5.5   Exponential and Logarithmic Equations

Chapter 5 Project

Chapter 5 Summary

Chapter 5 Review

Chapter 5 Test

By the end of this chapter you should be able to:

What if you had to predict the amount of time it will take for the earth's population to reach 20 billion (ignoring the effect of limited resources on population growth)? How would you perform this calculation?

By the end of this chapter, you'll be able to apply the skills regarding exponential and logarithmic functions to estimate the behavior of growing populations, interest-earning accounts, and decaying elements. On page 448 you'll encounter a problem like the population problem given above. You'll master this type of problem using tools such as the Summary of Logarithmic Properties, found on page 441.

Introduction

This chapter introduces two entirely new classes of functions, both of which are of enormous importance in many natural and man-made contexts. As we will see, the two classes of functions are inverses of one another, though historically exponential and logarithmic functions were developed independently and for unrelated reasons.

We will begin with a study of exponential functions. These are functions in which the variable appears in the exponent while the base is a constant, just the opposite of what we have seen so often in the individual terms of polynomials. As with many mathematical concepts, the argument can be made that exponential functions exist in the natural world independently of mankind and that consequently mathematicians have done nothing more than observe (and formalize) what there is to be seen. Exponential behavior is exhibited, for example, in the rate at which radioactive substances decay, in how the temperature of an object changes when placed in an environment held at a constant temperature, and in the fundamental principles of population growth. But exponential functions also arise in discussing such man-made phenomena as the growth of investment funds.

In fact, we will use the formula for compound interest to motivate the introduction of the most famous and useful base for exponential functions, the irrational constant e (the first few digits of which are 2.718281828459...). The Swiss mathematician Leonhard Euler (1707–1783), who identified many of this number's unique properties (such as the fact that $e = 1 + \frac{1}{1} + \frac{1}{1 \cdot 2} + \frac{1}{1 \cdot 2 \cdot 3} + \ldots$), was one of the first to recognize the fundamental importance of e and in fact is responsible for the choice of the letter e as its symbol. The constant e also arises very naturally in the context of calculus, but that discussion must wait for a later course.

Napier

Logarithms are inverses, in the function sense, of exponentials, but historically their development was for very different reasons. Much of the development of logarithms is due to John Napier (1550–1617), a Scottish writer of political and religious tracts and an amateur mathematician and scientist. It was the goal of simplifying computations of large numbers that led him to devise early versions of logarithmic functions, to construct what today would be called tables of logarithms, and to design a prototype of what would eventually become a slide rule. Of course, today it is not necessary to resort to logarithms in order to carry out difficult computations, but the properties of logarithms make them invaluable in solving certain equations. Further, the fact that they are inverses of exponential functions means they have just as much a place in the natural world.

Exponential Functions and Their Graphs

TOPICS

1. Definition and classification of exponential functions

2. Graphing exponential functions

3. Solving elementary exponential equations

T. Inputting exponents, graphing exponential functions

TOPIC

Definition and Classification of Exponential Functions

We have studied many functions in which a variable is raised to a constant power, including polynomial functions such as $f(x) = x^3 + 2x^2 - 1$, radical functions like $g(x) = x^{\frac{1}{3}}$, and rational functions such as $h(x) = x^{-2}$.

An *exponential function* is a function in which a constant is raised to a variable power. Exponential functions are extremely important because of the large number of natural situations in which they arise. Examples include radioactive decay, population growth, compound interest, spread of epidemics, and rates of temperature change.

DEFINITION

Exponential Functions

Let a be a fixed, positive real number not equal to 1. The **exponential function with base a** is the function

$$f(x) = a^x.$$

Why do we have the restrictions $a > 0$, $a \neq 1$? The base of the exponent can't be negative, since a^x would not be real for many values of x. For example, if $a = -1$ and $x = \dfrac{1}{2}$, a^x is not real.

If we let $a = 1$, we don't have a problem producing real numbers. Instead, the "exponential" function turns out to be constant. Recall that for all values of x, $1^x = 1$. For this reason, 1^x is considered a constant function, not an exponential function, and should always be written in its simplified form, 1.

Note that for any positive constant a, a^x is defined for all real numbers x. Consequently, the domain of $f(x) = a^x$ is the set of real numbers. What about the range of f? Since a is positive, we know that a^x must be positive. We will see that the range of all exponential functions is the set of all positive real numbers.

Recall that if a is any nonzero number, a^0 is defined to be 1. This means the y-intercept of any exponential function, regardless of the base, is the point $(0,1)$. Beyond this, exponential functions fall into two classes, depending on whether a lies between 0 and 1 or if a is larger than 1.

Consider the following calculations for two sample exponential functions:

x	$f(x) = \left(\dfrac{1}{3}\right)^x$	$g(x) = 2^x$
-2	$f(-2) = 9$	$g(-2) = \dfrac{1}{4}$
-1	$f(-1) = 3$	$g(-1) = \dfrac{1}{2}$
0	$f(0) = 1$	$g(0) = 1$
1	$f(1) = \dfrac{1}{3}$	$g(1) = 2$
2	$f(2) = \dfrac{1}{9}$	$g(2) = 4$

Note that the values of f decrease as x increases while the values of g do just the opposite. We say that f is an example of a *decreasing* function and g is an example of an *increasing* function. If we plot these points and then fill in the gaps with a smooth curve, we get the graphs of f and g that appear in Figure 1.

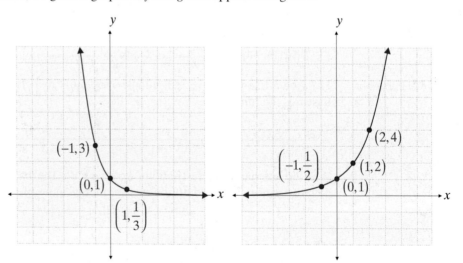

Figure 1: Two Exponential Functions

The above graphs suggest that the range of an exponential function is $(0, \infty)$, and that is indeed the case. Note that the base does not matter: the range of a^x is the positive real numbers for any allowable base (that is, for any positive a not equal to 1).

PROPERTIES

Behavior of
Exponential Functions

Given a positive real number a not equal to 1, the function $f(x) = a^x$ is:

- a **decreasing function** if $0 < a < 1$, with $f(x) \to \infty$ as $x \to -\infty$ and $f(x) \to 0$ as $x \to \infty$.

- an **increasing function** if $a > 1$, with $f(x) \to 0$ as $x \to -\infty$ and $f(x) \to \infty$ as $x \to \infty$.

In either case, the point $(0, 1)$ lies on the graph of f, the domain of f is the set of real numbers, and the range of f is the set of positive real numbers.

TOPIC 2   Graphing Exponential Functions

Given that all exponential functions take one of two basic shapes (depending on whether a is less than 1 or greater than 1), they are relatively easy to graph. Plotting a few points, including the y-intercept of $(0,1)$, will provide an accurate sketch.

EXAMPLE 1

Graphing Exponential
Functions

Sketch the graphs of the following exponential functions:

a. $f(x) = 3^x$
b. $g(x) = \left(\dfrac{1}{2}\right)^x$

Note:
Plugging in $x = -1$, $x = 0$, and $x = 1$ will produce a good idea of the shape of the graph.

Solutions:

In both cases, we plot 3 points by plugging in $x = -1$, $x = 0$, and $x = 1$. We then connect the points with a smooth curve.

a.

b.
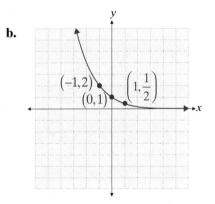

An exponential function, like any function, can be transformed in ways that result in the graph being shifted, reflected, stretched, or compressed. It often helps to graph the base function before trying to graph the transformed one.

EXAMPLE 2

Graphing Exponential Functions

Sketch the graphs of each of the following functions.

a. $f(x) = \left(\dfrac{1}{2}\right)^{x+3}$ **b.** $g(x) = -3^x + 1$ **c.** $h(x) = 2^{-x}$

Note:
For each example, first graph the base function, then apply any transformations.

Solutions:

a.

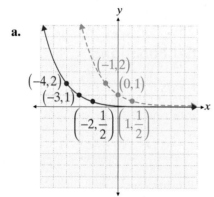

First, draw the graph of the function $\left(\dfrac{1}{2}\right)^x$, as in Example 1b.

Then, since x has been replaced by $x + 3$, shift the graph to the left by 3 units.

b.

Begin with the graph of 3^x shown in green.

The effect of multiplying a function by -1 is to reflect the graph with respect to the x-axis.

Following this, the second transformation of adding 1 to a function causes a vertical shift of the graph. The purple curve at left is the graph of $g(x) = -3^x + 1$.

c.

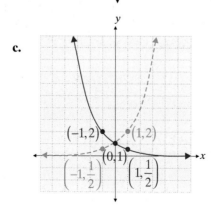

Begin by graphing the base function 2^x, shown in green.

For h, x has been replaced by $-x$, so we reflect the graph of 2^x across the y-axis to obtain the graph of h, shown in purple.

Note that this graph is also the graph of $\left(\dfrac{1}{2}\right)^x$. Using properties of exponents is another way to think about this problem, as

$$2^{-x} = \left(2^{-1}\right)^x = \left(\frac{1}{2}\right)^x.$$

TOPIC 3

Solving Elementary Exponential Equations

As you might expect, an equation in which the variable appears as an exponent is called an *exponential equation*. We are not yet ready to tackle exponential equations in full generality. Even something as simple as

$$2^x = 5$$

currently stumps us. We know that $2^2 = 4$ and $2^3 = 8$, so the answer must be between 2 and 3, but beyond this we don't have a method to proceed. This must wait until we have discussed a class of functions called *logarithms* (which will happen in Section 5.3).

However, we *are* ready to solve exponential equations that can be written in the form

$$a^x = a^b,$$

where a is an exponential base (positive and not equal to 1) and b is a constant.

You might guess, just from the form of the equation $a^x = a^b$ that the solution is $x = b$. This guess is correct, but we need to investigate why it is true.

The reason that the single value b is the solution of an exponential equation of the form $a^x = a^b$ is that the exponential function $f(x) = a^x$ is one-to-one (its graph passes the horizontal line test). Recall that if g is a one-to-one function, then the only way for $g(x_1)$ to equal $g(x_2)$ is if $x_1 = x_2$.

In the case of the function $f(x) = a^x$, the equation $a^x = a^b$ is equivalent to the statement $f(x) = f(b)$, and this implies $x = b$, since f is one-to-one.

An exponential equation may not appear in the simple form $a^x = a^b$ initially. The procedure below describes the steps you may need to take to solve an elementary exponential equation.

PROCEDURE

Solving Elementary
Exponential
Equations

To solve an elementary exponential equation,

Step 1: Isolate the exponential. Move the exponential containing x to one side of the equation and any constants or other variables in the expression to the other side. Simplify, if necessary.

Step 2: Find a base that can be used to rewrite both sides of the equation.

Step 3: Equate the powers, and solve the resulting equation.

EXAMPLE 3

Solving Elementary
Exponential
Equations

Solve the following exponential equations.

a. $25^x - 125 = 0$ **b.** $8^{y-1} = \dfrac{1}{2}$ **c.** $\left(\dfrac{2}{3}\right)^x = \dfrac{9}{4}$

Note:
As always, it is good
practice to check
your solution in the
original equation.

Solutions:

a. $25^x - 125 = 0$

$\quad\quad 25^x = 125$ — Begin by isolating the term with the variable on one side.

$\quad\quad \left(5^2\right)^x = 5^3$ — We can write both sides with the same base since 25 and 125 are both powers of 5.

$\quad\quad 5^{2x} = 5^3$ — Simplify using properties of exponents.

$\quad\quad 2x = 3$ — We then equate the power, resulting in a linear equation which we can easily solve.

$\quad\quad x = \dfrac{3}{2}$

b. $\quad 8^{y-1} = \dfrac{1}{2}$ — Again, we need to rewrite both sides using the same base.

$\quad\quad \left(2^3\right)^{y-1} = 2^{-1}$ — We see that 8 and $\dfrac{1}{2}$ can both be written as powers of 2.

$\quad\quad 2^{3y-3} = 2^{-1}$ — Simplify using properties of exponents.

$\quad\quad 3y - 3 = -1$ — Set the exponents equal to each other.

$\quad\quad 3y = 2$ — Solve the resulting linear equation.

$\quad\quad y = \dfrac{2}{3}$

c. $\left(\dfrac{2}{3}\right)^x = \dfrac{9}{4}$ — Sometimes, the choice of base is not obvious.

$\quad\quad \left(\dfrac{2}{3}\right)^x = \left(\dfrac{3}{2}\right)^2$ — Initially, we write the right-hand side as shown.

$\quad\quad \left(\dfrac{2}{3}\right)^x = \left(\dfrac{2}{3}\right)^{-2}$ — Then, we can make the two bases equal by using properties of exponents.

$\quad\quad x = -2$ — After making both bases the same, we equate the exponents to find the solution.

TOPIC

1.

2.

3.

4.

5.

6.

7.

8.

9.

Inputting Exponents, Graphing Exponential Functions

To input exponents into a graphing calculator, we can use one of two methods. If the exponent is a 2, we can type the base and then press [x^2]. Otherwise, we need to use the caret symbol, [^]. For example, to calculate 6^4, we would type the base, 6, then [^] followed by the exponent, 4.

CAUTION!

Regardless of which method is used, be careful with your negative signs. We know that $-3 \cdot -3 = 9$ and not -9 because a negative times a negative is a positive. However, if we type -3^2 into a calculator, the output is -9. The number that we are squaring is -3, not 3, so we need to type $(-3)^2$ or $(-3)\wedge 2$ into the calculator to get the correct answer, 9.

To graph an exponential function, where the exponent is the variable, we use the technique that incorporates the caret symbol. Consider the graph of the function $f(x) = 3\left(\dfrac{1}{2}\right)^x$. To graph this in a calculator, press [Y=] and type in the following:

Notice that only the fraction $\dfrac{1}{2}$ is being raised to the exponent, so we put it in parentheses. The graph looks like

In the next section, we will learn about the irrational number, e, which is used often in exponential equations. To calculate or input a value such as $e^{0.25}$ into a graphing calculator, type [2ND] [LN]. Then, type in the exponent, 0.25, and close the parentheses by pressing [ )]. Then press ENTER:

10.

11.

12.

13.

14.

15.

Exercises

Sketch the graphs of the following functions. See Examples 1 and 2.

1. $f(x) = 4^x$

2. $g(x) = (0.5)^x$

3. $s(x) = 3^{x-2}$

4. $f(x) = \left(\dfrac{1}{3}\right)^{x+1}$

5. $r(x) = 5^{x-2} + 3$

6. $h(x) = 1 - 2^{x+1}$

7. $f(x) = 2^{-x}$

8. $r(x) = 3^{2-x}$

9. $g(x) = 3\left(2^{-x}\right)$

10. $h(x) = 2^{2x}$

11. $s(x) = (0.2)^{-x}$

12. $f(x) = \dfrac{1}{2^x} + 1$

13. $g(x) = 3 - 2^{-x}$

14. $r(x) = \dfrac{1}{2^{3-x}}$

15. $h(x) = \left(\dfrac{1}{2}\right)^{5-x}$

16. $m(x) = 3^{2x+1}$

17. $p(x) = 2 - 4^{2-x}$

18. $q(x) = 5^{3-2x}$

19. $r(x) = \left(\dfrac{9}{2}\right)^{-x}$

20. $p(x) = \left(\dfrac{1}{3}\right)^{2-x}$

21. $r(x) = 1 - \left(\dfrac{15}{4}\right)^x$

Solve the following exponential equations. See Example 3.

22. $5^x = 125$

23. $3^{2x-1} = 27$

24. $9^{2x-5} = 27^{x-2}$

25. $10^x = 0.01$

26. $4^{-x} = 16$

27. $2^x = \left(\dfrac{1}{2}\right)^{13}$

28. $2^{x+1} = 64^3$

29. $\left(\dfrac{2}{3}\right)^{x+3} = \left(\dfrac{9}{4}\right)^{-x}$

30. $\left(\dfrac{1}{5}\right)^{x-4} = 625^{\frac{1}{2}}$

31. $4^{3x+2} = \left(\dfrac{1}{4}\right)^{-2x}$

32. $5^x = 0.2$

33. $7^{x^2+3x} = \dfrac{1}{49}$

34. $3^{x^2+4x} = 81^{-1}$

35. $\left(\dfrac{1}{2}\right)^{x-3} = \left(\dfrac{1}{4}\right)^{x-5}$

36. $64^{x+\frac{7}{6}} = 2$

37. $6^{2x} = 36^{2x-3}$

38. $4^{2x-5} = 8^{\frac{x}{2}}$

39. $\left(\dfrac{2}{5}\right)^{2x+4} = \left(\dfrac{4}{25}\right)^{11}$

40. $4^{4x-7} = \dfrac{1}{64}$

41. $-10^x = -0.001$

42. $3^x = 27^{x+4}$

43. $1000^{-x} = 10^{x-8}$

44. $1^{3x-7} = 4^{2-x}$

45. $5^{3x-1} = 625^x$

46. $\left(e^{x+2}\right)^3 = \left(e^x\right)\dfrac{1}{\left(e^{3x}\right)}$

47. $3^{2x-7} = 81^{\frac{x}{2}}$

20.

21.

22. $\{3\}$

23. $\{2\}$

24. $\{4\}$

25. $\{-2\}$

26. $\{-2\}$

27. $\{-13\}$

28. $\{17\}$

29. $\{3\}$

30. $\{2\}$

31. $\{-2\}$

32. $\{-1\}$

33. $\{-2, -1\}$

34. $\{-2\}$

35. $\{7\}$

36. $\{-1\}$

37. $\{3\}$

38. $\{4\}$

39. $\{9\}$

40. $\{1\}$

41. $\{-3\}$

42. $\{-6\}$

43. $\{2\}$

44. $\{2\}$

45. $\{-1\}$

46. $\left\{\dfrac{-6}{5}\right\}$

Match the graphs of the following functions to the appropriate equation.

48. $f(x) = 2^{3x}$

49. $h(x) = 5^{x} - 1$

50. $g(x) = 2\left(4^{x-1}\right)$

51. $p(x) = 1 - 2^{-x}$

52. $f(x) = 6^{4-x}$

53. $r(x) = \dfrac{1}{3^{x}}$

54. $m(x) = -2 + 2^{-3x}$

55. $g(x) = \left(\dfrac{1}{4}\right)^{1+x}$

56. $h(x) = 3^{\frac{1}{2}x}$

57. $s(x) = 1^{x} - 4$

a.

b.

c.

d.

e.

f.

g.

h.

i.

j.

47. No solution

48. a

49. g

50. i

51. b

52. d

53. f

54. e

55. c

56. h

57. j

Applications of Exponential Functions

TOPICS

1. Models of population growth

2. Radioactive decay

3. Compound interest and the number *e*

TOPIC   1

Models of Population Growth

Exponential functions arise naturally in a wide array of situations. In this section, we will study a few such situations in some detail, beginning with population models.

Many people working in such areas as mathematics, biology, and sociology study mathematical models of population. The models represent many types of populations, such as people in a given city or country, wolves in a wildlife habitat, or number of bacteria in a Petri dish. While such models can be quite complex, depending on factors like availability of food, space constraints, and effects of disease and predation, at their core, many population models assume that population growth displays exponential behavior.

The reason for this is that the growth of a population usually depends to a large extent on the number of members capable of producing more members. This assumes an abundant food supply and no constraints on population from lack of space, but at least initially this is often the case. In any situation where the rate of growth of a population is proportional to the size of the population, the population will grow exponentially, so we can write the function

$$P(t) = P_0 a^t.$$

This function tells us the size of a population $P(t)$ at time t. What do the other terms in the function represent?

Consider what happens if we substitute $t = 0$. $P(0) = P_0 a^0$, and since $a^0 = 1$ for all a, we have $P(0) = P_0$. Thus, P_0 is the *initial population* (population at time 0).

Recall from our work in graphing exponential equations that when $a > 1$, the larger the value of a is, the faster the exponential function grows. In situations of population growth, a is greater than 1 (otherwise we would have population decay). For this reason, the value of a is the *growth rate* of the population.

Often, we will need to use information about a population to determine the values of P_0 and a. Once we have the function for population growth, we can answer other questions about the population.

Example 1 illustrates this with a specific model of bacterial population growth.

EXAMPLE 1

Population Growth

A biologist is culturing bacteria in a Petri dish. She begins with 1000 bacteria, and supplies sufficient food so that for the first five hours the bacteria population grows exponentially, doubling every hour.

a. Find a function that models the population growth of this bacteria culture.

b. Determine when the population reaches 16,000 bacteria.

c. Calculate the population two and a half hours after the scientist begins.

Solutions:

a. We know that we seek a function of the form $P(t) = P_0 a^t$. Since the scientist starts with 1000 bacteria, the initial population, $P_0 = 1000$.

To solve for a, we use the fact that the population doubles every hour.

$2000 = P(1)$	Substitute, using the fact that the population after one hour equals 2000 bacteria.
$2000 = 1000a^1$	
$2000 = 1000a$	Simplify, then solve for a.
$a = 2$	

Thus, a function that models the population growth of the bacteria culture is $P(t) = 1000(2)^t$, where t is measured in hours.

b. We wish to find the time t for which $P(t) = 16,000$.

$16,000 = 1000(2)^t$	Substitute the desired population value.
$16 = 2^t$	Divide both sides by 1000.
$2^4 = 2^t$	Rewrite both sides with the same base, 2.
$t = 4$	Equate the exponents and solve.

Thus, the bacteria culture reaches a population of 16,000 in 4 hours.

c. To calculate the population two and half hours after growth begins, we substitute $t = 2.5$ into the population model function.

$$P(2.5) = 1000(2)^{2.5}$$

While we could rewrite this using rational exponents, and try to find an exact value, frequently this will be very tedious or impossible, so we use a calculator to evaluate.

$$P(2.5) \approx 5657 \text{ bacteria.}$$

TOPIC 2

Radioactive Decay

In contrast to populations (at least healthy populations), radioactive substances diminish with time. To be exact, the mass of a radioactive element decreases over time as it decays into other elements. Since exponential functions with a base between 0 and 1 are decreasing functions, this suggests that radioactive decay is modeled by

$$A(t) = A_0 a^t,$$

where $A(t)$ represents the amount of the substance at time t, A_0 is the amount at time $t = 0$, and a is a number between 0 and 1.

Radioactive decay is so predictable that archaeologists and anthropologists can use it to clock how long ago an organism died. The method of *radiocarbon dating* depends on the fact that living organisms constantly absorb molecules of the radioactive substance carbon-14 while alive, but the intake of carbon-14 ceases once the organism dies.

The percentage of carbon-14 on Earth (relative to other isotopes of carbon) has been relatively constant over time, so living organisms provide an accurate reference. By comparing the smaller percentage of carbon-14 in a dead organism to the percentage found in living tissue, an estimate of when the organism died can then be made.

Frequently, the decay rate of a radioactive substance is described by how long it takes for half of the material to decay into other elements. This is known as the *half-life* of the substance. The half-life of radioactive element ranges widely; Francium exhibits a half-life of about 22 minutes, while Bismuth has a half life of 1.9×10^{19} years, about one billion times the estimated age of the universe!

In the case of carbon-14, its half-life of 5728 years means that if an organism contained 12 grams of carbon-14 at death, it would contain 6 grams after 5728 years, 3 grams after another 5728 years, and so on.

EXAMPLE 2

Radioactive Decay

Determine the base, a, so that the function $A(t) = A_0 a^t$ accurately describes the decay of carbon-14 as a function of t years.

Solution:

Note that $A(0) = A_0 a^0 = A_0$, so A_0 represents the amount of carbon-14 at time $t = 0$. Since we are seeking a general formula, we don't know the value of A_0, that is, the value of A_0 will vary depending on the details of the situation. But we can still determine the base constant, a.

What we know is that half of the original amount of carbon-14 decays over a period of 5728 years, so $A(5728)$ will be half of A_0. This gives us the equation

$$A(5728) = \frac{A_0}{2} = A_0 (a)^{5728} .$$

To solve this equation for a, we can first divide both sides by A_0; the fact that A_0 then cancels from the equation emphasizes that its exact value is irrelevant in determining the value of a.

We are left with the equation

$$a^{5728} = \frac{1}{2}.$$

At this point we need a calculator, as we need to take the 5728th root of both sides:

$$a = \left(\frac{1}{2}\right)^{\frac{1}{5728}} \approx 0.999879$$

The function is thus: $A(t) = A_0(0.999879)^t$ (using our approximate value for a).

TOPIC 3

Compound Interest and the Number *e*

One of the most commonly encountered applications of exponential functions is in compounding interest. We run into compound interest when earning money (by interest on a savings account or investment) and when spending money (on car loans, mortgages and credit cards).

The basic compound interest formula can be understood by considering what happens when money is invested in a savings account. Typically, a savings account is set up to pay interest at an annual rate of r (which we will write in decimal form) compounded n times a year. For instance, a bank may offer an annual interest rate of 5% (meaning $r = 0.05$) compounded monthly (so $n = 12$).

Compounding is the act of calculating the interest earned on an investment and adding that amount to the investment. An investment in a monthly compounded account will have interest added to it twelve times over the course of a year, once each month.

Suppose an amount of P (for principal) dollars is invested in a savings account at an annual rate of r compounded n times per year. We want a formula for the amount of money $A(t)$ in the account after t years.

If we say that a period is the length of time between compoundings, interest is calculated at the rate of $\frac{r}{n}$ per period (for instance, if $r = 0.05$ and $n = 12$, interest is earned at a rate of $\frac{0.05}{12} \approx 0.00417$ per month). Table 1 illustrates how compounding increases the amount in the account over the course of several periods.

Period	Amount
0	$A = P$
1	$A = P\left(1 + \dfrac{r}{n}\right)$
2	$A = P\left(1 + \dfrac{r}{n}\right)\left(1 + \dfrac{r}{n}\right) = P\left(1 + \dfrac{r}{n}\right)^2$
3	$A = P\left(1 + \dfrac{r}{n}\right)^2\left(1 + \dfrac{r}{n}\right) = P\left(1 + \dfrac{r}{n}\right)^3$
k	$A = P\left(1 + \dfrac{r}{n}\right)^{k-1}\left(1 + \dfrac{r}{n}\right) = P\left(1 + \dfrac{r}{n}\right)^k$

Table 1: Effect of Compounding Interest on an Investment of P Dollars

Since there are nt compounding periods in t years, we obtain the following formula.

DEFINITION

Compound Interest
Formula

An investment of P dollars, compounded n times per year at an annual interest rate of r, has a value after t years of

$$A(t) = P\left(1 + \frac{r}{n}\right)^{nt}.$$

EXAMPLE 3

Compound Interest
Formula

Sandy invests \$10,000 in a savings account earning 4.5% annual interest compounded quarterly. What is the value of her investment after three and a half years?

Solution:

We know that $P = 10,000$, $r = 0.045$ (remember to express the interest rate in decimal form), $n = 4$ (since the account is compounded four times a year), and $t = 3.5$. Now we substitute and evaluate.

$$A(3.5) = 10,000\left(1 + \frac{0.045}{4}\right)^{(4)(3.5)}$$

$$= 10,000(1.01125)^{14}$$

$$\approx \$11,695.52$$

Thus, after three and a half years, Sandy's investment grows to \$11,695.52.

The compound interest formula can also be used to determine the interest rate of an existing savings account, as shown in Example 4.

EXAMPLE 4

Compound Interest Formula

Nine months after depositing $520.00 in a monthly compounded savings account, Frank checks his balance and finds the account has $528.84. Being the forgetful type, he can't remember what the annual interest rate for his account is, and sees the bank is advertising a rate of 2.5% for new accounts. Should he close out his existing account and open a new one?

Solution:

As in Example 3, we will begin by identifying the known quantities in the compound interest formula: $P = 520$, $n = 12$ (12 compoundings per year), and $t = 0.75$ (nine months is three-quarters of a year).

Further, the amount in the account, A, at this time is $528.84. This gives us the equation

$$528.84 = 520 \left(1 + \frac{r}{12} \right)^{(12)(0.75)}$$

to solve for r, the annual interest rate.

$$528.84 = 520 \left(1 + \frac{r}{12} \right)^{9} \qquad \text{Simplify the exponent.}$$

$$1.017 = \left(1 + \frac{r}{12} \right)^{9} \qquad \text{Divide both sides by 520.}$$

$$1.001875 \approx 1 + \frac{r}{12} \qquad \text{Take the ninth root of both sides.}$$

$$0.001875 \approx \frac{r}{12} \qquad \text{Simplify to solve for } r.$$

$$0.0225 \approx r$$

Thus Frank's current savings account is paying an annual interest rate of 2.25%, so he would gain a slight advantage by switching to a new account.

Even though the interest rate is divided by n, the number of periods per year, increasing the frequency of compounding always increases the total interest earned on the investment. This is because interest is always calculated on the current balance. For example, if you invest $1000 in an account with 5% interest, compounded once per year, you earn $50 in interest. However, if the interest is compounded *twice* per year, you earn 2.5%, or $25 for the first period, then 2.5% *of the new balance* of $1025. This brings the total interest earned to $50.63! Table 2 shows the interest earned in one year on an investment of $1000 in an account with a 5% annual interest rate, compounded n times per year.

n	Calculation	Value After 1 Year
1 (annually)	$A = 1000(1 + 0.05)$	$1050.00
2 (biannually)	$A = 1000\left(1 + \dfrac{0.05}{2}\right)^2$	$1050.63
4 (quarterly)	$A = 1000\left(1 + \dfrac{0.05}{4}\right)^4$	$1050.95
12 (monthly)	$A = 1000\left(1 + \dfrac{0.05}{12}\right)^{12}$	$1051.16
52 (weekly)	$A = 1000\left(1 + \dfrac{0.05}{52}\right)^{52}$	$1051.25
365 (daily)	$A = 1000\left(1 + \dfrac{0.05}{365}\right)^{365}$	$1051.27
8760 (hourly)	$A = 1000\left(1 + \dfrac{0.05}{8760}\right)^{8760}$	$1051.27

Table 2: Value of an Investment with Interest Compounded n Times per Year

Looking at the table, we see that the amount of interest keeps growing as we divide the year into more compounding periods, but that this growth slows dramatically. It looks as if there is a *limit* to how much interest we can earn, and this is in fact the case.

Let's examine the formula as $n \to \infty$. In order to do this we need to perform some algebraic manipulation of the formula:

$$A(t) = P\left(1 + \frac{r}{n}\right)^{nt}$$

$$= P\left(1 + \frac{1}{m}\right)^{rmt} \qquad \text{Substitute } m = \frac{n}{r}, \text{ so } \frac{1}{m} = \frac{r}{n}.$$

$$= P\left(\left(1 + \frac{1}{m}\right)^{m}\right)^{rt} \qquad \begin{array}{l}\text{Bring together the instances of the}\\\text{variable } m \text{ using the properties of}\\\text{exponents.}\end{array}$$

Although the manipulation may appear strange, it has accomplished the important task of isolating the part of the formula that changes as $n \to \infty$. Looking at the change of variables $m = \dfrac{n}{r}$, we see that letting $n \to \infty$ means that $m \to \infty$ as well. Since every other quantity remains fixed, we only have to understand what happens to

$$\left(1 + \frac{1}{m}\right)^{m}$$

as m grows without bound.

This is not a trivial undertaking. We might think that letting m get larger and larger would make the expression grow larger and larger, as m is the exponent in the expression and the base is larger than 1. But at the same time, the base approaches 1 as m increases without bound, and 1 raised to any power is simply 1. It turns out that these two effects balance one another out, as we can see in the table below:

m	$\left(1+\dfrac{1}{m}\right)^m$
10	2.59374
100	2.70481
1000	2.71692
10,000	2.71815
100,000	2.71827

Table 3: Values of $\left(1+\dfrac{1}{m}\right)^m$ as m Increases

As m gets larger and larger, we find that the expression approaches a fixed number:

$$\left(1+\frac{1}{m}\right)^m \rightarrow 2.71828182846...$$

This value is a very important number in mathematics, so important that it gets its own symbol, the letter e (which highlights its connection to exponential functions).

DEFINITION

The Number e

The number e is defined as the value of $\left(1+\dfrac{1}{m}\right)^m$ as $m \rightarrow \infty$.

$$e \approx 2.71828182846$$

In most applications, the approximation $e \approx 2.7183$ is sufficiently accurate.

Let's look back at the compound interest formula from before.

$$A(t) = P\left(1+\frac{r}{n}\right)^{nt} = P\left(\left(1+\frac{1}{m}\right)^m\right)^{rt}$$

This means that if P dollars is invested in an account that is **compounded continuously**, that is, with $n \rightarrow \infty$, then the amount in the account is determined by the following formula, which results from substituting the definition of e.

DEFINITION

Continuous
Compounding
Formula

An investment of P dollars compounded continuously at an annual interest rate of r has a value after t years of:

$$A(t) = Pe^{rt}.$$

EXAMPLE 5

Continuous
Compounding
Formula

If Sandy (last seen in Example 3) has the option of investing her $10,000 in a continuously compounded account earning 4.5% annual interest, what will be the value of her account in three and a half years?

Solution:

Again, the solution boils down to substituting the correct values and evaluating the result. Here, $P = 10,000$, $r = 0.045$, and $t = 3.5$.

$$A(3.5) = 10,000e^{(0.045)(3.5)}$$

$$= 10,000e^{0.1575}$$

$$\approx \$11,705.81$$

This account earns $10.29 more than the quarterly compounded account in Example 3.

As we will see in Section 5.4, all exponential functions can be expressed with the base e (or any other base, for that matter). The base e is so commonly used for exponential functions that it is often called the *natural base*. For instance, the formula for the radioactive decay of carbon-14, using the base e, is

$$A(t) = A_0 e^{-0.000121t}.$$

You should verify that this version of the decay formula does indeed give the same values for $A(t)$ as the version derived in Example 2.

Exercises

The following problems all involve working with exponential functions similar to those encountered in this section.

1. $V \approx 178$ people

2. $w \approx 6.6$ watts

1. A new virus has broken out in isolated parts of Africa and is spreading exponentially through tribal villages. The growth of this new virus can be mapped using the following formula where P stands for the number of people in a village and d stands for the number of days since the virus first appeared. According to this equation, how many people in a tribe of 300 will be infected after 5 days?

$$V = P\left(1 - e^{-0.18d}\right)$$

2. A new hybrid car is equipped with a battery that is meant to improve gas mileage by making the car run on electric power as well as gasoline. The batteries in these new cars must be changed out every so often to ensure proper operation. The power in the battery decreases according to the exponential equation below. After 30 days (d), how much power in watts (w) is left in the battery?

$$w(d) = 40e^{-0.06d}$$

3. $C \approx \$8526.20$

4. $P \approx 325{,}510$ rabbits

5. $W \approx 93$ computers

6. $F \approx 32$ apartments

7. a. 0.999567

   b. 0.958 grams

   c. 0.648 grams

8. Approximately 16.4%

---

3. A young economics student has come across a very profitable investment scheme in which his money will accrue interest according to the equation listed below. If this student invests $1250 into this lucrative endeavor, how much money will he have after 24 months? I represents the investment and m represents the number of months the money has been invested for.

$$C = Ie^{0.08m}$$

4. A family releases a couple of pet rabbits into the wild. Upon being released the rabbits begin to reproduce at an exponential rate, as shown in the formula below. After 2 years how large is the rabbit population where n stands for the initial rabbit population (2) and m stands for the number of months?

$$P = ne^{0.5m}$$

5. Inside a business network, an e-mail worm was downloaded by an employee. This worm goes through the infected computer's address book and sends itself to all the listed e-mail addresses. This worm very rapidly works its way through the network following the equation below where C is the number of computers in the network and W is the number of computers infected h hours after its discovery. After only 8 hours, how many computers has the worm infected if there are 150 computers in the network?

$$W = C\left(1 - e^{-0.12h}\right)$$

6. A construction crew has been assigned to build an apartment complex. The work of the crew can be modeled using the exponential formula below where A is the total number of apartments to be built, w is the number of weeks, and F is the number of finished apartments. Out of a total of 100 apartments, how many apartments have been finished after 4 weeks of work?

$$F = A\left(1 - e^{-0.1w}\right)$$

7. The half-life of radium is approximately 1600 years.

   a. Determine a so that $A(t) = A_0 a^t$ describes the amount of radium left after t years, where A_0 is the amount at time $t = 0$.

   b. How much of a 1-gram sample of radium would remain after 100 years?

   c. How much of a 1-gram sample of radium would remain after 1000 years?

8. The radioactive element Polonium-210 has a relatively short half-life of 140 days, and one way to model the amount of Polonium-210 remaining after t days is with the function $A(t) = A_0 e^{-0.004951t}$, where A_0 is the mass at time $t = 0$ (note that $A(140) = \dfrac{A_0}{2}$.) What percentage of the original mass of a sample of Polonium-210 remains after one year?

9. a. 3 years

   b. 9 years

10. 197,959 people

11. 1118 rabbits

12. Approximately

   a. 7.24 billion

   b. 15.38 billion

13. The bank offering 2.75% and monthly compounding.

14. 30 cents

15. Approximately 3.18%

16. Approximately 13.9%

17. $134,392

---

9. A certain species of fish is to be introduced into a new man-made lake, and wildlife experts estimate the population will grow according to

$$P(t) = (1000)2^{\frac{t}{3}},$$ where t represents the number of years from the time of introduction.

a. What is the doubling time for this population of fish?

b. How long will it take for the population to reach 8000 fish, according to this model?

10. The population of a certain inner-city area is estimated to be declining according to the model $P(t) = 237,000e^{-0.018t}$, where t is the number of years from the present. What does this model predict the population will be in ten years?

11. In an effort to control vegetation overgrowth, 100 rabbits are released in an isolated area that is free of predators. After one year, it is estimated that the rabbit population has increased to 500. Assuming exponential population growth, what will the population be after another 6 months?

12. Assuming a current world population of 6 billion people, an annual growth rate of 1.9% per year, and a worst-case scenario of exponential growth, what will the world population be in:
a. 10 years? **b.** 50 years?

13. Madiha has $3500 that she wants to invest in a simple savings account for two and a half years, at which time she plans to close out the account and use the money as a down payment on a car. She finds one local bank offering an annual interest rate of 2.75% compounded monthly, and another bank offering an annual interest rate of 2.7% compounded daily (365 times per year). Which bank should she choose?

14. Madiha, from the last problem, does some more searching and finds an online bank offering an annual rate of 2.75% compounded continuously. How much more money will she earn over two and a half years if she chooses this bank rather than the local bank offering the same rate compounded monthly?

15. Tom hopes to earn $1000 in interest in three years time from $10,000 that he has available to invest. To decide if it's feasible to do this by investing in a simple monthly compounded savings account, he needs to determine the annual interest rate such an account would have to offer for him to meet his goal. What would the annual rate of interest have to be?

16. An investment firm claims that its clients usually double their principal in five years time. What annual rate of interest would a savings account, compounded monthly, have to offer in order to match this claim?

17. The function $C(t) = C(1 + r)^t$ models the rise in the cost of a product that has a cost of C today, subject to an average yearly inflation rate of r for t years. If the average annual rate of inflation over the next decade is assumed to be 3%, what will the inflation-adjusted cost of a $100,000 house be in 10 years? Round your answer to the nearest dollar.

18. Approximately 2.5%

19. a. 10

  b. 7490 people

  c. The function approaches 10,000 as time goes on.

20. Approximately 0.043

21. a. 0.965936

  b. 0.707 kg

  c. 7.628 mg

22. Although the difference is minor in the short term, the account with the continuously compounded rate of 3.15% will pay more over time.

23. a. $1521.74

  b. $271.74

24. a. $P(t) =$ $375(1.026^t)$

  b. 8160 bacteria

25. $9459.48; $9942.41

---

18. Given the inflation model $C(t) = C(1 + r)^t$ (see the previous problem), and given that a loaf of bread that currently sells for $3.60 sold for $3.10 six years ago, what has the average annual rate of inflation been for the past six years?

19. The function $N(t) = \dfrac{10{,}000}{1 + 999e^{-t}}$ models the number of people in a small town who have caught the flu t weeks after the initial outbreak.

  a. How many people were ill initially?
  b. How many people have caught the flu after 8 weeks?
  c. Describe what happens to the function $N(t)$ as $t \to \infty$.

20. The concentration, $C(t)$, of a certain drug in the bloodstream after t minutes is given by the formula $C(t) = 0.05\left(1 - e^{-0.2t}\right)$. What is the concentration after 10 minutes?

21. Carbon-11 has a radioactive half-life of approximately 20 minutes, and is useful as a diagnostic tool in certain medical applications. Because of the relatively short half-life, time is a crucial factor when conducting experiments with this element.

  a. Determine a so that $A(t) = A_0 a^t$ describes the amount of carbon-11 left after t minutes, where A_0 is the amount at time $t = 0$.
  b. How much of a 2-kg sample of carbon-11 would be left after 30 minutes?
  c. How many milligrams of a 2-kg sample of carbon-11 would be left after 6 hours?

22. Charles has recently inherited $8000 that he wants to deposit into a savings account. He has determined that his two best bets are an account that compounds annually at a rate of 3.20% and an account that compounds continuously at an annual rate of 3.15%. Which account would pay Charles more interest?

23. Marshall invests $1250 in a mutual fund which boasts a 5.7% annual return compounded semiannually (twice a year). After three and a half years, Marshall decides to withdraw his money.

  a. How much is in his account?
  b. How much has he made in interest from his investment?

24. Adam is working in a lab testing bacteria populations. After starting out with a population of 375 bacteria, he observes the change in population and notices that the population doubles every 27 minutes. Find
  a. the equation for the population P in terms of time t in minutes, rounding a to the nearest thousandth, and
  b. the population after 2 hours.

25. Your credit union offers a special interest rate of 10% compounded monthly for the first year for a student savings account opened in August if the student deposits $5000 or more. You received a total of $9000 for graduation, and you decide to deposit all of it in this special account. Assuming you open your account in August and make no withdrawals for the first year, how much money will you have in your account at the end of February (after 6 months)? How much will you have at the end of the following July (after one full year)?

26. a. $421.87
 b. $429.28
 c. Second case; explanations will vary.

27. $2835.71

28. Yes; $202.02 remaining

29. $20,000

30. a. $70,664.27
 b. $3.33
 c. $6.32
 d. $63,225.93

31. $7318.71

32. $8795.91

33. a. $7647.95
 b. $7647.57
 c. Yes; explanations will vary.

26. You have a savings account of $3000 with an interest rate of 6.8%.

a. How much interest would be earned in 2 years if the interest is compounded annually?

b. How much interest would be earned in 2 years if the interest is compounded semiannually?

c. In which case do you make more money on interest? Explain why this is so.

27. If $2500 is invested in a continuously compounded certificate of deposit with an annual interest rate of 4.2%, what would be the account balance at the end of 3 years?

28. The new furniture store in town boasts a special in which you can buy any set of furniture in their store and make no monthly payments for the first year. However, the fine print says that the interest rate of 7.25% is compounded quarterly beginning when you buy the furniture. You are considering buying a set of living room furniture for $4000, but know you cannot save up more than $4500 in one year's time. Can you fully pay off your furniture on the one year anniversary of having bought the furniture? If so, how much money will you have left over? If not, how much more money will you need?

29. When Nicole was born, her grandmother was so excited about her birth that she opened a certificate of deposit in Nicole's honor to help send her to college. Now at age 18, Nicole's account has $81,262.93. How much did her grandmother originally invest if the interest rate has been 8.1% compounded annually?

30. Inflation is a relative measure of your purchasing power over time. The formula for inflation is the same as the compound interest formula, but with $n = 1$. Given the current values below, what will the values of the following items be 10 years from now if inflation is at 6.4%?

a. an SUV: $38,000
 b. a loaf of bread: $1.79
 c. a gallon of milk: $3.40
 d. your salary: $34,000

31. Depreciation is the decrease of an item's value and can be determined using a formula similar to that for compound interest:
$$V = P(1-r)^t,$$
where V is the new value.

If the particular car you buy upon graduation from college costs $17,500 and depreciates at a rate of 16% per year, what will the value of the car be in 5 years when you pay it off?

32. Assume the interest on your credit card is compounded continuously with an APR (annual percentage rate) of 19.8%. If you put your first term bill of $3984 on your credit card, but do not have to make payments until you graduate (4 years later), how much will you owe when you start making payments?

33. Suppose you deposit $5000 in an account for 5 years at a rate of 8.5%.

a. What would be the ending account balance if the interest is continuously compounded?

b. What would be the ending account balance if the interest is compounded daily?

c. Are these two answers similar? Why or why not?

Logarithmic Functions and Their Graphs

TOPICS

1. Definition of logarithmic functions

2. Graphing logarithmic functions

3. Evaluating elementary logarithmic expressions

4. Common and natural logarithms

T. Inputting logarithms, graphing logarithmic functions

TOPIC 1

Definition of Logarithmic Functions

Currently, we are only able to solve a small subset of possible exponential equations.

Solvable	Not Easily Solvable Yet
We can solve the equation $$2^x = 8$$ by writing 8 as 2^3 and equating exponents. This is an example of an elementary exponential equation we learned to solve in Section 5.1.	We cannot solve the equation $$2^x = 9$$ in the same way, although this equation is only slightly different. All we can say at the moment is that x must be a bit larger than 3.
If we know $A, P, n,$ and $t,$ we can solve the compound interest equation $$A = P\left(1 + \frac{r}{n}\right)^{nt}$$ for the annual interest rate $r,$ as we did in Example 4 of Section 5.2.	If we know $A, P,$ and $t,$ it is not so easy to determine the annual interest rate r from the equation $$A = Pe^{rt},$$ for calculating interest that is continuously compounded.

In both of the first two equations, the variable x appears in the exponent (making them exponential equations), and we aren't able to rewrite the equation with the variable outside the exponent. We can only solve the first equation because we can rewrite 8 as a power of 2, and even this doesn't remove the variable from the exponent.

In the second pair of equations, we are able to solve for r in the first case, but it is difficult in the continuous-compounding case: once again the variable is inconveniently "stuck" in the exponent. We need a way of undoing exponentiation.

This might seem familiar. We have "undone" functions before (whether we realized it at the time or not) by finding inverses of functions. Therefore, we need to find the inverse of the general exponential function $f(x) = a^x$. Fortunately, we know before we start that $f(x) = a^x$ has an inverse, because the graph of f passes the horizontal line test (regardless of whether $0 < a < 1$ or $a > 1$).

If we apply the algorithm for finding inverses of functions, we have the following:

$f(x) = a^x$ Rewrite the function as an equation by replacing $f(x)$ with y.

$y = a^x$

$x = a^y$ Then interchange x and y and proceed to solve for y.

$y = ?$ At this point, we are stuck again. What is y?

No concept or notation that we have encountered up to this point allows us to solve the equation $x = a^y$ for the variable y, and this is the reason for introducing a new class of functions called *logarithms*.

DEFINITION

Logarithmic Functions

Let a be a fixed positive real number not equal to 1. The **logarithmic function with base a** is defined to be the inverse of the exponential function with base a, and is denoted $\log_a x$. In symbols, if $f(x) = a^x$, then $f^{-1}(x) = \log_a x$.

In equation form, the definition of logarithm means that the equations

$$x = a^y \text{ and } y = \log_a x$$

are equivalent. Note that a is the base in both equations: either the base of the exponential function or the base of the logarithmic function.

EXAMPLE 1

Exponential and Logarithmic Equations

Use the definition of logarithmic functions to rewrite the following exponential equations as logarithmic equations.

a. $8 = 2^3$ **b.** $5^4 = 625$ **c.** $7^x = z$

Then rewrite the following logarithmic equations as exponential equations.

d. $\log_3 9 = 2$ **e.** $3 = \log_8 512$ **f.** $y = \log_2 x$

Solutions:

a. $3 = \log_2 8$ Note that in each case, the base of the exponential
b. $4 = \log_5 625$ equation is the base of the logarithmic equation.

c. $x = \log_7 z$

d. $3^2 = 9$

e. $8^3 = 512$

f. $2^y = x$

TOPIC 2

Graphing Logarithmic Functions

Since logarithmic functions are inverses of exponential functions, we can learn a great deal about the graphs of logarithmic functions by recalling the graphs of exponential functions. For instance, the domain of $\log_a x$ (for any allowable a) is the positive real numbers, because the positive real numbers constitute the range of a^x (the domain of f^{-1} is the range of f). Similarly, the range of $\log_a x$ is the entire set of real numbers, because the real numbers make up the domain of a^x.

Recall that the graphs of a function and its inverse are reflections of one another with respect to the line $y = x$. Since exponential functions come in two forms based on the value of a ($0 < a < 1$ and $a > 1$), logarithmic functions also fall into two categories. In both of the graphs in Figure 1, the green curve is the graph of an exponential function and the purple curve is the corresponding logarithmic function.

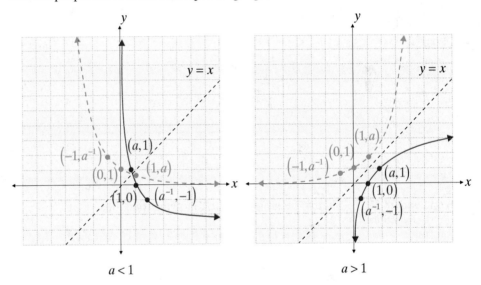

Figure 1: The Two Classes of Logarithmic Functions

In all cases, the points $(1,0)$, $(a,1)$, and $\left(a^{-1},-1\right)$ lie on the graph of a logarithmic function with base a. Frequently, these points will be enough to get a good idea of the shape of the graph.

Note that the domain of each of the logarithmic functions in Figure 1 is indeed $(0, \infty)$, and that the range in each case is $(-\infty, \infty)$. Also note that the y-axis is a vertical asymptote for both, and that neither has a horizontal asymptote.

CAUTION!

While it may appear so on the graph, logarithmic functions do **not** have a horizontal asymptote. The reason that logarithmic functions often look like they have a horizontal asymptote is because they are among the slowest growing functions in mathematics! Consider the function $f(x) = \log_2 x$. We have $f(1024) = 10$, $f(2048) = 11$, and $f(4096) = 12$. While the function may increase its value very slowly as x increases, it never approaches an asymptote.

EXAMPLE 2

Graphing Logarithmic Functions

Sketch the graphs of the following logarithmic functions.

a. $f(x) = \log_3 x$

b. $g(x) = \log_{\frac{1}{2}} x$

Note:
Once again, plotting a few key points will provide a good idea of the shape of the function.

Solutions:

a.

b.

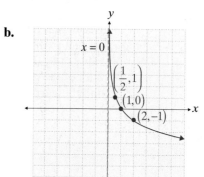

EXAMPLE 3

Graphing Logarithmic Functions

Sketch the graphs of the following functions.

a. $f(x) = \log_3 (x + 2) + 1$

b. $g(x) = \log_2 (-x - 1)$

c. $h(x) = \log_{\frac{1}{2}} (x) - 2$

Note:
Begin by graphing the base function, then apply any transformations.

Solutions:

a.

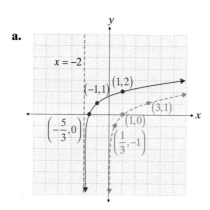

Begin by graphing the base function, which is $\log_3 x$. This is shown in green.

Since x has been replaced by $x + 2$, we shift the graph 2 units to the left.

To find the graph of f, we shift the result up 1 unit, since 1 has been added to the function. This graph is shown in purple.

Note that the asymptote has also shifted to the left.

b.

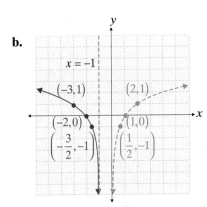

The basic shape of the graph of g is the same as the shape of $y = \log_2 x$, drawn in green.

To obtain g from $\log_2 x$, the variable x is replaced with $x - 1$, which shifts the graph 1 unit to the right, and then x is replaced by $-x$, which reflects the graph with respect to the y-axis.

c.

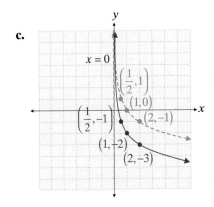

We begin with the green curve, which is the graph of $\log_{\frac{1}{2}} x$.

We then shift the graph 2 units down to obtain the graph of the function h.

TOPIC 3

Evaluating Elementary Logarithmic Expressions

Now that we have graphed logarithmic functions, we can augment our understanding of their behavior with a few algebraic observations. These will enable us to evaluate some logarithmic expressions and solve some elementary logarithmic equations.

First, our work in Example 2 suggests that the point $(1, 0)$ is always on the graph of $\log_a x$ for any allowable base a, and this is indeed the case. A similar observation is that $(a, 1)$ is always on the graph of $\log_a x$. These two facts are actually just restatements of two corresponding facts about exponential functions, a consequence of the definition of logarithms:

$$\log_a 1 = 0 \text{, because } a^0 = 1$$

$$\log_a a = 1 \text{, because } a^1 = a$$

More generally, we can use the fact that the functions $\log_a x$ and a^x are inverses of one another to write:

$$\log_a \left(a^x \right) = x \text{ and } a^{\log_a x} = x$$

In Example 4, we use the first statement to evaluate logarithmic expressions. Note the similarity to solving exponential equations; we rewrite the argument of the logarithm as a power of the base, allowing us to simplify the expression.

━━━━━━ **EXAMPLE 4** ━━━━━━

Logarithmic Expressions

Evaluate the following logarithmic expressions.

a. $\log_5 25$ **b.** $\log_{\frac{1}{2}} 2$ **c.** $\log_\pi \sqrt{\pi}$

d. $\log_{17} 1$ **e.** $\log_{16} 4$ **f.** $\log_{10}\left(\dfrac{1}{100}\right)$

Note:
We write each of the equivalent exponential equations as a reference.

Solutions:

a. $\log_5 25 = \log_5 5^2$ Rewrite 25 as a power of 5.

$\qquad\quad = 2$ Equivalent exponential equation: $25 = 5^2$

b. $\log_{\frac{1}{2}} 2 = \log_{\frac{1}{2}}\left(\dfrac{1}{2}\right)^{-1}$ Rewrite 2 as a power of $\dfrac{1}{2}$.

$\qquad\quad = -1$ Equivalent exponential equation: $2 = \left(\dfrac{1}{2}\right)^{-1}$

c. $\log_\pi \sqrt{\pi} = \log_\pi \pi^{\frac{1}{2}}$ Rewrite the radical as a rational exponent.

$\qquad\quad = \dfrac{1}{2}$ Equivalent exponential equation: $\sqrt{\pi} = \pi^{\frac{1}{2}}$

d. $\log_{17} 1 = 0$ Equivalent exponential equation: $1 = 17^0$

e. $\log_{16} 4 = \log_{16} 16^{\frac{1}{2}}$ Rewrite 4 as a power of 16.

$\qquad\quad = \dfrac{1}{2}$ Equivalent exponential equation: $4 = 16^{\frac{1}{2}}$

f. $\log_{10}\left(\dfrac{1}{100}\right) = \log_{10} 10^{-2}$ Rewrite $\dfrac{1}{100}$ as a power of 10.

$\qquad\quad = -2$ Equivalent exponential equation: $\dfrac{1}{100} = 10^{-2}$

In Example 5, we use both statements to solve logarithmic and exponential equations. Observe how converting between logarithmic and exponential form can make an equation easier to solve.

EXAMPLE 5

Solving Logarithmic
Equations

Solve the following equations involving logarithms.

a. $\log_6(2x) = -1$ **b.** $3^{\log_{3x} 2} = 2$ **c.** $\log_2 8^x = 5$

Solutions:

a. $\log_6(2x) = -1$

$2x = 6^{-1}$ Convert the equation to exponential form.

$2x = \dfrac{1}{6}$ Simplify the exponent, then solve for x.

$x = \dfrac{1}{12}$

b. $3^{\log_{3x} 2} = 2$ This time, we convert from the exponential form to the logarithmic form.

$\log_{3x} 2 = \log_3 2$

$3x = 3$ We can equate the bases, just like we equate the exponents in an elementary exponential equation.

$x = 1$

c. $\log_2 8^x = 5$

$8^x = 2^5$ Rewrite the equation in exponential form.

$\left(2^3\right)^x = 2^5$ Rewrite 8 as a power of 2.

$2^{3x} = 2^5$ Simplify using properties of exponents.

$3x = 5$ Set the exponents equal to each other, then solve.

$x = \dfrac{5}{3}$

TOPIC Common and Natural Logarithms

We have already mentioned the fact that the number e, called the natural base, plays a fundamental role in many important real-world situations and in higher mathematics, so it is not surprising that the logarithmic function with base e is worthy of special attention. For historical reasons, namely the fact that our number system is based on powers of 10, the logarithmic function with base 10 is also singled out.

DEFINITION

Common and Natural
Logarithms

- The function $\log_{10} x$ is called the **common logarithm**, and is usually written $\log x$.

- The function $\log_e x$ is called the **natural logarithm**, and is usually written $\ln x$.

Another way in which these particular logarithms are special is that most calculators, if they are capable of calculating logarithms at all, are only equipped to evaluate common and natural logarithms. Such calculators normally have a button labeled "LOG" for the common logarithm and a button labeled "LN" for the natural logarithm.

PROPERTIES

Properties of Natural Logarithms

$\ln x = y \Leftrightarrow e^y = x$

Properties	**Reasons**
1. $\ln 1 = 0$	Raise e to the power 0 to get 1.
2. $\ln e = 1$	Raise e to the power 1 to get e.
3. $\ln e^x = x$	Raise e to the power x to get e^x.
4. $e^{\ln x} = x$	$\ln x$ is the power to which e must be raised to get x.

EXAMPLE 6

Evaluating Logarithmic Expressions

Evaluate the following logarithmic expressions.

a. $\ln \sqrt[3]{e}$ **b.** $\log 1000$

c. $\ln(4.78)$ **d.** $\log(10.5)$

Solutions:

a. $\ln \sqrt[3]{e} = \ln e^{\frac{1}{3}} = \dfrac{1}{3}$

No calculator is necessary for this problem, just an application of an elementary property of logarithms.

b. $\log 1000 = \log 10^3 = 3$

Again, no calculator is required.

c. $\ln(4.78) \approx 1.564$

This time, a calculator is needed, and only an approximate answer can be given. Be sure to use the correct logarithm.

d. $\log(10.5) \approx 1.021$

Again, we must use a calculator, though we can say beforehand that the answer should be only slightly larger than 1, as $\log 10 = 1$ and 10.5 is only slightly larger than 10.

TOPIC   T   Inputting Logarithms, Graphing Logarithmic Functions

Common and natural logarithms can be entered into a graphing calculator using the **LOG** and **LN** buttons, respectively. For instance, to graph the function $f(x) = \log(3-x)$, press **Y=** and **LOG**. The opening parenthesis appears automatically, so we just have to type in the argument, $3-x$, and the right-hand parenthesis.

1. $4 = \log_5 625$

2. $3 = \log_6 216$

3. $3 = \log_x 27$

4. $2 = \log_b 3.2$

5. $3 = \log_{4.2} C$

6. $2 = \log_{1.3} V$

7. $x = \log_4 31$

8. $2x = \log_{16} 215$

9. $\sqrt{3} = \log_{4x} 13$

10. $x = \log_e \pi$

11. $e^x = \log_2 11$

12. $e = \log_4 N$

13. $81 = 3^4$

14. $\frac{1}{8} = 2^{-3}$

15. $4 = b^{\frac{1}{2}}$

16. $9 = y^2$

17. $15 = 2^b$

18. $8 = 5^d$

19. $W = 5^{12}$

20. $T = 7^6$

21. $2x = \pi^4$

22. $2\pi = \left(\sqrt{3}\right)^x$

23. $e^x = 2$

24. $e^3 = 5x$

25.

26.

27.

The following graph should appear:

CAUTION!

Notice that the graph appears to cut off around the point where $x = 3$. We know, however, that the graph has a vertical asymptote at $x = 3$, and approaches that asymptote even though it does not show on the graphing calculator.

Exercises

Write the following equations in logarithmic terms.

1. $625 = 5^4$ **2.** $216 = 6^3$ **3.** $x^3 = 27$

4. $b^2 = 3.2$ **5.** $4.2^3 = C$ **6.** $1.3^2 = V$

7. $4^x = 31$ **8.** $16^{2x} = 215$ **9.** $(4x)^{\sqrt{3}} = 13$

10. $e^x = \pi$ **11.** $2^{e^x} = 11$ **12.** $4^e = N$

Write the following logarithmic equations as exponential equations.

13. $\log_3 81 = 4$ **14.** $\log_2 \frac{1}{8} = -3$ **15.** $\log_b 4 = \frac{1}{2}$

16. $\log_y 9 = 2$ **17.** $\log_2 15 = b$ **18.** $\log_5 8 = d$

19. $\log_5 W = 12$ **20.** $\log_7 T = 6$ **21.** $\log_\pi 2x = 4$

22. $\log_{\sqrt{3}} 2\pi = x$ **23.** $\ln 2 = x$ **24.** $\ln 5x = 3$

28.

29.

30.

31.

32.

33.

34.

35.

36.

37. e

38. g

39. b

40. f

41. h

Sketch the graphs of the following functions. See Examples 2 and 3.

25. $f(x) = \log_3(x - 1)$

26. $g(x) = \log_5(x + 2) - 1$

27. $r(x) = \log_{\frac{1}{2}}(x - 3)$

28. $p(x) = 3 - \log_2(x + 1)$

29. $q(x) = \log_3(2 - x)$

30. $s(x) = \log_{\frac{1}{3}}(5 - x)$

31. $h(x) = \log_7(x - 3) + 3$

32. $m(x) = \log_{\frac{1}{2}}(1 - x)$

33. $f(x) = \log_3(6 - x)$

34. $p(x) = 4 - \log(x + 3)$

35. $s(x) = -\log_{\frac{1}{3}}(-x)$

36. $g(x) = \log_5(2x) - 1$

Match the graph of the appropriate equation to the logarithmic function.

37. $f(x) = \log_2 x - 1$ **38.** $f(x) = \log_2(2 - x)$ **39.** $f(x) = \log_2(-x)$

40. $f(x) = \log_2(x - 3)$ **41.** $f(x) = 1 - \log_2 x$ **42.** $f(x) = -\log_2 x$

43. $f(x) = -\log_2(-x)$ **44.** $f(x) = \log_2 x$ **45.** $f(x) = \log_2 x + 3$

a.

b.

c.

d.

e.

f.

g.

h.

i.

42. c

43. d

44. a

45. i

46. $\dfrac{1}{2}$

47. -2

48. -2

49. 3

50. $\dfrac{1}{3}$

51. -1

52. 2.89

53. -4

54. $\dfrac{5}{3}$

55. -1

56. 1

57. 0

58. $\dfrac{1}{5}$

59. $-\dfrac{1}{2}$

60. 2

61. $\{8\}$

62. $\{64\}$

63. $\left\{\dfrac{1}{8}\right\}$

64. $\{9\}$

65. $\{1\}$

66. $\left\{-\dfrac{1}{2}\right\}$

67. $\{-2\}$

68. $\left\{\dfrac{1}{21}\right\}$

69. No solution

70. $\{10\}$

71. $\{2\}$

72. $\{36\}$

73. $\{41.96\}$

Evaluate the following logarithmic expressions without the use of a calculator. See Examples 4 and 6.

46. $\log_7 \sqrt{7}$

47. $\log_{\frac{1}{2}} 4$

48. $\log_9\left(\dfrac{1}{81}\right)$

49. $\log_3 27$

50. $\log_{27} 3$

51. $\log_3\left(\log_{27} 3\right)$

52. $\ln e^{2.89}$

53. $\log 0.0001$

54. $\log_a a^{\frac{5}{3}}$

55. $\ln\left(\dfrac{1}{e}\right)$

56. $\log\left(\log\left(10^{10}\right)\right)$

57. $\log_3 1$

58. $\ln \sqrt[5]{e}$

59. $\log_{\frac{1}{16}} 4$

60. $\log_8 4^{\log 1000}$

Use the elementary properties of logarithms to solve the following equations. See Example 5.

61. $\log_{16} x = \dfrac{3}{4}$

62. $\log_{16} x^{\frac{1}{2}} = \dfrac{3}{4}$

63. $\log_{16} x = -\dfrac{3}{4}$

64. $\log_5 5^{\log_3 x} = 2$

65. $\log_a a^{\log_b x} = 0$

66. $\log_3 9^{2x} = -2$

67. $\log_{\frac{1}{3}} 3^x = 2$

68. $\log_7(3x) = -1$

69. $4^{\log_3 x} = 0$

70. $\log x^{10} = 10$

71. $\log_x\left(\log_{\frac{1}{2}} \dfrac{1}{4}\right) = 1$ **72.** $6^{\log_x e^2} = e$

Solve the following logarithmic equations, using a calculator if necessary to evaluate the logarithms. See Examples 5 and 6. Express your answer either as a fraction or a decimal rounded to the nearest hundredth.

73. $\log(3x) = 2.1$

74. $\log x^2 = -2$

75. $\ln(x+1) = 3$

76. $\ln 2x = -1$

77. $\ln e^x = 5.6$

78. $\ln\left(\ln x^2\right) = 0$

79. $\log 19 = 3x$

80. $\log e^x = 5.6$

81. $\log 300^{\log x} = 9$

82. $\log_9(2x-1) = 2$ **83.** $\log\left(\log(x-2)\right) = 1$

74. $\left\{\pm\dfrac{1}{10}\right\}$

75. $\{19.09\}$

76. $\{0.18\}$

77. $\{5.60\}$

78. $\left\{\pm\sqrt{e}\right\}$ or $\{\pm 1.65\}$

79. $\{0.43\}$

80. $\{12.89\}$

81. $\{4297.83\}$

82. $\{41\}$

83. $\{10,000,000,002\}$

Properties and Applications of Logarithms

TOPICS

1. Properties of logarithms

2. The change of base formula

3. Applications of logarithmic functions

TOPIC [1]

Properties of Logarithms

In Section 5.3, we introduced logarithmic functions and studied some of their elementary properties. The motivation was our inability, at that time, to solve certain exponential equations. Let us reconsider the two sample problems that initiated our discussion of logarithms and see if we have made progress. We begin with the continuously compounding interest problem.

EXAMPLE 1

Continuously Compounded Interest

Anne reads an ad in the paper for a new bank in town. The bank is advertising "continuously compounded savings accounts" in an attempt to attract customers, but fails to mention the annual interest rate. Curious, she goes to the bank and is told by an account agent that if she were to invest, $10,000 in an account, her money would grow to $10,202.01 in one year's time. But, strangely, the agent also refuses to divulge the yearly interest rate. What rate is the bank offering?

Solution:

We need to solve the equation $A = Pe^{rt}$ for r, given that $A = 10,202.01$, $P = 10,000$, and $t = 1$.

$$10,202.01 = 10,000e^{r(1)}$$ Substitute the given values.

$$1.020201 = e^r$$ Divide both sides by 10,000.

$$\ln(1.020201) = r$$ Convert to logarithmic form.

$$r \approx 0.02$$ Evaluate using a calculator.

Note that we use the natural logarithm since the base of the exponential function is e. While we must use a calculator, we can now solve the equation for r.

EXAMPLE 2

Solving Exponential Equations

Solve the equation $2^x = 9$.

Solution:

We convert the equation to logarithmic form to obtain the solution $x = \log_2 9$. Unfortunately, this answer still doesn't tell us anything about x in decimal form, other than that it is bound to be slightly more than 3. Further, we can't use a calculator to evaluate $\log_2 9$ since the base is neither 10 nor e.

As Example 2 shows, our ability to work with logarithms is still incomplete. In this section we will derive some important properties of logarithms that allow us to solve more complicated equations, as well as provide a decimal approximation to the solution of $2^x = 9$.

The following properties of logarithmic functions are analogs of corresponding properties of exponential functions, a consequence of how logarithms are defined.

PROPERTIES

Properties of Logarithms

Let a (the logarithmic base) be a positive real number not equal to 1, let x and y be positive real numbers, and let r be any real number.

1. $\log_a(xy) = \log_a x + \log_a y$ ("the log of a product is the sum of the logs")

2. $\log_a\left(\dfrac{x}{y}\right) = \log_a x - \log_a y$ ("the log of a quotient is the difference of the logs")

3. $\log_a(x^r) = r\log_a x$ ("the log of something raised to a power is the power times the log")

We illustrate the link between these properties and the related properties of exponents by proving the first one. Try proving the second and third as further practice.

Proof: Let $m = \log_a x$ and $n = \log_a y$. The equivalent exponential forms of these two equations are $x = a^m$ and $y = a^n$.

Since we are interested in the product xy, note that

$$xy = a^m a^n = a^{m+n}.$$

The statement $xy = a^{m+n}$ can then be converted to logarithmic form, giving us

$$\log_a(xy) = m + n.$$

Referring back to the definition of m and n, we have $\log_a(xy) = \log_a x + \log_a y$.

If the properties of logarithms appear strange at first, remember that they are just the properties of exponents restated in logarithmic form.

CAUTION!

Errors in working with logarithms often arise from incorrect recall of the logarithmic properties. The table below highlights some common mistakes.

Incorrect Statements	Correct Statements
$\log_a (x + y) = \log_a x + \log_a y$	$\log_a (xy) = \log_a x + \log_a y$
$\log_a (xy) = (\log_a x)(\log_a y)$	$\log_a (xy) = \log_a x + \log_a y$
$\dfrac{\log_a x}{\log_a y} = \log_a x - \log_a y$	$\log_a \left(\dfrac{x}{y}\right) = \log_a x - \log_a y$
$\dfrac{\log_a x}{\log_a y} = \log_a \left(\dfrac{x}{y}\right)$	$\log_a \left(\dfrac{x}{y}\right) = \log_a x - \log_a y$
$\dfrac{\log_a (xz)}{\log_a (yz)} = \dfrac{\log_a x}{\log_a y}$	$\dfrac{\log_a (xz)}{\log_a (yz)} = \dfrac{\log_a x + \log_a z}{\log_a y + \log_a z}$

In some situations, we will find it useful to use properties of logarithms to decompose a complicated expression into a sum or difference of simpler expressions, while in other situations we will do the reverse, combining a sum or a difference of logarithms into one logarithm. Examples 3 and 4 illustrate these processes.

EXAMPLE 3

Expanding Logarithmic Expressions

Use the properties of logarithms to expand the following expressions as much as possible (that is, decompose the expressions into sums or differences of the simplest possible terms).

a. $\log_4 \left(64x^3 \sqrt{y}\right)$ **b.** $\log_a \sqrt[3]{\dfrac{xy^2}{z^4}}$ **c.** $\log\left(\dfrac{2.7 \times 10^4}{x^{-2}}\right)$

Solutions:

Note:
As long as the base is the same for each term, its value does not affect the use of the properties.

a. $\log_4 \left(64x^3 \sqrt{y}\right) = \log_4 64 + \log_4 x^3 + \log_4 \sqrt{y}$ Use the first property to rewrite the expression as three terms.

$$= \log_4 4^3 + \log_4 x^3 + \log_4 y^{\frac{1}{2}}$$

$$= 3 + 3\log_4 x + \frac{1}{2}\log_4 y$$

We can evaluate the first term and rewrite the second and third terms using the third property.

b. $\log_a \sqrt[3]{\dfrac{xy^2}{z^4}} = \log_a \left(\dfrac{xy^2}{z^4} \right)^{\frac{1}{3}}$ Rewrite the radical as an exponent.

$= \dfrac{1}{3} \log_a \left(\dfrac{xy^2}{z^4} \right)$ Bring the exponent in front of the logarithm using the third property.

$= \dfrac{1}{3} \left(\log_a x + \log_a y^2 - \log_a z^4 \right)$ Expand the expression using the first two properties.

$= \dfrac{1}{3} \left(\log_a x + 2 \log_a y - 4 \log_a z \right)$ Apply the third property to the terms that result.

c. Recall that if a base is not explicitly written, it is assumed to be 10. This base is convenient when working with numbers in scientific notation.

$\log \left(\dfrac{2.7 \times 10^4}{x^{-2}} \right) = \log(2.7) + \log(10^4) - \log x^{-2}$ Expand using the first and second properties.

$= \log(2.7) + 4 + 2 \log x$ Evaluate the first two terms and use the third property on the last term.

$\approx 4.43 + 2 \log x$

It is appropriate to either evaluate $\log(2.7)$ or leave it in exact form. Use the context of the problem to decide which form is more convenient.

EXAMPLE 4

Condensing Logarithmic Expressions

Use the properties of logarithms to condense the following expressions as much as possible (that is, rewrite the expressions as a sum or difference of as few logarithms as possible).

a. $2 \log_3 \left(\dfrac{x}{3} \right) - \log_3 \left(\dfrac{1}{y} \right)$

b. $\ln x^2 - \dfrac{1}{2} \ln y + \ln 2$

c. $\log_b 5 + 2 \log_b x^{-1}$

Solutions:

Note:
Often, there will be multiple orders in which we can apply the properties to find the final result.

a. $2 \log_3 \left(\dfrac{x}{3} \right) - \log_3 \left(\dfrac{1}{y} \right) = \log_3 \left(\dfrac{x}{3} \right)^2 + \log_3 \left(\dfrac{1}{y} \right)^{-1}$ Use the third property to make the coefficients appear as exponents.

$= \log_3 \left(\dfrac{x^2}{9} \right) + \log_3 y$ Evaluate the exponents.

$= \log_3 \left(\dfrac{x^2 y}{9} \right)$ Combine terms using the first property.

b. $\ln x^2 - \dfrac{1}{2}\ln y + \ln 2 = \ln x^2 - \ln y^{\frac{1}{2}} + \ln 2$

Rewrite each term to have a coefficient of 1 or −1 using the third property.

$$= \ln\left(\dfrac{x^2}{y^{\frac{1}{2}}}\right) + \ln 2$$

We can then combine the terms using the second property.

$$= \ln\left(\dfrac{2x^2}{y^{\frac{1}{2}}}\right) \text{ or } \ln\left(\dfrac{2x^2}{\sqrt{y}}\right)$$

The final answer can be written in several different ways, two of which are shown.

c. $\log_b 5 + 2\log_b x^{-1} = \log_b 5 + \log_b x^{-2}$

Rewrite the coefficient as an exponent, then combine terms.

$$= \log_b 5x^{-2} \text{ or } \log_b\left(\dfrac{5}{x^2}\right)$$

TOPIC 2

The Change of Base Formula

The properties we just derived can be used to provide an answer to a question about logarithms that has been left unanswered thus far. A specific illustration of the question arose in Example 2: how do we determine the decimal form of a number like $\log_2 9$?

Surprisingly, to answer this question we will undo our work in Example 2. We assign a variable to the result $\log_2 9$, convert the resulting logarithmic equation into exponential form, take the natural logarithm of both sides, and then solve for the variable.

$x = \log_2 9$ — Let x equal the result from before, $\log_2 9$.

$2^x = 9$ — Convert the equation to exponential form.

$\ln\left(2^x\right) = \ln 9$ — Take the natural logarithm of both sides.

$x\ln 2 = \ln 9$ — Move the variable out of the exponent using the third property of logarithms.

$x = \dfrac{\ln 9}{\ln 2}$ — Simplify.

$x \approx 3.17$ — Evaluate with a calculator.

While using a logarithm with any base will give the correct solution to this problem, if a calculator is to be used to approximate the number $\log_2 9$, there are (for most calculators) only two good choices: the natural log and the common log. If we had done the work above with the common logarithm, the final answer would have been the same. That is,

$$\dfrac{\log 9}{\log 2} \approx 3.17.$$

And even though it would not be easy to evaluate,

$$\dfrac{\log_a 9}{\log_a 2} \approx 3.17$$

for any allowable logarithmic base a.

More generally, a logarithm with base b can be converted to a logarithm with base a through the same reasoning. This allows us to evaluate all logarithmic expressions.

THEOREM

Change of Base Formula

Let a and b both be positive real numbers, neither of them equal to 1, and let x be a positive real number. Then

$$\log_b x = \frac{\log_a x}{\log_a b}.$$

EXAMPLE 5

Change of Base Formula

Evaluate the following logarithmic expressions, using the base of your choice.

a. $\log_7 15$ **b.** $\log_{\frac{1}{2}} 3$ **c.** $\log_\pi 5$

Note:
Both the common and natural logarithm work in solving these problems.

Solutions:

a. $\log_7 15 = \dfrac{\ln 15}{\ln 7}$ Apply the change of base formula.

≈ 1.392 Evaluate using a calculator.

b. $\log_{\frac{1}{2}} 3 = \dfrac{\log 3}{\log\left(\dfrac{1}{2}\right)}$ Apply the change of base formula. This time we use the common logarithm.

≈ -1.585 Since the base of the logarithm is a fraction, we should expect a negative answer.

c. $\log_\pi 5 = \dfrac{\log 5}{\log \pi}$ Once again, we apply the change of base formula, then evaluate using a calculator.

≈ 1.406

TOPIC 3

Applications of Logarithmic Functions

Logarithms appear in many different contexts and have a wide variety of uses. This is due partly to the fact that logarithmic functions are the inverses of exponential functions, and partly to the logarithmic properties we have discussed. In fact, the mathematician who can be most credited for "inventing" logarithms, John Napier (1550–1617) of Scotland, was inspired in his work by the convenience of what we now call logarithmic properties.

Computationally, logarithms are useful because they relocate exponents as coefficients, thus making them easier to work with. Consider, for example, a very large number such as 3×10^{17} or a very small number such as 6×10^{-9}. The common logarithm (used because it has a base of 10) expresses these numbers on a more comfortable scale:

$$\log\left(3\times10^{17}\right) = \log 3 + \log\left(10^{17}\right) = 17 + \log 3 \approx 17.477$$

$$\log\left(6\times10^{-9}\right) = \log 6 + \log\left(10^{-9}\right) = -9 + \log 6 \approx -8.222$$

Napier, working long before the advent of electronic calculating devices, devised logarithms in order to take advantage of this property.

In chemistry, the concentration of hydronium ions in a solution determines its acidity. Since concentrations are small numbers that vary over many orders of magnitude, it is convenient to express acidity in terms of the pH scale, as follows.

DEFINITION

The pH Scale

The **pH** of a solution is defined to be $-\log\left[H_3O^+\right]$, where $\left[H_3O^+\right]$ is the concentration of hydronium ions in units of moles/liter. Solutions with a pH less than 7 are said to be *acidic*, while those with a pH greater than 7 are *basic*.

pH = 0   Battery Acid

pH = 1   Hydrochloric Acid Secreted by Stomach Lining

pH = 2   Lemon Juice, Gastric Acid, Vinegar

pH = 3   ⟵——— Grapefruit, Orange Juice

pH = 4   Tomato Juice

pH = 5   Soft Drinking Water

pH = 6   Urine, Saliva

pH = 7   Pure Water

pH = 8   Sea Water

pH = 9   Baking Soda

pH = 10   Great Salt Lake

pH = 11   Ammonia Solution

pH = 12   Soapy Water

pH = 13   Bleaches

pH = 14   Liquid Drain Cleaner

Figure 1: pH of Common Substances

EXAMPLE 6

The pH Scale

If a sample of orange juice is determined to have a $\left[H_3O^+ \right]$ concentration of 1.58×10^{-4} moles/liter, what is its pH?

Solution:

Applying the above formula (and using a calculator), the pH is equal to

$$pH = -\log\left(1.58 \times 10^{-4}\right) \approx -(-3.80) = 3.8.$$

After doing this calculation, the reason for the minus sign in the formula is more apparent. By multiplying the log of the concentration by −1, the pH of a solution is positive, which is convenient for comparative purposes.

The energy released during earthquakes can vary greatly, but logarithms provide a convenient way to analyze and compare the intensity of earthquakes.

DEFINITION

The Richter Scale

Earthquake intensity is measured on the **Richter scale** (named for the American seismologist Charles Richter, 1900–1985). In the formula that follows, I_0 is the intensity of a just-discernible earthquake, I is the intensity of an earthquake being analyzed, and R is its ranking on the Richter scale.

$$R = \log\left(\frac{I}{I_0}\right)$$

By this measure, earthquakes range from a classification of small ($R < 4.5$), to moderate ($4.5 \le R < 5.5$), to large ($5.5 \le R < 6.5$), to major ($6.5 \le R < 7.5$), and finally to greatest ($7.5 \le R$).

The base 10 logarithm means that every increase of 1 unit on the Richter scale corresponds to an increase by a factor of 10 in the intensity. This is a characteristic of all logarithmic scales. Also, note that a barely discernible earthquake has a rank of 0, since $\log 1 = 0$.

EXAMPLE 7

The Richter Scale

1. $3 + 3\log_5 x$

2. $2\ln x + \ln y - \ln 3$

3. $2 + \ln p - 3\ln q$

4. $2 + \log x$

5. $1 + \log_9 x$

   $-3\log_9 y$

6. $\dfrac{2}{3}\log_6 p$

   $-\dfrac{1}{3}\log_6 q$

7. $\dfrac{3}{2}\ln x + \ln p$

   $+5\ln q - 7$

8. $\dfrac{4}{5} + \dfrac{1}{5}\log_a b$

   $-\dfrac{2}{5}\log_a c$

9. $\log(2 + 3\log x)$

10. $2 + \log_3(x + 3y)$

11. $1 - \dfrac{1}{2}\log(x + y)$

12. $1 + \ln x$

The January 2001 earthquake in the state of Gujarat in India was $80,000,000$ times as intense as a 0-level earthquake. What was the Richter ranking of this devastating event?

Solution:

If we let I denote the intensity of the Gujarat earthquake, then $I = 80,000,000I_0$, so

$$R = \log\left(\frac{80,000,000I_0}{I_0}\right)$$
$$= \log\left(8 \times 10^7\right)$$
$$= \log(8) + \log\left(10^7\right)$$
$$= \log(8) + 7$$
$$\approx 7.9.$$

The Gujarat earthquake thus fell in the category of greatest on the Richter scale.

Sound intensity is another quantity that varies greatly, and the measure of how the human ear perceives intensity, in units called decibels, is very similar to the measure of earthquake intensity.

DEFINITION

The Decibel Scale

In the **decibel scale**, I_0 is the intensity of a just-discernible sound, I is the intensity of the sound being analyzed, and D is its decibel level:

$$D = 10\log\left(\frac{I}{I_0}\right)$$

Decibel levels range from 0 for a barely discernible sound, to 40 for the level of normal conversation, to 80 for heavy traffic, to 120 for a loud rock concert, and finally (as far as humans are concerned) to around 160, at which point the eardrum is likely to rupture.

EXAMPLE 8

The Decibel Scale

13. $\log_2 (y^2 + z) - 4\log_2 x - 4$

14. $1 + \log x$

15. $2\log_b x + \dfrac{1}{2}\log_b y - \log_b z$

16. $\ln 7 + 2\ln(x-3)$

17. $2 + \log_b a + b\log_b c$

18. x

19. $\log \dfrac{x}{y}$

20. $\log_5 \left(\dfrac{x}{y^2}\right)$

21. $\log_5 (x+5)$

22. $\ln x$

23. $\log_2 \left(x^{\frac{4}{3}} + 3x^{\frac{1}{3}}\right)$

24. $\log_7 \dfrac{x^{\frac{2}{5}}}{p^{\frac{1}{5}}q^{\frac{1}{5}}}$

25. $\ln\left(\dfrac{3p}{q^2}\right)$

26. $\log_5 \left(\dfrac{x}{y^2}\right)$

27. $\log\left(\dfrac{x-10}{x}\right)$

28. $\log a^3 b^3$

29. $\ln\left(\dfrac{z^2}{x^3 y^3}\right)$

30. $\log_2 4 = 2$

31. $\log_5 4$

32. $\log 3$

Given that $I_0 = 10^{-12}$ watts/meter^2, what is the decibel level of jet airliner's engines at a distance of 45 meters, for which the sound intensity is 50 watts/meter^2?

Solution:

$$D = 10\log\left(\dfrac{50}{10^{-12}}\right)$$

$$= 10\log\left(5 \times 10^{13}\right)$$

$$= 10\left(\log 5 + 13\right)$$

$$\approx 137$$

In other words, the sound level would probably not be literally ear-splitting, but it would be very painful.

Exercises

Use the properties of logarithms to expand the following expressions as much as possible. Simplify any numerical expressions that can be evaluated without a calculator. See Example 3.

1. $\log_5 \left(125x^3\right)$

2. $\ln\left(\dfrac{x^2 y}{3}\right)$

3. $\ln\left(\dfrac{e^2 p}{q^3}\right)$

4. $\log(100x)$

5. $\log_9 9xy^{-3}$

6. $\log_6 \sqrt[3]{\dfrac{p^2}{q}}$

7. $\ln\left(\dfrac{\sqrt{x^3}\, pq^5}{e^7}\right)$

8. $\log_a \sqrt[5]{\dfrac{a^4 b}{c^2}}$

9. $\log\left(\log\left(100x^3\right)\right)$

10. $\log_3 (9x + 27y)$

11. $\log\left(\dfrac{10}{\sqrt{x+y}}\right)$

12. $\ln\left(\ln\left(e^{ex}\right)\right)$

13. $\log_2 \left(\dfrac{y^2 + z}{16x^4}\right)$

14. $\log\left(\log\left(100{,}000^{2x}\right)\right)$

15. $\log_b \sqrt{\dfrac{x^4 y}{z^2}}$

16. $\ln\left(7x^2 - 42x + 63\right)$

17. $\log_b ab^2 c^b$

18. $\ln\left(\ln\left(e^{e^x}\right)\right)$

33. $\ln 45$

34. $\ln 6$

35. $\log_3 1 = 0$

36. $\log_7 \dfrac{1}{2}$

37. $\ln 12$

38. $\log 64$

39. $\log 11$

40. $\log_4 x^{12}$

41. $\log_8 \left(x^2 - y\right)$

42. $\log_{3x} 3x^2$

43. x^2

44. $\dfrac{y^2}{x^3}$

45. $\dfrac{e^2 p}{x}$

46. $3x^5$

47. $\dfrac{x^3}{y^4}$

48. $a^2 b$

49. x^2

50. x^2

51. 4

52. $e^2 x^3$

53. $12x^2$

54. 16

55. 2.04

56. −2.93

57. 0.95

58. −6.64

59. 0.95

60. −0.16

61. 2.45

62. 1.37

Use the properties of logarithms to condense the following expressions as much as possible, writing each answer as a single term with a coefficient of 1. See Example 4.

19. $\log x - \log y$

20. $\log_5 x - 2\log_5 y$

21. $\log_5 \left(x^2 - 25\right) - \log_5 \left(x - 5\right)$

22. $\ln \left(x^2 y\right) - \ln y - \ln x$

23. $\dfrac{1}{3}\log_2 x + \log_2 \left(x + 3\right)$

24. $\dfrac{1}{5}\left(\log_7 \left(x^2\right) - \log_7 \left(pq\right)\right)$

25. $\ln 3 + \ln p - 2\ln q$

26. $2\left(\log_5 \sqrt{x} - \log_5 y\right)$

27. $\log\left(x - 10\right) - \log x$

28. $2\log a^2 b - \log \dfrac{1}{b} + \log \dfrac{1}{a}$

29. $3\left(\ln \sqrt[3]{z^2} - \ln xy\right)$

30. $\log_2 \left(4x\right) - \log_2 x$

31. $\log_5 20 - \log_5 5$

32. $\log 30 - \log 2 - \log 5$

33. $\ln 15 + \ln 3$

34. $\ln 8 - \ln 4 + \ln 3$

35. $0.5\log_3 16 - \log_3 4$

36. $3\log_7 2 - 2\log_7 4$

37. $0.25\ln 81 + \ln 4$

38. $2\left(\log 4 - \log 1 + \log 2\right)$

39. $\log 11 + 0.5\log 9 - \log 3$

40. $3\log_4 \left(x^2\right) + \log_4 \left(x^6\right)$

41. $\log_8 \left(2x^2 - 2y\right) - 0.25\log_8 16$

42. $\log_{3x} x^2 + \log_{3x} 18 - \log_{3x} 6$

Use the properties of logarithms to write each of the following as a single term that does not contain a logarithm.

43. $5^{2\log_5 x}$

44. $10^{\log y^2 - 3\log x}$

45. $e^{2 - \ln x + \ln p}$

46. $e^{5\left(\ln \sqrt[5]{3} + \ln x\right)}$

47. $10^{\log x^3 - 4\log y}$

48. $a^{\log_a b + 4\log_a \sqrt{a}}$

49. $10^{2\log x}$

50. $10^{4\log x - 2\log x}$

51. $\log_4 16 \cdot \log_x x^2$

52. $e^{\ln x + 2 + \ln x^2}$

53. $4^{\log_4 (3x) + 0.5\log_4 \left(16x^2\right)}$

54. $4^{2\log_2 6 - \log_2 9}$

Evaluate the following logarithmic expressions. See Example 5.

55. $\log_4 17$

56. $2\log_{\frac{1}{3}} 5$

57. $\log_9 8$

58. $\log_2 0.01$

59. $\log_{12} 10.5$

60. $\log\left(\ln 2\right)$

61. $\log_6 3^4$

62. $\log_7 14.3$

63. $\log_{\frac{1}{2}} \pi^{-2}$

64. $\log_{\frac{1}{5}} 626$

65. $\ln\left(\log 123\right)$

66. $\log_{17} 0.041$

63. 3.30

64. –4.00

65. 0.74

66. –1.13

67. 1.20

68. 2

69. 1.86

70. 1.57

71. –1

72. 3.74

73. 3.85

74. –3.32

75. 0.77

76. –2.88

77. –1.76

78. –0.56

79. 2

80. 6

81. 7

82. –3

83. 1

84. 6

85. $4\sqrt{2} \approx 5.66$

86. 3.68

87. 9.05

88. 4.64

89. 2.08

90. 65,536

91. 12

92. 4.01

93. 1,048,576

67. $\log 16$

70. $\log_8 26$

73. $\log_{2.5} 34$

76. $\log_{0.4} 14$

68. $\log_3 9$

71. $\log_4 0.25$

74. $\log_{0.5} 10$

77. $\log_{0.2} 17$

69. $\log_5 20$

72. $\log_{1.8} 9$

75. $\log_4 2.9$

78. $\log_{0.16} 2.8$

Without using a calculator, evaluate the following expressions.

79. $\log_4 16$

82. $\log_4 \dfrac{1}{64}$

80. $\log_5 25^3$

83. $\ln e^{1.5} - \log_4 2$

81. $\ln e^4 + \ln e^3$

84. $\log_2 8^{(2\log_2 4 - \log_2 4)}$

Find the value of x in each of the following equations. Express your answer as exact as possible, or as a decimal rounded to the nearest hundredth.

85. $\log_x 1024 = 4$

88. $\log_4 625 = x$

91. $\log_{12} x = 1$

86. $\log_6 729 = x$

89. $\log_x 729 = 9$

92. $\log_x 16,807 = 7$

87. $\log_2 529 = x$

90. $\log_4 x = 8$

93. $\log_4 x = 10$

Solve the following application problems. See Examples 6, 7, and 8.

94. A certain brand of tomato juice has a $\left[H_3O^+\right]$ concentration of 3.16×10^{-6} moles/liter. What is the pH of this brand?

95. One type of detergent, when added to neutral water with a pH of 7, results in a solution with a $\left[H_3O^+\right]$ concentration that is 5.62×10^{-4} times weaker than that of the water. What is the pH of the solution?

96. What is the concentration of $\left[H_3O^+\right]$ in lemon juice with a pH of 3.2?

97. The 1994 Northridge, California earthquake measured 6.7 on the Richter scale. What was the intensity, relative to a 0-level earthquake, of this event?

98. How much stronger was the 2001 Gujarat earthquake (7.9 on the Richter scale) than the 1994 Northridge earthquake described in Exercise 97?

99. A construction worker operating a jackhammer would experience noise with an intensity of 20 watts/meter² if it weren't for ear protection. Given that $I_0 = 10^{-12}$ watts/meter², what is the decibel level for such noise?

100. A microphone picks up the sound of a thunderclap and measures its decibel level as 105. Given that $I_0 = 10^{-12}$ watts/meter², with what sound intensity did the thunderclap reach the microphone?

94. 5.5

95. 10.25

96. 6.31×10^{-4} moles/liter

97. 5,011,872 times stronger

98. a factor of 15.8

99. 133 decibels

100. 0.03 watts/meter^2

101. 7.62, yes

102. 13.4 decibels

103. a. 15.05 minutes

   b. 7:00 p.m.

   c. 112°F; no

101. Matt, a lifeguard, has to make sure that the pH of the swimming pool stays between 7.2 and 7.6. If the pH is out of this range, he has to add chemicals that alter the pH level of the pool. If Matt measures the $\left[H_3O^+ \right]$ concentration in the swimming pool to be 2.40×10^{-8} moles/liter, what is the pH? Does he need to change the pH by adding chemicals to the water?

102. The intensity of a cat's soft purring is measured to be 2.19×10^{-11}. Given that $I_0 = 10^{-12}$ watts/meter^2, what is the decibel level of this noise?

103. Newton's Law of Cooling states that the rate at which an object cools is proportional to the difference between the temperature of the object and the surrounding temperature. If C denotes the surrounding temperature and T_0 denotes the temperature at time $t = 0$, the temperature of an object at time t is given by $T(t) = C + (T_0 - C) e^{-kt}$, where k is a constant that depends on the particular object under discussion.

a. You are having friends over for tea and want to know how long after boiling the water it will be drinkable. If the temperature of your kitchen stays around 74°F and you found online that the constant k for tea is approximately 0.049, how many minutes after boiling the water will the tea be drinkable (you prefer your tea no warmer than 140°F)? Recall that water boils at 212°F.

b. As you intern for your local crime scene investigation department, you are asked to determine at what time a victim died. If you are told k is approximately 0.1947 for a human body and the body's temperature was 72°F at 1:00 a.m., and the body has been in a storage building at a constant 60°F, approximately what time did the victim die? Recall the average temperature for a human body is 98.6°F. Note in this situation, t is measured in hours.

c. When helping your father cook a turkey, you were told to remove the turkey when the thickest part had reached 180°F. If you remove the turkey and place it on the table in a room that is 72°F, and it cools to 155°F in 20 minutes, what will the temperature of the turkey be at lunch time (an hour and 15 minutes after the turkey is removed from the oven)? Should you warm the turkey before eating?

Exponential and Logarithmic Equations

TOPICS

1. Converting between exponential and logarithmic forms

2. Further applications of exponential and logarithmic equations

---

TOPIC 1

Converting Between Exponential and Logarithmic Forms

At this point, we have all the tools we need to solve the most common sorts of exponential and logarithmic equations. All that is left is to develop our skill in using the tools.

We have already solved many exponential and logarithmic equations, using elementary facts about exponential and logarithmic functions to obtain solutions. However, many equations require a bit more work to solve. While there is no algorithm to follow in dealing with more complicated equations, if a given equation doesn't yield a solution easily, try converting it from exponential form to logarithmic form or vice versa.

All of the properties of exponents and their logarithmic counterparts are of great use as well. For reference, the logarithmic properties that we have noted throughout Sections 5.3 and 5.4 are restated here.

PROPERTIES

Summary of Logarithmic Properties

1. The equations $x = a^y$ and $y = \log_a x$ are equivalent, and are, respectively, the exponential form and the logarithmic form of the same statement.

2. The inverse of the function $f(x) = a^x$ is $f^{-1}(x) = \log_a x$, and vice versa.

3. A consequence of the last point is that $\log_a (a^x) = x$ and $a^{\log_a x} = x$. In particular, $\log_a 1 = 0$ and $\log_a a = 1$.

4. $\log_a (xy) = \log_a x + \log_a y$ ("the log of a product is the sum of the logs")

5. $\log_a \left(\dfrac{x}{y} \right) = \log_a x - \log_a y$ ("the log of a quotient is the difference of the logs")

6. $\log_a (x^r) = r \log_a x$ ("the log of something raised to a power is the power times the log")

The next several examples illustrate typical uses of the properties, and how converting between the exponential and logarithmic forms of an equation can lead to a solution.

EXAMPLE 1

Solving Exponential Equations

Note:
This equation is not easily solved in exponential form, since the two sides of the equation do not have the same base.

Solve the equation $3^{2-5x} = 11$. Express the answer exactly and as a decimal approximation.

Solution:

There are two ways to convert the equation into logarithmic form. We will explore both, and see that they lead to the same answer.

The first method is to take the natural (or common) logarithm of both sides.

$$3^{2-5x} = 11$$

$$\ln\left(3^{2-5x}\right) = \ln 11 \qquad \text{Take the natural logarithm of both sides.}$$

$$(2-5x)\ln 3 = \ln 11 \qquad \text{Use properties of logarithms to bring the variable out of the exponent.}$$

$$2-5x = \frac{\ln 11}{\ln 3} \qquad \text{Divide both sides by ln 3.}$$

$$-5x = \frac{\ln 11}{\ln 3} - 2 \qquad \text{Simplify.}$$

$$x = -\frac{\ln 11}{5\ln 3} + \frac{2}{5} \qquad \text{An exact form of the answer.}$$

The second method is to rewrite the equation using the definition of logarithms, then apply the change of base formula to work with natural (or common) logarithms.

$$3^{2-5x} = 11 \qquad \text{Rewrite the equation using the definition of logarithms.}$$

$$2-5x = \log_3 11$$

$$2-5x = \frac{\ln 11}{\ln 3} \qquad \text{Rewrite the logarithmic term using the change of base formula.}$$

$$x = -\frac{\ln 11}{5\ln 3} + \frac{2}{5} \qquad \text{Applying the same algebra as above leads to the same exact answer.}$$

The key step to finding the exact answer is to remove the variable from the exponent, which is achieved by converting to logarithmic form.

The key step to finding a decimal approximation is to change the base (of the exponent and logarithm) to either e or 10, allowing the use of a calculator.

$$x = -\frac{\ln 11}{5\ln 3} + \frac{2}{5} \approx -0.037 \qquad \text{An approximate form of the answer.}$$

We can also use a calculator to verify this solution in the original equation.

================

EXAMPLE 2

================

Solving Exponential
Equations

Solve the equation $5^{3x-1} = 2^{x+3}$. Express the answer exactly and as a decimal approximation.

Solution:

As in the first example, taking a logarithm of both sides is the key. We will use the common logarithm this time, but the natural logarithm would work just as well.

$$5^{3x-1} = 2^{x+3}$$
$$\log\left(5^{3x-1}\right) = \log\left(2^{x+3}\right)$$ Take the logarithm of both sides.
$$(3x-1)\log 5 = (x+3)\log 2$$ Bring the exponents down using a property of logarithms, then multiply the terms out.
$$3x\log 5 - \log 5 = x\log 2 + 3\log 2$$
$$3x\log 5 - x\log 2 = 3\log 2 + \log 5$$
$$x(3\log 5 - \log 2) = 3\log 2 + \log 5$$ Collect the terms with x on one side, then factor out x.
$$x = \frac{3\log 2 + \log 5}{3\log 5 - \log 2} \approx 0.892$$ Simplify and evaluate with a calculator.

The exact answer could appear in many different forms, depending on the base of the logarithm chosen and the order of logarithmic properties used in simplifying the answer. We could simplify it further as follows:

$$x = \frac{3\log 2 + \log 5}{3\log 5 - \log 2} = \frac{\log 8 + \log 5}{\log 125 - \log 2} = \frac{\log 40}{\log\left(\frac{125}{2}\right)}.$$

================

EXAMPLE 3

================

Solving Logarithmic
Equations

Solve the equation $\log_7(3x - 2) = 2$.

Solution:

Note:
Since calculators can evaluate exponents of any base, it often does not matter what the base is when we convert logarithmic equations into their exponential forms.

Note that rewriting this equation using the change of base formula does not help, since the variable would still be trapped inside the logarithm. Instead, we use the definition of logarithms to rewrite the equation in exponential form.

$$\log_7(3x-2) = 2$$
$$3x - 2 = 7^2$$ Rewrite the equation in exponential form.
$$3x = 51$$ Simplify and solve for x.
$$x = 17$$

Solving Logarithmic
Equations

Note:
We need to check
for extraneous
solutions when
solving logarithmic
equations.

In particular,
remember that
logarithms of
negative numbers are
undefined.

Solve the equation $\log_5 x = \log_5 (2x + 3) - \log_5 (2x - 3)$.

Solution:

This is an example of a logarithmic equation that is not easily solved in logarithmic form. Once a few properties of logarithms have been utilized, the equation can be rewritten in a very familiar form.

$$\log_5 x = \log_5 (2x+3) - \log_5 (2x-3)$$

$$\log_5 x = \log_5 \left(\frac{2x+3}{2x-3}\right)$$ Combine terms using a property of logarithms.

$$x = \frac{2x+3}{2x-3}$$ Equate the arguments since each term has the same base.

$$x(2x-3) = 2x+3$$ Multiply both sides by $2x - 3$.

$$2x^2 - 3x = 2x+3$$ The result is a quadratic equation.

$$2x^2 - 5x - 3 = 0$$ Rewrite the equation with 0 on one side.

$$(2x+1)(x-3) = 0$$ Factor.

$$x = -\frac{1}{2}, 3$$ Solve using the Zero-Factor Property.

A crucial step remains! While these two solutions definitely solve the quadratic equation, we must check that they solve the initial logarithmic equation, as the process of solving logarithmic equations can introduce extraneous solutions. If we check our two potential solutions in the original equation, we quickly discover that only one of them is valid.

$$\log_5 \left(-\frac{1}{2}\right) = \log_5 \left(2\left(-\frac{1}{2}\right)+3\right) - \log_5 \left(2\left(-\frac{1}{2}\right)-3\right)$$

We can already see that $-\frac{1}{2}$ is not a solution to the equation, because logarithms of negative numbers are undefined. We move on to the second solution.

$$\log_5 3 = \log_5 (2(3)+3) - \log_5 (2(3)-3)$$ Substitute into the equation.

$$\log_5 3 = \log_5 9 - \log_5 3$$ Simplify.

$$\log_5 3 = \log_5 \left(\frac{9}{3}\right)$$ Combine the terms using a property of logarithms.

$$\log_5 3 = \log_5 3$$ A true statement.

Thus, the solution to the equation is the single value $x = 3$.

Further Applications of Exponential and Logarithmic Equations

We will conclude our discussion of exponential and logarithmic equations by revisiting some important applications.

EXAMPLE 5

Compounding Interest

Rita is saving up money for a down payment on a new car. She currently has $5500 but she knows she can get a loan at a lower interest rate if she can put down $6000. If she invests her $5500 in a money market account that earns an annual interest rate of 4.8% compounded monthly, how long will it take her to accumulate the $6000?

Solution:

We need to solve the compound interest formula for t, the amount of time Rita invests her money. Given that $P = 5500$, $A(t) = 6000$, $r = 0.048$, and $n = 12$, we have

$$6000 = 5500\left(1 + \frac{0.048}{12}\right)^{12t}.$$

Our solution needs to be a decimal approximation to be of practical use, so we need to use the natural or common logarithm to rewrite the equation in logarithmic form.

$$6000 = 5500\left(1 + \frac{0.048}{12}\right)^{12t}$$

$$\frac{6000}{5500} = \left(1 + \frac{0.048}{12}\right)^{12t} \qquad \text{Divide both sides by 5500.}$$

$$\ln\left(\frac{6000}{5500}\right) = \ln\left(\left(1 + \frac{0.048}{12}\right)^{12t}\right) \qquad \text{Take the natural logarithm of both sides.}$$

$$\ln\left(\frac{6000}{5500}\right) = 12t\ln(1.004) \qquad \text{Bring the variable out of the exponent using a property of logarithms.}$$

$$\frac{\ln\left(\frac{6000}{5500}\right)}{12\ln(1.004)} = t \qquad \text{Solve for } t.$$

$$t \approx 1.82$$

Since t is measured in years, the solution tells us that it will take a bit less than a year and 10 months for the $5500 to grow to $6000.

EXAMPLE 6

Radiocarbon Dating

We have already discussed how the radioactive decay of carbon-14 can be used to arrive at age estimates of carbon-based fossils, and we constructed an exponential function describing the rate of decay. A much more common form of the function is

$$A(t) = A_0 e^{-0.000121t},$$

where $A(t)$ is the mass of carbon-14 remaining after t years, and A_0 is the mass of carbon-14 initially. Use this formula to determine:

a. The half-life of carbon-14.

b. The age of a fossilized organism containing 1.5 grams of carbon-14, given that a living organism of the same size contains 2.3 grams of carbon-14.

Solutions:

a. We are looking for the value of t for which $A(t)$ is half of A_0, or $\frac{A_0}{2}$.

$$\frac{A_0}{2} = A_0 e^{-0.000121t}$$

$$\frac{1}{2} = e^{-0.000121t} \qquad \text{Divide both sides by } A_0.$$

$$\ln\left(\frac{1}{2}\right) = \ln\left(e^{-0.000121t}\right) \qquad \text{Since the base involved is } e, \text{ using the natural logarithm makes calculation much simpler.}$$

$$\ln\left(\frac{1}{2}\right) = -0.000121t \qquad \text{Simplify using a property of logarithms.}$$

$$\frac{\ln\left(\frac{1}{2}\right)}{-0.000121} = t \qquad \text{Solve for } t.$$

$$t \approx 5728 \text{ years} \qquad \text{Evaluate.}$$

Thus, as we saw before, the half-life of carbon-14 is approximately 5728 years.

b. We simply plug the given information into the formula and solve for t.

$$1.5 = 2.3 e^{-0.000121t}$$

$$\frac{1.5}{2.3} = e^{-0.000121t} \qquad \text{Divide both sides by 2.3.}$$

$$\ln\left(\frac{1.5}{2.3}\right) = -0.000121t \qquad \text{Take the natural logarithm of both sides and simplify using a property of logarithms.}$$

$$\frac{\ln\left(\frac{1.5}{2.3}\right)}{-0.000121} = t \qquad \text{Solve for } t.$$

$$t \approx 3533 \text{ years} \qquad \text{Evaluate.}$$

According to the radiocarbon dating, the fossil is about 3533 years old.

1. $x \approx 0.26$

2. $x \approx 0.80$

3. $x \approx 3.12$

4. $x \approx 0.72$

5. $x \approx 3.89$

6. $x \approx 2.30$

7. $x \approx -2.28$

8. $x \approx 5.93$

9. $x \approx 8.09$

10. $x \approx 1.43$

11. $x \approx 2.68$

12. $x \approx 15.21$

13. $x \approx -1.12$

14. $x \approx 12.30$

15. $x \approx 52.77$

16. $x \approx 3.86$

17. $x \approx \pm 0.71$

18. $x \approx 1.62$

19. $x = -12$

20. $x = 8$

21. $x \approx 1.32$

22. $x = 14$

23. $x \approx 3.27$

24. $x \approx 0.42$

25. $x = 125$

26. $x = 16$

27. $x = 5$

28. $x = 16$

29. $x \approx 40.17$

30. $x \approx 80.69$

31. $x = 35$

32. $x \approx 6.63$

33. $x \approx 9.38$

34. $x = 2$

35. $x = 4$

36. $x = \pm 2$

37. $x = \dfrac{1}{162}$

Exercises

Solve the following exponential and logarithmic equations (round decimal approximations to the nearest hundredth). See Examples 1 through 4.

1. $3e^{5x} = 11$

2. $4^{3-2x} = 7$

3. $11^{\frac{3}{x}} = 10$

4. $8^{3x+2} = 7^{2x+3}$

5. $e^{15-3x} = 28$

6. $10^{\frac{5}{x}} = 150$

7. $10^{2x+5} = e$

8. $e^{8x+e} = 8^{ex+8}$

9. $6^{x-7} = 7$

10. $2e^{3x} = 145$

11. $2^{6-x} = 10$

12. $e^{3x-6} = 10^{x+2}$

13. $e^{-4x-2} = 12$

14. $10^{x-9} = 2001$

15. $e^{x-4} = 4^{\frac{2x}{3}}$

16. $5^{x-2} = 20$

17. $8^{x^2+1} = 23$

18. $3^{\frac{4}{x}} = 15$

19. $6^{3x-4} = 36^{2x+4}$

20. $81^x = 3^{2x+16}$

21. $e^{2x} = 14$

22. $e^{4x} = e^{3x+14}$

23. $5^{5x-7} = 10^{2x}$

24. $10^{6x} = 3^{3x+4}$

25. $\log_5 x = 3$

26. $\log_2 x = 4$

27. $\log x + \log(4x) = 2$

28. $\log_4(x^2) - \log_4 x = 2$

29. $\ln(2x) - \ln 4 = 3$

30. $\ln(15x) - \ln 3 = 6$

31. $\log_4(x-3) + \log_4 2 = 3$

32. $\log_3 24 - \log_3 4 = x - 5$

33. $\log_5(8x) - \log_5 3 = 2$

34. $e^{2\ln x} = 4\log 10$

35. $9^{\log_3 x} = 16$

36. $3\log_8\left(512^{x^2}\right) = 36$

37. $\log_3(6x) - 2\log_3(6x) = 3$

38. $\ln(3e) = \log x$

39. $\ln\left(2^{4e^x}\right) = \ln\left(16^e\right)$

40. $\log(x-2) + \log(x+2) = 2$

41. $\log(x-3) + \log(x+3) = 4$

42. $\log_2(7x-4) = \log_2(16-3x)$

43. $\log_\pi(x-5) + \log_\pi(x+3) = \log_\pi(1-2x)$

44. $\log_3(x+3) + \log_3(x-5) = 2$

45. $\log x + \log(x-3) = 1$

38. $x \approx 125.49$

39. $x = 1$

40. $x \approx 10.20$

41. $x \approx 100.04$

42. $x = 2$

43. No solution

44. $x = 6$

45. $x = 5$

46. $x = 2$

47. $x = 8$

48. $x = 5$

49. $x = \dfrac{37}{8}$

50. $x = 4$

51. $x = 5$

52. $x = 3$

53. No solution

54. $x = 7$

55. $x = 6$

56. $x = 8$

57. $x = \sqrt{2} - 1$

58. $x = \ln 5$

59. $x = 2, 3$

60. $x = \ln 2$

61. $x = 1, 2$

62. $f(x) = \ln x$

63. $f(x) = \log 2x^2$

64. $f(x) = \ln 25x^2$

65. $f(x) = \ln 9x^2$

66. $f(x) = 3x - 3$

67. $f(x) = 256^x$

68. $f(x) = \ln x^{12}$

69. $f(x) = \ln 1 = 0$

70. $f(x) = \ln(x+1)$

71. $f(x) = \ln 5^x$

72. $f(x) = \log x^e - 1$

73. $f(x) = \ln 5$

46. $\log_7 (3x + 2) - \log_7 x = \log_7 4$

48. $\log_{12} (x - 2) + \log_{12} (x - 1) = 1$

50. $\ln(x + 1) + \ln(x - 2) = \ln(x + 6)$

51. $\log_4 (x - 3) + \log_4 (x - 2) = \log_4 (x + 1)$

52. $2\ln(x + 3) = \ln(12x)$

53. $\log_5 (x - 1) + \log_5 (x + 4) = \log_5 (x - 5)$

54. $\log_{255} (2x + 3) + \log_{255} (2x + 1) = 1$

56. $\log_6 (x + 1) + \log_6 (x - 4) = 2$

58. $e^{2x} - 3e^x - 10 = 0$ (**Hint:** First solve for e^x.)

59. $2^{2x} - 12(2^x) + 32 = 0$ (**Hint:** First solve for 2^x.)

60. $e^{2x} + 2e^x - 8 = 0$

47. $\log_2 x + \log_2 (x - 7) = 3$

49. $\log_3 (x + 1) - \log_3 (x - 4) = 2$

55. $\log_2 (x - 5) + \log_2 (x + 2) = 3$

57. $\ln(x + 2) + \ln(x) = 0$

61. $3^{2x} - 12(3^x) + 27 = 0$

Using the properties of logarithmic functions, simplify the following functions as much as possible. Write each function as a single term with a coefficient of 1, if possible.

62. $f(x) = 0.5\ln(x^2)$

64. $f(x) = 4\ln(\sqrt{5x})$

66. $f(x) = 3\ln(e^x) - 3$

68. $f(x) = 2\ln(x^3) + \ln(x^6)$

70. $f(x) = \ln(x^2 + x) - \ln x$

72. $f(x) = e^{\ln(\log x^e - 1)}$

63. $f(x) = 0.25\log(16x^8)$

65. $f(x) = 8\ln(\sqrt[4]{3x})$

67. $f(x) = 10^{2x\log 16}$

69. $f(x) = 2\ln(x^3) - \ln(x^6)$

71. $f(x) = 2\ln\left(5^{x\log_{20}(2\sqrt{5})}\right)$

73. $f(x) = 2\ln\left(5^{\log_4 2}\right)$

Solve the following application problems. See Examples 5 and 6.

74. Assuming that there are currently 6 billion people on Earth and a growth rate of 1.9% per year, how long will it take for the Earth's population to reach 20 billion?

75. How long does it take for an investment to double in value if:

    a. the investment is in a monthly compounded savings account earning 4% a year?

    b. the investment is in a savings account earning 7% a year that is continuously compounded?

74. About 64 years

75. a. 17.36 years

   b. 9.90 years

76. About 9079 years

77. 4.98 hours

78. $k \approx 0.1666$, so the coffee reaches 100°F when $t \approx 9.16$ minutes.

79. 4.99 years

80. a. 2.84 years

   b. 23.90 years

   c. $28,062.19

81. 0.271 years (about 99 days)

82. 216.9 days

76. Assuming a half-life of 5728 years, how long would it take for 3 grams of carbon-14 to decay to 1 gram?

77. Suppose a population of bacteria in a Petri dish has a doubling time of one and a half hours. How long will it take for an initial population of 10,000 bacteria to reach 100,000?

78. According to Newton's Law of Cooling, the temperature $T(t)$ of a hot object, at time t after being placed in an environment with a constant temperature C, is given by $T(t) = C + (T_0 - C)e^{-kt}$, where T_0 is the temperature of the object at time $t = 0$ and k is a constant that depends on the object. If a hot cup of coffee, initially at 190°F, cools to 125°F in 5 minutes when placed in a room with a constant temperature of 75°F, how long will it take for the coffee to reach 100°F?

79. Wayne has $12,500 in a high interest savings account at 3.66% annual interest compounded monthly. Assuming he makes no deposits or withdrawals, how long will it take for his investment to grow to $15,000?

80. Ben and Casey both open money market accounts with 4.9% annual interest compounded continuously. Ben opens his account with $8700 while Casey opens her account with $3100.

   a. How long will it take Ben's account to reach $10,000?

   b. How long will it take Casey's account to reach $10,000?

   c. How much money will be in Ben's account after the time found in part b?

81. Cesium-137 has a half-life of approximately 30 years. How long would it take for 160 grams of cesium-137 to decay to 159 grams?

82. A chemist, running tests on an unknown sample from an illegal waste dump, isolates 50 grams of what he suspects is a radioactive element. In order to help identify the element, he would like to know its half-life. He determines that after 40 days only 44 grams of the original element remains. What is the half-life of this mystery element?

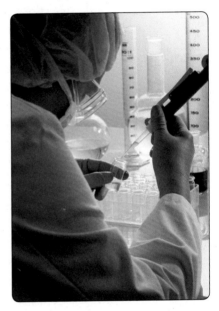

Chapter 5 Project

Exponential Functions

Computer viruses have cost U.S. companies billions of dollars in damages and lost revenues over the last few years. One factor that makes computer viruses so devastating is the rate at which they spread. A virus can potentially spread across the world in a matter of hours depending on its characteristics and whom it attacks.

Consider the growth of the following virus. A new virus has been created and is distributed to 100 computers in a company via a corporate e-mail. From these workstations the virus continues to spread. Let $t = 0$ be the time of the first 100 infections, and at $t = 17$ minutes the population of infected computers grows to 200. Assume the antivirus companies are not able to identify the virus or slow its progress for 24 hours, allowing the virus to grow exponentially.

1. What will the population of the infected computers be after 1 hour?
2. What will the population be after 1 hour 30 minutes?
3. What will the population be after a full 24 hours?

Suppose another virus is developed and released on the same 100 computers. This virus grows according to $P(t) = (100)2^{\frac{t}{2}}$, where t represents the number of hours from the time of introduction.

4. What is the doubling time for this virus?
5. How long will it take for the virus to infect 2000 computers, according to this model?

Chapter Summary

A summary of concepts and skills follows each chapter. Refer to these summaries to make sure you feel comfortable with the material in the chapter. The concepts and skills are organized according to the section title and topic title in which the material is first discussed.

5.1: Exponential Functions and Their Graphs

Definition and Classification of Exponential Functions
- The definition of *exponential functions*
- The two classes of exponential functions and their behavior

Graphing Exponential Functions
- Graphing transformations of exponential functions

Solving Elementary Exponential Equations
- The definition of an exponential equation
- Using the one-to-one nature of exponential functions to solve simple exponential equations

5.2: Applications of Exponential Functions

Models of Population Growth
- Modeling population growth with exponential functions
- Using population information to construct a corresponding exponential function

Radioactive Decay
- Modeling radioactive decay with exponential functions
- The use of exponential functions in *radiocarbon* dating
- Using half-life measurements and other data to construct exponential functions that model radioactive decay

Compound Interest and the Number *e*
- The meaning of *compounding* in investment contexts
- The use of the basic *compound interest formula*
- The *continuously compounding interest formula*
- The meaning of the *natural base e*, and its approximate value

5.3: Logarithmic Functions and Their Graphs

Definition of Logarithmic Functions
- Understanding the need for *logarithmic functions*
- The connection between a^x and $\log_a x$

Graphing Logarithmic Functions
- The two classes of logarithmic functions and their behavior

Evaluating Elementary Logarithmic Expressions
- Evaluating simple logarithmic expressions manually

Common and Natural Logarithms
- The definition of the *common* and *natural* logarithms
- Using a calculator to evaluate common and natural logarithms

5.4: Properties and Applications of Logarithms

Properties of Logarithms
- Basic properties of logarithmic functions, and how they are derived from properties of exponents
- Using the properties of logarithms to expand a single complicated logarithmic expression into a sum of simpler logarithmic expressions
- Using the properties of logarithms to combine sums of logarithmic expressions into a single logarithmic expression

The Change of Base Formula
- The derivation of the *change of base formula* and its use

Applications of Logarithmic Functions
- The historical use of logarithms in simplifying computation
- The use of logarithms in defining pH, earthquake intensity, and noise levels

5.5: Exponential and Logarithmic Equations

Converting Between Exponential and Logarithmic Forms
- Using the definition of logarithms and their properties to convert between the exponential and logarithmic forms of a given equation

Further Applications of Exponential and Logarithmic Equations
- Using logarithms to solve exponential equations

1.

2.

3.

4.

5.

6.

7.

8.

9.

10.

Chapter Review

Section 5.1

Sketch the graphs of the following functions.

1. $f(x) = \left(\dfrac{1}{2}\right)^{x-1} + 3$

2. $r(x) = 2^{-x+4} - 2$

3. $h(x) = 3^x$

4. $f(x) = 1 - 2^{-x}$

5. $p(x) = \left(\dfrac{1}{4}\right)^x$

6. $s(x) = (0.2)^{x-2}$

7. $g(x) = 4 - 2^x$

8. $m(x) = \dfrac{1}{2^x} - 3$

9. $f(x) = \dfrac{1}{2^{4-x}}$

10. $r(x) = \left(\dfrac{9}{2}\right)^{3-x}$

Solve the following elementary exponential equations.

11. $3^x = 243$

12. $2^{-x} = 16$

13. $0.5^x = 0.25$

14. $3^{3x-5} = 81$

15. $\left(\dfrac{2}{5}\right)^{-4x} = \left(\dfrac{25}{4}\right)^{x-1}$

16. $10{,}000^x = 10^{-2x-1}$

17. $9^{x-1} = 27^{-x+2}$

18. $\left(\dfrac{1}{3}\right)^{x-1} = 81^{\frac{1}{2}}$

19. $5^{3x-6} = 1$

Section 5.2

Solve the following problems involving exponential functions.

20. Melissa has recently inherited $15,000 that she wants to deposit into a savings account for 10 years. She has determined that her two best bets are an account that compounds annually at a rate of 3.95% and an account that compounds continuously at an annual rate of 3.85%. Which account would pay Melissa more interest?

21. Bill has come upon a 37 gram sample of iodine-131. He isolates the sample and waits for 2 weeks. After this time period, only 11 grams of iodine-131 remains. What is the half-life of this isotope?

11. $x = 5$

12. $x = -4$

13. $x = 2$

14. $x = 3$

15. $x = -1$

16. $x = -\dfrac{1}{6}$

17. $x = \dfrac{8}{5}$

18. $x = -1$

19. $x = 2$

20. The account at 3.95%

21. 8 days

22. a. $P(t) = 870(1.032^t)$
b. Approximately 68.39 minutes

23. a. 173 flies
b. 20 flies

24. 0.58 pounds

25. $x = \log_3 8$

26. $\sqrt{2} \log 3a = 1$

27. $\log_4 4096 = 3a$

28. $4^x = 64$

29. $3^{-3} = \dfrac{1}{27}$

30. $8^3 = 2A$

31.

32.

33.

34. 2

35. -2

36. -3

37. -3

38. -2

39. $\dfrac{2}{3}$

40. $x = 125$

---

22. Katherine is working in a lab testing bacteria populations. Starting out with a population of 870 bacteria, she notices that the population doubles every 22 minutes. Find **a.** the equation for the population P in terms of time t in minutes, and **b.** the time it would take for the population to reach 7500 bacteria.

23. The number of fruit flies in an experimental population after t hours is given by $Q(t) = 20e^{0.03t}$, $t \geq 0$.

a. How large is the population of fruit flies after 72 hours?
b. Find the initial number of fruit flies in the population.

24. In 1986, a nuclear reactor accident occurred in Chernobyl in what was then the Soviet Union. The explosion spread radioactive chemicals over hundreds of square miles and the government evacuated the city and surrounding areas. To see why the city is now uninhabited, consider the following model.

$$P = 10e^{-0.00002845t}$$

This model represents the amount of plutonium that remains (from an initial amount of 10 pounds) after t years. Sketch the graph of this function over the interval from $t = 0$ to $t = 100{,}000$. How much of the 10 pounds will remain after 100,000 years?

Section 5.3

Write the following equations in logarithmic terms.

25. $3^x = 8$ **26.** $(3a)^{\sqrt{2}} = 10$ **27.** $4^{3a} = 4096$

Write the following logarithmic equations as exponential equations.

28. $\log_4 64 = x$ **29.** $\log_3 \dfrac{1}{27} = -3$ **30.** $\log_8 2A = 3$

Sketch the graphs of the following functions.

31. $f(x) = -\log_3(-x)$ **32.** $f(x) = \log_{\frac{1}{2}} x$ **33.** $m(x) = \log_{\frac{1}{2}}(x+2)$

Evaluate the following logarithmic expressions without the use of a calculator.

34. $\log_{27} 9^{\log 1000}$ **35.** $\log_{\frac{1}{3}} 9$ **36.** $\log_4 \left(\dfrac{1}{64} \right)$

37. $\log_{\frac{1}{2}} 8$ **38.** $\log_{\sqrt{3}} \dfrac{1}{3}$ **39.** $\ln \sqrt[3]{e^2}$

Use the elementary properties of logarithms to solve the following equations.

40. $\log_6 6^{\log_5 x} = 3$ **41.** $\log_9 x^{\frac{1}{2}} = \dfrac{3}{4}$

42. $\log_x \left(\log_{\frac{1}{2}} \dfrac{1}{16} \right) = 2$ **43.** $\log_4(2x-1) = 2$

41. $x = 27$

42. $x = 2$

43. $x = \dfrac{17}{2}$

44. $x \approx 6.13$

45. $x \approx 155.41$

46. $x = 601$

47. $\dfrac{3}{2}\log x -$
$\dfrac{5}{2}\log \pi - \log 2$

48. $\dfrac{5}{2}\ln a + \ln m +$
$2\ln n - 5$

49. $3 + 3\log_3 a$

50. $1 + \ln 2 + \ln x$

51. $\log_2\left(\dfrac{a^{\frac{5}{3}}}{b^{\frac{1}{3}}c}\right)$

52. $\ln\left(\dfrac{4}{x^2 y^7}\right)$

53. $\log_2 (x - 3)$

54. $\log \dfrac{a^2 b^3}{d\sqrt{c}}$

55. $\log_3 \dfrac{x^2 - 2x}{x^2 + 4}$

56. 2.58

57. 6.18

58. -6.34

59. x^3

60. $\dfrac{x}{y^2}$

61. 0

62. 6

63. a. $251{,}188{,}643$

  b. $7{,}079{,}458$

  c. $1{,}258{,}925{,}412$

Solve the following logarithmic equations, using a calculator if necessary to evaluate the logarithms. Express your answer either as a fraction or a decimal rounded to the nearest hundredth.

44. $\ln(4x) = 3.2$ **45.** $\ln(x - 7) = 5$ **46.** $\log_7(4x - 3) = 4$

Section 5.4

Use properties of logarithms to expand the following expressions as much as possible. Simplify any numerical expressions that can be evaluated without a calculator.

47. $\log \sqrt{\dfrac{x^3}{4\pi^5}}$ **48.** $\ln\left(\dfrac{\sqrt{a^5}\,mn^2}{e^5}\right)$

49. $\log_3\left(27a^3\right)$ **50.** $\ln\left(\ln e^{2ex}\right)$

Use properties of logarithms to condense the following expressions as much as possible, writing each answer as a single term with a coefficient of 1.

51. $\dfrac{1}{3}\left(\log_2\left(a^5\right) - \log_2\left(bc^3\right)\right)$ **52.** $\ln 4 - \ln x^2 - 7\ln y$

53. $\log_2\left(x^2 - 9\right) - \log_2\left(x + 3\right)$ **54.** $2\log a + 3\log b - \dfrac{1}{2}\log c - \log d$

55. $\log_3(x - 2) + \log_3 x - \log_3\left(x^2 + 4\right)$

Evaluate the following logarithmic expressions

56. $\log_3 17$ **57.** $\log_{1.4} 8$ **58.** $4\log_{\frac{1}{2}} 3$

Use the properties of logarithms to write each of the following as a single term that does not contain a logarithm.

59. $6^{3\log_6 x}$ **60.** $5^{\log_5 x - 2\log_5 y}$

Without using a calculator, evaluate the following expressions.

61. $\ln \dfrac{1}{e^2} + \ln e^2$ **62.** $\log_4 64^2$

63. On the Richter scale, the magnitude R on a earthquake of intensity I is given by

$R = \log \dfrac{I}{I_0}$, where $I_0 = 1$ is the minimum intensity used for comparison. Find the intensity per unit of area for the following values of R.

  a. $R = 8.4$ **b.** $R = 6.85$ **c.** $R = 9.1$

64. $\dfrac{8-\ln 16}{5} \approx 1.05$

65. $\dfrac{6}{\log 321} \approx 2.39$

66. $\dfrac{3\ln 19}{\ln 7} + 12 \approx 16.54$

67. $\dfrac{\ln 5}{4-3\ln 5} \approx -1.94$

68. $\ln 8 - 2 \approx 0.08$

69. $\dfrac{\log 12}{\log 18} \approx 0.86$

70. 3

71. -2

72. 16

73. 73.5

74. 4

75. $f(x) = \ln x^3$

76. $f(x) = \log\left(8x^3\right)$

77. $f(x) = \log x^{10}$

78. $f(x) = \ln\left(3x^3\right)$

79. $f(x) = \log 7$

80. $f(x) = \ln 27$

81. 20.4 months (1.7 years)

82. 73.0 hours (about 3 days)

Section 5.5

Solve the following exponential and logarithmic equations. When appropriate, write the answer as both an exact expression and as a decimal approximation.

64. $e^{8-5x} = 16$

65. $10^{\frac{6}{x}} = 321$

66. $7^{\frac{x}{3}-4} = 19$

67. $e^{4x} = 5^{3x+1}$

68. $24 = 3e^{x+2}$

69. $3^{2x-1} = 2^{2-x}$

70. $\ln(x+1) + \ln(x-1) = \ln(x+5)$

71. $\log_2(x+3) + \log_2(x+4) = \log_2(3x+8)$

72. $\log_5(8x-3) = 3$

73. $\log_7(4x) - \log_7 6 = 2$

74. $\ln(5x+8) = \ln(40-3x)$

Using properties of logarithmic functions, simplify the following functions as much as possible. Write each function as a single term with a coefficient of 1, if possible.

75. $f(x) = 0.75\ln x^4$

76. $f(x) = 6\log\sqrt{2x}$

77. $f(x) = 4\log x^3 - \log x^2$

78. $f(x) = 0.5\ln\left(9x^6\right)$

79. $f(x) = 2\log 7^{\log_9 3}$

80. $f(x) = 2\ln 3^{\log_4 8}$

Use your knowledge of exponential and logarithmic functions to answer the following questions.

81. Rick puts $6500 in a high interest money market account at 4.36% annual interest compounded monthly. Assuming he makes no deposits or withdrawals, how long will it take for his investment to grow to $7000?

82. Sodium-24 has a half-life of approximately 15 hours. How long would it take for 350 grams of sodium-24 to decay to 12 grams?

1.

2.

3. $x = -4$

4. $x = -4$

5. $x = \dfrac{1}{3}$

6. a. $18,726

  b. $18,820

7. $\log_5 125 = x$

8. $x = \ln 3\pi$

9. $3^{-3} = \dfrac{1}{27}$

10. $a = 5^{10}$

11.

12.

13. 2

14. $\dfrac{2}{3}$

15. $x = \dfrac{1}{\sqrt{3}}$

16. $x = 3^{\frac{1}{3}}$

17. $\ln 2 + 2\ln a +$

  $\ln b - \ln 3 - \ln c$

18. $1 + \log_4 (3x + 5y)$

═══════════ **Chapter Test** ═══════════

Sketch the graphs of the following functions.

1. $g(x) = 2^{-x}$ **2.** $f(x) = 4^{1-2x}$

Solve the following exponential equations.

3. $5^x = \dfrac{1}{625}$ **4.** $2^{-x} + 1 = 17$ **5.** $\left(\dfrac{2}{3}\right)^{x-3} = \left(\dfrac{9}{4}\right)^{x+1}$

6. Suppose you invest a total of $12,000 at an annual interest rate of 9%. Find the balance to the nearest dollar after 5 years if it is compounded:

  a. Quarterly   **b.** Continuously

Write the following equations in logarithmic terms.

7. $125 = 5^x$ **8.** $e^x = 3\pi$

Write the following logarithmic equations as exponential equations.

9. $\log_3 \dfrac{1}{27} = -3$ **10.** $\log_5 a = 10$

Sketch the graphs of the following functions.

11. $f(x) = \log_2 (x - 1)$ **12.** $f(x) = -\log_2 x + 1$

Evaluate the following logarithmic expressions or equations.

13. $\log_2 (\log_3 81)$ **14.** $\ln \sqrt[3]{e^2}$

15. $\log_3 x = \dfrac{-1}{2}$ **16.** $\log_b b^{\log_3 x} = \dfrac{1}{3}$

Use the properties of logarithms to expand the following expressions as much as possible. Simplify any numerical expressions.

17. $\ln\left(\dfrac{2a^2 b}{3c}\right)$ **18.** $\log_4 (12x + 20y)$

Use the properties of logarithms to condense the following expressions as much as possible, writing each answer as a single term with a coefficient of 1.

19. $\log x + \log y - \log c$ **20.** $\log_3 (x - 2) + \log_3 x - \log_3 (x^2 + 4)$

19. $\log \dfrac{xy}{c}$

20. $\log_3 \dfrac{x^2 - 2x}{x^2 + 4}$

21. $x = \dfrac{1 - \log_5 7}{2}$

22. $x = \dfrac{2 + \ln 40}{5}$

23. $x = \dfrac{3}{\log_2 5}$

24. $x = \dfrac{9}{2}$

25. $x = \dfrac{3}{\sqrt{2}}$

26. $x = 16,\ x = \dfrac{1}{4}$

27. $x = 10$

28. $x = 9$

29. $I = 3.16$ watts/m²

30. $f^{-1}(x) = e^{\frac{x+1}{2}} - 1$

Solve the following exponential and logarithmic equations, if possible.

21. $5^{1-2x} = 7$

22. $e^{5x-2} = 40$

23. $2^{\frac{3}{x}} = 5$

24. $\log_5(x-2) + \log_5 2 = 1$

25. $\log_{\sqrt{3}} x + \log_{\sqrt{3}} 2x = 4$

26. $(\log_2 x)^2 - 2(\log_2 x) - 8 = 0$

27. $3^{2\log_3(x-1)} = 81$

28. $1 - 2\log(x+1) = -1$

29. The relationship between the number of decibels D and the intensity of sound I in watts per square meter is given by $D = 10 \cdot \log\left(\dfrac{I}{10^{-12}}\right)$. Determine the intensity of sound in watts per square meter if the decibel level is 125.

Find the inverse of the following function.

30. $f(x) = 2\ln(x+1) - 1$

Trigonometric Functions

6.1 Radian and Degree Measure of Angles

6.2 Trigonometric Functions of Acute Angles

6.3 Trigonometric Functions of Any Angle

6.4 Graphs of Trigonometric Functions

6.5 Inverse Trigonometric Functions

Chapter 6 Project

Chapter 6 Summary

Chapter 6 Review

Chapter 6 Test

By the end of this chapter you should be able to:

What if you were playing a game of pickup basketball with your friends? How would you model the path of the ball as you dribbled it in preparation for making a game-winning shot?

By the end of this chapter, you'll be able to evaluate and graph trigonometric functions. This type of function can be used to describe the behavior of many naturally occurring phenomena, such as waves and harmonic motion. On page 521, you will use a trigonometric function to model the displacement of a basketball. You will master this type of problem using the definition of simple harmonic motion and frequency on page 514.

Introduction

This chapter is the first of three dealing almost exclusively with an area of mathematics called *trigonometry*, an area which is big enough to study on its own (as it often is) but which is also intrinsically related to the algebraic and geometric concepts that make up the rest of this text. This close association will become even more apparent when you study calculus, so its inclusion in a book preparing you for calculus is very much appropriate.

Babylonian
Numbers

The early history of trigonometry is not quite so well-documented as that of geometry, but archaeologists have unearthed clay tablets indicating that Babylonian mathematicians around 2000 BC were already developing ideas that we would classify today as trigonometric. And that Babylonian heritage is of more than academic interest. We owe to the Babylonians of the 1st millennium BC our *degree* unit of angle measure. Although many competing arguments have been proposed as to why the Babylonians fixed on 360° as being the measure of one full rotation, there is no doubt that the convention began with them. (Some of the competing arguments are built around such things as connections between a full circle and the calendar, the fact that 360 can be factored many different ways, and the fact that Babylonians apparently divided the day into 12 "hours" of 30 parts each.)

The word "trigonometry" itself is Greek, and translates roughly as "measurement of triangles." From the start, trigonometry found important applications in astronomy, navigation, and surveying, and those applications have only grown in importance over time. Initially, trigonometry focused on ratios of side lengths of triangles, and that perspective is alive and strong still. But with the development of calculus in the 17th century, mathematicians also began to view the trigonometric relations as functions of real numbers. This secondary perspective lends itself to applications involving rotations or oscillations, and trigonometry quickly became an indispensable tool in engineering, explanations of wave propagation, and modern signal processing.

As with every topic in this text, try to keep the historical background in mind as you learn the material. Mathematics is not immune to societal and other pressures, and many aspects of trigonometry's history demonstrate this. The presentation in this text draws upon more than 2000 years of development, and would be very unfamiliar to early users of trigonometry. Relatively recent developments in technology have also had a profound effect on the way trigonometry is taught and learned—if you find yourself tiring at some point while studying this chapter, comfort yourself with the thought that dreaded "trig tables" are a thing of the past! Learning how to use them once constituted a large part of trigonometry, but calculators and computer software make such tedium unnecessary now. (If you have no idea what a "trig table" is, and want to subject yourself to a lecture on how kids today have it too easy, ask someone who learned trigonometry prior to the mid-1970s for an explanation.)

Radian and Degree Measure of Angles

TOPICS

1. The unit circle and angle measure

2. Conversion between degrees and radians

3. Commonly encountered angles

4. Arc length and angular speed

5. Area of a circular sector

TOPIC **1**

The Unit Circle and Angle Measure

Trigonometry is, at heart, the study of angles. Although much of trigonometry can be discussed without reference to angle measure (and in fact early Greek mathematicians did just that), we will find it very useful to have a method of describing the size of angles. As it turns out, there are two common ways to measure angles (as well as a number of less common ways). The method of measuring angles in terms of *degrees* is one that you are probably very familiar with—references to degree measure occur in all sorts of nonmathematical contexts. But, as mentioned in the introduction to this chapter, the definition of degree has more of a cultural basis than a mathematical basis, so it shouldn't be too surprising that there is a way of measuring angles that makes more mathematical sense. That more mathematically useful way is called *radian measure*.

DEFINITION

Radian Measure

Let θ (the Greek letter *theta*) be an angle at the center of a circle of radius 1 (the unit circle), as shown in the diagram. The measure of θ in **radians** (abbreviated as **rad**) is the length of that portion of the circle *subtended* by θ (that is, the portion of the circumference shown in green). Note that the unit of length measurement is immaterial. As long as the circle has a radius of 1 (unit), the length of the subtended portion of the circle (in the same units) is defined to be the radian measure of the angle.

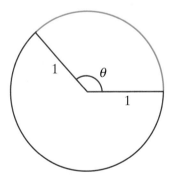

Of course, in general an angle θ is defined by any two rays R_1 and R_2 sharing a common origin, as shown in Figure 1. We can associate a sign with the measure of θ by designating one ray, say R_1, as the **initial side** and the other ray, R_2, as the **terminal side**. If θ is defined by a counterclockwise rotation from the initial side to the terminal side, we say θ has **positive measure**, and if θ is defined by a clockwise rotation we say it has **negative measure**. In Figure 1, the green angle has positive measure while the purple angle is negative.

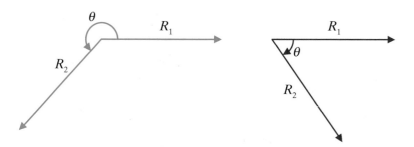

Figure 1: Positive and Negative Angle Measure

In order to use our definition of radian to measure an angle defined by two rays, as in Figure 1, we place the vertex of the angle at the center of a **unit circle** (a circle of radius 1) and measure the length of the arc between the initial and terminal sides of the angle. Further, we say the angle is in **standard position** if its vertex is located at the origin of the Cartesian plane and its initial side lies along the positive x-axis. In this case, the unit circle is then the graph of the equation $x^2 + y^2 = 1$. Figure 2 illustrates the second angle from Figure 1 placed in standard position.

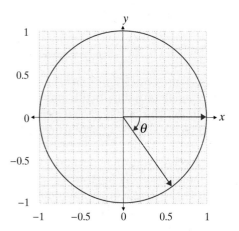

Figure 2: Standard Position of an Angle

TOPIC 2   Conversion Between Degrees and Radians

Once radian measure has been defined, the next order of business is to acquire a degree of familiarity with its use. To begin with, we want to be able to translate between degree measure and radian measure of an angle.

This is easily done if we recall the formula for the circumference C of a circle of radius r: $C = 2\pi r$. For the unit circle under discussion, $r = 1$ so $C = 2\pi$. If we think of the entire circumference as being the portion of the unit circle subtended by an angle of $360°$ (that is, an angle whose terminal side and initial side coincide), we have just determined that an angle of $360°$ corresponds to 2π radians. Using this as a starting point, we see that an angle of $180°$ corresponds to π radians (such an angle subtends half the circumference of the unit circle) and an angle of $90°$ corresponds to $\frac{\pi}{2}$ radians (that is, a right angle has a measure of $\frac{\pi}{2}$ radians). From the equation $180° = \pi$ rad, we can derive the conversion formulas that follow.

Degrees	Radians
$360°$	2π
$180°$	π
$90°$	$\frac{\pi}{2}$

DEFINITION

Conversion Formulas

Since $180° = \pi$ rad, we know that $1° = \dfrac{\pi}{180}$ rad and $\left(\dfrac{180}{\pi}\right)^{\circ} = 1$ rad. Multiplying both sides of these equations by an arbitrary quantity x, we have:

1. $x° = x\left(\dfrac{\pi}{180}\right)$ rad, and

2. x rad $= x\left(\dfrac{180}{\pi}\right)^{\circ}$.

In particular, note that 1 rad $\approx 57.296°$, so an angle of 1 rad cuts off a bit less than one-sixth of a circle (an angle of $60°$ cuts off exactly one-sixth of a circle).

Converting Between
Degrees and Radians

Convert the following angle measures as directed.

a. Express $\dfrac{\pi}{3}$ rad in degrees.   **b.** Express 270° in radians.

c. Express −2 rad in degrees.

Solutions:

a. $\dfrac{\pi}{3}$ rad $= \left(\dfrac{\pi}{3}\right)\left(\dfrac{180}{\pi}\right)^{\circ} = 60°.$ **b.** $270° = \left(270\right)\left(\dfrac{\pi}{180}\right)$ rad $= \dfrac{3\pi}{2}$ rad.

c. -2 rad $= \left(-2\right)\left(\dfrac{180}{\pi}\right)^{\circ} \approx -114.592°.$

Before continuing, a note on terminology: Whenever an angle is measured in degrees, its measure will appear followed by the degree symbol (°); angles measured in radians will either appear with the abbreviation "rad" afterward or, more commonly, with no notation at all. It is a reflection of the importance of radian measure in mathematics that if no indication of the method of measurement appears, we are to assume the angle is measured in radians.

TOPIC 3   Commonly Encountered Angles

It is tempting, when teaching or learning a new area of mathematics, to restrict attention to examples that are artificially "nice." That is, examples in which complicated terms in the accompanying equations either never appear or else conveniently cancel, and examples in which the final answer is suspiciously devoid of ugly fractions and approximations. With this in mind, you might dismiss the angles in the following discussion as unrealistically pleasant to work with. But the justification for studying the angles of 30°, 45°, and 60° is that, first, they actually do appear fairly frequently in real life and, second, they are undeniably useful in building an understanding of trigonometric functions. We will encounter them repeatedly in the sections that follow.

At the moment, we are primarily interested in determining the radian measures that correspond to these common angles, but that is a simple matter of applying the appropriate conversion formula. While we have them before us, therefore, we will also note how these angles relate to one another in the context of triangles. This knowledge will prove to be useful very soon.

First, the radian equivalents:

$$30° = (30)\left(\frac{\pi}{180}\right) = \frac{\pi}{6}$$

$$45° = (45)\left(\frac{\pi}{180}\right) = \frac{\pi}{4}$$

$$60° = (60)\left(\frac{\pi}{180}\right) = \frac{\pi}{3}$$

and, while we're at it:

$$90° = (90)\left(\frac{\pi}{180}\right) = \frac{\pi}{2}.$$

(Remember that the absence of notation following an angle means the angle is measured in radians.)

The triangle connection comes from an application of the Pythagorean Theorem. Recall that if a and b are the lengths of the two legs of a right triangle and if c is the length of the hypotenuse, then $a^2 + b^2 = c^2$. Recall also that the sum of the angles of a triangle is always 180°, or π radians. So a triangle with one vertex of measure $\frac{\pi}{6}$ and a second vertex of measure $\frac{\pi}{3}$ must have a right angle for the third vertex, and similarly for a triangle with two vertices of measure $\frac{\pi}{4}$. These observations are illustrated in Figure 3.

Figure 3: Common Triangles

Now, suppose the triangle on the left in Figure 3 has legs of length 1. The Pythagorean Theorem tells us then that the length of the hypotenuse is $\sqrt{1^2 + 1^2} = \sqrt{2}$. In general, any triangle with two angles of measure $\frac{\pi}{4}$ will be a right triangle with two legs of equal length, and the ratios of the lengths of the sides will be $1:1:\sqrt{2}$.

We have to work slightly harder to figure out the ratios of the lengths of the sides of the triangle on the right (the 30°-60°-90° triangle, in degree terms). Note that if we join the triangle with its mirror image, we obtain an equilateral triangle (since all the angles will measure 60°) as shown in Figure 4. This means that the length of the shorter leg of the original triangle must be half of the length of the hypotenuse. So if we assume the shorter leg has a length of 1, the hypotenuse has a length of two and the Pythagorean Theorem tells us that the longer leg has a length of $\sqrt{2^2 - 1^2} = \sqrt{3}$. In general, the ratio of the short leg to the long leg to the hypotenuse of such a triangle is $1 : \sqrt{3} : 2$.

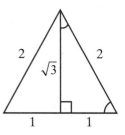

Figure 4: Doubling the $\dfrac{\pi}{6}$-$\dfrac{\pi}{3}$-$\dfrac{\pi}{2}$ Triangle

EXAMPLE 2

Determining the Measure of an Angle

Use the information in each diagram to determine the radian measure of the indicated angle.

a.

b.

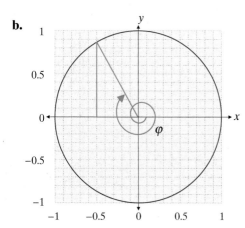

Solutions:

a. The angle θ is in standard position, and is positive. It can also be seen that the measure of the angle is π radians plus a bit more, where the "bit more" comes from the angle whose initial side is the negative x-axis and whose terminal side contains the hypotenuse of the green $\dfrac{\pi}{4}$-$\dfrac{\pi}{4}$-$\dfrac{\pi}{2}$ triangle. So the "bit more" must be $\dfrac{\pi}{4}$ radians, and the angle θ has measure $\pi + \dfrac{\pi}{4} = \dfrac{5\pi}{4}$.

b. The angle φ (the Greek letter *phi*) is also in standard position, but its measure is negative. It is defined by beginning at the positive *x*-axis and rotating -2π radians (that is, going full circle in the clockwise direction), continuing for another $-\pi$ radians (another half circle), and then continuing on for a bit more. This time, the angle corresponding to the "bit more" has its initial side on the negative *x*-axis and terminal side on the hypotenuse of a $\frac{\pi}{6}$-$\frac{\pi}{3}$-$\frac{\pi}{2}$ triangle. We know the triangle must be of this sort because its hypotenuse has length 1 (do you see why?) and its shorter leg has length $\frac{1}{2}$, so the ratio of the shorter leg to the hypotenuse is $1:2$. Hence the "bit more" must have measure $-\frac{\pi}{3}$ and altogether the measure of φ is

$$-2\pi - \pi - \frac{\pi}{3} = -\frac{10\pi}{3}.$$

TOPIC **4**

Arc Length and Angular Speed

The advantages of radian measure over degree measure will appear repeatedly over the next several chapters (and later in calculus). The first advantage is actually just a restatement of the definition of radian. Recall that the radian measure of an angle is related to that portion of the unit circle cut off (or subtended) by the angle when the angle is placed at the center of the circle; the length of the subtended arc is an example of *arc length*. In other words, if θ is a central angle of a unit circle, then $\frac{\theta}{2\pi}$ is the fraction of the circle's circumference subtended by θ. More generally, if θ is a central angle of a circle of radius *r*, the length *s* of the portion of the circle subtended by θ is the same fraction multiplied by the circumference.

FORMULA

Arc Length Formula

Given a circle of radius *r*, the length *s* of the arc subtended by a central angle θ is given by:

$$s = \left(\frac{\theta}{2\pi}\right)(2\pi r)$$

$$= r\theta$$

(In the figure on the right, the unit circle is drawn in black, while the larger circle has radius *r*.)

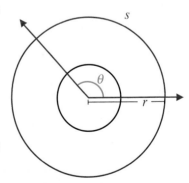

EXAMPLE 3

Applying the Arc Length Formula

The Galapagos Islands lie almost exactly on the equator, and are located at 90° West longitude. Suppose a ship sails along the equator from the Galapagos to the International Date Line (at 180° longitude). How far does the ship travel? (Assume the Earth's radius is 6370 kilometers.)

Solution:

The critical observation is that the ship travels one-quarter of the Earth's circumference (from 90° West of 0° longitude to 180° from 0°), so the angle (at the center of the Earth) described by the ship's path is $\frac{\pi}{2}$.

Using the arc length formula, the distance traveled is

$$s = (6370)\left(\frac{\pi}{2}\right)$$

$$\approx 10{,}006 \text{ km}.$$

Now that we have a convenient formula for arc length, we can easily determine the speed with which an object traverses a given arc. For instance, if we are told that the ship in Example 3 takes 15 days to make its journey, we can calculate that its average speed is

$$\frac{10{,}006}{(15)(24)} \approx 27.8 \frac{\text{km}}{\text{hr}}.$$

Often, information about rate of travel along a circle's circumference is given in terms of *angular speed*, which is a measure of the angle traversed over time.

DEFINITION

Angular Speed and Linear Speed

If an object moves along an arc of a circle defined by a central angle θ in time t, the object is said to have an **angular speed** ω (the Greek letter "omega") given by:

$$\omega = \frac{\theta}{t}.$$

If the circle has a radius of r, the distance traveled in time t is the arc length s, and the **linear speed** v is given by:

$$v = \frac{s}{t} = \frac{r\theta}{t} = r\omega.$$

EXAMPLE 4

Finding the Angular and Linear Speeds

Suppose an ant crawls along the rim of a circular glass with radius 2 inches, and traverses the arc indicated in red below in 20 seconds. What are the angular and linear speeds of the ant, and how far does it travel?

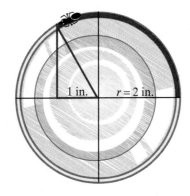

Solution:

As in Example 2b, we can determine that the triangle shown in purple is a $30°\text{-}60°\text{-}90°$ right triangle, and its vertex at the origin must have measure $\frac{\pi}{3}$. This means the ant describes an angle of $\theta = \frac{\pi}{2} + \frac{\pi}{6} = \frac{2\pi}{3}$ as it walks, so its angular speed is

$$\omega = \frac{\theta}{t} = \frac{\frac{2\pi}{3}}{20} = \frac{\pi}{30} \frac{\text{rad}}{\text{s}}.$$

Given that the radius of the glass is 2 inches, the linear speed of the ant is

$$v = r\omega = 2\left(\frac{\pi}{30}\right) = \frac{\pi}{15} \approx 0.21 \frac{\text{in.}}{\text{s}}.$$

Finally, the distance the ant travels is $s = r\theta = (2)\left(\frac{2\pi}{3}\right) = \frac{4\pi}{3}$ in. or approximately 4.19 in.

CAUTION!

The arc length and angular speed formulas, as well as the area formula that follows, are only true for angles measured in radians. Equivalent but less convenient formulas can be derived for angles measured in terms of degrees.

TOPIC 5   Area of a Circular Sector

We will close this section with one last example of the value of measuring angles in radians.

A **sector** of a circle is the portion of a circle between two radii. The area of a sector, then, can range from 0 to πr^2 square units, where the radius of the circle is assumed to be r units. Of course, the two radii defining a given sector can also be taken to be the initial and terminal sides of a central angle θ. Just as $\dfrac{\theta}{2\pi}$ represents the fraction of a circle's circumference subtended by θ, this same ratio represents the portion of a circle's area contained in the sector of angular size θ. This gives us the following formula.

FORMULA

Sector Area Formula

The area A of a sector with a central angle of θ in a circle of radius r is

$$A = \left(\frac{\theta}{2\pi}\right)\left(\pi r^2\right) = \frac{r^2\theta}{2}.$$

EXAMPLE 5

Finding the Area of a Sector

Determine the areas of the sectors defined by the given radii and angles.

a. Circle of radius 3 cm, central angle of $52°$

b. Circle of radius $\dfrac{1}{2}$ ft, central angle of $\dfrac{4\pi}{3}$

Solutions:

a. In order to use the above sector area formula, the first step is to convert the angle measure to radians:

$$52° = (52)\left(\frac{\pi}{180}\right) = \frac{13\pi}{45}.$$

Now, the formula is easily applied:

$$A = \frac{\left(3^2\right)\left(\dfrac{13\pi}{45}\right)}{2} = \frac{13\pi}{10} \approx 4.08 \text{ cm}^2.$$

b. Since the angle is given in radians, we have immediately:

$$A = \frac{\left(\dfrac{1}{2}\right)^2\left(\dfrac{4\pi}{3}\right)}{2} = \frac{\pi}{6} \approx 0.52 \text{ ft}^2.$$

1. 225°   2. 1°

3. −67.5°   4. −210°

5. 120°   6. 63°

7. 150°   8. 198°

9. −405°   10. −300°

11. $\dfrac{47\pi}{180}$ 12. $\dfrac{31\pi}{60}$

13. $\dfrac{11\pi}{15}$ 14. $\dfrac{77\pi}{90}$

15. $\dfrac{37\pi}{45}$ 16. $\dfrac{2\pi}{3}$

17. $\dfrac{8\pi}{3}$ 18. $\dfrac{26\pi}{9}$

19. $\dfrac{25\pi}{36}$ 20. $\dfrac{\pi}{2}$

21. 270°   22. −405°

23. 540°   24. 15°

25. −72°   26. 120°

27. $\dfrac{\pi}{9}$ 28. $\dfrac{17\pi}{9}$

29. $-\dfrac{4\pi}{5}$ 30. $\dfrac{11\pi}{30}$

31. $\dfrac{\pi}{6}$ 32. π

33. $\dfrac{83\pi}{90}$ 34. $\dfrac{11\pi}{15}$

35. 115°   36. $\dfrac{5\pi}{12}$

37. 50°   38. $\dfrac{\pi}{15}$

39. 345°   40. $\dfrac{26\pi}{15}$

41. 295°   42. 260°

43. $\dfrac{5\pi}{4}$ 44. 198°

Exercises

In questions 1–10, convert the radian measure to degrees. See Example 1.

1. $\dfrac{5\pi}{4}$ 2. $\dfrac{\pi}{180}$ 3. $\dfrac{-3\pi}{8}$ 4. $\dfrac{-7\pi}{6}$ 5. $\dfrac{2\pi}{3}$

6. $\dfrac{7\pi}{20}$ 7. $\dfrac{5\pi}{6}$ 8. $\dfrac{11\pi}{10}$ 9. $\dfrac{-9\pi}{4}$ 10. $\dfrac{-5\pi}{3}$

In questions 11–20, convert the degree measure to radians. See Example 1.

11. 47°   12. 93°   13. 132°   14. 154°   15. 148°

16. 120°   17. 480°   18. 520°   19. 125°   20. 90°

Convert the following angle measures as directed. See Example 1.

21. Express $\dfrac{3\pi}{2}$ in degrees.   22. Express $-\dfrac{9\pi}{4}$ in degrees.

23. Express 3π in degrees.   24. Express $\dfrac{\pi}{12}$ in degrees.

25. Express $-\dfrac{2\pi}{5}$ in degrees.   26. Express $\dfrac{2\pi}{3}$ in degrees.

27. Express 20° in radians.   28. Express 340° in radians.

29. Express −144° in radians.   30. Express 66° in radians.

31. Express 30° in radians.   32. Express 180° in radians.

The unit circle shown below shows several angles in radians or degrees. Fill in the corresponding radian or degree for questions 33–44.

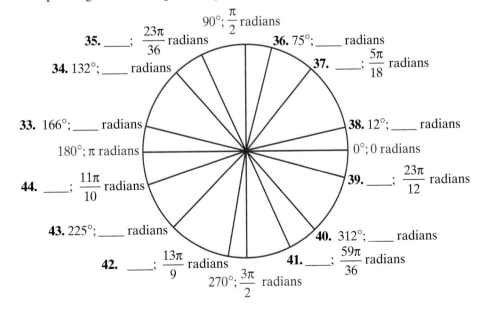

33. 166°; ____ radians

34. 132°; ____ radians

35. ____; $\dfrac{23\pi}{36}$ radians

36. 75°; ____ radians

37. ____; $\dfrac{5\pi}{18}$ radians

38. 12°; ____ radians

39. ____; $\dfrac{23\pi}{12}$ radians

40. 312°; ____ radians

41. ____; $\dfrac{59\pi}{36}$ radians

42. ____; $\dfrac{13\pi}{9}$ radians

43. 225°; ____ radians

44. ____; $\dfrac{11\pi}{10}$ radians

90°; $\dfrac{\pi}{2}$ radians

180°; π radians

0°; 0 radians

270°; $\dfrac{3\pi}{2}$ radians

45. $\dfrac{\pi}{4}$ **46.** $-\dfrac{5\pi}{6}$

47. $\dfrac{4\pi}{3}$ **48.** $-\dfrac{7\pi}{2}$

49. $\dfrac{13\pi}{4}$ **50.** $\dfrac{2\pi}{3}$

Use the information in each diagram to determine the radian measure of the indicated angle. See Example 2.

45.

46.

47.

48.

51.

52.

53.

54.

55.

49.

50.

56.

Sketch the indicated angles. See Example 2.

51. $\dfrac{5\pi}{2}$ **52.** $-60°$ **53.** $210°$

54. $-\dfrac{\pi}{3}$ **55.** $\dfrac{7\pi}{4}$ **56.** $120°$

57. 4 inches

58. 14.14 cm

59. 11.78 feet

60. 251.33 km

61. 8.64 m

62. 11.00 feet

63. 1.48 inches

64. 212.06 cm

65. 3.56 feet

66. 19.24 m

67. 528.10 km

68. 1.45°

69. 43.0°

70. 445.79 miles

71. 611.48 km

72. 1335.24 km

73. 3004.02 km

74. 7853.58 km

Find the arc length of a circle with the given radius r and central angle θ. Give the answer in the given unit of measure and round decimals to the nearest hundredth.

57. $r = 4$ in.; $\theta = 1$ **58.** $r = 9$ cm; $\theta = \dfrac{\pi}{2}$ **59.** $r = 15$ feet; $\theta = \dfrac{\pi}{4}$

60. $r = 80$ km; $\theta = 180°$ **61.** $r = 16.5$ m; $\theta = 30°$ **62.** $r = 7$ feet; $\theta = 90°$

Find the indicated arc length in each of the following problems. See Example 3. (Round your answers to the nearest hundredth.)

63. Given a circle of radius 5 inches, find the length of the arc subtended by a central angle of 17° (**Hint:** Convert to radians first).

64. Given a circle of radius 22.5 cm, find the length of the arc subtended by a central angle of 3π.

65. Given a circle with a diameter of 6 feet, find the length of the arc subtended by a central angle of 68° (**Hint:** Convert to radians first).

66. Given a circle of radius 7 meters, find the length of the arc subtended by a central angle of $\dfrac{7\pi}{8}$.

67. Assuming that Columbia, SC and Daytona Beach, FL have the same longitude (81° W), use a radius of 6370 kilometers for the Earth and the following to find the distance between the two cities.

City	Latitude
Columbia, SC	34° N
Daytona Beach, FL	29.25° N

68. Given that two cities on the equator are 100 miles apart and have the same latitude (that is, one is due west of the other), what is the difference in their longitudes? Use a value of 3960 miles for the radius of the Earth.

69. Using a radius of 1.2 cm for the average eyeball, find the central angle formed to meet the edges of an iris (the colored portion of the eye) with an arc length of 9 mm.

70. Find the distance between Denver, CO and Roswell, NM which lie on the same longitude. The latitude of Denver is 39.75° N and the latitude of Roswell is 33.3° N. Use a radius of 3960 miles for the Earth.

71. Find the distance between Atlanta, Georgia and Cincinnati, Ohio which lie on the same longitude. The latitude of Atlanta is 33.67° N and the latitude of Cincinnati is 39.17° N. Assume the Earth's radius is 6370 kilometers.

72. Find the distance between Greenwich, England and Valencia, Spain which lie on the same longitude. The latitude of Greenwich is 51.48° N and the latitude of Valencia is 39.47° N. Assume the Earth's radius is 6370 kilometers.

73. Find the distance between La Paz, Bolivia and Caracas, Venezuela which lie on the same longitude. The latitude of La Paz is 16.50° S and the latitude of Caracas is 10.52° N. Assume the Earth's radius is 6370 kilometers.

74. Find the distance between Bucharest, Romania and Johannesburg, South Africa which lie on the same longitude. The latitude of Bucharest is 44.43° N and the latitude of Johannesburg is 26.21° S. Assume the Earth's radius is 6370 kilometers.

75. $\dfrac{9}{2}$ **76.** $\dfrac{3}{8}$

77. $\dfrac{21}{47}$ **78.** 10

79. $\dfrac{45}{4}$ **80.** $\dfrac{1}{3}$

81. a. 2000π rad/min
b. 87.27 ft/s

82. 1039.61 mph

83. a. 100π rad/min
b. 366.52 ft/min

84. 17.14 mph

Find the radian measure of the central angle θ given the radius r and the arc length s transcribed by θ. Leave all answers in fraction form.

75. $r = 14$ ft; $s = 63$ ft   **76.** $r = 16$ in.; $s = 6$ in.   **77.** $r = 23.5$ dm; $s = 10.5$ dm

78. $r = 13$ cm; $s = 130$ cm   **79.** $r = 2$ km; $s = 22.5$ km   **80.** $r = 33$ ft; $s = 11$ ft

The following problems ask you to determine the angular and/or linear speeds of various objects. See Example 4. (Round your answers to the nearest hundredth.)

81. An industrial circular saw blade has a 10-inch radius and spins at 1000 rpm. Find **a.** the angular speed of a tooth of the blade in radians per minute, and **b.** the linear speed of the tooth in feet per second.

82. The Earth takes roughly 23 hours and 56 minutes to rotate once about its axis. Using a radius for the Earth of 3960 miles, what is the linear speed in miles per hour (relative to the center of the Earth) of a person standing on the equator? (Ignore, for the purposes of this problem, such motion as the rotation of the Earth about the Sun.)

83. A stationary exercise bike is ridden at a constant speed, causing the wheel to spin at a rate of 50 revolutions per minute. If a tack becomes lodged in the tire of radius 14 inches, find **a.** the angular speed of the tack in radians per minute, and **b.** the linear speed of the tack in feet per minute.

84. A horse is tethered and urged to trot such that it completes a circular path every 5 seconds. If the rope which tethers it is 20 feet long, what is the linear speed of the horse in miles per hour?

85. $\dfrac{5\pi}{12} \approx 1.31$

86. $\dfrac{5\pi}{8} \approx 1.96$

87. 29.93 cm^2

88. 2.89 ft^2

89. 15.08 m^2

90. 308.29 in.^{2}

91. 314.16 ft^2

92. 16.73 km^2

93. 9.82 in.^{2}

94. 4.19 in.^{2}

95. 26.18 cm

96. a. 10π rad/s
  b. 122.84 in./s

The following problems ask you to calculate the area of a sector of a circle. See Example 5.

85. Find the area of the green shaded portion of the circle below:

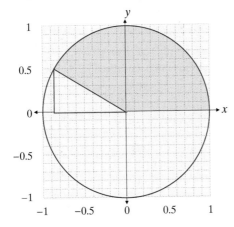

86. Find the area of the green shaded portion of the circle below:

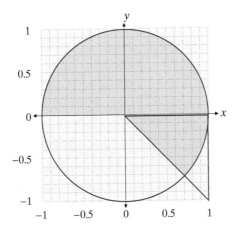

87. Find the area of the sector of a circle of radius 7 cm with a central angle of 70°.

88. Find the area of the sector of a circle of radius 3.5 ft with a central angle of 27°.

89. Find the area of the sector of a circle of radius 4 m with a central angle of $\dfrac{3\pi}{5}$.

90. Find the area of the sector of a circle of radius 16 inches with a central angle of 138°.

91. Find the area of the sector of a circle of radius 20 ft with a central angle of $\dfrac{\pi}{2}$.

92. Find the area of the sector of a circle of radius 19 km with a central angle of 5.31°.

93. A pie of radius 5 inches is cut into 8 equal pieces. What is the area of each piece?

94. The minute hand of a clock extends out to the edge of the clock's face, which is a circle of radius 2 inches. What area does the minute hand sweep out between 9:05 and 9:25?

95. The circular spinner for a board game is divided into 6 equal wedges, each of a different color. If the radius is 5 cm, what area is encompassed by 2 wedges?

96. The floppy disk drive (FDD) was invented in 1967 to store information for computer users. The first floppy drive used an 8-inch disk and had a radius of 3.91 inches. The drive motor would spin at 300 rotations per minute (RPM).

   a. Find the angular speed of the 8-inch disk in radians per second.

   b. Find the linear speed of a particular point on the circumference of the 8-inch disk in inches per second.

97. a. 12π rad/s
   b. 95.38 in./s

98. 11.7 in.

99. 38.5°

97. The 8-inch floppy disk drive evolved into a smaller 5.25-inch disk that was used in the personal computers (PC) in the early 1980s. The 5.25-inch disk had a radius of 2.53 inches. The usual drive motor for the 5.25-inch disk would spin at 360 rotations per minute.

   a. Find the angular speed of the 5.25-inch disk in radians per second.

   b. Find the linear speed of a particular point on the circumference of the 5.25-inch disk in inches per second.

98. Two gears are rotating to turn a conveyor belt. The smaller gear rotates 80° as the larger gear rotates 50°. If the larger gear has a radius of 18.7 in., what is the radius of the smaller gear?

99. Two water mills are on display at a local museum. The smaller water mill rotates counterclockwise and turns the larger water mill in a clockwise direction. If the smaller water mill has a radius of 5.23 ft and the larger water mill has a radius of 8.16 ft, what is the degree of rotation of the larger wheel when the smaller rotates 60°?

6.2

Trigonometric Functions of Acute Angles

TOPICS

1. The six basic trigonometric functions

2. Evaluation of trigonometric functions

3. Applications of trigonometric functions

TOPIC 1

The Six Basic Trigonometric Functions

Now that we are equipped with a useful way of measuring angles, we can proceed to define the basic trigonometric functions that will dominate the discussion in this and the next two chapters.

It is worthwhile to reflect again on the fact that the material we are studying was developed by countless individuals over the span of several thousand years, and that the cultures they lived in and the problems they hoped to solve varied greatly. Much of the early impetus in developing trigonometry came from astronomy, but by the time of the Renaissance the utility of trigonometry in navigation, surveying, and engineering was thoroughly well-recognized. That utility has only increased with time, along with the fields in which trigonometric skill is necessary. We will find trigonometry particularly useful, for example, in solving plane geometry problems.

With the benefit of thousands of years of development behind us, we will take an approach to defining the six basic trigonometric functions that calls upon the work of many different eras. That is, our treatment of trigonometry would not be immediately recognizable to, say, early Greek mathematicians or to Italian mathematicians of the 15th and 16th centuries. Instead, our definitions will be motivated by the desire to make the trigonometric functions most readily useful in a wide variety of applications.

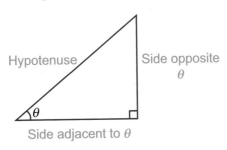

Figure 1: Legs Labeled Relative to θ

To begin, consider a right triangle such as the one shown in the figure to the left. We will define six functions which are functions of the angle θ. In order to do so, we label the two legs of the triangle as **adjacent** to and **opposite** θ, as shown.

DEFINITION

Assume θ is one of the acute (less than a right angle) angles in a right triangle, as in Figure 1, and let *adj* and *opp* stand for, respectively, the lengths of the sides adjacent to and opposite θ. Let *hyp* stand for the length of the hypotenuse of the right triangle. Then the **sine**, **cosine**, and **tangent** of θ, abbreviated sin θ, cos θ and tan θ, are the ratios:

$$\sin\theta = \frac{\text{opp}}{\text{hyp}}, \quad \cos\theta = \frac{\text{adj}}{\text{hyp}}, \quad \tan\theta = \frac{\text{opp}}{\text{adj}}.$$

(Note that sine, cosine, and tangent are indeed functions of the angle θ, and to be consistent with our functional notation we would expect to see, for example, $\sin(\theta) = \frac{\text{opp}}{\text{hyp}}$. By convention, though, the parentheses around the argument are omitted unless called for to make the meaning clear.)

Incidentally, the name sine appears to have evolved through a complicated history of abbreviations and mistranslations, beginning with an Arabic word for "half-chord." Our name for the ratio comes from the Latin word *sinus*, which means "bay," but the reference to water is entirely accidental. On the other hand, *cosine* and *tangent*, along with the three functions still to be defined, have meaningful names. More on the subject of names will appear soon.

The remaining three basic trigonometric functions are reciprocals of the first three, as follows:

DEFINITION

Again, assume θ is one of the acute angles in a right triangle, as in Figure 1. Then the **cosecant**, **secant**, and **cotangent** of θ, abbreviated csc θ, sec θ, and cot θ, are the reciprocals, respectively, of sin θ, cos θ, and tan θ. That is,

$$\csc\theta = \frac{1}{\sin\theta} = \frac{\text{hyp}}{\text{opp}}, \quad \sec\theta = \frac{1}{\cos\theta} = \frac{\text{hyp}}{\text{adj}}, \quad \cot\theta = \frac{1}{\tan\theta} = \frac{\text{adj}}{\text{opp}}.$$

EXAMPLE 1

Finding the Values of the Six Trigonometric Functions

Use the information contained in the two figures below to determine the values of the six trigonometric functions of θ.

a.

b.

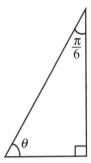

Solutions:

a. With the information given, we can determine $\tan\theta$ and $\cot\theta$ without effort (make sure you see why this is so). In order to evaluate the remaining four trigonometric functions at θ, all we need to do is determine the length of the hypotenuse. By the Pythagorean Theorem,

$$\left(\text{hyp}\right)^2 = \left(\text{adj}\right)^2 + \left(\text{opp}\right)^2$$

$$= 3^2 + 4^2$$

$$= 25$$

so hyp $= 5$. Thus:

$$\sin\theta = \frac{\text{opp}}{\text{hyp}} = \frac{4}{5}, \quad \cos\theta = \frac{\text{adj}}{\text{hyp}} = \frac{3}{5}, \quad \tan\theta = \frac{\text{opp}}{\text{adj}} = \frac{4}{3}$$

and

$$\csc\theta = \frac{1}{\sin\theta} = \frac{5}{4}, \quad \sec\theta = \frac{1}{\cos\theta} = \frac{5}{3}, \quad \cot\theta = \frac{1}{\tan\theta} = \frac{3}{4}.$$

b. Since the triangle pictured contains a right angle and an angle of $\frac{\pi}{6}$ radians, the angle θ must have measure $\frac{\pi}{3}$; in degree terms, this is a 30°-60°-90° triangle. Such triangles always have sides in the ratio $1:\sqrt{3}:2$, so even though we are not given the length of any side, we can still evaluate all six trigonometric functions at θ. For example, if the shorter leg (the leg adjacent to θ) is assumed to have a length of a, then the other leg must have a length of $a\sqrt{3}$ and the hypotenuse must have a length of $2a$. So,

$$\sin\theta = \frac{\text{opp}}{\text{hyp}} = \frac{a\sqrt{3}}{2a} = \frac{\sqrt{3}}{2}.$$

Similarly,

$$\cos\theta = \frac{\text{adj}}{\text{hyp}} = \frac{1}{2}, \tan\theta = \frac{\text{opp}}{\text{adj}} = \sqrt{3},$$

$$\csc\theta = \frac{1}{\sin\theta} = \frac{2}{\sqrt{3}}, \sec\theta = \frac{1}{\cos\theta} = 2, \cot\theta = \frac{1}{\tan\theta} = \frac{1}{\sqrt{3}}.$$

TOPIC 2 — Evaluation of Trigonometric Functions

The last example introduces the sort of reasoning we can use to evaluate trigonometric functions of many angles, and we will employ similar methods often in what follows. In other cases we will want a numerical approximation of the value of some trigonometric function, and a calculator will prove to be very useful.

EXAMPLE 2

Evaluating Trigonometric Functions

Evaluate the tangent and secant of $\theta = \frac{\pi}{4}$.

Solution:

With practice, you'll be able to determine $\tan\frac{\pi}{4}$ and $\sec\frac{\pi}{4}$ mentally, but initially it's very useful to draw a picture in order to visualize what is being asked. Since we are working with an angle of $\frac{\pi}{4}$ radians (45°), the remaining angle of our right triangle must be the same size. We've already noted that the sides of such a triangle have lengths in the ratio $1:1:\sqrt{2}$, so we can draw a triangle such as:

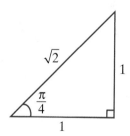

Note that the lengths of the sides have been arbitrarily chosen to be 1, 1, and $\sqrt{2}$. Any three numbers in the ratio of $1:1:\sqrt{2}$ could be used (the common factor will cancel out in the evaluation of any trigonometric function), so we may as well use these relatively simple numbers.

Now it is straightforward to note that

$$\tan\frac{\pi}{4} = \frac{1}{1} = 1$$

and

$$\sec\frac{\pi}{4} = \frac{\sqrt{2}}{1} = \sqrt{2}.$$

Up to this point in this chapter, a calculator has not been needed to perform any of the evaluations. But if we are asked to evaluate an expression such as sin 56.4° (we may encounter such an expression while solving a real-world application), some technological assistance is called for.

Fortunately, calculators (and computer software) that are equipped to handle trigonometric functions are readily available and easily used. However, it is important to remember that angles can be measured in terms of either degrees or radians, and *you* are responsible for putting the calculator in the correct mode (degree or radian) before performing the evaluation. This warning deserves to be repeated:

CAUTION!

Before using a calculator to evaluate a given trigonometric expression, determine whether the angle in the expression is measured in degrees or radians. Then put the calculator in the appropriate mode prior to the evaluation.

EXAMPLE 3

Evaluating Trigonometric Functions Using a Calculator

Use a calculator to evaluate the following expressions.

a. sin 56.4°

b. $\cot \dfrac{5\pi}{11}$

Solutions:

a. Refer to the user's manual to determine how to put your calculator in degree mode. Typically, there is a button labeled "mode" and pressing it leads to the option of choosing either "degree" or "radian." Once the calculator is in the correct mode, press the "sin" button, enter 56.4 on the number pad, and press the "=" or "Enter" button. The answer, rounded to 4 decimal places, is 0.8329. The exact number of digits on your display will depend on the calculator and its current settings. If your display reads –0.1481, your calculator is in the incorrect (radian) mode for this problem.

b. The absence of the degree symbol in the expression $\dfrac{5\pi}{11}$ tells us that the angle is measured in radians, so the first step is to place your calculator in radian mode. Next, recall that cotangent is the reciprocal of tangent. Most calculators don't have buttons specifically for the cotangent, secant, and cosecant functions; if you need to evaluate an expression containing one of these functions, take the reciprocal of, respectively, the tangent, cosine, or sine of the given angle.

In this case, use your calculator to confirm that

$$\tan \frac{5\pi}{11} \approx 6.9552,$$

and therefore

$$\cot \frac{5\pi}{11} = \frac{1}{\tan \dfrac{5\pi}{11}} \approx 0.1438.$$

As mentioned at the start of this section, trigonometry has been around for several thousand years. For all but the last few decades of that very long history, users of trigonometry did not have the option of being able to punch a few buttons on a calculator in order to perform their calculations. In the not too distant past, a large part of trigonometry consisted of teaching students how to use tables of predetermined evaluations (so-called "trig tables"). Thankfully, we are past the need for such instruction; however, one legacy of the precalculator days of trigonometry lives on and needs to be discussed. That legacy concerns notation.

Today, decimal notation most naturally suits the use of calculators. When calculations were done by hand, however, angles were more commonly expressed in the "degrees, minutes, seconds" (DMS) notation, and we still encounter this notation frequently in some contexts (surveying and astronomy, to name two). We need, therefore, to be able to convert from the DMS notation to decimal notation.

DEFINITION

Degree, Minute,
Second Notation

In the context of angle measure,

$$1' = \text{one minute} = \left(\frac{1}{60}\right)(1°)$$

and

$$1'' = \text{one second} = \left(\frac{1}{60}\right)(1') = \left(\frac{1}{3600}\right)(1°).$$

For instance, an angle given as $14°37'23''$ ("14 degrees, 37 minutes, 23 seconds") can also be written in decimal form (rounded to four decimal places) as $14.6231°$, since

$$14°37'23'' = 14 + \frac{37}{60} + \frac{23}{3600} \approx 14.6231°.$$

TOPIC 3   Applications of Trigonometric Functions

One way or another, most applications of trigonometry involve using given information about a triangle to determine something else about the triangle. At this point, the triangles we work with are all right triangles, and a basic knowledge of the trigonometric functions and the Pythagorean Theorem suffice as tools. The process of determining unknown angles and/or dimensions from known data in such cases is often termed **solving right triangles**. In later sections, we will consider arbitrary triangles and will enlarge our collection of tools with a variety of trigonometric identities and theorems.

EXAMPLE 4

Using Trigonometric Functions

Before cutting down a dead tree in your yard, you very sensibly decide to determine its height. Backing up 40 feet from the tree (which rises straight up from level ground), you use a *theodolite* (a surveyor's instrument that accurately measures angles) and note that the angle between the ground and the top of the tree is $61°55'39''$. How tall is the tree?

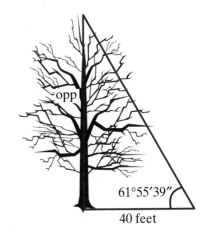

Solution:

As in so many problems, a picture is of great help. Note that this problem indeed involves solving a right triangle: we know the measure of an angle and the length of the angle's adjacent leg, and we want the length of the opposite leg. This observation gives us the best clue to which trigonometric function to use; since tangent and cotangent are the two which don't depend on the length of the hypotenuse, chances are one of these is a good choice.

Note also that we have to convert the theodolite's reading into decimal form before using a calculator:

$$61°55'39'' = 61 + \frac{55}{60} + \frac{39}{3600} \approx 61.9275°.$$

Now we can use the figure to see that

$$\tan 61.9275° = \frac{\text{opp}}{40},$$

so

$$\text{opp} = 40 \tan 61.9275° \approx 75.$$

That is, the tree is approximately 75 feet tall.

EXAMPLE 5

Using Trigonometric
Functions

The manufacturer of a certain brand of 16-foot ladder recommends that, when in use, the angle between the ground and the ladder should equal 75°. What distance should the foot of the ladder be from the base of the wall it is leaning against?

Solution:

Since we are given information about an angle, its adjacent side, and the hypotenuse of a right triangle, cosine is the logical trigonometric function to use in solving this problem (equivalently, secant could be used, but calculators are equipped with a "cos" and not a "sec" button so our current technology tends to lead to the use of cosine).

We want to determine the length of the adjacent side when the ladder is resting against the wall with its recommended angle of 75°, and we note that

$$\cos 75° = \frac{\text{adj}}{16}.$$

This gives us

$$\text{adj} = 16 \cos 75° \approx 4.14 \text{ feet},$$

or a bit less than 4 feet, 2 inches.

In many surveying problems, it is frequently necessary to determine the height of some distant object when it is impossible or impractical to measure how far away the object is. One way to determine the height anyway begins with the diagram in Figure 2. In the diagram, assume that distance d and angles α and β can be measured, but that distance x is unknown. How can we determine height h?

Figure 2: Determining h from Two Angles

There are two right triangles apparent in the diagram, and since we know or desire to know something about an angle and its opposite and adjacent sides in each triangle, the tangent function looks promising. We'll begin with the two trigonometric relations

$$\tan \alpha = \frac{h}{x}$$

and

$$\tan \beta = \frac{h}{x+d}.$$

The premise is that we want to find a formula for h in terms of α, β, and d alone, since we don't know the length x. We can start by solving the two equations above for h to get $h = x \tan \alpha$ and $h = (x+d) \tan \beta$. These two equations actually are an example of a *system of two equations* in the two unknowns x and h, and we will study such systems in detail in Chapter 10. For our present purposes, we'll note that if we can solve one equation for x and use the result in the other equation, we'll obtain a single equation in which h is the only unknown (this is called the *method of substitution*). For instance, if we solve the first equation for x,

$$x = \frac{h}{\tan \alpha},$$

and make that substitution for x in the second equation, we obtain

$$h = \left(\frac{h}{\tan \alpha} + d \right) \tan \beta.$$

Now we can solve this equation for h:

$$h = \left(\frac{h}{\tan \alpha} + d \right) \tan \beta$$

$$h = \frac{h \tan \beta}{\tan \alpha} + d \tan \beta \qquad \text{Multiply out.}$$

$$h \tan \alpha = h \tan \beta + d \tan \alpha \tan \beta \qquad \text{Clear fractions by multiplying by } \tan \alpha.$$

$$h \tan \alpha - h \tan \beta = d \tan \alpha \tan \beta \qquad \text{Isolate terms with } h \text{ on one side.}$$

$$h (\tan \alpha - \tan \beta) = d \tan \alpha \tan \beta \qquad \text{Factor out } h.$$

$$h = \frac{d \tan \alpha \tan \beta}{\tan \alpha - \tan \beta}. \qquad \text{Divide to solve for } h.$$

This formula is well-known to surveyors, though it often appears in the slightly more appealing form obtained by dividing the numerator and denominator of the fraction on the right by $\tan\alpha \tan\beta$:

$$h = \frac{d \tan\alpha \tan\beta}{\tan\alpha - \tan\beta}$$

$$= \frac{d}{\dfrac{\tan\alpha}{\tan\alpha \tan\beta} - \dfrac{\tan\beta}{\tan\alpha \tan\beta}}$$

$$= \frac{d}{\dfrac{1}{\tan\beta} - \dfrac{1}{\tan\alpha}}$$

$$= \frac{d}{\cot\beta - \cot\alpha}.$$

EXAMPLE 6

Using Trigonometric Functions

Approached from one direction, Mt. Baldy rises out of a perfectly level desert plain. A surveyor standing in the desert some distance from the mountain measures the angle of elevation between the desert floor and the top of the mountain to be $60°1'6''$. She then backs up 1000 feet and determines the new angle of elevation to be $56°3'23''$. How high above the desert plain does Mt. Baldy rise?

Solution:

Using the notation of the derivation above, we are given $\alpha = 60°1'6''$, $\beta = 56°3'23''$, and $d = 1000$. Converting to decimal notation, $\alpha \approx 60.0183°$ and $\beta \approx 56.0564°$, so

$$h = \frac{1000}{\cot 56.0564° - \cot 60.0183°}$$

$$\approx \frac{1000}{0.6731 - 0.5769}$$

$$\approx 10{,}395 \text{ feet.}$$

1. $\sin\theta = \cos\theta = \dfrac{\sqrt{2}}{2}$

$\tan\theta = \cot\theta = 1$

$\csc\theta = \sec\theta = \sqrt{2}$

2. $\sin\theta = \dfrac{12}{13}$, $\cos\theta = \dfrac{5}{13}$

$\tan\theta = \dfrac{12}{5}$, $\csc\theta = \dfrac{13}{12}$

$\sec\theta = \dfrac{13}{5}$, $\cot\theta = \dfrac{5}{12}$

3. $\sin\theta = \dfrac{1}{3}$, $\cos\theta = \dfrac{2\sqrt{2}}{3}$

$\tan\theta = \dfrac{\sqrt{2}}{4}$, $\csc\theta = 3$

$\sec\theta = \dfrac{3\sqrt{2}}{4}$, $\cot\theta = 2\sqrt{2}$

4. $\sin\theta = \dfrac{1}{2}$, $\cos\theta = \dfrac{\sqrt{3}}{2}$

$\tan\theta = \dfrac{\sqrt{3}}{3}$, $\csc\theta = 2$

$\sec\theta = \dfrac{2\sqrt{3}}{3}$, $\cot\theta = \sqrt{3}$

5. $\sin\theta = \cos\theta = \dfrac{\sqrt{2}}{2}$

$\tan\theta = \cot\theta = 1$

$\csc\theta = \sec\theta = \sqrt{2}$

6. $\sin\theta = \dfrac{\sqrt{3}}{2}$, $\cos\theta = \dfrac{1}{2}$

$\tan\theta = \sqrt{3}$, $\csc\theta = \dfrac{2\sqrt{3}}{3}$

$\sec\theta = 2$, $\cot\theta = \dfrac{\sqrt{3}}{3}$

7. $\sin\theta = \dfrac{5}{7}$, $\cos\theta = \dfrac{2\sqrt{6}}{7}$

$\tan\theta = \dfrac{5\sqrt{6}}{12}$, $\csc\theta = \dfrac{7}{5}$

$\sec\theta = \dfrac{7\sqrt{6}}{12}$,

$\cot\theta = \dfrac{2\sqrt{6}}{5}$

8. $\sin\theta = \cos\theta = \dfrac{\sqrt{2}}{2}$

$\tan\theta = \cot\theta = 1$

$\csc\theta = \sec\theta = \sqrt{2}$

Exercises

Use the information contained in the figures to determine the values of the six trigonometric functions of θ. Rationalize all denominators. See Example 1.

1.

2.

3.

4.

5.

6.

7.

8.

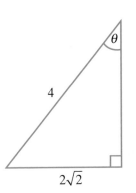

9. $\sin\theta = \dfrac{12}{13},\ \cos\theta = \dfrac{5}{13}$

$\tan\theta = \dfrac{12}{5},\ \csc\theta = \dfrac{13}{12}$

$\sec\theta = \dfrac{13}{5},\ \cot\theta = \dfrac{5}{12}$

9.

10.

10. $\sin\theta = \dfrac{11}{61},\ \cos\theta = \dfrac{60}{61}$

$\tan\theta = \dfrac{11}{60},\ \csc\theta = \dfrac{61}{11}$

$\sec\theta = \dfrac{61}{60},\ \cot\theta = \dfrac{60}{11}$

11.

12.

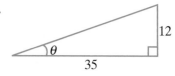

11. $\sin\theta = \dfrac{33}{65},\ \cos\theta = \dfrac{56}{65}$

$\tan\theta = \dfrac{33}{56},\ \csc\theta = \dfrac{65}{33}$

$\sec\theta = \dfrac{65}{56},\ \cot\theta = \dfrac{56}{33}$

12. $\sin\theta = \dfrac{12}{37},\ \cos\theta = \dfrac{35}{37}$

$\tan\theta = \dfrac{12}{35},\ \csc\theta = \dfrac{37}{12}$

$\sec\theta = \dfrac{37}{35},\ \cot\theta = \dfrac{35}{12}$

13.

14.

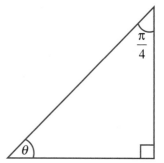

13. $\sin\theta = \dfrac{1}{2},\ \cos\theta = \dfrac{\sqrt{3}}{2}$

$\tan\theta = \dfrac{\sqrt{3}}{3},\ \csc\theta = 2$

$\sec\theta = \dfrac{2\sqrt{3}}{3},\ \cot\theta = \sqrt{3}$

14. $\sin\theta = \cos\theta = \dfrac{\sqrt{2}}{2}$

$\tan\theta = \cot\theta = 1$

$\csc\theta = \sec\theta = \sqrt{2}$

15. $\sin\dfrac{\pi}{4} = \dfrac{\sqrt{2}}{2},$

$\csc\dfrac{\pi}{4} = \sqrt{2}$

16. $\cos\dfrac{\pi}{7} \approx 0.9010$

$\tan\dfrac{\pi}{7} \approx 0.4816$

17. $\sec 60° = 2$

Evaluate the expressions, using a calculator if necessary. See Examples 2 and 3.

15. sine and cosecant of $\dfrac{\pi}{4}$

16. cosine and tangent of $\dfrac{\pi}{7}$

17. sec 60°

18. tan 71° and cot 71°

19. csc $\dfrac{\pi}{6}$

20. sine of $\dfrac{3\pi}{7}$

21. secant and tangent of 5°

22. cosine of 28.37°

23. cotangent of $\dfrac{\pi}{3}$

24. $\sin\dfrac{2\pi}{5}$ and $\cos\dfrac{2\pi}{5}$

25. tan 87.2°

26. csc 54°

18. $\tan 71° \approx 2.9042$;
$\cot 71° \approx 0.3443$

19. $\csc \dfrac{\pi}{6} = 2$

20. $\sin \dfrac{3\pi}{7} \approx 0.9749$

21. $\sec 5° \approx 1.0038$;
$\tan 5° \approx 0.0875$

22. $\cos 28.37° \approx 0.8799$

23. $\cot \dfrac{\pi}{3} = \dfrac{\sqrt{3}}{3}$

24. $\sin \dfrac{2\pi}{5} \approx 0.9511$

$\cos \dfrac{2\pi}{5} \approx 0.3090$

25. $\tan 87.2° \approx 20.4465$

26. $\csc 54° \approx 1.2361$

27. 0.9945

28. 0.3090

29. 1.0355

30. 3.4203

31. 28.6537

32. 0.6009

33. 3.0777

34. 0.7071

35. 0.3827

36. 0.7975

37. 2

38. 1.0103

39. 38.9053°

40. 56.2003°

41. 25.325°

42. 6.1472°

43. 21.6656°

44. 88.6667°

45. 0.1149

46. −0.5714

47. 0.7746   **48.** 0.4

49. 5   **50.** −5

51. True   **52.** True

53. True   **54.** False

55. False   **56.** True

Use a graphing calculator to evaluate the following expressions. Round answers to four decimal places. See Example 3.

27. $\sin 84°$

28. $\cos 72°$

29. $\tan 46°$

30. $\csc 17°$

31. $\sec 88°$

32. $\cot 59°$

33. $\tan \dfrac{2\pi}{5}$

34. $\cos \dfrac{\pi}{4}$

35. $\sin \dfrac{\pi}{8}$

36. $\cot \dfrac{2\pi}{7}$

37. $\sec \dfrac{\pi}{3}$

38. $\csc \dfrac{5\pi}{11}$

Convert each expression from degrees, minutes, seconds (DMS) notation to decimal notation. Round answers to four decimal places.

39. $38°54'19''$ **40.** $56°12'1''$ **41.** $25°18'90''$ **42.** $6°8'50''$

43. $21°39'56''$ **44.** $88°30'600''$

Determine the value of the given trigonometric expression given the value of another trigonometric expression. Round answers to four decimal places.

45. Find $\sin \theta$, if $\csc \theta = 8.7$.

46. Find $\cos \theta$, if $\sec \theta = -\dfrac{7}{4}$.

47. Find $\tan \theta$, if $\cot \theta = \dfrac{\sqrt{15}}{3}$.

48. Find $\cot \theta$, if $\tan \theta = 2.5$.

49. Find $\sec \theta$, if $\cos \theta = 0.2$.

50. Find $\csc \theta$, if $\sin \theta = -\dfrac{1}{5}$.

Determine whether the following statements are true or false. Use a graphing calculator when necessary.

51. If $\sin \theta = 0.8$, then $\csc \theta = 1.25$.

52. If $\cos \theta = 0.96$, then $\sec \theta = 1\dfrac{1}{24}$.

53. If $\tan \theta = 4\dfrac{4}{9}$, then $\cot \theta = 0.225$.

54. If $\sin \theta = 0.5625$, then $\csc \theta = 2.48$.

55. If $\cos \theta = 0.75$, then $\sec \theta = \dfrac{8}{3}$.

56. If $\tan \theta = 0.2540$, then $\cot \theta = 3.937$.

Use an appropriate trigonometric function and a calculator if necessary to solve the following problems. See Examples 4 and 5. Round answers to two decimal places.

57. A hang glider wants to determine if a certain vertical cliff is a suitable height for her liftoff. From a distance of 40 yards, she measures the angle from the ground to its tip as $80°55'24''$. How high is the cliff in feet?

58. A mahimahi is hooked on 70 feet of fishing line, 10 feet of which is above the surface of the water. The angle of depression from the water's surface to the line is 40°. How deep is the fish?

59. A filing cabinet is 3 feet and 4 inches tall from the floor. If a piece of string is stretched from the top of the cabinet to a point on the floor and the angle between the string and the floor is 11°, what is the length of the string?

60. A tree being cut down makes a 70° angle with the ground when the tip of the tree is directly above a spot that is 40 feet from the base of the tree. Find the height of the tree.

57. 751.19 feet

58. 38.57 feet

59. 17.47 feet

60. 116.95 feet

61. 48.54 yards

62. 11.47 feet

63. 12.04 m

64. 19.91 m

65. 314.57 feet

66. 21.67 feet

67. 13.86 feet

68. 35.03 feet

69. 6.86 feet

70. 9.98 feet

71. 20 m

61. Stephen is standing 15 yards from a stream, but instead of walking directly towards the stream, he decides to take a more scenic (though straight-line) path to the stream. If the angle between the scenic route and the stream is 18°, how far did Stephen walk?

62. The builder of a parking garage wants to build a ramp at an angle of 16° that covers a horizontal span of 40 feet. What is the vertical rise of the ramp?

63. A kitesurfer's lines are 20 m long and make an angle of 37° with the ocean while heading away from the beach under current wind conditions. How high above the water is the kite flying?

64. An anthropologist studying a tribe of indigenous people wants to know the dimensions of their stone-hewn temple. After walking 15 meters from the structure, she measures the angle to its top to be 53°. What is the height of the temple?

65. A radio tower has a 64-foot shadow cast by the Sun. If the angle from the tip of the shadow to the top of the tower is 78.5°, what is the height of the radio tower?

66. A ladder is propped up to a barn at an 80° angle. If the ladder is 22 feet long, what is the approximate height where the top of the ladder touches the barn?

67. The ramp of a moving truck touches the ground 12 feet away from the end of the truck. If the ramp makes an angle of 30° relative to the ground, what is the length of the ramp?

68. The angle of elevation of a flying kite is 61°7′21″. If the other end of the 40-foot string attached to the kite is tied to the ground, what is the approximate height of the kite?

69. A length of rope is attached from the top of a dock to the rope tie device located on the underneath of the boat at the water's surface. The rope is 33 feet in length and has an angle of elevation relative to the surface of the water of 12°. How high above the water does the dock sit?

Use the formula from Example 6 to solve the following problems.

70. A surveyor wants to find the width of a river without crossing it. He sights an abandoned tire on the opposite bank (the banks are straight and parallel) and measures the angle from where he stands relative to the shore to be 31°. After walking precisely 15 feet away from the tire, he measures the same angle to be 13.5°. How wide is the river?

71. A drawbridge operator in a control room observes a sailboat approaching and finds the angle of depression to the boat to be 9°. Twenty minutes later, the angle to the same boat is 19°. If the sailboat has traveled 68.2 m, how high above water is the control room?

72. 58.9 feet

73. 3196.80 feet

74. 2066.79 feet

72. A birdwatcher discovers a hawk's nest in a tree some distance away. She wants to determine its height, so she measures the angle from the level ground to the nest at 40°. After approaching 25 feet closer to the tree, she finds the same angle to be 52.5°. How high does the nest sit, in feet?

73. A surveyor standing some distance from a plateau measures the angle of elevation from the ground to the top of the plateau to be 46°57′12″. The surveyor then walks forward 800 feet and measures the angle of elevation to be 55°37′70″. What is the height of the plateau?

74. A surveyor standing some distance from a hill measures the angle of elevation from the ground to the top of the hill to be 83°45′97″. The surveyor then steps back 300 feet and measures the angle of elevation to be 75°44′16″. What is the height of the hill?

Trigonometric Functions of Any Angle

TOPICS

1. Extending the domains of the trigonometric functions
2. Evaluation using reference angles
3. Relationships between trigonometric functions

TOPIC 1 Extending the Domains of the Trigonometric Functions

The definitions given in the last section of the six trigonometric functions implicitly assumed that the angle under discussion was greater than 0 and less than $\frac{\pi}{2}$ radians. With just a little bit of extrapolation, the domain of definition of each function can be extended slightly as indicated in the table below:

Function	Initial Extended Domain (interval from which θ can be chosen)
$\sin \theta$	$\left[0, \frac{\pi}{2}\right]$ or $0 \leq \theta \leq \frac{\pi}{2}$
$\cos \theta$	$\left[0, \frac{\pi}{2}\right]$ or $0 \leq \theta \leq \frac{\pi}{2}$
$\tan \theta$	$\left[0, \frac{\pi}{2}\right)$ or $0 \leq \theta < \frac{\pi}{2}$
$\csc \theta$	$\left(0, \frac{\pi}{2}\right]$ or $0 < \theta \leq \frac{\pi}{2}$
$\sec \theta$	$\left[0, \frac{\pi}{2}\right)$ or $0 \leq \theta < \frac{\pi}{2}$
$\cot \theta$	$\left(0, \frac{\pi}{2}\right]$ or $0 < \theta \leq \frac{\pi}{2}$

The reasoning behind the extrapolation is as follows. Suppose that (x, y) is a point anywhere in the first quadrant of the Cartesian plane (including possibly on the x-axis or the y-axis). By drawing a line segment from the origin to (x, y) and another line segment from (x, y) vertically down to the x-axis (that is, the point $(x, 0)$), we obtain a right triangle with θ defined as pictured in Figure 1, where $r = \sqrt{x^2 + y^2}$.

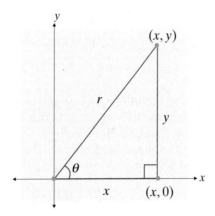

Figure 1: Trigonometric Functions in the First Quadrant

If either x or y is 0 (that is, if the point (x, y) lies on one of the coordinate axes), the triangle formed is *degenerate*. This is not a moral condemnation—it just means that two of the edges coincide so the figure doesn't appear to be a triangle. But as we will see, degenerate triangles won't affect our extended definitions of the trigonometric functions.

Recall that the original definitions of the functions were in terms of the hypotenuse and opposite and adjacent sides of a right triangle. Referring to Figure 1, we can rephrase and extend the definitions as follows:

$\sin \theta = \dfrac{y}{r}$	This now defines $\sin \theta$ for $0 \le \theta \le \dfrac{\pi}{2}$.
$\cos \theta = \dfrac{x}{r}$	This now defines $\cos \theta$ for $0 \le \theta \le \dfrac{\pi}{2}$.
$\tan \theta = \dfrac{y}{x}$ $\left(\text{for } x \ne 0\right)$	Note that the restriction $x \ne 0$ means $\theta \ne \dfrac{\pi}{2}$.
$\csc \theta = \dfrac{r}{y}$ $\left(\text{for } y \ne 0\right)$	Note that the restriction $y \ne 0$ means $\theta \ne 0$.
$\sec \theta = \dfrac{r}{x}$ $\left(\text{for } x \ne 0\right)$	Note that the restriction $x \ne 0$ means $\theta \ne \dfrac{\pi}{2}$.
$\cot \theta = \dfrac{x}{y}$ $\left(\text{for } y \ne 0\right)$	Note that the restriction $y \ne 0$ means $\theta \ne 0$.

Note that for a given angle θ in one of the intervals on the previous page, there are an infinite number of points (x, y) in the first quadrant that could be used in order to complete the picture in Figure 1 (any point lying along the ray rotated θ from the positive x-axis will work). But because all of the possible triangles thus formed are similar, any one of them suffices to define a given trigonometric function.

EXAMPLE 1

Evaluate all six trigonometric functions at $\theta = 0$, if possible.

Solution:

To use the definitions on the previous page, we need to pick a point along the ray defined by the angle $\theta = 0$. This ray lies along the positive x-axis, so any point on the positive x-axis will suffice (note that the angle $\theta = 0$ leads to one of the two degenerate triangles mentioned above). The first point on the positive x-axis that may come to mind is $(1, 0)$, and this will certainly work. For this point, $x = 1$, $y = 0$, and $r = 1$. Now:

$$\sin 0 = \frac{0}{1} = 0, \ \cos 0 = \frac{1}{1} = 1, \ \tan 0 = \frac{0}{1} = 0$$

and

$$\csc 0 \text{ is undefined}, \ \sec 0 = \frac{1}{1} = 1, \ \cot 0 \text{ is undefined}.$$

The association between an angle θ and a point (x, y) on the ray defined by θ actually allows us to extend the domains of the trigonometric functions to a far greater extent. The table below simply repeats, for reference, the definitions on the previous page.

DEFINITION

Given an angle θ in standard position, let (x, y) be any point (other than the origin) on the terminal side of the angle. Letting $r = \sqrt{x^2 + y^2}$ we define:

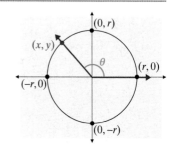

$\sin \theta = \dfrac{y}{r}$	$\cos \theta = \dfrac{x}{r}$	$\tan \theta = \dfrac{y}{x}$ (for $x \neq 0$)
$\csc \theta = \dfrac{r}{y}$ (for $y \neq 0$)	$\sec \theta = \dfrac{r}{x}$ (for $x \neq 0$)	$\cot \theta = \dfrac{x}{y}$ (for $y \neq 0$)

In other words,

- sin and cos are defined for all real numbers;
- tan and sec are defined for all real numbers except $\dfrac{\pi}{2} + n\pi$; and
- cot and csc are defined for all real numbers except $n\pi$

for some integer n.

We have now defined $\sin\theta$ and $\cos\theta$ for any real number θ, and the other four trigonometric functions have been defined for *nearly* any real number θ. The exact meaning of this last sentence will become clear as you study the remaining examples in this chapter.

EXAMPLE 2

Finding the Values of the Six Trigonometric Functions

Determine the values of the six trigonometric functions of the angle θ.

a. $\theta = -\dfrac{5\pi}{2}$ 　　　　　　　　　**b.** $\theta = 120°$

Solutions:

a. Recall that a negative angle corresponds to a clockwise rotation from the positive x-axis. The angle $\theta = -\dfrac{5\pi}{2}$ indicates one full revolution clockwise (-2π radians) plus another quarter revolution clockwise ($-\dfrac{\pi}{2}$ radians), resulting in the terminal side of the angle pointing straight down along the negative y-axis as shown. To evaluate the six trigonometric functions, we just need to choose a point on the terminal side of the angle; although it's a somewhat boring choice, the point $(0, -1)$ is probably easiest to work with (note that this gives us $r = 1$). Now:

$$\sin\left(-\frac{5\pi}{2}\right) = \frac{-1}{1} = -1, \ \cos\left(-\frac{5\pi}{2}\right) = \frac{0}{1} = 0, \ \tan\left(-\frac{5\pi}{2}\right) = \frac{-1}{0} \text{ is undefined}$$

and

$$\csc\left(-\frac{5\pi}{2}\right) = \frac{1}{-1} = -1, \ \sec\left(-\frac{5\pi}{2}\right) = \frac{1}{0} \text{ is undefined}, \ \cot\left(-\frac{5\pi}{2}\right) = \frac{0}{-1} = 0.$$

b. The angle $\theta = 120°$ is $30°$ more than a (counterclockwise-oriented) right angle, leading to the triangle shown at right. Once we realize we are dealing with a $30°$-$60°$-$90°$ triangle, we know that the ratios of the side lengths must be $1 : \sqrt{3} : 2$ and we can easily locate the point $\left(-1, \sqrt{3}\right)$ on the terminal side of the angle.

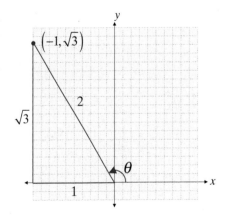

The rest is straightforward. Using the point $\left(-1, \sqrt{3}\right)$ and the value $r = 2$ we determine that

$$\sin 120° = \frac{\sqrt{3}}{2}, \quad \cos 120° = \frac{-1}{2}, \quad \tan 120° = \frac{\sqrt{3}}{-1} = -\sqrt{3}$$

and

$$\csc 120° = \frac{2}{\sqrt{3}}, \quad \sec 120° = \frac{2}{-1} = -2, \quad \cot 120° = \frac{-1}{\sqrt{3}}.$$

TOPIC 2 — Evaluation Using Reference Angles

The last example hinted at the fact that the evaluation of a trigonometric function at a nonacute angle can be related to its evaluation at an angle in the interval $\left[0, \dfrac{\pi}{2}\right]$. Such angles are called *reference angles*, and the precise definition is as follows.

DEFINITION

Reference Angle

Given an angle θ in standard position, the **reference angle** θ' associated with it is the angle formed by the x-axis and the terminal side of θ. Reference angles are always greater than or equal to 0 and less than or equal to $\dfrac{\pi}{2}$ radians ($0 \le \theta' \le \dfrac{\pi}{2}$).

Figure 2 illustrates four ways in which a given angle θ can relate to its reference angle θ'.

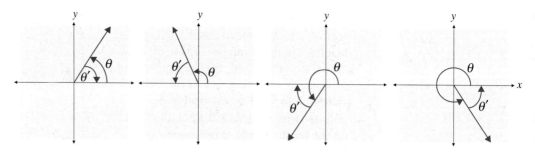

Figure 2: Angles and Associated Reference Angles

EXAMPLE 3

Finding the
Reference Angle

Determine the reference angle associated with each of the following angles.

a. $\theta = \dfrac{9\pi}{8}$

b. $\varphi = -655°$

Solutions:

a. The terminal side of $\theta = \dfrac{9\pi}{8}$ lies in the third quadrant, so the reference angle is determined by it and the negative x-axis. Specifically,

$$\theta' = \frac{9\pi}{8} - \pi = \frac{\pi}{8}.$$

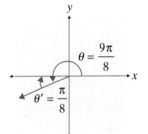

b. The terminal side of $\varphi = -655°$ lies in the first quadrant (note that the angle describes one complete revolution and more than three-quarters of a second clockwise revolution around the origin). An additional clockwise rotation of 65° would result in two full revolutions, so that is the reference angle. That is,

$$2(360°) - 655° = 65°.$$

The value of reference angles lies in the fact that a trigonometric function evaluated at a given angle will be the same as the function evaluated at the reference angle, except possibly for sign. We implicitly used this fact in Example 2, and the reason why it's true is not hard to see. Since we have defined the trigonometric functions in terms of the coordinates of a point chosen on the terminal side of an angle, the value doesn't depend at all on how many revolutions around the origin (or in which direction) the angle describes. In Example 2b, for instance, the key step lay in determining that the reference angle for 120° is 60°. This led to the construction of a 30°-60°-90° triangle and the easy evaluation of all six trigonometric functions.

Sign is the one thing that may differ between the value of a trigonometric function at θ and the value of the same function at the reference angle θ'. By thinking about the signs of the x and y coordinates of points in the four quadrants, it's easy to see that all trig functions are positive in the first quadrant, that sine (and its reciprocal cosecant) are positive in the second quadrant, that tangent (and its reciprocal cotangent) are positive in the third quadrant, and that cosine (and its reciprocal secant) are positive in the fourth quadrant. A bit of propaganda has evolved as a mnemonic to help students remember this: "All Students Take Calculus" may remind you that, beginning in the first quadrant, the functions that are positive are:

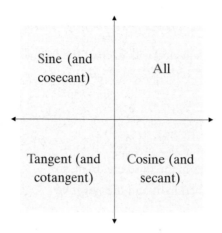

Signs of the Trigonometric Functions		
Quadrant	Positive	Negative
I	all	none
II	sin, csc	cos, sec, tan, cot
III	tan, cot	sin, csc, cos, sec
IV	cos, sec	sin, csc, tan, cot

EXAMPLE 4

Evaluating Trigonometric Functions

Evaluate the following:

a. $\cos \dfrac{7\pi}{6}$ **b.** $\tan(-225°)$

Solutions:

a. The terminal side of $\dfrac{7\pi}{6}$ lies in the third quadrant and its reference angle is $\dfrac{\pi}{6}$. $\cos \dfrac{\pi}{6} = \dfrac{\sqrt{3}}{2}$, but cosine is negative in the third quadrant, so $\cos \dfrac{7\pi}{6} = -\dfrac{\sqrt{3}}{2}$.

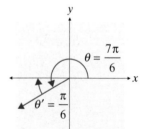

b. The terminal side of $-225°$ lies in the second quadrant and its reference angle is $45°$. Since tangent is negative in the second quadrant, $\tan(-225°) = -\tan 45° = -1$.

TOPIC **3**

Relationships Between Trigonometric Functions

If you have been studying the definitions and examples in this chapter carefully, you may be starting to develop the feeling that there is a great deal of redundancy in trigonometry. This is no illusion; as specific examples, we know now that there are two ways to measure angles and that three of the trigonometric functions are just reciprocals of the other three. There are other redundancies that are a bit more subtle, but you may have developed a sense of them already. We will end this section with a discussion of the qualities of trigonometric functions that lead to this feeling.

The preceding paragraph should not be taken to mean, however, that some of what you are learning in this chapter is pointless. Take, for example, the fact that cosecant, secant, and cotangent are simply reciprocals of sine, cosine, and tangent, and therefore seem unnecessary. In calculus, you'll encounter problems that are more easily stated and solved in terms of, for example, secant rather than cosine. This argument (that a problem is easier to solve with the choice of one function over another) is potent, and can't be disregarded. Why discard something if its existence makes life easier?

Another reason, if not a justification, for the existence of all six trigonometric functions is historical. This history is worth spending a paragraph or two on just for the light it sheds on the nomenclature of trigonometry, but as it turns out the nomenclature in turn leads to some useful facts.

Starting with a circle of radius 1, construct the lines as shown in Figure 3. The words *secant* and *tangent* are derived from Latin names for, respectively, the lines $\overline{OB}$ and $\overline{AB}$ (we use "tangent" in everyday language to describe a situation where one object is just touching another). If the line passing through C and D were continued down until it intersected the circle again, it would form a *chord*; as it is, $\overline{CD}$ forms a half-chord and *sine* is the word that evolved to denote its length.

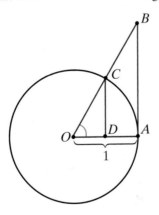

Figure 3: Etymology of the Trigonometric Functions

Make sure you see how the lengths of the line segments $\overline{OB}$, $\overline{AB}$, and $\overline{CD}$ represent, respectively, the secant, tangent, and sine of the central angle shown in green. (**Hint:** It's important to remember that $\overline{OA}$ and $\overline{OC}$ have length 1.) The next step is to see how these three functions lead to the remaining three trigonometric functions (cosine, cosecant, and cotangent.)

Complementary angles are two angles which, when put together, form a right angle. That is, the sum of their measures is $\dfrac{\pi}{2}$ (or 90° in terms of degree measure). The cosine, cosecant, and cotangent of an angle θ are, respectively, the sine, secant, and tangent of the *complement* of θ. In terms of formulas, this gives us our first set of identities.

IDENTITY

Cofunction Identities

Given an angle θ (measured in radians), $\dfrac{\pi}{2} - \theta$ is the measure of its complement. Thus:

$$\cos\theta = \sin\left(\frac{\pi}{2} - \theta\right), \quad \csc\theta = \sec\left(\frac{\pi}{2} - \theta\right), \text{ and } \cot\theta = \tan\left(\frac{\pi}{2} - \theta\right).$$

We will encounter many more identities in the sections to come. In general, trigonometric identities are equations that are useful in simplifying or evaluating expressions. For reference purposes, we list here another set of three identities that you know well by now:

IDENTITY

Reciprocal Identities

For a given angle θ for which both sides of the equation make sense,

$$\csc\theta = \frac{1}{\sin\theta}, \quad \sec\theta = \frac{1}{\cos\theta}, \text{ and } \cot\theta = \frac{1}{\tan\theta}.$$

We'll finish this initial list of identities with two that you may have already noted. Recall our definitions of tangent and cotangent in terms of a point (x, y) on the terminal side of an angle θ:

$$\tan\theta = \frac{y}{x} \text{ and } \cot\theta = \frac{x}{y}.$$

Since $\sin\theta = \dfrac{y}{r}$ and $\cos\theta = \dfrac{x}{r}$, the following identities are apparent.

IDENTITY

Quotient Identities

For a given angle θ for which both sides of the equation make sense,

$$\tan\theta = \frac{\sin\theta}{\cos\theta} \text{ and } \cot\theta = \frac{\cos\theta}{\sin\theta}.$$

EXAMPLE 5

Using Cofunction
Identities

Express each of the following in terms of the appropriate cofunction, and verify the equivalence of the two expressions.

a. $\cos\left(-\dfrac{5\pi}{11}\right)$

b. $\cot 195°$

Solutions:

a. $\cos\left(-\dfrac{5\pi}{11}\right) = \sin\left(\dfrac{\pi}{2} - \left(-\dfrac{5\pi}{11}\right)\right)$ The cosine of an angle is equal to the sine of the complement of the angle.

$= \sin\left(\dfrac{11\pi}{22} + \dfrac{10\pi}{22}\right)$ Simplify the argument.

$= \sin\dfrac{21\pi}{22}.$ Now use a calculator to verify that $\cos\left(-\dfrac{5\pi}{11}\right)$ and $\sin\dfrac{21\pi}{22}$ are both approximately 0.1423.

b. $\cot 195° = \tan\left(90° - 195°\right)$ The same relation between cotangent and tangent applies, though the angles are measured in degrees in this example. Verify that both sides are approximately 3.7321.

$= \tan\left(-105°\right).$

We'll conclude this section with examples that illustrate another way to use the relationships between trigonometric functions.

Using the
Relationships
Between
Trigonometric
Functions

EXAMPLE 6

Given that $\cos\theta = -\dfrac{\sqrt{3}}{2}$ and $\tan\theta$ is negative, determine θ and $\tan\theta$.

Solution:

Since $\cos\theta$ is negative, we know the terminal side of θ must lie in either the second or third quadrant (remember: "All Students Take Calculus" reminds us the cosine is only positive in the first and fourth quadrants). Given that $\tan\theta$ is also negative, the terminal side of θ must lie in the second quadrant (tangent is positive in the third). A diagram is helpful at this point:

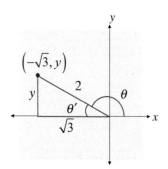

In the diagram, we've drawn θ with its terminal side in Quadrant II and we've drawn the reference angle θ'. We've also noted the relative magnitudes of the adjacent leg and hypotenuse of the right triangle, as indicated by the fact that $\cos\theta = -\dfrac{\sqrt{3}}{2}$. Of course, the actual lengths don't have to be $\sqrt{3}$ and 2, respectively, but they do have to be some multiple of $\sqrt{3}$ and 2, so we may as well use the simplest choice.

What we were not given initially is the length labeled "y" and the actual angle θ. But from the diagram, we can now recognize that the triangle is a familiar one and that $\theta' = \dfrac{\pi}{6}$, so $\theta = \dfrac{5\pi}{6}$. Finally, it must be the case that $y = 1$ and therefore $\tan\theta = -\dfrac{1}{\sqrt{3}}$.

Using the
Relationships
Between
Trigonometric
Functions

EXAMPLE 7

Given that $\cot\theta = 0.4$ and that θ lies in the first quadrant, determine $\sin\theta$.

Solution:

All trigonometric functions are ratios, so it will probably be useful to express cotangent as a fraction. The result will help us construct a right triangle that relates to the given information. To that end, note that $\cot\theta = 0.4 = \dfrac{4}{10} = \dfrac{2}{5}$. If we take the numerator and denominator as the lengths of the adjacent and opposite sides of a right triangle, we are led to the diagram on the next page.

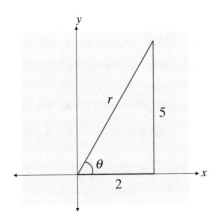

Now we can use the Pythagorean Theorem to determine that $r = \sqrt{4+25} = \sqrt{29}$, and so $\sin\theta = \dfrac{5}{\sqrt{29}}$.

Exercises

1. $\sin 45° = \cos 45° = \dfrac{\sqrt{2}}{2}$

$\tan 45° = \cot 45° = 1$

$\csc 45° = \sec 45° = \sqrt{2}$

2. $\sin \dfrac{\pi}{2} = \csc \dfrac{\pi}{2} = 1$

$\cos \dfrac{\pi}{2} = \cot \dfrac{\pi}{2} = 0$

$\tan \dfrac{\pi}{2} = \sec \dfrac{\pi}{2}$
$= $ undefined

3. $\sin 60° = \dfrac{\sqrt{3}}{2}$

$\cos 60° = \dfrac{1}{2}$

$\tan 60° = \sqrt{3}$

$\csc 60° = \dfrac{2\sqrt{3}}{3}$

$\sec 60° = 2$

$\cot 60° = \dfrac{\sqrt{3}}{3}$

4. $\sin \dfrac{3\pi}{4} = \dfrac{\sqrt{2}}{2}$,

$\cos \dfrac{3\pi}{4} = -\dfrac{\sqrt{2}}{2}$,

$\tan \dfrac{3\pi}{4} = \cot \dfrac{3\pi}{4} = -1,\ \csc \dfrac{3\pi}{4} = \sqrt{2},\ \sec \dfrac{3\pi}{4} = -\sqrt{2}$

5. $\sin \dfrac{5\pi}{2} = \csc \dfrac{5\pi}{2} = 1,\ \cos \dfrac{5\pi}{2} = \cot \dfrac{5\pi}{2} = 0,\ \tan \dfrac{5\pi}{2} = \sec \dfrac{5\pi}{2} =$ undefined

6. $\sin -520° \approx -0.3420, \cos -520° \approx -0.9397, \tan -520° \approx 0.3640, \csc -520° \approx -2.9238, \sec -520° \approx -1.0642, \cot -520° \approx 2.7475$

Evaluate all six trigonometric functions at the given θ, using a calculator if necessary. See Examples 1 and 2.

1. $\theta = 45°$ **2.** $\theta = \dfrac{\pi}{2}$ **3.** $\theta = 60°$ **4.** $\theta = \dfrac{3\pi}{4}$ **5.** $\theta = \dfrac{5\pi}{2}$

6. $\theta = -520°$ **7.** $\theta = 305°$ **8.** $\theta = -1105°$ **9.** $\theta = 6\pi$ **10.** $\theta = 670°$

11. $\theta = \dfrac{3\pi}{2}$ **12.** $\theta = -215°$ **13.** $\theta = \dfrac{5\pi}{4}$ **14.** $\theta = 780°$ **15.** $\theta = -445°$

Determine the reference angle associated with the given angle. See Example 3.

16. $\theta = 98°$ **17.** $\theta = \dfrac{9\pi}{2}$ **18.** $\theta = -60°$ **19.** $\theta = \dfrac{5\pi}{4}$ **20.** $\theta = \dfrac{5\pi}{2}$

21. $\theta = 313°$ **22.** $\theta = \dfrac{7\pi}{6}$ **23.** $\theta = -168°$ **24.** $\theta = \dfrac{6\pi}{5}$ **25.** $\theta = 216°$

26. $\theta = \dfrac{3\pi}{2}$ **27.** $\theta = -330°$ **28.** $\theta = \dfrac{7\pi}{4}$ **29.** $\theta = 718°$ **30.** $\theta = 105°$

Determine which of the four quadrants angle θ is located in.

31. $\sin \theta > 0$ and $\tan \theta < 0$ **32.** $\sin \theta < 0$ and $\cos \theta > 0$ **33.** $\tan \theta > 0$ and $\sec \theta > 0$

34. $\cos \theta > 0$ and $\cot \theta < 0$ **35.** $\sec \theta < 0$ and $\csc \theta < 0$ **36.** $\cot \theta > 0$ and $\csc \theta > 0$

37. $\cot \theta > 0$ and $\cos \theta < 0$ **38.** $\sin \theta > 0$ and $\sec \theta < 0$

7. $\sin 305° \approx -0.8192$
$\cos 305° \approx 0.5736$
$\tan 305° \approx -1.4281$
$\csc 305° \approx -1.2208$
$\sec 305° \approx 1.7434$
$\cot 305° \approx -0.7002$

8. $\sin -1105° \approx -0.4226$
$\cos -1105° \approx 0.9063$
$\tan -1105° \approx -0.4663$
$\csc -1105° \approx -2.3662$
$\sec -1105° \approx 1.1034$
$\cot -1105° \approx -2.1445$

9. $\sin 6\pi = \tan 6\pi = 0$
$\cos 6\pi = \sec 6\pi = 1$
$\csc 6\pi = \cot 6\pi$
$= $ undefined

10. $\sin 670° \approx -0.7660$
$\cos 670° \approx 0.6428$
$\tan 670° \approx -1.1918$
$\csc 670° \approx -1.3054$
$\sec 670° \approx 1.5557$
$\cot 670° \approx -0.8391$

11. $\sin \dfrac{3\pi}{2} = \csc \dfrac{3\pi}{2} = -1$
$\cos \dfrac{3\pi}{2} = \cot \dfrac{3\pi}{2} = 0$
$\tan \dfrac{3\pi}{2} = \sec \dfrac{3\pi}{2}$
$\qquad = $ undefined

12. $\sin -215° \approx 0.5736,$
$\cos -215° \approx -0.8192,$
$\tan -215° \approx -0.7002,$
$\csc -215° \approx 1.7434,$
$\sec -215° \approx -1.2208,$
$\cot -215° \approx -1.4281$

13. $\sin \dfrac{5\pi}{4} = -\dfrac{\sqrt{2}}{2}$
$\cos \dfrac{5\pi}{4} = -\dfrac{\sqrt{2}}{2}$
$\csc \dfrac{5\pi}{4} = -\sqrt{2}$
$\sec \dfrac{5\pi}{4} = -\sqrt{2}$
$\tan \dfrac{5\pi}{4} = \cot \dfrac{5\pi}{4} = 1$

Match the angle θ in questions 39–48 with the correct reference angle θ' in choices a.–c. Answers will be used more than once.

a. $\theta' = 30°$ **b.** $\theta' = 45°$ **c.** $\theta' = 60°$

39. $\theta = 300°$ **40.** $\theta = 150°$ **41.** $\theta = -135°$ **42.** $\theta = 210°$ **43.** $\theta = -120°$

44. $\theta = 315°$ **45.** $\theta = 510°$ **46.** $\theta = 600°$ **47.** $\theta = 855°$ **48.** $\theta = 480°$

In each of the following problems, **a.** rewrite the expression in terms of the given angle's reference angle, and then **b.** evaluate the result, using a calculator if necessary. See Example 4.

49. $\tan 98°$ **50.** $\sin \dfrac{9\pi}{2}$ **51.** $\cos (-60°)$ **52.** $\tan \dfrac{5\pi}{4}$ **53.** $\cos \dfrac{5\pi}{2}$

54. $\sin 313°$ **55.** $\cos \dfrac{7\pi}{6}$ **56.** $\tan (-168°)$ **57.** $\cos \dfrac{6\pi}{5}$ **58.** $\sin 216°$

59. $\tan \dfrac{3\pi}{2}$ **60.** $\cos (-330°)$ **61.** $\sin \dfrac{7\pi}{4}$ **62.** $\tan 718°$ **63.** $\sin 105°$

Use the appropriate identity to answer the following questions. Choose only one answer per question.

64. Which choice is equivalent to $\sin 18°$?

a. $\tan 72°$ **b.** $\cos 72°$ **c.** $\csc 72°$ **d.** $\sec 162°$ **e.** $\cos 162°$

65. Which choice is equivalent to $\sec \dfrac{\pi}{6}$?

a. $\csc \dfrac{\pi}{3}$ **b.** $\cos \dfrac{\pi}{2}$ **c.** $\sin \dfrac{\pi}{6}$ **d.** $\cos \dfrac{\pi}{3}$ **e.** $\tan \dfrac{\pi}{6}$

66. Which choice is equivalent to $\tan \dfrac{\pi}{12}$?

a. $\sin \dfrac{\pi}{2}$ **b.** $\cos \dfrac{\pi}{12}$ **c.** $\cot \dfrac{\pi}{2}$ **d.** $\cot \dfrac{\pi}{12}$ **e.** $\cot \dfrac{5\pi}{12}$

67. Which choice is equivalent to $\cos 87°$?

a. $\sin 93°$ **b.** $\cos 93°$ **c.** $\sin 273°$ **d.** $\sec 3°$ **e.** $\sin 3°$

Express each of the following in terms of the appropriate cofunction, and verify the equivalence of the two expressions. See Example 5.

68. $\cot 135°$ **69.** $\sec \dfrac{\pi}{2}$ **70.** $\sin (-60°)$ **71.** $\cos \left(-\dfrac{3\pi}{4}\right)$ **72.** $\csc \dfrac{5\pi}{6}$

73. $\cot 313°$ **74.** $\cos \left(\dfrac{-3\pi}{6}\right)$ **75.** $\csc (-168°)$ **76.** $\sin \left(\dfrac{-4\pi}{5}\right)$ **77.** $\sec 216°$

78. $\csc \dfrac{3\pi}{2}$ **79.** $\cos (-15°)$ **80.** $\cot \dfrac{\pi}{4}$ **81.** $\tan (-105°)$ **82.** $\sec 105°$

14. $\sin 780° = \dfrac{\sqrt{3}}{2}$, $\cos 780° = \dfrac{1}{2}$, $\tan 780° = \sqrt{3}$, $\csc 780° = \dfrac{2\sqrt{3}}{3}$, $\sec 780° = 2$, $\cot 780° = \dfrac{\sqrt{3}}{3}$

15. $\sin -445° \approx -0.9962$
$\cos -445° \approx 0.0872$
$\tan -445° \approx -11.4301$
$\csc -445° \approx -1.0038$
$\sec -445° \approx 11.4737$
$\cot -445° \approx -0.0875$

16. $\theta' = 82°$ **17.** $\theta' = \dfrac{\pi}{2}$

18. $\theta' = 60°$ **19.** $\theta' = \dfrac{\pi}{4}$

20. $\theta' = \dfrac{\pi}{2}$ **21.** $\theta' = 47°$

22. $\theta' = \dfrac{\pi}{6}$ **23.** $\theta' = 12°$

24. $\theta' = \dfrac{\pi}{5}$ **25.** $\theta' = 36°$

26. $\theta' = \dfrac{\pi}{2}$ **27.** $\theta' = 30°$

28. $\theta' = \dfrac{\pi}{4}$ **29.** $\theta' = 2°$

30. $\theta' = 75°$ **31.** II

32. IV **33.** I **34.** IV

35. III **36.** I **37.** III

38. II **39.** c **40.** a **41.** b
42. a **43.** c **44.** b **45.** a

46. c **47.** b **48.** c

49. $\tan 98° = -\tan 82°$
≈ -7.1154

50. $\sin \dfrac{9\pi}{2} = \sin \dfrac{\pi}{2} = 1$

51. $\cos -60° = \cos 60°$
$= 0.5$

52. $\tan \dfrac{5\pi}{4} = \tan \dfrac{\pi}{4} = 1$

53. $\cos \dfrac{5\pi}{2} = \cos \dfrac{\pi}{2} = 0$

54. $\sin 313° = -\sin 47°$
≈ -0.7314

Using a graphing calculator, determine the tangent and cotangent for each question. Round each answer to three decimal places.

83. $\sin\theta = 0.978$ and $\cos\theta = 0.208$ **84.** $\sin\theta = 0.588$ and $\cos\theta = -0.809$

85. $\sin\theta = -0.966$ and $\cos\theta = -0.259$ **86.** $\sin\theta = -0.866$ and $\cos\theta = -0.5$

87. $\sin\theta = -0.699$ and $\cos\theta = 0.743$ **88.** $\sin\theta = -0.995$ and $\cos\theta = -0.105$

The three cofunction identities presented in this section have three companion identities, as follows:

$$\sin\theta = \cos\left(\dfrac{\pi}{2} - \theta\right), \ \sec\theta = \csc\left(\dfrac{\pi}{2} - \theta\right), \text{ and } \tan\theta = \cot\left(\dfrac{\pi}{2} - \theta\right).$$

89. Prove these three identities. (**Hint:** For the first identity, begin with the observation that $\sin\theta = \sin\left(\dfrac{\pi}{2} - \left(\dfrac{\pi}{2} - \theta\right)\right)$ and then apply one of the three original cofunction identities.)

Use the given information about each angle to evaluate the expressions. See Examples 6 and 7.

90. Given that $\cos\theta = \dfrac{\sqrt{12}}{4}$ and $\tan\theta$ is negative, determine θ and $\tan\theta$.

91. Given that $\csc\theta = 1.25$ and θ lies in the second quadrant, determine $\cot\theta$.

92. Given that $\tan\theta = \dfrac{\sqrt{3}}{3}$ and $\sin\theta$ is positive, determine θ and $\sin\theta$.

93. Given that $\sec\theta = 0.3$ and θ lies in the fourth quadrant, determine $\csc\theta$.

Express each of the following in terms of the appropriate cofunction. Check the term given and the cofunction found with your calculator to verify the expressions are equivalent.

94. $\sin\dfrac{7\pi}{4}$ **95.** $\csc\dfrac{8\pi}{3}$ **96.** $\cot\dfrac{3\pi}{4}$

97. $\cos\left(-\dfrac{5\pi}{3}\right)$ **98.** $\tan 15°$ **99.** $\sec(-315°)$

55. $\cos\dfrac{7\pi}{6} = -\cos\dfrac{\pi}{6} = \dfrac{-\sqrt{3}}{2}$ **56.** $\tan -168° = \tan 12° \approx 0.2126$ **57.** $\cos\dfrac{6\pi}{5} = -\cos\dfrac{\pi}{5} \approx -0.8090$

58. $\sin 216° = -\sin 36° \approx -0.5878$ **59.** $\tan\dfrac{3\pi}{2} = \tan\dfrac{\pi}{2} = $ undefined   **60.** $\cos -330° = \cos 30° = \dfrac{\sqrt{3}}{2}$

61. $\sin\dfrac{7\pi}{4} = -\sin\dfrac{\pi}{4} = \dfrac{-\sqrt{2}}{2}$ **62.** $\tan 718° = -\tan 2° \approx -0.0349$ **63.** $\sin 105° = \sin 75° \approx 0.9659$ **64.** b **65.** a

66. e **67.** e **68.** $\cot 135° = \tan -45° = -1$ **69.** $\sec\dfrac{\pi}{2} = \csc 0 = $ undefined   **70.** $\sin -60° = \cos 150° = -\dfrac{\sqrt{3}}{2}$

71. $\cos -\dfrac{3\pi}{4} = \sin\dfrac{5\pi}{4} = \dfrac{-\sqrt{2}}{2}$ **72.** $\csc\dfrac{5\pi}{6} = \sec -\dfrac{\pi}{3} = 2$ **73.** $\cot 313° = \tan -223° \approx -0.9325$ **74.** $\cos -\dfrac{3\pi}{6} = \sin\pi = 0$

75. $\csc -168° = \sec 258° \approx -4.8097$ **76.** $\sin -\dfrac{4\pi}{5} = \cos\dfrac{13\pi}{10} \approx -0.5878$ **77.** $\sec 216° = \csc -126° \approx -1.2361$

6.4 Graphs of Trigonometric Functions

TOPICS

1. Graphing the basic trigonometric functions
2. Periodicity and other observations
3. Graphing transformed trigonometric functions
4. Damped harmonic motion

TOPIC 1   Graphing the Basic Trigonometric Functions

In the preceding sections, you have been exposed to some of the highlights of more than 2000 years of thought regarding trigonometry and its uses. With the fundamentals out of the way, it's time to reflect on a subtle but important point.

Most of the applications of the trigonometric functions implicitly view them as either functions of (acute) angles or as functions of real numbers. (Incidentally, the trigonometric functions can be extended even further to be functions of complex numbers, but that discussion will have to wait for a later course.) We've had quite a bit of experience with applications of the first sort; chances are, if you sketch a triangle in the course of solving a problem, you're using the "angle" point of view. In this section, we will concentrate on the second point of view.

To emphasize the theme of this section, we will now frequently use x to represent the argument of a given trigonometric function, and x will be a variable standing for (nearly) any real number. To start with, we know from Section 6.3 that sine and cosine are functions of any real number x, so the first order of business will be to graph them as functions defined on the real line. The process will be familiar to you—we have taken similar steps to graph the functions we encountered in Chapters 3, 4, and 5.

Recall that the most elementary approach to graphing a function for the first time is to plot points. That is, evaluate a given function $f(x)$ for a sufficient number of values x so that, when all the calculated points of the form $(x, f(x))$ are plotted, a reasonable guess for the graph of f can be constructed. We'll do this for the two functions sine and cosine at the same time.

78. $\csc \dfrac{3\pi}{2} = \sec -\pi$
$$= -1$$

79. $\cos(-15°) = \sin 105°$
$$\approx 0.9659$$

80. $\cot \dfrac{\pi}{4} = \tan \dfrac{\pi}{4}$
$$= 1$$

81. $\tan(-105°) = \cot 195°$
$$\approx 3.7321$$

82. $\sec 105° = \csc(-15°)$
$$\approx -3.8637$$

83. $\tan \theta \approx 4.702,$
$\cot \theta \approx 0.213$

84. $\tan \theta \approx -0.727,$
$\cot \theta \approx -1.376$

85. $\tan \theta \approx 3.730,$
$\cot \theta \approx 0.268$

86. $\tan \theta = 1.732,$
$\cot \theta \approx 0.577$

87. $\tan \theta \approx -0.941$, $\cot \theta \approx -1.063$ **88.** $\tan \theta \approx 9.476$, $\cot \theta \approx 0.106$ **89.** Prove using identities on pages 500 and 501.

90. $\tan \theta = -\dfrac{\sqrt{3}}{3}$ **91.** $\csc \theta = \dfrac{125}{100} = \dfrac{5}{4}$ **92.** $\sin \theta = \dfrac{1}{2}$, $\theta = 30°$ **93.** $\sec \theta = \dfrac{3}{10}$, hyp = 3, adj = 10, does not exist.
$\theta = -30°$
$\qquad\qquad\qquad \cot \theta = -\dfrac{3}{4}$

94. $\cos \dfrac{-5\pi}{4}$, -0.7071 **95.** $\sec \dfrac{-13\pi}{6}$, 1.1547 **96.** $\tan \dfrac{-\pi}{4}$, -1 **97.** $\sin \dfrac{13\pi}{6}$, $\dfrac{1}{2}$ **98.** $\cot 75°$, 0.2679 **99.** $\csc 405°$, 1.4142

For instance, for $x = 0$ we know that $\sin 0 = 0$ and $\cos 0 = 1$. We could use a calculator to confirm these facts, but it's better to *understand* why the values are what they are. In this context, letting $x = 0$ tells us to consider a degenerate triangle whose initial and terminal sides both lie along the positive x-axis, and the definitions of sine and cosine in Section 6.3 then lead to

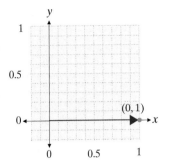

$$\sin 0 = \frac{0}{1} = 0 \text{ and } \cos 0 = \frac{1}{1} = 1.$$

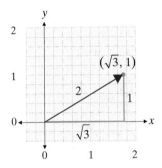

Similarly, for $x = \frac{\pi}{6}$ we know

$$\sin \frac{\pi}{6} = \frac{1}{2} \text{ and } \cos \frac{\pi}{6} = \frac{\sqrt{3}}{2}.$$

If we let x take on other values between 0 and 2π that are easy to work with, we obtain the following table:

x	$\sin x$	$\cos x$	x	$\sin x$	$\cos x$
0	0	1	π	0	-1
$\dfrac{\pi}{6}$	$\dfrac{1}{2}$	$\dfrac{\sqrt{3}}{2}$	$\dfrac{7\pi}{6}$	$-\dfrac{1}{2}$	$-\dfrac{\sqrt{3}}{2}$
$\dfrac{\pi}{4}$	$\dfrac{1}{\sqrt{2}}$	$\dfrac{1}{\sqrt{2}}$	$\dfrac{5\pi}{4}$	$-\dfrac{1}{\sqrt{2}}$	$-\dfrac{1}{\sqrt{2}}$
$\dfrac{\pi}{3}$	$\dfrac{\sqrt{3}}{2}$	$\dfrac{1}{2}$	$\dfrac{4\pi}{3}$	$-\dfrac{\sqrt{3}}{2}$	$-\dfrac{1}{2}$
$\dfrac{\pi}{2}$	1	0	$\dfrac{3\pi}{2}$	-1	0
$\dfrac{2\pi}{3}$	$\dfrac{\sqrt{3}}{2}$	$-\dfrac{1}{2}$	$\dfrac{5\pi}{3}$	$-\dfrac{\sqrt{3}}{2}$	$\dfrac{1}{2}$
$\dfrac{3\pi}{4}$	$\dfrac{1}{\sqrt{2}}$	$-\dfrac{1}{\sqrt{2}}$	$\dfrac{7\pi}{4}$	$-\dfrac{1}{\sqrt{2}}$	$\dfrac{1}{\sqrt{2}}$
$\dfrac{5\pi}{6}$	$\dfrac{1}{2}$	$-\dfrac{\sqrt{3}}{2}$	$\dfrac{11\pi}{6}$	$-\dfrac{1}{2}$	$\dfrac{\sqrt{3}}{2}$

Figure 1: Selected Values of Sine and Cosine

This table represents the values of sine and cosine as x assumes the value of convenient angles, beginning with $x = 0$ and rotating around counterclockwise for one full circle. Of course, there's no need to let x take on higher values, or negative numbers for that matter, since the above table will simply repeat. For instance, $\sin 2\pi = \sin 0$ and $\sin\left(-\dfrac{\pi}{6}\right) = \sin\dfrac{11\pi}{6}$. This observation is a recognition of the *periodicity* of sine and cosine; we say that these two functions both have a period of 2π. We will discuss the periodicity of all trigonometric functions more thoroughly soon.

Figure 2 shows the result of plotting the calculated points for sine, with the purple curve drawn to smoothly pass through the points.

Figure 2: Graph of Sine Between 0 and 2π

Similarly, Figure 3 contains the result of plotting the calculated points for cosine.

Figure 3: Graph of Cosine Between 0 and 2π

One fact that leaps out from a glance at the graphs is that both sine and cosine take on values only between –1 and 1. This makes perfect sense, of course, since the length of the hypotenuse of a right triangle is always greater than or equal to the length of either leg, but the graphs drive this point home in a way that tables of figures don't. Another fact that is starting to appear is that sine and cosine seem to have similar shapes, one shifted horizontally with respect to the other. This is clearer if we extend the graphs to more of the real line, making use of each function's periodicity.

In Figure 4, both functions are graphed over a longer interval of the x-axis, with sine in purple and cosine in green.

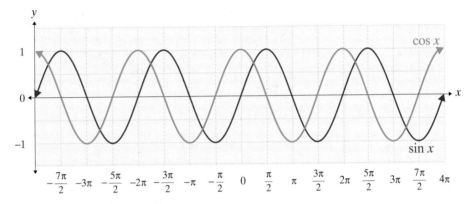

Figure 4: Graphs of Sine and Cosine

In the examples that follow, we will take similar steps to construct sketches of graphs of various functions. As with nearly all of our graphing, however, our intent is not to compete with graphing calculators and computer software for the most accurate pictures. Instead, our goal is to understand the behavior of functions and be able to pick out important qualities like periodicity, symmetry, the existence of asymptotes, and so on.

EXAMPLE 1

Graphing the Cosecant Function

Sketch the graph of the cosecant function.

Solution:

Since cosecant is the reciprocal of sine, the data in Figure 1 and the graphs in Figure 4 are sufficient to give us the understanding we seek. First, we note that $\sin x = 0$ when x is any multiple of π, so $\csc x$ is undefined for the same values of x. That is, the domain of the cosecant function is all real numbers *except* $\pm\{0, \pi, 2\pi, 3\pi,...\}$. We knew this already—this corresponds to the restriction in the general definition of cosecant given in Section 6.3. Second, in the interval $[0, \pi]$, the graph of sine is symmetric with respect to the line $x = \dfrac{\pi}{2}$, so the graph of cosecant will have the same property. Similar statements hold for other intervals. And since sine is nonnegative (and less than or equal to 1) for all x in $[0, \pi]$, the reciprocal of sine will also be nonnegative (but greater than or equal to 1) for the same x.

Since $\sin x \to 0$ as x approaches any multiple of π, we know that cosecant will have vertical asymptotes at these points. Extending the reasoning in the last paragraph, we also know that $\csc x \geq 1$ on the intervals $\ldots, \left(-4\pi, -3\pi\right), \left(-2\pi, -\pi\right), \left(0, \pi\right), \left(2\pi, 3\pi\right), \ldots$ and that $\csc x \leq -1$ on the intervals $\ldots, \left(-3\pi, -2\pi\right), \left(-\pi, 0\right), \left(\pi, 2\pi\right), \left(3\pi, 4\pi\right), \ldots.$ Putting all this together, we obtain the following sketch:

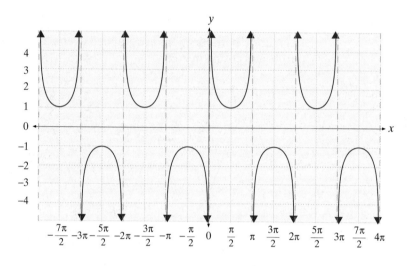

TOPIC 2

Periodicity and Other Observations

With a few graphs to look back on, we'll now formally define what we mean by periodicity.

DEFINITION

Period of a Function

A function f is said to be **periodic** if there is a positive number p such that

$$f\left(x+p\right) = f\left(x\right)$$

for all x in the domain of f. The smallest such number p is called the **period** of f.

For instance, we know that $\sin\left(x+2\pi\right) = \sin x$, $\cos\left(x+2\pi\right) = \cos x$, and $\csc\left(x+2\pi\right) = \csc x$. It's also true that $\sin\left(x+2n\pi\right) = \sin x$ for any integer n, but 2π is the smallest positive constant p for which $\sin\left(x+p\right) = \sin x$, so the period of sine (and cosine and cosecant) is 2π.

EXAMPLE 2

Finding the Period
of Trigonometric
Functions

Determine the period of the secant, tangent, and cotangent functions.

Solution:

As you no doubt expect after studying cosecant, the period of secant is the same as the period of cosine, since secant is the reciprocal of cosine. We can prove this algebraically as follows:

$$\sec(x+2\pi) = \frac{1}{\cos(x+2\pi)} = \frac{1}{\cos x} = \sec x,$$

so the period of secant is no larger than 2π. And if there were a smaller positive number p for which $\sec(x+p) = \sec x$, then cosine would necessarily have the same period p, contradicting what we know about the period of cosine. Thus the period of secant is 2π. Note, however, that secant is not defined for all real numbers; the domain of secant is all real numbers except where cosine $= 0$, namely $\pm\left\{\dfrac{\pi}{2}, \dfrac{3\pi}{2}, \dfrac{5\pi}{2}, ...\right\}$.

We'll have to work a bit harder to determine the period of tangent. A reasonable guess might be 2π, since four of the trigonometric functions have period 2π, but we'll see that this is wrong. Recall the general definition of the tangent function: given an angle θ and any point (x, y) on the terminal side of θ (other than the origin, of course),

$$\tan\theta = \frac{y}{x}.$$

If $\theta \in \left[0, \dfrac{\pi}{2}\right)$, $\tan\theta$ is positive because both y and x are positive. If $\theta \in \left[\pi, \dfrac{3\pi}{2}\right)$, $\tan\theta$ is again positive because y and x are both negative. If the terminal side of θ lies in the second or fourth quadrant, $\tan\theta$ is negative. This alone is enough to tell us that the period of tangent can't be less than π (do you see why?). But more precisely, consider what the tangent function does to a given angle θ and to $\theta + \pi$, as illustrated in Figure 5.

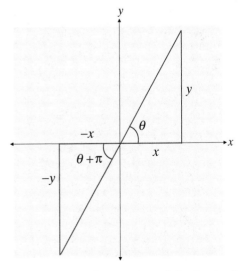

Figure 5: Tangent of an Angle and the Angle plus π

Since $\dfrac{-y}{-x} = \dfrac{y}{x}$ (with the restriction $x \neq 0$), we see that $\tan(\theta + \pi) = \tan\theta$. This, along with the observation in the preceding paragraph, tells us that the period of tangent is π, and consequently the period of cotangent is also π.

With the information in the last examples for inspiration, we'll proceed to get a better understanding of tangent.

EXAMPLE 3

Graphing the Tangent Function

Sketch the graph of the tangent function.

Solution:

First, we'll recap what we know: tangent is not defined for multiples of $\dfrac{\pi}{2}$, its period is π, and it is nonnegative on the interval $\left[0, \dfrac{\pi}{2}\right)$ and nonpositive on the interval $\left(-\dfrac{\pi}{2}, 0\right]$. And we can either draw some triangles or use the identity $\tan x = \dfrac{\sin x}{\cos x}$ to easily calculate some values:

x	$-\dfrac{\pi}{3}$	$-\dfrac{\pi}{4}$	$-\dfrac{\pi}{6}$	0	$\dfrac{\pi}{6}$	$\dfrac{\pi}{4}$	$\dfrac{\pi}{3}$
$\tan x$	$-\sqrt{3}$	-1	$-\dfrac{1}{\sqrt{3}}$	0	$\dfrac{1}{\sqrt{3}}$	1	$\sqrt{3}$

Now we can proceed to plot the above points to sketch the graph of tangent in the interval $\left(-\dfrac{\pi}{2}, \dfrac{\pi}{2}\right)$, and we can use its periodicity to extend the graph over a larger interval:

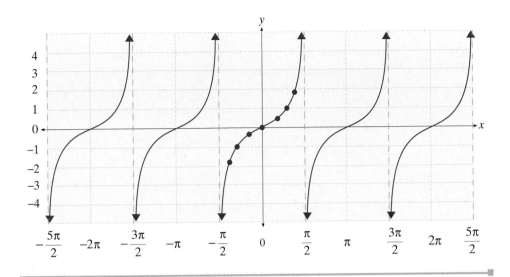

In Chapter 3 we defined the terms *even* and *odd* as applied to functions. Recall that a function f is *odd* if $f(-x) = -f(x)$ for all x in the domain, and *even* if $f(-x) = f(x)$ for all x in the domain. Geometrically, this means that the graph of f is either symmetric with respect to the origin or the y-axis, respectively. We've constructed enough graphs now to note that cosine and its reciprocal, secant, are even functions, while sine, cosecant, tangent, and cotangent are all odd. These facts are summarized below.

IDENTITY

Even/Odd Identities

$\sin(-x) = -\sin x$	$\cos(-x) = \cos x$	$\tan(-x) = -\tan x$
$\csc(-x) = -\csc x$	$\sec(-x) = \sec x$	$\cot(-x) = -\cot x$

TOPIC 3

Graphing Transformed Trigonometric Functions

In actual use, a trigonometric function is unlikely to appear in its pristine form. That is, in order to be useful in solving a problem, a given function will probably have to be modified somewhat. Geometrically, this means that the graph of a function will appear stretched, compressed, reflected, or shifted relative to the graph of its fundamental form. We will finish out this section with a discussion of the more common transformations.

We have had quite a bit of experience with transformations of functions in general, but in the context of trigonometry some additional nomenclature is useful. We'll begin with a definition of one particularly useful term.

DEFINITION

Amplitude of Sine and Cosine Curves

Given a fixed real number a, the **amplitude** of the function $f(x) = a\sin x$ or the function $g(x) = a\cos x$ is the value $|a|$. As we know, the multiplication of $\sin x$ or $\cos x$ by a stretches (or compresses, if $-1 < a < 1$) the graph vertically by a factor of $|a|$, so the amplitude represents the distance between the x-axis and the maximum value of the function.

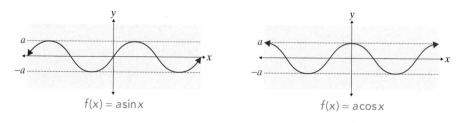

$$f(x) = a\sin x \qquad\qquad f(x) = a\cos x$$

As we have seen, the shapes of the sine and cosine curves are identical; one is merely shifted horizontally with respect to the other. For this reason, both graphs are said to be *sinusoidal*, so we have now defined the amplitude of a sinusoidal curve. Such curves arise in numerous physical situations, often when an object is displaying *simple harmonic motion (SHM)*.

DEFINITION

Simple Harmonic Motion and Frequency

If an object is oscillating and its displacement from some midpoint at time t can be described by either $f(t) = a\sin\omega t$ or $g(t) = a\cos\omega t$, the object is said to be in **simple harmonic motion**. In either case, a is a real number and ω ("omega") is a positive real number. The amplitude of the object (the maximum displacement from rest) is $|a|$ and the **frequency** of the object's motion is $\dfrac{\omega}{2\pi}$.

Frequency is a measure of the number of times an object goes through one complete cycle of motion in one unit of time. If time t is being measured in seconds, the frequency is measured in terms of *cycles per second*, or **Hertz (Hz)**.

EXAMPLE 4

Using Simple Harmonic Motion

A first approximation to the motion of an object suspended at the end of a spring and set into vertical oscillation is given by $y = a\cos\omega t$, where y is the displacement of the object above or below its rest position and a and ω are as defined on the previous page. Suppose several potatoes are dumped into the basket of a grocer's scale, which then proceeds to bounce up and down with a frequency of 3 Hz. Given that the distance traveled between peaks of the basket's oscillation is 8 centimeters, find a mathematical model for the basket's motion.

Solution:

From the given information, we know that our model must describe an object traveling 4 centimeters above and 4 centimeters below its rest position; in other words, the amplitude must be 4. We have a choice to make, however: we can define our coordinate system so that a positive displacement is either up or down. The most natural choice in this problem is probably to choose positive displacement as being in the upward direction, so we would like to arrange it so that $y = -4$ when $t = 0$. That is, a good model of the basket's oscillation will have the basket starting out 4 centimeters below its midpoint position at the moment $(t = 0)$ when the potatoes are dumped in. This means we want to set $a = -4$.

A frequency of 3 Hz means that the basket makes 3 complete up-and-down cycles every second. This means that $\omega = 6\pi$, and so our model is $y = -4\cos 6\pi t$. To see how our model relates to the physical situation, consider the following graph:

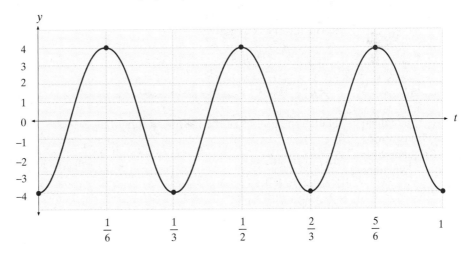

From the graph, it's clear that the basket really does make 3 complete cycles during the course of the first second, that it starts out 4 centimeters below the midpoint position, and that its maximum displacement above midpoint is also 4 centimeters.

Of course, our experience tells us that the simple model seen in Example 4 isn't very accurate over a long period of time—eventually the basket settles down and ceases its oscillatory motion. We'll return to this problem at the end of this section, but first we'll define a few more useful terms.

DEFINITION

Period Revisited

We know that sine and cosine (and their reciprocals) go through one complete cycle over an interval of length 2π. The modified functions

$$f(x) = \sin bx \text{ and } g(x) = \cos bx$$

both have a period of $\dfrac{2\pi}{b}$. Note then that the period is the reciprocal of the frequency. This makes sense, since the frequency measures how many cycles occur over an interval of length 1, while the period is the length of the interval required for one complete cycle.

Notice in Example 4 that $b = 6\pi$, so the period is $\dfrac{2\pi}{6\pi} = \dfrac{1}{3}$.

Replacing the variable x with bx thus has the effect of stretching or compressing the graph of $\sin x$ or $\cos x$ horizontally, just as multiplying either function by a constant a has the same effect vertically. One further effect we need to be able to achieve is a horizontal shifting to the left or right. We've actually already learned how to do this in general: replacing the variable x with $x - c$ shifts a graph to the right if c is positive and to the left if c is negative.

EXAMPLE 5

Graphing Transformed Trigonometric Functions

Sketch the graphs of the following functions.

a. $f(x) = \sin 2\pi x$ **b.** $g(x) = \sin(x - \pi)$

Solutions:

a.

The period of $f(x) = \sin 2\pi x$ is $\dfrac{2\pi}{2\pi} = 1$ but this is the only difference between the function f and $\sin x$.

b.

The graph of $g(x) = \sin(x - \pi)$ is simply the usual sine shape shifted π units to the right.

In the context of trigonometry, shifting a function to the left or right is called a **phase shift**, and in general it may occur in combination with a change in period and amplitude. For instance, the function $f(x) = 3\sin(5\pi x - 4\pi)$ has an amplitude of 3, a phase shift, and an altered period. One way to determine the phase shift and period precisely is to rewrite the function as follows:

$$f(x) = 3\sin(5\pi x - 4\pi)$$

$$= 3\sin\left[5\pi\left(x - \frac{4}{5}\right)\right].$$

The function $3\sin 5\pi x$ has a period of $\frac{2}{5}$, but when x is replaced with $x - \frac{4}{5}$, we know the graph of $3\sin 5\pi x$ gets shifted to the right by $\frac{4}{5}$ units.

Another way to determine details about the period and phase shift of $f(x) = 3\sin(5\pi x - 4\pi)$ is to relate the beginning and end of one cycle of the function to the beginning and end of one cycle of $\sin x$. That is, we know that a cycle of f will begin when $x = 0$ and will end when $x = 2\pi$.

Solving these two equations for x gives us:

$$5\pi x - 4\pi = 0 \qquad\qquad 5\pi x - 4\pi = 2\pi$$

$$5\pi x = 4\pi \qquad\qquad 5\pi x = 6\pi$$

$$x = \frac{4}{5} \qquad\qquad x = \frac{6}{5}$$

So one complete cycle of f occurs over the interval between $x = \frac{4}{5}$ and $x = \frac{6}{5}$, telling us again that the period is $\frac{2}{5}$.

These observations are summarized on the next page for the functions sine and cosine. The same sort of analysis will prove to be just as useful for the remaining trigonometric functions, however.

DEFINITION

Amplitude, Period, and Phase Shift Combined

Given constants a, b (such that $b > 0$), and c, the functions

$$f(x) = a\sin(bx - c) \quad \text{and} \quad g(x) = a\cos(bx - c)$$

have **amplitude** $|a|$, **period** $\dfrac{2\pi}{b}$, and a **phase shift** of $\dfrac{c}{b}$. The left endpoint of one cycle of either function is $\dfrac{c}{b}$ and the right endpoint is $\dfrac{c}{b} + \dfrac{2\pi}{b}$.

$f(x) = a\sin(bx - c)$

$f(x) = a\cos(bx - c)$

EXAMPLE 6

Graphing Transformed Trigonometric Functions

Sketch the graph of $f(x) = -2\sec\left(\pi x + \dfrac{\pi}{2}\right)$.

Solution:

The starting point is the secant function. We know its period is 2π and that it has vertical asymptotes at $\pm\left\{\dfrac{\pi}{2}, \dfrac{3\pi}{2}, \dfrac{5\pi}{2}, \dots\right\}$. It completes one cycle between $\dfrac{\pi}{2}$ and $\dfrac{5\pi}{2}$, so these will be convenient values to use in determining the left and right endpoints of one cycle of f:

$$\pi x + \frac{\pi}{2} = \frac{\pi}{2} \qquad\qquad \pi x + \frac{\pi}{2} = \frac{5\pi}{2}$$

$$\pi x = 0 \qquad\qquad\qquad \pi x = 2\pi$$

$$x = 0 \qquad\qquad\qquad\quad x = 2$$

Note that this implicitly tells us that the function f has a period of 2.

The factor of -2 in front of secant stretches the graph vertically by a factor of 2 and reflects it with respect to the x-axis. For instance, where $\sec\left(\pi x + \dfrac{\pi}{2}\right)$ has values of 1 and -1, the function f will have values of, respectively, -2 and 2. Putting this all together, our sketch is as shown on the next page:

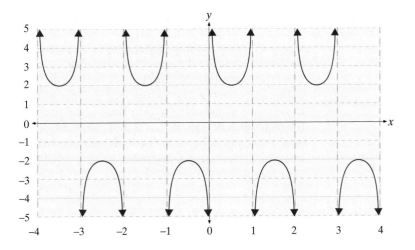

EXAMPLE 7

Graphing Transformed Trigonometric Functions

Sketch the graph of $g(x) = 1 + \sin\left(x - \dfrac{\pi}{4}\right)$.

Solution:

The graph of g is the graph of $\sin x$ shifted to the right by $\dfrac{\pi}{4}$ units and up by 1 (recall that adding a constant to a function merely shifts the graph up or down, according to whether the constant is positive or negative). Neither the amplitude nor period has changed, however. Our sketch is thus:

TOPIC

Damped Harmonic Motion

Let's now return briefly to our simple harmonic motion problem. If we want to more accurately describe the up-and-down motion of an object suspended from a spring, we need to account for the fact that oscillations are often *damped* as time progresses. That is, the amplitude of the oscillations decreases according to some rule. We can easily modify our sinusoidal wave in such a way by multiplying a sine or cosine function by an amplitude factor that is not constant. Remember: the fundamental behavior of a sinusoidal curve is to oscillate between values of −1 and 1. If we multiply such a wave by a decreasing amplitude, the result will be a wave whose oscillations diminish over time.

EXAMPLE 8

Modeling Damped Harmonic Motion

Sketch the graph of $f(t) = -4e^{-t} \cos 6\pi t$.

Solution:

We've already graphed the function $f(t) = -4\cos 6\pi t$ (this function was our simple model for the motion of the grocer's basket in Example 4). The factor of e^{-t} provides the desired damping effect. In the figure below, the green curves are the graphs of $4e^{-t}$ and $-4e^{-t}$, and are included to show how they describe the "envelope" of amplitude modulation. The result is that the magnitude of the displacement of the grocer's basket decreases over time. Notice, however, the period is unaffected.

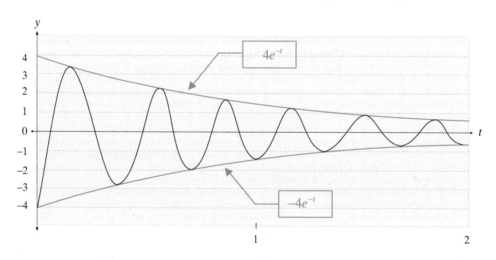

In the exercises to follow, you'll be asked to sketch graphs of similar products of damping factors and trigonometric functions.

1.

2.

3. A = 3 inches

P = 2 seconds

$g(t) = 3 \cos \pi t$

4. $P = \dfrac{8}{9}$ s

5. A = 1.5 ft

$P = \dfrac{3}{5}$ s

$g(t) = 1.5 \cos \dfrac{10}{3} \pi t$

6. A = 2, P = 2π,

no phase shift

7. $A = \dfrac{3}{2}$, P = 2π,

no phase shift

8. A = 4, P = 2π,

no phase shift

9. A = 1, P = 2π,

shifted right 5 units

10. A = 1, P = 2π,

no phase shift

11. $A = \dfrac{1}{2}$, P = 2π,

no phase shift

12. A = 3, P = 2π,

shifted left 7 units

13. $A = \dfrac{2}{3}$, P = 2π,

no phase shift

14. A = 2, P = π,

no phase shift

15. A = 3, P = 4π,

no phase shift

16. $A = \dfrac{3}{2}$, P = 2,

no phase shift

17. A = 1, P = $\dfrac{2}{3}$, shifted right $\dfrac{2}{3\pi}$ units

18. A = 0.5, P = $\dfrac{\pi}{4}$, shifted left $\dfrac{1}{8}$ units

19. A = 7, P = 4, shifted left $\dfrac{3}{\pi}$ units

Exercises

Given the information contained in Examples 1, 2, and 3, construct sketches of the following functions.

1. cotangent

2. secant

Use the given information to construct functions modeling the described behavior. See Example 4.

3. A baby is playing with a toy attached above his head on a coiled spring. The baby pulls the toy down a distance of 3 inches from its equilibrium position, and then releases it. The time for one oscillation is 2 seconds. Find the amplitude and period, then give the function for its displacement.

4. A pull cord for a lighted ceiling fan is swinging back and forth. The end of the cord swings a total distance of 4 inches from end to end and reaches a speed of 9 inches per second at the midpoint. Find the period of oscillation.

5. Marcel is bouncing a basketball at 10 ft/s. The distance from the ground to his waist is approximately 3 feet. Find the amplitude and period, then give the function for its displacement.

Determine the amplitude, period, and phase shift of each of the following trigonometric equations.

6. $y = 2 \cos x$

7. $y = \dfrac{3}{2} \sin x$

8. $y = 5 + 4 \cos x$

9. $y = \sin (x - 5)$

10. $y = -\sin x$

11. $2y = \cos x$

12. $y = -3 \cos (x + 7)$

13. $y = \dfrac{2}{3} \sin x$

14. $y = 2 \sin 2x$

15. $y = -3 \cos \dfrac{1}{2} x$

16. $2y = 3 \sin \pi\theta$

17. $y = \cos (3\pi\theta - 2)$

18. $y = 0.5 \sin (8x + 1)$

19. $y = 7 \cos \left(x \cdot \dfrac{\pi}{2} + \dfrac{3}{2} \right)$

20. $5y = 8 \cos (2\pi x + 4)$

21. $y = 2 - \dfrac{3}{4} \sin (-3 + x)$

20. $A = \dfrac{8}{5}$, $P = 1$,

shifted left $\dfrac{2}{\pi}$ units

21. $A = \dfrac{3}{4}$, $P = 2\pi$,

shifted up 2 units,
shifted right 3 units

22.

Sketch the graphs of the following functions. See Examples 5 and 6.

22. $f(x) = \cos \pi x$

23. $g(x) = -2\sin 5x$

24. $f(x) = \csc \dfrac{3\pi}{4} x$

25. $g(x) = 3\sin(x - 2\pi)$

26. $g(x) = \tan\left(3\pi x - \dfrac{\pi}{2}\right)$

27. $f(x) = -5\cot \pi x$

28. $g(x) = \sin\left(x - \dfrac{\pi}{4}\right)$

29. $f(x) = 4\cos\left(\dfrac{3x}{2} + \dfrac{\pi}{2}\right)$

30. $f(x) = \dfrac{1}{2}\tan\left(\dfrac{1}{2}x - \dfrac{\pi}{3}\right)$

31. $g(x) = 2\cos(4x - 2)$

32. $f(x) = \cos(x - \pi)$

33. $g(x) = 3\sin 4x$

34. $g(x) = \csc\left(\dfrac{3\pi}{2}x - \dfrac{1}{2}\right)$

35. $f(x) = -\sin 2\pi x$

36. $f(x) = 5\tan\left(3\pi - \dfrac{\pi}{2}x\right)$

Sketch the graphs of the following functions. See Example 7.

37. $g(x) = 1 + \sin(x - 2\pi)$

38. $f(x) = 2 - \cos 2\pi x$

39. $f(x) = 4 + \csc\left(1 - \dfrac{5\pi}{4}x\right)$

40. $g(x) = 5 - 2\sin\left(x - \dfrac{\pi}{2}\right)$

41. $g(x) = 1 + \tan\left(\pi x - \dfrac{\pi}{4}\right)$

42. $f(x) = -3 + 5\cos x$

43. $g(x) = 2 - \sin\left(2x - \dfrac{\pi}{4}\right)$

44. $f(x) = 4 + \tan\left(x + \dfrac{3\pi}{2}\right)$

45. $f(x) = \dfrac{1}{2} - 5\sin\left(\dfrac{1}{2}x - \dfrac{\pi}{2}\right)$

46. $g(x) = 1 - \dfrac{1}{4}\cos\left(\dfrac{1}{4}x - \dfrac{\pi}{2}\right)$

47. $g(x) = 2 + \dfrac{5}{6}\sec\left(\dfrac{1}{2}x - \pi x\right)$

48. $f(x) = \dfrac{1}{2}\tan\left(\dfrac{3}{4}x - 2\pi\right) + 3$

Sketch the following functions modeling damped harmonic motion. See Example 8.

49. $g(t) = -2e^{-t}\cos 5\pi t$

50. $f(t) = e^{-t}\sin \dfrac{3\pi}{4} t$

51. $g(t) = e^{t}\sin\left(3t - \dfrac{\pi}{2}\right)$

52. $g(t) = 3e^{-t}\cos\left(5t - \dfrac{\pi}{2}\right)$

23.

24.

25.

26.

27.

28.

53. $f(t) = -3 + 5e^{-t}\cos t$

54. $f(t) = -5e^{t}\cos\dfrac{3\pi}{2}t$

55. $f(t) = \dfrac{1}{2}e^{-t}\sin\left(\dfrac{5}{6}t - 4\pi\right) + 2$

56. $g(t) = 2 + e^{-t}\sin\left(t - \dfrac{\pi}{4}\right)$

29.

30.

31.

32.

33.

34.

35.

36.

37.

38.

39.

40.

41.

42.

43.

44.

45.

46.

47.

48.

49.

50.

51.

52.

53.

54.

55.

56.

Inverse Trigonometric Functions

TOPICS

1. The definitions of inverse trigonometric functions

2. Evaluation of inverse trigonometric functions

3. Applications of inverse trigonometric functions

T. Evaluating inverse trigonometric functions

TOPIC 1

The Definitions of Inverse Trigonometric Functions

The rationale for the inverse trigonometric functions is the rationale for inverses of functions in general. In many situations, we will want to find an angle having a certain specified property, and our method will be to "undo" the action of a given trigonometric function. As a simple example, suppose we need to find an acute angle θ for which $\sin\theta = \dfrac{1}{2}$. Our experience is sufficient for this task; we've worked with nice angles enough to recognize that $\sin\dfrac{\pi}{6} = \dfrac{1}{2}$, so it must be the case that $\theta = \dfrac{\pi}{6}$. But what if we seek an angle φ for which $\sin\varphi = 0.7$? The problem is similar, but we don't yet have a way to determine φ.

Recall from Chapter 3, however, that a function will have an inverse only if it is one-to-one. Recall also, if the graph of the function is available, this means the graph must pass the horizontal line test; this is something the trigonometric functions markedly fail to do. Fortunately, there is a way out. By restricting the domain of a trigonometric function wisely, we can make it one-to-one and thus invertible. We will go through the process step by step for the sine function and then briefly show how the other trigonometric functions are dealt with similarly.

There are many ways in which we could restrict the domain of sine in order to make it one-to-one, but we are guided also by the desire to not lose more than we have to in the restriction. For instance, we could specify that we will only define sine over the interval $\left[0, \dfrac{\pi}{2}\right]$, but by doing so we prevent the newly defined function from ever taking on a negative value (note that $0 \le \sin x \le 1$ for $x \in \left[0, \dfrac{\pi}{2}\right]$). Figure 1 indicates that $\left[-\dfrac{\pi}{2}, \dfrac{\pi}{2}\right]$ is the largest interval containing $\left[0, \dfrac{\pi}{2}\right]$ that we could choose for the restricted domain; the bold green portion of the graph is one-to-one and takes on all values between −1 and 1.

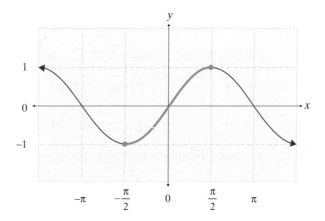

Figure 1: Restricting the Domain of Sine

In practice, context will tell us whether we want to think of sine as being defined on the entire real line or only over the interval $\left[-\dfrac{\pi}{2}, \dfrac{\pi}{2} \right]$, but the biggest hint will be whether we need to apply the inverse of the sine function. If so, the restricted domain for sine is called for.

Two notations are commonly used for the inverse trigonometric functions. In the case of sine, $y = \sin x$ is equivalent to the equations

$$x = \arcsin y \ \text{ and } \ x = \sin^{-1} y.$$

The arcsine notation derives from the fact that $\arcsin y$ is the length of the arc (on the unit circle) corresponding to the angle x. The $\sin^{-1} y$ notation is in keeping with our use of f^{-1} to stand for the inverse of the function f.

CAUTION!

But in using this notation, remember:

$$\sin^{-1} y \neq \frac{1}{\sin y}.$$

In order to avoid this possible source of confusion, some texts use only arcsine for the inverse sine function.

In summary,

DEFINITION

Arcsine

Given $x \in \left[-\dfrac{\pi}{2}, \dfrac{\pi}{2}\right]$, **arcsine** is defined by either of the following:

$$y = \sin x \Leftrightarrow x = \arcsin y \text{ and } y = \sin x \Leftrightarrow x = \sin^{-1} y.$$

In words, arcsin y is the angle whose sine is y. Since the (restricted) domain of sine is $\left[-\dfrac{\pi}{2}, \dfrac{\pi}{2}\right]$ and its range is $[-1, 1]$, the domain of arcsine is $[-1, 1]$ and its range is $\left[-\dfrac{\pi}{2}, \dfrac{\pi}{2}\right]$.

The best way to finish up this introduction to arcsine is with a graph of the function. In Chapter 3 we saw that the graphs of a function and its inverse are reflections of one another with respect to the line $y = x$, and this is all we need in order to generate Figure 2.

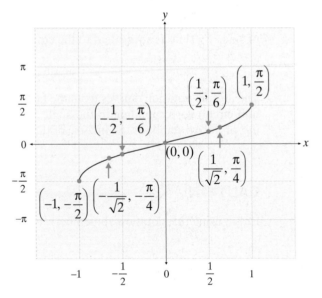

Figure 2: Graph of Arcsine

EXAMPLE 1

Defining and Plotting the Arccosine Function

With the derivation of arcsine as a guide, construct a definition of arccosine and plot the resulting function.

Solution:

As with sine, we first need to restrict the domain of cosine to an interval over which cosine is one-to-one. Picture the graph of cosine in your mind (or refer to Figure 4 in Section 6.4 for a refresher). Most people would probably say that the natural choice for the restricted domain is the interval $[0, \pi]$, and this is indeed the convention. This is all we need in order to make our definition: Given $x \in [0, \pi]$, **arccosine** (with its two notations) is defined by

$$y = \cos x \Leftrightarrow x = \arccos y \text{ and } y = \cos x \Leftrightarrow x = \cos^{-1} y.$$

To graph the arccosine, we simply reflect the restricted graph of cosine with respect to the line $y = x$. Since the (restricted) domain of cosine is $[0, \pi]$ and the range is $[-1, 1]$, we know that the domain of arccosine will be $[-1, 1]$ and its range will be $[0, \pi]$. This knowledge serves as a good way to double-check our graph of arccosine.

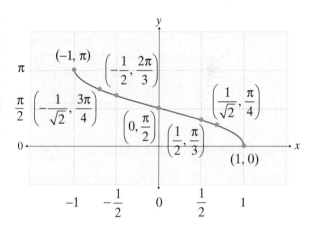

Arctangent is the third commonly encountered inverse function, and its definition and graph are arrived at in a similar manner. The box below summarizes facts about the definitions, domains, and ranges of arcsine, arccosine, and arctangent, and Figure 3 illustrates the graph of arctangent. Note the horizontal asymptotes in the graph of arctangent, corresponding to the vertical asymptotes in the graph of tangent.

DEFINITION

Arcsine, Arccosine, and Arctangent

Function	Notation 1	Notation 2	Domain	Range
Inverse Sine	$\arcsin y = x \Leftrightarrow y = \sin x$	$\sin^{-1} y = x \Leftrightarrow y = \sin x$	$[-1, 1]$	$\left[-\dfrac{\pi}{2}, \dfrac{\pi}{2}\right]$
Inverse Cosine	$\arccos y = x \Leftrightarrow y = \cos x$	$\cos^{-1} y = x \Leftrightarrow y = \cos x$	$[-1, 1]$	$[0, \pi]$
Inverse Tangent	$\arctan y = x \Leftrightarrow y = \tan x$	$\tan^{-1} y = x \Leftrightarrow y = \tan x$	$(-\infty, \infty)$	$\left(-\dfrac{\pi}{2}, \dfrac{\pi}{2}\right)$

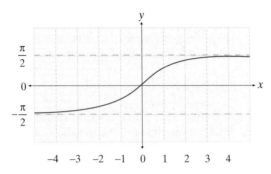

Figure 3: Graph of Arctangent

TOPIC 2   Evaluation of Inverse Trigonometric Functions

The evaluation of inverse trigonometric functions can take several forms, depending on context. One meaning is the actual numerical evaluation of an expression containing an inverse trig function; this may or may not require the use of a calculator (discussed in Topic T). Another meaning is the simplification of expressions containing inverse trig functions, using nothing more than our knowledge of how functions and their inverses behave relative to one another. We'll begin with some numerical examples.

EXAMPLE 2

Evaluating Inverse
Trigonometric
Functions

Evaluate the following expressions.

a. $\arctan(-1)$ **b.** $\sin^{-1} 2.3$

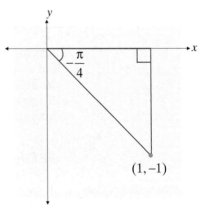

Solutions:

a. A glance at the graph of arctangent tells us that $\arctan(-1)$ is a negative number apparently halfway between 0 and $-\dfrac{\pi}{2}$ (that is, a negative angle whose terminal side is in the fourth quadrant). By drawing the appropriate right triangle, we can verify that it is indeed the case that $\tan\left(-\dfrac{\pi}{4}\right) = -1$, so $\arctan(-1) = -\dfrac{\pi}{4}$.

b. The domain of $\sin^{-1}$ is $[-1,1]$, so the short answer is that $\sin^{-1} 2.3$ cannot be evaluated. After all, how could there be an angle whose sine is more than 1? For the purposes of this text, the answer is indeed simply that $\sin^{-1} 2.3$ is not defined. As a teaser toward more advanced mathematics, however, you may recall an aside at the start of Section 6.4 that the domains of the trigonometric functions can be extended to include complex numbers. Under those conditions, $\sin^{-1} 2.3$ actually has a complex number value (with nonzero imaginary part).

The single most important attribute of the inverse trigonometric functions is that they reverse the action of the functions they are associated with; remember that, in general, $f^{-1}(f(x)) = x$ and $f(f^{-1}(x)) = x$. But these statements are only true if all of the expressions contained in them make sense. A solid understanding of domains and ranges prevents potential errors, as shown in the next example.

EXAMPLE 3

Evaluating Compositions of Trigonometric Functions

Evaluate the following expressions, if possible.

a. $\arcsin\left(\sin\dfrac{3\pi}{4}\right)$ **b.** $\cos\left(\cos^{-1}(-0.2)\right)$ **c.** $\tan^{-1}\left(\tan\dfrac{7\pi}{6}\right)$

Solutions:

a. The potential error in this problem is to assume that $\arcsin\left(\sin\dfrac{3\pi}{4}\right) = \dfrac{3\pi}{4}$, since arcsin and sin are inverse functions of one another. But $\dfrac{3\pi}{4}$ lies outside the range of arcsin, which is $\left[-\dfrac{\pi}{2}, \dfrac{\pi}{2}\right]$ so we know this can't be the answer. The key is to evaluate the expressions individually:

$$\sin\frac{3\pi}{4} = \frac{1}{\sqrt{2}}$$

and then

$$\arcsin\frac{1}{\sqrt{2}} = \frac{\pi}{4}.$$

b. The number -0.2 lies in the domain of arccosine, and all real numbers lie in the domain of cosine, so all the parts of the expression $\cos\left(\cos^{-1}(-0.2)\right)$ make sense and we are safe in stating $\cos\left(\cos^{-1}(-0.2)\right) = -0.2$. If we wanted to explore the expression a bit further, we could note that, from the graph of arccosine, $\cos^{-1}(-0.2)$, is some positive number, and further that $\cos^{-1}(-0.2)$ must be greater than $\dfrac{\pi}{2}$ since $\cos\left(\cos^{-1}(-0.2)\right)$ is negative. This is indeed the case: $\cos^{-1}(-0.2)$ is approximately $101.5°$.

c. We run into the same problem with $\tan^{-1}\left(\tan\dfrac{7\pi}{6}\right)$ as with $\arcsin\left(\sin\dfrac{3\pi}{4}\right)$, but we will present a slightly different way of thinking about the resolution here. Instead of evaluating $\tan\dfrac{7\pi}{6}$ literally, consider only the steps involved in doing so. The first step is to determine that the reference angle for $\dfrac{7\pi}{6}$ is $\dfrac{\pi}{6}$, and the second step is to note that $\tan\dfrac{7\pi}{6}$ and $\tan\dfrac{\pi}{6}$ have the same sign (the terminal side of $\dfrac{7\pi}{6}$ is in the third quadrant, and tangent is positive there). So $\tan^{-1}\left(\tan\dfrac{7\pi}{6}\right) = \tan^{-1}\left(\tan\dfrac{\pi}{6}\right) = \dfrac{\pi}{6}$, as $\dfrac{\pi}{6}$ lies in the range of $\tan^{-1}$.

The last example demonstrated the evaluation of compositions of trig functions with their inverses, but of course other compositions are possible. In many cases, a picture aids greatly in the computation.

EXAMPLE 4

Evaluating Compositions of Trigonometric Functions

Evaluate the following expressions.

a. $\tan\left(\sin^{-1}\left(-\dfrac{4}{5}\right)\right)$ **b.** $\cos(\arctan 0.4)$

Solutions:

a. Remember that the range of arcsin is $\left[-\dfrac{\pi}{2}, \dfrac{\pi}{2}\right]$, and in particular that $\sin^{-1}\left(-\dfrac{4}{5}\right)$ will lie between $-\dfrac{\pi}{2}$ and 0 (the graph tells us that arcsin of a negative number is negative). If we let $\theta = \sin^{-1}\left(-\dfrac{4}{5}\right)$ then $\sin\theta = -\dfrac{4}{5}$ and we can sketch the following triangle to illustrate the relationship between θ and the given numbers:

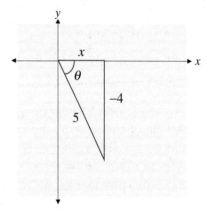

Of course, the Pythagorean Theorem allows us to calculate x:

$$x = \sqrt{5^2 - (-4)^2} = \sqrt{9} = 3.$$

Now we can see that $\tan\theta = -\dfrac{4}{3}$, so

$$\tan\left(\sin^{-1}\left(-\dfrac{4}{5}\right)\right) = -\dfrac{4}{3}.$$

b. We can employ the same method and let $\theta = \arctan 0.4$. This leads to

$$\tan\theta = 0.4 = \dfrac{4}{10} = \dfrac{2}{5}$$

and then to the sketch

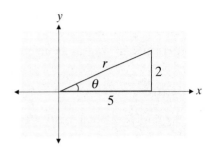

The Pythagorean Theorem gives us $r = \sqrt{5^2 + 2^2} = \sqrt{29}$, and so $\cos\left(\arctan 0.4\right) = \dfrac{5}{\sqrt{29}}$.

TOPIC 3   Applications of Inverse Trigonometric Functions

Many applications calling for the use of inverse trigonometric functions are dynamic, and feature an angle that is changing over time; you will encounter many such problems in calculus. The first step is often to determine a formula for a given angle in terms of other quantities, as illustrated in the next examples.

EXAMPLE 5

Using Inverse Trigonometric Functions

A lighthouse is to be constructed half a mile from a long, straight reef, as shown. In order to ensure the light illuminates certain portions of the reef within specified lengths of time, the engineer needs a formula for θ in terms of x. Find such a formula.

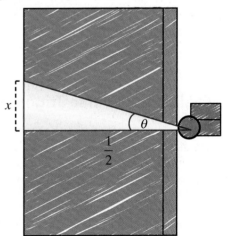

Solution:

From the diagram, we see that $\tan\theta = \dfrac{x}{\frac{1}{2}} = 2x,$ so the formula for θ is simply

$$\theta = \tan^{-1} 2x.$$

Using Inverse
Trigonometric
Functions

EXAMPLE 6

Express $\sin\left(\cos^{-1} 2x\right)$ as an algebraic function of x, assuming $-\dfrac{1}{2} \le x \le \dfrac{1}{2}$.

Solution:

Let $\theta = \cos^{-1} 2x$. Then $\cos\theta = 2x$ and we are led to consider a sketch like the one below:

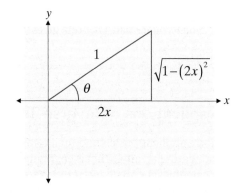

In the sketch, we have chosen the simplest lengths for the adjacent side and the hypotenuse that make $\cos\theta = 2x$, though of course any positive multiple of these lengths would also work. And as always, once the lengths of two sides of the right triangle have been determined, the Pythagorean Theorem provides the length of the third side. Now we can refer to the sketch to see that

$$\sin\left(\cos^{-1} 2x\right) = \sin\theta = \frac{\sqrt{1-4x^2}}{1} = \sqrt{1-4x^2}.$$

TOPIC Evaluating Inverse Trigonometric Functions

The evaluation of inverse trigonometric functions may at times require the use of a calculator. For example, consider the expression $\cos^{-1}(-0.3)$. None of the commonly encountered angles we have worked with has a cosine of –0.3, so we must use a calculator to evaluate this expression.

Remember that the angles are usually measured in radians unless directed otherwise, so make sure your calculator is in radian mode before proceeding. This can be done by pressing **MODE** and making sure RADIAN is selected.

The labels on calculators vary, but many calculators have a $\cos^{-1}$ function accessed by first pressing a button labeled "2ND" or "INV" and then the **COS** button. One possible sequence of keystrokes to evaluate $\cos^{-1}(-0.3)$ is:

The result you see should be approximately 1.8755 (and remember, this is 1.8755 radians).

```
cos⁻¹(-.3)
          1.875488981
```

Be careful when evaluating expressions with a calculator. If we look back to Example 2a and use a calculator to evaluate the expression $\tan^{-1}(-1)$, the answer we get is approximately –0.7854.

```
tan⁻¹(-1)
          -.7853981634
```

Although this result looks different than the one obtained in Example 2a, they are equivalent. It is up to us to recognize that –0.7854 is approximately $-\dfrac{\pi}{4}$.

```
tan⁻¹(-1)
          -.7853981634
-π/4
          -.7853981634
```

Exercises

1.

Construct sketches of the following functions. See Example 1.

1. arccosecant: Assume a domain of $\left[-\dfrac{\pi}{2}, 0\right) \cup \left(0, \dfrac{\pi}{2}\right]$ for cosecant.

2.

2. arcsecant: Assume a domain of $\left[0, \dfrac{\pi}{2}\right) \cup \left(\dfrac{\pi}{2}, \pi\right]$ for secant.

3. arccotangent: Assume a domain of $(0, \pi)$ for cotangent.

Solve each equation for y without the use of a calculator.

3.

4. $y = \sin^{-1}(-1)$ **5.** $y = \cos^{-1}\dfrac{\sqrt{2}}{2}$ **6.** $y = \tan^{-1} 1$ **7.** $y = \cot^{-1}\left(-\dfrac{\sqrt{3}}{3}\right)$

8. $y = \sec^{-1}\dfrac{2\sqrt{3}}{3}$ **9.** $y = \csc^{-1}(-2)$ **10.** $y = \arcsin 0$ **11.** $y = \arccos(-1)$

12. $y = \arctan\left(-\sqrt{3}\right)$ **13.** $y = \text{arccot}\left(-\sqrt{3}\right)$ **14.** $y = \text{arcsec } 2$ **15.** $y = \text{arccsc}\sqrt{2}$

16. $y = \text{arccot}(-1)$ **17.** $y = \tan^{-1}\dfrac{\sqrt{3}}{3}$ **18.** $y = \cos^{-1}\left(-\dfrac{1}{2}\right)$ **19.** $y = \csc^{-1} 2$

4. $y = -\dfrac{\pi}{2}$ **5.** $y = \dfrac{\pi}{4}$ **20.** $y = \arcsin\left(-\dfrac{1}{2}\right)$ **21.** $y = \sec^{-1}(-1)$ **22.** $y = \text{arccsc } 1$ **23.** $y = \arctan 0$

6. $y = \dfrac{\pi}{4}$ **7.** $y = \dfrac{2\pi}{3}$ **24.** $y = \sin^{-1}\dfrac{\sqrt{2}}{2}$ **25.** $y = \arccos\left(-\dfrac{\sqrt{2}}{2}\right)$ **26.** $y = \text{arcsec}(-2)$ **27.** $y = \cot^{-1}\left(-\sqrt{3}\right)$

8. $y = \dfrac{\pi}{6}$ **9.** $y = -\dfrac{\pi}{6}$

10. $y = 0$ **11.** $y = \pi$

12. $y = -\dfrac{\pi}{3}$ **13.** $y = \dfrac{5\pi}{6}$

14. $y = \dfrac{\pi}{3}$ **15.** $y = \dfrac{\pi}{4}$

16. $y = \dfrac{3\pi}{4}$ **17.** $y = \dfrac{\pi}{6}$

18. $y = \dfrac{2\pi}{3}$ **19.** $y = \dfrac{\pi}{6}$

20. $y = -\dfrac{\pi}{6}$ **21.** $y = \pi$

22. $y = \dfrac{\pi}{2}$ **23.** $y = 0$

24. $y = \dfrac{\pi}{4}$ **25.** $y = \dfrac{3\pi}{4}$

26. $y = \dfrac{2\pi}{3}$ **27.** $y = \dfrac{5\pi}{6}$

28. -0.2014 **29.** DNE

30. -1.1198 **31.** 1.3734

32. 1.9823 **33.** 0.6747

34. 0.2014 **35.** DNE

36. 0.0067 **37.** 1.3734

38. -0.1126 **39.** 1.0472

40. $\dfrac{\pi}{2}$ **41.** $-\dfrac{\pi}{2}$

42. $\dfrac{1}{2}$ **43.** $-\dfrac{\pi}{6}$

44. -0.8 **45.** $\dfrac{\pi}{4}$

46. 0.3714 **47.** 0

48. 0.8944 **49.** DNE

50. 0.75 **51.** 0.2764

52. $\dfrac{\sqrt{3}}{2}$ **53.** $-\dfrac{1}{2}$

54. 1 **55.** 2

56. -1 **57.** 2

Evaluate the following expressions, using a calculator if necessary. See Example 2.

28. $\sin^{-1}(-0.2)$ **29.** $\cos^{-1} 4$ **30.** $\sin^{-1}(-0.9)$

31. $\tan^{-1} 5$ **32.** $\cos^{-1}(-0.4)$ **33.** $\tan^{-1} 0.8$

Most calculators are not equipped with arccosecant, arcsecant, and arccotangent buttons, but expressions involving these functions can still be evaluated. To evaluate, for example, $\csc^{-1} x$, let $\theta = \csc^{-1} x$. Then

$$\csc \theta = x$$

$$\frac{1}{\sin \theta} = x$$

$$\sin \theta = \frac{1}{x}$$

$$\theta = \sin^{-1}\frac{1}{x}.$$

Use the above method to evaluate the following expressions.

34. $\csc^{-1} 5$ **35.** $\sec^{-1}(-0.5)$ **36.** $\cot^{-1} 150$

37. $\cot^{-1} 0.2$ **38.** $\csc^{-1}(-8.9)$ **39.** $\sec^{-1} 2$

Evaluate the following expressions, if possible. See Example 3.

40. $\cos^{-1}\left(\cos\dfrac{2\pi}{4}\right)$ **41.** $\sin^{-1}\left(\sin\dfrac{3\pi}{2}\right)$ **42.** $\tan(\tan^{-1} 0.5)$

43. $\sin^{-1}\left(\sin\dfrac{7\pi}{6}\right)$ **44.** $\cos(\cos^{-1}(-0.8))$ **45.** $\tan^{-1}\left(\tan\dfrac{5\pi}{4}\right)$

Evaluate the following expressions. See Example 4.

46. $\sin(\arctan 0.4)$ **47.** $\sin^{-1}\left(\cos\dfrac{3\pi}{2}\right)$ **48.** $\cos(\tan^{-1} 0.5)$

49. $\arcsin(\tan 1)$ **50.** $\tan(\cos^{-1}(-0.8))$ **51.** $\tan^{-1}(\cos 5)$

Find the value of each expression without using a calculator.

52. $\sin(\arctan \sqrt{3})$ **53.** $\cos(\sec^{-1}(-2))$ **54.** $\tan(\operatorname{arccot} 1)$

55. $\csc\left(\arccos\left(-\dfrac{\sqrt{3}}{2}\right)\right)$ **56.** $\tan\left(\sin^{-1}\left(-\dfrac{\sqrt{2}}{2}\right)\right)$ **57.** $\sec\left(\csc^{-1}\dfrac{2\sqrt{3}}{3}\right)$

58. $-\dfrac{\sqrt{2}}{2}$ **59.** $\dfrac{2\sqrt{3}}{3}$

60. 1 **61.** $-\sqrt{3}$

62. $\dfrac{\sqrt{2}}{2}$ **63.** $\dfrac{2\sqrt{3}}{3}$

64. $-\sqrt{2}$ **65.** $-\dfrac{\sqrt{3}}{3}$

66. $\dfrac{\sqrt{2}}{2}$ **67.** 2

68. -1 **69.** $\dfrac{\sqrt{3}}{2}$

70. a. 0.2450
   b. 0.7854
   c. 0.9505

71. a. 0.2898
   b. 0.4429
   c. 0.7956

72. $\dfrac{\sqrt{1-x^2}}{x}$

73. $\dfrac{\sqrt{x^2-4}}{2}$

74. $\sqrt{9x^2+1}$

75. $\dfrac{\sqrt{3}x}{3}$

76. $\left|\dfrac{1}{x}\right|\sqrt{x^2-1}$

77. $\dfrac{4}{\sqrt{x^2+16}}$

58. $\cos\left(\cot^{-1}(-1)\right)$

59. $\sec\left(\arcsin\left(-\dfrac{1}{2}\right)\right)$

60. $\cot\left(\operatorname{arcsec}\sqrt{2}\right)$

61. $\cot\left(\operatorname{arccsc}(-2)\right)$

62. $\sin\left(\cos^{-1}\dfrac{\sqrt{2}}{2}\right)$

63. $\sec\left(\tan^{-1}\left(-\dfrac{\sqrt{3}}{3}\right)\right)$

64. $\sec\left(\arccos\left(-\dfrac{\sqrt{2}}{2}\right)\right)$

65. $\tan\left(\csc^{-1}(-2)\right)$

66. $\sin\left(\operatorname{arcsec}\sqrt{2}\right)$

67. $\csc\left(\cot^{-1}\sqrt{3}\right)$

68. $\cot\left(\sin^{-1}\left(-\dfrac{\sqrt{2}}{2}\right)\right)$

69. $\cos\left(\arctan\left(-\dfrac{\sqrt{3}}{3}\right)\right)$

Use inverse trigonometric functions to solve the following problems. See Example 5.

70. Kim is watching a space shuttle launch from an observation spot two miles away from the launch pad. Find the angle of elevation to the shuttle for each of the following heights.
   a. 0.5 miles   **b.** 2 miles   **c.** 2.8 miles

71. Jesse is rowing in the men's singles race. The length of the oar from the side of the shell to the water is 7 feet. At what angle is the oar from the side of the boat when the blade is at the following distance from the boat?

7 ft

a. 2 feet   **b.** 3 feet   **c.** 5 feet

Express the following functions as purely algebraic functions. See Example 6.

72. $\tan\left(\cos^{-1}x\right)$

73. $\cot\left(\sin^{-1}\dfrac{2}{x}\right)$

74. $\sec\left(\tan^{-1}3x\right)$

75. $\tan\left(\sin^{-1}\dfrac{x}{\sqrt{x^2+3}}\right)$

76. $\sin\left(\sec^{-1}x\right)$

77. $\cos\left(\tan^{-1}\dfrac{x}{4}\right)$

78. $\theta = 47.88848865$

79. $\theta = -14.48917361$

80. $\theta = 32.76975166$

81. $\theta = 152.9872134$

82. $\theta = 60.10239082$

83. $\theta = 1.682588837$

84. $\theta = 0.5736213262$

85. $\theta = 1.405647622$

86. $\theta = -0.8480623625$

87. $\theta = 0.7199835305$

Using a graphing calculator, find the measure of θ in degrees. Remember to make sure your calculator is in the correct mode.

78. $\theta = \sin^{-1} 0.74184113$ **79.** $\theta = \arctan(-0.258416)$ **80.** $\theta = \text{arccsc } 1.847526$

81. $\theta = \sec^{-1}(-1.1224539)$ **82.** $\theta = \cot^{-1} 0.57496998$

Using a graphing calculator, find the value of θ in radians. Remember to make sure your calculator is in the correct mode.

83. $\theta = \arccos(-0.1115598)$ **84.** $\theta = \text{arccot } 1.547773$ **85.** $\theta = \tan^{-1} 5.999999$

86. $\theta = \csc^{-1}(-1.333333)$ **87.** $\theta = \arcsin 0.65937229$

Sketch a graph of the following functions. Then graph the functions using a graphing calculator to check your answer.

88. $f(x) = \sin^{-1}(x - 3)$ **89.** $f(x) = \sec^{-1} 2x$ **90.** $f(x) = \arctan \dfrac{x}{2}$

91. $f(x) = 2 \arccos x$

88.

89.

90.

91.

Chapter 6 Project

Trigonometric Applications

At the Paris Observatory in 1851, Jean Foucault used a long pendulum to prove that the Earth is rotating. As it swings, the pendulum appears to change its path. However, it is not the pendulum that changes path, but the room rotating underneath it. At the North Pole the Earth revolves 360° underneath the pendulum over 24 hours. The path of a pendulum at the equator does not revolve at all; instead, the pendulum travels in a huge circle

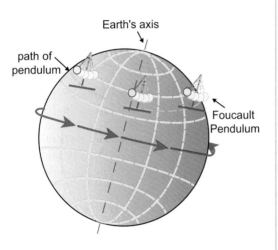

while the Earth spins. At points between the two, the pendulum cannot show how far it travels, but it can show how much the Earth is revolving underneath it. To calculate how much the Earth revolves in a particular location, use the following equation:

$$\text{degrees of revolution} = 360° \sin (\text{latitude of location})$$

Location	Latitude	Location	Latitude
United Nations, NY	40°44′58″	St. Isaac's Cathedral, Russia	59°53′02″
California Academy of Sciences	37°46′12″	Paris Observatory, France	48°48′58″
Smithsonian, Washington, DC	38°53′19″		

1. In a 24-hour period, how many degrees does the Earth revolve at the above locations?

Your university has decided to install a 50-foot Foucault Pendulum in your science building and asked you to make sure that there is enough room.

2. If the pendulum swings a total of 16°, how long is the arc traced in the air by the tip of the pendulum during one swing?

3. The school plans to build a small wall encircling the swinging pendulum. What should the diameter of the circle be if they want the tip of the pendulum to come within 6 inches of the wall?

4. When the pendulum reaches the farthest point from the center, how much higher will the tip be compared to when it is at the center?

5. If the science center only has room for a circular wall of diameter 12 feet, how many degrees can the pendulum swing and still stay 6 inches from the wall?

6. The Foucault Pendulum in the United Nations building has a length of 75 feet and a period of 10 seconds. Assuming simple harmonic motion and that at $t = 0$ the pendulum is at its farthest distance away (6 feet from the center of the circle), what function models the motion of the pendulum? Graph this function.

Chapter Summary

A summary of concepts and skills follows each chapter. Refer to these summaries to make sure you feel comfortable with the material in the chapter. The concepts and skills are organized according to the section title and topic title in which the material is first discussed.

6.1: Radian and Degree Measure of Angles

The Unit Circle and Angle Measure
- The definition of *radian* (length of the portion of the circle subtended by θ)
- Definition of initial side and terminal side of an angle

Conversion Between Degrees and Radians
- Converting from degrees to radians: $x° = x\left(\dfrac{\pi}{180}\right)$ rad

- Converting from radians to degrees: x rad $= x\left(\dfrac{180}{\pi}\right)°$

Commonly Encountered Angles
- 30, 45, 60, and 90 and the ratios of sides of triangles with common angles

Arc Length and Angular Speed
- The definition of (fraction of a circle's circumference subtended by θ) and formula for *arc length*, s ($s = r\theta$)
- The definition of (speed of an object moving along the arc of a circle defined by a central angle θ and time t) and formula for *angular speed*, $\omega \left(\omega = \dfrac{\theta}{t}\right)$

- The definition of (the distance traveled in time t around an arc length s of a circle with radius r) and formula for *linear speed*, $v \left(v = \dfrac{s}{t} = \dfrac{r\theta}{t} = r\omega\right)$

Area of a Circular Sector
- The definition of *sector* (the portion of the circle between two radii)
- Finding the area of a sector: $A = \left(\dfrac{\theta}{2\pi}\right)\left(\pi r^2\right) = \dfrac{r^2\theta}{2}$

6.2: Trigonometric Functions of Acute Angles

The Six Basic Trigonometric Functions
- Using the sides of a triangle to define the six basic trigonometric functions:

$$\sin\theta = \frac{\text{opp}}{\text{hyp}}, \quad \cos\theta = \frac{\text{adj}}{\text{hyp}}, \quad \tan\theta = \frac{\text{opp}}{\text{adj}}$$

$$\csc\theta = \frac{1}{\sin\theta} = \frac{\text{hyp}}{\text{opp}}, \quad \sec\theta = \frac{1}{\cos\theta} = \frac{\text{hyp}}{\text{adj}}, \quad \cot\theta = \frac{1}{\tan\theta} = \frac{\text{adj}}{\text{opp}}.$$

Evaluation of Trigonometric Functions
- Find the value of each of the trigonometric functions given the value of the angle
- Degree, minute, second notation: $1' = \text{one minute} = \left(\frac{1}{60}\right)(1°)$

$$1'' = \text{one second} = \left(\frac{1}{60}\right)(1') = \left(\frac{1}{3600}\right)(1°).$$

Applications of Trigonometric Functions
- Using trigonometric functions and the Pythagorean Theorem to solve application problems
- Solving surveying problems and the formula: $h = \dfrac{d}{\cot\beta - \cot\alpha}$

6.3: Trigonometric Functions of Any Angle

Extending the Domains of Trigonometric Functions
- Using $r = \sqrt{x^2 + y^2}$ where (x, y) is a point on the terminal side of an angle to extend the domains of the trig functions:

$$\sin\theta = \frac{y}{r}, \quad \cos\theta = \frac{x}{r}, \quad \tan\theta = \frac{y}{x} \ (\text{for } x \neq 0),$$

$$\csc\theta = \frac{r}{y} \ (\text{for } y \neq 0), \quad \sec\theta = \frac{r}{x} \ (\text{for } x \neq 0), \quad \cot\theta = \frac{x}{y} \ (\text{for } y \neq 0)$$

Evaluation Using Reference Angles
- The definition of *reference angle* (angle formed by the x-axis and the terminal side of the given angle)
- Using reference angles to evaluate a given angle

6.3: Trigonometric Functions of Any Angle (cont.)

Relationships Between Trigonometric Functions

- Cofunction identities: $\cos\theta = \sin\left(\dfrac{\pi}{2} - \theta\right)$, $\csc\theta = \sec\left(\dfrac{\pi}{2} - \theta\right)$,

  and $\cot\theta = \tan\left(\dfrac{\pi}{2} - \theta\right)$

- Reciprocal identities: $\csc\theta = \dfrac{1}{\sin\theta}$, $\sec\theta = \dfrac{1}{\cos\theta}$, and $\cot\theta = \dfrac{1}{\tan\theta}$

- Quotient identities: $\tan\theta = \dfrac{\sin\theta}{\cos\theta}$ and $\cot\theta = \dfrac{\cos\theta}{\sin\theta}$

- Using trig identities to find the value of an angle or another trig function

6.4: Graphs of Trigonometric Functions

Graphing the Basic Trigonometric Functions
- Plotting known points to determine the general shape of the trig graphs

Periodicity and Other Observations
- The definition of the *period* of a function (the smallest number p such that $f(x + p) = f(x)$)
- Even trig functions: $\cos x$ and $\sec x$
- Odd trig functions: $\sin x$, $\csc x$, $\tan x$, and $\cot x$

Graphing Transformed Trigonometric Functions
- The definition of the *amplitude* of sine and cosine
- The definition of *simple harmonic motion* (oscillating motion that can be described by $f(t) = a \sin \omega t$ or $g(t) = a \cos \omega t$)
- The definition of *frequency* (the number of times an object goes through one complete cycle of motion in one unit of time, $\dfrac{\omega}{2\pi}$)
- The definition of *period* for a function of the form $f(x) = \sin bx$ or $g(x) = \cos bx$: $\dfrac{2\pi}{b}$
- Sketching transformed graphs

Damped Harmonic Motion
- Graphing functions representing damped harmonic motion

6.5: Inverse Trigonometric Functions

The Definitions of Inverse Trigonometric Functions
- Restricting the domain of trig functions so the inverse is defined
- Inverse notation for trig functions
- Graphing inverse trig functions

Evaluation of Inverse Trigonometric Functions
- Numerical evaluations of inverse trig functions
- Simplifying expressions involving inverse trig functions
- Evaluating compositions of trig functions and inverse trig functions

Applications of Inverse Trigonometric Functions
- Determining a formula for a given angle in terms of other quantities

1. 4°   **2.** 216°

3. –315°   **4.** 54°

5. $\dfrac{7\pi}{30}$ **6.** $\dfrac{\pi}{3}$

7. $\dfrac{-79\pi}{180}$ **8.** $\dfrac{-5\pi}{36}$

9.

10.

11. 15.71 ft

12. 18.85 km

13. 30.72 meters

14. 10.05 inches

15. 1296.35 km

16. $\dfrac{26}{9}$ rad   **17.** 3 rad

18. a. 1200π
     b. 8.80 m/s

19. 108π ft²
     or 339.29 ft²

20. $\sin\theta = \dfrac{\sqrt{2}}{2}$,

$\cos\theta = \dfrac{\sqrt{2}}{2}$,

$\tan\theta = 1$,

$\csc\theta = \sqrt{2}$,

$\sec\theta = \sqrt{2}$,

$\cot\theta = 1$

Chapter Review

Section 6.1

Convert the following angle measures as directed.

1. Express $\dfrac{\pi}{45}$ in degrees.

2. Express $\dfrac{6\pi}{5}$ in degrees.

3. Express $\dfrac{-7\pi}{4}$ in degrees.

4. Express $\dfrac{-3\pi}{10}$ in degrees.

5. Express 42° in radians.

6. Express 60° in radians.

7. Express –79° in radians.

8. Express –25° in radians.

Sketch the indicated angles.

9. 300°

10. $\dfrac{-9\pi}{4}$

Find the arc length of a circle with the given radius r and central angle θ. Give the answer in the given unit of measure and round decimals to the nearest hundredth.

11. $r = 5$ ft; $\theta = 180°$

12. $r = 8$ km; $\theta = \dfrac{3\pi}{4}$

Find the indicated arc length in each of the following problems. Give the answer in the given unit of measure and round decimals to the nearest hundredth.

13. Given a circle of radius 16.3 meters, find the length of the arc subtended by a central angle of $\dfrac{3\pi}{5}$.

14. Given a circle of radius 8 inches, find the length of the arc subtended by a central angle of 72° (**Hint:** Convert to radians first).

15. Find the distance between Vancouver, British Columbia and San Francisco, California which have the same longitude. The latitude of Vancouver is 49.25° N and the latitude of San Francisco is 37.77° N. Assume the Earth's radius is 6470 km.

Find the radian measure of the central angle θ given the radius r and the arc length s transcribed by θ. Leave all answers in fraction form.

16. $r = 18$ ft; $s = 52$ ft

17. $r = 6.4$ dm; $s = 19.2$ dm

Solve the following problems. Round answers to the nearest hundredth.

18. A gear in a machine has a radius of 14 cm and is rotating at a constant speed of 600 rpm. Find **a.** the angular speed of a tooth of the gear in radians per minute, and **b.** the linear speed of the tooth in meters per second.

19. Find the area of the sector of a circle of radius 18 ft with a central angle of $\dfrac{2\pi}{3}$.

21. $\sin\theta = \dfrac{2\sqrt{5}}{5},$

$\cos\theta = \dfrac{\sqrt{5}}{5},$

$\tan\theta = 2,$

$\csc\theta = \dfrac{\sqrt{5}}{2},$

$\sec\theta = \sqrt{5},$

$\cot\theta = \dfrac{1}{2}$

22. 0.9903

23. 4.0108

24. 1.0353

25. 0.2225

26. 36.9372°

27. 15.2203°

28. True   **29.** False

30. $10\sqrt{3}$ feet or 17.32 feet

31. 129.13 feet

32. $\sin 90° = 1,$

$\cos 90° = 0,$

$\tan 90° = \text{undefined},$

$\csc 90° = 1,$

$\sec 90° = \text{undefined},$

$\cot 90° = 0$

33. $\sin(-460°) \approx -0.9848,$

$\cos(-460°) \approx -0.1736,$

$\tan(-460°) \approx 5.6713,$

$\csc(-460°) \approx -1.0154,$

$\sec(-460°) \approx -5.7588,$

$\cot(-460°) \approx 0.1763$

34. $\sin\dfrac{\pi}{4} = \cos\dfrac{\pi}{4} = \dfrac{\sqrt{2}}{2},$

$\tan\dfrac{\pi}{4} = \cot\dfrac{\pi}{4} = 1,$

$\csc\dfrac{\pi}{4} = \sec\dfrac{\pi}{4} = \sqrt{2}$

Section 6.2

Use the information contained in the figures to determine the values of the six trigonometric functions of θ. Rationalize all denominators.

20.

21.

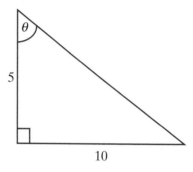

Use a graphing calculator to evaluate the following expressions. Round answers to four decimal places.

22. $\sin 82°$

23. $\cot 14°$

24. $\csc\dfrac{5\pi}{12}$

25. $\cos\dfrac{3\pi}{7}$

Convert each expression from degrees, minutes, seconds (DMS) notation to decimal notation. Round answers to four decimal places.

26. $36°56'14''$

27. $15°12'73''$

Determine whether the following statements are true or false. Use a graphing calculator when necessary.

28. If $\tan\theta = 1.6$, then $\cot\theta = 0.625$.

29. If $\csc\theta = 3.4$, then $\sin\theta = 1.7$.

Use an appropriate trigonometric function and a calculator if necessary to solve the following problems.

30. A wheelchair ramp touches the ground 15 feet away from the top of the steps. If the ramp makes an angle of 30° relative to the ground, how long is the ramp?

31. A building is 83 feet tall and a cable is stretched from the top of the building to the ground. If the angle between the cable and the ground is 40°, how long is the cable?

Section 6.3

Evaluate all six trigonometric functions at the given θ, using a calculator if necessary. Round answers to four decimal places.

32. $\theta = 90°$

33. $\theta = -460°$

34. $\theta = \dfrac{\pi}{4}$

35. $\theta = \dfrac{7\pi}{3}$

35. $\sin\dfrac{7\pi}{3}=\dfrac{\sqrt{3}}{2},$

$\cos\dfrac{7\pi}{3}=\dfrac{1}{2},$

$\tan\dfrac{7\pi}{3}=\sqrt{3},$

$\csc\dfrac{7\pi}{3}=\dfrac{2\sqrt{3}}{3},$

$\sec\dfrac{7\pi}{3}=2,$

$\cot\dfrac{7\pi}{3}=\dfrac{\sqrt{3}}{3}$

36. $\theta'=86°$ **37.** $\theta'=37°$

38. $\theta'=\dfrac{\pi}{2}$ **39.** $\theta'=\dfrac{\pi}{4}$

40. I   **41.** III

42. $\sin 290°=\sin 70°$

≈ -0.9397

43. $\tan\dfrac{4\pi}{3}=\tan\dfrac{\pi}{3}=\sqrt{3}$

44. $\csc 193°=\sec -103°$

≈ -4.4454

45. $\sin -42°=\cos 132°$

≈ -0.6691

46. $\cot\dfrac{3\pi}{4}=\tan\dfrac{-\pi}{4}=-1$

47. $\cos\dfrac{5\pi}{4}=\sin\dfrac{-3\pi}{4}$

$=\dfrac{-\sqrt{2}}{2}$

48. $\theta=\dfrac{3\pi}{4}$ and

$\tan\theta=-1$

49. $\sec\theta=\dfrac{13}{5}$

50. $A=3, P=\dfrac{\pi}{2}$, no

phase shift

51. $A=6, P=2\pi$, no

phase shift

52. $A=-\dfrac{1}{2}, P=\dfrac{2\pi}{3}$,

shifted right $\dfrac{\pi}{3}$ units

Determine the reference angle associated with the given angle.

36. $\theta=86°$ **37.** $\theta=-143°$

38. $\theta=\dfrac{3\pi}{2}$ **39.** $\theta=\dfrac{11\pi}{4}$

Determine which of the four quadrants angle θ is located in.

40. $\csc\theta>0$ and $\tan\theta>0$ **41.** $\sec\theta<0$ and $\cot\theta>0$

In each of the following problems, **a.** rewrite the expression in terms of the given angle's reference angle, and then **b.** evaluate the result, using a calculator if necessary. Round answers to four decimal places.

42. $\sin 290°$ **43.** $\tan\dfrac{4\pi}{3}$

Express each of the following in terms of the appropriate cofunction, and verify the equivalence of the two expressions.

44. $\csc 193°$ **45.** $\sin(-42°)$

46. $\cot\dfrac{3\pi}{4}$ **47.** $\cos\dfrac{5\pi}{4}$

Use the given information about each angle to evaluate the expressions.

48. Given that $\sin\theta=\dfrac{\sqrt{2}}{2}$ and $\tan\theta$ is negative, determine θ and $\tan\theta$.

49. Given that $\csc\theta=\dfrac{13}{12}$ and θ lies in the first quadrant, determine $\sec\theta$.

Section 6.4

Determine the amplitude, period, and phase shift of each of the following trigonometric equations.

50. $y=3\cos 4x$ **51.** $y=10+6\cos x$

52. $y=6-\dfrac{1}{2}\sin(3\theta-\pi)$ **53.** $y=-3+9\sin(2\theta+2\pi)$

Sketch the graphs of the following functions.

54. $f(x)=-2\tan 3x$ **55.** $g(x)=4\sin(2x-5)$

56. $f(\theta)=5\cos\left(\theta-\dfrac{\pi}{3}\right)$ **57.** $h(x)=2+\sin(x-\pi)$

58. $g(x)=1-\dfrac{1}{2}\sin\left(\dfrac{1}{2}x+\dfrac{\pi}{4}\right)$ **59.** $f(t)=\dfrac{1}{2}e^{-t}\cos(t+2\pi)-1$

53. $A = 9, P = \pi$, shifted left π units

54.

55.

56.

57.

58.

59.

60. $y = \dfrac{\pi}{6}$ **61.** $y = \dfrac{\pi}{2}$

62. $y = \dfrac{-\pi}{4}$ **63.** $y = \dfrac{\pi}{3}$

64. DNE

65. 0.4636 rad or 26.5651 degrees

Section 6.5

Solve each equation for y without the use of a calculator.

60. $y = \cos^{-1} \dfrac{\sqrt{3}}{2}$

61. $y = \cos^{-1} 0$

62. $y = \arctan(-1)$

63. $y = \csc^{-1} \dfrac{2\sqrt{3}}{3}$

Evaluate the following expressions, using a calculator if necessary.

64. $\sin^{-1} 2$

65. $\tan^{-1} 0.5$

Evaluate the following expressions, if possible.

66. $\arccos(\sin \pi)$

67. $\cos\left(\cos^{-1} 0.9\right)$

68. $\tan\left(\tan^{-1} 0.75\right)$

69. $\sin^{-1}\left(\cos \dfrac{3\pi}{4}\right)$

Find the value of each expression without using a calculator.

70. $\sin\left(\arctan \dfrac{\sqrt{3}}{3}\right)$

71. $\cot\left(\sec^{-1} 2\right)$

72. $\cos\left(\arcsin\left(\dfrac{-\sqrt{2}}{2}\right)\right)$

73. $\sec\left(\sin^{-1} \dfrac{\sqrt{3}}{2}\right)$

Using a graphing calculator, find the measure of θ in radians.

74. $\theta = \cos^{-1} 0.3492581$

75. $\theta = \tan^{-1}(-4.18249588)$

Using a graphing calculator, find the measure of θ in degrees.

76. $\theta = \arcsin(-0.66666667)$

77. $\theta = \arccos 0.56894372$

Express the following function as a purely algebraic function.

78. $\tan\left(\sin^{-1} \dfrac{x}{\sqrt{x^2 + 4}}\right)$

66. $\dfrac{\pi}{2}$

67. 0.9

68. 0.75

69. $\dfrac{-\pi}{4}$

70. $\dfrac{1}{2}$

71. $\dfrac{\sqrt{3}}{3}$

72. $\dfrac{\sqrt{2}}{2}$

73. 2

74. 1.2140171

75. −1.336110366

76. −41.81031515

77. 55.32339906

78. $\dfrac{x}{2}$

1. 240°

2. $-\dfrac{73\pi}{60}$

3.

4.

5. 1.204 m

6. 17.36°

7. a. 1200π rad/min
b. 125.66 in./s

8. 180π rad/min

9. 73.55 cm²

10. $\sin\theta = \dfrac{2\sqrt{10}}{7}$

$\cos\theta = \dfrac{3}{7}$

$\tan\theta = \dfrac{2\sqrt{10}}{3}$

$\csc\theta = \dfrac{7\sqrt{10}}{20}$

$\sec\theta = \dfrac{7}{3}$

$\cot\theta = \dfrac{3\sqrt{10}}{20}$

11. $\tan 63° \approx 1.9626$
$\cot 63° \approx 0.5095$

12. $\sin\dfrac{3\pi}{8} \approx 0.9239;$

$\csc\dfrac{3\pi}{8} \approx 1.0824$

Chapter Test

Convert the following angle measures as directed.

1. Express $\dfrac{4\pi}{3}$ in degrees.

2. Express −219° in radians.

Sketch the indicated angles.

3. $-\dfrac{5\pi}{3}$

4. 175°

Find the indicated arc length in each of the following problems.

5. Given a circle with a diameter of 1.5 m, find the length of the arc subtended by a central angle of 92° (**Hint:** Convert to radians first).

6. Given that two cities are 1200 miles apart and have the same longitude (that is, one is due south of the other), what is the difference in their latitudes? Use a value of 3960 miles for the radius of the Earth.

Solve the following word problems.

7. A DVD has a 2-inch radius and spins at 600 rpm. Find **a.** the angular speed of a mark on its edge in radians per minute, and **b.** the linear speed of the mark in inches per second.

8. A round metal gear has a radius of 3 inches and spins at 90 rpm. What is the angular speed of a notch on the edge of the gear?

9. If a circle of radius 7 cm contains a sector marked by a central angle of −172°, what is the area of the sector?

Use the information contained in the figure to determine the values of the six trigonometric functions of θ.

10.

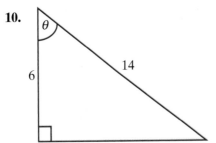

Evaluate the expressions, using a calculator if necessary.

11. $\tan 63°$ and $\cot 63°$

12. sine and cosecant of $\dfrac{3\pi}{8}$

13. 2966.02 feet

14. 15.16 feet

15. 176.99 feet

16. 56.93 feet

17. $\sin -125° \approx -0.8192$
$\cos -125° \approx -0.5736$
$\tan -125° \approx 1.4281$
$\csc -125° \approx -1.2208$
$\sec -125° \approx -1.7434$
$\cot -125° \approx 0.7002$

18. $\sin 5\pi = 0$
$\cos 5\pi = -1$
$\tan 5\pi = 0$
$\csc 5\pi = $ undefined
$\sec 5\pi = -1$
$\cot 5\pi = $ undefined

19. **a.** $\sin 179° = \sin 1°$
b. 0.0175

20. **a.** $\cos \dfrac{\pi}{4}$ **b.** $\dfrac{\sqrt{2}}{2}$

21. $\cot \dfrac{7\pi}{10}$
≈ -0.7265

22. $\sec 38°$
≈ 1.2690

23. $\sin\theta = \dfrac{-\sqrt{5}}{5}$

24. $\csc\theta = -\dfrac{2.6}{1} = -\dfrac{13}{5}$,
$\cot\theta = \dfrac{5}{12}$

25. $A = 1.5, P = 1$
$f(t) = 1.5 \cos 2\pi t$

Use an appropriate trigonometric function and a calculator, if necessary, to solve the following problems.

13. A tourist is going to ride a cable car up a mountain near Lake Tahoe. He knows that the elevation at the base of the mountain is 4530 feet, and that the top of the mountain reaches 6976 feet. If the angle of ascent for the cable car line is 55°33′20″, how far does each car travel?

14. A professor uses a laser pointer while giving a PowerPoint® lecture. If the laser makes an angle of 13° with the horizontal when she is pointing at a spot on screen that is 3.5 feet above the horizontal, how far is she standing from the screen?

Use the formula $h = \dfrac{d}{\cot \beta - \cot \alpha}$ to solve the following problems.

15. To determine the height of a skyscraper, a surveyor stands down the street and measures the angle to its highest point to be 52°. He then travels 45 feet closer to the building and finds the same angle to be 44°. How tall is the building?

16. A camera operator sits in a stationary crane, 40 feet above ground. He films a scene below of an automobile approaching. At the beginning of the scene, the angle of depression to the driver is 23°. After a direct approach towards the crane, the scene ends with the camera at an angle of 47°. How far did the vehicle travel?

Evaluate the six trigonometric functions for the angles given below, using a calculator if necessary.

17. $\theta = -125°$

18. $\theta = 5\pi$

In each of the following problems, **a.** rewrite the expression in terms of the given angle's reference angle, and then **b.** evaluate the result, using a calculator if necessary.

19. $\sin 179°$

20. $\cos \left(-\dfrac{9\pi}{4}\right)$

Express each of the following in terms of the appropriate cofunction, and verify the equivalence of the two expressions.

21. $\tan \left(-\dfrac{\pi}{5}\right)$

22. $\csc 52°$

Use the given information about each angle to evaluate the expressions.

23. Given that $\tan\theta = \dfrac{1}{2}$ and $\sin \theta$ is negative, determine $\sin \theta$.

24. Given that $\csc \theta = -2.6$ and θ lies in the third quadrant, determine $\cot \theta$.

Construct a function modeling the following behavior.

25. A clock's pendulum spans a distance of 3 inches from one end of its swing to the other. The time for one oscillation is one second.

26.

Sketch the graphs of the following functions.

26. $f(x) = \csc 2\pi x$

27. $f(x) = 1 - \cos 3\pi x$

27.

Sketch the following function modeling damped harmonic motion.

28. $g(t) = -e^{-t+2} \sin 3\pi t$

Evaluate the following expressions, using a calculator if necessary.

29. $\tan^{-1} 0.6$

30. $\cos^{-1}(-2)$

28.

Convert the following expressions to ones that can be entered in a calculator, then evaluate.

31. $\sec^{-1}(-5.3)$

32. $\cot^{-1} 127$

29. 0.5404

Evaluate the following expressions, if possible.

30. The cos function has a range of $[-1, 1]$, so the expression can't be evaluated.

33. $\cos^{-1}\left(\cos\dfrac{\pi}{2}\right)$

34. $\cos(\tan^{-1}(-0.4))$

Express the following function as a purely algebraic function.

31. $\cos^{-1}\left(-\dfrac{1}{5.3}\right)$

≈ 1.7606

35. $\sin(\sec^{-1} 2x)$

32. $\tan^{-1}\dfrac{1}{127} \approx 0.0079$

33. $\dfrac{\pi}{2}$

34. 0.9285

35. $\dfrac{\sqrt{4x^2 - 1}}{2x}$

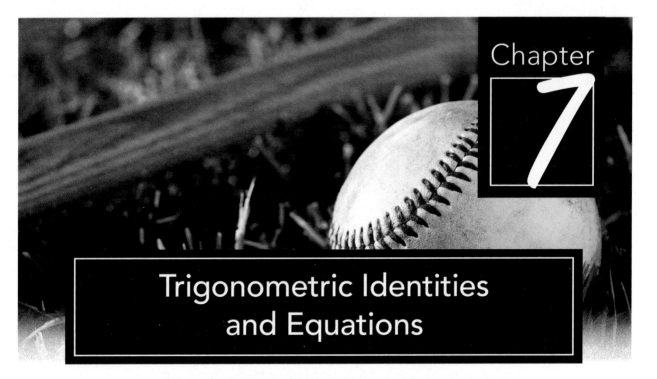

Trigonometric Identities and Equations

7.1   Fundamental Identities and Their Uses

7.2   Sum and Difference Identities

7.3   Product-Sum Identities

7.4   Trigonometric Equations

Chapter 7 Project

Chapter 7 Summary

Chapter 7 Review

Chapter 7 Test

By the end of this chapter you should be able to:

What if you were walking up to bat in a baseball game? At what angle from the ground should you hit the ball to make sure that it clears the back fence, giving you a home run?

By the end of this chapter, you will have been introduced to and know how to solve equations using several fundamental trigonometric identities. These identities are very useful for simplifying trigonometric expressions. On page 596, you will find a baseball problem similar to the one given above. You will need to use identities such as the Double-Angle Identities on page 576 and inverse trigonometric functions to find the angle to hit the ball in order to achieve the needed range.

Introduction

The last chapter introduced the trigonometric functions and their basic properties and uses. This chapter now delves deeper into how the functions relate to one another; this deeper understanding will allow us to simplify unwieldy expressions, solve trigonometric equations, and extend our grasp of trigonometry yet further.

The relationships between the trigonometric functions fall into a category of equations called *identities*, the formal definition of which was given in Chapter 1. Briefly, a trigonometric identity (also sometimes referred to as a trigonometric formula) is an equation that is always true. As such, there is no need to solve an identity—instead, identities are used as tools in accomplishing other tasks. Identities began to be used soon after the appearance of trigonometric functions. By the second century AD, Ptolemy, the Greek astronomer and geographer, was using identities in his work with the chords of a circle, the accepted form of trigonometry at the time. For his early work entitled *Almagest*, he used the sum and difference formulas and something that resembled the modern day half angle formula in order to update the current chord tables to an accuracy of three decimal places.

Identities must either be discovered or verified, and both processes call for the application of algebra. Further, the use of an identity usually calls for algebraic skill; some of the uses we will explore include simplifying expressions, evaluating trigonometric functions exactly, and solving *conditional* equations (which are equations that are *not* always true). In fact, the underlying theme of this chapter is the marriage of trigonometry and algebra. Whereas the last chapter focused almost exclusively on the basics of trigonometry, this chapter and the next bring algebra back into the discussion. The union of trigonometry and algebra culminates in the last section of the chapter, where you will see that such algebraic methods as factoring and solving for specific terms are critical in being able to solve trigonometric equations.

Viéte

The concept of mixing algebra and trigonometry did not happen overnight. For centuries mathematicians separated the realms of real numbers and geometric ideas. It was not until the sixteenth century that mathematicians finally began to use algebra for abstract quantities rather than simply for concrete values. This transition was thanks in part to François Viéte (1540–1603), who was the first to consistently apply algebraic methods to his work in trigonometry. He introduced the sum-to-product formulas, the law of tangents, and a recurrence formula that allows $\cos nx$ to be presented in terms of the cosine of lower multiples of x. Although Viéte's breakthrough application of algebra to trigonometry may seem remarkably basic, it is good to remember that many important mathematical discoveries come from examining a problem in a new, unexpected, and sometimes simple way.

Fundamental Identities and Their Uses

TOPICS

1. Previously encountered identities
2. Simplifying trigonometric expressions
3. Verifying trigonometric identities
4. Trigonometric substitutions

TOPIC 1

Previously Encountered Identities

Chapter 6 presented the foundations of trigonometry, and concentrated largely on the geometric and functional aspects of the material. This chapter focuses more on the relationships between trigonometric functions and on the marriage of algebra and trigonometry. It does so by introducing and then using statements of equality known as *trigonometric identities*. We begin with a review of the identities already seen, though three of the equations below have only been alluded to in passing.

IDENTITY

Identities Already Seen

Reciprocal Identities

$$\csc x = \frac{1}{\sin x} \qquad \sec x = \frac{1}{\cos x} \qquad \cot x = \frac{1}{\tan x}$$

$$\sin x = \frac{1}{\csc x} \qquad \cos x = \frac{1}{\sec x} \qquad \tan x = \frac{1}{\cot x}$$

Quotient Identities

$$\tan x = \frac{\sin x}{\cos x} \qquad \cot x = \frac{\cos x}{\sin x}$$

Cofunction Identities

$$\cos x = \sin\left(\frac{\pi}{2} - x\right) \qquad \csc x = \sec\left(\frac{\pi}{2} - x\right) \qquad \cot x = \tan\left(\frac{\pi}{2} - x\right)$$

$$\sin x = \cos\left(\frac{\pi}{2} - x\right) \qquad \sec x = \csc\left(\frac{\pi}{2} - x\right) \qquad \tan x = \cot\left(\frac{\pi}{2} - x\right)$$

Period Identities

$$\sin(x+2\pi)=\sin x \qquad \cos(x+2\pi)=\cos x$$

$$\csc(x+2\pi)=\csc x \qquad \sec(x+2\pi)=\sec x$$

$$\tan(x+\pi)=\tan x \qquad \cot(x+\pi)=\cot x$$

Even/Odd Identities

$$\sin(-x)=-\sin x \qquad \cos(-x)=\cos x \qquad \tan(-x)=-\tan x$$

$$\csc(-x)=-\csc x \qquad \sec(-x)=\sec x \qquad \cot(-x)=-\cot x$$

Pythagorean Identities

$$\sin^2 x+\cos^2 x=1 \qquad \tan^2 x+1=\sec^2 x \qquad 1+\cot^2 x=\csc^2 x$$

The Pythagorean Identities have not explicitly appeared yet, but they are based on an idea we have used frequently. Their names allude to the familiar Pythagorean Theorem, and the first identity follows from consideration of a diagram such as that in Figure 1.

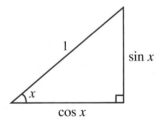

Figure 1: Derivation of $\sin^2 x+\cos^2 x=1$

The diagram assumes that x is a real number between 0 and $\dfrac{\pi}{2}$, but a similar diagram can be drawn given any real number (recall that the use of reference angles makes this possible). If a right triangle with a hypotenuse of length 1 is drawn, then the legs of the right triangle must be of length $\sin x$ and $\cos x$ (be sure you see how this follows from the definitions of sine and cosine). The Pythagorean Theorem then leads to the first Pythagorean Identity; dividing through by $\cos^2 x$ and $\sin^2 x$ leads to, respectively, the second and third Pythagorean Identities.

The ideas behind the identities can lead to statements that appear, superficially, to be different. For instance, we know that the 2π-periodicity of sine makes all of the following statements true:

$$\sin(x-6\pi)=\sin x,\ \ \sin(x+4\pi)=\sin(x+2\pi),\ \ \text{and}\ \ \sin(x+2\pi)=\sin(x-2\pi).$$

Deeper contemplation of the graph of sine leads to the similar identity

$$\sin x = -\sin(x+\pi).$$

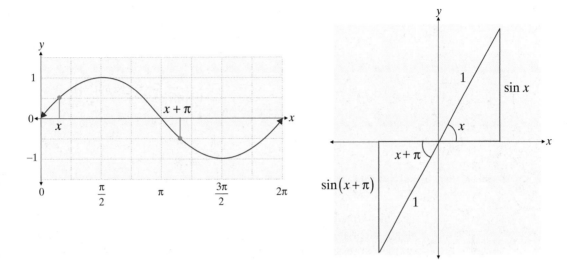

Similar statements based on periodicity can be deduced for the other trigonometric functions.

TOPIC 2

Simplifying Trigonometric Expressions

One common use of trigonometric identities is in simplifying expressions. Frequently, the first answer obtained in solving a trigonometry-related problem is unnecessarily complicated, and the judicious use of identities leads to a simpler form of the answer. This happens often, for instance, in calculus problems.

EXAMPLE 1

Simplifying
Trigonometric
Expressions

Simplify the expression $\cos\theta + \sin\theta\tan\theta$.

Solution:

Skill in using identities to simplify trigonometric expressions only comes with practice, but the guiding principle is to rewrite the expression in such a way that the use of one or more identities becomes apparent. In the case of the expression $\cos\theta + \sin\theta\tan\theta$, a good way to begin is to rewrite it in terms of only sine and cosine, since three distinct functions are unnecessary.

$$\cos\theta + \sin\theta\tan\theta = \cos\theta + \sin\theta\left(\frac{\sin\theta}{\cos\theta}\right)$$

$$= \cos\theta + \frac{\sin^2\theta}{\cos\theta}$$

Now, the presence of $\sin^2\theta$ should remind you of the first Pythagorean Identity (one that you'll find yourself using very frequently). Remember that $\sin^2\theta + \cos^2\theta = 1$, so if some other term can be rewritten in such a way that $\sin^2\theta$ is added to $\cos^2\theta$, there is a good chance that the identity applies. In the problem at hand,

$$\cos\theta + \frac{\sin^2\theta}{\cos\theta} = \frac{\cos^2\theta}{\cos\theta} + \frac{\sin^2\theta}{\cos\theta}$$

$$= \left(\frac{1}{\cos\theta}\right)\left(\cos^2\theta + \sin^2\theta\right)$$

$$= \frac{1}{\cos\theta}$$

$$= \sec\theta.$$

The fact that $\cos\theta + \sin\theta\tan\theta = \sec\theta$ is not at all obvious initially, but is also not at all surprising. Such equivalences between relatively complicated and relatively simple expressions occur frequently in trigonometry, and it's usually worth spending some amount of time to see if a complicated expression can be simplified. After all, if the expression $\cos\theta + \sin\theta\tan\theta$ were the answer to a real-world problem, and if the next step was to evaluate the expression for numerous values of θ, it would certainly be easier to simply evaluate $\sec\theta$ instead.

EXAMPLE 2

Simplifying Trigonometric Expressions

Simplify the expression $\cot\alpha + \dfrac{\sin\alpha}{1+\cos\alpha}$.

Solution:

As in Example 1, rewriting the expression in terms of only sine and cosine is a good way to start. And as in so many problems, finding a common denominator and combining the resulting fractions is a promising way to proceed. After that, the remaining steps suggest themselves clearly:

$$\cot\alpha + \frac{\sin\alpha}{1+\cos\alpha} = \frac{\cos\alpha}{\sin\alpha} + \frac{\sin\alpha}{1+\cos\alpha}$$

$$= \left(\frac{\cos\alpha}{\sin\alpha}\right)\left(\frac{1+\cos\alpha}{1+\cos\alpha}\right) + \left(\frac{\sin\alpha}{1+\cos\alpha}\right)\left(\frac{\sin\alpha}{\sin\alpha}\right) \quad \text{Multiply by appropriate factors to achieve a common denominator.}$$

$$= \frac{\cos\alpha + \cos^2\alpha + \sin^2\alpha}{\sin\alpha(1+\cos\alpha)} \quad \text{Apply the first Pythagorean Identity.}$$

$$= \frac{\cos\alpha + 1}{\sin\alpha(1+\cos\alpha)} \quad \text{Cancel the common factors.}$$

$$= \frac{1}{\sin\alpha}$$

$$= \csc\alpha.$$

TOPIC 3   Verifying Trigonometric Identities

The identities we have seen to this point are all very fundamental, and in fact some are merely restatements of definitions. Nevertheless, they are also very useful and serve as good examples of the class of equations known as identities. Recall from Section 1.5 that an identity is an equation that is true for any (allowable) value(s) of the variable(s); this would be a good time to review the list of trigonometric identities at the start of this section and verify that those equations indeed fit this description.

Of course, most equations that we encounter in algebra are not identities, and in fact the goal of much of our work is to determine exactly which values of the variable(s) make the equation true. In Section 1.5 equations that were not identities or contradictions were labeled *conditional*, but in practice the adjective is usually dropped since most equations we encounter are of this sort. We have already solved many simple conditional trigonometric equations, but we will study such equations in much greater depth in Section 7.4. At that time, we will see that trigonometric identities are very useful in determining the solutions of trigonometric equations.

Before we get to that point, though, we need to build up our repertoire of identities, and one step in doing so is to verify that a proposed identity really is true for all values of the variable. This is called *verifying an identity*, and the process is often very similar to using identities to simplify expressions. While there is no guaranteed method to use in such verification, there are some general guidelines to follow.

PROCEDURE

Guidelines
for Verifying
Trigonometric
Identities

1. **Work with one side at a time.** Choose one side of the equation to work with and simplify it. The more complicated side is usually the best choice. The goal is to transform it into the other side.

2. **Apply trigonometric identities as appropriate.** To do so, it will probably be necessary to combine fractions, add or subtract terms, factor expressions, or use other algebraic manipulations.

3. **Rewrite in terms of sine and cosine if necessary.** If you are stuck, expressing everything in terms of sine and cosine often leads to inspiration.

EXAMPLE 3

Verify the identity $2\csc^2 x = \dfrac{1}{1-\cos x} + \dfrac{1}{1+\cos x}$.

Solution:

The right-hand side is more complicated, so we'll begin with it. Clearly, combining the two fractions is a good way to begin, especially as we can see that a denominator of $1-\cos^2 x$ will eventually appear.

$$\frac{1}{1-\cos x} + \frac{1}{1+\cos x} = \frac{1+\cos x + 1 - \cos x}{(1-\cos x)(1+\cos x)}$$

Modify each fraction in order to obtain a common denominator, and combine.

$$= \frac{2}{1-\cos^2 x}$$

Multiply out the denominator.

$$= \frac{2}{\sin^2 x}$$

Apply the first Pythagorean Identity.

$$= 2\csc^2 x$$

In verifying some identities, it may be easiest to simplify both sides of the equation individually. The goal in this case is to achieve a single simpler expression which is equivalent to both sides of the original equation.

EXAMPLE 4

Verify the identity $\dfrac{\tan^2 x}{1+\sec x} = \dfrac{1-\cos x}{\cos x}$.

Solution:

It's not immediately clear that either side is more complicated than the other, so we can try simplifying both. Beginning with the left-hand side, we can use another of the Pythagorean Identities as follows:

$$\frac{\tan^2 x}{1+\sec x} = \frac{\sec^2 x - 1}{1+\sec x}$$

Use the second Pythagorean Identity to rewrite the numerator.

$$= \frac{(\sec x - 1)(\sec x + 1)}{1+\sec x}$$

Factor the difference of two squares.

$$= \sec x - 1.$$

Cancel the common factors.

The right-hand side is more easily dealt with:

$$\frac{1-\cos x}{\cos x} = \frac{1}{\cos x} - \frac{\cos x}{\cos x}$$

Break the single fraction into two and simplify.

$$= \sec x - 1.$$

Since both sides of the original equation are equivalent to $\sec x - 1$, the identity is true.

EXAMPLE 5

Verifying Trigonometric Identities

Verify the identity $\dfrac{\cos\varphi\cot\varphi}{1-\sin\varphi}-1=\csc\varphi$.

Solution:

The left-hand side is clearly more complicated, and there are several ways of beginning the process of simplifying it. First, rewriting the numerator of the fraction in terms of sine and cosine looks promising, as a factor of $\cos^2\varphi$ would then appear. Second, obtaining a denominator of $1-\sin^2\varphi$ in the fraction would be easily done and would allow one of the Pythagorean Identities to apply. We'll try both ideas:

$$\frac{\cos\varphi\cot\varphi}{1-\sin\varphi}-1=\frac{\cos\varphi\left(\dfrac{\cos\varphi}{\sin\varphi}\right)}{1-\sin\varphi}-1$$

Rewrite cotangent in terms of sine and cosine.

$$=\left(\frac{\cos^2\varphi}{\sin\varphi(1-\sin\varphi)}\right)\left(\frac{1+\sin\varphi}{1+\sin\varphi}\right)-1$$

Multiply appropriately in order to obtain $1-\sin^2\varphi$ in the denominator.

$$=\left(\frac{\cos^2\varphi}{\sin\varphi\left(1-\sin^2\varphi\right)}\right)(1+\sin\varphi)-1$$

$$=\left(\frac{\cos^2\varphi}{\sin\varphi\cos^2\varphi}\right)(1+\sin\varphi)-1$$

Apply the first Pythagorean Identity.

$$=\frac{1+\sin\varphi}{\sin\varphi}-1$$

Cancel common factors.

$$=\frac{1}{\sin\varphi}+1-1$$

Break apart fraction and simplify.

$$=\csc\varphi$$

TOPIC 4

Trigonometric Substitutions

In several classes of calculus problems, it is very convenient to be able to replace certain algebraic expressions with trigonometric expressions. Most often, these replacements depend on one of the Pythagorean Identities, and the work involved is reminiscent of that in earlier examples in this section. The example below illustrates a typical trigonometric substitution.

EXAMPLE 6

Using Trigonometric Substitutions

Use the substitution $\sin\theta = \dfrac{x}{2}$ to write $\sqrt{4-x^2}$ as a trigonometric expression. Assume $0 \le \theta \le \dfrac{\pi}{2}$.

Solution:

Although it is not necessary for the task at hand, a diagram motivating the substitution may be helpful. The triangle below illustrates the geometric relation between θ and the various algebraic expressions:

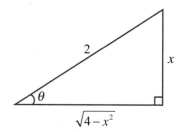

The suggested substitution can be rewritten as $x = 2\sin\theta$, and so we obtain

$$\sqrt{4-x^2} = \sqrt{4-\left(2\sin\theta\right)^2}$$
$$= \sqrt{4-4\sin^2\theta}$$
$$= 2\sqrt{1-\sin^2\theta}$$
$$= 2\cos\theta.$$

1. $\sec x$

2. $\cos^2 \theta$

3. $\sin t$

4. $\cos^2 x$

5. $\cos x - \sec x$

6. 1

7. $\tan \alpha$

8. $\cos^2 t$

9. $\sec y$

10. $-\tan x$

11. $\tan^2 x$

12. 1

13.–25. Answers will vary.

Exercises

Use trigonometric identities to simplify the expressions. There may be more than one correct answer. See Examples 1 and 2.

1. $\tan x \csc x$

2. $\dfrac{1}{\tan^2 \theta + 1}$

3. $\dfrac{\tan t}{\sec t}$

4. $\cot^2 x - \cot^2 x \cos^2 x$ **5.** $\sin(-x)\tan x$

6. $\dfrac{1}{\sec^2 x} + \sin x \cos\left(\dfrac{\pi}{2} - x\right)$

7. $\sin(\alpha + 2\pi)\sec \alpha$ **8.** $\sin t\,(\csc t - \sin t)$ **9.** $\cos y\,(1 + \tan^2 y)$

10. $\dfrac{1}{\cos x \csc(-x)}$ **11.** $\dfrac{1 - \tan^2 x}{\cot^2 x - 1}$ **12.** $\dfrac{\sin \beta \tan\left(\dfrac{\pi}{2} - \beta\right)}{\cos \beta}$

Verify the identities. See Examples 3, 4, and 5.

13. $(1 - \cos\theta)(1 + \cos\theta) = \sin^2 \theta$

14. $\csc x - \sin x = \cos x \cot x$

15. $\sec^2 y - \tan^2 y = \sec y \cos(-y)$

16. $\dfrac{\cos \beta \cot \beta}{1 - \sin \beta} - 1 = \csc \beta$

17. $\dfrac{\sin\left(\dfrac{\pi}{2} - x\right)}{\cos\left(\dfrac{\pi}{2} - x\right)} = \cot x$

18. $\dfrac{\sec^2 \theta}{\tan \theta} = \sec \theta \csc \theta$

19. $\dfrac{1}{\tan x} + \tan x = \dfrac{\sec^2 x}{\tan x}$

20. $\sin^2 t + \sin^2\left(\dfrac{\pi}{2} - t\right) = 1$

21. $\dfrac{1}{\sin(\theta + 2\pi) + 1} + \dfrac{1}{\csc(\theta + 2\pi) + 1} = 1$ **22.** $3 + \cot^2 \alpha = 2 + \csc^2 \alpha$

23. $\sin^2 x - \sin^4 x = \cos^2(-x) - \cos^4(-x)$

24. $\cot\left(\dfrac{\pi}{2} - \beta\right)\cot \beta = 1$

25. $\dfrac{\cos\left(\dfrac{\pi}{2} - \alpha\right)}{\csc \alpha} - 1 = \sin \alpha \cot(-\alpha)\cos(-\alpha)$

26. $\sec \theta$

27. $4 \tan \theta$

28. $3 \sin \theta$

29. $10 \csc \theta$

30. $8 \cos \theta$

31. $2 \cot \theta$

32. $5 \sec \theta$

33. $12 \sin \theta$

34. Answers will vary.

35. Answers will vary.

Use the suggested substitution to rewrite the given expression as a trigonometric expression. Assume $0 \le \theta \le \dfrac{\pi}{2}$. See Example 6.

26. $\sqrt{x^2 + 1}$, $x = \tan \theta$

27. $\sqrt{x^2 - 16}$, $x = 4 \sec \theta$

28. $\sqrt{9 - x^2}$, $\cos \theta = \dfrac{x}{3}$

29. $\sqrt{4x^2 + 100}$, $\cot \theta = \dfrac{x}{5}$

30. $\sqrt{64 - x^2}$, $x = 8 \sin \theta$

31. $\sqrt{x^2 - 4}$, $x = 2 \csc \theta$

32. $\sqrt{x^2 + 25}$, $\tan \theta = \dfrac{x}{5}$

33. $\sqrt{144 - 9x^2}$, $x = 4 \cos \theta$

Show how the identities below follow from the first Pythagorean Identity.

34. $\tan^2 x + 1 = \sec^2 x$

35. $1 + \cot^2 x = \csc^2 x$

Sum and Difference Identities

TOPICS

1. The identities

2. Using sum and difference identities for exact evaluation

3. Applications of sum and difference identities

TOPIC 1

The Identities

This section and the next will introduce new identities, selected proofs of the identities, and many examples of how the identities can be used. The first group of identities concerns functions acting on sums or differences of angles.

IDENTITY

Sum and Difference Identities

Sine Identities

$$\sin(u+v) = \sin u \cos v + \cos u \sin v \qquad \sin(u-v) = \sin u \cos v - \cos u \sin v$$

Cosine Identities

$$\cos(u+v) = \cos u \cos v - \sin u \sin v \qquad \cos(u-v) = \cos u \cos v + \sin u \sin v$$

Tangent Identities

$$\tan(u+v) = \frac{\tan u + \tan v}{1 - \tan u \tan v} \qquad \tan(u-v) = \frac{\tan u - \tan v}{1 + \tan u \tan v}$$

We will prove two of these identities here, and indicate in the exercises how the other four identities can then be derived.

Consider the unit circle in Figure 1 on the following page. In the figure, angle v has been placed in standard position and angle u has been positioned in such a way that angle $u + v$ is in standard position. In addition, the negative of angle u has also been drawn in standard position.

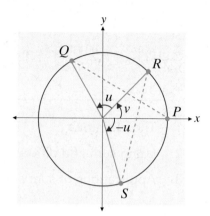

Figure 1: Derivation of Cosine Sum Identity

Using the fact that the radius of the circle is 1, we can easily identify the coordinates of the points $P, Q, R,$ and S:

$$P = (1, 0), \quad Q = \left(\cos(u+v), \sin(u+v)\right), \quad R = (\cos v, \sin v), \quad \text{and} \quad S = (\cos u, -\sin u).$$

Note that for point S, we have used the fact that $\sin(-u) = -\sin u$ and $\cos(-u) = \cos u$. Note also that the chord $\overline{PQ}$ has the same length as the chord $\overline{RS}$, since the subtended angles both have magnitude $u + v$. Since we know the coordinates of the endpoints of the chords, we can use the Distance Formula to obtain the equation

$$\sqrt{\left(\cos(u+v)-1\right)^2 + \left(\sin(u+v)-0\right)^2} = \sqrt{\left(\cos v - \cos u\right)^2 + \left(\sin v + \sin u\right)^2}.$$

The square roots are easily eliminated by squaring both sides, and we can then proceed to expand the squared terms:

$$\cos^2(u+v) - 2\cos(u+v) + 1 + \sin^2(u+v) =$$

$$\cos^2 v - 2\cos u \cos v + \cos^2 u + \sin^2 v + 2\sin u \sin v + \sin^2 u.$$

By now, you should be attuned to the presence of $\sin^2(\ )$ and $\cos^2(\ )$ terms; every pair of these having the same argument can be replaced with 1. Making that replacement three times in the preceding equation gives us

$$2 - 2\cos(u+v) = 2 - 2\cos u \cos v + 2\sin u \sin v,$$

and subtracting 2 from both sides and then dividing by -2 yields

$$\cos(u+v) = \cos u \cos v - \sin u \sin v.$$

Now that we have proved one of the identities, the rest follow relatively quickly. In particular, the difference identity for cosine is very easily proved by replacing v with $-v$ in the sum identity and making use of the fact that cosine is an even function and sine is odd:

$$\cos(u-v) = \cos u \cos(-v) - \sin u \sin(-v)$$

$$= \cos u \cos v + \sin u \sin v.$$

TOPIC 2

Using Sum and Difference Identities for Exact Evaluation

It was fairly easy, back in Chapter 6, to determine the exact values of the trigonometric functions acting on the angles $\dfrac{\pi}{6}$, $\dfrac{\pi}{4}$, and $\dfrac{\pi}{3}$. It may seem odd, therefore, that these are still the only (acute) angles for which we can perform exact evaluation. To this point, for instance, our only option for evaluating $\sin 75^\circ$ has been to use a calculator and note that $\sin 75^\circ \approx 0.9659$. The sum and difference identities extend our ability to obtain exact values greatly, as seen in the next example.

EXAMPLE 1

Using the Sum and Difference Identities for Exact Evaluation

Determine the exact value of $\sin 75^\circ$.

Solution:

All problems of this sort will call for us to express the given angle in terms of angles about which we know more. In this case, we'll use the fact that $75^\circ = 45^\circ + 30^\circ$:

$$\sin 75^\circ = \sin\left(45^\circ + 30^\circ\right)$$

$$= \sin 45^\circ \cos 30^\circ + \cos 45^\circ \sin 30^\circ$$

$$= \left(\frac{1}{\sqrt{2}}\right)\left(\frac{\sqrt{3}}{2}\right) + \left(\frac{1}{\sqrt{2}}\right)\left(\frac{1}{2}\right)$$

$$= \frac{\sqrt{3}+1}{2\sqrt{2}}.$$

You can now easily verify that this exact value is approximately 0.9659.

EXAMPLE 2

Using the Sum and Difference Identities for Exact Evaluation

Determine the exact value of $\cos 75^\circ$.

Solution:

The purpose of this example is twofold. The first is to point out that we are starting to build up a significant collection of tools and knowledge. The second is to emphasize that two identical answers may appear, superficially, to be different. We could certainly use the sum identity for cosine to evaluate $\cos 75^\circ$, taking steps very similar to those in Example 1. Using this approach, we would obtain

$$\cos 75^\circ = \frac{\sqrt{6}-\sqrt{2}}{4}.$$

But, coming immediately after the evaluation of $\sin 75°$, it would also make sense to use the identity $\cos^2 x + \sin^2 x = 1$ to obtain

$$\cos 75° = \sqrt{1 - \sin^2 75°}$$

$$= \sqrt{1 - \left(\frac{\sqrt{3}+1}{2\sqrt{2}}\right)^2}$$

$$= \sqrt{1 - \frac{4 + 2\sqrt{3}}{8}}$$

$$= \sqrt{\frac{2 - \sqrt{3}}{4}}$$

$$= \frac{\sqrt{2 - \sqrt{3}}}{2}.$$

Are these two answers actually the same?

You can use a calculator to determine that both are approximately 0.2588, but it is more convincing to prove the equivalence mathematically. A common way of doing this is to set the two answers equal to one another and to simplify the expressions in the resulting equation until either an obviously true or obviously false statement is achieved. Keep in mind, however, that the "equality" below is actually a question initially; we won't know that the two expressions actually are equal until we achieve an obviously true equation.

$$\frac{\sqrt{6} - \sqrt{2}}{4} \overset{?}{=} \frac{\sqrt{2 - \sqrt{3}}}{2} \qquad \text{Use the two expressions to form an "equation."}$$

$$\sqrt{6} - \sqrt{2} \overset{?}{=} 2\sqrt{2 - \sqrt{3}} \qquad \text{Multiply both sides by 4.}$$

$$6 - 2\sqrt{12} + 2 \overset{?}{=} 4\left(2 - \sqrt{3}\right) \qquad \text{Square both sides.}$$

$$8 - 4\sqrt{3} = 8 - 4\sqrt{3} \qquad \text{Simplify to obtain an obviously true statement.}$$

EXAMPLE 3

Using the Sum and
Difference Identities
for Exact Evaluation

Determine the exact value of $\tan \dfrac{\pi}{12}$.

Solution:

To make use of previous results, we need to note that $\dfrac{\pi}{12} = \dfrac{\pi}{3} - \dfrac{\pi}{4}$. So

$$\tan \frac{\pi}{12} = \tan\left(\frac{\pi}{3} - \frac{\pi}{4} \right)$$

$$= \frac{\tan \dfrac{\pi}{3} - \tan \dfrac{\pi}{4}}{1 + \tan \dfrac{\pi}{3} \tan \dfrac{\pi}{4}}$$

$$= \frac{\sqrt{3} - 1}{1 + \left(\sqrt{3} \right)(1)}$$

$$= \frac{\sqrt{3} - 1}{\sqrt{3} + 1}.$$

Although the need arises less often, the sum and difference identities can also be used in reverse.

EXAMPLE 4

Using the Sum and
Difference Identities
for Exact Evaluation

Determine the exact value of $\sin 80^\circ \cos 20^\circ - \cos 80^\circ \sin 20^\circ$.

Solution:

The key step is to recognize the expression as the difference identity for sine. We can then proceed as follows:

$$\sin 80^\circ \cos 20^\circ - \cos 80^\circ \sin 20^\circ = \sin\left(80^\circ - 20^\circ \right)$$

$$= \sin 60^\circ$$

$$= \frac{\sqrt{3}}{2}.$$

TOPIC 3   Applications of Sum and Difference Identities

The truth of the cofunction identities can be easily seen through the use of a unit circle diagram, as in the argument at the start of this section. But the difference identities can also be used to furnish a very quick algebraic verification, as in the next example.

EXAMPLE 5

Using the Sum and Difference Identities

Use a difference identity to verify that $\sin\left(\dfrac{\pi}{2}-x\right)=\cos x$.

Solution:

$$\sin\left(\frac{\pi}{2}-x\right)=\sin\frac{\pi}{2}\cos x-\cos\frac{\pi}{2}\sin x$$
$$=(1)(\cos x)-(0)(\sin x)$$
$$=\cos x$$

The identities in this section are useful in simplifying trigonometric functions of sums or differences no matter how unwieldy they may initially appear to be.

EXAMPLE 6

Using the Sum and Difference Identities

Express $\sin\left(\tan^{-1}x+\cos^{-1}x\right)$ as an algebraic function of x.

Solution:

From our previous experience with compositions of trig functions and inverse trig functions, we know the general approach to take; the only additional complication here is that the argument of sine is a sum. If we let $u=\tan^{-1}x$ and $v=\cos^{-1}x$, then $\tan u=x$ and $\cos v=x$ and we are led to consider two right triangles such as those below.

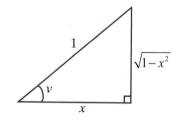

We can now apply the appropriate sum identity:

$$\sin\left(\tan^{-1} x + \cos^{-1} x\right) = \sin\left(u + v\right)$$

$$= \sin u \cos v + \cos u \sin v$$

$$= \left(\frac{x}{\sqrt{1+x^2}}\right)(x) + \left(\frac{1}{\sqrt{1+x^2}}\right)\left(\sqrt{1-x^2}\right)$$

$$= \frac{x^2}{\sqrt{1+x^2}} + \frac{\sqrt{1-x^2}}{\sqrt{1+x^2}}.$$

EXAMPLE 7

Using the Sum and Difference Identities

Show that $\dfrac{\sin(x+h) - \sin x}{h} = \sin x\left(\dfrac{\cos h - 1}{h}\right) + \cos x\left(\dfrac{\sin h}{h}\right).$

Solution:

$$\frac{\sin(x+h) - \sin x}{h} = \frac{\sin x \cos h + \cos x \sin h - \sin x}{h}$$

$$= \frac{\sin x \cos h - \sin x}{h} + \frac{\cos x \sin h}{h}$$

$$= \sin x\left(\frac{\cos h - 1}{h}\right) + \cos x\left(\frac{\sin h}{h}\right)$$

The last example of an application of the sum/difference identities leads to a surprising conclusion. As you know, the graph of cosine has the same shape as the graph of sine shifted to the left by $\dfrac{\pi}{2}$ (one of the cofunction identities is the algebraic form of this statement). But the connection between cosine and sine goes deeper; as it turns out, a *sum* of a sine term and a cosine term can be rephrased in terms of sine or cosine alone. This is true even if the sine and cosine functions have different amplitudes.

To see this, consider the expression $A\sin x + B\cos x$, a sum of a sine curve with amplitude A and a cosine curve with amplitude B. This has some resemblance to the right-hand side of the sum identity for sine, but only if A can be replaced with a cosine term and B with a sine term. That is the inspiration for the following:

$$A\sin x + B\cos x = \sqrt{A^2+B^2}\left(\frac{A}{\sqrt{A^2+B^2}}\sin x + \frac{B}{\sqrt{A^2+B^2}}\cos x\right).$$

If an angle φ exists for which

$$\cos\varphi = \frac{A}{\sqrt{A^2+B^2}} \quad \text{and} \quad \sin\varphi = \frac{B}{\sqrt{A^2+B^2}},$$

then the sum identity gives us

$$A\sin x + B\cos x = \sqrt{A^2+B^2}\left(\cos\varphi\sin x + \sin\varphi\cos x\right)$$
$$= \sqrt{A^2+B^2}\,\sin(x+\varphi).$$

Does such a φ exist? Figure 2 makes it clear that φ is easily found.

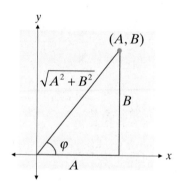

Figure 2: Determining the Phase Shift φ

THEOREM

Sum of Sines and Cosines

$$A\sin x + B\cos x = \sqrt{A^2+B^2}\left(\frac{A}{\sqrt{A^2+B^2}}\sin x + \frac{B}{\sqrt{A^2+B^2}}\cos x\right)$$

$$= \sqrt{A^2+B^2}\,\sin(x+\varphi)$$

$$\text{where } \cos\varphi = \frac{A}{\sqrt{A^2+B^2}} \quad \text{and} \quad \sin\varphi = \frac{B}{\sqrt{A^2+B^2}}.$$

EXAMPLE 8

Using the Sum of
Sines and Cosines
Identity

Express the function $f(x) = \sin x - \sqrt{3}\cos x$ in terms of a single sine function, and graph the result.

Solution:

Since $A = 1$ and $B = -\sqrt{3}$, $\sqrt{A^2 + B^2} = 2$ and φ must satisfy

$$\cos\varphi = \frac{1}{2} \text{ and } \sin\varphi = \frac{-\sqrt{3}}{2}.$$

This means φ lies in the fourth quadrant, and as a negative angle it can be expressed as $\varphi = -\frac{\pi}{3}$ (we could also write $\varphi = \frac{5\pi}{3}$). Using the above derivation,

$$f(x) = 2\sin\left(x - \frac{\pi}{3}\right),$$

and the graph of this function is shown below.

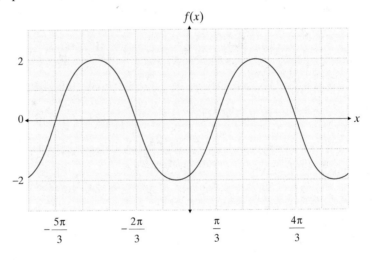

Exercises

Use the sum and difference identities to determine the exact value of each of the following expressions. See Examples 1, 2, and 3.

1. $-\dfrac{\sqrt{6}+\sqrt{2}}{4}$

2. $\dfrac{\sqrt{6}-\sqrt{2}}{4}$

3. $-\sqrt{3}-2$

4. $\dfrac{\sqrt{6}-\sqrt{2}}{4}$ 5. -1

6. $-\sqrt{3}-2$ 7. -1

8. $2-\sqrt{3}$ 9. 0

10. $\dfrac{-\sqrt{3}}{2}$ 11. 0

12. $\dfrac{\sqrt{3}}{2}$ 13. $\dfrac{\sqrt{6}+\sqrt{2}}{4}$

14. $\dfrac{\sqrt{6}-\sqrt{2}}{4}$

15. $\dfrac{\sqrt{2}+\sqrt{6}}{4}$

16. $\dfrac{\sqrt{2}+\sqrt{6}}{4}$

17. $\sqrt{3}+2$ 18. $-\sqrt{3}+2$

19. $\dfrac{\sqrt{6}-\sqrt{2}}{4}$ 20. $\dfrac{-\sqrt{3}}{3}$

21. $\dfrac{\sqrt{6}+\sqrt{2}}{4}$ 22. $\dfrac{-1}{2}$

23. $\sqrt{3}+2$ 24. $\dfrac{-\sqrt{2}}{2}$

25. $\dfrac{-\sqrt{6}-\sqrt{2}}{4}$ 26. -1

27. $\dfrac{-\sqrt{6}-\sqrt{2}}{4}$ 28. $\dfrac{-\sqrt{2}}{2}$

29. $\dfrac{\sqrt{6}-\sqrt{2}}{4}$

30. $\sqrt{3}+2$ 31. $\dfrac{-1}{2}$

32. $\dfrac{\sqrt{2}-\sqrt{6}}{4}$

33. $\dfrac{\sqrt{6}+\sqrt{2}}{4}$

1. $\cos\left(\dfrac{\pi}{4}+\dfrac{\pi}{3}\right)$

2. $\sin\left(\dfrac{\pi}{6}+\dfrac{3\pi}{4}\right)$

3. $\tan\left(\dfrac{4\pi}{3}+\dfrac{5\pi}{4}\right)$

4. $\sin\left(\dfrac{2\pi}{3}+\dfrac{\pi}{4}\right)$

5. $\cos\left(\dfrac{7\pi}{6}-\dfrac{\pi}{6}\right)$

6. $\tan\left(\dfrac{\pi}{3}-\dfrac{3\pi}{4}\right)$

7. $\cos\left(\dfrac{4\pi}{3}+\dfrac{5\pi}{3}\right)$

8. $\tan\left(\dfrac{4\pi}{3}-\dfrac{5\pi}{4}\right)$

9. $\cos\left(\dfrac{7\pi}{6}-\dfrac{5\pi}{3}\right)$

10. $\cos\left(\dfrac{7\pi}{6}+\dfrac{5\pi}{3}\right)$

11. $\sin\left(\dfrac{7\pi}{4}+\dfrac{5\pi}{4}\right)$

12. $\cos\left(\dfrac{\pi}{3}+\dfrac{11\pi}{6}\right)$

13. $\cos\left(\dfrac{\pi}{4}-\dfrac{\pi}{6}\right)$

14. $\sin\left(\dfrac{5\pi}{4}-\dfrac{\pi}{3}\right)$

15. $\sin\left(\dfrac{7\pi}{4}+\dfrac{2\pi}{3}\right)$

16. $\cos\left(\dfrac{\pi}{3}-\dfrac{\pi}{4}\right)$

17. $\tan 75°$

18. $\tan 15°$

19. $\sin 165°$

20. $\tan 150°$

21. $\cos(-15°)$

22. $\sin(-30°)$

23. $\tan 255°$

24. $\cos 135°$

25. $\cos 195°$

26. $\sin 270°$

27. $\cos 165°$

28. $\sin 315°$

29. $\sin\dfrac{\pi}{12}$

30. $\tan\dfrac{5\pi}{12}$

31. $\sin\dfrac{-5\pi}{6}$

32. $\cos\dfrac{7\pi}{12}$

33. $\cos\dfrac{25\pi}{12}$

34. $\sin\dfrac{13\pi}{12}$

35. $\sin\dfrac{11\pi}{12}$

36. $\tan\dfrac{7\pi}{12}$

37. $\cos\dfrac{-7\pi}{6}$

38. $\cos\dfrac{-\pi}{3}$

39. $\sin\dfrac{5\pi}{12}$

40. $\tan\dfrac{\pi}{12}$

Find the sum or difference for each given question.

41. $\sin\alpha=\dfrac{4}{5}$ and $\sin\beta=\dfrac{5}{13}$. Both α and β are in quadrant I. Find $\cos(\alpha-\beta)$.

42. $\sin\alpha=-\dfrac{15}{17}$ and $\cos\beta=-\dfrac{3}{5}$. Both α and β are in quadrant III. Find $\sin(\alpha-\beta)$.

34. $\dfrac{-\sqrt{6}+\sqrt{2}}{4}$

35. $\dfrac{\sqrt{6}-\sqrt{2}}{4}$

36. $-\sqrt{3}-2$ **37.** $\dfrac{-\sqrt{3}}{2}$

38. $\dfrac{1}{2}$ **39.** $\dfrac{\sqrt{6}+\sqrt{2}}{4}$

40. $2-\sqrt{3}$

41. $\dfrac{56}{65}$ **42.** $\dfrac{13}{85}$

43. $\dfrac{36}{85}$ **44.** $\dfrac{-253}{325}$

45. $\dfrac{4\sqrt{6}-\sqrt{21}}{25}$

46. $\dfrac{\sqrt{2}-\sqrt{5}}{2}$

47. $\sin 45° = \dfrac{\sqrt{2}}{2}$

48. $\cos\dfrac{-\pi}{4} = \dfrac{\sqrt{2}}{2}$

49. $\tan 135° = -1$

50. $\sin 90° = 1$

51. $\tan\dfrac{\pi}{4} = 1$

52. $\cos 30° = \dfrac{\sqrt{3}}{2}$

53. $\sin 150° = \dfrac{1}{2}$

54. $\cos\dfrac{\pi}{2} = 0$

55. $\cos 135° = \dfrac{-\sqrt{2}}{2}$

56. $\tan\dfrac{7\pi}{6} = \dfrac{\sqrt{3}}{3}$

57. $\tan 60° = \sqrt{3}$

58. $\sin\dfrac{\pi}{3} = \dfrac{\sqrt{3}}{2}$

59.–68. Answers will vary.

43. $\cos\alpha = -\dfrac{15}{17}$ and $\cos\beta = -\dfrac{3}{5}$. α is in quadrant II and β is in quadrant III. Find $\sin(\alpha+\beta)$.

44. $\cos\alpha = -\dfrac{24}{25}$ and $\sin\beta = \dfrac{5}{13}$. α is in quadrant III and β is in quadrant I. Find $\cos(\alpha+\beta)$.

Use the sum and difference identities to rewrite each of the following expressions as a trigonometric function of a single number, and then evaluate the result. See Example 4.

47. $\sin 15° \cos 30° + \cos 15° \sin 30°$

49. $\dfrac{\tan 100° + \tan 35°}{1-\tan 100° \tan 35°}$

51. $\dfrac{\tan\dfrac{5\pi}{16}-\tan\dfrac{\pi}{16}}{1+\tan\dfrac{5\pi}{16}\tan\dfrac{\pi}{16}}$

53. $\sin 70° \cos 80° + \cos 70° \sin 80°$

55. $\cos 182° \cos 47° + \sin 182° \sin 47°$

57. $\dfrac{\tan 70° - \tan 10°}{1+\tan 70° \tan 10°}$

45. $\cos\alpha = \dfrac{2}{5}$ and $\cos\beta = \dfrac{1}{5}$. Both α and β are in quadrant I. Find $\sin(\beta-\alpha)$.

46. $\cos\alpha = -\dfrac{2}{3}$ and $\sin\beta = -\dfrac{2\sqrt{2}}{3}$. α is in quadrant III and β is in quadrant IV. Find $\tan(\alpha+\beta)$.

48. $\cos\dfrac{5\pi}{12}\cos\dfrac{2\pi}{3} + \sin\dfrac{5\pi}{12}\sin\dfrac{2\pi}{3}$

50. $\sin 125° \cos 35° - \cos 125° \sin 35°$

52. $\cos 15° \cos 15° - \sin 15° \sin 15°$

54. $\cos\dfrac{\pi}{5}\cos\dfrac{3\pi}{10} - \sin\dfrac{\pi}{5}\sin\dfrac{3\pi}{10}$

56. $\dfrac{\tan\dfrac{5\pi}{12}+\tan\dfrac{3\pi}{4}}{1-\tan\dfrac{5\pi}{12}\tan\dfrac{3\pi}{4}}$

58. $\sin\dfrac{5\pi}{12}\cos\dfrac{\pi}{12} - \cos\dfrac{5\pi}{12}\sin\dfrac{\pi}{12}$

Use the sum and difference identities to verify the following identities. See Example 5.

59. $\tan\left(\dfrac{\pi}{2}-\theta\right) = \cot\theta$

   (**Hint:** Use sin and cos.)

61. $\cos\left(\dfrac{3\pi}{2}-\alpha\right) = -\sin\alpha$

60. $\cos^2 u - \sin^2 v = \cos(u+v)\cos(u-v)$

62. $\sin(\beta-\theta) + \sin(\beta+\theta) = 2\sin\beta\cos\theta$

69. True **70.** True

71. False **72.** True

73. True **74.** False

75. 1

76. $\dfrac{4x^2 - \sqrt{1-4x^2}}{\sqrt{1+4x^2}}$

77. $\dfrac{2x^2 + \sqrt{1-x^2}}{\sqrt{1+4x^2}}$

78. $2x\sqrt{1-x^2}$

79. $x\left(\sqrt{1-4x^2}\right.$

$\left. -2\sqrt{1-x^2}\right)$

80. $2x^2 - 1$

81. $\sqrt{2}\sin\left(x + \dfrac{\pi}{4}\right)$

82. $2\sin\left(x + \dfrac{\pi}{3}\right)$

83. $\sqrt{2}\sin\left(2\beta - \dfrac{\pi}{4}\right)$

84. $2\sin\left(\theta + \dfrac{5\pi}{6}\right)$

85. $13\sin(5u + 1.1760)$

86. $10\sin\left(\dfrac{v}{2} + 0.9273\right)$

63. $\tan\left(\alpha - \dfrac{5\pi}{4}\right) = \dfrac{\tan\alpha - 1}{1 + \tan\alpha}$

64. $\sec\left(\dfrac{\pi}{2} - u\right) = \csc u$

65. $\tan(\pi + 2\pi) = 0$

66. $\sin\left(\dfrac{5\pi}{6} + \theta\right) = \dfrac{1}{2}\left(\cos\theta - \sqrt{3}\sin\theta\right)$

67. $\sin(u+v)\sin(u-v) = \sin^2 u - \sin^2 v$

68. $\cos\left(\dfrac{7\pi}{4} - \beta\right) = \dfrac{\sqrt{2}}{2}\left(\cos\beta - \sin\beta\right)$

Using a graphing calculator, determine whether the following identities are true or false. (**Hint:** Graph both expressions on each side of the equality separately and determine if the graphs coincide.)

69. $\sin\left(\dfrac{\pi}{2} - \theta\right) = \cos\theta$

70. $\cos\left(\theta - \dfrac{3\pi}{2}\right) = -\sin\theta$

71. $\cot(\pi + \theta) = -\tan\theta$

72. $\sec\left(\dfrac{\pi}{2} - \theta\right) = \csc\theta$

73. $\sin\left(\dfrac{\pi}{6} + \theta\right) = \dfrac{1}{2}\left(\cos\theta + \sqrt{3}\sin\theta\right)$

74. $\dfrac{1 + \tan\theta}{1 - \tan\theta} = -\tan\theta$

Express each of the following as an algebraic function of x. See Example 6.

75. $\sin(\sin^{-1}2x + \cos^{-1}2x)$

76. $\sin(\arctan 2x - \arccos 2x)$

77. $\cos(\arctan 2x - \arcsin x)$

78. $\cos(\cos^{-1}x - \sin^{-1}x)$

79. $\cos(\arccos x + \arcsin 2x)$

80. $\sin(\sin^{-1}x - \cos^{-1}x)$

Express each of the following functions in terms of a single sine function, and graph the result. See Example 8.

81. $f(x) = \sin x + \cos x$

82. $g(x) = \sin x + \sqrt{3}\cos x$

83. $h(\beta) = \sin 2\beta - \cos 2\beta$

84. $f(\theta) = -\sqrt{3}\sin\theta + \cos\theta$

85. $g(u) = 5\sin 5u + 12\cos 5u$

86. $h(v) = 8\cos\dfrac{v}{2} + 6\sin\dfrac{v}{2}$

87. Use a cofunction identity to prove the sum and difference identities for sine.

(**Hint:** Note that $\sin(u+v) = \cos\left(\dfrac{\pi}{2} - (u+v)\right) = \cos\left(\left(\dfrac{\pi}{2} - u\right) - v\right)$ and apply the difference identity for cosine.)

87. Answers will vary.

88. Answers will vary.

89. True

90. False

91. False

92. True

93. Answers will vary.

88. Given sum identities for sine and cosine, prove the sum identity for tangent.

89. Prove or disprove that $\sin(u + v) + \sin(u - v) = 2 \sin u \cos v$.

90. Prove or disprove that $\dfrac{\cos(u+v)}{\cos u \cos v} = \tan u + \tan v$.

91. Prove or disprove that $\dfrac{\cos(u-v)}{\cos(u+v)} = 2\tan u \tan v$.

92. Prove or disprove that $\dfrac{\sin(u+v)}{\sin(u-v)} = \dfrac{\tan u + \tan v}{\tan u - \tan v}$.

93. Use the sine and cosine difference formulas to prove $\tan(u-v) = \dfrac{\tan u - \tan v}{1 + \tan u \tan v}$.

Product-Sum Identities

TOPICS

1. Double-angle identities
2. Power-reducing identities
3. Half-angle identities
4. Product-to-sum and sum-to-product identities

TOPIC

Double-Angle Identities

We will enlarge our repertoire greatly in this section, introducing and using four useful classes of identities. We will also prove a few selected identities in order to demonstrate typical methods of proof.

Our first class of identities contains those in which the argument of a given trigonometric function is twice an angle, and the identities are thus commonly called *double-angle identities*.

IDENTITY

Double-Angle
Identities

Sine Identity: $\sin 2u = 2\sin u \cos u$

Cosine Identities: $\cos 2u = \cos^2 u - \sin^2 u$

$= 2\cos^2 u - 1$

$= 1 - 2\sin^2 u$

Tangent Identity: $\tan 2u = \dfrac{2\tan u}{1 - \tan^2 u}$

These identities are easily verified using the sum identities of the previous section. We will prove two of the cosine identities here, and leave the remainder of the proofs as exercises. For the first, note that

$$\cos 2u = \cos(u + u)$$

$$= \cos u \cos u - \sin u \sin u$$

$$= \cos^2 u - \sin^2 u.$$

And for the second cosine identity, we can apply a Pythagorean Identity:

$$\cos 2u = \cos^2 u - \sin^2 u$$
$$= \cos^2 u - \left(1 - \cos^2 u\right)$$
$$= 2\cos^2 u - 1.$$

EXAMPLE 1

Using the Double-Angle Identities

Given that $\cos x = -\dfrac{2}{\sqrt{5}}$ and that $\sin x$ is positive, determine $\cos 2x$, $\sin 2x$, and $\tan 2x$.

Solution:

First, note that since $\cos x$ is negative and $\sin x$ is positive, we know that x lies in quadrant II. The Pythagorean Theorem (or the first Pythagorean Identity) then tells us that $\sin x = \dfrac{1}{\sqrt{5}}$, as follows:

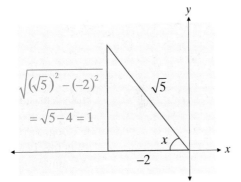

$$\sin x = \sqrt{1 - \left(-\frac{2}{\sqrt{5}}\right)^2} = \sqrt{1 - \frac{4}{5}} = \frac{1}{\sqrt{5}}.$$

We can now easily apply two of the double-angle identities:

$$\cos 2x = \cos^2 x - \sin^2 x = \left(-\frac{2}{\sqrt{5}}\right)^2 - \left(\frac{1}{\sqrt{5}}\right)^2 = \frac{4}{5} - \frac{1}{5} = \frac{3}{5}$$

and

$$\sin 2x = 2\sin x \cos x = 2\left(\frac{1}{\sqrt{5}}\right)\left(-\frac{2}{\sqrt{5}}\right) = -\frac{4}{5}.$$

We can now most easily determine $\tan 2x$ by noting that $\tan 2x = \dfrac{\sin 2x}{\cos 2x} = -\dfrac{4}{3}$, but we could also have determined that $\tan x = -\dfrac{1}{2}$ and used the tangent double-angle identity as follows:

$$\tan 2x = \frac{2\tan x}{1 - \tan^2 x} = \frac{2\left(-\frac{1}{2}\right)}{1 - \left(-\frac{1}{2}\right)^2} = \frac{-1}{\frac{3}{4}} = -\frac{4}{3}.$$

Of course, the double-angle identities (in conjunction with others) can be used to prove new identities, as shown in the next example.

EXAMPLE 2

Prove that $\sin 3x = 3\sin x - 4\sin^3 x.$

Solution:

$$\sin 3x = \sin(2x + x) \qquad\qquad \text{Rewrite the argument as a sum.}$$

$$= \sin 2x \cos x + \cos 2x \sin x \qquad \text{Apply the sum identity for sine.}$$

$$= 2\sin x \cos x \cos x + (1 - 2\sin^2 x)\sin x \qquad \text{Apply two of the double-angle identities.}$$

$$= 2\sin x \cos^2 x + \sin x - 2\sin^3 x$$

$$= 2\sin x(1 - \sin^2 x) + \sin x - 2\sin^3 x \qquad \text{Apply the first Pythagorean Identity.}$$

$$= 2\sin x - 2\sin^3 x + \sin x - 2\sin^3 x$$

$$= 3\sin x - 4\sin^3 x$$

Not surprisingly, the same sort of argument can be used on any expression of the form $\sin nx$ or $\cos nx$, where n is a positive integer. In fact, $\cos nx$ can be rewritten as an nth-degree polynomial in $\cos x$ (the polynomials are called *Chebyshev polynomials,* after the 19th-century Russian Pafnuty Lvovich Chebyshev, a mathematician famous for, among many other things, the variety of spellings of his last name). A similar statement is almost true for $\sin nx$; see the exercises for more on this topic.

TOPIC 2   Power-Reducing Identities

The double-angle identities just introduced are useful when $\sin x, \cos x,$ or $\tan x$ is known and an expression containing $\sin nx$, $\cos nx$, or $\tan nx$ (for some positive integer n) must be dealt with. But they are also useful when faced with an expression containing $\sin^2 x$, $\cos^2 x$, or $\tan^2 x$. Quite often in calculus, for instance, it's very helpful to be able to lower the power of a trigonometric expression—that is, to be able to replace, say, $\cos^2 x$, with an expression containing only cosine to the first power. The following three identities follow quickly from the double-angle identities and accomplish the power-reducing task.

IDENTITY

Sine Identity:	$\sin^2 x = \dfrac{1 - \cos 2x}{2}$
Cosine Identity:	$\cos^2 x = \dfrac{1 + \cos 2x}{2}$
Tangent Identity:	$\tan^2 x = \dfrac{1 - \cos 2x}{1 + \cos 2x}$

The first identity is easily proved by beginning with one of the double-angle identities for cosine:

$$\cos 2x = 1 - 2\sin^2 x$$

$$2\sin^2 x = 1 - \cos 2x$$

$$\sin^2 x = \frac{1 - \cos 2x}{2}$$

We begin with the one double-angle identity that contains $\sin^2 x$ and no other squared terms, and then proceed to solve for $\sin^2 x$.

Of course, the power-reducing identities can be used to reduce powers other than 2, as shown in the next example.

EXAMPLE 3

Using the Power-Reducing Identities

Express $\sin^5 x$ in terms containing only first powers of sine and cosine.

Solution:

The procedure is to break the expression down into pieces to which the power-reducing identities apply, repeating the process if necessary:

$$\sin^5 x = \sin^2 x \; \sin^2 x \; \sin x$$

$$= \left(\frac{1 - \cos 2x}{2} \right)\left(\frac{1 - \cos 2x}{2} \right)\sin x$$

$$= \left(\frac{1 - 2\cos 2x + \cos^2 2x}{4} \right)\sin x$$

$$= \left(\frac{1}{4} \right)\left(\sin x - 2\cos 2x \; \sin x + \cos^2 2x \; \sin x \right)$$

$$= \left(\frac{1}{4} \right)\left(\sin x - 2\cos 2x \; \sin x + \left(\frac{1 + \cos 4x}{2} \right)\sin x \right)$$

$$= \left(\frac{1}{4} \right)\left(\sin x - 2\cos 2x \; \sin x + \frac{\sin x}{2} + \frac{\cos 4x \; \sin x}{2} \right)$$

$$= \left(\frac{1}{4} \right)\left(\frac{3\sin x}{2} - 2\cos 2x \; \sin x + \frac{\cos 4x \; \sin x}{2} \right).$$

TOPIC 3   Half-Angle Identities

We have reached the point where we have enough identities to accomplish most of the common trigonometric tasks that may present themselves. For instance, if we needed the exact evaluation of $\sin\frac{\pi}{8}$, $\cos\frac{\pi}{8}$, or $\tan\frac{\pi}{8}$, we could start with an appropriate power-reducing identity and make use of the fact that twice $\frac{\pi}{8}$ is $\frac{\pi}{4}$, one of the "nice" angles. The box below contains general identities relating sine, cosine, and tangent of half an angle to expressions involving the whole angle. After proving one of the identities, we will proceed to determine the exact values of $\sin\frac{\pi}{8}$, $\cos\frac{\pi}{8}$, and $\tan\frac{\pi}{8}$.

IDENTITY

Half-Angle Identities

Sine Identity: $\sin\dfrac{x}{2} = \pm\sqrt{\dfrac{1-\cos x}{2}}$

Cosine Identity: $\cos\dfrac{x}{2} = \pm\sqrt{\dfrac{1+\cos x}{2}}$

Tangent Identities: $\tan\dfrac{x}{2} = \dfrac{1-\cos x}{\sin x} = \dfrac{\sin x}{1+\cos x}$

The proofs of the first two half-angle identities follow from the corresponding power-reducing identities, with the sign of the result being determined by the quadrant in which $\dfrac{x}{2}$ lies (see the exercises for guidance in the proofs). For the tangent half-angle identity, we begin by replacing x with $\dfrac{x}{2}$ in the power-reducing identity and proceed as follows:

$$\tan^2\frac{x}{2} = \frac{1-\cos 2\left(\dfrac{x}{2}\right)}{1+\cos 2\left(\dfrac{x}{2}\right)}$$

Apply the tangent power-reducing identity with an argument of $\dfrac{x}{2}$.

$$\tan\frac{x}{2} = \pm\sqrt{\frac{1-\cos x}{1+\cos x}}$$

Simplify the right-hand side and take the square root of both sides.

$$\tan\frac{x}{2} = \pm\sqrt{\left(\frac{1-\cos x}{1+\cos x}\right)\left(\frac{1-\cos x}{1-\cos x}\right)}$$

Multiply the numerator and denominator by $1 - \cos x$.

$$\tan\frac{x}{2} = \pm\sqrt{\frac{(1-\cos x)^2}{1-\cos^2 x}}$$

Expand the denominator of the right-hand side and apply the Pythagorean Identity.

$$\tan\frac{x}{2} = \pm\frac{|1-\cos x|}{|\sin x|}$$

Simplify the radical.

All that remains is to simplify the formula a bit. First, note that $1 - \cos x$ is always nonnegative, so the absolute value in the numerator is unnecessary. The next step is to note that $\tan \dfrac{x}{2}$ and $\sin x$ have the same sign for any value of x; the verification of this is left to the reader (consider individually the four cases of $0 \le \dfrac{x}{2} < \dfrac{\pi}{2}$, $\dfrac{\pi}{2} < \dfrac{x}{2} \le \pi$, $-\dfrac{\pi}{2} < \dfrac{x}{2} \le 0$ and $-\pi \le \dfrac{x}{2} < -\dfrac{\pi}{2}$). This means that if the absolute value in the denominator is removed, then the $\pm$ in the formula can also be discarded, since its only purpose is to remind us to choose the correct sign depending on the quadrant in which $\dfrac{x}{2}$ lies. We are thus led to the desired result; proving the second formula in the tangent half-angle identity is left as an exercise.

EXAMPLE 4

Using the Half-Angle Identities

Determine the exact values of $\sin \dfrac{\pi}{8}$, $\cos \dfrac{\pi}{8}$, and $\tan \dfrac{\pi}{8}$.

Solution:

Since $\cos \dfrac{\pi}{4} = \sin \dfrac{\pi}{4} = \dfrac{\sqrt{2}}{2}$, we can easily determine that

$$\sin \frac{\pi}{8} = \sqrt{\frac{1 - \cos \frac{\pi}{4}}{2}} = \sqrt{\frac{1 - \frac{\sqrt{2}}{2}}{2}} = \sqrt{\frac{2 - \sqrt{2}}{4}} = \frac{1}{2}\sqrt{2 - \sqrt{2}},$$

$$\cos \frac{\pi}{8} = \sqrt{\frac{1 + \cos \frac{\pi}{4}}{2}} = \sqrt{\frac{1 + \frac{\sqrt{2}}{2}}{2}} = \sqrt{\frac{2 + \sqrt{2}}{4}} = \frac{1}{2}\sqrt{2 + \sqrt{2}},$$

and

$$\tan \frac{\pi}{8} = \frac{1 - \cos \frac{\pi}{4}}{\sin \frac{\pi}{4}} = \frac{1 - \frac{\sqrt{2}}{2}}{\frac{\sqrt{2}}{2}} = \frac{2\sqrt{2} - 2}{2} = \sqrt{2} - 1.$$

Of course, we could also have used the fact that $\tan x = \dfrac{\sin x}{\cos x}$ for the last calculation.

Verifying that the two seemingly different answers obtained are actually equivalent is left to the reader.

EXAMPLE 5

Using the Half-Angle
Identities for Exact
Evaluation

Determine the exact value of $\cos 105°$.

Solution:

Once we note that $105°$ is half of $210°$, and recall that the reference angle of $210°$ is $30°$ the calculation is straightforward. We must also be careful, however, to insert a minus sign before the radical, as $105°$ lies in the second quadrant and $\cos 105°$ is therefore negative.

$$\cos 105° = -\sqrt{\frac{1+\cos 210°}{2}} = -\sqrt{\frac{1-\frac{\sqrt{3}}{2}}{2}} = -\frac{1}{2}\sqrt{2-\sqrt{3}}$$

TOPIC 4

Product-to-Sum and Sum-to-Product Identities

The final set of identities in this section is easily verified using the sum and difference identities, and the proofs are left to the reader. The first box below contains identities that allow us to rewrite certain products of trigonometric functions as sums or differences of functions.

IDENTITY

Product-to-Sum
Identities

$$\sin x \cos y = \frac{1}{2}\left[\sin(x+y)+\sin(x-y)\right] \qquad \cos x \sin y = \frac{1}{2}\left[\sin(x+y)-\sin(x-y)\right]$$

$$\sin x \sin y = \frac{1}{2}\left[\cos(x-y)-\cos(x+y)\right] \qquad \cos x \cos y = \frac{1}{2}\left[\cos(x+y)+\cos(x-y)\right]$$

EXAMPLE 6

Using the Product-to-Sum Identities

Express $\sin^5 x$ in terms containing only first powers of sine.

Solution:

In Example 3, we used a power-reducing identity to determine that

$$\sin^5 x = \left(\frac{1}{4}\right)\left(\frac{3\sin x}{2} - 2\cos 2x \sin x + \frac{\cos 4x \sin x}{2}\right).$$

We can now use one of the product-to-sum identities to rewrite two of the terms in this expression, as follows:

$$\sin^5 x = \left(\frac{1}{4}\right)\left[\frac{3\sin x}{2} - 2\cos 2x \sin x + \frac{\cos 4x \sin x}{2}\right]$$

$$= \left(\frac{1}{4}\right)\left[\frac{3\sin x}{2} - 2\left(\frac{1}{2}(\sin 3x - \sin x)\right) + \frac{1}{2}\left(\frac{1}{2}(\sin 5x - \sin 3x)\right)\right]$$

$$= \frac{3}{8}\sin x - \frac{1}{4}\sin 3x + \frac{1}{4}\sin x + \frac{1}{16}\sin 5x - \frac{1}{16}\sin 3x$$

$$= \frac{5}{8}\sin x - \frac{5}{16}\sin 3x + \frac{1}{16}\sin 5x.$$

Consider, now, the first product-to-sum identity, with u and v in place of x and y:

$$\sin u \cos v = \frac{1}{2}\left[\sin(u+v) + \sin(u-v)\right].$$

If we replace u with $\frac{x+y}{2}$ and v with $\frac{x-y}{2}$, we obtain:

$$\sin\left(\frac{x+y}{2}\right)\cos\left(\frac{x-y}{2}\right) = \frac{1}{2}(\sin x + \sin y).$$

Multiplying this last equation through by 2 gives us an identity that allows us to rewrite a sum of sine functions as a product of a sine and cosine function, and it is the first of the four sum-to-product identities in the box on the following page.

IDENTITY

Sum-to-Product
Identities

$$\sin x + \sin y = 2\sin\left(\frac{x+y}{2}\right)\cos\left(\frac{x-y}{2}\right) \qquad \sin x - \sin y = 2\cos\left(\frac{x+y}{2}\right)\sin\left(\frac{x-y}{2}\right)$$

$$\cos x + \cos y = 2\cos\left(\frac{x+y}{2}\right)\cos\left(\frac{x-y}{2}\right) \qquad \cos x - \cos y = -2\sin\left(\frac{x+y}{2}\right)\sin\left(\frac{x-y}{2}\right)$$

EXAMPLE 7

Using the Sum-to-
Product Identities

Verify the identity $\dfrac{\sin 2x + \sin 4x}{\cos 2x + \cos 4x} = \tan 3x$

Solution:

We can apply two of the sum-to-product identities to easily verify the statement:

$$\frac{\sin 2x + \sin 4x}{\cos 2x + \cos 4x} = \frac{2\sin\left(\dfrac{6x}{2}\right)\cos\left(\dfrac{-2x}{2}\right)}{2\cos\left(\dfrac{6x}{2}\right)\cos\left(\dfrac{-2x}{2}\right)}$$

$$= \frac{\sin 3x}{\cos 3x}$$

$$= \tan 3x$$

---

1. $\cos 2x = \dfrac{7}{25}$;

$\sin 2x = \dfrac{24}{25}$;

$\tan 2x = \dfrac{24}{7}$

2. $\cos 2x = -\dfrac{15}{17}$;

$\sin 2x = -\dfrac{8}{17}$;

$\tan 2x = \dfrac{8}{15}$

Exercises

Use the information given in each problem to determine $\cos 2x$, $\sin 2x$, and $\tan 2x$, if possible. See Example 1.

1. $\sin x = \dfrac{3}{5}$ and $\cos x$ is positive

2. $\tan x = -4$ and $\sin x$ is negative

3. $\cos x = -\dfrac{2}{\sqrt{6}}$ and $\sin x$ is positive

4. $\sin x = \dfrac{1}{\sqrt{5}}$ and $\tan x$ is positive

5. $\tan x = \dfrac{1}{\sqrt{3}}$ and $\cos x$ is negative

6. $\cos x = -3$ and $\tan x$ is negative

3. $\cos 2x = \dfrac{1}{3}$; $\sin 2x = \dfrac{-2\sqrt{2}}{3}$; $\tan 2x = -2\sqrt{2}$ **4.** $\cos 2x = \dfrac{3}{5}$; $\sin 2x = \dfrac{4}{5}$; $\tan 2x = \dfrac{4}{3}$

5. $\cos 2x = \dfrac{1}{2}$; $\sin 2x = \dfrac{\sqrt{3}}{2}$; $\tan 2x = \sqrt{3}$ **6.** Not possible

7.–12. Answers will vary.

13. $\dfrac{\sin x - \sin x \cos 2x}{2}$

14. $\dfrac{3 - 4\cos 2x + \cos 4x}{8}$

15. $\dfrac{1 - \cos 2x - \cos 4x}{16}$
$+ \dfrac{\cos 2x \cos 4x}{16}$

16. $\dfrac{\cos x - \cos x \cos 4x}{8}$

17. $\dfrac{3\sin x - 4\sin x \cos 2x}{3 + 4\cos 2x + \cos 4x}$
$+ \dfrac{\sin x \cos 4x}{3 + 4\cos 2x + \cos 4x}$

18. $\dfrac{35 - 48\cos 2x}{128}$
$+ \dfrac{28\cos 4x + \cos 8x}{128}$
$- \dfrac{16\cos 2x \cos 4x}{128}$

19. $\dfrac{\sqrt{2 + \sqrt{2}}}{2}$

20. $-\sqrt{2} - 1$

21. $\dfrac{\sqrt{2 + \sqrt{3}}}{2}$

22. $-\left(\sqrt{3} + 2\right)$

23. $\dfrac{\sqrt{2 + \sqrt{3}}}{2}$

24. $\dfrac{-\sqrt{2 + \sqrt{3}}}{2}$

25. $\dfrac{\sin 6x + \sin 0}{2}$

26. $\dfrac{\sin \dfrac{\pi}{2} - \sin 0}{2}$

27. $\dfrac{5(\sin 120° - \sin 90°)}{2}$

28. $\cos 30° + \cos 120°$

29. $\dfrac{\cos 2y - \cos 2x}{2}$

30. $\dfrac{\sin \pi + \sin \dfrac{2\pi}{3}}{2}$

31. $\dfrac{\cos \dfrac{7\pi}{12} - \cos \dfrac{23\pi}{12}}{2}$

32. $\dfrac{\cos 2\beta + \cos 4\beta}{2}$

33. $\sin \dfrac{\pi}{2} - \sin \dfrac{\pi}{6}$

Verify the following trigonometric identities. See Example 2.

7. $\tan 3x = \dfrac{3\tan x - \tan^3 x}{1 - 3\tan^2 x}$

8. $\sin 2x = \dfrac{2\tan x}{1 + \tan^2 x}$

9. $\dfrac{\sin 4x - \sin 2x}{\cos 4x + \cos 2x} = \tan x$

10. $\dfrac{\sin 3x}{\sin x} = 3 - 4\sin^2 x$

11. $2\sin^2 3x = 1 - \cos 6x$

12. $\sin 3x = 3\sin x \cos^2 x - \sin^3 x$

Use a power-reducing identity to rewrite the expressions as directed. See Example 3.

13. Rewrite $\sin^3 x$ in terms containing only first powers of sine and cosine.

14. Rewrite $\sin^4 x$ in terms containing only first powers of cosine.

15. Rewrite $\sin^4 x \cos^2 x$ in terms containing only first powers of cosine.

16. Rewrite $\cos^3 x \sin^2 x$ in terms containing only first powers of cosine.

17. Rewrite $\tan^4 x \sin x$ in terms containing only first powers of sine and cosine.

18. Rewrite $\sin^8 x$ in terms containing only first powers of cosine.

Determine the exact values of each of the following expressions. See Examples 4 and 5.

19. $\sin \dfrac{3\pi}{8}$

20. $\tan 112°30'$

21. $\cos\left(-\dfrac{\pi}{12}\right)$

22. $\tan \dfrac{7\pi}{12}$

23. $\sin 75°$

24. $\cos 165°$

Use the product-to-sum identities to rewrite the expressions as a sum or difference. See Example 6.

25. $\sin 3x \cos 3x$

26. $\cos \dfrac{\pi}{4} \sin \dfrac{\pi}{4}$

27. $5 \cos 105° \sin 15°$

28. $2 \cos 75° \cos 45°$

29. $\sin(x + y) \sin(x - y)$

30. $\sin \dfrac{5\pi}{6} \cos \dfrac{\pi}{6}$

31. $\sin \dfrac{5\pi}{4} \sin \dfrac{2\pi}{3}$

32. $\cos \beta \cos 3\beta$

33. $2\cos \dfrac{\pi}{3} \sin \dfrac{\pi}{6}$

34. $2\sin\dfrac{\pi}{2}\cos\dfrac{\pi}{4}$

35. $2\sin 4x\cos 2x$

36. $2\cos 45°\cos 15°$

37. $-2\sin 2\beta\sin\beta$

38. $2\cos\dfrac{3\pi}{4}\sin\dfrac{\pi}{4}$

39. $2\cos 75°\sin 60°$

40. $-2\sin 4x\sin 2x$

41. $-2\sin\dfrac{17\pi}{24}\sin\dfrac{11\pi}{24}$

42. $2\sin\pi\cos\theta$

43.–46. Answers will vary.

Use the sum-to-product identities to rewrite the expressions as a product. See Example 7.

34. $\sin\dfrac{\pi}{4}+\sin\dfrac{3\pi}{4}$ **35.** $\sin 6x+\sin 2x$ **36.** $\cos 60°+\cos 30°$

37. $\cos 3\beta-\cos\beta$ **38.** $\sin\pi-\sin\dfrac{\pi}{2}$ **39.** $\sin 135°-\sin 15°$

40. $\cos 6x-\cos 2x$ **41.** $\cos\dfrac{7\pi}{6}-\cos\dfrac{\pi}{4}$ **42.** $\sin(\pi+\theta)+\sin(\pi-\theta)$

43. Two of the double-angle identities were proved in this section. Prove the remaining three double-angle identities.

44. The power-reducing identity for sine was proved in this section. Prove the remaining two power-reducing identities.

45. Prove the half-angle identities for sine and cosine by replacing x with $\dfrac{x}{2}$ in an appropriately chosen identity.

46. One of the formulas in the tangent half-angle identity was proved in this section. Prove the second formula in the tangent half-angle identity.

47. As mentioned in this section, $\cos nx$ can be expressed as a polynomial of degree n in $\cos x$; such polynomials are called Chebyshev polynomials. For $\sin nx$, the equivalent rewriting is a product of $\sin x$ and a polynomial of degree $n-1$ in $\cos x$. Expand $\sin nx$ and $\cos nx$ for $n=2,3$, and 4 and compare the results.

47. $\cos 2x=2\cos^2 x-1$

$\cos 3x=4\cos^3 x-3\cos x$

$\cos 4x=8\cos^4 x-8\cos^2 x+1$

$\sin 2x=2\sin x\cos x$

$\sin 3x=\sin x\left(4\cos^2 x-1\right)$

$\sin 4x=\sin x\left(8\cos^3 x-4\cos x\right)$

7.4

Trigonometric Equations

TOPICS

1. Applying algebraic techniques to trigonometric equations
2. Using inverse trigonometric functions

TOPIC

Applying Algebraic Techniques to Trigonometric Equations

The majority of the trigonometric equations we have seen thus far in this chapter have been identities, and the goal has been to either verify the identity or use it in simplifying a trigonometric expression. In this section, we will study equations that are conditional—that is, equations which are not true for all allowable values of the variable. The goal, of course, is to identify exactly those values of the variable that *do* make a given equation true.

The identities we have learned in the first three sections of this chapter will prove to be very useful again, especially when combined with standard algebraic techniques covered in previous chapters. It is important to remember, however, that the solution to an equation consists of *all* values that make the equation true. Since trigonometric functions are periodic, it is not at all unusual for a trigonometric equation to have an infinite number of solutions, and the goal is to find and describe all of them.

One common approach in solving trigonometric equations, especially if only one trigonometric function is involved, is to isolate the trigonometric expression on one side of the equation. Our knowledge of the behavior of trigonometric functions will then often allow us to describe the complete solution set.

EXAMPLE 1

Solving Equations by Isolating the Trigonometric Function

Solve the equation $3 - 6\cos x = 0$.

Solution:

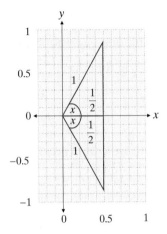

Isolating $\cos x$ is easily accomplished:

$$3 - 6\cos x = 0$$
$$3 = 6\cos x$$
$$\cos x = \frac{1}{2}.$$

This tells us several things. First, the fact that $\cos x$ is positive means that x lies in the first or fourth quadrant. This means that $0 \le x \le \dfrac{\pi}{2}$ or $-\dfrac{\pi}{2} \le x \le 0$ (we could also use the range $\dfrac{3\pi}{2} \le x \le 2\pi$ to describe fourth-quadrant angles). We can then sketch a triangle or two, if necessary, to remind ourselves that $\cos\dfrac{\pi}{3} = \dfrac{1}{2}$ and $\cos\left(-\dfrac{\pi}{3}\right) = \dfrac{1}{2}$.

Then, since cosine is 2π-periodic, we can describe the complete solution set with the two equations

$$x = \frac{\pi}{3} + 2n\pi \text{ and } x = -\frac{\pi}{3} + 2n\pi,$$

where n is any integer. Note that this set could be described, less precisely, as

$$\left\{ \ldots - \frac{7\pi}{3}, \ -\frac{5\pi}{3}, \ -\frac{\pi}{3}, \frac{\pi}{3}, \frac{5\pi}{3}, \frac{7\pi}{3}, \ldots \right\}.$$

EXAMPLE 2

Solving Equations
by Isolating the
Trigonometric
Function

Solve the equation $\tan^2 x - 1 = 2$.

Solution:

The trigonometric function can again be isolated fairly easily:

$$\tan^2 x - 1 = 2$$

$$\tan^2 x = 3$$

$$\tan x = \pm\sqrt{3}.$$

As in Example 1, the task now is to identify *all* the angles which solve the last equation (which, because of the plus-or-minus symbol, is actually two equations). Since tangent has a period of π, we'll be done if we find all the solutions in an interval of length π and add to them all integer multiples of π. For instance, in the interval $\left[-\dfrac{\pi}{2}, \dfrac{\pi}{2}\right]$, $\tan x = \sqrt{3}$ when $x = \dfrac{\pi}{3}$ and $\tan x = -\sqrt{3}$ when $x = -\dfrac{\pi}{3}$. So the complete solution set is described by

$$x = \frac{\pi}{3} + n\pi \ \text{ and } \ x = -\frac{\pi}{3} + n\pi.$$

The last example included a squared trigonometric term, but the equation was easily solved by taking the square root of both sides. More complicated quadratic-like trigonometric equations can be solved by factoring, by completing the square, or by the quadratic formula.

EXAMPLE 3

Solving Trigonometric
Equations by
Factoring

Solve the equation $\sin^2 x - \sin x = \sin x + 3$.

Solution:

The equation $\sin^2 x - \sin x = \sin x + 3$ is quadratic in $\sin x$; that is, if $\sin x$ were replaced by x, it would simply be a quadratic equation. This observation provides the key to its solution: we will use our previous experience with quadratic equations to solve for $\sin x$ and then use our knowledge of trigonometry to solve for x.

$$\sin^2 x - \sin x = \sin x + 3$$

$$\sin^2 x - 2\sin x - 3 = 0$$

$$(\sin x - 3)(\sin x + 1) = 0$$

Remember that the first step in solving quadratic equations by factoring or by the quadratic formula is to collect all the terms on one side. In this problem, we can easily factor the resulting left-hand side.

As in any quadratic equation solved by factoring, the two factors now lead to two equations:

$$\sin x = 3 \qquad \text{or} \qquad \sin x = -1$$

$$\text{no real solution} \qquad\qquad x = \frac{3\pi}{2} + 2n\pi$$

The first equation has no real number solution, as 3 lies outside the range of the sine function. The second equation has an infinite number of solutions.

The examples in this section so far have involved only one trigonometric function. If an equation initially appears with two or more functions, one approach to solving it is to use an identity to obtain a new equation in only one function.

EXAMPLE 4

Solving Equations Using Trigonometric Identities

Solve the equation $-2\cos^2 x + 1 = \sin x$.

Solution:

The presence of a $\cos^2 x$ or $\sin^2 x$ term is almost always an indication that the first Pythagorean Identity may be useful:

$$-2\cos^2 x + 1 = \sin x$$

$$-2\left(1 - \sin^2 x\right) + 1 = \sin x$$

Use the first Pythagorean Identity to rewrite the equation in terms of sine alone.

$$2\sin^2 x - \sin x - 1 = 0$$

$$\left(2\sin x + 1\right)\left(\sin x - 1\right) = 0.$$

Solve the resulting quadratic in sine.

As in the last example, we have two equations to solve:

$$2\sin x + 1 = 0 \qquad\qquad \text{or} \qquad\qquad \sin x - 1 = 0$$

$$\sin x = -\frac{1}{2} \qquad\qquad\qquad\qquad \sin x = 1$$

$$x = -\frac{5\pi}{6} + 2n\pi, \ -\frac{\pi}{6} + 2n\pi \qquad\qquad x = \frac{\pi}{2} + 2n\pi$$

The solutions to the first equation are angles lying in the third and fourth quadrants, while the solution to the second equation is a right angle. Three equations are thus necessary to describe the solution set.

There is another way to consider the equation $-2\cos^2 x + 1 = \sin x$ that is very instructive, though usually not sufficient to clearly identify the solutions. The equation is satisfied at those values of x for which the graphs of the functions $f(x) = -2\cos^2 x + 1$ and $g(x) = \sin x$ coincide. Figure 1 on the next page contains the graphs of these two functions over the interval $\left[-\pi, \pi\right]$ (f is shown in purple and g is shown in green); note that the two graphs intersect at exactly the points found above.

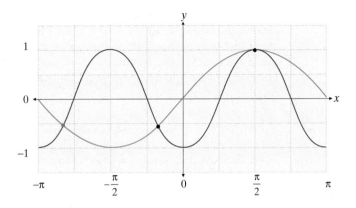

Figure 1: Graphical Representation of Solution to $-2\cos^2 x + 1 = \sin x$

EXAMPLE 5

Solving Trigonometric Equations by Graphing

Solve the equation $\sin x = \cos x$ in the interval $[0,\, 2\pi)$.

Solution:

We have used the graphs of sine and cosine to such an extent that the solutions to this equation should come as no surprise. Graphically, we are looking for those x's at which sine and cosine coincide. The graph below illustrates the two points in the interval $[0,\, 2\pi)$ where this happens:

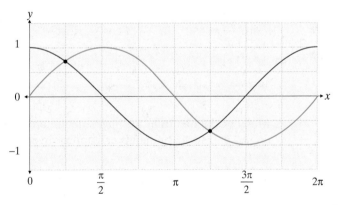

From the graph, it appears that the solution of the equation is the set $\left\{\dfrac{\pi}{4}, \dfrac{5\pi}{4}\right\}$, and this is indeed the case. Algebraically, we could also solve the equation $\sin x = \cos x$ by dividing both sides by $\cos x$ and making use of our knowledge of the tangent function. Note, though, that we can only do this after we determine that we won't lose any solutions in the process — dividing by $\cos x$ is only a legitimate step if $\cos x$ is nonzero. But when cosine *is* zero, sine is not, so we can safely discard those values of x as they can't possibly be part of the solution set. After the division, we have the equation $\tan x = 1$, and the solution set of this equation in the interval $[0,\, 2\pi)$ is $\left\{\dfrac{\pi}{4}, \dfrac{5\pi}{4}\right\}$.

In some nontrigonometric equations, such as those involving radicals or fractional exponents, raising both sides to a power is a useful technique. The same is true for some trigonometric equations. But just as in the nontrigonometric case, we must check our answers at the end to guard against extraneous solutions; remember that squaring both sides of an equation, for instance, often introduces solutions that do not solve the original problem.

EXAMPLE 6

Solving Equations Using Trigonometric Identities

Solve the equation $\sin x - 1 = \cos x$ in the interval $[0, 2\pi)$.

Solution:

$$\sin x - 1 = \cos x$$

$$(\sin x - 1)^2 = \cos^2 x$$

By squaring both sides of the equation, we can apply an identity and rewrite the equation in terms of a single trigonometric function.

$$\sin^2 x - 2\sin x + 1 - \cos^2 x = 0$$

$$\sin^2 x - 2\sin x + 1 - (1 - \sin^2 x) = 0$$

$$2\sin^2 x - 2\sin x = 0$$

$$2(\sin x)(\sin x - 1) = 0$$

The new equation is quadratic in sine, and can be solved by factoring.

$$\sin x = 0 \text{ and } \sin x = 1$$

$$x = 0,\ \pi \text{ or } x = \frac{\pi}{2}$$

$$\overset{?}{\sin 0 - 1 = \cos 0} \quad \overset{?}{\sin \frac{\pi}{2} - 1 = \cos \frac{\pi}{2}} \quad \overset{?}{\sin \pi - 1 = \cos \pi}$$

$$-1 \neq 1 \qquad\qquad 0 = 0 \qquad\qquad -1 = -1$$

At this point, there are three potential solutions. However, inserting the potential solutions into the original equation reveals that the solution set is actually $\left\{ \frac{\pi}{2},\ \pi \right\}$.

Many of the identities we have encountered contain arguments that are multiples of angles, and such arguments can appear in conditional equations as well.

EXAMPLE 7

Solving Equations by Isolating the Trigonometric Function

Solve the equation $\dfrac{\sec 3t}{2} - 1 = 0$.

Solution:

$$\frac{\sec 3t}{2} - 1 = 0$$

Begin by isolating the single trigonometric term in the equation.

$$\sec 3t = 2$$

$$3t = \frac{\pi}{3} + 2n\pi \quad \text{and} \quad 3t = -\frac{\pi}{3} + 2n\pi$$

There are two solutions of the equation in any interval of length 2π.

$$t = \frac{\pi}{9} + \frac{2n\pi}{3} \quad \text{and} \quad t = -\frac{\pi}{9} + \frac{2n\pi}{3}$$

Divide both sides (including the $2n\pi$ term) by 3.

TOPIC 2   Using Inverse Trigonometric Functions

It is, of course, no accident that the examples so far in this section have had "nice" solutions based on the commonly encountered angles of $\frac{\pi}{6}$, $\frac{\pi}{4}$, $\frac{\pi}{3}$, and $\frac{\pi}{2}$. The preceding examples have been undeniably convenient for demonstrating the sorts of techniques used in solving trigonometric equations. But real-world problems do not, necessarily, give rise to equations with such easily described solutions. It is in solving these more unwieldy problems that inverse trigonometric functions are most useful.

EXAMPLE 8

Solving Equations Using Inverse Trigonometric Functions

Solve the equation $\tan^2 x + 2\tan x = 3$ on the interval $\left(-\frac{\pi}{2}, \frac{\pi}{2}\right)$.

Solution:

$$\tan^2 x + 2\tan x = 3$$

$$\tan^2 x + 2\tan x - 3 = 0$$

$$(\tan x + 3)(\tan x - 1) = 0$$

$$\tan x = -3 \text{ or } \tan x = 1$$

$$x = \tan^{-1}(-3) \text{ and } x = \frac{\pi}{4}$$

The equation is quadratic in tangent, and can be solved by factoring.

At this point, one of the two resulting equations can be solved easily. The other solution is $\tan^{-1}(-3)$, which is approximately -1.2490.

EXAMPLE 9

Solving Equations Using Inverse Trigonometric Functions

Solve the equation $6\sin x - 2 = \sin x$ on the interval $[0, 2\pi)$.

Solution:

$$6\sin x - 2 = \sin x$$

$$5\sin x = 2$$

$$\sin x = 0.4$$

$$x = \sin^{-1} 0.4 \text{ and } x = \pi - \sin^{-1} 0.4$$

The trigonometric function can be easily isolated.

Note that since $\sin x$ is positive, x lies in the first or second quadrant. The solution in the first quadrant is $\sin^{-1} 0.4$, and the solution in the second quadrant has $\sin^{-1} 0.4$ as its reference angle.

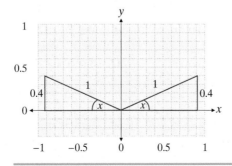

Exercises

1. $x = \dfrac{7\pi}{6} + 2n\pi$

$x = \dfrac{11\pi}{6} + 2n\pi$

2. $x = \dfrac{\pi}{6} + n\pi$

$x = \dfrac{5\pi}{6} + n\pi$

3. $x = \dfrac{\pi}{4} + 2n\pi$

$x = \dfrac{7\pi}{4} + 2n\pi$

4. $x = \dfrac{\pi}{3} + 2n\pi$

$x = \dfrac{5\pi}{3} + 2n\pi$

5. $x = \dfrac{\pi}{6} + 2n\pi$

$x = \dfrac{11\pi}{6} + 2n\pi$

6. $x = \dfrac{\pi}{6} + \dfrac{n\pi}{2}$

7. No solution

8. $x = \dfrac{\pi}{2} + 2n\pi$

9. $x = \dfrac{2\pi}{3} + n\pi$

10. $x = \dfrac{\pi}{6} + n\pi$

$x = \dfrac{5\pi}{6} + n\pi$

$x = \dfrac{\pi}{3} + n\pi$

$x = \dfrac{2\pi}{3} + n\pi$

11. $x = n\pi$

12. $x = \dfrac{\pi}{2} + n\pi$

13. $x = 2n\pi$

14. $x = \dfrac{\pi}{3} + 2n\pi$

$x = \dfrac{5\pi}{3} + 2n\pi$

Use trigonometric identities and algebraic methods, as necessary, to solve the following trigonometric equations. See Examples 1 through 7.

1. $2\sin x + 1 = 0$

2. $4\sin^2 x + 2 = 3$

3. $\sqrt{2} - 2\cos x = 0$

4. $4\cos x = 2$

5. $2\cos x - \sqrt{3} = 0$

6. $\sin 2x = \sqrt{3}\cos 2x$

7. $-\dfrac{1}{\sqrt{48}\sin x} = \dfrac{1}{8}$

8. $\sin^2 x - \sin x = 2\sin x - 2$

9. $\sqrt{3}\tan x + 1 = -2$

10. $(3\tan^2 x - 1)(\tan^2 x - 3) = 0$

11. $\sec^2 x - 1 = 0$

12. $\sin^2 x = \sin^2 x + \cos^2 x$

13. $\sec x + \tan x = 1$

14. $\cos x + \sin x \tan x = 2$

15. $\sin^2 x + \cos^2 x + \tan^2 x = 0$

16. $\cos^2 x = 3\sin^2 x$

17. $2\cos^2 x - 3 = 5\cos x$

18. $\sin^3 x = \sin x$

19. $\dfrac{\cot 2x}{\sqrt{3}} = -1$

20. $\cos x - 1 = \sin x$

Use trigonometric identities, algebraic methods, and inverse trigonometric functions, as necessary, to solve the following trigonometric equations on the interval $[0, 2\pi)$. See Examples 8 and 9.

21. $2\sin^2 x + 7\sin x = 4$

22. $\tan^2 x = \tan x + 6$

23. $2\cos^2 x - 1 = 0$

24. $\sec^2 x - 3 = -\tan x$

25. $0.05\sin^3 x = 0.1\sin x$

26. $12\sin x - 1 = 8\sin x$

27. $2\cos^2 x + 11\cos x = -5$

28. $10\sin x - 16 = -11 + 18\sin x$

29. $\sec^2 x - 2 + 8\tan x = 42$

30. $\sin(-x) = 5\sin x - 3$

Verify that the x-values given are solutions to the given equation.

31. $2\cos x + 1 = 0$;

    a. $x = \dfrac{2\pi}{3}$ **b.** $x = \dfrac{4\pi}{3}$

32. $3\sec^2 x - 4 = 0$

    a. $x = \dfrac{\pi}{6}$ **b.** $x = \dfrac{5\pi}{6}$

33. $2\sin^2 x - \sin x - 1 = 0$

    a. $x = \dfrac{\pi}{2}$ **b.** $x = \dfrac{7\pi}{6}$

34. $\tan^2 3x = 3$

    a. $x = \dfrac{\pi}{9}$ **b.** $x = \dfrac{2\pi}{9}$

15. No solution

16. $x = \dfrac{\pi}{6} + n\pi$

   $x = \dfrac{5\pi}{6} + n\pi$

17. $x = \dfrac{2\pi}{3} + 2n\pi$

   $x = \dfrac{4\pi}{3} + 2n\pi$

18. $x = n\pi$

   $x = \dfrac{\pi}{2} + n\pi$

19. $x = \dfrac{5\pi}{12} + \dfrac{n\pi}{2}$

20. $x = 2n\pi$

   $x = \dfrac{3\pi}{2} + 2n\pi$

21. $x = \dfrac{\pi}{6}, \dfrac{5\pi}{6}$

22. $x = \tan^{-1} 3,$

   $\tan^{-1} -2$

23. $x = \dfrac{\pi}{4}, \dfrac{3\pi}{4}, \dfrac{5\pi}{4}, \dfrac{7\pi}{4}$

24. $x = \dfrac{\pi}{4}, \dfrac{5\pi}{4}, \tan^{-1} -2$

25. $x = 0, \pi, 2\pi$

26. $x = \sin^{-1} \dfrac{1}{4}$

27. $x = \dfrac{2\pi}{3}, \dfrac{4\pi}{3}$

28. $x = \sin^{-1} \dfrac{-5}{8}$

29. $x = \tan^{-1}(-4 + \sqrt{59}),$

   $\tan^{-1}(-4 - \sqrt{59})$

30. $x = \dfrac{\pi}{6}, \dfrac{5\pi}{6}$

31.–39. Answers will vary.
41. 1.1925 and 3.8088
42. 2.6779 and 5.8195
43. 0.8959 and 2.2457
44. 0.6662, 2.4754,
   3.8078, and 5.6169
45. 0.9169 and 2.2247

35. $\csc^4 x - 4\csc^2 x = 0$

   a. $x = \dfrac{\pi}{6}$ **b.** $x = \dfrac{5\pi}{6}$

37. $2\cot x + 1 = -1$

   a. $x = \dfrac{3\pi}{4}$ **b.** $x = \dfrac{7\pi}{4}$

39. $2\sec x + 1 = \sec x + 3$

   a. $x = \dfrac{\pi}{3}$ **b.** $x = \dfrac{5\pi}{3}$

Use a graphing calculator to approximate the solutions of the given equation for the interval $[0, 2\pi)$. Round answers to four decimal places.

41. $x \tan x - 3 = 0$

43. $2\cos^2 x - \sin x = 0$

45. $2\sin x - \csc^2 x = 0$

47. $\log x = -\sin x$

36. $3\cot^2 x - 1 = 0$

   a. $x = \dfrac{\pi}{3}$ **b.** $x = \dfrac{2\pi}{3}$

38. $\csc^2 x = 2\cot x$

   a. $x = \dfrac{\pi}{4}$ **b.** $x = \dfrac{5\pi}{4}$

40. $2\sin^2 2x = 1$

   a. $x = \dfrac{\pi}{8}$ **b.** $x = \dfrac{3\pi}{8}$

42. $2\sin x + \cos x = 0$

44. $\cot^2 x - \sec^2 x = 0$

46. $2\sin x = 1 - 2\cos x$

48. $\sin \dfrac{x}{2} = 2\cos 2x$

Determine if the value given is a solution to the trigonometric function. If the value of x is not a solution, give all solutions to the equation.

49. $2\cos x = -1$; $x = \dfrac{4\pi}{3} + 2n\pi$

50. $\tan 3x (\tan x - 1) = 0$; $x = \dfrac{\pi}{4} + n\pi$

51. $3\sec^2 x = 4$; $x = \dfrac{\pi}{6} + n\pi$

52. $\sin^2 x - 3\cos^2 x = 0$; $x = \dfrac{\pi}{3} + 2n\pi$

53. $\sqrt{3}\csc x = 2$; $x = \dfrac{2\pi}{3} + 2n\pi$

54. $2\sin^2 x - 1 = 0$; $x = \dfrac{\pi}{4} + n\pi$

55. $\tan x = -\sqrt{3}$; $x = \dfrac{\pi}{6} + n\pi$

56. $\tan^2 3x - 3 = 0$; $x = \dfrac{2\pi}{9} + \dfrac{n\pi}{3}$

57. $3\cot^2 x = 1$; $x = \dfrac{\pi}{3} + n\pi$

58. $\cos 2x (2\cos x + 1) = 0$; $x = \dfrac{5\pi}{6} + n\pi$

Solve the following quadratic-like equations in the interval $[0, 2\pi)$.

59. $2\sin^2 x - \sin x - 1 = 0$

60. $2\sin^2 x + 3\cos x - 3 = 0$

61. $\sin x - \cos x - 1 = 0$

62. $\tan x + \sqrt{3} = \sec x$

46. 1.9948 and 5.8592 **47.** 0.4043, 3.7535, and 5.4549 **48.** 0.6993, 2.6078, 3.6754, and 5.5839 **49.** True **50.** True

51. True **52.** True **53.** True **54.** True **55.** $x = \dfrac{2\pi}{3} + n\pi$ **56.** True **57.** True **58.** $x = \dfrac{\pi}{4} + \dfrac{n\pi}{2}, \dfrac{2\pi}{3} + 2n\pi, \dfrac{4\pi}{3} + 2n\pi$

59. $x = \dfrac{\pi}{2}, \dfrac{7\pi}{6}, \dfrac{11\pi}{6}$

60. $x = 0, \dfrac{\pi}{3}, \dfrac{5\pi}{3}$

61. $x = \dfrac{\pi}{2}, \pi$

62. $x = \dfrac{11\pi}{6}$

63. $x = 0, \dfrac{2\pi}{3}, \dfrac{4\pi}{3}$

64. $x = \dfrac{\pi}{6}, \dfrac{\pi}{2}, \dfrac{3\pi}{2}, \dfrac{11\pi}{6}$

65. $x = \dfrac{\pi}{4}, \dfrac{5\pi}{4}$

66. $x = 90°, 270°$

67. $x = 45°, 135°, 225°, 315°$

68. $x = 45°, 135°, 225°, 315°$

69. $x = 90°, 210°, 330°$

70. $x = 71.6°, 135°, 251.6°, 315°$

71. No solution

72. $x = 68.5°, 291.5°$

73. $x = 30°, 210°, 240°, 300°$

74. $s = \dfrac{2}{3}, \dfrac{3}{2}$

$t = 0.8411, 5.4421$

75. $s = -4, 3$

t has no solution.

76. $s = -5, \dfrac{3}{2}$

$t = 0.9828, 1.7682, 4.1244, 4.9098$

77. $s = \dfrac{1 \pm \sqrt{2}}{2}$

$t = 1.7794, 4.5038$

78. They can have at most two real solutions.

79. They can have a maximum of four solutions.

63. $\cos 2x - \cos x = 0$

65. $\csc^2 x - 2 \cot x = 0$

Solve the following equations in the interval $[0°, 360°)$. Give the exact solutions when appropriate; otherwise, round answers to the nearest tenth.

66. $\sin^2 x \cos x - \cos x = 0$

68. $\tan x = \cot x$

70. $\sec^2 x - 2 \tan x = 4$

72. $2 \cos^2 x - 1 = -2 \cos x$

Solve the algebraic and trigonometric equations given. Restrict the solutions to the trigonometric equations to the interval $[0, 2\pi)$.

74. $6s^2 - 13s + 6 = 0; 6 \cos^2 t - 13 \cos t + 6 = 0$

75. $s^2 + s - 12 = 0; \sin^2 t + \sin t - 12 = 0$

76. $2s^2 + 7s - 15 = 0; 2 \tan^2 t + 7 \tan t - 15 = 0$

77. $4s^2 - 4s - 1 = 0; 4 \cos^2 t - 4 \cos t - 1 = 0$

78. While working Exercises 74–77, what did you observe as the maximum number of real solutions the algebraic equations can have?

79. While working Exercises 74–77, what did you observe as the maximum number of real solutions the trigonometric equations can have in the interval $[0, 2\pi)$?

80. Use your graphing calculator to solve the equation $2\sin 2x = \sqrt{3}$ on the interval $[0°, 360°)$. (Remember to set your MODE to DEGREE.)

81. Use your graphing calculator to solve the equation $\sin 3x - \dfrac{1}{2} = 0$ on the interval $[0°, 360°)$. (Remember to set your MODE to DEGREE.)

64. $2 \cos^2 x - \sqrt{3} \cos x = 0$

67. $\cos^2 x = \sin^2 x$

69. $2 \sin x = \csc x + 1$

71. $\sin^2 x = 2 \sin x - 3$

73. $2 \sin x \cot x + \sqrt{3} \cot x - 2\sqrt{3} \sin x - 3 = 0$

82. Use your graphing calculator to solve the equation $2 \sin 4x - 1 = 0$ on the interval $[0°, 360°)$. (Remember to set your MODE to DEGREE.)

83. While working with the trigonometric equations in Exercises 80–82, what did you observe about the solutions to the equation of the form $y = \sin ax$ in the interval $[0°, 360°)$?

84. An arrow is shot by an archer at an angle of θ in reference to the horizontal. The initial velocity is $v_0 = 100$ feet per second. If the arrow hits a target 300 feet from where the archer is standing, determine θ for the range given by $r = \dfrac{1}{32} v_0^2 \sin 2\theta$.

85. A baseball leaves a bat at an angle of θ in reference to the horizontal. The initial velocity is $v_0 = 95$ feet per second. The ball is caught 160 feet from where it is hit. What is the value of θ if the range is given by the projectile $r = \dfrac{1}{32} v_0^2 \sin 2\theta$?

80. $x = 30°, 60°, 210°, 240°$

81. $x = 10°, 50°, 130°, 170°, 250°, 290°$

82. $x = 7.5°, 37.5°, 97.5°, 127.5°, 187.5°, 217.5°, 277.5°, 307.5°$

83. They may have as many as $2a$ solutions from $0°$ to $360°$.

84. $36.9°$

85. $17.3°$

Trigonometric Identities

Lasers are used in everything from CD players to scanners at the check-out counter, to intricate medical surgeries, to weaponry. A beam of light from a flashlight shines for a couple hundred yards, but a laser's narrow band of light can be reflected off the moon and detected on Earth. A laser has this ability because its light is *coherent*. Coherent light means that each light wave has exactly the same amplitude, direction, and phase. Coherence reflects the superposition principle which states that when combining two waves, the resulting wave is the sum of the two individual waves.

Let's examine how the superposition principle works:

1. Consider two waves with a difference in displacement of $\dfrac{\pi}{2}$:

$$y_1 = 2\sin(kx - \omega t)$$
$$y_2 = 2\sin\left(kx - \omega t + \frac{\pi}{2}\right)$$

Using a trigonometric identity, add these two waves to find the equation of their superposition. What is the amplitude of the resulting wave?

2. For the set of equations:

$$y_1 = A\sin(kx - \omega t)$$
$$y_2 = A\sin\left(kx - \omega t + \delta\right)$$

a. For what values of δ would the amplitude be the largest? (This happens when two waves are coherent and it is called constructive interference.)

b. What is the smallest amplitude possible for $y_1 + y_2$?

c. For what values of δ would the smallest amplitude occur? (This is called destructive interference.)

3. Graph the following equations:

$$y_1 = 3\sin t$$
$$y_2 = 3\sin\left(t + \frac{\pi}{3}\right)$$

Now graph $y_1 + y_2$. Discuss the relationship between the three graphs.

Chapter Summary

A summary of concepts and skills follows each chapter. Refer to these summaries to make sure you feel comfortable with the material in the chapter. The concepts and skills are organized according to the section title and topic title in which the material is first discussed.

7.1: Fundamental Identities and Their Uses

Previously Encountered Identities
- Reciprocal identities, quotient identities, cofunction identities, period identities, even identities, odd identities, and Pythagorean Identities

Simplifying Trigonometric Expressions
- Using trigonometric identities to simplify more complicated expressions

Verifying Trigonometric Identities
- Guidelines for verifying trigonometric identities: **1.** Work with one side at a time; **2.** Apply trigonometric identities as appropriate; **3.** Rewrite in terms of sine and cosine if necessary

Trigonometric Substitutions
- Replacing algebraic expressions with trigonometric expressions

7.2: Sum and Difference Identities

The Identities
- Sine Identities: $\sin(u+v) = \sin u \cos v + \cos u \sin v$ and
$$\sin(u-v) = \sin u \cos v - \cos u \sin v$$
- Cosine Identities: $\cos(u+v) = \cos u \cos v - \sin u \sin v$ and
$$\cos(u-v) = \cos u \cos v + \sin u \sin v$$
- Tangent Identities: $\tan(u+v) = \dfrac{\tan u + \tan v}{1 - \tan u \tan v}$ and
$$\tan(u-v) = \dfrac{\tan u - \tan v}{1 + \tan u \tan v}$$

Using Sum and Difference Identities for Exact Evaluation

Applications of Sum and Difference Identities

7.3: Product-Sum Identities

Double-Angle Identities

- Sine Identity: $\sin 2u = 2\sin u \cos u$

- Cosine Identities: $\cos 2u = \cos^2 u - \sin^2 u$

$$= 2\cos^2 u - 1$$

$$= 1 - 2\sin^2 u$$

- Tangent Identity: $\tan 2u = \dfrac{2\tan u}{1 - \tan^2 u}$

Power-Reducing Identities

- Sine Identity: $\sin^2 x = \dfrac{1 - \cos 2x}{2}$

- Cosine Identity: $\cos^2 x = \dfrac{1 + \cos 2x}{2}$

- Tangent Identity: $\tan^2 x = \dfrac{1 - \cos 2x}{1 + \cos 2x}$

Half-Angle Identities

- Sine Identity: $\sin \dfrac{x}{2} = \pm\sqrt{\dfrac{1 - \cos x}{2}}$

- Cosine Identity: $\cos \dfrac{x}{2} = \pm\sqrt{\dfrac{1 + \cos x}{2}}$

- Tangent Identity: $\tan \dfrac{x}{2} = \dfrac{1 - \cos x}{\sin x} = \dfrac{\sin x}{1 + \cos x}$

Product-to-Sum and Sum-to-Product Identities

- Product-to-Sum Identities:

$$\sin x \cos y = \frac{1}{2}\left[\sin(x+y) + \sin(x-y)\right]$$

$$\cos x \sin y = \frac{1}{2}\left[\sin(x+y) - \sin(x-y)\right]$$

$$\sin x \sin y = \frac{1}{2}\left[\cos(x-y) - \cos(x+y)\right]$$

$$\cos x \cos y = \frac{1}{2}\left[\cos(x+y) + \cos(x-y)\right]$$

7.3: Product-Sum Identities (cont.)

- Sum-to-Product Identities:

$$\sin x + \sin y = 2\sin\left(\frac{x+y}{2}\right)\cos\left(\frac{x-y}{2}\right)$$

$$\sin x - \sin y = 2\cos\left(\frac{x+y}{2}\right)\sin\left(\frac{x-y}{2}\right)$$

$$\cos x + \cos y = 2\cos\left(\frac{x+y}{2}\right)\cos\left(\frac{x-y}{2}\right)$$

$$\cos x - \cos y = -2\sin\left(\frac{x+y}{2}\right)\sin\left(\frac{x-y}{2}\right)$$

7.4: Trigonometric Equations

Applying Algebraic Techniques to Trigonometric Equations
- Using trigonometric identities and algebraic techniques to solve trigonometric equations

Using Inverse Trigonometric Functions
- Using inverse trigonometric functions to solve trigonometric equations

1. $\csc x$

2. $\cos^2 x$

3. $-\tan^2 y$

4. 1

5. $-1 - \sin\theta$

6. $-\sin^2 x$

7.–12. Answers will vary.

13. $4\sec\theta$

14. $8\cos\theta$

15. $10\cot\theta$

16. $6\sec\theta$

17. $\dfrac{\sqrt{3}}{2}$

18. $\dfrac{\sqrt{2} - \sqrt{6}}{4}$

19. $\dfrac{\sqrt{2} - \sqrt{6}}{4}$

20. $\dfrac{-\sqrt{6} - \sqrt{2}}{4}$

21. $\sqrt{3}$

22. $-2 - \sqrt{3}$

23. $\dfrac{13}{84}$

24. $\dfrac{16}{65}$

25. $\sin 210°; -\dfrac{1}{2}$

26. $\tan\left(\dfrac{3\pi}{4}\right); -1$

27. Answers will vary.

28. Answers will vary.

29. $\dfrac{\sqrt{1 - x^2} - x^2}{\sqrt{1 + x^2}}$

30. $\dfrac{2x\left(1 - \sqrt{1 - 4x^2}\right)}{\sqrt{1 + 4x^2}}$

31. $2\sin\left(x + \dfrac{7\pi}{4}\right)$

32. $2\sin\left(4\alpha - \dfrac{\pi}{6}\right)$

Chapter Review

Section 7.1

Use trigonometric identities to simplify the expressions. There may be more than one correct answer.

1. $\cot x \sec x$

2. $\left(\csc^2 x - 1\right)\cos^2\left(\dfrac{\pi}{2} - x\right)$

3. $\dfrac{\tan(-y)}{\cot(\pi + y)}$

4. $\dfrac{\tan^2\alpha}{\csc^2\left(\dfrac{\pi}{2} - \alpha\right)} + \dfrac{\cos\alpha}{\sec(\alpha + 2\pi)}$

5. $\sin^2\theta\sec^2\theta - \csc^2\left(\dfrac{\pi}{2} - \theta\right) + \sin(-\theta)$

6. $\sin\left(\dfrac{\pi}{2} - x\right)\cos(x + 2\pi) + \sin(-x)\sec\left(\dfrac{\pi}{2} - x\right)$

Verify the identities.

7. $(\tan x + \sec x)(\sec x - \tan x) = 1$

8. $\cos^2 x\tan^2 x = 1 - \dfrac{1}{\sec^2 x}$

9. $\dfrac{\cos\left(\dfrac{\pi}{2} - t\right)}{\tan(-t)} = -\cos t$

10. $5 + \tan^2 y = 4 + \sec^2 y$

11. $\tan(\theta + \pi) = -\dfrac{\sec(\theta + 2\pi)}{\csc(-\theta)}$

12. $-\tan\left(\dfrac{\pi}{2} - x\right)\tan(-x) - \tan^2 x\sin^2\left(\dfrac{\pi}{2} - x\right) = \cos^2 x$

Use the suggested substitution to rewrite the given expression as a trigonometric expression.

13. $\sqrt{16 + x^2}$, $\tan\theta = \dfrac{x}{4}$

14. $\sqrt{64 - 16x^2}$, $2\sin\theta = x$

15. $\sqrt{25x^2 - 100}$, $\csc\theta = \dfrac{x}{2}$

16. $\sqrt{9x^2 + 36}$, $x = 2\tan\theta$

Section 7.2

Use the sum and difference identities to determine the exact value of each of the following expressions.

17. $\cos\left(\dfrac{\pi}{2} + \dfrac{5\pi}{3}\right)$

18. $\cos 255°$

33. $\cos 2x = \dfrac{-7}{25}$,

$\sin 2x = \dfrac{24}{25}$,

$\tan 2x = \dfrac{-24}{7}$

34. $\cos 2x = \dfrac{4}{5}$,

$\sin 2x = \dfrac{3}{5}$,

$\tan 2x = \dfrac{3}{4}$

35. Answers will vary.

36. Answers will vary.

37. $\dfrac{\sin x - \sin x \cos 4x}{8}$

38. $\dfrac{3\sin x - 4\cos 2x \sin x}{4 + 4\cos 2x}$

$\quad + \dfrac{\cos 4x \sin x}{4 + 4\cos 2x}$

39. $2 + \sqrt{3}$

40. $-\dfrac{1}{2}\sqrt{2 + \sqrt{2}}$

41. $2 - \sqrt{3}$

42. $-\dfrac{1}{2}\sqrt{2 + \sqrt{2}}$

43. $\dfrac{\sin 2x - \sin 2y}{2}$

44. $\dfrac{\cos \dfrac{11\pi}{2} + \cos \dfrac{7\pi}{2}}{2}$

45. $\dfrac{1}{2}\left(\sin 180° + \sin 150°\right)$

46. $\dfrac{1}{2}\left(\cos x - \cos 7x\right)$

47. $2\cos 4\alpha \sin \alpha$

48. $2\cos 120° \cos 105°$

49. $-2\sin \dfrac{17\pi}{24}\sin \dfrac{\pi}{24}$

50. $2\sin \dfrac{7\pi}{12}\cos \dfrac{\pi}{4}$

51. $x = \dfrac{\pi}{6} + 2n\pi$

$\quad x = \dfrac{-\pi}{6} + 2n\pi$

---

19. $\sin(-15°)$

20. $\sin\left(\dfrac{5\pi}{4} + \dfrac{\pi}{6}\right)$

21. $\tan\left(\pi - \dfrac{2\pi}{3}\right)$

22. $\tan 105°$

Find the sum or difference for each given question.

23. $\sin \alpha = \dfrac{8}{17}$ and $\sin \beta = \dfrac{3}{5}$. Both α and β are in quadrant II. Find $\tan(\alpha - \beta)$.

24. $\sin \alpha = \dfrac{5}{13}$ and $\cos \beta = \dfrac{4}{5}$. α is in quadrant II and β is in quadrant IV. Find $\sin(\alpha - \beta)$.

Use the sum and difference identities to rewrite each of the following expressions as a trigonometric function of a single number, and then evaluate the result.

25. $\sin 175° \cos 35° + \cos 175° \sin 35°$

26. $\dfrac{\tan \dfrac{9\pi}{8} - \tan \dfrac{3\pi}{8}}{1 + \tan \dfrac{9\pi}{8}\tan \dfrac{3\pi}{8}}$

Use the sum and difference identities to verify the following identities.

27. $\sin x = \cos\left(\dfrac{\pi}{2} - x\right)$

28. $\cos(\alpha + \beta) - \cos(\alpha - \beta) = -2\sin \alpha \sin \beta$

Express each of the following as an algebraic expression of x.

29. $\cos\left(\sin^{-1} x + \tan^{-1} x\right)$

30. $\cos\left(\cos^{-1} 2x + \tan^{-1} 2x\right)$

Express each of the following functions in terms of a single sine function.

31. $f(x) = \sqrt{2}\sin x - \sqrt{2}\cos x$

32. $h(\alpha) = \sqrt{3}\sin 4\alpha - \cos 4\alpha$

Section 7.3

Use the information given to determine $\cos 2x$, $\sin 2x$, and $\tan 2x$.

33. $\tan x = \dfrac{4}{3}$ and $\sin x$ is positive

34. $\sin x = \dfrac{-1}{\sqrt{10}}$ and $\tan x$ is positive

Verify the following trigonometric identities.

35. $\cos 3x = 4\cos^3 x - 3\cos x$

36. $\dfrac{\sin 4x}{4} = \sin x \cos x - 2\sin^3 x \cos x$

Use a power-reducing identity to rewrite the expressions as directed.

37. Rewrite $\sin^3 x \cos^2 x$ in terms containing only the first powers of sine and cosine.

38. Rewrite $\tan^2 x \sin^3 x$ in terms containing only the first powers of sine and cosine.

52. $x = \dfrac{\pi}{6} + 2n\pi$

$x = \dfrac{5\pi}{6} + 2n\pi$

$x = n\pi$

53. $x = \dfrac{\pi}{2} + 2n\pi$

$x = \dfrac{-\pi}{2} + 2n\pi$

54. $x = n\pi$

$x = \dfrac{\pi}{4} + n\pi$

55. $x = \dfrac{\pi}{3} + 2n\pi$

$x = \dfrac{5\pi}{3} + 2n\pi$

$x = \pi + 2n\pi$

56. $x = \dfrac{7\pi}{6} + 2n\pi$

$x = \dfrac{11\pi}{6} + 2n\pi$

57. $x = \dfrac{\pi}{6}, \dfrac{5\pi}{6}, \dfrac{7\pi}{6}, \dfrac{11\pi}{6}$

58. $x = \dfrac{\pi}{2}$

59. True

60. $x = \dfrac{\pi}{6} + 2n\pi$

$x = \dfrac{5\pi}{6} + 2n\pi$

61. $x = \dfrac{\pi}{3} + 2n\pi$

$x = \dfrac{2\pi}{3} + 2n\pi$

62. True

63. $x = 0°, 180°$

64. $x = 60°, 300°$

Determine the exact values of each of the following expressions.

39. $\tan \dfrac{5\pi}{12}$

40. $\cos 157.5°$

41. $\tan 15°$

42. $\sin\left(-\dfrac{5\pi}{8}\right)$

Use the product-to-sum identities to rewrite the expressions as a sum or difference.

43. $\cos(x + y)\sin(x - y)$

44. $\cos \dfrac{3\pi}{4}\cos \dfrac{\pi}{6}$

45. $\sin 165° \cos 15°$

46. $\sin 4x \sin 3x$

Use the sum-to-product identities to rewrite the expression as a product.

47. $\sin 5\alpha - \sin 3\alpha$

48. $\cos 225° + \cos 15°$

49. $\cos \dfrac{3\pi}{4} - \cos \dfrac{2\pi}{3}$

50. $\sin \dfrac{5\pi}{6} + \sin \dfrac{\pi}{3}$

Section 7.4

Use the trigonometric identities and algebraic methods, as necessary, to solve the following trigonometric equations.

51. $8\cos^2 x + 1 = 7$

52. $2\sin^2 x = \sin x$

53. $-\sin^2 x + 4\cos x + 1 = 0$

54. $\tan^3 x = \tan x$

55. $-2\sin^2 x = -\cos x - 1$

56. $\sin x + \cos x \cot x = -2$

Use trigonometric identities, algebraic methods, and inverse trigonometric functions, as necessary, to solve the following trigonometric equations on the interval $[0, 2\pi]$.

57. $3\tan^2 x + 9 = 10$

58. $\sin^2 x = 3 - 2\sin x$

Determine if the value given is a solution to the trigonometric function. If the value of x is not a solution, give all solutions to the equation.

59. $4\sin^2 x = 3;\ x = \dfrac{5\pi}{3} + 2n\pi$

60. $\dfrac{1}{2}\csc x + 1 = 2;\ x = \dfrac{3\pi}{4} + 2n\pi$

61. $\tan 2x \cos x = -\dfrac{\sqrt{3}}{2};\ x = \dfrac{\pi}{6} + 2n\pi$

62. $\sin x + \cos 2x = 1;\ x = \dfrac{5\pi}{6} + 2n\pi$

Solve the following equations in the interval $[0°, 360°)$. Give exact solutions when appropriate; otherwise, round answers to the nearest tenth.

63. $\cos^2 x \sin x = \sin x$

64. $2\cos^2 x + 7\cos x = 4$

Chapter Test

1. $\sin^2 x - 5 \csc^2 x$

2. -1

3. Answers will vary.

4. Answers will vary.

5. $13 \sec \theta$

6. $8\sqrt{2} \cos \theta$

7. Answers will vary.

8. $\dfrac{\sqrt{6} - \sqrt{2}}{4}$

9. $\sqrt{3} - 2$

10. $\cos 103°, -0.2250$

11. $\tan \dfrac{7\pi}{8}, -0.4142$

Use trigonometric identities to simplify the expressions. There may be more than one correct answer.

1. $\sin^2(-x) - 5 \sec^2 x \cot^2 x$

2. $\dfrac{\tan \beta \cos \beta}{\sin(-\beta)}$

Verify the identities.

3. $1 + \sin^2 \theta = \csc^2 \theta - \cot^2 \theta + \sin^2 \theta$

4. $(\sin^2 x)(\csc^2 x - 1) = \cos^2 x$

Use the suggested substitution to rewrite the given expression as a trigonometric expression. For all problems, assume $0 \le \theta \le \dfrac{\pi}{2}$.

5. $\sqrt{x^2 + 169}$, $\dfrac{x}{13} = \tan \theta$

6. $\sqrt{128 - 2x^2}$, $x = 8 \sin \theta$

Show how the identity below follows from the first Pythagorean Identity.

7. $\tan^2 x + 1 = \sec^2 x$

Use the sum and difference identities to determine the exact value of each of the following expressions.

8. $\sin \dfrac{25\pi}{12}$

9. $\tan 165°$

Use the sum and difference identities to rewrite each of the following expressions as a trigonometric function of a single number, and then evaluate the result.

10. $\cos 131° \cos 28° + \sin 131° \sin 28°$

11. $\dfrac{\tan \dfrac{5\pi}{8} + \tan \dfrac{\pi}{4}}{1 - \tan \dfrac{5\pi}{8} \tan \dfrac{\pi}{4}}$

12. Answers will vary.

13. Answers will vary.

14. $\dfrac{2x + x\sqrt{1 - 4x^2}}{\sqrt{1 + x^2}}$

15. $h(\beta) = 4\sin\left(4\beta + \dfrac{\pi}{6}\right)$

16. $\cos 2x = \dfrac{7}{9}$

$\sin 2x = \dfrac{-4\sqrt{2}}{9}$

$\tan 2x = \dfrac{-4\sqrt{2}}{7}$

17. Answers will vary.

18. $\dfrac{3\cos x + 4\cos x \cos 2x}{8}$

$+ \dfrac{\cos x \cos 4x}{8}$

19. $\dfrac{\sqrt{2 - \sqrt{3}}}{2}$

20. $\dfrac{\sqrt{2 + \sqrt{3}}}{2}$

21. $\dfrac{\sin 10x}{2}$

22. $\dfrac{\sin \dfrac{6\pi}{5}}{2}$

23. $2\sin\dfrac{x}{2}\cos\dfrac{x}{3}$

24. $2\cos 4x \cos x$

Use the sum and difference identities to verify the following identities.

12. $\csc\left(\dfrac{3\pi}{4} + \theta\right) = \sec\left(\dfrac{\pi}{4} + \theta\right)$

13. $\cos^2\dfrac{\pi}{2} - \sin^2\dfrac{3\pi}{4} = \cos\dfrac{5\pi}{4}\cos\left(-\dfrac{\pi}{4}\right)$

Express the following as an algebraic function of x.

14. $\sin(\tan^{-1} x + \sin^{-1} 2x)$

Express the following function in terms of sine functions only.

15. $h(\beta) = 2\sqrt{3}\sin 4\beta + 2\cos 4\beta$

Use the information given to determine $\cos 2x$, $\sin 2x$, and $\tan 2x$.

16. $\sin x = \dfrac{1}{3}$ and $\cos x$ is negative

Verify the following trigonometric identity.

17. $\tan 4x = \dfrac{4\tan x - 4\tan^3 x}{1 - 6\tan^2 x + \tan^4 x}$

Use a power-reducing identity to rewrite the expression as directed.

18. Rewrite $\cos^5 x$ in terms containing only first powers of sine and cosine.

Determine the exact values of each of the following expressions.

19. $\sin\dfrac{11\pi}{12}$

20. $\cos 15°$

Use the product-to-sum identities to rewrite each expression as a sum or difference.

21. $\sin 5x \cos 5x$

22. $\cos\dfrac{3\pi}{5}\sin\dfrac{3\pi}{5}$

Use the sum-to-product identities to rewrite each expression as a product.

23. $\sin\dfrac{5x}{6} + \sin\dfrac{x}{6}$

24. $\cos 5x + \cos 3x$

25. $x = \dfrac{\pi}{6} + n\pi$,

$x = \dfrac{\pi}{3} + n\pi$

26. $x = \dfrac{\pi}{6} + 2n\pi$,

$x = \dfrac{5\pi}{6} + 2n\pi$

$x = \dfrac{\pi}{2} + 2n\pi$

27. $x = \dfrac{n\pi}{2}$, where n is odd

28. $x = \dfrac{\pi}{3} + n\pi$,

$x = \dfrac{3\pi}{4} + n\pi$

29. $x = \sin^{-1}\left(\dfrac{-2+\sqrt{34}}{5}\right) + 2n\pi$

30. $x = \cos^{-1}\left(\dfrac{-2}{3}\right) + 2n\pi$

31. $x = \tan^{-1}\left(-2+\sqrt{10}\right) + n\pi$,

$x = \tan^{-1}\left(-2-\sqrt{10}\right) + n\pi$

32. $x = \dfrac{7\pi}{6} + 2n\pi$,

$x = \dfrac{11\pi}{6} + 2n\pi$

Use trigonometric identities and algebraic methods, as necessary, to solve the following trigonometric equations.

25. $4\sin x \cos x = \sqrt{3}$

26. $-\dfrac{2\cos^2 \theta}{3} = \sin\theta - 1$

27. $2\cos^8 x = 4\cos^6 x$

28. $\tan\theta + 1 = \sqrt{3} + \sqrt{3}\cot\theta$

Use trigonometric identities, algebraic methods, and inverse trigonometric functions, as necessary, to solve the following trigonometric equations.

29. $5\sin^2 x + 4\sin x = 6$

30. $3\cos^2 x + 8\cos x = -4$

31. $4\sec^2 x - 4 + 16\tan x = 24$

32. $8\sin x - 4 = 16\sin x$

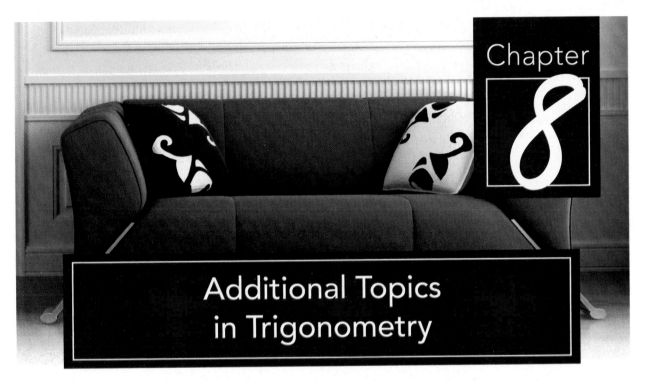

Chapter 8

Additional Topics in Trigonometry

8.1 The Law of Sines and the Law of Cosines

8.2 Polar Coordinates and Polar Equations

8.3 Parametric Equations

8.4 Trigonometric Form of Complex Numbers

8.5 Vectors in the Cartesian Plane

8.6 The Dot Product and Its Uses

Chapter 8 Project

Chapter 8 Summary

Chapter 8 Review

Chapter 8 Test

By the end of this chapter you should be able to:

What if you are moving to a new apartment? If you know the weight of your couch, how would you determine the force necessary to slide it from your living room to the front door?

By the end of this chapter, you will be able to perform many operations with vectors, including addition, scalar multiplication, and a special kind of vector multiplication called the dot product. Vectors have many uses in the physical sciences due to their ability to concisely describe and perform calculations about the forces acting around us. To find the force necessary to move your couch, you will need to solve problems similar to the ones on page 676. You will master these types of problems using tools such as Vector Operations Using Components on page 669.

Introduction

This is the last of three chapters devoted to the introduction and use of trigonometric concepts. Chapter 6 introduced the nomenclature of trigonometry and the six fundamental trigonometric functions, and Chapter 7 focused on a deeper understanding of how those functions behave. The goal of this chapter is to show how trigonometry relates to a surprisingly large number of topics seen elsewhere in this text, even in situations which would at first seem to have nothing in common with trigonometry.

The applications of trigonometry seen in this chapter vary widely. As a preview, here is a cursory list of the uses that will be developed:

- New theorems and relations that are exceedingly useful in surveying, navigation, and other practical concerns
- New ways to measure areas of triangles
- A second planar coordinate system that simplifies the graphing and solving of some kinds of equations
- A second way of describing curves in the plane that allows us to answer questions that have been awkward up to this point
- A second way of representing complex numbers that simplifies multiplication and division, and offers an easy method for the calculation and visualization of roots
- A very powerful method of representing *directed magnitudes*, allowing us to mathematically describe quantities in which the direction that a magnitude is applied is of critical importance
- A new mathematical operation that greatly simplifies the solving of problems involving, for example, force and work

As usual, the concepts presented come from many different eras and cultures. One of the formulas for triangular area is named for a 1st-century-AD mathematician from Alexandria, but was probably known by mathematicians from several different areas a century or two before. The method of representing directed magnitudes owes much to the work of the 19th-century Irish mathematician William Rowan Hamilton. And the use of trigonometry in handling complex numbers developed over the course of several centuries and through the work of many mathematicians working toward many widely differing goals.

Hamilton

The Law of Sines and the Law of Cosines

TOPICS

1. The Law of Sines and its use

2. The Law of Cosines and its use

3. Areas of triangles

TOPIC 1

The Law of Sines and Its Use

Chapter 6 introduced the trigonometric functions and demonstrated how they are used to solve a variety of problems involving triangles. Chapter 7 then explored the properties of the trigonometric functions and developed relationships and identities that greatly extended their usefulness. But the triangles analyzed to this point have all been right triangles, and even our more general treatment of the trigonometric functions has drawn, explicitly or implicitly, upon properties of right triangles. In this section we will expand the class of triangles we can analyze to include *oblique triangles*—those that have no right angle.

Most of the problems in this section contain a step in which a triangle must be *solved*; that is, given some information about a triangle's sides and/or angles, we will need to determine something else about the triangle. Note that there are six fundamental quantities that can be measured for any triangle: the lengths of the three sides, and the sizes of the three angles. We will see that if we know the length of any one side and any two of the other five quantities, we can determine the remaining three quantities.

In keeping with tradition, we will let A, B, and C denote the measures of the three angles of a given triangle and a, b, and c the lengths of the sides, with the side of length a opposite the angle of measure A and similarly for b and c. We will also follow the convention of letting a letter stand for both a quantity and the leg or angle it measures. Figure 1 illustrates a typical oblique triangle and the labeling of its sides and angles.

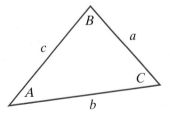

Figure 1: Oblique Triangle with Sides and Angles Labeled

In describing what we know about a triangle, we will let S represent knowledge about a leg and A knowledge about an angle, so for instance the notation SAS is shorthand for a situation in which we know the lengths of two legs and the measure of the angle between them. The table below categorizes the cases in which the two laws, known as the Law of Sines and the Law of Cosines, allow us to solve a given triangle.

DEFINITION

Use of the Law of Sines and the Law of Cosines

Law of Sines	Law of Cosines
Two angles and a side (AAS or ASA)	Two sides and the included angle (SAS)
Two sides and a nonincluded angle (SSA)	Three sides (SSS)

We are now ready to state and use the Law of Sines.

THEOREM

The Law of Sines

Given a triangle with sides and angles labeled according to the convention on the previous page,

$$\frac{\sin A}{a} = \frac{\sin B}{b} = \frac{\sin C}{c}.$$

Note that since A, B, and C represent measures of angles in a triangle, all three must lie between 0 and π radians; hence, $\sin A$, $\sin B$, and $\sin C$ are all nonzero and the Law of Sines can also be written in the form

$$\frac{a}{\sin A} = \frac{b}{\sin B} = \frac{c}{\sin C}.$$

While the Law of Sines is typically applied to oblique triangles, its truth relies on decomposing a triangle into two right triangles. For instance, if angles A and C are acute, we can construct the altitude h and determine its value two different ways, as shown in Figure 2.

$$c \sin A = h = a \sin C$$

Figure 2: The Law of Sines (Acute Case)

The exact same relation is true if one of the angles, say A, is obtuse, but this realization depends on recalling that $\sin A = \sin(\pi - A)$ (that is, the sine of an angle in the second quadrant is equal to the sine of the reference angle). Figure 3 illustrates this case.

$$c \sin A = h = a \sin C$$

Figure 3: The Law of Sines (Obtuse Case)

In either case, dividing both sides of the equation $c \sin A = a \sin C$ by ac leads to the Law of Sines as it relates to A, C, a, and c. Similar diagrams provide the rest of the law.

The next few examples demonstrate the use of the Law of Sines in the situations where it is applicable.

EXAMPLE 1

Using the Law of Sines in an *AAS* Situation

Sarah is piloting a hot-air balloon, and finds herself becalmed directly above a long straight road. She notices mile markers on the road, and determines the angle of depression to the two markers as shown below. How far is she from marker A? What is her altitude?

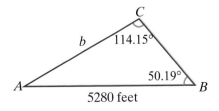

Solution:

To begin, note that the angle at mile marker B also has measure $50.19°$ (this follows from the equality of opposite angles formed by a line cutting two parallel lines). Further, the angle at Sarah's position has measure $180° - 15.66° - 50.19° = 114.15°$. Expressing the one length that we know in feet, we have the following information:

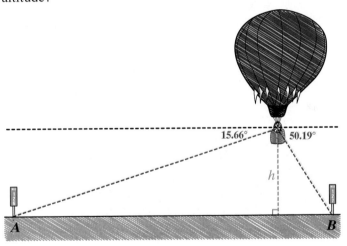

We can now use the Law of Sines to obtain the equation

$$\frac{b}{\sin 50.19°} = \frac{5280}{\sin 114.15°},$$

which we can easily solve for length b to get $b = (\sin 50.19°)\left(\dfrac{5280}{\sin 114.15°}\right) \approx 4445$ feet.

To determine Sarah's altitude h, we can use the fact that the measure of the angle at marker A is 15.66° and observe that

$$\sin 15.66° = \frac{h}{4445}.$$

Solving this for h, we obtain $h \approx 1200$ feet.

EXAMPLE 2

Using the Law of Sines in an *ASA* Situation

A surveyor has the task of determining the dimensions of the triangular plot of land shown below. He has already measured the length of the short edge of the plot, and has also determined the measure of two of the vertices. What are the lengths of the other two edges?

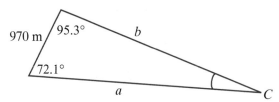

Solution:

The first step is to determine the measure of the third angle. Since the sum of the angles in a triangle is 180°, the third angle has measure $180° - 95.3° - 72.1°$, or 12.6°. The Law of Sines now tells us

$$\frac{970}{\sin 12.6°} = \frac{a}{\sin 95.3°} \quad \text{and} \quad \frac{970}{\sin 12.6°} = \frac{b}{\sin 72.1°}.$$

Solving for a and b, we obtain $a \approx 4428$ meters and $b \approx 4231$ meters.

As Examples 1 and 2 demonstrated, problems in which two angles and a side are known are readily dealt with. Usually, the application of some simple facts about triangles and angles, followed by the Law of Sines, allows us to completely determine the dimensions of a triangle in the *AAS* and *ASA* cases. Unfortunately, the same is not true for the *SSA* case. Given two sides a and b and the angle A opposite a, there may be no triangle, exactly one triangle, or two triangles that satisfy the conditions. To see why this is so, consider the four possibilities that may occur when A is acute:

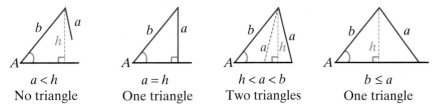

| $a < h$ | $a = h$ | $h < a < b$ | $b \leq a$ |
| No triangle | One triangle | Two triangles | One triangle |

Figure 4: *SSA* Case with Acute Angle *A*

As Figure 4 shows, the size of a in relation to h and b, where $h = b \sin A$, determines whether a triangle exists which fits the given information, and whether the triangle is unique if it does exist. There are fewer possibilities if A is obtuse, but the ambiguity is still present. Figure 5 illustrates the obtuse case.

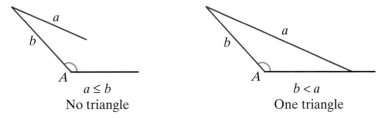

| $a \leq b$ | $b < a$ |
| No triangle | One triangle |

Figure 5: *SSA* Case with Obtuse Angle *A*

Because of the uncertainty over whether a unique triangle exists in the *SSA* case, it is sometimes called the **ambiguous case**. In practice, the ambiguity is resolved through consideration of additional information.

<table>
<tr><td>Determining If a
Triangle Exists</td></tr>
</table>

EXAMPLE 3

Determining If a
Triangle Exists

Construct a triangle, if possible, for which $A = \dfrac{\pi}{6}$, $b = 12$ cm, and $a = 7$ cm.

Solution:

A sketch is often useful in order to get an initial idea of whether such a triangle is possible. In this case, a rough sketch might look something like the following:

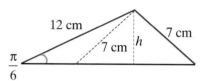

Such a sketch doesn't prove anything, but it certainly hints at the possibility of two triangles satisfying the given information, and the sketch also reminds us of the criterion to check: whether $h < a < b$. Since $h = 12 \sin\left(\dfrac{\pi}{6}\right) = 6$, and since $6 < 7 < 12$, we know two such triangles really do exist. The remaining task is to completely determine the dimensions of both triangles.

Triangle 1: We will arbitrarily designate the triangle in which B is acute as Triangle 1. The Law of Sines tells us that

$$\frac{\sin \dfrac{\pi}{6}}{7} = \frac{\sin B}{12},$$

and so $\sin B = \dfrac{6}{7}$. Thus, $B = \sin^{-1}\left(\dfrac{6}{7}\right) \approx 1.03$ (remember that this is in radians). This means that angle C has measure $\pi - \dfrac{\pi}{6} - \sin^{-1}\left(\dfrac{6}{7}\right)$, or approximately 1.59, and a second application of the Law of Sines gives us

$$\frac{c}{\sin 1.59} = \frac{7}{\sin \dfrac{\pi}{6}},$$

so $c \approx 14.0$ cm.

Triangle 2: In the second triangle, B is the obtuse angle for which $\sin B = \dfrac{6}{7}$; that is, the reference angle of B is 1.03, and hence $B \approx \pi - 1.03 = 2.11$ (a quick check with a calculator will verify that $\sin 2.11 \approx \dfrac{6}{7}$). In this case, $C \approx \pi - \dfrac{\pi}{6} - 2.11 = 0.51$, and hence

$$\frac{c}{\sin 0.51} = \frac{7}{\sin \dfrac{\pi}{6}}.$$

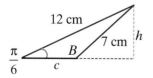

Solving this for c gives us $c \approx 6.8$ cm.

EXAMPLE 4

Determining If a Triangle Exists

EXAMPLE 4

Construct a triangle, if possible, for which $A = 75°$, $b = 15$ units, and $a = 10$ units.

Solution:

A sufficiently accurate sketch may lead you to conclude that no such triangle is possible. The proof of this fact follows from noting that $h = 15\sin 75° \approx 14.5$, and that $a < h$. In other words, a is too short to reach side c and form a triangle, so no triangle satisfies the given conditions.

TOPIC 2

The Law of Cosines and Its Use

The Law of Sines is of no use if we are given information about three sides of a triangle or two sides and the included angle of a triangle. Fortunately, the Law of Cosines handles the *SSS* and *SAS* cases easily.

THEOREM

The Law of Cosines

Given a triangle ABC, with sides labeled conventionally, the following are all true:

$$a^2 = b^2 + c^2 - 2bc\cos A$$

$$b^2 = a^2 + c^2 - 2ac\cos B$$

$$c^2 = a^2 + b^2 - 2ab\cos C$$

The similarity of these statements to the Pythagorean Theorem is no coincidence, and you should verify that the Law of Cosines, when applied to a right triangle, reduces to the Pythagorean Theorem. The Law of Cosines is thus an extension of the simpler theorem, but its proof depends on constructing an auxiliary right triangle to which we can apply the Pythagorean Theorem. To prove the first statement, for example, position the triangle under consideration so that angle A is in standard position in the plane, as shown in Figure 6.

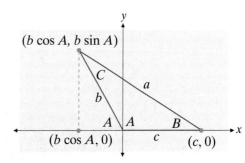

Figure 6: Proving the Law of Cosines

Applying the Pythagorean Theorem to the right triangle formed by the points $(c, 0)$, $(b \cos A, b \sin A)$ and $(b \cos A, 0)$ we obtain

$$a^2 = (c - b \cos A)^2 + (b \sin A)^2$$
$$= c^2 - 2bc \cos A + b^2 \cos^2 A + b^2 \sin^2 A$$
$$= b^2 (\cos^2 A + \sin^2 A) + c^2 - 2bc \cos A$$
$$= b^2 + c^2 - 2bc \cos A.$$

The other two parts of the Law of Cosines are proved similarly.

In solving triangles, it's useful to remember a simple fact: the longest side of a triangle is opposite the largest angle. This is important especially when working backward to determine an angle from the sine of the angle, as seen in Example 5 below. And to avoid possible confusion when applying the Law of Cosines to solve for an angle, it's a good idea to solve for the largest angle first. If the largest angle is obtuse, the other two angles must be acute (a triangle can have at most one obtuse angle). Of course, if the largest angle is acute, this also means the other two angles (which are by definition smaller) must be acute.

EXAMPLE 5

Using the Law of Cosines in a SSS Situation

Determine the three angles for a triangle in which $a = 3$ inches, $b = 5$ inches, and $c = 7$ inches.

Solution:

Guided by the observations above, we'll solve for C, the largest angle, first. By the Law of Cosines,

$$7^2 = 3^2 + 5^2 - 2(3)(5) \cos C$$
$$49 = 34 - 30 \cos C$$

and so $\cos C = \dfrac{(49 - 34)}{(-30)} = -\dfrac{1}{2}$. Since $\cos C$ is negative, we know C is obtuse, and a calculator tells us that $C = \cos^{-1}(-0.5) = 120°$. (We can also determine C without a calculator by first determining C's reference angle; since $\cos 60° = 0.5$, we know that $C = 180° - 60° = 120°$.)

We can now use either law to determine another angle. Using the Law of Sines to determine A, we have

$$\frac{\sin A}{3} = \frac{\sin 120°}{7}$$

and so $\sin A \approx 0.37$. Using a calculator, this means $A \approx 21.79°$ (and we know that A is not $180° - 21.79°$ because we have already determined that A is acute). We can now easily determine that $B \approx 180° - 120° - 21.79° = 38.21°$.

EXAMPLE 6

Using the Law of Cosines in an SAS Situation

The course of a sailboat race instructs the sailors to head due east 11 kilometers to the first buoy. They are then to veer off to port (to the left) by 20° and proceed for another 15 kilometers, at which point they should find the second buoy. What bearing would a sailor take to go from the starting point directly to the second buoy?

Solution:

A picture is definitely in order:

We wish to determine A, but neither law allows us to do so directly. However, the Law of Cosines will allow us to determine the third side, which we have labeled b in the diagram:

$$b^2 = 11^2 + 15^2 - 2(11)(15)\cos 160°$$
$$= 121 + 225 - 330 \cos 160°$$
$$\approx 656.10.$$

This gives us $b \approx 25.61$ kilometers, and we can now apply the Law of Sines to determine $\sin A$:

$$\frac{\sin A}{15} = \frac{\sin 160°}{25.61}.$$

This gives us $\sin A \approx 0.20$ and thus $A \approx 11.56°$ North of East.

TOPIC 3

Areas of Triangles

Look again at the proof of the Law of Sines. In that discussion, we determined the height of a given triangle two different ways, both of which involved the sine of an angle. The two formulas in one particular case are depicted again for reference in Figure 7.

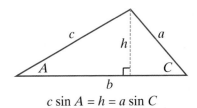

$c \sin A = h = a \sin C$

Figure 7: Two Formulas for Height

Recall that the area of a triangle is given by the formula $\left(\dfrac{1}{2}\right)(\text{base})(\text{height})$; in Figure 7, the length of the base is b, and we have two ways of determining the height. However, multiplying the base b by either expression for height results in a product of the lengths of two sides and the sine of the included angle. This same sort of product occurs if we instead consider one of the other sides of the triangle to be the base. In summary, we have derived the following.

THEOREM

Area of a Triangle
(Sine Formula)

The area of a triangle is one-half the product of the lengths of any two sides and the sine of their included angle. That is,

$$\text{Area} = \frac{1}{2}ab\sin C = \frac{1}{2}bc\sin A = \frac{1}{2}ac\sin B.$$

Remarkably, there is also a formula for triangular area that does not depend on knowing *any* of the three angles. The formula is usually called Heron's Formula (Heron was an Alexandrian mathematician of the first century AD), but there is evidence that Archimedes (287 BC–212 BC) and mathematicians of other cultures had also discovered it.

THEOREM

Area of a Triangle
(Heron's Formula)

Given a triangle with sides $a, b,$ and c, let $s = \dfrac{a+b+c}{2}$. Then

$$\text{Area} = \sqrt{s(s-a)(s-b)(s-c)}.$$

Our knowledge of trigonometry and the two laws of this section allows us to derive Heron's Formula. Beginning with the Sine Formula for area, followed by one of the Pythagorean Identities and some factoring, we know

$$\text{Area}^2 = \frac{1}{4}a^2b^2\sin^2 C$$

$$= \frac{1}{4}a^2b^2\left(1-\cos^2 C\right)$$

$$= \frac{1}{4}a^2b^2\left(1-\cos C\right)\left(1+\cos C\right).$$

By the Law of Cosines, we can rewrite the last two factors as follows.

$$1 - \cos C = 1 - \frac{a^2 + b^2 - c^2}{2ab}$$

$$= \frac{2ab - a^2 - b^2 + c^2}{2ab}$$

$$= \frac{c^2 - \left(a^2 - 2ab + b^2\right)}{2ab}$$

$$= \frac{c^2 - \left(a - b\right)^2}{2ab}$$

$$= \frac{\left(c + a - b\right)\left(c - a + b\right)}{2ab}$$

$$1 + \cos C = 1 + \frac{a^2 + b^2 - c^2}{2ab}$$

$$= \frac{2ab + a^2 + b^2 - c^2}{2ab}$$

$$= \frac{\left(a^2 + 2ab + b^2\right) - c^2}{2ab}$$

$$= \frac{\left(a + b\right)^2 - c^2}{2ab}$$

$$= \frac{\left(a + b + c\right)\left(a + b - c\right)}{2ab}$$

Replacing the above expressions for the original factors, we now have

$$\text{Area}^2 = \frac{1}{4} a^2 b^2 \left(\frac{\left(c + a - b\right)\left(c - a + b\right)}{2ab} \right)\left(\frac{\left(a + b + c\right)\left(a + b - c\right)}{2ab} \right)$$

$$= \frac{1}{16} \left(a + b + c\right)\left(c - a + b\right)\left(c + a - b\right)\left(a + b - c\right)$$

$$= \left(\frac{a + b + c}{2} \right)\left(\frac{-a + b + c}{2} \right)\left(\frac{a - b + c}{2} \right)\left(\frac{a + b - c}{2} \right)$$

$$= s(s - a)(s - b)(s - c).$$

Taking the square root of both sides completes the process.

EXAMPLE 7

Using the Sine Formula to Find the Area of a Triangle

A businessman has an opportunity to buy a triangular plot of land in a part of town known colloquially as "Five Points," as five roads intersect there to form five equal angles. The property has 147 feet of frontage on one road, and 207 feet of frontage on the other. What is the square footage of the property?

Solution:

Since the roads intersect to form five equal angles, the included angle of the lot must be $72°$. The Sine Formula for area then quickly gives us

$$\text{Area} = \frac{1}{2}(147)(207)(\sin 72°)$$

$$\approx 14{,}470 \text{ ft}^2.$$

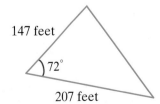

EXAMPLE 8

Using Heron's Formula to Find the Area of a Triangle

A set designer is putting together a backdrop for a play, and one element of the scene is a large triangular piece of wood. The edges of the triangle are of lengths 4 meters, 7 meters, and 9 meters. She wants to know the square area of the triangle in order to estimate the amount of paint needed to cover it.

Solution:

Heron's Formula is easily applied:

$$s = \frac{4+7+9}{2} = 10$$

4 meters   9 meters   7 meters

and so

$$\text{Area} = \sqrt{(10)(10-4)(10-7)(10-9)}$$

$$= \sqrt{(10)(6)(3)(1)}$$

$$= \sqrt{180}$$

$$= 6\sqrt{5}$$

$$\approx 13.4 \text{ m}^2.$$

1. 934.2 miles

2. 21 feet

3. 31.6 feet

4. 442.3 feet

5. 28.3 feet

6. 9.4 feet

7. 156.8 feet

8. 26.3 feet

9. 2835.6 feet

10. 32.6 feet, 26.5 feet

11. 98.1 feet

12. 2.6 feet

13. 105.5 miles

14. 4.4 feet

15. $C = 105°$, $b \approx 4.2426$, $c \approx 5.7956$

Exercises

Solve the following problems. See Examples 1 and 2.

1. A plane flies 730 miles from Charleston, SC to Cleveland, OH with a bearing of N 30° W (30° West of North). The plane then flies from Cleveland to Dallas, TX at a S 42° W bearing (42° West of South). How far is Dallas from Charleston (assume Dallas and Charleston are at the same latitude)?

2. Jack wants to build a tree house. His parents worry that he is building it too high. If Jack's dad is looking at the tree house location from a 70° angle and then moves back 10 feet so he can see it at a 50° angle, how high is the tree house location?

3. A telephone pole was recently hit by a car and now leans 6° from the vertical. A point 40 feet away from the base of the pole has an angle of elevation of 36° to the top of the pole. How tall is the pole?

6°

36°

40 feet

16. $C = 80°$,
    $b \approx 1.4845$,
    $c \approx 2.2743$

17. $C = 60°$,
    $a \approx 4.9067$,
    $c \approx 4.5221$

18. $C = 60°$,
    $a \approx 2.8794$,
    $c \approx 2.5321$

19. $A = 80°$,
    $a \approx 3.9392$,
    $b \approx 3.7588$

20. $A = 20°$,
    $a = 2.3696$
    $c = 4.4534$

21. $C = 150°$,
    $b \approx 1.0154$,
    $c \approx 2.9238$

22. $A = 50°$,
    $b \approx 3.8567$,
    $c \approx 1.9581$

23. No triangle

24. $B = 40°$,
    $C = 100°$,
    $c \approx 6.1284$

25. $A \approx 20.7048°$,
    $B \approx 114.2952°$,
    $b \approx 5.1559$

26. $A \approx 35.9044°$,
    $B \approx 4.0956°$,
    $a \approx 8.2110$

27. No triangle

28. $A \approx 19.1638°$,
    $C \approx 80.8362°$,
    $c \approx 6.0148$

29. No triangle

30. $A \approx 6.7456°$,
    $C \approx 63.2544°$,
    $c \approx 7.6026$

31. $c \approx 11.54$,
    $B \approx 31.31°$,
    $C \approx 88.69°$

32. $h \approx 6.02$, $c < h$, so triangle cannot be constructed.

4. Brandy is flying an airplane and is descending at a 10° angle towards a runway. If she can see a lake behind her that is 500 feet away from the runway at a 50° angle, how much farther does she have to fly till she lands?

5. John's lizard ran out of the house. It ran 20 feet, turned and ran 30 feet, and then turned 140° to face the house. How far away from the house is John's lizard?

6. Kristin is playing miniature golf. She hits the ball and it bounces off a brick, making a 110° angle. Her ball comes to a stop 8 feet away at a 20° angle from where it started. How far did the ball travel?

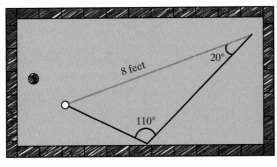

7. Janet is racing her friend Susan. Susan runs 10° away from Janet for 2000 feet. If she has to turn at a 50° angle to get back to Janet's path, how much shorter was Janet's run?

8. Two pieces of mail blew out into the yard. When Bob went to pick them up, he walked 10 feet to the first piece, turned 40° and walked to the next piece, and finally turned 150° to walk back to where he started. How far did Bob walk?

9. A bridge is suspended over a gorge shaped like an upside down isosceles triangle. If the bridge makes an 80° angle with the side of the gorge and the bridge is 1000 feet long, how deep is the gorge?

10. Brittany and Jim are playing catch. They are standing 30 feet away from each other. Ryan wants to join them and stands at a 50° angle away from Jim and at a 70° angle away from Brittany. How far away is Ryan from Jim and Ryan from Brittany?

11. Alan is golfing and sets up for a long drive. He slices it and hits a tree 80 feet away. The ball ricochets off of it at a 110° angle and comes to a stop 20° away from the direction he hit it. How far from Alan did the ball land?

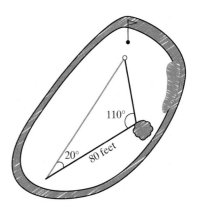

12. A ping pong net has become bent at a 70° angle instead of a 90° angle. The bottom of the net is 4.5 feet away from the end of the table. If the top of the net is 4.35 feet away from the end of the table, how high is the net?

13. An airplane has to fly between 3 airports. The trip from the 1st to the 2nd is 120 miles. After landing at the 2nd airport, the airplane must turn 140° to head toward the 3rd airport. At the 3rd airport it must turn 100° to head to the first airport. How far does the airplane have to travel from the 2nd airport to the 3rd?

33. $C = 103.1°$,
$a \approx 83.23, c \approx 90.42$

34. $C = 109°22'$,
$a \approx 0.29, b \approx 2.39$

35. $b \approx 6.78$,
$A \approx 60.90°$,
$B \approx 19.10°$

36. $c \approx 0.31$,
$B \approx 70.13°$,
$C \approx 1.87°$

37. $h \approx 0.98$. Since
$0.98 < 1.5 < 2.4$,
there can be two
triangles.
Triangle 1: $a \approx 3.33$,
$A \approx 115.40°$,
$B \approx 40.60°$
Triangle 2:
$a \approx 1.05, A \approx 16.60°$,
$B \approx 139.40°$

14. An extremely limber girl can bend over backwards at her waist making a 60° with her back. If her legs are 3.5 feet long and the distance between her hands and feet is 4 feet, what is the distance between her waist and hands?

Solve for the remaining angle and sides of the triangles. See Examples 1 and 2.

15. $A = 30°, B = 45°, a = 3$ **16.** $A = 60°, B = 40°, a = 2$

17. $A = 70°, B = 50°, b = 4$ **18.** $A = 100°, B = 20°, b = 1$

19. $B = 70°, C = 30°, c = 2$ **20.** $B = 120°, C = 40°, b = 6$

21. $A = 20°, B = 10°, a = 2$ **22.** $B = 100°, C = 30°, a = 3$

Solve for the remaining angles and side of any triangle that can be created. See Examples 3 and 4.

23. $A = 40°, a = 2, b = 4$ **24.** $A = 40°, a = 4, b = 4$

25. $C = 45°, a = 2, c = 4$ **26.** $C = 140°, b = 1, c = 9$

27. $A = 60°, a = 5, c = 6$ **28.** $B = 80°, a = 2, b = 6$

29. $B = 50°, b = 2, c = 5$ **30.** $B = 110°, a = 1, b = 8$

Construct a triangle, if possible, using the following information. Round the missing lengths and angles to two decimal places. See Examples 3 and 4.

31. $A = 60°, a = 10, b = 6$ **32.** $C = 42°, b = 9, c = 3$

33. $B = 13.2°, A = 63.7°, b = 21.2$ **34.** $A = 6°23', B = 64°15', c = 2.5$

35. $C = 100°, a = 18.1, c = 20.4$ **36.** $A = 108°, a = 9, b = 8.9$

37. $C = 24°, b = 2.4, c = 1.5$ **38.** $B = 16.9°, A = 29.7°, b = 17.8$

39. $A = 46°53', B = 74°13', c = 3.1$ **40.** $C = 116°, a = 24.1, c = 25$

41. $A = 30°, a = 15, b = 13$ **42.** $C = 74°, b = 4.5, c = 23$

Solve for the remaining angles and side of the triangles. See Examples 5 and 6.

43. $A = 60°, b = 3, c = 7$ **44.** $A = 40°, b = 2, c = 3$

45. $B = 50°, a = 4, c = 6$ **46.** $B = 45°, a = 5, c = 4$

47. $C = 30°, a = 8, b = 6$ **48.** $A = 110°, b = 2, c = 1$

49. $C = 70°, a = 5, b = 7$ **50.** $B = 100°, a = 1, c = 3$

38. $C = 133.4°$,
$a \approx 30.34, c \approx 44.49$

39. $C = 58°54'$,
$a \approx 2.64, b \approx 3.48$

40. $b \approx 1.92$,
$A \approx 60.05°$,
$B \approx 3.95°$

41. $c \approx 24.78$,
$C \approx 124.32°$,
$B \approx 25.68°$

42. $a \approx 23.83$,
$A \approx 95.16°$,
$B \approx 10.84°$

43. $a = \sqrt{37}$,
$B \approx 25.2850°$,
$C \approx 94.7150°$

44. $a \approx 1.9513$,
$B \approx 41.2114°$,
$C \approx 98.7886°$

45. $b \approx 4.5985$,
$A \approx 41.7854°$,
$C \approx 88.2146°$

46. $b \approx 3.5659$,
$A \approx 82.5159°$,
$C \approx 52.4841°$

47. $c \approx 4.1063$,
$A \approx 103.0643°$,
$B \approx 46.9357°$

48. $a \approx 2.5235$,
$B \approx 48.1378°$,
$C \approx 21.8622°$

49. $c \approx 7.0752$,
$A \approx 41.6113°$,
$B \approx 68.3887°$

50. $b \approx 3.3229$,
$A \approx 17.2395°$,
$C \approx 62.7605°$

51. $A \approx 46.5675°$,
$B \approx 104.4775°$,
$C \approx 28.9550°$

52. $A \approx 51.3178°$,
$B \approx 18.1949°$,
$C \approx 110.4873°$

53. $A \approx 121.8554°$
$B \approx 39.5712°$,
$C \approx 18.5734°$

54. $A \approx 106.6015°$,
$B \approx 25.2088°$,
$C \approx 48.1897°$

55. $A = B = C = 60°$

Solve for the angles of the given triangles. See Examples 5 and 6.

51. $a = 3, b = 4, c = 2$

52. $a = 5, b = 2, c = 6$

53. $a = 8, b = 6, c = 3$

54. $a = 9, b = 4, c = 7$

55. $a = 5, b = 5, c = 5$

56. $a = 6, b = 4, c = 7$

57. $a = 5, b = 3, c = 4$

58. $a = 7, b = 2, c = 8$

Solve the following problems. See Examples 5 and 6.

59. A log is seen floating down a stream. The log is first spotted 10 feet away. Ten seconds later the log is 70 feet away making a 60° angle between the two sightings. How far did the log travel?

60. A bullet is fired and ricochets off a metal sign 100 feet away making an 80° angle as it speeds toward a tree where it embeds itself. If the sign and tree are 60 feet apart, how far did the bullet stop from where it was fired?

61. Astronomers once thought the Sun revolved around the Earth. The Sun is 9.3×10^7 miles away and moves 15° across the sky in an hour. Assuming the Sun travels in a straight line, how far would it have had to travel?

62. A pitcher 60 feet away throws a baseball to Joey. Joey bunts the ball at a 20° angle away from the pitcher. If the ball travels 50 feet, how far does the pitcher have to run to pick up the ball?

63. Nick is surfing a wave that carries him for 20 feet. He executes a sharp spray making a 100° angle. He rides the wave for 5 feet more before he topples into the water. How far is Nick from where he started?

64. A farmer puts a piece of fence across an inside corner of his barn to make a pen for his chickens. The lengths of the sides of the pen are 7 feet, 5 feet, and 8 feet. What are the respective angles?

65. Teresa wants to make a picture frame with two 5-inch and two 12-inch pieces of wood. If the diagonal length is 13 inches, what do the inside angles have to be for the two imaginary triangles?

66. Brian is up to bat. He hits the ball straight at the pitcher 60 feet away. The ball ricochets off the pitcher's head at a 100° angle and comes to rest 40 feet away from the pitcher. How far did the ball travel away from Brian?

67. A plane took off and ascended for 1000 feet before leveling off. Once level, the plane flew for 500 feet which put it 1480 feet directly away from where it started. After leveling off, what is the angle between the plane's current horizontal flight path and its ascending flight path?

56. $A \approx 58.8114°$,
$\quad B \approx 34.7719°$,
$\quad C \approx 86.4167°$

57. $A = 90°$,
$\quad B \approx 36.8699°$,
$\quad C \approx 53.1301°$

58. $A \approx 53.5764°$,
$\quad B \approx 13.2912°$,
$\quad C \approx 113.1324°$

59. 65.5744 feet

60. 107.3137 feet

61. 2.4×10^7 miles

62. 21.4906 feet

63. 21.4413 feet

64. $60°, 38.2132°, 81.7868°$

65. $22.6199°, 67.3801°$

66. 77.6757 feet

67. 160.1188°

68. $a \approx 11.88$,
$\quad B \approx 32.28°$,
$\quad C \approx 82.72°$

69. $c \approx 8.05$,
$\quad A \approx 86.21°$,
$\quad B \approx 58.79°$

70. $b \approx 10.58$,
$\quad A \approx 31.01°$,
$\quad C \approx 124.79°$

71. $c \approx 20.04$,
$\quad B \approx 44°15'$,
$\quad A \approx 89°38'$

Construct a triangle, if possible, using the following information and the Law of Cosines. See Examples 5 and 6.

68. $A = 65°, c = 13, b = 7$

69. $C = 35°, b = 12, a = 14$

70. $B = 24.2°, a = 13.3, c = 21.2$

71. $C = 46°7', a = 27.8, b = 19.4$

72. $A = 103°, c = 8, b = 6.3$

73. $C = 75°4', b = 15.4, a = 16.8$

74. $b = 12, c = 9, a = 15$

75. $c = 4.78, b = 16.46, a = 16.54$

76. $b = 4.2, a = 7.6, c = 9.2$

77. $b = 6.84, c = 10.87, a = 7.37$

78. $a = 76.45, b = 94.45, c = 84.42$

79. $a = 5, b = 10, c = 7$

Find the area of the triangle using the following information. See Examples 7 and 8.

80. $A = 131°, b = 10, c = 25$

81. $B = 60°7', c = 18, a = 6$

82. $C = 103°, a = 10, b = 2$

83. $B = 54°, a = 10, c = 7$

84. $A = 67°49', c = 4.2, b = 9.5$

85. $C = 46°, b = 20, a = 19$

86. $A = 86°, b = 24, c = 28$

87. $b = 12, c = 18, a = 15$

Solve the following problems. See Examples 7 and 8.

88. Bob wants to build an ice skating rink in his backyard, but his wife says he can only use the part beyond the wood-chipped path running through their yard. How large would his rink be if it is triangular-shaped with sides of length 20 feet, 23 feet, and 32 feet?

89. Nancy wants to plant wildflowers between the two intersecting paths in her garden. If the paths intersect at a 72° angle and she wants the flowers to extend 12 feet down one path and 15 feet down the other, how large is the area she wants to plant?

90. The U.S.S. Cyclops mysteriously disappeared somewhere in the Bermuda Triangle in 1910. Miami, Florida; San Juan, Puerto Rico; and the Bermudas are generally accepted as the three points of the triangle. The distances from Miami to San Juan and from Miami to the Bermudas are both 908.2 nautical miles and the distance from San Juan to the Bermudas is 839.1 nautical miles. How large an area must be searched to look for the remains of the missing ship?

91. An A-frame house overlooking the Atlantic Ocean has windows entirely covering one end. If the roof intersects at a 54° angle and the roof is 21 feet long from peak to ground, how much area do the windows cover?

72. $a \approx 11.24, B \approx 33.1°$,
$\quad C \approx 43.9°$

73. $c \approx 19.65, A \approx 55°42'$,
$\quad B \approx 49°14'$

74. $A \approx 90°$, $B \approx 53.13°$,
$C \approx 36.87°$

75. $A \approx 82.62°$,
$B \approx 80.72°$, $C \approx 16.66°$

76. $A \approx 54.82°$, $B \approx 26.85°$,
$C \approx 98.33°$

77. $A \approx 41.93°$, $B \approx 38.33°$,
$C \approx 99.74°$

78. $A \approx 50.22°$,
$B \approx 71.71°$,
$C \approx 58.07°$

79. $A \approx 27.66°$,
$B \approx 111.80°$,
$C \approx 40.54°$

92. Brian just bought a used sailboat, but it needs a new sail. The dimensions of the sail are 11 feet × 12 feet × 7 feet.

a. What are the measures of the three angles of the sail?

b. How much fabric would a sail of this size require?

c. Suppose he plans to make the sail three different colors by dividing the largest angle so the 12-foot side is split into three sections of 5 feet (blue), 4 feet (yellow), and 3 feet (red), respectively. How much fabric of each color would he need?

93. Any regular (all sides are equal) *n*-sided polygon can be divided into *n* equal triangles by drawing a line from each vertex to the center of the polygon. A pentagon would be divided as shown in the figure.

a. If each side has a length of 6 inches, what would the area of the given pentagon be?

b. Using a similar method, what would be the area of an octagon with sides of length 11 inches?

c. What would be the area of a five-pointed star where each line segment has a length of 8 inches?

80. 94.34
81. 46.82
82. 9.74
83. 28.32
84. 18.47
85. 136.67
86. 335.18
87. 89.29
88. 228.7705 ft²
89. 85.5951 ft²

90. 337,940.9589 square nautical miles
91. 178.3882 ft²
92. a. 80.2801°, 64.6231°, 35.0968°
   b. 37.9473 ft²
   c. 9.4868 ft² (red),
   15.8114 ft² (blue),
   12.6491 ft² (yellow)

93. a. 61.9372 in.²
   b. 584.2397 in.²
   c. 136.1041 in.²

Polar Coordinates and Polar Equations

TOPICS

1. The polar coordinate system
2. Coordinate conversion
3. The form of polar equations
4. Graphing polar equations
T. Graphing polar equations

TOPIC 1

The Polar Coordinate System

The Cartesian coordinate system was introduced in Chapter 2, and it has served us well to this point. But there are many situations for which a rectangular coordinate system is not the most natural choice. Some planar images and some equations in two variables have a symmetry which is awkward to express in terms of the familiar x and y coordinates. Polar coordinates provide an alternative framework for these cases.

The **polar coordinate system**, like the Cartesian coordinate system, serves as a means of locating points in the plane, and both systems are centered at a point O called the **origin**, sometimes referred to as the **pole** in the polar system. Starting from the origin O, a ray (or half-line) called the **polar axis** is drawn; in practice, the polar axis is usually drawn extending horizontally to the right, so that it corresponds with the positive x-axis in the Cartesian system. Now, given any point P in the plane other than the origin, the line segment $\overline{OP}$ has a unique positive length; we will label this length r (as in *radius*). Finally, we let θ denote the angle, measured counterclockwise, between the polar axis and the segment $\overline{OP}$, and we say (r, θ) are **polar coordinates** of the point P. The origin is the unique point for which $r = 0$ and for which the angle θ is irrelevant; the coordinates $(0, \theta)$ refer to O for any angle θ. Figure 1 illustrates the process of determining r and θ for several points.

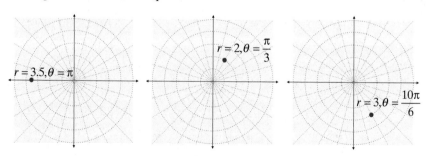

Figure 1: The Polar Coordinate System

Before proceeding further, it is important to recognize one very critical difference between Cartesian and polar coordinates. A given point P corresponds to unique Cartesian coordinates (x, y), but the polar coordinates (r, θ) of P as described above are only one of an infinite number of ways of specifying P in the polar system. Our familiarity with trigonometric functions indicates one reason for this: $(r, \theta + 2n\pi)$ also represents P, since θ and $\theta + 2n\pi$ have the same terminal sides for any integer n. Further, $(-r, \theta + (2n + 1)\pi)$ also represents P for any integer n, given the interpretation that $-r$ indicates travel in the opposite direction through the origin. These observations are illustrated in Figure 2 with alternate descriptions of the points from Figure 1.

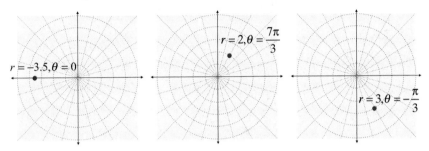

Figure 2: Alternate Polar Coordinates

EXAMPLE 1

Plotting in Polar
Coordinates

Plot the points given by the following polar coordinates.

a. $\left(2, \dfrac{3\pi}{4} \right)$
   b. $\left(3.5, -\dfrac{5\pi}{2} \right)$
   c. $\left(-1, \dfrac{4\pi}{3} \right)$

Solutions:

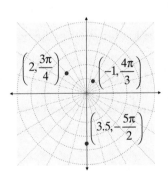

TOPIC 2 Coordinate Conversion

Since we have two systems by which to specify points in the plane, it should come as no surprise that we will occasionally need to be able to translate information from one system to the other. As we will soon see, this will be especially useful when faced with an equation which is awkward to graph in one coordinate system, but straightforward in the other. Fortunately, converting from Cartesian coordinates to polar coordinates, and vice versa, is easily accomplished. To do so, we will assume the polar axis is aligned with the positive x-axis, and that a fixed point P has Cartesian coordinates (x, y) and polar coordinates (r, θ). Then, as seen in Figure 3, $r^2 = x^2 + y^2$ and

$$\cos\theta = \frac{x}{r}, \;\; \sin\theta = \frac{y}{r}, \;\; \text{and} \;\; \tan\theta = \frac{y}{x}.$$

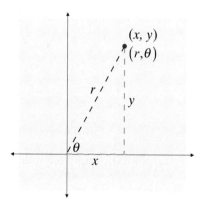

Figure 3: Coordinate Conversion

These relations, and the diagram in Figure 3, should seem very familiar—we encountered them first in defining the trigonometric functions. They are restated as conversion formulas in the box below.

DEFINITION

Coordinate Conversion

Converting from polar to Cartesian coordinates: Given (r, θ), x and y are defined by

$$x = r\cos\theta \;\; \text{and} \;\; y = r\sin\theta.$$

Converting from Cartesian to polar coordinates: Given (x, y), r and θ are defined by

$$r^2 = x^2 + y^2 \;\; \text{and} \;\; \tan\theta = \frac{y}{x} \;\; (x \neq 0).$$

Make note of the quadrant of the original Cartesian coordinate when converting, to be sure the polar coordinates fall in the same quadrant. (See Example 3.)

EXAMPLE 2

Converting from Polar to Cartesian Coordinates

Convert the following points from polar to Cartesian coordinates.

a. $\left(2, -\dfrac{\pi}{3}\right)$

b. $\left(-3, \dfrac{\pi}{4}\right)$

Solutions:

a. $x = 2\cos\left(-\dfrac{\pi}{3}\right) = 1$ and $y = 2\sin\left(-\dfrac{\pi}{3}\right) = -\sqrt{3}$, so the Cartesian coordinates are $\left(1, -\sqrt{3}\right)$.

b. $x = -3\cos\left(\dfrac{\pi}{4}\right) = -\dfrac{3}{\sqrt{2}}$ and $y = -3\sin\left(\dfrac{\pi}{4}\right) = -\dfrac{3}{\sqrt{2}}$, so the Cartesian coordinates are $\left(-\dfrac{3}{\sqrt{2}}, -\dfrac{3}{\sqrt{2}}\right)$.

EXAMPLE 3

Converting from Cartesian to Polar Coordinates

Convert the following points from Cartesian to polar coordinates.

a. $(-3, 2)$

b. $\left(\sqrt{3}, -1\right)$

Solutions:

a. To avoid careless error, make sure you have some rough idea of what the answer should be before doing any calculations. Since $(-3, 2)$ is in the second quadrant, one possible conversion will lead to $\dfrac{\pi}{2} < \theta < \pi$. To get the exact angle, we use the fact that $\tan\theta = -\dfrac{2}{3}$, so $\theta = \tan^{-1}\left(-\dfrac{2}{3}\right) \approx 2.55$ (remember, this is in radians). Depending on your calculator, though, you may have found $\tan^{-1}\left(-\dfrac{2}{3}\right) \approx -0.59$, an angle in the fourth quadrant; if so, it is up to you to remember to either add π in order to get an angle in the second quadrant, or to use a negative value for r. The radius is more easily determined: $r = \sqrt{(-3)^2 + (2)^2} = \sqrt{13}$. The two answers we have found are thus $\left(\sqrt{13}, 2.55\right)$ and $\left(-\sqrt{13}, -0.59\right)$.

b. Polar coordinates of $\left(\sqrt{3},-1\right)$ are more easily determined. The point lies in the fourth quadrant, so the two most obvious conversions will lead to either $\dfrac{3\pi}{2}<\theta<2\pi$ or $-\dfrac{\pi}{2}<\theta<0$. The coordinates $\left(\sqrt{3},-1\right)$ give rise to a familiar triangle with convenient angles, and it's probably easiest to use $\theta=-\dfrac{\pi}{6}$. The radius is $r=2$, so one set of polar coordinates for the point is $\left(2,-\dfrac{\pi}{6}\right)$.

TOPIC 3

The Form of Polar Equations

In the most general terms, a *polar equation* is an equation in the variables r and θ that defines a relationship between these polar coordinates (just as equations in x and y define a relationship between rectangular coordinates). A solution to a polar equation is thus an ordered pair (r,θ) that makes the equation true, and the graph of a polar equation consists of all such ordered pairs. Many polar equations can be written in the form $r=f(\theta)$; in such a form, the distance from the origin is expressed as a function of the angle θ and the graph of the equation is usually fairly easy to determine. We will soon study many examples of polar equations, but first we will examine the effect of translating an equation from one coordinate system to another.

The basic tools for such translation are the formulas for coordinate conversion used in Examples 2 and 3. Translating from rectangular coordinates to polar coordinates is particularly straightforward: simply replace every occurrence of x with $r\cos\theta$ and every occurrence of y with $r\sin\theta$. Translation in the opposite direction may require significantly more effort.

EXAMPLE 4

Rewriting an Equation in Polar Coordinates

Rewrite the equation $x^2-2x+y^2=0$ in polar coordinates.

Solution:

Making the appropriate substitutions in the equation $x^2-2x+y^2=0$ and simplifying, we obtain

$$\left(r\cos\theta\right)^2-2\left(r\cos\theta\right)+\left(r\sin\theta\right)^2=0$$
$$r^2\cos^2\theta-2r\cos\theta+r^2\sin^2\theta=0$$
$$r^2\left(\cos^2\theta+\sin^2\theta\right)-2r\cos\theta=0$$
$$r^2-2r\cos\theta=0.$$

EXAMPLE 5

Rewriting an Equation
in Rectangular
Coordinates

Rewrite the equation $2r = \sec \theta$ in rectangular coordinates.

Solution:

One good way to begin is to rewrite the equation as follows:

$$2r = \sec\theta$$

$$2r = \frac{1}{\cos\theta}$$

$$r\cos\theta = \frac{1}{2}.$$

Now we recognize the term on the left-hand side as x, and we see the equation is $x = \dfrac{1}{2}$.

TOPIC **Graphing Polar Equations**

Gaining a sense of confidence in your ability to graph polar equations calls for nothing more than familiarity with a number of examples. As with equations in rectangular coordinates, we will begin with some very simple, but illustrative, examples.

EXAMPLE 6

Graphing Polar
Equations

Sketch the graphs of the following polar equations, and then convert the equations to rectangular coordinates.

a. $r = 3$ **b.** $\theta = \dfrac{2\pi}{3}$

Solutions:

a. Since θ doesn't appear in the equation $r = 3$, θ is allowed to take on any value. The equation thus describes all points (r, θ) for which $r = 3$; that is, a circle of radius 3. The equation in rectangular coordinates is $x^2 + y^2 = 9$.

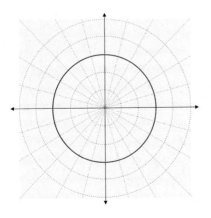

b. In this equation, r is allowed to take on any value. But every point (r, θ) that satisfies the equation must have $\theta = \dfrac{2\pi}{3}$. The graph is thus a straight line passing through the origin, and $\dfrac{y}{x} = \tan\theta = \tan\left(\dfrac{2\pi}{3}\right) = -\sqrt{3}$, so $y = -\sqrt{3}x$.

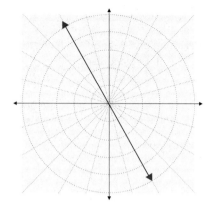

Of course, in general we expect an equation in polar coordinates to contain both r's and θ's. As with rectangular coordinates, the most basic approach to sketching the graph of a polar equation is to plot some representative points and to connect the points as seems appropriate. This method, applied judiciously and perhaps in combination with some algebra, will take us far.

EXAMPLE 7

Graphing Polar
Equations

Sketch the graph of the equation $r = 2\cos\theta$.

Solution:

We can begin by calculating some values of r for given θ's:

θ	0	$\dfrac{\pi}{6}$	$\dfrac{\pi}{4}$	$\dfrac{\pi}{3}$	$\dfrac{\pi}{2}$	$\dfrac{2\pi}{3}$	$\dfrac{3\pi}{4}$	$\dfrac{5\pi}{6}$
r	2	$\sqrt{3}$	$\sqrt{2}$	1	0	-1	$-\sqrt{2}$	$-\sqrt{3}$

Now if we plot the pairs (r, θ) from the table, we obtain the points in green shown at right. These certainly appear to lie along the circumference of a circle, so the points have been connected with the curve in purple. If we convert the equation into rectangular coordinates as shown on the next page, we see that the graph indeed is a circle of radius 1 centered at $x = 1, y = 0$.

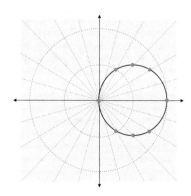

$$r = 2\cos\theta$$
$$r^2 = 2r\cos\theta$$
$$x^2 + y^2 = 2x$$
$$x^2 - 2x + y^2 = 0$$
$$x^2 - 2x + 1 + y^2 = 1$$
$$(x-1)^2 + y^2 = 1$$

We have used symmetry in the past as an aid in graphing functions and equations, and the concept is no less useful in polar coordinates. Consider the three types of symmetry illustrated in Figure 4.

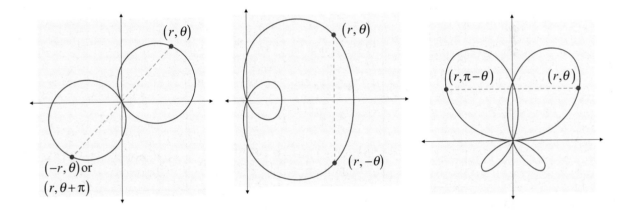

Figure 4: Symmetry in Polar Coordinates

Algebraically, symmetry can be recognized with the following tests.

DEFINITION

Symmetry of Polar Equations

An equation in r and θ is symmetric with respect to:

1. **The pole** if replacing r with $-r$ (or replacing θ with $\theta + \pi$) results in an equivalent equation.

2. **The polar axis** if replacing θ with $-\theta$ results in an equivalent equation.

3. **The line** $\theta = \dfrac{\pi}{2}$ if replacing θ with $\pi - \theta$ results in an equivalent equation.

We will conclude this section with a catalog of some polar equations that arise frequently enough to have been given names. Exploring a few of these further will give us the opportunity to apply the above symmetry tests and gain more familiarity with graphing in polar coordinates.

DEFINITION

Common Polar Equations and Graphs

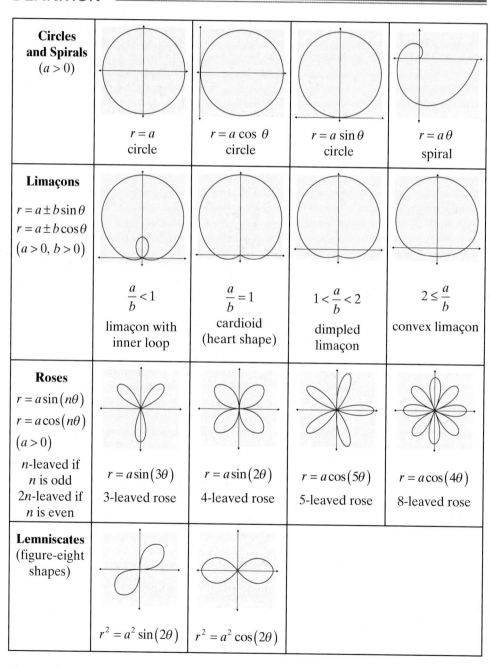

Circles and Spirals $(a > 0)$	$r = a$ circle	$r = a\cos\theta$ circle	$r = a\sin\theta$ circle	$r = a\theta$ spiral
Limaçons $r = a \pm b\sin\theta$ $r = a \pm b\cos\theta$ $(a > 0, b > 0)$	$\dfrac{a}{b} < 1$ limaçon with inner loop	$\dfrac{a}{b} = 1$ cardioid (heart shape)	$1 < \dfrac{a}{b} < 2$ dimpled limaçon	$2 \le \dfrac{a}{b}$ convex limaçon
Roses $r = a\sin(n\theta)$ $r = a\cos(n\theta)$ $(a > 0)$ n-leaved if n is odd $2n$-leaved if n is even	$r = a\sin(3\theta)$ 3-leaved rose	$r = a\sin(2\theta)$ 4-leaved rose	$r = a\cos(5\theta)$ 5-leaved rose	$r = a\cos(4\theta)$ 8-leaved rose
Lemniscates (figure-eight shapes)	$r^2 = a^2 \sin(2\theta)$	$r^2 = a^2 \cos(2\theta)$		

EXAMPLE 8

Graphing Common
Polar Equations Using
Symmetry

Use symmetry and the table on page 634 to sketch the graph of $r = 2\sin(3\theta)$.

Solution:

The graph of $r = 2\sin(3\theta)$ is a 3-leaved rose. Since $a = 2$, the maximum distance between the origin and the points on the graph is 2; this follows from the fact that $-1 \le \sin(3\theta) \le 1$ for all θ. The graph and some points on the graph from $0 \le \theta \le \pi$ in increments of $\dfrac{\pi}{12}$ are as follows.

θ	r
0	0
$\dfrac{\pi}{12}$	$\sqrt{2}$
$\dfrac{\pi}{6}$	2
$\dfrac{\pi}{4}$	$\sqrt{2}$
$\dfrac{\pi}{3}$	0
$\dfrac{5\pi}{12}$	$-\sqrt{2}$
$\dfrac{\pi}{2}$	-2

θ	r
$\dfrac{7\pi}{12}$	$-\sqrt{2}$
$\dfrac{2\pi}{3}$	0
$\dfrac{3\pi}{4}$	$\sqrt{2}$
$\dfrac{5\pi}{6}$	2
$\dfrac{11\pi}{12}$	$\sqrt{2}$
π	0

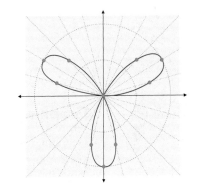

EXAMPLE 9

Graphing Common
Polar Equations
Using Symmetry

Use symmetry and the table on page 634 to sketch the graph of $r = 1 + 2\cos\theta$.

Solution:

The graph of $r = 1 + 2\cos\theta$ is a limaçon with an inner loop. Because

$$1 + 2\cos\theta = 1 + 2\cos(-\theta)$$

(since cosine is an even function), the graph is symmetric with respect to the polar axis. The graph and some points on the graph are as follows:

θ	r
0	3
$\dfrac{\pi}{3}$	2
$\dfrac{\pi}{2}$	1
$\dfrac{2\pi}{3}$	0
π	-1

θ	r
$\dfrac{4\pi}{3}$	0
$\dfrac{3\pi}{2}$	1
$\dfrac{5\pi}{3}$	2
2π	3

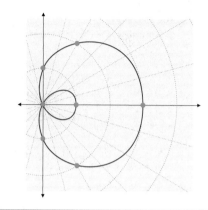

TOPIC   T   Graphing Polar Equations

Graphing calculators can be very useful in constructing accurate graphs of polar equations, although we must keep in mind that the capabilities of graphing technologies in general are limited and vary considerably.

Use the following method to graph the limaçon $r = 1 + 2\cos\theta$ (from Example 9) on a graphing calculator.

To begin, press **MODE** and make sure POL is selected on the 4ᵗʰ line. Next, press **Y=** and type in the following. (**Note:** To enter θ, press the **X,T,θ,n** button.)

Then press **WINDOW** and set the parameters as:

$\theta\,\text{min} = \emptyset,$

$\theta\,\text{max} = 2\pi,$

$\text{Xmin} = -3,$

$\text{Xmax} = 3,$

$\text{Ymin} = -2, \text{and}$

$\text{Ymax} = 2.$

Finally, press **GRAPH** and the screen should appear as shown below.

We follow the same steps to graph the 8-leaved rose $r = 3\cos(4\theta)$, however the window parameters must be adjusted to fit the entire rose on the screen.

Again, make sure mode is set to polar coordinates. Press ⬤ ⬤ and type in the following.

Then press ⬤ WINDOW and set the parameters as:

θ min = 0,

θ max = 2π,

Xmin = −3,

Xmax = 3,

Ymin = −3, and

Ymax = 3.

Finally, press ⬤ GRAPH and the screen should appear as shown below.

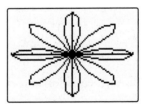

Note that a polar equation must be solved for r in order to use the graphing feature of a calculator. This means that, in order to graph a lemniscate, we solve the equation for r and then instruct the graphing calculator to graph both the positive and negative square roots that result.

Exercises

Plot the point given by the polar coordinates. See Example 1.

1. $\left(-1, \dfrac{5\pi}{4}\right)$ **2.** $\left(-5, \dfrac{3\pi}{2}\right)$ **3.** $\left(\dfrac{1}{4}, \dfrac{-7\pi}{6}\right)$

4. $\left(\sqrt{3}, \dfrac{-\pi}{3}\right)$ **5.** $\left(4\dfrac{8}{9}, -\pi\right)$ **6.** $\left(\dfrac{7}{\sqrt{2}}, \dfrac{\pi}{2}\right)$

1. $\left(-1, \dfrac{5\pi}{4}\right)$

2. $\left(-5, \dfrac{3\pi}{2}\right)$

3. $\left(\dfrac{1}{4}, -\dfrac{7\pi}{6}\right)$

4. $\left(\sqrt{3}, -\dfrac{\pi}{3}\right)$

5. $\left(4\dfrac{8}{9}, -\pi\right)$

6. $\left(\dfrac{7}{\sqrt{2}}, \dfrac{\pi}{2}\right)$

7. $(3.54, -3.54)$

8. $(0, 0)$

9. $(-4.42, -4.42)$

10. $(-1.59, -1.59)$

11. $(-2.6, -1.5)$

12. $(9.53, -5.5)$

13. $(-3, 0)$ and $(3, \pi)$

14. $\left(\sqrt{39}, 2.86\right)$ and $\left(-\sqrt{39}, -0.28\right)$

15. $\left(\sqrt{145}, -0.08\right)$ and $\left(-\sqrt{145}, 3.06\right)$

16. $(8, 0)$ and $(-8, \pi)$

17. $\left(2\sqrt{21}, 1.76\right)$ and $\left(-2\sqrt{21}, -1.38\right)$

18. $\left(5\sqrt{2}, \dfrac{5\pi}{4}\right)$ and $\left(-5\sqrt{2}, \dfrac{\pi}{4}\right)$

19. $r^2 = 25$

20. $r^2 = 81$

21. $r\cos\theta = 12$

22. $r\sin\theta = 16$

23. $\sin\theta = \cos\theta$

24. $r\sin\theta = b$

25. $r\cos\theta = 16a$

26. $r^2 = a$

27. $r^2 - 4ar\cos\theta = 0$

28. $r^2 - 4ar\sin\theta = 0$

29. $r^2\sin^2\theta - 4r\cos\theta - 4 = 0$

30. $r^2 = 36a^2$

31. $x^2 + y^2 = 5x$

32. $x^2 + y^2 = 8y$

33. $x^2 + y^2 = 49$

Convert each point from polar to Cartesian coordinates. See Example 2.

7. $\left(5, \dfrac{7\pi}{4}\right)$ **8.** $(0, 2\pi)$ **9.** $\left(6.25, \dfrac{-3\pi}{4}\right)$

10. $\left(-2.25, \dfrac{\pi}{4}\right)$ **11.** $\left(3, \dfrac{-5\pi}{6}\right)$ **12.** $\left(-11, \dfrac{10\pi}{12}\right)$

Convert each point from Cartesian to polar coordinates. See Example 3.

13. $(-3, 0)$ **14.** $\left(-6, \sqrt{3}\right)$ **15.** $(12, -1)$

16. $(8, 0)$ **17.** $\left(-\sqrt{3}, 9\right)$ **18.** $(-5, -5)$

Rewrite each equation in polar coordinates. See Example 4.

19. $x^2 + y^2 = 25$ **20.** $x^2 + y^2 = 81$ **21.** $x = 12$

22. $y = 16$ **23.** $y = x$ **24.** $y = b$

25. $x = 16a$ **26.** $x^2 + y^2 = a$ **27.** $x^2 + y^2 = 4ax$

28. $x^2 + y^2 = 4ay$ **29.** $y^2 - 4 = 4x$ **30.** $x^2 + y^2 = 36a^2$

Rewrite each equation in rectangular coordinates. See Example 5.

31. $r = 5\cos\theta$ **32.** $r = 8\sin\theta$ **33.** $r = 7$

34. $\theta = \dfrac{\pi}{6}$ **35.** $18r = 9\csc\theta$ **36.** $r = 2\sec\theta$

37. $r^2 = \sin 2\theta$ **38.** $r = \dfrac{2}{1 - \cos\theta}$ **39.** $r = \dfrac{12}{4\sin\theta + 7\cos\theta}$

40. $r = \dfrac{16}{4 + 4\sin\theta}$

Rewrite each equation in rectangular coordinates and sketch the graph. See Examples 6 and 7.

41. $r = 3$ **42.** $r = 6$ **43.** $\theta = \dfrac{5\pi}{6}$

44. $\theta = \dfrac{\pi}{4}$ **45.** $r = 7\sec\theta$ **46.** $r = 2\csc\theta$

34. $y = \dfrac{x}{\sqrt{3}}$ **35.** $y = \dfrac{1}{2}$ **36.** $x = 2$ **37.** $x^4 + y^4 + 2x^2y^2 = 2xy$ **38.** $y^2 = 4 + 4x$

39. $4y + 7x = 12$ **40.** $x^2 = 16 - 8y$

41. $x^2 + y^2 = 9$

Sketch a graph of the polar equation. See Examples 8 and 9.

47. $r = 4$ **48.** $r = 5$ **49.** $\theta = \dfrac{4\pi}{3}$

50. $\theta = \dfrac{-\pi}{3}$ **51.** $r = 6\cos\theta$ **52.** $r = 2\sin\theta$

53. $r = 3 - 3\sin\theta$ **54.** $r = 6 + 5\cos\theta$ **55.** $r = 7(1 + \cos\theta)$

42. $x^2 + y^2 = 36$

56. $r = 2(1 - 2\sin\theta)$ **57.** $r = 4 - 3\sin\theta$ **58.** $r = 3 + 4\sin\theta$

59. $r = 3\sin 3\theta$ **60.** $r = 5\sin 3\theta$ **61.** $r = 2\sin 2\theta$

62. $r = 4\sin 2\theta$ **63.** $r = 5\cos 5\theta$ **64.** $r = 4\cos 5\theta$

65. $r = 4\cos 4\theta$ **66.** $r = 3\cos 4\theta$ **67.** $r^2 = 16\sin 2\theta$

68. $r^2 = 9\cos 2\theta$

43. $y = -\dfrac{x}{\sqrt{3}}$

44. $y = x$ **45.** $x = 7$ **46.** $y = 2$

47.   **48.**   **49.**   **50.**   **51.**

52.   **53.**   **54.**   **55.**   **56.**

57.   **58.**   **59.**   **60.**   **61.**

62.   **63.**   **64.**   **65.**   **66.**

67.   **68.**

Parametric Equations

TOPICS

1. Sketching parametric curves

2. Eliminating the parameter

3. Constructing parametric equations

Sketching Parametric Curves

Many curves in the plane are most naturally defined in terms of a variable called a **parameter**, a variable different from those representing the coordinates. For instance, if (x, y) denotes the position of a thrown object, both x and y may be thought of as functions of a third variable t, representing time. In fact, thinking of the position (x, y) as a function of time t provides more information than a simple equation relating x and y: by calculating (x, y) for various values of t, we can determine not just the shape of the object's flight, but *where* the object is at any given time. If we have determined exactly how x and y depend on t, so that

$$x = f(t) \text{ and } y = g(t)$$

for some functions f and g, we have determined **parametric equations** for the curve traced out by (x, y).

The example provided by a thrown object is important enough to warrant further study, and a thorough analysis will lead to our first illustration of how parametric curves are sketched. In Section 1.7, we saw that the height of an object, moving only under the influence of the force of gravity g, is given by the formula

$$h(t) = -\frac{1}{2} g t^2 + v_0 t + h_0,$$

where v_0 is the initial (vertical) velocity and h_0 is the initial height of the object. In this formula, t represents the variable time; in the derivation that follows, t will retain this meaning, but we will now also consider t as the parameter describing the object's travel.

Only strictly vertical travel was considered in Section 1.7, but we now have the tools to analyze more realistic scenarios. We will set up our coordinate system so that the object starts off at position $(0, h_0)$ at time $t = 0$, and we will assume that it has an initial velocity of v_0 at an angle θ, as shown in Figure 1.

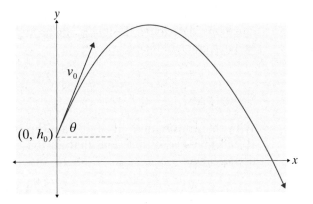

Figure 1: Describing an Object's Travel Parametrically

It is important to realize that v_0 is the size of the initial velocity in the direction θ, and that therefore the initial *vertical* velocity is $v_0 \sin \theta$ and the initial *horizontal* velocity is $v_0 \cos \theta$. If we let (x, y) stand for the position of the object at time t, the formula from Section 1.7 tells us

$$y = -\frac{1}{2} g t^2 + \left(v_0 \sin \theta \right) t + h_0,$$

so we have succeeded in writing y as a function of t. The situation in the horizontal direction is different. Because there is no force affecting the object's horizontal velocity over time (we are assuming there is no frictional force present), the horizontal displacement is simply the product of the horizontal velocity and time (remember, distance = rate $\times$ time). That is,

$$x = \left(v_0 \cos \theta \right) t.$$

We now have parametric equations defining the path of the object's travel. A specific application of these general equations follows.

EXAMPLE 1

Suppose a baseball is hit 3 feet above the ground, and that it leaves the bat at a speed of 100 miles an hour at an angle of 20° from the horizontal. Construct a set of parametric equations describing the ball's flight, and sketch a graph of the ball's travel.

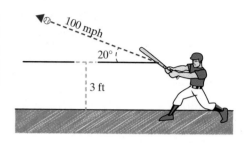

Solution:

With the information given, it's appropriate to work in units of feet and seconds. Recall that $g = 32$ ft/s², and that all the quantities must be expressed in the same units. Hence:

$$h_0 = 3 \text{ feet}$$

and

$$v_0 = 100 \ \frac{\text{miles}}{\text{hour}} = \left(\frac{100 \text{ miles}}{\text{hour}} \right) \left(\frac{5280 \text{ feet}}{1 \text{ mile}} \right) \left(\frac{1 \text{ hour}}{3600 \text{ seconds}} \right) \approx 146.7 \ \frac{\text{feet}}{\text{second}}.$$

We can now determine the parametric equations:

$$y \approx -\frac{1}{2}(32)t^2 + (146.7)\left(\sin 20° \right)t + 3$$

$$= -16t^2 + 50.2t + 3$$

and

$$x \approx (146.7)\left(\cos 20° \right)t$$

$$= 137.9t.$$

For the purposes of this application, the part of the curve we are interested in begins with $t = 0$, the moment the bat hits the ball. Note that at this time, $x = 0$ and $y = 3$; that is, the ball starts off at the point $(0, 3)$, as we intended. As time moves on, the x-coordinate increases and the y-coordinate at first increases and then decreases. Considering the physical situation giving rise to the equations, we are probably only interested in the curve up to the point where $y = 0$ (when the ball hits the ground). If we calculate (x, y) for values of t from 0 to 3.5, in half-second increments, we obtain the following table:

t	0	0.5	1.0	1.5
(x, y)	$(0, 3)$	$(69.0, 24.1)$	$(137.9, 37.2)$	$(206.9, 42.3)$

t	2.0	2.5	3.0	3.5
(x, y)	$(275.8, 39.4)$	$(344.8, 28.5)$	$(413.7, 9.6)$	$(482.7, -17.3)$

We can now plot these points and connect them, as usual, with a smooth curve. The resulting graph is:

We have completed the tasks set in Example 1, but there are still some advantages to parametric curves to point out. First, curves described by parametric equations automatically possess an **orientation**, which is the direction traveled along the curve as the parameter increases. If the parameter represents time, as in Example 1, the orientation has an obvious physical meaning: it is the direction the ball travels as it traces out the curve. Second, the parametric description of the curve allows many questions to be easily answered. Consider, for example, the following question.

EXAMPLE 2

Using Parametric
Equations

Suppose the ball in Example 1 has been hit toward a 10-foot-high fence that is 400 feet from home plate. Will the ball clear the fence?

Solution:

The graph of the ball's flight makes it appear that the answer may be yes, but pictures can be deceptive and a graph is certainly not a proof. However, we can easily resolve the situation in two steps: first, determine the time when the ball is 400 feet from home plate, and then calculate the height of the ball at that time. We determine the time by setting the equation for x equal to 400 and solving for t:

$$137.9t = 400$$

$$t = 2.9 \text{ seconds.}$$

Then we evaluate the height y at this time:

$$y = -16(2.9)^2 + 50.2(2.9) + 3$$

$$= 14.0 \text{ feet.}$$

So the ball does indeed clear the fence.

TOPIC   2   Eliminating the Parameter

While the parametric form of a curve is very useful in answering some questions and for some graphing purposes, it should not be assumed that one form or the other is better for *all* purposes. In some cases, it is convenient to convert parametric equations into a single equation relating x and y; we will call such equations rectangular equations, to make the distinction with parametric equations.

The process of converting from parametric form to rectangular form is called, appropriately enough, **eliminating the parameter**. The most common method is to solve one of the equations for the parameter and to substitute the result for the parameter in the other equation.

EXAMPLE 3

Eliminating the
Parameter

Sketch the graph described by the parametric equations $x = \dfrac{1}{\sqrt{t-1}} + 1$ and $y = \dfrac{1}{t-1}$ by eliminating the parameter.

Solution:

If we solve the second equation for t, the result is as follows:

$$y = \frac{1}{t-1}$$

$$t - 1 = \frac{1}{y}$$

$$t = \frac{1}{y} + 1.$$

Substituting this into the first equation and simplifying, we obtain an equation in x and y.

$$x = \frac{1}{\sqrt{\left(\dfrac{1}{y}+1\right)-1}} + 1$$

$$= \frac{1}{\sqrt{\dfrac{1}{y}}} + 1$$

$$= \sqrt{y} + 1$$

We are more accustomed to graphing such equations in x and y if they are solved for y, and doing so in this case leads to a curve we know well:

$$x = \sqrt{y} + 1$$

$$x - 1 = \sqrt{y}$$

$$(x-1)^2 = y.$$

As we know, the graph of this last equation is an upward-opening parabola with vertex at $(1,0)$. But before sketching this curve and proclaiming it as the answer, look again at the original parametric equations. The formulas in t are only valid for $t > 1$ (because of the presence of $\sqrt{t-1}$), and this restriction on t leads to corresponding restrictions on x and y. If we consider all possible values of t bigger than 1, we see that the restriction on y is simply $y > 0$. However, as $t \to \infty$, $x \to 1$ and as $t \to 1$, $x \to \infty$. In particular, note that there is no value of t for which $x = 1$ (or anything smaller than 1). The graph defined by the original parametric equations is thus only half of the parabola, as shown above.

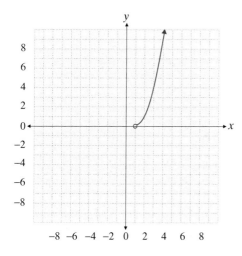

In addition to time t, angle measure θ is often used as a parameter. In fact, we have encountered parametric equations in θ already, though this was not clear at the time. Consider the following familiar example.

EXAMPLE 4

Eliminating the
Parameter

Sketch the graph described by the parametric equations $x = 5\cos\theta$ and $y = 5\sin\theta$ for $0 \le \theta \le 2\pi$.

Solution:

We have many ways of determining the shape of this curve. One way is to recognize the equations $x = 5\cos\theta$ and $y = 5\sin\theta$ as specific instances of the formulas $x = r\cos\theta$ and $y = r\sin\theta$ that relate rectangular and polar coordinates. If we make this connection, we see immediately that $r = 5$ and the curve is thus a circle of radius 5 centered at the origin.

But we don't have to rely on this insight in order to obtain the answer. Another approach is to eliminate the parameter as in the last example, and again we have several options. One is to solve for θ in one equation and substitute the result in the other equation:

$$y = 5\sin\theta$$

$$\sin\theta = \frac{y}{5}$$

$$\theta = \sin^{-1}\left(\frac{y}{5}\right)$$

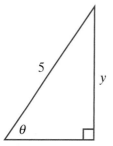

and so

$$x = 5\cos\left(\sin^{-1}\left(\frac{y}{5}\right)\right)$$

$$= 5\left(\frac{\sqrt{25-y^2}}{5}\right)$$

$$= \sqrt{25-y^2}.$$

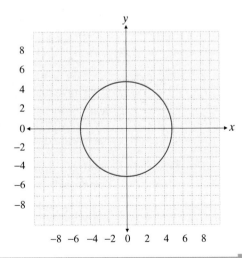

This last equation is more meaningful if we square both sides and rearrange terms slightly:

$$x = \sqrt{25-y^2}$$

$$x^2 = 25 - y^2$$

$$x^2 + y^2 = 25.$$

Using this method, we arrive at the rectangular form for the equation of a circle of radius 5 centered at the origin.

Another method of eliminating the parameter is indirect, but very useful. If we square both sides of each of the parametric equations, we can apply the Pythagorean Theorem as follows:

$$x^2 = 25\cos^2\theta \ \text{ and } \ y^2 = 25\sin^2\theta$$

so

$$x^2 + y^2 = 25\cos^2\theta + 25\sin^2\theta$$

$$= 25\left(\cos^2\theta + \sin^2\theta\right)$$

$$= 25.$$

Using any method, we arrive at the graph sketched at right.

TOPIC 3   Constructing Parametric Equations

As we saw in Examples 1 and 2, the parametric form of an equation can prove to be very useful in answering some questions. This naturally leads us to the question of *how* parametric equations describing a given curve can be constructed. We will consider several methods.

EXAMPLE 5

Constructing Parametric Equations

Construct parametric equations describing the graph of $y = x^2 + 3$.

Solution:

One hallmark of parametric descriptions of graphs is that they are not unique. Geometrically, the reason for this is that a given curve can be traced out by two parameters at different "speeds" or even in opposite directions. We will illustrate this point with two solutions to the problem in this example.

Solution 1: One parameterization of the graph is achieved by letting $x = t$ and thereby defining $y = t^2 + 3$. As t increases, the parabola defined by the parametric equations is traced out from left to right. Plot a few points to verify this claim.

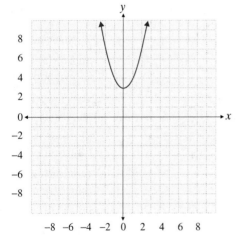

Solution 2: One of an infinite number of alternative solutions, and one that traces out the graph in the opposite direction, is achieved by letting $x = 5 - t$ (there is no special significance to the number 5 here). Now, as t increases, x decreases and so the points on the parabola will be traced out from right to left. We know this to be true even before determining that $y = x^2 + 3 = (5-t)^2 + 3 = t^2 - 10t + 28$. Again, plot a few points to verify the claim.

EXAMPLE 6

A **cycloid** is the curve (in red below) traced out by a point on a circle as it rolls along a straight line. Fix a point P on a circle of radius a, and assume that P lies initially at the origin and that the circle rolls to the right as shown. Find parametric equations describing the cycloid traced out by P.

P

Solution:

The first step is to make a wise choice of parameter. To this end, let θ measure the angle between the ray extending straight down from the circle's center and the ray extending from the circle's center through the point P. Then when the circle is in the initial position, the two rays coincide and $\theta = 0$. As the circle rolls to the right, θ increases; when the circle is in the position of the second circle shown above, $\theta = \dfrac{\pi}{2}$, and $\theta = \pi$ when the point P reaches its topmost position the first time (the third circle shown above).

Consider the enlarged diagram at right. If we let (x,y) denote the coordinates of the point P, our goal is to write x and y as functions of θ. Our first observation is that the line segment $\overline{OA}$ must have the same length as the arc $\overparen{PA}$ (since the circle rolls without slipping); the arc-length formula tells us the length of $\overparen{PA}$ is $a\theta$, so this is the length of $\overline{OA}$ as well. This means the coordinates of point C are $(a\theta, a)$ and that the coordinates of point A are $(a\theta, 0)$. The dashed line in the diagram has length $a\sin\theta$, and the other leg of the right triangle has length $a\cos\theta$. Putting this all together, we have

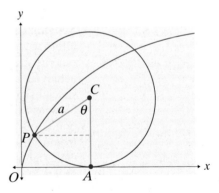

$$x + a\sin\theta = a\theta \text{ and } y + a\cos\theta = a.$$

Solving for x and y, we obtain the desired parametric equations as shown below (verifying that the equations above remain true as θ takes on larger values is left to the reader):

$$x = a(\theta - \sin\theta) \text{ and } y = a(1 - \cos\theta).$$

1.

t	x	y
0	5	0
1	6	-1
2	7	undefined
3	8	$\sqrt{3}$
4	9	1
5	10	$\dfrac{\sqrt{5}}{3}$
6	11	$\dfrac{\sqrt{6}}{4}$

2.

t	x	y
0	0	4
$\dfrac{\pi}{6}$	$\dfrac{\sqrt{3}}{6}$	$\dfrac{15}{4}$
$\dfrac{\pi}{3}$	$\dfrac{\sqrt{3}}{2}$	$\dfrac{13}{4}$
$\dfrac{\pi}{2}$	undefined	3
$\dfrac{2\pi}{3}$	$-\dfrac{\sqrt{3}}{2}$	$\dfrac{13}{4}$
$\dfrac{5\pi}{6}$	$-\dfrac{\sqrt{3}}{6}$	$\dfrac{15}{4}$
π	0	4

3. a. $x = 14.05t$

$y = -16t^2 + 15.61t + 7$

b.

c. No, he won't

Exercises

Solve the following problems concerning parametric equations. See Examples 1 and 2.

1. Given the parametric equations $x = 5 + t$ and $y = \dfrac{\sqrt{t}}{t-2}$, construct a table of the points (x, y) that result from t values from zero to six, and then sketch the curve.

2. Given the parametric equations $x = \dfrac{\tan\theta}{2}$ and $y = \cos^2\theta + 3$, construct a table of the points (x, y) that result from the values $\theta = 0, \dfrac{\pi}{6}, \dfrac{\pi}{3}, \dfrac{\pi}{2}, \dfrac{2\pi}{3}, \dfrac{5\pi}{6},$ and π. Using these values, sketch the graph of the equations.

3. François shoots a basketball at an angle of 48° from the horizontal. It leaves his hands 7 feet from the ground with a velocity of 21 ft/s.

a. Construct a set of parametric equations describing the shot.

b. Sketch a graph of the basketball's flight.

c. If the goal is 15 ft away and 11 ft high, will he make the shot?

4. On his morning paper route, John throws a newspaper from his car window 3.5 feet from the ground. The paper has an initial velocity of 10 ft/s, and is tossed at an angle of 10° from the horizontal.

a. Construct a set of parametric equations modeling the path of the newspaper.

b. Sketch a graph of the paper's path.

4. a. $x = 9.85t$

$y = -16t^2 + 1.74t + 3.5$

b.

5. Suppose that a circus performer is shot from a cannon at a rate of 80 mph, at an angle of 60° from the horizontal. The cannon sits on a platform 10 feet above ground.

a. Construct a set of parametric equations describing the performer's stunt.

b. Sketch a graph of his flight.

c. How high is the acrobat 1.5 seconds after leaving the cannon?

d. How far from the cannon should a landing net be placed, if it is placed at ground level?

e. At what time t will the performer land in the net?

f. If a 12-foot-high wall of flames is placed 70 feet from the cannon, will he clear it unharmed?

5. a. $x = 58.67t$

$y = -16t^2 + 101.61t + 10$

b.

c. 126.42 ft
d. 378.42 ft
e. 6.45 s   **f.** Yes

6. $2x = 3y + 6$

7. $y = 3x^2 + 4,$

$x \geq 0$

8. $x = 2y + 4$

9. $x = |y + 8|$

10. $y = 16x^2$

Sketch the graphs of the following parametric equations by eliminating the parameter. See Examples 3 and 4.

6. $x = 3(t + 1)$ and $y = 2t$

7. $x = \sqrt{t - 2}$ and $y = 3t - 2$

8. $x = 1 + t$ and $y = \dfrac{t - 3}{2}$

9. $x = |t + 3|$ and $y = t - 5$

10. $x = \dfrac{t}{4}$ and $y = t^2$

11. $x = \dfrac{t}{t + 2}$ and $y = \sqrt{t}$

12. $x = \sqrt{t + 3}$ and $y = t + 3$

13. $x = \dfrac{2}{|t - 3|}$ and $y = 2t - 1$

14. $x = \cos\theta$ and $y = 2\sin\theta$

15. $x = 3\sin\theta - 1$ and $y = \dfrac{\cos\theta}{2}$

16. $x = 1 - \sin\theta$ and $y = \sin\theta - 1$

17. $x = 2\cos\theta$ and $y = 3\cos\theta$

18. $x = 2\sin\theta + 2$ and $y = 2\cos\theta + 2$

19. $x = \sin\theta$ and $y = 4 - 3\cos\theta$

Construct parametric equations describing the graphs of the following equations. See Example 5.

20. $y = (x + 1)^2$

21. $y = 5x - 2$

22. $y = -x^2 - 5$

23. $x^2 + \dfrac{y^2}{4} = 1$

24. $x = y^2 + 4$

25. $y = x^2 + 1$

26. $y = \dfrac{1}{x}$

27. $x = 4y - 6$

28. $y = |x - 1|$

29. $x = 2(y - 3)$

30. $y^2 = 1 - x^2$

31. $x = \dfrac{1}{3y}$

32. $y = x^2 - x - 6$

Construct parametric equations for the line with the given attributes.

33. Slope -2, passing through $(-5, -2)$

34. Slope $\dfrac{1}{4}$, passing through $(10, 12)$

35. Slope 3, passing through $(7, 2)$

36. Passing through $(0, 0)$ and $(7, 4)$

37. Passing through $(6, -3)$ and $(2, 3)$

38. Passing through $(12, 3)$ and $(-4, -5)$

Using the given values for x, construct parametric equations describing the graph of each of the following equations.

39. $y = 3x + 1$, given that $x = 2 + t$

40. $y = 2 - |x|$, given that $x = t - 5$

41. $y = 5 - x^2$, given that $x = t + 1$

42. $5 = 2y + x$, given that $x = 4t$

43. $\dfrac{x^2}{2} + 6 = y$, given that $x = t - 4$

44. $(x - 2)^2 = y$, given that $x = 5t + 1$

11. $y = \sqrt{\dfrac{2x}{1-x}}$

Construct parametric equations for the circle with the given attributes.

45. Center $(0, 0)$, radius 1

46. Center $(-4, 2)$, radius 3

47. Center $(7, -5)$, radius 4

48. Center $(0, -2)$, radius 6

Solve the following application problems.

12. $x = \sqrt{y}$

49. A wheel of radius 12 inches rolls along a flat surface in a straight line. There is a fixed point P that initially lies at the point $(0, 0)$. Find parametric equations describing the cycloid traced out by P.

50. A ball is rolled on the floor in a straight line from one person to another person. The ball has a radius of 3 cm and there is a fixed point P located on the ball. Let the person rolling the ball represent the origin. Find parametric equations describing the cycloid traced out by P.

13. $x = \dfrac{4}{|y - 5|}$

17. $y = \dfrac{3}{2}x, -2 \le x \le 2$

14. $y = \pm 2\sqrt{1 - x^2}$

18. $y = \pm\sqrt{4x - x^2} + 2$

15. $y = \pm\dfrac{\sqrt{8 - 2x - x^2}}{6}$

19. $y = 4 \pm 3\sqrt{1 - x^2}$

16. $y = -x, 0 \le x \le 2$

20. $x = t, y = t^2 + 2t + 1$

21. $x = t, y = 5t - 2$

22. $x = t, y = -t^2 - 5$

23. $x = t, y = \pm 2\sqrt{1 - t^2}$

24. $x = t, y = \pm\sqrt{t - 4}$

25. $x = t, y = t^2 + 1$

26. $x = t, y = \dfrac{1}{t}$

27. $x = t, y = \dfrac{6 + t}{4}$

28. $x = t, y = |t - 1|$

29. $x = t, y = \dfrac{t + 6}{2}$

30. $x = t, y = \pm\sqrt{1 - t^2}$

31. $x = t, y = \dfrac{1}{3t}$

32. $x = t, y = t^2 - t - 6$

33. $x = t, y = -2t - 12$

34. $x = t, y = \dfrac{1}{4}t + \dfrac{19}{2}$

35. $x = t, y = 3t - 19$

36. $x = t, y = \dfrac{4}{7}t$

37. $x = t, y = -\dfrac{3}{2}t + 6$

38. $x = t, y = \dfrac{1}{2}t - 3$

39. $x = 2 + t, y = 3t + 7$

40. $x = t - 5, y = 2 - |t - 5|$

41. $x = t + 1, y = -t^2 - 2t + 4$

42. $x = 4t, y = -2t + \dfrac{5}{2}$

43. $x = t - 4, y = \dfrac{t^2}{2} - 4t + 14$

44. $x = 5t + 1, y = 25t^2 - 10t + 1$

45. $x = \cos\theta, y = \sin\theta$

46. $x = -4 + 3\cos\theta, y = 2 + 3\sin\theta$

47. $x = 7 + 4\cos\theta, y = -5 + 4\sin\theta$

48. $x = 6\cos\theta, y = -2 + 6\sin\theta$

49. $x = 12(\theta - \sin\theta), y = 12(1 - \cos\theta)$

50. $x = 3(\theta - \sin\theta), y = 3(1 - \cos\theta)$

8.4 Trigonometric Form of Complex Numbers

TOPICS

1. The complex plane
2. Complex numbers in trigonometric form
3. Multiplication and division of complex numbers
4. Powers of complex numbers
5. Roots of complex numbers

TOPIC 1   The Complex Plane

Complex numbers were introduced in Section 1.4, and at that point the most basic arithmetic operations were extended from the field of real numbers to the field of complex numbers. Since then you have been exposed to many algebraic and trigonometric concepts, some of which can now be put to use to enlarge our understanding of complex numbers and expand our ability to work with them.

The first idea we can put to use is that of a two-dimensional plane. We will use the plane as a way to visualize complex numbers, exactly as we use the real line to visualize real numbers; this use was foreshadowed in Topic 4 on recursive graphics in Section 3.6. We need a two-dimensional framework in order to plot complex numbers since complex numbers consist of two independent components—the real part and the imaginary part. In fact, for the purposes of graphing, we identify a given complex number $z = a + bi$ with the ordered pair (a, b). In this context, we label the horizontal axis of a rectangular coordinate system the **real axis**, and we label the vertical axis the **imaginary axis**; the plane as a whole is called the **complex plane**. Figure 1 illustrates how several complex numbers are graphed.

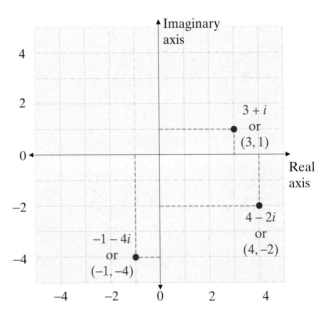

Figure 1: The Complex Plane

The association of a complex number $z = a + bi$ with the ordered pair (a, b) in the plane gives rise immediately to a way to measure the size of z.

DEFINITION

Magnitude of a Complex Number

We say the **magnitude** of a complex number $z = a + bi$, also known as the **modulus** or **absolute value** of z, is the real number

$$|z| = \sqrt{a^2 + b^2},$$

and we use $|z|$ to denote this nonnegative quantity.

The formula for $|z|$ is, of course, nothing more than the formula for the distance between (a, b) and the origin of the plane, so the magnitude of a complex number is a measure of how distant it is from the complex number $0 + 0i$ (which we usually just write as 0). Note also that if z is real, $|z|$ has the same meaning as the absolute value of a real number.

EXAMPLE 1

Finding the Magnitude

Determine the magnitudes of the following complex numbers.

a. $-2 + 5i$ **b.** $1 - 3i$

Solutions:

a. $|-2 + 5i| = \sqrt{(-2)^2 + (5)^2} = \sqrt{29}$

b. $|1 - 3i| = \sqrt{(1)^2 + (-3)^2} = \sqrt{10}$

EXAMPLE 2

Graphing Regions in the Complex Plane

Graph the regions of the complex plane defined by the following.

a. $\left\{ z \mid |z| \leq 1 \right\}$ **b.** $\left\{ z = a + bi \mid |z| \leq 1 \text{ and } b \geq 0 \right\}$

Solutions:

a. In words, $\left\{ z \mid |z| \leq 1 \right\}$ is the set of all complex numbers with magnitude less than or equal to 1. The region appears at right.

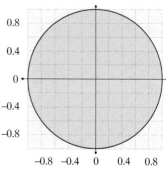

b. The region $\left\{ z = a + bi \mid |z| \leq 1 \text{ and } b \geq 0 \right\}$ consists of only those complex numbers with magnitude less than or equal to 1 and nonnegative coefficients of i. The graph of the region is a half circle as shown.

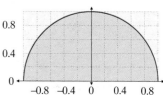

TOPIC 2   Complex Numbers in Trigonometric Form

We can also put our knowledge of trigonometry to use in describing complex numbers, with the result that many operations, such as multiplication, division, and the taking of roots, are made much easier.

To see how trigonometry applies, consider how the graph of a complex number gives rise to an angle θ in a very natural way, as shown below. Given $z = a + bi$, we know

$$\sin\theta = \frac{b}{|z|} \text{ and } \cos\theta = \frac{a}{|z|},$$

so $a = |z|\cos\theta$ and $b = |z|\sin\theta$. This leads to the following definition.

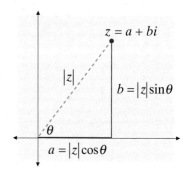

DEFINITION

Trigonometric
Form of $z = a + bi$

Given $z = a + bi$, the **trigonometric form** of z is given by

$$z = |z|\cos\theta + i|z|\sin\theta$$
$$= |z|(\cos\theta + i\sin\theta),$$

where $|z| = \sqrt{a^2 + b^2}$ and θ satisfies $\tan\theta = \dfrac{b}{a}$. The angle θ is called the **argument** of z.

In this way, we also relate the graphing of complex numbers to the polar coordinates discussed in Section 8.2. And as mentioned in that section, the angle θ corresponding to a specific point in the plane is not unique; however, any two arguments for a given complex number will differ by a multiple of 2π.

EXAMPLE 3

Writing Complex
Numbers in
Trigonometric Form

Write each of the following complex numbers in trigonometric form.

a. $-1 + i$ **b.** $3 - 4i$

Solutions:

a. $|-1 + i| = \sqrt{1+1} = \sqrt{2}$ and the argument of $-1 + i$ is $\dfrac{3\pi}{4}$, so in trigonometric form $-1 + i$ is written $\sqrt{2}\left(\cos\dfrac{3\pi}{4} + i\sin\dfrac{3\pi}{4}\right)$.

b. The magnitude of $3 - 4i$ is 5, but the argument is not so neatly expressed. In degrees, $\theta = \tan^{-1}\left(-\dfrac{4}{3}\right) \approx -53.1°$. So $3 - 4i \approx 5\left(\cos\left(-53.1°\right) + i\sin\left(-53.1°\right)\right)$.

EXAMPLE 4

Writing Complex
Numbers in
Standard Form

Write the complex number $2\left(\cos\dfrac{\pi}{6} + i\sin\dfrac{\pi}{6}\right)$ in standard form.

Solution:

$$2\left(\cos\frac{\pi}{6} + i\sin\frac{\pi}{6}\right) = 2\left(\frac{\sqrt{3}}{2} + i\frac{1}{2}\right) = \sqrt{3} + i.$$

Although it is beyond the scope of this text to prove, there is a very famous and useful identity that relates exponential functions and arguments of complex numbers. The identity is called **Euler's Formula**, and states: $e^{i\theta} = \cos\theta + i\sin\theta$. In the context of the trigonometric form of complex numbers, we can use this to write

$$z = |z|(\cos\theta + i\sin\theta) = |z|e^{i\theta}.$$

We will return to this elegant formula soon.

TOPIC 3

Multiplication and Division of Complex Numbers

The remainder of this section will show how, as promised, the trigonometric form of complex numbers simplifies certain computations. We start with multiplication and division of complex numbers.

Given two complex numbers $z_1 = |z_1|(\cos\theta_1 + i\sin\theta_1)$ and $z_2 = |z_2|(\cos\theta_2 + i\sin\theta_2)$, we can use our basic knowledge of complex multiplication to write the product as

$$z_1 z_2 = |z_1|(\cos\theta_1 + i\sin\theta_1)|z_2|(\cos\theta_2 + i\sin\theta_2)$$
$$= |z_1||z_2|\left[(\cos\theta_1\cos\theta_2 - \sin\theta_1\sin\theta_2) + i(\sin\theta_1\cos\theta_2 + \cos\theta_1\sin\theta_2)\right].$$

The combinations of sines and cosines that result should look familiar: they appear in the sum and difference trigonometric identities of Section 7.2. Applying these, we can simplify the above product to obtain

$$z_1 z_2 = |z_1||z_2|\left[\cos(\theta_1 + \theta_2) + i\sin(\theta_1 + \theta_2)\right].$$

A similar result can be obtained for the division of complex numbers, but the verification will be left to the reader. The box below summarizes the observations.

DEFINITION

Product and Quotient of Complex Numbers

Given $z_1 = |z_1|(\cos\theta_1 + i\sin\theta_1)$ and $z_2 = |z_2|(\cos\theta_2 + i\sin\theta_2)$,

$$z_1 z_2 = |z_1||z_2|\left[\cos(\theta_1 + \theta_2) + i\sin(\theta_1 + \theta_2)\right]$$

and

$$\frac{z_1}{z_2} = \frac{|z_1|}{|z_2|}\left[\cos(\theta_1 - \theta_2) + i\sin(\theta_1 - \theta_2)\right] \quad (z_2 \neq 0).$$

EXAMPLE 5

Multiplying Complex Numbers

Use the product formula on the previous page to find the product of $z_1 = 3\left(\cos\dfrac{5\pi}{6} + i\sin\dfrac{5\pi}{6}\right)$ and $z_2 = 2\left(\cos\dfrac{2\pi}{3} + i\sin\dfrac{2\pi}{3}\right)$.

Solution:

$$z_1 z_2 = (3)(2)\left(\cos\left(\frac{5\pi}{6} + \frac{2\pi}{3}\right) + i\sin\left(\frac{5\pi}{6} + \frac{2\pi}{3}\right)\right)$$

$$= 6\left(\cos\left(\frac{9\pi}{6}\right) + i\sin\left(\frac{9\pi}{6}\right)\right)$$

$$= 6\left(\cos\left(\frac{3\pi}{2}\right) + i\sin\left(\frac{3\pi}{2}\right)\right)$$

$$= -6i.$$

EXAMPLE 6

Dividing Complex Numbers

Use the quotient formula on the previous page to divide $z_1 = 6\left(\cos 117° + i\sin 117°\right)$ by $z_2 = 2\left(\cos 72° + i\sin 72°\right)$.

Solution:

$$\frac{z_1}{z_2} = \frac{6\left(\cos 117° + i\sin 117°\right)}{2\left(\cos 72° + i\sin 72°\right)}$$

$$= 3\left(\cos\left(117° - 72°\right) + i\sin\left(117° - 72°\right)\right)$$

$$= 3\left(\cos 45° + i\sin 45°\right)$$

$$= \frac{3}{\sqrt{2}} + i\frac{3}{\sqrt{2}}.$$

To remember the product and quotient formulas just derived, it's useful to note that multiplication of complex numbers corresponds to addition of the arguments, while division of complex numbers corresponds to subtraction of the arguments (along with a respective product or quotient of the magnitudes). This correspondence should seem familiar: it's exactly the behavior of multiplication and division of exponential functions. This similarity is no accident, as we see if we rephrase multiplication and division using Euler's Formula as follows.

$$z_1 z_2 = \left(|z_1| e^{i\theta_1}\right)\left(|z_2| e^{i\theta_2}\right) = |z_1||z_2| e^{i\theta_1} e^{i\theta_2} = |z_1||z_2| e^{i(\theta_1 + \theta_2)}$$

and

$$\frac{z_1}{z_2} = \frac{|z_1| e^{i\theta_1}}{|z_2| e^{i\theta_2}} = \left(\frac{|z_1|}{|z_2|}\right)\left(\frac{e^{i\theta_1}}{e^{i\theta_2}}\right) = \frac{|z_1|}{|z_2|} e^{i(\theta_1 - \theta_2)}.$$

This behavior of multiplication and division of complex numbers also manifests itself graphically, as seen in the next example.

EXAMPLE 7

Graphing Complex Numbers and Their Product

Plot the two complex numbers $2e^{i\left(\frac{\pi}{3}\right)}$ and $3e^{i\left(\frac{5\pi}{6}\right)}$ and then, on the same graph, plot their product.

Solution:

First note that, by the above formula, $\left(2e^{i\left(\frac{\pi}{3}\right)}\right)\left(3e^{i\left(\frac{5\pi}{6}\right)}\right) = 6e^{i\left(\frac{7\pi}{6}\right)}$. We can now use the magnitudes and arguments of the three complex numbers as their polar coordinates and construct the graph below. Note in particular how the argument of the product is the sum of the arguments of the two given complex numbers.

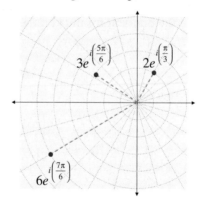

TOPIC 4   Powers of Complex Numbers

The product formula can be used to greatly speed up the process of calculating powers of complex numbers. For instance, if $z = |z|(\cos\theta + i\sin\theta)$, then

$$z^2 = \left[|z|(\cos\theta + i\sin\theta)\right]^2$$

$$= |z|^2 (\cos 2\theta + i\sin 2\theta).$$

Similarly,

$$z^3 = z^2 z$$

$$= |z|^2 \left(\cos 2\theta + i\sin 2\theta\right)|z|\left(\cos\theta + i\sin\theta\right)$$

$$= |z|^3 \left(\cos 3\theta + i\sin 3\theta\right).$$

The pattern that is developing is even more apparent if we use Euler's Formula:

$$z^n = \left(|z|e^{i\theta}\right)^n$$

$$= |z|^n e^{in\theta}.$$

These observations are named after the early 18th-century mathematician Abraham DeMoivre.

THEOREM

Powers of Complex Numbers (DeMoivre's Theorem)

Given $z = |z|\left(\cos\theta + i\sin\theta\right)$ and a positive integer n,

$$z^n = |z|^n \left(\cos n\theta + i\sin n\theta\right).$$

Using Euler's Formula, this appears in the form

$$z^n = |z|^n e^{in\theta}.$$

EXAMPLE 8

Using DeMoivre's Theorem

Use DeMoivre's Theorem to calculate $\left(\sqrt{3} - i\right)^7$.

Solution:

In one form,

$$\left(\sqrt{3} - i\right)^7 = 2^7 e^{7i\left(-\frac{\pi}{6}\right)}$$

$$= 128 e^{i\left(-\frac{7\pi}{6}\right)}$$

$$= 128 e^{i\left(\frac{5\pi}{6}\right)}.$$

In the alternative form, $\left(\sqrt{3} - i\right)^7 = 128\left(\cos\dfrac{5\pi}{6} + i\sin\dfrac{5\pi}{6}\right)$.

TOPIC **5**

Roots of Complex Numbers

DeMoivre's Theorem can be used to determine roots of complex numbers. The first step is to realize that if $z = |z|(\cos\theta + i\sin\theta)$ is a nonzero complex number and if n is a positive integer, z has n distinct n^{th} roots. This follows from an application of the Fundamental Theorem of Algebra: the nth-degree polynomial equation $w^n = z$ has n solutions (here, w is a complex variable).

One n^{th} root of z is easily determined. If we let $w_0 = |z|^{\frac{1}{n}}\left(\cos\dfrac{\theta}{n} + i\sin\dfrac{\theta}{n}\right)$, then by DeMoivre's Theorem

$$w_0{}^n = \left(|z|^{\frac{1}{n}}\right)^n\left(\cos\left(n\frac{\theta}{n}\right) + i\sin\left(n\frac{\theta}{n}\right)\right)$$

$$= |z|(\cos\theta + i\sin\theta)$$

$$= z,$$

so w_0 is indeed an nth root. But as we know, replacing θ with $\theta + 2k\pi$ results in an equivalent way of writing z. This observation leads to the following:

DEFINITION

Roots of Complex
Numbers

Given the nonzero complex number $z = |z|(\cos\theta + i\sin\theta)$ and the positive integer n,

$$w_k = |z|^{\frac{1}{n}}\left[\cos\left(\frac{\theta + 2k\pi}{n}\right) + i\sin\left(\frac{\theta + 2k\pi}{n}\right)\right]$$

defines n distinct n^{th} roots of z for $k = 0, 1, \ldots, n-1$. Alternatively,

$$w_k = |z|^{\frac{1}{n}} e^{i\left(\frac{\theta + 2k\pi}{n}\right)}.$$

EXAMPLE 9

Finding Roots of Complex Numbers

Find all the fifth roots of 1, and graph their locations in the complex plane.

Solution:

The easiest way to describe the five fifth roots uses the Euler formulation. Since 1 can be written as e^{0i} we know $\theta = 0$. Thus we have:

$$\left\{ (1)^{\frac{1}{5}} e^{i\left(\frac{0+2(0)\pi}{5}\right)}, (1)^{\frac{1}{5}} e^{i\left(\frac{0+2(1)\pi}{5}\right)}, (1)^{\frac{1}{5}} e^{i\left(\frac{0+2(2)\pi}{5}\right)}, (1)^{\frac{1}{5}} e^{i\left(\frac{0+2(3)\pi}{5}\right)}, (1)^{\frac{1}{5}} e^{i\left(\frac{0+2(4)\pi}{5}\right)} \right\}$$

which simplifies to

$$\left\{ 1, e^{i\left(\frac{2\pi}{5}\right)}, e^{i\left(\frac{4\pi}{5}\right)}, e^{i\left(\frac{6\pi}{5}\right)}, e^{i\left(\frac{8\pi}{5}\right)} \right\}.$$

The graph of the five roots appears below. Note that they are evenly spaced around a circle of radius 1.

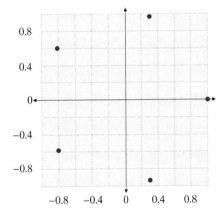

EXAMPLE 10

Finding Roots of Complex Numbers

Find all the fourth roots of $-1 - i\sqrt{3}$.

Solution:

Again, the roots are most easily written using Euler's Formula, and we begin by writing $-1 - i\sqrt{3}$ as $2e^{i\left(\frac{4\pi}{3}\right)}$. Then the four fourth roots are:

$$\left\{ 2^{\frac{1}{4}} e^{i\left(\frac{\pi}{3}\right)}, 2^{\frac{1}{4}} e^{i\left(\frac{5\pi}{6}\right)}, 2^{\frac{1}{4}} e^{i\left(\frac{4\pi}{3}\right)}, 2^{\frac{1}{4}} e^{i\left(\frac{11\pi}{6}\right)} \right\}.$$

1. $\sqrt{34}$

2. $\sqrt{10}$

3. $\sqrt{20} = 2\sqrt{5}$

4. $\sqrt{37}$

5. $\sqrt{32} = 4\sqrt{2}$

6. $\sqrt{29}$

7.

Exercises

Graph and determine the magnitudes of the following complex numbers. See Example 1.

1. $3 + 5i$ **2.** $-1 + 3i$ **3.** $2 - 4i$

4. $-6 - i$ **5.** $4 + 4i$ **6.** $5 + 2i$

Sketch z_1, z_2, $z_1 + z_2$, and $z_1 z_2$ on the same complex plane.

7. $z_1 = 7 + 2i, z_2 = -2 + 3i$ **8.** $z_1 = -1 - 2i, z_2 = 4 + 4i$

9. $z_1 = 3 + i, z_2 = 5 - i$ **10.** $z_1 = 5 - 2i, z_1 = -1 + 5i$

Graph the regions of the complex plane defined by the following. See Example 2.

11. $\left\{ z \,\middle|\, |z| < 3 \right\}$ **12.** $\left\{ z \,\middle|\, 1 \le |z| \le 4 \right\}$

13. $\left\{ z \,\middle|\, |z| \ge 1 \right\}$ **14.** $\left\{ z = a + bi \,\middle|\, a > 1,\ b > 2 \right\}$

15. $\left\{ z = a + bi \,\middle|\, a \ge b \right\}$ **16.** $\left\{ z \,\middle|\, |z| = 4 \right\}$

Write each of the following complex numbers in trigonometric form. See Example 3.

17. $-3 - i$ **18.** $5 - 3i$ **19.** $1 + 2i$

20. $3 + \sqrt{3}i$ **21.** $4 + 2i$ **22.** $5 - \sqrt{2}i$

23. $\sqrt{2} - \sqrt{2}i$ **24.** $2\sqrt{3} - 2i$ **25.** $3 + 4i$

26. $1 + i$ **27.** $4 - 4\sqrt{3}i$ **28.** $2\sqrt{2} - i$

Write each of the complex numbers in standard form. See Example 4.

29. $3\left(\cos\dfrac{5\pi}{6} + i\sin\dfrac{5\pi}{6} \right)$ **30.** $\cos\dfrac{\pi}{3} + i\sin\dfrac{\pi}{3}$

31. $2\left(\cos\dfrac{4\pi}{3} + i\sin\dfrac{4\pi}{3} \right)$ **32.** $6\left(\cos\pi + i\sin\pi \right)$

33. $5\left(\cos\dfrac{3\pi}{4} + i\sin\dfrac{3\pi}{4} \right)$ **34.** $\cos\dfrac{5\pi}{4} + i\sin\dfrac{5\pi}{4}$

35. $\dfrac{3}{2}\left(\cos 150° + i\sin 150° \right)$ **36.** $4\left(\cos 210° + i\sin 210° \right)$

37. $5\left[\cos\left(78°20' \right) + i\sin\left(78°20' \right) \right]$ **38.** $3\left[\cos\left(121°40' \right) + i\sin\left(121°40' \right) \right]$

8.

9.

10.

11.

12.

13.

14.

15.

16.

Perform the following operations and show the answer in both trigonometric form and standard form. See Examples 5 and 6.

39. $\left[4\left(\cos 60^\circ + i \sin 60^\circ\right)\right]\left[4\left(\cos 330^\circ + i \sin 330^\circ\right)\right]$

40. $\left[3\left(\cos 180^\circ + i \sin 180^\circ\right)\right]\left[4\left(\cos 30^\circ + i \sin 30^\circ\right)\right]$

41. $\left[\sqrt{2}\left(\cos \dfrac{5\pi}{4} + i \sin \dfrac{5\pi}{4}\right)\right]\left[3\sqrt{3}\left(\cos \dfrac{\pi}{6} + i \sin \dfrac{\pi}{6}\right)\right]$

42. $\left[10\left(\cos \dfrac{5\pi}{6} + i \sin \dfrac{5\pi}{6}\right)\right]\left[6\left(\cos \dfrac{2\pi}{3} + i \sin \dfrac{2\pi}{3}\right)\right]$

43. $(-1+3i)\left(\sqrt{3}+i\right)$ **44.** $2i(4+5i)$

45. $\dfrac{6\left(\cos 225^\circ + i \sin 225^\circ\right)}{3\left(\cos 45^\circ + i \sin 45^\circ\right)}$ **46.** $\dfrac{8\left(\cos 135^\circ + i \sin 135^\circ\right)}{4\left(\cos 30^\circ + i \sin 30^\circ\right)}$

47. $\dfrac{10\left(\cos \dfrac{5\pi}{3} + i \sin \dfrac{5\pi}{3}\right)}{3\left(\cos \dfrac{7\pi}{6} + i \sin \dfrac{7\pi}{6}\right)}$ **48.** $\dfrac{12\left(\cos \dfrac{10\pi}{3} + i \sin \dfrac{10\pi}{3}\right)}{6\left(\cos 2\pi + i \sin 2\pi\right)}$

49. $\dfrac{-i}{1+i}$ **50.** $\dfrac{-2-2i}{4+3i}$

51. $\dfrac{2e^{\frac{2\pi}{3}i}}{e^{\frac{\pi}{4}i}}$ **52.** $\left(e^{210^\circ i}\right)\left(e^{90^\circ i}\right)$

Plot both complex numbers and, on the same graph, plot their product. See Example 7.

53. $\left[4\left(\cos \dfrac{5\pi}{6} + i \sin \dfrac{5\pi}{6}\right)\right]\left[2\left(\cos \pi + i \sin \pi\right)\right]$

54. $\left[5\left(\cos \dfrac{3\pi}{2} + i \sin \dfrac{3\pi}{2}\right)\right]\left[3\left(\cos \dfrac{5\pi}{6} + i \sin \dfrac{5\pi}{6}\right)\right]$

55. $(2-5i)\left(\sqrt{2}+2i\right)$ **56.** $(-4-2i)\left(-\sqrt{3}+4i\right)$

57. $\left(2e^{\frac{\pi}{3}i}\right)\left(3e^{\frac{5\pi}{4}i}\right)$ **58.** $\left(5e^{\frac{5\pi}{3}i}\right)\left(e^{\pi i}\right)$

17. $\sqrt{10}\left(\cos(3.46) + i \sin(3.46)\right)$ **18.** $\sqrt{34}\left(\cos(-0.54) + i \sin(-0.54)\right)$ **19.** $\sqrt{5}\left(\cos(1.11) + i \sin(1.11)\right)$

20. $2\sqrt{3}\left(\cos\left(\dfrac{\pi}{6}\right) + i \sin\left(\dfrac{\pi}{6}\right)\right)$ **21.** $2\sqrt{5}\left(\cos(0.46) + i \sin(0.46)\right)$ **22.** $3\sqrt{3}\left(\cos(-0.28) + i \sin(-0.28)\right)$

23. $2\left(\cos\left(-\dfrac{\pi}{4}\right) + i \sin\left(-\dfrac{\pi}{4}\right)\right)$ **24.** $4\left(\cos\left(-\dfrac{\pi}{6}\right) + i \sin\left(-\dfrac{\pi}{6}\right)\right)$

25. $5(\cos 0.93 + i\sin 0.93)$

26. $\sqrt{2}\left(\cos\dfrac{\pi}{4} + i\sin\dfrac{\pi}{4}\right)$

Use DeMoivre's Theorem to calculate the following. See Example 8.

27. $8\left(\cos\left(-\dfrac{\pi}{3}\right) + i\sin\left(-\dfrac{\pi}{3}\right)\right)$

59. $\left(1-\sqrt{3}i\right)^5$

60. $\left(\dfrac{\sqrt{2}}{2}+\dfrac{\sqrt{2}}{2}i\right)^{22}$

61. $(5+3i)^{17}$

28. $3(\cos(-0.34) + i\sin(-0.34))$

62. $\left(-\sqrt{3}+i\right)^{13}$

63. $\left(\cos\dfrac{\pi}{4}+i\sin\dfrac{\pi}{4}\right)^8$

64. $\left[2\left(\cos 135° + i\sin 135°\right)\right]^4$

29. $\dfrac{-3\sqrt{3}}{2}+\dfrac{3i}{2}$

Find the indicated roots of the following and graphically represent each set in the complex plane. See Examples 9 and 10.

30. $\dfrac{1}{2}+\dfrac{i\sqrt{3}}{2}$ **31.** $-1-i\sqrt{3}$ **65.** The fourth roots of -1.

66. The cube roots of $64i$.

32. -6 **33.** $-\dfrac{5}{\sqrt{2}}+\dfrac{5i}{\sqrt{2}}$

67. The square roots of $2\sqrt{3}+2i$.

68. The fourth roots of $-1-i$.

34. $-\dfrac{1}{\sqrt{2}}-\dfrac{i}{\sqrt{2}}$

69. The fourth roots of 256.

70. The fourth roots of $16\left(\cos\dfrac{4\pi}{3}+i\sin\dfrac{4\pi}{3}\right)$.

35. $\dfrac{-3\sqrt{3}}{4}+\dfrac{3i}{4}$

71. The square roots of $4\left(\cos 120° + i\sin 120°\right)$.

36. $-2\sqrt{3}-2i$

37. $1.01+4.9i$

Solve the following equations. See Examples 9 and 10.

72. $z^3 - i = 0$

73. $z^2 - 4\sqrt{3} - 4i = 0$

74. $z^4 + 81i = 0$

38. $-1.57+2.55i$

75. $z^5 + 32 = 0$

76. $z^3 + 4\sqrt{2} - i = 0$

77. $z^2 + 25i = 0$

39. $16\left(\cos 30° + i\sin 30°\right) = 8\sqrt{3}+8i$ **40.** $12\left(\cos 210° + i\sin 210°\right) = -6\sqrt{3}-6i$ **41.** $3\sqrt{6}\left(\cos\dfrac{17\pi}{12}+i\sin\dfrac{17\pi}{12}\right) = -1.9-7.1i$

42. $60\left(\cos\dfrac{3\pi}{2}+i\sin\dfrac{3\pi}{2}\right) = -60i$ **43.** $2\sqrt{10}\left(\cos 2.42 + i\sin 2.42\right) = \left(-3-\sqrt{3}\right)+i\left(3\sqrt{3}-1\right)$

44. $2\sqrt{41}\left(\cos 2.47 + i\sin 2.47\right) = -10+8i$ **45.** $2\left(\cos 180° + i\sin 180°\right) = -2$ **46.** $2\left(\cos 105° + i\sin 105°\right) = -0.52+1.93i$

47. $\dfrac{10}{3}\left(\cos\dfrac{\pi}{2}+i\sin\dfrac{\pi}{2}\right) = \dfrac{10i}{3}$ **48.** $2\left(\cos\dfrac{4\pi}{3}+i\sin\dfrac{4\pi}{3}\right) = -1-i\sqrt{3}$ **49.** $\dfrac{1}{\sqrt{2}}\left(\cos\left(-\dfrac{3\pi}{4}\right)+i\sin\left(-\dfrac{3\pi}{4}\right)\right) = -\dfrac{1}{2}-\dfrac{i}{2}$

50. $\dfrac{2\sqrt{2}}{5}\left(\cos 3.28 + i\sin 3.28\right) = -0.56-0.08i$ **51.** $2\left(\cos\dfrac{5\pi}{12}+i\sin\dfrac{5\pi}{12}\right) = 0.52+1.93i$ **52.** $\cos 300° + i\sin 300° = \dfrac{1}{2}-\dfrac{i\sqrt{3}}{2}$

53.   **54.**   **55.**   **56.**   **57.**

58.   **59.** $32e^{\frac{\pi}{3}i}$ **60.** $e^{\frac{3\pi}{2}i}$ **61.** $1.04\times 10^{13}\,e^{2.9i}$ **62.** $8192e^{\frac{5\pi}{6}i}$ **63.** $e^{2\pi i}$ **64.** $16e^{180°i}$

65. $e^{\frac{\pi i}{4}}, e^{\frac{3\pi i}{4}}, e^{\frac{5\pi i}{4}}, e^{\frac{7\pi i}{4}}$ **66.** $4e^{\frac{\pi i}{6}}, 4e^{\frac{5\pi i}{6}}, 4e^{\frac{3\pi i}{2}}$ **67.** $2e^{\frac{\pi i}{12}}, 2e^{\frac{13\pi i}{12}}$ **68.** $2^{\frac{1}{8}}e^{\frac{5\pi i}{16}}, 2^{\frac{1}{8}}e^{\frac{13\pi i}{16}}, 2^{\frac{1}{8}}e^{\frac{21\pi i}{16}}, 2^{\frac{1}{8}}e^{\frac{29\pi i}{16}}$

8.5

Vectors in the Cartesian Plane

TOPICS

1. Vector terminology
2. Basic vector operations
3. Component form of a vector
4. Vector applications

69. $4, 4e^{\frac{\pi i}{2}}, 4e^{\pi i}, 4e^{\frac{3\pi i}{2}}$

70. $2e^{\frac{\pi i}{3}}, 2e^{\frac{5\pi i}{6}}, 2e^{\frac{4\pi i}{3}},$ $2e^{\frac{11\pi i}{6}}$

71. $2e^{60°i}, 2e^{240°i}$

72. $e^{\frac{\pi i}{6}}, e^{\frac{5\pi i}{6}}, e^{\frac{3\pi i}{2}}$

73. $2\sqrt{2}e^{\frac{\pi i}{12}}, 2\sqrt{2}e^{\frac{13\pi i}{12}}$

74. $3e^{-\frac{\pi i}{8}}, 3e^{\frac{3\pi i}{8}}, 3e^{\frac{7\pi i}{8}},$ $3e^{\frac{11\pi i}{8}}$

75. $2e^{\frac{\pi i}{5}}, 2e^{\frac{3\pi i}{5}}, 2e^{\pi i},$ $2e^{\frac{7\pi i}{5}}, 2e^{\frac{9\pi i}{5}}$

76. $1.79e^{0.99i}, 1.79e^{3.08i},$ $1.79e^{5.18i}$

77. $5e^{-\frac{\pi i}{4}}, 5e^{\frac{3\pi i}{4}}$

TOPIC 1

Vector Terminology

Many quantities are defined by their size. For example, length, area, mass, price, and temperature are fully determined by a single number; such numbers, representing only magnitude, are called **scalars**. Other quantities, however, cannot be adequately described by a single number. Force and velocity, for instance, possess both a *magnitude* and a *direction*, and a complete description of these quantities must somehow include both. Such *directed magnitudes* are called **vectors**.

We will introduce vectors in the setting of the two-dimensional plane, but the study of vectors in general constitutes an enormous area of mathematics, and vectors are easily extended into spaces of any dimension. In the plane, vectors are often represented as directed line segments (informally known as "arrows"). Such a directed line segment begins at a point P called the **initial point** and ends at a point Q called the **terminal point**, and the notation $\overrightarrow{PQ}$ is used to refer to the directed line segment. A subtle but very important point, though, is that a vector is characterized *entirely* by its direction and its magnitude, not by its initial and terminal points. That is, for a specific pair of points P and Q, $\overrightarrow{PQ}$ is only one way of depicting the vector it represents. We will use bold lowercase letters to denote vectors in general, and an expression such as $\mathbf{u} = \overrightarrow{PQ}$ means that $\mathbf{u}$ is a vector whose length is the same as the length of the line segment $\overrightarrow{PQ}$ and whose direction is defined by the initial point P and the terminal point Q. To make this important point clear, Figure 1 illustrates five different ways of depicting the one vector $\mathbf{u}$.

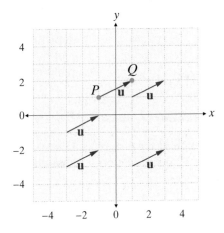

Figure 1: Five Depictions of **u**

Since vectors are characterized entirely by length and direction, we say that two vectors **u** and **v** with the same length and the same direction are **equal** (no matter where they might actually be depicted, if they are graphed) and we write **u** = **v**. The **length** of a vector may also be referred to as its **magnitude** or **norm**, and is denoted $\|\mathbf{u}\|$. In Figure 1, the length of **u** is the same as the length of the line segment from $(-1, 1)$ to $(1, 2)$, and the familiar distance formula tells us

$$\|\mathbf{u}\| = \left\|\overline{PQ}\right\| = \sqrt{\left(1-(-1)\right)^2 + \left(2-1\right)^2} = \sqrt{5}.$$

DEFINITION

Magnitude of a Vector

The length of a vector is called its **magnitude** and is denoted $\|\mathbf{u}\|$.

$$\|\mathbf{u}\| = \sqrt{a^2 + b^2}$$

where a is the horizontal displacement and b is the vertical displacement of the vector.

TOPIC 2

Basic Vector Operations

The applications of vectors point the way toward the basic ways that vectors can be combined. Consider the following scenario: wind with a velocity of **v** is affecting the flight of a plane that would be flying, if the wind were absent, with a velocity of **u**. What is the actual velocity of the plane given the presence of the wind? Both the plane and the wind possess, individually, a speed and a direction; the combination of the plane's velocity and the wind's velocity results in a third velocity that represents the actual progress of the plane.

The use of vectors facilitates such calculations. **Vector addition** is the operation called for in the above scenario; the plane's actual velocity is given by **u** + **v**, and this expression is referred to as the **sum** of the two vectors. Graphically, **u** + **v** can be represented by depicting **u**, placing **v** so that its initial point coincides with the terminal point of **u**, and then drawing a directed line segment from the initial point of **u** to the terminal point of **v**.

EXAMPLE 1

Graphing the Sum of Two Vectors

Given the two vectors **u** and **v** depicted at right, construct a graphical representation of **u** + **v**.

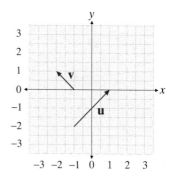

Solution:

The simplest way to proceed is to leave **u** alone and to redraw **v** so that its initial point is the terminal point of **u**. If we then construct a directed line segment from **u**'s initial point to **v**'s terminal point, the result will depict **u** + **v**. This has been done in the graph to the right, with **u** + **v** shown in green.

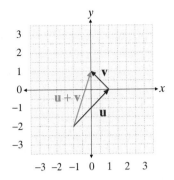

Now consider a second scenario. A person is attempting to paddle a kayak upstream, and her velocity would be **u** if the water wasn't flowing. But suppose the water *is* flowing, with a magnitude equal to the kayaker's but the exact opposite direction. If we let **v** denote the water's velocity, then **u** + **v** = **0**, where **0** is the **zero vector**. The zero vector is the unique vector which has a magnitude of 0, and in the context of vectors its function is equivalent to the number 0 on the real line (or the number 0 on the complex plane). Further, it makes sense to write **v** = −**u** and to say that −**u** is the *additive inverse* of the vector **u**.

Consider one last scenario. Suppose you are walking along a hallway with a velocity **u**, and realize you are late for a meeting. You break into a slow run, moving three times faster than you were before. Your direction is unchanged, but your magnitude is three times as large; a convenient way to denote your new velocity is 3**u**, i.e., the scalar 3 times the vector **u**. Negative numbers can multiply vectors too: (−1)**u** means the same as −**u** and represents a vector with the same magnitude as **u** but the opposite direction.

In all, the set of vectors we are describing possess properties that are very familiar to us from our study of real and complex numbers. The complete set of such properties appears below.

PROPERTIES

Properties of Vector Addition and Scalar Multiplication

Assume **u**, **v**, and **w** represent vectors, while a and b represent scalars. Then the following hold:

Vector Addition Properties	Scalar Multiplication Properties
u + **v** = **v** + **u**	$a(\mathbf{u} + \mathbf{v}) = a\mathbf{u} + a\mathbf{v}$
u + (**v** + **w**) = (**u** + **v**) + **w**	$(a + b)\mathbf{u} = a\mathbf{u} + b\mathbf{u}$
u + **0** = **u**	$(ab)\mathbf{u} = a(b\mathbf{u}) = b(a\mathbf{u})$
u + (−**u**) = **0**	1**u** = **u**, 0**u** = **0**, and $a\mathbf{0} = \mathbf{0}$
	$\|a\mathbf{u}\| = |a|\|\mathbf{u}\|$

EXAMPLE 2

Graphing Vectors

Given the two vectors **u** and **v** depicted at right, construct graphical representations of $-2\mathbf{u}$, $3\mathbf{u} + \mathbf{v}$, and $\mathbf{u} - \mathbf{v}$.

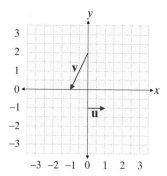

Solution:

Remember that the vectors can be placed anywhere on the plane, as long as they have the correct magnitude and direction. With a bit of planning, all three answers can be graphed in the same image, as shown to the right. Note that $\mathbf{u} - \mathbf{v}$ means the same thing as $\mathbf{u} + (-\mathbf{v})$.

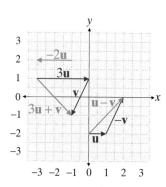

TOPIC 3   Component Form of a Vector

Our graphical work with vectors to this point has illustrated the important differences between scalars and vectors, and it has also begun to show how vectors can be used to solve certain problems. But in order to be really useful, we need a way to treat vectors in a more analytical fashion; that is, we need a form that can be used in equations.

The **component form** of a vector is one such form. Consider again the two vectors **u** and **v** from the last example. One way of completely characterizing **u** is to note that the horizontal displacement between the initial and terminal points is 1, while the vertical displacement is 0. The notation for this observation is $\mathbf{u} = \langle 1, 0 \rangle$; in a similar fashion, we would write $\mathbf{v} = \langle -1, -2 \rangle$. The similarity between this notation and ordered pair notation is no accident, as the order in which the components appear is critical in both cases. Note, however, the very important distinction between vectors and points in the plane: $(-1, -2)$ refers to

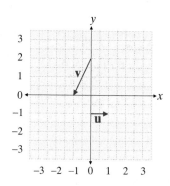

the one single point that is 1 unit to the left and 2 units down from the origin, while $\langle -1, -2 \rangle$ refers to the vector whose horizontal displacement is -1 and whose vertical displacement is -2. A depiction of $\mathbf{v} = \langle -1, -2 \rangle$ can be placed anywhere in the plane, as long as the displacement between the initial point and the terminal point is correct.

Determining the components of a vector from its initial point P and its terminal point Q is simply a matter of calculating the displacements. If $P = (x_1, y_1)$ and $Q = (x_2, y_2)$, then

$$\overrightarrow{PQ} = \langle x_2 - x_1, y_2 - y_1 \rangle.$$

The component form of a vector provides a simple and precise way of performing vector operations, as shown below.

DEFINITION

Vector Operations Using Components

Given two vectors $\mathbf{u} = \langle u_1, u_2 \rangle$ and $\mathbf{v} = \langle v_1, v_2 \rangle$ and a scalar a,

$$\mathbf{u} + \mathbf{v} = \langle u_1 + v_1, u_2 + v_2 \rangle$$

$$a\mathbf{u} = \langle au_1, au_2 \rangle$$

and

$$\|\mathbf{u}\| = \sqrt{u_1^2 + u_2^2}$$

EXAMPLE 3

Using Vector Operations

Given the two vectors $\mathbf{u}$ and $\mathbf{v}$ depicted at right, determine the component forms of $-2\mathbf{u}$, $3\mathbf{u} + \mathbf{v}$, and $\mathbf{u} - \mathbf{v}$, and then find the magnitudes of $\mathbf{u}$ and $\mathbf{v}$.

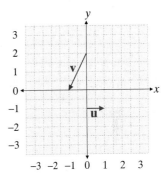

Solution:

In the previous discussion, we determined that $\mathbf{u} = \langle 1, 0 \rangle$ and $\mathbf{v} = \langle -1, -2 \rangle$; note how these components can be determined by identifying $(0, -1)$ as the initial point of $\mathbf{u}$, $(1, -1)$ as the terminal point of $\mathbf{u}$, and $(0, 2)$ and $(-1, 0)$ as the initial and terminal points of $\mathbf{v}$.

The remainder of the work is straightforward:

$$-2\mathbf{u} = \langle (-2)(1), (-2)(0) \rangle = \langle -2, 0 \rangle,$$

$$3\mathbf{u} + \mathbf{v} = \langle 3, 0 \rangle + \langle -1, -2 \rangle = \langle 2, -2 \rangle,$$

$$\mathbf{u} - \mathbf{v} = \langle 1, 0 \rangle - \langle -1, -2 \rangle = \langle 2, 2 \rangle,$$

$$\|\mathbf{u}\| = \sqrt{1^2 + 0^2} = 1,$$

$$\|\mathbf{v}\| = \sqrt{(-1)^2 + (-2)^2} = \sqrt{5}.$$

The vector **u** in the preceding two examples is just one example of a **unit vector**, which is any vector with a length of 1. There are, of course, an infinite number of unit vectors (think of any arrow of length 1 pointing in an arbitrary direction), but two of them are given special names:

$$\mathbf{i} = \langle 1, 0 \rangle \text{ and } \mathbf{j} = \langle 0, 1 \rangle.$$

In other words, **i** and **j** are the two unit vectors parallel to the coordinate axes. These two unit vectors are useful in some contexts because *any* vector in the plane can be written in terms of **i** and **j**, as follows:

$$\langle a, b \rangle = a \langle 1, 0 \rangle + b \langle 0, 1 \rangle = a\mathbf{i} + b\mathbf{j}$$

The last expression in the equation above is an example of a **linear combination**; any sum of scalar multiples of vectors constitutes a linear combination of the vectors, and the properties of linear combinations turn out to be very important in the study of vectors.

EXAMPLE 4

Finding the Unit Vector and Linear Combination of a Vector

Let $\mathbf{u} = \langle -5, 3 \rangle$. Find **a.** a unit vector pointing in the same direction as **u**, and **b.** the linear combination of **i** and **j** that is equivalent to **u**.

Solutions:

a. The key in constructing a unit vector with the same direction as **u** is to multiply **u** by an appropriate scalar a such that $\|a\mathbf{u}\| = 1$. We can solve this for a as follows:

$$\|a\mathbf{u}\| = 1$$
$$|a|\|\mathbf{u}\| = 1$$
$$a = \frac{1}{\|\mathbf{u}\|}.$$

Note that the absolute value sign around a has been dropped in the third equation, since we know in advance that we want to multiply **u** by a positive number. In our case, we have $\|\mathbf{u}\| = \sqrt{25 + 9} = \sqrt{34}$, so the desired unit vector is

$$\left(\frac{1}{\sqrt{34}} \right) \langle -5, 3 \rangle = \left\langle -\frac{5}{\sqrt{34}}, \frac{3}{\sqrt{34}} \right\rangle.$$

b. The second task is fairly straightforward: $\mathbf{u} = \langle -5, 3 \rangle = -5\mathbf{i} + 3\mathbf{j}$.

In some cases, information about a vector's magnitude and its angle with respect to the positive x-axis may be given to us. If we need to work instead with the components of the vector, our knowledge of trigonometry comes to the rescue. The right triangle in Figure 2 is familiar from our study of sine and cosine.

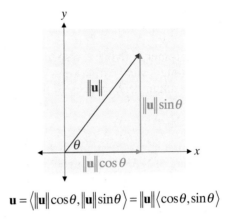

$$\mathbf{u} = \left\langle \|\mathbf{u}\| \cos\theta, \|\mathbf{u}\| \sin\theta \right\rangle = \|\mathbf{u}\| \left\langle \cos\theta, \sin\theta \right\rangle$$

Figure 2: Components Derived from Magnitude and Angle

EXAMPLE 5

Finding the Vector Form of the Velocity

A baseball is hit, and leaves the bat at a speed of 100 mph and at an angle of $20°$ from the horizontal. Express this velocity in vector form.

Solution:

The speed is the vector's magnitude, and $\theta = 20°$. So the ball's velocity as it leaves the bat is the vector

$$100\left\langle \cos 20°,\ \sin 20° \right\rangle \approx \left\langle 93.97,\ 34.20 \right\rangle.$$

TOPIC Vector Applications

Let us return to an application that served as the inspiration for vector addition.

EXAMPLE 6

Applying Vector Operations

An airplane is flying at a speed of 200 miles per hour at a bearing of N 33° E when it encounters wind with a velocity of 35 miles per hour at a bearing of N 47° W. What is the resultant true velocity of the airplane?

Solution:

The first step is to express the plane's velocity and the wind's velocity as vectors. A bearing of N 33° E means 33° East of North, but it is more convenient to think of this as 57° North of East (in keeping with our convention of measuring angles relative to the positive x-axis). Similarly, the wind's bearing of N 47° W equates to a bearing of 137° as measured from due East. If we let **p** denote the plane's velocity and **w** the wind's velocity, we have:

$$\mathbf{p} = 200\langle \cos 57°, \sin 57° \rangle \approx \langle 108.9, 167.7 \rangle$$

and

$$\mathbf{w} = 35\langle \cos 137°, \sin 137° \rangle \approx \langle -25.6, 23.9 \rangle.$$

The plane's true velocity is now simply $\mathbf{p} + \mathbf{w} = \langle 83.3, 191.6 \rangle$. It may also be useful to determine that the speed of the plane is now $\|\mathbf{p} + \mathbf{w}\| = \sqrt{(83.3)^2 + (191.6)^2} \approx 208.9$ miles per hour, and that its bearing is 66.5° North of East. This last angle is derived from the fact that $\tan\theta = \dfrac{191.6}{83.3}$, so $\theta = \tan^{-1}\dfrac{191.6}{83.3} \approx 66.5°$. The diagram below illustrates the three vectors in this problem.

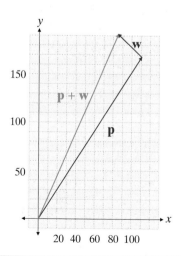

EXAMPLE 7

Applying Vector
Operations

A cat is slowly pushing a 5-pound plant across a table, with the intention of knocking it off the edge (determining why cats feel the need to do so is beyond the scope of this text). The cat is pushing with a force of 1 pound. What is the total force being applied to the plant?

Solution:

Weight is itself a force—it is the force due to gravity that the Earth exerts on an object. Forces exerted on an object are added as vectors, and the result is the total applied force. If we let $\mathbf{F}_1$ denote the weight of the plant and $\mathbf{F}_2$ the force exerted by the cat, the force on the plant is

$$\mathbf{F}_1 + \mathbf{F}_2 = \langle 0, -5 \rangle + \langle 1, 0 \rangle = \langle 1, -5 \rangle.$$

The magnitude of this total force is $\sqrt{1+25} \approx 5.1$ pounds, and the diagram below illustrates the situation (note the difference in scale between the axes).

4.

5.

6.

Exercises

Use the figure to sketch a graph for the specified vector. See Examples 1 and 2.

1. −**u**

2. 2**u** + **v**

3. 3**v**

4. $-\dfrac{1}{2}\mathbf{u} - \mathbf{v}$

5. 2**u** − 2**v**

6. **u** + 3**v**

1.

2.

3.

---

7. $\mathbf{v} = \langle 3, -3 \rangle$,
$\|\mathbf{v}\| = 3\sqrt{2}$

8. $\mathbf{v} = \langle -2, 3 \rangle$,
$\|\mathbf{v}\| = \sqrt{13}$

9. $\mathbf{v} = \langle 5, 3 \rangle$,
$\|\mathbf{v}\| = \sqrt{34}$

10. $\mathbf{v} = \langle -3, -5 \rangle$,
$\|\mathbf{v}\| = \sqrt{34}$

11. $\mathbf{v} = \langle 5, -1 \rangle$,
$\|\mathbf{v}\| = \sqrt{26}$

12. $\mathbf{v} = \langle 1, -3 \rangle$,
$\|\mathbf{v}\| = \sqrt{10}$

13. $\mathbf{v} = \langle -7, 7 \rangle$,
$\|\mathbf{v}\| = 7\sqrt{2}$

14. $\mathbf{v} = \langle -5, 7 \rangle$,
$\|\mathbf{v}\| = \sqrt{74}$

15. $\mathbf{v} = \langle -4, -6 \rangle$,
$\|\mathbf{v}\| = 2\sqrt{13}$

16. $\mathbf{v} = \langle -1, 6 \rangle$,
$\|\mathbf{v}\| = \sqrt{37}$

Find the component form and the magnitude of vector **v** for each of the following. See Example 3.

7.

8.

9.

10.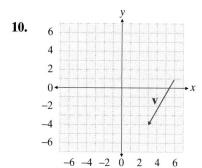

Find the component form and the magnitude of the vectors defined by the points given below. Assume the first point given is the initial point and the second point given is the terminal point. See Example 3.

11. $(-2, 4), (3, 3)$

12. $(1, 6), (2, 3)$

13. $(5, -2), (-2, 5)$

14. $(4, 0), (-1, 7)$

15. $(3, 4), (-1, -2)$

16. $(1, -6), (0, 0)$

17. a. $\langle -2, 8 \rangle$ **b.** $\langle 8, -4 \rangle$
c. $\langle -4, 0 \rangle$

18. a. $\langle 10, 7 \rangle$ **b.** $\langle 2, 14 \rangle$
c. $\langle -4, -10 \rangle$

19. a. $\langle 1, 4 \rangle$ **b.** $\langle -11, 12 \rangle$
c. $\langle 6, -8 \rangle$

20. a. $\langle 6, 10 \rangle$ **b.** $\langle 11, 9 \rangle$
c. $\langle -8, -8 \rangle$

21. a. $\langle -5, -10 \rangle$
b. $\langle -8, -2 \rangle$ **c.** $\langle 6, 4 \rangle$

22. a. $\langle -1, -8 \rangle$
b. $\langle -3, 11 \rangle$ **c.** $\langle 2, -4 \rangle$

23. $-\mathbf{u} = \langle -1, -1 \rangle$,
$2\mathbf{u} - \mathbf{v} = \langle -1, 5 \rangle$,
$\mathbf{u} + \mathbf{v} = \langle 4, -2 \rangle$,
$\|\mathbf{u}\| = \sqrt{2}$,
$\|\mathbf{v}\| = 3\sqrt{2}$

24. $-\mathbf{u} = \langle 4, -4 \rangle$,
$2\mathbf{u} - \mathbf{v} = \langle -14, 11 \rangle$,
$\mathbf{u} + \mathbf{v} = \langle 2, 1 \rangle$,
$\|\mathbf{u}\| = 4\sqrt{2}$,
$\|\mathbf{v}\| = 3\sqrt{5}$

25. $-\mathbf{u} = \langle 4, -4 \rangle$,
$2\mathbf{u} - \mathbf{v} = \langle -12, 12 \rangle$,
$\mathbf{u} + \mathbf{v} = \langle 0, 0 \rangle$,
$\|\mathbf{u}\| = 4\sqrt{2}$,
$\|\mathbf{v}\| = 4\sqrt{2}$

26. $-\mathbf{u} = \langle 3, 0 \rangle$,
$2\mathbf{u} - \mathbf{v} = \langle -6, 4 \rangle$,
$\mathbf{u} + \mathbf{v} = \langle -3, -4 \rangle$,
$\|\mathbf{u}\| = 3$,
$\|\mathbf{v}\| = 4$

For each of the following, calculate **a.** $2\mathbf{u} + \mathbf{v}$, **b.** $-\mathbf{u} + 3\mathbf{v}$, and **c.** $-2\mathbf{v}$. See Example 3.

17. $\mathbf{u} = \langle -2, 4 \rangle$, $\mathbf{v} = \langle 2, 0 \rangle$

18. $\mathbf{u} = \langle 4, 1 \rangle$, $\mathbf{v} = \langle 2, 5 \rangle$

19. $\mathbf{u} = \langle 2, 0 \rangle$, $\mathbf{v} = \langle -3, 4 \rangle$

20. $\mathbf{u} = \langle 1, 3 \rangle$, $\mathbf{v} = \langle 4, 4 \rangle$

21. $\mathbf{u} = \langle -1, -4 \rangle$, $\mathbf{v} = \langle -3, -2 \rangle$

22. $\mathbf{u} = \langle 0, -5 \rangle$, $\mathbf{v} = \langle -1, 2 \rangle$

For each of the following graphs, determine the component forms of $-\mathbf{u}$, $2\mathbf{u} - \mathbf{v}$, and $\mathbf{u} + \mathbf{v}$ and find the magnitudes of $\mathbf{u}$ and $\mathbf{v}$. See Example 3.

23.

24.

25.

26.
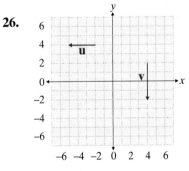

Given the vector $\mathbf{u}$, find **a.** a unit vector pointing in the same direction as $\mathbf{u}$, and **b.** the linear combination of $\mathbf{i}$ and $\mathbf{j}$ that is equivalent to $\mathbf{u}$. See Example 4.

27. $\mathbf{u} = \langle 6, -3 \rangle$

28. $\mathbf{u} = \langle 1, 4 \rangle$

29. $\mathbf{u} = \langle -5, -1 \rangle$

30. $\mathbf{u} = \langle -4, 3 \rangle$

31. $\mathbf{u} = \langle 2, 3 \rangle$

32. $\mathbf{u} = \langle 5, 2 \rangle$

Find the magnitude and direction angle of the vector $\mathbf{v}$.

33. $\mathbf{v} = 5\left(\cos 30° \mathbf{i} + \sin 30° \mathbf{j} \right)$

34. $\mathbf{v} = 7\left(\cos 45° \mathbf{i} + \sin 45° \mathbf{j} \right)$

35. $\mathbf{v} = 4\mathbf{i} + 3\mathbf{j}$

36. $\mathbf{v} = -2\mathbf{i} - 2\mathbf{j}$

27. a. $\left\langle \dfrac{2}{\sqrt{5}}, -\dfrac{1}{\sqrt{5}} \right\rangle$

    b. $\mathbf{u} = 6\mathbf{i} - 3\mathbf{j}$

28. a. $\left\langle \dfrac{1}{\sqrt{17}}, \dfrac{4}{\sqrt{17}} \right\rangle$

    b. $\mathbf{u} = \mathbf{i} + 4\mathbf{j}$

29. a. $\left\langle \dfrac{-5}{\sqrt{26}}, -\dfrac{1}{\sqrt{26}} \right\rangle$

    b. $\mathbf{u} = -5\mathbf{i} - \mathbf{j}$

30. a. $\left\langle -\dfrac{4}{5}, \dfrac{3}{5} \right\rangle$

    b. $\mathbf{u} = -4\mathbf{i} + 3\mathbf{j}$

31. a. $\left\langle \dfrac{2}{\sqrt{13}}, \dfrac{3}{\sqrt{13}} \right\rangle$

    b. $\mathbf{u} = 2\mathbf{i} + 3\mathbf{j}$

32. a. $\left\langle \dfrac{5}{\sqrt{29}}, \dfrac{2}{\sqrt{29}} \right\rangle$

    b. $\mathbf{u} = 5\mathbf{i} + 2\mathbf{j}$

33. $\|\mathbf{v}\| = 5, \theta = 30°$

34. $\|\mathbf{v}\| = 7, \theta = 45°$

35. $\|\mathbf{v}\| = 5, \theta = 36.9°$

36. $\|\mathbf{v}\| = 2\sqrt{2}, \theta = 225°$

37. $\langle 3\sqrt{3}, 3 \rangle$

38. $\left\langle \dfrac{5}{2}, 0 \right\rangle$

39. $\langle -9\sqrt{2}, 9\sqrt{2} \rangle$

40. $\langle 0, 3\sqrt{3} \rangle$

41. $\left\langle -\dfrac{1}{2}, \dfrac{\sqrt{3}}{2} \right\rangle$

42. $\langle 4, 4 \rangle$

43. $\left\langle \dfrac{8}{\sqrt{13}}, \dfrac{12}{\sqrt{13}} \right\rangle$

44. $\left\langle \dfrac{7}{\sqrt{17}}, \dfrac{28}{\sqrt{17}} \right\rangle$

45. $\langle 2\sqrt{3}, 2 \rangle$

46. $\langle 48.21, 57.45 \rangle$

Find the component form of **v** given its magnitude and the angle it makes with the positive *x*-axis. See Example 5.

37. $\|\mathbf{v}\| = 6, \theta = 30°$

38. $\|\mathbf{v}\| = \dfrac{5}{2}, \theta = 0°$

39. $\|\mathbf{v}\| = 18, \theta = 135°$

40. $\|\mathbf{v}\| = 3\sqrt{3}, \theta = 90°$

41. $\|\mathbf{v}\| = 1, \theta = 120°$

42. $\|\mathbf{v}\| = 4\sqrt{2}, \theta = 45°$

43. $\|\mathbf{v}\| = 4$, **v** in the direction of $2\mathbf{i} + 3\mathbf{j}$

44. $\|\mathbf{v}\| = 7$, **v** in the direction of $\mathbf{i} + 4\mathbf{j}$

Solve the following problems. See Examples 6 and 7.

45. A spitball is fired into the air at a speed of 4 ft/s and at an angle of 30° from the horizontal. Express this velocity in vector form.

46. A golf ball is driven into the air at a speed of 75 miles per hour and at an angle of 50° from the horizontal. Express this velocity in vector form.

47. A sailboat is traveling at a speed of 45 miles per hour with a bearing of N 59° W, when it encounters a front with winds blowing at 15 miles per hour with a bearing of S 3° E. What is the resultant true velocity of the sailboat?

48. An underwater missile is traveling at a speed of 350 miles per hour and bearing of S 17° W, when it meets a current traveling at 44 miles per hour in the direction of N 61° W. What is the resultant true velocity of the underwater missile?

49. Prometheus is slowly pushing a 1235-pound boulder across a flat plain with a force of 150 pounds. What is the total force being applied to the boulder?

50. A boy is pushing a toy truck across the floor. If the toy weighs 3 pounds and the boy is exerting half a pound of pressure on the toy, what is the total force being applied to the toy truck?

350 mph
S 17° W

44 mph
N 61° W

47. 38.67 mph, N 77.76° W

48. 343.56 mph, S 24.2° W

49. 1244.08 pounds

50. 3.04 pounds

The Dot Product and Its Uses

TOPICS

1. The dot product
2. Projections of vectors
3. Applications of the dot product

TOPIC

The Dot Product

This last section introduces a vector operation that is distinct from the vector addition and scalar multiplication that you learned in Section 8.5. The reason for introducing this third operation is simply its usefulness in performing a variety of tasks.

The operation goes by a variety of names; it is often referred to as the **dot product** of two vectors, due to the notation used, but it is also often called the **scalar product** because it is an operation that turns two vectors into a scalar. More advanced texts often refer to the operation as the **inner product**, to differentiate it from a number of other products that can be defined on vectors (including one called, as you might have guessed, the *outer product*).

DEFINITION

The Dot Product in the Plane

Given two vectors $\mathbf{u} = \langle u_1, u_2 \rangle$ and $\mathbf{v} = \langle v_1, v_2 \rangle$, the **dot product $\mathbf{u} \cdot \mathbf{v}$** of the two vectors is the scalar defined by

$$\mathbf{u} \cdot \mathbf{v} = u_1 v_1 + u_2 v_2.$$

Note that the dot product of two vectors is, indeed, a scalar, and that it can be positive, 0, or negative. Its calculation is simple, assuming the components of the two vectors are known.

EXAMPLE 1

Calculating the Dot
Product

Calculate each of the following dot products.

a. $\langle -5, 2 \rangle \cdot \langle 3, -1 \rangle$ **b.** $\langle -5, 2 \rangle \cdot \langle 2, 5 \rangle$ **c.** $\langle -5, 2 \rangle \cdot \langle -5, 2 \rangle$

Solutions:

a. $\langle -5, 2 \rangle \cdot \langle 3, -1 \rangle = (-5)(3) + (2)(-1) = -15 - 2 = -17$

b. $\langle -5, 2 \rangle \cdot \langle 2, 5 \rangle = (-5)(2) + (2)(5) = -10 + 10 = 0$

c. $\langle -5, 2 \rangle \cdot \langle -5, 2 \rangle = (-5)(-5) + (2)(2) = 25 + 4 = 29$

A number of properties of the dot product can be proven. The verifications of the following statements are left to the reader.

PROPERTIES

Elementary
Properties of the
Dot Product

Given two vectors $\mathbf{u} = \langle u_1, u_2 \rangle$ and $\mathbf{v} = \langle v_1, v_2 \rangle$ and a scalar a, the following hold:

1. $\mathbf{u} \cdot \mathbf{v} = \mathbf{v} \cdot \mathbf{u}$ **2.** $\mathbf{0} \cdot \mathbf{u} = 0$

3. $\mathbf{u} \cdot (\mathbf{v} + \mathbf{w}) = \mathbf{u} \cdot \mathbf{v} + \mathbf{u} \cdot \mathbf{w}$ **4.** $a(\mathbf{u} \cdot \mathbf{v}) = (a\mathbf{u}) \cdot \mathbf{v} = \mathbf{u} \cdot (a\mathbf{v})$

5. $\mathbf{u} \cdot \mathbf{u} = \|\mathbf{u}\|^2$

EXAMPLE 2

Using the Properties
of the Dot Product

Given that the dot product of a vector $\mathbf{u}$ with itself is 8, determine the magnitude of $\mathbf{u}$.

Solution:

Since $\|\mathbf{u}\|^2 = \mathbf{u} \cdot \mathbf{u} = 8$, it follows that $\|\mathbf{u}\| = 2\sqrt{2}$.

There is another property, however, whose truth is not so immediately clear. It often goes by the name of the Dot Product Theorem.

THEOREM

The Dot Product Theorem

Let two nonzero vectors **u** and **v** be depicted so that their initial points coincide, and let θ be the smaller of the two angles formed by **u** and **v** (so $0 \leq \theta \leq \pi$). Then

$$\mathbf{u} \cdot \mathbf{v} = \|\mathbf{u}\|\|\mathbf{v}\|\cos\theta.$$

The proof of this statement makes good use of the Law of Cosines and several of the elementary properties listed above. Starting with the two vectors **u** and **v** drawn as directed, the third side of a triangle is formed by **u** − **v** as shown in Figure 1.

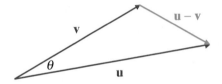

Figure 1: Proving the Dot Product Theorem

By the Law of Cosines, then, $\|\mathbf{u}-\mathbf{v}\|^2 = \|\mathbf{u}\|^2 + \|\mathbf{v}\|^2 - 2\|\mathbf{u}\|\|\mathbf{v}\|\cos\theta$. But we have a second way of expressing $\|\mathbf{u}-\mathbf{v}\|^2$:

$$\|\mathbf{u}-\mathbf{v}\|^2 = (\mathbf{u}-\mathbf{v})\cdot(\mathbf{u}-\mathbf{v})$$
$$= \mathbf{u}\cdot\mathbf{u} - \mathbf{u}\cdot\mathbf{v} - \mathbf{v}\cdot\mathbf{u} + \mathbf{v}\cdot\mathbf{v}$$
$$= \|\mathbf{u}\|^2 - 2(\mathbf{u}\cdot\mathbf{v}) + \|\mathbf{v}\|^2.$$

Equating these two different ways of expressing $\|\mathbf{u}-\mathbf{v}\|^2$ and canceling like terms, we obtain

$$-2\|\mathbf{u}\|\|\mathbf{v}\|\cos\theta = -2(\mathbf{u}\cdot\mathbf{v}),$$

and dividing both sides by −2 then gives us the desired result.

While the Dot Product Theorem is occasionally useful in calculating the dot product of two given vectors, it is more often used to determine the angle between two vectors.

EXAMPLE 3

Applying the Dot
Product Theorem

Find the angle between the vector $\mathbf{u} = \langle -3, 1 \rangle$ and the vector $\mathbf{v} = \langle 2, 5 \rangle$.

Solution:

By the Dot Product Theorem,

$$\cos\theta = \frac{\mathbf{u} \cdot \mathbf{v}}{\|\mathbf{u}\|\|\mathbf{v}\|} = \frac{\langle -3, 1 \rangle \cdot \langle 2, 5 \rangle}{\left(\sqrt{10}\right)\left(\sqrt{29}\right)} = \frac{-1}{\sqrt{290}}$$

and so

$$\theta = \cos^{-1}\left(-\frac{1}{\sqrt{290}}\right) \approx 93.4°.$$

The Dot Product Theorem also makes it clear that pairs of vectors for which θ is a right angle are easily identified by the dot product. If $\theta = \frac{\pi}{2}$ for two given vectors $\mathbf{u}$ and $\mathbf{v}$, then $\cos\theta = 0$ and hence the dot product is 0. We turn this into a definition as follows:

DEFINITION

Orthogonal Vectors

Two nonzero vectors $\mathbf{u}$ and $\mathbf{v}$ are said to be **orthogonal** (or **perpendicular**) if

$$\mathbf{u} \cdot \mathbf{v} = 0.$$

This is equivalent to saying that the angle θ defined by the two vectors is a right angle.

EXAMPLE 4

Finding an
Orthogonal Vector

Find a vector orthogonal to the vector $\mathbf{u} = \langle 7, -2 \rangle$.

Solution:

There are an infinite number of vectors $\mathbf{v}$ for which $\mathbf{u} \cdot \mathbf{v} = 0$, and hence an infinite number of correct answers. One such is $\mathbf{v} = \langle 2, 7 \rangle$, and another is $\mathbf{v} = \langle -4, -14 \rangle$. Any nonzero multiple of $\langle 2, 7 \rangle$ is orthogonal to $\mathbf{u}$.

TOPIC [2]

Projections of Vectors

The solution of many problems depends on being able to decompose a vector into a sum of two vectors in a specific way. Usually, the task amounts to writing a vector $\mathbf{u}$ as a sum $\mathbf{w}_1 + \mathbf{w}_2$ where $\mathbf{w}_1$ is parallel to a second vector $\mathbf{v}$ and $\mathbf{w}_2$ is orthogonal to $\mathbf{v}$.

In the language of vectors, we say $\mathbf{w}_1$ is the **projection of u onto v**, and write $\mathbf{w}_1 = \text{proj}_{\mathbf{v}}\mathbf{u}$. If we can determine $\text{proj}_{\mathbf{v}}\mathbf{u}$, then we can easily find $\mathbf{w}_2$ by noting that

$$\mathbf{w}_2 = \mathbf{u} - \text{proj}_{\mathbf{v}}\mathbf{u},$$

and since $\mathbf{w}_2$ is perpendicular to $\mathbf{v}$, some books refer to $\mathbf{w}_2$ as $\text{perp}_{\mathbf{v}}\mathbf{u}$.

The first step is to find a formula for $\text{proj}_{\mathbf{v}}\mathbf{u}$ in terms of $\mathbf{u}$ and $\mathbf{v}$. From our study of trigonometry and the right triangle in the diagram, we know that

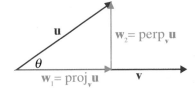

$$\cos\theta = \frac{\|\mathbf{w}_1\|}{\|\mathbf{u}\|},$$

so

$$\|\mathbf{w}_1\| = \|\mathbf{u}\|\cos\theta.$$

We can now apply the Dot Product Theorem and write

$$\|\mathbf{w}_1\| = \|\mathbf{u}\|\cos\theta = \|\mathbf{u}\|\left(\frac{\mathbf{u}\cdot\mathbf{v}}{\|\mathbf{u}\|\|\mathbf{v}\|}\right) = \frac{\mathbf{u}\cdot\mathbf{v}}{\|\mathbf{v}\|}.$$

This, however, is only the *magnitude* of the projection of $\mathbf{u}$ onto $\mathbf{v}$. In order to construct the actual vector $\mathbf{w}_1$, we need to multiply this magnitude by a unit vector pointing in the same direction as $\mathbf{v}$:

$$\mathbf{w}_1 = \left(\frac{\mathbf{u}\cdot\mathbf{v}}{\|\mathbf{v}\|}\right)\left(\frac{\mathbf{v}}{\|\mathbf{v}\|}\right) = \left(\frac{\mathbf{u}\cdot\mathbf{v}}{\|\mathbf{v}\|^2}\right)\mathbf{v}.$$

This is summarized in the following box.

DEFINITION

Projection of **u** onto **v**

Let $\mathbf{u}$ and $\mathbf{v}$ be nonzero vectors. The **projection of u onto v** is the vector

$$\text{proj}_{\mathbf{v}}\mathbf{u} = \left(\frac{\mathbf{u}\cdot\mathbf{v}}{\|\mathbf{v}\|^2}\right)\mathbf{v}.$$

EXAMPLE 5

Finding the Projection of u onto v

Find the projection of $\mathbf{u} = \langle 2, 4 \rangle$ onto $\mathbf{v} = \langle 7, -1 \rangle$, and then write $\mathbf{u}$ as a sum of two orthogonal vectors, one of which is $\text{proj}_\mathbf{v}\mathbf{u}$.

Solution:

First,

$$\text{proj}_\mathbf{v}\mathbf{u} = \left(\frac{\langle 2, 4 \rangle \cdot \langle 7, -1 \rangle}{\left\|\langle 7, -1 \rangle\right\|^2}\right)\langle 7, -1 \rangle$$

$$= \left(\frac{14-4}{\left(\sqrt{49+1}\right)^2}\right)\langle 7, -1 \rangle$$

$$= \left(\frac{1}{5}\right)\langle 7, -1 \rangle$$

$$= \left\langle \frac{7}{5}, -\frac{1}{5}\right\rangle.$$

Now, we can determine that $\text{perp}_\mathbf{v}\mathbf{u} = \langle 2, 4 \rangle - \left\langle \frac{7}{5}, -\frac{1}{5}\right\rangle = \left\langle \frac{3}{5}, \frac{21}{5}\right\rangle.$

So, we can write $\mathbf{u}$ as the sum of the two orthogonal vectors $\text{proj}_\mathbf{v}\mathbf{u}$ and $\text{perp}_\mathbf{v}\mathbf{u}$.

$$\mathbf{u} = \text{proj}_\mathbf{v}\mathbf{u} + \text{perp}_\mathbf{v}\mathbf{u} = \left\langle \frac{7}{5}, -\frac{1}{5}\right\rangle + \left\langle \frac{3}{5}, \frac{21}{5}\right\rangle = \langle 2, 4 \rangle = \mathbf{u}$$

The graphs of $\mathbf{u}$ and $\mathbf{v}$ are shown here in purple, while the orthogonal decomposition of $\mathbf{u}$ appears in green.

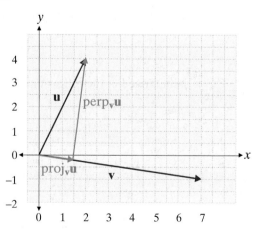

TOPIC 3

Applications of the Dot Product

The dot product and associated concepts are very useful in answering some common questions. We will close this section with a few examples.

EXAMPLE 6

Applying the Dot Product

A boat and trailer, which together weigh 650 pounds, are to be pulled up a boat ramp that has an incline of 30°. What force is required to merely prevent the boat and trailer from rolling down the ramp?

Solution:

As seen in Example 7 of Section 8.5, weight is simply an example of force: it is the force that the Earth exerts on an object through gravity. To gain some insight into this problem, suppose the question were "What is the force needed to lift the boat and trailer straight up?" The answer would be that 650 pounds of force is needed to counter the pull of gravity, so anything greater than 650 pounds of force would result in the boat and trailer being lifted.

But the actual situation isn't so extreme. Only the portion of the combined weight that is parallel to the boat ramp must be countered. That is, we are looking for the component of the 650 pounds that is directed at an angle of 30° from the horizontal. The picture at right illustrates the vectors under discussion.

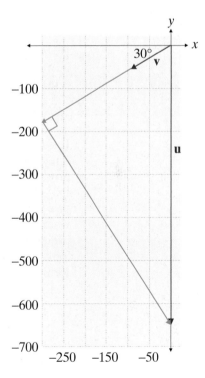

In the diagram, the long vector of magnitude 650 pointing straight down represents the weight of the boat and trailer, and the shorter purple vector points down the boat ramp. We seek the component of the weight vector in the direction of the boat ramp. The weight vector is simply $\mathbf{u} = \langle 0, -650 \rangle$. A vector of any length pointing down the boat ramp can be used; one that is easy to work with is $\mathbf{v} = \langle -\sqrt{3}, -1 \rangle$ (make sure you see how these components follow from the known angle of 30°).

Now,

$$\text{proj}_v \mathbf{u} = \left(\frac{\langle 0, -650 \rangle \cdot \langle -\sqrt{3}, -1 \rangle}{\left\| \langle -\sqrt{3}, -1 \rangle \right\|^2} \right) \langle -\sqrt{3}, -1 \rangle$$

$$= \left(\frac{(-650)(-1)}{\left(\sqrt{3}+1\right)^2} \right) \langle -\sqrt{3}, -1 \rangle$$

$$= \left(\frac{650}{4} \right) \langle -\sqrt{3}, -1 \rangle$$

$$= \left\langle -\frac{325\sqrt{3}}{2}, \frac{-325}{2} \right\rangle.$$

The magnitude of this projection, which is 325 pounds, is thus the force that must be exerted to keep the boat and trailer stationary. Anything more than this will allow the boat and trailer to be pulled up the ramp.

Work has a technical meaning in physics and engineering, and the dot product is very useful in calculating how much work is required to accomplish a task. Technically, **work** is the application of a force through a certain distance. For instance, applying a force of 375 pounds to pull the boat and trailer of Example 6 a distance of 20 feet up the ramp results in a certain amount of work being done.

It is important to realize, though, that only the component of the force applied in the direction of motion contributes to the work done. Fortunately, we can use the dot product to make this calculation. If we let $\mathbf{D}$ represent the vector of the motion (so that $\|\mathbf{D}\|$ is the distance traveled) and $\mathbf{F}$ the force applied (not necessarily in the same direction as $\mathbf{D}$), then the component of the force in the direction of motion is $\|\mathbf{F}\| \cos\theta$, where θ is, as usual, the angle between the two vectors. The work done is then the product of $\|\mathbf{D}\|$ and $\|\mathbf{F}\| \cos\theta$, that is, $W = \|\mathbf{D}\| \|\mathbf{F}\| \cos\theta$. But this last expression should look familiar—it appears in the Dot Product Theorem, allowing us to write the simpler formula below:

$$W = \mathbf{F} \cdot \mathbf{D}.$$

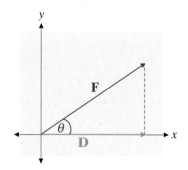

EXAMPLE 7

Applying the Dot Product

A child pulls a wagon along a sidewalk, exerting a force of 15 pounds on the handle of the wagon. The handle is at an angle of 40° to the horizontal. If the child pulls the wagon a distance of 50 feet, what work has been done?

Solution:

We start by defining the force and distance vectors:

$$\mathbf{F} = 15\langle \cos 40°, \sin 40° \rangle \text{ and } \mathbf{D} = \langle 50, 0 \rangle.$$

The calculation is now straightforward:

$$W = 15\langle \cos 40°, \sin 40° \rangle \cdot \langle 50, 0 \rangle \approx (15)(0.766)(50) + (15)(0.643)(0) = 574.5 \text{ foot-pounds.}$$

Exercises

Calculate each of the following dot products. See Example 1.

1. $\langle 4, 3 \rangle \cdot \langle 5, -1 \rangle$ 　　　**2.** $\langle 2, 4 \rangle \cdot \langle -1, -1 \rangle$ 　　　**3.** $\langle 3, 5 \rangle \cdot \langle 2, 0 \rangle$

4. $\langle -1, 6 \rangle \cdot \langle 6, 1 \rangle$ 　　　**5.** $\langle 2, 2 \rangle \cdot \langle 2, 2 \rangle$ 　　　**6.** $\langle 1, 2 \rangle \cdot \langle 3, 4 \rangle$

7. $\langle -4, 3 \rangle \cdot \langle 2, 3 \rangle$ 　　　**8.** $\langle -2, -4 \rangle \cdot \langle 6, 2 \rangle$ 　　　**9.** $\mathbf{u} = 5\mathbf{i} + \mathbf{j}, \mathbf{v} = -2\mathbf{i} + 3\mathbf{j}$

10. $\mathbf{u} = \mathbf{i} - 5\mathbf{j}, \mathbf{v} = -2\mathbf{i} - 4\mathbf{j}$

Find the indicated quantity given $\mathbf{u} = \langle -2, 3 \rangle$ and $\mathbf{v} = \langle 4, 4 \rangle$. See Example 1.

11. $\mathbf{v} \cdot \mathbf{v}$ 　　　**12.** $4\mathbf{u} \cdot \mathbf{v}$ 　　　**13.** $\|\mathbf{u}\| + 2$ 　　　**14.** $(\mathbf{u} \cdot \mathbf{v})2\mathbf{v}$

Find the magnitude of $\mathbf{u}$ using the dot product. See Example 2.

15. $\mathbf{u} = \langle 6, -1 \rangle$ 　　**16.** $\mathbf{u} = \langle 10, 3 \rangle$ 　　**17.** $\mathbf{u} = 2\mathbf{i} + 7\mathbf{j}$ 　　**18.** $\mathbf{u} = -3\mathbf{i} + 4\mathbf{j}$

Find the angle between the given vectors. See Example 3.

19. $\mathbf{u} = \langle -2, 3 \rangle, \mathbf{v} = \langle 1, 0 \rangle$ 　　　　**20.** $\mathbf{u} = \langle 5, 4 \rangle, \mathbf{v} = \langle 3, 2 \rangle$

21. $\mathbf{u} = \langle 3, 5 \rangle, \mathbf{v} = \langle 4, 4 \rangle$ 　　　　**22.** $\mathbf{u} = \langle -4, 2 \rangle, \mathbf{v} = \langle 1, 5 \rangle$

23. $\mathbf{u} = -\mathbf{i} + 2\mathbf{j}, \mathbf{v} = 3\mathbf{i} - 3\mathbf{j}$ 　　　　**24.** $\mathbf{u} = 5\mathbf{i} + 2\mathbf{j}, \mathbf{v} = 4\mathbf{i} + \mathbf{j}$

25. $\dfrac{\pi}{4}$

26. $\dfrac{7\pi}{12} \approx 1.8326$

27. $8°, 69°, 103°$

28. $26.6°, 45°, 108.4°$

29. $57.1°, 60.8°, 62.1°$

30. $4.3°, 35.5°, 140.2°$

31. -62.5 **32.** $128\sqrt{3}$

33. $-32\sqrt{2}$ **34.** -45

35. $\langle 1, 1 \rangle, \langle 5, 5 \rangle$

36. $\langle -1, 4 \rangle, \langle 1, -4 \rangle$

37. $\langle 3, 1 \rangle, \langle -6, -2 \rangle$

38. $\langle -4, 5 \rangle, \langle 4, -5 \rangle$

39. Neither

40. Parallel

41. Orthogonal

42. Neither

43. $\text{proj}_v \mathbf{u} = \langle 2, 1 \rangle,$
$\text{perp}_v \mathbf{u} = \langle -1, 2 \rangle$

44. $\text{proj}_v \mathbf{u} = \left\langle -\dfrac{6}{25}, \dfrac{42}{25} \right\rangle,$
$\text{perp}_v \mathbf{u} = \left\langle \dfrac{56}{25}, \dfrac{8}{25} \right\rangle$

45. $\text{proj}_v \mathbf{u} = \left\langle \dfrac{6}{5}, \dfrac{2}{5} \right\rangle,$
$\text{perp}_v \mathbf{u} = \left\langle \dfrac{9}{5}, -\dfrac{27}{5} \right\rangle$

46. $\text{proj}_v \mathbf{u} = \left\langle \dfrac{9}{10}, \dfrac{27}{10} \right\rangle,$
$\text{perp}_v \mathbf{u} = \left\langle -\dfrac{9}{10}, \dfrac{3}{10} \right\rangle$

47. $\text{proj}_v \mathbf{u} = \left\langle -\dfrac{60}{17}, -\dfrac{15}{17} \right\rangle,$
$\text{perp}_v \mathbf{u} = \left\langle \dfrac{9}{17}, -\dfrac{36}{17} \right\rangle$

48. $\text{proj}_v \mathbf{u} = \left\langle \dfrac{7}{13}, \dfrac{35}{13} \right\rangle,$
$\text{perp}_v \mathbf{u} = \left\langle \dfrac{45}{13}, -\dfrac{9}{13} \right\rangle$

49. 14 **50.** 18 **51.** 3 **52.** 0

25. $\mathbf{u} = \cos\left(\dfrac{3\pi}{4}\right)\mathbf{i} + \sin\left(\dfrac{3\pi}{4}\right)\mathbf{j}, \ \mathbf{v} = \cos\left(\dfrac{\pi}{2}\right)\mathbf{i} + \sin\left(\dfrac{\pi}{2}\right)\mathbf{j}$

26. $\mathbf{u} = \cos\left(\dfrac{\pi}{4}\right)\mathbf{i} + \sin\left(\dfrac{\pi}{4}\right)\mathbf{j}, \ \mathbf{v} = \cos\left(\dfrac{5\pi}{6}\right)\mathbf{i} + \sin\left(\dfrac{5\pi}{6}\right)\mathbf{j}$

Use vectors to find the interior angles of the triangles given the following sets of vertices.

27. $(3, 3), (4, 2), (-1, -6)$

28. $(0, 0), (0, 5), (3, 6)$

29. $(-2, -1), (2, 4), (-4, 5)$

30. $(6, 3), (-5, 2), (-6, 1)$

Find $\mathbf{u} \cdot \mathbf{v}$ where θ is the angle between $\mathbf{u}$ and $\mathbf{v}$. See Example 3.

31. $\|\mathbf{u}\| = 25, \|\mathbf{v}\| = 5, \theta = 120°$

32. $\|\mathbf{u}\| = 4, \|\mathbf{v}\| = 64, \theta = \dfrac{\pi}{6}$

33. $\|\mathbf{u}\| = 16, \|\mathbf{v}\| = 4, \theta = \dfrac{3\pi}{4}$

34. $\|\mathbf{u}\| = 9, \|\mathbf{v}\| = 10, \theta = \dfrac{2\pi}{3}$

Find <u>two</u> vectors orthogonal to the given vector. See Example 4. Answers may vary.

35. $\mathbf{u} = \langle 3, -3 \rangle$ **36.** $\mathbf{u} = \langle 4, 1 \rangle$ **37.** $\mathbf{u} = \langle 2, -6 \rangle$ **38.** $\mathbf{u} = \langle 5, 4 \rangle$

Determine whether $\mathbf{u}$ and $\mathbf{v}$ are orthogonal, parallel, or neither. See Example 4.

39. $\mathbf{u} = \langle 2, -3 \rangle, \mathbf{v} = \langle 1, 6 \rangle$

40. $\mathbf{u} = \langle -12, 30 \rangle, \mathbf{v} = \left\langle \dfrac{1}{2}, -\dfrac{5}{4} \right\rangle$

41. $\mathbf{u} = 2\mathbf{i} - 2\mathbf{j}, \mathbf{v} = -\mathbf{i} - \mathbf{j}$

42. $\mathbf{u} = \mathbf{i}, \mathbf{v} = -2\mathbf{i} + 2\mathbf{j}$

Find the projection of $\mathbf{u}$ onto $\mathbf{v}$, and then write $\mathbf{u}$ as a sum of two orthogonal vectors, one of which is $\text{proj}_v \mathbf{u}$. See Example 5.

43. $\mathbf{u} = \langle 1, 3 \rangle, \mathbf{v} = \langle 4, 2 \rangle$

44. $\mathbf{u} = \langle 2, 2 \rangle, \mathbf{v} = \langle 1, -7 \rangle$

45. $\mathbf{u} = \langle 3, -5 \rangle, \mathbf{v} = \langle 6, 2 \rangle$

46. $\mathbf{u} = \langle 0, 3 \rangle, \mathbf{v} = \langle 2, 6 \rangle$

47. $\mathbf{u} = \langle -3, -3 \rangle, \mathbf{v} = \langle -4, -1 \rangle$

48. $\mathbf{u} = \langle 4, 2 \rangle, \mathbf{v} = \langle 1, 5 \rangle$

Find the work done on a particle moving from J to K if the magnitude and direction of the force are given by $\mathbf{v}$. See Example 7.

49. $J = (1, 4), K = (5, 6), \mathbf{v} = \langle 2, 3 \rangle$

50. $J = (-3, 2), K = (0, 5), \mathbf{v} = \langle 4, 2 \rangle$

51. $J = (3, 0), K = (-4, -2), \mathbf{v} = -\mathbf{i} + 2\mathbf{j}$

52. $J = (3, -3), K = (5, 1), \mathbf{v} = 6\mathbf{i} - 3\mathbf{j}$

53. 3479.3 pounds

54. 13.1 pounds

55. 109.6 pounds

56. 1638.3 ft-lb

57. 579,555.5 ft-lb

58. 11,431,535.3 ft-lb

Solve the following problems. See Examples 6 and 7.

53. A truck with a gross weight of 25,000 pounds is parked on an 8° slope. What force is required to prevent the truck from rolling down the hill?

54. A child sits in his go-cart at the start position of a race atop a hill. If the hill has a slope of 3°, and the child and go-cart have a total weight of 250 pounds, what force is required to keep them stationary at the start position?

55. A woman on skis holds herself stationary, with the use of her ski poles, on a ski slope that is 45° from the horizontal. If the woman and her skis have a total weight of 155 pounds, what is the force required to prevent her from sliding down the slope?

56. A child pulls a sled over the snow, exerting a force of 25 pounds on the attached rope. The rope is 35° from the horizontal. If the child pulls the sled a distance of 80 feet, what work has been done?

57. The world's strongest man pulls a log 200 feet, and the tension in the cable connecting the man and log is 3000 pounds. What is the work being done if the cable is being held 15° from the horizontal?

58. A recreational vehicle pulls a passenger car behind it, exerting 1250 pounds on the attach point. The point of attachment is 30° from the horizontal. If the RV pulls the car a distance of 2 miles, what work has been done?

Chapter 8 Project

Trigonometric Applications

Built by King Khufu from 2589–2566 BC to serve as his tomb, the Great Pyramid of Giza covers 13 acres and weighs more than 6.5 million tons. The Great Pyramid is the oldest and the only remaining wonder of the 7 Wonders of the Ancient World. Even using modern technology, engineers in the 21ˢᵗ century would have difficulties recreating this impressive structure. Today, we can only put forth theories as to how this amazing structure was created.

When first built, the Egyptians called the Great Pyramid Ikhet, which means Glorious Light, for the sides of the pyramid were covered in highly polished limestone which would have shone brightly under the hot Egyptian Sun.

1.   When built, the length of each side of the base was 754 ft and the distance from each corner of the base to the peak was 718 ft. What was the surface area of the four sides?

The Egyptians quarried most of the stone locally, but they also floated huge granite blocks down the Nile River from Aswan.

2.   Your barge with the latest shipment of granite for King Khufu is quickly approaching Giza. The river is flowing at 195 yards per minute. You command your oarsmen to start rowing towards shore at a 65° angle from the direction of the current. They can row 260 yards per minute.

   a. What is the resultant true velocity of the barge?
   b. The Nile River is 840 yards wide near Giza. If the boat is in the center of the river and the dock is 750 yards ahead, will they hit the bank before or after it? By how much will they miss it? (**Hint:** (velocity)(time) = distance.)

3.   Once the boat reaches the shore, the granite stones need to be moved into place.

   a. If a granite block weighs 8300 pounds and 12 of your men are each pulling on it (in the same direction) with a force of 115 pounds, what is the total force being applied?

b. In order to get the stone to the necessary spot, the stone must be pulled up a ramp that has been built around the pyramid. If the ramp has a 9° grade, what force is necessary to keep the block from sliding?

c. If the top of the pyramid is currently 320 feet above the desert, how long does the ramp have to be?

d. How much work is it for the 12 men to drag the stone to the top of the pyramid? (Remember, in this case work is done in both the horizontal and vertical directions.)

Chapter Summary

A summary of concepts and skills follows each chapter. Refer to these summaries to make sure you feel comfortable with the material in the chapter. The concepts and skills are organized according to the section title and topic title in which the material is first discussed.

8.1: The Law of Sines and the Law of Cosines

The Law of Sines and Its Use
- Use of the Law of Sines: *AAS, ASA,* or *SSA*
- Use of the Law of Cosines: *SAS* or *SSS*
- The Law of Sines: $\dfrac{\sin A}{a} = \dfrac{\sin B}{b} = \dfrac{\sin C}{c}$
- Solving applications using the Law of Sines
- *SSA* Case: If *A* is acute and $a < h$, then no triangle exists;
  if *A* is acute and $a = h$, then one triangle exists;
  if *A* is acute and $h < a < b$, then two triangles exist;
  if *A* is acute and $b \le a$, then one triangle exists;
  if *A* is obtuse and $a \le b$, then no triangle exists;
  if *A* is obtuse and $b < a$, then one triangle exists.

The Law of Cosines and Its Use
- The Law of Cosines: $a^2 = b^2 + c^2 - 2bc \cos A$
  $$b^2 = a^2 + c^2 - 2ac \cos B$$
  $$c^2 = a^2 + b^2 - 2ab \cos C$$
- Solving applications using the Law of Cosines

Areas of Triangles
- Area of a triangle (Sine Formula):
  $$\text{Area} = \frac{1}{2} ab \sin C = \frac{1}{2} bc \sin A = \frac{1}{2} ac \sin B.$$
- Area of a triangle (Heron's Formula):
  $$\text{Area} = \sqrt{s(s-a)(s-b)(s-c)} \text{ where } s = \frac{a+b+c}{2}.$$

8.2: Polar Coordinates and Polar Equations

The Polar Coordinate System
- Polar coordinates: (r, θ) where *r* is the distance from the origin to the given point and θ is the angle, measured counterclockwise from the polar axis to a line segment connecting the origin to the given point.

8.2: Polar Coordinates and Polar Equations (cont.)

Coordinate Conversion
- Converting from polar to Cartesian coordinates:
$$x = r\cos\theta \ \text{ and } \ y = r\sin\theta.$$
- Converting from Cartesian to polar coordinates:
$$r^2 = x^2 + y^2 \ \text{ and } \ \tan\theta = \frac{y}{x} \ \ (x \neq 0).$$

The Form of Polar Equations
- Polar equation: an equation in the variables r and θ that defines a relationship between these coordinates
- Solution to a polar equation: an ordered pair (r, θ) that makes the equation true

Graphing Polar Equations
- Symmetry of polar equations: pole symmetry, polar axis symmetry, and symmetry with respect to the line $\theta = \dfrac{\pi}{2}$
- Common polar equations and their graphs

8.3: Parametric Equations

Sketching Parametric Curves
- Describing an object's travel parametrically

Eliminating the Parameter
- Eliminating the parameter: the process of converting from parametric form to rectangular form

Constructing Parametric Equations
- How parametric equations describing a given curve are constructed

8.4: Trigonometric Form of Complex Numbers

The Complex Plane
- Coordinates of a complex number $z = a + bi$: (a, b)
- Horizontal axis = real axis; vertical axis = imaginary axis
- Complex plane: plane created by the real and imaginary axes.
- Magnitude of a complex number: $|z| = \sqrt{a^2 + b^2}$

Complex Numbers in Trigonometric Form
- Trigonometric form of $z = a + bi$: $|z|(\cos\theta + i\sin\theta)$ where $|z| = \sqrt{a^2 + b^2}$ and $\tan\theta = \dfrac{b}{a}$.
- Complex number written using Euler's Formula: $z = |z|e^{i\theta}$

Multiplication and Division of Complex Numbers
- Multiplication: $z_1 z_2 = |z_1||z_2|\left[\cos(\theta_1 + \theta_2) + i\sin(\theta_1 + \theta_2)\right]$
- Division: $\dfrac{z_1}{z_2} = \dfrac{|z_1|}{|z_2|}\left[\cos(\theta_1 - \theta_2) + i\sin(\theta_1 - \theta_2)\right]$

Powers of Complex Numbers
- DeMoivre's Theorem: $z^n = |z|^n(\cos n\theta + i\sin n\theta) = |z|^n e^{in\theta}$

Roots of Complex Numbers
- Roots of complex numbers:

$$w_k = |z|^{\frac{1}{n}}\left[\cos\left(\frac{\theta + 2k\pi}{n}\right) + i\sin\left(\frac{\theta + 2k\pi}{n}\right)\right] = |z|^{\frac{1}{n}} e^{i\left(\frac{\theta + 2k\pi}{n}\right)}$$

8.5: Vectors in the Cartesian Plane

Vector Terminology
- Vector: quantity possessing magnitude and direction, often represented by a directed line segment. The point where the directed line segment begins is called the initial point and the point where it ends is called the terminal point.
- Magnitude: length of the vector, denoted $\|\mathbf{u}\|$.

8.5: Vectors in the Cartesian Plane (cont.)

Basic Vector Operations
- Properties of vector addition
- Properties of scalar multiplication

Component Form of a Vector
- Component form: $\mathbf{v} = \langle x_2 - x_1, y_2 - y_1 \rangle$
- Vector operations using components
- Unit vector: any vector with length 1 such as $\mathbf{i} = \langle 1, 0 \rangle$ and $\mathbf{j} = \langle 0, 1 \rangle$
- $\langle a, b \rangle = a\mathbf{i} + b\mathbf{j}$
- Linear combinations: any sum of scalar multiples of vectors
- $\mathbf{u} = \|\mathbf{u}\| \langle \cos\theta, \sin\theta \rangle$

Vector Applications

8.6: The Dot Product and Its Uses

The Dot Product
- Dot product: $\mathbf{u} \cdot \mathbf{v} = u_1 v_1 + u_2 v_2$.
- Properties of the dot product
- The Dot Product Theorem: $\mathbf{u} \cdot \mathbf{v} = \|\mathbf{u}\|\|\mathbf{v}\|\cos\theta$.
- Orthogonal vectors: two vectors are orthogonal if $\mathbf{u} \cdot \mathbf{v} = 0$.

Projections of Vectors
- Projection of $\mathbf{u}$ onto $\mathbf{v}$: $\text{proj}_{\mathbf{v}}\mathbf{u} = \left(\dfrac{\mathbf{u} \cdot \mathbf{v}}{\|\mathbf{v}\|^2} \right)\mathbf{v}$.

Applications of the Dot Product
- Work $= \mathbf{F} \cdot \mathbf{D}$

1. Height ≈ 22.5762 ft,
   angle $\approx 15.8886°$

2. $C = 105°$, $c \approx 5.46$,
   $a \approx 2.83$

3. $A \approx 75.40°$,
   $B \approx 47.60°$, $b \approx 11.45$
   or $A \approx 104.60°$,
   $B \approx 18.40°$, $b \approx 4.89$

4. $B = 68°40'$,
   $a \approx 36.80$, $b \approx 35.60$

5. $A \approx 64.99°$,
   $B \approx 37.01°$, $a \approx 12.04$

6. $a \approx 9.43$, $B \approx 48.52°$,
   $C \approx 69.48°$

7. $A \approx 22°31'$,
   $C \approx 63°22'$, $b \approx 15.62$

8. $A \approx 126.01°$,
   $B \approx 12.99°$, $C \approx 41°$

Chapter Review

Section 8.1

Solve the following problem.

1. The base of a 25-ft ladder is positioned 7 ft away from an office building situated on a slight hill, and the ladder and ground form a 62° angle. At what angle and at what height does the ladder touch the building?

Construct a triangle, if possible, using the following information.

2. $A = 30°$, $B = 45°$, $b = 4$

3. $a = 15$, $c = 13$, $C = 57°$

4. $A = 74°20'$, $C = 37°$, $c = 23$

5. $b = 8$, $c = 13$, $C = 78°$

Construct a triangle, if possible, using the following information and the Law of Cosines.

6. $A = 62°$, $b = 8$, $c = 10$

7. $B = 94°7'$, $a = 6$, $c = 14$

8. $a = 9$, $b = 2.5$, $c = 7.3$

9. $a = 10.8$, $b = 13.4$, $c = 6$

Section 8.2

Convert the points from polar to Cartesian coordinates.

10. $\left(-3.45, \dfrac{\pi}{3} \right)$

11. $\left(7, \dfrac{7\pi}{6} \right)$

Convert the points from Cartesian to polar coordinates.

12. $\left(-\sqrt{3}, -1 \right)$

13. $(10, 12)$

Rewrite each equation in polar coordinates.

14. $x^2 + y^2 = 16a^2$

15. $x^2 + y^2 = 9ax$

Rewrite each equation in rectangular coordinates.

16. $r = 2\cos\theta$

17. $r = \dfrac{16}{4\cos\theta + 4\sin\theta}$

Sketch a graph of each polar equation.

18. $r = 4\sin 3\theta$

19. $r^2 = 25\cos 2\theta$

9. $A \approx 52.04°,$
$B \approx 101.99°,$
$C \approx 25.98°$

10. $(-1.725, -2.99)$

11. $(-6.06, -3.5)$

12. $\left(2, \dfrac{7\pi}{6}\right)$ and $\left(-2, \dfrac{\pi}{6}\right)$

13. $(15.62, 0.88)$ and $(-15.62, -2.26)$

14. $r^2 = 16a^2$

15. $r^2 - 9ar\cos\theta = 0$

16. $x^2 - 2x + y^2 = 0$

17. $x + y = 4$

18.

19.

20. $y = \dfrac{1}{1296x^2} \; (t \neq 0)$

21. $y = |x - 7|$

Section 8.3

Sketch the graphs of the following parametric equations by eliminating the parameter.

20. $x = \dfrac{1}{36t}$ and $y = t^2$

21. $x = t + 5$ and $y = |t - 2|$

22. $x = \dfrac{3}{4t - 2}$ and $y = 2t - 2$

23. $x = 4\sin\theta$ and $y = \cos\theta + 1$

Construct parametric equations describing the graphs of the following equations.

24. $y^2 = x^2 + 4$

25. $6x = 2 - y$

Construct parametric equations for the line or conic with the given attributes.

26. Line: Passing through $(14, 4)$ and $(-3, -8)$

27. Circle: Center $(1, 1)$, radius 1

Section 8.4

Graph and determine the magnitudes of the following complex numbers.

28. $5 + 2i$

29. $-3 + 3i$

Sketch $z_1, z_2, z_1 + z_2$, and $z_1 z_2$ on the same complex plane.

30. $z_1 = -2 - 3i, z_2 = 6 + 3i$

31. $z_1 = 4 + 2i, z_2 = -5 + i$

Graph the regions of the complex plane defined by the following.

32. $\left\{z \,\middle|\, 2 \leq |z| \leq 3\right\}$

33. $\left\{z = a + bi \,\middle|\, a > 2, \, b > 3\right\}$

Write each of the following complex numbers in trigonometric form.

34. $2\sqrt{3} - 3i$

35. $1 + 4i$

Write each of the following complex numbers in standard form.

36. $4\left(\cos\dfrac{7\pi}{4} + i\sin\dfrac{7\pi}{4}\right)$

37. $3\left(\cos 60° + i\sin 60°\right)$

22. $y = \dfrac{3}{2x} - 1 \; \left(t \neq \dfrac{1}{2}\right)$

23. $\dfrac{x^2}{16} + (y - 1)^2 = 1$

24. $x = t, \; y^2 = t^2 + 4$

25. $x = t, \; y = 2 - 6t$

26. $x = t,$

$$y = \frac{12}{17}t - \frac{100}{17}$$

27. $x = 1 + \cos\theta,$
$y = 1 - \sin\theta$

28. $\sqrt{29}$

(5, 2)

29. $3\sqrt{2}$

(−3, 3)

30.

$z_2 = (6, 3)$
$z_1 = (-2, -3)$ $z_1 + z_2 = (4, 0)$
$z_1 z_2 = (-3, -24)$

31.

$z_1 + z_2 = (-1, 3)$
$z_1 = (-5, 1)$ $z_2 = (4, 2)$
$z_1 z_2 = (-22, -6)$

32.

Perform the following operations and show the answer in both trigonometric form and standard form.

38. $\left[\sqrt{3}\left(\cos\frac{2\pi}{3} + i\sin\frac{2\pi}{3}\right)\right]\left[4\sqrt{3}\left(\cos\frac{7\pi}{6} + i\sin\frac{7\pi}{6}\right)\right]$

39. $\dfrac{5\left(\cos 240° + i\sin 240°\right)}{\left(\cos 120° + i\sin 120°\right)}$

40. $\dfrac{-\sqrt{3} + i}{1 - i\sqrt{3}}$

41. $\left(12e^{35°i}\right)\left(2e^{280°i}\right)$

Use DeMoivre's Theorem to calculate the following.

42. $\left(1 + \sqrt{3}i\right)^6$

43. $\left[3\left(\cos 240° + i\sin 240°\right)\right]^{11}$

Find the indicated roots of the following and graphically represent each set in the complex plane.

44. The square roots of $-144i$.

45. The cube roots of $125\left(\cos\frac{7\pi}{4} + i\sin\frac{7\pi}{4}\right)$.

Solve the following equations.

46. $z^4 - 1 + i = 0$

47. $z^3 + 4\sqrt{2} - 4i\sqrt{2} = 0$

Section 8.5

Find the component form and the magnitude of the vector $\mathbf{v}$ for each of the following, assuming the first point given is the initial point and the second point given is the terminal point.

48. $(-1, 0), (4, -5)$

49. $(6, 5), (-4, -1)$

For each of the following, calculate and graph **a.** $2\mathbf{u} + \mathbf{v}$, **b.** $-\mathbf{u} + 3\mathbf{v}$, and **c.** $-2\mathbf{v}$.

50. $\mathbf{u} = \langle 1, 3 \rangle$, $\mathbf{v} = \langle -5, 2 \rangle$

51. $\mathbf{u} = \langle 1, -1 \rangle$, $\mathbf{v} = \langle 4, -3 \rangle$

33.

34. $\sqrt{21}\left(\cos(-0.71)\right.$
$\left.+i\sin(-0.71)\right)$

35. $\sqrt{17}\left(\cos 1.33\right.$
$\left.+i\sin 1.33\right)$

36. $2\sqrt{2}-2\sqrt{2}i$

37. $\dfrac{3}{2}+\dfrac{3i\sqrt{3}}{2}$

38. $12\left(\cos\left(\dfrac{11\pi}{6}\right)\right.$
$\left.+i\sin\left(\dfrac{11\pi}{6}\right)\right),$
$6\sqrt{3}-6i$

39. $5(\cos 120°$
$+i\sin 120°),$
$-\dfrac{5}{2}+\dfrac{5i\sqrt{3}}{2}$

40. $\cos\dfrac{7\pi}{6}+i\sin\dfrac{7\pi}{6},$
$-\dfrac{\sqrt{3}}{2}-\dfrac{i}{2}$

41. $24(\cos 315°$
$+i\sin 315°),$
$12\sqrt{2}-12i\sqrt{2}$

42. 64

43. $177,147e^{120°i}$

44. $12e^{i\left(-\frac{\pi}{4}\right)},12e^{i\left(\frac{3\pi}{4}\right)}$

For each of the following graphs, determine the component forms of $-\mathbf{u}$, $2\mathbf{u}-\mathbf{v}$, and $\mathbf{u}+\mathbf{v}$ and find the magnitudes of $\mathbf{u}$ and $\mathbf{v}$.

52.

53.

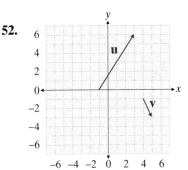

Given the vector $\mathbf{u}$, find **a.** a unit vector pointing in the same direction as $\mathbf{u}$, and **b.** the linear combination of $\mathbf{i}$ and $\mathbf{j}$ that is equivalent to $\mathbf{u}$.

54. $\mathbf{u}=\langle-4,5\rangle$

55. $\mathbf{u}=\langle 6,3\rangle$

Find the magnitude and direction angle of the vector $\mathbf{v}$.

56. $\mathbf{v}=4\left(\cos 135°\,\mathbf{i}+\sin 135°\,\mathbf{j}\right)$

57. $\mathbf{v}=5\mathbf{i}-\mathbf{j}$

Find the component form of $\mathbf{v}$ given its magnitude and the angle it makes with the positive x-axis.

58. $\|\mathbf{v}\|=2\sqrt{2},\ \theta=60°$

59. $\|\mathbf{v}\|=6,$ $\mathbf{v}$ in the direction of $3\mathbf{i}-4\mathbf{j}$

Solve the following problems.

60. A golf ball is driven into the air at a speed of 90 miles per hour and an angle of 45° from the horizontal. Express this velocity in vector form.

61. A sailboat is traveling at a speed of 55 miles per hour with a bearing of W 66° N, when it encounters a front with winds blowing at 20 miles per hour with a bearing of S 10° W. What is the resultant true velocity of the sailboat?

45. $5e^{i\left(\frac{7\pi}{12}\right)},\ 5e^{i\left(\frac{5\pi}{4}\right)},$
$5e^{i\left(\frac{23\pi}{12}\right)}$

46. $2^{\frac{1}{8}}e^{i\left(-\frac{\pi}{16}\right)},\ 2^{\frac{1}{8}}e^{i\left(\frac{7\pi}{16}\right)},$
$2^{\frac{1}{8}}e^{i\left(\frac{15\pi}{16}\right)},\ 2^{\frac{1}{8}}e^{i\left(\frac{23\pi}{16}\right)}$

47. $2e^{i\left(\frac{\pi}{4}\right)},\ 2e^{i\left(\frac{11\pi}{12}\right)},$
$2e^{i\left(\frac{19\pi}{12}\right)}$

48. $\mathbf{v}=\langle 5,-5\rangle$
$\|\mathbf{v}\|=5\sqrt{2}$

49. $\mathbf{v}=\langle-10,-6\rangle$
$\|\mathbf{v}\|=2\sqrt{34}$

50.

51.

Section 8.6

Find the indicated quantity given $\mathbf{u} = \langle 1, -4 \rangle$ and $\mathbf{v} = \langle 2, 5 \rangle$.

62. $3\mathbf{u} \cdot \mathbf{v}$ **63.** $(\mathbf{u} \cdot \mathbf{v})3\mathbf{v}$

Find the magnitude of $\mathbf{u}$ using the dot product.

64. $\mathbf{u} = \langle -2, -3 \rangle$ **65.** $\mathbf{u} = -\mathbf{i} - 3\mathbf{j}$

Find the angle between the given vectors.

66. $\mathbf{u} = \langle 5, -5 \rangle, \mathbf{v} = \langle 1, 4 \rangle$

52. $-\mathbf{u} = \langle -4, -6 \rangle$
$2\mathbf{u} - \mathbf{v} = \langle 7, 14 \rangle$
$\mathbf{u} + \mathbf{v} = \langle 5, 4 \rangle$
$\|\mathbf{u}\| = 2\sqrt{13}$
$\|\mathbf{v}\| = \sqrt{5}$

67. $\mathbf{u} = \cos\left(\dfrac{\pi}{4}\right)\mathbf{i} + \sin\left(\dfrac{\pi}{4}\right)\mathbf{j}, \mathbf{v} = \cos\left(\dfrac{2\pi}{3}\right)\mathbf{i} + \sin\left(\dfrac{2\pi}{3}\right)\mathbf{j}$

Find $\mathbf{u} \cdot \mathbf{v}$ where θ is the angle between $\mathbf{u}$ and $\mathbf{v}$.

53. $-\mathbf{u} = \langle -5, -1 \rangle$
$2\mathbf{u} - \mathbf{v} = \langle 7, 1 \rangle$
$\mathbf{u} + \mathbf{v} = \langle 8, 2 \rangle$
$\|\mathbf{u}\| = \sqrt{26}$
$\|\mathbf{v}\| = \sqrt{10}$

68. $\|\mathbf{u}\| = 16, \|\mathbf{v}\| = 2, \theta = 60°$ **69.** $\|\mathbf{u}\| = 8, \|\mathbf{v}\| = 9, \theta = \dfrac{2\pi}{3}$

Find the projection of $\mathbf{u}$ onto $\mathbf{v}$, and then write $\mathbf{u}$ as a sum of two orthogonal vectors, one of which is $\text{proj}_\mathbf{v}\mathbf{u}$.

70. $\mathbf{u} = \langle 2, 3 \rangle, \mathbf{v} = \langle -1, 5 \rangle$ **71.** $\mathbf{u} = \langle 4, -1 \rangle, \mathbf{v} = \langle 2, 2 \rangle$

54. a. $\left\langle \dfrac{-4}{\sqrt{41}}, \dfrac{5}{\sqrt{41}} \right\rangle$

b. $-4\mathbf{i} + 5\mathbf{j}$

Find the work done in a particle moving from J to K if the magnitude and direction of the force are given by $\mathbf{v}$.

55. a. $\left\langle \dfrac{2}{\sqrt{5}}, \dfrac{1}{\sqrt{5}} \right\rangle$

b. $6\mathbf{i} + 3\mathbf{j}$

72. $J = (2, 4), K = (3, 6), \mathbf{v} = \langle 1, 3 \rangle$ **73.** $J = (-5, 3), K = (0, 4), \mathbf{v} = \langle 5, 6 \rangle$

74. A truck with a gross weight of 33,000 pounds is parked on a 6° slope. What force is required to prevent the truck from rolling down the hill?

56. $\|\mathbf{v}\| = 4, \theta = 135°$

57. $\|\mathbf{v}\| = \sqrt{26}, \theta = -11.3°$

58. $\left(\sqrt{2}, \sqrt{6} \right)$

75. The world's strongest man pulls a log 160 feet, and the tension in the cable connecting the man and log is 2650 pounds. What is the work being done if the cable is being held 10° from the horizontal?

59. $\langle 3.6, -4.8 \rangle$ **60.** $\left\langle 45\sqrt{2}, 45\sqrt{2} \right\rangle$ **61.** 40.01 mph, W 49.77° N   **62.** -54 **63.** $\langle -108, -270 \rangle$ **64.** $\sqrt{13}$

65. $\sqrt{10}$ **66.** $120.96°$ **67.** $\dfrac{5\pi}{12}$ **68.** 16   **69.** -36 **70.** $\text{proj}_\mathbf{v}\mathbf{u} = \left\langle \dfrac{-1}{2}, \dfrac{5}{2} \right\rangle, \text{perp}_\mathbf{v}\mathbf{u} = \left\langle \dfrac{5}{2}, \dfrac{1}{2} \right\rangle$

71. $\text{proj}_\mathbf{v}\mathbf{u} = \left\langle \dfrac{3}{2}, \dfrac{3}{2} \right\rangle, \text{perp}_\mathbf{v}\mathbf{u} = \left\langle \dfrac{5}{2}, \dfrac{-5}{2} \right\rangle$ **72.** 7   **73.** 31   **74.** 3449.4 lb   **75.** 417,558.5 ft-lb

1. 6.5 feet

2. $B \approx 20.05°$,
$C \approx 119.95°, c \approx 20.22$

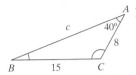

3. $h \approx 13.93, a < h$, so triangle cannot be constructed.

4. $A \approx 104.48°$,
$B \approx 28.96°, C \approx 46.57°$

5. $A \approx 18.19°$,
$B \approx 128.68°$,
$C \approx 33.12°$

6. 11.62

7. 38.57

8. $\left(-3\sqrt{2}, 3\sqrt{2}\right)$

9. $\left(-\sqrt{5}, -1.11\right)$ and $\left(\sqrt{5}, 2.03\right)$

10. $r^2 = 49$

11. $y = \dfrac{1}{4}$

12.

13.

Chapter Test

1. Bill is practicing billiards. He hits the cue ball so that it bounces off of the rail cushion, making an angle of 100°. The ball comes to a stop 5.0 feet away from where it started at a 40° angle. How far did the ball travel? Round your answer to the nearest tenth.

Construct a triangle, if possible, using the following information.

2. $A = 40°, a = 15, b = 8$

3. $A = 55°, a = 10, b = 17$

Construct a triangle, if possible, using the following information and the Law of Cosines.

4. $a = 8, b = 4, c = 6$

5. $a = 4, b = 10, c = 7$

Find the area of the triangle using the following information.

6. $a = 8, b = 4, c = 6$

7. $A = 40°, b = 15, c = 8$

Convert the point as indicated.

8. Polar to Cartesian: $\left(6, \dfrac{3\pi}{4}\right)$

9. Cartesian to polar: $(-1, 2)$

Rewrite the equation as indicated.

10. Rectangular to polar: $x^2 + y^2 = 49$

11. Polar to rectangular: $4r = \csc\theta$

Sketch a graph of the polar equation.

12. $r = 3\sin 2\theta$

Sketch the graphs of the following parametric equations by eliminating the parameter.

13. $x = t - 2, y = 2t + 1$

14. $x = 2 + 2\sin\theta, y = 5 + 2\cos\theta$

14.

15. $x = t,\ y = 8t - 3$

16. $x = t,\ y = \pm\sqrt{t - 3}$

17. $x = t,\ y = -3t + 13$

18. $x = 1 + 3\cos\theta,$

$y = 2 + 3\sin\theta$

19.

20. $2\left(\cos\dfrac{11\pi}{6} + i\sin\dfrac{11\pi}{6}\right)$

21. $2\sqrt{10}\left(\cos 4.12 + i\sin 4.12\right)$

22. $3 - 3i\sqrt{3}$

23. $\dfrac{-5\sqrt{2}}{2} - \dfrac{-5i\sqrt{2}}{2}$

24. $10\left(\cos\dfrac{5\pi}{6} + i\sin\dfrac{5\pi}{6}\right)$

$= -5\sqrt{3} + 5i$

25. $\dfrac{1}{2}\left(\cos 45° + i\sin 45°\right)$

$= \dfrac{\sqrt{2}}{4} + i\dfrac{\sqrt{2}}{4}$

26. $8e^{\frac{\pi}{2}i}$

27. $\mathbf{v} = \langle 1, -5\rangle, \|\mathbf{v}\| = \sqrt{26}$

28. $\mathbf{v} = \langle 2, 2\rangle, \|\mathbf{v}\| = 2\sqrt{2}$

29. a. $\langle 6, 12\rangle$ **b.** $\langle 5, -4\rangle$

c. $\langle -6, 0\rangle$

30. a. $\left\langle \dfrac{1}{\sqrt{5}}, \dfrac{2}{\sqrt{5}}\right\rangle$

b. $\mathbf{u} = 2\mathbf{i} + 4\mathbf{j}$

31. $\|\mathbf{v}\| = 6,\ \theta = 210°$

32. $\langle 65.11, 54.64\rangle$

Construct parametric equations describing the graphs of the following equations.

15. $y = 8x - 3$

16. $x = y^2 + 3$

Construct parametric equations with the given attributes for the specified curve.

17. A line with slope -3 passing through $(2, 7)$

18. A circle with center $(1, 2)$ and radius 3

19. Graph the region of the complex plane defined by $\{z\,|\,3 \le |z| \le 6\}$.

Write each of the following complex numbers in trigonometric form.

20. $\sqrt{3} - i$

21. $-4 - 6i$

Write each of the complex numbers in standard form.

22. $6\left(\cos\dfrac{5\pi}{3} + i\sin\dfrac{5\pi}{3}\right)$

23. $5\left(\cos 225° + i\sin 225°\right)$

Perform the following operation and show the answer in both trigonometric form and standard form.

24. $\left[2\left(\cos\dfrac{\pi}{6} + i\sin\dfrac{\pi}{6}\right)\right]\left[5\left(\cos\dfrac{2\pi}{3} + i\sin\dfrac{2\pi}{3}\right)\right]$

25. $\dfrac{4\left(\cos 315° + i\sin 315°\right)}{8\left(\cos 270° + i\sin 270°\right)}$

26. Use DeMoivre's Theorem to calculate $(1 - i)^6$.

Find the component form and the magnitude of the vectors defined by the points given below. Assume the first point given is the initial point and the second point given is the terminal point.

27. $(4, 2), (5, -3)$

28. $(-6, 0), (-4, 2)$

29. For $\mathbf{u} = \langle 1, 4\rangle$ and $\mathbf{v} = \langle 3, 0\rangle$, calculate **a.** $3\mathbf{u} + \mathbf{v}$, **b.** $-\mathbf{u} + 2\mathbf{v}$, and **c.** $-2\mathbf{v}$.

30. Given the vector $\mathbf{u} = \langle 2, 4\rangle$, find **a.** a unit vector pointing in the same direction as $\mathbf{u}$, and **b.** the linear combination of $\mathbf{i}$ and $\mathbf{j}$ that is equivalent to $\mathbf{u}$.

31. Find the magnitude and direction angle of vector $\mathbf{v} = 6\left(\cos 210° + i\sin 210°\right)$.

32. A golf ball is driven into the air at a speed of 85 miles per hour and at an angle of $40°$ from the horizontal. Express this velocity in vector form.

33. $\langle 18,6 \rangle$

34. $3\sqrt{10} + 1$

35. $33.7°$

36. $63.4°$

37. 12

38. $\text{proj}_v \mathbf{u} = \left\langle \dfrac{16}{13}, \dfrac{24}{13} \right\rangle$,

$\text{perp}_v \mathbf{u} = \left\langle \dfrac{-3}{13}, \dfrac{2}{13} \right\rangle$

Find the indicated quantity given $\mathbf{u} = \langle 2, -4 \rangle$ and $\mathbf{v} = \langle 3, 1 \rangle$.

33. $(\mathbf{u} \cdot \mathbf{v})3\mathbf{v}$

34. $\|3\mathbf{v}\| + 1$

Find the angle between the given vectors.

35. $\mathbf{u} = \langle 2, 3 \rangle$, $\mathbf{v} = \langle 0, 1 \rangle$

36. $\mathbf{u} = -\mathbf{i} + 2\mathbf{j}$, $\mathbf{v} = 3\mathbf{i} + 4\mathbf{j}$

Find $\mathbf{u} \cdot \mathbf{v}$ where θ is the angle between $\mathbf{u}$ and $\mathbf{v}$.

37. $\|\mathbf{u}\| = 6$, $\|\mathbf{v}\| = 4$, $\theta = \dfrac{\pi}{3}$

Find the projection of $\mathbf{u}$ onto $\mathbf{v}$, and then write $\mathbf{u}$ as a sum of two orthogonal vectors, one of which is $\text{proj}_v \mathbf{u}$.

38. $\mathbf{u} = \langle 1, 2 \rangle$, $\mathbf{v} = \langle 2, 3 \rangle$

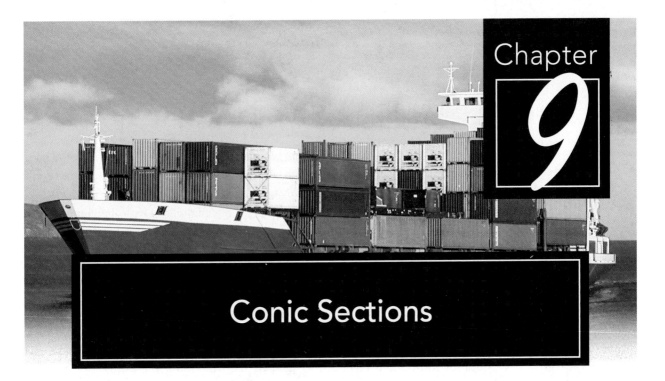

Chapter 9

Conic Sections

9.1   The Ellipse

9.2   The Parabola

9.3   The Hyperbola

9.4   Rotation of Conics

9.5   Polar Form of Conic Sections

Chapter 9 Project

Chapter 9 Summary

Chapter 9 Review

Chapter 9 Test

By the end of this chapter you should be able to:

What if you were sailing a ship and didn't know your exact location in the sea but could receive signals from two radio transmitters? How would you determine your possible locations?

By the end of this chapter, you'll be able to graph conic sections and describe them with equations. On page 736, you'll find that the answer to the ship lost at sea involves a hyperbola. You'll master this type of problem using tools such as the Standard Form of a Hyperbola, found on page 731.

Introduction

This chapter is tightly focused on a family of plane curves called *conic sections*. We have actually studied simple examples of such curves already (circles and parabolas are conic sections), but the more comprehensive study presented in this chapter makes use of the advanced tools we now possess. For instance, our familiarity with the language and concepts of trigonometry will prove to be very useful.

The study of conic sections, the curves obtained by intersecting a plane with a cone, has a history extending back to Greek mathematics of the third century BC and continuing very much to the present. In fact, few branches of mathematics display such a long-lived vitality. Such famous early mathematical figures as Euclid and Archimedes studied conics and discovered many of their properties, while later mathematicians like Isaac Newton, René Descartes, and Carl Friedrich Gauss continued the tradition and made significant contributions of their own. But perhaps the most interesting aspect of this history is the long span of time that lies between the early (almost purely intellectual) formulation of the theory and its modern (thoroughly pragmatic) applications.

Kepler

The Greek mathematician Apollonius (c. 262–190 BC) is largely remembered today for his eight volume book entitled *Conic Sections*. Apollonius improved upon the slightly earlier work of Euclid and Archimedes and proceeded to develop almost all of the theory of conic sections you will encounter in this chapter. Apparently, he is also responsible for the names of the three varieties of conic sections: ellipses, parabolas, and hyperbolas. The names adhere to the Pythagorean tradition of identifying mathematical objects by their geometric properties, and the Greek words refer to the behavior of certain projections of the curves. (Incidentally, the three figures of speech known as *ellipsis*, *parabole*, and *hyperbole* have a similar root.) And while the mathematicians of Apollonius' time did have a few worldly uses for their knowledge of conics, it must be stressed that their driving force was the joy of intellectual accomplishment.

Contrast this with the applications of the theory that were discovered several millennia later. The German astronomer and mathematician Johannes Kepler (1571–1630) made the observation that planets orbit the Sun in elliptical paths, and proceeded to quantify those paths. The unique focusing property of parabolas leads to their use in modern lighting, in the design of satellite dishes, and in telescopes. And an analysis of the problems of long-range navigation leads naturally to a use of hyperbolas in such navigational systems as LORAN. All of these applications, along with many others, make use of geometric properties identified by people with no foreknowledge of their eventual use.

9.1

The Ellipse

TOPICS

1. Overview of conic sections
2. The standard form of an ellipse
3. Planetary orbits

TOPIC 1   Overview of Conic Sections

The three types of conic sections—ellipses, parabolas, and hyperbolas—are so named because all three types of curves arise from intersecting a plane with a circular cone. As shown in Figure 1, an **ellipse** is a closed curve resulting from the intersection of a cone with a plane that intersects only one *nappe* of the cone (the part of the cone on one side of the vertex). A **parabola** results from intersecting a cone with a plane that is parallel to a line on the surface of the cone passing through the vertex (a parabola also intersects only one nappe). Finally, a **hyperbola** is the intersection of a cone with a plane that intersects both nappes. In each case, we specify that the intersecting plane does not contain the vertex. A figure that results when the plane *does* contain the vertex is called a *degenerate* conic section and is a point, a line, or a pair of intersecting lines.

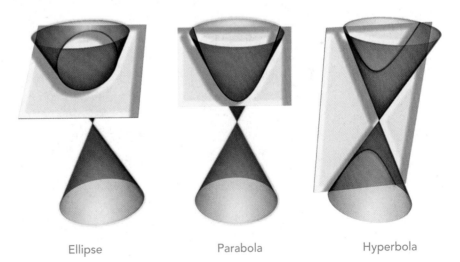

Ellipse   Parabola   Hyperbola

Figure 1: Conic Sections

These conic sections share similarities in their algebraic definitions as well. As curves in the Cartesian plane, every conic section is the graph of an equation that can be written in the form $Ax^2 + Bxy + Cy^2 + Dx + Ey + F = 0$, where $A, B, C, D, E,$ and F are real constants. In the first three sections of this chapter, we will study only those conic sections for which $B = 0$; we will then see the effect of allowing B to be nonzero in Section 9.4.

DEFINITION

Algebraic Definition
of Conics

A conic section described by an equation of the form $Ax^2 + Cy^2 + Dx + Ey + F = 0$, where at least one of the two coefficients A and C is not equal to 0, is:

1. an **ellipse** if the product AC is positive

2. a **parabola** if the product AC is 0

3. a **hyperbola** if the product AC is negative

Finally, the three conic sections are characterized by certain geometric properties that are similar in nature. We will make use of these properties in the work to follow.

DEFINITION

Plane Geometric
Definition of Conics

1. An ellipse consists of the set of points in the plane for which the sum of the distances d_1 and d_2 to two foci (plural of focus) is a fixed constant. See the first diagram in Figure 2.

2. A parabola consists of the set of points that are the same distance d from a line (called the directrix), and a point not on the line (called the focus). See the second diagram in Figure 2.

3. A hyperbola consists of the set of points for which the magnitude of the difference of the distances d_1 and d_2 to two foci is fixed. See the third diagram in Figure 2.

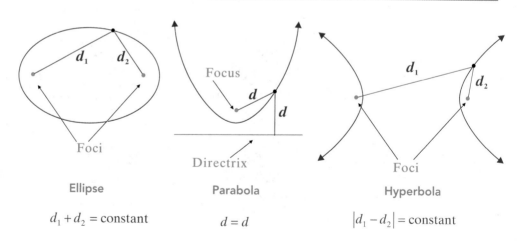

Figure 2: Plane Geometric Definition of Conics

TOPIC 2

The Standard Form of an Ellipse

We would like to use the properties of the conic sections to derive useful forms of their equations. Specifically, we want to be able to construct the graph of a conic section from its equation and, reversing the process, be able to find the equation for a conic section with known properties. We begin with the ellipse.

Initially, assume we have an ellipse centered at the origin with foci at $(-c, 0)$ and $(c, 0)$. We want an equation in x and y that identifies all points (x, y) on the ellipse. The only property we need is that the sum $d_1 + d_2$ shown in Figure 3 is fixed. To make our algebra easier, we will denote this sum as $2a$; that is, $d_1 + d_2 = 2a$.

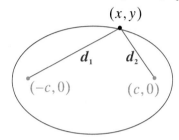

Figure 3: The Geometric Property of the Ellipse

Using the distance formula to find d_1 and d_2, we know that

$$\sqrt{(x+c)^2 + y^2} + \sqrt{(x-c)^2 + y^2} = 2a.$$

While technically we have found an equation in x and y that describes all the points on the ellipse, it does not appear particularly easy to work with. We can make the equation easier to interpret by eliminating the radicals.

$$\left(\sqrt{(x+c)^2 + y^2}\right)^2 = \left(2a - \sqrt{(x-c)^2 + y^2}\right)^2$$

$$(x+c)^2 + y^2 = 4a^2 - 4a\sqrt{(x-c)^2 + y^2} + (x-c)^2 + y^2$$

$$\cancel{x^2} + 2cx + \cancel{c^2} + \cancel{y^2} = 4a^2 - 4a\sqrt{(x-c)^2 + y^2} + \cancel{x^2} - 2cx + \cancel{c^2} + \cancel{y^2}$$

There is still one radical left, so we isolate it and repeat the process:

$$4cx - 4a^2 = -4a\sqrt{(x-c)^2 + y^2}$$

$$a^2 - cx = a\sqrt{(x-c)^2 + y^2}$$

$$\left(a^2 - cx\right)^2 = a^2\left((x-c)^2 + y^2\right)$$

$$a^4 - \cancel{2a^2cx} + c^2x^2 = a^2x^2 - \cancel{2a^2cx} + a^2c^2 + a^2y^2$$

$$a^4 - a^2c^2 = a^2x^2 - c^2x^2 + a^2y^2$$

A change of variables will improve the appearance even more. From the diagram in Figure 3, we can see that $2a$ is larger than $2c$. Even in the extreme case of very skinny ellipses, the sum $d_1 + d_2$ (which we have called $2a$) will approach $2c$ but will always be larger than $2c$.

We can conclude that $a > c$, so $a^2 - c^2$ is a positive number, so we can safely rename $a^2 - c^2$ as b^2. Making this substitution, we simplify further.

$$a^4 - a^2c^2 = a^2x^2 - c^2x^2 + a^2y^2$$
$$a^2\left(a^2 - c^2\right) = \left(a^2 - c^2\right)x^2 + a^2y^2$$
$$a^2b^2 = b^2x^2 + a^2y^2$$

One last cosmetic adjustment is to divide the equation by a^2b^2:

$$b^2x^2 + a^2y^2 = a^2b^2$$
$$\frac{x^2}{a^2} + \frac{y^2}{b^2} = 1$$

Note that this equation satisfies the algebraic definition for an ellipse. If we solve the equation for zero, we have

$$\frac{1}{a^2}x^2 + \frac{1}{b^2}y^2 - 1 = 0.$$

Highlighting the coefficients A and C of the general conic section form, we calculate $AC = \frac{1}{a^2b^2}$, which must be positive. This tells us that the above equation indeed describes an ellipse in the plane.

EXAMPLE 1

Determine the foci of the ellipse $\frac{x^2}{16} + \frac{y^2}{9} = 1$, and then graph the ellipse.

Solution:

The given equation is of the form $\frac{x^2}{a^2} + \frac{y^2}{b^2} = 1$, so it represents an ellipse centered at the origin with foci on the x-axis.

We see that $a = 4$ and $b = 3$, so using the fact that $a^2 - c^2 = b^2$ (by definition), we can calculate the coordinates of the foci.

$$a^2 - c^2 = b^2$$
$$c^2 = a^2 - b^2$$
$$c^2 = 16 - 9 = 7$$
$$c = \pm\sqrt{7}$$

Thus, the foci are located at $\left(-\sqrt{7}, 0\right)$ and $\left(\sqrt{7}, 0\right)$.

Again, we know the graph of this equation is an ellipse. To graph it, we will compute the x- and y-intercepts.

Setting x and y equal to zero, we have:

$$\frac{(0)^2}{16}+\frac{y^2}{9}=1 \quad \text{and} \quad \frac{x^2}{16}+\frac{(0)^2}{9}=1$$

$$\frac{y^2}{9}=1 \qquad\qquad \frac{x^2}{16}=1$$

$$y^2=9 \qquad\qquad x^2=16$$

$$y=\pm 3 \qquad\qquad x=\pm 4$$

This shows that the x-intercepts are $(-4,0)$ and $(4,0)$ and the y-intercepts are $(0,-3)$ and $(0,3)$. Note that these four points are extremes of the graph, as any x-value larger than 4 in magnitude leads to imaginary values for y, and any y-value larger than 3 in magnitude leads to imaginary values for x.

Connecting the four extreme points with an elliptically shaped curve, we obtain the picture below.

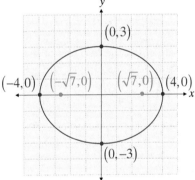

At this point, we know that

$$\frac{x^2}{a^2}+\frac{y^2}{b^2}=1$$

describes an ellipse centered at the origin that is elongated horizontally. If the foci of an origin-centered ellipse are instead at $(0,c)$ and $(0,-c)$, so that the ellipse is elongated vertically, then the equation is

$$\frac{x^2}{b^2}+\frac{y^2}{a^2}=1,$$

where $a>b>0$ and $c^2=a^2-b^2$.

The last step is to come up with the equation for an ellipse that is centered at the point (h,k). We already know how to do this: replacing x with $x-h$ in an equation shifts the graph of the equation h units horizontally, and replacing y with $y-k$ shifts the graph k units vertically. We introduce a few terms before writing the standard form.

DEFINITION

Axes and Vertices of an Ellipse

The **major axis** of an ellipse is the line segment extending from one extreme point of the ellipse to the other and passing through the two foci and the center. The major axis has a length of $2a$. The two ends of the major axis are the **vertices** of the ellipse.

The **minor axis** is the line segment that also passes through the center and is perpendicular to the major axis; it spans the distance between the other two extremes of the ellipse and has a length of $2b$.

DEFINITION

Standard Form of an Ellipse

Let a and b be positive constants with $a > b$. The **standard form of the equation for the ellipse** centered at (h, k) with major axis of length $2a$ and minor axis of length $2b$ is:

- $\dfrac{(x-h)^2}{a^2} + \dfrac{(y-k)^2}{b^2} = 1$ if the major axis is horizontal

- $\dfrac{(x-h)^2}{b^2} + \dfrac{(y-k)^2}{a^2} = 1$ if the major axis is vertical

In both cases, the two foci of the ellipse are located on the major axis c units away from the center of the ellipse, where $c^2 = a^2 - b^2$.

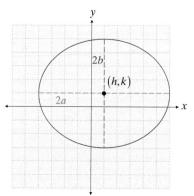

EXAMPLE 2

Graphing an Ellipse in Standard Form

Graph the ellipse $\dfrac{(x-1)^2}{25} + \dfrac{(y+4)^2}{9} = 1$. Include the foci on the graph.

Solution:

First, we note that the denominator on the x term is larger than the denominator on the y term. This means the major axis is horizontal. We can now collect information about the ellipse from the standard form equation:

$$\frac{(x-1)^2}{25} + \frac{(y-(-4))^2}{9} = 1$$

The ellipse is centered at $(1,-4)$, and we have $a = 5$ and $b = 3$. This means the major axis extends from $(-4,-4)$ to $(6,-4)$ and the minor axis runs from $(1,-7)$ to $(1,-1)$. Connecting these points with an elliptical curve will provide a good graph.

Before graphing, we locate the foci.

$$c^2 = a^2 - b^2$$
$$c^2 = 25 - 9 = 16$$
$$c = \pm 4$$

Thus, the foci are located 4 units from the center, along the major axis. This places them at $(-3,-4)$ and $(5,-4)$. Putting all of this together, we graph the ellipse.

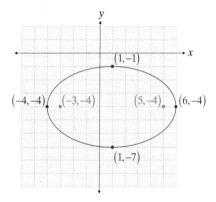

EXAMPLE 3

Graphing an Ellipse Not in Standard Form

Graph the ellipse $25x^2 + 4y^2 + 100x - 24y + 36 = 0$.

Solution:

Note:
Even though it is not in standard form, we know this equation represents an ellipse because the product of the coefficients of x^2 and y^2 is positive.

As written, the equation doesn't provide any useful information about the ellipse. To rewrite it in standard form, we need to complete the square for both variables.

$$25x^2 + 4y^2 + 100x - 24y + 36 = 0$$

$$25x^2 + 100x + 4y^2 - 24y = -36 \qquad \text{Arrange the terms to pair like variables.}$$

$$25(x^2 + 4x) + 4(y^2 - 6y) = -36 \qquad \text{Factor to obtain a coefficient of 1 on each of the squared terms.}$$

$$25(x^2 + 4x + 4) + 4(y^2 - 6y + 9) = -36 + 100 + 36 \qquad \text{Complete the square. Note that we have to balance the equation.}$$

$$25(x+2)^2 + 4(y-3)^2 = 100 \qquad \text{Rewrite the perfect square trinomials as squared binomials.}$$

$$\frac{(x+2)^2}{4} + \frac{(y-3)^2}{25} = 1 \qquad \text{Divide both sides by 100.}$$

At this point, we have an equivalent equation that is in the standard form for an ellipse. Using this form, we can find all the information we need to construct a graph.

$$\frac{\left(x-(-2)\right)^2}{4}+\frac{(y-3)^2}{25}=1$$

Reading the equation, we have the center at $(-2,3)$ with $a=5$ and $b=2$. Note that a^2 appears in the y term, so this ellipse is elongated vertically. This means that the extreme points of the ellipse lie at $(-4,3)$, $(0,3)$, $(-2,-2)$ and $(-2,8)$.

Using $c^2=a^2-b^2=21$, we have $c=\pm\sqrt{21}\approx\pm4.6$. Thus, the foci lie about 4.6 units below and above the center of the ellipse, at $(-2,-1.6)$ and $(-2,7.6)$. Putting all this information together, we have the graph below.

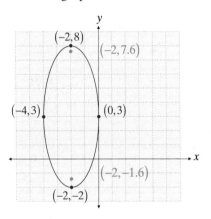

EXAMPLE 4

Constructing the Equation for an Ellipse

Construct the standard form equation for the ellipse whose graph is given below.

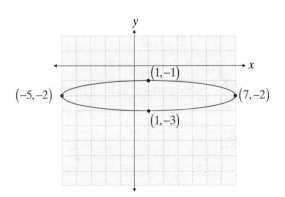

Solution:

In order to construct the standard form equation, we need four pieces of information: the center of the ellipse, whether the major axis is horizontal or vertical, and the values of a and b.

Examining the graph, we see that the center is $(1,-2)$, the major axis is horizontal with length 12, and the minor axis has length 2. This gives us $a = 6$ and $b = 1$. Plugging this information into the standard form, we arrive at the equation for this ellipse:

$$\frac{(x-1)^2}{6^2} + \frac{(y-(-2))^2}{1^2} = 1, \text{ which we rewrite as } \frac{(x-1)^2}{36} + (y+2)^2 = 1.$$

Note that the technique of graphing ellipses is similar to that of graphing circles, including the need to complete the square to obtain standard form, as in Example 3. In fact, circles are a special type of ellipse. Consider what happens if the numbers a and b in the standard form of an ellipse happen to be the same. If $a = b = r$, then the standard form is

$$\frac{(x-h)^2}{r^2} + \frac{(y-k)^2}{r^2} = 1.$$

If we multiply through by r^2, we obtain $(x-h)^2 + (y-k)^2 = r^2$, the standard form of a circle. So a circle is an ellipse whose major and minor axes are the same length.

A circle, viewed as a type of ellipse, has only one focus, and that focus coincides with the center of the circle. Ellipses that are nearly circular have two foci, but they are relatively close to the center, while narrow ellipses have foci far away from the center. This gives us a convenient way to measure the relative "skinniness" of an ellipse.

DEFINITION

Eccentricity of an Ellipse

Given an ellipse with major and minor axes of lengths $2a$ and $2b$, respectively, the **eccentricity** of the ellipse, denoted by the symbol e, is defined by

$$e = \frac{c}{a} = \frac{\sqrt{a^2 - b^2}}{a}.$$

If $e = 0$, then $c = 0$ and the ellipse is a circle. At the other extreme, c may be close to (but cannot equal) a, in which case e is close to 1. An eccentricity close to 1 indicates a relatively skinny ellipse.

TOPIC 3

Planetary Orbits

Johannes Kepler (1571–1630) was the German-born astronomer and mathematician who first demonstrated that the planets in our solar system follow elliptical orbits. He did this by laboring over a period of twenty-one years to mathematically model the astronomical observations of his predecessor Tycho Brahe. The ultimate result was Kepler's Three Laws of Planetary Motion:

THEOREM

Kepler's Laws of Planetary Motion

1. The planets orbit the Sun in elliptical paths, with the Sun at one focus of each orbit.

2. A line segment between the Sun and a given planet sweeps out equal areas of space in equal intervals of time.

3. The square of the time needed for a planet to complete a full revolution about the Sun is proportional to the cube of the orbit's semimajor axis (half of the major axis).

The shapes of the various planetary orbits in our solar system vary widely, and these differences greatly influence seasons on the planets. For instance, Mars has a much more eccentric orbit than Earth, in both the informal and formal meaning of the term. The eccentricity of Mars' orbit is approximately 0.093, while the eccentricity of Earth's orbit is approximately 0.017. As a result, seasonal climate differences on Mars are much more dramatic than on Earth, including a large change in atmospheric pressure. Orbital eccentricities can be easily calculated from astronomical data, and we can use the eccentricity of a planet's orbit to answer particular questions.

EXAMPLE 5

Planetary Orbits

Given that the furthest Earth gets from the Sun is approximately 94.56 million miles, and that the eccentricity of Earth's orbit is approximately 0.017, estimate the closest approach of the Earth to the Sun.

Solution:

Kepler's first law states that each planet follows an elliptical orbit, and that the Sun is positioned at one focus of the ellipse. Thus the Earth is furthest from the Sun when it is at the end of the ellipse's major axis on the other side of the center from the Sun.

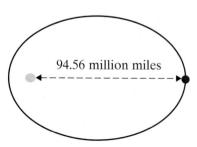

94.56 million miles

1. Center: $(5, 2)$

  Foci: $\left(5, 2 \pm \sqrt{21}\right)$

  Vertices: $(5, 7)$, $(5, -3)$

2. Center: $(-3, -1)$

  Foci: $\left(-3, -1 \pm \sqrt{7}\right)$

  Vertices: $(-3, 3)$, $(-3, -5)$

3. Center: $(-2, -5)$

  Foci: $\left(-2 \pm \sqrt{6}, -5\right)$

  Vertices: $(1, -5)$, $(-5, -5)$

4. Center: $(4, 2)$

  Foci: $\left(4, 2 \pm \sqrt{6}\right)$

  Vertices: $\left(4, 2 \pm 2\sqrt{2}\right)$

5. Center: $(-3, 2)$

  Foci: $\left(-3 \pm \sqrt{2}, 2\right)$

  Vertices: $(-1, 2)$, $(-5, 2)$

6. Center: $(1, -2)$

  Foci: $\left(1, -2 \pm 2\sqrt{2}\right)$

  Vertices: $(1, 2)$, $(1, -6)$

7. Center: $(-5, 1)$

  Foci: $\left(-5, 1 \pm 2\sqrt{3}\right)$

  Vertices: $(-5, 5)$, $(-5, -3)$

8. Center: $(3, -4)$

  Foci: $\left(3 \pm \sqrt{2}, -4\right)$

  Vertices: $(5, -4)$, $(1, -4)$

9. Center: $(-4, 2)$

  Foci: $\left(-4 \pm 3\sqrt{2}, 2\right)$

  Vertices: $\left(-4 \pm 3\sqrt{3}, 2\right)$

10. Center: $(1, -3)$

  Foci: $(1, -2)$, $(1, -4)$

  Vertices: $(1, -1)$, $(1, -5)$

11. Center: $(2, 0)$

  Foci: $(4, 0)$, $(0, 0)$

  Vertices: $\left(2 \pm \sqrt{5}, 0\right)$

12. Center: $(0, -3)$

  Foci: $\left(\pm \sqrt{6}, -3\right)$

  Vertices: $\left(\pm 2\sqrt{2}, -3\right)$

13. e

14. h

15. f

16. a

17. c

18. g

19. b

20. d

Using our standard terminology, and units of millions of miles, this means that

$$a + c = 94.56.$$

By the definition of eccentricity, we also know that

$$0.017 = \frac{c}{a}.$$

We can use these two equations to find the values of a and c.

$$a + c = 94.56$$
$$a + 0.017a = 94.56 \qquad \text{Substitute } c = 0.017a.$$
$$1.017a = 94.56 \qquad \text{Simplify.}$$
$$a \approx 92.98$$

Substituting back into our first equation, we have $c \approx 94.56 - 92.98$, so $c \approx 1.58$. The closest approach to the Sun must be a distance of $a - c$, which is approximately $92.98 - 1.58 = 91.40$ million miles.

Exercises

Find the center, foci, and vertices of each ellipse that the equation describes. See Examples 1, 2 and 3.

1. $\dfrac{(x-5)^2}{4} + \dfrac{(y-2)^2}{25} = 1$

2. $\dfrac{(x+3)^2}{9} + \dfrac{(y+1)^2}{16} = 1$

3. $(x+2)^2 + 3(y+5)^2 = 9$

4. $4(x-4)^2 + (y-2)^2 = 8$

5. $x^2 + 6x + 2y^2 - 8y + 13 = 0$

6. $2x^2 + y^2 - 4x + 4y - 10 = 0$

7. $4x^2 + y^2 + 40x - 2y + 85 = 0$

8. $x^2 + 2y^2 - 6x + 16y + 37 = 0$

9. $x^2 + 3y^2 + 8x - 12y + 1 = 0$

10. $4x^2 + 3y^2 - 8x + 18y + 19 = 0$

11. $x^2 - 4x + 5y^2 - 1 = 0$

12. $x^2 + 4y^2 + 24y + 28 = 0$

21. $\left(3 \pm 2\sqrt{2}, -1\right)$

22. $\left(-5, -2 \pm 2\sqrt{3}\right)$

23. $\left(3 \pm \sqrt{5}, 4\right)$

24. $(\pm 3, 3)$

25.

$$\left(1, 4 \pm \sqrt{3}\right)$$

26.

$$\left(4 \pm 2\sqrt{3}, 4\right)$$

27.

$$\left(-1 \pm \sqrt{21}, -5\right)$$

28.

$$\left(2, -1\right)$$

29.

$$\left(-2 \pm \sqrt{7}, -1\right)$$

30.

$$\left(\pm 2\sqrt{6}, -2\right)$$

31.

$$\left(-1 \pm \sqrt{7}, 2\right)$$

Match the following equations to their graphs.

13. $\dfrac{(x-1)^2}{4} + \dfrac{y^2}{81} = 1$

14. $\dfrac{(x-3)^2}{49} + \dfrac{(y-2)^2}{25} = 1$

15. $\dfrac{(x+4)^2}{9} + \dfrac{(y-3)^2}{4} = 1$

16. $\dfrac{(x+3)^2}{36} + \dfrac{(y-1)^2}{64} = 1$

17. $\dfrac{(x+3)^2}{4} + \dfrac{(y+3)^2}{9} = 1$

18. $\dfrac{x^2}{16} + \dfrac{(y+5)^2}{4} = 1$

19. $\dfrac{(x-4)^2}{9} + \dfrac{(y+3)^2}{49} = 1$

20. $\dfrac{(x-5)^2}{16} + \dfrac{(y+1)^2}{9} = 1$

a.

b.

c.

d.

e.

f.

g.

h.

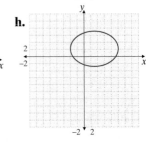

Sketch the graphs of the following ellipses and determine the coordinates of the foci. See Examples 1, 2, and 3.

21. $\dfrac{(x-3)^2}{9} + \dfrac{(y+1)^2}{1} = 1$

22. $\dfrac{(x+5)^2}{4} + \dfrac{(y+2)^2}{16} = 1$

23. $\dfrac{(x-3)^2}{9} + \dfrac{(y-4)^2}{4} = 1$

24. $\dfrac{x^2}{25} + \dfrac{(y-3)^2}{16} = 1$

25. $(x-1)^2 + \dfrac{(y-4)^2}{4} = 1$

26. $\dfrac{(x-4)^2}{16} + \dfrac{(y-4)^2}{4} = 1$

27. $\dfrac{(x+1)^2}{25} + \dfrac{(y+5)^2}{4} = 1$

28. $\dfrac{(x-2)^2}{9} + \dfrac{(y+1)^2}{9} = 1$

32.

$$\left(2, 3 \pm \sqrt{5}\right)$$

33.

$$\left(-5, 3 \pm \sqrt{15}\right)$$

34.

$$\left(2, -1 \pm \sqrt{21}\right)$$

35.

$$\left(-5 \pm \sqrt{5}, -5\right)$$

36.

$$\left(2, -3 \pm \sqrt{15}\right)$$

37.

$$\left(0, -2 \pm \sqrt{3}\right)$$

38.

$$\left(-6, 4 \pm \sqrt{5}\right)$$

29. $\dfrac{(x+2)^2}{16} + \dfrac{(y+1)^2}{9} = 1$

30. $\dfrac{x^2}{25} + (y+2)^2 = 1$

31. $9x^2 + 16y^2 + 18x - 64y = 71$

32. $9x^2 + 4y^2 - 36x - 24y + 36 = 0$

33. $16x^2 + y^2 + 160x - 6y = -393$

34. $25x^2 + 4y^2 - 100x + 8y + 4 = 0$

35. $4x^2 + 9y^2 + 40x + 90y + 289 = 0$

36. $16x^2 + y^2 - 64x + 6y + 57 = 0$

37. $4x^2 + y^2 + 4y = 0$

38. $9x^2 + 4y^2 + 108x - 32y = -352$

In each of the following problems, an ellipse is described either by picture or by properties it possesses. Find the equation, in standard form, for each ellipse. See Example 4.

39. Center at the origin, major axis of length 10 on the y-axis, foci 3 units from the center.

40. Center at $(-2, 3)$, major axis of length 8 oriented horizontally, minor axis of length 4.

41. Vertices at $(1, 4)$ and $(1, -2)$, foci $2\sqrt{2}$ units from the center.

42. Vertices at $(5, -1)$ and $(1, -1)$, minor axis of length 2.

43. Foci at $(0, 0)$ and $(6, 0)$, $e = \dfrac{1}{2}$.

44. Vertices at $(-1, 4)$ and $(-1, 0)$, $e = 0$.

45. Vertices at $(-2, -1)$ and $(-2, -5)$, minor axis of length 2.

46. Vertices at $(-4, 6)$ and $(-14, 6)$, $e = \dfrac{2}{5}$.

47. Vertices at $(1, 3)$ and $(9, 3)$, one of the foci at $(6, 3)$.

48. Foci at $(2, -4)$ and $(2, -8)$, minor axis of length 6.

49.

50.

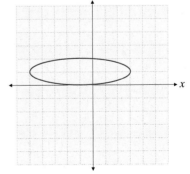

39. $\dfrac{x^2}{16} + \dfrac{y^2}{25} = 1$

40. $\dfrac{(x+2)^2}{16} + \dfrac{(y-3)^2}{4} = 1$

41. $(x-1)^2 + \dfrac{(y-1)^2}{9} = 1$

42. $\dfrac{(x-3)^2}{4} + (y+1)^2 = 1$

43. $\dfrac{(x-3)^2}{36} + \dfrac{y^2}{27} = 1$

44. $\dfrac{(x+1)^2}{4} + \dfrac{(y-2)^2}{4} = 1$

45. $(x+2)^2 + \dfrac{(y+3)^2}{4} = 1$

46. $\dfrac{(x+9)^2}{25} + \dfrac{(y-6)^2}{21} = 1$

47. $\dfrac{(x-5)^2}{16} + \dfrac{(y-3)^2}{15} = 1$

48. $\dfrac{(x-2)^2}{9} + \dfrac{(y+6)^2}{13} = 1$

49. $\dfrac{(x-2)^2}{4} + \dfrac{(y+2)^2}{9} = 1$

51.

52.

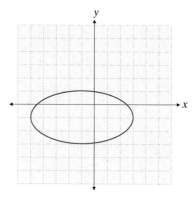

Find the eccentricity and the lengths of the minor and major axes of the following ellipses.

53. $\dfrac{x^2}{100} + \dfrac{y^2}{144} = 1$ **54.** $\dfrac{x^2}{64} + \dfrac{y^2}{9} = 1$ **55.** $x^2 + 9y^2 = 36$

56. $25x^2 + 4y^2 = 100$ **57.** $4x^2 + 16y^2 = 16$ **58.** $5x^2 + 8y^2 = 40$

59. $20x^2 + 10y^2 = 40$ **60.** $\dfrac{1}{4}x^2 + \dfrac{1}{12}y^2 = \dfrac{1}{2}$ **61.** $x^2 = 49 - 7y^2$

Solve the following application problems. See Example 5.

62. The orbit of Halley's Comet is an ellipse with an eccentricity of 0.967. Its closest approach to the Sun is approximately 54,591,000 miles. What is the farthest Halley's Comet ever gets from the Sun?

63. Pluto's closest approach to the Sun is approximately 4.43×10^9 kilometers, and its maximum distance from the Sun is approximately 7.37×10^9 kilometers. What is the eccentricity of Pluto's orbit?

64. Use the information given in Example 5 to determine the length of the minor axis of the ellipse formed by Earth's orbit around the Sun.

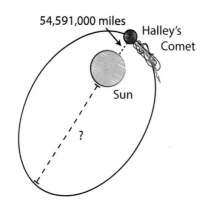

54,591,000 miles

Halley's Comet

Sun

?

50. $\dfrac{(x+1)^2}{16} +$
$(y-1)^2 = 1$

51. $\dfrac{(x-1)^2}{9} + \dfrac{y^2}{16} = 1$

52. $\dfrac{(x+1)^2}{16} +$
$\dfrac{(y+1)^2}{4} = 1$

53. $e = \dfrac{\sqrt{11}}{6}$
major $= 24$
minor $= 20$

54. $e = \dfrac{\sqrt{55}}{8}$
major $= 16$
minor $= 6$

55. $e = \dfrac{2\sqrt{2}}{3}$
major $= 12$
minor $= 4$

56. $e = \dfrac{\sqrt{21}}{5}$
major $= 10$
minor $= 4$

57. $e = \dfrac{\sqrt{3}}{2}$
major $= 4$
minor $= 2$

58. $e = \dfrac{\sqrt{6}}{4}$
major $= 4\sqrt{2}$
minor $= 2\sqrt{5}$

59. $e = \dfrac{\sqrt{2}}{2}$
major $= 4$
minor $= 2\sqrt{2}$

60. $e = \dfrac{\sqrt{6}}{3}$
major $= 2\sqrt{6}$
minor $= 2\sqrt{2}$

65. The archway supporting a bridge over a river is in the shape of half an ellipse. The archway is 60 feet wide and is 15 feet tall at the middle. A boat is 10 feet wide and 14 feet, 9 inches tall. Is the boat capable of passing under the archway?

15 ft
60 ft

66. Since the sum of the distances from each of the two foci to any point on an ellipse is constant, we can draw an ellipse using the following method. Tack the ends of a length of string at two points (the foci) and, keeping the string taut by pulling outward with the tip of a pencil, trace around the foci to form an ellipse (the total length of the string remains constant). If you want to create an ellipse with a major axis of length 5 cm and a minor axis of length 3 cm, how long should your string be and how far apart should you place the tacks? Use the relationships of distances and formulas that you have learned in this section.

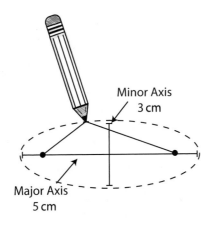

Minor Axis
3 cm

Major Axis
5 cm

67. Using the method described in exercise 66, describe the change in your ellipse when you move the two foci closer together. What happens when you move them farther apart?

68. *The Whispering Gallery* in Chicago's Museum of Science and Industry is a giant ellipsoid that transmits the slightest whisper from one focus to the other focus. This giant ellipse is known to have a length of about 568 inches and a width of about 162 inches. Find the eccentricity of *The Whispering Gallery*. About how far apart are two whisperers when communicating in this gallery? Round your answers to four decimal places.

61. $e = \dfrac{\sqrt{42}}{7}$
major $= 14$
minor $= 2\sqrt{7}$

62. 3,253,954,454 miles

63. $e \approx 0.249$

64. 185.93 million miles

65. Yes, just barely, if the boat is centered on the river.

66. The string should be 5 cm long, and the tacks should be 4 cm apart.

67. Moving the foci closer together creates a more circular ellipse. Moving them further apart creates a more elongated ellipse.

68. $e = 0.9585$; 544.4079 inches

The Parabola

TOPICS

1. The standard form of a parabola

2. Parabolic mirrors

TOPIC 1

The Standard Form of a Parabola

We have already studied parabolas in the context of quadratic functions. Specifically, any function of the form $f(x) = a(x-h)^2 + k$ describes a vertically oriented parabola in the plane (opening upward or downward, depending on the sign of a) whose vertex is at (h,k).

The material in this section does not replace what we have learned. Instead, viewing parabolas as conic sections broadens our understanding of them, and allows us to work with parabolic curves that are not defined by functions.

Just as with ellipses, we want to derive a useful form of the equation that describes parabolas from their characteristic geometric property. Recall that the plane geometric definition of a parabola is that each point on a given parabola is equidistant from a fixed point called the **focus** and a fixed line called the **directrix**. Let p denote the distance between the vertex of the parabola and the focus, and hence also the distance between the vertex and the directrix, as shown in Figure 1.

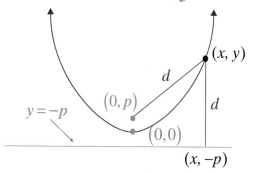

Figure 1: The Geometric Property of the Parabola

We begin our derivation by assuming the parabola is oriented vertically and that the vertex is at the origin. As we did with ellipses, we generalize this basic result later.

Given this, we know that the focus of our parabola is at $(0,p)$ and the equation for the directrix is $y = -p$. The equation below uses the distance formula to characterize all those points that are of equal distance d from the focus and the directrix:

$$\sqrt{x^2 + (y-p)^2} = \sqrt{(y+p)^2}$$

Note that the x-coordinate does not appear in the distance formula on the right side of the equation. This is because the shortest path from this parabola to the directrix is always a vertical line.

One way to make the equation look nicer is to square both sides. We then proceed to eliminate some terms and combine others to arrive at a useful form:

$$x^2 + (y-p)^2 = (y+p)^2$$
$$x^2 + y^2 - 2py + p^2 = y^2 + 2py + p^2$$
$$x^2 = 4py$$

The conclusion is that any equation that can be written in the above form describes a parabola with vertex at the origin, focus at $(0, p)$ and directrix the line $y = -p$. This is true even if p is negative; in this case the parabola opens downward.

To generalize, recall that replacing x with y and vice versa reflects a graph about the line $y = x$, so the equation $y^2 = 4px$ describes a horizontally oriented parabola with vertex at $(0, 0)$. Swapping x and y reflects the focus and directrix as well, so the focus of $y^2 = 4px$ is at $(p, 0)$ and the directrix is the vertical line $x = -p$. As always, replacing x with $x - h$ and y with $y - k$ shifts a graph h units horizontally and k units vertically.

Summarizing all of this information, we arrive at the standard form.

DEFINITION

Standard Form of a Parabola

Let p be a nonzero real constant. The **standard form of the equation for the parabola** with vertex at (h, k) is:

- $(x-h)^2 = 4p(y-k)$ if the parabola is vertically oriented. In this case, the focus is at $(h, k+p)$ and the equation of the directrix is $y = k - p$.

- $(y-k)^2 = 4p(x-h)$ if the parabola is horizontally oriented. In this case, the focus is at $(h+p, k)$ and the equation of the directrix is $x = h - p$.

As we have seen, parabolas can be relatively flat or relatively skinny. With quadratic functions, we saw that the coefficient a governed the flatness of a parabola. In standard form, the parameter p determines the level of flatness. To see how this effect works, consider the standard form for a vertically opening parabola with vertex at the origin.

$$x^2 = 4py$$

Solving this equation for y, we return to the quadratic equation form, with $a = \dfrac{1}{4p}$.

$$y = \frac{1}{4p}x^2$$

Thus, we see that if the magnitude of p increases, the magnitude of a decreases, producing a flatter parabola, while the magnitude of p decreases, the magnitude of a increases, producing a skinnier parabola. This relationship holds even if the vertex moves or if the parabola opens horizontally.

EXAMPLE 1

Graphing a Parabola in Standard Form

Graph the parabola $(y+2)^2 = 8(x-4)$ and determine its focus and directrix.

Solution:

This equation is in standard form, so we can quickly determine the vertex, focus and directrix.

$$(y-(-2))^2 = 8(x-4)$$

We can see the vertex is at $(4,-2)$. To find the focus and directrix we first calculate p. $8 = 4p$, so $p = 2$. This means the focus lies at $(6,-2)$ and the directrix is the vertical line $x = 2$.

To get an idea of the shape of the parabola, we need to find a few more points on its graph. Here we make use of the geometric property of the parabola.

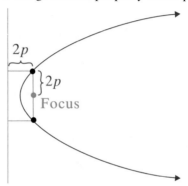

Figure 2: Two Easily Plotted Points

Recall that every point on the parabola must be equidistant from the focus and directrix. With our horizontally opening parabola, we know the focus is a distance of $2p$ from the directrix. Drawing the line parallel to the directrix that passes through the focus, we see that there are two points on the parabola that are $2p$ from the focus. By definition, these points must lie $2p$ above and below the focus.

Thus, two points on our parabola are $(6,-2-4) = (6,-6)$ and $(6,-2+4) = (6,2)$. Plotting these along with our vertex, we have the graph below.

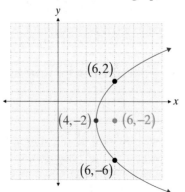

EXAMPLE 2

**Graphing a Parabola
Not in Standard Form**

Graph the parabola $-y^2 + 2x + 2y + 5 = 0$ and determine its focus and directrix.

Solution:

Note:
Even though it is not in standard form, we know this equation represents a parabola because the product of the coefficients of x^2 and y^2 is zero.

Written in this form, it is difficult to graph the parabola. We can rewrite the equation in standard form by completing the square with respect to y (the squared variable).

$$-y^2 + 2x + 2y + 5 = 0$$

Begin by rearranging the terms to obtain a coefficient of 1 on the y^2 term.

$$y^2 - 2y = 2x + 5$$

$$y^2 - 2y + 1 = 2x + 5 + 1$$

In order to complete the square, we have to add 1 to both sides.

$$(y-1)^2 = 2(x+3)$$

Rewrite the trinomial as a binomial squared.

$$(y-1)^2 = 4\left(\frac{1}{2}\right)(x+3)$$

Put the right-hand side into the form $4p(x-h)$ to make the value of p easy to see.

Now that the equation is in standard form, by inspection we can tell that the vertex is at $(-3,1)$ and that $p = \dfrac{1}{2}$. Since the focus is p units to the right of the vertex and the directrix is a vertical line p units to the left of the vertex, we obtain:

$$\text{focus: } \left(-\frac{5}{2}, 1\right), \text{ directrix: } x = -\frac{7}{2}$$

To sketch the graph, we begin by plotting the vertex at $(-3,1)$. Using the same process as in Example 1, note that the two points with x-coordinates of $-\dfrac{5}{2}$ are 1 unit away from the directrix; therefore they must be 1 unit away from the focus as well. Thus, $\left(-\dfrac{5}{2}, 0\right)$ and $\left(-\dfrac{5}{2}, 2\right)$ must also lie on the parabola. This gives us some idea of the "flatness" of the parabola.

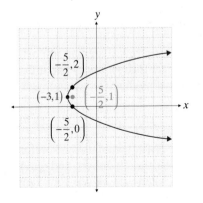

───────────────────────────

EXAMPLE 3

Standard Form of a
Parabola

Given that the directrix of a parabola is the line $y = 2$, that $p = -1$, and that the parabola is symmetric with respect to the line $x = 2$, find its standard form equation.

Solution:

Because the directrix $y = 2$ is a horizontal line, the parabola opens vertically. This means we need to use the form

$$(x - h)^2 = 4p(y - k).$$

From the equation for the line of symmetry we know the x-coordinate of the vertex (and the focus) must be 2. Moving down 1 unit from the directrix (since p is negative) puts the vertex at $(2, 1)$. Plugging this information into the standard form, we have:

$$(x - 2)^2 = 4(-1)(y - 1)$$
$$(x - 2)^2 = -4(y - 1)$$

We can verify our equation with a graph of the parabola.

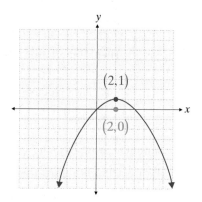

───────────────────────────

TOPIC 2   Parabolic Mirrors

We have already seen an important algebraic application of the parabola: the path of an object thrown under the influence of gravity is a parabola. Algebraically, we know that the height $h(t)$ of an object with initial velocity v_0 and initial height h_0 is

$$h(t) = -\frac{1}{2}gt^2 + v_0 t + h_0,$$

where g, a constant, is the acceleration due to gravity (see Section 1.7 for details). This is a quadratic function, and we experience the fact that a parabola is the shape of its graph whenever we see, for example, a thrown baseball.

In some applications, the geometric properties of parabolas are the key issue, and the standard form may be more helpful than a quadratic equation. One useful geometric property of the parabola is that if rays of light are emanated from the focus of the parabola, they are reflected from the inner surface of the parabola in such a way that the reflected rays are parallel to the axis of symmetry, as shown in Figure 3.

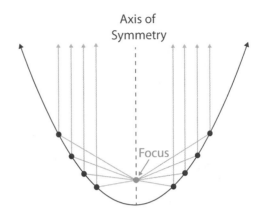

Figure 3: Parallel Rays of Reflection in a Parabola

This property is used in a variety of real-world applications. If a parabola is rotated about its axis of symmetry, a three-dimensional shape called a *paraboloid* is the result. A *parabolic mirror* is then made by coating the inner surface of a paraboloid with a reflecting material. Parabolic mirrors are the basis of searchlights and vehicle headlights, with a light source placed at the focus of the paraboloid, and are also the basis of one design of telescope, in which incoming (parallel) starlight is reflected to an eyepiece at the focus.

EXAMPLE 4

Parabolic Mirrors

The Hale Telescope at the Mount Palomar observatory in California is a very large reflecting telescope. The paraboloid is the top surface of a large cylinder of Pyrex glass 200 inches in diameter. Along the outer rim, the cylinder is 26.8 inches thick, while at the center, it is 23 inches thick. Where is the focus of the parabolic mirror located?

Note:
The same concept can be used to focus sunlight, intensely heating a small area at the focus. This is called a parabolic furnace.

Solution:

First, we need to draw a picture of the situation. In order to make the math as easy as possible, we can locate the origin of our coordinate system at the vertex of a parabolic cross section of the mirror, and we can assume the parabola opens upward.

Since we placed the vertex at $(0,0)$, we know the equation $x^2 = 4py$ describes the shape of the cross section for some value p. If we can determine p, we can find the focus of the parabola.

1.

$(-1, 4),\ y = 2$

2.

$(1, 2),\ x = -3$

3.

$\left(\dfrac{1}{2}, 4\right),\ x = \dfrac{3}{2}$

4.

$(-1, 1), x = -5$

5.

$(2, 0), y = -2$

6.

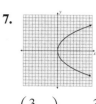

$(-4, -1), x = 2$

7.

$\left(\dfrac{3}{2}, 0\right), x = -\dfrac{3}{2}$

To find p, we need the coordinates of another point on the parabola. The difference in thickness of the mirror between the center and the outer rim is 3.8 inches, and the mirror has a diameter of 200 inches, so the two points $(-100, 3.8)$ and $(100, 3.8)$ must lie on the graph. Plugging a point into the equation $x^2 = 4py$, we can solve for p:

$$\left(100\right)^2 = 4p(3.8)$$
$$10000 = 15.2p$$
$$p \approx 657.9 \text{ inches } (54.8 \text{ feet})$$

We know that the focus of a parabola is p units from the vertex, so the focus of the Hale Telescope is nearly 55 feet from the mirror.

Exercises

Graph the following parabolas and determine the focus and directrix of each. See Examples 1 and 2.

1. $(x+1)^2 = 4(y-3)$ **2.** $y^2 - 4y = 8x + 4$ **3.** $(y-4)^2 = -2(x-1)$

4. $(y-1)^2 = 8(x+3)$ **5.** $(x-2)^2 = 4(y+1)$ **6.** $(y+1)^2 = -12(x+1)$

7. $y^2 = 6x$ **8.** $x^2 = 2y$ **9.** $x^2 = 7y$

10. $x^2 = -5y$ **11.** $y = -12x^2$ **12.** $x = -4y^2$

13. $x = \dfrac{1}{6}y^2$ **14.** $\dfrac{1}{5}x = -y^2$ **15.** $y^2 + 16x = 0$

16. $-6x - 2y^2 = 0$ **17.** $4y + 2x^2 = 4$ **18.** $2y^2 - 10x = 10$

19. $y^2 + 2y + 12x + 37 = 0$ **20.** $x^2 - 8y = 6x - 1$ **21.** $x^2 + 6x + 8y = -17$

22. $x^2 + 2x + 8y = 31$ **23.** $y^2 + 6y - 2x + 13 = 0$ **24.** $y^2 - 2y - 4x + 13 = 0$

8.

$\left(0, \dfrac{1}{2}\right), y = -\dfrac{1}{2}$

9.

$\left(0, \dfrac{7}{4}\right), y = -\dfrac{7}{4}$

10.

$\left(0, -\dfrac{5}{4}\right), y = \dfrac{5}{4}$

11.

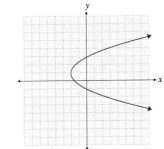

$\left(0, -\dfrac{1}{48}\right), y = \dfrac{1}{48}$

12.

$\left(-\dfrac{1}{16}, 0\right), x = \dfrac{1}{16}$

13.

$\left(\dfrac{3}{2}, 0\right), x = -\dfrac{3}{2}$

Match the following equations to the appropriate graph on the page.

25. $(x+2)^2 = 3(y-1)$

26. $(y-1)^2 = 2(x+2)$

27. $y^2 = 4(x+1)$

28. $x^2 = 2(y+1)$

29. $(x-1)^2 = -(y-2)$

30. $(y+2)^2 = 3x$

31. $(x-2)^2 = 4y$

32. $y^2 = -2(x+1)$

a.

b.

c.

d.

e.

f.

g.

h.

14.

$\left(-\dfrac{1}{20}, 0\right), x = \dfrac{1}{20}$

15.

$(-4, 0), x = 4$

16.

$\left(-\dfrac{3}{4}, 0\right), x = \dfrac{3}{4}$

17.

$\left(0, \dfrac{1}{2}\right), y = \dfrac{3}{2}$

18.

$\left(\dfrac{1}{4}, 0\right), x = -\dfrac{9}{4}$

19.

$(-6, -1), x = 0$

20.

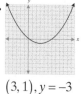

$(3, 1), y = -3$

Find the equation, in standard form, for the parabola with the given properties or with the given graph. See Example 3.

33. Focus at $(-2, 1)$, directrix is the y-axis.

34. Focus at $(-2, 1)$, directrix is the x-axis.

35. Vertex at $(3, -1)$, focus at $(3, 1)$.

36. Symmetric with respect to the line $y = 1$, directrix is the line $x = 2$, and $p = -3$.

37. Vertex at $(3, -2)$, directrix is the line $x = -3$.

38. Vertex at $(7, 8)$, directrix is the line $x = \dfrac{27}{4}$.

39. Focus at $\left(-3, -\dfrac{3}{2}\right)$, directrix is the line $y = -\dfrac{1}{2}$.

40. Vertex at $(3, 16)$, focus at $(3, 11)$.

41. Vertex at $(-4, 3)$, focus at $\left(-\dfrac{3}{2}, 3\right)$.

42. Symmetric with respect to the x-axis, focus at $(-3, 0)$, and $p = 2$.

43.

44.

21.

$(-3,-3), y=1$

22.

$(-1,2), y=6$

23.

$\left(\dfrac{5}{2},-3\right), x=\dfrac{3}{2}$

24.

$(4,1), x=2$

25. g

26. c

27. b

28. f

29. e

30. a

31. d

32. h

33. $(y-1)^2 = -4(x+1)$

34. $(x+2)^2 = 2\left(y-\dfrac{1}{2}\right)$

35. $(x-3)^2 = 8(y+1)$

36. $(y-1)^2 = -12(x+1)$

37. $(y+2)^2 = 24(x-3)$

38. $(y-8)^2 = (x-7)$

Solve the following application problems.

45. One design for a solar furnace is based on the paraboloid formed by rotating the parabola $x^2 = 8y$ around its axis of symmetry. The object to be heated in the furnace is then placed at the focus of the paraboloid (assume that x and y are in units of feet). How far from the vertex of the paraboloid is the hottest part of the furnace?

46. A certain brand of satellite dish antenna is a paraboloid with a diameter of 6 feet and a depth of 1 foot. How far from the vertex of the dish should the receiver of the antenna be placed given that the receiver should be located at the focus of the paraboloid?

47. A spotlight is made by placing a strong light bulb inside a reflective paraboloid formed by rotating the parabola $x^2 = 6y$ around its axis of symmetry (assume that x and y are in units of inches). In order to have the brightest, most concentrated light beam, how far from the vertex should the bulb be placed?

39. $(x+3)^2 = -2(y+1)$

40. $(x-3)^2 = -20(y-16)$

41. $(y-3)^2 = 10(x+4)$

42. $y^2 = 8(x+5)$

43. $(y+1)^2 = -8(x-2)$

44. $(x-2)^2 = 2(y-1)$

45. 2 feet

46. $\dfrac{9}{4}$ feet

47. 1.5 inches

The Hyperbola

TOPICS

1. The standard form of a hyperbola

2. Guidance systems

TOPIC

The Standard Form of a Hyperbola

In studying hyperbolas, we once again begin with the characteristic geometric property and use this to derive a useful form of the equation for a hyperbola which we call the standard form.

The characteristic geometric property of hyperbolas is similar to the one for ellipses. Recall that the points of an ellipse are those for which the sum of the distances to two foci is a fixed constant. For the points on a hyperbola, the magnitude of the *difference* of the distances to two foci is a fixed constant. This results in two disjoint pieces, called *branches*, of a hyperbola. The point halfway between the two branches is the *center* of the hyperbola, and the two points on the hyperbola closest to the center are the two *vertices*. Figure 1 illustrates how these parts of a hyperbola relate to one another.

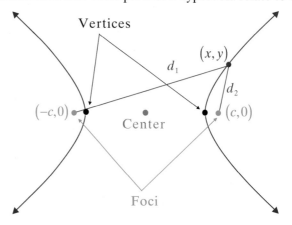

Figure 1: The Geometric Property of the Hyperbola

To begin, we will work with a hyperbola oriented as in Figure 1, with the center at the origin and the two foci at $(-c, 0)$ and $(c, 0)$.

We know that for every point (x, y) on the hyperbola, the quantity $|d_1 - d_2|$ is fixed; in order to simplify the algebra, let $2a$ denote this fixed quantity. Using the distance formula to express d_1 and d_2, we have:

$$\left| \sqrt{(x+c)^2 + (y-0)^2} - \sqrt{(x-c)^2 + (y-0)^2} \right| = 2a$$

Note that if we find $\left|d_1 - d_2\right|$ for one of the vertices, it is equal to the distance between the two vertices, which is twice the distance between each vertex and the center. This means that the distance from the vertex to the center is equal to a.

We wish to manipulate this equation into a more useful form. Because of the absolute value symbols, this one equation actually represents two equations. The left-hand side represents either $d_1 - d_2$ or $d_2 - d_1$ depending on which quantity is nonnegative.

Fortunately, the two cases result in the same equation after we isolate one of the radicals and square both sides. Here, d_1 has been isolated and both sides squared:

$$\left(\sqrt{(x+c)^2 + (y-0)^2} \right)^2 = \left(2a + \sqrt{(x-c)^2 + (y-0)^2} \right)^2$$

$$(x+c)^2 + y^2 = 4a^2 + 4a\sqrt{(x-c)^2 + y^2} + (x-c)^2 + y^2$$

This work is familiar, as it closely parallels the derivation of the standard form for an ellipse. In fact, if we follow the same simplification steps, we arrive at the form

$$\frac{x^2}{a^2} - \frac{y^2}{c^2 - a^2} = 1,$$

and as with ellipses we make a change of variables to improve the appearance. Note that a (the distance of either vertex from the center) is less than c (the distance of either focus from the center), so $c^2 - a^2$ is positive, so we can rename $c^2 - a^2$ as b^2. This changes the above equation to

$$\frac{x^2}{a^2} - \frac{y^2}{b^2} = 1,$$

with $b^2 = c^2 - a^2$ (we will write this as $c^2 = a^2 + b^2$ later so that we can solve for c).

To generalize this equation, we make the same observations as with ellipses and parabolas. Swapping x and y reflects the graph in Figure 1 with respect to the line $y = x$ (giving us a hyperbola with foci on the y-axis) and replacing x with $x - h$ and y with $y - k$ moves the center to (h, k).

DEFINITION

Standard Form of
a Hyperbola

Let a and b be positive constants. The **standard form of the equation for the hyperbola** with center at (h, k) is:

- $\dfrac{(x-h)^2}{a^2} - \dfrac{(y-k)^2}{b^2} = 1$ if the foci are aligned horizontally (where the y-values of the foci are equal).

- $\dfrac{(y-k)^2}{a^2} - \dfrac{(x-h)^2}{b^2} = 1$ if the foci are aligned vertically (where the x-values of the foci are equal).

In either case, the foci are located c units away from the center, where $c^2 = a^2 + b^2$, and the vertices are located a units away from the center.

The standard form for a hyperbola is useful, as it tells us by inspection where the center is and hence where the two vertices are (as well as the foci if we wish). Unfortunately, this knowledge alone leaves a lot of uncertainty about the shape of the hyperbola.

We need a reliable way to understand the "flatness" of the branches of a hyperbola, and it turns out that the branches of a hyperbola approach two *oblique asymptotes* far away from the center.

Consider a hyperbola centered at the origin with foci aligned horizontally, as in Figure 1. Then for some pair of constants a and b, the hyperbola is described by

$$\frac{x^2}{a^2} - \frac{y^2}{b^2} = 1.$$

To understand how this hyperbola behaves far away from the center, we can solve the equation for y:

$$y^2 = b^2 \left(\frac{x^2}{a^2} - 1 \right)$$

We are wondering how y behaves when x is large in magnitude, but the answer is not clear from the equation in this form. As is often the case, a little algebraic manipulation sheds light on the issue. Factoring out the fraction from the parentheses gives us the equation

$$y^2 = \frac{b^2 x^2}{a^2} \left(1 - \frac{a^2}{x^2} \right),$$

and taking the square root of both sides leads to

$$y = \pm \frac{b}{a} x \sqrt{1 - \frac{a^2}{x^2}} \; .$$

The advantage of this last form is that as x goes to ∞ or $-\infty$, the radicand approaches 1. This means that y gets closer and closer to the value

$$\frac{b}{a} x \quad \text{or} \quad -\frac{b}{a} x$$

for values of x that are large in magnitude. In other words, the two straight lines

$$y = \frac{b}{a} x \quad \text{and} \quad y = -\frac{b}{a} x$$

(which intersect at the center) are the asymptotes of the hyperbola.

For hyperbolas whose foci are aligned vertically, the equations for the asymptotes are

$$x = \frac{b}{a} y \quad \text{and} \quad x = -\frac{b}{a} y;$$

that is, x and y exchange places. If we solve these last two equations for y, and also consider hyperbolas not centered at the origin by adding translations, we obtain the equations for asymptotes of all hyperbolas. This information allows us to sketch actual graphs of hyperbolas.

THEOREM

Asymptotes of
Hyperbolas

- The asymptotes of the hyperbola $\dfrac{(x-h)^2}{a^2} - \dfrac{(y-k)^2}{b^2} = 1$ are the two lines

  $y - k = \dfrac{b}{a}(x-h)$ and $y - k = -\dfrac{b}{a}(x-h)$.

- The asymptotes of the hyperbola $\dfrac{(y-k)^2}{a^2} - \dfrac{(x-h)^2}{b^2} = 1$ are the two lines

  $y - k = \dfrac{a}{b}(x-h)$ and $y - k = -\dfrac{a}{b}(x-h)$.

As shown in the figure below, these asymptotes are the diagonals of a rectangle centered at the center of the hyperbola with sides of length $2a$ and $2b$.

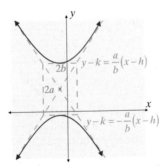

Figure 2: Asymptotes of Horizontally and Vertically Oriented Hyperbolas

EXAMPLE 1

Graphing a Hyperbola
in Standard Form

Graph the hyperbola $\dfrac{(x-2)^2}{25} - \dfrac{(y-1)^2}{4} = 1,$ indicating its asymptotes.

Solution:

Since the hyperbola is written in standard form, the first step is to gather all of the information contained in the equation.

$$\frac{(x-2)^2}{25} - \frac{(y-1)^2}{4} = 1$$

The positive term contains the x variable, so the foci of this hyperbola are aligned horizontally. We can also read that the center is $(2,1)$, $a=5$, and $b=2$. Using the center and the value of a, we know the vertices must lie at

$$(2-5, 1) = (-3, 1) \text{ and } (2+5, 1) = (7, 1).$$

Using $c^2 = a^2 + b^2$, we know that $c = \sqrt{29}$, so we know the foci must lie at

$$(2-\sqrt{29}, 1) \approx (-3.4, 1) \text{ and } (2+\sqrt{29}, 1) \approx (7.4, 1).$$

The best tool for graphing the hyperbola is its set of asymptotes. According to our formula, the asymptotes are the lines $y-1=\frac{2}{5}(x-2)$ and $y-1=-\frac{2}{5}(x-2)$. Using these lines and our vertices, we plot the graph of the hyperbola.

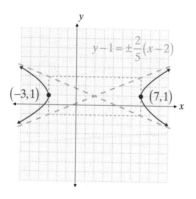

EXAMPLE 2

Graphing a Hyperbola Not in Standard Form

Graph the hyperbola $-16x^2 + 9y^2 + 96x + 18y = 279$, indicating its asymptotes.

Solution:

Note:
Even though it is not in standard form, we know this equation represents a hyperbola because the product of the coefficients of x^2 and y^2 is negative.

In its current form, the equation tells us nothing about the graph of the hyperbola it represents. We need to complete the square (twice) to find the standard form.

$$-16x^2 + 9y^2 + 96x + 18y = 279$$

Factor to obtain a coefficient of 1 on each of the squared terms.

$$-16(x^2 - 6x) + 9(y^2 + 2y) = 279$$

$$-16(x^2 - 6x + 9) + 9(y^2 + 2y + 1) = 279 - 144 + 9$$

Add the needed constant to complete each perfect square trinomial, and compensate by adding the appropriate numbers to the right-hand side as well.

$$-16(x-3)^2 + 9(y+1)^2 = 144$$

$$\frac{(y+1)^2}{16} - \frac{(x-3)^2}{9} = 1$$

Finally, divide by 144 to arrive at the standard form.

With the equation in standard form, we now know that the foci of the hyperbola are aligned vertically (since the positive fraction on the left is in the variable y), that $a = 4$ and $b = 3$, and that the center of the hyperbola is at $(3, -1)$. This tells us that the two vertices of the hyperbola must lie at

$$(3, -1-4) = (3, -5) \text{ and } (3, -1+4) = (3, 3).$$

Using $c^2 = a^2 + b^2 = 25$, we know that $c = 5$, so the two foci are at

$$(3, -1-5) = (3, -6) \text{ and } (3, -1+5) = (3, 4).$$

For this example, we will use a shortcut to graph the asymptotes. We know the asymptotes pass through the center of the hyperbola, and we know that one has a slope of $\dfrac{4}{3}$ and the other a slope of $-\dfrac{4}{3}$. The rectangle drawn on the graph is centered at $(3, -1)$, with the bottom and top edges 4 units away from the center and the left and right edges 3 units away. The asymptotes are the diagonals of this rectangle.

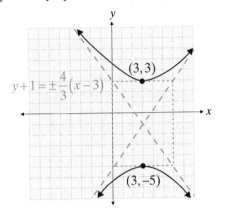

EXAMPLE 3

Standard Form of a Hyperbola

Given that the asymptotes of a hyperbola have slopes of $\dfrac{1}{2}$ and $-\dfrac{1}{2}$ and that the vertices of the hyperbola are at $(-1, 0)$ and $(7, 0)$, find its standard form equation.

Solution:

The fact that the vertices (and hence the foci) are aligned horizontally tells us that we need to construct an equation of the form

$$\frac{(x-h)^2}{a^2} - \frac{(y-k)^2}{b^2} = 1.$$

We need to determine $h, k, a,$ and b. Since the vertices are 8 units apart, we know that $2a = 8$, or $a = 4$. We also know the center lies halfway between the vertices, at the point $(3, 0)$. All that remains is to determine b.

Since the foci are aligned horizontally, we know the asymptotes are of the form

$$y - k = \frac{b}{a}(x-h) \text{ and } y - k = -\frac{b}{a}(x-h),$$

and we already know a, h, and k. Specifically, we know that the two given slopes must correspond to the fractions $\dfrac{b}{a}$ and $-\dfrac{b}{a}$. That is, $\dfrac{1}{2} = \dfrac{b}{a}$, so $\dfrac{1}{2} = \dfrac{b}{4}$, and so $b = 2$. This means that the equation of the hyperbola is

$$\frac{(x-3)^2}{16} - \frac{y^2}{4} = 1,$$

and the equations for the asymptotes are

$$y = \pm\frac{1}{2}(x-3).$$

A sketch of the hyperbola, along with its asymptotes, appears below.

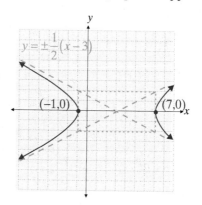

TOPIC ❷ Guidance Systems

Hyperbolas, like ellipses and parabolas, arise in a wide variety of contexts. Hyperbolas are seen in architecture and structural engineering (for example, the shape of a nuclear power plant's cooling towers) and in astronomy (comets that make a single pass through our solar system don't have elliptical orbits, but instead trace one branch of a hyperbola).

One important application of hyperbolas concerns guidance systems, such as LORAN (**LO**ng **RA**nge **N**avigation). LORAN is a radio-communication system that can be used to determine the location of a ship at sea, and the basis of LORAN is an understanding of hyperbolic curves.

Consider a situation in which two land-based radio transmitters, located at sites A and B in Figure 3, send out a signal simultaneously. A receiver on a ship, located at C, would receive the two signals at slightly different times due to the difference in the distances the signals must travel. Since the times for signal travel are proportional to the respective distances d_1 and d_2, the difference in time between receipt of the two signals is proportional to $|d_1 - d_2|$. In other words, a person on the ship can determine $|d_1 - d_2|$ by measuring the time difference in receiving the two signals.

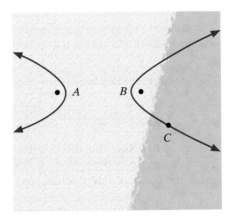

Figure 3: Possible Ship Locations

As Figure 3 indicates, knowing $|d_1 - d_2|$ and the locations of the radio transmitters defines a hyperbola of possible locations for the ship. This hyperbola provides a convenient means of navigation; by maintaining the same time difference between the signals, the captain ensures that the ship stays on a known hyperbolic path. Often LORAN stations are placed very close to the shore; the resulting hyperbolas have their vertices on the shore as well. Thus, different time differences of the signals correspond to different landing locations on the shore.

EXAMPLE 4

LORAN

A ship measures a time difference of 0.000108 seconds between the signals of two LORAN signals sent from stations 100 miles apart along a coastline. If the ship maintains this time difference, where will it land on the coastline? Assume that the signal moves at the speed of light $(186{,}000$ miles per second$)$.

Note:
As usual, setting up a convenient system of coordinates can make the resulting equations simpler.

Solution:

Begin by graphing the situation in the Cartesian plane. We place the two stations (which are the foci of the hyperbola) at $(-50,0)$ and $(50,0)$. The ship must lie on the hyperbola (and hopefully is in the water).

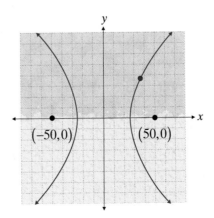

Note that the hyperbola intersects the shore at its vertices. Thus, if the ship maintains its path on the hyperbola, it will land at a vertex. We use the time difference to calculate the position of the vertex using the rate equation $d = rt$.

Substituting $r = 186{,}000\dfrac{\text{miles}}{\text{second}}$ and $t = 0.000108$ seconds, we have

$$d = \left(186{,}000\frac{\text{miles}}{\text{second}}\right)(0.000108\text{ seconds}) \approx 20\text{ miles}.$$

Recall from before that this distance $|d_1 - d_2| = 2a$, where a is the distance of the vertex from the center of the hyperbola. Thus, we know that the ship will reach shore 10 miles from the origin of our coordinate system.

Whether the ship is heading towards $(-10,0)$ or $(10,0)$ depends on which signal reached the ship's receiver first.

LORAN can actually determine the location of the ship by performing the same computations for another pair of simultaneous signals sent out from two additional transmitters, located at A' and B'. This defines a second hyperbola, and the ship must be at a point where the two hyperbolas intersect, as shown in Figure 4.

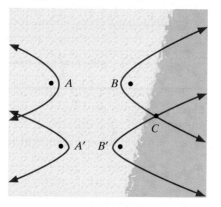

Figure 4: Two Sets of Transmitters

1.

$\left(-3\pm\sqrt{13},-1\right)$

2.

$\left(-2,2\pm\sqrt{34}\right)$

3.

$\left(1,3\pm2\sqrt{5}\right)$

4.

$\left(-2\pm\sqrt{10},1\right)$

5.

$\left(1\pm\sqrt{34},1\right)$

6.

$\left(2,3\right),\left(2,-7\right)$

7.

$\left(\pm2\sqrt{5},2\right)$

8.

$\left(-3,1\pm\sqrt{10}\right)$

Exercises

Sketch the graphs of the following hyperbolas, using asymptotes as guides. Determine the coordinates of the foci in each case. See Examples 1 and 2.

1. $\dfrac{(x+3)^2}{4}-\dfrac{(y+1)^2}{9}=1$

2. $\dfrac{(y-2)^2}{25}-\dfrac{(x+2)^2}{9}=1$

3. $4y^2-x^2-24y+2x=-19$

4. $x^2-9y^2+4x+18y-14=0$

5. $9x^2-25y^2=18x-50y+241$

6. $9x^2-16y^2+116=36x+64y$

7. $\dfrac{x^2}{16}-\dfrac{(y-2)^2}{4}=1$

8. $\dfrac{(y-1)^2}{9}-(x+3)^2=1$

9. $9y^2-25x^2-36y-100x=289$

10. $9x^2+18x=4y^2+27$

11. $9x^2-16y^2-36x+32y-124=0$

12. $x^2-y^2+6x-6y=4$

13. $\dfrac{(y-2)^2}{64}-\dfrac{(x+7)^2}{49}=1$

14. $\dfrac{(y-4)^2}{49}-\dfrac{(x+2)^2}{16}=1$

15. $\dfrac{(x+1)^2}{64}-\dfrac{(y+7)^2}{4}=1$

16. $\dfrac{(x+10)^2}{16}-\dfrac{(y+8)^2}{25}=1$

Find the center, foci, and vertices of each hyperbola that the equation describes.

17. $\dfrac{(x+3)^2}{4}-\dfrac{(y-2)^2}{9}=1$

18. $\dfrac{(y-2)^2}{16}-\dfrac{(x+1)^2}{9}=1$

19. $3(x-1)^2-(y+4)^2=9$

20. $(y-2)^2-2(x-4)^2=4$

21. $(x+2)^2-5(y-1)^2=25$

22. $6(y+2)^2-(x+1)^2=12$

23. $2x^2+12x-y^2-2y+9=0$

24. $y^2-9x^2+6y+72x-144=0$

25. $x^2-4y^2-2x=0$

26. $4y^2-x^2+32y+2x+47=0$

27. $4x^2-y^2-64x+10y+167=0$

28. $4x^2-9y^2-36y-72=0$

9.

$\left(-2, 2 \pm \sqrt{34}\right)$

10.

$\left(-1 \pm \sqrt{13}, 0\right)$

11.

$(7, 1), (-3, 1)$

12.

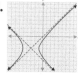

$\left(-3 \pm 2\sqrt{2}, -3\right)$

13.

$\left(-7, 2 \pm \sqrt{113}\right)$

14.

$\left(-2, 4 \pm \sqrt{65}\right)$

15.

$\left(-1 \pm 2\sqrt{17}, -7\right)$

16.

$\left(-10 \pm \sqrt{41}, -8\right)$

Match the corresponding equation to the appropriate graph.

29. $\dfrac{x^2}{9} - y^2 = 1$

30. $y^2 - \dfrac{x^2}{4} = 1$

31. $x^2 - \dfrac{(y-3)^2}{4} = 1$

32. $\dfrac{(x-3)^2}{4} - \dfrac{(y+1)^2}{9} = 1$

33. $(y+2)^2 - \dfrac{(x-2)^2}{4} = 1$

34. $\dfrac{x^2}{9} - \dfrac{(y+2)^2}{4} = 1$

35. $\dfrac{y^2}{4} - (x-1)^2 = 1$

36. $x^2 - y^2 = 1$

a.

b.

c.

d.

e.

f.

g.

h.

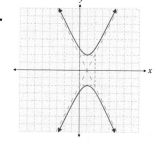

17. Center: $(-3, 2)$
Foci: $\left(-3 \pm \sqrt{13}, 2\right)$
Vertices: $(-1, 2)$, $(-5, 2)$

18. Center: $(-1, 2)$
Foci: $(-1, 7), (-1, -3)$
Vertices: $(-1, 6)$, $(-1, -2)$

19. Center: $(1, -4)$
Foci: $\left(1 \pm 2\sqrt{3}, -4\right)$
Vertices: $\left(1 \pm \sqrt{3}, -4\right)$

20. Center: $(4, 2)$
Foci: $\left(4, 2 \pm \sqrt{6}\right)$
Vertices: $(4, 4), (4, 0)$

21. Center: $(-2, 1)$
Foci: $\left(-2 \pm \sqrt{30}, 1\right)$
Vertices: $(3, 1)$, $(-7, 1)$

22. Center: $(-1, -2)$
Foci:
$\left(-1, -2 \pm \sqrt{14}\right)$
Vertices:
$\left(-1, -2 \pm \sqrt{2}\right)$

23. Center: $(-3, -1)$
Foci:
$\left(-3 \pm 2\sqrt{3}, -1\right)$
Vertices: $(-1, -1)$, $(-5, -1)$

24. Center: $(4, -3)$
Foci: $\left(4, -3 \pm \sqrt{10}\right)$
Vertices: $(4, 0)$, $(4, -6)$

Find the equation, in standard form, for the hyperbola with the given properties or with the given graph. See Example 3.

37. Foci at $(-3, 0)$ and $(3, 0)$ and vertices at $(-2, 0)$ and $(2, 0)$.

38. Foci at $(1, 5)$ and $(1, -1)$ and vertices at $(1, 3)$ and $(1, 1)$.

39. Asymptotes of $y = \pm 2x$ and vertices at $(0, -1)$ and $(0, 1)$.

40. Asymptotes of $y = \pm(x - 2) + 1$ and vertices at $(-1, 1)$ and $(5, 1)$.

41. Foci at $(2, 4)$ and $(-2, 4)$ and asymptotes of $y = \pm 3x + 4$.

42. Foci at $(-1, 3)$ and $(-1, -1)$ and asymptotes of $y = \pm(x + 1) + 1$.

43. Foci at $(2, 5)$ and $(10, 5)$ and vertices at $(3, 5)$ and $(9, 5)$.

44. Foci at $(7, 4)$ and $(7, -4)$ and vertices at $(7, 1)$ and $(7, -1)$.

45. Asymptotes of $y = \pm(2x + 8) + 3$ and vertices at $(-6, 3)$ and $(-2, 3)$.

46. Asymptotes of $y = \pm \dfrac{4}{3}x - 3$ and vertices at $(0, -7)$ and $(0, 1)$.

47.

48.

49.

50.

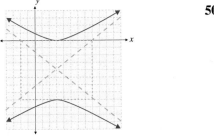

25. Center: $(1, 0)$
Foci: $\left(1 \pm \dfrac{\sqrt{5}}{2}, 0\right)$
Vertices: $(2, 0), (0, 0)$

26. Center: $(1, -4)$
Foci: $\left(1, -4 \pm 2\sqrt{5}\right)$
Vertices: $(1, -2)$, $(1, -6)$

27. Center: $(8, 5)$
Foci: $\left(8 \pm 4\sqrt{5}, 5\right)$
Vertices: $(12, 5)$, $(4, 5)$

28. Center: $(0, -2)$
Foci: $\left(\pm \sqrt{13}, -2\right)$
Vertices: $(3, -2)$, $(-3, -2)$

29. a

30. e

31. b

32. f

33. g

34. c

35. h

36. d

37. $\dfrac{x^2}{4} - \dfrac{y^2}{5} = 1$

38. $(y-2)^2 - \dfrac{(x-1)^2}{8} = 1$

39. $y^2 - 4x^2 = 1$

40. $\dfrac{(x-2)^2}{9} - \dfrac{(y-1)^2}{9} = 1$

41. $\dfrac{5x^2}{2} - \dfrac{5(y-4)^2}{18} = 1$

42. $\dfrac{(y-1)^2}{2} - \dfrac{(x+1)^2}{2} = 1$

43. $\dfrac{(x-6)^2}{9} - \dfrac{(y-5)^2}{7} = 1$

44. $y^2 - \dfrac{(x-7)^2}{15} = 1$

45. $\dfrac{(x+4)^2}{4} - \dfrac{(y-3)^2}{16} = 1$

46. $\dfrac{(y+3)^2}{16} - \dfrac{x^2}{9} = 1$

Solve the following application problems. See Example 4.

51. As mentioned in this section, some comets trace one branch of a hyperbola through the solar system, with the Sun at one focus. Suppose a comet is spotted that appears to be headed straight for Earth, as shown in the figure. As the comet gets closer, however, it becomes apparent that it will pass between the Earth, which lies at the center of the hyperbolic path of the comet, and the Sun. In the end, the closest the comet comes to Earth is 60,000,000 miles. Using a figure of 94,000,000 miles for the distance from the Earth to the Sun, and positioning the Earth at the origin of a coordinate system, find the equation for the path of the comet.

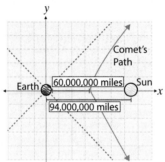

52. Suppose two LORAN radio transmitters are 26 miles apart. A ship at sea receives signals sent simultaneously from the two transmitters and is able to determine that the difference between the distances from the ship to each of the transmitters is 24 miles. By positioning the two transmitters on the y-axis, each 13 miles from the origin, find the equation for the hyperbola that describes the set of possible locations for the ship.

47. $\dfrac{(x-1)^2}{9} - \dfrac{(y+1)^2}{4} = 1$

48. $\dfrac{(y+2)^2}{16} - \dfrac{x^2}{9} = 1$

49. $\dfrac{(y+4)^2}{16} - \dfrac{(x-3)^2}{25} = 1$

50. $\dfrac{(x+7)^2}{16} - \dfrac{(y+3)^2}{9} = 1$

51. $\dfrac{x^2}{\left(6 \times 10^7\right)^2} - \dfrac{y^2}{\left(7.2 \times 10^7\right)^2} = 1$

52. $\dfrac{y^2}{144} - \dfrac{x^2}{25} = 1$

9.4

Rotation of Conics

TOPICS

1.   Rotation formulas

2.   Rotation invariants

TOPIC 1   Rotation Formulas

In the first three sections of this chapter, we studied equations of the form

$$Ax^2 + Cy^2 + Dx + Ey + F = 0.$$

The choice of letters for the coefficients is traditional, and certainly hints at a missing term. As mentioned briefly in Section 9.1, the most general form of a conic in Cartesian coordinates is

$$Ax^2 + Bxy + Cy^2 + Dx + Ey + F = 0$$

and we are now ready to explore the consequences of adding a nonzero xy term to the equation.

Geometrically, the condition $B \neq 0$ leads to a conic that is not oriented vertically or horizontally. The ellipses, parabolas, and hyperbolas graphed so far have all been aligned so that the axis (or axes in the case of ellipses and hyperbolas) of symmetry is parallel to the x-axis or y-axis. If $B \neq 0$, the graph is a conic section that has been rotated through some angle θ. We can exploit this fact to make the graphing of a rotated conic section relatively easy. Our method will be to introduce a new set of coordinate axes (i.e., a second Cartesian plane) rotated by an acute angle θ with respect to the original coordinate axes, and then to graph the conic in the new coordinate system. Algebraically, the goal is to begin with an equation of the form

$$Ax^2 + Bxy + Cy^2 + Dx + Ey + F = 0$$

and define a new set of coordinate axes x' and y' in which the equation has the form

$$A'x'^2 + C'y'^2 + D'x' + E'y' + F' = 0.$$

That is, the coefficient B' in the new coordinate system is 0, and hence the graphing techniques of the first three sections of this chapter apply.

We begin with a picture. Figure 1 is an illustration of two rectangular coordinate systems, with the new system rotated by an acute angle θ with respect to the original.

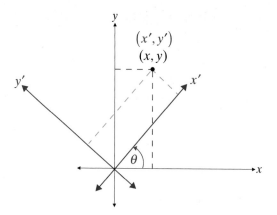

Figure 1: Two Coordinate Systems

The point in the figure has two sets of coordinates, corresponding to the two coordinate planes. However, the distance r between the origin and the point is the same in both, and this fact and the introduction of the angle θ' shown in Figure 2 allow us to begin to relate the two sets of coordinates.

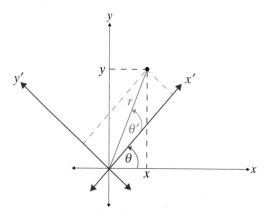

$$x = r\cos(\theta + \theta') \qquad x' = r\cos\theta'$$
$$y = r\sin(\theta + \theta') \qquad y' = r\sin\theta'$$

Figure 2: Relation Between x', y', x, and y

We can now apply one of the trigonometric identities from Chapter 7:

$$x = r\cos(\theta + \theta')$$
$$= r(\cos\theta\cos\theta' - \sin\theta\sin\theta')$$
$$= r\cos\theta'\cos\theta - r\sin\theta'\sin\theta$$
$$= x'\cos\theta - y'\sin\theta.$$

Similarly, $y = x'\sin\theta + y'\cos\theta$. We will also need to be able to express x' and y' in terms of x and y, and we can do so through clever application of another trigonometric identity. By multiplying the two equations we have just derived by $\cos\theta$ and $\sin\theta$, respectively, and then adding the results, we obtain:

$$\left(x\cos\theta = x'\cos^2\theta - y'\sin\theta\cos\theta\right)$$

$$+\left(y\sin\theta = x'\sin^2\theta + y'\sin\theta\cos\theta\right)$$

$$\overline{x\cos\theta + y\sin\theta = x'\left(\cos^2\theta + \sin^2\theta\right)}$$

$$= x'.$$

A similar manipulation allows us to express y' in terms of x and y. All four relations are summarized below.

DEFINITION

Rotation Relations

Given a rectangular coordinate system with axes x' and y' rotated by angle θ with respect to axes x and y, as in Figure 1, the two sets of coordinates (x', y') and (x, y) for the same point are related by:

$$x = x'\cos\theta - y'\sin\theta \qquad x' = x\cos\theta + y\sin\theta$$
$$y = x'\sin\theta + y'\cos\theta \qquad y' = -x\sin\theta + y\cos\theta$$

EXAMPLE 1

Finding
$x'y'$-Coordinates

Given that $\theta = \dfrac{\pi}{6}$, find the $x'y'$-coordinates of the point with xy-coordinates $(-1, 5)$.

Solution:

Using the above relations,

$$x' = (-1)\cos\frac{\pi}{6} + 5\sin\frac{\pi}{6}$$

$$= (-1)\left(\frac{\sqrt{3}}{2}\right) + (5)\left(\frac{1}{2}\right)$$

$$= \frac{5 - \sqrt{3}}{2}$$

and

$$y' = -(-1)\sin\frac{\pi}{6} + 5\cos\frac{\pi}{6}$$

$$= \left(\frac{1}{2}\right) + (5)\left(\frac{\sqrt{3}}{2}\right)$$

$$= \frac{1+5\sqrt{3}}{2}.$$

The $x'y'$-coordinates are thus $\left(\dfrac{5-\sqrt{3}}{2}, \dfrac{1+5\sqrt{3}}{2}\right)$.

We are familiar with the graph of $y = \dfrac{1}{x}$ from our work in Chapters 3 and 4. The next example gives us another perspective on this equation, written in the form $xy = 1$.

EXAMPLE 2

Using the Rotation
Relations

Use the rotation $\theta = 45°$ to show that the graph of $xy = 1$ is a hyperbola.

Solution:

Using the angle $\theta = 45°$ in the rotation relations, we convert the equation as follows:

$$xy = 1$$

$$\left(x'\cos 45° - y'\sin 45°\right)\left(x'\sin 45° + y'\cos 45°\right) = 1$$

$$\left(\frac{x'}{\sqrt{2}} - \frac{y'}{\sqrt{2}}\right)\left(\frac{x'}{\sqrt{2}} + \frac{y'}{\sqrt{2}}\right) = 1$$

$$\frac{x'^2}{2} - \frac{x'y'}{2} + \frac{x'y'}{2} - \frac{y'^2}{2} = 1$$

$$\frac{x'^2}{2} - \frac{y'^2}{2} = 1.$$

We recognize this last equation as a hyperbola in the $x'y'$-plane, with center at the origin and vertices $\sqrt{2}$ away. The asymptotes are $y' = \pm x'$, which correspond to the x- and y-axes.

Remember that the goal in general is to determine θ so that the conversion of the equation

$$Ax^2 + Bxy + Cy^2 + Dx + Ey + F = 0,$$

has no $x'y'$ term. Example 2 will serve as the inspiration; by replacing x and y with the corresponding x' and y' expressions and simplifying the result, we can derive a formula for the appropriate angle θ. We begin with the replacements:

$$A\left(x'\cos\theta - y'\sin\theta\right)^2 + B\left(x'\cos\theta - y'\sin\theta\right)\left(x'\sin\theta + y'\cos\theta\right)$$

$$+C\left(x'\sin\theta + y'\cos\theta\right)^2 + D\left(x'\cos\theta - y'\sin\theta\right)$$

$$+E\left(x'\sin\theta + y'\cos\theta\right) + F = 0.$$

When the left-hand side of this equation is expanded and like terms collected, the result is an equation of the form

$$A'x'^2 + B'x'y' + C'y'^2 + D'x' + E'y' + F' = 0$$

where

$$A' = A\cos^2\theta + B\cos\theta\sin\theta + C\sin^2\theta$$

$$B' = 2(C - A)\cos\theta\sin\theta + B\left(\cos^2\theta - \sin^2\theta\right)$$

$$C' = A\sin^2\theta - B\cos\theta\sin\theta + C\cos^2\theta$$

$$D' = D\cos\theta + E\sin\theta$$

$$E' = -D\sin\theta + E\cos\theta$$

$$F' = F.$$

Since we want $B' = 0$, this gives us an equation in θ to solve. To do so, we will use the double-angle formulas for both sine and cosine in reverse:

$$2(C - A)\cos\theta\sin\theta + B\left(\cos^2\theta - \sin^2\theta\right) = 0$$

$$(C - A)\sin 2\theta + B\cos 2\theta = 0$$

$$B\cos 2\theta = (A - C)\sin 2\theta$$

$$\frac{\cos 2\theta}{\sin 2\theta} = \frac{A - C}{B}$$

$$\cot 2\theta = \frac{A - C}{B}.$$

This result is summarized on the following page.

DEFINITION

The graph of the equation $Ax^2 + Bxy + Cy^2 + Dx + Ey + F = 0$ in the xy-plane is the same as the graph of the equation $A'x'^2 + C'y'^2 + D'x' + E'y' + F' = 0$ in the $x'y'$-plane, where the angle of rotation θ between the two coordinate systems satisfies

$$\cot 2\theta = \frac{A-C}{B}.$$

Although formulas relating the primed coefficients to the unprimed coefficients were derived in the preceding discussion, in practice it is easier to simply determine θ and use the rotation relations to convert equations, as shown in this next example.

EXAMPLE 3

Graph the conic section $x^2 + 2\sqrt{3}xy + 3y^2 + \sqrt{3}x - y = 0$ by first determining the appropriate angle θ by which to rotate the axes.

Solution:

By the formula above,

$$\cot 2\theta = \frac{1-3}{2\sqrt{3}} = -\frac{1}{\sqrt{3}}.$$

Since the angle θ is to be acute, it must be the case that $2\theta = \frac{2\pi}{3}$ and hence $\theta = \frac{\pi}{3}$. By the rotation relations, then,

$$x = x'\cos\frac{\pi}{3} - y'\sin\frac{\pi}{3}$$

$$= \frac{1}{2}x' - \frac{\sqrt{3}}{2}y'$$

and

$$y = x'\sin\frac{\pi}{3} + y'\cos\frac{\pi}{3}$$

$$= \frac{\sqrt{3}}{2}x' + \frac{1}{2}y'.$$

Making these substitutions into the equation, we obtain:

$$\left(\frac{1}{2}x' - \frac{\sqrt{3}}{2}y'\right)^2 + 2\sqrt{3}\left(\frac{1}{2}x' - \frac{\sqrt{3}}{2}y'\right)\left(\frac{\sqrt{3}}{2}x' + \frac{1}{2}y'\right) + 3\left(\frac{\sqrt{3}}{2}x' + \frac{1}{2}y'\right)^2$$

$$+\sqrt{3}\left(\frac{1}{2}x' - \frac{\sqrt{3}}{2}y'\right) - \left(\frac{\sqrt{3}}{2}x' + \frac{1}{2}y'\right) = 0.$$

Multiplying out and collecting like terms in this equation results in

$$y' = 2x'^2,$$

a much simpler equation. We recognize this as a parabola with vertex at the origin. The graph of this equation in the $x'y'$-plane is easily sketched:

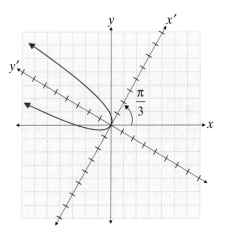

TOPIC 2

Rotation Invariants

The relationships we derived between A, B, C, D, E, F and $A', B', C', D', E',$ and F' lead to some interesting observations, one of which we have an immediate use for. Since $F' = F$, we say the constant term is *invariant* under rotation. Slightly less obviously, $A' + C' = A + C$, so $A + C$ is also said to be an invariant.

Less obvious still is the very important fact that $B'^2 - 4A'C' = B^2 - 4AC$. This invariant is called the **discriminant** of the conic section, and its sign (except in the case of degenerate conics) identifies the graph of the equation as an ellipse, a parabola, or a hyperbola. The discriminant thus generalizes the role that the product AC played in Sections 9.1, 9.2, and 9.3.

DEFINITION

Classifying Conics

Assuming the graph of the equation $Ax^2 + Bxy + Cy^2 + Dx + Ey + F = 0$ is a nondegenerate conic section, it is classified by its discriminant as follows:

1. **Ellipse** if $B^2 - 4AC < 0$

2. **Parabola** if $B^2 - 4AC = 0$

3. **Hyperbola** if $B^2 - 4AC > 0$

EXAMPLE 4

Classifying and Graphing Conic Sections

Classify the conic section $13x^2 + 10xy + 13y^2 + 42\sqrt{2}x - 6\sqrt{2}y + 18 = 0$ and then sketch its graph.

Solution:

The discriminant is $10^2 - 4(13)(13) = -576$. Since this result is negative, the conic section is an ellipse. The next step is to determine the rotation angle θ:

$$\cot 2\theta = \frac{13 - 13}{10} = 0$$

$$\Rightarrow 2\theta = \frac{\pi}{2}$$

$$\Rightarrow \theta = \frac{\pi}{4}.$$

The rotation relations are thus

$$x = x' \cos\frac{\pi}{4} - y' \sin\frac{\pi}{4}$$

$$= \frac{x'}{\sqrt{2}} - \frac{y'}{\sqrt{2}}$$

and

$$y = x' \sin\frac{\pi}{4} + y' \cos\frac{\pi}{4}$$

$$= \frac{x'}{\sqrt{2}} + \frac{y'}{\sqrt{2}}.$$

Making these substitutions in the original equation gives us

$$13\left(\frac{x'}{\sqrt{2}}-\frac{y'}{\sqrt{2}}\right)^2+10\left(\frac{x'}{\sqrt{2}}-\frac{y'}{\sqrt{2}}\right)\left(\frac{x'}{\sqrt{2}}+\frac{y'}{\sqrt{2}}\right)+13\left(\frac{x'}{\sqrt{2}}+\frac{y'}{\sqrt{2}}\right)^2$$

$$+42\sqrt{2}\left(\frac{x'}{\sqrt{2}}-\frac{y'}{\sqrt{2}}\right)-6\sqrt{2}\left(\frac{x'}{\sqrt{2}}+\frac{y'}{\sqrt{2}}\right)+18=0$$

which, when multiplied out and simplified, reduces to

$$9x'^2+4y'^2+18x'-24y'+9=0.$$

In order to easily graph this ellipse in the $x'y'$-plane, we follow the usual completing-the-square process:

$$9\left(x'^2+2x'\right)+4\left(y'^2-6y'\right)=-9$$

$$9\left(x'^2+2x'+1\right)+4\left(y'^2-6y'+9\right)=-9+9+36$$

$$9\left(x'+1\right)^2+4\left(y'-3\right)^2=36$$

$$\frac{\left(x'+1\right)^2}{4}+\frac{\left(y'-3\right)^2}{9}=1.$$

We can now easily graph this ellipse whose center, in the $x'y'$-plane, is $(-1, 3)$ and whose minor and major axes have lengths 4 and 6, respectively.

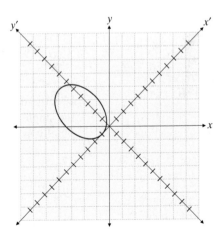

1. $\left(4\sqrt{3}+3,-4+3\sqrt{3}\right)$

2. $\left(\dfrac{-5+\sqrt{3}}{2},\dfrac{5\sqrt{3}+1}{2}\right)$

3. $\left(\dfrac{-5\sqrt{2}}{16},\dfrac{3\sqrt{2}}{16}\right)$

4. $(3.9021,4.1308)$

5. $(-1.2097,-13.5476)$

6. $(7,3)$

7. $(3.3485,-1.5002)$

8. $(1,1)$

9. Ellipse

10. Ellipse

11. Hyperbola

12. Hyperbola

13. Hyperbola

14. Parabola

15. Ellipse

16. Parabola

17. Hyperbola

$\theta=\dfrac{\pi}{4},\ x'^{2}-y'^{2}=4$

18. Hyperbola

$\theta=\dfrac{\pi}{4},\ x'^{2}-y'^{2}=8$

19. Parabola

$\theta=\dfrac{\pi}{4},\ y'=-\sqrt{2}x'^{2}$

Exercises

In the exercises below, the coordinates of a point are given. Find the $x'y'$-coordinates of each point for the given rotation angle θ. See Example 1.

1. $(8,6),\ \theta=30°$

2. $(-5,1),\ \theta=\dfrac{\pi}{3}$

3. $\left(\dfrac{-1}{2},\dfrac{-1}{8}\right),\ \theta=\dfrac{\pi}{4}$

4. $(2.7,5),\ \theta=15°$

5. $(13,-4),\ \theta=78°$

6. $\left(\dfrac{12}{\sqrt{18}},\dfrac{240}{\sqrt{1152}}\right),\ \theta=45°$

7. $\left(3.65,\dfrac{3}{8}\right),\ \theta=\dfrac{\pi}{6}$

8. $\left(\dfrac{3+\sqrt{48}}{-\left(2\sqrt{12}+3\right)},\dfrac{\sqrt{4096}}{8\sqrt{25}\cdot\dfrac{1}{5}\sqrt{64}}\right),\ \theta=\dfrac{\pi}{2}$

Use the discriminant to determine whether the equation of the given conic represents an ellipse, a parabola, or a hyperbola. See Example 4.

9. $2x^{2}-3xy+2y^{2}-2x=0$

10. $3x^{2}+7xy+5y^{2}-6x+7y+15=0$

11. $3x^{2}+8xy+4y^{2}-7=0$

12. $5x^{2}+6xy-3y^{2}-9=0$

13. $-2x^{2}-8xy+2y^{2}+2y+5=0$

14. $3x^{2}-6xy+3y^{2}+3x-9=0$

15. $x^{2}-xy+4y^{2}+2x-3y+1=0$

16. $x^{2}-4xy+4y^{2}+2x+3y-1=0$

Use the discriminant to classify each of the following conic sections. Then determine the angle θ that will allow you to convert the equation and eliminate the xy term. Finally, sketch the conic section. See Examples 2, 3, and 4.

17. $xy=2$

18. $xy-4=0$

19. $x^{2}+2xy+y^{2}-x+y=0$

20. $7x^{2}+5\sqrt{3}xy+2y^{2}=14$

21. $22x^{2}+6\sqrt{3}xy+16y^{2}-49=276$

22. $2\sqrt{3}x^{2}-6xy+\sqrt{3}x-9y=0$

23. $34x^{2}+8\sqrt{3}xy+42y^{2}=1380$

24. $xy+x-4y=6$

Sketch the graphs of the following conic sections. See Examples 2, 3, and 4.

25. $x^{2}+6xy+y^{2}=18$

26. $x^{2}-4xy+3y^{2}=12$

27. $9x^{2}+14xy-9y^{2}=15$

28. $36x^{2}-19xy+8y^{2}=72$

29. $40x^{2}+20xy+10y^{2}+(2\sqrt{2}-6)x-(4\sqrt{2}+8)y=90$

30. $72x^{2}+19xy+4y^{2}=20$

31. $48x^{2}+15xy+7y^{2}=28$

32. $72x^{2}+18xy-9y^{2}=14$

20. Hyperbola

$\theta = \dfrac{\pi}{6}, 19x'^2 - y'^2 = 28$

21. Ellipse

$\theta = \dfrac{\pi}{6}, \dfrac{x'^2}{13} + \dfrac{y'^2}{25} = 1$

22. Hyperbola

$\theta = \dfrac{\pi}{3},$

$\dfrac{(x'+2)^2}{3} - \left(y' - \dfrac{1}{\sqrt{3}}\right)^2 = 1$

23. Ellipse

$\theta = \dfrac{\pi}{3}, \dfrac{x'^2}{30} + \dfrac{y'^2}{46} = 1$

24. Hyperbola

$\theta = \dfrac{\pi}{4},$

$\dfrac{\left(x' - \dfrac{3}{\sqrt{2}}\right)^2}{4} - \dfrac{\left(y' + \dfrac{5}{\sqrt{2}}\right)^2}{4} = 1$

Match the equation with its corresponding graph.

33. $3x^2 + 2xy + y^2 - 10 = 0$

34. $x^2 - 4xy + 4y^2 + 5\sqrt{5}y + 1 = 0$

35. $xy - 1 = 0$

36. $x^2 + y^2 - 16x + 39 = 0$

37. $x^2 - y^2 - 16 = 0$

38. $3x^2 + 8xy + 4y^2 - 7 = 0$

39. $4xy - 9 = 0$

40. $x^2 - 6xy + 9y^2 - 2y + 1 = 0$

a.

b.

c.

d.

e.

f.

25.

26.

27.

28.

29.

30.

g.

h.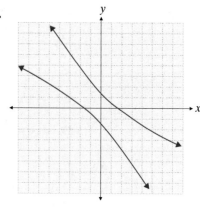

Use your knowledge of rotation of conics to answer the following questions.

41. You have just used the rotation of axes to rotate the x- and y-axes until they were parallel to the axes of the conic. The resulting equation in the $x'y'$-plane is of the form

$$A'(x')^2 + Bx'y' + E'y' + F' = 0.$$

What is wrong with the resulting equation?

42. What must the angle of rotation (θ) be if the coefficients of x^2 and y^2 are equal and $B \neq 0$? Support your answer.

43. Using the equation

$$7x^2 - 6\sqrt{3}xy + 13y^2 - 16 = 0:$$

a. Show that the rotation invariant $F = F'$ is true.

b. Show that the rotation invariant $A + C = A' + C'$ is true.

c. Show that the rotation invariant $B^2 - 4AC = (B')^2 - 4A'C'$ is true.

31. **32.**

33. c **34.** d
35. a **36.** b
37. g **38.** h
39. e **40.** f

41. The objective of the rotation of axes is to eliminate the $x'y'$ term. If your final equation contains an $x'y'$ term, you know that a mistake has occurred.

42. θ would be $45°$. Support: Let $A = C$. Then $\cot 2\theta = \dfrac{A-C}{B} = \dfrac{A-A}{B} = \dfrac{0}{B} = 0.$ Therefore, $2\theta = 90°$ so $\theta = 45°$.

43. a. Use the rotation of axes procedure to obtain the equation $4(x')^2 + 16(y')^2 - 16 = 0$. Now we know $F = -16$ and $F' = -16$. We can plug these values in $F = F'$ and obtain $-16 = -16$, which is true.

b. Use the rotation of axes procedure to obtain the equation $4(x')^2 + 16(y')^2 - 16 = 0$. Now we know $A = 7, C = 13, A' = 4,$ and $C' = 16$. We can plug these values in $A + C = A' + C'$, and obtain $7 + 13 = 4 + 16$, or $20 = 20$, which is true.

c. Use the rotation of axes procedure to obtain the equation $4(x')^2 + 16(y')^2 - 16 = 0$. Now we know $A = 7, B = -6\sqrt{3}, C = 13, A' = 4, B' = 0,$ and $C' = 16$. We can plug these values in $B^2 - 4AC = (B')^2 - 4A'C'$ and obtain $(-6\sqrt{3})^2 - 4(7)(13) = (0)^2 - 4(4)(16)$, or $-256 = -256$, which is true.

Polar Form of Conic Sections

TOPICS

1. The focus/directrix definition of conic sections

2. Using the polar form of conic sections

The first three sections of this chapter dealt with the three varieties of conic sections individually, describing their geometric properties and the formulation of their associated equations in rectangular coordinates. This individual attention is useful when introducing conic sections, and rectangular coordinates are undeniably useful in working with conics to solve certain applications. But polar coordinates provide us with an alternative approach to the study of conic sections, one which possesses important advantages of its own.

TOPIC `1`

The Focus/Directrix Definition of Conic Sections

Probably the most striking virtue of the polar coordinate formulation of conics is that the equations for the three types of conics all have the same form. The equations are easily identified, and the magnitude of a parameter e, called the *eccentricity*, determines whether the conic is an ellipse, a parabola, or a hyperbola. You first encountered eccentricity in the discussion of ellipses—the use of e in this section is an extension of the original use. An additional characteristic of the polar form of conics is that all three varieties are defined in terms of a *focus* and a *directrix*; previously, the directrix only made an appearance in the discussion of parabolas.

Throughout this section, we will assume a point F, called the **focus**, lies at the origin of the plane and that a line L, called the **directrix**, lies d units away from the focus. Until we discuss rotated conics in polar form, the directrix will be oriented either vertically or horizontally, so the equation for the directrix will be $x = -d, x = d, y = -d,$ or $y = d$. We will let $D(P, F)$ denote the distance between a variable point P and the fixed focus F, and we will let $D(P, L)$ denote the shortest distance between a variable point P and the fixed directrix L. The **eccentricity** e will be a fixed positive number for any given conic, and we will make frequent use of the polar coordinates (r, θ) of points in the plane.

DEFINITION

Let $D(P, F)$ denote the distance between a variable point P and the fixed focus F; and let $D(P, L)$ denote the shortest distance between P and the fixed directrix L. A conic section consists of all points P in the plane that satisfy the equation

$$\frac{D(P,F)}{D(P,L)} = e,$$

where e is a fixed positive constant.

The conic is:

an ellipse if $0 < e < 1$,
a parabola if $e = 1$, and
a hyperbola if $e > 1$.

In words, a conic section consists of all those points for which the ratio of the distance to the focus and the distance to the directrix is a fixed constant e. This common definition for all three types of conic sections is pleasantly unified. Figure 1 illustrates three conic sections that share the directrix $x = -1$ but with three different eccentricities. In all three graphs, the directrix and the focus appear in green.

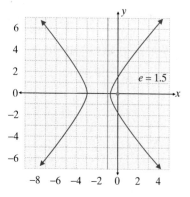

Figure 1: Common Directrix and Focus, Three Eccentricities

The next step is to use the focus/directrix description to develop the form of the polar equations for conics. If we let (r,θ) denote a point P on a given conic section, then $D(P,F) = r$. To determine $D(P,L)$, consider the diagram in Figure 2, which depicts an ellipse and a directrix of the form $x = d$.

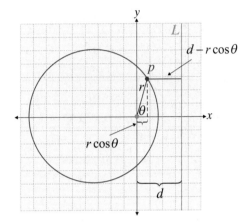

Figure 2: Determining $D(P, L)$

From Figure 2, we see that $D(P,L) = d - r\cos\theta$, so the equation in polar coordinates is

$$\frac{r}{d - r\cos\theta} = e.$$

This equation will be more useful if we solve for r:

$$r = e(d - r\cos\theta)$$
$$r = ed - er\cos\theta$$
$$r + er\cos\theta = ed$$
$$r(1 + e\cos\theta) = ed$$
$$r = \frac{ed}{1 + e\cos\theta}.$$

The derivations of the equations for a directrix of the form $x = -d$, $y = -d$, or $y = d$ are very similar, and will be left to the reader. The four possible forms and their geometric meanings are described in the following box.

DEFINITION

Polar Form of
Conic Sections

In polar coordinates, a conic section with its focus at the origin and either a horizontal or vertical directrix has one of the following forms:

Conic Section Equation	Directrix
$r = \dfrac{ed}{1+e\cos\theta}$	Vertical Directrix $x = d$
$r = \dfrac{ed}{1-e\cos\theta}$	Vertical Directrix $x = -d$
$r = \dfrac{ed}{1+e\sin\theta}$	Horizontal Directrix $y = d$
$r = \dfrac{ed}{1-e\sin\theta}$	Horizontal Directrix $y = -d$

EXAMPLE 1

Identifying the Type
and Directrix
of a Conic

Identify the variety of each of the following conics, and determine the equation for the directrix in each case.

a. $r = \dfrac{15}{5-3\cos\theta}$
   b. $r = \dfrac{2}{3+5\sin\theta}$

Solutions:

a. All we need to do is determine the two constants e and d, and we can accomplish this with simple algebraic manipulation. In order to have the correct form, the constant term in the denominator must be a 1. Thus:

$$r = \frac{15}{5-3\cos\theta}$$

$$= \frac{15\left(\dfrac{1}{5}\right)}{\left(5-3\cos\theta\right)\left(\dfrac{1}{5}\right)}$$

$$= \frac{3}{1-\left(\dfrac{3}{5}\right)\cos\theta}.$$

This tells us that $e = \dfrac{3}{5}$, so the conic is an ellipse. To determine the constant d, we need to write the numerator as a product of e and d. Since the numerator is 3 and since we now know $e = \dfrac{3}{5}$, we can do this as follows:

$$r = \dfrac{3}{1 - \left(\dfrac{3}{5}\right)\cos\theta}$$

$$= \dfrac{\left(\dfrac{3}{5}\right)(5)}{1 - \left(\dfrac{3}{5}\right)\cos\theta}.$$

Hence, $d = 5$. The last observation is that the trigonometric function in the denominator is cosine and that the sign between the two terms in the denominator is negative. By the guidelines for conics in polar form, this tells us that $x = -5$ is the equation for the directrix.

b.   We use the same methods as above to determine first e and then d:

$$r = \dfrac{2}{3 + 5\sin\theta}$$

$$= \dfrac{\dfrac{2}{3}}{1 + \left(\dfrac{5}{3}\right)\sin\theta}$$

$$= \dfrac{\left(\dfrac{5}{3}\right)\left(\dfrac{2}{5}\right)}{1 + \left(\dfrac{5}{3}\right)\sin\theta}.$$

From this form, we can easily see that $e = \dfrac{5}{3}$ and that the directrix is $y = \dfrac{2}{5}$. Thus, the conic section is a hyperbola and the directrix is horizontal and above the x-axis.

EXAMPLE 2

Constructing a
Polar Equation

Construct a polar equation for a leftward-opening parabola with focus at the origin and directrix 2 units from the focus.

Solution:

Since we are discussing a parabola, $e = 1$. If the parabola is to open to the left, the directrix must be oriented vertically and must lie to the right of the focus, so $d = 2$ and the trigonometric function in the denominator must be cosine. Thus, the equation is

$$r = \frac{2}{1 + \cos\theta}.$$

The graph of the parabola appears below, with the focus and directrix shown in green:

TOPIC

Using the Polar Form of Conic Sections

Example 1 illustrated how the two constants e and d can be easily determined from the equation for a conic in polar form. In order to get an even better understanding of a given conic section, it may be useful to determine the constants a, b, and c as well. These constants have the same meaning in this context as they did in Sections 9.1 and 9.3; that is, $2a$ is the distance between vertices of an ellipse or a hyperbola, $2b$ is the length of the minor axis of an ellipse or the width of the auxiliary rectangle that aids in sketching the asymptotes of a hyperbola, and $2c$ is the distance between foci of an ellipse or a hyperbola.

We begin with a. In Section 9.1, eccentricity was defined as $e = \frac{c}{a}$, and this relation still holds for the expanded definition of eccentricity that we now have. In originally discussing ellipses, we noted that $b^2 = a^2 - c^2$, and in originally discussing hyperbolas we noted that $b^2 = c^2 - a^2$; we can combine these two statements by noting that b is nonnegative and that $b^2 = |a^2 - c^2|$. Note that if $e = 1, a = c$ and, hence, $b = 0$; this should not be surprising, as b played no role in the original discussion of parabolas.

When graphing an ellipse or a hyperbola described in polar form, it is computationally easy to determine a. Using the above observations, it is then easy to determine c and b. This process is illustrated in the next several examples.

EXAMPLE 3

Graphing Conic Sections in Polar Form

Sketch the graph of the conic section $r = \dfrac{15}{5 - 3\cos\theta}$.

Solution:

This is the first equation from Example 1, and we have already seen that the equation can be written in the form

$$r = \frac{\left(\dfrac{3}{5}\right)(5)}{1 - \left(\dfrac{3}{5}\right)\cos\theta}.$$

So $e = \dfrac{3}{5}$, $d = 5$, and the directrix is the line $x = -5$. This tells us that the graph is an ellipse and that the major axis is oriented horizontally (perpendicular to the directrix). The entire graph is traced out as θ increases from 0 to 2π, with the right vertex corresponding to $\theta = 0$, and the left vertex corresponding to $\theta = \pi$, (halfway around the ellipse). When $\theta = 0$, $r = \dfrac{15}{2}$ and when $\theta = \pi$, $r = \dfrac{15}{8}$. In rectangular coordinates, the coordinates of the right vertex are $\left(\dfrac{15}{2}, 0\right)$ and the coordinates of the left vertex are $\left(-\dfrac{15}{8}, 0\right)$ (remember that $\theta = \pi$, and a positive r corresponds to a point left of the origin).

Now that we know the coordinates of the two vertices, we can determine that

$$2a = \frac{15}{2} + \frac{15}{8} = \frac{75}{8}$$

and so $a = \dfrac{75}{16}$. Since $e = \dfrac{3}{5}$ and $c = ea$, this means

$$c = \left(\frac{3}{5}\right)\left(\frac{75}{16}\right) = \frac{45}{16}.$$

Finally, from the relation $b^2 = |a^2 - c^2|$, we can determine that $b = \dfrac{15}{4}$.

With this knowledge of a and b, we can sketch the graph below:

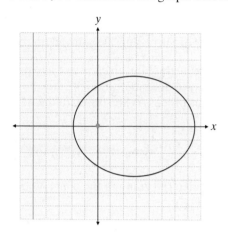

EXAMPLE 4

Graphing Conic
Sections in
Polar Form

Sketch the graph of the conic section $r = \dfrac{2}{3+5\sin\theta}$.

Solution:

This equation is the second from Example 1, and we already know $e = \dfrac{5}{3}$ and that $y = \dfrac{2}{5}$ is the equation for the directrix. This tells us that the graph is a hyperbola and that the vertices are aligned vertically (perpendicular again to the directrix). One vertex corresponds to $\theta = \dfrac{\pi}{2}$ while the other vertex corresponds to $\theta = \dfrac{3\pi}{2}$ (it might also be useful to note that the lower branch of the hyperbola crosses the x-axis at $\theta = 0$ and $\theta = \pi$). When $\theta = \dfrac{\pi}{2}$, $r = \dfrac{1}{4}$ and when $\theta = \dfrac{3\pi}{2}$, $r = -1$ (make note of the negative sign on r). This means that the rectangular coordinates of the two vertices are $\left(0, \dfrac{1}{4}\right)$ and $(0, 1)$, so $2a = \dfrac{3}{4}$ and $a = \dfrac{3}{8}$. From this, $c = ea = \dfrac{5}{8}$ and the relation $b^2 = \left|a^2 - c^2\right|$ leads to $b = \dfrac{1}{2}$.

Our work in Section 9.3 tells us the equations for the asymptotes of the hyperbola are $y - \dfrac{5}{8} = \pm\dfrac{3}{4}x$. Now, using the vertices and asymptotes we calculated, we can sketch the following graph:

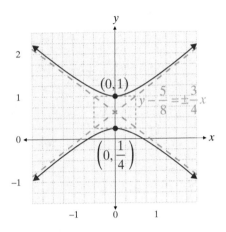

We will conclude this section with an example of rotation of conics in polar form. In general, the graph of an equation $r = f(\theta - \varphi)$ is the rotation of the graph of $r = f(\theta)$ by the angle φ counterclockwise. This makes rotation in polar coordinates particularly easy to handle.

EXAMPLE 5

Graphing Conic Sections Using Rotation in Polar Coordinates

Sketch the graph of the conic section $r = \dfrac{2}{1 + \cos\left(\theta - \dfrac{\pi}{6}\right)}$.

Solution:

We constructed the equation $r = \dfrac{2}{1 + \cos\theta}$ in Example 2, so we know its graph is a parabola opening to the left with directrix $x = 2$. The graph of $r = \dfrac{2}{1 + \cos\theta}$ is repeated to the right.

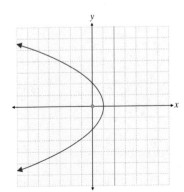

The graph of $r = \dfrac{2}{1+\cos\left(\theta - \dfrac{\pi}{6}\right)}$ is simply the same shape rotated $\dfrac{\pi}{6}$ radians counterclockwise:

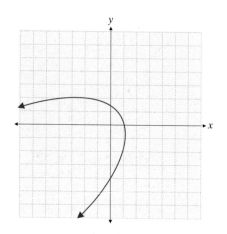

1. c

2. a

3. f

4. e

5. b

6. d

7. Hyperbola, $y = \dfrac{7}{6}$

8. Parabola, $y = -2$

9. Ellipse, $x = -3$

10. Parabola, $x = -2$

11. Hyperbola, $x = \dfrac{1}{3}$

12. Ellipse, $y = \dfrac{7}{2}$

13. Ellipse, $x = 5$

14. Ellipse, $y = -1$

15. Hyperbola, $x = -\dfrac{6}{5}$

16. Hyperbola, $y = -\dfrac{4}{3}$

17. Parabola, $y = \dfrac{3}{2}$

18. Hyperbola, $x = -\dfrac{1}{4}$

Exercises

Match the graph with its corresponding polar equation. See Examples 2 and 3.

1. $r = \dfrac{3}{4-\cos\theta}$

2. $r = \dfrac{9}{6-2\sin\theta}$

3. $r = \dfrac{3}{3+4\sin\theta}$

4. $r = \dfrac{1}{2+2\cos\theta}$

5. $r = \dfrac{6}{1+3\sin\theta}$

6. $r = \dfrac{6}{1+3\cos\theta}$

a.

b.

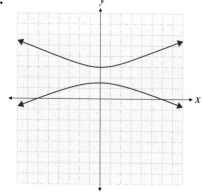

19. Hyperbola,

$$x = -\frac{4}{7}$$

20. Ellipse, $y = -\frac{9}{4}$

21. $r = \dfrac{2}{1 - \cos\theta}$

22. $r = \dfrac{6}{1 - 2\cos\theta}$

23. $r = \dfrac{3}{1 - 4\sin\theta}$

24. $r = \dfrac{2}{1 + \cos\theta}$

25. $r = \dfrac{3}{1 + \frac{1}{4}\cos\theta}$

26. $r = \dfrac{4}{1 + \frac{1}{2}\sin\theta}$

27.

28.

29.

30.

c.

d.

e.

f.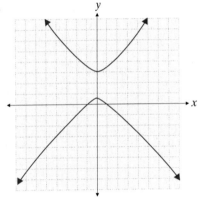

Identify each conic section and find the equation for its directrix. See Example 1.

7. $r = \dfrac{7}{1 + 6\sin\theta}$

8. $r = \dfrac{2}{1 - \sin\theta}$

9. $r = \dfrac{3}{4 - \cos\theta}$

10. $r = \dfrac{4}{2 - 2\cos\theta}$

11. $r = \dfrac{1}{1 + 3\cos\theta}$

12. $r = \dfrac{7}{3 + 2\sin\theta}$

13. $r = \dfrac{5}{2 + \cos\theta}$

14. $r = \dfrac{3}{4 - 3\sin\theta}$

15. $r = \dfrac{6}{3 - 5\cos\theta}$

16. $r = \dfrac{8}{5 - 6\sin\theta}$

17. $r = \dfrac{3}{2 + 2\sin\theta}$

18. $r = \dfrac{-1}{3 + 4\cos\theta}$

19. $r = \dfrac{4}{6 - 7\cos\theta}$

20. $r = \dfrac{9}{5 - 4\sin\theta}$

31.

32.

33.

34.

35.

36.

37.

38.

Construct polar equations for the conic sections described below. See Example 2.

	Conic	Eccentricity	Directrix
21.	Parabola	$e = 1$	$x = -2$
22.	Hyperbola	$e = 2$	$x = -3$
23.	Hyperbola	$e = 4$	$y = -\dfrac{3}{4}$
24.	Parabola	$e = 1$	$x = 2$
25.	Ellipse	$e = \dfrac{1}{4}$	$x = 12$
26.	Ellipse	$e = \dfrac{1}{2}$	$y = 8$

Sketch the graphs of the following conic sections. See Examples 3, 4, and 5.

27. $r = \dfrac{5}{1 + 3\cos\theta}$ **28.** $r = \dfrac{3}{2 + \sin\theta}$

29. $r = \dfrac{4}{1 - 2\sin\theta}$ **30.** $r = \dfrac{6}{2 - 4\cos\theta}$

31. $r = \dfrac{9}{3 - 2\cos\theta}$ **32.** $r = \dfrac{5}{3 + \sin\theta}$

33. $r = \dfrac{4}{1 + 2\cos\theta}$ **34.** $r = \dfrac{4}{2 + 2\sin\theta}$

35. $r = \dfrac{-3}{4 - 9\cos\theta}$ **36.** $r = \dfrac{9}{-4 + \dfrac{3}{2}\sin\theta}$

37. $r = \dfrac{-11}{3 - \cos\theta}$ **38.** $r = \dfrac{2}{10 + 4\sin\theta}$

39. $r = \dfrac{3}{7 + 3\cos\theta}$ **40.** $r = \dfrac{2}{2 + 3\cos\left(\theta - \dfrac{\pi}{4}\right)}$

39.

40.

41.

42.

43.

44.

45.

46.

41. $r = \dfrac{-7}{5 + 3\sin\left(\theta - \dfrac{\pi}{6}\right)}$

42. $r = \dfrac{5}{-2 - 4\sin\left(\theta + \dfrac{2\pi}{3}\right)}$

43. $r = \dfrac{4}{-3 - 2\cos\left(\theta + \dfrac{\pi}{3}\right)}$

44. $r = \dfrac{1}{1 + 4\sin\left(\theta + \dfrac{\pi}{6}\right)}$

45. $r = \dfrac{2}{1 + \cos\left(\theta - \dfrac{\pi}{4}\right)}$

46. $r = \dfrac{4}{2 + 2\sin\left(\theta - \dfrac{\pi}{3}\right)}$

Use your knowledge of conics to answer the following questions.

47. The planets of our solar system follow elliptical orbits with the Sun located at one of the foci. If we assume the Sun is located at the pole and the major axes of these elliptical orbits lie along the polar axis, the orbits of the planets can be expressed by the polar equation

$$r = \frac{\left(1 - e^2\right)a}{1 - e\cos\theta}$$

where e is the eccentricity. Verify the above equation.

48. Using the equation from Exercise 47, answer the following:

a. Show that the shortest distance from the Sun to a planet, called the *perihelion*, is $r = a(1 - e)$.

b. Show that the longest distance from the Sun to a planet, called the *aphelion*, is $r = a(1 + e)$.

c. Uranus is approximately 2.74×10^9 km away from the Sun at perihelion and 3.00×10^9 km at aphelion. Find the eccentricity of Uranus' orbit.

d. The eccentricity of Neptune's path is 0.0113 and $a = 4.495 \times 10^9$. Determine the perihelion and aphelion distances for Neptune.

47. Answers will vary.

48. a. Answers will vary.

   b. Answers will vary.

   c. $e \approx 0.0453$

   d. 4.44×10^9 km,

   4.55×10^9 km

Constructing a Bridge

Plans are in process to develop an uninhabited coastal island into a new resort. Before development can begin, a bridge must be constructed joining the island to the mainland.

Two possibilities are being considered for the support structure of the bridge. The archway could be built as a parabola, or in the shape of a semiellipse.

Assume all measurements that follow refer to dimensions at high tide. The county building inspector has deemed that in order to establish a solid foundation, the space between supports must be at most 300 feet and the height at the center of the arch should be 80 feet. There is a commercial fishing dock located on the mainland whose fishing vessels travel constantly along this intracoastal waterway. The tallest of these ships requires a 60 ft clearance to pass comfortably beneath the bridge. With these restrictions, the width of a channel with a minimum height of 60 ft has to be determined for both possible shapes of the bridge to confirm that it will be suitable for the water traffic beneath it.

80 ft

? ft

60 ft

300 ft

1. Find the equation of a parabola that will fit these constraints.

2. How wide is the channel with a minimum of 60 ft vertical clearance for the parabola in question 1?

3. Find the equation of a semiellipse that will fit these constraints.

4. How wide is the channel with a minimum of 60 ft vertical clearance for the semiellipse in question 3?

5. Which of these bridge designs would you choose, and why?

6. Suppose the tallest fishing ship installs a new antenna which raises the center height by 12 ft. How far off of center (to the left or right) can the ship now travel and still pass under the bridge without damage to the antenna:
   a. For the parabola?
   b. For the semiellipse?

Chapter Summary

A summary of concepts and skills follows each chapter. Refer to these summaries to make sure you feel comfortable with the material in the chapter. The concepts and skills are organized according to the section title and topic title in which the material is first discussed.

9.1: The Ellipse

Overview of Conic Sections
- The geometric definitions of the three types of conic sections: *ellipses*, *parabolas*, and *hyperbolas*
- How to identify, algebraically, equations whose graphs are ellipses, parabolas, or hyperbolas
- The meaning of the *foci* (plural of *focus*) and *vertices* of an ellipse

The Standard Form of an Ellipse
- How the standard form of the equation for an ellipse relates to its geometric properties
- The meaning of an ellipse's *major* and *minor axes*, and *vertices*
- Using the standard form of an ellipse to sketch its graph
- How to construct the equation for an ellipse with prescribed properties
- The geometric and algebraic meaning of elliptical *eccentricity*

Planetary Orbits
- Using knowledge of ellipses to answer questions about planetary orbits

9.2: The Parabola

The Standard Form of a Parabola
- How the standard form of the equation for a parabola relates to its geometric properties
- The meaning of the *focus*, *vertex*, and *directrix* of a parabola
- Using the standard form of a hyperbola to sketch its graph
- How to construct the equation for a hyperbola with prescribed properties

Parabolic Mirrors
- Using knowledge of parabolas to answer questions about parabolic mirrors

9.3: The Hyperbola

The Standard Form of a Hyperbola
- How the standard form of the equation for a hyperbola relates to its geometric properties
- The meaning of the *foci, vertices,* and *asymptotes* of a hyperbola
- Using the standard form of a hyperbola to sketch its graph
- How to construct the equation for a hyperbola with prescribed properties

Guidance Systems
- How properties of hyperbolas are used in designing guidance systems

9.4: Rotation of Conics

Rotation Formulas
- How to relate two sets of coordinates
- How to eliminate the xy term

Rotation Invariants
- Classifying conics by their invariants
- The meaning of the *discriminant*

9.5: Polar Form of Conic Sections

The Focus/Directrix Definition of Conic Sections
- How to determine the polar form of conic sections
- The meaning of the *focus, directrix,* and *eccentricity* of the polar form of conic sections

Using the Polar Form of Conic Sections
- Graphing conic sections in polar form

1. Center: $(3,-1)$
Vertices: $(7,-1)$,
$(-1,-1)$
Foci: $\left(3\pm2\sqrt{3},-1\right)$

2. Center: $(-1,2)$
Vertices: $(-1,0)$,
$(-1,4)$
Foci: $\left(-1,2\pm\dfrac{2\sqrt{5}}{3}\right)$

3.

$\left(-1\pm\sqrt{7},2\right)$

4.

$\left(3\pm2\sqrt{2},-1\right)$

5.

$\left(0,\pm3\sqrt{2}\right)$

6.

$\left(4,\pm\sqrt{21}\right)$

7. $\dfrac{(x+1)^2}{9}+$
$\dfrac{(y-4)^2}{16}=1$

8. $\dfrac{(x-4)^2}{36}+$
$\dfrac{(y-2)^2}{27}=1$

Chapter Review

Section 9.1

Find the center, foci, and vertices of the ellipse that each equation describes.

1. $(x-3)^2+4(y+1)^2=16$ **2.** $9x^2+4y^2+18x-16y+9=0$

Sketch the graphs of the following ellipses, and determine the coordinates of the foci.

3. $\dfrac{(x+1)^2}{16}+\dfrac{(y-2)^2}{9}=1$ **4.** $x^2+9y^2-6x+18y=-9$

5. $3x^2+y^2=27$ **6.** $25x^2+4y^2-200x+300=0$

In each of the following problems, an ellipse is described by properties it possesses. Find the equation, in standard form, for each ellipse.

7. Center at $(-1,4)$, major axis is vertical and of length 8, foci $\sqrt{7}$ units from the center.

8. Foci at $(1,2)$ and $(7,2)$, $e=\dfrac{1}{2}$.

9. Vertices at $\left(\dfrac{7}{2},-1\right)$ and $\left(\dfrac{1}{2},-1\right)$, $e=0$.

10. Vertices at $(1,-8)$ and $(1,2)$, minor axis of length 6.

11. Foci at $(0,0)$ and $(4,0)$, major axis of length 8.

12. Center at $(0,4)$, $a=2c$, and vertices at $(-4,4)$ and $(4,4)$.

13.

14.
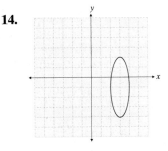

Given that the area A of the ellipse $\dfrac{x^2}{a^2}+\dfrac{y^2}{b^2}=1$ is $A=\pi\cdot a\cdot b$ and $a+b=30$, find the equation or function to satisfy each of the following.

15. Write the area of the ellipse as a function of a.

16. Find the equation of an ellipse with an area of 200π square inches.

9. $\dfrac{4(x-2)^2}{9} +$

$\dfrac{4(y+1)^2}{9} = 1$

10. $\dfrac{(x-1)^2}{9} +$

$\dfrac{(y+3)^2}{25} = 1$

11. $\dfrac{(x-2)^2}{16} + \dfrac{y^2}{12} = 1$

12. $\dfrac{x^2}{16} + \dfrac{(y-4)^2}{12} = 1$

13. $\dfrac{(x+1)^2}{4} +$

$\dfrac{(y-1)^2}{25} = 1$

14. $(x-3)^2 +$

$\dfrac{(y+1)^2}{9} = 1$

15. $30\pi a - \pi a^2$

16. $\dfrac{x^2}{400} + \dfrac{y^2}{100} = 1$ or

$\dfrac{x^2}{100} + \dfrac{y^2}{400} = 1$

17.

$(-6,-1), x = 0$

18.

Section 9.2

Graph the following parabolas and determine the focus and directrix of each.

17. $(y+1)^2 = -12(x+3)$

18. $y^2 - 8y + 2x + 14 = 0$

19. $y^2 + 2y = 4x - 1$

20. $x + \dfrac{1}{4}y^2 = 0$

21. $2y + 4x^2 = 8$

22. $y^2 - 4y + 2x + 24 = 0$

Find the equation, in standard form, for the parabola with the given properties.

23. Vertex at $(-2, 3)$, directrix is the line $y = 2$.

24. Vertex at $(5, -3)$, focus at $(5, 1)$.

25. Focus at $(3, -1)$, directrix is the line $x = 2$.

26. Focus at $(1, -2)$, directrix is the x-axis.

27. Vertex at $(2, -1)$, directrix is the line $x = -2$.

28. Symmetric with respect to the x-axis, focus at $(-3, 0)$, and $p = 4$.

29.

30.

31. A motorcycle headlight is made by placing a strong light bulb inside a reflective paraboloid formed by rotating the parabola $x^2 = 5y$ around its axis of symmetry (assume that x and y are in units of inches). In order to have the brightest, most concentrated light beam, how far from the vertex should the bulb be placed?

19.

20.

21.

22.

$\left(\dfrac{1}{2}, 4\right), x = \dfrac{3}{2}$

$(1, -1), \ x = -1$

$(-1, 0), \ x = 1$

$\left(0, \dfrac{31}{8}\right), \ y = \dfrac{33}{8}$

$\left(\dfrac{-21}{2}, 2\right), x = \dfrac{-19}{2}$

23. $(x+2)^2 = 4(y-3)$

24. $(x-5)^2 = 16(y+3)$

25. $(y+1)^2 = 2\left(x - \dfrac{5}{2}\right)$

26. $(x-1)^2 = -4(y+1)$

27. $(y+1)^2 = 16(x-2)$

28. $y^2 = 16(x+7)$

29. $(y+1)^2 = -16(x-5)$

30. $(x-2)^2 = 20(y+4)$

31. $\dfrac{5}{4}$ inches

32.

$(2,3),(2,-7)$

33.

$\left(-3\pm\sqrt{13},-1\right)$

34.

$\left(\pm\sqrt{2},0\right)$

35.

$(0,\pm13)$

36. Center: $(-1,2)$
Foci: $\left(-1\pm3\sqrt{5},2\right)$
Vertices: $(-7,2),(5,2)$

37. Center: $(0,2)$
Foci: $\left(\pm2\sqrt{10},2\right)$
Vertices: $(\pm6,2)$

Section 9.3

Sketch the graphs of the following hyperbolas, using asymptotes as guides. Determine the coordinates of the foci in each case.

32. $\dfrac{(y+2)^2}{9} - \dfrac{(x-2)^2}{16} = 1$

33. $9x^2 - 4y^2 + 54x - 8y + 41 = 0$

34. $x^2 - y^2 = 1$

35. $\dfrac{y^2}{25} - \dfrac{x^2}{144} = 1$

Find the center, foci, and vertices of each hyperbola that the equation describes.

36. $(x+1)^2 - 4(y-2)^2 = 36$

37. $x^2 - 9y^2 + 36y - 72 = 0$

38. $y^2 - 4x^2 - 2y - 32x = 67$

39. $\dfrac{(y-3)^2}{4} - \dfrac{(x-3)^2}{49} = 1$

Find the equation, in standard form, for the hyperbola with the given properties.

40. Vertices at $(4,-1)$ and $(-2,-1)$ and foci at $(5,-1)$ and $(-3,-1)$.

41. Asymptotes of $y = \pm\dfrac{5}{2}(x+1) - 2$ and vertices at $(-3,-2)$ and $(1,-2)$.

42. Foci at $(-1,-2)$ and $(-1,8)$ and asymptotes of $y = \pm\left(\dfrac{3}{4}x + \dfrac{3}{4}\right) + 3$.

43. Asymptotes of $y = \pm(3x-6) + 2$ and vertices at $(2,-1)$ and $(2,5)$.

44. Vertices at $(\pm3,0)$ and foci at $(\pm5,0)$.

45. Foci at $\left(-1, 7\pm\sqrt{13}\right)$ and asymptotes of $y = \pm\dfrac{2}{3}(x+1) + 7$.

46.

47.
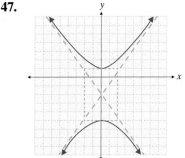

38. Center: $(-4,1)$
Foci: $\left(-4, 1\pm\sqrt{5}\right)$
Vertices: $(-4,3),(-4,-1)$

39. Center: $(3,3)$
Foci: $\left(3, 3\pm\sqrt{53}\right)$
Vertices: $(3,1),(3,5)$

40. $\dfrac{(x-1)^2}{9} - \dfrac{(y+1)^2}{7} = 1$

41. $\dfrac{(x+1)^2}{4} - \dfrac{(y+2)^2}{25} = 1$

42. $\dfrac{(y-3)^2}{9} - \dfrac{(x+1)^2}{16} = 1$

43. $\dfrac{(y-2)^2}{9} - \dfrac{(x-2)^2}{1} = 1$

44. $\dfrac{x^2}{9} - \dfrac{y^2}{16} = 1$

45. $\dfrac{(y-7)^2}{4} - \dfrac{(x+1)^2}{9} = 1$

46. $\dfrac{(x-1)^2}{36} - \dfrac{(y+1)^2}{16} = 1$

47. $\dfrac{(y+2)^2}{9} - \dfrac{x^2}{4} = 1$

48. $\left(\dfrac{-\sqrt{2}}{2}, \dfrac{15\sqrt{2}}{2}\right)$

49. $\left(\begin{array}{c} 11 + 43\sqrt{3}, \\ -11\sqrt{3} + 43 \end{array}\right)$

50. $(-4.3395, -9.0299)$

51. $\left(3 + 3\sqrt{3}, 9 - \sqrt{3}\right)$

52. Hyperbola,

$\theta = \dfrac{\pi}{4}, x'^2 - y'^2 = 12$

Section 9.4

In the exercises below, the xy-coordinates of a point are given. Find the $x'y'$-coordinates of each point for the given rotation angle θ.

48. $(-8, 7), \theta = \dfrac{\pi}{4}$

49. $(22, 86), \theta = \dfrac{\pi}{3}$

50. $(4.6, -8.9), \theta = 53°$

51. $\left(2\sqrt{3}, 6\sqrt{3}\right), \theta = 30°$

Use the discriminant to classify each of the following conic sections. Then determine the angle θ that will allow you to convert the equation and eliminate the xy term. Finally, sketch the conic section.

52. $xy - 6 = 0$

53. $10x^2 + 2\sqrt{3}xy + 12y^2 - y = 0$

54. $10\sqrt{3}x^2 + 42xy - 4\sqrt{3}y^2 = 187\sqrt{3}$

55. $x^2 + 2xy + y^2 + x - y = 0$

Section 9.5

Identify each conic section and find the equation for its directrix.

56. $r = \dfrac{5}{4 - 8\sin\theta}$

57. $r = \dfrac{7}{4 + 4\sin\theta}$

58. $r = \dfrac{4}{6 - 3\cos\theta}$

59. $r = \dfrac{7}{5 + 2\cos\theta}$

53. Ellipse,

$\theta = \dfrac{\pi}{3}, \dfrac{\left(x' - \dfrac{\sqrt{3}}{52}\right)^2}{9} + \dfrac{\left(y' - \dfrac{1}{36}\right)^2}{13} = \dfrac{5}{27,378}$

54. Hyperbola,

$\theta = \dfrac{\pi}{6}, \dfrac{x'^2}{11} - \dfrac{y'^2}{17} = 1$

55. Parabola,

$\theta = \dfrac{\pi}{4}, y' = \sqrt{2}x'^2$

56. Hyperbola, $y = \dfrac{-5}{8}$

57. Parabola, $y = \dfrac{7}{4}$

58. Ellipse, $x = \dfrac{-4}{3}$

59. Ellipse, $x = \dfrac{7}{2}$

60. $r = \dfrac{12}{1 + 4\sin\theta}$

61. $r = \dfrac{4}{1 + \dfrac{1}{4}\cos\theta}$

62. $r = \dfrac{7}{1 - \sin\theta}$

63. $r = \dfrac{3}{1 + 9\cos\theta}$

Construct polar equations for the conic sections described below.

	Eccentricity	**Directrix**
60.	$e = 4$	$y = 3$
61.	$e = \dfrac{1}{4}$	$x = 16$
62.	$e = 1$	$y = -7$
63.	$e = 9$	$x = \dfrac{1}{3}$

1. Center: $(2, -1)$
   Vertices: $(2, 8)$, $(2, -10)$
   Foci: $\left(2, -1 \pm \sqrt{65}\right)$

2. Center: $(-3, 1)$
   Vertices: $(0, 1)$, $(-6, 1)$
   Foci: $\left(-3 \pm \dfrac{3\sqrt{3}}{2}, 1\right)$

3. $\dfrac{(x-2)^2}{4} + (y-2)^2 = 1$

4. $\dfrac{(x-3)^2}{36} + \dfrac{(y-2)^2}{32} = 1$

5. $\dfrac{x^2}{48} + \dfrac{y^2}{64} = 1$

6. Approx. 97.98 yards apart

7.

   Focus: $\left(2, \dfrac{-1}{2}\right)$;

   Directrix: $y = \dfrac{-3}{2}$

8.

   Focus: $\left(-1, \dfrac{-25}{6}\right)$;

   Directrix: $y = \dfrac{-7}{6}$

9. $(x-4)^2 = 16(y+2)$

10. $(y-2)^2 = -20(x+1)$

11. $8\sqrt{6}$ m ≈ 19.6 m

Chapter Test

Find the center, foci, and vertices of the ellipse that each equation describes.

1. $\dfrac{(x-2)^2}{16} + \dfrac{(y+1)^2}{81} = 1$

2. $x^2 + 4y^2 + 6x - 8y + 4 = 0$

Find the equation, in standard form, for the ellipse with the specified properties.

3. Vertices at $(0, 2)$ and $(4, 2)$, minor axis of length 2.

4. Center at $(3, 2)$, $a = 3c$, and foci at $(1, 2)$ and $(5, 2)$.

5. Vertices at $(0, \pm 8)$, $e = \dfrac{1}{2}$.

6. Stan, an avid runner and math enthusiast designs an elliptical track for a race. The length of the major axis is 100 yards and the track is 20 yards wide. He has decided to set up two cameras to document the event and wants to place them at the foci of the ellipse. How far apart will the cameras be?

Graph the following parabolas and determine the focus and directrix of each.

7. $(x-2)^2 = 2(y+1)$

8. $x^2 + 2x + 6y + 17 = 0$

Find the equation, in standard form, for the parabola with the given properties.

9. Vertex at $(4, -2)$, focus at $(4, 2)$.

10. Symmetric with respect to the line $y = 2$, directrix is the line $x = 4$, and $p = -5$.

Use your knowledge of parabolas to answer the following question.

11. A parabolic archway is 12 meters high at the vertex. At a height of 10 meters the width of the archway is 8 meters. How wide is the archway at ground level?

12.

Center: $(-1,-2)$

Foci: $\left(-1\pm\sqrt{145},-2\right)$

Vertices: $(-13,-2)$, $(11,-2)$

13.

Center: $(2,-3)$

Foci: $\left(2\pm\sqrt{10},-3\right)$

Vertices: $(3,-3)$, $(1,-3)$

14. Center: $(-1,2)$

Foci: $\left(-1\pm\sqrt{29},2\right)$

Vertices: $(-3,2),(1,2)$

15. Center: $(0,0)$

Foci: $(\pm 5,0)$

Vertices: $(\pm 3,0)$

16. $\dfrac{y^2}{4}-\dfrac{x^2}{12}=1$

17. $x^2 - \dfrac{y^2}{9}=1$

18. $(3.4823, -32.2999)$

19. $\left(2\sqrt{3},0\right)$

Find the center, foci, and vertices of each of the following hyperbolas, and sketch the graph, using asymptotes as guides.

12. $\dfrac{(x+1)^2}{144}-\dfrac{(y+2)^2}{1}=1$

13. $9x^2 - y^2 - 36x - 6y + 18 = 0$

Find the center, foci, and vertices of each hyperbola that the equation describes.

14. $\dfrac{(x+1)^2}{4}-\dfrac{(y-2)^2}{25}=1$

15. $16x^2 - 9y^2 = 144$

Find the equation, in standard form, for the hyperbola with the given properties.

16. Vertices at $(0,\pm 2)$ and foci at $(0,\pm 4)$.

17. Vertices at $(\pm 1,0)$ and asymptotes at $y=\pm 3x$.

In the exercises below, the xy-coordinates of a point are given. Find the $x'y'$-coordinates of each point for the given rotation angle θ.

18. $(5, -32.1)$, $\theta = 2.7°$

19. $\left(3,\sqrt{3}\right)$, $\theta = \dfrac{\pi}{6}$

20. Ellipse,

$$\theta = \frac{\pi}{6}, \frac{x'^2}{26} + \frac{y'^2}{50} = 1$$

21. Parabola,

$$\theta = \frac{\pi}{4}, y' = \frac{\sqrt{2}}{3}x'^2 - 2$$

22. Hyperbola, $y = 4$

23. Ellipse, $y = \frac{1}{2}$

24. Parabola, $x = \frac{-2}{3}$

25. $r = \frac{3}{1 - \cos\theta}$

26. $r = \dfrac{3}{1 - \dfrac{1}{5}\sin\theta}$

Use the discriminant to classify each of the following conic sections. Then determine the angle θ that will allow you to convert the equation and eliminate the xy term. Finally, sketch the conic section.

20. $44x^2 + 12\sqrt{3}xy + 32y^2 - 582 = 718$ **21.** $3x^2 + 6xy + 3y^2 + 9x - 9y = \frac{36}{\sqrt{2}}$

Identify each conic section and find the equation for its directrix.

22. $r = \dfrac{8}{1 + 2\sin\theta}$ **23.** $r = \dfrac{3}{7 + 6\sin\theta}$ **24.** $r = \dfrac{6}{9 - 9\cos\theta}$

Construct polar equations for the conic sections described below.

	Eccentricity	Directrix
25.	$e = 1$	$x = -3$
26.	$e = \dfrac{1}{5}$	$y = -15$

Chapter 10

Systems of Equations

10.1   Solving Systems by Substitution and Elimination

10.2   Matrix Notation and Gaussian Elimination

10.3   Determinants and Cramer's Rule

10.4   The Algebra of Matrices

10.5   Inverses of Matrices

10.6   Partial Fraction Decomposition

10.7   Linear Programming

10.8   Nonlinear Systems of Equations and Inequalities

        Chapter 10 Project

        Chapter 10 Summary

        Chapter 10 Review

        Chapter 10 Test

By the end of this chapter you should be able to:

What if you were given a large handful of nickels and pennies and were told their total value and the number of coins? Without counting, would you know how many of each coin you have?

By the end of this chapter, you'll be able to solve both linear and nonlinear systems of equations. Given two or more equations in the same variables, you will find solutions for the variables that solve all the equations simultaneously. You'll encounter the coin problem on page 793. You'll master this type of problem using tools such as the Method for Solving Systems by Elimination, found on page 784.

Introduction

In this chapter, we return to the study of linear equations, arguably the simplest class of equations. But as we will soon see, our understanding of linear equations has room for growth in many directions. In fact, the material in this chapter serves as a good illustration of how, given incentive and opportunity, mathematicians extend ideas and techniques beyond the familiar.

In the particular case of linear equations, this extension has taken the form of:

1. Considering equations containing more variables (we have already seen the first step in this process when we moved from linear equations in one variable to linear equations in two).

2. Trying to find solutions that satisfy more than one linear equation at a time; such sets of equations are called *systems*.

3. Making use of elementary methods to solve systems if possible, and developing entirely new methods if necessary or desirable.

The third point above constitutes the bulk of this chapter. We start off, in our quest to solve systems of linear equations, by using some intuitive methods that work admirably if the equations contain only a few variables and if the system has only a few equations. Several centuries ago, however, mathematicians began to realize the limitations of such simple methods. As the problems that people wanted to solve led to systems of many variables and equations, refinements were made to the elementary methods of solution, and eventually, entirely new techniques were developed. The two refinements we will study are called *Gaussian elimination* and *Gauss-Jordan elimination*. Some of the new techniques that were developed for solving large systems of equations make use of *determinants of matrices* and *matrix inverses*, two concepts that can serve as an introduction to higher mathematics.

Gauss

The notion of the inverse of a matrix (a *matrix* is a rectangular array of numbers) closes out our discussion of solution methods, but in fact it brings us full circle and allows us to write systems of linear equations as single matrix equations. This illustrates how mathematicians, in extending our reach in terms of problem solving and in constructing new mathematics, are guided by the achievements of the past. Sections 10.6 and 10.7 then demonstrate two important uses of linear systems of equations, and the chapter ends with an introduction to systems of nonlinear equations and inequalities.

Solving Systems by Substitution and Elimination

TOPICS

1. Definition and classification of linear systems of equations
2. Solving systems by substitution
3. Solving systems by elimination
4. Larger systems of equations
5. Applications of systems of equations
T. Solving systems of equations using technology

TOPIC 1 Definitions and Classification of Linear Systems of Equations

Many problems in mathematics can be described by two or more equations in two or more variables. When the equations are all linear, such a collection of equations is called a *linear system of equations*, or sometimes *simultaneous linear equations*. The word *simultaneous* refers to the goal of identifying the values of the variables (if there are any) that solve all of the equations simultaneously.

In the case of two linear equations in two variables, it turns out that there are only three possible configurations of solutions, which we see in Figure 1.

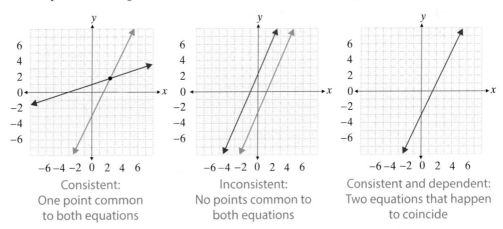

Consistent:
One point common
to both equations

Inconsistent:
No points common to
both equations

Consistent and dependent:
Two equations that happen
to coincide

Figure 1: Solutions to Systems of Two Linear Equations

Now, each equation individually has an infinite number of solutions. What we are interested in are the points that solve both equations. These are *solutions to the system of equations*.

In the first graph of Figure 1, the two lines intersect in exactly one point. In the second graph, the two lines are parallel, so the system of equations has no solution since there is no point lying on both lines. In the final graph, the two lines actually coincide and appear as one, meaning that any ordered pair that solves one of the equations in the system solves the other as well, so the system has an infinite number of solutions.

We will encounter systems consisting of more than two equations and/or more than two variables, but larger systems have one important similarity to the two-variable, two-equation systems in Figure 1: every linear system will have exactly one solution, no solution, or an infinite number of solutions.

DEFINITION

Solutions to Linear Systems of Equations

- A linear system of equations with no solution is called **inconsistent**.

- A linear system of equations that has at least one solution is called **consistent**.

- Any linear system with more than one solution must have an infinite number of solutions and is called **dependent**.

Our goal is to develop a systematic, effective method of solving linear systems. Because linear systems of equations arise in so many different contexts and are of great importance both theoretically and practically, many solution methods have been devised. In this chapter, we will use solution methods that fall into four broad classes. The two we cover in this section are fairly easy to apply when the number of equations and variables is small.

TOPIC 2   Solving Systems by Substitution

The solution method of substitution hinges on solving one equation in a system for one of the variables, and substituting the result for that variable in the remaining equations. This can be a time-consuming process if the system is large (meaning more than a few equations and more than a few variables), and in fact the task may have to be repeated many times. But it is a very natural method to use for some small systems, as illustrated in Examples 1 and 2.

EXAMPLE 1

Solving a System of Equations by Substitution

Use the method of substitution to solve the system $\begin{cases} 2x - y = 1 \\ x + y = 5 \end{cases}$

Solution:

Note:
Solving for a variable with a coefficient of 1 will make the process of substitution a bit easier.

Either equation can be solved for either variable, and the choice doesn't affect the final answer. We will solve the second equation for x, and then substitute the result in the first equation, giving us one equation in the variable y. Once we have solved for y, we can substitute the value of y into either original equation to find x.

$$x = -y + 5 \qquad \text{Solve the second equation for x.}$$

$$2(-y+5) - y = 1 \qquad \text{Substitute the result in the first equation.}$$

$$-2y + 10 - y = 1 \qquad \text{Simplify and solve for y.}$$

$$-3y = -9$$

$$y = 3$$

We then substitute our answer for y into the original second equation.

$$x + 3 = 5 \qquad \text{Substitute } y = 3 \text{ in the second equation.}$$

$$x = 2 \qquad \text{Solve for x.}$$

Thus, the solution to the system of equations is $(2,3)$.

Note how the following graph corresponds to the system and the ordered pair solution that we have found.

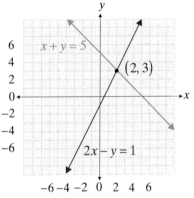

EXAMPLE 2

Solving a System
of Equations by
Substitution

Use the method of substitution to solve the system $\begin{cases} -2x + 6y = 6 \\ x + 3 = 3y \end{cases}$

Solution:

$$x = 3y - 3 \qquad \text{Solve the second equation for x.}$$

$$-2(3y - 3) + 6y = 6 \qquad \text{Substitute the result in the first equation.}$$

$$-6y + 6 + 6y = 6 \qquad \text{Simplify.}$$

$$0 = 0$$

The resulting equation is always true. This means that for any value of y, letting $x = 3y - 3$ results in an ordered pair $(x,y) = (3y-3, y)$ that solves both equations. Since there are an infinite number of solutions, the system is dependent.

Graphically, this means the graphs of the two equations are exactly the same. In fact, the first equation is simply a rearranged multiple of the second equation.

Algebraically, we can describe the solution set as $\{(3y-3,y)|y \in \mathbb{R}\}$. If we had solved either equation for y instead of x, we would have obtained the alternative but equivalent solution $\left\{\left(x, \dfrac{x+3}{3}\right) \middle| x \in \mathbb{R}\right\}$.

TOPIC 3 Solving Systems by Elimination

In some systems, the expressions obtained by solving one equation for one variable are difficult to work with, no matter which variable we choose. In such cases, the solution method of elimination may be a more efficient choice. The elimination method is also often a better choice for larger systems.

The method of elimination is based on the goal of eliminating one variable in one equation by adding two equations together, and in fact, the elimination method is often called the *addition method*.

The method works by making sure that the resulting equation has fewer variables than either of the original two, simplifying the system. In fact, if the system is of the two-variable, two-equation variety, the new equation is ready to be solved for its remaining variable, and the solution of the system is then straightforward to find.

EXAMPLE 3

Solving a System
of Equations by
Elimination

Use the method of elimination to solve the system $\begin{cases} 5x+3y=-7 \\ 7x-6y=-20 \end{cases}$

Solution:

The coefficient of y in the second equation is -6, while the coefficient of y in the first equation is 3. This means that if we multiply all the terms in the first equation by 2, the coefficients of y will be negatives of one another, so adding the two equations will eliminate the y variable.

In order to keep track of these steps, we annotate our work with labeled arrows. When we modify the system, we are not changing the solutions, we are just writing an equivalent system that is easier to solve. The ultimate goal is to rewrite the system so that we can "read off" the answer.

The notation above the arrow indicates that we have modified the system by multiplying each term in equation 1 by the constant 2.

$$\begin{cases} 5x+3y=-7 \\ 7x-6y=-20 \end{cases} \xrightarrow{2E_1} \begin{cases} 10x+6y=-14 \\ 7x-6y=-20 \end{cases}$$

$$17x=-34 \qquad \text{Add the equations.}$$

$$x=-2 \qquad \text{Solve for } x.$$

We can then substitute $x=-2$ into either of the original equations to determine y. Here, we substitute into the first equation of the original system.

$$5(-2) + 3y = -7$$

$$3y = 3$$

$$y = 1$$

The ordered pair $(-2, 1)$ is thus the solution of the system. Note that using the second equation gives the same y-value and is a good way to check our work.

EXAMPLE 4

Solving a System
of Equations by
Elimination

Use the method of elimination to solve the system $\begin{cases} 2x - 3y = 3 \\ 3x - \dfrac{9}{2}y = 5 \end{cases}$

Solution:

Eliminate the variable x by multiplying both equations by a constant:

$$\begin{cases} 2x - 3y = 3 \\ 3x - \dfrac{9}{2}y = 5 \end{cases} \xrightarrow[-2E_2]{3E_1} \begin{cases} 6x - 9y = 9 \\ -6x + 9y = -10 \end{cases}$$
$$0 = -1$$

Although the intent was to obtain coefficients of x that were negatives of one another, we have achieved the same thing for y.

When we add the equations, the result is $0 = -1$, a false statement. This means that no ordered pair solves both equations, and the system is inconsistent. Graphically, the two lines defined by the equations are parallel.

TOPIC Larger Systems of Equations

Algebraically, larger systems of equations can be dealt with in the same way as the two-variable, two-equation systems that we have studied (though the number of steps needed to obtain a solution might increase). Geometrically, however, larger systems can mean something quite different.

For example, if an equation contains three variables, say x, y, and z, a given solution of the equation must consist of an *ordered triple* of numbers, not an ordered pair. A graphical representation of the ordered triple requires three coordinate axes, as opposed to two. This leads to the concept of three-dimensional space, and a coordinate system with three axes. Figure 2 is an illustration of the way in which the positive x, y, and z axes are typically represented on a two-dimensional surface, such as a piece of paper, a computer monitor, or a blackboard. The negative portion of each of the axes is not drawn, and the three axes meet at the origin (the point with $(0, 0, 0)$ as its coordinates) at right angles. As an illustration of how ordered triples appear plotted in Cartesian *space*, the point $(1, 2, 4)$ is plotted (the thin colored lines are drawn merely for reference and are not part of the plot).

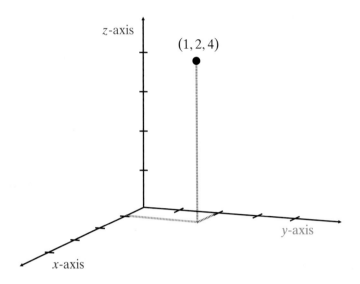

Figure 2: Plotting the Point (1, 2, 4)

We are interested in the graph of a linear equation in three variables. It turns out that any equation of the form

$$Ax + By + Cz = D,$$

where not all of A, B, and C are 0, depicts a plane in three-dimensional space. A linear system of equations in three variables will thus describe a collection of planes, one per equation. If a linear system of three variables contains three equations, it is possible for the three planes to intersect in exactly one point. It is also possible, however, for two of the planes to intersect in a line while the third plane contains no point of that line, for the three planes to intersect in a common line, for two or three of the planes to be parallel, or for two or three of the planes to coincide. The addition of another variable and another equation leads to many more possibilities than we have seen thus far. However, it is still the case that a linear system has no solution, exactly one solution, or an infinite number of solutions. Figure 3 illustrates some of the possible configurations of a three-variable, three-equation linear system.

Figure 3: Three-Variable, Three-Equation Systems

EXAMPLE 5

Solving a System of Equations by Elimination

Solve the system $\begin{cases} 2x + y + z = 6 \\ 2x + 3y - z = -2 \\ -3x + 2y - z = 5 \end{cases}$

Solution:

We will follow the same type of approach we used in Example 4. If we add the first equation to the second equation or the third equation, we will eliminate z, resulting in a two-equation system in the variables x and y.

Equation 1: $\begin{cases} 2x + y + z = 6 \\ 2x + 3y - z = -2 \end{cases}$
Equation 2:
$$4x + 4y = 4$$

Equation 1: $\begin{cases} 2x + y + z = 6 \\ -3x + 2y - z = 5 \end{cases}$
Equation 3:
$$-x + 3y = 11$$

Putting these two equations together, we have the system

$$\begin{cases} 4x + 4y = 4 \\ -x + 3y = 11 \end{cases}$$

We use elimination once more, multiplying the second equation by 4 to eliminate x.

$$\begin{cases} 4x + 4y = 4 \\ -x + 3y = 11 \end{cases} \xrightarrow{\ 4E_2\ } \begin{cases} 4x + 4y = 4 \\ -4x + 12y = 44 \end{cases}$$
$$16y = 48$$
$$y = 3$$

Now that we have solved for y, we can plug this back into one of the equations from the two variable system to solve for x.

$$-x + 3(3) = 11$$
$$-x + 9 = 11$$
$$x = -2$$

Finally, we substitute both x and y into one of the equations from the original system.

$$2(-2) + 3 + z = 6$$
$$-4 + 3 + z = 6$$
$$z = 7$$

Thus, the solution to the system of equations is the ordered triple $(-2, 3, 7)$.

Solving a System
of Equations by
Elimination

EXAMPLE 6

Solve the system $\begin{cases} 3x - 5y + z = -10 \\ -x + 2y - 3z = -7 \\ x - y - 5z = -24 \end{cases}$

Solution:

In general, a good approach to solving a large system of equations is to try to eliminate a variable and obtain a smaller system.

There are many possible ways to proceed. One option is to use the second equation (or a multiple of it) to eliminate x when we add it to the first and third equations. The result will be a two equation system in the variables y and z.

$$\text{Equation 1: } \begin{cases} 3x - 5y + z = -10 \\ \text{Equation 2: } \end{cases} \!\! \begin{cases} -x + 2y - 3z = -7 \end{cases} \xrightarrow{3E_2} \begin{cases} 3x - 5y + z = -10 \\ -3x + 6y - 9z = -21 \\ \hline \qquad\qquad y - 8z = -31 \end{cases}$$

$$\text{Equation 2: } \begin{cases} -x + 2y - 3z = -7 \\ \text{Equation 3: } \end{cases} \!\! \begin{cases} x - y - 5z = -24 \\ \hline \qquad\quad y - 8z = -31 \end{cases}$$

The two resulting equations are identical, and tell us that $y = 8z - 31$. We can now use any equation that contains x to determine the relation between x and z. For instance, the third equation in the system tells us that $x = y + 5z - 24$, or

$$x = (8z - 31) + 5z - 24 = 13z - 55.$$

One description of the solution set is thus $\{(13z - 55,\ 8z - 31,\ z) \mid z \in \mathbb{R}\}$. Geometrically, the three planes described by the equations of the system intersect in a line.

TOPIC 5   Applications of Systems of Equations

Many applications that we have previously analyzed using a single equation are more naturally stated in terms of two or more equations. Consider, for example, the following mixture problem.

EXAMPLE 7

Mixing Alloys

A foundry needs to produce 75 tons of an alloy that is 34% copper. It has supplies of 9% copper alloy and 84% copper alloy. How many tons of each alloy must be mixed to obtain the desired result?

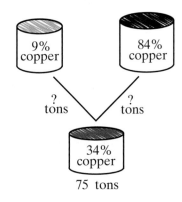

Solution:

Let x represent the number of tons of 9% copper alloy needed, and y the number of tons of 84% copper alloy needed. We have two variables, so we will need two equations to find the solution.

Since we need 75 total tons of alloy, one equation is $x + y = 75$. We also know that 9% of the x tons and 84% of the y tons represent the total amount of copper, and this amount must equal 34% of 75 tons. The second equation is thus $0.09x + 0.84y = 0.34(75)$. This gives us a system that can be solved by elimination:

$$\begin{cases} x + y = 75 \\ 0.09x + 0.84y = 0.34(75) \end{cases} \xrightarrow[100E_2]{-9E_1} \begin{cases} -9x - 9y = -675 \\ 9x + 84y = 2550 \end{cases}$$

$$75y = 1875$$

$$y = 25$$

Substituting back, we have $x + 25 = 75$, so $x = 50$. Thus, 50 tons of 9% alloy and 25 tons of 84% alloy are needed.

EXAMPLE 8

Determining Ages

If the ages of three girls, Xenia, Yolanda, and Zsa Zsa, are added, the result is 30. The sum of Xenia's and Yolanda's ages is Zsa Zsa's age, while Xenia's age subtracted from Yolanda's is half of Zsa Zsa's age a year ago. How old is each girl?

Solution:

Let x, y, and z represent Xenia's, Yolanda's, and Zsa Zsa's ages, respectively. The first sentence tells us that $x + y + z = 30$. The second sentence tells us that $x + y = z$, and that $y - x = \dfrac{z-1}{2}$. To make the work easier, these equations can be rewritten as shown:

$$\begin{cases} x+y+z = 30 \\ x+y = z \\ y-x = \dfrac{z-1}{2} \end{cases} \xrightarrow{\;2E_3\;} \begin{cases} x+y+z = 30 \\ x+y-z = 0 \\ -2x+2y-z = -1 \end{cases}$$

From this, we see that the sum of the first and second equations results in $2x + 2y = 30$, or $x + y = 15$, and the sum of the first and third equations is $-x + 3y = 29$. The method of elimination is the best choice for solving the new system

$$\begin{cases} x+y = 15 \\ -x+3y = 29 \end{cases}$$

as the sum of the two equations gives us $4y = 44$, or $y = 11$. We can use this value to determine that $x = 4$, and then use these two values to determine that $z = 15$. Geometrically, it means that the ordered triple $(4, 11, 15)$ is the point of intersection of the three planes described by these equations. In context, it means that Xenia is 4, Yolanda is 11, and Zsa Zsa is 15.

TOPIC

Solving Systems of Equations Using Technology

The solution to a consistent pair of linear equations is the point common to both equations. Graphically speaking, the solution is the point where the graphs of the two equations intersect. We can use a graphing calculator to find this point. Consider the following system of equations: $\begin{cases} 2x - 3y = -13 \\ x = y - 6 \end{cases}$. One way to solve this system using a

calculator is to graph each equation. Remember to solve for y before entering the equation in ⬤ Y= and selecting ⬤ GRAPH.

Once the graph of the two lines is displayed, press ⬤ 2ND ⬤ TRACE to access the CALC menu and select 5:intersect. The phrase "First curve?" should appear. Use the arrows to move the cursor along the first line to where it appears to intersect the other line and press ENTER. When the phrase "Second curve?" appears, press ENTER again (as the cursor should now be on the second line, still near the point of intersection). Now the word "Guess?" should appear. Press ENTER a final time and the x- and y-values of the point of intersection will appear at the bottom.

1. $(-5, 2)$

2. $(-1, 1)$

3. $(5, 3)$

4. $\left\{\left(\dfrac{-y-2}{3}, y\right)\middle|\right.$

$\left.y \in \mathbb{R}\right\}$

5. $\varnothing$

6. $(-4, 1)$

7. $\left\{\left(\dfrac{y-3}{2}, y\right)\middle|\right.$

$\left.y \in \mathbb{R}\right\}$

8. $\left\{(-2y-4, y)\middle|\right.$

$\left.y \in \mathbb{R}\right\}$

9. $(-1, 7)$

10. $(-4, 14)$

11. $(3, 11)$

12. $(1, 1)$

13. $\left\{(x, 4x+1)\middle|\right.$

$\left.x \in \mathbb{R}\right\}$

14. $\varnothing$

15. $(2, 19)$

16. $(1, -2)$

17. $(-5, 1)$

18. $(3, -2)$

19. $(5, 6)$

20. $\varnothing$

21. $\left\{(-y-2, y)\middle|\, y \in \mathbb{R}\right\}$

22. $(2, 3)$

23. $(-5, 4)$

24. $\varnothing$

So the point where the lines intersect, and thus the solution to this system of equations, is $(-5, 1)$. This method works with any system of equations that can be graphed on a calculator, not just linear ones.

Exercises

Use the method of substitution to solve the following systems of equations. If a system is dependent, express the solution set in terms of one of the variables. See Examples 1 and 2.

1. $\begin{cases} 2x - y = -12 \\ 3x + y = -13 \end{cases}$ **2.** $\begin{cases} 2x - 4y = -6 \\ 3x - y = -4 \end{cases}$ **3.** $\begin{cases} 3y = 9 \\ x + 2y = 11 \end{cases}$

4. $\begin{cases} -3x - y = 2 \\ 9x + 3y = -6 \end{cases}$ **5.** $\begin{cases} 2x + y = -2 \\ -4x - 2y = 5 \end{cases}$ **6.** $\begin{cases} 5x - y = -21 \\ 9x + 2y = -34 \end{cases}$

7. $\begin{cases} 2x - y = -3 \\ -4x + 2y = 6 \end{cases}$ **8.** $\begin{cases} 3x + 6y = -12 \\ 2x + 4y = -8 \end{cases}$ **9.** $\begin{cases} 2x + 5y = 33 \\ 3x = -3 \end{cases}$

10. $\begin{cases} 5x + 2y = 8 \\ 2x + y = 6 \end{cases}$ **11.** $\begin{cases} -2x + y = 5 \\ 9x - 2y = 5 \end{cases}$ **12.** $\begin{cases} 3x + y = 4 \\ -2x + 3y = 1 \end{cases}$

13. $\begin{cases} 4x - y = -1 \\ -8x + 2y = 2 \end{cases}$ **14.** $\begin{cases} 4x - 2y = 3 \\ -2x + y = -7 \end{cases}$ **15.** $\begin{cases} 9x - y = -1 \\ 3x + 2y = 44 \end{cases}$

Use the method of elimination to solve the following systems of equations. If a system is dependent, express the solution set in terms of one of the variables. See Examples 3 and 4.

16. $\begin{cases} 2x - 3y = 8 \\ 8x + 5y = -2 \end{cases}$ **17.** $\begin{cases} -2x + 3y = 13 \\ 4x + 2y = -18 \end{cases}$ **18.** $\begin{cases} 5x + 7y = 1 \\ -2x + 3y = -12 \end{cases}$

19. $\begin{cases} x + 2y = 17 \\ 3x + 4y = 39 \end{cases}$ **20.** $\begin{cases} 5x - 10y = 9 \\ -x + 2y = -3 \end{cases}$ **21.** $\begin{cases} -2x - 2y = 4 \\ 3x + 3y = -6 \end{cases}$

22. $\begin{cases} 4x + y = 11 \\ 3x - 2y = 0 \end{cases}$ **23.** $\begin{cases} 7x + 8y = -3 \\ -5x - 4y = 9 \end{cases}$ **24.** $\begin{cases} -2x - y = 9 \\ 4x + 2y = 1 \end{cases}$

25. $(-1, 1)$

26. $(7, 6)$

27. $(3, -5)$

28. $(4, 3)$

29. $\varnothing$

30. $\varnothing$

31. $(-1, 3, 0)$

32. $(3, -2, 3)$

33. $(2, 2, -1)$

34. $\left\{(-3z-5, \right.$

$-6z-10,$

$\left. z)\,\middle|\, z \in \mathbb{R}\right\}$

35. $\left\{\left(\dfrac{y-z+2}{3}, y, z\right)\,\middle|\, \right.$

$\left. y \in \mathbb{R}, z \in \mathbb{R}\right\}$

36. $(2, 2, 2)$

37. $\varnothing$

38. $\left\{\left(-\dfrac{5z}{2}+4, \right.\right.$

$-5z+10, z)\Big|$

$\left. z \in \mathbb{R}\right\}$

39. $(1, 1, 0)$

40. $(1, 2, 3)$

41. $(9, 1, 1)$

42. $\varnothing$

43. $(3, 1, -2)$

44. $\left\{\left(\dfrac{10z+39}{9}, \right.\right.$

$\dfrac{7z+12}{9}, z\Big)\Big|$

$\left. z \in \mathbb{R}\right\}$

25. $\begin{cases} -2x+4y=6 \\ 3x-y=-4 \end{cases}$ **26.** $\begin{cases} 5x-6y=-1 \\ -4x+3y=-10 \end{cases}$ **27.** $\begin{cases} \dfrac{2}{3}x+y=-3 \\ 3x+\dfrac{5}{2}y=-\dfrac{7}{2} \end{cases}$

28. $\begin{cases} \dfrac{x}{5}-y=-\dfrac{11}{5} \\ \dfrac{x}{4}+y=4 \end{cases}$ **29.** $\begin{cases} \dfrac{2}{3}x+2y=1 \\ x+3y=0 \end{cases}$ **30.** $\begin{cases} -x-5y=-6 \\ \dfrac{3}{5}x+3y=1 \end{cases}$

Use any convenient method to solve the following systems of equations. If a system is dependent, express the solution set in terms of one or more of the variables, as appropriate. See Examples 5 and 6.

31. $\begin{cases} x-y+4z=-4 \\ 4x+y-2z=-1 \\ -y+2z=-3 \end{cases}$ **32.** $\begin{cases} x+2y=-1 \\ y+3z=7 \\ 2x+5z=21 \end{cases}$

33. $\begin{cases} x+y=4 \\ y+3z=-1 \\ 2x-2y+5z=-5 \end{cases}$ **34.** $\begin{cases} 2x-y=0 \\ 5x-3y-3z=5 \\ 2x+6z=-10 \end{cases}$

35. $\begin{cases} 3x-y+z=2 \\ -6x+2y-2z=-4 \\ -3x+y-z=-2 \end{cases}$ **36.** $\begin{cases} 2x-3y=-2 \\ x-4y+3z=0 \\ -2x+7y-5z=0 \end{cases}$

37. $\begin{cases} 3x-y+z=2 \\ -6x+2y-2z=1 \\ 5x+2y-3z=2 \end{cases}$ **38.** $\begin{cases} 4x-y+5z=6 \\ 4x-3y-5z=-14 \\ -2x-5z=-8 \end{cases}$

39. $\begin{cases} 3x+8z=3 \\ -3x+y-7z=-2 \\ x+2y+3z=3 \end{cases}$ **40.** $\begin{cases} x+2y+z=8 \\ 2x-3y-4z=-16 \\ x-5y+5z=6 \end{cases}$

41. $\begin{cases} 2x-7y-4z=7 \\ -x+4y+2z=-3 \\ 3y-4z=-1 \end{cases}$ **42.** $\begin{cases} 4x+4y-2z=6 \\ x-5y+3z=-2 \\ -2x-2y+z=3 \end{cases}$

43. $\begin{cases} 2x+3y+4z=1 \\ 3x-4y+5z=-5 \\ 4x+5y+6z=5 \end{cases}$ **44.** $\begin{cases} x-4y+2z=-1 \\ 2x+y-3z=10 \\ -3x+12y-6z=3 \end{cases}$

45. $(4, 5, 5)$

46. $(2, -1, 2)$

47. $\left(\dfrac{49}{3}, \dfrac{-16}{3}, \dfrac{5}{4}\right)$

48. $(-2, -1, 4)$

49. $(0, 3, 2)$

50. $(-2, 3, 4)$

51. 22 pennies,
23 nickels

52. $a = -\dfrac{1}{2}, b = -\dfrac{3}{2},$

$c = 5$

53. 25 people

54. 40 ounces of
12%,

20 ounces of
30%

55. Eliza is 15 years
old.

56. $7000 at 8% and
$5000 at 15%

57. 7 shirts and
4 pairs of shorts

45. $\begin{cases} x + 2y + 3z = 29 \\ 2x - y - z = -2 \\ 3x + 2y - 6z = -8 \end{cases}$

46. $\begin{cases} 5x - 2y + z = 14 \\ 8x + 4y = 12 \\ 9x = 18 \end{cases}$

47. $\begin{cases} 2x + 5y = 6 \\ 3y + 8z = -6 \\ x + 4y = -5 \end{cases}$

48. $\begin{cases} 4x + 3y + 4z = 5 \\ 5x - 6y - 2z = -12 \\ 5z = 20 \end{cases}$

49. $\begin{cases} 9x + 4y - 8z = -4 \\ -6x + 3y - 9z = -9 \\ 8y - 3z = 18 \end{cases}$

50. $\begin{cases} 21x - 7y + 51z = 141 \\ 13x + 9y - 5z = -19 \\ 19x - 8y + 23z = 30 \end{cases}$

Solve the following application problems. See Examples 7 and 8.

51. Karen empties out her purse and finds 45 loose coins, consisting entirely of nickels and pennies. If the total value of the coins is $1.37, how many nickels and how many pennies does she have?

52. What choice of a, b, and c will force the graph of the polynomial $f(x) = ax^2 + bx + c$ to have a y-intercept of 5 and to pass through the points $(1, 3)$ and $(2, 0)$?

53. A tour organizer is planning on taking a group of 40 people to a musical. Balcony tickets cost $29.95 and regular tickets cost $19.95. The organizer collects a total of $1048.00 from her group to buy the tickets. How many people chose to sit in the balcony?

54. How many ounces each of a 12% alcohol solution and a 30% alcohol solution must be combined to obtain 60 ounces of an 18% solution?

55. Eliza's mother is 20 years older than Eliza, but 3 years younger than Eliza's father. Eliza's father is 7 years younger than three times Eliza's age. How old is Eliza?

56. An investor decides at the beginning of the year to invest some of his cash in an account paying 8% annual interest, and to put the rest in a stock fund that ends up earning 15% over the course of the year. He puts $2000 more in the first account than in the stock fund, and at the end of the year he finds he has earned $1310 in interest. How much money was invested at each of the two rates?

57. Jack and Tyler went shopping for summer clothes. Shirts were $12.47 each, including tax, and shorts were $17.23 per pair, including tax. Jack and Tyler spent a total of $156.21 on 11 items. How many shirts and pairs of shorts did they buy?

58. Bob is 51 years old.

59. 3 quarters, 11 dimes, and 28 pennies

60. $800 in stocks

61. Jim is 28 years old.

62. 50 tickets

63. 3 thumb screws

64. Certificate of deposit: $3000, Stocks: $6000, Bonds: $1000

65. Apples: $0.78, Oranges: $0.93, Mangos: $1.05

58. Three years ago, Bob was twice as old as Marla. Fifteen years ago, Bob was three times as old as Marla. How old is Bob?

59. Deyanira empties her pockets and finds 42 coins consisting of quarters, dimes, and pennies. There are twice as many pennies as dimes and quarters total. If the total value of the coins is $2.13, how many coins of each denomination does she have?

60. If an investor has invested $1000 in stocks and bonds, how much has he invested in stocks if he invested four times more in stocks than in bonds?

61. Twelve years ago, Jim was twice as old as Kristin. Sixteen years ago, Jim was three times older than Kristin. How old is Jim?

62. A movie brought in $740 in ticket sales in one day. Tickets during the day cost $5 and tickets at night cost $7. If 120 tickets were sold, how many were sold during the day?

63. A computer has 24 screws in its case. If there are 7 times more slotted screws than thumb screws, how many thumb screws are in the computer?

64. Jael has $10,000 she would like to invest. She has narrowed her options down to a certificate of deposit paying 5% annually, bonds paying 4% annually, and stocks with an expected annual rate of return of 13.5%. If she wants to invest twice as much in the stocks as in the certificate of deposit and she wants to earn $1000 in interest by the end of the year, how much should she invest in each type of investment?

65. Lea ordered fruit baskets for three of her coworkers. One contained 5 apples, 2 oranges, and 1 mango and cost $6.81. Another contained 2 mangos, 8 oranges, and 3 apples and cost $11.88. The third contained 4 apples, 4 oranges, and 4 mangos and cost $11.04. How much did each type of fruit cost?

Matrix Notation and Gaussian Elimination

TOPICS

1. Linear systems, matrices, and augmented matrices
2. Gaussian elimination and row echelon form
3. Gauss-Jordan elimination and reduced row echelon form
T. Matrices

TOPIC **1**

Linear Systems, Matrices, and Augmented Matrices

Many applications of linear systems involve *thousands* of variables, including the scheduling of planes and passengers at an airport and the modeling of materials at the atomic level. The solution methods of substitution and elimination, learned in the last section, are inadequate for dealing with such large systems. Further, it can become impractical to even *write down* these systems of equations. In this section, we explore the matrix, a mathematical tool that can be very helpful in understanding linear systems, both large and small.

DEFINITION

Matrices and Matrix Notation

A **matrix** is a rectangular array of numbers, called **elements** or **entries** of the matrix. As the numbers are in a rectangular array, they naturally form **rows** and **columns**. It is often important to determine the size of a given matrix; we say that a matrix with m rows and n columns is an $m \times n$ matrix (read "m by n"), or of **order** $m \times n$. By convention, the number of rows is always stated first.

$$
\text{rows} \begin{array}{c} \rightarrow \\ \rightarrow \\ \rightarrow \end{array} \overset{\text{columns}}{\begin{bmatrix} 4 & 0 & 1 & -3 \\ 12 & -5 & -5 & 8 \\ 0 & -7 & 25 & 2 \end{bmatrix}}
$$

This matrix has 3 rows and 4 columns, so it is a 3×4 matrix.

Matrices are often labeled with capital letters. The same letter in lowercase, with a pair of subscripts attached, is usually used to refer to its individual elements. For instance, if A is a matrix, a_{ij} refers to the element in the ith row and the jth column.

If we call the above matrix A, then we have $a_{23} = -5$ and $a_{32} = -7$.

======== EXAMPLE 1 ========

Matrices and Matrix Notation

Given the matrix $A = \begin{bmatrix} -27 & 0 & 1 \\ 5 & -\pi & 13 \end{bmatrix}$, determine:

a. The order of A **b.** The value of a_{13} **c.** The value of a_{21}

Solutions:

a. A has 2 rows and 3 columns, and is thus a 2×3 matrix.

b. The value of the entry in the first row and third column is $a_{13} = 1$.

c. The value of the entry in the second row and first column is $a_{21} = 5$.

DEFINITION

Standard Form of a System of Linear Equations

A linear system of equations is in **standard form** when each equation has been simplified with its variables on the left-hand side and its constant term on the right-hand side. Each equation should have its variable terms listed in the same order.

For example, the system $\begin{cases} 3x + 4 = 7y \\ -2x + 8y = 18 \end{cases}$ is written in standard form as $\begin{cases} 3x - 7y = -4 \\ -x + 4y = 9 \end{cases}$

Now that we have a standard way to organize the terms in a system of equations, we can put the matrix to use as a way to represent a system of equations concisely.

DEFINITION

Augmented Matrices

Given a linear system of equations written in standard form, the **augmented matrix** of that system is a matrix consisting of the coefficients of the variables listed in their relative positions with an adjoined column of the constants of the system. The matrix of coefficients and the column of constants are customarily separated by a vertical bar.

For example, the augmented matrix for the system $\begin{cases} 3x - 7y = -4 \\ -x + 4y = 9 \end{cases}$ is $\left[\begin{array}{cc|c} 3 & -7 & -4 \\ -1 & 4 & 9 \end{array} \right]$

The augmented matrix will have as many rows as there are equations in the system, and one more column than there are variables.

=== EXAMPLE 2 ===

Augmented Matrices

Construct the augmented matrix for the linear system $\begin{cases} \dfrac{2x-6y}{2} = 3-z \\ z-x+5y = 12 \\ x+3y-2 = 2z \end{cases}$

Solution:

Write each equation in standard form, then read off the coefficients and constants to construct the augmented matrix.

Linear System	Standard Form	Augmented Matrix
$\dfrac{2x-6y}{2} = 3-z$ $z-x+5y = 12$ $x+3y-2 = 2z$	$x-3y+z = 3$ $-x+5y+z = 12$ $x+3y-2z = 2$	$\begin{bmatrix} 1 & -3 & 1 & 3 \\ -1 & 5 & 1 & 12 \\ 1 & 3 & -2 & 2 \end{bmatrix}$

TOPIC 2

Gaussian Elimination and Row Echelon Form

Consider the following augmented matrix:

$$\begin{bmatrix} 1 & 2 & -2 & 11 \\ 0 & 1 & -1 & 3 \\ 0 & 0 & 1 & -1 \end{bmatrix}$$

We can translate this back into system form to obtain

$$\begin{cases} x+2y-2z = 11 \\ y-z = 3 \\ z = -1 \end{cases}$$

and we note that this system can be solved when we substitute the last equation $z = -1$ into the second equation to obtain $y-(-1) = 3$, or $y = 2$, and then substitute again in the first equation to obtain $x+2(2)-2(-1) = 11$, or $x = 5$.

Typically, systems of equations aren't so straightforward to solve, at least not at first. Recall that the methods of substitution and elimination made systems of equations simpler to solve by reducing the number of equations and/or variables present.

Matrices, a powerful organizational tool, make it easier to transform complicated systems into ones like the example above. The process of *Gaussian elimination* (named after the mathematician Carl Friedrich Gauss) can transform any augmented matrix into a form like the one above, and the solution of the corresponding system then solves the original system as well. The technical name for an augmented matrix in the form shown above is *row echelon form*.

DEFINITION

Row Echelon Form

A matrix is in **row echelon form** if:

1. The first nonzero entry in each row is 1. We call this a **leading 1**.

2. Every entry below a leading 1 is 0, and each leading 1 appears farther to the right than the leading 1's in the rows above it.

3. All rows consisting entirely of 0's (if there are any) appear at the bottom.

EXAMPLE 3

Row Echelon Form

Determine if the following matrices are in row echelon form. If not, explain why not.

a. $\begin{bmatrix} 2 & 0 & 13 & | & -1 \\ 0 & 5 & 1 & | & 1 \\ 0 & 0 & 1 & | & -4 \end{bmatrix}$

   b. $\begin{bmatrix} 1 & 2 & -8 & | & 0 \\ 0 & 1 & 1 & | & 3 \\ 0 & 0 & 1 & | & 10 \end{bmatrix}$

c. $\begin{bmatrix} 1 & 0 & 4 & | & -2 \\ 1 & 2 & 0 & | & 7 \\ 1 & -5 & -3 & | & 6 \end{bmatrix}$

   d. $\begin{bmatrix} 1 & -3 & 1 & | & 6 \\ 0 & 0 & 0 & | & 0 \\ 0 & 1 & -8 & | & -10 \end{bmatrix}$

Solutions:

a. This matrix fails the first condition, as the first nonzero entry is not always 1.

$\begin{bmatrix} 2 & 0 & 13 & | & -1 \\ 0 & 5 & 1 & | & 1 \\ 0 & 0 & 1 & | & -4 \end{bmatrix}$

b. This matrix meets each of the three conditions, and is in row echelon form.

c. This matrix fails the second condition, as nonzero values are present beneath a leading 1.

$\begin{bmatrix} 1 & 0 & 4 & | & -2 \\ 1 & 2 & 0 & | & 7 \\ 1 & -5 & -3 & | & 6 \end{bmatrix}$

d. This matrix fails the third condition, since there is a row of all zeros that lies above a row with nonzero entries.

$\begin{bmatrix} 1 & -3 & 1 & | & 6 \\ 0 & 0 & 0 & | & 0 \\ 0 & 1 & -8 & | & -10 \end{bmatrix}$

Since the real goal of Gaussian elimination is to solve systems of equations that are represented by augmented matrices, we need to make sure any changes we apply to the matrix do not change the solution of the system.

These "legal" operations are called *elementary row operations*. There are three such matrix operations, based on the fact that the following actions have no effect on the solutions of a system of equations.

- Interchanging the order of two equations in a system

- Multiplying all the terms in an equation by a nonzero constant

- Adding one equation to another (or a nonzero multiple of an equation)

DEFINITION

Elementary Row Operations

Let A be an augmented matrix corresponding to a system of equations. Each of the following operations on A results in the augmented matrix of an equivalent system. In the notation, R_i refers to row i of the matrix A.

1. Rows i and j can be interchanged. (Denoted as $R_i \leftrightarrow R_j$)

2. Each entry in row i can be multiplied by a nonzero constant c. (Denoted as cR_i)

3. Row j can be replaced with the sum of itself and a constant multiple of row i. (Denoted as $cR_i + R_j$)

Typically, we write that an operation has been performed by connecting the original matrix to the new matrix with an arrow, writing the type of operation above it. For example, the matrix from Example 3d can be placed in row echelon form as follows.

$$\left[\begin{array}{ccc|c} 1 & -3 & 1 & 6 \\ 0 & 0 & 0 & 0 \\ 0 & 1 & -8 & -10 \end{array}\right] \xrightarrow{R_2 \leftrightarrow R_3} \left[\begin{array}{ccc|c} 1 & -3 & 1 & 6 \\ 0 & 1 & -8 & -10 \\ 0 & 0 & 0 & 0 \end{array}\right]$$

━━━━━━ **EXAMPLE 4** ━━━━━━

Elementary Row
Operations

Perform the indicated elementary row operation.

$$\begin{bmatrix} 1 & 2 & 3 & | & 0 \\ 5 & 4 & 1 & | & -7 \\ 6 & 1 & 8 & | & -9 \end{bmatrix} \xrightarrow{-5R_1+R_2} ?$$

Solution:

The row operation indicates that we should add -5 times Row 1 to Row 2.

$$\begin{bmatrix} 1 & 2 & 3 & | & 0 \\ 5 & 4 & 1 & | & -7 \\ 6 & 1 & 8 & | & -9 \end{bmatrix} \xrightarrow{-5R_1+R_2} \begin{bmatrix} 1 & 2 & 3 & | & 0 \\ -5(1)+5 & -5(2)+4 & -5(3)+1 & | & -5(0)+-7 \\ 6 & 1 & 8 & | & -9 \end{bmatrix}$$

$$\begin{bmatrix} 1 & 2 & 3 & | & 0 \\ 0 & -6 & -14 & | & -7 \\ 6 & 1 & 8 & | & -9 \end{bmatrix}$$

Note that the first row now begins with 1 and the second row now begins with 0. The row operation has begun to change this matrix towards row echelon form.

────────────────────────────────

In summary, Gaussian elimination refines the method of elimination by removing the need to write variables at each step and by providing a framework for systematic application of row operations. Just as with substitution and elimination, there is no one correct method to apply Gaussian elimination. With the upcoming examples, we will see some general guidelines to follow.

━━━━━━ **EXAMPLE 5** ━━━━━━

Gaussian Elimination

Use Gaussian elimination to solve the system $\begin{cases} -2x+y-5z=-6 \\ x+2y-z=-8 \\ 3x-y+2z=2 \end{cases}$

Solution:

First, we read off the augmented matrix corresponding to this system.

$$\begin{bmatrix} -2 & 1 & -5 & | & -6 \\ 1 & 2 & -1 & | & -8 \\ 3 & -1 & 2 & | & 2 \end{bmatrix}$$

Now, we transform it into row echelon form. It is usually easiest to work one column at a time. After getting a leading 1 in the first row, use row operations to obtains 0's below it. Repeat this process with each successive column.

$$\begin{bmatrix} -2 & 1 & -5 & | & -6 \\ 1 & 2 & -1 & | & -8 \\ 3 & -1 & 2 & | & 2 \end{bmatrix} \xrightarrow{R_1 \leftrightarrow R_2} \begin{bmatrix} 1 & 2 & -1 & | & -8 \\ -2 & 1 & -5 & | & -6 \\ 3 & -1 & 2 & | & 2 \end{bmatrix}$$

Exchange Rows 1 and 2 to make 1 the first entry of the first row.

$$\xrightarrow{2R_1+R_2} \begin{bmatrix} 1 & 2 & -1 & | & -8 \\ 0 & 5 & -7 & | & -22 \\ 3 & -1 & 2 & | & 2 \end{bmatrix}$$

Add 2 times Row 1 to Row 2 to get a 0 as the first entry of Row 2.

$$\xrightarrow{-3R_1+R_3} \begin{bmatrix} 1 & 2 & -1 & | & -8 \\ 0 & 5 & -7 & | & -22 \\ 0 & -7 & 5 & | & 26 \end{bmatrix}$$

Add −3 times Row 1 to Row 3 to get a 0 as the first entry of Row 3.

$$\xrightarrow{\frac{1}{5}R_2} \begin{bmatrix} 1 & 2 & -1 & | & -8 \\ 0 & 1 & -\frac{7}{5} & | & -\frac{22}{5} \\ 0 & -7 & 5 & | & 26 \end{bmatrix}$$

Multiply Row 2 by $\frac{1}{5}$ to make its first nonzero entry 1.

$$\xrightarrow{7R_2+R_3} \begin{bmatrix} 1 & 2 & -1 & | & -8 \\ 0 & 1 & -\frac{7}{5} & | & -\frac{22}{5} \\ 0 & 0 & -\frac{24}{5} & | & -\frac{24}{5} \end{bmatrix}$$

Add 7 times Row 2 to Row 3 to get a 0 as the second entry of Row 3.

$$\xrightarrow{-\frac{5}{24}R_3} \begin{bmatrix} 1 & 2 & -1 & | & -8 \\ 0 & 1 & -\frac{7}{5} & | & -\frac{22}{5} \\ 0 & 0 & 1 & | & 1 \end{bmatrix}$$

Multiply Row 3 by $-\frac{5}{24}$ to make its first nonzero entry a 1.

The third row in the last matrix tells us that $z = 1$. When we substitute this in the second equation, we obtain

$$y - \frac{7}{5}(1) = -\frac{22}{5},$$

so $y = -3$. From the first row equation, we then obtain

$$x + 2(-3) - 1(1) = -8,$$

or $x = -1$. Thus, the ordered triple $(-1, -3, 1)$ solves the system of equations.

TOPIC   # Gauss-Jordan Elimination and Reduced Row Echelon Form

Once a matrix has been put in row echelon form, we can back-substitute, beginning with the last equation and working backward to solve the system of equations. We reached the row echelon form with matrix operations, and it turns out that we can perform the remaining steps in matrix form as well.

Just as Gaussian elimination is a refinement of the method of elimination, Gauss-Jordan elimination, co-named for Wilhelm Jordan (1842–1899), is a refinement of Gaussian elimination. The goal in the method is to put a given matrix into *reduced row echelon form*.

DEFINITION

Reduced Row
Echelon Form

A matrix is said to be in **reduced row echelon form** if:

1. It is in row echelon form.

2. Each entry *above* a leading 1 is also 0.

Consider, for instance, the last matrix obtained in Example 5. If we begin with that matrix and eliminate entries (that is, convert to 0) above the leading 1 in the third column, and then do the same in the second column, we obtain a matrix in reduced row echelon form:

$$
\begin{bmatrix} 1 & 2 & -1 & -8 \\ 0 & 1 & -\dfrac{7}{5} & -\dfrac{22}{5} \\ 0 & 0 & 1 & 1 \end{bmatrix}
\xrightarrow[R_3+R_1]{\frac{7}{5}R_3+R_2}
\begin{bmatrix} 1 & 2 & 0 & -7 \\ 0 & 1 & 0 & -3 \\ 0 & 0 & 1 & 1 \end{bmatrix}
\xrightarrow{-2R_2+R_1}
\begin{bmatrix} 1 & 0 & 0 & -1 \\ 0 & 1 & 0 & -3 \\ 0 & 0 & 1 & 1 \end{bmatrix}
$$

If we now write this in system form, we have

$$
\begin{cases} x = -1 \\ y = -3 \\ z = 1 \end{cases}
$$

which is equivalent to the original system, but in a form that directly tells us the solution of the system.

EXAMPLE 6

Reduced Row Echelon Form

Use Gauss-Jordan elimination to solve the system $\begin{cases} x - 2y + 3z = -5 \\ 2x + 3y - z = 1 \\ -x - 5y + 4z = -6 \end{cases}$

Solution:

$$\begin{bmatrix} 1 & -2 & 3 & | & -5 \\ 2 & 3 & -1 & | & 1 \\ -1 & -5 & 4 & | & -6 \end{bmatrix} \xrightarrow[R_1 + R_3]{-2R_1 + R_2} \begin{bmatrix} 1 & -2 & 3 & | & -5 \\ 0 & 7 & -7 & | & 11 \\ 0 & -7 & 7 & | & -11 \end{bmatrix} \xrightarrow{R_2 + R_3} \begin{bmatrix} 1 & -2 & 3 & | & -5 \\ 0 & 7 & -7 & | & 11 \\ 0 & 0 & 0 & | & 0 \end{bmatrix}$$

Note that the row of 0's at the bottom corresponds to the true statement $0 = 0$, indicating that this system has an infinite number of solutions. In order to describe the solution set algebraically, we continue the row operations until the matrix is in reduced row echelon form.

$$\begin{bmatrix} 1 & -2 & 3 & | & -5 \\ 0 & 7 & -7 & | & 11 \\ 0 & 0 & 0 & | & 0 \end{bmatrix} \xrightarrow{\frac{1}{7}R_2} \begin{bmatrix} 1 & -2 & 3 & | & -5 \\ 0 & 1 & -1 & | & \frac{11}{7} \\ 0 & 0 & 0 & | & 0 \end{bmatrix} \xrightarrow{2R_2 + R_1} \begin{bmatrix} 1 & 0 & 1 & | & -\frac{13}{7} \\ 0 & 1 & -1 & | & \frac{11}{7} \\ 0 & 0 & 0 & | & 0 \end{bmatrix}$$

The last matrix is in reduced row echelon form, as every leading 1 has 0's both below and above it. The corresponding system of equations is

$$\begin{cases} x + z = -\frac{13}{7} \\ y - z = \frac{11}{7} \end{cases}, \text{ which is equivalent to } \begin{cases} x = -z - \frac{13}{7} \\ y = z + \frac{11}{7} \end{cases}.$$

Thus, we can describe the solution set as $\left\{ \left(-z - \frac{13}{7}, z + \frac{11}{7}, z \right) \middle| z \in \mathbb{R} \right\}$.

TOPIC T

Matrices

Operations with matrices can also be done on a graphing calculator. Suppose we wanted to find the row reduced form of the matrix $\begin{bmatrix} 8 & -1 & 2 & 30 \\ -3 & 4 & 5 & -20 \\ 1 & 2 & -1 & -6 \end{bmatrix}$. First, we need to enter in this matrix by pressing to access the **MATRIX** menu and use the arrows to highlight **EDIT**. Select **1:[A]** for matrix A and enter the dimensions, 3×4, and the values for each entry: If there are already numbers present on the display, type over them and the calculator will adjust automatically. To enter the values for each entry, move the cursor to the upper left entry position and type the value that goes in that position. Press **ENTER** and the cursor will move to the next position in the matrix. The double subscripts appear at the bottom of the display as each value is entered.

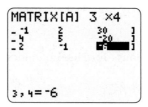

Return to the main screen by pressing 2ND MODE. Now that the matrix is stored in the calculator as matrix A, we can perform operations. Press 2ND x^{-1} to access the MATRIX menu again, but this time select MATH. Scroll down to select A:ref(. When you press ENTER, the following will appear on the main screen:

```
ref(
```

Now we need to select which matrix we want to put in row echelon form, matrix A. To do this, press 2ND x^{-1} once more. NAMES and 1:[A] should already be highlighted, so press ENTER. Add the right-hand parenthesis and press ENTER.

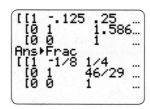

To view this matrix with fractional entries, press MATH and ENTER, since 1:Frac is already highlighted. Press ENTER again on the main screen.

```
[[1  -.125 .25 ...
 [0  1     1.586...
 [0  0     1     ...
Ans▶Frac
[[1  -1/8  1/4  ...
 [0  1     46/29 ...
 [0  0     1     ...
```

Since the entire matrix does not fit on the screen, use the arrow keys to view the right side of the matrix.

Finding the row-reduced echelon form of this matrix is similar. Press 2ND x^{-1} and select MATH but this time arrow down farther to highlight B:rref(and press ENTER. Select matrix A again, add the right-hand parenthesis, and press ENTER.

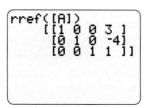

1. a. 3×2
  b. -1
  c. None

2. a. 1×3
  b. 2
  c. None

3. a. 5×2
  b. None
  c. 10

4. a. 3×3
  b. 3
  c. 0

5. a. 3×4
  b. None
  c. 286

6. a. 2×3
  b. 4
  c. 0

7. a. 3×2
  b. 1
  c. None

8. a. 3×4
  b. None
  c. 75

9. a. 2×5
  b. 5
  c. 2

10. $\begin{bmatrix} 4 & 5 & -3 & 8 \\ 7 & -2 & 0 & -6 \\ 5 & -6 & 3 & 0 \end{bmatrix}$

11. $\begin{bmatrix} -3 & 1 & -2 & -4 \\ \frac{1}{2} & -4 & -1 & 1 \\ 0 & -3 & 3 & 1 \end{bmatrix}$

12. $\begin{bmatrix} 5 & \frac{1}{2} & \frac{-1}{2} & 3 \\ -7 & 1 & 7 & 2 \\ 1 & -1 & 1 & 4 \end{bmatrix}$

13. $\begin{bmatrix} \frac{-3}{2} & -1 & 0 & -1 \\ 2 & 2 & 3 & 0 \\ 0 & -1 & 6 & 0 \end{bmatrix}$

14. $\begin{bmatrix} -1 & 1 & 1 & -6 \\ -3 & 6 & -5 & 1 \\ \frac{-2}{3} & -1 & 2 & 0 \end{bmatrix}$

Exercises

1. Let $A = \begin{bmatrix} 4 & -1 \\ 0 & 3 \\ 9 & -5 \end{bmatrix}$. Determine the following, if possible:

  a. The order of A **b.** The value of a_{12} **c.** The value of a_{23}

2. Let $B = \begin{bmatrix} -7 & 2 & 11 \end{bmatrix}$. Determine the following, if possible:

  a. The order of B **b.** The value of b_{12} **c.** The value of b_{31}

3. Let $C = \begin{bmatrix} 1 & 0 \\ 5 & -3 \\ 2 & 9 \\ \pi & e \\ 10 & -7 \end{bmatrix}$. Determine the following, if possible:

  a. The order of C **b.** The value of c_{23} **c.** The value of c_{51}

4. Let $D = \begin{bmatrix} -8 & 13 & -1 \\ 0 & 6 & 3 \\ 0 & -9 & 0 \end{bmatrix}$. Determine the following, if possible:

  a. The order of D **b.** The value of d_{23} **c.** The value of d_{33}

5. Let $E = \begin{bmatrix} -443 & 951 & 165 & 274 \\ 286 & -653 & 812 & -330 \\ 909 & 377 & 429 & -298 \end{bmatrix}$. Determine the following, if possible:

  a. The order of E **b.** The value of e_{42} **c.** The value of e_{21}

6. Let $A = \begin{bmatrix} 9 & 5 & 0 \\ 7 & 4 & 2 \end{bmatrix}$. Determine the following, if possible:

  a. The order of A **b.** The value of a_{22} **c.** The value of a_{13}

7. Let $B = \begin{bmatrix} 8 & 1 \\ 3 & 0 \\ 6 & 7 \end{bmatrix}$. Determine the following, if possible:

  a. The order of B **b.** The value of b_{12} **c.** The value of b_{13}

15. $\begin{bmatrix} 12 & 1 & -\frac{3}{2} & | & 1 \\ 5 & 2 & \frac{}{} & | & \frac{1}{5} \\ 1 & 0 & 3 & | & 1 \\ 5 & 2 & 1 & | & -2 \end{bmatrix}$

16. $\begin{bmatrix} \frac{3}{2} & 2 & -3 & | & 6 \\ 3 & -6 & 27 & | & 0 \\ 2 & 6 & 1 & | & 3 \end{bmatrix}$

17. $\begin{bmatrix} \frac{2}{3} & \frac{-4}{3} & -2 & | & 0 \\ 8 & -2 & 6 & | & 7 \\ 3 & -2 & 0 & | & 0 \end{bmatrix}$

18. $\begin{bmatrix} \frac{4}{3} & \frac{-2}{3} & 1 & | & 7 \\ -4 & 4 & 12 & | & 3 \\ -5 & -8 & 0 & | & -4 \end{bmatrix}$

19. $\begin{bmatrix} \frac{1}{2} & -14 & \frac{-1}{4} & | & -8 \\ \frac{1}{5} & \frac{-7}{6} & \frac{1}{4} & | & -3 \\ 5 & -5 & \frac{8}{3} & | & -5 \end{bmatrix}$

20. $\begin{cases} 5x + 3y = 9 \\ x + 4y = 12 \end{cases}$

21. $\begin{cases} x = 8 \\ y = 3 \end{cases}$

22. $\begin{cases} 14x + z = 16 \\ 3x + 6y + 4z = 0 \\ 8x + 2y + 5z = 21 \end{cases}$

23. $\begin{cases} x + 3y + 6z = 16 \\ y + 2z = 9 \\ z = 4 \end{cases}$

24. $\begin{cases} 2x + y + z = 22 \\ x + 3y + z = 17 \\ x + y + 4z = 8 \end{cases}$

25. $\begin{cases} 9y + 13z = 27 \\ 2x + 21z = 19 \\ 7x + 18y = 32 \end{cases}$

26. $\begin{bmatrix} 0 & -7 & | & -22 \\ 1 & 3 & | & 5 \end{bmatrix}$

8. Let $C = \begin{bmatrix} 65 & 32 & 91 & 45 \\ 23 & 18 & 75 & 47 \\ 8 & 63 & 28 & 31 \end{bmatrix}$. Determine the following, if possible:

    a. The order of C **b.** The value of c_{43} **c.** The value of c_{23}

9. Let $D = \begin{bmatrix} 4 & 9 & 7 & 1 & 8 \\ 5 & 3 & 0 & 2 & 6 \end{bmatrix}$. Determine the following, if possible:

    a. The order of D **b.** The value of d_{21} **c.** The value of d_{24}

Construct the augmented matrix that corresponds to each of the following systems of equations. See Example 2. (Answers may appear in slightly different, but equivalent, form.)

10. $\begin{cases} 4x + 5y - 3z = 8 \\ 7x - 2y + 9 = 3 \\ 5x - 6y + 3z = 0 \end{cases}$

11. $\begin{cases} y - 2z + 4 = 3x \\ \dfrac{x}{2} - 4y - 1 = z \\ 3(-y + z) - 1 = 0 \end{cases}$

12. $\begin{cases} 5x + \dfrac{y-z}{2} = 3 \\ 7(z - x) + y - 2 = 0 \\ x - (4 - z) = y \end{cases}$

13. $\begin{cases} \dfrac{2 - 3x}{2} = y \\ 3z + 2(x + y) = 0 \\ 2x - y = 2(x - 3z) \end{cases}$

14. $\begin{cases} 2(z + 3) - x + y = z \\ -3(x - 2y) - 1 = 5z \\ \dfrac{x}{3} - (y - 2z) = x \end{cases}$

15. $\begin{cases} \dfrac{12x - 1}{5} + \dfrac{y}{2} = \dfrac{3z}{2} \\ y - (x + 3z) = -(1 - y) \\ 2x - 2 - z - 2y = 7x \end{cases}$

16. $\begin{cases} \dfrac{3x + 4y}{2} - 3z = 6 \\ 3(x - 2y + 9z) = 0 \\ 2x + 6y = 3 - z \end{cases}$

17. $\begin{cases} \dfrac{2x - 4y}{3} = 2z \\ 8x = 2(y - 3z) + 7 \\ 3x = 2y \end{cases}$

18. $\begin{cases} \dfrac{2(2x - y)}{3} + z = 7 \\ 4 = \dfrac{3}{-x + y + 3z} \\ 4x - 8y + 4 = 9x \end{cases}$

19. $\begin{cases} 0.5x - 14y = \dfrac{z}{4} - 8 \\ \dfrac{x}{5} - y + \dfrac{z}{4} = \dfrac{y}{6} - 3 \\ \dfrac{2}{3}\left(\dfrac{4}{y - x - 1}\right) = \dfrac{5}{z} \end{cases}$

27. $\begin{bmatrix} 2 & -5 & | & 3 \\ 0 & -7 & | & 5 \end{bmatrix}$

28. $\begin{bmatrix} 2 & 1 & | & -4 \\ 1 & -3 & | & 0 \end{bmatrix}$

29. $\begin{bmatrix} 1 & 3 & | & -2 \\ 9 & -2 & | & 7 \end{bmatrix}$

30. $\begin{bmatrix} 8 & 2 & | & 10 \\ 3 & 6 & | & 0 \end{bmatrix}$

31. $\begin{bmatrix} 8 & -2 & | & -4 \\ -6 & 2 & | & -14 \end{bmatrix}$

32. $\begin{bmatrix} -3 & -4 & | & 2 \\ 15 & -3 & | & 0 \end{bmatrix}$

33. $\begin{bmatrix} 4 & 12 & | & -6 \\ 9 & 9 & | & 6 \end{bmatrix}$

34. $\begin{bmatrix} 3 & 0 & | & 1 \\ 14 & 7 & | & 1 \end{bmatrix}$

35. $\begin{bmatrix} 4 & -1 & | & 5 \\ -6 & 2 & | & 0 \end{bmatrix}$

36. $\begin{bmatrix} 5 & 2 & 9 & | & 7 \\ 2 & 6 & -10 & | & 0 \\ -3 & -6 & -8 & | & 1 \end{bmatrix}$

37. $\begin{bmatrix} 18 & -6 & 15 & | & 42 \\ -7 & 19 & 2 & | & 3 \\ -4.5 & 5.5 & -2 & | & 3.5 \end{bmatrix}$

38. $\begin{bmatrix} 5 & 3 & 13 & | & 15 \\ 17 & 9 & -8 & | & -14 \\ -30 & -29 & 35 & | & 36 \end{bmatrix}$

39. $\begin{bmatrix} 5 & 18 & 22 & | & 5 \\ 32 & -9 & -27 & | & -23 \\ -9 & 21 & 12 & | & 9 \end{bmatrix}$

40. $\begin{bmatrix} 1 & 3 & -2 & | & 4 \\ 0 & -10 & 14 & | & -10 \\ 0 & 15 & -8 & | & 27 \end{bmatrix}$

41. $\begin{bmatrix} 0 & 1 & -9 & | & -3 \\ 1 & 1 & 3 & | & 4 \\ 0 & 0 & 0 & | & 0 \end{bmatrix}$

42. $\begin{bmatrix} -3 & 2 & | & 2 \\ -1 & 0 & | & 5 \end{bmatrix}$

Construct the system of equations that corresponds to each of the following matrices.

20. $\begin{bmatrix} 5 & 3 & | & 9 \\ 1 & 4 & | & 12 \end{bmatrix}$

21. $\begin{bmatrix} 1 & 0 & | & 8 \\ 0 & 1 & | & 3 \end{bmatrix}$

22. $\begin{bmatrix} 14 & 0 & 1 & | & 16 \\ 3 & 6 & 4 & | & 0 \\ 8 & 2 & 5 & | & 21 \end{bmatrix}$

23. $\begin{bmatrix} 1 & 3 & 6 & | & 16 \\ 0 & 1 & 2 & | & 9 \\ 0 & 0 & 1 & | & 4 \end{bmatrix}$

24. $\begin{bmatrix} 2 & 1 & 1 & | & 22 \\ 1 & 3 & 1 & | & 17 \\ 1 & 1 & 4 & | & 8 \end{bmatrix}$

25. $\begin{bmatrix} 0 & 9 & 13 & | & 27 \\ 2 & 0 & 21 & | & 19 \\ 7 & 18 & 0 & | & 32 \end{bmatrix}$

Fill in the blanks by performing the indicated row operations. See Example 4.

26. $\begin{bmatrix} 3 & 2 & | & -7 \\ 1 & 3 & | & 5 \end{bmatrix} \xrightarrow{-3R_2+R_1} \underline{\ ?\ }$

27. $\begin{bmatrix} 2 & -5 & | & 3 \\ -4 & 3 & | & -1 \end{bmatrix} \xrightarrow{2R_1+R_2} \underline{\ ?\ }$

28. $\begin{bmatrix} 4 & 2 & | & -8 \\ 3 & -9 & | & 0 \end{bmatrix} \xrightarrow[\frac{1}{3}R_2]{\frac{1}{2}R_1} \underline{\ ?\ }$

29. $\begin{bmatrix} 9 & -2 & | & 7 \\ 1 & 3 & | & -2 \end{bmatrix} \xrightarrow{R_1\leftrightarrow R_2} \underline{\ ?\ }$

30. $\begin{bmatrix} 4 & 1 & | & 5 \\ 3 & 6 & | & 0 \end{bmatrix} \xrightarrow{2R_1} \underline{\ ?\ }$

31. $\begin{bmatrix} 8 & -2 & | & -4 \\ 3 & -1 & | & 7 \end{bmatrix} \xrightarrow{-2R_2} \underline{\ ?\ }$

32. $\begin{bmatrix} 9 & 12 & | & -6 \\ 15 & -3 & | & 0 \end{bmatrix} \xrightarrow{-\frac{1}{3}R_1} \underline{\ ?\ }$

33. $\begin{bmatrix} 4 & 12 & | & -6 \\ 7 & 3 & | & 9 \end{bmatrix} \xrightarrow{\frac{1}{2}R_1+R_2} \underline{\ ?\ }$

34. $\begin{bmatrix} 3 & 0 & | & 1 \\ 5 & 7 & | & -2 \end{bmatrix} \xrightarrow{3R_1+R_2} \underline{\ ?\ }$

35. $\begin{bmatrix} 8 & -2 & | & 10 \\ 9 & -3 & | & 0 \end{bmatrix} \xrightarrow[-\frac{2}{3}R_2]{\frac{1}{2}R_1} \underline{\ ?\ }$

36. $\begin{bmatrix} 5 & 2 & 9 & | & 7 \\ 1 & 3 & -5 & | & 0 \\ 2 & -4 & 1 & | & 8 \end{bmatrix} \xrightarrow[-R_1+R_3]{2R_2} \underline{\ ?\ }$

37. $\begin{bmatrix} 6 & -2 & 5 & | & 14 \\ -7 & 19 & 2 & | & 3 \\ -9 & 11 & -4 & | & 7 \end{bmatrix} \xrightarrow[0.5R_3]{3R_1} \underline{\ ?\ }$

38. $\begin{bmatrix} 5 & 3 & 13 & | & 15 \\ 17 & 9 & -8 & | & -14 \\ 4 & -11 & 19 & | & 8 \end{bmatrix} \xrightarrow{-2R_2+R_3} \underline{\ ?\ }$

39. $\begin{bmatrix} 8 & 11 & 18 & | & 2 \\ 14 & 33 & -3 & | & -5 \\ -9 & 21 & 12 & | & 9 \end{bmatrix} \xrightarrow[-2R_3+R_2]{\frac{1}{3}R_3+R_1} \underline{\ ?\ }$

40. $\begin{bmatrix} 1 & 3 & -2 & | & 4 \\ 3 & -1 & 8 & | & 2 \\ -5 & 0 & 2 & | & 7 \end{bmatrix} \xrightarrow[5R_1+R_3]{-3R_1+R_2} \underline{\ ?\ }$

41. $\begin{bmatrix} 2 & 3 & -3 & | & 5 \\ 1 & 1 & 3 & | & 4 \\ 3 & 3 & 9 & | & 12 \end{bmatrix} \xrightarrow[-3R_2+R_3]{-2R_2+R_1} \underline{\ ?\ }$

42. $\begin{bmatrix} -3 & 2 & | & 2 \\ 5 & -4 & | & 1 \end{bmatrix} \xrightarrow{2R_1+R_2} \underline{\ ?\ }$

43. $\begin{bmatrix} -5 & 20 & | & -15 \\ 2 & -12 & | & 5 \end{bmatrix} \xrightarrow[\frac{1}{2}R_2]{\frac{1}{5}R_1} \underline{\ ?\ }$

44. $\begin{bmatrix} 2 & 2 & 3 & | & 7 \\ -3 & 2 & 8 & | & -2 \\ 1 & 5 & 2 & | & 6 \end{bmatrix} \xrightarrow[3R_3+R_2]{-2R_3+R_1} \underline{\ ?\ }$

45. $\begin{bmatrix} 1 & 5 & -9 & | & 11 \\ 1 & 4 & -1 & | & 4 \\ 4 & 3 & 5 & | & 45 \end{bmatrix} \xrightarrow[-4R_1+R_3]{-R_1+R_2} \underline{\ ?\ }$

43. $\begin{bmatrix} -1 & 4 & | & -3 \\ 1 & -6 & | & \frac{5}{2} \end{bmatrix}$

44. $\begin{bmatrix} 0 & -8 & -1 & | & -5 \\ 0 & 17 & 14 & | & 16 \\ 1 & 5 & 2 & | & 6 \end{bmatrix}$

45. $\begin{bmatrix} 1 & 5 & -9 & | & 11 \\ 0 & -1 & 8 & | & -7 \\ 0 & -17 & 41 & | & 1 \end{bmatrix}$

46. Row echelon form

47. Neither

48. Reduced row echelon form

49. Neither

50. Row echelon form

51. Neither

52. $(-1, 1)$

53. $(3, -1)$

54. $(-4, 1)$

55. $(1, 3)$

56. $(5, 6)$

57. $(-7, 3)$

58. $(7, 8)$

59. $\varnothing$

60. $\varnothing$

61. $(3, 2)$

62. $(-2, 5)$

63. $\{(-2y - 4, y)\ |\ y \in \mathbb{R}\}$

64. $(-4, 5)$

65. $\varnothing$

66. $(5, 4)$

67. $(4, 0, 3)$

68. $(-2, 7, 3)$

69. $(15, -21, 8)$

70. $(1, -2)$

71. $(3, -5)$

72. $(5, 3)$

73. $\{(x, -3x - 2)\ |\ x \in \mathbb{R}\}$

For each matrix below, determine if it is in row echelon form, reduced row echelon form, or neither.

46. $\begin{bmatrix} 1 & 5 & | & 4 \\ 0 & 1 & | & 3 \end{bmatrix}$

47. $\begin{bmatrix} 1 & 2 & 0 & | & 9 \\ 0 & 1 & 3 & | & 4 \\ 0 & 1 & 1 & | & 12 \end{bmatrix}$

48. $\begin{bmatrix} 1 & 0 & 0 & | & 4 \\ 0 & 1 & 0 & | & 1 \\ 0 & 0 & 1 & | & 8 \end{bmatrix}$

49. $\begin{bmatrix} 1 & 0 & 0 & | & 7 \\ 5 & 1 & 0 & | & 14 \\ 3 & 4 & 1 & | & -16 \end{bmatrix}$

50. $\begin{bmatrix} 1 & 2 & 5 & | & 0 \\ 0 & 1 & 9 & | & 3 \\ 0 & 0 & 0 & | & 1 \end{bmatrix}$

51. $\begin{bmatrix} 0 & 1 & | & 3 \\ 1 & 0 & | & 6 \end{bmatrix}$

Use Gaussian elimination and back-substitution to solve the following systems of equations. See Example 5.

52. $\begin{cases} 2x - 4y = -6 \\ 3x - y = -4 \end{cases}$

53. $\begin{cases} 2x - 5y = 11 \\ 3x + 2y = 7 \end{cases}$

54. $\begin{cases} 5x - y = -21 \\ 9x + 2y = -34 \end{cases}$

55. $\begin{cases} x - 4y = -11 \\ 7x - y = 4 \end{cases}$

56. $\begin{cases} x + 2y = 17 \\ 3x + 4y = 39 \end{cases}$

57. $\begin{cases} 2x + 6y = 4 \\ -4x - 7y = 7 \end{cases}$

58. $\begin{cases} 3x - 2y = 5 \\ -5x + 4y = -3 \end{cases}$

59. $\begin{cases} 2x + y = -2 \\ -4x - 2y = 5 \end{cases}$

60. $\begin{cases} 6x - 16y = 10 \\ -3x + 8y = 4 \end{cases}$

61. $\begin{cases} 2x - 3y = 0 \\ 5x + y = 17 \end{cases}$

62. $\begin{cases} 6x + 3y = 3 \\ x + y = 3 \end{cases}$

63. $\begin{cases} 3x + 6y = -12 \\ 2x + 4y = -8 \end{cases}$

64. $\begin{cases} 4x + 5y = 9 \\ 8x + 3y = -17 \end{cases}$

65. $\begin{cases} \frac{2}{3}x + 2y = 1 \\ x + 3y = 0 \end{cases}$

66. $\begin{cases} 13x - 17y = -3 \\ -19x + 15y = -35 \end{cases}$

67. $\begin{cases} 3x - 9y - 7z = -9 \\ 5x + 11y - z = 17 \\ -4x - 8y + 7z = 5 \end{cases}$

68. $\begin{cases} 8x - y + 5z = -8 \\ 11x - 2y + 9z = -9 \\ 7x - 3y + 13z = 4 \end{cases}$

69. $\begin{cases} 17x + 13y + 8z = 46 \\ -12x + 3y + 28z = -19 \\ 14x + 5y - 15z = -15 \end{cases}$

Use Gauss-Jordan elimination to solve the following systems of equations. See Example 6.

70. $\begin{cases} 2x - 3y = 8 \\ 8x + 5y = -2 \end{cases}$

71. $\begin{cases} \frac{2}{3}x + y = -3 \\ 3x + \frac{5}{2}y = -\frac{7}{2} \end{cases}$

72. $\begin{cases} 3y = 9 \\ x + 2y = 11 \end{cases}$

73. $\begin{cases} 6x + 2y = -4 \\ -9x - 3y = 6 \end{cases}$

74. $\begin{cases} 3y = 6 \\ 5x + 2y = 4 \end{cases}$

75. $\begin{cases} 3x + 8y = -4 \\ x + 2y = -2 \end{cases}$

76. $\begin{cases} -3x + 2y = 5 \\ 5x - 2y = 1 \end{cases}$

77. $\begin{cases} 9x - 11y = 10 \\ -4x + 3y = -12 \end{cases}$

78. $\begin{cases} 9x - 15y = -6 \\ -3x + 11y = -10 \end{cases}$

74. $(0, 2)$

75. $(-4, 1)$

76. $(3, 7)$

77. $(6, 4)$

78. $(-4, -2)$

79. $(-11, -5)$

80. $(1, 8, 7)$

81. $(7, 3, 3)$

82. $\{(-3z - 5,$

$-6z - 10,$

$z)\big| z \in \mathbb{R}\}$

83. $(2, 2, -1)$

84. $(2, 2, 2)$

85. $\{(1, y, 0)\big| y \in \mathbb{R}\}$

86. $\varnothing$

87. $(3, -2, 3)$

88. $(3, -1, 3)$

89. $(2, 3, 4)$

90. $(5, -2, 3)$

91. $(9, -19, 7)$

92. $(-15, 3, 9)$

93. $(1, -2, -1, 3)$

94. $\left\{ \left(\dfrac{-5 - z - 5y}{2}, \right. \right.$

$\dfrac{-19 + 3z - 5y}{2},$

$\left. y, z \right) \big| y \in \mathbb{R}, z \in \mathbb{R} \}$

95. $42, 26, 87$

96. Eggs $\$1.03$,
Bread $\$2.12$,
Bacon $\$4.27$

97. Small: 10,
Medium: 24,
Large: 48

79. $\begin{cases} 3x - 8y = 7 \\ 18x - 35y = -23 \end{cases}$

80. $\begin{cases} 4x + y - 3z = -9 \\ 2x - 3z = -19 \\ 7x - y - 4z = -29 \end{cases}$

81. $\begin{cases} -5x + 9y + 3z = 1 \\ 3x + 2y - 6z = 9 \\ x + 4y - z = 16 \end{cases}$

82. $\begin{cases} 2x - y = 0 \\ 5x - 3y - 3z = 5 \\ 2x + 6z = -10 \end{cases}$

83. $\begin{cases} x + y = 4 \\ y + 3z = -1 \\ 2x - 2y + 5z = -5 \end{cases}$

84. $\begin{cases} 2x - 3y = -2 \\ x - 4y + 3z = 0 \\ -2x + 7y - 5z = 0 \end{cases}$

85. $\begin{cases} 3x + 8z = 3 \\ -3x - 7z = -3 \\ x + 3z = 1 \end{cases}$

86. $\begin{cases} 3x - y + z = 2 \\ -6x + 2y - 2z = 1 \\ 5x + 2y - 3z = 2 \end{cases}$

87. $\begin{cases} x + 2y = -1 \\ y + 3z = 7 \\ 2x + 5z = 21 \end{cases}$

88. $\begin{cases} 2x + 8y - z = -5 \\ -5x + 3y + 4z = -6 \\ x - 4y - 5z = -8 \end{cases}$

89. $\begin{cases} 7x - 8y + 2z = -2 \\ 5x - 3y - z = -3 \\ 8x + y - 3z = 7 \end{cases}$

90. $\begin{cases} 8x + 14y - 3z = 3 \\ -6x + 2y + 7z = -13 \\ 8x + 19y + 3z = 11 \end{cases}$

91. $\begin{cases} 8x + 5y + 3z = -2 \\ 12x - y - 18z = 1 \\ 7x + 6y + 10z = 19 \end{cases}$

92. $\begin{cases} 4x + 8y + 7z = 27 \\ -2x + 9y - 8z = -15 \\ 9x + 13y + 7z = -33 \end{cases}$

93. $\begin{cases} w - x + 2z = 9 \\ 2w + 3y = -1 \\ -2w - 5y - z = 0 \\ x + 2y = -4 \end{cases}$

94. $\begin{cases} 3w - x + 5y + 3z = 2 \\ -4w - 10y - 2z = 10 \\ w - x + 2z = 7 \\ 4w - 2x + 5y + 5z = 9 \end{cases}$

Solve the following application problems.

95. The sum of three integers is 155. The first integer is sixteen more than the second. The third integer is seven less than the sum of the first integer and twice the second. What are the three integers?

96. Mario bought a pound of bacon, a dozen eggs, and a loaf of bread to make breakfast for his family. The total cost was $\$7.42$. The bacon cost $\$0.03$ more than twice the price of the bread and the eggs cost $\$0.03$ less than half the price of the bread. Find the price of each item.

97. The Pizza House sells three sizes of pizzas: small, medium, large. The prices of the pizzas are $\$9.00$, $\$12.00$, and $\$15.00$, respectively. In one day, they sold 82 pizzas for a total of $\$1098.00$. If the number of large pizzas sold was twice the number of medium pizzas sold, how many of each size pizza did the Pizza House sell?

Determinants and Cramer's Rule

TOPICS

1. Determinants and their evaluation
2. Using Cramer's rule to solve linear systems
T. Evaluating determinants

TOPIC

Determinants and Their Evaluation

While Gaussian elimination and Gauss-Jordan elimination both provide significant improvements in ease and speed when solving systems of equations, they are not truly new methods. Each simply uses the matrix as a powerful, agile organizer to make the substitution and elimination methods simpler. Cramer's rule, named after the Swiss mathematician Gabriel Cramer (1704–1752), is a solution method that truly brings something new to the discussion.

Cramer's rule relies on the computation of a number, called the *determinant*, associated with every *square* matrix (a matrix with the same number of rows as columns).

DEFINITION

Determinant of a
2×2 Matrix

The **determinant** of the matrix $A = \begin{bmatrix} a_{11} & a_{12} \\ a_{21} & a_{22} \end{bmatrix}$, denoted $|A|$, is $a_{11}a_{22} - a_{21}a_{12}$.

CAUTION!

If A is a matrix, $|A|$ stands for the determinant of A, not the absolute value of A. In fact, the concept of absolute value does not apply to matrices.

Of course, this only defines determinants for matrices of one size, and unfortunately the simplicity of the definition does not carry over to larger sizes. The good news, however, is that determinants of 3×3, 4×4, and larger matrices can all be related, ultimately, to determinants of 2×2 matrices. We will do a few calculations before considering larger matrices.

EXAMPLE 1

Determinant of a
2×2 Matrix

Evaluate the determinants of each of the following matrices:

a. $A = \begin{bmatrix} -2 & 3 \\ -1 & 3 \end{bmatrix}$

b. $B = \begin{bmatrix} 2 & -3 \\ -4 & 6 \end{bmatrix}$

Solutions:

a. $|A| = a_{11}a_{22} - a_{21}a_{12}$

$= (-2)(3) - (-1)(3) = -6 + 3 = -3$

b. $|B| = b_{11}b_{22} - b_{21}b_{12}$

$= (2)(6) - (-4)(-3) = 12 - 12 = 0$

Calculating the determinant of a 3×3 matrix involves calculating three specific 2×2 determinants. Similarly, finding the determinant of a 4×4 matrix requires evaluating four specific 3×3 determinants.

These specific smaller determinants are called the *cofactors* of a matrix.

DEFINITION

Minors and Cofactors

Let A be an $n \times n$ matrix, and let i and j be numbers between 1 and n, so that a_{ij} is an element of A.

- The **minor** of the element a_{ij} is the determinant of the $(n-1) \times (n-1)$ matrix formed from A by deleting its ith row and j^{th} column.

- The **cofactor** of the element a_{ij} is $(-1)^{i+j}$ times the minor of a_{ij}. Thus, if $i+j$ is even, the cofactor of a_{ij} is the same as the minor of a_{ij}; if $i+j$ is odd, the cofactor is the negative of the minor.

The following matrix of signs may help you remember which minors to multiply by -1: if a_{ij} is in a position occupied by a $-$ sign, change the sign of the minor of a_{ij} to obtain the cofactor.

$$\begin{bmatrix} + & - & + & - & \cdots \\ - & + & - & + & \cdots \\ + & - & + & - & \cdots \\ - & + & - & + & \cdots \\ \vdots & \vdots & \vdots & \vdots & \ddots \end{bmatrix}$$

Figure 1: The Cofactor Sign Matrix

EXAMPLE 2

Minors and Cofactors   For the matrix $A = \begin{bmatrix} -5 & 3 & 2 \\ 1 & 0 & -1 \\ -3 & 1 & 0 \end{bmatrix}$,

a. Evaluate the minor of a_{12}.   **b.** Evaluate the cofactor of a_{23}.

Solutions:

a. Finding the minor of a_{12} requires deleting the first row and second column of the matrix A. This gives us the following.

$$\begin{bmatrix} -5 & 3 & 2 \\ 1 & 0 & -1 \\ -3 & 1 & 0 \end{bmatrix}$$

Thus, the minor of $a_{12} = \begin{vmatrix} 1 & -1 \\ -3 & 0 \end{vmatrix} = (1)(0)-(-3)(-1) = 0-3 = -3$.

b. Similarly, the cofactor of $a_{23} = (-1)^{2+3}\begin{vmatrix} -5 & 3 \\ -3 & 1 \end{vmatrix} = (-1)^5\left[(-5)(1)-(-3)(3)\right]$

$$= (-1)[-5+9] = (-1)(4) = -4.$$

PROCEDURE

Determinant of an
$n \times n$ Matrix

Evaluation of an $n \times n$ determinant is accomplished by **expansion** along a fixed row or column. The result does not depend on which row or column is chosen.

- To expand along the ith row, each element of that row is multiplied by its cofactor, and the n products are then added.

- To expand along the jth column, each element of that column is multiplied by its cofactor, and the n products are then added.

For example, if we expand along the first column of a 3×3 matrix, we get the following. Note the minus sign in front of a_{21}.

$$\begin{vmatrix} a_{11} & a_{12} & a_{13} \\ a_{21} & a_{22} & a_{23} \\ a_{31} & a_{32} & a_{33} \end{vmatrix} = a_{11}\begin{vmatrix} a_{22} & a_{23} \\ a_{32} & a_{33} \end{vmatrix} - a_{21}\begin{vmatrix} a_{12} & a_{13} \\ a_{32} & a_{33} \end{vmatrix} + a_{31}\begin{vmatrix} a_{12} & a_{13} \\ a_{22} & a_{23} \end{vmatrix}$$

We could expand along a row or a different column in a similar manner.

EXAMPLE 3

Determinant of an
$n \times n$ Matrix

Evaluate the determinant of the matrix $A = \begin{bmatrix} -1 & 3 & 2 \\ -2 & 0 & 0 \\ 4 & 1 & 5 \end{bmatrix}$.

Note:
Minimize the number of computations by choosing which row or column to expand along carefully.

Solution:

First, we decide which row or column to expand along. A row or column with many zeros is generally a good choice, since it makes the multiplication much easier. In this case, Row 2 has the most zeros, so expand along it.

$$\begin{vmatrix} -1 & 3 & 2 \\ -2 & 0 & 0 \\ 4 & 1 & 5 \end{vmatrix} = -(-2)\begin{vmatrix} 3 & 2 \\ 1 & 5 \end{vmatrix} + (0)\begin{vmatrix} -1 & 2 \\ 4 & 5 \end{vmatrix} - (0)\begin{vmatrix} -1 & 3 \\ 4 & 1 \end{vmatrix}$$

$$= -(-2)(13) + 0 - 0$$

$$= 26$$

Thus, $|A| = 26$. We get the same answer if we expand along a different row or column.

$$\begin{vmatrix} -1 & 3 & 2 \\ -2 & 0 & 0 \\ 4 & 1 & 5 \end{vmatrix} = (-1)\begin{vmatrix} 0 & 0 \\ 1 & 5 \end{vmatrix} - (-2)\begin{vmatrix} 3 & 2 \\ 1 & 5 \end{vmatrix} + (4)\begin{vmatrix} 3 & 2 \\ 0 & 0 \end{vmatrix}$$

$$= (-1)(0) - (-2)(13) + (4)(0)$$

$$= 26$$

As we saw in the previous example, even 3×3 determinants can involve a large number of calculations, but by taking advantage of zeros, we are able to reduce the amount of work. We can take this a step farther by applying a few properties of determinants.

PROPERTIES

Properties of
Determinants

1. A constant can be factored out of each of the terms in a given row or column when computing determinants. For example,

$$\begin{vmatrix} 2 & -1 \\ 15 & 5 \end{vmatrix} = 5\begin{vmatrix} 2 & -1 \\ 3 & 1 \end{vmatrix} \quad \text{and} \quad \begin{vmatrix} 4 & 7 \\ 12 & 9 \end{vmatrix} = 4\begin{vmatrix} 1 & 7 \\ 3 & 9 \end{vmatrix}.$$

2. Interchanging two rows or two columns changes the determinant by a factor of -1. For example,

$$\begin{vmatrix} 2 & -1 \\ 15 & 5 \end{vmatrix} = -\begin{vmatrix} 15 & 5 \\ 2 & -1 \end{vmatrix} \quad \text{and} \quad \begin{vmatrix} 3 & -2 \\ 7 & 1 \end{vmatrix} = -\begin{vmatrix} -2 & 3 \\ 1 & 7 \end{vmatrix}.$$

3. The determinant is unchanged by adding a multiple of one row (or column) to another row (or column). For example,

$$\begin{vmatrix} 3 & -2 \\ 1 & -1 \end{vmatrix} \overset{-3R_2 + R_1}{=} \begin{vmatrix} 0 & 1 \\ 1 & -1 \end{vmatrix}.$$

EXAMPLE 4

Properties of
Determinants

Evaluate the determinant of the matrix $B = \begin{bmatrix} 4 & -2 & 3 & 0 \\ 2 & 1 & -1 & 3 \\ 3 & 0 & 1 & 1 \\ 2 & -2 & 0 & 0 \end{bmatrix}$.

Solution:

Use the properties of determinants to try to obtain rows or columns with as many zeros as possible.

$$\begin{vmatrix} 4 & -2 & 3 & 0 \\ 2 & 1 & -1 & 3 \\ 3 & 0 & 1 & 1 \\ 2 & -2 & 0 & 0 \end{vmatrix} \overset{-3R_3+R_2}{=} \begin{vmatrix} 4 & -2 & 3 & 0 \\ -7 & 1 & -4 & 0 \\ 3 & 0 & 1 & 1 \\ 2 & -2 & 0 & 0 \end{vmatrix}$$

Applying the third property makes the fourth column have only one nonzero entry.

Now expand along the fourth column (remembering that the minus sign is part of the cofactor):

$$\begin{vmatrix} 4 & -2 & 3 & 0 \\ -7 & 1 & -4 & 0 \\ 3 & 0 & 1 & 1 \\ 2 & -2 & 0 & 0 \end{vmatrix} = -(1)\begin{vmatrix} 4 & -2 & 3 \\ -7 & 1 & -4 \\ 2 & -2 & 0 \end{vmatrix}$$

We can continue to apply the third property of determinants to simplify the evaluation of the 3×3 determinant.

$$|B| = -\begin{vmatrix} 4 & -2 & 3 \\ -7 & 1 & -4 \\ 2 & -2 & 0 \end{vmatrix} \overset{C_1+C_2}{=} -\begin{vmatrix} 4 & 2 & 3 \\ -7 & -6 & -4 \\ 2 & 0 & 0 \end{vmatrix}$$

Add the first column to the second.

$$= -(2)\begin{vmatrix} 2 & 3 \\ -6 & -4 \end{vmatrix}$$

Expand along the third row, which now has two zeros.

$$= (-2)(10)$$

$$= -20$$

Notice that we have reduced the work to the evaluation of only one 3×3 determinant, which in turn involved evaluating only one 2×2 determinant.

TOPIC 2

Using Cramer's Rule to Solve Linear Systems

To understand the form of Cramer's rule, we will solve the general two-variable, two-equation linear system by elimination. To do this, we note that any such system can be put into the form

$$\begin{cases} ax + by = e \\ cx + dy = f \end{cases}$$

where $a, b, c, d, e,$ and f are all constants. If we can solve this system for x and y, then we will have a formula for the solution of any such system.

Using the method of elimination, we can obtain an equation in x alone by multiplying the first equation by d and the second equation by $-b$:

$$\begin{cases} ax + by = e \\ cx + dy = f \end{cases} \xrightarrow[\substack{dE_1 \\ -bE_2}]{} \begin{cases} adx + bdy = ed \\ -bcx - bdy = -bf \end{cases}$$
$$\overline{(ad - bc)x = ed - bf}$$

This equation can then be solved for x to obtain

$$x = \frac{ed - bf}{ad - bc}.$$

Similarly, the system can be solved for y to obtain

$$y = \frac{af - ce}{ad - bc}.$$

Note that these formulas only make sense if the denominator is not zero. We will deal with this possibility shortly.

These formulas are worthwhile on their own, but they are a bit complex and would be difficult to memorize. Note that each term in the two fractions appears in the form of a 2×2 determinant. In fact, the above formulas are equivalent to:

$$x = \frac{\begin{vmatrix} e & b \\ f & d \end{vmatrix}}{\begin{vmatrix} a & b \\ c & d \end{vmatrix}} \text{ and } y = \frac{\begin{vmatrix} a & e \\ c & f \end{vmatrix}}{\begin{vmatrix} a & b \\ c & d \end{vmatrix}}$$

The denominator D in both formulas is the determinant of the coefficient matrix, the square matrix consisting of the coefficients of the variables. If we let D_x and D_y represent the numerators of the formulas for x and y, respectively, then D_x is the determinant of the coefficient matrix with the first column (the x-column) replaced by the column of constants, and D_y is the determinant of the coefficient matrix with the second column replaced by the column of constants. Putting these observations together, we obtain Cramer's rule for the two-variable, two-equation case:

THEOREM

The solution of a two-variable, two-equation linear system $\begin{cases} ax + by = e \\ cx + dy = f \end{cases}$, is given by

$$x = \frac{D_x}{D} \text{ and } y = \frac{D_y}{D}$$

where D is the determinant of the coefficient matrix $\begin{vmatrix} a & b \\ c & d \end{vmatrix}$, D_x is the determinant of the matrix formed by replacing the column of x-coefficients with the column of constant terms $\begin{vmatrix} e & b \\ f & d \end{vmatrix}$, and D_y is the determinant of the matrix formed by replacing the column of y-coefficients with the column of constant terms $\begin{vmatrix} a & e \\ c & f \end{vmatrix}$.

CAUTION!

Whenever a fraction appears in our work, we need to ask if the expression in the denominator can ever be zero, and what it means if this happens. In Cramer's rule, the determinant D can equal 0, which prevents us from using the given formulas.

If $D = 0$, the system is either dependent or inconsistent. If both D_x and D_y are also zero, the system is dependent. If at least one of D_x and D_y is nonzero, the system has no solution.

EXAMPLE 5

Cramer's Rule

Use Cramer's rule to solve the following systems.

a. $\begin{cases} 4x - 5y = 3 \\ -3x + 7y = 1 \end{cases}$

b. $\begin{cases} -x + 2y = -1 \\ 3x - 6y = 3 \end{cases}$

Solutions:

a. $D = \begin{vmatrix} 4 & -5 \\ -3 & 7 \end{vmatrix} = 28 - 15 = 13$ Calculate D first. Since $D \neq 0$, we know there is a single solution to the system.

$D_x = \begin{vmatrix} 3 & -5 \\ 1 & 7 \end{vmatrix} = 21 - (-5) = 26$ Calculate D_x and D_y.

$D_y = \begin{vmatrix} 4 & 3 \\ -3 & 1 \end{vmatrix} = 4 - (-9) = 13$

Applying Cramer's rule, we have $x = \dfrac{D_x}{D} = \dfrac{26}{13} = 2$ and $y = \dfrac{D_y}{D} = \dfrac{13}{13} = 1$, so the solution is $(2, 1)$.

b. $D = \begin{vmatrix} -1 & 2 \\ 3 & -6 \end{vmatrix} = 6 - 6 = 0$

Again we calculate D first. Since $D = 0$ either the system has no solution or it has an infinite number of solutions.

$D_x = \begin{vmatrix} -1 & 2 \\ 3 & -6 \end{vmatrix} = 6 - 6 = 0$

Since D_x and D_y both equal zero, the system is dependent.

$D_y = \begin{vmatrix} -1 & -1 \\ 3 & 3 \end{vmatrix} = -3 - (-3) = 0$

The solution set can be found by solving either equation for either variable: $\{(2y+1, y) \mid y \in \mathbb{R}\}$.

Cramer's rule can be extended to solve any linear system of n equations in n variables. While Cramer's rule is an extremely powerful method and remarkable for its succinctness, keep in mind that using Cramer's rule to solve an n-equation, n-variable system entails calculating $(n+1)$ $n \times n$ determinants, so it is important to make use of the labor-saving properties of determinants.

THEOREM

Cramer's Rule   A linear system of n equations in the n variables x_1, x_2, ..., x_n can be written in the form

$$\begin{cases} a_{11}x_1 + a_{12}x_2 + \ldots + a_{1n}x_n = b_1 \\ a_{21}x_1 + a_{22}x_2 + \ldots + a_{2n}x_n = b_2 \\ \vdots \\ a_{n1}x_1 + a_{n2}x_2 + \ldots + a_{nn}x_n = b_n \end{cases}$$

The solution of the system is given by the formulas $x_1 = \dfrac{D_{x_1}}{D}$, $x_2 = \dfrac{D_{x_2}}{D}$, ..., $x_n = \dfrac{D_{x_n}}{D}$, where D is the determinant of the coefficient matrix and D_{x_i} is the determinant of the same matrix with the ith column replaced by the column of constants b_1, b_2, ..., b_n.

If $D = 0$ and if each $D_{x_i} = 0$ as well, the system is dependent and has an infinite number of solutions. If $D = 0$ and at least one of the D_{x_i}'s is nonzero, the system has no solution.

━━━━━━━━━━━━━ **EXAMPLE 6** ━━━━━━━━━━━━━

Cramer's Rule   Use Cramer's rule to solve the system $\begin{cases} 3x - 2y - 2z = -1 \\ 3y + z = -7 \\ x + y + 2z = 0 \end{cases}$

Solution:

Note how the properties of determinants are used to simplify each calculation. In each case, the row or column used for expansion is written in blue.

$$D = \begin{vmatrix} 3 & -2 & -2 \\ 0 & 3 & 1 \\ 1 & 1 & 2 \end{vmatrix} \overset{-3R_3 + R_1}{=} \begin{vmatrix} 0 & -5 & -8 \\ 0 & 3 & 1 \\ 1 & 1 & 2 \end{vmatrix} = (1)\begin{vmatrix} -5 & -8 \\ 3 & 1 \end{vmatrix} = 19$$

Since $D \neq 0$, the system has a unique solution.

$$D_x = \begin{vmatrix} -1 & -2 & -2 \\ -7 & 3 & 1 \\ 0 & 1 & 2 \end{vmatrix} \overset{-2C_2 + C_3}{=} \begin{vmatrix} -1 & -2 & 2 \\ -7 & 3 & -5 \\ 0 & 1 & 0 \end{vmatrix} = -(1)\begin{vmatrix} -1 & 2 \\ -7 & -5 \end{vmatrix} = -19$$

$$D_y = \begin{vmatrix} 3 & -1 & -2 \\ 0 & -7 & 1 \\ 1 & 0 & 2 \end{vmatrix} \overset{-2C_1 + C_3}{=} \begin{vmatrix} 3 & -1 & -8 \\ 0 & -7 & 1 \\ 1 & 0 & 0 \end{vmatrix} = (1)\begin{vmatrix} -1 & -8 \\ -7 & 1 \end{vmatrix} = -57$$

$$D_z = \begin{vmatrix} 3 & -2 & -1 \\ 0 & 3 & -7 \\ 1 & 1 & 0 \end{vmatrix} \overset{-C_1 + C_2}{=} \begin{vmatrix} 3 & -5 & -1 \\ 0 & 3 & -7 \\ 1 & 0 & 0 \end{vmatrix} = (1)\begin{vmatrix} -5 & -1 \\ 3 & -7 \end{vmatrix} = 38$$

After evaluating D_x, D_y, and D_z, we know the solution is the single ordered triple

$$(x, y, z) = \left(\frac{D_x}{D}, \frac{D_y}{D}, \frac{D_z}{D} \right) = \left(\frac{-19}{19}, \frac{-57}{19}, \frac{38}{19} \right) = (-1, -3, 2).$$

TOPIC T   Evaluating Determinants

To find the determinant of a matrix using the calculator, we must first edit a matrix and enter in the dimensions and elements of the matrix whose determinant we wish to find, for example, the matrix $\begin{bmatrix} 2 & 4 & -8 \\ 1 & 3 & 6 \\ -7 & 5 & 1 \end{bmatrix}$. With that saved as matrix A, we press

 and select MATH. Press ENTER since 1:det (is already highlighted. We then press `2ND` `x⁻¹`, select 1:[A] under NAMES, add the right-hand parenthesis and press ENTER.

```
MATRIX[A] 3 ×3
[ 2      4      -8   ]
[ 1      3      6    ]
[ -7     5      A    ]



3,3=1
```

```
det([A])
                   -434
```

Notice that if we try to find the determinant of a 3×4 matrix, for instance, we get the following error message:

```
ERR:INVALID DIM
1■Quit
2:Goto
```

Remember that the definition of determinant only applies to square matrices; matrices that are not square do not have determinants.

1. 11

2. 0

3. 15

4. 0

5. $ab - x^2$

6. $7x$

7. 8

8. $-abcd$

9. -10

10. $wz - xy$

11. -39

12. $2x^2y - 3xy + 3x$

Exercises

Evaluate the following determinants. See Example 1.

1. $\begin{vmatrix} 4 & -3 \\ 1 & 2 \end{vmatrix}$ **2.** $\begin{vmatrix} 5 & -2 \\ 5 & -2 \end{vmatrix}$ **3.** $\begin{vmatrix} 0 & 3 \\ -5 & 2 \end{vmatrix}$ **4.** $\begin{vmatrix} 34 & -2 \\ 17 & -1 \end{vmatrix}$ **5.** $\begin{vmatrix} a & x \\ x & b \end{vmatrix}$

6. $\begin{vmatrix} 5x & 2 \\ -x & 1 \end{vmatrix}$ **7.** $\begin{vmatrix} -2 & 2 \\ -2 & -2 \end{vmatrix}$ **8.** $\begin{vmatrix} ac & 2ad \\ bc & db \end{vmatrix}$ **9.** $\begin{vmatrix} -1 & 2 \\ 3 & 4 \end{vmatrix}$ **10.** $\begin{vmatrix} w & x \\ y & z \end{vmatrix}$

11. $\begin{vmatrix} -2 & 9 \\ 5 & -3 \end{vmatrix}$ **12.** $\begin{vmatrix} 2y & 3x \\ y-1 & x^2 \end{vmatrix}$

13. $\{-2, 3\}$

14. $\{-4, -1\}$

15. $\{-5, 1\}$

16. $\{-3, 6\}$

17. $\{-5, -4\}$

18. $\{-8, 3\}$

19. $\{-6, 4\}$

20. $\left\{-5 \pm \sqrt{34}\right\}$

21. $\{2, 5\}$

22. -3

23. 3

24. -9

25. -9

26. -6

27. 2

28. -1

29. -2

30. -8

31. 159

32. 66

33. 78

34. 444

35. -254

36. -84

37. 404

38. -1080

Solve for x by calculating the determinant.

13. $\begin{vmatrix} x-2 & 2 \\ 2 & x+1 \end{vmatrix} = 0$
14. $\begin{vmatrix} x+7 & -2 \\ 9 & x-2 \end{vmatrix} = 0$
15. $\begin{vmatrix} x+1 & 8 \\ 1 & x+3 \end{vmatrix} = 0$

16. $\begin{vmatrix} x-8 & 11 \\ -2 & x+5 \end{vmatrix} = 0$
17. $\begin{vmatrix} x+6 & 2 \\ -1 & x+3 \end{vmatrix} = 0$
18. $\begin{vmatrix} x-4 & -4 \\ 3 & x+9 \end{vmatrix} = 0$

19. $\begin{vmatrix} x+5 & 3 \\ 3 & x-3 \end{vmatrix} = 0$
20. $\begin{vmatrix} x+3 & 6 \\ 5 & x+7 \end{vmatrix} = 0$
21. $\begin{vmatrix} x-3 & 2 \\ 1 & x-4 \end{vmatrix} = 0$

Use the matrix $A = \begin{bmatrix} 2 & -1 & 5 \\ 0 & 1 & 3 \\ 1 & 0 & -2 \end{bmatrix}$ to evaluate the following. See Example 2.

22. The minor of a_{12}
23. The cofactor of a_{12}

24. The minor of a_{22}
25. The cofactor of a_{22}

26. The cofactor of a_{32}
27. The cofactor of a_{33}

28. The minor of a_{13}
29. The cofactor of a_{21}

30. The cofactor of a_{31}

Find the determinant by the method of expansion by cofactors along the given row or column. See Example 3.

31. $\begin{vmatrix} 4 & 5 & 3 \\ -1 & 2 & 7 \\ 11 & 6 & 2 \end{vmatrix}$ Expand along Row 3
32. $\begin{vmatrix} 8 & 2 & 0 \\ 3 & 4 & 7 \\ 1 & 0 & 2 \end{vmatrix}$ Expand along Column 3

33. $\begin{vmatrix} 5 & 8 & 5 \\ 0 & -6 & 3 \\ 2 & 4 & -1 \end{vmatrix}$ Expand along Row 2
34. $\begin{vmatrix} -4 & 2 & 1 \\ 9 & 12 & 8 \\ 0 & 6 & -3 \end{vmatrix}$ Expand along Column 1

35. $\begin{vmatrix} 13 & 0 & -7 \\ 4 & 2 & 3 \\ 1 & 4 & 0 \end{vmatrix}$ Expand along Row 2
36. $\begin{vmatrix} 7 & 0 & 1 \\ 2 & 5 & 3 \\ 8 & 6 & 2 \end{vmatrix}$ Expand along Column 3

37. $\begin{vmatrix} 8 & 0 & -7 & 5 \\ 4 & -2 & 3 & 3 \\ -1 & 1 & 0 & 2 \\ 2 & 0 & 6 & 0 \end{vmatrix}$ Expand along Row 4
38. $\begin{vmatrix} 4 & -2 & 9 & 2 \\ 7 & 0 & 1 & 7 \\ -6 & 3 & 0 & 1 \\ 3 & 1 & 2 & 0 \end{vmatrix}$ Expand along Column 2

39. 4

40. 26

41. 120

42. 0

43. 10

44. x^4

45. x^4

46. 12

47. x^8

48. $(1, -2)$

49. $(76, -53)$

50. $\varnothing$

51. $\left\{(-y-2, y)\,\middle|\, y \in \mathbb{R}\right\}$

52. $(3, -5)$

53. $\varnothing$

54. $(3, -2, 3)$

55. $\left\{(-3z-5,\right.$
$-6z-10,$
$\left. z)\,\middle|\, z \in \mathbb{R}\right\}$

56. $\left\{(1, y, 0)\,\middle|\, y \in \mathbb{R}\right\}$

57. $\left\{\left(\dfrac{-5y-z-5}{2},\right.\right.$
$\dfrac{-5y+3z-19}{2}, y,$
$\left.\left. z\right)\,\middle|\, y \in \mathbb{R}, z \in \mathbb{R}\right\}$

58. $\varnothing$

59. $\left\{(-z+8, -5z+31,\right.$
$\left. -2z+37, z)\,\middle|\, z \in \mathbb{R}\right\}$

Evaluate the following determinants. In each case, minimize the required number of computations by carefully choosing a row or column to expand along, and use the properties of determinants to simplify the process. See Examples 3 and 4.

39. $\begin{vmatrix} 2 & 0 & 1 \\ -5 & 1 & 0 \\ 3 & -1 & 1 \end{vmatrix}$

40. $\begin{vmatrix} 12 & 3 & 1 \\ 1 & 1 & -1 \\ 0 & 2 & 0 \end{vmatrix}$

41. $\begin{vmatrix} 12 & 3 & 6 \\ 2 & 2 & -4 \\ 0 & 2 & 0 \end{vmatrix}$

42. $\begin{vmatrix} 1 & 2 & 3 \\ 4 & 5 & 6 \\ 7 & 8 & 9 \end{vmatrix}$

43. $\begin{vmatrix} 2 & 1 & -3 & 0 \\ 1 & -2 & 1 & 0 \\ 0 & 1 & 0 & 1 \\ 2 & 0 & 1 & 1 \end{vmatrix}$

44. $\begin{vmatrix} x & 0 & 0 & 0 \\ 0 & x & 0 & 0 \\ 0 & 0 & x & 0 \\ 0 & 0 & 0 & x \end{vmatrix}$

45. $\begin{vmatrix} x & x & x & x \\ 0 & x & x & x \\ 0 & 0 & x & x \\ 0 & 0 & 0 & x \end{vmatrix}$

46. $\begin{vmatrix} 0 & 2 & 0 & 0 \\ -2 & -4 & 5 & 9 \\ 1 & 3 & -1 & 1 \\ 0 & 7 & 0 & 2 \end{vmatrix}$

47. $\begin{vmatrix} x & x & 0 & 0 \\ yz & x^3 & z & x^4 \\ z & xy & x & 0 \\ x^2 & 0 & 0 & 0 \end{vmatrix}$

Use Cramer's rule to solve the following systems. See Examples 5 and 6.

48. $\begin{cases} 2x - 3y = 8 \\ 8x + 5y = -2 \end{cases}$

49. $\begin{cases} 5x + 7y = 9 \\ 2x + 3y = -7 \end{cases}$

50. $\begin{cases} 5x - 10y = 9 \\ -x + 2y = -3 \end{cases}$

51. $\begin{cases} -2x - 2y = 4 \\ 3x + 3y = -6 \end{cases}$

52. $\begin{cases} \dfrac{2}{3}x + y = -3 \\ 3x + \dfrac{5}{2}y = -\dfrac{7}{2} \end{cases}$

53. $\begin{cases} \dfrac{2}{3}x + 2y = 1 \\ x + 3y = 0 \end{cases}$

54. $\begin{cases} x + 2y = -1 \\ y + 3z = 7 \\ 2x + 5z = 21 \end{cases}$

55. $\begin{cases} 2x - y = 0 \\ 5x - 3y - 3z = 5 \\ 2x + 6z = -10 \end{cases}$

56. $\begin{cases} 3x + 8z = 3 \\ -3x - 7z = -3 \\ x + 3z = 1 \end{cases}$

57. $\begin{cases} 3w - x + 5y + 3z = 2 \\ -4w - 10y - 2z = 10 \\ w - x + 2z = 7 \\ 4w - 2x + 5y + 5z = 9 \end{cases}$

58. $\begin{cases} 2w + x - 3y = 3 \\ w - 2x + y = 1 \\ x + z = -2 \\ y + z = 0 \end{cases}$

59. $\begin{cases} 3w - 2x + y - 5z = -1 \\ w + x - y + 4z = 2 \\ 4w - x - z = 1 \\ 5w - x = 9 \end{cases}$

60. $(27, 109)$

61. $(1647, 2071)$

62. $(7, -1, 2, 4)$

63. $\varnothing$

64. $(-2, 2)$

65. $(-3, -1, 0, -4)$

66. 6 ft, 17 ft, 20 ft

67. Candy bars: 5,
Ice cream: 6

68. Soybeans: 147.5 acres,
Corn: 295 acres,
Wheat: 57.5 acres

60. $\begin{cases} -4x + y = 1 \\ 7x + 2y = 407 \end{cases}$

61. $\begin{cases} 5x - 4y = -49 \\ 24x - 19y = 179 \end{cases}$

62. $\begin{cases} 2w - 3x + 4y - z = 21 \\ w + 5x = 2 \\ -2x + 3y + z = 12 \\ -3w + 4z = -5 \end{cases}$

63. $\begin{cases} -5x + 10y = 3 \\ \dfrac{7}{2}x - 7y = 20 \end{cases}$

64. $\begin{cases} 23x + 21y = -4 \\ x - 3y = -8 \end{cases}$

65. $\begin{cases} w - x + y - z = 2 \\ 2w - x + 3y = -5 \\ x - 2z = 7 \\ 3w + 4x = -13 \end{cases}$

Solve the following application problems.

66. The three sides of a triangle are related as follows: the perimeter is 43 feet, the second side is 5 feet more than twice the first side, and the third side is 3 feet less than the sum of the other two sides. Find the lengths of the three sides of the triangle.

67. Eric's favorite candy bar and ice cream flavor have fat and calorie contents as follows: each candy bar has 5 grams of fat and 280 calories; each serving of ice cream has 10 grams of fat and 150 calories. How many candy bars and servings of ice cream did he eat during the weekend he consumed 85 grams of fat and 2300 calories from these two treats?

68. A farmer plants soybeans, corn, and wheat and rotates the planting each year on her 500-acre farm. In a particular year, the profits from her crops were $120 per acre of soybeans, $100 per acre of corn, and $80 per acre of wheat. She planted twice as many acres of corn as soybeans. How many acres did she plant with each crop the year she made a total profit of $51,800?

10.4

The Algebra of Matrices

TOPICS

1. Matrix addition
2. Scalar multiplication
3. Matrix multiplication
T. Matrix algebra

TOPIC 1

Matrix Addition

We've already seen once in this book that the mathematical operations of addition, subtraction, multiplication and division can be defined on objects besides real numbers. Specifically, we saw how these operations can be defined for real-valued functions. In the next two sections, we will explore how to define these operations for matrices.

DEFINITION

Matrix Addition

Two matrices A and B can be added to form the new matrix $A + B$ only if A and B are of the same order. The addition is performed by adding corresponding entries of the two matrices together; that is, the element in the ith row and the jth column of $A + B$ is given by $a_{ij} + b_{ij}$.

CAUTION!

It is very important to note the restriction on the order of the two matrices in the above definition; a matrix with m rows and n columns can only be added to another matrix with m rows and n columns. In all aspects of matrix algebra, the orders of the matrices involved must be considered.

DEFINITION

Matrix Equality

Two matrices A and B are **equal**, denoted $A = B$, if they are of the same order and all corresponding entries of A and B are equal.

EXAMPLE 1

Matrix Addition

Perform the indicated addition, if possible.

a. $\begin{bmatrix} -3 & 2 \\ 0 & -5 \\ 11 & -9 \end{bmatrix} + \begin{bmatrix} 3 & 17 \\ 5 & 4 \\ -10 & 4 \end{bmatrix}$

b. $\begin{bmatrix} 2 & -5 \\ 1 & 0 \\ 0 & 3 \\ -7 & 10 \end{bmatrix} + \begin{bmatrix} 2 & 1 & 0 & -7 \\ -5 & 0 & 3 & 10 \end{bmatrix}$

Solutions:

a. Both matrices are 3×2, so the sum is defined and

$\begin{bmatrix} -3 & 2 \\ 0 & -5 \\ 11 & -9 \end{bmatrix} + \begin{bmatrix} 3 & 17 \\ 5 & 4 \\ -10 & 4 \end{bmatrix} = \begin{bmatrix} 0 & 19 \\ 5 & -1 \\ 1 & -5 \end{bmatrix}$

Each entry in the first matrix is added to its corresponding entry in the second matrix.

b. The first matrix is 4×2 and the second is 2×4, so the addition cannot be performed.

The entries in matrices do not all have to be constants. In many applications of matrices, it is convenient to represent some of the entries initially as variables, with the intent of eventually solving for the variables. We can solve some examples of such *matrix equations* now, using only matrix addition and matrix equality.

EXAMPLE 2

Matrix Equations

Determine the values of the variables that will make each of the following statements true.

a. $\begin{bmatrix} -3 & a & b \\ -2 & a+b & 5 \end{bmatrix} = \begin{bmatrix} c & 3 & 7 \\ -2 & d & 5 \end{bmatrix}$

b. $\begin{bmatrix} 3x \\ 4 \end{bmatrix} + \begin{bmatrix} -y \\ 2x \end{bmatrix} = \begin{bmatrix} 13 \\ 7y \end{bmatrix}$

Solutions:

a. Solving for the four variables $a, b, c,$ and d is just a matter of comparing the entries one-by-one.

$\begin{bmatrix} -3 & a & b \\ -2 & a+b & 5 \end{bmatrix} = \begin{bmatrix} c & 3 & 7 \\ -2 & d & 5 \end{bmatrix}$

Comparing the top rows tells us that $a = 3$, $b = 7$, and $c = -3$.

In the bottom rows, note the -2 in the left corner of each matrix and the 5 in each right corner. If these constants were not equal we would have a contradiction, and there would be no way to make the matrix equation true.

The only variable left to solve for is d, which we see is equal to $a+b$. So, $d = 3+7 = 10$.

b. Each matrix consists of only two entries, and after performing the matrix addition on the left and comparing corresponding entries, we arrive at the system of equations

$$\begin{cases} 3x - y = 13 \\ 4 + 2x = 7y \end{cases}$$

We know a number of ways of solving such a system. If we choose to use Cramer's rule, we first rewrite the system as

$$\begin{cases} 3x - y = 13 \\ 2x - 7y = -4 \end{cases}$$

From here we can determine

$$D = \begin{vmatrix} 3 & -1 \\ 2 & -7 \end{vmatrix} = (3)(-7) - (2)(-1) = -19,$$

$$D_x = \begin{vmatrix} 13 & -1 \\ -4 & -7 \end{vmatrix} = (13)(-7) - (-4)(-1) = -95, \text{ and}$$

$$D_y = \begin{vmatrix} 3 & 13 \\ 2 & -4 \end{vmatrix} = (3)(-4) - (2)(13) = -38,$$

so $x = \dfrac{D_x}{D} = \dfrac{-95}{-19} = 5$ and $y = \dfrac{D_y}{D} = \dfrac{-38}{-19} = 2.$

TOPIC 2 — Scalar Multiplication

In the context of matrix algebra, a *scalar* is a real number, and *scalar multiplication* refers to the product of a real number and a matrix. It may seem strange to combine two very different sorts of objects (a scalar and a matrix), but consider the following matrix sum

$$\begin{bmatrix} -5 & 2 \\ 1 & -3 \\ -2 & 7 \end{bmatrix} + \begin{bmatrix} -5 & 2 \\ 1 & -3 \\ -2 & 7 \end{bmatrix} = \begin{bmatrix} -10 & 4 \\ 2 & -6 \\ -4 & 14 \end{bmatrix}.$$

Since the result of adding a matrix to itself has the effect of doubling each entry, it makes sense to write

$$2\begin{bmatrix} -5 & 2 \\ 1 & -3 \\ -2 & 7 \end{bmatrix} = \begin{bmatrix} -10 & 4 \\ 2 & -6 \\ -4 & 14 \end{bmatrix}.$$

Extending the idea to all scalars (real numbers) leads to the following definition.

DEFINITION

Scalar Multiplication

If A is an $m \times n$ matrix and c is a scalar, cA stands for the $m \times n$ matrix for which each entry is c times the corresponding entry of A. In other words, the entry in the ith row and j^{th} column of cA is ca_{ij}.

EXAMPLE 3

Scalar Multiplication

Given the matrices $A = \begin{bmatrix} -1 & 6 & 2 \\ -8 & 0 & 1 \end{bmatrix}$ and $B = \begin{bmatrix} 0 & -3 & 4 \\ 1 & -2 & 6 \end{bmatrix}$, write $-3A + 2B$ as a single matrix.

Solution:

Before calculating, note that the operation can be performed since both matrices are of the same order, 2×3.

This problem has two steps. First, multiply each entry of A and B by its corresponding scalar. Second, add the resulting matrices together:

$$-3A + 2B = -3\begin{bmatrix} -1 & 6 & 2 \\ -8 & 0 & 1 \end{bmatrix} + 2\begin{bmatrix} 0 & -3 & 4 \\ 1 & -2 & 6 \end{bmatrix}$$ Multiply each matrix by its scalar.

$$= \begin{bmatrix} 3 & -18 & -6 \\ 24 & 0 & -3 \end{bmatrix} + \begin{bmatrix} 0 & -6 & 8 \\ 2 & -4 & 12 \end{bmatrix}$$ Add the resulting matrices.

$$= \begin{bmatrix} 3 & -24 & 2 \\ 26 & -4 & 9 \end{bmatrix}$$

As we try to understand algebraic operations on matrices, we can use real numbers as a model. Subtraction of real numbers is defined in terms of addition, and we define matrix subtraction in the same way.

DEFINITION

Matrix Subtraction

Let A and B be two matrices of the same order. The difference $A - B$ is defined by

$$A - B = A + (-B).$$

EXAMPLE 4

Matrix Subtraction

Perform the indicated subtraction: $\begin{bmatrix} 3 & -5 & 2 \end{bmatrix} - \begin{bmatrix} -2 & -5 & 3 \end{bmatrix}$.

Solution:

Since both matrices are of order 1×3, we know the subtraction is possible. Subtract each entry in the second matrix from the corresponding entry in the first:

$$\begin{bmatrix} 3 & -5 & 2 \end{bmatrix} - \begin{bmatrix} -2 & -5 & 3 \end{bmatrix} = \begin{bmatrix} 3-(-2) & -5-(-5) & 2-3 \end{bmatrix}$$

$$= \begin{bmatrix} 5 & 0 & -1 \end{bmatrix}$$

TOPIC 3

Matrix Multiplication

The definition of matrix multiplication is not as straightforward as that of matrix addition or subtraction. Simply put, matrix multiplication does not refer to multiplying the corresponding entries of two matrices together. To understand the process of matrix multiplication, it helps to first think about matrices as functions.

Note that the system of equations

$$\begin{cases} x' = ax + by \\ y' = cx + dy \end{cases}$$

can be thought of as a function that transforms the ordered pair (x, y) into the ordered pair (x', y'), and that the function is characterized entirely by the matrix

$$A = \begin{bmatrix} a & b \\ c & d \end{bmatrix}.$$

Similarly, the system

$$\begin{cases} x' = ex + fy \\ y' = gx + hy \end{cases}$$

is a function that transforms ordered pairs into ordered pairs, and it is characterized by the matrix

$$B = \begin{bmatrix} e & f \\ g & h \end{bmatrix}.$$

Now we can look at the result of plugging the output of the first function into the second function (*composing* the two functions so that B acts on the result of A).

If we let (x', y') denote the output of the first function, we obtain the new ordered pair (x'', y'') given by

$$\begin{cases} x'' = ex' + fy' \\ y'' = gx' + hy' \end{cases} \text{ or } \begin{cases} x'' = e(ax + by) + f(cx + dy) \\ y'' = g(ax + by) + h(cx + dy) \end{cases}$$

We can change the way the last system is written to obtain

$$\begin{cases} x'' = (ea + fc)x + (eb + fd)y \\ y'' = (ga + hc)x + (gb + hd)y \end{cases}$$

so the composition of the two functions, in the order BA, is characterized by the matrix

$$BA = \begin{bmatrix} ea + fc & eb + fd \\ ga + hc & gb + hd \end{bmatrix}.$$

Note that the entries of this last matrix are the sums of products, where the products in each sum are between elements of the rows of B with elements of the columns of A. This pattern is the basis of our formal definition of matrix multiplication.

DEFINITION

Matrix Multiplication

Two matrices A and B can be multiplied together, resulting in a new matrix denoted AB, only if the number of columns of A (the matrix on the left) is the same as the number of rows of B (the matrix on the right). Thus, if A is of order $m \times n$, the product AB is only defined if B is of order $n \times p$. The order of AB will be $m \times p$.

If we let c_{ij} denote the entry in the ith row and j^{th} column of AB, c_{ij} is obtained from the ith row of A and the jth column of B by the formula

$$c_{ij} = a_{i1}b_{1j} + a_{i2}b_{2j} + \ldots + a_{in}b_{nj}.$$

In words, c_{ij} equals the product of the first element of row i of matrix A and the first element of column j of matrix B, plus the product of the second element of row i and the second element of column j, and so on.

CAUTION!

Unlike numerical multiplication, matrix multiplication is not commutative. That is, given two matrices A and B, AB *in general* is not equal to BA. As an illustration of this fact, suppose A is a 3×4 matrix and B is a 4×2 matrix. Then AB is defined (and is of order 3×2), but BA doesn't even exist. Even when both AB and BA are defined, they are generally not equal.

EXAMPLE 5

Matrix Multiplication   Given the matrices $A = \begin{bmatrix} 2 & 0 \\ -5 & 1 \end{bmatrix}$ and $B = \begin{bmatrix} 7 & -2 \\ 3 & 1 \end{bmatrix}$, find AB.

Solution:

Since both matrices are 2×2, they can be multiplied, and the result will also be a 2×2 matrix. For this example, we will follow the procedure one entry at a time.

$$AB = \begin{bmatrix} 2 & 0 \\ -5 & 1 \end{bmatrix}\begin{bmatrix} 7 & -2 \\ 3 & 1 \end{bmatrix} = \begin{bmatrix} 2(7)+0(3) & AB_{12} \\ AB_{21} & AB_{22} \end{bmatrix}$$ Sum the product of entries in the first row of A and first column of B.

$$= \begin{bmatrix} 2 & 0 \\ -5 & 1 \end{bmatrix}\begin{bmatrix} 7 & -2 \\ 3 & 1 \end{bmatrix} = \begin{bmatrix} 14 & 2(-2)+0(1) \\ AB_{21} & AB_{22} \end{bmatrix}$$ Then the first row of A and second column of B.

$$= \begin{bmatrix} 2 & 0 \\ -5 & 1 \end{bmatrix}\begin{bmatrix} 7 & -2 \\ 3 & 1 \end{bmatrix} = \begin{bmatrix} 14 & -4 \\ -5(7)+1(3) & AB_{22} \end{bmatrix}$$ Then the second row of A and first column of B.

$$= \begin{bmatrix} 2 & 0 \\ -5 & 1 \end{bmatrix}\begin{bmatrix} 7 & -2 \\ 3 & 1 \end{bmatrix} = \begin{bmatrix} 14 & -4 \\ -32 & -5(-2)+1(1) \end{bmatrix}$$ Finally, the second row of A and second column of B.

$$= \begin{bmatrix} 14 & -4 \\ -32 & 11 \end{bmatrix}$$

EXAMPLE 6

Matrix Multiplication   Given the matrices $A = \begin{bmatrix} 2 & -3 \\ 4 & -1 \\ 1 & 0 \end{bmatrix}$ and $B = \begin{bmatrix} 5 & 0 & -2 \\ -4 & 1 & 3 \end{bmatrix}$, find the following products.

a. AB **b.** BA

Solutions:

a. A is of order 3×2 and B is of order 2×3, so AB is defined and is of order 3×3. Each entry of AB is formed from a row of A and a column of B.

$$AB = \begin{bmatrix} 2 & -3 \\ 4 & -1 \\ 1 & 0 \end{bmatrix}\begin{bmatrix} 5 & 0 & -2 \\ -4 & 1 & 3 \end{bmatrix} = \begin{bmatrix} 2(5)+(-3)(-4) & 2(0)+(-3)(1) & 2(-2)+(-3)(3) \\ 4(5)+(-1)(-4) & 4(0)+(-1)(1) & 4(-2)+(-1)(3) \\ 1(5)+0(-4) & 1(0)+0(1) & 1(-2)+0(3) \end{bmatrix}$$

$$= \begin{bmatrix} 22 & -3 & -13 \\ 24 & -1 & -11 \\ 5 & 0 & -2 \end{bmatrix}$$

b. From the orders of the two matrices, we know BA exists and will be a 2×2 matrix. Note that each entry of BA is a sum of three products.

$$BA = \begin{bmatrix} 5 & 0 & -2 \\ -4 & 1 & 3 \end{bmatrix} \begin{bmatrix} 2 & -3 \\ 4 & -1 \\ 1 & 0 \end{bmatrix} = \begin{bmatrix} 5(2)+0(4)+(-2)(1) & 5(-3)+0(-1)+(-2)(0) \\ (-4)(2)+1(4)+3(1) & (-4)(-3)+1(-1)+3(0) \end{bmatrix}$$

$$= \begin{bmatrix} 8 & -15 \\ -1 & 11 \end{bmatrix}$$

Now that matrix multiplication has been defined, it can be used in another way to illustrate the fact that matrices can be considered as functions. As already mentioned, the matrix

$$A = \begin{bmatrix} a & b \\ c & d \end{bmatrix}$$

characterizes the function that transforms (x, y) into (x', y') in the system

$$\begin{cases} x' = ax + by \\ y' = cx + dy \end{cases}$$

This is even more clear if we associate the ordered pair (x, y) with the 2×1 matrix

$$\begin{bmatrix} x \\ y \end{bmatrix}$$

Then the matrix product

$$\begin{bmatrix} a & b \\ c & d \end{bmatrix} \begin{bmatrix} x \\ y \end{bmatrix}$$

which has the "look" of a function acting on its argument, gives us the expressions in the system. Verify for yourself that

$$\begin{bmatrix} a & b \\ c & d \end{bmatrix} \begin{bmatrix} x \\ y \end{bmatrix} = \begin{bmatrix} ax + by \\ cx + dy \end{bmatrix}$$

TOPIC Matrix Algebra

To perform algebra on matrices using the calculator, we must first create the matrices we wish to use by defining them in the MATRIX menu under EDIT. Once matrices have been defined, we can select them in the MATRIX menu under NAMES and use them in the operations we wish to perform. For instance, suppose we have two matrices,

$A = \begin{bmatrix} -1 & 4 \\ 9 & -2 \end{bmatrix}$ and $B = \begin{bmatrix} 2 & 10 \\ -1 & 4 \end{bmatrix}$, and we want to find $A + 3B$.

1. $\begin{bmatrix} 5 & -1 \\ 0 & 0 \\ 2 & 13 \end{bmatrix}$

2. Not possible

3. $\begin{bmatrix} 6 & -3 \\ 18 & 30 \\ -9 & 21 \end{bmatrix}$

4. $\begin{bmatrix} \dfrac{3}{2} & 1 & \dfrac{5}{2} \\ -1 & -2 & \dfrac{1}{2} \end{bmatrix}$

5. Not possible

6. $\begin{bmatrix} 9 & -8 \\ 10 & 10 \\ -5 & 14 \end{bmatrix}$

7. $\begin{bmatrix} 14 & -14 \\ 8 & 0 \\ -4 & 14 \end{bmatrix}$

8. $\begin{bmatrix} 7 & -8 \\ \dfrac{15}{2} & 5 \\ -\dfrac{9}{2} & \dfrac{13}{2} \end{bmatrix}$

9. $\begin{bmatrix} -7 & 5 \\ 3 & 10 \\ -3 & -8 \end{bmatrix}$

10. $\begin{bmatrix} 3 & -1 \\ 17 & 30 \\ -9 & 16 \end{bmatrix}$

11. Not possible

12. $\begin{bmatrix} -18 & 26 \\ -16 & 0 \\ 12 & -2 \end{bmatrix}$

13. $a = 3, b = -1,$
  $c = 10$

14. $x = 8, y = -3,$
  $z = 6$

15. $a = 2, b = -2,$
  $c = -1$

After creating matrices A and B, we would enter in the following:

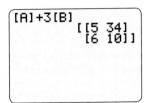

Notice that if we try to perform an operation that isn't possible due to dimension size, we will get an error.

Exercises

Given $A = \begin{bmatrix} 3 & -2 \\ 1 & 0 \\ 0 & 5 \end{bmatrix}$, $B = \begin{bmatrix} 4 & -5 \\ 3 & 0 \\ -2 & 2 \end{bmatrix}$, $C = \begin{bmatrix} 2 & -1 \\ 6 & 10 \\ -3 & 7 \end{bmatrix}$, and $D = \begin{bmatrix} 3 & 2 & 5 \\ -2 & -4 & 1 \end{bmatrix}$, determine

the following, if possible. See Examples 1, 3, and 4.

1. $3A - B$ **2.** $B - 2D$ **3.** $3C$ **4.** $\dfrac{1}{2}D$

5. $3D + C$ **6.** $A + B + C$ **7.** $2A + 2B$ **8.** $\dfrac{3}{2}B + \dfrac{1}{2}C$

9. $C - 3A$ **10.** $3C - A$ **11.** $4A - 3D$ **12.** $2(A - 3B)$

Determine values of the variables that will make the following equations true, if possible. See Examples 1, 3, and 4.

13. $\begin{bmatrix} 2a & b & 3 \\ -5 & 9 & 7 \end{bmatrix} = \begin{bmatrix} 6 & -1 & 3 \\ -5 & 9 & c-3 \end{bmatrix}$ **14.** $\begin{bmatrix} x \\ -9 \\ -1+z \end{bmatrix} = \begin{bmatrix} 8 \\ 3y \\ 5 \end{bmatrix}$

15. $\begin{bmatrix} a & 2b & c \end{bmatrix} + 3\begin{bmatrix} a & 2 & -c \end{bmatrix} = \begin{bmatrix} 8 & 2 & 2 \end{bmatrix}$ **16.** $\begin{bmatrix} w & 5x \\ 2y & z \end{bmatrix} - 5\begin{bmatrix} w & x \\ y & -z \end{bmatrix} = \begin{bmatrix} w+5 & 0 \\ 6 & 1 \end{bmatrix}$

17. $\begin{bmatrix} 3x \\ 2y \end{bmatrix} + \begin{bmatrix} x \\ -y \\ z \end{bmatrix} = \begin{bmatrix} 4 \\ 0 \\ 2 \end{bmatrix}$ **18.** $\begin{bmatrix} 2a & 3b & c \end{bmatrix} = \begin{bmatrix} 4 \\ 3 \\ 0 \end{bmatrix}$

19. $\begin{bmatrix} x \\ 3x \end{bmatrix} - \begin{bmatrix} y \\ 2y \end{bmatrix} = \begin{bmatrix} 5 \\ 20 \end{bmatrix}$ **20.** $7\begin{bmatrix} -1 \\ y \end{bmatrix} = \begin{bmatrix} 2x \\ 5x \end{bmatrix} + 3\begin{bmatrix} y \\ 1 \end{bmatrix}$

21. $2\begin{bmatrix} x \\ 2y \end{bmatrix} - 3\begin{bmatrix} 5y \\ -3x \end{bmatrix} = \begin{bmatrix} -9 \\ 31 \end{bmatrix}$ **22.** $2\begin{bmatrix} 3r & s & 2t \end{bmatrix} - \begin{bmatrix} r & s & t \end{bmatrix} = \begin{bmatrix} 15 & 3 & 9 \end{bmatrix}$

16. $w = -1, y = -2,$

$z = \dfrac{1}{6},$

No restriction on x

17. Not possible

18. Not possible

19. $x = 10, y = 5$

20. $x = -2, y = -1$

21. $x = 3, y = 1$

22. $r = 3, s = 3, t = 3$

23. Not possible

24. $x = -4, y = 3$

25. $a = 8, b = 5$

26. $r = 1, s = 6$

27. $[24 \quad -5]$

28. Not possible

29. $[35 \quad 18]$

30. $[38]$

31. Not possible

32. $\begin{bmatrix} 20 & 0 & -12 \\ 10 & 0 & -6 \\ -30 & 0 & 18 \end{bmatrix}$

33. $[-30 \quad -3]$

34. Not possible

35. $\begin{bmatrix} 15 & -3 & -24 \\ 25 & -5 & -40 \\ 30 & -6 & -48 \end{bmatrix}$

36. Not possible

37. $[-34 \quad -7]$

38. $[-26 \quad -12]$

39. $\begin{bmatrix} 11 & 0 \\ 0 & 11 \end{bmatrix}$

23. $2\begin{bmatrix} 2x^2 & x \\ 7x & 4 \end{bmatrix} - \begin{bmatrix} 5x \\ x-2 \end{bmatrix} = \begin{bmatrix} 2x & 0 \\ 6 & x^2 \end{bmatrix}$

25. $3\begin{bmatrix} 2a \\ -a \end{bmatrix} - 3\begin{bmatrix} 3b \\ 2b \end{bmatrix} = \begin{bmatrix} 3 \\ -54 \end{bmatrix}$

Evaluate the following matrix products, if possible. See Examples 5 and 6.

27. $[3 \quad -2 \quad 1]\begin{bmatrix} 5 & -1 \\ 0 & 3 \\ 9 & 4 \end{bmatrix}$

29. $[3 \quad 7]\begin{bmatrix} 0 & -8 \\ 5 & 6 \end{bmatrix}$

31. $\begin{bmatrix} 3 & 9 & -4 \\ 0 & 0 & 2 \\ 5 & -2 & 7 \end{bmatrix}\begin{bmatrix} 3 & 2 \\ 2 & 1 \end{bmatrix}$

33. $[-3 \quad -6 \quad -3]\begin{bmatrix} 6 & 9 \\ 6 & -8 \\ -8 & 8 \end{bmatrix}$

35. $\begin{bmatrix} -3 \\ -5 \\ -6 \end{bmatrix}[-5 \quad 1 \quad 8]$

24. $\begin{bmatrix} -x \\ 3 \end{bmatrix} - 5\begin{bmatrix} 2 \\ y \end{bmatrix} = \begin{bmatrix} -2y \\ 3x \end{bmatrix}$

26. $2\begin{bmatrix} -s \\ -7 \end{bmatrix} + 2\begin{bmatrix} -2r \\ r \end{bmatrix} = -2\begin{bmatrix} 8 \\ s \end{bmatrix}$

28. $\begin{bmatrix} 0 & -8 \\ 5 & 6 \end{bmatrix}[3 \quad 7]$

30. $[5 \quad 0 \quad -3]\begin{bmatrix} 4 \\ 2 \\ -6 \end{bmatrix}$

32. $\begin{bmatrix} 4 \\ 2 \\ -6 \end{bmatrix}[5 \quad 0 \quad -3]$

34. $\begin{bmatrix} 4 & -5 \\ 7 & -9 \end{bmatrix}[-8 \quad 3]$

Given $A = \begin{bmatrix} -3 & 1 \\ 2 & 3 \end{bmatrix}, B = [8 \quad -5], C = \begin{bmatrix} 4 \\ 7 \\ -2 \end{bmatrix}$, and $D = \begin{bmatrix} -5 & 4 \\ -1 & -1 \end{bmatrix}$,

determine the following, if possible. See Examples 5 and 6.

36. AB **37.** BA **38.** $BA + B$ **39.** A^2

40. C^2 **41.** CB **42.** D^2 **43.** $CD + C$

44. DA **45.** AD **46.** DB **47.** $(BD)A$

40. Not possible

41. $\begin{bmatrix} 32 & -20 \\ 56 & -35 \\ -16 & 10 \end{bmatrix}$

42. $\begin{bmatrix} 21 & -24 \\ 6 & -3 \end{bmatrix}$

43. Not possible

44. $\begin{bmatrix} 23 & 7 \\ 1 & -4 \end{bmatrix}$

45. $\begin{bmatrix} 14 & -13 \\ -13 & 5 \end{bmatrix}$

46. Not possible

47. $[179 \quad 76]$

Inverses of Matrices

TOPICS

1. The matrix form of a linear system

2. Finding the inverse of a matrix

3. Using matrix inverses to solve linear systems

T. Inverting matrices

TOPIC 1

The Matrix Form of a Linear System

As we saw in the last section, if we express the ordered pair (x, y) as a 2×1 matrix, then the linear system

$$\begin{cases} ax + by = e \\ cx + dy = f \end{cases}$$

can be written as

$$\begin{bmatrix} a & b \\ c & d \end{bmatrix} \begin{bmatrix} x \\ y \end{bmatrix} = \begin{bmatrix} e \\ f \end{bmatrix}.$$

The fact that the matrix equation is equivalent to the system of equations above it is a great leap in efficiency: it converts a system of any number of equations into a single matrix equation. More importantly, the function interpretation of a matrix allows us to express a *whole system* of equations in a form like that of a *single linear equation* of a single variable.

Since the generic linear equation $ax = b$ can be solved by dividing both sides by a,

$$ax = b \Leftrightarrow x = \frac{b}{a} \ \left(\text{assuming } a \neq 0 \right)$$

it is tempting to solve

$$\begin{bmatrix} a & b \\ c & d \end{bmatrix} \begin{bmatrix} x \\ y \end{bmatrix} = \begin{bmatrix} e \\ f \end{bmatrix}$$

by "dividing" both sides by the 2×2 matrix of coefficients. Unfortunately, we don't yet have a way of making sense of "matrix division."

We will return to this thought soon, but first we will see how some specific linear systems appear in matrix form.

EXAMPLE 1

Matrix Equation

Write each linear system as a matrix equation.

a. $\begin{cases} -3x + 5y = 2 \\ x - 4y = -1 \end{cases}$

b. $\begin{cases} 3y - x = -2 \\ 4 - z + y = 5 \\ z - 3x + 3 = y - x \end{cases}$

Solutions:

a. Since the system is in standard form, we can just read off the coefficients of x and y to form the equation

$$\begin{bmatrix} -3 & 5 \\ 1 & -4 \end{bmatrix} \begin{bmatrix} x \\ y \end{bmatrix} = \begin{bmatrix} 2 \\ -1 \end{bmatrix}.$$

b. First, we write each equation in standard form:

$$\begin{cases} 3y - x = -2 \\ 4 - z + y = 5 \\ z - 3x + 3 = y - x \end{cases} \longrightarrow \begin{cases} -x + 3y = -2 \\ y - z = 1 \\ -2x - y + z = -3 \end{cases}$$

Now we can read off the coefficients to form the matrix equation

$$\begin{bmatrix} -1 & 3 & 0 \\ 0 & 1 & -1 \\ -2 & -1 & 1 \end{bmatrix} \begin{bmatrix} x \\ y \\ z \end{bmatrix} = \begin{bmatrix} -2 \\ 1 \\ -3 \end{bmatrix}.$$

TOPIC 2 — Finding the Inverse of a Matrix

In order to solve matrix equations like the two we obtained in Example 1, we need a way to "undo" the matrix of coefficients that appears in front of the column of variables.

In order to figure out how to "undo" a matrix, we will first need to understand how to do *nothing* to a matrix. Consider the following matrix products:

$$\begin{bmatrix} 1 & 0 \\ 0 & 1 \end{bmatrix} \begin{bmatrix} x \\ y \end{bmatrix} \quad \text{and} \quad \begin{bmatrix} 1 & 0 & 0 \\ 0 & 1 & 0 \\ 0 & 0 & 1 \end{bmatrix} \begin{bmatrix} x \\ y \\ z \end{bmatrix}$$

Evaluating these products, we have:

$$\begin{bmatrix} 1 & 0 \\ 0 & 1 \end{bmatrix} \begin{bmatrix} x \\ y \end{bmatrix} = \begin{bmatrix} 1x + 0y \\ 0x + 1y \end{bmatrix} \quad \text{and} \quad \begin{bmatrix} 1 & 0 & 0 \\ 0 & 1 & 0 \\ 0 & 0 & 1 \end{bmatrix} \begin{bmatrix} x \\ y \\ z \end{bmatrix} = \begin{bmatrix} 1x + 0y + 0z \\ 0x + 1y + 0z \\ 0x + 0y + 1z \end{bmatrix}$$

$$= \begin{bmatrix} x \\ y \end{bmatrix} \qquad\qquad\qquad\qquad = \begin{bmatrix} x \\ y \\ z \end{bmatrix}$$

If these matrix products appear as the left-hand side of matrix equations, the equations would correspond to solutions of linear systems:

$$\begin{bmatrix} 1 & 0 \\ 0 & 1 \end{bmatrix} \begin{bmatrix} x \\ y \end{bmatrix} = \begin{bmatrix} a \\ b \end{bmatrix} \text{ corresponds to } \begin{cases} x = a \\ y = b \end{cases}$$

and

$$\begin{bmatrix} 1 & 0 & 0 \\ 0 & 1 & 0 \\ 0 & 0 & 1 \end{bmatrix} \begin{bmatrix} x \\ y \\ z \end{bmatrix} = \begin{bmatrix} a \\ b \\ c \end{bmatrix} \text{ corresponds to } \begin{cases} x = a \\ y = b \\ z = c \end{cases}$$

When we multiply any matrix by one of these matrices, the original matrix is *unchanged*. This fact is very useful in solving matrix equations.

DEFINITION

Identity Matrices

The $n \times n$ **identity matrix**, denoted I_n (just I when there is no possibility of confusion), is the $n \times n$ matrix consisting of 1's on the *main diagonal* and 0's everywhere else. The **main diagonal** consists of those entries in the first row-first column, the second row-second column, and so on down to the nth row-n^{th} column. Every identity matrix has the form

$$I = \begin{bmatrix} 1 & 0 & 0 & \cdots & 0 \\ 0 & 1 & 0 & \cdots & 0 \\ 0 & 0 & 1 & \cdots & 0 \\ \vdots & \vdots & \vdots & \ddots & \vdots \\ 0 & 0 & 0 & \cdots & 1 \end{bmatrix}.$$

If the matrices A and B have appropriate order, so that the matrix products are defined, then $AI = A$ and $IB = B$. Thus, the identity matrix serves as the multiplicative identity on the set of appropriately sized matrices. In this sense, I serves the same purpose as the number 1 in the set of real numbers.

We know that a linear system of n equations and n variables can be expressed as a matrix equation $AX = B$, where A is an $n \times n$ matrix of coefficients, X is an $n \times 1$ matrix containing the n variables, and B is an $n \times 1$ matrix of the constants from the right-hand sides of the equations. If we could find a matrix, which we call A^{-1}, with the property that $A^{-1}A = I$, then we could use A^{-1} to "undo" the matrix A. We call the matrix A^{-1}, if it exists, the *inverse* of A. This is analogous to the fact that $\dfrac{1}{a}$, sometimes denoted a^{-1}, is the (multiplicative) inverse of the real number a.

DEFINITION

The Inverse of a Matrix

Let A be an $n \times n$ matrix. If there exists an $n \times n$ matrix A^{-1} such that

$$A^{-1}A = I_n \text{ and } AA^{-1} = I_n,$$

we call A^{-1} the **inverse** of A.

EXAMPLE 2

Finding the Inverse of
a Matrix

Find the inverse of the matrix $A = \begin{bmatrix} 2 & -3 \\ -1 & 2 \end{bmatrix}$.

Solution:

If we let $A^{-1} = \begin{bmatrix} w & x \\ y & z \end{bmatrix}$, we can use the equation $AA^{-1} = I$ to find $w, x, y,$ and z.

$$\begin{bmatrix} 2 & -3 \\ -1 & 2 \end{bmatrix}\begin{bmatrix} w & x \\ y & z \end{bmatrix} = \begin{bmatrix} 1 & 0 \\ 0 & 1 \end{bmatrix}$$

Multiplying the left-hand side out, we see that we need to solve the equation

$$\begin{bmatrix} 2w - 3y & 2x - 3z \\ -w + 2y & -x + 2z \end{bmatrix} = \begin{bmatrix} 1 & 0 \\ 0 & 1 \end{bmatrix}$$

which, if we equate columns on each side, means we need to solve the two linear systems

$$\begin{cases} 2w - 3y = 1 \\ -w + 2y = 0 \end{cases} \text{ and } \begin{cases} 2x - 3z = 0 \\ -x + 2z = 1 \end{cases}$$

We have covered many methods for solving such systems. If we write the augmented matrix for each system, we get

$$\begin{bmatrix} 2 & -3 & | & 1 \\ -1 & 2 & | & 0 \end{bmatrix} \text{ and } \begin{bmatrix} 2 & -3 & | & 0 \\ -1 & 2 & | & 1 \end{bmatrix}.$$

Note that the left-hand sides of these matrices are the same. This allows us to combine them into a new kind of augmented matrix so we can use Gauss-Jordan elimination to solve the systems at the same time. Combining the matrices, we get

$$\begin{bmatrix} 2 & -3 & | & 1 & 0 \\ -1 & 2 & | & 0 & 1 \end{bmatrix}.$$

When we change this new matrix into reduced row echelon form, we will have solved the first system with the numbers in the third column and the second system with the numbers in the fourth column.

$$\begin{bmatrix} 2 & -3 & | & 1 & 0 \\ -1 & 2 & | & 0 & 1 \end{bmatrix} \xrightarrow{R_1 \leftrightarrow R_2} \begin{bmatrix} -1 & 2 & | & 0 & 1 \\ 2 & -3 & | & 1 & 0 \end{bmatrix} \xrightarrow{2R_1 + R_2} \begin{bmatrix} -1 & 2 & | & 0 & 1 \\ 0 & 1 & | & 1 & 2 \end{bmatrix}$$

$$\xrightarrow{-R_1} \begin{bmatrix} 1 & -2 & | & 0 & -1 \\ 0 & 1 & | & 1 & 2 \end{bmatrix} \xrightarrow{2R_2 + R_1} \begin{bmatrix} 1 & 0 & | & 2 & 3 \\ 0 & 1 & | & 1 & 2 \end{bmatrix}$$

This tells us that $w = 2$ and $y = 1$ (from the third column) and $x = 3$ and $z = 2$ (from the fourth column). So

$$A^{-1} = \begin{bmatrix} 2 & 3 \\ 1 & 2 \end{bmatrix}.$$

We can now verify that

$$\begin{bmatrix} 2 & -3 \\ -1 & 2 \end{bmatrix}\begin{bmatrix} 2 & 3 \\ 1 & 2 \end{bmatrix} = \begin{bmatrix} 1 & 0 \\ 0 & 1 \end{bmatrix} \text{ and also } \begin{bmatrix} 2 & 3 \\ 1 & 2 \end{bmatrix}\begin{bmatrix} 2 & -3 \\ -1 & 2 \end{bmatrix} = \begin{bmatrix} 1 & 0 \\ 0 & 1 \end{bmatrix}$$

so we have indeed found A^{-1}.

Note how, during the solution process, the identity matrix passed from the right side of the matrix to the left, resulting in reduced row echelon form. With this observation, we can omit the intermediate step of constructing the systems of equations, and skip to the process of putting the appropriate augmented matrix into reduced row echelon form.

PROCEDURE

Finding the Inverse of a Matrix

Let A be an $n \times n$ matrix. The inverse of A can be found by:

Step 1: Forming the augmented matrix $[A \mid I]$, where I is the $n \times n$ identity matrix.

Step 2: Using Gauss-Jordan elimination to put $[A \mid I]$ into the form $[I \mid B]$, if possible.

Step 3: Defining A^{-1} to be B.

If it is not possible to put $[A \mid I]$ into reduced row echelon form, then A doesn't have an inverse, and we say A is **not invertible**.

If the coefficient matrix of a system of equations is not invertible, it means that the system either has an infinite number of solutions or has no solution.

There is a shortcut for finding inverses of 2×2 matrices that can save you some time. This shortcut also quickly identifies those 2×2 matrices that are not invertible.

THEOREM

Inverse of a 2 × 2 Matrix

Let $A = \begin{bmatrix} a & b \\ c & d \end{bmatrix}$. Then $A^{-1} = \dfrac{1}{|A|}\begin{bmatrix} d & -b \\ -c & a \end{bmatrix}$, where $|A| = ad - bc$ is the determinant of A. Since $|A|$ appears in the denominator of a fraction, A^{-1} fails to exist if $|A| = 0$.

In fact, we can extend this last observation to all square matrices.

THEOREM

Invertible Matrices

A square matrix A is **invertible** if and only if $|A| \neq 0$.

EXAMPLE 3

Finding the Inverse of
a Matrix

Find the inverses of the following matrices, if possible.

a. $A = \begin{bmatrix} 3 & -5 \\ 2 & 1 \end{bmatrix}$
                                                b. $B = \begin{bmatrix} 2 & 4 & 2 \\ -1 & 5 & -1 \\ 3 & 1 & 3 \end{bmatrix}$

Solutions:

a. Since A is 2×2, we can use the shortcut,

$$A^{-1} = \frac{1}{3-(-10)}\begin{bmatrix} 1 & 5 \\ -2 & 3 \end{bmatrix} = \begin{bmatrix} \dfrac{1}{13} & \dfrac{5}{13} \\ \dfrac{-2}{13} & \dfrac{3}{13} \end{bmatrix}.$$

We can verify our work as follows:

$$\begin{bmatrix} 3 & -5 \\ 2 & 1 \end{bmatrix}\begin{bmatrix} \dfrac{1}{13} & \dfrac{5}{13} \\ \dfrac{-2}{13} & \dfrac{3}{13} \end{bmatrix} = \begin{bmatrix} 1 & 0 \\ 0 & 1 \end{bmatrix} = \begin{bmatrix} \dfrac{1}{13} & \dfrac{5}{13} \\ \dfrac{-2}{13} & \dfrac{3}{13} \end{bmatrix}\begin{bmatrix} 3 & -5 \\ 2 & 1 \end{bmatrix}$$

b. Since B is a 3×3 matrix, we use the augmented identity matrix process.

$$\left[\begin{array}{ccc|ccc} 2 & 4 & 2 & 1 & 0 & 0 \\ -1 & 5 & -1 & 0 & 1 & 0 \\ 3 & 1 & 3 & 0 & 0 & 1 \end{array}\right] \xrightarrow[3R_2+R_3]{2R_2+R_1} \left[\begin{array}{ccc|ccc} 0 & 14 & 0 & 1 & 2 & 0 \\ -1 & 5 & -1 & 0 & 1 & 0 \\ 0 & 16 & 0 & 0 & 3 & 1 \end{array}\right]$$

$$\xrightarrow{R_1 \leftrightarrow R_2} \left[\begin{array}{ccc|ccc} -1 & 5 & -1 & 0 & 1 & 0 \\ 0 & 14 & 0 & 1 & 2 & 0 \\ 0 & 16 & 0 & 0 & 3 & 1 \end{array}\right] \xrightarrow[\frac{1}{14}R_2]{-R_1} \left[\begin{array}{ccc|ccc} 1 & -5 & 1 & 0 & -1 & 0 \\ 0 & 1 & 0 & \dfrac{1}{14} & \dfrac{1}{7} & 0 \\ 0 & 16 & 0 & 0 & 3 & 1 \end{array}\right]$$

$$\xrightarrow{-16R_2+R_3} \left[\begin{array}{ccc|ccc} 1 & -5 & 1 & 0 & -1 & 0 \\ 0 & 1 & 0 & \dfrac{1}{14} & \dfrac{1}{7} & 0 \\ 0 & 0 & 0 & -\dfrac{8}{7} & \dfrac{5}{7} & 1 \end{array}\right]$$

At this point, we can stop. Once the first three entries of any row are 0 in the 3×3 matrix, there is no way to put the matrix into reduced row echelon form. Thus, B has no inverse.

We could have seen this fact earlier by considering the determinant of B, which is 0, meaning B has no inverse. To quickly see that $|B| = 0$, note that B has two identical columns. Recall that switching two columns changes the sign of the determinant, but if we switch identical columns, the determinant must remain the same! The only way this is possible is if $|B| = 0$.

TOPIC 3

Using Matrix Inverses to Solve Linear Systems

We have now assembled all the tools we need to solve linear systems by the inverse matrix method.

PROCEDURE

The Inverse Matrix Method

To solve a linear system of n equations in n variables:

Step 1: Write the system in matrix form as $AX = B$, where A is the $n \times n$ matrix of coefficients, X is the $n \times 1$ matrix of variables, and B is the $n \times 1$ matrix of constants.

Step 2: Calculate A^{-1}, if it exists. If A^{-1} does not exist, the system either has an infinite number of solutions, or no solution, and a different method is required.

Step 3: Multiply both sides of the equation $AX = B$ by A^{-1}. We obtain

$$A^{-1}AX = A^{-1}B$$

$$IX = A^{-1}B$$

$$X = A^{-1}B$$

Step 4: The entries in the $n \times 1$ matrix $A^{-1}B$ are the solutions for the variables listed in the $n \times 1$ matrix X.

CAUTION!

It is crucial to multiply both sides of the equation $AX = B$ by A^{-1} on the *left-hand side*. Recall that matrix multiplication is not commutative, so failing to do this can result in an incorrect answer, and often the multiplication will not even be defined.

EXAMPLE 4

Inverse Matrix Method

Solve the following systems by the inverse matrix method.

a. $\begin{cases} 4x - 5y = 3 \\ -3x + 7y = 1 \end{cases}$

b. $\begin{cases} -x + 2y = 3 \\ 3x - 6y = -5 \end{cases}$

Solutions:

a. $\begin{cases} 4x - 5y = 3 \\ -3x + 7y = 1 \end{cases}$

$$\begin{bmatrix} 4 & -5 \\ -3 & 7 \end{bmatrix} \begin{bmatrix} x \\ y \end{bmatrix} = \begin{bmatrix} 3 \\ 1 \end{bmatrix}$$

Write the system in matrix form.

$$\begin{bmatrix} 4 & -5 \\ -3 & 7 \end{bmatrix}^{-1} = \frac{1}{28-15}\begin{bmatrix} 7 & 5 \\ 3 & 4 \end{bmatrix}$$

Find the inverse of the coefficient matrix. Since the matrix is of order 2×2, we can use the shortcut to obtain the inverse quickly.

$$= \begin{bmatrix} \dfrac{7}{13} & \dfrac{5}{13} \\ \dfrac{3}{13} & \dfrac{4}{13} \end{bmatrix}$$

$$\begin{bmatrix} x \\ y \end{bmatrix} = \begin{bmatrix} \dfrac{7}{13} & \dfrac{5}{13} \\ \dfrac{3}{13} & \dfrac{4}{13} \end{bmatrix}\begin{bmatrix} 3 \\ 1 \end{bmatrix}$$

It is a good idea to check your work at this stage by making sure that the product of the original matrix and its inverse is the identity matrix.

$$= \begin{bmatrix} \dfrac{21}{13}+\dfrac{5}{13} \\ \dfrac{9}{13}+\dfrac{4}{13} \end{bmatrix}$$

Rewrite the equation $AX = B$ in the form $X = A^{-1}B$ and carry out the matrix multiplication.

$$= \begin{bmatrix} 2 \\ 1 \end{bmatrix}$$

Thus, the solution to the system is $(2,1)$.

b. $\begin{cases} -x+2y=3 \\ 3x-6y=-5 \end{cases}$

$$\begin{bmatrix} -1 & 2 \\ 3 & -6 \end{bmatrix}\begin{bmatrix} x \\ y \end{bmatrix} = \begin{bmatrix} 3 \\ -5 \end{bmatrix}$$

Again, write the system in matrix form.

$$\begin{bmatrix} -1 & 2 \\ 3 & -6 \end{bmatrix}^{-1} = \frac{1}{6-6}\begin{bmatrix} -6 & -2 \\ -3 & -1 \end{bmatrix}$$

The coefficient matrix has no inverse (since its determinant is 0), so the system has no solution or an infinite number of solutions.

$$D_x = \begin{vmatrix} 3 & 2 \\ -5 & -6 \end{vmatrix} = -18-(-10) \neq 0$$

Since $D_x \neq 0$, Cramer's rule tells us the system has no solution.

This system has no solution.

We have now reached the end of our list of solution methods for linear systems. One important question has not yet been addressed. How should we choose a method of solution, given a system of equations?

Unfortunately, there is no simple answer. If the system is small, or looks especially simple, then the methods of substitution or elimination may be the quickest route to a solution. Generally, the larger and/or the more complicated the system, the more likely we are to prefer a matrix method of solution, such as Gauss-Jordan elimination, Cramer's rule, or the inverse matrix method. If you find that your first choice leads to a computational headache, don't be reluctant to stop and try another method.

TOPIC T

Inverting Matrices

We can also use a graphing calculator to find the inverse of a matrix by first defining the matrix whose inverse we want to find. Then, enter the matrix and press [x⁻¹]. To show the answer in fraction form, press **MATH** and select 1:Frac. If we defined matrix A to be $\begin{bmatrix} 7 & 4 \\ 1 & 2 \end{bmatrix}$, we would find the following to be its inverse:

```
[A]⁻¹
        [[.2  -.4]
         [-.1  .7]]
Ans▶Frac
        [[1/5   -2/5]
         [-1/10 7/10]]
```

Exercises

1. $\begin{bmatrix} 14 & -5 \\ 1 & 9 \end{bmatrix}\begin{bmatrix} x \\ y \end{bmatrix} = \begin{bmatrix} 7 \\ 2 \end{bmatrix}$

2. $\begin{bmatrix} 1 & -9 \\ -2 & 3 \end{bmatrix}\begin{bmatrix} x \\ y \end{bmatrix} = \begin{bmatrix} 5 \\ 8 \end{bmatrix}$

3.
$\begin{bmatrix} 1 & 2 \\ 9 & -3 \end{bmatrix}\begin{bmatrix} x \\ y \end{bmatrix} = \begin{bmatrix} -6 \\ -14 \end{bmatrix}$

4.
$\begin{bmatrix} 1 & -1 & 0 \\ -1 & 0 & -1 \\ 0 & -3 & 1 \end{bmatrix}\begin{bmatrix} x \\ y \\ z \end{bmatrix} = \begin{bmatrix} 5 \\ -2 \\ 4 \end{bmatrix}$

5.
$\begin{bmatrix} 3 & -7 & 1 \\ 1 & -1 & 0 \\ 0 & 8 & 5 \end{bmatrix}\begin{bmatrix} x_1 \\ x_2 \\ x_3 \end{bmatrix} = \begin{bmatrix} -4 \\ 2 \\ -3 \end{bmatrix}$

6.
$\begin{bmatrix} 0 & -1 & 1 \\ -1 & 1 & 0 \\ 1 & 0 & -1 \end{bmatrix}\begin{bmatrix} x_1 \\ x_2 \\ x_3 \end{bmatrix} = \begin{bmatrix} 0 \\ 0 \\ 0 \end{bmatrix}$

7. $\begin{bmatrix} \dfrac{3}{5} & -\dfrac{8}{5} \\ 0 & 1 \end{bmatrix}\begin{bmatrix} x \\ y \end{bmatrix} = \begin{bmatrix} 2 \\ 2 \end{bmatrix}$

8.
$\begin{bmatrix} 1 & -7 \\ 1 & -6 \end{bmatrix}\begin{bmatrix} x \\ y \end{bmatrix} = \begin{bmatrix} 5 \\ -10 \end{bmatrix}$

Write each of the following systems of equations as a single matrix equation. See Example 1.

1. $\begin{cases} 14x - 5y = 7 \\ x + 9y = 2 \end{cases}$

2. $\begin{cases} x - 5 = 9y \\ 3y - 2x = 8 \end{cases}$

3. $\begin{cases} -6 - 2y = x \\ 9x + 14 = 3y \end{cases}$

4. $\begin{cases} x - y = 5 \\ 2 - z = x \\ z - 3y = 4 \end{cases}$

5. $\begin{cases} 3x_1 - 7x_2 + x_3 = -4 \\ x_1 - x_2 = 2 \\ 8x_2 + 5x_3 = -3 \end{cases}$

6. $\begin{cases} x_3 = x_2 \\ x_2 = x_1 \\ x_1 = x_3 \end{cases}$

7. $\begin{cases} \dfrac{3x - 8y}{5} = 2 \\ y - 2 = 0 \end{cases}$

8. $\begin{cases} x - 7y = 5 \\ \dfrac{6 + x}{2} = 3y - 2 \end{cases}$

9. $\begin{cases} 4x = 3y - 9 \\ 13 - 2x = -4y \end{cases}$

10. $\begin{cases} -\dfrac{7}{3}y = \dfrac{5 - x}{6} \\ x - 5(y - 3) = -2 \end{cases}$

11. $\begin{cases} 2x - y = -3z \\ y - x = 17 \\ 2 + z + 4x = 5y \end{cases}$

12. $\begin{cases} 2x_1 - 3x_3 = 7 \\ x_2 - 10x_3 = 0 \\ 2x_1 - x_2 + x_3 = 1 \end{cases}$

Find the inverse of each of the following matrices, if possible. See Examples 2 and 3.

13. $\begin{bmatrix} 0 & 4 \\ -5 & -1 \end{bmatrix}$

14. $\begin{bmatrix} -2 & -2 \\ -1 & 2 \end{bmatrix}$

15. $\begin{bmatrix} 3 & 4 \\ -4 & -5 \end{bmatrix}$

16. $\begin{bmatrix} -1 & -1 \\ -\dfrac{1}{4} & \dfrac{1}{2} \end{bmatrix}$

17. $\begin{bmatrix} -\dfrac{1}{5} & 0 \\ \dfrac{1}{5} & \dfrac{1}{2} \end{bmatrix}$

18. $\begin{bmatrix} -7 & 2 \\ 7 & -2 \end{bmatrix}$

9.
$$\begin{bmatrix} 4 & -3 \\ 2 & -4 \end{bmatrix} \begin{bmatrix} x \\ y \end{bmatrix} = \begin{bmatrix} -9 \\ 13 \end{bmatrix}$$

10.
$$\begin{bmatrix} 1 & -14 \\ 1 & -5 \end{bmatrix} \begin{bmatrix} x \\ y \end{bmatrix} = \begin{bmatrix} 5 \\ -17 \end{bmatrix}$$

11.
$$\begin{bmatrix} 2 & -1 & 3 \\ -1 & 1 & 0 \\ 4 & -5 & 1 \end{bmatrix} \begin{bmatrix} x \\ y \\ z \end{bmatrix} = \begin{bmatrix} 0 \\ 17 \\ -2 \end{bmatrix}$$

12.
$$\begin{bmatrix} 2 & 0 & -3 \\ 0 & 1 & -10 \\ 2 & -1 & 1 \end{bmatrix} \begin{bmatrix} x_1 \\ x_2 \\ x_3 \end{bmatrix} = \begin{bmatrix} 7 \\ 0 \\ 1 \end{bmatrix}$$

13. $\begin{bmatrix} -\dfrac{1}{20} & -\dfrac{1}{5} \\ \dfrac{1}{4} & 0 \end{bmatrix}$

14. $\begin{bmatrix} -\dfrac{1}{3} & -\dfrac{1}{3} \\ -\dfrac{1}{6} & \dfrac{1}{3} \end{bmatrix}$

15. $\begin{bmatrix} -5 & -4 \\ 4 & 3 \end{bmatrix}$

16. $\begin{bmatrix} -2 & 4 \\ 1 & -4 \end{bmatrix}$

17. $\begin{bmatrix} -5 & 0 \\ 2 & 2 \end{bmatrix}$

18. Not invertible

19. Not invertible

20.
$$\begin{bmatrix} \dfrac{1}{3} & \dfrac{2}{3} & \dfrac{2}{3} \\ \dfrac{4}{15} & \dfrac{1}{3} & \dfrac{2}{15} \\ -\dfrac{1}{2} & -\dfrac{1}{2} & -\dfrac{1}{2} \end{bmatrix}$$

19. $\begin{bmatrix} -2 & -4 & -2 \\ 1 & -4 & 1 \\ 4 & -3 & 4 \end{bmatrix}$

20. $\begin{bmatrix} -3 & 0 & -4 \\ 2 & 5 & 4 \\ 1 & -5 & -2 \end{bmatrix}$

21. $\begin{bmatrix} -\dfrac{5}{11} & -\dfrac{8}{11} & 1 \\ \dfrac{13}{11} & \dfrac{12}{11} & -2 \\ -\dfrac{2}{11} & -\dfrac{1}{11} & 0 \end{bmatrix}$

22. $-\dfrac{1}{31}\begin{bmatrix} 17 & -8 & -2 \\ 1 & 5 & 9 \\ -6 & 1 & 8 \end{bmatrix}$

23. $\begin{bmatrix} -1 & 2 & -1 \\ 0 & 3 & -1 \\ 0 & 4 & -1 \end{bmatrix}$

24. $\begin{bmatrix} -1 & 0 & -1 \\ -\dfrac{3}{2} & \dfrac{1}{2} & -\dfrac{3}{2} \\ -\dfrac{1}{2} & 0 & -\dfrac{1}{4} \end{bmatrix}$

25. $\begin{bmatrix} -\dfrac{6}{5} & -\dfrac{2}{5} & -1 \\ \dfrac{3}{5} & \dfrac{1}{5} & 1 \\ 1 & 0 & 1 \end{bmatrix}$

26. $\begin{bmatrix} 2 & -2 & 1 \\ -2 & 2 & -3 \\ 1 & 0 & 2 \end{bmatrix}$

27. $\begin{bmatrix} 0 & 1 & 1 \\ 1 & 1 & 0 \\ 0 & 1 & 2 \end{bmatrix}$

28. $\begin{bmatrix} 9 & 8 & 7 \\ 6 & 5 & 4 \\ 3 & 2 & 1 \end{bmatrix}$

29. $\begin{bmatrix} \dfrac{2}{3} & \dfrac{8}{9} & \dfrac{1}{9} \\ -\dfrac{1}{3} & \dfrac{2}{9} & -\dfrac{2}{9} \\ -\dfrac{1}{3} & -\dfrac{7}{9} & -\dfrac{2}{9} \end{bmatrix}$

30. $\begin{bmatrix} -3 & -3 & -4 \\ 0 & \dfrac{1}{4} & \dfrac{1}{2} \\ 2 & 2 & 3 \end{bmatrix}$

For each set of matrices, determine if either matrix is the inverse of the other.

31. $\begin{bmatrix} -5 & -2 \\ -7 & 4 \end{bmatrix}, \begin{bmatrix} 10 & 4 \\ 14 & -8 \end{bmatrix}$

32. $\begin{bmatrix} 9 & -18 \\ 3 & 12 \end{bmatrix}, \begin{bmatrix} -3 & -6 \\ -1 & -4 \end{bmatrix}$

33. $\begin{bmatrix} -6 & -1 & 1 \\ 4 & -1 & -2 \\ 1 & -1 & -1 \end{bmatrix}, \begin{bmatrix} -1 & -2 & 3 \\ 2 & 5 & -8 \\ -3 & -7 & 10 \end{bmatrix}$

34. $\begin{bmatrix} -1 & 4 & 5 \\ 3 & -11 & -17 \\ 4 & -17 & -19 \end{bmatrix}, \begin{bmatrix} -80 & -9 & -13 \\ -11 & -1 & -2 \\ -7 & -1 & -1 \end{bmatrix}$

35. $\begin{bmatrix} 2 & 0 & -1 \\ 3 & 4 & 2 \\ 1 & 1 & -3 \end{bmatrix}, \begin{bmatrix} 4 & 0 & -2 \\ 6 & 8 & 4 \\ 2 & 2 & -6 \end{bmatrix}$

36. $\begin{bmatrix} -7 & 0 & -2 \\ -10 & -1 & -2 \\ -7 & -1 & -1 \end{bmatrix}, \begin{bmatrix} -1 & 2 & -2 \\ 4 & -7 & 6 \\ 3 & -7 & 7 \end{bmatrix}$

21. $\begin{bmatrix} 2 & 1 & -4 \\ -4 & -2 & -3 \\ -1 & -1 & -4 \end{bmatrix}$

22. $\begin{bmatrix} -1 & -2 & 2 \\ 2 & -4 & 5 \\ -1 & -1 & -3 \end{bmatrix}$

23. $\begin{bmatrix} -1 & 2 & -1 \\ 0 & -1 & 1 \\ 0 & -4 & 3 \end{bmatrix}$

24. $\begin{bmatrix} 1 & 0 & -4 \\ -3 & 2 & 0 \\ -2 & 0 & 4 \end{bmatrix}$

25. $\begin{bmatrix} -1 & -2 & 1 \\ -2 & 1 & -3 \\ 1 & 2 & 0 \end{bmatrix}$

26. $\begin{bmatrix} 1 & 1 & 1 \\ \dfrac{1}{4} & \dfrac{3}{4} & 1 \\ -\dfrac{1}{2} & -\dfrac{1}{2} & 0 \end{bmatrix}$

27. $\begin{bmatrix} -2 & 1 & 1 \\ 2 & 0 & -1 \\ -1 & 0 & 1 \end{bmatrix}$

28. Not invertible

29. $\begin{bmatrix} 2 & -1 & 2 \\ 0 & 1 & -1 \\ -3 & -2 & -4 \end{bmatrix}$

30. $\begin{bmatrix} 1 & -4 & 2 \\ -4 & 4 & -6 \\ 2 & 0 & 3 \end{bmatrix}$

31. No

32. No

33. Yes

34. Yes

35. No

36. Yes

37. $\begin{bmatrix} \dfrac{-2}{11} & \dfrac{3}{11} \\ \dfrac{-1}{11} & \dfrac{7}{11} \end{bmatrix}$

Using a graphing calculator, find the inverse of each of the following matrices, if possible. Round answers to the nearest thousandth, when necessary.

37. $\begin{bmatrix} -7 & 3 \\ -1 & 2 \end{bmatrix}$

38. $\begin{bmatrix} -6 & 2 \\ -5 & 5 \end{bmatrix}$

39. $\begin{bmatrix} -2 & 0 & 2 \\ 2 & -3 & 1 \\ 1 & -2 & 3 \end{bmatrix}$

40. $\begin{bmatrix} 2.3 & 7.8 \\ -3.4 & 1.6 \end{bmatrix}$

41. $\begin{bmatrix} 4.5 & -9.4 & 6.9 \\ 8.6 & -2.8 & 1.2 \\ 3.1 & 0.3 & -7.0 \end{bmatrix}$

42. $\begin{bmatrix} 38 & -44 & 72 \\ -93 & 16 & 29 \\ 65 & 23 & -19 \end{bmatrix}$

Solve the following systems by the inverse matrix method, if possible. If the inverse matrix method doesn't apply, use any other method to determine if the system is inconsistent or dependent. See Example 4.

43. $\begin{cases} -2x - 2y = 9 \\ -x + 2y = -3 \end{cases}$

44. $\begin{cases} 3x + 4y = -2 \\ -4x - 5y = 9 \end{cases}$

45. $\begin{cases} -2x + 3y = 1 \\ 4x - 6y = -2 \end{cases}$

46. $\begin{cases} -2x + 4y = 5 \\ x - 4y = -3 \end{cases}$

47. $\begin{cases} -5x = 10 \\ 2x + 2y = -4 \end{cases}$

48. $\begin{cases} -3x + y = 2 \\ 9x - 3y = 5 \end{cases}$

49. $\begin{cases} 8x + 2y = 26 \\ -16x - 2y = -90 \end{cases}$

50. $\begin{cases} 3x - 7y = -2 \\ -6x + 14y = 4 \end{cases}$

51. $\begin{cases} 3y = 15 \\ 8x + 4y = 20 \end{cases}$

52. $\begin{cases} 4y + 3z = -254 \\ 2x - 2y - z = 100 \\ -x + y - 2z = 155 \end{cases}$

53. $\begin{cases} 2x - y - 3z = -10 \\ 2y - z = 11 \\ -x + 4z = 0 \end{cases}$

54. $\begin{cases} 3y - 4z = 15 \\ x + 2y - 3z = 9 \\ -x - y + 2z = -5 \end{cases}$

38. $\begin{bmatrix} \dfrac{-1}{4} & \dfrac{1}{10} \\ \dfrac{-1}{4} & \dfrac{3}{10} \end{bmatrix}$

39. $\begin{bmatrix} \dfrac{-7}{12} & \dfrac{-1}{3} & \dfrac{1}{2} \\ \dfrac{-5}{12} & \dfrac{-2}{3} & \dfrac{1}{2} \\ \dfrac{-1}{12} & \dfrac{-1}{3} & \dfrac{1}{2} \end{bmatrix}$

40. $\begin{bmatrix} 0.053 & -0.258 \\ 0.113 & 0.076 \end{bmatrix}$

41. $\begin{bmatrix} -0.044 & 0.146 & -0.018 \\ -0.146 & 0.121 & -0.124 \\ -0.026 & 0.070 & -0.156 \end{bmatrix}$

42. $\begin{bmatrix} 0.004 & -0.003 & 0.009 \\ 0 & 0.020 & 0.029 \\ 0.012 & 0.014 & 0.013 \end{bmatrix}$

43. $\left(-2, -\dfrac{5}{2}\right)$

44. $(-26, 19)$

45. $\left\{ \left(\dfrac{3y - 1}{2}, y \right) \middle| y \in \mathbb{R} \right\}$

46. $\left(-2, \dfrac{1}{4}\right)$

47. $(-2, 0)$

48. $\varnothing$

49. $(8, -19)$

50. $\left\{ \left(\dfrac{7y - 2}{3}, y \right) \middle| y \in \mathbb{R} \right\}$

51. $(0, 5)$

52. $(7, -2, -82)$

53. $(-4, 5, -1)$

54. $(-2, 1, -3)$

Partial Fraction Decomposition

TOPICS

1. The pattern of decompositions
2. Completing the decomposition process

TOPIC 1

The Pattern of Decompositions

Throughout this text, we have frequently found it useful or necessary to combine fractions. You have done this so often by now, and for such a variety of reasons, that you may not even be consciously aware of the process—the act of finding a common denominator and combining fractions may be second nature. There are occasions, however, when it is helpful to be able to reverse the process. In performing certain operations in calculus, for instance, the ability to write a single fraction as a sum of simpler fractions comes in very handy. The process of doing so is called **partial fraction decomposition**, and we will find the methods of this chapter useful in the execution of the process.

To be specific, the fractions we will want to decompose are proper rational functions; that is, fractions of the form

$$f(x) = \frac{p(x)}{q(x)}$$

where p and q are polynomials and the degree of p is less than the degree of q (recall that we already know how to perform polynomial division on fractions where the degree of p is greater than or equal to the degree of q). As a consequence of the Fundamental Theorem of Algebra and the Conjugate Roots Theorem (Chapter 4), if $q(x)$ has only real coefficients then it can be written as a product of factors of the form $(ax+b)^m$ and $(ax^2+bx+c)^n$, where m and n are positive integers, a, b, and c are real numbers, and ax^2+bx+c cannot be factored further without resorting to complex coefficients (we say ax^2+bx+c is irreducible). The appearance of such factors tells us how the rational function can be decomposed, as outlined on the next page.

DEFINITION

Decomposition Pattern

Given the proper rational function $f(x) = \dfrac{p(x)}{q(x)}$, assume $q(x)$ has been completely factored as a product of factors of the form $(ax+b)^m$ and $\left(ax^2 + bx + c\right)^n$, where a, b, and c are real numbers, $ax^2 + bx + c$ is irreducible over the real numbers, and m and n are positive integers. Then $f(x)$ can be decomposed as a sum of simpler rational functions, where

1. Each factor of the form $(ax+b)^m$ leads to a sum of the form

$$\frac{A_1}{ax+b} + \frac{A_2}{(ax+b)^2} + \cdots + \frac{A_m}{(ax+b)^m}$$

2. Each factor of the form $\left(ax^2 + bx + c\right)^n$ leads to a sum of the form

$$\frac{A_1 x + B_1}{ax^2 + bx + c} + \frac{A_2 x + B_2}{\left(ax^2 + bx + c\right)^2} + \cdots + \frac{A_n x + B_n}{\left(ax^2 + bx + c\right)^n}$$

EXAMPLE 1

Finding the Partial Fraction Decomposition of a Function

Write the form of the partial fraction decomposition of the rational function

$$f(x) = \frac{p(x)}{x^3 + 6x^2 + 12x + 8}.$$

Assume that the degree of p is 2 or smaller.

Solution:

The primary task in this problem is to factor the denominator. The Rational Zero Theorem (Section 4.3) tells us that the potential rational zeros of the denominator are $\pm\{1, 2, 4, 8\}$ (remember that the potential rational zeros of a polynomial are the factors of the constant term divided by the factors of the leading coefficient). Synthetic division or long division can then be used to test these potential zeros one by one; the work following uses synthetic division to show that -2 is a zero.

$$\begin{array}{r|rrrr} -2 & 1 & 6 & 12 & 8 \\ & & -2 & -8 & -8 \\ \hline & 1 & 4 & 4 & 0 \end{array}$$

From this, we conclude that $x^3 + 6x^2 + 12x + 8 = (x+2)(x^2 + 4x + 4) = (x+2)^3$. Following the guidelines of the decomposition pattern, we now know that $f(x)$ can be written as a sum of rational functions of the form

$$f(x) = \frac{A_1}{x+2} + \frac{A_2}{(x+2)^2} + \frac{A_3}{(x+2)^3}.$$

Of course, we can go no further with the problem above since we were not given a specific polynomial in the numerator. Note, though, that the numerator played no role in our work—any rational function with the denominator $(x+2)^3$ can be decomposed as in Example 1, as long as the degree of the numerator is 2 or smaller.

EXAMPLE 2

Finding the Partial Fraction Decomposition of a Function

Write the form of the partial fraction decomposition of the rational function

$$f(x) = \frac{p(x)}{(x-5)^2(x^2+4)^2}.$$

Assume that the degree of p is 5 or smaller.

Solution:

In this example, the denominator is already appropriately factored; while it is true that $x^2 + 4$ can be factored as $(x - 2i)(x + 2i)$, partial fraction decomposition calls for leaving irreducible quadratics such as $x^2 + 4$ in their unfactored state. All that remains is following the decomposition guidelines to write $f(x)$ as:

$$f(x) = \frac{A_1}{x-5} + \frac{A_2}{(x-5)^2} + \frac{B_1x + C_1}{x^2+4} + \frac{B_2x + C_2}{(x^2+4)^2}.$$

Note the choice of letters in the numerators of the decomposition. The names of the unknown coefficients do not matter, of course, but it is important to realize that there are a total of six such coefficients and so six different symbols must be used to denote them.

TOPIC 2

Completing the Decomposition Process

Assuming the degree of the numerator is smaller than the degree of the denominator, the numerator plays no role in determining the form of the partial fraction decomposition of a given rational function. But of course it must be considered when we need to actually solve for the unknown coefficients appearing in the decomposition. It is at this stage that one of the methods of solving systems of equations may be called for.

EXAMPLE 3

Finding the Partial Fraction Decomposition of a Function

Find the partial fraction decomposition of the rational function

$$f(x) = \frac{-2x+14}{x^2+2x-3}.$$

Solution:

Since $x^2 + 2x - 3 = (x+3)(x-1)$, we know

$$\frac{-2x+14}{x^2+2x-3} = \frac{A_1}{x+3} + \frac{A_2}{x-1}.$$

There are many ways we can go about solving for A_1 and A_2, but most begin with the step of eliminating the fractions. Multiplying through by $(x+3)(x-1)$ leads immediately, upon canceling common factors, to the equation

$$-2x+14 = A_1(x-1) + A_2(x+3).$$

This particular partial fraction decomposition can be accomplished simply—no advanced methods of solving systems of equations will be necessary. The key observation is that the equation above is true for *all* values of x; thus, we can substitute well-chosen values for x into the equation and quickly solve for the two coefficients. The values of x that are useful are those that make either $x - 1$ or $x + 3$ zero; that is, we will first let $x = 1$ and then let $x = -3$:

$$\text{For } x = 1: \ -2(1)+14 = A_1(1-1) + A_2(1+3)$$

$$12 = 4A_2$$

$$A_2 = 3$$

and

$$\text{For } x = -3: \ -2(-3)+14 = A_1(-3-1) + A_2(-3+3)$$

$$20 = -4A_1$$

$$A_1 = -5.$$

Thus, $\dfrac{-2x+14}{x^2+2x-3} = \dfrac{-5}{x+3} + \dfrac{3}{x-1}.$

Our answer can be checked by combining these two fractions.

The method shown in Example 3 will not always be sufficient, by itself, to solve for the unknown coefficients. Consider the next example.

EXAMPLE 4

Finding the Partial Fraction Decomposition of a Function

Find the partial fraction decomposition of the rational function

$$f(x) = \frac{3x+1}{(x+2)^3}.$$

Solution:

We already considered rational functions with this particular denominator in Example 1, and we know that

$$\frac{3x+1}{(x+2)^3} = \frac{A_1}{x+2} + \frac{A_2}{(x+2)^2} + \frac{A_3}{(x+2)^3}.$$

Eliminating fractions is again a good first step:

$$3x+1 = A_1(x+2)^2 + A_2(x+2) + A_3.$$

The obvious convenient value for x is -2, and making this substitution allows us to quickly solve for A_3:

$$3(-2)+1 = A_1(-2+2)^2 + A_2(-2+2) + A_3$$
$$-5 = A_3.$$

However, there are no other values for x that allow us to immediately determine A_1 and A_2. There are still several other ways we can complete the decomposition, though. One is to simply substitute two other values for x and so obtain two equations in two unknowns. Since we now know that $A_3 = -5$, we will make the substitutions for x in the equation

$$3x+6 = A_1(x+2)^2 + A_2(x+2).$$

In the work below, we use $x = 0$ and $x = -1$, with these values chosen simply because they make the calculations easy:

$$3(0)+6 = A_1(0+2)^2 + A_2(0+2)$$
$$6 = 4A_1 + 2A_2$$

and

$$3(-1)+6 = A_1(-1+2)^2 + A_2(-1+2)$$
$$3 = A_1 + A_2.$$

We recognize this as a linear system in the two variables A_1 and A_2, and we can use any of the methods of this chapter to solve it. Below, we have used the method of elimination.

$$\begin{cases} 4A_1 + 2A_2 = 6 \\ A_1 + A_2 = 3 \end{cases} \xrightarrow{-2E_2} \begin{cases} 4A_1 + 2A_2 = 6 \\ -2A_1 - 2A_2 = -6 \end{cases}$$
$$2A_1 \qquad = 0$$

So $A_1 = 0$ and hence $A_2 = 3$. This gives us the partial fraction decomposition

$$\frac{3x+1}{(x+2)^3} = \frac{3}{(x+2)^2} + \frac{-5}{(x+2)^3}.$$

There is another way of thinking about the decomposition process that is very important conceptually. After eliminating fractions, the decomposition in Example 4 has the form

$$3x + 1 = A_1 (x+2)^2 + A_2 (x+2) + A_3.$$

This is an equation relating two polynomials, a fact that is explicitly clear if we multiply out the right-hand side and collect powers of x:

$$3x + 1 = A_1 x^2 + (4A_1 + A_2) x + (4A_1 + 2A_2 + A_3)$$

As we know, two polynomials are equal only if corresponding coefficients are equal. Equating the coefficients in this example gives us the following three equations.

$$\begin{cases} 0 = A_1 \\ 3 = 4A_1 + A_2 \\ 1 = 4A_1 + 2A_2 + A_3 \end{cases} \qquad \text{Note that there is no } x^2 \text{ term on the left.}$$

This is a linear system of three equations in three variables, and again any method from this chapter could be used to solve it. The verification that the solution of this system leads to the answer in Example 4 is left to the reader.

We conclude this section with an example that calls for the application of several skills learned in this text.

EXAMPLE 5

Finding the Partial Fraction Decomposition of a Function

Find the partial fraction decomposition of the rational function

$$f(x) = \frac{3x^5 + 6x^4 + 9x^3 + 7x^2 + 4x - 1}{(x^2 + x + 1)^2}.$$

Solution:

First, note that this rational function is improper — the degree of the numerator is greater than the degree of the denominator, so we must perform some polynomial division before the partial fraction decomposition. To do so, we first expand the denominator:

$$\left(x^2 + x + 1\right)^2 = x^4 + 2x^3 + 3x^2 + 2x + 1.$$

Polynomial long division (Section 4.2) then gives us the result

$$\frac{3x^5 + 6x^4 + 9x^3 + 7x^2 + 4x - 1}{\left(x^2 + x + 1\right)^2} = 3x + \frac{x^2 + x - 1}{\left(x^2 + x + 1\right)^2},$$

and it is the fractional part of this that we will decompose.

Next, note that $x^2 + x + 1$ is irreducible. This follows from the use of the quadratic formula to solve the equation $x^2 + x + 1 = 0$; since the solutions of this equation are complex numbers, the linear factors of $x^2 + x + 1$ contain complex coefficients. The partial fraction decomposition thus has the form

$$\frac{x^2 + x - 1}{\left(x^2 + x + 1\right)^2} = \frac{Ax + B}{x^2 + x + 1} + \frac{Cx + D}{\left(x^2 + x + 1\right)^2}.$$

(For the sake of variety, and to avoid the necessity for subscripts, we will use different letters to stand for the unknown coefficients.)

Clearing the equation of fractions gives us

$$x^2 + x - 1 = \left(Ax + B\right)\left(x^2 + x + 1\right) + \left(Cx + D\right)$$

and hence

$$x^2 + x - 1 = Ax^3 + \left(A + B\right)x^2 + \left(A + B + C\right)x + \left(B + D\right).$$

From this polynomial equation we derive the following system.

$$\begin{cases} 0 = A \\ 1 = A + B \\ 1 = A + B + C \\ -1 = B + D \end{cases}$$

This system is easily solved, considering the equations in the order presented, and gives us $A = 0$, $B = 1$, $C = 0$, and $D = -2$. So our final answer is

$$\frac{3x^5 + 6x^4 + 9x^3 + 7x^2 + 4x - 1}{\left(x^2 + x + 1\right)^2} = 3x + \frac{1}{x^2 + x + 1} + \frac{-2}{\left(x^2 + x + 1\right)^2}.$$

1. $\dfrac{A_1}{x-3}+\dfrac{A_2}{x+2}$

2. $\dfrac{A_1}{x-6}+\dfrac{A_2}{x+4}$

3. $\dfrac{A_1}{x+3}+\dfrac{A_2}{x+4}$
$+\dfrac{A_3}{(x+4)^2}$

4. $\dfrac{A_1}{x+5}+\dfrac{A_2}{(x+5)^2}+$
$\dfrac{B_1 x+C_1}{x^2+3}+\dfrac{B_2 x+C_2}{(x^2+3)^2}$

5. $\dfrac{A_1}{x+3}+\dfrac{A_2}{x-2}+\dfrac{A_3}{x+2}$

6. $\dfrac{A_1}{x+4}+\dfrac{A_2}{x-1}$
$+\dfrac{B_1 x+C_1}{x^2+5}$

7. d   **8.** b

9. h   **10.** e

11. a   **12.** f

13. c   **14.** g

15. $\dfrac{-1}{x}+\dfrac{2}{x-2}+\dfrac{2}{x+2}$

16. $\dfrac{-2}{3(x+6)}+\dfrac{4}{7(x+4)}$
$+\dfrac{2}{21(x-3)}$

17. $\dfrac{5}{72(x-2)}-\dfrac{1}{32x}$
$-\dfrac{7}{288(x+4)}$
$-\dfrac{x+17}{72(x^2+8)}$

18. $\dfrac{1}{x-1}-\dfrac{1}{x+4}$

Exercises

Write the form of the partial fraction decomposition of each of the following rational functions. In each case, assume the degree of the numerator is less than the degree of the denominator. See Examples 1 and 2.

1. $f(x)=\dfrac{p(x)}{x^2-x-6}$ **2.** $f(x)=\dfrac{p(x)}{x^2-2x-24}$

3. $f(x)=\dfrac{p(x)}{x^3+11x^2+40x+48}$ **4.** $f(x)=\dfrac{p(x)}{(x+5)^2(x^2+3)^2}$

5. $f(x)=\dfrac{p(x)}{(x+3)(x^2-4)}$ **6.** $f(x)=\dfrac{p(x)}{(x^2+5)(x^2+3x-4)}$

Match each rational expression with the form of its decomposition. The decompositions are labeled a.–h.

7. $\dfrac{2x-1}{(x+2)^3(x-2)}$ **8.** $\dfrac{2x-1}{x^2(x^2-4)}$ **9.** $\dfrac{2x-1}{x^3-4x^2+4x}$

10. $\dfrac{2x-1}{x^4-16}$ **11.** $\dfrac{2x-1}{x^3(x-2)^2}$ **12.** $\dfrac{2x-1}{x^5+2x^4+4x^3+8x^2}$

13. $\dfrac{2x-1}{x^3(x-2)}$ **14.** $\dfrac{2x-1}{x^5+6x^4+12x^3+8x^2}$

a. $\dfrac{A}{x}+\dfrac{B}{x^2}+\dfrac{C}{x^3}+\dfrac{D}{x-2}+\dfrac{E}{(x-2)^2}$

b. $\dfrac{A}{x}+\dfrac{B}{x^2}+\dfrac{C}{x+2}+\dfrac{D}{x-2}$

c. $\dfrac{A}{x}+\dfrac{B}{x^2}+\dfrac{C}{x^3}+\dfrac{D}{x-2}$

d. $\dfrac{A}{x+2}+\dfrac{B}{(x+2)^2}+\dfrac{C}{(x+2)^3}+\dfrac{D}{x-2}$

e. $\dfrac{Ax+B}{x^2+4}+\dfrac{C}{x+2}+\dfrac{D}{x-2}$

f. $\dfrac{A}{x}+\dfrac{B}{x^2}+\dfrac{C}{x+2}+\dfrac{Dx+E}{x^2+4}$

g. $\dfrac{A}{x+2}+\dfrac{B}{(x+2)^2}+\dfrac{C}{(x+2)^3}+\dfrac{D}{x}+\dfrac{E}{x^2}$

h. $\dfrac{A}{x}+\dfrac{B}{x-2}+\dfrac{C}{(x-2)^2}$

19. $\dfrac{10}{x-4} - \dfrac{5}{x-2}$

20. $\dfrac{121x}{546\left(x^2+3\right)}$

$-\dfrac{99}{182\left(x^2+3\right)}$

$-\dfrac{212}{39(x+6)}$

$+\dfrac{73}{14(x+5)}$

21. $\dfrac{3}{2(x+3)} + \dfrac{3}{14(x-1)}$

$-\dfrac{12}{7(x+6)}$

22. $\dfrac{26}{5(x+3)}$

$+\dfrac{7}{15(x-2)}$

$-\dfrac{21}{4(x+5)}$

$-\dfrac{5}{12(x+1)}$

23. $\dfrac{1}{2(x-1)} - \dfrac{1}{2(x+1)}$

24. $\dfrac{1}{7(x-2)} - \dfrac{x}{7(x^2+3)}$

$-\dfrac{2}{7(x^2+3)}$

25. $\dfrac{1}{48(x-2)}$

$-\dfrac{x}{80(x^2+4)}$

$-\dfrac{3}{40(x^2+4)}$

$-\dfrac{1}{120(x+4)}$

26. $\dfrac{1}{x-1} - \dfrac{1}{x}$

27. $\dfrac{5}{4(x-2)} - \dfrac{1}{4(x+2)}$

Find the partial fraction decomposition of each of the following rational functions. See Examples 3, 4, and 5.

15. $f(x) = \dfrac{3x^2+4}{x^3-4x}$

16. $f(x) = \dfrac{2x}{x^3+7x^2-6x-72}$

17. $f(x) = \dfrac{4x+2}{(x^3+8x)(x^2+2x-8)}$

18. $f(x) = \dfrac{5}{x^2+3x-4}$

19. $f(x) = \dfrac{5x}{x^2-6x+8}$

20. $f(x) = \dfrac{6x^2-4}{(x^2+3)(x+6)(x+5)}$

21. $f(x) = \dfrac{6x}{x^3+8x^2+9x-18}$

22. $f(x) = \dfrac{12x^2+x-1}{x^4+7x^3+5x^2-31x-30}$

23. $f(x) = \dfrac{1}{x^2-1}$

24. $f(x) = \dfrac{x+3}{(x^2+3)(x^2+x-6)}$

25. $f(x) = \dfrac{x^2-4}{(x^4-16)(x^2+2x-8)}$

26. $f(x) = \dfrac{x+1}{x^3-x}$

27. $f(x) = \dfrac{x+3}{x^2-4}$

28. $f(x) = \dfrac{x}{x^3+6x^2+11x+6}$

29. $f(x) = \dfrac{x}{x^4-16}$

30. $f(x) = \dfrac{5}{x^2-6x+8}$

31. $f(x) = \dfrac{2x+3}{(x^2-9)(x^2+4x-12)}$

32. $f(x) = \dfrac{x}{x^2-7x+12}$

33. $f(x) = \dfrac{x^2}{x^3+5x^2+3x-9}$

34. $f(x) = \dfrac{2x}{(x+4)(x^2-2x-3)}$

35. $f(x) = \dfrac{2x}{x^2-9}$

36. $f(x) = \dfrac{4x+3}{(x^2-9)(x^2-2x-24)}$

37. $f(x) = \dfrac{x^2}{x^3+4x^2-12x}$

38. $f(x) = \dfrac{2}{x^3+7x^2-8x}$

28. $\dfrac{2}{x+2}-\dfrac{3}{2(x+3)}$

$-\dfrac{1}{2(x+1)}$

29. $\dfrac{1}{16(x+2)}+\dfrac{1}{16(x-2)}$

$-\dfrac{x}{8(x^2+4)}$

30. $\dfrac{5}{2(x-4)}-\dfrac{5}{2(x-2)}$

31. $\dfrac{1}{24(x+6)}-\dfrac{1}{30(x+3)}$

$-\dfrac{7}{40(x-2)}$

$+\dfrac{1}{6(x-3)}$

32. $\dfrac{4}{x-4}-\dfrac{3}{x-3}$

33. $\dfrac{15}{16(x+3)}+\dfrac{1}{16(x-1)}$

$-\dfrac{9}{4(x+3)^2}$

34. $\dfrac{1}{6(x+1)}+\dfrac{3}{14(x-3)}$

$-\dfrac{8}{21(x+4)}$

35. $\dfrac{1}{x+3}+\dfrac{1}{x-3}$

36. $\dfrac{13}{70(x+4)}-\dfrac{1}{6(x+3)}$

$-\dfrac{5}{42(x-3)}$

$+\dfrac{1}{10(x-6)}$

37. $\dfrac{1}{4(x-2)}+\dfrac{3}{4(x+6)}$

38. $\dfrac{1}{36(x+8)}+\dfrac{2}{9(x-1)}-\dfrac{1}{4x}$

Using your graphing calculator, determine whether the partial fraction decomposition is true or false by graphing the left and right side of the equation on the same coordinate plane.

39. $\dfrac{x+7}{x^2-x-6}=\dfrac{2}{x-3}-\dfrac{1}{x+2}$

40. $\dfrac{5x^2+20x+6}{x^3+2x^2+x}=\dfrac{6}{x}-\dfrac{1}{x+1}+\dfrac{9}{(x+1)^2}$

41. $\dfrac{4x^2-3x-4}{x^3+x^2-2x}=\dfrac{2}{x}-\dfrac{1}{x-1}+\dfrac{3}{x+2}$

42. $\dfrac{2x+4}{x^3-2x^2}=-\dfrac{2}{x}-\dfrac{2}{x^2}+\dfrac{2}{x-2}$

43. $\dfrac{1}{x^2+x-2}=\dfrac{1}{x-1}+\dfrac{1}{x+2}$

44. $\dfrac{6x^2+2}{x^2(x-3)^3}=\dfrac{4}{x}+\dfrac{2}{x^2}+\dfrac{3x}{x-3}+\dfrac{4}{(x-3)^2}+\dfrac{6x}{(x-3)^3}$

Write the partial fraction decomposition for the rational expression. Then assign a value to the constant, a, and check the result graphically.

45. $\dfrac{1}{x(x+a)}$

46. $\dfrac{1}{x(a-x)}$

47. $\dfrac{1}{a^2-x^2}$

48. $\dfrac{1}{(x+1)(a-x)}$

49. $\dfrac{1}{(x+a)(x+1)}$

39. True

40. True

41. True

42. True

43. False

44. False

45. $\dfrac{1}{a}\left(\dfrac{1}{x}-\dfrac{1}{x+a}\right)$

46. $\dfrac{1}{a}\left(\dfrac{1}{x}+\dfrac{1}{a-x}\right)$

47. $\dfrac{1}{2a}\left(\dfrac{1}{a+x}+\dfrac{1}{a-x}\right)$

48. $\dfrac{1}{a+1}\left(\dfrac{1}{x+1}+\dfrac{1}{a-x}\right)$

49. $\dfrac{1}{a-1}\left(\dfrac{1}{x+1}-\dfrac{1}{x+a}\right)$

Linear Programming

TOPICS

1. Planar feasible regions

2. Linear programming in two variables

TOPIC 1   Planar Feasible Regions

Many applications of mathematics require identifying the values that a set of variables is allowed to take. If an application imposes limitations on the values that two variables can assume, the set of all allowable values forms a portion of the plane called the **feasible region** or **region of constraint**. For instance, the constraints $1 \leq x \leq 5$ and $2 \leq y \leq 4$ define the rectangular feasible region shown below.

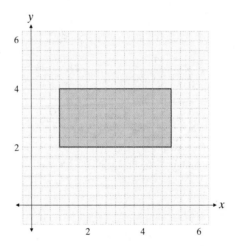

Figure 1: A Rectangular Feasible Region

More often, constraint inequalities involve both variables and the corresponding feasible region is a bit more interesting than a rectangle. The collection of all constraints in a problem comprises a system of linear *inequalities*, and the techniques learned in this chapter can be used to extract information about the feasible region described by such systems.

EXAMPLE 1

Feasible Regions

A family orchard sells peaches and nectarines. To prevent a pest infestation, the number of nectarine trees in the orchard cannot exceed the number of peach trees. Also, because of the space requirements of each type of tree, the number of nectarine trees plus twice the number of peach trees cannot exceed 100 trees. Construct the constraints and graph the feasible region for this situation.

Solution:

Let p be the number of peach trees and n be the number of nectarine trees. The constraints given in the problem can now be written in terms of p and n.

The second sentence tells us that $n \le p$, and the third sentence tells us that $n + 2p \le 100$. Since we can't have a negative number of trees, we also know that $n \ge 0$ and $p \ge 0$.

Letting the horizontal axis represent p and the vertical axis represent n, we can graph the feasible region defined by these constraints.

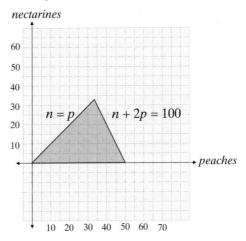

Any point (with integer coordinates) in the shaded feasible region corresponds to a certain number of peach trees and nectarine trees, and all such points meet the stated conditions.

TOPIC 2   Linear Programming in Two Variables

Often, given a feasible region, we want to find the point in the region such that the associated values of the variables maximize or minimize some quantity. Two important examples come from business: profit is a quantity that people usually wish to maximize, and cost is a quantity best kept to a minimum. When the quantity to be optimized (that is, either maximized or minimized) is a linear function of the variables concerned, the problem is known as a **linear programming** problem. Linear programming is used in a wide variety of applications, some involving thousands of variables, and a complete study of its techniques is beyond the scope of this text. However, the basic concepts are simple and easily applied, especially when working with applications involving just two variables.

DEFINITION

Bounded and
Unbounded Regions

A **bounded** region is one in which all the points lie within some finite distance of the origin. An **unbounded** region has points that are arbitrarily far away from the origin.

PROCEDURE

Linear Programming

Step 1: Identify the variables to be considered in the problem, determine all the constraints on the variables imposed by the problem, and sketch the feasible region described by the constraints.

Step 2: Determine the function that is to be either maximized or minimized. This function is called the **objective function**.

Step 3: Evaluate the function at each of the vertices of the feasible region and compare the values found. If the feasible region is bounded, the optimum value of the function will occur at a vertex. If the feasible region is unbounded, the optimum value of the function *may not exist*, but if it does exist, it will occur at a vertex.

EXAMPLE 2

Linear Programming

Find the maximum and minimum values of $f(x,y) = 3x + 2y$ subject to the constraints

$$\begin{cases} x \geq 0 \\ y \geq 0 \\ y \geq 3 - 3x \\ 2x + y \leq 10 \\ x + y \leq 7 \\ y \geq x - 2 \end{cases}$$

Solution:

The two variables are given to us, as are the constraints; all we must do to complete the first step is sketch the region defined by the constraints. The following graph contains the sketch, with selected boundaries identified by their corresponding equations.

The function to be optimized is also given to us in this problem, so step two is complete. According to the linear programming method, the only remaining task is to evaluate the function at the vertices of the feasible region. Why should we expect the maximum and minimum values of f to occur at vertices, and not in the interior of the region?

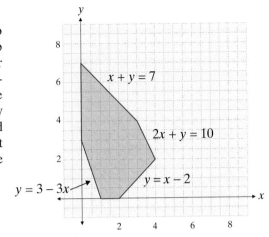

Another diagram will answer this question. The function $f(x, y)$ takes on a variety of values at different ordered pairs (x, y). But for any particular value, say k, there will be a line of ordered pairs in the plane such that $f(x, y) = k$. Such lines for different values of k will be parallel, as shown here. Two particular values for k are labeled in the diagram, and these values are the minimum and maximum values of f.

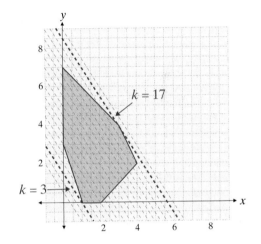

The extreme values will thus occur where a line of constant value *just touches* the region. This will be at a vertex or one of the edges of the region, if the edge is parallel to the line of constant value. Thus, evaluating the function at the vertices of the feasible region will suffice to identify the extreme values.

Each vertex of a feasible region is the intersection of two edges, so the coordinates of each vertex are found by solving a system of two equations. The six vertices of the region in this problem are determined by the following six systems:

$$\begin{cases} x + y = 7 \\ 2x + y = 10 \end{cases} \quad \begin{cases} 2x + y = 10 \\ y = x - 2 \end{cases} \quad \begin{cases} y = x - 2 \\ y = 0 \end{cases} \quad \begin{cases} y = 0 \\ y = 3 - 3x \end{cases} \quad \begin{cases} y = 3 - 3x \\ x = 0 \end{cases} \quad \begin{cases} x = 0 \\ x + y = 7 \end{cases}$$

Solving each of these systems, we find the vertices to be $(3, 4), (4, 2), (2, 0), (1, 0), (0, 3)$, and $(0, 7)$. Substituting each of these ordered pairs into $f(x, y) = 3x + 2y$, we find the maximum value of 17 occurs at the vertex $(3, 4)$, and the minimum value of 3 occurs at the vertex $(1, 0)$.

EXAMPLE 3

Linear Programming

Find the maximum and minimum values of $f(x,y) = x + 2y$ subject to the constraints

$$\begin{cases} x \geq 0 \\ y \geq 3 - x \\ x - y \leq 1 \\ y \leq 2x + 4 \end{cases}$$

Solution:

The feasible region defined by the constraints is unbounded; a portion of its sketch appears to the right.

Because the region is unbounded, the function f may not have a maximum or a minimum value over the region. In this case, f has no maximum value. One way to see this is to note that ordered pairs of the form (x, x) lie in the region for all $x \geq \dfrac{3}{2}$, and $f(x, x) = x + 2x = 3x$ grows without bound as x increases.

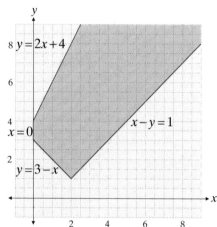

On the other hand, f does have a minimum value over the feasible region. The three vertices are $(0, 4)$, $(0, 3)$, and $(2, 1)$. Evaluating the function at these points, we obtain $f(0, 4) = 8, f(0, 3) = 6$, and $f(2, 1) = 4$, so the minimum value is 4.

Our last example illustrates how linear programming can be applied in a typical business application.

EXAMPLE 4

Linear Programming

A manufacturer makes two models of MP3 players. Model A requires 3 minutes to assemble, 4 minutes to test, and 1 minute to package. Model B requires 4 minutes to assemble, 3 minutes to test, and 6 minutes to package. The manufacturer can allot 7400 minutes total for assembly, 8000 minutes for testing, and 9000 minutes for packaging. Model A generates a profit of $7.00 and model B generates a profit of $8.00. How many of each model should be made in order to maximize profit?

Solution:

If we let x denote the number of Model A players made and y the number of Model B players made, the constraint inequalities are

$$\begin{cases} x \geq 0, y \geq 0 \\ 3x + 4y \leq 7400 \\ 4x + 3y \leq 8000 \\ x + 6y \leq 9000 \end{cases}$$

and the sketch of the feasible region appears below.

The profit function in this situation is $f(x, y) = 7x + 8y$, and our goal is to maximize this function. The vertices of the feasible region are: $(0, 1500)$, $(600, 1400)$, $(1400, 800)$, $(2000, 0)$, and $(0, 0)$. Of course, manufacturing 0 units of Model A and Model B is not going to be profitable, so we really only need to evaluate f at four vertices:

$f(0, 1500) = 12,000$

$f(600, 1400) = 15,400$

$f(1400, 800) = 16,200$

$f(2000, 0) = 14,000$

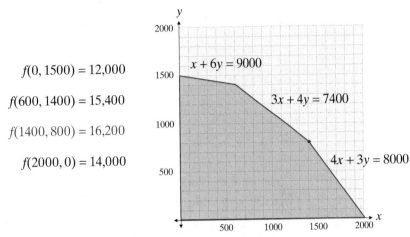

From these calculations, we conclude that the maximum profit is generated from making 1400 units of Model A and 800 units of Model B. The maximum profit would be \$16,200.

Exercises

Construct the constraints and graph the feasible regions for the following situations. See Example 1.

1.

$60x + 50y \le 50,000;$

$x + 10y \le 6000;$

$x \ge 0; \ y \ge 0$

2.

$40x + 60y \le 200;$

$8x + 5y \le 25;$

$x \ge 0; \ y \ge 0$

1. A plane carrying relief food and water can carry a maximum of 50,000 pounds, and is limited in space to carrying no more than 6000 cubic feet. Each container of water weighs 60 pounds and takes up 1 cubic foot, and each container of food weighs 50 pounds and takes up 10 cubic feet. What is the region of constraint for the number of containers of food and water that the plane can carry?

2. A furniture company makes two kinds of sofas, the Standard model and the Deluxe model. The Standard model requires 40 hours of labor to build, and the Deluxe model requires 60 hours of labor to build. The finish of the Deluxe model uses both teak and fabric, while the Standard uses only fabric, with the result that each Deluxe sofa requires 5 square yards of fabric and each Standard sofa requires 8 square yards of fabric. Given that the company can use 200 hours of labor and 25 square yards of fabric per week building sofas, what is the region of constraint for the number of Deluxe and Standard sofas the company can make per week?

3.

$12x + 32y \le 80$;

$2x + 3y \le 10$;

$x \ge 0;\ y \ge 0$

4.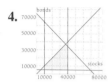

$x + y \le 75,000$;

$x - y \ge 0$;

$x \le 40,000$;

$x \ge 10,000$;

$y \ge 5000$

5. Min $= 0$ at $(0,0)$

Max $= 21$ at $(0,7)$

6. Min $= 0$ at $(0,0)$

Max $= 12$ at $(3,0)$

7. Min $= 0$ at $(0,0)$

Max $= 35$ at $(0,7)$

8. Min $= 0$ at $(0,0)$

Max $= 12$ at $(0,3)$

9. Min $= 0$ at $(0,0)$

Max $= 48$ at $(0,8)$

10. Min $= 0$ at $(0,0)$

Max $= \dfrac{575}{3}$ at

$\left(20, \dfrac{5}{3}\right)$

11. Min $= 0$ at $(0,0)$

Max $= 15$ at $(2.5, 0)$

12. Min $= 0$ at $(0,0)$

Max $= 35$ at $(0,5)$

3. Sarah is looking through a clothing catalog and she is willing to spend up to \$80 on clothes and \$10 for shipping. Shirts cost \$12 each plus \$2 shipping, and a pair of pants costs \$32 plus \$3 shipping. What is the region of constraint for the number of shirts and pairs of pants Sarah can buy?

4. Suppose you inherit \$75,000 from a previously unknown (and highly eccentric) uncle, and that the inheritance comes with certain stipulations regarding investments. First, the dollar amount invested in bonds must not exceed the dollar amount invested in stocks. Second, a minimum of \$10,000 must be invested in stocks and a minimum of \$5000 must be invested in bonds. Finally, a maximum of \$40,000 can be invested in stocks. What is the region of constraint for the dollar amount that can be invested in the two categories of stocks and bonds?

Find the minimum and maximum values of the given functions, subject to the given constraints. See Examples 2 and 3.

5. Objective Function

$f(x, y) = 2x + 3y$

Constraints

$x \ge 0$

$y \ge 0$

$x + y \le 7$

6. Objective Function

$f(x, y) = 4x + y$

Constraints

$x \ge 0$

$y \ge 0$

$x + y \le 3$

7. Objective Function

$f(x, y) = 2x + 5y$

Constraints

$x \ge 0$

$y \ge 0$

$x + y \le 7$

8. Objective Function

$f(x, y) = 7x + 4y$

Constraints

$x \ge 0$

$y \ge 0$

$3x + y \le 3$

9. Objective Function

$f(x, y) = 5x + 6y$

Constraints

$0 \le x \le 7$

$0 \le y \le 10$

$8x + 5y \le 40$

10. Objective Function

$f(x, y) = 9x + 7y$

Constraints

$0 \le x \le 20$

$0 \le y \le 10$

$6x + 12y \le 140$

11. Objective Function

$f(x, y) = 6x + 4y$

Constraints

$0 \le x \le 4$

$0 \le y \le 5$

$4x + 3y \le 10$

12. Objective Function

$f(x, y) = 3x + 7y$

Constraints

$0 \le x \le 8$

$0 \le y \le 6$

$7x + 10y \le 50$

13. Min = 0 at $(0, 0)$
    Max = 50 at $(3, 4)$

14. Min = 0 at $(0, 0)$
    Max = 20.625 at
    $(13.875, 3.375)$

15. Min = 0 at $(0, 0)$
    Max = 48 at $(8, 0)$

16. Min = 0 at $(0, 0)$
    Max = 300 at $(0, 10)$

17. Min = $\dfrac{172}{11}$ at
    $\left(\dfrac{24}{11}, \dfrac{10}{11}\right)$
    Max = $\dfrac{400}{3}$ at
    $\left(0, \dfrac{40}{3}\right)$

18. Min = $\dfrac{680}{7}$ at
    $\left(\dfrac{88}{7}, \dfrac{-36}{7}\right)$
    Max = $\dfrac{1980}{7}$ at
    $\left(\dfrac{108}{7}, \dfrac{-6}{7}\right)$

19. Model X: 1000
    units, Model Y: 500
    units, Maximum
    Profit: $76,000

20. Type X: 75 units;
    Type Y: 100 units;
    Maximum Profit:
    $712.50

13. Objective Function
    $f(x, y) = 6x + 8y$
    Constraints
    $x \geq 0; y \geq 0$
    $4x + y \leq 16$
    $x + 3y \leq 15$

15. Objective Function
    $f(x, y) = 6x + y$
    Constraints
    $x \geq 0; y \geq 0$
    $3x + 4y \leq 24$
    $3x + 4y \leq 48$

17. Objective Function
    $f(x, y) = 3x + 10y$
    Constraints
    $x \geq 0$
    $2x + 4y \geq 8$
    $5x - y \leq 10$
    $x + 3y \leq 40$

14. Objective Function
    $f(x, y) = x + 2y$
    Constraints
    $x \geq 0; y \geq 0$
    $3x + y \leq 45$
    $x + 3y \leq 24$

16. Objective Function
    $f(x, y) = 15x + 30y$
    Constraints
    $x \geq 0; y \geq 0$
    $5x + 7y \leq 70$
    $5x + 7y \leq 140$

18. Objective Function
    $f(x, y) = 20x + 30y$
    Constraints
    $x \geq 0$
    $12x + 6y \geq 120$
    $9x - 6y \leq 144$
    $x + 4y \leq 12$

Solve the following application problems. See Example 4.

19. A manufacturer produces two models of computers. The times (in hours) required for assembling, testing, and packaging each model are listed in the table below.

Process	Model X	Model Y
Assemble	2.5	3
Test	2	1
Package	0.75	1.25

The total times available for assembling, testing, and packaging are 4000 hours, 2500 hours, and 1500 hours, respectively. The profits per unit are $50 for Model X and $52 for Model Y. How many of each type should be produced to maximize profit? What is the maximum profit?

20. A manufacturer produces two types of fans. The times (in minutes) required for assembling, packaging, and shipping each type are listed in the table below.

Process	Type X	Type Y
Assemble	20	25
Package	40	10
Ship	10	7.5

The total times available for assembling, packaging, and shipping are 4000 minutes, 4800 minutes, and 1500 minutes, respectively. The profits per unit are $4.50 for Type X and $3.75 for Type Y. How many of each type should be produced to maximize profit? What is the maximum profit?

21. Ashley should buy 3 olive bundles and 2 cranberry bundles. The total cost to make the curtains is $70.

22. The volunteer could choose any point on the line $3x + 2y = 45$ from $(7.5, 11.25)$ to $(15, 0)$, but note the answers must be whole numbers. The minimum cost is $22.50 for all values in this range.

23. $20,000 should be placed in municipal bonds, $5000 should be placed in Treasury bills.

24. a. Flip phone: manufacture 120 phones, Camera phone: manufacture 80 phones, Minimum cost: $1160

b. Flip phone: manufacture 100 phones, Camera phone: manufacture 170 phones, Maximum profit: $650

21. Ashley is making a set of patchwork curtains for her apartment. She needs a minimum of 16 yards of the solid material, at least 5 yards of the striped material, and at least 20 yards of the flowered material. She can choose between two sets of precut bundles. The olive-based bundle costs $10 per bundle and contains 8 yards of the solid material, 1 yard of the striped material, and 2 yards of the flowered material. The cranberry-based bundle costs $20 per bundle and includes 2 yards of the solid material, 1 yard of the striped material, and 7 yards of the flowered material. How many of each bundle should Ashley buy to minimize her cost and yet buy enough material to complete the curtains? What is her minimum cost?

23. On your birthday your grandmother gave you $25,000, but told you she would like you to invest the money for 10 years before you use any of it. Since you wish to respect your grandmother's wishes, you seek out the advice of a financial adviser. She suggests you invest at least $15,000 in municipal bonds yielding 6% and no more than $5,000 in Treasury bills yielding 9%. How much should be placed in each investment so that income is maximized?

22. A volunteer has been asked to drop off some supplies at a facility housing victims of a hurricane evacuation. The volunteer would like to bring at least 60 bottles of water, 45 first aid kits, and 30 security blankets on his visit. The relief organization has a standing agreement with two companies that provide victim packages. Company A can provide packages of 5 water bottles, 3 first aid kits, and 4 security blankets at a cost of $1.50. Company B can provide packages of 2 water bottles, 2 first aid kits, and 1 security blanket at a cost of $1.00. How many of each package should the volunteer pick up to minimize the cost? What total amount does the relief organization pay?

24. An independent cell phone manufacturer produces two models: a flip phone and a camera phone. The manufacturer's quota per day is to produce at least 100 flip phones and 80 camera phones. No more than 200 flip phones and 170 camera phones can be produced per day due to limitations on production. A total of at least 200 phones must be shipped every day.

a. If the production costs are $5 for a flip phone and $7 for a camera phone, how many of each model should be produced on a daily basis to minimize cost? What is the minimum cost to produce the minimum amount?

b. If each flip phone results in a $2 loss but each camera phone results in a $5 gain, how many of each model should be manufactured daily to maximize profit? What is the maximum profit if this number of phones is produced?

10.8 Nonlinear Systems of Equations and Inequalities

TOPICS

1. Approximating solutions by graphing
2. Solving nonlinear systems algebraically
3. Solving nonlinear systems of inequalities

TOPIC 1   Approximating Solutions by Graphing

Solving Nonlinear Systems by Graphing

In this last section of the chapter, we deal with systems of equations in which one or more of the equations are nonlinear. Because the systems are nonlinear, the matrix-based methods that we have learned *do not apply*. The theory of nonlinear systems is much more complex than that of linear systems, so we will only use general guiding principles to solve them.

Since we have learned how to graph many different classes of equations, we can put our graphing knowledge to use in solving nonlinear systems. Keep in mind, however, that graphing will only help us identify the *real* solutions; as we will see, some nonlinear systems also have complex number solutions (with nonzero imaginary parts).

EXAMPLE 1

Use graphing to guess the real solution(s) of the following system, and then verify that your answer is correct.

$$\begin{cases} x + y = 2 \\ x^2 + y^2 = 4 \end{cases}$$

Solution:

The first equation describes a line in the plane with an x-intercept at $(2,0)$ and a y-intercept of $(0,2)$. The second equation describes a circle of radius 2 centered at the origin.

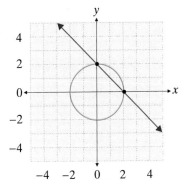

We are interested in whether any ordered pairs of numbers satisfy both equations simultaneously. Graphically, we are looking for the points of intersection of the two graphs.

The graph suggests that $(2,0)$ and $(0,2)$ make up the real solutions of the system. We already know these two points lie on the line, and it can be verified that they also solve the second equation:

$$2^2 + 0^2 = 4 \text{ and } 0^2 + 2^2 = 4.$$

This example points out another very significant difference between linear and nonlinear systems. Recall that a linear system is guaranteed to have no solution, one solution, or an infinite number of solutions. Nonlinear systems may have any number of solutions.

EXAMPLE 2

Solving Nonlinear Systems by Graphing

Use graphing to guess the real solution(s) of the following system, and then verify that your answer is correct.

$$\begin{cases} x + 2 = y^2 \\ x^2 + y^2 = 4 \end{cases}$$

Solution:

The first equation represents a horizontally oriented parabola opening to the right, with vertex at $(-2, 0)$. The second equation is again a circle of radius 2 centered at the origin.

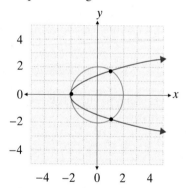

This time, the system appears to have three real solutions. The point $(-2, 0)$ is easily verified to be one of them, but the other two are harder to determine just by looking at the graph. Noting that the x-coordinate of the other two points appears to be 1, we can replace x with 1 in the two equations to obtain the equation $y^2 = 3$ in both cases. Solving for y yields $y = \pm\sqrt{3}$, which means that $\left(1, \sqrt{3}\right)$ and $\left(1, -\sqrt{3}\right)$ are the other two solutions of the system. Checking these solutions verifies that both ordered pairs solve both equations.

TOPIC 2   Solving Nonlinear Systems Algebraically

In Example 2, we needed to combine graphical knowledge with some algebra in order to solve the system. The picture corresponding to the system suggested that there were three real solutions, and we can be fairly confident that we found all solutions to the system. But graphs can be misleading, especially if they are not drawn accurately. Consider, for example, the following very similar system.

EXAMPLE 3

Solving Nonlinear Systems Algebraically

Solve the following nonlinear system.

$$\begin{cases} x+4 = y^2 \\ x^2 + y^2 = 4 \end{cases}$$

Solution:

The only difference between this system and the one in Example 2 is that the vertex of the parabola is at $(-4, 0)$, so the graph of the system is as shown.

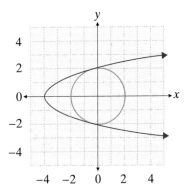

Looking at the graphs, it isn't clear exactly how many intersections exist or where those intersections are. A reasonable guess might be the two points $(0, 2)$ and $(0, -2)$, and it can be verified that these two points do indeed lie on both equations. But if we were to conclude that these are the only two solutions, we would be wrong.

We need a more reliable method for this example, and algebra comes to the rescue. While we can't apply the matrix-based solution methods to nonlinear systems, we can use the methods of substitution and elimination. For the system

$$\begin{cases} x+4 = y^2 \\ x^2 + y^2 = 4 \end{cases}$$

the method of substitution works well if we solve both equations for y^2.

$$\begin{cases} x+4 = y^2 \\ y^2 = 4 - x^2 \end{cases}$$

The two expressions equal to y^2 both contain only the variable x, so if we equate the expressions we will have a single equation in one variable. The resulting equation is solved as shown.

1. $\{(0,-3),(2,0)\}$

2. $\{(0,1),(2,0)\}$

3. $\{(-1,-1),(-1,1),$
$(1,-1),(1,1)\}$

4. $\{(1,-1),(1,1)\}$

Solving Nonlinear Systems Algebraically

5. $\{(-1,1),(1,1)\}$

6. $\{(2,1)\}$

$$x+4=4-x^2$$
$$x=-x^2$$
$$x^2+x=0$$
$$x(x+1)=0$$
$$x=0,\ -1$$

We already know that 0 is one possible value for x, as we have verified that $(0,2)$ and $(0,-2)$ are solutions of the system. But the value of -1 is new, and if we substitute $x=-1$ into either equation, we find that $\left(-1,\sqrt{3}\right)$ and $\left(-1,-\sqrt{3}\right)$ are also solutions. So the system has a total of four solutions, which we now indicate on the graph.

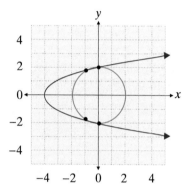

The next example of a nonlinear system can be solved by the method of elimination.

EXAMPLE 4

Solve the following nonlinear system.

$$\begin{cases} y+1=(x-2)^2 \\ 3-y=(x-1)^2 \end{cases}$$

Solution:

Before solving the system algebraically, you may want to graph these two vertically oriented parabolas to get some idea of what to expect.

We can eliminate the variable y by adding the two equations, leaving the quadratic equation $4=(x-2)^2+(x-1)^2$.

7. $\{(-1, 2), (1, 2)\}$

8. $\{(-1, -3), (1, -3)\}$

9. No solution

10. $\{(1, 3)\}$

11. $\{(-3, 1), (1, 1)\}$

We can now solve the single-variable equation:

$$4 = (x-2)^2 + (x-1)^2$$

$$4 = x^2 - 4x + 4 + x^2 - 2x + 1$$

$$0 = 2x^2 - 6x + 1$$

$$x = \frac{6 \pm \sqrt{36-8}}{4}$$

$$x = \frac{3 \pm \sqrt{7}}{2}$$

For each of the two values of x, we need to determine the corresponding values of y, and we can use either equation to do so. The work below uses the first equation.

For $x = \dfrac{3+\sqrt{7}}{2}$:

$$y = \left(\frac{3+\sqrt{7}}{2} - 2\right)^2 - 1$$

$$y = \left(\frac{-1+\sqrt{7}}{2}\right)^2 - 1$$

$$y = \frac{8 - 2\sqrt{7}}{4} - 1$$

$$y = 2 - \frac{\sqrt{7}}{2} - 1$$

$$y = 1 - \frac{\sqrt{7}}{2}$$

For $x = \dfrac{3-\sqrt{7}}{2}$:

$$y = \left(\frac{3-\sqrt{7}}{2} - 2\right)^2 - 1$$

$$y = \left(\frac{-1-\sqrt{7}}{2}\right)^2 - 1$$

$$y = \frac{8 + 2\sqrt{7}}{4} - 1$$

$$y = 2 + \frac{\sqrt{7}}{2} - 1$$

$$y = 1 + \frac{\sqrt{7}}{2}$$

Thus, the two solutions of the system are

$$\left(\frac{3+\sqrt{7}}{2}, 1-\frac{\sqrt{7}}{2}\right) \text{ and } \left(\frac{3-\sqrt{7}}{2}, 1+\frac{\sqrt{7}}{2}\right).$$

EXAMPLE 5

Solving Nonlinear Systems Algebraically

Solve the following nonlinear system.

$$\begin{cases} \dfrac{(x+1)^2}{9} + y^2 = 1 \\ y^2 = x + 2 \end{cases}$$

Solution:

The first equation describes an ellipse, and the second describes a parabola opening to the right. The graph of the system indicates that we should expect two real solutions:

12. $\{(-1, 0), (0, -1),$

$\quad (1, 0)\}$

13. $\{(0, 0)\}$

14. $\left\{\left(2.5, \pm\sqrt{2.75}\right)\right\}$

15. $\{(\pm3, 0)\}$

16. $\left\{\left(\pm\sqrt{3}, -1\right)\right\}$

17. $\{(\pm1, 0)\}$

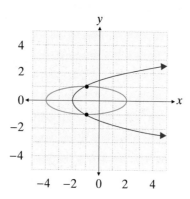

If we solve this system algebraically, we discover that there is more to this problem.

Since the second equation is already solved for y^2, the method of substitution is a good option. Substituting in the first equation, we obtain

$$\frac{(x+1)^2}{9} + x + 2 = 1.$$

We solve this second-degree equation in x the usual way:

$$\frac{(x+1)^2}{9} + x + 2 = 1$$

$$\frac{(x+1)^2}{9} + x + 1 = 0$$

$$(x+1)^2 + 9x + 9 = 0$$

$$x^2 + 11x + 10 = 0$$

$$(x+10)(x+1) = 0$$

$$x = -10, \ -1$$

The solution $x = -1$ isn't surprising, since our graph certainly suggested this value, but the solution $x = -10$ doesn't seem to relate to the picture at all. But remember, these are only the x-coordinates of the ordered pairs that solve the system, and we still need to determine the corresponding y-coordinates. We will use the second equation to do this.

For $x = -10$:   For $x = -1$:

$$y^2 = -10 + 2 \qquad y^2 = -1 + 2$$

$$y^2 = -8 \qquad\qquad y^2 = 1$$

$$y = \pm 2i\sqrt{2} \qquad y = \pm 1$$

Now we know that the two real solutions are $(-1, -1)$ and $(-1, 1)$, as was suggested by our graph, and that the two ordered pairs $\left(-10, -2i\sqrt{2}\right)$ and $\left(-10, 2i\sqrt{2}\right)$ also solve the system. These last two solutions don't appear on the graph because of their nonreal second coordinates.

TOPIC 3

Solving Nonlinear Systems of Inequalities

The feasible regions of Section 10.7 are portions of the plane that simultaneously satisfy a collection of linear inequalities, and thus represent solutions of linear systems of inequalities. Combining the skills learned there with our newly acquired knowledge of how to solve nonlinear systems of equations, we can now graph solutions of nonlinear systems of inequalities.

Recall that a linear inequality in two variables divides the Cartesian plane into two regions separated by a line: one region consists of all those points that satisfy the inequality, and the other region consists of those that do not. The boundary line is either a part of the solution region or not, depending on whether the inequality is nonstrict or strict. In general, a nonlinear inequality in two variables also separates the plane into regions, with the difference (as we would expect) that the boundary between the regions is not linear. To illustrate, the graphs in Figure 1 below represent the solutions of, respectively, the inequalities $x^2 + y^2 < 1$ and $y \geq x^2 + 1$.

18. $\{(0, 1),(1, 2)\}$

19. $\{(2, \pm 3), (-2, \pm 3)\}$

20. $\{(-1,-1),(0,0),(1,1)\}$

21. $\{(5, 3)\}$

22. $\{(\pm 1, -1)\}$

23. $\{(\pm 2, 0),(0, -2)\}$

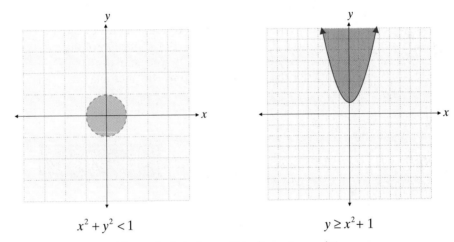

$$x^2 + y^2 < 1 \qquad\qquad y \geq x^2 + 1$$

Figure 1: Solutions of Nonlinear Inequalities

As with linear inequalities, the solution regions can often be identified by considering the geometric meaning of the given nonlinear inequality (e.g., the solutions of $x^2 + y^2 < 1$ are those points lying within a radius of 1 of the origin, and the solutions of $y \geq x^2 + 1$ are those points for which the y-coordinate lies on or above the parabola $y = x^2 + 1$). Alternatively, a "test point" within a given region can be substituted into the inequality; if the point satisfies the inequality, that region is part of the solution set. All that remains in order to solve nonlinear systems of inequalities is to note that the solution of a system consists of the intersection of the solutions of the individual inequalities. We will illustrate this point with a variation of Example 5.

EXAMPLE 6

Solving Nonlinear Systems of Inequalities by Graphing

Graph the solution of the following nonlinear system of inequalities.

$$\begin{cases} \dfrac{(x+1)^2}{9} + y^2 < 1 \\ y^2 \le x + 2 \end{cases}$$

24. $\{(3, \pm 2), (-3, \pm 2)\}$

Solution:

The region solving the first inequality is the interior of an ellipse (with the boundary excluded), while the region solving the second inequality consists of those points on and to the right of a rightward-opening parabola. Thus, the solution of the system consists of all those points in the plane lying in both regions; that is, the intersection of the regions. The graph of this intersection appears shaded in the figure to the right (note the dashed line for the portion of the boundary defined by the ellipse and the solid line for the portion defined by the parabola).

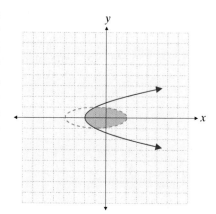

25. $\left\{\left(-i\sqrt{6}, -6\right), \left(i\sqrt{6}, -6\right), \left(-\sqrt{5}, 5\right), \left(\sqrt{5}, 5\right)\right\}$

26. $\{(-2, 0), (2, 0)\}$

27. $\{(-2, 3)\}$

28. $\left\{\left(-2i\sqrt{2}, -2\sqrt{3}\right), \left(-2i\sqrt{2}, 2\sqrt{3}\right), \left(2i\sqrt{2}, -2\sqrt{3}\right), \left(2i\sqrt{2}, 2\sqrt{3}\right)\right\}$

29. $\left\{(-1, -4), (1, 4), \left(-2\sqrt{2}, -\sqrt{2}\right), \left(2\sqrt{2}, \sqrt{2}\right)\right\}$

30. $\left\{\left(-\sqrt{5}, -\sqrt{5}\right), \left(\sqrt{5}, \sqrt{5}\right)\right\}$

31. $\left\{(-3i, -5), (3i, -5), (0, 4)\right\}$

32. $\{(-2, -8), (0, 0)\}$

33. $\left\{\left(-\sqrt{3}, 0\right), \left(\sqrt{3}, 0\right)\right\}$

34. $\{(-2, 0), (2, 0)\}$

Exercises

Use graphing to guess the real solution(s) of the following systems, and then verify your answer algebraically. See Examples 1 and 2.

1. $\begin{cases} 3x - 2y = 6 \\ \dfrac{x^2}{4} + \dfrac{y^2}{9} = 1 \end{cases}$

2. $\begin{cases} x + 2y = 2 \\ \dfrac{x^2}{4} + y^2 = 1 \end{cases}$

3. $\begin{cases} x^2 + 4y^2 = 5 \\ x^2 + y^2 = 2 \end{cases}$

4. $\begin{cases} 4x^2 + y^2 = 5 \\ 4(x-2)^2 + y^2 = 5 \end{cases}$

5. $\begin{cases} y = x^2 \\ 2 - y = x^2 \end{cases}$

6. $\begin{cases} x^2 + (y-1)^2 = 4 \\ (x-3)^2 + (y-1)^2 = 1 \end{cases}$

7. $\begin{cases} y - x^2 = 1 \\ y + 2 = 4x^2 \end{cases}$

8. $\begin{cases} x^2 + y^2 = 10 \\ x^2 + y = -2 \end{cases}$

9. $\begin{cases} x = y^2 - 3 \\ x^2 + 4y^2 = 4 \end{cases}$

10. $\begin{cases} (x-1)^2 + (y-6)^2 = 9 \\ (x-1)^2 + (y+1)^2 = 16 \end{cases}$

35. $\left\{\begin{array}{l}(-i,-6),(i,-6),\\ \left(-\sqrt{2},3\right),\left(\sqrt{2},3\right)\end{array}\right\}$

36. $\left\{\left(1,\dfrac{1}{4}\right)\right\}$

37. $\left\{\begin{array}{l}(1,1),(1,-1),\\ \left(-\dfrac{3}{2},\dfrac{\sqrt{14}}{2}\right),\\ \left(-\dfrac{3}{2},-\dfrac{\sqrt{14}}{2}\right)\end{array}\right\}$

38. $\left\{\begin{array}{l}\left(3-\sqrt{13},51-12\sqrt{13}\right),\\ \left(3+\sqrt{13},51+12\sqrt{13}\right)\end{array}\right\}$

39. $\left\{\left(\dfrac{7}{2},\dfrac{17}{4}\right)\right\}$

40. $\left\{\begin{array}{l}\left(2i,-i\sqrt{10}\right),\\ \left(2i,i\sqrt{10}\right),\\ \left(-2i,-i\sqrt{10}\right),\\ \left(-2i,i\sqrt{10}\right)\end{array}\right\}$

41. $\left\{\begin{array}{l}\left(-6,-i\sqrt{15}\right),\\ \left(-6,i\sqrt{15}\right),\\ (2,-1),(2,1)\end{array}\right\}$

42. $\left\{\begin{array}{l}\left(-\sqrt{6},2\right),\left(\sqrt{6},2\right),\\ (3,-1),(-3,-1)\end{array}\right\}$

43. $\left\{\begin{array}{l}(0,1),\\ \left(-\dfrac{100}{101},-\dfrac{99}{101}\right)\end{array}\right\}$

44. $\left\{\begin{array}{l}\left(\dfrac{3}{4},-\dfrac{\sqrt{46}}{4}\right),\\ \left(\dfrac{3}{4},\dfrac{\sqrt{46}}{4}\right)\end{array}\right\}$

11. $\begin{cases}(x+1)^2+(y-1)^2=4\\(x+1)^2+4(y-1)^2=4\end{cases}$

13. $\begin{cases}(x-2)^2+y^2=4\\(x+2)^2+y^2=4\end{cases}$

15. $\begin{cases}x^2+y^2=9\\\dfrac{x^2}{9}+\dfrac{y^2}{25}=1\end{cases}$

17. $\begin{cases}x^2+\dfrac{y^2}{4}=1\\y=0\end{cases}$

19. $\begin{cases}2y^2-3x^2=6\\2y^2+x^2=22\end{cases}$

21. $\begin{cases}x=y^2-4\\x+13=6y\end{cases}$

23. $\begin{cases}2y=x^2-4\\x^2+y^2=4\end{cases}$

12. $\begin{cases}x^2+y^2=1\\y=x^2-1\end{cases}$

14. $\begin{cases}x^2+y^2=9\\x^2+y^2-2x-3=1\end{cases}$

16. $\begin{cases}x^2+y^2=4\\-x^2=2y-1\end{cases}$

18. $\begin{cases}y=x^2+1\\y-1=x^3\end{cases}$

20. $\begin{cases}y=x^3\\y=\sqrt[3]{x}\end{cases}$

22. $\begin{cases}y=2x^2-3\\y=-x^2\end{cases}$

24. $\begin{cases}x^2-y^2=5\\\dfrac{x^2}{25}+\dfrac{4y^2}{25}=1\end{cases}$

Solve the following nonlinear systems algebraically. Be sure to check for nonreal solutions. See Examples 3, 4, and 5.

25. $\begin{cases}x^2+y^2=30\\x^2=y\end{cases}$

27. $\begin{cases}x^2-1=y\\4x+y=-5\end{cases}$

29. $\begin{cases}y=\dfrac{4}{x}\\2x^2+y^2=18\end{cases}$

31. $\begin{cases}y-x^2=4\\x^2+y^2=16\end{cases}$

33. $\begin{cases}2x^2+3y^2=6\\x^2+3y^2=3\end{cases}$

35. $\begin{cases}3x^2-y=3\\9x^2+y^2=27\end{cases}$

26. $\begin{cases}3x^2+2y^2=12\\x^2+2y^2=4\end{cases}$

28. $\begin{cases}x^2+y^2=4\\3x^2+4y^2=24\end{cases}$

30. $\begin{cases}xy=5\\x^2+y^2=10\end{cases}$

32. $\begin{cases}y-x^2=6x\\y=4x\end{cases}$

34. $\begin{cases}x^2+y^2=4\\\dfrac{x^2}{4}-\dfrac{y^2}{8}=1\end{cases}$

36. $\begin{cases}\dfrac{1}{x}+\dfrac{1}{y}=5\\\dfrac{1}{x}-\dfrac{1}{y}=-3\end{cases}$

45. $\left\{\left(-\dfrac{\sqrt{42}}{6},-\dfrac{\sqrt{66}}{6}\right),\right.$
$\left(-\dfrac{\sqrt{42}}{6},\dfrac{\sqrt{66}}{6}\right),$
$\left(\dfrac{\sqrt{42}}{6},-\dfrac{\sqrt{66}}{6}\right),$
$\left.\left(\dfrac{\sqrt{42}}{6},\dfrac{\sqrt{66}}{6}\right)\right\}$

46. $\left\{(5,0),\left(-6,\pm i\sqrt{11}\right)\right\}$

47. $\left\{(-5,0),(-2,0),(-1,0)\right\}$

48. $\{(0,\pm 4)\}$

49. $\left\{(3,2),(2,3),\left(\dfrac{-1\pm i\sqrt{23}}{2},\dfrac{-1\pm i\sqrt{23}}{2}\right)\right\}$

50. $\left\{\left(\dfrac{5}{3},\pm\dfrac{2\sqrt{6}}{3}\right),\left(-\dfrac{5}{2},\pm\dfrac{i\sqrt{6}}{2}\right)\right\}$

51. $\left\{(0,-1),\left(\dfrac{\sqrt{6}}{3},\dfrac{2\sqrt{6}}{9}-1\right)\right\}$

52. $\{(2,4)\}$

53. $\left\{\left(\sqrt{2}+1,2\sqrt{2}\right),\left(-\sqrt{2}+1,-2\sqrt{2}\right),\left(2\sqrt{2}+1,\sqrt{2}\right),\left(-2\sqrt{2}+1,-\sqrt{2}\right)\right\}$

37. $\begin{cases} x+y^2=2 \\ 2x^2-y^2=1 \end{cases}$

39. $\begin{cases} y-2=(x-2)^2 \\ y+2=(x-1)^2 \end{cases}$

41. $\begin{cases} (x+1)^2+y^2=10 \\ \dfrac{(x-2)^2}{4}+y^2=1 \end{cases}$

43. $\begin{cases} 2x=y-1 \\ \dfrac{x^2}{25}+y^2=1 \end{cases}$

45. $\begin{cases} x^2+7y^2=14 \\ x^2+y^2=3 \end{cases}$

47. $\begin{cases} y=x^3+8x^2+17x+10 \\ -y=x^3+8x^2+17x+10 \end{cases}$

49. $\begin{cases} xy=6 \\ (x-2)^2+(y-2)^2=1 \end{cases}$

51. $\begin{cases} y=x^3-1 \\ 3y=2x-3 \end{cases}$

53. $\begin{cases} xy-y=4 \\ (x-1)^2+y^2=10 \end{cases}$

55. $\begin{cases} y=\sqrt{x-4}+1 \\ (x-3)^2+(y-1)^2=1 \end{cases}$

57. $\begin{cases} y^2-y-12=x-x^2 \\ y-1+\dfrac{2x-12}{y}=0 \end{cases}$

59. $\begin{cases} \dfrac{(y+2)^2}{(x+y)}=1 \\ x=y^2+5y+4 \end{cases}$

38. $\begin{cases} y-2=(x+3)^2 \\ \dfrac{1}{3}y=(x-1)^2 \end{cases}$

40. $\begin{cases} y^2+2=2x^2 \\ y^2=x^2-6 \end{cases}$

42. $\begin{cases} x^2+y^2=10 \\ x^2+y=8 \end{cases}$

44. $\begin{cases} 2x^2+y^2=4 \\ 2(x-1)^2+y^2=3 \end{cases}$

46. $\begin{cases} x^2+y^2=25 \\ y^2=x-5 \end{cases}$

48. $\begin{cases} \dfrac{x^2}{25}+\dfrac{y^2}{16}=1 \\ x^2+y^2=16 \end{cases}$

50. $\begin{cases} y^2=x+1 \\ \dfrac{x^2}{5}+\dfrac{y^2}{6}=1 \end{cases}$

52. $\begin{cases} y+5=(x+1)^2 \\ y-3=(x-3)^2 \end{cases}$

54. $\begin{cases} 2x^2+5y^2=16 \\ 4x^2+3y^2=4 \end{cases}$

56. $\begin{cases} y=\sqrt[3]{x} \\ \sqrt{y}=x \end{cases}$

58. $\begin{cases} y=7x^2+1 \\ x^2+y^2=1 \end{cases}$

60. $\begin{cases} x=\sqrt{6y+1} \\ y=\sqrt{\dfrac{x^2+7}{2}} \end{cases}$

54. $\left\{ \left(\pm i\sqrt{2}, 2 \right), \right.$
$\left. \left(\pm i\sqrt{2}, -2 \right) \right\}$

55. $\left\{ (4, 1), \left(1, 1 + i\sqrt{3} \right) \right\}$

56. $\left\{ (0, 0), (1, 1) \right\}$

57. $\left\{ (0, 4), (0, -3), \right.$
$\left. (3, -2), (3, 3) \right\}$

58. $\left\{ \left(\pm \dfrac{\sqrt{15}}{7} i, \dfrac{-8}{7} \right), \right.$
$\left. (0, 1) \right\}$

59. $\left\{ (4, 0) \right\}$

60. $\left\{ (5, 4) \right\}$

61. $\left\{ \left(\pm \dfrac{1}{2}, \pm \dfrac{1}{4} \right) \right\}$

62. $\left\{ \left(3, \pm \dfrac{\sqrt{26}}{2} \right), \right.$
$\left. \left(6, \pm \dfrac{7\sqrt{2}}{2} \right) \right\}$

63. b

64. a

65. b, d

61. $\begin{cases} \dfrac{-2}{x^2} + \dfrac{1}{y^2} = 8 \\ \dfrac{9}{x^2} - \dfrac{2}{y^2} = 4 \end{cases}$

62. $\begin{cases} x^2 + 3x - 2y^2 = 5 \\ -4x^2 + 6y^2 = 3 \end{cases}$

Draw the graph and determine whether the ordered pairs are solutions to the system of inequalities.

63. $\begin{cases} x \ge 3 \\ y > 4 \end{cases}$
a. $(2, 5)$ **b.** $(7, 8)$ **c.** $(5, 0)$ **d.** $(3, 4)$

64. $\begin{cases} y \le 2x + 1 \\ y < 4 \\ y > x \end{cases}$
a. $(1, 2)$ **b.** $(3, 4)$ **c.** $(-1, -1)$ **d.** $(3, 3)$

65. $\begin{cases} y \ge x^2 \\ y < x^3 \\ y \le 4x \end{cases}$
a. $(2, 2)$ **b.** $(2, 4)$ **c.** $(2, 8)$ **d.** $(3, 9)$

66. $\begin{cases} y \ge x^2 - 2 \\ y \le (x - 2)^2 \\ 3y > 2x + 12 \end{cases}$
a. $(2, 5)$ **b.** $(7, 8)$ **c.** $(5, 0)$ **d.** $(3, 4)$

67. $\begin{cases} x < 4 \\ y \ge \sqrt{x} \\ 2y > -x \end{cases}$
a. $(2, 5)$ **b.** $(7, 8)$ **c.** $(5, 0)$ **d.** $(3, 4)$

Graph the following systems of inequalities. See Example 6.

68. $\begin{cases} y < 2x \\ y > x^2 \end{cases}$
69. $\begin{cases} y \le 2x + 3 \\ y \ge 0 \\ x \ge 0 \end{cases}$
70. $\begin{cases} y > x^2 \\ -3y \le x - 9 \end{cases}$

71. $\begin{cases} y \le x \\ 2y > -x \\ x < 4 \end{cases}$
72. $\begin{cases} y \le \sqrt{x} \\ 2y > (x - 1)^2 - 4 \end{cases}$
73. $\begin{cases} y > x^3 \\ y \le \sqrt[3]{x} \\ y > 0 \end{cases}$

74. $\begin{cases} y \ge x^3 \\ y \ge -x^3 \\ y < 2(x + 1) \end{cases}$
75. $\begin{cases} x^2 + y^2 < 9 \\ -4y \ge x - 12 \end{cases}$
76. $\begin{cases} y \le \sin x \\ x \ge 0 \end{cases}$

66. None

67. a, d

68.

69.

70.

71.

Solve the following application problems.

77. The area of a certain rectangle is 45 square inches, and its perimeter is 28 inches. Find the dimensions of the rectangle.

78. The product of two positive integers is 88, and their sum is 19. What are the integers?

79. Jack takes half an hour longer than his wife does to make the 210-mile drive between two cities. His wife drives 10 miles an hour faster. How fast do the two drive?

80. To construct the two garden beds shown below, 48.5 meters of fencing are needed. The combined area of the beds is 95 square meters. There are two possibilities for the overall dimensions of the two beds. What are they?

81. The product of two integers is −84, and their sum is −5. What are the integers?

82. Paul and Maria were driving the same 24-mile route, and they departed at the same time. After 20 minutes, Maria was 4 miles ahead of Paul. If it took Paul 10 minutes longer to reach their destination, how fast were they each driving?

83. The surface area of a certain right circular cylinder is $54\pi \ \text{cm}^2$ and the volume is $54\pi \ \text{cm}^3$. Find the height h and the radius r of this cylinder.
(**Hint:** The formulas for the volume and surface area of a right circular cylinder are as follows: $V = \pi r^2 h$ and $A = 2\pi rh + 2\pi r^2$.)

72.

73.

74.

75.

76.

77. 9 inches by 5 inches

78. 8 and 11

79. 60 mph and 70 mph

80. $6\frac{2}{3} \times 14\frac{1}{4}$ meters

or $9\frac{1}{2} \times 10$ meters

81. −12 and 7

82. 36 mph and 48 mph

83. $h = 6$ cm, $r = 3$ cm

Chapter 10 Project

Market Share Matrix

Assume you are the sales and marketing director for Joe's Java, a coffee shop located on a crowded city street corner. There are two competing coffee shops on this block—Buck's Café and Tweak's Coffee. The management has asked you to develop a marketing campaign to increase your market share from 25% to at least 35% within 6 months. With the resulting plan to meet this goal, you predict that each month:

a. You will retain 93% of your customers, 4% will go to Buck's Café, and 3% will go to Tweak's Coffee.

b. Buck's Café will retain 91% of their customers, 6% will come to Joe's Java, and 3% will go to Tweak's Coffee.

c. Tweak's Coffee will retain 92% of their customers, 3% will come to Joe's Java and 5% will go to Buck's Café.

The current percentage of the market is shown in this matrix:

$$x_0 = \begin{bmatrix} 0.25 \\ 0.45 \\ 0.30 \end{bmatrix} \begin{matrix} Joe's \\ Buck's \\ Tweak's \end{matrix}$$

After one month the shares of the coffee shops are:

$$x_1 = Px_0 = \begin{bmatrix} 0.93 & 0.06 & 0.03 \\ 0.04 & 0.91 & 0.05 \\ 0.03 & 0.03 & 0.92 \end{bmatrix} \begin{bmatrix} 0.25 \\ 0.45 \\ 0.30 \end{bmatrix} = \begin{bmatrix} 0.2685 \\ 0.4345 \\ 0.2970 \end{bmatrix}$$

1. Construct a table that lists the market share for all of the coffee shops at the end of each of the first 6 months.

2. Will your campaign be successful based on this model (will you reach 35% market share in 6 months)?

3. What actions do you think Buck's Café and Tweak's Coffee will take as your market share changes?

4. What effect could their actions have on the market?

Chapter Summary

A summary of concepts and skills follows each chapter. Refer to these summaries to make sure you feel comfortable with the material in the chapter. The concepts and skills are organized according to the section title and topic title in which the material is first discussed.

10.1: Solving Systems by Substitution and Elimination

Definition and Classification of Linear Systems of Equations
- The geometric and algebraic meaning of *linear* (or *simultaneous*) *systems of equations*
- The varieties of linear systems of equations, and the meaning of the terms *inconsistent, consistent,* and *dependent*

Solving Systems by Substitution
- The use of the *substitution method* to solve linear systems of equations
- Alternatives for describing solutions that consist of more than one point

Solving Systems by Elimination
- The use of the *elimination method* to solve linear systems of equations
- The use of arrow notation to keep track of the steps used in solving a system

Larger Systems of Equations
- The geometric and algebraic meaning of larger linear systems of equations
- How the solution process of larger systems relates to that of smaller systems

Applications of Systems of Equations
- Using systems of equations to solve application problems

10.2: Matrix Notation and Gaussian Elimination

Linear Systems, Matrices, and Augmented Matrices
- The definition of a *matrix*, the meaning of *rows* and *columns*, and the definition of *matrix order*
- Notation commonly used to define and refer to matrices and their elements
- The meaning of an *augmented matrix*

10.2: Matrix Notation and Gaussian Elimination (cont.)

Gaussian Elimination and Row Echelon Form
- The meaning of *Gaussian elimination* as a solution strategy
- The definition of *row echelon form*, and how it relates to solving systems of equations
- The definition, notation, and use of the three *elementary row operations*

Gauss-Jordan Elimination and Reduced Row Echelon Form
- The meaning of *Gauss-Jordan elimination* as a solution strategy
- The definition of *reduced row echelon form*

10.3: Determinants and Cramer's Rule

Determinants and Their Evaluation
- The definition of the *determinant* of a matrix
- The definition of a matrix element's *minor* and *cofactor*, and the use of minors and cofactors in evaluating determinants
- The use of properties of determinants and elementary row and column operations to simplify the computation of determinants

Using Cramer's Rule to Solve Linear Systems
- The formulas that go by the name Cramer's rule, and how these formulas can be used to solve linear systems of equations

10.4: The Algebra of Matrices

Matrix Addition
- The definition of *matrix equality*
- The meaning of *matrix addition*, and how to perform it

Scalar Multiplication
- The meaning of *scalar multiplication*, and how to perform it

Matrix Multiplication
- The meaning of *matrix multiplication*, how it relates to composition of functions, and how to perform it

10.5: Inverses of Matrices

The Matrix Form of a Linear System
- How to write a linear system of equations as a single matrix equation

Finding the Inverse of a Matrix
- The meaning of the *identity matrix*, and the effect of multiplying a given square matrix by the appropriate identity
- The meaning of the *inverse* of a matrix, and why the inverse of a matrix is useful in solving a given system
- The algorithm used to find the inverse of a matrix

Using Matrix Inverses to Solve Linear Systems
- How to solve a system of equation using the inverse matrix method

10.6: Partial Fraction Decomposition

The Pattern of Decompositions
- The definition of *partial fraction decomposition*
- How to write a rational function in the form of a partial fraction decomposition

Completing the Decomposition Process
- Finding the partial fraction decomposition of a rational function

10.7: Linear Programming

Planar Feasible Regions
- The definition of feasible region or region of constraint
- How to construct graphs of the feasible region

Linear Programming in Two Variables
- How to use the linear programming method
- How to find the objective function

10.8: Nonlinear Systems of Equations and Inequalities

Approximating Solutions by Graphing
- Using graphs of systems of nonlinear equations to gain an approximate understanding of the real-number solutions to expect

Solving Nonlinear Systems Algebraically
- Using algebra to determine all the solutions of nonlinear systems of equations

Solving Nonlinear Systems of Inequalities
- Using algebra to determine the graphs and solutions of nonlinear systems of inequalities

Chapter Review

Section 10.1

1. $\varnothing$

2. $(4,-5,5)$

3. $(3,0,2)$

4. $(2,-1)$

5. $\varnothing$

6. $(-2,2)$

7. $\left\{(3-3y,y)\big| y \in \mathbb{R}\right\}$

8. $(0,-3)$

9. $(3,2)$

10. $(-1,1,2)$

11. $(8,12,10)$

12. $\varnothing$

13. $y = \dfrac{4}{15}x^2 - x + \dfrac{11}{15}$

14. a. 3×3

  b. -8

  c. 7

15. a. 1×4

  b. 8

  c. None

Use any convenient method to solve the following systems of equations. If a system is dependent, express the solution set in terms of one or more of the variables, as appropriate.

1. $\begin{cases} 3x - y + z = 2 \\ -x + y - 2z = -4 \\ -6x + 2y - 2z = -7 \end{cases}$

2. $\begin{cases} 2x - y = 13 \\ 5x - 2y - z = 25 \\ 7x - 6z = -2 \end{cases}$

3. $\begin{cases} x + y - z = 1 \\ 3x - 4y - 5z = -1 \\ 6x - 3y + z = 20 \end{cases}$

4. $\begin{cases} 6x - 5y = 17 \\ -4x + 9y = -17 \end{cases}$

5. $\begin{cases} 3x - y = 2 \\ -6x + 2y = 5 \end{cases}$

6. $\begin{cases} 3x - 2y = -10 \\ x + 2y = 2 \end{cases}$

7. $\begin{cases} \dfrac{x}{3} + y - 1 = 0 \\ x + 3y = 3 \end{cases}$

8. $\begin{cases} \dfrac{x}{3} - \dfrac{y+1}{2} = 1 \\ \dfrac{x}{2} - \dfrac{y}{4} = \dfrac{3}{4} \end{cases}$

9. $\begin{cases} x + y = 5 \\ 2x - y = 4 \\ 5x + y = 17 \end{cases}$

10. $\begin{cases} 2x + 3y - 4z = -7 \\ x - y + 4z = 6 \\ x + y + z = 2 \end{cases}$

11. $\begin{cases} 3x - 2y + z = 10 \\ x + y + z = 30 \\ 2x - y - z = -6 \end{cases}$

12. $\begin{cases} 3x - 2y - 2z = -8 \\ x - y - z = -5 \\ x + y + z = -3 \end{cases}$

13. Find the equation of a parabola $y = ax^2 + bx + c$, passing through the points $(1,0)$, $(-4,9)$, and $(-1,2)$.

Section 10.2

14. Let $A = \begin{bmatrix} 2 & -8 & 9 \\ 7 & 3 & 0 \\ 11 & 6 & 1 \end{bmatrix}$. Determine the following, if possible:

  a. The order of A
  b. The value of a_{12}
  c. The value of a_{21}

15. Let $B = \begin{bmatrix} 13 & 8 & 20 & 5 \end{bmatrix}$. Determine the following, if possible:

  a. The order of B
  b. The value of b_{12}
  c. The value of b_{31}

16. $\begin{bmatrix} 2 & 1 & -1 & | & 3 \\ -2 & 3 & -1 & | & 2 \\ 6 & -8 & 1 & | & 3 \end{bmatrix}$

17. $\begin{bmatrix} 4 & 5 & -1 & | & 0 \\ 1 & 3 & 2 & | & 3 \\ 10 & -1 & -6 & | & 0 \end{bmatrix}$

18. $\begin{cases} 8x - 2y = 2 \\ -x + 5y = 3 \end{cases}$

19. $\begin{cases} 8x + 7z = 5 \\ -3y + 4z = 16 \\ 16x - 2y + z = 2 \end{cases}$

20. $\begin{cases} 3x - 7y + 6z = 9 \\ -11x + 3z = -14 \\ 8z = 2 \end{cases}$

21. $\begin{bmatrix} 0 & -5 & | & -11 \\ 1 & 2 & | & 3 \end{bmatrix}$

22. $\begin{bmatrix} 2 & 3 & | & 5 \\ 0 & 5 & | & 12 \end{bmatrix}$

23. $\begin{bmatrix} 1 & -4 & | & -4 \\ 1 & 7 & | & 11 \end{bmatrix}$

24. $\begin{bmatrix} -1 & 0 & 2 & | & -6 \\ 2 & -6 & 8 & | & 2 \\ -1 & -1 & -5 & | & 6 \end{bmatrix}$

25. $(2,-1)$

26. $\left(\dfrac{-50}{3}, -16\right)$

27. $(3,-5)$

28. $\varnothing$

29. $2x^4$

30. 5

31. 7

32. x^8

Construct the augmented matrix that corresponds to each of the following systems of equations.

16. $\begin{cases} 2x + (y - z) = 3 \\ 2(y - x) + y - 2 = z \\ 3x - \dfrac{3 - z}{2} = 4y \end{cases}$

17. $\begin{cases} z - 4x = 5y \\ 14z + 7(x + 3y) = 21 \\ 8x - y = -2(x - 3z) \end{cases}$

Construct the system of equations that corresponds to each of the following matrices.

18. $\begin{bmatrix} 8 & -2 & | & 2 \\ -1 & 5 & | & 3 \end{bmatrix}$

19. $\begin{bmatrix} 8 & 0 & 7 & | & 5 \\ 0 & -3 & 4 & | & 16 \\ 16 & -2 & 1 & | & 2 \end{bmatrix}$

20. $\begin{bmatrix} 3 & -7 & 6 & | & 9 \\ -11 & 0 & 3 & | & -14 \\ 0 & 0 & 8 & | & 2 \end{bmatrix}$

Fill in the blanks by performing the indicated elementary row operations.

21. $\begin{bmatrix} 3 & 1 & | & -2 \\ 1 & 2 & | & 3 \end{bmatrix} \xrightarrow{-3R_2 + R_1} \underline{\quad ? \quad}$

22. $\begin{bmatrix} 2 & 3 & | & 5 \\ -4 & -1 & | & 2 \end{bmatrix} \xrightarrow{2R_1 + R_2} \underline{\quad ? \quad}$

23. $\begin{bmatrix} 1 & -4 & | & -4 \\ 3 & -1 & | & 3 \end{bmatrix} \xrightarrow{-2R_1 + R_2} \underline{\quad ? \quad}$

24. $\begin{bmatrix} -1 & 0 & 2 & | & -6 \\ 1 & -3 & 4 & | & 1 \\ -2 & -1 & -3 & | & 0 \end{bmatrix} \xrightarrow[-R_1 + R_3]{2R_2} \underline{\quad ? \quad}$

Use Gaussian elimination and back-substitution to solve the following systems of equations.

25. $\begin{cases} 3x - y = 7 \\ x - 4y = 6 \end{cases}$

26. $\begin{cases} \dfrac{x}{5} - \dfrac{y}{3} = 2 \\ -6x + 5y = 20 \end{cases}$

Use Gauss-Jordan elimination to solve the following systems of equations.

27. $\begin{cases} 5x - 4y = 35 \\ 25x - 18y = 165 \end{cases}$

28. $\begin{cases} x - 3y - 4z = -5 \\ -x + 7y + 8z = 17 \\ 2x - 10y - 12z = -10 \end{cases}$

Section 10.3

Evaluate the following determinants.

29. $\begin{vmatrix} x^3 & -x^2 \\ x^2 & x \end{vmatrix}$

30. $\begin{vmatrix} -1 & 3 & 1 \\ 1 & -4 & 0 \\ 0 & 2 & 3 \end{vmatrix}$

31. $\begin{vmatrix} -2 & -1 & -3 & 0 \\ 3 & 3 & 1 & 5 \\ 4 & 0 & 0 & 1 \\ 2 & 0 & 1 & 0 \end{vmatrix}$

32. $\begin{vmatrix} x^4 & x & x & 2x \\ 0 & x & x^3 & x \\ 0 & 0 & x & x \\ 0 & 0 & 0 & x^2 \end{vmatrix}$

33. $9, -9$

34. $-15, -15$

35. $(-4, 1)$

36. $(2, -1, 1)$

37. $\varnothing$

38. $(4, 2, 0)$

39. $\begin{bmatrix} 4 & -16 & 6 \\ -5 & 8 & 12 \end{bmatrix}$

40. $\begin{bmatrix} 4 & 0 \\ -5 & 9 \end{bmatrix}$

41. Not possible

42. Not possible

43. $\begin{bmatrix} 9 & -23 & 3 \\ 5 & -3 & 8 \end{bmatrix}$

44. $\begin{bmatrix} 47 & -77 \\ 34 & 4 \end{bmatrix}$

45. $w = -2, x = 1,$
$\quad y = 3, z = -4$

46. Not possible

47. $x = 2, y = -3$

48. $x = 1, y = -1$

49. $\begin{bmatrix} 12 & 46 \end{bmatrix}$

50. $\begin{bmatrix} -12 & 8 & 12 \\ -15 & 10 & 15 \\ -18 & 12 & 18 \end{bmatrix}$

Use the matrix $A = \begin{bmatrix} 0 & -3 & 1 \\ 2 & 0 & 5 \\ -1 & 3 & 2 \end{bmatrix}$ to evaluate the minor and cofactor of the following elements.

33. a_{12}

34. a_{31}

Use Cramer's rule to solve the following systems.

35. $\begin{cases} x + 6y = 2 \\ 3x - y = -13 \end{cases}$

36. $\begin{cases} x + 2y - 3z = -3 \\ -5x - y + 4z = -5 \\ 3x + y + z = 6 \end{cases}$

37. $\begin{cases} -4x + 2y = 3 \\ 2x - y = 4 \end{cases}$

38. $\begin{cases} x - 2y = 0 \\ x + y + z = 6 \\ 3x - y - 4z = 10 \end{cases}$

Section 10.4

Given $A = \begin{bmatrix} 2 & -8 & 3 \\ -1 & 0 & 5 \end{bmatrix}$, $B = \begin{bmatrix} 2 & 0 \\ -1 & 3 \end{bmatrix}$, $C = \begin{bmatrix} 3 & 1 & -6 \\ 8 & -3 & -7 \end{bmatrix}$, and $D = \begin{bmatrix} 0 & 4 \\ -3 & 11 \\ 7 & 1 \end{bmatrix}$,

determine the following, if possible.

39. BA

40. B^2

41. $CD + C$

42. BD

43. $3A + C$

44. $AD + B$

Determine values of the variables that will make the following equations true, if possible.

45. $\begin{bmatrix} w & 5x \\ 2y & z \end{bmatrix} - 3\begin{bmatrix} w & x \\ 2 & -z \end{bmatrix} = \begin{bmatrix} 4 & 2 \\ y-3 & -16 \end{bmatrix}$

46. $\begin{bmatrix} 4x & 2y^2 & z \end{bmatrix} = \begin{bmatrix} 12 \\ 18 \\ -2 \end{bmatrix}$

47. $2\begin{bmatrix} x \\ -3y \end{bmatrix} - \begin{bmatrix} y \\ 2x \end{bmatrix} = \begin{bmatrix} 7 \\ 14 \end{bmatrix}$

48. $\begin{bmatrix} 3x \\ 5y \end{bmatrix} - \begin{bmatrix} y \\ -2x \end{bmatrix} = \begin{bmatrix} 4 \\ -3 \end{bmatrix}$

Evaluate the following matrix products, if possible.

49. $\begin{bmatrix} 7 & 1 & -1 \end{bmatrix}\begin{bmatrix} 1 & 6 \\ 2 & 1 \\ -3 & -3 \end{bmatrix}$

50. $\begin{bmatrix} 4 \\ 5 \\ 6 \end{bmatrix}\begin{bmatrix} -3 & 2 & 3 \end{bmatrix}$

51. $\begin{bmatrix} 1 & -1 & 2 \\ 2 & -3 & -1 \\ -3 & 0 & 6 \end{bmatrix}\begin{bmatrix} x_1 \\ x_2 \\ x_3 \end{bmatrix} = \begin{bmatrix} -4 \\ 1 \\ 5 \end{bmatrix}$

52. $\begin{bmatrix} 3 & -1 & 1 \\ 2 & 0 & -5 \\ 4 & 3 & 0 \end{bmatrix}\begin{bmatrix} x \\ y \\ z \end{bmatrix} = \begin{bmatrix} 4 \\ 1 \\ 6 \end{bmatrix}$

53. $\begin{bmatrix} \dfrac{3}{16} & \dfrac{1}{8} \\ -\dfrac{1}{8} & \dfrac{1}{4} \end{bmatrix}$

54. $\begin{bmatrix} 1 & -2 \\ -\dfrac{1}{2} & 2 \end{bmatrix}$

55. Not possible

56. $\begin{bmatrix} \dfrac{1}{12} & \dfrac{1}{6} & \dfrac{11}{24} \\ \dfrac{5}{12} & -\dfrac{1}{6} & \dfrac{7}{24} \\ \dfrac{1}{12} & \dfrac{1}{6} & -\dfrac{1}{24} \end{bmatrix}$

57. No

58. Yes

59. Yes

60. No

61. $\left(1, -\dfrac{1}{3}\right)$

62. $(0, -9, 7)$

63. $(-15, 20, -33), (-2, 3, -4), (-9, 11, -17)$

Section 10.5

Write each of the following systems of equations as a single matrix equation.

51. $\begin{cases} x_1 - x_2 + 2x_3 = -4 \\ 2x_1 - 3x_2 - x_3 = 1 \\ -3x_1 + 6x_3 = 5 \end{cases}$

52. $\begin{cases} 3x - y + z = 4 \\ 2x - 5z = 1 \\ 4x + 3y - 6 = 0 \end{cases}$

Find the inverse of each of the following matrices, if possible.

53. $\begin{bmatrix} 4 & -2 \\ 2 & 3 \end{bmatrix}$

54. $\begin{bmatrix} 2 & 2 \\ \dfrac{1}{2} & 1 \end{bmatrix}$

55. $\begin{bmatrix} 4 & 12 \\ 3 & 9 \end{bmatrix}$

56. $\begin{bmatrix} -1 & 2 & 3 \\ 1 & -1 & 4 \\ 2 & 0 & -2 \end{bmatrix}$

For each set of matrices, determine if either matrix is the inverse of the other.

57. $\begin{bmatrix} 3 & 12 \\ 2 & 9 \end{bmatrix}, \begin{bmatrix} 1 & 4 \\ \dfrac{2}{3} & 3 \end{bmatrix}$

58. $\begin{bmatrix} 1 & -2 \\ -3 & 4 \end{bmatrix}, \begin{bmatrix} -2 & -1 \\ -\dfrac{3}{2} & -\dfrac{1}{2} \end{bmatrix}$

59. $\begin{bmatrix} -2 & 4 & -3 \\ 0 & 6 & -3 \\ 0 & 8 & -3 \end{bmatrix}, \begin{bmatrix} -\dfrac{1}{2} & 1 & -\dfrac{1}{2} \\ 0 & -\dfrac{1}{2} & \dfrac{1}{2} \\ 0 & -\dfrac{4}{3} & 1 \end{bmatrix}$

60. $\begin{bmatrix} 5 & -3 & 7 \\ 6 & 0 & 2 \\ -9 & 1 & 0 \end{bmatrix}, \begin{bmatrix} -9 & 2 & 1 \\ 7 & 0 & -3 \\ 1 & 8 & 2 \end{bmatrix}$

Solve the following systems by the inverse matrix method, if possible. If the inverse matrix method doesn't apply, use any other method to determine if the system is inconsistent or dependent.

61. $\begin{cases} 5x + 9y = 2 \\ -2x - 3y = -1 \end{cases}$

62. $\begin{cases} 2y + 3z = 3 \\ -2x = 0 \\ 8x + 4y + 5z = -1 \end{cases}$

Solve the following set of systems by the inverse matrix method.

63. $\begin{cases} 2x - z = 3 \\ x + 4y + 2z = -1 \\ x + y = 5 \end{cases}$ $\begin{cases} 2x - z = 0 \\ x + 4y + 2z = 2 \\ x + y = 1 \end{cases}$ $\begin{cases} 2x - z = -1 \\ x + 4y + 2z = 1 \\ x + y = 2 \end{cases}$

64. $\dfrac{A_1}{x}+\dfrac{A_2}{x^2}+\dfrac{A_3}{3x-1}$

$+\dfrac{A_4}{(3x-1)^2}$

65. $\dfrac{A_1}{x+4}+\dfrac{A_2}{x-1}$

66. $\dfrac{1}{x-1}+\dfrac{1}{x+1}$

67. $-\dfrac{3}{2(2x-5)^2}$

$+\dfrac{1}{2(2x-5)}$

68. $\dfrac{x}{(4+x^2)^2}+\dfrac{2}{(4+x^2)}$

69. $x\ge 20,\ y\ge 10,\ x\le 40$

$y\le 40,\ x+y\le 60$

70. $x\ge 6,\ y\ge 6,\ x\le 10,$

$y\le 8,\ x+y\ge 15$

71. Min $=0$ at $(0,0)$
Max $=30$ at $(5,0)$

72. Min $=20$ at $(0,10)$
Max $=50$ at $(10,0)$

73. Min $=8$ at $(0,2)$
Max $=24$ at
$\left(\dfrac{24}{7},\dfrac{12}{7}\right)$

74. Min $=350$ at $(5,0)$
Max $=1028$ at $(10,4)$

Section 10.6

Write the form of the partial fraction decomposition of each of the following rational functions. In each case, assume the degree of the numerator is less than the degree of the denominator.

64. $f(x)=\dfrac{p(x)}{9x^4-6x^3+x^2}$

65. $f(x)=\dfrac{p(x)}{x^2+3x-4}$

Find the partial fraction decomposition of each of the following rational functions.

66. $f(x)=\dfrac{2x}{(x-1)(x+1)}$

67. $f(x)=\dfrac{x-4}{(2x-5)^2}$

68. $f(x)=\dfrac{2x^2+x+8}{(x^2+4)^2}$

Section 10.7

Construct the constraints and graph the feasible regions for the following situations.

69. Each bag of nuts contains peanuts and cashews. The total number of nuts in the bag cannot exceed 60. There must be at least 20 peanuts and 10 cashews per bag. There can be no more than 40 peanuts or 40 cashews per bag. What is the region of constraint for the number of nuts per bag?

70. You wish to study at least 15 hours (over a 4-day span) for your upcoming statistics and biology tests. You need to study a minimum of 6 hours for each test. The maximum you wish to study for statistics is 10 hours and for biology is 8 hours. What is the region of constraint for the number of hours you should study for each test?

Find the minimum and maximum values of the given functions, subject to the given constraints.

71. Objective Function
$f(x,\ y)=6x+10y$
Constraints
$x\ge 0,\ y\ge 0,\ 2x+5y\le 10$

72. Objective Function
$f(x,\ y)=5x+2y$
Constraints
$x\ge 0,\ y\ge 0,\ x+y\le 10,$
$x+2y\ge 10,\ 2x+y\ge 10$

73. Objective Function
$f(x,\ y)=5x+4y$
Constraints
$x\ge 0,\ y\ge 0,\ 2x+3y\le 12,$
$3x+y\le 12,\ x+y\ge 2$

74. Objective Function
$f(x,\ y)=70x+82y$
Constraints
$x\ge 0,\ y\ge 0,\ x\le 10,\ y\le 20,$
$x+y\ge 5,\ x+2y\le 18$

75. 12 vases should be produced, 12 pitchers should be produced, Max profit: $660

76. 35 bionic arms, 15 bionic legs, Min cost: $24,000

77. $\{(0, -2), (3, 1)\}$

78. $\left\{ (i, -1), (-i, -1), \right.$

$\left(\dfrac{\sqrt{2}}{2}, \dfrac{1}{2} \right),$

$\left. \left(-\dfrac{\sqrt{2}}{2}, \dfrac{1}{2} \right) \right\}$

79. $\{(4, 3), (-4, 3),$ $(-4, -3), (4, -3)\}$

80. $\{(2, 1)\}$

81. 9 and 16

82. 36 mph and 24 mph

Use linear programming to answer the following questions.

75. Krueger's Pottery manufactures two kinds of hand-painted pottery: a vase and a pitcher. There are three processes to create the pottery: throwing (forming the pottery on the potter's wheel), baking, and painting. No more than 90 hours are available per day for throwing, only 120 hours are available per day for baking, and no more than 60 hours per day are available for painting. The vase requires 3 hours for throwing, 6 hours for baking, and 2 hours for painting. The pitcher requires 3 hours for throwing, 4 hours for baking, and 3 hours for painting. The profit for each vase is $25 and the profit for each pitcher is $30. How many of each piece of pottery should be produced a day to maximize profit? What would the maximum profit be if Krueger's produced this amount?

76. Pranas produces bionic arms and legs for those that are missing a limb. Pranas can produce at least 20, but no more than 60 arms in a week due to the lab limitations. They can produce at least 15, but no more than 40 legs in a week. To keep their research grant, the company must produce at least 50 limbs per week. It costs $450 to produce the bionic arm and $550 to produce the bionic leg. How many of each should be produced per week to minimize the cost? What would the minimum cost be if Pranas produced this amount?

Section 10.8

Use graphing to guess the real solution(s) of the following system, and then verify your answer algebraically.

77. $\begin{cases} (x-2)^2 + y = 2 \\ x - y = 2 \end{cases}$

Solve the following nonlinear systems algebraically. Be sure to check for nonreal solutions.

78. $\begin{cases} x^2 + 2y^2 = 1 \\ x^2 = y \end{cases}$ **79.** $\begin{cases} x^2 + y^2 = 25 \\ 2x^2 - y^2 = 23 \end{cases}$ **80.** $\begin{cases} y = (x-1)^2 \\ y + 8 = (x+1)^2 \end{cases}$

Solve the following problems involving nonlinear systems of equations.

81. The product of two positive integers is 144, and their sum is 25. What are the integers?

82. Stephen and Scott were driving the same 72-mile route, and they departed at the same time. After 30 minutes, Stephen was 6 miles ahead of Scott. If it took Scott one more hour than Stephen to reach their destination, how fast were they each driving?

83. None

84.

85.

86.

Draw the graph and determine whether the ordered pairs are solutions to the system of inequalities.

83. $\begin{cases} y^2 \le 9 - x^2 \\ y < |x| \\ y > -|x| \end{cases}$ **a.** $(2,5)$ **b.** $(7,8)$ **c.** $(5,0)$ **d.** $(3,4)$

Graph the following systems of inequalities.

84. $\begin{cases} y \le \sin x \\ y > -\sin x \end{cases}$ **85.** $\begin{cases} y \le \sqrt{x+1} \\ y > x^2 - 1 \end{cases}$ **86.** $\begin{cases} x^2 y \le 1 \\ 2y \le x^2 + 2 \\ y < 16x^2 \end{cases}$

Chapter Test

1. $(2,1)$

2. $\varnothing$

3. $\left(\dfrac{3}{2}, \dfrac{4}{3}, \dfrac{19}{6}\right)$

4. 15

5. $\begin{cases} 3x - y = 4 \\ 6x + 2y = 0 \end{cases}$

6. a. 4×2

   b. 3

   c. None

7. $\begin{bmatrix} 0 & 1 & -2 & | & 1 \\ 3 & 1 & -1 & | & 4 \\ 0 & -4 & 6 & | & -1 \end{bmatrix}$

8. $\begin{bmatrix} 2 & -5 & -1 & | & 4 \\ 1 & -1 & 1 & | & 6 \\ -4 & -3 & 1 & | & 4 \end{bmatrix}$

9. $\begin{bmatrix} 4 & -3 & | & -2 \\ 1 & -2 & | & 3 \end{bmatrix}$

10. $\begin{bmatrix} -9 & 6 & 10 & | & 5 \\ 5 & -10 & -8 & | & -7 \\ -1 & 2 & 4 & | & 3 \end{bmatrix}$

11. $(0, -3)$

12. $(1, 1, 1)$

Use any convenient method to solve the following systems of equations. If a system is dependent, express the solution set in terms of one or more of the variables, as appropriate.

1. $\begin{cases} x + 2y = 4 \\ 2x - y = 3 \end{cases}$

2. $\begin{cases} 3x - 2y = \dfrac{1}{2} \\ 6x - 4y = 3 \end{cases}$

3. $\begin{cases} x - 2y + z = 2 \\ 2x + y - 2z = -2 \\ -x + 3z = 8 \end{cases}$

4. Rose's age is half of what John's will be in 3 years and half of what Bob's was 3 years ago. If Bob's age is $\dfrac{11}{9}$ of John's age, how old is Rose?

5. Find a system of linear equations that has the solution $\left(\dfrac{2}{3}, -2\right)$. Answers may vary.

6. Let $X = \begin{bmatrix} 3 & -2 \\ -1 & 0 \\ -4 & 1 \\ 2 & 3 \end{bmatrix}$. Determine the following, if possible.

   a. The order of X **b.** The value of x_{42} **c.** The value of x_{23}

Construct the augmented matrix that corresponds to each of the following systems of equations.

7. $\begin{cases} y - 2z = 1 \\ 3x + y - z = 4 \\ 1 - 4y + 6z = 0 \end{cases}$

8. $\begin{cases} \dfrac{x}{5} - \dfrac{y}{2} = \dfrac{z+4}{10} \\ \dfrac{2}{x - y + z} = \dfrac{1}{3} \\ -x = 1 - \dfrac{z - 3y}{4} \end{cases}$

Fill in the blanks by performing the indicated row operations.

9. $\begin{bmatrix} 8 & -6 & | & -4 \\ 5 & -10 & | & 15 \end{bmatrix} \xrightarrow[\frac{1}{5}R_2]{\frac{1}{2}R_1} \; ?$

10. $\begin{bmatrix} -8 & 4 & 6 & | & 2 \\ 3 & -6 & 0 & | & -1 \\ -1 & 2 & 4 & | & 3 \end{bmatrix} \xrightarrow[-2R_3 + R_2]{R_3 + R_1} \; ?$

Use Gaussian elimination and back-substitution to solve the following systems of equations.

11. $\begin{cases} \dfrac{x}{2} - \dfrac{y}{3} = 1 \\ \dfrac{2x}{5} - \dfrac{5y}{3} = 5 \end{cases}$

12. $\begin{cases} \dfrac{x}{2} - \dfrac{3z}{2} = -1 \\ 2y + \dfrac{5z - 2}{2} = \dfrac{7}{2} \\ -6x + y - 2z = -7 \end{cases}$

13. -6

14. 0

15. $\{\pm 4\}$

16. $\{5, -4\}$

17. $(2, -2)$

18. $(3, -1, 1)$

19. Not possible

20. $\{(3, -4, t) \mid t \in \mathbb{R}\}$

21. $[-6 \quad 18]$

22. $\begin{bmatrix} 0 & -1 & -5 \\ 6 & -3 & 3 \\ 8 & -6 & -6 \end{bmatrix}$

23. Not possible

24. $\begin{bmatrix} 0 & 1 \\ -1 & 3 \end{bmatrix}$

25. $\begin{bmatrix} \dfrac{1}{3} & -\dfrac{1}{3} \\ 1 & 0 \end{bmatrix} \begin{bmatrix} x \\ y \end{bmatrix} = \begin{bmatrix} 4 \\ -1 \end{bmatrix}$

26. $\begin{bmatrix} 3 & -2 & 1 \\ 1 & -1 & 0 \\ 0 & 1 & -1 \end{bmatrix} \begin{bmatrix} x \\ y \\ z \end{bmatrix} = \begin{bmatrix} 1 \\ 6 \\ 4 \end{bmatrix}$

Evaluate the following determinants.

13. $\begin{vmatrix} 4 & -2 \\ 3 & -3 \end{vmatrix}$

14. $\begin{vmatrix} xy & -axy \\ bx & -xab \end{vmatrix}$

Solve for x by finding the determinant.

15. $\begin{vmatrix} x-1 & 5 \\ 3 & x+1 \end{vmatrix} = 0$

16. $\begin{vmatrix} x+2 & 2 \\ 7 & x-3 \end{vmatrix} = 0$

Use Cramer's rule to solve the following systems.

17. $\begin{cases} 3x - 4y = 14 \\ x + 3y = -4 \end{cases}$

18. $\begin{cases} x + y - 3z = -1 \\ 2x + 2y + z = 5 \\ -x + 3y - 2z = -8 \end{cases}$

Determine the values of the variables that will make the following equations true, if possible. Answers may vary.

19. $\begin{bmatrix} x \\ 2y \end{bmatrix} + \begin{bmatrix} x \\ -y \\ z \end{bmatrix} = \begin{bmatrix} 1 \\ 2 \\ -1 \end{bmatrix}$

20. $3 \begin{bmatrix} 2r & -s & t \end{bmatrix} - \begin{bmatrix} -r & 2s & 3t \end{bmatrix} = \begin{bmatrix} 21 & 20 & 0 \end{bmatrix}$

Evaluate the following matrix products, if possible.

21. $\begin{bmatrix} 1 & -2 & 3 & 4 \end{bmatrix} \begin{bmatrix} -2 & 0 \\ 3 & -1 \\ 2 & 4 \\ -1 & 1 \end{bmatrix}$

22. $\begin{bmatrix} -2 & 1 \\ 0 & 3 \\ -4 & 6 \end{bmatrix} \begin{bmatrix} 1 & 0 & 3 \\ 2 & -1 & 1 \end{bmatrix}$

Given $A = \begin{bmatrix} 4 & -1 \\ -2 & 3 \end{bmatrix}$, $B = \begin{bmatrix} -4 & 6 \end{bmatrix}$, $C = \begin{bmatrix} 4 \\ -1 \\ 3 \end{bmatrix}$, and $D = \begin{bmatrix} -4 & 2 \\ 1 & 0 \end{bmatrix}$, determine the following, if possible.

23. AB

24. $D + A$

Write each of the following systems as a single matrix equation.

25. $\begin{cases} \dfrac{x-y}{3} = 4 \\ x + 1 = 0 \end{cases}$

26. $\begin{cases} 3x - 2y + z = 1 \\ x - y = 6 \\ y - z = 4 \end{cases}$

27. $\begin{bmatrix} \dfrac{7}{11} & \dfrac{-3}{11} \\ \dfrac{-5}{11} & \dfrac{-1}{11} \end{bmatrix}$

28. Not possible

29. $\dfrac{A_1}{x+3} + \dfrac{A_2}{x+1}$

30. $\dfrac{A_1}{x-3} + \dfrac{B_1 x + C_1}{x^2+4}$

31. $\dfrac{1}{x+1} +$

$\dfrac{2}{x^2 - 2x + 3}$

32. $\dfrac{5}{x-2} - \dfrac{3}{x} - \dfrac{1}{x+2}$

33.

$900x + 1500y \le 40,000;$

$28x + 16y \le 1000$

34. $\{(2,1),(-2,1),$

$(-\sqrt{5},0),(\sqrt{5},0)\}$

35. $\{(i,3),(i,-3),$

$(-i,3),(-i,-3)\}$

36.

Find the inverse of each of the following matrices, if possible.

27. $\begin{bmatrix} \dfrac{1}{2} & -\dfrac{3}{2} \\ -\dfrac{5}{2} & -\dfrac{7}{2} \end{bmatrix}$

28. $\begin{bmatrix} 3 & -1 & 4 \\ 0 & 2 & -2 \\ -3 & 1 & -4 \end{bmatrix}$

Write the form of the partial fraction decomposition of each of the following rational functions. In each case, assume the degree of the numerator is less than the degree of the denominator.

29. $f(x) = \dfrac{p(x)}{x^2 + 4x + 3}$

30. $f(x) = \dfrac{p(x)}{x^3 - 3x^2 + 4x - 12}$

Find the partial fraction decomposition of each of the following rational functions.

31. $f(x) = \dfrac{x^2 + 5}{(x+1)(x^2 - 2x + 3)}$

32. $f(x) = \dfrac{x^2 + 12x + 12}{x^3 - 4x}$

33. An appliance store is stocking up for a sale featuring two types of refrigerators. Model A costs the store \$900 and is 28 cubic feet. Model B, which has new energy-saving technology, costs \$1500 and is 16 cubic feet. Given that the store has budgeted \$40,000 to spend on refrigerators for this sale and that there is only room on the sales floor for 1000 cubic feet of display, what is the region of constraint for the number of each type of refrigerator the store can purchase?

Solve the following nonlinear systems algebraically.

34. $\begin{cases} x^2 + y^2 = 5 \\ x^2 + y = 5 \end{cases}$

35. $\begin{cases} x^2 + y^2 = 8 \\ 2x^2 + 3y^2 = 25 \end{cases}$

Graph the following system of inequalities.

36. $\begin{cases} 2y - 3x < 4 \\ \qquad y \ge (x-1)^2 \end{cases}$

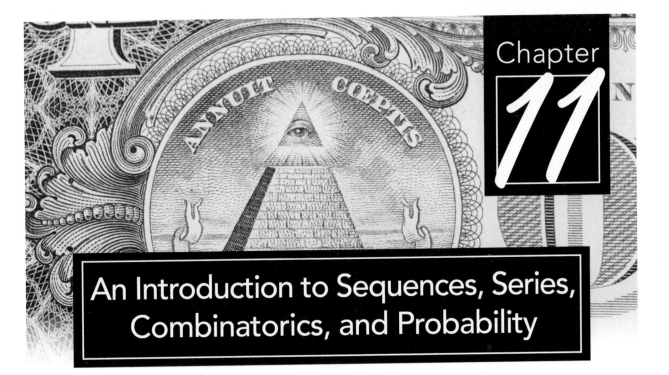

Chapter 11

An Introduction to Sequences, Series, Combinatorics, and Probability

11.1 Sequences and Series

11.2 Arithmetic Sequences and Series

11.3 Geometric Sequences and Series

11.4 Mathematical Induction

11.5 An Introduction to Combinatorics

11.6 An Introduction to Probability

Chapter 11 Project

Chapter 11 Summary

Chapter 11 Review

Chapter 11 Test

By the end of this chapter you should be able to:

What if your boss offered you a $1 raise this year, a $2 raise next year, a $4 raise the following year, and promised to continue doubling in this manner. How much of a raise would you receive after 20 years? (Hint: Your boss is not likely to agree to this deal!)

By the end of this chapter, you'll be able to find individual terms in sequences and series as well as their general formulas. You'll solve application problems similar to the one above, including one involving a king, grains of wheat, and a chess board on page 926. You'll master this type of problem using tools such as the formula for the Partial Sum of a Geometric Sequence, found on page 919.

Introduction

This last chapter introduces a variety of mathematical topics that, depending on your major, you are likely to encounter and study in greater detail in later courses.

Nearly everyone will, at some point, find it convenient to know the basic facts about sequences and series presented here. (This is not to say that everyone *will* use this material, but those who don't will be forced to find someone who does when faced with a sequence or series problem.) For instance, sequences and series arise in the financial world in the guise of retirement planning, portfolio analysis, and loan repayment; in sociology in the construction of population models; in the study of computer algorithms; and in medical models of epidemic spread. The counting techniques of *combinatorics*, a branch of mathematics, are immensely useful in a large number of surprisingly diverse situations; the fields of statistics and computer science, to name just two, make heavy use of the methods studied here. Finally, the science of *probability* finds applications ranging from the deadly serious (actuarial mortality tables) to the purely entertaining (the modern "gaming" industry).

Pascal

Probability, in particular, has a long, convoluted, and fascinating history. People have pondered questions of chance, and the possibility of analyzing or predicting fate, for all of recorded history (and longer, of course). Mathematicians, as a subset of the general population, have been no exception to this rule, but their understanding of probability has had no particularly mathematical aspect until relatively recently. While Italian mathematicians of the Renaissance made some small progress in the area, mathematical probability is usually said to have been born in France in the year 1654. It was in that year that a member of the French nobility, Antoine Gombaud, Chevalier de Méré, posed several questions born from his gambling experience to the mathematician Blaise Pascal (1623–1662). Pascal, intrigued by the questions, consulted with Pierre de Fermat (1601–1665), another mathematician famous today largely for his so-called "Last Theorem." In a series of letters back and forth, Pascal and Fermat laid the foundation of modern mathematical probability, a field with profound applications inside and outside of the world of gambling.

The material in this chapter can be viewed as a bridge between algebra and further study in mathematics. All of the topics introduced in this chapter rely on the mathematical maturity you have gained by studying algebra, but of course there are many more areas of math that are not explored at all in this book. Calculus, in particular, is one such area; many of you will soon be taking at least one semester of calculus, a field of mathematics that extends the static world of algebra to the dynamic world in which we live.

Sequences and Series

TOPICS

1. Recursively and explicitly defined sequences
2. Summation notation and a few formulas
3. Partial sums and series
4. Fibonacci sequences
T. Sequences and sums

TOPIC

Recursively and Explicitly Defined Sequences

Many natural and mathematical phenomena can be described with a simple list of numbers. For instance, the growth of an isolated population of rabbits over the course of many months might be described with a list of the number of rabbits born each month. When we write out a list of the sum of the first n odd numbers, we can recognize it as another familiar list, the squares of the integers:

$$1 = 1$$
$$1 + 3 = 4$$
$$1 + 3 + 5 = 9$$
$$1 + 3 + 5 + 7 = 16$$
$$1 + 3 + 5 + 7 + 9 = 25$$

These lists are called *sequences*, and we can think of them as a special type of function.

DEFINITION

Sequences

An **infinite sequence** (or just **sequence**) is a function whose domain is the set of natural numbers $\{1, 2, 3, \ldots\}$. Instead of giving the function a name such as f and referring to the values of the sequence as $f(1), f(2), f(3), \ldots$, we commonly use subscripts and refer to the **terms** of the sequence as $a_1, a_2, a_3, \ldots$. In other words, a_n stands for the nth term in the sequence.

Finite sequences are defined similarly, but the domain of a finite sequence is a set of the form $\{1, 2, 3, \ldots, k\}$ for some positive integer k.

A consequence of sequence notation is that we don't have a label, such as f, to attach to a sequence; instead, a given sequence is normally identified by some formula for determining a_n, the nth term of the sequence, and the entire sequence may be referred to as $\{a_n\}$. If it is necessary to discuss several sequences at once, use different letters to denote the different terms. For instance, a problem may refer to two sequences $\{a_n\}$ and $\{b_n\}$.

━━━━━━━━━━━━━━ EXAMPLE 1 ━━━━━━━━━━━━━━

Sequences   Determine the first five terms of the following sequences:

a. $a_n = 5n - 2$ **b.** $b_n = \dfrac{(-1)^n + 1}{2}$ **c.** $c_n = \dfrac{n}{n+1}$

Note:
Sequences often have distinct patterns. As you write out the terms for each sequence, see if you can predict the next term before calculating it.

Solutions:

a. Replacing n in the formula for the general term with the first five positive integers, we obtain

$$a_1 = 5(1) - 2, \quad a_2 = 5(2) - 2, \quad a_3 = 5(3) - 2, \quad a_4 = 5(4) - 2, \quad a_5 = 5(5) - 2,$$

so the sequence starts out as $3, 8, 13, 18, 23, \ldots$

b. Again, we replace n with the first five positive integers to determine the first five terms of the sequence.

$$b_1 = \frac{(-1)^1 + 1}{2} = \frac{0}{2} = 0$$

$$b_2 = \frac{(-1)^2 + 1}{2} = \frac{2}{2} = 1$$

$$b_3 = \frac{(-1)^3 + 1}{2} = \frac{0}{2} = 0$$

$$b_4 = \frac{(-1)^4 + 1}{2} = \frac{2}{2} = 1$$

$$b_5 = \frac{(-1)^5 + 1}{2} = \frac{0}{2} = 0$$

Note that $(-1)^n$ is equal to -1 if n is odd and is equal to 1 if n is even.

When we add 1 to $(-1)^n$ and divide the result by 2, we obtain either 0 or 1, so the sequence begins 0, 1, 0, 1, 0, ...

This example illustrates the fact that a given value may appear more than once in a sequence.

c. Replacing n in the formula for the general term with the first five positive integers, we obtain

$$c_1 = \frac{1}{1+1}, \quad c_2 = \frac{2}{2+1}, \quad c_3 = \frac{3}{3+1}, \quad c_4 = \frac{4}{4+1}, \quad c_5 = \frac{5}{5+1},$$

so the sequence starts out as $\dfrac{1}{2}, \dfrac{2}{3}, \dfrac{3}{4}, \dfrac{4}{5}, \dfrac{5}{6}, \ldots$

━━━━━━━━━━━━━━━━━━━━━━━━━━━━━━━━

The formulas for the general n^{th} terms in Example 1 are all examples of **explicit** formulas, named because they provide a direct rule for calculating any term in the sequence. In many cases, though, an explicit formula for the general term cannot be found, or is not as easily determined.

In such cases, the terms of a sequence are often defined *recursively*. A **recursive** formula is one that refers to one or more of the terms preceding a_n in the definition for a_n. For instance, if the first term of a sequence is -5 and if it is known that each of the remaining terms of the sequence is 7 more than the term preceding it, the sequence can be defined by the rules $a_1 = -5$ and $a_n = a_{n-1} + 7$ for $n \geq 2$.

EXAMPLE 2

Recursively Defined
Sequences

Determine the first five terms of the following recursively defined sequences.

a. $a_1 = 3$ and $a_n = a_{n-1} + 5$ for $n \geq 2$ **b.** $a_1 = 2$ and $a_n = 3a_{n-1} + 1$ for $n \geq 2$

Solutions:

a. We find the first five terms by replacing n with the first five positive integers, just as in Example 1. Note that in using the recursive definition we must determine the elements of the sequence in order; that is, to determine a_5, we need to first know a_4. And to determine a_4, we need to first know a_3, and so on back to a_1.

$$a_1 = 3$$
$$a_2 = a_1 + 5 = 3 + 5 = 8$$
$$a_3 = a_2 + 5 = 8 + 5 = 13$$
$$a_4 = a_3 + 5 = 13 + 5 = 18$$
$$a_5 = a_4 + 5 = 18 + 5 = 23$$

Thus, the sequence starts out as $3, 8, 13, 18, 23, \ldots$

The sequence defined by this recursive definition appears to be the same as the first sequence in Example 1, defined explicitly by $a_n = 5n - 2$. This illustrates the fact that a given sequence can often be defined several different ways.

b. Using the recursive formula $a_1 = 2$ and $a_n = 3a_{n-1} + 1$ for $n \geq 2$, we obtain

$$a_1 = 2$$
$$a_2 = 3a_1 + 1 = 3(2) + 1 = 7$$
$$a_3 = 3a_2 + 1 = 3(7) + 1 = 22$$
$$a_4 = 3a_3 + 1 = 3(22) + 1 = 67$$
$$a_5 = 3a_4 + 1 = 3(67) + 1 = 202$$

Thus, the sequence starts out as $2, 7, 22, 67, 202, \ldots$

In general, an explicit formula is more useful when finding the terms of a sequence than a recursive formula. Consider the task of calculating a_{100} based on the formula $a_n = 5n - 2$, versus the same task given the formula $a_1 = 3$ and $a_n = a_{n-1} + 5$ for $n \geq 2$.

To an extent, the problems in Examples 1 and 2 can be turned around, so that the question is finding a formula for the general n^{th} term of a sequence given its first few terms. This is often the challenge in modeling situations, where the goal is to extrapolate the behavior of a sequence of numbers by finding a formula that produces the first few terms of the sequence. The catch is that there is always more than one formula that will produce identical terms of a sequence up to a certain point and then differ beyond that. Consider the following two explicit formulas:

$$a_n = 3n \quad \text{and} \quad b_n = 3n + (n-1)(n-2)(n-3)(n-4)(n-5)n^2$$

These two formulas will produce identical results for the first five terms, but different results from then on. For this reason, the instructions in Example 3 ask for *possible* formulas for the given sequences.

EXAMPLE 3

Finding the Formula
for a Sequence

Find a possible formula for the general n^{th} term of the sequences that begin as follows.

a. $-3, 9, -27, 81, -243, \ldots$ **b.** $1, 3, 6, 10, 15, \ldots$

Solutions:

a. There is no general method to find a formula for the terms of a sequence. Observation is usually the best tool. If a formula for a given sequence does not come to mind quickly, it may help to associate each term of the sequence with its place in the sequence, as shown below:

$$\begin{array}{cccccc}
1 & 2 & 3 & 4 & 5 & \ldots \\
\updownarrow & \updownarrow & \updownarrow & \updownarrow & \updownarrow & \updownarrow \\
-3 & 9 & -27 & 81 & -243 & \ldots
\end{array}$$

If a pattern is still not apparent, try rewriting the terms of the sequence in different ways. In this case, factoring the terms leads to:

$$\begin{array}{cccccc}
1 & 2 & 3 & 4 & 5 & \ldots \\
\updownarrow & \updownarrow & \updownarrow & \updownarrow & \updownarrow & \updownarrow \\
-3^{1} & 3^{2} & -3^{3} & 3^{4} & -3^{5} & \ldots
\end{array}$$

The nth term of the sequence is the nth power of 3, multiplied by -1 if n is odd. One way to express alternating signs in a sequence is to multiply the nth term by $(-1)^{n}$ (if the odd terms are negative) or by $(-1)^{n+1}$ (if the even terms are negative). In this case, a possible formula for the general n^{th} term is $a_n = (-1)^{n}(3)^{n}$, or $a_n = (-3)^{n}$.

Note that we might also have come up with the recursive formula $a_1 = -3$ and $a_n = -3a_{n-1}$ for $n \geq 2$.

b. Associate each term with its place in the sequence, as follows:

$$\begin{array}{cccccc}
1 & 2 & 3 & 4 & 5 & \ldots \\
\updownarrow & \updownarrow & \updownarrow & \updownarrow & \updownarrow & \updownarrow \\
1 & 3 & 6 & 10 & 15 & \ldots
\end{array}$$

In this case, factoring the terms does not seem to help in identifying a pattern, but thinking about the difference between successive terms does:

$$\begin{array}{cccccc}
1 & 2 & 3 & 4 & 5 & \ldots \\
\updownarrow & \updownarrow & \updownarrow & \updownarrow & \updownarrow & \updownarrow \\
1 & 3=1+2 & 6=3+3 & 10=6+4 & 15=10+5 & \ldots
\end{array}$$

This observation leads to the recursive formula $a_1 = 1$ and $a_n = a_{n-1} + n$ for $n \geq 2$.

TOPIC 2

Summation Notation and a Few Formulas

One very common use of sequence notation is to define terms that are to be added together. If the first n terms of a given sequence are to be added, we can write the sum as $a_1 + a_2 + \ldots + a_n$, but this can be confusing and unwieldy. Further, this notation doesn't describe the terms being added.

Summation notation (also known as *sigma notation*) provides a better option. Summation notation borrows the capital Greek letter Σ ("sigma") to denote the operation of summation, as described below.

DEFINITION

Summation Notation

The sum $a_1 + a_2 + \ldots + a_n$ is expressed in **summation notation** as $\sum\limits_{i=1}^{n} a_i$.

When this notation is used, the letter i is called the **index of summation**, and a_i often appears as the formula for the ith term of a sequence. In the sum above, all the terms of the sequence beginning with a_1 and ending with a_n are to be added. The notation can be modified to indicate a different first or last term of the sum.

EXAMPLE 4

Evaluating Sums

Rewrite the following sums in expanded form, then evaluate them.

a. $\sum\limits_{i=1}^{4}(3i-2)$ **b.** $\sum\limits_{i=3}^{5} i^2$

Solutions:

a. $\sum\limits_{i=1}^{4}(3i-2) = (3\cdot 1-2)+(3\cdot 2-2)+(3\cdot 3-2)+(3\cdot 4-2)$ Replace the index i with the numbers 1 through 4.

$\qquad = 1+4+7+10$ Evaluate each term.

$\qquad = 22$ Sum.

b. $\sum\limits_{i=3}^{5} i^2 = 3^2+4^2+5^2$ Replace the index i with the numbers 3, 4, and 5.

$\qquad = 9+16+25$ Evaluate each term.

$\qquad = 50$ Sum.

Keep in mind that sigma notation is just a more concise way of representing addition. The following properties of sigma notation are just restatements of familiar properties of addition.

PROPERTIES

Properties of Sigma Notation

Let $\{a_n\}$ and $\{b_n\}$ be two sequences, and let c be a constant.

1. $\displaystyle\sum_{i=1}^{n}\left(a_i + b_i\right) = \sum_{i=1}^{n} a_i + \sum_{i=1}^{n} b_i$ (the terms of a sum can be rearranged)

2. $\displaystyle\sum_{i=1}^{n} c a_i = c \sum_{i=1}^{n} a_i$ (constants can be factored out of a sum)

3. $\displaystyle\sum_{i=1}^{n} a_i = \sum_{i=1}^{k} a_i + \sum_{i=k+1}^{n} a_i$ for any $1 \le k \le n-1$, (a sum can be broken apart into two smaller sums)

CAUTION!

There are many similar looking statements that are not true. For instance, it is not generally true that $\displaystyle\sum_{i=1}^{n} a_i b_i = \sum_{i=1}^{n} a_i \cdot \sum_{i=1}^{n} b_i$.

In addition to the arithmetic properties listed above, formulas for sums that occur frequently can be very useful when calculating more complicated sums. We will cover four such formulas in this section and prove two of them. Proofs of the other two formulas will be covered in Section 11.4.

THEOREM

Four Summation Formulas

1. $\displaystyle\sum_{i=1}^{n} 1 = n$

2. $\displaystyle\sum_{i=1}^{n} i = \frac{n(n+1)}{2}$

3. $\displaystyle\sum_{i=1}^{n} i^2 = \frac{n(n+1)(2n+1)}{6}$

4. $\displaystyle\sum_{i=1}^{n} i^3 = \frac{n^2(n+1)^2}{4}$

The first formula really requires no proof at all. We can see that it is true just by writing the sum in expanded form:

$$\sum_{i=1}^{n} 1 = \underbrace{1 + 1 + \ldots + 1}_{n \text{ terms}} = n.$$

One proof of the second formula begins by letting S stand for the sum; that is,

$$S = \sum_{i=1}^{n} i = 1+2+\ldots+n.$$

Since the addition can be performed in any order, it is also true that

$$S = n+(n-1)+\ldots+1.$$

Note, now, the result of adding these two equations:

$$
\begin{array}{ccccccc}
S & = & 1 & +2 & +\ldots & +n \\
S & = & n & +(n-1) & +\ldots & +1 \\
\hline
2S & = & (n+1) & +(n+1) & +\ldots & +(n+1)
\end{array}
$$

Since the term $n+1$ appears n times on the right-hand side of the bottom equation, we have $2S = n(n+1)$ or, after dividing both sides by 2,

$$\sum_{i=1}^{n} i = \frac{n(n+1)}{2}.$$

Note that this provides an explicit formula for the nth term of the sequence in Example 3b. Since the nth term is the sum $1+2+\ldots+n$, the formula is $a_n = \dfrac{n(n+1)}{2}$.

EXAMPLE 5

Evaluating Sums

Use the above properties and formulas to evaluate the following sums.

a. $\displaystyle\sum_{i=1}^{9}(7i-3)$

b. $\displaystyle\sum_{i=4}^{6}3i^2$

Note:
A good strategy is to break the sum into the simplest parts possible using the properties of sigma notation, then apply the known formulas.

Solutions:

a. $\displaystyle\sum_{i=1}^{9}(7i-3) = \sum_{i=1}^{9}7i - \sum_{i=1}^{9}3$

Apply the first property to split the given sum into two sums.

$\displaystyle = 7\sum_{i=1}^{9}i - 3\sum_{i=1}^{9}1$

Apply the second property to factor the constants out of each sum.

$\displaystyle = 7\left(\frac{9\cdot10}{2}\right) - 3\cdot9$

Use the summation formulas.

$= 315 - 27$

Simplify.

$= 288$

b. $\displaystyle\sum_{i=4}^{6} 3i^2 = 3\sum_{i=4}^{6} i^2$ Use the second property to factor out the 3.

$\displaystyle = 3\sum_{i=1}^{6} i^2 - 3\sum_{i=1}^{3} i^2$ Apply the third property of sigma notation to obtain two sums that begin with $i = 1$.

$\displaystyle = 3\left(\frac{6 \cdot 7 \cdot 13}{6}\right) - 3\left(\frac{3 \cdot 4 \cdot 7}{6}\right)$ Use a summation formula.

$= 273 - 42$ Simplify.

$= 231$

TOPIC 3 — Partial Sums and Series

Given a sequence $\{a_n\}$, we may need to calculate sums of the form $a_1 + a_2 + \ldots + a_n$ or the sum of *all* the terms, which we might express as $a_1 + a_2 + a_3 + \ldots$. The following definition formalizes these two ideas.

DEFINITION

Partial Sums and Series

Given an infinite sequence $\{a_n\}$, the **n^{th} partial sum** is $S_n = \displaystyle\sum_{i=1}^{n} a_i$. S_n is an example of a **finite series**. The **infinite series** associated with $\{a_n\}$ is the sum $\displaystyle\sum_{i=1}^{\infty} a_i$. Note that the adjective *infinite* refers to the infinite number of terms that appear in the sum and does not imply that the sum itself is infinite.

It may seem that the sum of an infinite number of terms can't possibly be a finite number, especially if all of the terms are positive. But in fact, many simple examples of infinite series with finite sums exist. Consider the fractions of an inch marked on a typical ruler.

Figure 1: Visualizing an Infinite Series

If half of an inch and a quarter of an inch are added, the result is three-quarters of an inch. If an additional eighth of an inch is added, the result is seven-eighths of an inch. If the ruler is of sufficient precision, a sixteenth of an inch can be added, resulting in a total of fifteen-sixteenths of an inch.

In principle, we can continue this process indefinitely, with the result that the sum of the fractions at every stage is closer to, but never exceeds, one inch. In the language of partial sums and series, we can say

$$S_n = \frac{1}{2} + \frac{1}{4} + \ldots + \frac{1}{2^n} = \frac{2^n - 1}{2^n}$$

and

$$S = \frac{1}{2} + \frac{1}{4} + \frac{1}{8} + \ldots = \sum_{i=1}^{\infty} \frac{1}{2^i} = 1.$$

An infinite series $\sum_{i=1}^{\infty} a_i$ **converges** if the sequence of partial sums $S_n = \sum_{i=1}^{n} a_i$ approaches a fixed real number S, and in this case we write $S = \sum_{i=1}^{\infty} a_i$.

If the sequence of partial sums does not approach some fixed real number (either by getting larger and larger in magnitude or by "bouncing around"), we say the series **diverges**. The next example illustrates both possible outcomes.

EXAMPLE 6

Evaluating Partial Sums and Series

Examine the partial sums associated with each infinite series to determine if the series converges or diverges.

a. $\displaystyle\sum_{i=1}^{\infty} \left(\frac{1}{i} - \frac{1}{i+1} \right)$ **b.** $\displaystyle\sum_{i=1}^{\infty} (-1)^i$

Solutions:

a. $S_1 = 1 - \dfrac{1}{2} = 1 - \dfrac{1}{1+1}$

Begin by evaluating the first few partial sums.

$S_2 = \left(1 - \dfrac{1}{2}\right) + \left(\dfrac{1}{2} - \dfrac{1}{3}\right)$

$\quad = 1 - \dfrac{\cancel{1}}{\cancel{2}} + \dfrac{\cancel{1}}{\cancel{2}} - \dfrac{1}{3} = 1 - \dfrac{1}{3} = 1 - \dfrac{1}{2+1}$

Writing the terms out, we see that all of the terms except the first and last one cancel out. Since the length of the expression collapses down, this is an example of a *telescoping series*.

$S_3 = \left(1 - \dfrac{1}{2}\right) + \left(\dfrac{1}{2} - \dfrac{1}{3}\right) + \left(\dfrac{1}{3} - \dfrac{1}{4}\right)$

$\quad = 1 - \dfrac{\cancel{1}}{\cancel{2}} + \dfrac{\cancel{1}}{\cancel{2}} - \dfrac{\cancel{1}}{\cancel{3}} + \dfrac{\cancel{1}}{\cancel{3}} - \dfrac{1}{4} = 1 - \dfrac{1}{4} = 1 - \dfrac{1}{3+1}$

$\vdots$

$S_n = 1 - \dfrac{1}{n+1}$

We can write a formula for the nth partial sum by observing the pattern.

The partial sums approach 1 as n gets larger and larger, therefore the series converges. In fact, $\displaystyle\sum_{i=1}^{\infty} \left(\frac{1}{i} - \frac{1}{i+1} \right) = 1.$

b. $S_1 = (-1)^1 = -1$

$S_2 = (-1)^1 + (-1)^2$

$= -1 + 1 = 0$

$S_3 = (-1)^1 + (-1)^2 + (-1)^3$

$= -1 + 1 + (-1) = -1$

$S_4 = (-1)^1 + (-1)^2 + (-1)^3 + (-1)^4$

$= -1 + 1 + (-1) + 1 = 0$

$\vdots$

$S_n = \begin{cases} -1 & \text{if } n \text{ is odd} \\ 0 & \text{if } n \text{ is even} \end{cases}$

Again, begin by calculating the first few partial sums to look for a pattern.

A pattern emerges based on whether n is even or odd.

The partial sums of this series oscillate between −1 and 0 and do not approach a fixed number as n gets large. Therefore, this series diverges.

TOPIC 4

Fibonacci Sequences

One of the most famous sequences in mathematics has a long and colorful history. Leonardo Fibonacci (meaning "Leonardo, son of Bonaccio") was born in Pisa, in approximately the year 1175, and was one of the first of many famous Italian mathematicians of the Middle Ages. As a child and young adult, he traveled with his father (a merchant) to ports in northern Africa and the Middle East where Arab scholars had collected, preserved, and expanded the mathematics of many different cultures. Among other achievements, Leonardo is known for the book he wrote after returning to Italy, the *Liber abaci*, which exposed European scholars to Hindu-Arabic notation and the mathematics he had learned, as well as to his own contributions.

One of the problems in the *Liber abaci* gives rise to what we now call the Fibonacci sequence: $1, 1, 2, 3, 5, 8, 13, 21, \ldots$. The problem, loosely translated, asks

> *How many pairs of rabbits can a single pair produce, if every month each pair of rabbits can beget a new pair and if each pair becomes productive in the second month of existence?*

We can solve this problem using the sequence notation from this lesson. In the first month, we have just the initial pair of rabbits, so

$$a_1 = 1.$$

In the second month, this pair is still too young to reproduce, so again there is just one pair, meaning

$$a_1 = 1, \ a_2 = 1.$$

In the third month, the pair produces a new pair of rabbits, giving a total of 2:

$$a_1 = 1, \ a_2 = 1, \ a_3 = 2.$$

In the fourth month, we have 2 existing pairs, and one new pair (since the rabbits born in the third month are not yet old enough to reproduce), thus

$$a_1 = 1, \; a_2 = 1, \; a_3 = 2, \; a_4 = 3.$$

Can we now deduce a recursive formula for the number of pairs in the next month? For the fifth month $(n = 5)$, the number of pairs will be equal to the number of existing pairs, which is equal to a_4, plus the number of new pairs. Since every pair that is at least 2 months old produces 1 new pair, this is equal to a_3. Thus, we have

$$a_5 = a_4 + a_3$$
$$= 3 + 2 = 5$$

This logic applies to any month beginning with the third, so we can define the Fibonacci sequence recursively.

$$a_1 = 1, \; a_2 = 1 \text{ and } a_n = a_{n-1} + a_{n-2} \text{ for } n \ge 3.$$

Generalized Fibonacci sequences are defined similarly, with the first two terms given and each successive term defined as the sum of the previous two.

TOPIC

Sequences and Sums

If we are given an explicit formula for a sequence, like $a_n = 3n^2 - 8$, we can use a graphing calculator to find specific terms in that sequence. First, press **Y=** and enter the sequence as Y1, using **X,T,θ,n** for "n."

```
Plot1  Plot2  Plot3
\Y1■3X^2-8
\Y2=
\Y3=
\Y4=
\Y5=
\Y6=
\Y7=
```

QUIT by pressing **2ND** **MODE** and select **VARS** . Use the right arrow to highlight Y-VARS and press ENTER to select Function. Since we entered the sequence as Y1, press ENTER to select Y1. Then, use parentheses and enter the desired value for n.

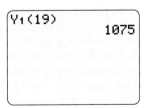

This tells us that the 19th term of the sequence, a_{19}, is 1075.

1. Infinite

2. Infinite

3. Finite

4. Infinite

5. Finite

6. Infinite

7. Infinite

8. Finite

9. Infinite

10. $4, 11, 18, 25, 32$

11. $2, -1, -4, -7, -10$

12. $-2, 4, -8, 16, -32$

13. $1, \dfrac{3}{2}, \dfrac{9}{5}, 2, \dfrac{15}{7}$

14. $-1, \dfrac{1}{4}, -\dfrac{1}{9}, \dfrac{1}{16}, -\dfrac{1}{25}$

15. $\dfrac{2}{3}, -\dfrac{4}{9}, \dfrac{8}{27},$
$-\dfrac{16}{81}, \dfrac{32}{243}$

16. $1, -\dfrac{1}{3}, \dfrac{1}{9}, -\dfrac{1}{27}, \dfrac{1}{81}$

17. $\dfrac{1}{2}, \dfrac{4}{3}, \dfrac{9}{4}, \dfrac{16}{5}, \dfrac{25}{6}$

18. $0, \dfrac{1}{9}, \dfrac{1}{4}, \dfrac{9}{25}, \dfrac{4}{9}$

19. $-1, 6, -5, 20, -27$

20. $26, 7, 0, -1, -2$

21. $2, 2, \dfrac{18}{7}, \dfrac{16}{5}, \dfrac{50}{13}$

22. $-1, \sqrt{2}, -\sqrt{3}, 2, -\sqrt{5}$

23. $2, 1, \dfrac{8}{9}, 1, \dfrac{32}{25}$

24. $1, 5, 9, 13, 17$

25. $10, 5, 0, -5, -10$

26. $\dfrac{1}{2}, 1, 2, 4, 8$

We can use graphing calculators to evaluate partial sums as well as finding terms in a sequence. Suppose we wanted to evaluate $\displaystyle\sum_{i=2}^{7}\left(3i^2 - 8\right)$.

To find a sum, press **2ND** **STAT** to access the LIST menu. Use the arrow key to highlight MATH and select 5 : sum (and press ENTER).

In the parentheses for "sum" we need to enter the sequence $3i^2 - 8$. To do so, press **2ND** **STAT** to access the LIST menu again, but this time highlight OPS and select 5 : seq (and press ENTER).

In the parentheses for "sequence," type the sequence, using **X,T,θ,n** for the variable. Use the **,** button to insert a comma and then type the name of the variable, again using **X,T,θ,n**, the beginning value, 2, the ending value, 7, and the increment, which is 1. Each of these should be separated by a comma. We need to close the parentheses for both the sequence and the sum, so type two right-hand parentheses and press ENTER.

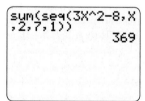

So $\displaystyle\sum_{i=2}^{7}\left(3i^2 - 8\right) = 369$.

27. $\dfrac{1}{27}, \dfrac{1}{81}, \dfrac{1}{243},$
    $\dfrac{1}{729}, \dfrac{1}{2187}$

28. $3, 36, 729, 20,736,$
    $759,375$

29. $8, 2^{\frac{3}{2}}, 2^{\frac{3}{4}}, 2^{\frac{3}{8}}, 2^{\frac{3}{16}}$

30. $\dfrac{5}{4}, 2, \dfrac{5}{2}, \dfrac{20}{7}, \dfrac{25}{8}$

31. $\dfrac{1}{3}, 1, \dfrac{9}{5}, \dfrac{8}{3}, \dfrac{25}{7}$

32. $1, 3, 6, 10, 15$

33. $-1, 2, -3, 4, -5$

34. Undefined, $9, 4,$
    $\dfrac{25}{9}, \dfrac{9}{4}$

35. $2, 6, 12, 20, 30$

36. $\dfrac{1}{3}, \dfrac{1}{2}, \dfrac{5}{9}, \dfrac{7}{12}, \dfrac{3}{5}$

37. $\sqrt{3}+1, \sqrt{6}+1, 4,$
    $2\sqrt{3}+1, \sqrt{15}+1$

38. $0, -1, -4, -9, -16$

39. $0, -4, 0, 18, 56$

40. $2, 4, 16, 256,$
    $65,536$

41. $-2, -11, -74,$
    $-515, -3602$

42. $1, 2, 6, 24, 120$

43. $-1, 0, -1, 0, -1$

44. $2, \sqrt{5}, \sqrt{6},$
    $\sqrt{7}, 2\sqrt{2}$

45. $3, 1, 3, 1, 3$

46. $a_n = 7n - 2$

47. $a_n = (-2)^n$

48. $a_1 = -1,$
    $a_n = -na_{n-1}, n \geq 2$

Exercises

Determine if each of the sequences given is finite or infinite.

1. The sequence of odd numbers

2. $2, 4, 6, 8, 10, 12\ldots$

3. $1, 3, 5, 7, 9, 11$

4. $3, 9, 27, 81, 243\ldots$

5. The sequence of the days of the week

6. $1, 0, 1, 0, 1, 0, 1, 0\ldots$

7. $1, 2, 3, 4, 5, 6, 7\ldots$

8. The sequence of letters in the alphabet

9. The sequence of the number of ants in a colony recorded daily

Determine the first five terms of the sequences whose n^{th} terms are defined as follows. See Examples 1 and 2.

10. $a_n = 7n - 3$

11. $a_n = -3n + 5$

12. $a_n = (-2)^n$

13. $a_n = \dfrac{3n}{n+2}$

14. $a_n = \dfrac{(-1)^n}{n^2}$

15. $a_n = \dfrac{(-1)^{n+1} 2^n}{3^n}$

16. $a_n = \left(-\dfrac{1}{3}\right)^{n-1}$

17. $a_n = \dfrac{n^2}{n+1}$

18. $a_n = \dfrac{(n-1)^2}{(n+1)^2}$

19. $a_n = (-2)^n + n$

20. $a_n = (-n+4)^3 - 1$

21. $a_n = \dfrac{2n^2}{3n-2}$

22. $a_n = (-1)^n \sqrt{n}$

23. $a_n = \dfrac{2^n}{n^2}$

24. $a_n = 4n - 3$

25. $a_n = -5n + 15$

26. $a_n = 2^{n-2}$

27. $a_n = 3^{-n-2}$

28. $a_n = (3n)^n$

29. $a_n = \sqrt[2^n]{64}$

30. $a_n = \dfrac{5n}{n+3}$

31. $a_n = \dfrac{n^2}{n+2}$

32. $a_n = \dfrac{n^2+n}{2}$

33. $a_n = (-1)^n n$

34. $a_n = \dfrac{(n+1)^2}{(n-1)^2}$

35. $a_n = n^2 + n$

36. $a_n = \dfrac{2n-1}{3n}$

37. $a_n = \sqrt{3n} + 1$

38. $a_n = -(n-1)^2$

39. $a_n = (n-1)(n+2)(n-3)$

40. $a_1 = 2$ and $a_n = (a_{n-1})^2$ for $n \geq 2$

41. $a_1 = -2$ and $a_n = 7a_{n-1} + 3$ for $n \geq 2$

42. $a_1 = 1$ and $a_n = na_{n-1}$ for $n \geq 2$

43. $a_1 = -1$ and $a_n = -a_{n-1} - 1$ for $n \geq 2$

44. $a_1 = 2$ and $a_n = \sqrt{(a_{n-1})^2 + 1}$ for $n \geq 2$

45. $a_1 = 3$, $a_2 = 1$, and $a_n = (a_{n-2})^{a_{n-1}}$ for $n \geq 3$

49. $a_n = \dfrac{n}{n+2}$

50. $a_n = \left(\dfrac{1}{n}\right)^2$

51. $a_1 = 1$,

$a_n = \dfrac{1}{n}a_{n-1}, n \ge 2$

52. $a_n = 9n - 43$

53. $a_n = \dfrac{4-n}{n+13}$

54. $a_n = 2^{n-3}$

55. $a_n = 4 - 5n$

56. $a_n = \dfrac{n}{2^n}$

57. $a_1 = 1$, $a_n = n + n\left(a_{n-1}\right), n \ge 2$

58. $-2 + 1 + 4 + 7 + 10 + 13 + 16 = 49$

59. $-3 - 12 - 27 - 48 - 75 = -165$

60. $\displaystyle\sum_{i=1}^{6} i^3 = 441$

61. $\displaystyle\sum_{i=0}^{7}(1 + 3i) = 92$

62. $45 + 80 + 125 + 180 + 245 + 320 + 405 + 500 = 1900$

63. $\displaystyle\sum_{i=3}^{9} i^2 = 280$

64. $-6 - 12 - 24 - 48 - 96 - 192 = -378$

65. $-36 - 30 - 22 - 12 + 0 + 14 + 30 + 48 = -8$

Find a possible formula for the general n^{th} term of the sequences that begin as follows. Answers may vary. See Example 3.

46. $5, 12, 19, 26, 33, \ldots$

47. $-2, 4, -8, 16, -32, \ldots$

48. $-1, 2, -6, 24, -120, \ldots$

49. $\dfrac{1}{3}, \dfrac{2}{4}, \dfrac{3}{5}, \dfrac{4}{6}, \dfrac{5}{7}, \ldots$

50. $1, \dfrac{1}{4}, \dfrac{1}{9}, \dfrac{1}{16}, \dfrac{1}{25}, \ldots$

51. $1, \dfrac{1}{2}, \dfrac{1}{6}, \dfrac{1}{24}, \dfrac{1}{120}, \ldots$

52. $-34, -25, -16, -7, 2, \ldots$

53. $\dfrac{3}{14}, \dfrac{2}{15}, \dfrac{1}{16}, 0, -\dfrac{1}{18}, \ldots$

54. $\dfrac{1}{4}, \dfrac{1}{2}, 1, 2, 4, \ldots$

55. $-1, -6, -11, -16, -21, \ldots$

56. $\dfrac{1}{2}, \dfrac{1}{2}, \dfrac{3}{8}, \dfrac{1}{4}, \dfrac{5}{32}, \ldots$

57. $1, 4, 15, 64, 325, \ldots$

Translate each expanded sum that follows into summation notation, and vice versa. Then use the formulas and properties from this section to evaluate the sums. See Examples 4 and 5.

58. $\displaystyle\sum_{i=1}^{7}(3i - 5)$

59. $\displaystyle\sum_{i=1}^{5} -3i^2$

60. $1 + 8 + 27 + \ldots + 216$

61. $1 + 4 + 7 + \ldots + 22$

62. $\displaystyle\sum_{i=3}^{10} 5i^2$

63. $9 + 16 + 25 + \ldots + 81$

64. $\displaystyle\sum_{i=1}^{6} -3(2)^i$

65. $\displaystyle\sum_{i=6}^{13}(i + 3)(i - 10)$

66. $9 + 27 + 81 + \ldots + 19{,}683$

Find a formula for the nth partial sum S_n of each of the following series. If the series is finite, determine the sum. If the series is infinite, determine if it converges or diverges, and if it converges, determine the sum. See Example 6.

67. $\displaystyle\sum_{i=1}^{100}\left(\dfrac{1}{i+3} - \dfrac{1}{i+4}\right)$

68. $\displaystyle\sum_{i=1}^{\infty}\left(\dfrac{1}{i+3} - \dfrac{1}{i+4}\right)$

69. $\displaystyle\sum_{i=1}^{\infty}\left(2^i - 2^{i-1}\right)$

70. $\displaystyle\sum_{i=1}^{15}\left(2^i - 2^{i-1}\right)$

71. $\displaystyle\sum_{i=1}^{49}\left(\dfrac{1}{2i} - \dfrac{1}{2i+2}\right)$

72. $\displaystyle\sum_{i=1}^{\infty}\left(\dfrac{1}{2i} - \dfrac{1}{2i+2}\right)$

73. $\displaystyle\sum_{i=1}^{100} \ln\left(\dfrac{i}{i+1}\right)$ (**Hint:** Make use of a property of logarithms to rewrite the sum.)

74. $\displaystyle\sum_{i=1}^{\infty} \ln\left(\dfrac{i}{i+1}\right)$ (**Hint:** Make use of a property of logarithms to rewrite the sum.)

75. $\displaystyle\sum_{i=1}^{30}\left(\dfrac{1}{2i+5} - \dfrac{1}{2i+7}\right)$ **76.** $\displaystyle\sum_{i=1}^{\infty}\left(\dfrac{1}{3i+1} - \dfrac{1}{3i+4}\right)$ **77.** $\displaystyle\sum_{i=1}^{65} \ln\left(\dfrac{i}{i+1}\right)$

66. $\displaystyle\sum_{i=2}^{9} 3^i = 29,520$

67. $S_n = \dfrac{n}{4(n+4)}$,

$S_{100} = \dfrac{25}{104}$

68. $S_n = \dfrac{n}{4(n+4)}$,

$S = \dfrac{1}{4}$

69. $S_n = 2^n - 1$,
series diverges

70. $S_n = 2^n - 1$,

$S_{15} = 32,767$

71. $S_n = \dfrac{n}{2(n+1)}$,

$S_{49} = \dfrac{49}{100}$

72. $S_n = \dfrac{n}{2(n+1)}$, $S = \dfrac{1}{2}$

73. $S_n = -\ln(n+1)$,

$S_{100} = -\ln 101$

74. $S_n = -\ln(n+1)$,
series diverges

75. $S_n = \dfrac{2n}{7(2n+7)}$,

$S_{30} = \dfrac{60}{469}$

76. $S_n = \dfrac{3n}{4(3n+4)}$,

$S = \dfrac{1}{4}$

77. $S_n = -\ln(n+1)$,
$S_{65} = -\ln 66$

78. $4, 7, 11, 18, 29$

Determine the first five terms of the following generalized Fibonacci sequences.

78. $a_1 = 4$, $a_2 = 7$, and $a_n = a_{n-2} + a_{n-1}$ for $n \geq 3$

79. $a_1 = -9$, $a_2 = 1$, and $a_n = a_{n-2} + a_{n-1}$ for $n \geq 3$

80. $a_1 = 10$, $a_2 = 20$, and $a_n = a_{n-2} + a_{n-1}$ for $n \geq 3$

81. $a_1 = -17$, $a_2 = 13$, and $a_n = a_{n-2} + a_{n-1}$ for $n \geq 3$

82. $a_1 = 13$, $a_2 = -17$, and $a_n = a_{n-2} + a_{n-1}$ for $n \geq 3$

Determine the first five terms of the following recursively defined sequences.

83. $a_1 = 2$, $a_2 = -3$, and $a_n = 3a_{n-1} + a_{n-2}$ for $n \geq 3$

84. $a_1 = 1$, $a_2 = -3$, and $a_n = a_{n-1}a_{n-2}$ for $n \geq 3$

Solve the following application problems.

85. Suppose you buy one cow and a number of bulls. In year one, your cow gives birth to a female calf and continues to bear another female calf every year for the rest of her life. Assuming that every calf born is female, that each cow begins calving in her third year (at age two), and that your cows never die, determine the number of cows (do not count the bulls) you will have at the end of the 14th year.

86. Beginning with yourself, create a sequence describing the number of biological predecessors you have in each of the past 7 generations of your family.

87. You borrow $638 to buy a new car stereo. You plan to pay this sum back with monthly payments of $74. The interest rate on your loan is 6% compounded monthly (recall that's 0.5% per month). Let A_n be the amount you owe at the end of the nth month. Find a recursive sequence to represent A_n. Use this sequence to find the amount owed after 4 months and the amount owed after 6 months. How many months will it take to pay off your loan?

88. The Fibonacci sequence is quite prevalent in nature. Do some research on your own to find an occurrence in nature (other than population growth) of the Fibonacci sequence.

79. $-9, 1, -8, -7, -15$

80. $10, 20, 30, 50, 80$

81. $-17, 13, -4, 9, 5$

82. $13, -17, -4, -21, -25$

83. $2, -3, -7, -24, -79$

84. $1, -3, -3, 9, -27$

85. 987

86. $1, 2, 4, 8, 16, 32, 64, 128$

87. $A_n = A_{n-1}(1.005) - 74$; $A_4 = 352.63$;

$A_6 = 207.79$; 9 months

88. Answers will vary.

11.2

Arithmetic Sequences and Series

TOPICS

1. Characteristics of arithmetic sequences and series
2. The formula for the general term of an arithmetic sequence
3. Evaluating partial sums of arithmetic sequences

TOPIC 1

Characteristics of Arithmetic Sequences and Series

Suppose that the parents of a ten-year-old child decide to increase her $1.00/week allowance by $0.50/week with the start of each new year. The sequence describing her weekly allowance, beginning with age ten, is then

$$1.00, \ 1.50, \ 2.00, \ 2.50, \ 3.00, \ 3.50, \ldots$$

This type of sequence, in which the difference between any two consecutive terms is constant, is called an *arithmetic sequence*.

DEFINITION

Arithmetic Sequences

A sequence $\{a_n\}$ is an **arithmetic sequence** (also called an **arithmetic progression**) if there is a constant d such that $a_{n+1} - a_n = d$ for each $n = 1, 2, 3, \ldots$. The constant d is called the **common difference** of the sequence.

Since every sequence $\{a_n\}$ can be used to determine an associated series $a_1 + a_2 + a_3 + \ldots$, arithmetic series follow naturally from arithmetic sequences. We can prove that any nontrivial arithmetic series diverges.

If we denote the first term of the sequence a_1, the second term is then $a_1 + d$ (where d is the common difference), the third term is $(a_1 + d) + d = a_1 + 2d$, and so on.

So if we add up the first n terms of the sequence (that is, if we find the nth partial sum), we have

$$S_n = a_1 + (a_1 + d) + (a_1 + 2d) + \ldots + (a_1 + (n-1)d)$$

$$= \sum_{i=1}^{n} (a_1 + (i-1)d)$$

$$= \sum_{i=1}^{n} a_1 + \sum_{i=1}^{n} (i-1)d \qquad \text{Note the use of a property of } \sum.$$

This may be enough to convince you that the partial sums are getting larger and larger in magnitude as n grows, since S_n consists of a_1 added to itself n times, plus a sum of multiples of d. However, having an explicit formula for the nth partial sum is useful, so we will simplify further.

$$S_n = \sum_{i=1}^{n} a_1 + \sum_{i=1}^{n} (i-1)d$$

$$= a_1 \sum_{i=1}^{n} 1 + d \sum_{i=1}^{n} (i-1)$$

Factor out constants using a property of sigma notation.

$$= na_1 + d\left(0+1+2+\ldots+(n-1)\right)$$

Evaluate the first term using a summation formula. Writing out the second term, we see it can be written using a different index of summation.

$$= na_1 + d\sum_{i=1}^{n-1} i$$

$$= na_1 + d\left(\frac{(n-1)n}{2}\right)$$

Apply another summation formula.

It is clear that the sequence of partial sums does not approach a fixed number S, except in the trivial case when $a_1 = 0$ and $d = 0$. Thus, except for the series $0 + 0 + 0\ldots$, every arithmetic series diverges.

TOPIC   The Formula for the General Term of an Arithmetic Sequence

As with all sequences, an explicit formula for the general n^{th} term of an arithmetic sequence is very useful. We have already noted that if a_1 is the first term of an arithmetic sequence, and if the common difference is d, then $a_1 + d$ is the second term, $a_1 + 2d$ is the third term, and so on. This pattern is summarized below.

THEOREM

General Term of an Arithmetic Sequence

The explicit formula for the **general n^{th} term of an arithmetic sequence** is

$$a_n = a_1 + (n-1)d,$$

where d is the common difference for the sequence.

EXAMPLE 1

Find the formula for the general n^{th} term of each arithmetic sequence.

a. $-3, 2, 7, 12, \ldots$ **b.** $a_1 = \dfrac{1}{3}$ and $a_4 = \dfrac{11}{6}$ **c.** $a_7 = -8$ and $d = -3$

Note:
An arithmetic
sequence can be
defined by two
terms, or by one term
and the common
difference.

Solutions:

a. First, find d, the difference, by calculating the difference between any two consecutive terms.

$$d = 2 - (-3) = 5$$

Since a_1 is listed in the sequence $(a_1 = -3)$, we have all the information we need.

$$
\begin{aligned}
a_n &= a_1 + (n-1)d \\
&= -3 + (n-1)(5) \qquad &&\text{Substitute the known values.} \\
&= 5n - 8 &&\text{Simplify.}
\end{aligned}
$$

b. Here, two nonconsecutive terms are given. Since $a_1 = \dfrac{1}{3}$, we can use the formula for the nth term to find d.

$$a_4 = a_1 + (4-1)d \qquad \text{Write the formula for } n = 4.$$

$$\frac{11}{6} = \frac{1}{3} + 3d \qquad \text{Substitute the known values.}$$

$$\frac{3}{2} = 3d \qquad \text{Simplify.}$$

$$d = \frac{1}{2}$$

Now, substitute back in to the formula to find the general n^{th} term.

$$a_n = \frac{1}{3} + (n-1)\left(\frac{1}{2}\right) \qquad \text{Substitute the known values.}$$

$$= \frac{1}{2}n - \frac{1}{6} \qquad \text{Simplify.}$$

c. Similarly, we have enough information to use the formula to find a_1.

$$
\begin{aligned}
a_7 &= a_1 + (7-1)d \\
-8 &= a_1 + 6(-3) \qquad &&\text{Substitute the known values.} \\
-8 &= a_1 - 18 &&\text{Simplify.} \\
a_1 &= 10
\end{aligned}
$$

Substitute back in to the formula to find a_n.

$$
\begin{aligned}
a_n &= 10 + (n-1)(-3) \\
&= -3n + 13
\end{aligned}
$$

============ EXAMPLE 2 ============

Modeling Population
Growth

A demographer models the population growth of a small town as an arithmetic progression. He knows that the population in 2002 was 12,790 and that in 2005 the population was 13,150. He wants to treat the population in 2000 as the first term of the arithmetic progression. What is the sought-after formula?

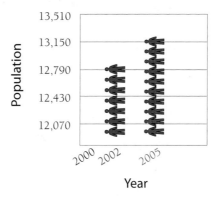

Solution:

We know that a_1 represents the population in 2000. Similarly, a_2 represents the population in 2001 and a_3 represents the population in 2002.

The known information gives us $a_3 = 12,790$ and $a_6 = 13,150$ (since the population in 2005 is 13,150). If we let d represent the common difference, the population increases by $3d$ between 2002 and 2005

$$3d = 13,150 - 12,790 = 360,$$

so $d = 120$.

Now, we can use the fact that $a_3 = 12,790$ to determine a_1,

$$a_3 = a_1 + 2d$$
$$12,790 = a_1 + 2(120)$$
$$12,790 - 240 = a_1$$
$$a_1 = 12,550$$

Putting this information together, we have the desired formula:

$$a_n = 12,550 + 120(n-1)$$
$$= 12,430 + 120n$$

Note that to apply the formula, we need to remember $n = 1$ corresponds to 2000, $n = 2$ corresponds to 2001, and so on.

TOPIC 3   Evaluating Partial Sums of Arithmetic Sequences

To begin our study of partial sums of arithmetic sequences, we will return to the ten-year-old girl whose weekly allowance grows in an arithmetic progression.

EXAMPLE 3

Partial Sums of Arithmetic Sequences

Melinda, the ten-year-old, knows that her parents plan to increase her weekly allowance by \$0.50/week with the start of each new year, and she currently receives \$1.00/week. Assuming that Melinda receives an allowance every week from the current year up through the year she turns 18, what is her total take?

Solution:

In the current year, Melinda will receive

$$(52)(\$1.00) = \$52.00.$$

The next year, when Melinda is 11, she will receive

$$(52)(\$1.50) = \$78.00.$$

The sum of these two amounts, \$130.00, is her total allowance during her 10th and 11th years. Extending this reasoning through Melinda's 18th year, we make use of the partial sum formula we already derived:

$$S_n = na_1 + d\left(\frac{(n-1)n}{2}\right)$$

We know that $a_1 = 52$ since her total allowance during the first year is \$52.00. Since her weekly allowance increases \$0.50 every year, Melinda's yearly allowance increases by $52(\$0.50) = \$26.00$ each year, so $d = 26$.

We then use the formula to calculate S_9, since Melinda's 18th year will be her 9th year of receiving allowance.

$$S_9 = 9 \cdot 52 + (26)\left(\frac{(9-1) \cdot 9}{2}\right) = 468 + 936 = 1404.$$

Melinda will thus receive a total of \$1404 in allowance money from age 10 through age 18.

We can rewrite the partial sum formula in a way that is useful when d has not yet been found. Recall that the formula for the general term of an arithmetic sequence is

$$a_n = a_1 + (n-1)d.$$

Rewriting this, we have $(n-1)d = a_n - a_1$, and we can use this fact in our partial sum formula to eliminate d, as follows:

$$S_n = na_1 + d\left(\frac{(n-1)n}{2}\right)$$ The known partial sum formula.

$$= na_1 + \left(\frac{n}{2}\right)(n-1)d$$ Rewrite the second term to isolate $(n-1)d$.

$$= na_1 + \left(\frac{n}{2}\right)(a_n - a_1)$$ Substitute.

$$= \left(\frac{n}{2}\right)(a_1 + a_n)$$ Simplify.

THEOREM

Partial Sums of
Arithmetic Sequences

Let $\{a_n\}$ be an arithmetic sequence with common difference d. The **sum of the first n terms** (the nth partial sum) is given by both

$$S_n = na_1 + d\left(\frac{(n-1)n}{2}\right)$$

and

$$S_n = \left(\frac{n}{2}\right)(a_1 + a_n).$$

In practice, the choice of formula is based on whether it is easier to determine d or the last term, a_n, in the sum to be evaluated.

EXAMPLE 4

Partial Sums of
Arithmetic Sequences

Determine the sum of the first 100 positive odd integers.

Solution:

Listing the first few positive odd integers, 1, 3, 5, 7, ..., we can see that the nth odd is equal to $2n-1$.

Thus, we have $a_1 = 1$, $d = 2$, $n = 100$, and $a_n = a_{100} = 2(100)-1 = 199$. Given this information, we can use either partial sum formula to find the answer.

Using the first formula, $S_{100} = 100(1) + 2\left(\frac{(100-1)(100)}{2}\right) = 100 + 2\left(\frac{99 \cdot 100}{2}\right) = 10,000$.

Using the second formula, $S_{100} = \left(\frac{100}{2}\right)(1 + 199) = 50 \cdot 200 = 10,000$.

1. $a_n = 3n - 5$

2. $a_n = 2n + 3$

3. $a_n = -2n + 9$

4. $a_n = 5n + 4$

5. $a_n = 9n - 4$

6. $a_n = 4n + 5$

7. $a_n = -6n + 9$

8. $a_n = 3n + 7$

9. $a_n = 19n + 5$

10. $a_n = 17n - 54$

11. $a_n = n + \dfrac{5}{2}$

12. $a_n = -4n + 19$

13. $a_n = -\dfrac{19}{2}n + \dfrac{43}{2}$

14. $a_n = 3n + 5$

15. $a_n = -2n + 1$

16. $a_n = \dfrac{3}{2}n - 10$

17. $a_n = -4n + 33$

18. $a_n = -2n - 3$

19. $d = 2$

20. No

21. $d = 1$

22. No

23. No

24. $d = -4$

25. No

26. $d = 0$

27. 13

28. −14

29. 2

30. $6\pi + 3$

31. 2

32. $\dfrac{37}{2}$

Exercises

Find the explicit formula for the general n^{th} term of each arithmetic sequence described below. See Examples 1 and 2.

1. $-2, 1, 4, 7, 10, \ldots$

2. $5, 7, 9, 11, 13, \ldots$

3. $7, 5, 3, 1, -1, \ldots$

4. $a_2 = 14$ and $a_3 = 19$

5. $a_1 = 5$ and $a_5 = 41$

6. $a_2 = 13$ and $a_4 = 21$

7. $a_3 = -9$ and $d = -6$

8. $a_{12} = 43$ and $d = 3$

9. $a_5 = 100$ and $d = 19$

10. $-37, -20, -3, 14, 31, \ldots$

11. $\dfrac{7}{2}, \dfrac{9}{2}, \dfrac{11}{2}, \dfrac{13}{2}, \dfrac{15}{2}, \ldots$

12. $15, 11, 7, 3, -1, \ldots$

13. $a_1 = 12$ and $a_3 = -7$

14. $a_{73} = 224$ and $a_{75} = 230$

15. $a_1 = -1$ and $a_6 = -11$

16. $a_5 = -\dfrac{5}{2}$ and $d = \dfrac{3}{2}$

17. $a_4 = 17$ and $d = -4$

18. $a_{34} = -71$ and $d = -2$

Determine if each of the following sequences is arithmetic. If so, find the common difference.

19. The sequence of even numbers

20. $1, 2, 4, 7, 11, 16, \ldots$

21. $1, 2, 3, 4, 5, 6, 7, \ldots$

22. The Fibonacci sequence

23. $1, 2, 4, 8, 16, 32, \ldots$

24. $42, 38, 34, 30, 26, 22, \ldots$

25. $0, 1, 0, 2, 0, 3, 0, 4, \ldots$

26. $12, 12, 12, 12, 12, \ldots$

Given the initial term and the common difference, find the value of the 7th term of the arithmetic sequences described below. See Examples 1 and 2.

27. $a_1 = 1$ and $d = 2$ **28.** $a_1 = 4$ and $d = -3$ **29.** $a_1 = 0$ and $d = \dfrac{1}{3}$

30. $a_1 = 3$ and $d = \pi$ **31.** $a_1 = 8$ and $d = -1$ **32.** $a_1 = \dfrac{1}{2}$ and $d = 3$

33. $d = 2.5; 5, 7.5, 10,$
$12.5, 15$

34. $d = 5; 2, 7, 12, 17, 22$

35. $d = 7; 7, 14, 21, 28, 35$

36. $d = 20; 16, 36,$
$56, 76, 96$

37. $d = 9; -62, -53,$
$-44, -35, -26$

38. $d = 4; 17, 21,$
$25, 29, 33$

39. $d = 5$

40. $d = 3$

41. $d = 1$

42. $d = -4$

43. $d = -2$

44. $d = \sqrt{3}$

45. 195

46. a_{50}

47. a_{73}

48. $\dfrac{109}{3}$

49. 117

50. a_{23}

51. 26

52. 11

53. -8

54. 11

55. 13

56. -18

57. 55

58. $14\sqrt{2}$

Given two terms of an arithmetic sequence, find the common difference and the first five terms of the sequence.

33. $a_1 = 5$ and $a_2 = 7.5$

34. $a_6 = 27$ and $a_9 = 42$

35. $a_7 = 49$ and $a_{11} = 77$

36. $a_4 = 76$ and $a_8 = 156$

37. $a_5 = -26$ and $a_9 = 10$

38. $a_8 = 45$ and $a_{10} = 53$

Find the common difference of the given sequence. See Examples 1 and 2.

39. $\{5n - 3\}$

40. $\left\{3n - \dfrac{1}{2}\right\}$

41. $\{n + 6\}$

42. $\{1 - 4n\}$

43. $\{\sqrt{2} - 2n\}$

44. $\{n\sqrt{3} + 5\}$

Use the given information about each arithmetic sequence to answer the question. See Examples 1 and 2.

45. Given that $a_1 = -3$ and $a_5 = 5$, what is a_{100}?

46. In the sequence $24, 43, 62, \ldots$, which term is 955?

47. In the sequence $1, \dfrac{4}{3}, \dfrac{5}{3}, \ldots$, which term is 25?

48. Given that $a_5 = -\dfrac{5}{3}$ and $a_9 = 1$, what is a_{62}?

49. In the sequence $-16, -9, -2, \ldots$, what is a_{20}?

50. In the sequence $\dfrac{1}{4}, \dfrac{7}{16}, \dfrac{5}{8}, \ldots$, which term is $\dfrac{35}{8}$?

51. In the sequence $2, 5, 8, 11, \ldots$, what is the 9th term?

52. In the sequence $1, 3, 5, 7, \ldots$, what is the 6th term?

53. In the sequence $16, 12, 8, 4, \ldots$, what is the 7th term?

54. In the sequence $\dfrac{1}{2}, 2, \dfrac{7}{2}, 5, \ldots$, what is the 8th term?

55. In the sequence $-2, 1, 4, 7, \ldots$, what is the 6th term?

56. In the sequence $9, 6, 3, 0, \ldots$, what is the 10th term?

57. In the sequence $5, 10, 15, 20, \ldots$, what is the 11th term?

58. In the sequence $2\sqrt{2}, 4\sqrt{2}, 6\sqrt{2}, 8\sqrt{2}, \ldots$, what is the 7th term?

59. 14,350

60. −2300

61. 17,114

62. 39,900

63. −1475

64. 16,550

65. $-\dfrac{3219}{5}$

66. 51,207

67. 902

68. 15

69. −1316

70. 8,100,292

71. 6 years

72. 108 in the last row, 2760 altogether

73. 1620 pounds

74. 20 rows

75. $625; $8100

76. $3290

Each of the following sums is a partial sum of an arithmetic sequence; use either formula to determine its value. See Examples 3 and 4.

59. $\displaystyle\sum_{i=1}^{100}(3i-8)$ **60.** $\displaystyle\sum_{i=1}^{50}(-2i+5)$ **61.** $\displaystyle\sum_{i=5}^{90}(4i+9)$

62. $3+11+\ldots+795$ **63.** $25+18+\ldots+(-143)$ **64.** $-12+2+\ldots+674$

65. $\displaystyle\sum_{i=1}^{37}\left(-\dfrac{3}{5}i-6\right)$ **66.** $\displaystyle\sum_{i=100}^{200}(3i+57)$ **67.** $\displaystyle\sum_{i=2}^{42}(2i-22)$

68. $-90+(-77)+\ldots+92$ **69.** $7+3+\ldots+(-101)$ **70.** $4+\dfrac{81}{20}+\ldots+900$

Solve the following application problems. See Examples 3 and 4.

71. Cynthia borrows $21,000, interest-free, from her parents to help pay for her college education, and promises that upon graduation she will pay back the sum beginning with $1000 the first year and increasing the amount by $1000 with each successive year. How many years will it take for her to repay the entire $21,000?

72. A certain theatre is shaped so that the first row has 30 seats, and, moving toward the back, each successive row has two seats more than the previous one. If there are 40 rows, how many seats does the last row contain? How many seats are there altogether?

73. A brick mason spends a morning moving a pile of bricks from his truck to the work site by wheelbarrow. Each brick weighs two pounds, and on his first trip he transports 100 pounds. On each successive trip, as he tires, he decides to move one less brick. How many pounds of bricks has he transported after 20 trips?

74. The manager of a grocery store decides to create a display of soup cans by placing cans in a row on the floor and then stacking successive rows so that each level of the tower has one less can than the one below it. The manager wants the top row to have 5 cans, and the store has 290 cans that can be used for the display. If all of the cans are used, how many rows will the display have?

75. A man decides to lease a car and is told that his payment to the car dealership will be $50 in the first month. He is also told that every month thereafter, for the next 60 months, his payments will increase by $25. How much is his monthly payment after two years? How much has he paid in total after the first two years?

76. Your grandmother doesn't trust banks, so she decided to save for your college education by periodically adding money to a mason jar buried in her flower garden. She began the practice with $65 and added $15 every time she got her monthly paycheck. If she continued this routine for 18 years, how much money did she manage to save for you?

11.3

Geometric Sequences and Series

TOPICS

1. Characteristics of geometric sequences
2. The formula for the general term of a geometric sequence
3. Evaluating partial sums of geometric sequences
4. Evaluating infinite geometric series

TOPIC 1

Characteristics of Geometric Sequences

Suppose Marilyn is nearing the end of a job hunt, and has a choice of two positions. Both offer a starting salary of $40,000 per year, but they differ in their projected future salaries. Employer A offers a yearly salary increase of $1250/year. Employer B offers an increase of 3%/year for all employees. Assuming the positions are equally desirable in all other respects, which one should Marilyn choose?

This question doesn't have a simple answer. One important consideration is the number of years Marilyn anticipates working for her next employer. A table comparing the two projected salaries over the next decade will help us determine an answer. To construct this table, we need a formula for the salary of each company each year.

We know how to find such a formula for Employer A, as the sequence of yearly salaries forms an arithmetic progression with a common difference of $1250; this means the nth term of the sequence is given by

$$a_n = 40{,}000 + (n-1)1250.$$

The nth term of the sequence of salaries for Employer B is of a different form. Since each successive salary is 3% greater than the preceding one, the recursive formula

$$b_1 = 40{,}000 \text{ and } b_n = b_{n-1} + (0.03)b_{n-1} \text{ for } n \geq 2$$

describes the sequence of salaries. We can then simplify the formula by writing $b_n = (1.03)b_{n-1}$ for $n \geq 2$. With these two formulas, we can construct the table comparing salaries seen in Table 1 (all salaries are rounded off to the nearest dollar).

	1	2	3	4	5	6	7	8	9	10
a_n	40,000	41,250	42,500	43,750	45,000	46,250	47,500	48,750	50,000	51,250
b_n	40,000	41,200	42,436	43,709	45,020	46,371	47,762	49,195	50,671	52,191

Table 1: Comparison of Salaries of Employers A and B in Year n

The first thing we notice is that Employer A offers a higher yearly salary up through the fourth year, but that thereafter Employer B pays more. So if Marilyn anticipates staying in her next job for five years or more, she may want to go with Employer B.

This reasoning doesn't consider the *accumulated* salary through year n; for a more accurate comparison of the two jobs, we need to compare the *partial sums*. We will return to the question of partial sums later in this section.

The salary sequence for Employer B is an example of a *geometric sequence*, and its identifying characteristic is that the *ratio* of consecutive terms in the sequence is a fixed constant. As we noted above, the ratio of any term in the sequence to the preceding term is 1.03. (Remember that for arithmetic sequences, the *difference* between consecutive terms is fixed.)

DEFINITION

Geometric Sequences

A sequence $\{a_n\}$ is a **geometric sequence** (also called a **geometric progression**) if there is a constant $r \neq 0$ so that $\dfrac{a_{n+1}}{a_n} = r$ for each $n = 1, 2, 3, \ldots$. The constant r is called the **common ratio** of the sequence.

TOPIC 2 The Formula for the General Term of a Geometric Sequence

We can develop an explicit formula for the general n^{th} term of a geometric sequence by relating each term to a_1, the first term of the sequence, just as we did for arithmetic sequences.

Since the sequence is geometric, there is some fixed number r such that $\dfrac{a_{n+1}}{a_n} = r$ for each $n = 1, 2, 3, \ldots$. In particular,

$$\frac{a_2}{a_1} = r, \text{ or } a_2 = a_1 r.$$

Similarly,

$$\frac{a_3}{a_2} = r, \text{ so } a_3 = a_2 r = \left(a_1 r\right) r = a_1 r^2.$$

Extending this pattern, we arrive at an explicit formula for a_n.

THEOREM

General Term of a Geometric Sequence

The explicit formula for the **general n^{th} term of a geometric sequence** is

$$a_n = a_1 r^{n-1},$$

where r is the common ratio for the sequence.

Given this information, we now know that the yearly salary offered by Employer B above is given by the formula

$$b_n = (40,000)(1.03)^{n-1}.$$

EXAMPLE 1

General Term of a
Geometric Sequence

Find the formula for the general n^{th} term of each geometric sequence.

a. $\dfrac{1}{3}, \dfrac{1}{9}, \dfrac{1}{27}, \ldots$ **b.** $3, -6, 12, \ldots$ **c.** $a_4 = \dfrac{5}{16}$ and $r = \dfrac{1}{2}$

Solutions:

a. First, use any two consecutive terms to determine the common ratio, r.

$$r = \frac{\left(\dfrac{1}{9}\right)}{\left(\dfrac{1}{3}\right)} = \frac{3}{9} = \frac{1}{3}$$

Since we know that $a_1 = \dfrac{1}{3}$, the explicit formula is as follows

$$a_n = a_1 (r)^{n-1}$$
$$= \frac{1}{3}\left(\frac{1}{3}\right)^{n-1} = \left(\frac{1}{3}\right)^{n} \qquad \text{Substitute the known values and simplify.}$$

b. Using the same process as before, we know $a_1 = 3$ and can calculate r.

$$r = \frac{-6}{3} = -2$$

Applying the formula, we have

$$a_n = 3(-2)^{n-1}$$

c. Here we are given r and another term. We can use the formula to find a_1.

$$a_4 = a_1 (r)^{4-1} \qquad \text{Write out the explicit formula for } n = 4.$$

$$\frac{5}{16} = a_1 \left(\frac{1}{2}\right)^{3} \qquad \text{Substitute the known values.}$$

$$\frac{5}{16} = a_1 \left(\frac{1}{8}\right) \qquad \text{Simplify and solve for } a_1.$$

$$a_1 = \frac{5}{2}$$

We then substitute a_1 and r to find the explicit formula.

$$a_n = \frac{5}{2}\left(\frac{1}{2}\right)^{n-1}$$

━━━━━━━━━━━ EXAMPLE 2 ━━━━━━━━━━━

General Term of a
Geometric Sequence

Given that the second term of a geometric sequence is –6 and the fifth term is 162, what is the fourth term?

Solution:

Given that the sequence is geometric, we know that

$$-6 = a_1 r \text{ and } 162 = a_1 r^4.$$

We can eliminate a_1 from this pair of equations by dividing:

$$\frac{162}{-6} = \frac{a_1 r^4}{a_1 r}, \text{ so } r^3 = -27.$$

Given r, we can find a_1 using one of the equations above.

$$-6 = a_1 r$$

$$a_1 = \frac{-6}{-3} = 2$$

So the general term is given by $a_n = 2(-3)^{n-1}$, and we find the fourth term by plugging in $n = 4$:

$$a_4 = 2(-3)^{4-1} = -54.$$

TOPIC 3

Evaluating Partial Sums of Geometric Sequences

Given a geometric sequence, many applications require us to compute the partial sum $S_n = a_1 + a_2 + \ldots + a_n$ for some value of n. We will now derive an explicit formula for the nth partial sum.

First, we use the formula for the nth term of a geometric sequence to rewrite S_n.

$$S_n = a_1 + a_2 + \ldots + a_n$$

$$= a_1 + a_1 r + a_1 r^2 + \ldots + a_1 r^{n-1}.$$

If we multiply both sides of this equation by r, we obtain

$$r S_n = a_1 r + a_1 r^2 + a_1 r^3 + \ldots + a_1 r^n.$$

If we now subtract the second equation from the first, most of the terms from the right-hand sides of the equations cancel, as shown below:

$$S_n - r S_n = \left(a_1 + \cancel{a_1 r} + \cancel{\ldots} + \cancel{a_1 r^{n-1}} \right) - \left(\cancel{a_1 r} + \cancel{a_1 r^2} + \cancel{\ldots} + a_1 r^n \right)$$

$$= a_1 - a_1 r^n$$

The goal is to solve for S_n, and we can do this by factoring the left and right sides of the equation and dividing:

$$S_n\left(1-r\right) = a_1\left(1-r^n\right)$$

$$S_n = \frac{a_1\left(1-r^n\right)}{1-r}$$

THEOREM

Partial Sum of a Geometric Sequence

Let $\{a_n\}$ be a geometric sequence with common ratio r, and assume r is neither 0 nor 1. Then **the nth partial sum of the sequence**, $S_n = a_1 + a_1 r + a_1 r^2 + \ldots + a_1 r^{n-1}$, is given by

$$S_n = \frac{a_1\left(1-r^n\right)}{1-r}.$$

The condition $r \neq 1$ is necessary to prevent division by zero in the formula, but it isn't difficult to determine the partial sum without the formula if r happens to be 1. In this case, $S_n = a_1 + a_1 + \ldots + a_1 = na_1$.

With this formula in hand, along with one of the two partial sum formulas for arithmetic sequences, let us return to the question with which we began this section.

EXAMPLE 3

Comparing Salary Offers

Recall that Marilyn has the option of taking a job with Employer A, who offers a yearly salary increase of \$1250, or Employer B, who offers a yearly salary increase of 3%. Given that the starting salary is \$40,000 for both, in what year does the accumulated salary paid by Employer B overtake the accumulated salary paid by Employer A?

Solution:

We have already determined that Employer B pays a higher salary beginning with the fifth year of employment, but we need to compare the partial sums of the salaries paid by the two employers up through year n to answer the question above. Recall that the sequence of yearly salaries paid by Employer A is defined by

$$a_n = 40,000 + \left(n-1\right)1250$$

and that the sequence of yearly salaries paid by Employer B is defined by

$$b_n = 40,000\left(1.03\right)^{n-1}.$$

Recall also that one partial sum formula for an arithmetic sequence is

$$S_n = na_1 + d\left(\frac{\left(n-1\right)n}{2}\right),$$

where d is the common difference for the sequence.

In order to keep the partial sums straight, let A_n denote the sum of the salaries paid by Employer A through year n, and let B_n be the same for Employer B. Then

$$A_n = n(40{,}000) + 1250\left(\frac{(n-1)n}{2}\right) = 40{,}000n + 625(n^2 - n) = 625n^2 + 39{,}375n$$

and

$$B_n = \frac{40{,}000(1 - 1.03^n)}{1 - 1.03} = \left(\frac{40{,}000}{0.03}\right)(1.03^n - 1).$$

We can now use these formulas to compute the accumulated salaries paid by the two employers up through year n, as shown in Table 2.

	1	2	3	4	5	6	7	8
A_n	40,000	81,250	123,750	167,500	212,500	258,750	306,250	355,000
B_n	40,000	81,200	123,636	167,345	212,365	258,736	306,498	355,693

Table 2: Accumulated Salaries through Year 8

Comparing values in the table indicates that the accumulated salary paid by Employer B overtakes that of Employer A in the seventh year.

We can also use the geometric partial sum formula to evaluate certain expressions defined with the sigma notation, as shown in the next example.

EXAMPLE 4

Partial Sum of a Geometric Sequence

Evaluate $\displaystyle\sum_{i=2}^{7} 5\left(-\frac{1}{2}\right)^i$.

Solution:

This is a partial sum of a geometric sequence, but as it is written the first term and the common ratio of the sequence are not apparent. A good way to begin is to write the sum in expanded form:

$$\sum_{i=2}^{7} 5\left(-\frac{1}{2}\right)^i = 5\left(-\frac{1}{2}\right)^2 + 5\left(-\frac{1}{2}\right)^3 + \ldots + 5\left(-\frac{1}{2}\right)^7$$

$$= \frac{5}{4} - \frac{5}{8} + \ldots - \frac{5}{128}$$

We can see that $a_1 = \frac{5}{4}$ and that the common ratio is $r = -\frac{1}{2}$. Since there are six terms in the sum, we let $n = 6$ in the partial sum formula. Putting this all together, we have

$$S_6 = \frac{a_1\left(1-r^6\right)}{1-r}$$ Write the formula for the 6th partial sum.

$$= \frac{\left(\dfrac{5}{4}\right)\left(1-\left(-\dfrac{1}{2}\right)^6\right)}{1-\left(-\dfrac{1}{2}\right)}$$ Substitute the known values.

$$= \frac{\left(\dfrac{5}{4}\right)\left(1-\dfrac{1}{64}\right)}{\left(\dfrac{3}{2}\right)}$$ Simplify.

$$= \left(\frac{5}{4}\right)\left(\frac{63}{64}\right)\left(\frac{2}{3}\right) = \frac{105}{128}$$

TOPIC Evaluating Infinite Geometric Series

In Section 11.1, we used an intuitive approach to think about the result of adding half an inch to a quarter of an inch to an eighth of an inch, and so on. We came to the conclusion that if the process were continued indefinitely, the sum of the fractions would be 1. That is,

$$\frac{1}{2}+\frac{1}{4}+\frac{1}{8}+\ldots = 1, \text{ or } \sum_{i=1}^{\infty}\left(\frac{1}{2}\right)^i = 1.$$

Our reasoning was that the sequence of partial sums of the sequence

$$\left\{\left(\frac{1}{2}\right)^n\right\}$$

never exceeded 1, yet got closer and closer to 1 as more terms were added.

We now have the machinery to analyze infinite series like the one above more rigorously. For one thing, we know that the sequence

$$\frac{1}{2}, \frac{1}{4}, \frac{1}{8}, \ldots$$

is a geometric sequence, and that both the first term and the common ratio are $\dfrac{1}{2}$. We can use the partial sum formula, substituting $a_1 = \dfrac{1}{2}$ and $r = \dfrac{1}{2}$.

$$S_n = \frac{1}{2}+\frac{1}{4}+\ldots+\frac{1}{2^n} = \frac{\left(\cancel{\dfrac{1}{2}}\right)\left(1-\left(\dfrac{1}{2}\right)^n\right)}{\left(1-\cancel{\dfrac{1}{2}}\right)} = 1-\left(\frac{1}{2}\right)^n$$

If we write S_n as a single fraction, we obtain

$$S_n = \frac{2^n - 1}{2^n},$$

the formula we intuitively derived in Section 11.1. In either form, we can see that S_n approaches 1 as $n \to \infty$, but we can now generalize this observation for all convergent geometric series.

THEOREM

Sum of an Infinite Geometric Series

If $|r| < 1$, the infinite geometric series

$$\sum_{n=1}^{\infty} a_1 r^{n-1} = a_1 + a_1 r + a_1 r^2 + a_1 r^3 + \ldots$$

converges, and the sum of the series is given by $S = \dfrac{a_1}{1-r}$. We can use sigma notation with the index beginning at 0 to write the same series; in this form,

$$\sum_{n=0}^{\infty} a_1 r^n = a_1 + a_1 r + a_1 r^2 + a_1 r^3 + \ldots = \frac{a_1}{1-r}.$$

The proof of this result follows from the fact that if $|r| < 1$, then $r^n \to 0$ as $n \to \infty$, so

$$S_n = \frac{a_1 \left(1 - r^n\right)}{1-r} \to \frac{a_1}{1-r} \text{ as } n \to \infty.$$

In the case of the series $\dfrac{1}{2} + \dfrac{1}{4} + \dfrac{1}{8} + \ldots$, where $a_1 = \dfrac{1}{2}$ and $r = \dfrac{1}{2}$, the sum is

$$S = \frac{\dfrac{1}{2}}{1 - \dfrac{1}{2}} = \frac{\dfrac{1}{2}}{\dfrac{1}{2}} = 1.$$

EXAMPLE 5

Infinite Geometric Series

Find the sums of the following series.

a. $\displaystyle\sum_{n=1}^{\infty} 5\left(-\frac{1}{2}\right)^{n-1}$ **b.** $\displaystyle\sum_{n=1}^{\infty} 3\left(\frac{1}{10}\right)^n$

Solutions:

a. First, we identify the values of a_1 and r. Plugging in $n = 1$, we see that $a_1 = 5$, and from the summation formula, we know $r = -\dfrac{1}{2}$.

Since the common ratio r is less than 1 in magnitude, we can apply the formula.

$$S = \frac{a_1}{1-r}$$

$$= \frac{5}{1-\left(-\dfrac{1}{2}\right)} = \frac{5}{\left(\dfrac{3}{2}\right)} = \frac{10}{3} \qquad \text{Substitute the known values, then simplify.}$$

b. It is important to note exactly how a series is written. The formula $S = \dfrac{a_1}{1-r}$ applies directly to two forms of summation notation:

$$\sum_{n=1}^{\infty} a_1 r^{n-1} \quad \text{or} \quad \sum_{n=0}^{\infty} a_1 r^{n}.$$

In this case, the form of the series doesn't quite match either of these forms, so we write out the first few terms to identify the first term of the series and the common ratio:

$$\sum_{n=1}^{\infty} 3\left(\frac{1}{10}\right)^{n} = \frac{3}{10} + \frac{3}{100} + \frac{3}{1000} + \ldots$$

From this, we can see that $a_1 = \dfrac{3}{10}$ and $r = \dfrac{1}{10}$. Plugging these values in, we have

$$S = \frac{a_1}{1-r}$$

$$= \frac{\left(\dfrac{3}{10}\right)}{1-\dfrac{1}{10}} = \frac{\left(\dfrac{3}{10}\right)}{\left(\dfrac{9}{10}\right)} = \frac{1}{3}.$$

The last example illustrates an important relation between geometric series and the decimal system we use to write real numbers. Note that we have shown that

$$\sum_{n=1}^{\infty} 3\left(\frac{1}{10}\right)^{n} = \frac{3}{10} + \frac{3}{100} + \frac{3}{1000} + \ldots = 0.33\overline{3}$$

is the decimal representation of the fraction $\dfrac{1}{3}$. Similarly,

$$0.\overline{52} = \frac{52}{100} + \frac{52}{10,000} + \frac{52}{1,000,000} + \ldots = \sum_{n=1}^{\infty} 52\left(\frac{1}{100}\right)^{n}$$

can be written in fractional form by noting that for this series $a_1 = \dfrac{52}{100}$ and $r = \dfrac{1}{100}$, so

$$0.\overline{52} = \frac{\dfrac{52}{100}}{1-\dfrac{1}{100}} = \frac{52}{99}.$$

1. $a_n = -3(2)^{n-1}$

2. $a_n = 7\left(\dfrac{1}{2}\right)^{n-1}$

3. $a_n = 2\left(\dfrac{-1}{3}\right)^{n-1}$

4. $a_n = 5(2)^{n-1}$

5. $a_n = \left(\dfrac{-1}{4}\right)^{n-1}$

6. $a_n = (-0.1)^{n-1}$

7. $a_n = \left(\dfrac{1}{7}\right)^{n-1}$

8. $a_n = \left(-\dfrac{3}{4}\right)^{n-1}$

9. $a_n = (-3)^{n-1}$

10. $a_n = (-3)^{n}$

11. $a_n = 3\left(\dfrac{2}{3}\right)^{n-1}$

12. $a_n = -5\left(-\dfrac{1}{4}\right)^{n-1}$

13. $a_n = 7(-2)^{n-1}$

14. $a_n = -16\left(\dfrac{3}{2}\right)^{n-1}$

15. $a_n = \dfrac{1}{16}(2)^{n-1}$

16. $a_n = \dfrac{1}{3}(7)^{n-1}$

17. $a_n = \dfrac{39}{68}\left(\dfrac{4}{3}\right)^{n-1}$

18. $a_n = (2)^{n-1}$

19. No

20. $r = 1$

21. $r = \dfrac{1}{2}$

22. No

Exercises

Find the explicit formula for the general n^{th} term of each geometric sequence described below. See Examples 1 and 2.

1. $-3, -6, -12, -24, -48, \ldots$

2. $7, \dfrac{7}{2}, \dfrac{7}{4}, \dfrac{7}{8}, \dfrac{7}{16}, \ldots$

3. $2, -\dfrac{2}{3}, \dfrac{2}{9}, -\dfrac{2}{27}, \dfrac{2}{81}, \ldots$

4. $a_1 = 5$ and $a_4 = 40$

5. $a_2 = -\dfrac{1}{4}$ and $a_5 = \dfrac{1}{256}$

6. $a_1 = 1$ and $a_4 = -0.001$

7. $a_2 = \dfrac{1}{7}$ and $r = \dfrac{1}{7}$

8. $a_3 = \dfrac{9}{16}$ and $r = -\dfrac{3}{4}$

9. $a_3 = 9$, $a_5 = 81$, and $r < 0$

10. $-3, 9, -27, 81, -243, \ldots$

11. $3, 2, \dfrac{4}{3}, \dfrac{8}{9}, \dfrac{16}{27}, \ldots$

12. $-5, \dfrac{5}{4}, -\dfrac{5}{16}, \dfrac{5}{64}, -\dfrac{5}{256}, \ldots$

13. $a_3 = 28$ and $a_6 = -224$

14. $a_2 = -24$ and $a_5 = -81$

15. $a_5 = 1$ and $a_6 = 2$

16. $a_4 = \dfrac{343}{3}$ and $r = 7$

17. $a_2 = \dfrac{13}{17}$ and $r = \dfrac{4}{3}$

18. $a_4 = 8$, $a_8 = 128$, and $r > 0$

Determine if each of the following sequences is geometric. If so, find the common ratio.

19. The sequence of odd numbers

20. $4, 4, 4, 4, 4, 4, \ldots$

21. $100, 50, 25, 12.5, 6.25, \ldots$

22. $2, 5, 11, 23, 47, \ldots$

23. $\dfrac{7}{8}, \dfrac{7}{4}, \dfrac{7}{2}, 7, 14, \ldots$

24. The sequence of numbers called out at a Bingo game

25. $7, 49, 343, 2401, \ldots$

26. $10, 15, 22.5, 33.75, \ldots$

Given the two terms of a geometric sequence, find the common ratio and first five terms of the sequence.

27. $a_1 = 8$ and $a_2 = 24$

28. $a_6 = \dfrac{1}{2}$ and $a_9 = \dfrac{1}{54}$

29. $a_7 = 16$ and $a_{11} = 256$

30. $a_4 = 108$ and $a_8 = 8748$

31. $a_5 = 100$ and $a_9 = \dfrac{4}{25}$

32. $a_8 = 100$ and $a_{10} = 1$

23. $r = 2$

24. No

25. $r = 7$

26. $r = 1.5$

27. $r = 3; 8, 24, 72, 216,$
     648

28. $r = \dfrac{1}{3}; \dfrac{243}{2}, \dfrac{81}{2},$
     $\dfrac{27}{2}, \dfrac{9}{2}, \dfrac{3}{2}$

29. $r = 2; \dfrac{1}{4}, \dfrac{1}{2}, 1, 2, 4$

30. $r = 3; 4, 12, 36, 108,$
     324

31. $r = \dfrac{1}{5}; 62,500,$
     $12,500, 2500, 500,$
     100

32. $r = \dfrac{1}{10}; 10^9, 10^8,$
     $10^7, 10^6, 10^5$

33. $\dfrac{5}{16,384}$

34. $\dfrac{2}{3}$

35. $-2,147,483,648$

36. $\dfrac{5}{2}$

37. $r = \pm 2$

38. $r = -6$

39. $r = \pm \dfrac{1}{5}$

40. $-\dfrac{1023}{1024}$

41. $\dfrac{52,222,139,775}{1,048,576}$
     $\approx 49,802.9$

42. $2,199,023,254,528$

43. $\dfrac{10,923}{16,384}$
     ≈ 0.666687

Use the given information about each geometric sequence to answer the question.

33. Given that $a_2 = -\dfrac{5}{2}$ and $a_5 = \dfrac{5}{16}$, what is a_{15}?

34. Given that $a_1 = 1$ and $a_4 = \dfrac{8}{27}$, what is the common ratio r?

35. Given that $a_3 = -2$ and $a_4 = -16$, what is a_{13}?

36. Given that $a_2 = 24$ and $a_5 = 375$, what is the common ratio r?

37. Given that $a_1 = -1$ and $a_3 = -4$, what is the common ratio r?

38. Given that $a_3 = 108$ and $a_4 = -648$, what is the common ratio r?

39. Given that $a_3 = -\dfrac{4}{25}$ and $a_7 = -\dfrac{4}{15,625}$, what is the common ratio r?

Each of the following sums is a partial sum of a geometric sequence. Use this fact to evaluate the sums. See Examples 3 and 4.

40. $\displaystyle\sum_{i=1}^{10} 3\left(-\dfrac{1}{2}\right)^i$ **41.** $\displaystyle\sum_{i=5}^{20} 5\left(\dfrac{3}{2}\right)^i$ **42.** $\displaystyle\sum_{i=10}^{40} 2^i$

43. $1 - \dfrac{1}{2} + \ldots + \dfrac{1}{16,384}$ **44.** $2 + 6 + \ldots + 39,366$ **45.** $5 - \dfrac{5}{3} + \ldots - \dfrac{5}{19,683}$

46. $1 - 3 + \ldots + 59,049$ **47.** $\displaystyle\sum_{i=4}^{15} 5(-2)^i$ **48.** $1 + \dfrac{3}{5} + \ldots + \dfrac{243}{3125}$

Determine if the following infinite geometric series converge. If a given series converges, find the sum. See Example 5.

49. $\displaystyle\sum_{i=0}^{\infty} -\dfrac{1}{2}\left(\dfrac{2}{3}\right)^i$ **50.** $\displaystyle\sum_{i=1}^{\infty} \left(\dfrac{4}{5}\right)^i$ **51.** $\displaystyle\sum_{i=0}^{\infty} \left(-\dfrac{9}{8}\right)^i$

52. $\displaystyle\sum_{i=0}^{\infty} \left(-\dfrac{8}{9}\right)^i$ **53.** $\displaystyle\sum_{i=5}^{\infty} \left(\dfrac{19}{20}\right)^i$ **54.** $\displaystyle\sum_{i=0}^{\infty} (-1)^i$

55. $\displaystyle\sum_{i=1}^{\infty} \dfrac{1}{3}(2)^{i-1}$ **56.** $\displaystyle\sum_{i=0}^{\infty} 5\left(\dfrac{6}{11}\right)^i$ **57.** $\displaystyle\sum_{i=4}^{\infty} \left(\dfrac{13}{24}\right)^i$

Write each of the following repeating decimal numbers as fractions. See Example 5b.

58. $1.\overline{65}$ **59.** $0.\overline{123}$ **60.** $-0.\overline{5}$

61. $-3.\overline{8}$ **62.** $0.0\overline{29}$ **63.** $9.\overline{98}$

44. 59,048

45. $\dfrac{73{,}810}{19{,}683}$

≈ 3.749936

46. 44,287

47. $-109{,}200$

48. $\dfrac{7448}{3125}$

$= 2.38336$

49. $-\dfrac{3}{2}$

50. 4

51. Series diverges

52. $\dfrac{9}{17}$

53. $\dfrac{2{,}476{,}099}{160{,}000}$

≈ 15.475619

54. Series diverges

55. Series diverges

56. 11

57. $\dfrac{28{,}561}{152{,}064}$

≈ 0.187822

58. $\dfrac{164}{99}$

59. $\dfrac{123}{999}$

60. $-\dfrac{5}{9}$

61. $-\dfrac{35}{9}$

62. $\dfrac{29}{999}$

63. $\dfrac{989}{99}$

64. 79.26 feet, 90 feet

65. $14,802.44

66. $14,908.33

67. 1.845×10^{19}

Solve the following application problems.

64. A rubber ball is dropped from a height of 10 feet, and on each bounce it rebounds up to 80% of its previous height. How far has it traveled vertically at the moment when it hits the ground for the tenth time? If we assume it bounces indefinitely, what is the total vertical distance traveled?

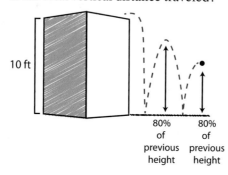

10 ft

80% of previous height   80% of previous height

65. If $10,000 is invested in a simple savings account with an annual interest rate of 4% compounded once a year, what is the value of the account after ten years?

66. If $10,000 is invested in a simple savings account with an annual interest rate of 4% compounded once a month, what is the value of the account after ten years?

67. An ancient story about the game of chess tells of a king who offered to grant the inventor of the game a wish. The inventor replied, "Place a grain of wheat on the first square of the board, 2 grains on the second square, 4 grains on the third, and so on. The wheat will be my reward." How many grains of wheat would the king have had to come up with? (There are 64 squares on a chessboard.)

68. An isosceles right triangle is divided into two similar triangles, one of the new triangles is divided into two similar triangles, and this process is continued without end. If the shading pattern seen below is continued indefinitely, what fraction of the original triangle is shaded?

68. $\dfrac{2}{3}$

69. Approximately 13,778 students

70. \$50,489.36, $r = 1.07$

71. $S_{30} = 1.1 \times 10^{30}$; $r = 10$

72. 0.140625 ft

73. Yes; explanations will vary (any example such that $r = 1$ and $d = 0$).

69. Each year the university admissions committee accepts 3% more students than they accepted in the previous year. If 2130 students were admitted in the first year of this trend, how many total students will have been admitted after 6 years?

70. On the day you were born, your parents deposited \$15,000 in a simple savings account for your college education. If the annual interest rate is 6.8%, compounded quarterly, how much money will be in the account when you begin college at the age of 18? What is the common ratio of this series?

71. Spamway, an internet advertising agency, uses e-mail forwards to collect addresses to which to send their advertisements. They begin with an e-mail chain letter that they send to 10 people. According to the letter, each of those 10 people has to forward the e-mail to 10 more people or they will have 7 years of bad luck. Assuming the e-mails are received and forwarded only once a day and all the recipients are superstitious and follow the rules, how many e-mail addresses will Spamway have collected after 30 days? Is this series geometric? If so, find the common ratio.

72. Last summer you were a camp counselor for Sunny Days day camp. An arts and crafts project required you to distribute pieces of string (of different lengths) among your kids. If you began with a piece of string 18 feet long and gave each child exactly half of your remaining string as you distributed, how long (in inches) was the seventh child's string?

73. Is it possible for a geometric sequence to also be an arithmetic sequence? If so, give an example and if not, explain your reasoning.

11.4 Mathematical Induction

TOPICS

1. The role of induction
2. Proofs by mathematical induction

TOPIC   The Role of Induction

The first three sections of this chapter focused on concepts built on the natural numbers: sequences are functions whose domain is the natural numbers, and series are sums based on sequences. This section is similarly focused. While the word "induction" has many meanings in everyday language, the *Principle of Mathematical Induction* has a precise mathematical meaning that depends on the properties of natural numbers.

Mathematical induction is most often used to prove statements involving natural numbers. More precisely, if $P(n)$ is a statement about each natural number n, mathematical induction can be used to prove that the statement is true for all n. As an example, consider the statement

$$P(n)\text{: The sum of the first } n \text{ positive integers is } \frac{n(n+1)}{2}.$$

We already know this statement is true for each n, as we proved it in Section 11.1. However, induction provides a more powerful and more general method of proof, and we will see how it can be used to prove similar statements. In particular, we can use induction to prove the other summation formulas from Section 11.1.

It is important to see how induction relates to some of the other concepts in this chapter. We have seen several examples of recursive formulas, for instance, $a_1 = 1$ and $a_n = a_{n-1} + n$ for $n \geq 2$ is a recursive formula that generates the sequence of sums $1 + 2 + \cdots + n$. As we know, this recursive formula is inconvenient for calculating, say, a_{100}, as we would need to first calculate a_{99}, which would in turn require us to calculate a_{98}, and so on back to a_1. In contrast, the explicit formula

$$a_n = \frac{n(n+1)}{2}$$

allows us to quickly determine a_{100} or any other such sum. As we will see, the Principle of Mathematical Induction is a useful tool for making the transition from a recursive formula to an explicit formula.

TOPIC 2

Proofs by Mathematical Induction

Consider again the statement

$$P(n): \text{The sum of the first } n \text{ positive integers is } \frac{n(n+1)}{2}.$$

Suppose we demonstrate that **if** the statement is true for an integer k, **then** the statement is also true for the next integer, $k + 1$. We can easily show that $P(1)$ is true,

as the sum of the first 1 positive integers, namely 1, is indeed $\frac{(1)(1+1)}{2} = 1.$

Since $P(1)$ is true, we can conclude from our demonstration that $P(2)$ is also true. And since $P(2)$ is true, it follows that $P(3)$ is true. We don't have to explicitly show that $P(2)$ being true implies that $P(3)$ is true, because we've already demonstrated that, in general, the validity of $P(k)$ implies the validity of $P(k + 1)$. This is the essence of the Principle of Mathematical Induction.

THEOREM

Principle of
Mathematical
Induction

Assume that $P(n)$ is a statement about each natural number n. Suppose that the following two conditions hold:

1. $P(1)$ is true.

2. For each natural number k, if $P(k)$ is true, then $P(k + 1)$ is true.

Then the statement $P(n)$ is true for every natural number n.

An inductive proof thus consists of two steps. The first task, which we will call the **basic step**, is generally very easy; we simply verify that $P(1)$ is true. The second task, called the **induction step**, may require more work and algebraic manipulation.

It is important to realize that the induction step begins with a very powerful assumption: under the *assumption* that $P(k)$ is true, our goal is to prove that $P(k + 1)$ is also true. The single most important step, and the step that is most unfamiliar initially, is the one in which the inductive assumption is used to simplify part of the statement $P(k + 1)$ in order to prove that $P(k + 1)$ is true. Be sure to note how the inductive assumption is used in the following examples.

EXAMPLE 1

Proof by
Mathematical
Induction

Prove that for each natural number n, $1 + 3 + 5 + \ldots + (2n - 1) = n^2$.

Solution:

Basic Step: The statement $P(1)$ is the equation $1 = 1^2$, which is clearly true.

Induction Step: As always, we begin this step with the assumption that $P(k)$ is true.

That is, $1+3+5+\ldots+(2k-1)=k^2$. We want to prove that $P(k+1)$ is true, so we need to show that $1+3+5+\ldots+(2(k+1)-1)=(k+1)^2$. To do this, we rewrite the left-hand side using the inductive assumption.

$$1+3+5+\ldots+(2(k+1)-1)=1+3+5+\ldots+(2k-1)+(2(k+1)-1)$$

$$= k^2+(2(k+1)-1) \qquad \text{Use the inductive assumption to rewrite all but the last term.}$$

$$= k^2+2k+1 \qquad \text{Simplify.}$$

$$= (k+1)^2 \qquad \text{Factor.}$$

Since we've completed the basic step and the induction step, the proof is complete.

Induction can be used to prove statements other than equations. The next two examples prove an inequality and a divisibility fact.

EXAMPLE 2

Proof by Mathematical Induction

Prove that for each natural number n, $2n \le 2^n$.

Solution:

Basic Step: The statement $P(1)$ is the inequality $2(1) \le 2^1$, or $2 \le 2$. This is true.

Induction Step: Assume that $2k \le 2^k$. Then:

$$2(k+1) = 2k+2 \qquad \text{Begin by writing the left-hand side of } P(k+1).$$

$$\le 2^k + 2 \qquad \text{Use the inductive assumption.}$$

$$\le 2^k + 2^k \qquad \text{Use the fact that } k \ge 1 \text{ implies } 2 \le 2^k.$$

$$= 2(2^k)$$

$$= 2^{k+1} \qquad \text{Simplify until the right-hand side of } P(k+1) \text{ is reached.}$$

EXAMPLE 3

Proof by Mathematical Induction

Prove that for each natural number n, $8^n - 3^n$ is divisible by 5.

Solution:

Basic Step: Clearly, $8^1 - 3^1 = 5$ is divisible by 5, so $P(1)$ is true.

Induction Step: Assume that $8^k - 3^k$ is divisible by 5. Then there is some integer p for which $8^k - 3^k = 5p$. Note that

$$8^{k+1} - 3^{k+1} = 8^k \cdot 8 - 3^k \cdot 3 \qquad \text{Rewrite using properties of exponents.}$$

$$= 8^k \cdot (5+3) - 3^k \cdot 3 \qquad \text{Rewrite 8 as } 5 + 3.$$

$$= 5 \cdot 8^k + 3 \cdot 8^k - 3 \cdot 3^k \qquad \text{Apply the distributive property.}$$

$$= 5 \cdot 8^k + 3 \cdot \left(8^k - 3^k\right) \qquad \text{Factor the last two terms.}$$

$$= 5 \cdot 8^k + 3 \cdot 5p \qquad \text{Apply the inductive assumption.}$$

$$= 5\left(8^k + 3p\right). \qquad \text{Factor once more.}$$

This shows that $8^{k+1} - 3^{k+1}$ is a product of 5 and the integer $8^k + 3p$, so we have shown that $8^{k+1} - 3^{k+1}$ is divisible by 5, completing the inductive step, and thus the proof.

Finally, we will prove the formula for the summation of the first n squares. The final summation formula from Section 11.1 is left as an exercise (Exercise 22).

EXAMPLE 4

Proof by
Mathematical
Induction

Prove that for each natural number n, $\displaystyle\sum_{i=1}^{n} i^2 = \frac{n(n+1)(2n+1)}{6}$.

Solution:

Basic Step: Note that $P(1)$ is the statement $1^2 = \dfrac{(1)(1+1)(2+1)}{6}$ which is true.

Induction Step: The inductive assumption is $1^2 + 2^2 + \ldots + k^2 = \dfrac{k(k+1)(2k+1)}{6}$.

Consider the left-hand side of $P(k + 1)$:

$$1^2 + 2^2 + \ldots + k^2 + \left(k+1\right)^2$$

To show that $P(k + 1)$ is true, we need to demonstrate that the sum above is equal to the right-hand side of $P(k + 1)$, namely

$$\frac{(k+1)\left((k+1)+1\right)\left(2(k+1)+1\right)}{6}, \text{ better written as } \frac{(k+1)(k+2)(2k+3)}{6}.$$

We use the inductive assumption to rewrite all but the last term of the left-hand side; some algebraic manipulation then leads to the desired result.

$$1^2 + 2^2 + \cdots + (k+1)^2 = 1^2 + 2^2 + \cdots + k^2 + (k+1)^2$$

$$= \frac{k(k+1)(2k+1)}{6} + (k+1)^2 \qquad \text{Use the inductive assumption to rewrite all but the last term.}$$

$$= \frac{k(k+1)(2k+1)}{6} + \frac{6(k+1)^2}{6} \qquad \text{Rewrite the second term using the LCD.}$$

$$= \frac{(k+1)}{6}\left[k(2k+1) + 6(k+1)\right] \qquad \text{Factor out } \frac{k+1}{6}.$$

$$= \frac{(k+1)}{6}\left[2k^2 + k + 6k + 6\right] \qquad \text{Expand the remaining terms.}$$

$$= \frac{(k+1)}{6}\left[2k^2 + 7k + 6\right] \qquad \text{Combine like terms.}$$

$$= \frac{(k+1)}{6}\left[(k+2)(2k+3)\right] \qquad \text{Factor the resulting quadratic.}$$

$$= \frac{(k+1)(k+2)(2k+3)}{6} \qquad \text{Rewrite in the desired form.}$$

This completes the induction step, and thus the entire proof.

Exercises

1. $S_{k+1} = \dfrac{1}{3k+9}$

2. $S_{k+1} = \dfrac{k+1}{k}$

3. $S_{k+1} = \dfrac{(k+1)(k+2)(2k+3)}{4}$

4. $S_{k+1} = \dfrac{1}{(2k+1)(2k+3)}$

Find S_{k+1} for the given S_k.

1. $S_k = \dfrac{1}{3(k+2)}$

2. $S_k = \dfrac{k^2}{k(k-1)}$

3. $S_k = \dfrac{k(k+1)(2k+1)}{4}$

4. $S_k = \dfrac{1}{(2k-1)(2k+1)}$

Use the Principle of Mathematical Induction to prove the following statements. See Examples 1 through 4.

5. $1 + 2 + 3 + 4 + \ldots + n = \dfrac{n(n+1)}{2}$

5. Basic Step: $n = 1,\ 1 = 1$ and $\dfrac{1(1+1)}{2} = 1$;

Induction Step: If $1 + 2 + 3 + \ldots + k = \dfrac{k(k+1)}{2}$, then $(1 + 2 + 3 + \ldots + k) + (k+1) = \dfrac{k(k+1)}{2} + (k+1) = \dfrac{k^2 + k + 2k + 2}{2} = \dfrac{(k+1)(k+2)}{2}$

6. $\dfrac{1}{2} + \dfrac{1}{2^2} + \dfrac{1}{2^3} + \ldots + \dfrac{1}{2^n} = 1 - \dfrac{1}{2^n}$

6. Basic Step: $n = 1$, $\dfrac{1}{2^1} = \dfrac{1}{2}$ and $1 - \dfrac{1}{2^1} = \dfrac{1}{2}$;

Induction Step: If $\dfrac{1}{2} + \dfrac{1}{2^2} + \dfrac{1}{2^3} + \ldots + \dfrac{1}{2^k} = 1 - \dfrac{1}{2^k}$, then $\left[\dfrac{1}{2} + \dfrac{1}{2^2} + \dfrac{1}{2^3} + \ldots + \dfrac{1}{2^k}\right] + \dfrac{1}{2^{k+1}} = \left[1 - \dfrac{1}{2^k}\right] + \dfrac{1}{2^{k+1}} = 1 - \dfrac{2}{2^{k+1}} + \dfrac{1}{2^{k+1}} = 1 - \dfrac{1}{2^{k+1}}$

7. $2+4+6+8+\ldots+2n = n(n+1)$

7. Basic Step: $n=1$, $2(1)=2$ and $1(1+1)=2$;

Induction Step: If $2+4+6+\ldots+2k = k(k+1)$, then $(2+4+6+\ldots+2k)+2(k+1) = k^2+k+2k+2 = (k+1)(k+2)$

8. Basic Step: $n=1$, $\dfrac{1}{(2(1)-1)(2(1)+1)} = \dfrac{1}{3}$ and $\dfrac{1}{2(1)+1} = \dfrac{1}{3}$;

8. $\displaystyle\sum_{i=1}^{n} \frac{1}{(2i-1)(2i+1)} = \frac{n}{2n+1}$

Induction Step: If $\displaystyle\sum_{i=1}^{k} \frac{1}{(2i-1)(2i+1)} = \frac{k}{2k+1}$, then $\displaystyle\sum_{i=1}^{k+1} \frac{1}{(2i-1)(2i+1)}$

$= \left[\displaystyle\sum_{i=1}^{k} \frac{1}{(2i-1)(2i+1)}\right] + \frac{1}{(2(k+1)-1)(2(k+1)+1)} = \frac{k}{2k+1} + \frac{1}{(2k+1)(2k+3)} = \frac{2k^2+3k+1}{(2k+1)(2k+3)} = \frac{(2k+1)(k+1)}{(2k+1)(2(k+1)+1)} = \frac{(k+1)}{(2(k+1)+1)}$

9. $4^0 + 4^1 + 4^2 + \ldots + 4^{n-1} = \dfrac{4^n - 1}{3}$

9. Basic Step: $n=1$, $4^{1-1} = 1$ and $\dfrac{4^1-1}{3} = 1$;

Induction Step: If $4^0 + 4^1 + 4^2 + \ldots + 4^{k-1} = \dfrac{4^k-1}{3}$, then $4^0 + 4^1 + 4^2 + \ldots + 4^{k-1} + 4^{k+1-1} = \dfrac{4^k-1}{3} + 4^k = \dfrac{4^k-1+3\cdot4^k}{3} = \dfrac{4\cdot4^k-1}{3} = \dfrac{4^{k+1}-1}{3}$

10. $2^n > n^2$ for all $n \ge 5$.

10. Basic Step: $n=5$, $2^5 = 32$ and $5^2 = 25$, so $2^5 > 5^2$;

Induction Step: If $2^k > k^2$, then $2^{k+1} = 2^k \cdot 2^1 > k^2 \cdot 2 = k^2 + k^2 \ge k^2 + 5k > k^2 + 3k = k^2 + 2k + k \ge k^2 + 2k + 5 > k^2 + (2k+1) = (k+1)^2$

11. Basic Step: $n=1$,

11. $\dfrac{1}{1\cdot4} + \dfrac{1}{4\cdot7} + \dfrac{1}{7\cdot10} + \ldots + \dfrac{1}{(3n-2)(3n+1)} = \dfrac{n}{3n+1}$

$\dfrac{1}{(3(1)-2)(3(1)+1)} = \dfrac{1}{4}$ and $\dfrac{1}{3(1)+1} = \dfrac{1}{4}$;

Induction Step: If $\dfrac{1}{1\cdot4} + \dfrac{1}{4\cdot7} + \dfrac{1}{7\cdot10} + \ldots + \dfrac{1}{(3k-2)(3k+1)} = \dfrac{k}{3k+1}$, then $\dfrac{1}{1\cdot4} + \dfrac{1}{4\cdot7} + \dfrac{1}{7\cdot10} + \ldots + \dfrac{1}{(3k-2)(3k+1)} + \dfrac{1}{(3(k+1)-2)(3(k+1)+1)}$

$= \left[\dfrac{1}{1\cdot4} + \dfrac{1}{4\cdot7} + \dfrac{1}{7\cdot10} + \ldots + \dfrac{1}{(3k-2)(3k+1)}\right] + \dfrac{1}{(3k+1)(3k+4)} = \dfrac{k}{3k+1} + \dfrac{1}{(3k+1)(3k+4)} = \dfrac{3k^2+4k+1}{(3k+1)(3k+4)} = \dfrac{(3k+1)(k+1)}{(3k+1)(3k+4)} = \dfrac{(k+1)}{(3(k+1)+1)}$

12. $5^0 + 5^1 + 5^2 + \ldots + 5^{n-1} = \dfrac{5^n - 1}{4}$

12. Basic Step: $n=1$, $5^{1-1} = 1$ and $\dfrac{5^1-1}{4} = 1$;

Induction Step: If $5^0 + 5^1 + 5^2 + \ldots + 5^{k-1} = \dfrac{5^k-1}{4}$, then $5^0 + 5^1 + 5^2 + \ldots + 5^{k-1} + 5^{k+1-1} = \dfrac{5^k-1}{4} + 5^k = \dfrac{5^k-1+4\cdot5^k}{4} = \dfrac{5\cdot5^k-1}{4} = \dfrac{5^{k+1}-1}{4}$

13. $5 + 10 + 15 + \ldots + 5n = \dfrac{5n(n+1)}{2}$

13. Basic Step: $n=1$, $5(1)=5$ and $\dfrac{5(1)(1+1)}{2} = 5$;

Induction Step: If $5+10+15+\ldots+5k = \dfrac{5k(k+1)}{2}$, then $5+10+15+\ldots+5k+5(k+1) = (5+10+15+\ldots+5k)+5k+5 = \dfrac{5k(k+1)}{2} + 5k + 5$

$= \dfrac{5k^2+15k+10}{2} = \dfrac{5(k+1)(k+2)}{2} = \dfrac{5(k+1)[(k+1)+1]}{2}$

14. $n^2 \geq 100n$ for all $n \geq 100$.

14. Basic Step: $n = 100$, $100^2 = 10,000$ and $100 \cdot 100 = 10,000$, so $100^2 \geq 100 \cdot 100$; Induction Step: If $k^2 \geq 100k$, then $(k+1)^2 = k^2 + 2k + 1 \geq 100k + 2k + 1 \geq 100k + 200 + 1 > 100k + 100 = 100(k+1)$

15. $\left(1 + \dfrac{1}{1}\right)\left(1 + \dfrac{1}{2}\right)\left(1 + \dfrac{1}{3}\right)\ldots\left(1 + \dfrac{1}{n}\right) = n + 1$

15. Basic Step: $n = 1$, $1 + \dfrac{1}{1} = 2$ and $1 + 1 = 2$

Induction Step: If $\left(1 + \dfrac{1}{1}\right)\left(1 + \dfrac{1}{2}\right)\left(1 + \dfrac{1}{3}\right)\ldots\left(1 + \dfrac{1}{k}\right) = k + 1$, then $\left(1 + \dfrac{1}{1}\right)\left(1 + \dfrac{1}{2}\right)\left(1 + \dfrac{1}{3}\right)\ldots\left(1 + \dfrac{1}{k}\right)\left(1 + \dfrac{1}{k+1}\right) = (k+1)\left(1 + \dfrac{1}{k+1}\right)$

$$= k + 1 + \dfrac{k+1}{k+1} = (k+1) + 1$$

16. $3 + 5 + 7 + \ldots + (2n+1) = n(n+2)$

16. Basic Step: $n = 1$, $2(1) + 1 = 3$ and $1(1+2) = 3$ Induction Step: If $3 + 5 + 7 + \ldots + (2k+1) = k(k+2)$, then $\left[3 + 5 + 7 + \ldots + (2k+1)\right] + \left[2(k+1) + 1\right] = k(k+2) + (2k+3) = k^2 + 4k + 3 = (k+1)(k+3) = (k+1)\left[(k+1) + 2\right]$

17. $1 + 4 + 7 + 10 + \ldots + (3n-2) = \dfrac{n}{2}(3n-1)$

17. Basic Step: $n = 1$, $3(1) - 2 = 1$ and $\dfrac{1}{2}(3(1) - 1) = 1$;

Induction Step: If $1 + 4 + 7 + 10 + \ldots + (3k-2) = \dfrac{k}{2}(3k-1)$, then $\left[1 + 4 + 7 + 10 + \ldots + (3k-2)\right] + \left[3(k+1) - 2\right] = \dfrac{k}{2}(3k-1) + (3k+1)$

$$= \dfrac{k(3k-1) + 2(3k+1)}{2} = \dfrac{3k^2 + 5k + 2}{2} = \dfrac{(k+1)(3k+2)}{2} = \dfrac{k+1}{2}(3(k+1) - 1)$$

18. $-2 - 3 - 4 - \ldots - (n+1) = -\dfrac{1}{2}(n^2 + 3n)$

18. Basic Step: $n = 1$, $-(1+1) = -2$ and $-\dfrac{1}{2}(1^2 + 3(1)) = -2$; Induction Step: If $-2 - 3 - 4 - \cdots - (k+1) = -\dfrac{1}{2}(k^2 + 3k)$,

then $-2 - 3 - 4 - \cdots - (k+1) - ((k+1) + 1) = -\dfrac{1}{2}(k^2 + 3k) - (k+2) = -\dfrac{k^2 + 3k + 2k + 4}{2} = -\dfrac{(k^2 + 2k + 1) + (3k + 3)}{2} = -\dfrac{1}{2}((k+1)^2 + 3(k+1))$

19. $3^n > 2n + 1$ for all $n \geq 2$.

19. Basic Step: $n = 2$, $3^2 = 9$ and $2(2) + 1 = 5$, so $3^2 > 2(2) + 1$;

Induction Step: If $3^k > 2k + 1$, then $3^{k+1} = 3^1 \cdot 3^k > 3(2k+1) = 6k + 3 > 2k + 3 = 2k + 2 + 1 = 2(k+1) + 1$

20. For all natural numbers n, $2^n > n$.

20. Basic Step: $n = 1$, since $2^1 = 2$, $2^1 > 1$;

Induction Step: If $2^k > k$, then $2^{k+1} = 2^1 \cdot 2^k > 2k = k + k \geq k + 1$

21. $1^3 + 2^3 + 3^3 + 4^3 + \ldots + n^3 = \dfrac{n^2(n+1)^2}{4}$

21. Basic Step: $n = 1$, $1^3 = 1$ and $\dfrac{1^2(1+1)^2}{4} = 1$;

Induction Step: If $1^3 + 2^3 + 3^3 + 4^3 + \ldots + k^3 = \dfrac{k^2(k+1)^2}{4}$,

then $\left(1^3 + 2^3 + 3^3 + 4^3 + \ldots + k^3\right) + (k+1)^3 = \dfrac{k^2(k+1)^2}{4} + (k+1)^3 = \dfrac{k^2(k+1)^2 + 4(k+1)^3}{4} = \dfrac{(k+1)^2(k+2)^2}{4} = \dfrac{(k+1)^2((k+1) + 1)^2}{4}$

22. $1\cdot2+2\cdot3+3\cdot4+\ldots+n(n+1)=\dfrac{n(n+1)(n+2)}{3}$

22. Basic Step: $n=1$, $1(1+1)=2$ and $\dfrac{1(1+1)(1+2)}{3}=2$;

Induction Step: If $1\cdot2+2\cdot3+3\cdot4+\ldots+k(k+1)=\dfrac{k(k+1)(k+2)}{3}$, then $\left[1\cdot2+2\cdot3+3\cdot4+\ldots+k(k+1)\right]+(k+1)\big((k+1)+1\big)$

$=\dfrac{k(k+1)(k+2)}{3}+(k+1)(k+2)=\dfrac{k(k+1)(k+2)+3(k+1)(k+2)}{3}=\dfrac{(k+1)(k+2)(k+3)}{3}=\dfrac{(k+1)\big((k+1)+1\big)\big((k+1)+2\big)}{3}$

23. If $a>1$, then $a^n>1$.

23. Basic Step: $n=1$, $a^1=a$ so $a^1>1$, when $a>1$;

Induction Step: If $a^k>1$, then $a^{k+1}=a^k\cdot a^1>1\cdot a=a>1$

24. $2^n>4n$ for all $n\geq5$.

24. Basic Step: $n=5$, $2^5=32$ and $4\cdot5=20$, so $2^5>4\cdot5$;

Induction Step: If $2^k>4k$, then $2^{k+1}=2\cdot2^k=2^k+2^k>4k+4k>4k+4=4(k+1)$

25. $1^4+2^4+3^4+4^4+\ldots+n^4=\dfrac{n(n+1)(2n+1)(3n^2+3n-1)}{30}$

25. Basic Step: $n=1$, $1^4=1$

and $\dfrac{1(1+1)(2(1)+1)(3(1)^2+3(1)-1)}{30}=1$;

Induction Step: If $1^4+2^4+3^4+\ldots+k^4=\dfrac{k(k+1)(2k+1)(3k^2+3k-1)}{30}$, then $\big(1^4+2^4+3^4+\ldots+k^4\big)+(k+1)^4=\dfrac{k(k+1)(2k+1)(3k^2+3k-1)}{30}$

$+(k+1)^4=\dfrac{6k^5+45k^4+130k^3+180k^2+119k+30}{30}=\dfrac{(k+1)(k+2)(2k+3)(3k^2+9k+5)}{30}=\dfrac{(k+1)(k+2)(2(k+1)+1)(3(k+1)^2+3(k+1)-1)}{30}$

26. $1^5+2^5+3^5+4^5+\ldots+n^5=\dfrac{n^2(n+1)^2(2n^2+2n-1)}{12}$

26. Basic Step: $n=1$, $1^5=1$ and

$\dfrac{1^2(1+1)^2(2(1)^2+2(1)-1)}{12}=1$; Induction Step: If $1^5+2^5+3^5+\ldots+k^5=\dfrac{k^2(k+1)^2(2k^2+2k-1)}{12}$, then $\big(1^5+2^5+3^5+\ldots+k^5\big)+(k+1)^5$

$=\dfrac{k^2(k+1)^2(2k^2+2k-1)}{12}+(k+1)^5=\dfrac{(k+1)^2\left[k^2(2k^2+2k-1)+12(k+1)^3\right]}{12}=\dfrac{(k+1)^2(2k^4+14k^3+35k^2+36k+12)}{12}$

$=\dfrac{(k+1)^2(k+2)^2(2k^2+6k+3)}{12}=\dfrac{(k+1)^2(k+2)^2(2(k+1)^2+2(k+1)-1)}{12}$

27. $\dfrac{1}{\sqrt{1}}+\dfrac{1}{\sqrt{2}}+\dfrac{1}{\sqrt{3}}+\ldots+\dfrac{1}{\sqrt{n}}>\sqrt{n}$ for all $n \geq 2$. **27.** Basic Step: $n \geq 2$, $\dfrac{1}{\sqrt{1}}+\dfrac{1}{\sqrt{2}}=1+\dfrac{\sqrt{2}}{2}$ and $1+\dfrac{\sqrt{2}}{2}>\sqrt{}$

Induction Step: If $\dfrac{1}{\sqrt{1}}+\dfrac{1}{\sqrt{2}}+\dfrac{1}{\sqrt{3}}+\ldots+\dfrac{1}{\sqrt{k}}>\sqrt{k}$, then $\left[\dfrac{1}{\sqrt{1}}+\dfrac{1}{\sqrt{2}}+\dfrac{1}{\sqrt{3}}+\ldots+\dfrac{1}{\sqrt{k}}\right]+\dfrac{1}{\sqrt{k+1}}>\sqrt{k}+\dfrac{1}{\sqrt{k+1}}=\dfrac{\sqrt{k}\left(\sqrt{k+1}\right)-}{\sqrt{k+1}}$

$$=\dfrac{\sqrt{k^2+k}+1}{\sqrt{k+1}}>\dfrac{\sqrt{k^2}+1}{\sqrt{k+1}}=\dfrac{k+1}{\sqrt{k+1}}=\sqrt{k}$$

28. $1+3+5+7+\ldots+(2n-1)=n^2$

28. Basic Step: $n=1$, $2(1)-1=1$ and $1^2=1$;

Induction Step: If $1+3+5+7+\ldots+2k-1=k^2$, then $(1+3+5+7+\ldots+2k-1)+(2(k+1)-1)=k^2+(2k+1)=(k+1)^2$

Use the Principle of Mathematical Induction to prove the given properties.

29. $(ab)^n = a^n b^n$, for all positive integers n. (Assume a and b are constants.)

29. Basic Step: $n=1$, $(ab)^1=ab$ and $a^1 b^1=ab$;

Induction Step: If $(ab)^k=a^k b^k$, then $(ab)^{k+1}=(ab)^k \cdot (ab)=a^k b^k \cdot ab = \left(a \cdot a^k\right)\left(b \cdot b^k\right)=a^{k+1}b^{k+1}$

30. $\left(a^m\right)^n = a^{mn}$, for all positive integers m and n. (Assume a and m are constants.)

30. Basic Step: $n=1$, $\left(a^m\right)^1=a^m$ and $a^{m\cdot 1}=a^m$; Induction Step: If $\left(a^m\right)^k=a^{mk}$, then $\left(a^m\right)^{k+1}=\left(a^m\right)^k\cdot\left(a^m\right)^1=a^{mk}\cdot a^m=a^{mk+m}=a^{m(k+1)}$

31. If $x_1>0$, $x_2>0$, ..., $x_n>0$, then
$$\ln\left(x_1\cdot x_2\cdot x_3\cdot\ldots\cdot x_n\right)=\ln x_1+\ln x_2+\ln x_3+\ldots+\ln x_n.$$

31. Basic Step: $n=1$, $\ln\left(x_1\right)=\ln x_1$; Induction Step: If $\ln\left(x_1\cdot x_2\cdot x_3\cdot\ldots\cdot x_k\right)=\ln x_1+\ln x_2+\ln x_3+\ldots+\ln x_k$ when $x_1>0$, $x_2>0$, ..., $x_n>0$,

then $\ln\left(x_1\cdot x_2\cdot x_3\cdot\ldots\cdot x_k\cdot x_{k+1}\right)=\ln\left(x_1\cdot x_2\cdot x_3\cdot\ldots\cdot x_k\right)+\ln\left(x_{k+1}\right)=\left(\ln x_1+\ln x_2+\ln x_3+\ldots+\ln x_k\right)+\ln x_{k+1}$

32. For all natural numbers n, 5 is a factor of $\left(2^{2n-1}+3^{2n-1}\right)$.

32. Basic Step: $n=1$, $\left(2^{2(1)-1}+3^{2(1)-1}\right)=5$, so 5 is a factor;

Induction Step: If $\left(2^{2(k)-1}+3^{2(k)-1}\right)=5p$ for some integer p, then $\left(2^{2(k+1)-1}+3^{2(k+1)-1}\right)=\left(2^{2k+1}+3^{2k+1}\right)=2^2\cdot 2^{2k-1}+3^2\cdot 3^{2k-1}=4\cdot 2^{2k-1}+9\cdot 3^{2k-1}$

$$=4\cdot 2^{2k-1}+(4+5)\cdot 3^{2k-1}=4\cdot 2^{2k-1}+4\cdot 3^{2k-1}+5\cdot 3^{2k-1}=4\left(2^{2k-1}+3^{2k-1}\right)+5\cdot 3^{2k-1}=4(5p)+5\cdot 3^{2k-1}=5\left(4p+3^{2k-1}\right)$$

33. 64 is a factor of $\left(9^n-8n-1\right)$ for all $n\geq 2$.

33. Basic Step: $n=2$, $\left(9^2-8(2)-1\right)=64$ of which 64 is a factor;   Induction Step: If $\left(9^k-8k-1\right)=64p$ for some integer p,

then $\left(9^{k+1}-8(k+1)-1\right)=9\cdot 9^k-8k-9=9\cdot 9^k-9\cdot 8k+8\cdot 8k-9=9\left(9^k-8k-1\right)+64k=9(64p)+64k=64(9p+k)$

34. For all natural numbers n, 3 is a factor of $\left(n^3 + 3n^2 + 2n\right)$.

34. Basic Step: $n = 1$, $\left(1^3 + 3(1)^2 + 2(1)\right) = 6 = 3 \cdot 2$, so 3 is a factor;

Induction Step: If $\left(k^3 + 3k^2 + 2k\right) = 3p$ for some integer p, then $\left(k+1\right)^3 + 3\left(k+1\right)^2 + 2\left(k+1\right) = k^3 + 6k^2 + 11k + 6$

$= \left(k^3 + 3k^2 + 2k\right) + \left(3k^2 + 9k + 6\right) = 3p + 3\left(k^2 + 3k + 2\right) = 3\left(p + k^2 + 3k + 2\right)$

35. For all natural numbers n, $n^3 - n + 3$ is divisible by 3.

35. Basic Step: $n = 1$, $\left(1^3 - 1 + 3\right) = 3$, which is divisible by 3; Induction Step: If $\dfrac{k^3 - k + 3}{3} = p$ or $k^3 - k + 3 = 3p$ for some integer p,

then $\left(k+1\right)^3 - \left(k+1\right) + 3 = k^3 + 3k^2 + 2k + 3 = \left(k^3 - k + 3\right) + \left(3k^2 + 3k\right) = 3p + 3\left(k^2 + k\right) = 3\left(p + k^2 + k\right)$

36. For all natural numbers n, $5^n - 1$ is divisible by 4.

36. Basic Step: $n = 1$, $5^1 - 1 = 4$, which is divisible by 4;   Induction Step: If $\dfrac{5^k - 1}{4} = p$ or $5^k - 1 = 4p$ for some integer p,

then $5^{k+1} - 1 = 5 \cdot 5^k - 5 + 4 = 5\left(5^k - 1\right) + 4 = 5\left(4p\right) + 4 = 4\left(5p + 1\right)$

37. Basic Step: $n = 1$, $1\left(1+1\right)\left(1+2\right) = 6$, which is divisible by 6; Induction Step: If $\dfrac{k\left(k+1\right)\left(k+2\right)}{6} = p$ or $k\left(k+1\right)\left(k+2\right) = 6p$ for

some integer p, then   **37.** For all natural numbers n, $n\left(n+1\right)\left(n+2\right)$ is divisible by 6.

$\left(k+1\right)\left(k+2\right)\left(k+3\right) = k^3 + 6k^2 + 11k + 6 = \left(k^3 + 3k^2 + 2k\right) + \left(3k^2 + 9k + 6\right) = k\left(k+1\right)\left(k+2\right) + 3\left(k+1\right)\left(k+2\right) = 6p + 3\left(k+1\right)\left(k+2\right)$.

$6p$ is clearly divisible by 6. In order for $3\left(k+1\right)\left(k+2\right)$ to be divisible by 6, it must be divisible by 2 and 3. It is clearly divisible by 3.

If k is odd, then the term $\left(k+1\right)$ must be even, making it divisible by 2. If k is even, then the term $\left(k+2\right)$ is even, making it divisible by 2.

Therefore, $3\left(k+1\right)\left(k+2\right)$ is divisible by 6.

38. In the 19th century a mathematical puzzle was published telling of a mythical monastery in Benares, India with three crystal towers holding 64 disks made of gold. The disks are each of a different size and have holes in the middle so that they slide over the towers and sit in a stack with the largest on the bottom and the smallest on the top. The monks of the monastery were instructed to move all of the disks to the third tower following these three rules:

1. Each disk sits over a tower except when it is being moved.

2. No disk may ever rest on a smaller disk.

3. Only one disk at a time may be moved.

According to the puzzle, when the monks complete their task, the world would end! To move n disks requires $H(n) = 2^n - 1$ moves. Prove this is true through mathematical induction.

38. $1 + 2 + 4 + 8 + \ldots + 2^{n-1} = 2^n - 1$;   Basic Step: $n = 1$, $2^{1-1} = 1$ and $2^1 - 1 = 1$;

Induction Step: If $1 + 2 + 4 + 8 + \ldots + 2^{k-1} = 2^k - 1$, then $\left(1 + 2 + 4 + 8 + \cdots + 2^{k-1}\right) + 2^{k+1-1} = \left(2^k - 1\right) + 2^k = 2 \cdot 2^k - 1 = 2^{k+1} - 1$

39. If there are n people in a room, and every person shakes hands with every other person exactly once, then exactly $\dfrac{n(n-1)}{2}$ handshakes will occur. Prove this is true through mathematical induction.

39. $0+1+2+3+\ldots+(n-1)=\dfrac{n(n-1)}{2}$; Basic Step: $n=1$, $(1-1)=0$ and $\dfrac{1(1-1)}{2}=0$; Induction Step: If $0+1+2+\ldots+(k-1)=\dfrac{k(k-1)}{2}$,

then $\left[0+1+2+\ldots+(k-1)\right]+(k+1-1)=\dfrac{k(k-1)}{2}+k=\dfrac{k^2-k+2k}{2}=\dfrac{k(k+1)}{2}=\dfrac{(k+1)\left((k+1)-1\right)}{2}$

40. Any monetary value of 4 cents or higher can be composed of twopence (a British two-cent coin) and nickels. Your basic step would be 4 cents = twopence + twopence. Use the fact that $k = 2t + 5n$ where k is the total monetary value, t is the number of twopence, and n is the number of nickels, to prove $P(k + 1)$. (**Hint:** There are 3 induction steps to prove.)   **40.** Induction Steps: If $k = 2t + 5n$, then…

Case 1 $(t \geq 1, \ n \geq 1, \ \text{and} \ k = 2t+5n)$: $\ k+1 = 2t+5n+1 = 2t+5n+(6-5) = (2t+6)+(5n-5) = 2(t+3)+5(n-1)$

Case 2 $(t = 0, \ n \geq 1, \ \text{and} \ k = 5n)$: $\ k+1 = 5n+1 = 5n+(6-5) = 6+(5n-5) = 2(3)+5(n-1)$

Case 3 $(t \geq 2, \ n = 0, \ \text{and} \ k = 2t)$: $\ k+1 = 2t+1 = 2t+(5-4) = (2t-4)+5 = 2(t-2)+5(1)$

41. What is wrong with this "proof" by induction?

Proposition: All horses are the same color. (In any set of n horses, all horses are the same color.)

Basic Step: If you have only one horse in a group, then all of the horses in that group have the same color.

Induction Step: Assume that in any group of n horses, all horses are the same color. Now take any group of $n + 1$ horses. Remove the first horse from this group and the remaining n horses must be of the same color because of the hypothesis. Now replace the first horse and remove the last horse. Once again, the remaining n horses must be the same color because of the hypothesis. Since the two groups overlap, all $n + 1$ horses must be the same color.

Thus by induction, any group of n horses are the same color.

41. The Induction Step does not work for $n = 1$. In the case of $n = 1$, $n + 1 = 2$ and the groups formed by removing the first horse and then the last horse do not overlap.

11.5

An Introduction to Combinatorics

TOPICS

1. The Multiplication Principle of Counting

2. Permutations

3. Combinations

4. The Binomial Theorem

T. Permutations and combinations

TOPIC 1

The Multiplication Principle of Counting

Combinatorics can be informally defined as "the science of counting." More precisely, combinatorics is "the study of techniques used to determine the sizes of sets." Even this may sound deceptively simple. As we will see, there are many occasions when the cardinality (the number of elements) of a well-defined set may be difficult to determine at first glance. We will also see that in many cases the size of a set is of more importance than the actual elements of the set.

Many problems in combinatorics can be solved with just a few fairly intuitive ideas, the most basic of which is the *Multiplication Principle of Counting*. Before formally stating the principle, we will use it to solve a problem.

EXAMPLE 1

Counting Phone Numbers

In the United States, telephone numbers consist of a 3-digit area code followed by a 7-digit local number. Neither the first digit of the area code nor the first digit of the local number can be 0 or 1. How many such phone numbers are there?

Note:
This problem asks about the cardinality of a set (how many phone numbers are there) using informal language. This is common in combinatorics problems.

Solution:

We will count all the possible phone numbers by thinking about how we could construct them.

Every such phone number consists of a string of ten digits, with the restriction that the first and fourth digits (reading from left to right) can't be either 0 or 1. Since there are ten digits in all (0 through 9), this means there are eight possible ways to choose a digit for the first and fourth "slots," and ten possible ways to choose digits for all the remaining slots. This is illustrated below.

8 possible digits	10 possible digits	10 possible digits	8 possible digits	10 possible digits	10 possible digits	10 possible digits	10 possible digits	10 possible digits	10 possible digits

For the moment, consider how just the first two slots can be filled. Any of the eight allowable digits for the first slot can be paired with any of the ten allowable digits for the second slot, meaning that there are $8 \cdot 10$ ways of filling the first two slots altogether.

Now, any of these eighty possible choices for the first two slots can be matched with any of the ten possible digits for the third slot, giving us a total of 800 ways of filling the first three slots. This pattern continues, so that the total number of phone numbers of the required form is

$$8 \cdot 10 \cdot 10 \cdot 8 \cdot 10 \cdot 10 \cdot 10 \cdot 10 \cdot 10 \cdot 10 = 6,400,000,000.$$

The generalization of the reasoning we used in Example 1 is often stated in terms of a sequence of events.

THEOREM

The Multiplication Principle of Counting

Suppose $E_1, E_2, \ldots, E_n$ is a sequence of events, each of which has a certain number of possible outcomes. Suppose event E_1 has m_1 possible outcomes, and that after event E_1 has occurred, event E_2 has m_2 possible outcomes. Similarly, after event E_2, event E_3 has m_3 possible outcomes, and so on.

The **total number of ways that all n events can occur** is the product $m_1 \cdot m_2 \cdot \ldots \cdot m_n$.

An alternative interpretation of the principle is to think of the events as a sequence of tasks to be completed. In Example 1, each task consists of selecting a digit for a given slot. There are ten tasks in all, and the product of the number of ways each task can be performed gives us the total number of phone numbers.

EXAMPLE 2

The Multiplication Principle of Counting

A certain state specifies that all nonpersonalized license plates consist of two letters followed by four digits, and that the letter O (which could be mistaken for the digit 0) cannot be used. How many such license plates are there?

Solution:

Generating such a license plate is a matter of choosing six characters: the first two can be any of 25 letters and the last four can be any of 10 digits.

Applying the Multiplication Principle of Counting, there are 25 outcomes for each of the first two events and 10 outcomes for each of the last four events, thus the total number of such license plates is

$$25 \cdot 25 \cdot 10 \cdot 10 \cdot 10 \cdot 10 = 6,250,000.$$

TOPIC 2

Permutations

Permutations

A **permutation** of a set of objects is an ordering of the objects. In any such ordering, one of the objects is first, another is second, and so on, so the construction of a permutation of n objects consists of "filling in" n slots. This means we can use the Multiplication Principle of Counting to determine the number of permutations of a given set of objects.

EXAMPLE 3

Permutations

One brand of combination lock for a door consists of five buttons labeled A, B, C, D, and E, and the installer of the lock can set the combination to be any permutation of the five letters. How many such permutations are there?

Solution:

The difference between this problem and the first two examples is that once a letter has been chosen for a given slot, it can't be reused. So in constructing a combination code, there are five choices for the first letter but only four choices for the second letter, since whatever letter was used first cannot be used again.

Similarly, there are only three choices for the third slot and only two choices for the fourth slot. Finally, whichever of the five letters is left *must* be used for the fifth slot, so there is only one choice. The figure below illustrates the slot-filling process.

$$\underbrace{}_{5\ choices} \quad \underbrace{}_{4\ choices} \quad \underbrace{}_{3\ choices} \quad \underbrace{}_{2\ choices} \quad \underbrace{}_{1\ choice}$$

We then use the Multiplication Principle of Counting to determine that there are $5 \cdot 4 \cdot 3 \cdot 2 \cdot 1 = 120$ such combinations.

Products of the form $n \cdot (n-1) \cdot (n-2) \cdot \ldots \cdot 2 \cdot 1$ occur so frequently that there is a shorthand notation for them, defined as follows.

DEFINITION

Factorial Notation

If n is a positive integer, the notation $n!$ (which is read "**n factorial**") stands for the product of all the integers from 1 to n. That is,

$$n! = n \cdot (n-1) \cdot (n-2) \cdot \ldots \cdot 2 \cdot 1.$$

In addition, $0!$ is defined to be 1.

If all permutation problems involved nothing more than an application of the factorial operation, as in Example 3, there would be little left to say. But typically, the solution of a permutation problem requires counting the number of ways that a linear arrangement of k objects can be made from a collection of n objects, where $k \leq n$, so we need to be more careful in applying the Multiplication Principle of Counting.

EXAMPLE 4

Permutations

How many different five-letter combination codes are possible if every letter of the alphabet can be used, but no letter may be repeated?

Solution:

The difference between this problem and the one in Example 3 is that there are now 26 choices for the first letter, 25 choices for the second, and so on. The corresponding "slot diagram" describing the number of ways each slot can be filled appears below.

26 choices   25 choices   24 choices   23 choices   22 choices

The total number of such combination codes is thus $26 \cdot 25 \cdot 24 \cdot 23 \cdot 22 = 7,893,600$.

Products like the one in Example 4 are also very common. Note that we can state the answer to Example 4 in terms of factorials as follows:

$$26 \cdot 25 \cdot 24 \cdot 23 \cdot 22 = \frac{(26 \cdot 25 \cdot 24 \cdot 23 \cdot 22) \cdot 21!}{21!} = \frac{26!}{21!}$$

Generalizing this, we obtain a formula for the number of permutations of length k that can be formed from a collection of n objects, usually expressed as the number of permutations of n objects, taken k at a time.

THEOREM

Permutation Formula

The **number of permutations of n objects taken k at a time** is

$$_nP_k = \frac{n!}{(n-k)!}.$$

This is a simpler way of expressing the product found by the Multiplication Principle of Counting

$$_nP_k = n \cdot (n-1) \cdot \ldots \cdot (n-k+1).$$

The formula for $_nP_k$ is especially useful when the number of factors to be multiplied is large, as in the next example, assuming you have access to a calculator or computer with the factorial function.

EXAMPLE 5

Permutation Formula

A magician is preparing to demonstrate a card trick that involves 20 cards chosen at random from a standard deck of cards. Once chosen, the 20 cards are arranged in a row, and the order of the cards plays a role in the trick. How many such orderings are possible?

Solution:

A standard deck of cards contains 52 distinct cards, and this problem asks for the number of permutations of 52 cards taken 20 at a time. One way to determine this would be to evaluate $52 \cdot 51 \cdot \ldots \cdot 33$ (a product of 20 numbers), but a far faster way is to use the permutation formula:

$$_{52}P_{20} = \frac{52!}{(52-20)!} = \frac{52!}{32!} \approx \frac{8.07 \times 10^{67}}{2.63 \times 10^{35}} \approx 3.07 \times 10^{32}$$

The two factorials that appear in the formula above have been determined by a calculator. Note that 52!, 32!, and the final answer are all very large numbers, so it is convenient to use scientific notation.

TOPIC

Combinations

In contrast to permutations, where the order of the objects is important, combinations are simply collections of objects with no ordering. To be specific, a **combination** of n objects taken k at a time is one of the ways of forming a subset of size k from a set of size n, where again $k \leq n$. Combination problems typically ask us to determine the number of different subsets of size k that can be formed.

To emphasize the difference between permutations and combinations, and to understand the combination formula that we will derive, consider this problem.

EXAMPLE 6

Permutations and
Combinations

Let $S = \{a, b, c, d\}$.

a. How many permutations of size 3 can be formed from the set S?

b. How many combinations of size 3 can be formed from the set S?

Solutions:

a. The number of permutations of 4 objects taken 3 at a time is $_4P_3 = \frac{4!}{(4-3)!} = \frac{4!}{1!} = 24.$

b. We have already determined there are 24 permutations of size 3 that can be formed. For the purpose of determining the corresponding number of combinations, it is useful to actually list out the permutations:

$$
\begin{array}{cccccc}
abc & acb & bac & bca & cab & cba \\
abd & adb & bad & bda & dab & dba \\
acd & adc & cad & cda & dac & dca \\
bcd & bdc & cbd & cdb & dbc & dcb
\end{array}
$$

If we now view these collections of objects as sets, all six permutations in the first row describe the single set $\{a,b,c\}$. Similarly, the six permutations in the second row are simply six different ways of describing the set $\{a,b,d\}$. The third row describes the set $\{a,c,d\}$ and the fourth row describes the set $\{b,c,d\}$. In all, there are only four combinations of 4 objects taken 3 at a time.

Let $_nC_k$ denote the number of combinations of n objects taken k at a time. Taking a cue from Example 6, note that each of the size k combinations formed from S gives rise to $k!$ permutations of size k, and we know that there are $_nP_k$ permutations taken k at a time overall. This means that

$$
(k!)\left(_nC_k\right) = {_nP_k}, \text{ or } _nC_k = \frac{_nP_k}{k!}.
$$

If we now replace $_nP_k$ with $\dfrac{n!}{(n-k)!}$, we have the following formula.

THEOREM

Combination Formula

The **number of combinations of n objects taken k at a time** is

$$
_nC_k = \frac{n!}{k!(n-k)!}.
$$

At this point, we have seen all of the counting techniques we will need. There is much more to the subject of combinatorics, but the three ideas we have discussed—the Multiplication Principle of Counting, the permutation formula, and the combination formula—are sufficient to answer many, many questions. This is especially true when the techniques are combined, as the next few examples show.

EXAMPLE 7

Forming Committees

Suppose a Senate committee consists of 11 Democrats, 10 Republicans, and 1 Independent. The chair of the committee wants to form a subcommittee to be charged with researching a particular issue, and decides to appoint 3 Democrats, 2 Republicans, and the 1 Independent to the subcommittee. How many different subcommittees are possible?

Solution:

The chair needs to form a subset of size 3 from the 11 Democrats, a subset of size 2 from the 10 Republicans, and a subset of size 1 from the 1 Independent. Since the order of those chosen is irrelevant, this is a combination problem and not a permutation problem.

The combination formula tells us these tasks can be done in, respectively, $_{11}C_3$, $_{10}C_2$, and $_1C_1$ ways (of course, there is only 1 way to choose 1 member from a set of 1). These numbers are:

$$_{11}C_3 = \frac{11!}{3!8!} = \frac{11 \cdot 10 \cdot 9 \cdot 8!}{3! \, 8!} = \frac{990}{6} = 165$$

$$_{10}C_2 = \frac{10!}{2!8!} = \frac{10 \cdot 9 \cdot 8!}{2! \, 8!} = \frac{90}{2} = 45$$

$$_1C_1 = \frac{1!}{1!0!} = \frac{1}{1} = 1 \ \ \text{(Remember that } 0! = 1.)$$

Once the appropriate number of people from each party have been chosen, any of the 165 possible groups of 3 Democrats can be matched up with any of the 45 possible groups of 2 Republicans, and then further matched up with the 1 Independent. The Multiplication Principle thus tells us there are $165 \cdot 45 \cdot 1 = 7425$ ways of forming the desired subcommittee.

EXAMPLE 8

Forming "Words"

How many different arrangements are there of the letters in the following words?

a. STIPEND   **b.** SALAAM   **c.** MISSISSIPPI

Solutions:

a. We want to count the number of "words" that can be formed from the letters in the word STIPEND, using each letter once and only once (most of the arrangements will not actually be legitimate English words). Since the order of the letters is important, and since STIPEND contains 7 distinct letters, the answer is:

$$_7P_7 = \frac{7!}{0!} = 7! = 5040.$$

b. The word SALAAM contains 6 letters, but only 4 distinct letters. That is, the 3 A's are indistinguishable, so the answer is not simply $_6P_6$. In fact, 6! overcounts the total number of arrangements by a factor of 3!, because any one arrangement of the 6 letters in SALAAM is equivalent to 5 more arrangements. This is because the 3 A's can be permuted in $3! = 6$ ways, which we can see by coloring the A's differently:

MAALSA, MAALSA, MAALSA, MAALSA, MAALSA, MAALSA

This means the total number of arrangements is actually

$$\frac{6!}{3!} = \frac{720}{6} = 120.$$

c. MISSISSIPPI contains 11 characters, but 4 of them are S's, 4 of them are I's, 2 of them are P's, and the remaining 1 character is M. If the 11 characters were all distinct, the total number of arrangements of the letters would be 11!, but we need to divide this number by 4! to account for the indistinguishable S's, and then divide again by 4! to account for the I's, and then again by 2! to account for the P's. This gives us a total of

$$\frac{11!}{4!4!2!} = \frac{39,916,800}{(24)(24)(2)} = 34,650.$$

TOPIC 4

The Binomial Theorem

Consider the expression $(x+y)^7$, which is a binomial raised to a power. We often need to expand expressions like this in order to work toward the solution of an algebra problem.

We can expand $(x+y)^7$ using elementary methods as follows:

$$(x+y)^7 = (x+y)(x+y)(x+y)^5$$

$$= (x^2 + 2xy + y^2)(x+y)^5$$

$$= (x^2 + 2xy + y^2)(x+y)(x+y)^4$$

$$= \dots$$

This can be extremely tedious and error-prone if the exponent is larger than, say, 3 or 4. We can use the combinatorics methods we have seen to develop two formulas that greatly simplify the process.

When we expand an expression like $(x+y)^7$, we are really making sure we account for all the possible products that result from taking one term from each factor. For instance, when we multiply the terms that are boxed below, we obtain x^3y^4:

$$(\boxed{x}+y)(x+\boxed{y})(x+\boxed{y})(\boxed{x}+y)(x+\boxed{y})(\boxed{x}+y)(x+\boxed{y})$$

But there are many other choices of terms that also lead to x^3y^4, such as:

$$(x+\boxed{y})(x+\boxed{y})(\boxed{x}+y)(\boxed{x}+y)(\boxed{x}+y)(x+\boxed{y})(x+\boxed{y})$$

In all, there are 35 different ways of boxing 3 x's and 4 y's, meaning that there is a coefficient of 35 in front of x^3y^4 in the expansion of $(x+y)^7$.

Where did this number of 35 come from? We can relate these expansions to the "word" problems we studied in Example 8. For instance, the coefficient of x^3y^4 is the number of different arrangements of the letters $xxxyyyy$. Since these 7 letters consist of 3 x's and 4 y's, the total number of such arrangements is

$$\frac{7!}{3!4!} = \frac{5040}{(6)(24)} = 35.$$

Note that the boxed letters above correspond to the "words" $xyyxyxy$ and $yyxxxyy$.

DEFINITION

Binomial Coefficients

Given nonnegative integers n and k, with $k \leq n$, we define

$$\binom{n}{k} = \frac{n!}{k!(n-k)!}.$$

The expression $\binom{n}{k}$ is called a **binomial coefficient**, as it corresponds to the coefficient of $x^{n-k}y^k$ in the expansion of $(x+y)^n$.

We have already seen the formula for binomial coefficients in another context. Note that

$$\binom{n}{k} = {}_nC_k.$$

The last step in expanding a binomial, now that we know how to find the coefficients, is to put all the pieces together. Consider again the expression $(x+y)^7$. When this is expanded, there will be terms containing x^7, x^6y, x^5y^2, x^4y^3, x^3y^4, x^2y^5, xy^6, and y^7, each multiplied by the appropriate binomial coefficient. Note that the sum of the exponents in each term is 7, as this is the total number of x's and y's in each term.

EXAMPLE 9

Binomial Expansion

Expand the expression $(x+y)^7$.

Solution:

Based on the above reasoning, the expansion is

$$\binom{7}{0}x^7y^0 + \binom{7}{1}x^6y^1 + \binom{7}{2}x^5y^2 + \binom{7}{3}x^4y^3 + \binom{7}{4}x^3y^4 + \binom{7}{5}x^2y^5 + \binom{7}{6}x^1y^6 + \binom{7}{7}x^0y^7.$$

We evaluate the binomial coefficients using the binomial coefficient formula, simplifying the expression as follows:

$$(x+y)^7 = x^7 + 7x^6y + 21x^5y^2 + 35x^4y^3 + 35x^3y^4 + 21x^2y^5 + 7xy^6 + y^7$$

THEOREM

The Binomial Theorem

Given the positive integer n, and any two expressions A and B,

$$(A+B)^n = \sum_{k=0}^{n} \binom{n}{k} A^{n-k} B^k$$

$$= \binom{n}{0}A^nB^0 + \binom{n}{1}A^{n-1}B^1 + \binom{n}{2}A^{n-2}B^2 + \ldots + \binom{n}{n-1}A^1B^{n-1} + \binom{n}{n}A^0B^n.$$

Note that A and B can be more complicated expressions than just x and y.

EXAMPLE 10

The Binomial Theorem

Expand the expression $(2x - y)^4$.

Solution:

We use the Binomial Theorem with $A = 2x$ and $B = -y$ to generate the expansion

$$\binom{4}{0}(2x)^4(-y)^0 + \binom{4}{1}(2x)^3(-y)^1 + \binom{4}{2}(2x)^2(-y)^2 + \binom{4}{3}(2x)^1(-y)^3 + \binom{4}{4}(2x)^0(-y)^4.$$

Simplifying this expression, we have

$$(2x - y)^4 = (2x)^4 + 4(2x)^3(-y) + 6(2x)^2(-y)^2 + 4(2x)(-y)^3 + (-y)^4$$
$$= 16x^4 - 32x^3 y + 24x^2 y^2 - 8xy^3 + y^4.$$

TOPIC T

Permutations and Combinations

We can now calculate expressions like $_{18}P_7$ using the formula for permutations. We can also, however, use a graphing calculator. Start by typing the first number, 18. Then, press and use the arrow keys to highlight PRB, select 2:nPr, and press ENTER.

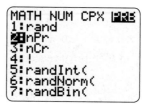

Finally, type in the second number, 7, and press ENTER.

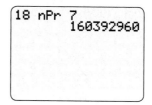

The process for computing $_{18}C_7$ is the same, except after pressing **MATH** we will need to select 3:nCr under the PRB menu instead of 2:nPr.

1. Combination

2. Combination

3. Combination

4. Permutation

5. $10^3 = 1000$

6. $5^7 = 78,125$

7. $9^7 = 4,782,969$

8. $6^5 = 7776$

9. $15! \approx 1.308 \times 10^{12}$

10. $2 \cdot 26^3 = 35,152$

11. $3! = 6$

12. $6 \cdot 3 \cdot 2 \cdot 2 = 72$

13. $5^{10} = 9,765,625$

14. $12! = 479,001,600$

15. $36^6 =$

     $2,176,782,336$

16. $36 \cdot 35 \cdot 34$

     $\cdot 33 \cdot 32 \cdot 31 =$

     $1,402,410,240$

17. $26 \cdot 25 \cdot 24 \cdot 10 \cdot 9 \cdot 8$

     $= 11,232,000$

Exercises

Consider each of the following situations and determine if each is a combination or permutation.

1. Double scoop options from 29 ice cream flavors

2. A board committee chosen from 15 candidates

3. A poker hand from a standard deck

4. A seating chart for 24 students

Use the Multiplication Principle of Counting to answer the following questions. See Examples 1 and 2.

5. Suppose you write down someone's phone number on a piece of paper, but then accidentally wash it along with your laundry. Upon drying the paper, all you can make out of the number is 42?-3?7?. How many different phone numbers fit this pattern?

6. How many different 7-digit phone numbers contain no odd digits? (Ignore the fact that certain 7-digit sequences are disallowed as phone numbers.)

7. How many different 7-digit phone numbers do not contain the digit 9? (Ignore the fact that certain 7-digit sequences are disallowed as phone numbers.)

8. A computer security system allows the buyer to set any password of five letters, with repetition allowed, but each of the letters must be A, B, C, D, E, or F. How many passwords are possible?

9. In how many different orders can 15 runners finish a race, assuming there are no ties?

10. How many different 4-letter radio station names can be made, assuming the first letter must be a K or a W? Assume repetition of letters is allowed.

11. How many different 4-letter radio station names can be made from the call letters K, N, I, T, assuming the letter K must appear first? Each of the four letters can be used only once.

12. Three men and three women line up in a row for a photograph, and decide men and women should alternate. In how many different ways can this be done? (Don't forget that the leftmost person can be a man or a woman.)

13. How many different ways can a 10-question multiple choice test be answered, assuming every question has 5 possible answers?

14. How many different ways can your 12 favorite novels be arranged in a row?

15. How many different 6-character license plates can be formed if all 26 letters and 10 numerical digits can be used with repetition?

16. How many different 6-character license plates can be formed if all 26 letters and 10 numerical digits can be used without repetition?

17. How many different 6-character license plates can be formed if the first 3 characters must be letters and the last 3 characters must be numerical digits? (Assume repetition is not allowed.)

18. $8! = 40,320$

19. $_{30}P_{12} \approx 4.143 \times 10^{16}$

20. $_{15}P_3 = 2730$

21. $_{36}P_8 \approx 1.220 \times 10^{12}$

22. $_7P_5 = 2520$

23. $_7P_6 = 5040$

$_7P_7 = 5040$ as well. (Having a child remain standing is numerically equivalent to putting a seventh chair in the room.)

24. $_{17}P_4 = 57,120$

25. $_{26}P_3 = 15,600$

26. 12

27. 210

28. 720

29. 6.0823×10^{16}

30. 15

31. 6

32. 792

33. 116,280

18. A box of crayons comes with 8 different colored crayons arranged in a single row. How many different ways can the crayons be ordered in the box?

Express the answers to the following permutation problems using permutation notation $\left(_nP_k\right)$ and numerically. See Examples 3, 4, and 5.

19. Suppose you have a collection of 30 cherished math books. How many different ways can you choose 12 of them to arrange in a row?

20. In how many different ways can first place, second place, and third place be decided in a 15-person race?

21. Suppose you need to select a user ID for a computer account, and the system administrator requires that each ID consist of 8 characters with no repetition allowed. The characters you may choose from are the 26 letters of the alphabet (with no distinction between uppercase and lowercase) and the 10 digits. How many choices for a user ID do you have?

22. How many different 5-letter "words" (they don't have to be actual English words) can be formed from the letters in the word PLASTIC?

23. Seven children rush into a room in which six chairs are lined up in a row. How many different ways can six of the seven children choose a chair to sit in? (The seventh remains standing.) How does the answer differ if there are seven chairs in the room?

24. At a meeting of 17 people, a president, vice president, secretary, and treasurer are to be chosen. How many different ways can these positions be filled?

25. Given 26 building blocks, each with a different letter of the alphabet printed on it, how many different 3-letter "words" can be formed?

Evaluate the following permutation expressions. See Examples 5 and 6.

26. $_4P_2$

27. $_{15}P_2$

28. $_6P_5$

29. $_{19}P_{17}$

Evaluate the following combination expressions. See Example 7.

30. $_6C_4$

31. $_4C_2$

32. $_{12}C_5$

33. $_{21}C_{14}$

34. $_{10}C_3 \cdot _{13}C_4$

  $\cdot _6C_2 \cdot _4C_1$

  $= 5,148,000$

35. $_7C_3 = 35$

36. $_{49}C_6 = 13,983,816$

37. $_9C_2 = 36$

38. $_{10}C_5 = 252$

39. $_{75}C_5 = 17,259,390$

40. $\dfrac{5!}{2!} = 60$

41. $\dfrac{6!}{2!3!} = 60$

42. $\dfrac{7!}{2!2!} = 1260$

43. $\dfrac{7!}{2!} = 2520$

44. $\dfrac{11!}{2!2!2!} = 4,989,600$

45. $\dfrac{9!}{2!2!} = 90,720$

46. $_{52}P_5$

  $= 311,875,200,$

  $_{52}C_5 = 2,598,960$

47. $_{10}C_4 \cdot _8C_4 \cdot _{13}C_4$

  $= 10,510,500$

Express the answers to the following combination problems using combination notation $\left(_nC_k\right)$ and numerically. See Examples 7 and 8.

34. In many countries, it is not uncommon for quite a few political parties to have their representatives in power. Suppose a committee composed of 10 Conservatives, 13 Liberals, 6 Greens, and 4 Socialists decides to form a subcommittee consisting of 3 Conservatives, 4 Liberals, 2 Greens, and 1 Socialist. How many different such subcommittees can be formed?

35. A trade union asks its members to select 3 people, from a slate of 7, to serve as representatives at a national meeting. How many different sets of 3 can be chosen?

36. Many lottery games are set up so that players select a subset of numbers from a larger set and the winner is the person whose selection matches that chosen by some random mechanism. The order of the numbers is irrelevant. How many choices of six numbers can be made from the numbers 1 through 49?

37. How many different lines can be drawn through a set of nine points in the plane, assuming that no three of the points are collinear? (Points are said to be *collinear* if a single line containing them can be drawn.)

38. Suppose you are taking a 10-question True-False test, and you are guessing that the professor has arranged it so that five of the answers are True and five are False. How many different ways are there of marking the test with five True answers and five False answers?

39. A caller in a Bingo game draws 5 marked ping pong balls from a basket of 75 and calls the numbers out to the players. How many different combinations are possible assuming that the order is irrelevant?

Determine the number of distinguishable different "words" that can be formed by rearranging the letters in each of the following words. See Example 8.

40. ABYSS

41. BANANA

42. COLLEGE

43. ALGEBRA

44. MATHEMATICS

45. FIBONACCI

Use a combination of techniques seen in this section to answer the following questions. See Examples 1 through 8.

46. How many different ways are there of choosing five cards from a standard 52-card deck and arranging them in a row? How many different five-card hands can be dealt from a standard 52-card deck?

47. Suppose you have 10 physics texts, 8 computer science texts, and 13 math texts. How many different ways can you select 4 of each to take with you on vacation?

48. $_{10}P_4 \cdot _8P_4 \cdot _{13}P_4 \cdot 3!$
$= 871,782,912,000$

49. 112 cones

50. 30 pizzas

51. 96 outfits

52. 8 birth orders

53. 288 schedules

54. 54 lineups

55. 120 5-letter strings

56. 504 finishes

57. 303,600 ways

58. 40,320 orders

59. 495 pizzas

60. 120 numbers

61. 752,538,150 groups

62. 1,816,214,400 lineups

63. 420 ways

64. $243x^5 + 405x^4y +$
$270x^3y^2 + 90x^2y^3$
$+15xy^4 + y^5$

65. $x^7 - 14x^6y +$
$84x^5y^2 - 280x^4y^3$
$+560x^3y^4 -$
$672x^2y^5$
$+448xy^6 - 128y^7$

66. $x^4 - 12x^3 + 54x^2$
$-108x + 81$

48. Suppose you have 10 physics texts, 8 computer science texts, and 13 math texts. How many different ways can you select 4 of each and then arrange them in a row on a shelf, so that the books are grouped by discipline?

49. A certain ice cream store has four different kinds of cones and 28 different flavors of ice cream. How many different single-scoop ice cream cones is it possible to order at this ice cream store?

50. If a local pizza shop has three different types of crust, two different kinds of sauce, and five different toppings, how many different one topping pizzas can be ordered?

51. A man has 8 different shirts, 4 different shorts, and 3 different pairs of shoes. How many different outfits can the man choose from?

52. A couple wants to have three children. They want to know the different possible gender outcomes for birth order. How many different birth orders are possible?

53. A student has to make out his schedule for classes next fall. He has to take a math class, a science class, an elective, a history class, and an English class. There are three math classes to choose from, two science classes, four electives, three history classes, and four English classes. How many different schedules could the student have?

54. A basketball team has 12 different people on the team. The team consists of three point guards, two shooting guards, one weak forward, three power forwards, and three centers. How many different starting lineups are possible? (The starting lineup will consist of one player in each of the 5 positions.)

55. How many 5-letter strings can be formed using the letters V, W, X, Y, and Z, if the same letter cannot be repeated?

56. If at the racetrack nine greyhounds are racing against each other, how many different first, second, and third place finishes are possible?

57. A basketball tryout has four distinct positions available on the team. If 25 people show up for tryouts, how many different ways can the four positions be filled?

58. If a trumpet player is practicing eight different pieces of music, in how many different orders can he play his pieces of music?

59. A pizza place has 12 total toppings to choose from. How many different 4-topping pizzas can be ordered?

60. How many different 5-digit numbers can be formed using each of the numbers 6, 8, 1, 9, and 4?

61. If eight cards are chosen randomly from a deck of 52, how many possible groups of eight can be chosen?

67. $x^8 - 4x^6 y^3$

$+ 6x^4 y^6$

$- 4x^2 y^9 + y^{12}$

68. $7776x^{10} + 6480x^8 y$

$+ 2160x^6 y^2$

$+ 360x^4 y^3 + 30x^2 y^4$

$+ y^5$

69. $4096x^6 + 30{,}720x^5 y^2$

$+ 96{,}000x^4 y^4$

$+ 160{,}000x^3 y^6$

$+ 150{,}000x^2 y^8$

$+ 75{,}000xy^{10}$

$+ 15{,}625y^{12}$

70. $2401x^8 + 10{,}976x^6 y^2$

$+ 18{,}816x^4 y^4$

$+ 14{,}336x^2 y^6$

$+ 4096y^8$

71.

$x^{15} - 5x^{12} y^2 + 10x^9 y^4$

$- 10x^6 y^6 + 5x^3 y^8 - y^{10}$

72. 32

73. -80

74. $x^{16} + 48x^{15} y$

$+ 1080x^{14} y^2$

$+ 15{,}120x^{13} y^3$

75.

$8192x^{13} + 159{,}744x^{12}$

$+ 1{,}437{,}696x^{11}$

76. $129{,}140{,}163x^{\frac{17}{4}}$

$+ 3{,}658{,}971{,}285x^4 y$

77. $2{,}008{,}326{,}144x^{14}$

78. $651{,}168x^5$

79. $831{,}409{,}920x^4 y^8$

62. A baseball team has 15 players on the roster and a batting lineup consists of 9 players. How many different batting lineups are possible?

63. How many different ways can two red balls, one orange ball, one black ball, and three yellow balls be arranged?

Use the Binomial Theorem in the following problems. See Examples 9 and 10.

64. Expand the expression $(3x + y)^5$.

65. Expand the expression $(x - 2y)^7$.

66. Expand the expression $(x - 3)^4$.

67. Expand the expression $(x^2 - y^3)^4$.

68. Expand the expression $(6x^2 + y)^5$.

69. Expand the expression $(4x + 5y^2)^6$.

70. Expand the expression $(7x^2 + 8y^2)^4$.

71. Expand the expression $(x^3 - y^2)^5$.

72. What is the coefficient of the term containing $x^3 y$ in the expansion of $(2x + y)^4$?

73. What is the coefficient of the term containing $x^4 y^3$ in the expansion of $(x^2 - 2y)^5$?

74. Find the first four terms in the expansion of $(x + 3y)^{16}$.

75. Find the first three terms in the expansion of $(2x + 3)^{13}$.

76. Find the first two terms in the expansion of $\left(3x^{\frac{1}{4}} + 5y\right)^{17}$.

77. Find the 11th term in the expansion of $(x + 2)^{24}$.

78. Find the 17th term in the expansion of $(2x + 1)^{21}$.

79. Find the 9th term in the expansion of $(x - 6y)^{12}$.

80. $\binom{n-1}{k-1} + \binom{n-1}{k} = \dfrac{(n-1)!}{(k-1)!(n-1-(k-1))!} + \dfrac{(n-1)!}{k!(n-1-k)!}$

$= \dfrac{k(n-1)!}{k(k-1)!(n-k)!} + \dfrac{(n-1)!(n-k)}{k!(n-k-1)!(n-k)}$

$= \dfrac{k(n-1)!}{k!(n-k)!} + \dfrac{n(n-1)! - k(n-1)!}{k!(n-k)!}$

$= \dfrac{n(n-1)!}{k!(n-k)!} = \dfrac{n!}{k!(n-k)!} = \binom{n}{k}$

Pascal's triangle is a triangular arrangement of binomial coefficients, the first few rows of which appear as follows:

$$1$$
$$1 \quad 1$$
$$1 \quad 2 \quad 1$$
$$1 \quad 3 \quad 3 \quad 1$$
$$1 \quad 4 \quad 6 \quad 4 \quad 1$$
$$\vdots$$

Each number (aside from those on the perimeter of the triangle) is the sum of the two numbers diagonally adjacent to it in the previous row. Pascal's triangle is a useful way of generating binomial coefficients, with the nth row containing the coefficients of a binomial raised to the $(n-1)^{\text{th}}$ power. It can also be used to suggest useful relationships between binomial coefficients. Prove each of the following such relationships algebraically.

80. $\dbinom{n}{k} = \dbinom{n-1}{k-1} + \dbinom{n-1}{k}$

(Note that this is a restatement of how Pascal's triangle is formed.)

81. $\dbinom{n}{k} = \dbinom{n}{n-k}$ **82.** $\dbinom{n}{0} = \dbinom{n}{n} = 1$

83. $\dbinom{n}{0} + \dbinom{n}{1} + \ldots + \dbinom{n}{n} = 2^n$

(**Hint:** Use the Binomial Theorem on $(x+y)^n$ for a convenient choice of x and y.)

81. $\dbinom{n}{n-k} = \dfrac{n!}{(n-k)!(n-(n-k))!} = \dfrac{n!}{(n-k)!(n-n+k)!} = \dfrac{n!}{(n-k)!k!} = \dbinom{n}{k}$

82. $\dbinom{n}{0} = \dfrac{n!}{0!(n-0)!} = \dfrac{n!}{(1)(n)!} = \dfrac{n!}{n!} = 1$

$\dbinom{n}{n} = \dfrac{\cancel{n!}}{\cancel{n!}(n-n)!} = \dfrac{1}{(0)!} = 1$

83. $2^n = (1+1)^n = \displaystyle\sum_{k=0}^{n} \dbinom{n}{k}(1)^k (1)^{n-k} = \sum_{k=0}^{n} \dbinom{n}{k} = \dbinom{n}{0} + \dbinom{n}{1} + \ldots + \dbinom{n}{n}$

An Introduction to Probability

TOPICS

1. The language of probability

2. Using combinatorics to compute probabilities

3. Unions, intersections, and independent events

TOPIC 1

The Language of Probability

e make use of the concept of *probability* nearly every day in a wide variety of ways. Most of the time, the use is informal; we might wonder what the probability of rain is during a game of tennis, or what the chances are of striking it rich by winning the state lottery. In mathematics, probability is much more rigorously defined, and the results of probabilistic analysis are used in many important applications.

DEFINITION

Terminology of Probability

An **experiment** is any activity that results in well-defined **outcomes**. In a given problem, we are usually concerned with finding the probability that one or more of the outcomes will occur, and we use the word **event** to refer to a set of outcomes.

The set of all possible outcomes of a given experiment is called the **sample space** of the experiment. This means that in the language of sets, an event is any subset of the sample space, including the empty set and the entire sample space.

For example, rolling a standard die is an experiment that has outcomes of the numbers 1, 2, 3, 4, 5, and 6. The sample space of this experiment is the set $S = \{\, 1, 2, 3, 4, 5, 6 \,\}$. The events consist of all the possible subsets of S. We might, for instance, be interested in the event E defined as "the number rolled is even." In terms of sets, $E = \{\, 2, 4, 6 \,\}$. Our intuition tells us that, if the die is a fair one, the probability of E occurring is one-half, as half the numbers are even and half are odd. We denote this by writing

$P(E) = \dfrac{1}{2}$, which is read "the probability of E is one-half."

This intuition points the way toward the main tool we have for calculating probabilities.

THEOREM

We say that the outcomes of an experiment are *equally likely* if they all have the same probability of occurring. If E is an event of such an experiment, and if S is the sample space of the experiment, then the **probability of E** is given by

$$P(E) = \frac{n(E)}{n(S)}$$

where $n(E)$ and $n(S)$ are, respectively, the cardinalities of the sets E and S.

The formula tells us that probabilities are always going to be real numbers between 0 and 1, inclusive. The probability of an event E is 0 if E is the empty set, and 1 if E is the entire sample space S. In all other cases, the size of E is going to be between 0 and the size of S, so the fraction will yield a number between 0 and 1.

CAUTION!

This formula has two important restrictions. First, it only applies in equally likely situations, so we can't use it to analyze weighted coins, crooked roulette tables, tampered decks of cards, and so on. Secondly, and more subtly, it assumes that the size of the sample space is finite (otherwise, the formula makes no sense), so we can't use it to analyze experiments based on, say, choosing a real number at random.

The complement of an event E, denoted E^C, is the set of all outcomes in the sample space that are not in E. Thus the probability of the complement of an event can be derived as:

$$P(E^C) = \frac{n(S) - n(E)}{n(S)} = \frac{n(S)}{n(S)} - \frac{n(E)}{n(S)} = 1 - P(E).$$

EXAMPLE 1

A (fair) die is rolled once. Find the probability that the number rolled is:

a. Prime   **b.** Divisible by 5   **c.** 7

Solutions:

The sample space $S = \{1,2,3,4,5,6\}$ is the same for all three questions, and we can see that $n(S) = 6$.

a. Let E be the event that the number is prime. Then $E = \{2, 3, 5\}$ since these are the the prime numbers between 1 and 6, so $n(E) = 3$.

Then, $P(E) = \dfrac{n(E)}{n(S)} = \dfrac{3}{6} = \dfrac{1}{2}$.

b. Let F be the event that the number is divisible by 5. Then $F = \{5\}$, as this is the only integer from 1 to 6 that is divisible by 5.

So $P(F) = \dfrac{n(F)}{n(S)} = \dfrac{1}{6}$.

c. Let G be the event that the number is 7. In this case, $G = \emptyset$, the empty set, as there is no way for the top face to show a 7. This means $P(G) = 0$.

<table>
<tr><td>**TOPIC**</td><td>2</td></tr>
</table>

Using Combinatorics to Compute Probabilities

Most probability questions are not as basic as those in Example 1, and what makes a probability problem more complex is finding the size of an event and/or the sample space. This is a major application of the combinatorics techniques we covered in the last section.

EXAMPLE 2

Computing Probabilities

A pair of dice is rolled, and the sum of the top faces noted. What is the probability that the sum is:

a. 2   **b.** 5   **c.** 7 or 11

Solutions:

The size of the sample space is the same for all three questions, so it makes sense to determine this first. In order to use the one probability formula we have, we need to make sure that all outcomes of the sample space are equally likely.

For this reason, we do *not* want to define the sample space to be all integers between 2 and 12. This is because these sums are not all equally likely. For instance, there is only one way for the sum to be 2: both top faces must show a 1. But there are two ways for the sum to be 3: one die (call it die A) shows a 1 and the other (die B) shows a 2, or else die A shows a 2 and die B shows a 1. Similarly, there are *three* ways for the sum of the top faces to be 4.

In order to define the sample space properly, we construct a table of ordered pairs. In the table below, the first number in each pair corresponds to the number showing on die A and the second number corresponds to the number showing on die B.

	1	2	3	4	5	6
1	(1,1)	(1,2)	(1,3)	(1,4)	(1,5)	(1,6)
2	(2,1)	(2,2)	(2,3)	(2,4)	(2,5)	(2,6)
3	(3,1)	(3,2)	(3,3)	(3,4)	(3,5)	(3,6)
4	(4,1)	(4,2)	(4,3)	(4,4)	(4,5)	(4,6)
5	(5,1)	(5,2)	(5,3)	(5,4)	(5,5)	(5,6)
6	(6,1)	(6,2)	(6,3)	(6,4)	(6,5)	(6,6)

Each of these ordered pairs is equally likely to come up, as any of the numbers 1 through 6 are equally likely for the first slot (die A) and similarly for the second slot (die B). Referring to the positions in the ordered pairs as "slots" points to a quick way of determining the size of the sample space. Since there are 6 choices for each slot, the Multiplication Principle of Counting tells us there are 36 possible outcomes of this experiment.

We can now proceed to answer the three specific questions.

a. There is only one ordered pair corresponding to a sum of 2, namely (1, 1), so the probability of this event is $\dfrac{1}{36}$.

b. There are four ordered pairs corresponding to a sum of 5:

$$\{(1, 4), (2, 3), (3, 2), (4, 1)\},$$

so the probability of this event is $\frac{4}{36}$, or $\frac{1}{9}$.

c. A sum of 7 or 11 comes from any of the following ordered pairs:

$$\{(1, 6), (2, 5), (3, 4), (4, 3), (5, 2), (6, 1), (5, 6), (6, 5)\}$$

Since there are eight elements in this event, the probability of rolling a sum of 7 or 11 is $\frac{8}{36}$, or $\frac{2}{9}$.

EXAMPLE 3

Computing
Probabilities

Suppose you are taking a 10-question True-False test, and you are completely unprepared for it. If you decide to guess on each question, what is the probability of getting 8 or more questions right?

Solution:

The sample space for this problem consists of all the possible sequences of 10 answers, each of which is True or False. Combinatorics tells us that there are $2^{10} = 1024$ such sequences (2 choices for each of 10 "slots"), and these 1024 possible sequences are equally likely if your choice of True or False on each question is random.

The probability of getting 8 or more questions right can be broken up into three possibilities: getting 8 right, 9 right, or all 10 right.

There is only one sequence of 10 answers that are all correct. There are more ways of correctly answering exactly 9 questions, and we can use another tool from combinatorics to find out exactly how many ways. We need to select 9 "objects" from a set of 10 to be correctly answered questions, and the number of ways to do this is

$$_{10}C_9 = \frac{10!}{9!1!} = 10.$$

Similarly, there are $_{10}C_8$ ways of selecting 8 of the 10 questions, so there are

$$_{10}C_8 = \frac{10!}{8!2!} = \frac{10 \cdot 9}{2} = 45$$

ways of getting exactly 8 questions right.

Altogether, there are $45 + 10 + 1 = 56$ ways of getting 8, 9, or 10 questions right, so the probability of this happening is

$$\frac{56}{1024} \approx 0.055.$$

This means you only have a 5.5% chance of scoring 80% or better by guessing, so studying ahead of time is clearly advantageous!

TOPIC 3   Unions, Intersections, and Independent Events

Since sample spaces and events are defined in terms of sets, it makes sense to use the terms *union* and *intersection* to describe ways of combining events. The notion of independence of events, on the other hand, is unique to probability. We will look first at probabilities of unions and intersections of events. For a review of set operations like union and intersection, see Section 1.1.

If E and F are two subsets of the same sample space S, then $E \cup F$ and $E \cap F$ are also subsets and constitute events in their own right. It makes sense, then, to talk about the probability of the events $E \cup F$ and $E \cap F$, and it's reasonable to suspect that they bear some relation to $P(E)$ and $P(F)$.

First, note that if all the outcomes in S are equally likely (as is the case with the problems we study in this section), then

$$P(E \cup F) = \frac{n(E \cup F)}{n(S)} \text{ and } P(E \cap F) = \frac{n(E \cap F)}{n(S)},$$

so what we are really after are the relations between $n(E \cup F)$, $n(E \cap F)$, $n(E)$, and $n(F)$.

Consider again the experiment of rolling a die. Let E be the event "the number rolled is divisible by 2" and let F be the event "the number rolled is divisible by 3." For this small experiment, we can list the elements of each of the four events we are interested in, and then determine each event's cardinality:

$$E = \{2, 4, 6\};\ n(E) = 3$$

$$F = \{3, 6\};\ n(F) = 2$$

$$E \cup F = \{2, 3, 4, 6\};\ n(E \cup F) = 4$$

$$E \cap F = \{6\};\ n(E \cap F) = 1$$

Note that even though union is in some ways the set equivalent of numerical addition, the cardinality of the union of E and F in this example is not the sum of the cardinalities of E and F individually. The reason: both E and F contain the element 6, so if we simply add the sizes of E and F together to get the size of $E \cup F$, we wind up counting 6 twice. This happens in general when we try to determine the cardinality of a union of two sets, and we remedy the situation by subtracting the cardinality of the intersection from the sum of the individual cardinalities. By doing this, we count those elements that lie in both sets just once.

THEOREM

Cardinality of a
Union of Sets

Let E and F be two finite sets. Then

$$n(E \cup F) = n(E) + n(F) - n(E \cap F).$$

Note that if E and F are *disjoint*, meaning they have no elements in common, then $n(E \cap F) = 0$, so $n(E \cup F) = n(E) + n(F)$.

We can use this fact to find a formula for the probability of the union of two events. Assuming E and F are two subsets of the same sample space S,

$$P(E \cup F) = \frac{n(E \cup F)}{n(S)}$$

$$= \frac{n(E) + n(F) - n(E \cap F)}{n(S)}$$

$$= \frac{n(E)}{n(S)} + \frac{n(F)}{n(S)} - \frac{n(E \cap F)}{n(S)}$$

$$= P(E) + P(F) - P(E \cap F).$$

THEOREM

Probability of a Union Of Two Events

Let E and F be two subsets of the same sample space. Then the **probability of the event "E or F,"** denoted $P(E \cup F)$, is given by the formula

$$P(E \cup F) = P(E) + P(F) - P(E \cap F).$$

The term $P(E \cap F)$ represents the probability of both events E and F happening. If, as sets, events E and F are disjoint (so $E \cap F = \varnothing$), then E and F are said to be *mutually exclusive*. In this case, $E \cap F = \varnothing$ and $P(E \cup F) = P(E) + P(F)$.

EXAMPLE 4

Probability of a Union of Two Events

Assume a single die has been rolled and the number showing on top noted. Let E be the event "the number is divisible by 2" and let F be the event "the number is divisible by 3." Find the following probabilities:

a. $P(E \cap F)$ **b.** $P(E \cup F)$

Solutions:

We have already determined the sizes of all the relevant sets in the discussion above, so we are ready to apply the appropriate formulas.

a. Since $n(E \cap F) = 1$, and since the sample space has six elements altogether,

$$P(E \cap F) = \frac{1}{6}.$$

b. Since $n(E) = 3$ and $n(F) = 2$,

$$P(E \cup F) = P(E) + P(F) - P(E \cap F)$$

$$= \frac{3}{6} + \frac{2}{6} - \frac{1}{6}$$

$$= \frac{2}{3}.$$

The formula for the probability of the union of three or more events can be complicated in general, and will be left for a later course. But if no two events have any elements in common, the formula is less complex. We say that events $E_1, E_2, ..., E_n$ are **pairwise disjoint** if $E_i \cap E_j = \varnothing$ whenever $i \neq j$. Another way of saying this is that every possible pair of events in the collection $E_1, E_2, ..., E_n$ is a mutually exclusive pair.

THEOREM

Unions of Mutually Exclusive Events

If $E_1, E_2, ..., E_n$ are pairwise disjoint, then

$$P(E_1 \cup E_2 \cup ... \cup E_n) = P(E_1) + P(E_2) + ... + P(E_n).$$

EXAMPLE 5

Unions of Mutually Exclusive Events

Suppose Jim has chosen a PIN of 8736 for his bank's ATM, and that all PINs at the bank consist of four digits from 0 to 9. Using the ATM one day, he senses someone looking over his shoulder as he enters the first two digits, and he decides to cancel the operation and leave. The next day Jim discovers his ATM card is missing. In the worst case scenario, a stranger now has Jim's ATM card and the first two digits of his PIN. As he calls the bank to cancel the card, he wonders what the chances are the unknown someone can guess the remaining digits in the three tries the ATM allows. What is the probability of this event?

Solution:

If the first two digits are indeed known, the stranger has the task of filling in the last two digits, and there are 10 choices for each:

$$\underline{\quad 8 \quad} \quad \underline{\quad 7 \quad} \quad \underbrace{\underline{\qquad}}_{10 \text{ choices}} \; \underbrace{\underline{\qquad}}_{10 \text{ choices}}$$

The size of the sample space is thus 100, and any guess at completing the PIN correctly has a probability equal to $\dfrac{1}{100}$ of being correct. Assuming the stranger tries three different ways of completing the PIN (so that the three events are pairwise disjoint), the probability of Jim's account being broken into is

$$\frac{1}{100} + \frac{1}{100} + \frac{1}{100} = \frac{3}{100}.$$

CAUTION!

A common error often made in finding the answer to Example 5 gives an answer of $\dfrac{1}{100} + \dfrac{1}{99} + \dfrac{1}{98} = \dfrac{14,701}{485,100} \approx .0303$. This is often due to confusion about an area of statistics called conditional probability. The probability of guessing right on the second guess is not $\dfrac{1}{99}$, but is actually $\left(\dfrac{1}{99}\right)\left(\dfrac{99}{100}\right)$, where $\dfrac{99}{100}$ is the probability that the first guess is wrong. Of course, $\left(\dfrac{1}{99}\right)\left(\dfrac{99}{100}\right) = \dfrac{1}{100}$.

The last probability idea we will consider is that of *independence*. Informally, we say that two events are independent if the occurrence of one of them has no effect on the occurrence of the other. Formally, independence of events is related to the probability of their intersection, as follows.

DEFINITION

Independent Events

Given two events E_1 and E_2 in the same sample space, we say E_1 and E_2 are **independent** if $P(E_1 \cap E_2) = P(E_1)P(E_2)$.

More generally, a collection of events $E_1, E_2, \ldots, E_n$ is **independent** if for any subcollection $E_{n_1}, E_{n_2}, \ldots, E_{n_k}$ of $E_1, E_2, \ldots, E_n$, it is true that

$$P\left(E_{n_1} \cap E_{n_2} \cap \ldots \cap E_{n_k}\right) = P\left(E_{n_1}\right)P\left(E_{n_2}\right)\ldots P\left(E_{n_k}\right).$$

EXAMPLE 6

Independent Events

If a coin is flipped three times, what is the probability of it coming up tails all three times?

Solution:

We actually have two good ways of answering this question, one using the notion of independence and one not.

Let E_i be the event "the coin comes up tails on the ith flip." We are interested, then, in the probability of $E_1 \cap E_2 \cap E_3$, the probability that we get tails each time. Since $P(E_i) = \frac{1}{2}$ for each $i = 1, 2, 3$ (remember, the coin is assumed fair), then

$$P(E_1 \cap E_2 \cap E_3) = P(E_1)P(E_2)P(E_3)$$
$$= \frac{1}{2} \cdot \frac{1}{2} \cdot \frac{1}{2}$$
$$= \frac{1}{8}.$$

The other way of obtaining the same answer is to consider the sample space made up of all possible three-toss sequences. Since each flip of the coin results in one of two possibilities, the Multiplication Principle of Counting tells us that there are $2^3 = 8$ possible sequences. Only one of these is the sequence consisting of three tails, so the probability of this event is $\frac{1}{8}$.

1. $\dfrac{3}{5}$

2. 0.28

3. $\dfrac{9}{13}$

4. 0.85

5. $\dfrac{1}{3}$

6. 0.51

7. a. 0 **b.** $\dfrac{5}{8}$

8. a. $\dfrac{3}{10}$ **b.** $\dfrac{2}{5}$

9. a. 0 **b.** $\dfrac{3}{5}$

10. a. 0 **b.** $\dfrac{5}{8}$

11. a. 0 **b.** 1

12. a. $\dfrac{1}{12}$ **b.** $\dfrac{5}{6}$

13. a. $\dfrac{1}{8}$ **b.** $\dfrac{9}{16}$

14. a. $\dfrac{3}{11}$ **b.** $\dfrac{8}{11}$

15. The set of all ordered 4-tuples made up of H's and T's. There are 16 such 4-tuples.

16. The set of 13 cards in the suit of hearts.

17. The set of all ordered pairs that have either an H or a T in the first slot and one of the 13 hearts in the second slot. There are 26 such ordered pairs.

18. The set $\{\text{I}, \text{II}, \text{III}, \text{IV}\}$

Exercises

Below is the given probability that an event will occur; find the probability that it will not occur.

1. $P(E) = \dfrac{2}{5}$

2. $P(E) = 0.72$

3. $P(E) = \dfrac{4}{13}$

4. $P(E) = 0.15$

5. $P(E) = \dfrac{2}{3}$

6. $P(E) = 0.49$

Apply the formulas for the probabilities of intersection and union to the following sets and determine **a.** $P(E \cap F)$ and **b.** $P(E \cup F)$. Let $n(S)$ equal the size of the sample space.

7. $n(S) = 8,\ E = \{2,5\},\ F = \{3,7,9\}$

8. $n(S) = 10,\ E = \{1,2,5\},\ F = \{1,2,3,5\}$

9. $n(S) = 5,\ E = \{4,B\},\ F = \{3\}$

10. $n(S) = 8,\ E = \{A\},\ F = \{B,C,D,E\}$

11. $n(S) = 4,\ E = \{1,\beta\},\ F = \{\alpha,2\}$

12. $n(S) = 12,\ E = \{A,C,g,5,n,7,8,t,L\},\ F = \{n,6\}$

13. $n(S) = 16,\ E = \{1,2,A,m,13,Y,8\},\ F = \{1,9,11,m\}$

14. $n(S) = 11,\ E = \{m,7,D,4,\theta\},\ F = \{\phi,D,3,7,m,\Sigma\}$

Determine the sample space of each of the following experiments.

15. A coin is flipped four times and the result recorded after each flip.

16. A card is drawn at random from the 13 hearts.

17. A coin is flipped and a card is drawn at random from the 13 hearts.

18. A quadrant of the Cartesian plane is chosen at random.

19. A slot machine lever is pulled; there are 3 slots, each of which can hold 6 different values.

19. The set of all ordered triples with any of the 6 values in each slot. There are 216 such triples.

20. The set of all ordered pairs that have an integer from 1 to 6 in each slot. There are 36 pairs.

21. The set of the 38 pockets.

22. The set of all ordered 6-tuples with numbers from 1 to 20 in each slot. There are 64 million such 6-tuples.

23. a. $\dfrac{2}{3}$ **b.** $\dfrac{1}{3}$

24. a. $\dfrac{3}{52}$ **b.** $\dfrac{12}{13}$

   c. $\dfrac{5}{13}$

25. a. $\dfrac{3}{8}$ **b.** $\dfrac{1}{8}$

   c. $\dfrac{1}{2}$

26. $\dfrac{1}{13,983,816}$

27. $\dfrac{3}{10}$

28. $\dfrac{5}{16}$

29. $\dfrac{387,420,489}{1,000,000,000}$

   ≈ 0.3874

30. a. $\dfrac{1}{361}$ **b.** $\dfrac{261}{361}$

   c. $\dfrac{180}{361}$

20. An individual die is rolled twice and each of the two results is recorded.

21. At a casino, a roulette wheel spins until a ball comes to rest in one of the 38 pockets.

Answer the following probability questions. Be careful to properly identify the sample space and the appropriate event in each case. See Examples 1, 2, and 3.

23. An ordinary die is rolled. Find the probability of rolling:
   a. A 3 or higher.
   b. An even composite number.

24. A card is drawn from a standard 52-card deck. Find the probability of drawing:
   a. A face card (jack, queen, or king) in the suit of hearts.
   b. Anything but an ace.
   c. A black (clubs or spades) card that is not a face card.

25. A coin is flipped three times. Find the probability of getting:
   a. Heads exactly twice.
   b. The sequence Heads, Tails, Heads.
   c. Two or more Heads.

26. A state lottery game is won by choosing the same six numbers (without repetition) as those selected by a mechanical device. The numbers are picked from the set $\{1, 2, \ldots, 49\}$, and the order of the numbers chosen is immaterial. What is the probability of winning?

22. A lottery drawing consists of 6 randomly drawn numbers from 1 to 20; the order of the numbers matters in this case, and repetition is possible.

27. What is the probability that a four-digit ATM PIN chosen at random ends in 7, 8, or 9?

28. Assume the probability of a newborn being male is one-half. What is the probability that a family with five children has exactly three boys?

29. What is the probability that a 9-digit driver's license number chosen at random will not have an 8 as a digit?

30. A roulette wheel in a casino has 38 pockets: 18 red, 18 black, and 2 green. Spinning the wheel causes a small ball to randomly drop into one of the pockets. All of the pockets are equally likely. The wheel is spun twice. Find the probability of getting:
   a. Green both times.
   b. Black at least once.
   c. Red exactly once.

31. There is a 25% chance of rain for each of the next 2 days. What is the probability that it will rain on one of the days but not the other?

The following problems all involve unions or intersections of events, or the notion of independent events. See Examples 4, 5, and 6.

32. A pair of dice is rolled. Find the probability that the sum of the top faces is:
   a. Seven.
   b. Seven or eleven.
   c. An even number or a number divisible by three.
   d. Ten or higher.

33. A card is drawn from a standard 52-card deck. Find the probability of drawing:
   a. A face card or a diamond.
   b. A face card but not a diamond.
   c. A red face card or a king.

31. $\frac{3}{8}$

32. a. $\frac{1}{6}$ **b.** $\frac{2}{9}$

   c. $\frac{2}{3}$ **d.** $\frac{1}{6}$

33. a. $\frac{11}{26}$ **b.** $\frac{9}{52}$

   c. $\frac{2}{13}$

34. $\frac{125}{1,747,977}$;

   $6,991,908$

35. a. $\frac{1}{169}$ **b.** $\frac{1}{221}$

36. $\frac{1}{4165}$

37. 18.75%

38. $\frac{11}{1105}$

39. a. $\frac{1}{6}$ **b.** $\frac{5}{18}$

   c. $\frac{1}{9}$

40. $\frac{4}{11}$

41. $\frac{1}{20}$

42. $\frac{1}{3}$

43. $\frac{2}{9}$

44. $\frac{1}{7}$

45. $\frac{3}{10}$

46. $\frac{2}{5}$

47. 27 tickets

48. $\frac{1}{14}$

34. A state lottery game is won by choosing the same six numbers (without repetition) as those selected by a mechanical device. The numbers are picked from the set $\{1, 2,...,49\}$, and the order of the numbers chosen is immaterial. What is the probability of winning if someone buys 1000 tickets? (Of course, no two of the tickets have the same set of six numbers.) How many tickets would have to be bought to raise the probability of winning to one-half?

35. Two cards are drawn at random from a standard 52-card deck. What is the probability of them both being aces if:
   a. The first card is drawn, looked at, placed back in the deck, and the deck is then shuffled before the second card is drawn?
   b. The two cards are drawn at the same time?

36. What is the probability of being dealt a five-card hand (from a standard 52-card deck) that has four cards of the same rank?

37. The probability of rain today is 75%, and the probability that Bob will forget to put the top up on his convertible is 25%. What is the probability of the inside of his car getting wet?

38. What is the probability of drawing 3 face cards in a row, without replacement, from a 52-card deck?

39. Two dice are rolled, and the difference is calculated by subtracting the smaller value from the larger value. Therefore, the difference may range from 0 to 5. Find the probability of each of the following differences:
   a. 0
   b. 1
   c. 4

40. A letter is randomly chosen from the word MISSISSIPPI. What is the probability of the letter being an S?

41. Mike works in a company of about 100 employees. If this year five people in the company are going to be randomly laid off, what is the probability that Mike will get laid off?

42. A pack of M&M's contains 10 yellow, 15 green, and 20 red pieces. What is the probability of choosing a green M&M out of the pack?

43. A jar of cookies has 3 sugar cookies, 4 chocolate chip cookies, and 2 peanut butter cookies. What is the probability of randomly choosing a peanut butter cookie out of the jar?

44. A big box of crayons contains 4 different blues, 3 different reds, 5 different greens, and 2 different yellows. What is the probability of randomly choosing a yellow crayon out of the box?

45. A bag of marbles contains 3 blue marbles, 2 red marbles, and 5 orange marbles. What is the probability of randomly picking a blue marble out of the bag?

46. Every week a teacher of a class of 25 randomly chooses a student to wash the blackboards. If there are 15 girls in the class, what is the probability that the student selected will be a boy?

47. If in a raffle 135 tickets are sold, how many tickets must be purchased for an individual to have a 20% chance of winning?

48. Zach is running for student council. At Zach's school the student council is chosen randomly from all qualified candidates. If there are 42 candidates running, including Zach, and a total of three positions, what is the probability that Zach will be selected for the council?

Chapter 11 Project

Probability

You may be familiar with the casino game of roulette. But have you ever tried to compute the probability of winning on a given bet?

The roulette wheel has 38 total slots. The wheel turns in one direction and a ball is rolled in the opposite direction around the wheel until it comes to rest in one of the 38 slots. The slots are numbered 00, and 0–36. Eighteen of the slots between 1 and 36 are colored black and eighteen are colored red. The 0 and 00 slots are colored green and are considered neither even nor odd, and neither red nor black. These slots are the key to the house's advantage.

The following are some common bets in roulette:
A gambler may bet that the ball will land on a particular number, or a red slot, or a black slot, or an odd number, or an even number (not including 0 or 00). He or she could wager instead that the ball will land on a column (one of 12 specific numbers between 1 and 36), or on a street (one of 3 specific numbers between 1 and 36).

The payoffs for winning bets are:
1 to 1 on odd, even, red, and black
2 to 1 on a column
11 to 1 on a street
35 to 1 any one number

1. Compute the probability of the ball landing on:
   a. A red slot.
   b. An odd number.
   c. The number 0.
   d. A street (any of 3 specific numbers).
   e. The number 2.
2. Based on playing each of the scenarios above (**a.–e.**) compute the winnings for each bet individually, if $5 is bet each time and all 5 scenarios lead to winnings.
3. If $1 is bet on hitting just one number, what would be the expected payoff? (**Hint:** Expected payoff is [(*probability of winning*) · (*payment for a win*)] − [(*probability of losing*) · (*payout for a loss*)].)
4. Given the information in question 3, would you like to play roulette on a regular basis? Why or why not? Why will the casino acquire more money in the long run?

Chapter Summary

A summary of concepts and skills follows each chapter. Refer to these summaries to make sure you feel comfortable with the material in the chapter. The concepts and skills are organized according to the section title and topic title in which the material is first discussed.

11.1: Sequences and Series

Recursively and Explicitly Defined Sequences
- The definition of *infinite* and *finite sequences*, and the notation used to define sequences
- The meaning of *explicit* and *recursive* formulas, and how to use them
- Finding a formula that reproduces a given number of terms of a sequence

Summation Notation and a Few Formulas
- The meaning of *summation notation* using the Greek letter sigma
- Converting between summation notation and expanded form
- Properties of summations
- The use of specific summation formulas

Partial Sums and Series
- The meaning of a *series*, and how series are related to sequences
- The meaning of a *partial sum* of a series
- *Finite series* and *infinite series*
- *Convergence* and *divergence* of series

Fibonacci Sequences
- Using recursive formulas to define Fibonacci sequences

11.2: Arithmetic Sequences and Series

Characteristics of Arithmetic Sequences and Series
- The definition of an *arithmetic sequence* and the meaning of the *common difference* of an arithmetic sequence

The Formula for the General Term of an Arithmetic Sequence
- The formula for the general term of an arithmetic sequence, and how to determine it

Evaluating Partial Sums of Arithmetic Sequences
- The two formulas for partial sums of arithmetic sequences

11.3: Geometric Sequences and Series

Characteristics of Geometric Sequences
- The definition of a *geometric sequence*, and the meaning of the *common ratio* of a geometric sequence

The Formula for the General Term of a Geometric Sequence
- The formula for the general term of a geometric sequence, and how to use it

Evaluating Partial Sums of Geometric Sequences
- The formula for the partial sum of a geometric sequence

Evaluating Infinite Geometric Series
- The extension of the partial sum formula to the case of infinite geometric series
- How to determine convergence or divergence of infinite geometric series
- The correspondence between geometric series and decimal notation

11.4: Mathematical Induction

The Role of Induction
- The mathematical meaning of *induction*

Proofs by Mathematical Induction
- Using mathematical induction to prove statements
- The conditions for applying proof by mathematical induction
- The two steps—*basic step* and *induction step*—for constructing an inductive proof

11.5: An Introduction to Combinatorics

The Multiplication Principle of Counting
- The meaning of the *Multiplication Principle of Counting* and its use in counting problems

Permutations
- The application of the *Multiplication Principle of Counting* to count the number of *permutations* of a set of objects
- The *Permutation Formula* and the meaning and use of the notation $_nP_k$

Combinations
- The connection between the *Permutation Formula* and the *Combination Formula*
- The meaning and use of the notation $_nC_k$

The Binomial Theorem
- The formulas for *binomial coefficients*
- The *Binomial Theorem* and its use

11.6: An Introduction to Probability

The Language of Probability
- The meaning of *experiment, outcome, event*, and *sample space* in probability theory
- The calculation of the probability of an event when all outcomes are *equally likely*

Using Combinatorics to Compute Probabilities
- The use of combinatorics in calculating probabilities

Unions, Intersections, and Independent Events
- The meaning of *independence* of events
- The meaning of *unions* and *intersections* of events, and the calculation of probabilities of unions of events
- The meaning of *mutually exclusive events*

1. $-3, 9, -27, 81, -243$

2.
$-1, \sqrt[3]{2}, -\sqrt[3]{3}, \sqrt[3]{4}, -\sqrt[3]{5}$

3. $-3, -4, -5, -6, -7$

4. $1, \dfrac{1}{2}, \dfrac{2}{9}, \dfrac{3}{32}, \dfrac{24}{625}$

5. $a_n = 6n - 13$

6. $a_n = \dfrac{3^{n-1}}{2^n}$

7. $a_n = n^2 - 1$

8. $a_n = \dfrac{2n+1}{n+1}$

9. $a_1 = -2,$

$a_n = n(a_{n-1})$

for $n \geq 2$

10. $a_n = n^2 + n$

11. $-3 - 5 - 7 - 9$

$-11 - 13 = -48$

12. $-2 + 4 - 8 + 16$

$-32 + 64 = 42$

13. $\displaystyle\sum_{i=2}^{7} i^3 = 783$

14. $-1 + 1 + 3 + 5 + 7$

$+ 9 = 24$

15. $-8 - 16 - 32 - 64$

$- 128 = -248$

16. $\displaystyle\sum_{i=2}^{10} 2i^2 = 768$

17.

$S_n = \dfrac{n}{2(n+2)}, \ S_{80} = \dfrac{20}{41}$

18.

$S_n = \dfrac{n}{2(n+2)}, \ S = \dfrac{1}{2}$

19. $S_n = 3 - 3^{n+1}$,

series diverges

20. $-2, \ 5, \ 3, \ 8, \ 11$

21. $-10, -12, -22,$

$-34, -56$

Chapter Review

Section 11.1

Determine the first five terms of the sequences whose n^{th} terms are defined as follows.

1. $a_n = (-3)^n$

2. $a_n = (-1)^n \sqrt[3]{n}$

3. $a_1 = -3, a_{n-1} = a_n + 1$ for $n \geq 2$

4. $a_n = \dfrac{n!}{n^n}$

Find a possible formula for the general n^{th} term of the sequences that begin as follows.

5. $-7, -1, 5, 11, 17, \ldots$

6. $\dfrac{1}{2}, \dfrac{3}{4}, \dfrac{9}{8}, \dfrac{27}{16}, \dfrac{81}{32}, \ldots$

7. $0, 3, 8, 15, 24, 35, \ldots$

8. $\dfrac{3}{2}, \dfrac{5}{3}, \dfrac{7}{4}, \dfrac{9}{5}, \dfrac{11}{6}, \ldots$

9. $-2, -4, -12, -48, -240, \ldots$

10. $2, 6, 12, 20, 30, \ldots$

Translate each expanded sum that follows into summation notation, and vice versa. Then use the formulas and properties from Section 11.1 to evaluate the sums.

11. $\displaystyle\sum_{i=3}^{8} (-2i + 3)$

12. $\displaystyle\sum_{i=2}^{7} (-2)^{i-1}$

13. $8 + 27 + 64 + \ldots + 343$

14. $\displaystyle\sum_{k=1}^{6} (2k - 3)$

15. $\displaystyle\sum_{i=1}^{5} -4(2^i)$

16. $8 + 18 + 32 + \ldots + 200$

Find a formula for the nth partial sum S_n of each of the following series. If the series is finite, determine the sum. If the series is infinite, determine if it converges or diverges, and if it converges, determine the sum.

17. $\displaystyle\sum_{i=1}^{80} \left(\dfrac{1}{i+1} - \dfrac{1}{i+2} \right)$

18. $\displaystyle\sum_{i=1}^{\infty} \left(\dfrac{1}{i+1} - \dfrac{1}{i+2} \right)$

19. $\displaystyle\sum_{i=1}^{\infty} (3^i - 3^{i+1})$

Determine the first five terms of the following generalized Fibonacci sequences.

20. $a_1 = -2$, $a_2 = 5$, and $a_n = a_{n-2} + a_{n-1}$ for $n \geq 3$

21. $a_1 = -10$, $a_2 = -12$, and $a_n = a_{n-2} + a_{n-1}$ for $n \geq 3$

22. $a_n = -3n + 8$

23. $a_n = \dfrac{5}{2}n + 9$

24. $a_n = -9n + 20$

25. $a_n = 3n - 1$

26. $a_n = \dfrac{3}{4}n - \dfrac{1}{4}$

27. $a_n = 9n - 14$

28. -146

29. 275

30. a_{61}

31. a_{36}

32. a_{25}

33. 8827

34. -7140

35. 66

36. $a_n = 2(4)^{n-1}$

37. $a_n = 3\left(\dfrac{1}{5}\right)^{n-1}$

38. $a_n = 18\left(-\dfrac{1}{3}\right)^{n-1}$

39. $a_n = 6(4)^{n-1}$

40. $a_n = 10(2)^{n-1}$

41. $a_n = 8\left(\dfrac{1}{4}\right)^{n-1}$

Section 11.2

Find the explicit formula for the general n^{th} term of each arithmetic sequence described below.

22. $5, 2, -1, -4, -7, \ldots$

23. $a_2 = 14$ and $a_4 = 19$

24. $a_7 = -43$ and $d = -9$

25. $a_1 = 2, a_4 = 11$

26. $a_9 = \dfrac{13}{2}, d = \dfrac{3}{4}$

27. $-5, 4, 13, 22, 31, \ldots$

Use the given information about each arithmetic sequence to answer the question.

28. Given that $a_1 = -2$ and $a_4 = -20$, what is a_{25}?

29. Given that $a_3 = 17$ and $a_7 = 29$, what is a_{89}?

30. In the sequence $8, 19, 30, \ldots$, which term is 668?

31. In the sequence $6, 1, -4, \ldots$, which term is -169?

32. In the sequence $\dfrac{8}{3}, \dfrac{10}{3}, 4, \ldots$, which term is $\dfrac{56}{3}$?

Each of the following sums is a partial sum of an arithmetic sequence; use either formula to determine its value.

33. $\displaystyle\sum_{i=1}^{97} (2i - 7)$

34. $\displaystyle\sum_{i=1}^{60} (-4i + 3)$

Solve the following application problem.

35. Sylvia suspects that she has an ant infestation in her apartment. The first day she noticed them, she saw 10 ants in her kitchen. Each day she notices 4 more ants than on the previous day. If she doesn't call an exterminator, how many ants would she see on the fifteenth day?

Section 11.3

Find the explicit formula for the general n^{th} term of each geometric sequence described below.

36. $2, 8, 32, 128, 512, \ldots$

37. $3, \dfrac{3}{5}, \dfrac{3}{25}, \dfrac{3}{125}, \dfrac{3}{625}, \ldots$

38. $18, -6, 2, -\dfrac{2}{3}, \dfrac{2}{9}, \ldots$

39. $a_1 = 6$ and $a_4 = 384$

40. $a_2 = 20$ and $a_6 = 320$

41. $a_1 = 8$ and $a_4 = \dfrac{1}{8}$

42. $r = 3$;

    $4, 12, 36, 108, 324$

43.
$r = \pm\dfrac{2}{3}$;

$\pm\dfrac{45}{8}, \dfrac{15}{4}, \pm\dfrac{5}{2}, \dfrac{5}{3}, \pm\dfrac{10}{9}$

44. $\pm\dfrac{1}{3}$

45. -2

46. $-\dfrac{1}{256}$

47. $\dfrac{381}{512}$

48. $40,955$

49. -12

50. Series diverges

51. 1

Given the two terms of a geometric sequence, find the common ratio and first five terms of the sequence.

42. $a_1 = 4$ and $a_4 = 108$ **43.** $a_4 = \dfrac{5}{3}$ and $a_6 = \dfrac{20}{27}$

Use the given information about each geometric sequence to answer the question.

44. Given that $a_2 = \dfrac{3}{5}$ and $a_4 = \dfrac{1}{15}$, what is the common ratio r?

45. Given that $a_1 = 3$ and $a_4 = -24$, what is the common ratio r?

46. Given that $a_5 = -16$ and $a_6 = -4$, what is a_{11}?

Each of the following sums is a partial sum of a geometric sequence. Use this fact to evaluate the sums.

47. $\displaystyle\sum_{i=3}^{9} 3\left(\dfrac{1}{2}\right)^i$ **48.** $5 + 10 + \ldots + 20,480$

Determine if the following infinite geometric series converge. If a given series converges, find the sum.

49. $\displaystyle\sum_{i=0}^{\infty} -3\left(\dfrac{3}{4}\right)^i$ **50.** $\displaystyle\sum_{i=1}^{\infty} \left(-\dfrac{5}{4}\right)^i$ **51.** $\displaystyle\sum_{i=1}^{\infty} \dfrac{2}{5}\left(\dfrac{5}{7}\right)^i$

Section 11.4

Use the Principle of Mathematical Induction to prove the following statements.

52. $1 + 4 + 9 + \ldots + n^2 = \dfrac{n(n+1)(2n+1)}{6}$

52. Basic Step: $n = 1$, $1^2 = 1$ and $\dfrac{1(1+1)(2(1)+1)}{6} = 1$;

Induction Step: If $1 + 4 + 9 + \ldots + k^2 = \dfrac{k(k+1)(2k+1)}{6}$, then $\left(1 + 4 + 9 + \ldots + k^2\right) + (k+1)^2 = \dfrac{k(k+1)(2k+1)}{6} + (k+1)^2$

$= \dfrac{(k+1)\left[2k^2 + k + 6(k+1)\right]}{6} = \dfrac{(k+1)\left[2k^2 + 7k + 6\right]}{6} = \dfrac{(k+1)(k+2)(2k+3)}{6} = \dfrac{(k+1)((k+1)+1)(2(k+1)+1)}{6}$

53. $\dfrac{1}{1\cdot 3} + \dfrac{1}{3\cdot 5} + \dfrac{1}{5\cdot 7} + \ldots + \dfrac{1}{(2n-1)(2n+1)} = \dfrac{n}{2n+1}$

53. Basic Step: $n = 1$, $\dfrac{1}{(2(1)-1)(2(1)+1)} = \dfrac{1}{3}$

and $\dfrac{1}{2(1)+1} = \dfrac{1}{3}$; Induction Step: If $\dfrac{1}{1\cdot 3} + \dfrac{1}{3\cdot 5} + \dfrac{1}{5\cdot 7} + \ldots + \dfrac{1}{(2k-1)(2k+1)} = \dfrac{k}{2k+1}$, then $\dfrac{1}{1\cdot 3} + \dfrac{1}{3\cdot 5} + \dfrac{1}{5\cdot 7} + \ldots + \dfrac{1}{(2k-1)(2k+1)}$

$+ \dfrac{1}{(2(k+1)-1)(2(k+1)+1)} = \dfrac{k}{2k+1} + \dfrac{1}{(2k+1)(2k+3)} = \dfrac{2k^2 + 3k + 1}{(2k+1)(2k+3)} = \dfrac{(2k+1)(k+1)}{(2k+1)(2(k+1)+1)} = \dfrac{(k+1)}{(2(k+1)+1)}$

54. $5+8+11+\cdots+(3n+2)=\dfrac{n(3n+7)}{2}$ **54.** Basic Step: $n=1$, $(3(1)+2)=5$ and $\dfrac{1(3(1)+7)}{2}=5$;

Induction Step: If $5+8+11+\cdots+(3k+2)=\dfrac{k(3k+7)}{2}$, then $5+8+11+\cdots+(3k+2)+(3(k+1)+2)=\dfrac{k(3k+7)}{2}+(3k+5)$

$=\dfrac{3k^2+7k+6k+10}{2}=\dfrac{(k+1)(3k+10)}{2}=\dfrac{(k+1)(3(k+1)+7)}{2}$

55. $1\cdot3+2\cdot4+3\cdot5+\cdots+n(n+2)=\dfrac{n(n+1)(2n+7)}{6}$ **55.** Basic Step: $n=1$, $1(1+2)=3$ and

$\dfrac{1(1+1)(2(1)+7)}{6}=3$;   Induction Step: If $1\cdot3+2\cdot4+3\cdot5+\cdots+k(k+2)=\dfrac{k(k+1)(2k+7)}{6}$, then $1\cdot3+2\cdot4+3\cdot5+\cdots+k(k+2)$

$+(k+1)(k+1+2)=\dfrac{k(k+1)(2k+7)}{6}+(k+1)(k+3)=\dfrac{(k+1)[k(2k+7)+6(k+3)]}{6}=\dfrac{(k+1)(k+2)(2k+9)}{6}=\dfrac{(k+1)((k+1)+1)(2(k+1)+7)}{6}$

56. For all natural numbers n, 11^n-7^n is divisible by 4.

56. Basic Step: $n=1$, $11^1-7^1=4$, which is divisible by 4;   Induction Step: If $\dfrac{11^k-7^k}{4}=p$ or $11^k-7^k=4p$ for some integer p,

then $11^{k+1}-7^{k+1}=11\cdot11^k-7\cdot7^k=4\cdot11^k+7\cdot11^k-7\cdot7^k=4\cdot11^k+7(11^k-7^k)=4\cdot11^k+7(4p)=4(11^k+7p)$

57. For all natural numbers n, 7^n-1 is divisible by 3.

57. Basic Step: $n=1$, $7^1-1=6$, and $2\cdot3=6$;   Induction Step: If $\dfrac{7^k-1}{3}=p$ or $7^k-1=3p$ for some integer p,

then $7^{k+1}-1=7\cdot7^k-7+6=7(7^k-1)+6=7(3p)+6=21p+6=3(7p+2)$

Section 11.5

Use the Multiplication Principle of Counting and the Permutation and Combination formulas to answer the following questions.

58. $8\cdot10^3\cdot26^3=$
   $140,608,000$

59. $8^7=2,097,152$

60. $\dfrac{8!}{3!2!}=3360$

61. $_{10}P_3=720$

62. $_{21}P_5=2,441,880$

63. $_{12}C_4=495$

64.
$-32y^5+80y^4-80y^3$
$+40y^2-10y+1$

65.
$x^7+14x^6+84x^5+$
$280x^4+560x^3+$
$672x^2+448x+128$

58. A license plate must contain 4 numerical digits followed by 3 letters. If the first digit cannot be 0 or 1, how many different license plates can be created?

59. How many different 7-digit phone numbers do not contain the digits 6 or 7?

60. In how many different orders can the letters in the word "aardvark" be arranged?

61. In how many different ways can first place, second place, and third place be awarded in a 10-person shot put competition?

62. At a meeting of 21 people, a president, vice president, secretary, treasurer, and recruitment officer are to be chosen. How many different ways can these positions be filled?

63. A college admissions committee selects 4 out of 12 scholarship finalists to receive merit-based financial aid. How many different sets of 4 recipients can be chosen?

64. Expand the expression $(1-2y)^5$.

65. Expand the expression $(x+2)^7$.

66. Expand the expression $(5x^2-2y)^5$.

66.

$3125x^{10} - 6250x^8 y +$

$5000x^6 y^2 - 2000x^4 y^3 +$

$400x^2 y^4 - 32y^5$

67. a. $\dfrac{1}{9}$　**b.** $\dfrac{2}{3}$

68. a. 0　**b.** $\dfrac{5}{6}$

69. a. $\dfrac{2}{7}$　**b.** $\dfrac{4}{7}$

70. a. $\dfrac{1}{8}$　**b.** $\dfrac{3}{4}$

71. a. $\dfrac{1}{10}$　**b.** $\dfrac{4}{5}$

72. a. $\dfrac{4}{13}$

b. $\dfrac{5}{26}$

c. $\dfrac{7}{26}$

73. $\dfrac{33}{108,290}$

74. Approximately 52.17%

Section 11.6

Apply the formulas for the probabilities of intersection and union to the following sets and determine **a.** $P(E \cap F)$ and **b.** $P(E \cup F)$. Let $n(S)$ equal the size of the sample space.

67. $n(S) = 9$, $E = \{3, 5, 7\}$, $F = \{1, 2, 3, 4\}$

68. $n(S) = 6$, $E = \{A, B\}$, $F = \{X, Y, Z\}$

69. $n(S) = 7$, $E = \{\alpha, 13\}$, $F = \{\alpha, \beta, 13, 14\}$

70. $n(S) = 8$, $E = \{a, 4, m, 7\}$, $F = \{m, 3, s\}$

71. $n(S) = 10$, $E = \{3, 4, X, Y, 5, Z\}$, $F = \{5, 6, 7\}$

Answer the following probability questions. Be careful to properly identify the sample space and the appropriate event in each case.

72. A card is drawn from a standard 52-card deck. Find the probability of drawing:

　a. A seven or a club.

　b. A face card but not a red queen.

　c. A black three or a spade.

73. What is the probability of being dealt a five-card hand (from a standard 52-card deck) that contains only face cards?

74. There is a 10% chance of rain each individual day for an entire week. What is the probability that it will rain at least once during this seven-day period?

1. $-3, -5, -7, -9, -11$

2. $-1, -4, -9, -16, -25$

3. $1, 3, 6, 10, 15$

4. $1, \dfrac{1}{2}, \dfrac{1}{6}, \dfrac{1}{24}, \dfrac{1}{120}$

5. $a_n = n^2 - 2$

6. $a_n = \dfrac{(-1)^n}{n^2}$

7. $-2 - 8 - 18 - 32 - 50$
$- 72 - 98 - 128 - 162$
$= -570$

8. $-2 - 9 - 20 - 35 -$
$54 - 77 - 104 - 135$
$- 170 - 209 - 252 -$
$299 = -1366$

9. $S_n = \dfrac{n}{2n+4}$

$S_{100} = \dfrac{25}{51}$

10. $S_n = \ln \dfrac{2}{n+2}$

$S_{80} = \ln \dfrac{1}{41}$

11. $3, 6, 9, 15, 24$

12. $4, -2, 0, -2, -4$

13. $a_n = -4n + 3$

14. $a_n = 8n - 12$

15. $a_n = -3 \cdot 2^{n-1}$

16. $a_n = \dfrac{9}{5}\left(\dfrac{2}{3}\right)^{n-1}$

17. Yes, $S = \dfrac{1}{2}$

18. No

19. $\displaystyle\sum_{k=1}^{6} a_k = 54$

20. Basic Step: $n = 4$, $4! = 24$, and $2^4 = 16$; so $4! > 2^4$;

Induction Step: If $k! > 2^k$, then

$(k+1)! = k!(k+1) > 2^k (k+1) \geq 2^k (4+1) > 2^k \cdot 2 = 2^{k+1}$

Chapter Test

Determine the first five terms of the sequences whose n^{th} terms are defined as follows.

1. $a_n = -2n - 1$

2. $a_n = -n^2$

3. $a_n = 1 + 2 + 3 + \ldots + n$

4. $a_1 = 1$ and $a_n = \dfrac{a_{n-1}}{n}$, for $n \geq 2$

Find a possible formula for the general n^{th} term of the sequences that begin as follows. There may be more than one correct answer.

5. $-1, 2, 7, 14, 23, \ldots$

6. $-1, \dfrac{1}{4}, -\dfrac{1}{9}, \dfrac{1}{16}, -\dfrac{1}{25}, \ldots$

Translate each summation that follows into an expanded sum. Then use the formulas to evaluate the sums.

7. $\displaystyle\sum_{k=1}^{9} -2k^2$

8. $\displaystyle\sum_{i=1}^{12} (1+i)(1-2i)$

Find a formula for the nth partial sum S_n of each of the following series, and determine the sum if possible.

9. $\displaystyle\sum_{n=1}^{100} \left(\dfrac{1}{n+1} - \dfrac{1}{n+2}\right)$

10. $\displaystyle\sum_{n=1}^{80} \ln\left(\dfrac{n+1}{n+2}\right)$

11. Determine the first five terms of the following generalized Fibonacci sequence:
$a_1 = 3$, $a_2 = 6$, and $a_n = a_{n-1} + a_{n-2}$ for $n \geq 3$

12. Determine the first terms of the following recursively defined sequence:
$a_1 = 4$, $a_2 = -2$, and $a_n = 2a_{n-1} + a_{n-2}$ for $n \geq 3$

Find the explicit formula for the general n^{th} term of each arithmetic sequence described below.

13. $-1, -5, -9, -13, -17, \ldots$

14. $a_2 = 4, d = 8$

Find the explicit formula for the general n^{th} term of each geometric sequence described below.

15. $a_1 = -3, r = 2$

16. $a_2 = \dfrac{6}{5}, r = \dfrac{2}{3}$

Determine if the following geometric series converge. If a series converges, find the sum.

17. $\displaystyle\sum_{n=0}^{\infty} \frac{1}{3}\left(\frac{1}{3}\right)^n$

18. $\displaystyle\sum_{n=0}^{\infty} \left(\frac{5}{2}\right)^n$

19. Given $\displaystyle\sum_{k=1}^{n} a_k = n^2 + 3n$, let $n = 6$ and solve.

Use the Principle of Mathematical Induction to prove the following statements.

20. $n! > 2^n$ for all $n \geq 4$.

21. $2 + 7 + 12 + 17 + \ldots + (5n-3) = \dfrac{n}{2}(5n-1)$

Use the Multiplication Principle of Counting to answer the following questions.

22. How many different 7-digit telephone numbers are possible within each area code in the USA? (Ignore the fact that certain 7-digit sequences are disallowed as phone numbers.)

23. How many different 6-character license plates can be formed if the first 2 places must be letters and last 4 places must be digits? (Assume repetition is not allowed.)

24. A restaurant serves 3 different salads, 10 different entrees, and 6 different desserts. How many different meals could be created, assuming each meal consists of all three courses?

25. Expand the expression $(a - 3b)^6$.

26. Expand the expression $(2b - 2c)^3$.

27. A man has five pairs of socks of which no two pairs are the same color. If he randomly selects two socks from a drawer, what is the probability that he gets a matched pair?

21.
Basic Step: $n = 1$, $5(1) - 3 = 2$ and $\dfrac{1}{2}(5(1) - 1) = 2$; Induction Step: If $2 + 7 + 12 + 17 + \ldots + (5k-3) = \dfrac{k}{2}(5k-1)$ then
$2 + 7 + 12 + 17 + \ldots + (5k-3) + (5(k+1) - 3) = \dfrac{k}{2}(5k-1) + (5(k+1) - 3) = \dfrac{5k^2 - k}{2} + 5k + 5 - 3 = \dfrac{5k^2 - k}{2} + \dfrac{10k + 4}{2}$
$= \dfrac{5k^2 + 9k + 4}{2} = \dfrac{(k+1)}{2}(5k + 4) = \left(\dfrac{k+1}{2}\right)(5k + 5 - 1) = \left(\dfrac{k+1}{2}\right)(5(k+1) - 1)$.

22. 10,000,000

23. 3,276,000

24. 180

25. $a^6 - 18a^5b + 135a^4b^2$

$- 540a^3b^3 + 1215a^2b^4$

$- 1458ab^5 + 729b^6$

26. $8b^3 - 24b^2c + 24bc^2$

$- 8c^3$

27. $\dfrac{1}{9}$

28. a. $\dfrac{52}{125}$

b. $\dfrac{4}{5}$

c. $\dfrac{37}{500}$

29. a. $\dfrac{3}{11}$

b. $\dfrac{3}{44}$

c. $\dfrac{1}{22}$

d. $\dfrac{3}{55}$

28. A sample of college students, faculty, and administration were asked whether they favored a proposed increase in the annual activity fee to enhance student life on campus. The results of the study are given in the following table.

	Students	Faculty	Admin	Total
Favor	237	37	18	292
Oppose	163	38	7	208
Total	400	75	25	500

A person is selected at random from the sample. Find each of the following specified probabilities.

a. The person is not in favor of the proposal.

b. The person is a student.

c. The person is a faculty member and is in favor of the proposal.

29. There are 5 red, 4 black, and 3 yellow pencils in a box. Three pencils are selected without replacement at random from the box. Find each of the following specified probabilities.

a. Each one is a different color.

b. All three are the same color.

c. All three are red.

d. 2 are yellow, and 1 is black.

Answer Key

1 ## Number Systems and Equations and Inequalities of One Variable

Section 1.1 Real Numbers and Algebraic Expressions

1. a. $19, 2^5$ **b.** $19, \dfrac{0}{15}, 2^5$ **c.** $19, \dfrac{0}{15}, 2^5, -33$ **d.** $19, -4.3, \dfrac{0}{15}, 2^5, -33$ **e.** $-\sqrt{3}$ **f.** All

3. **5.** **7.** $<, \leq$ **9.** $\leq, \geq$ **11.** $>, \geq$ **13.** $2a + b > c$ **15.** $2c \leq 3d$

17. $\{n \mid n \text{ is an integer and } 5 \leq n \leq 105\}$ **19.** $\{2^n \mid n \text{ is a whole number}\}$ **21.** $\left\{\dfrac{1}{n} \mid n \text{ is an odd integer}\right\}$

23. $[-3, 19)$ **25.** $(-\infty, 15)$ **27.** $(0, \infty)$ **29.** **31.** **33.**

35. $[-7, 7)$ **37.** $\varnothing$ **39.** $(-\infty, \infty)$ **41.** $\mathbb{Z}$ **43.** $\mathbb{Z}$ **45.** -11 **47.** $\sqrt{5} - \sqrt{3}$ **49.** 1 **51.** -15 **53.** 5 **55.** 6

57. $3x^2 y^3, -2\sqrt{x + y}, 7z$ **59.** $1, 8.5, -14$ **61.** $\dfrac{-5x}{2yz}, -8x^5 y^3, 6.9z$ **63.** -98 **65.** $6\pi + 8$ **67.** -20

69. Commutative **71.** Distributive **73.** Associative **75.** Multiplicative cancellation; $\dfrac{1}{5}$ **77.** Additive cancellation; x

79. Multiplicative cancellation; 6 **81.** Additive cancellation; $-2x + y$

83. **85.** 20 miles **87.** If sugar $= s$, $\{s \mid 3 \leq s \leq 4\} = [3, 4]$; If walnuts $= w$, $\left\{w \mid \dfrac{1}{2} \leq w \leq \dfrac{2}{3}\right\} = \left[\dfrac{1}{2}, \dfrac{2}{3}\right]$

89. $131.11°C$ **91.** Derek's BMI is 23.7 and his weight status is normal. **93.** 187 ft **95.** The union contains the citizens with brown hair or blue eyes and the intersection contains the citizens with brown hair and blue eyes. **97.** Yes, all whole numbers are also integers, but the negative integers are not whole numbers, so all integers are not whole numbers. **99.** Answers will vary.

Section 1.2 Properties of Exponents and Radicals

1. 16 **3.** 81 **5.** x^3 **7.** $\dfrac{3}{t^5}$ **9.** $2n^8$ **11.** $\dfrac{y^2}{s^3 z^{11}}$ **13.** $\dfrac{1}{9m^2 - 8n^6}$ **15.** $\dfrac{25m^8}{n^4}$ **17.** $27x^9$ **19.** $\dfrac{1}{5z^6 - 81x^{12}}$

21. -9.12×10^8 **23.** $32{,}000{,}000$ **25.** 3.1536×10^7 **27.** 0.0003937 **29.** 2.8×10^{-3} **31.** -11 **33.** 1×10^{-8}

35.–37. Answers will vary. **39.** -3 **41.** $-\dfrac{3}{5}$ **43.** 2 **45.** $\dfrac{2}{5}$ **47.** $\left|x^3\right| \sqrt{2y}$ **49.** $\dfrac{1}{2\left|xy^3\right|}$ **51.** $\dfrac{\left|x^3\right| y^2}{2}$

53. $2xy^2 \sqrt[5]{x^2}$ **55.** $-\dfrac{|a|\sqrt{2}}{2}$ **57.** $\sqrt{6} + \sqrt{3}$ **59.** $\dfrac{x + 2\sqrt{xy} + y}{x - y}$ **61.** $-|y|\sqrt{3x}$ **63.** 0 **65.** $4z\sqrt[3]{2z}$

67. $\dfrac{1}{8}$ **69.** 27 **71.** $\dfrac{1}{125}$ **73.** $(a - b)^{\frac{4}{3}}$ **75.** $x^{\frac{1}{4}}$ **77.** $6n^3 \sqrt[3]{36n}$ **79.** x^3 **81.** $\sqrt[16]{16{,}807}$

83. Answers will vary. **85.** $\dfrac{1}{2}bhl$ **87.** 180π in.^{3} **89.** $81\pi d$ ft^3 **91.** $2\pi r^2$ **93.** $3d^2 \sqrt{3}$; 3.326 cm^2

95. 2.998×10^8 m/s **97.** Answers will vary. **99.** Because a positive number squared is positive and a negative number squared is positive.

Section 1.3 Polynomials and Factoring

1. Not a polynomial   **3.** Degree 11; polynomial of four terms   **5.** Degree 0 monomial   **7.** Degree 4 binomial

9. Degree 2 trinomial   **11.** Degree 5 binomial   **13.** $-x^{13}+7x^{11}-4x^{10}+9$ **a.** 13   **b.** -1

15. $2s^6-10s^5+4s^3$ **a.** 6   **b.** 2   **17.** $9y^6-3y^5+y-2$ **a.** 6   **b.** 9   **19.** πz^5+8z^2-2z+1 **a.** 5   **b.** π

21. $-4x^3y-6y-x^2z$ **23.** $x^2y+xy^2+6x-6y$ **25.** $-3ab$ **27.** xy^2-x^2y-y

29. $3a^3b^3+21a^3b^2+2a^2b^2+14a^2b-3ab^3-21ab^2$ **31.** $3a^2-2ab-8b^2$ **33.** $6x^2+33xy-18y^2$

35. $7y^4-34xy^2-5x^2$ **37.** $6x^3y^3-3x^3y+36x^2y^3+4x^2y^2-18x^2y+24xy^2$ **39.** $m\left(4mn+16m^2+7\right)$

41. $6\left(a-b^2\right)$ **43.** $2x\left(x^5-7x^2+4\right)$ **45.** $\left(x^3-y\right)\left(x^3-y-1\right)$ **47.** $4y^2\left(3y^4-2-4y^3\right)$ **49.** $\left(a^2+b\right)(a-b)$

51. $z(1+z)\left(1+z^2\right)$ **53.** $(n-2)\left(x^2+y\right)$ **55.** $(a-5b)(x+5y)$ **57.** $(2x-11)(2x+11)$

59. $(7a-12b)(7a+12b)$ **61.** $\left(5x^2y-3\right)\left(5x^2y+3\right)$ **63.** $(x-10y)\left(x^2+10xy+100y^2\right)$

65. $\left(m^2+5n^3\right)\left(m^4-5m^2n^3+25n^6\right)$ **67.** $\left(3x^2-2y^4z\right)\left(9x^4+6x^2y^4z+4y^8z^2\right)$ **69.** $\left(4y^2z-3x^4\right)\left(4y^2z+3x^4\right)$

71. $\left(7y^3+3xz^2\right)\left(49y^6+21xy^3z^2+9x^2z^4\right)$ **73.** $(x+5)(x-3)$ **75.** $(x-1)^2$ **77.** $(x-2)^2$ **79.** $(y+7)^2$

81. $(x+11)(x+2)$ **83.** $(y-8)(y-1)$ **85.** $(5a+3)(a-8)$ **87.** $(x+6)(5x-3)$ **89.** $(16y-9)(y-1)$

91. $(4a-3)(2a+1)$ **93.** $(4y-5)(3y-1)$ **95.** $2x(2x-1)^{\frac{-3}{2}}$ **97.** $a^{-3}\left(7a^2-2b\right)$ **99.** $2y^{-5}\left(5y^3-x\right)$

101. $(5x+7)^{\frac{4}{3}}(5x+6)$ **103.** $y^{-4}\left(7y^3+5\right)$ **105.** No; a variable in the denominator is equivalent to a variable

with a negative exponent.   **107. a.** Yes; degree = 4; leading coefficient = 2; terms = 4   **b.** Yes; degree = 3; leading

coefficient = 2; terms = 3

Section 1.4 The Complex Number System

1. $5i$ **3.** $-3i\sqrt{3}$ **5.** $4i\sqrt{2x}$ **7.** $i\sqrt{29}$ **9.** $1-3i$ **11.** $8-6i$ **13.** $-5+6i$ **15.** $16-30i$ **17.** i **19.** -11

21. $40-42i$ **23.** -9 **25.** $1+5i$ **27.** $-1-4i$ **29.** $7i$ **31.** $3+i$ **33.** $-i$ **35.** $-i$ **37.** $10-2i$ **39.** $\dfrac{14}{37}+\dfrac{10}{37}i$

41. $\dfrac{21}{17}-\dfrac{1}{17}i$ **43.** $-5+2i\sqrt{6}$ **45.** 8 **47.** $-\dfrac{7}{3}i$ **49.** $22+10i\sqrt{3}$ **51.** $6+3j$ ohms   **53.** $11-2j$ ohms

Section 1.5 Linear Equations in One Variable

1. $\mathbb{R}$ (Identity)   **3.** $x=1$ **5.** $w=-3$ **7.** $\mathbb{R}$ (Identity)   **9.** $\varnothing$ (Contradiction)   **11.** $m=7$ **13.** $x=3.7$

15. $x=1.05$ **17.** $y=-5$ **19.** $\mathbb{R}$ (Identity)   **21.** $\mathbb{R}$ (Identity)   **23.** $x=3$ **25.** $\varnothing$ (Contradiction)

27. $y=-\dfrac{1}{3},-3$ **29.** $x=\dfrac{1}{3}$ **31.** $x=-311,420$ **33.** $x=-\dfrac{4}{5},2$ **35.** $\varnothing$ (Contradiction)   **37.** $x=-2,2$

39. $x=0$ **41.** $x=-99$ **43.** $x=\dfrac{1}{4}$ **45.** $x=\dfrac{1}{7}$ **47.** $r=\dfrac{C}{2\pi}$ **49.** $a=\dfrac{v^2-v_0^2}{2x}$ **51.** $F=\dfrac{9}{5}C+32$

53. $h = \dfrac{A - 2lw}{2w + 2l}$　**55.** $m = \dfrac{2K}{v^2}$　**57.** $\dfrac{19}{3}$ hours, or 6 hours and 20 minutes　**59.** 13.5 miles　**61.** \$390

63. 7.5%　**65.** 26 feet by 26 feet　**67.** 53, 55, and 57　**69.** 36.4%

Section 1.6 Linear Inequalities in One Variable

1. $\{-9, 3.14, -2.83, 1, -3, 4\}$　**3.** $\{-2.83, 1, -3\}$　**5.** $(-\infty, -3]$ ⟵|⟶ $_{-3}$　**7.** $(-\infty, 4.8)$ ⟵|⟶ $_{4.8}$

9. $(-\infty, 2.25)$ ⟵|⟶ $_{2.25}$　**11.** $\left(-\infty, \dfrac{3}{2}\right)$ ⟵|⟶ $_{\frac{3}{2}}$　**13.** $\left(-\infty, -\dfrac{3}{11}\right]$ ⟵|⟶ $_{-\frac{3}{11}}$

15. $(7, \infty)$ ⟵|⟶ $_{7}$　**17.** $(35, \infty)$ ⟵|⟶ $_{35}$　**19.** $(-3, \infty)$ ⟵|⟶ $_{-3}$

21. $(-0.11, \infty)$ ⟵|⟶ $_{-0.11}$　**23.** $(1, 5]$ ⟵|—|⟶ $_{1\ \ 5}$　**25.** $(-10, 6]$ ⟵|—|⟶ $_{-10\ \ 6}$

27. $[-8, -2)$ ⟵|—|⟶ $_{-8\ \ -2}$　**29.** $(21, 69]$ ⟵|—|⟶ $_{21\ \ 69}$　**31.** $\left(\dfrac{23}{7}, \dfrac{25}{7}\right)$ ⟵|—|⟶ $_{\frac{23}{7}\ \ \frac{25}{7}}$

33. $\left[\dfrac{13}{2}, 16\right)$ ⟵|—|⟶ $_{\frac{13}{2}\ \ 16}$　**35.** $\left(-\dfrac{5}{3}, 1\right]$ ⟵|—|⟶ $_{-\frac{5}{3}\ \ 1}$　**37.** $\left(-\infty, -\dfrac{7}{2}\right) \cup \left(\dfrac{15}{2}, \infty\right)$ ⟵|—|⟶ $_{-\frac{7}{2}\ \ \frac{15}{2}}$

39. $\left(-\infty, \dfrac{1}{2}\right) \cup \left(\dfrac{5}{2}, \infty\right)$ ⟵|—|⟶ $_{\frac{1}{2}\ \ \frac{5}{2}}$　**41.** $\varnothing$　**43.** $(-\infty, 2) \cup (6, \infty)$ ⟵|—|⟶ $_{2\ \ 6}$　**45.** $\varnothing$　**47.** $\varnothing$

49. $[-4, 0]$ ⟵|—|⟶ $_{-4\ \ 0}$　**51.** $(-\infty, \infty)$ ⟵⟶　**53.** $[73, 113]$ for an A, $(113, 115]$ for an A+.

55. $(1140, 1600]$

Section 1.7 Quadratic Equations

1. $\left\{-1, \dfrac{3}{2}\right\}$　**3.** $\{7\}$　**5.** $\left\{-1, \dfrac{8}{3}\right\}$　**7.** $\{3, 11\}$　**9.** $\{3, 11\}$　**11.** $\{0, 6\}$　**13.** $\{17, 19\}$　**15.** $\left\{\dfrac{1}{2} \pm \sqrt{2}\right\}$　**17.** $\{-5, -3\}$

19. $\left\{-5, \dfrac{3}{2}\right\}$　**21.** $\left\{\dfrac{13}{2}, \dfrac{15}{2}\right\}$　**23.** $\left\{-\dfrac{4}{3}, 1\right\}$　**25.** $\{-1\}$　**27.** $\left\{-\dfrac{3}{2}, 5\right\}$　**29.** $\{-4.5 \pm 4.5i\}$　**31.** $\left\{\pm \dfrac{9}{8}\right\}$

33. $\left\{\dfrac{-1 \pm \sqrt{7}}{2}\right\}$　**35.** $\{-3, 9\}$　**37.** $\{-21, 11\}$　**39.** $\{-1, 2\}$　**41.** $(x - 3 - 2i)(x - 3 + 2i)$　**43.** $\left(2x + 3 - 2\sqrt{2}\right)\left(2x + 3 + 2\sqrt{2}\right)$

45. $b = -5$ and $c = -24$　**47.** $\{8, 13\}$　**49.** $\{\pm 4, \pm 3\}$　**51.** $\left\{-1, \dfrac{1}{8}\right\}$　**53.** $\left\{1, 5, 3 \pm \sqrt{10}\right\}$　**55.** $\left\{\pm i\sqrt{13}, \pm 3i\right\}$

57. $\left\{\dfrac{1}{16}, 16\right\}$　**59.** $\left\{-216, -\dfrac{27}{125}\right\}$　**61.** $\left\{-\dfrac{1}{2}, \pm i\right\}$　**63.** $\left\{-2, 1 \pm i\sqrt{3}\right\}$　**65.** $\left\{\dfrac{3}{2}, \dfrac{-3 \pm 3i\sqrt{3}}{4}\right\}$　**67.** $\left\{-\dfrac{7}{3}, 0, 1\right\}$

69. $\left\{-7, 0, \dfrac{7}{5}\right\}$　**71.** $\{4, \pm i\}$　**73.** $\{0, 2, 3\}$　**75.** $\left\{\dfrac{41}{7}\right\}$　**77.** $\{4\}$　**79.** $\left\{-1, 0, \dfrac{2}{5}\right\}$　**81.** $\left\{-3, -\dfrac{13}{4}\right\}$

83. $b = -4$, $c = -12$, and $d = 0$　**85.** $a = 1$, $c = -36$, and $d = -144$　**87.** $a = 15$, $b = -16$, and $c = -5$

89. 4.5 seconds　**91.** 4.8 seconds

Section 1.8 Rational and Radical Equations

1. $\dfrac{2x+1}{x-5}; x \neq -3, 5$ **3.** $x(x-1); x \neq -3$ **5.** $\dfrac{2x-3}{x-2}; x \neq -5, 2$ **7.** $2x-3; x \neq -7$ **9.** $\dfrac{x^3+9x^2+11x+19}{(x-3)(x+5)}$

11. $\dfrac{13x}{(x-3)(x+5)}$ **13.** $\dfrac{x^3+4x^2-7x+18}{(x+3)(x-3)}$ **15.** $\dfrac{x+2}{x-6}$ **17.** $y-1$ **19.** $z-3$ **21.** $y-3$ **23.** -6 **25.** $\dfrac{x^2+9}{6x-3}$

27. $\dfrac{2x^2}{x+1}$ **29.** $\dfrac{-xy-y^2}{x}$ **31.** $\dfrac{x}{y}$ **33.** x^2y^2 **35.** $\dfrac{11x}{7y}$ **37.** $\left\{-2, -\dfrac{3}{2}\right\}$ **39.** $\left\{3 \pm \sqrt{10}\right\}$ **41.** $\{\pm i\}$ **43.** $\{-2\}$

45. $\varnothing$ **47.** $\dfrac{x-2}{x+2}$ **49.** $\dfrac{(z^2-11z+54)(z-9)}{(z-2)}$ **51.** $\dfrac{2y^2+5y-4}{y+1}$ **53.** $\{-1\}$ **55.** $\{-7, -2\}$ **57.** $\varnothing$ **59.** $\{4, 44\}$

61. $\varnothing$ **63.** $\{-2, 1\}$ **65.** $\{0, 5\}$ **67.** $\varnothing$ **69.** $\left\{\dfrac{1}{4}, 1\right\}$ **71.** $\left\{\pm\dfrac{8}{27}\right\}$ **73.** $\{-2, 5\}$ **75.** $\{7, 10\}$ **77.** $a = \pm\sqrt{c^2-b^2}$

79. $m = \dfrac{k}{\omega^2}$ **81.** $v = \pm\sqrt{\dfrac{Fr}{m}}$ **83.** $h = \pm\sqrt{\dfrac{w}{23}}$ **85.** $c = \pm\sqrt{\dfrac{2gm}{r}}$ **87.** $b = \pm\sqrt{c^2-a^2}$ **89.** $a = \sqrt[3]{\dfrac{uP^2}{4\pi^2}}$

91. $\dfrac{4}{3}$ minutes, or 1 minute and 20 seconds   **93.** 4 hours and 12 hours   **95.** 90 minutes   **97.** 9.1 hours

Chapter 1 Review

1. a. 2^3 **b.** $2^3, 0$ **c.** $-\sqrt{4}, 2^3, 0$ **d.** All except $\sqrt{17}$ **e.** $\sqrt{17}$ **f.** All   **2.** $\left\{\dfrac{1}{2n}\,\middle|\, n \text{ is a natural number}\right\}$

3. $[4, 17)$ **4.** $[-8, -1]$ **5.** -7 **6.** -1 **7.** $\dfrac{x^2}{2y}$, $12.1x$, $-\sqrt{y+5}$ **8.** $\dfrac{1}{2}$, 12.1, -1 **9.** $\dfrac{4\pi}{3} - 36$

10. -66 **11.** Commutative property   **12.** Multiplicative cancellation; $\dfrac{1}{4}$ **13.** $(-4, 13]$ **14.** $[5, 8)$

15. Melissa, Monica, Peter, Liz, James   **16.** $\dfrac{-t^9}{2s^7}$ **17.** $\dfrac{27z^3}{y^6}$ **18.** -0.0003005 **19.** 6.952×10^7

20. 4.152×10^{12} **21.** 2.0×10^{-8} **22.** 5 **23.** $\sqrt[6]{3}$ **24.** $5x^{10}$ **25.** $\dfrac{3\sqrt{x}-3\sqrt{2}}{x-2}$ **26.** $\dfrac{2y\sqrt[3]{9x^2y}}{3}$

27. $3|x|\sqrt{2xy} - 2x\sqrt[3]{2xy}$ **28.** $62 - 20\sqrt{6}$ **29.** $\dfrac{1}{x^{\frac{7}{4}}}$ **30.** $56x^{11}$ **31.** $\dfrac{4000\pi}{3}$ in.^{3} **32.** $m^4 - 5m^3 + 3m^2 + 2$

33. $-8x^2y + 8xy + y$ **34.** $3x^3 - 4x^2y^3 + 3xy - 4y^4$ **35.** $5a^2 - 7a^2b + 27ab - 35ab^2 + 10b^2$

36. $4xy(2x^2y + x^2 - 3y)$ **37.** $(2x-5y)(x+3)$ **38.** $(2a+1)(3a-5)$ **39.** $(2a+3b^2)(2a-3b^2)$

40. $(3x-2y)^{\frac{2}{3}}\left[(3x-2y)^{\frac{2}{3}} - 1\right]$ **41.** $x^{-2}(8+5x)$ **42.** $-2i\sqrt{2x}$ **43.** 3 **44.** $2+7i$ **45.** $-30+10i$

46. $-\dfrac{7}{25} + \dfrac{24}{25}i$ **47.** $-4\sqrt{3}$ **48.** $62 - 16i\sqrt{2}$ **49.** $\dfrac{-3i\sqrt{3}}{2}$ **50.** No solution   **51.** All real numbers   **52.** 6.25

53. $\{3, 4\}$ **54.** $\{1, 3\}$ **55.** $\{-3, 1\}$ **56.** $\{-3, 4\}$ **57.** $\left\{\dfrac{-3}{2}\right\}$ **58.** $b_2 = \dfrac{2A}{h} - b_1$ **59.** $C = \dfrac{5}{9}(F - 32)$

60. 246.7 miles   **61.** \$85   **62.** $[7,\infty)$ ◄────┼────► 7   **63.** $(4,\infty)$ ◄───┼──► 4

64. $[4,\infty)$ ◄───┼───► 4   **65.** $(1,\infty)$ ◄───┼───► 1   **66.** $[-7,4)$ ◄─┼───┼─► −7 4

67. $(7,27]$ ◄─┼───┼─► 7 27   **68.** $(-5,-1)$ ◄─┼───┼─► −5 −1   **69.** No solution

70. $(-\infty,-6]\cup[8,\infty)$ ◄─┼───┼─► −6 8   **71.** $(-\infty,-26]\cup[-7,\infty)$ ◄─┼──┼─► −26 −7   **72.** $\left\{-\dfrac{2}{5},3\right\}$ **73.** $\{2\pm3i\}$

74. $\left\{4\pm\sqrt{2}\right\}$ **75.** $\left\{\dfrac{19\pm\sqrt{701}}{17}\right\}$ **76.** $\left\{\pm1,\pm\sqrt{2}\right\}$ **77.** $\{-27,8\}$ **78.** $\{\pm2,\pm3\}$ **79.** $\left\{\pm\sqrt{2},4\right\}$ **80.** $\left\{0,\dfrac{1}{2},2\right\}$

81. $\{-1,0,4\}$ **82.** $\left\{\dfrac{3}{4}\right\}$ **83.** $b=-2$ and $c=-8$ **84.** 10   **85.** $\dfrac{x+3}{x-3}; x\neq0,\pm3$ **86.** $\dfrac{x+3}{x^2+3x+9}; x\neq3$

87. $\dfrac{-2}{x}$ **88.** $\dfrac{1}{a+1}$ **89.** $\dfrac{x+3}{3}$ **90.** $-x-y$ **91.** $\{\pm i\}$ **92.** $\left\{-\dfrac{1}{3}\right\}$ **93.** $\dfrac{x^2}{x+1}$ **94.** 2   **95.** $\{-4\}$

96. $\{13\}$ **97.** $\{-1,8\}$ **98.** $r=\sqrt{\dfrac{3V}{\pi h}}$ **99.** $\dfrac{24}{7}$

Chapter 1 Test

1. a. $\sqrt{1}, |-21|$ **b.** $\sqrt{1}, |-21|$ **c.** $\sqrt{1}, -1, |-21|$ **d.** $\sqrt{1}, 7.\overline{6}, -1, \dfrac{2}{9}, |-21|$ **e.** $2\sqrt{3}, 5\pi$

f. $2\sqrt{3}, 5\pi, \sqrt{1}, 7.\overline{6}, -1, \dfrac{2}{9}, |-21|$ **2.** $(-7,9]$ **3.** $[14,\infty)$ **4.** -9 **5.** $\sqrt{11}-\sqrt{5}$ **6.** 6   **7.** -0.0002004

8. 5.224×10^7 **9.** $\dfrac{18y^2}{x^4z^5}$ **10.** $\dfrac{|a|}{2}$ **11.** $\dfrac{-4y}{x^3}$ **12.** $3x^2y+7xy-2z+4xz$ **13.** $20x^3y+10x^2y^2+8x^2y+4xy^2-12x-6y$

14. $(2a+1)(3a-5)$ **15.** $\left(6x^3+y\right)\left(6x^3-y\right)$ **16.** $5i-5$ **17.** $-\dfrac{1}{2}$ **18.** $4+i$ **19.** $-8i$ **20.** $y=\dfrac{14}{9}$; conditional

21. $16=126$; contradiction   **22.** $\left\{-\dfrac{10}{7},0\right\}$ **23.** $v_0=\dfrac{h+16t^2}{t}$ **24.** $(-1,7]$ ◄─(┼────┼─► −1 7

25. $(-3,13]$ ◄─(┼────┼─► −3 13   **26.** Under \$420   **27.** $\left\{-4,\dfrac{5}{2}\right\}$ **28.** $\{-3,-2\}$ **29.** $\{-6,4\}$ **30.** $\{\pm3i,\pm\sqrt{2}\}$

31. $\left\{0,\dfrac{1\pm\sqrt{13}}{4}\right\}$ **32.** $\left\{-\dfrac{1}{3}\right\}$ **33.** $\dfrac{3t^2+8t}{3t+12}; t\neq-4,-5$ **34.** $\dfrac{3a^3+a}{5a+1}$ **35.** $\dfrac{2x+1}{(2x-1)(x-8)}$ **36.** $\dfrac{b-a}{4a+4b}$

37. $\{-5\}$ **38.** $\{2\}$ **39.** $\{4,5\}$

2 Introduction to Equations and Inequalities of Two Variables

Section 2.1 The Cartesian Coordinate System

1.

3.

5.

7.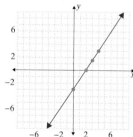

9. III   **11.** IV   **13.** Positive x-axis   **15.** III   **17.** IV   **19.** II   **21.** IV   **23.** I

25. Negative y-axis   **27.** $X = [-5, 6]; Y = [-8, 9]$ **29.** $X = [-3, 6]; Y = [-4, 5]$

31. $X = [-6, 8]; Y = [-9, 7]$

33. $\left\{ (0, -3), (2, 0), \left(3, \dfrac{3}{2} \right), (4, 3) \right\}$ **35.** $\{(0, 0), (1, \pm 1), (4, \pm 2),$ $(9, \pm 3), (2, -\sqrt{2})\}$ **37.** $\{(0, \pm 3), (\pm 3, 0), (-1, \pm 2\sqrt{2}),$ $(1, \pm 2\sqrt{2}), (\pm \sqrt{5}, 2)\}$

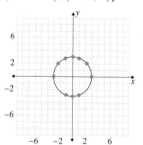

39. $\sqrt{34}, \left(\dfrac{-7}{2}, \dfrac{1}{2} \right)$ **41.** $\sqrt{58}, \left(\dfrac{3}{2}, \dfrac{7}{2} \right)$ **43.** $2\sqrt{2}, (-1, -1)$ **45.** $4\sqrt{34}, (3, -8)$ **47.** $10, (1, -6)$

49. $3\sqrt{13}, \left(2, \dfrac{1}{2} \right)$ **51.** $10\sqrt{2}, (3, 3)$ **53.** $x = 2$ or 18 **55.** $x = 10, y = 1$ **57.** 12 **59.** $2\sqrt{29} + \sqrt{26} + 5\sqrt{2}$

61. 54 **63.** Area $= \dfrac{25}{2}$ **65.** Area $= \dfrac{5}{2}$ **67.** Area $= 6$ **69.** Area $= 30$ **71.** 1.25 kilometers

73. a. 249.19 meters   **b.** $\left(\dfrac{133}{2}, \dfrac{709}{2} \right)$

Section 2.2 Linear Equations in Two Variables

1. Yes   **3.** No   **5.** No   **7.** No   **9.** Yes   **11.** Yes   **13.** Yes   **15.** No   **17.** No   **19.** No   **21.** No   **23.** Yes

25.   **27.**   **29.**

31.   **33.**   **35.**

37. 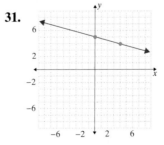   **39.**   **41.** e   **43.** c   **45.** f   **47.** $a = P - b - c$

49. $j = 24{,}000 + 9b;\ b = \dfrac{j - 24{,}000}{9};$ Yes

Section 2.3 Forms of Linear Equations

1. -4 **3.** 0 **5.** Undefined   **7.** $\dfrac{2}{3}$ **9.** $\dfrac{1}{6}$ **11.** -7 **13.** -3 **15.** $-\dfrac{9}{13}$ **17.** $-\dfrac{1}{4}$ **19.** 0 **21.** Undefined

23. 2 **25.** $\dfrac{7}{6}$ **27.** $-\dfrac{5}{2}$

29.   **31.**   **33.**

35.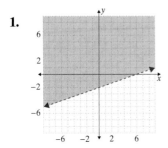

37. $y = \frac{3}{4}x - 3$ **39.** $y = -\frac{5}{2}x - 7$ **41.** $y = -5x - 9$ **43.** $3x - 2y = 3$

45. $y = 5$ **47.** $10x - y = 31$ **49.** $3x + y = 26$ **51.** $4x + 3y = 5$

53. $x = 2$ **55.** $y = -1$ **57.** $2x + 7y = 52$ **59.** $y = 5$ **61.** $15x - 8y = 0$

63. c   **65.** e   **67.** d   **69. a.** $2225 **b.** $2100 **c.** $0.25   **71.** $325

Section 2.4 Parallel and Perpendicular Lines

1. $y = 4x + 9$ **3.** $y = 3x - 11$ **5.** $y = -9$ **7.** $y = x$ **9.** $y = \frac{7}{6}x + \frac{53}{6}$ **11.** Yes   **13.** Yes   **15.** Yes   **17.** No

19. No   **21.** No   **23.** No   **25.** No   **27.** Yes   **29.** No   **31.** $y = -\frac{1}{3}x - 1$ **33.** $y = 7$ **35.** $y = -\frac{1}{4}x - \frac{3}{4}$

37. $y = x + 3$ **39.** $y = -3x + 28$ **41.** No   **43.** No   **45.** No   **47.** No   **49.** Yes   **51.** No   **53.** No   **55.** No

57. Yes   **59.** $41\frac{2}{3}$ ft

Section 2.5 Linear Inequalities in Two Variables

1.

3.

5.

7.

9.

11.

13.

15.

17.

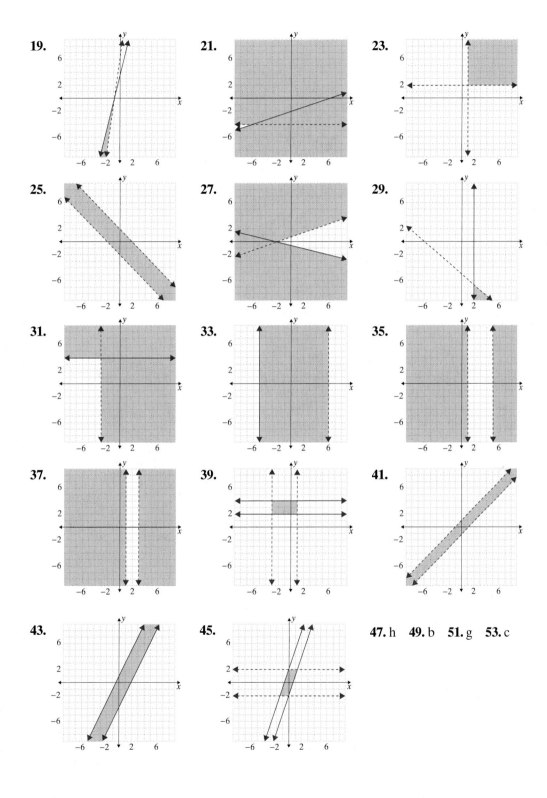

47. h   **49.** b   **51.** g   **53.** c

55. $12x + 22y < 150$ **57.** $73x + 46y < 1750$

Section 2.6 Introduction to Circles

1. $(x + 4)^2 + (y + 3)^2 = 25$ **3.** $(x - 7)^2 + (y + 9)^2 = 9$ **5.** $x^2 + y^2 = 6$ **7.** $\left(x - \sqrt{5}\right)^2 + \left(y - \sqrt{3}\right)^2 = 16$

9. $(x - 7)^2 + (y - 2)^2 = 4$ **11.** $(x + 3)^2 + (y - 8)^2 = 2$ **13.** $(x - 4)^2 + (y - 8)^2 = 10$ **15.** $x^2 + y^2 = 85$

17. $\left(x + \dfrac{7}{2}\right)^2 + \left(y - \dfrac{17}{2}\right)^2 = \dfrac{53}{2}$ **19.** $(x + 6)^2 + \left(y - \dfrac{3}{2}\right)^2 = \dfrac{125}{4}$ **21.** $\left(x + \dfrac{13}{2}\right)^2 + (y + 7)^2 = \dfrac{365}{4}$

23. $(x - 4)^2 + (y - 3)^2 = 25$ **25.** $(x - 2)^2 + y^2 = 4$ **27.** $(x - 2)^2 + (y - 4)^2 = 49$ **29.** $(x + 3)^2 + (y + 2)^2 = 64$

31. $(0, 0), r = 6$ **33.** $(0, 8), r = 3$ **35.** $(8, 0), r = 2\sqrt{2}$

37. $(-5, -4), r = 2$ **39.** $(5, -5), r = \sqrt{5}$ **41.** $(2, -2), r = 4$

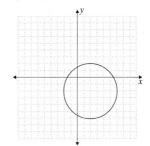

43. $(0, -5), r = 4$

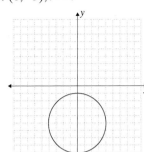

45. $(1, -3), r = 2\sqrt{2}$

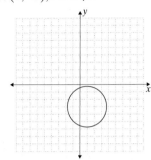

47. $(0, 0), r = 8$

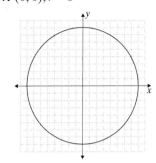

49. $(3, -2), r = 4$

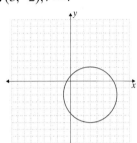

51. $(1, 0), r = 3$

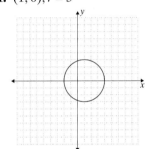

53. $(2, -4), r = 6$

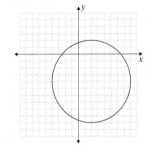

55. $(3, -3), r = 5$

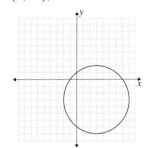

Chapter 2 Review

1.

2.

3.

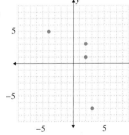

4. Origin (both axes)

5. Positive x-axis

6. Quadrant IV

7. $(2,0),(0,-3),$
$\left(-1,-\dfrac{9}{2}\right),\left(\dfrac{2}{3},-2\right),$
$(-2,-6)$

8. $(0,\pm2),\left(-\dfrac{4}{3},0\right),$
$\left(1,-\sqrt{7}\right),(-1,\pm1),$
$\left(\dfrac{5}{3},3\right)$

9. a. $\sqrt{2}$ **b.** $\left(\dfrac{5}{2},-\dfrac{13}{2}\right)$

10. a. 10 **b.** $(0,-6)$

11. a. $2\sqrt{13}$ **b.** $(-5,3)$

12. a. $\sqrt{97}$ **b.** $\left(\dfrac{1}{2},1\right)$

13. 2

14. $2+3\sqrt{2}+\sqrt{34}$

15. 24

16. $\left(\sqrt{5}\right)^2+\left(\sqrt{80}\right)^2=\left(\sqrt{85}\right)^2$; Area $=\dfrac{1}{2}\sqrt{5}\sqrt{80}=10$

17. No   **18.** Yes   **19.** Yes   **20.** Yes   **21.** No   **22.** No

23.

24.

25.

26.

27.

28.

29. 12 **30.** -4 **31.** Undefined

32.

33.

34.

35. $x - y = 5$ **36.** $3x - 2y = -12$ **37.** $y = \dfrac{5}{9}x - 2$ **38.** $y = -\dfrac{7}{3}x + 9$ **39.** $9x - 2y = 31$ **40.** $2x + 6y = 9$

41. $W = 0.08s + 2800$ **42.** Neither   **43.** Perpendicular   **44.** Neither   **45.** $y = 3x + 10$ **46.** $y = \dfrac{1}{6}x + 4$

47. $y = 2x - 3$ **48.** $y = -\dfrac{10}{3}x + \dfrac{5}{3}$ **49.** $y = -\dfrac{4}{3}x + 6$ **50.** $y = -3x - 19$ **51.** $x = 7$ **52.** $y = \dfrac{7}{5}x - \dfrac{7}{5}$

53. Yes   **54.** No

55.

56.

57.

58.

59.

60.

61.

62.

63.

64.

65.

66.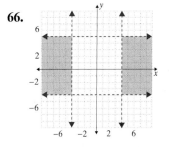

67. $3x + 4y \geq 1500$ **68.** $\left(x - \sqrt{5}\right)^2 + \left(y + \sqrt{2}\right)^2 = 16$ **69.** $\left(x + 2\right)^2 + y^2 = 18$ **70.** $\left(x - 2\right)^2 + \left(y + 1\right)^2 = 20$

71. $\left(x + 2\right)^2 + \left(y - 5\right)^2 = 18$ **72.** Center: $(-3, 1)$, Radius: $2\sqrt{2}$ **73.** ± 3

74. $r = 4, (h, k) = (-5, 2)$ **75.** $r = \sqrt{10}, (h, k) = (0, 3)$ **76.** $r = 3, (h, k) = (1, -4)$

                    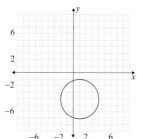

77. $r = \sqrt{29}, (h, k) = (-3, 5)$

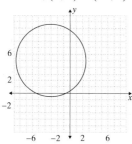

Chapter 2 Test

1. Quadrant III   **3.** $(\pm 2, 0), (0, \pm 2), \left(-1, \pm\sqrt{3}\right), \left(1, \pm\sqrt{3}\right), (0, 2)$ **4.** $5, -1$ **9.** $(0, -2), (6, 0)$

2. Negative y-axis

5. $\sqrt{26} + 2\sqrt{5} + \sqrt{10}$

6. 1

7. No

8. Yes, $y \neq -1$

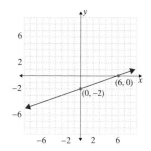

10. $-\dfrac{3}{4}$ **14.**

11. Undefined

12. $-\dfrac{16}{3}$

13. $2y - x = -8$

15. \$55,000   **16.** $y = 3x - 11$ **17.** $4, -2$

18. $y = 2x + 11$ **19.** Parallel   **20.** Neither

21. Answers will vary.

22. Answers will vary.

23. $m = 4$; y-intercept = 1

24. m is undefined; no y-intercept

25. a. $x + y = -1$ **b.** $-x + y = -2$

26. a. $x = 2$ **b.** $y = 5$

27.

28.

29.

30.

31.

32.

33. $(x + 3)^2 + (y + 2)^2 = 5$

34. $x^2 + y^2 = 169$

35. Center: $(-1, 3)$; radius: 4

36.

37. $B(6, 3)$; equation: $(x - 2)^2 + (y - 2)^2 = 17$

Relations, Functions, and Their Graphs

Section 3.1 Relations and Functions

1. Dom = {−2}; Ran = {5, 3, 0, −9} **3.** Dom = $\{\pi, -2\pi, 3, 1\}$; Ran = {2, 4, 0, 7} **5.** Dom = $\mathbb{Z}$;

Ran = even integers **7.** Dom = $\mathbb{Z}$; Ran = $\{..., -2, 1, 4, ...\}$ **9.** Dom = Ran = $\mathbb{R}$ **11.** Dom = $[0, \infty)$; Ran = $\mathbb{R}$

13. Dom = $\mathbb{R}$; Ran = {−1} **15.** Dom = {0}; Ran = $\mathbb{R}$ **17.** Dom = [−3, 1]; Ran = [0, 4] **19.** Dom = [0, 3];

Ran = [1, 5] **21.** Dom = [−1, 3]; Ran = [−4, 3] **23.** Dom = All males with siblings; Ran = All people who have

brothers **25.** Not a function; $(-2, 5)$ and $(-2, 3)$ **27.** Function **29.** Not a function; $(6, -1)$ and $(6, 4)$

31. Not a function; $(-1, 0)$ and $(-1, 4)$ **33.** Function **35.** Function **37.** Function; Dom = $(-\infty, 0) \cup (0, \infty)$

39. Not a function **41.** Function; Dom = $(-\infty, -2) \cup (-2, \infty)$ **43.** Function; Dom = $\mathbb{R}$ **45.** Not a function

47. $f(x) = -6x^2 + 2x$; $f(-1) = -8$ **49.** $f(x) = \dfrac{-x + 10}{3}$; $f(-1) = \dfrac{11}{3}$ **51.** $f(x) = -2x - 10$; $f(-1) = -8$

53. a. 10 **b.** $x^2 + x - 2$ **c.** $2ax + 3a + a^2$ **d.** $x^4 + 3x^2$ **55. a.** 8 **b.** $3x - 1$ **c.** $3a$ **d.** $3x^2 + 2$

57. a. -2 **b.** $-6x + 16$ **c.** $-6a$ **d.** $-6x^2 + 10$ **59. a.** $i - 3$ **b.** $\sqrt{2 - x} - 3$ **c.** $\sqrt{1 - x - a} - \sqrt{1 - x}$ **d.** $\sqrt{1 - x^2} - 3$

61. $2x + h - 5$ **63.** $\dfrac{-1}{(x + h + 2)(x + 2)}$ **65.** $5(2x + h)$ **67.** 2 **69.** $\dfrac{\sqrt{x + h} - \sqrt{x}}{h}$ **71.** Dom = Cod = Ran = $\mathbb{R}$

73. Dom = Cod = Ran = $\mathbb{Z}$ **75.** Dom = Cod = $\mathbb{N}$; Ran = {6, 7, 8, ...} **77.** $[1, \infty)$ **79.** $(-\infty, -2) \cup (-2, 3) \cup (3, \infty)$

81. $\mathbb{R}$ **83.** $\left(-\infty, \dfrac{1}{3}\right) \cup \left(\dfrac{1}{3}, \infty\right)$ **85.** $(-\infty, 2) \cup (2, \infty)$ **87.** $[-6, \infty)$ **89.** $(-\infty, 0) \cup (0, \infty)$

Section 3.2 Linear and Quadratic Functions

1.   **3.**   **5.**

7.

9.

11.

13.

15.

17. Vertex: $(-2, -1)$; no x-int.

19. Vertex: $(0, 2)$; no x-int.

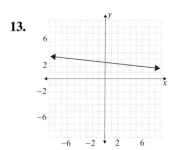

21. Vertex: $\left(\dfrac{1}{2}, \dfrac{25}{2}\right)$; x-int.: $x = -2, 3$

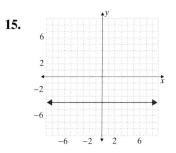

23. Vertex: $(0, -1)$; no x-int.

25. Vertex: $(-1, 3)$; no x-int.

27. Vertex: $(1, -4)$; no x-int.

29. Vertex: $(1, -2)$; x-int.: $x = 0, 2$

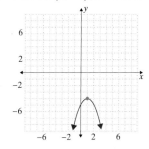

31. g **33.** a **35.** c **37.** h **39.** Width of 50 feet, length of 100 feet **41.** Width and length are 5 **43.** $(8, 4)$

45. 8 and 8 **47.** 11,250 square feet **49.** 500 rooms **51.** 1500 cars **53.** 6050 square feet **55.** 180 feet

57. Vertex: $(4,-1)$; x-int.: $x = \dfrac{8 \pm \sqrt{2}}{2}$ **59.** Vertex: $(4,-36)$; x-int.: $x = -2, 10$ **61.** Vertex: $(0, 25)$; x-int.: $x = -5, 5$

        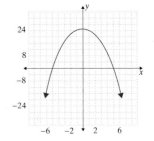

63. Vertex: $(-1, 0)$; x-int.: $x = -1$ **65.** Vertex: $(5, 21)$; x-int.: $x = 5 \pm \sqrt{21}$

    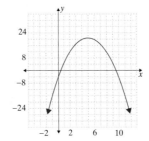

Section 3.3 Other Common Functions

1.   **3.**   **5.**

7.   **9.**   **11.**

13.

15.

17.

19.

21.

23.

25.

27.

29.

31.

33.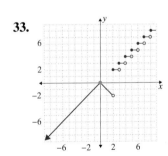

35. f   **37.** g   **39.** h   **41.** b

Section 3.4 Variation and Multivariable Functions

1. $A = kbh$ **3.** $W = \dfrac{k}{d^2}$ **5.** $r = \dfrac{k}{t}$ **7.** $x = ky^3z^2$ **9.** $y = 18\sqrt{5}$ **11.** $y = 60\sqrt[3]{2}$ **13.** $y = 0.75$ **15.** $z = 48$

17. $z = 112$ **19.** 256 feet   **21.** 20.60   **23.** 6.7 meters   **25.** 1.25 centimeters   **27.** 34.54 inches   **29.** 164.7872 in.^{2}

31. 9 watts   **33.** $P(\sigma, \varepsilon) = \dfrac{\sigma^2}{2\varepsilon}$ **35.** $10\sqrt{3}$ **37.** $a = \dfrac{9b^2}{4}$; 36   **39.** $a = 3bc$; 108   **41.** $P = 2.15g$; \$43

43. $F = \dfrac{15d}{9}$; 12 cm   **45.** $V = \dfrac{800}{P}$; 200 cm^3 **47.** $R = \dfrac{0.000009l}{d^2}$; 17.28 ohms

Section 3.5 Transformations of Functions

1. $f(x) = x^2$ **3.** $f(x) = \sqrt[3]{x}$ **5.** $f(x) = \sqrt{x}$ **7.** $f(x) = \sqrt{x}$ **9.** $f(x) = x^2$ **11.** $f(x) = |x|$

13. Dom = Ran = $\mathbb{R}$

15. Dom = $\mathbb{R}$; Ran = $(-\infty, 2]$

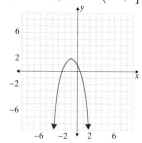

17. Dom = $\mathbb{R}$; Ran = $[0, \infty)$

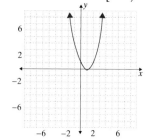

19. Dom = $(-\infty, 2]$; Ran = $[0, \infty)$

21. Dom = $(-\infty, 3) \cup (3, \infty)$; Ran = $(0, \infty)$

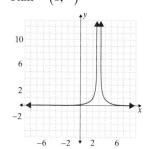

23. Dom = $(-\infty, 2) \cup (2, \infty)$; Ran = $(-\infty, 0) \cup (0, \infty)$

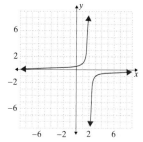

25. Dom = Ran = $\mathbb{R}$

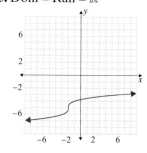

27. Dom = Ran = $\mathbb{R}$

29. Dom = $\mathbb{R}$; Ran = $[-3, \infty)$

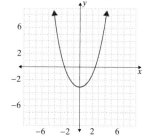

31. Dom = $\mathbb{R}$; Ran = $(-\infty, 0]$

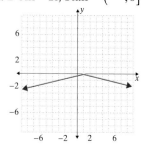

33. Dom = $[1, \infty)$; Ran = $(-\infty, 2]$

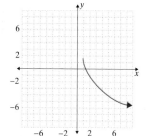

35. $f(x) = (x+3)^2 - 4$

37. $f(x) = -x^2 + 6$

39. $f(x) = (-x+1)^3$

41. $f(x) = -\sqrt{x} + 5$

43. $f(x) = -|-x+7|$

45. Even function;
y-axis symmetry

47. Neither; No symmetry

49. Not a function;
x-axis symmetry

51. Neither; No symmetry

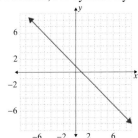

53. Not a function;
x-axis symmetry

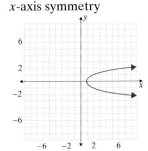

55. Even function;
y-axis symmetry

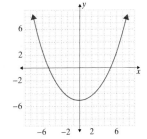

57. Odd function;
Origin symmetry

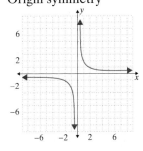

Section 3.6 Combining Functions

1. a. 2 **b.** −8 **c.** −15 **d.** $-\dfrac{3}{5}$ **3. a.** −3 **b.** −1 **c.** 2 **d.** 2 **5. a.** 12 **b.** 18 **c.** −45 **d.** −5 **7. a.** 3 **b.** 1 **c.** 2 **d.** 2

9. a. 6 **b.** 0 **c.** 9 **d.** 1 **11. a.** 5 **b.** −1 **c.** 6 **d.** $\dfrac{2}{3}$ **13. a.** 3 **b.** 5 **c.** −4 **d.** −4 **15. a.** $|x| + \sqrt{x}$; Dom $= [0, \infty)$

b. $\dfrac{|x|}{\sqrt{x}}$; Dom $= (0, \infty)$ **17. a.** $x^2 + x - 2$; Dom $= \mathbb{R}$ **b.** $\dfrac{1}{x+1}$; Dom $= (-\infty, -1) \cup (-1, 1) \cup (1, \infty)$

19. a. $x^3 + 3x - 8$; Dom $= \mathbb{R}$ **b.** $\dfrac{3x}{x^3 - 8}$; Dom $= (-\infty, 2) \cup (2, \infty)$ **21. a.** $-2x^2 + |x+4|$; Dom $= \mathbb{R}$ **b.** $\dfrac{-2x^2}{|x+4|}$;

Dom $= (-\infty, -4) \cup (-4, \infty)$ **23.** 2 **25.** 0 **27.** 8 **29.** 3 **31.** 1 **33.** $\dfrac{1}{3}$ **35. a.** $\dfrac{1}{x-1}$; Dom $= (-\infty, 1) \cup (1, \infty)$

b. $\dfrac{1}{x} - 1$; Dom $= (-\infty, 0) \cup (0, \infty)$ **37. a.** $1 - \sqrt{x}$; Dom $= [0, \infty)$ **b.** $\sqrt{1-x}$; Dom $= (-\infty, 1]$

39. a. $x^2 - 4x + 3;\ \text{Dom} = \mathbb{R}$ **b.** $x^2 + 2x - 3;\ \text{Dom} = \mathbb{R}$ **41. a.** $|x|^3 + |x|^2 - 5|x| + 3;\ \text{Dom} = \mathbb{R}$ **b.** $\left|x^3 + 4x^2\right| - 1;$

$\text{Dom} = \mathbb{R}$ **43. a.** $\dfrac{x^2 + 7}{2};\ \text{Dom} = \mathbb{R}$ **b.** $\dfrac{x^2 + 4x + 7}{2};\ \text{Dom} = \mathbb{R}$ **45.** $g(x) = \dfrac{2}{x},\ h(x) = 5x - 1,\ f(x) = g(h(x))$

47. $g(x) = x + \sqrt{x} - 5,\ h(x) = x + 2,\ f(x) = g(h(x))$ **49.** $g(x) = \dfrac{\sqrt{x}}{x^2},\ h(x) = x - 3,\ f(x) = g(h(x))$

51. $g(x) = x - 3,\ h(x) = \left|x^2 + 3x\right|,\ f(x) = g(h(x))$ **53.** $g(x) = \sqrt{x + 5}$ **55.** $g(x) = -x^3 - 7$ **57.** $V = 3\pi r^3$

59. $V = \dfrac{1}{12}\pi r^2 t^2$ **61.** $(f \circ g)(x) = \sqrt[3]{\dfrac{-x^3}{3x^2 - 9}},\ (f \circ g)(-x) = \sqrt[3]{\dfrac{x^3}{3x^2 - 9}} = -(f \circ g)(x)$ **63.** Yes **65.** Yes **67.** No

69. No **71.** No

Section 3.7 Inverses of Functions

1. Dom = {2, −1, −2};
  Ran = {−4, 3, 0}

3. Dom = Ran = $\mathbb{R}$

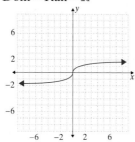

5. Dom = $\mathbb{R}$; Ran = $[0, \infty)$

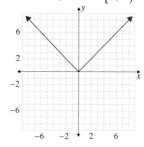

7. Dom = Ran = $\mathbb{R}$

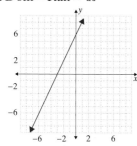

9. Dom = $[2, \infty)$; Ran = $[0, \infty)$

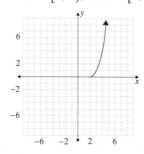

11. Dom = $\mathbb{R}$; Ran = $[-2, \infty)$

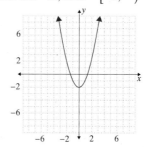

13. Dom = Ran = $\mathbb{R}$

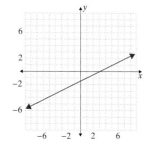

15. Not a one-to-one function; $f(-1) = f(1) = 1$ **17.** Restrict to $[0, \infty)$

19. Inverse exists **21.** Inverse exists **23.** Inverse exists

25. Restrict to $[2, \infty)$ **27.** Restrict to $[12, \infty)$ **29.** $f^{-1}(x) = (x + 2)^3$

31. $r^{-1}(x) = \dfrac{-2x - 1}{3x - 1}$ **33.** $F^{-1}(x) = (x - 2)^{\frac{1}{3}} + 5$ **35.** $V^{-1}(x) = 2x - 5$

37. $h^{-1}(x) = (x + 2)^{\frac{5}{3}}$ **39.** $J^{-1}(x) = \dfrac{x - 2}{3x}$ **41.** $h^{-1}(x) = (x - 6)^{\frac{1}{7}}$

43. $r^{-1}(x) = \dfrac{x^5}{2}$ **45.** $f^{-1}(x) = \dfrac{x^3}{54}$ **47.–55.** Answers will vary.   **57.** b   **59.** e   **61.** a

63. 73 1 53 13 97 73 29 57 17 73   **65.** FRISBEE VOLLEYBALL AND HORSESHOES

67. CATCH A WAVE

Chapter 3 Review

1. Dom $= \{-3,-1,0,3,4\}$; Ran $= \{-1,0,3,4\}$; Yes   **2.** Dom $= \mathbb{R}$; Ran $= \{2\}$; Yes   **3.** Dom $= \mathbb{Z}$;

Ran $= \{\ldots,-4,0,4,8,\ldots\}$; Yes   **4.** Dom $= \mathbb{R}$; Ran $= \mathbb{R}$; Yes   **5.** Dom $= [-6,\infty)$; Ran $= \mathbb{R}$; No

6. Dom $= [0,\infty)$; Ran $= [4,\infty)$; Yes   **7.** Dom $= \mathbb{R}$; Ran $= \{-5\}$; Yes   **8.** Dom $= \{-2,-4\}$; Ran $= \{-1,5\}$; Yes

9. Dom $= \mathbb{R}$; Ran $= (-\infty,3]$; Yes   **10.** $f(x) = 3\sqrt{x+11} - 4; f(-2) = 5$ **11.** $f(x) = -x^2 + 5x; f(-2) = -14$

12. $\sqrt{x+h}$ **13.** $\dfrac{\sqrt{x+h} - \sqrt{x}}{x+h}$ **14.** $\sqrt[3]{(x+h)^2}$ **15.** $\dfrac{\sqrt[3]{(x+h)^2} - \sqrt[3]{x^2}}{h}$ **16.** Dom $= \mathbb{N}$; Cod $= \mathbb{R}$;

Ran $= \left\{\dfrac{3}{4}, \dfrac{3}{2}, \dfrac{9}{4}, \ldots\right\}$ **17.** Dom $= \mathbb{R}$; Cod $= \mathbb{R}$; Ran $= \mathbb{R}$ **18.** $\mathbb{R}$ **19.** $(-\infty,1)\cup(1,\infty)$

20.

21.

22.

23.

24.

25.

26.

27.

28.

29.

30. 125   **31.**

32.

33.

34.

35.

36.

37.

38.

39.

40.

41.

42.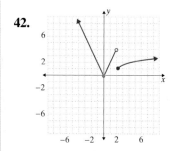

43. $V = khr^2$ **44.** $y = \dfrac{ka^3}{\sqrt{b}}$ **45.** $y = 112$ **49.** Dom = Ran = $\mathbb{R}$

46. $y = 72$ **47.** About 1226 videos per month

48. 7.44×10^7 meters

50. Dom $= \mathbb{R}$; Ran $= [0, \infty)$

51. Dom $= (-\infty, -2) \cup (-2, \infty)$; Ran $= (0, \infty)$

52. Dom $=$ Ran $= \mathbb{R}$

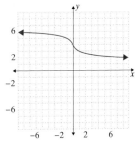

53. Dom $= (-\infty, 2) \cup (2, \infty)$; Ran $= (-\infty, -3) \cup (-3, \infty)$

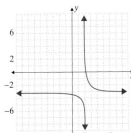

54. Dom $= [1, \infty)$; Ran $= [3, \infty)$

55. $f(x) = (x-1)^2 - 2$ **56.** $f(x) = -|x-3|$

57. $f(x) = -\sqrt{x} + 4$ **58.** Even function; y-axis symmetry **59.** Not a function; x-axis symmetry **60. a.** $-\dfrac{3}{2}$ **b.** $-\dfrac{5}{2}$ **c.** -1 **d.** -4

61. a. 7 **b.** -3 **c.** 10 **d.** $\dfrac{2}{5}$ **62. a.** -2 **b.** 18 **c.** -80 **d.** $-\dfrac{4}{5}$ **63. a.** $x^2 + \sqrt{x}$; Dom $= [0, \infty)$ **b.** $x^{\frac{3}{2}}$; Dom $= (0, \infty)$

64. a. $\dfrac{1}{x-2} + \sqrt[3]{x}$; Dom $= (-\infty, 2) \cup (2, \infty)$ **b.** $\dfrac{1}{\sqrt[3]{x}(x-2)}$; Dom $= (-\infty, 0) \cup (0, 2) \cup (2, \infty)$ **65. a.** $x^2 + x + 1$;

Dom $= \mathbb{R}$ **b.** $\dfrac{3x}{(x-1)^2}$; Dom $= (-\infty, 1) \cup (1, \infty)$ **66. a.** $x^2 + \sqrt[3]{x} - 5$; Dom $= \mathbb{R}$ **b.** $\dfrac{x^2 - 4}{\sqrt[3]{x} - 1}$; Dom $= (-\infty, 1) \cup (1, \infty)$

67. 5 **68.** $-\dfrac{9}{2}$ **69.** 4 **70.** $-\dfrac{2}{3}$ **71. a.** $4x^3 + 7$; Dom $= \mathbb{R}$ **b.** $(4x - 1)^3 + 2$; Dom $= \mathbb{R}$

72. a. $\dfrac{1}{\sqrt{x-2}}$; Dom $= (2, \infty)$ **b.** $\dfrac{1}{\sqrt{x-4}} + 2$; Dom $= (4, \infty)$ **73. a.** $2x^2 - 16x + 33$; Dom $= \mathbb{R}$ **b.** $2x^2 - 3$;

Dom $= \mathbb{R}$ **74. a.** $3\sqrt{x-3}$; Dom $= [3, \infty)$ **b.** $\sqrt{3x - 3}$; Dom $= [1, \infty)$ **75.** $g(x) = \dfrac{3}{x}$, $h(x) = 3x^2 + 1$, $f(x) = (g \circ h)(x)$

76. $g(x) = \dfrac{\sqrt{x}}{x^2}$, $h(x) = x + 2$, $f(x) = (g \circ h)(x)$ **77.** $g(x) = \dfrac{x+4}{6}$ **78.** $g(x) = \dfrac{2}{x} + 1$

79. Dom $= \{-1, 2, 4, 5\}$; Ran $= \{-6, -1, 0, 3\}$

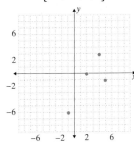

80. Dom $=$ Ran $= \mathbb{R}$

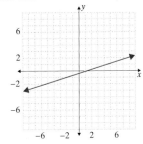

81. Dom $= [0, \infty)$; Ran $= [0, \infty)$:

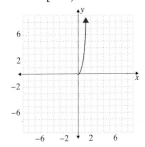

82. $r^{-1}(x) = \dfrac{x+2}{7x}$ **83.** $g^{-1}(x) = \dfrac{3}{4-x}$ **84.** $f^{-1}(x) = (x+6)^5$ **85.** $p^{-1}(x) = \dfrac{(x-3)^2}{4} + 1$ **86.** $f^{-1}(x) = \dfrac{-x-3}{x-2}$

87. $f^{-1}(x) = (x+1)^3 - 2$ **88.** $f^{-1}(x) = \sqrt{x+3} + 1,\ x \geq -3$ **89.** Answers will vary.

Chapter 3 Test

1. $-x^2 + 2(x-h) - h^2 + 2xh$ **2.** $-2x - h + 2$ **3.** $f(x) = 3x;\ -6$ **4.** $f(x) = \dfrac{8x}{4-3x};\ \dfrac{-8}{5}$ **5.** $\left(-\infty, \dfrac{1}{2}\right]$

6. $(-\infty, -2) \cup (-2, 2) \cup (2, \infty)$

7.   **8.**   **9.** $-\dfrac{25}{4}$ **10.**

11. $y = 48$ **12.** 64 feet   **13.** $f(x) = (x+2)^2 + 4$ **14. a.** -2 **b.** -6 **c.** -8 **d.** -2 **15.** $-x^4 - 2x^2 - 2$ **16.** $0, -2$

17. $g(x) = \sqrt{x},\ h(x) = 2x^2 + 1,\ f(x) = (g \circ h)(x)$ **18.** $g(x) = |x| - 2,\ h(x) = x^2 - x,\ f(x) = (g \circ h)(x)$

19. $f^{-1}(x) = (x-3)^4,\ x \geq 3$ **20.** $f^{-1}(x) = \sqrt{4-x}$ **21.** $f^{-1}(x) = \dfrac{2}{x} + 1$ **22.** $f^{-1}(x) = \dfrac{x^7}{3}$ **23.** -3 **24.** -3

25. 7   **26.** y-axis   **27.** Origin   **28.** $\dfrac{-dx+b}{cx-a}$ **29.** $\dfrac{4}{5}$ **30. a.** -64 **b.** 1

---

4   Polynomial Functions

Section 4.1 Introduction to Polynomial Equations and Graphs

1.–17. Answers will vary.   **19.** Yes   **21.** Yes   **23.** Yes   **25.** $1 \pm 2i$ **27.** $-3, \dfrac{1}{2}$ **29.** $\pm\sqrt{3}, \pm\sqrt{5}$ **31.** $-\dfrac{5}{2}$

33. $0, 4 \pm 3i$ **35.** $\pm 1, \pm 2i\sqrt{2}$ **37.** 7th-degree; lead coef $= 4$; $j(x) \to -\infty$ as $x \to -\infty$; $j(x) \to \infty$ as $x \to \infty$

39. 5th-degree; lead coef $= -6$; $h(x) \to \infty$ as $x \to -\infty$; $h(x) \to -\infty$ as $x \to \infty$

41. 4th-degree; lead coef $= -2$; $f(x) \to -\infty$ as $x \to -\infty$ and $x \to \infty$

43. *x*-int: $(-4, 0), (-2, 0),$
$(3, 0)$; *y*-int: $(0, 24)$
$g(x) \to \infty$ as $x \to -\infty$;
$g(x) \to -\infty$ as $x \to \infty$

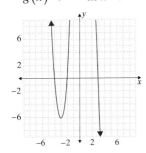

45. *x*-int: $(-2, 0)$;
y-int: $(0, -8)$
$h(x) \to \infty$ as $x \to -\infty$;
$h(x) \to -\infty$ as $x \to \infty$

47. *x*-int: $(-2, 0), (-1, 0),$
$(0, 0)$; *y*-int: $(0, 0)$
$s(x) \to -\infty$ as $x \to -\infty$;
$s(x) \to \infty$ as $x \to \infty$

49. *x*-int: $(3, 0)$;
y-int: $(0, -243)$
$g(x) \to -\infty$ as $x \to -\infty$;
$g(x) \to \infty$ as $x \to \infty$

51. e **53.** a **55.** f **57.** d **59.** f **61.** b **63.** $(-\infty, -2) \cup (3, \infty)$

65. $(-\infty, -2) \cup (-1, 0)$ **67.** $[-2, 1] \cup [3, \infty)$ **69.** $[-5, -1] \cup [1, 4]$

71. $\left(-\dfrac{1}{2}, 2\right)$ **73.** $(-\infty, -4) \cup (2, 3)$ **75.** All integers between 5 and 27,
inclusive **77.** All integers between 11 and 23, inclusive

79. About 17.9 months **81.** About 141.4 weeks

Section 4.2 Polynomial Division and the Division Algorithm

1. $3x^2 - x + 1 + \dfrac{5x - 1}{2x^2 + 2}$ **3.** $x - 2 + \dfrac{-2}{x^2 - 4x + 4}$ **5.** $4x^2 - 14x + 29 - \dfrac{65}{x + 2}$ **7.** $x^3 + 6x^2 - 2x + 5 + \dfrac{2x + 5}{3x^2 - 1}$

9. $2x^3 - 3x^2 + 2x - 5$ **11.** $x^3 + 3x^2 + 10x + 10 + \dfrac{22}{x - 3}$ **13.** $3x^2 + 5x + 9 + \dfrac{45}{3x - 5}$ **15.** $2x - 5 + \dfrac{7}{x + 3}$

17. $x^2 - ix + 6 + \dfrac{1 + i}{2x - i}$ **19.** $x^2 + 3$ **21.** $p(1) = 4$ **23.** k is a zero **25.** k is a zero **27.** $p(1) = 12$

29. k is a zero **31.** k is a zero **33.** k is a zero **35.** k is a zero **37.** $p(5) = -2$ **39.** k is a zero

41. $x^2 - 4x + 2 + \dfrac{-1}{x + 5}$ **43.** $x^7 - 3x^2 + \dfrac{3}{x + 1}$ **45.** $4x^2 - 4x + 2$ **47.** $x^4 - x^3 - x^2 - 7x - 14 - \dfrac{10}{x - 2}$

49. $x^3 - x^2 + x$ **51.** $2x^2 - 4ix + 17 + \dfrac{8 + 48i}{x - 3i}$ **53.** $f(x) = -x^2 - x + 12$ **55.** $f(x) = -x^2 + 4x - 13$

57. $f(x) = x^4 - 12x^3 + 54x^2 - 108x + 81$ **59.** $f(x) = 3x^4 + 9x^3 - 9x^2 - 21x + 18$

61. $SA = (x + 5)(x + 2) = x^2 + 7x + 10$

Section 4.3 Locating Real Zeros of Polynomials

1. $\pm\left\{\dfrac{1}{3},\dfrac{2}{3},1,\dfrac{4}{3},2,\dfrac{8}{3},4,8\right\},\left\{-4,\dfrac{1}{3},2\right\}$ **3.** $\pm\{1,2,3,4,6,8,12,24\},\{-2i,2i,2,3\}$ **5.** $\pm\{1,2,7,14\},\{1,2,7\}$

7. $\pm\left\{\dfrac{1}{2},1,\dfrac{3}{2},\dfrac{5}{2},3,5,\dfrac{15}{2},15\right\},\left\{-1,\dfrac{5}{2},3\right\}$ **9.** $\pm\left\{\dfrac{1}{3},1,3\right\},\{-1,1,-i,i\}$ **11.** $\pm\{1,11\},\{-11,-1,1\}$ **13.** $\{-1,1,-i,i\}$

15. $\{-1,2-3i,2+3i\}$ **17.** $\{-2i,2i,2,3\}$ **19.** $\{4,1-2i,1+2i\}$ **21.** $\{-5i,5i,-2,1\}$ **23.** $\{-11,-1,1\}$

25. 0 pos., 3 or 1 neg.   **27.** 2 or 0 pos., 1 neg.   **29.** 3 or 1 pos., 1 neg.   **31.** 1 pos., 1 neg.   **33.** 3 or 1 pos., 0 neg.

35. 0 pos., 0 neg.   **37.** $[-5,1]$ **39.** $[-1,6]$ **41.** $[-3,6]$ **43.** $[-3,3]$ **45.** $[-3,6]$ **47.** $\{-4,-1,1\}$ **49.** $\{-1,2,5\}$

51. $\left\{2,3,\pm 2\sqrt{2}\right\}$ **53.** $\pm\left\{\sqrt{5},i\sqrt{5}\right\}$ **55.** $\left\{-2,-\dfrac{1}{2},6\right\}$ **57.** $f(-3)=-84, f(-1)=16$ **59.** $f(2)=-15, f(3)=24$

61. $f(2)=15, f(3)=-24$ **63.** Answers will vary.   **65.** $\left\{-6,\dfrac{1}{4},1\right\}$ **67.** $\left\{-3,\pm\sqrt{3},\dfrac{1}{2}\right\}$ **69.** $\left\{\dfrac{-9\pm\sqrt{41}}{2},\dfrac{1}{2},2\right\}$

71. $\{-4,-3,1,2\}$ **73.** $\{\pm 5,\pm 6,\pm 8\}$ **75.** $\{-3,4,5\}$ **77.** $\left\{-\dfrac{2}{3},\dfrac{1}{2},7\right\}$ **79.** $\{-4,-3,2\}$ **81.** $\left\{\pm i\sqrt{7},0,3\right\}$

83. $\left\{\sqrt[3]{-5},-\dfrac{3}{4},2\right\}$

Section 4.4 The Fundamental Theorem of Algebra

1.

3.

5.

7.

9. $f(x)=(x+2)^3(x-1)^2$

11. $s(x)=-(x-3)(x+1)\left(x-\sqrt{5}\right)\left(x+\sqrt{5}\right)$

13. $H(x) = (x-2)(x+1) \times$
$(x-\sqrt{3})(x+\sqrt{3})$

15. $f(x) = (x+2)^2(2x+3)$

17. $\{-2,1\}$ **19.** $\{-2,1,\pm i\sqrt{5}\}$

21. $\{-3,1,\pm 2i\}$ **23.** $\{\pm 1,\pm 4i\}$

25. $(x-3+2i)(x-3-2i)(x-4)(x+1)$

27. $(x-1-3i)(x-2)(x+1)$

29. $(x+\sqrt{7})(x-\sqrt{7})(x-2+3i)(x-2-3i)$

31. $(x-2)(x+1)(x-1+2i)(x-1-2i)$ **33.** $x(x-3)(x-1)(x+11)$ **35.** $(x-2)(x-2+4i)(x-2-4i)$

37. $f(x) = -2x^3 + 18x^2 - 32x - 52$ **39.** $f(x) = 2x^5 + 2x^4 - 10x^3 - 2x^2 + 16x - 8$

41. $f(x) = 3x^4 - 18x^3 + 12x^2 - 72x$ **43.** $f(x) = -x^3 + 2x^2 + 14x - 40$ **45.** $f(x) = -x^3 + 4x^2 + 15x - 68$

47. a. $V(x) = 2x(15-2x)(5-x)$ **b.** $x=0, x=5, x=\dfrac{15}{2}$ **c.** $x=0$ and $x=5$ **49.** Answers will vary.

Section 4.5 Rational Functions and Rational Inequalities

1. $x=1$ **3.** No vertical asymptote   **5.** $x=2$ **7.** $x=0$ **9.** $x=-\dfrac{1}{2}$ **11.** No vertical asymptote   **13.** $x=7$

15. $x=-2$ **17.** $x=-2, x=2$ **19.** $y=0$ **21.** No horizontal or oblique asymptote   **23.** $y=0$ **25.** $y=2$

27. $y=3x+6$ **29.** $y=0$ **31.** $y=0$ **33.** $y=x-11$ **35.** $y=5x+4$

37.

39.

41.

43.

45.

47.

49. a. $x = -2$ **b.** $y = 0$ **c.** None **d.** None **e.** $(0,5)$ **51. a.** $x = 9$ **b.** $y = 0$ **c.** None **d.** None **e.** $\left(0, -\dfrac{1}{3}\right)$

53. a. $x = -1, x = 1$ **b.** None **c.** $y = x$ **d.** $\left(\sqrt[3]{3}, 0\right)$ **e.** $(0,3)$ **55. a.** $x = 1$ **b.** None **c.** $y = 3x$ **d.** None **e.** $(0,-3)$

57. $(-\infty,-2) \cup (-1,1)$ **59.** $(-8,-2) \cup (2,\infty)$ **61.** $(-\infty,-2) \cup (-2,3)$ **63.** $(0,3)$ **65.** $(-2,-1) \cup (1,\infty)$

67. $(-\infty,-1) \cup \left[-\dfrac{1}{2}, 0\right)$ **69. a.**

71. a.

b. Joan's rabbit population reaches a maximum of 200 rabbits.

b. The concentration of the drug disappears in the long run.

Chapter 4 Review

1.–4. Answers will vary. **5.** $\pm\sqrt{2}, \pm\sqrt{5}$ **6.** $0, \pm\sqrt{2}, \pm i$ **7.** $\pm\sqrt{2}$ **8.** $-4, -2, 0$ **9.** $0, \dfrac{-1 \pm \sqrt{5}}{2}$ **10.** $-2 \pm i\sqrt{3}$

11. x-int: $-2, 1, 3$; y-int: 6
$f(x) \to -\infty$ as $x \to -\infty$;
$f(x) \to \infty$ as $x \to \infty$

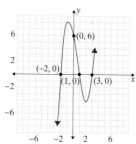

12. x-int: $2, -1$; y-int: 4
$f(x) \to \infty$ as $x \to \pm\infty$

13. x-int: $1, 4$; y-int: 4
$g(x) \to \infty$ as $x \to \pm\infty$

14. x-int: $0, -2, -5$; y-int: 0
$h(x) \to \infty$ as $x \to -\infty$;
$h(x) \to -\infty$ as $x \to \infty$

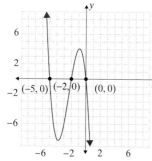

15. $\left[\dfrac{5}{2}, 3\right]$ **16.** $(-\infty,-1) \cup (-1,3) \cup (3,\infty)$ **17.** $[-2,-1] \cup [1,4]$

18. $[-2,0] \cup [4,\infty)$ **19.** $(-\infty,0) \cup (0,1) \cup (2,\infty)$ **20.** $\left(\dfrac{1}{3}, 2\right)$ **21.** All integers between 4 and 30, inclusive **22.** $4x^2 - 3x + 3 + \dfrac{7}{2x^2 - 1}$

23. $11x + 35 + \dfrac{100}{x - 3}$ **24.** $x^2 - 3x + 4 - \dfrac{5x + 16}{x^2 + 3x + 2}$

25. $2x^3 + 2x^2 - 2x - 3 + \dfrac{-2x - 2}{x^2 - x}$ **26.** $x^2 - 6 + \dfrac{-4 + 7i}{2x + i}$ **27.** $p(1) = 90$

28. k is a zero **29.** $p\left(\dfrac{2}{3}\right) = -\dfrac{7}{3}$ **30.** $x^3 + x^2 + 2x + 7$

31. $-x^3 + 2x^2 - 7x + 23$ **32.** $x^4 + 3x^2 - 5 + \dfrac{23}{x + 2}$

33. $-x^3 + 7x^2 + x - 3 + \dfrac{-1}{x-1}$ **34.** $x^3 + 4x^2 - x + 3$ **35.** $f(x) = x^2 - 4x - 12$ **36.** $y = \dfrac{1}{2}(x+4)^4$

37. $y = 2(x^2 - 4)(x - 3)$ **38.** $\pm\{1,2,3,6\}, \{-3,-1,2\}$ **39.** $\pm\left\{\dfrac{1}{2}, 1, \dfrac{3}{2}, 3, \dfrac{9}{2}, 9\right\}, \left\{1, \dfrac{3}{2}, 3\right\}$

40. $\pm\left\{\dfrac{1}{2}, 1, \dfrac{3}{2}, 3, \dfrac{9}{2}, 9\right\}, \left\{-3, \dfrac{2 \pm i\sqrt{2}}{2}\right\}$ **41.** $\pm\{1,3,9\}, \{-3,1\}$ **42.** $\{-3,-1,2\}$ **43.** $\left\{1, \dfrac{3}{2}, 3\right\}$ **44.** $\left\{-3, \dfrac{2 \pm i\sqrt{2}}{2}\right\}$

45. $\{-3,-1\}$ **46.** 2 or 0 pos., 2 or 0 neg.   **47.** 4, 2 or 0 pos., 2 or 0 neg.   **48.** $[-2,6]$ **49.** $[-3,7]$ **50.** $\left\{-\dfrac{3}{2}, 3, 4\right\}$

51. $\left\{-\dfrac{5}{2}, -\dfrac{1}{2}, 7\right\}$ **52.** $f(-2) = 73; f(0) = -5$ **53.** $f(2) = 3; f(4) = -15$ **54.** $\{\pm 1, 2, 3\}$ **55.** $\{\pm 3i, 4\}$

56. $\{-3,-2,-1\}$ **57.** $\left\{3, 2 \pm \sqrt{3}\right\}$ **58.** $\{1, \pm 3i\}$ **59.** $\left\{-\dfrac{1}{2}, 2\right\}$ **60.** $\left\{-\dfrac{4}{3}, \pm i\sqrt{2}\right\}$

61.

62.

63. $(x^2 + 1)(x - 3)$

64. $(x+1)(x-1)(x-2)(x^2 + x + 1)$ **65.** $\left\{-1, \dfrac{5}{3}, \pm 2i\right\}$ **66.** $\left\{\pm\sqrt{3}, -\dfrac{1}{2}\right\}$ **67.** $\{\pm 2i, -3, \pm 1\}$

68. $(x - 2 + i)(x - 2 - i)(7x - 2)(2x - 7)$

69. $(x - 5i)(x + 5i)(x - 6)(x + 1)$ **70.** $(x - 1 - i)(x - 1 + i)(2x + 1)(x + 3)$

71. $(x + 3)\left(x - \dfrac{1 + i\sqrt{19}}{4}\right)\left(x - \dfrac{1 - i\sqrt{19}}{4}\right)$

72. $f(x) = 2x^4 + 7x^3 - 18x^2 + 67x - 30$

73. $f(x) = x^5 + 3x^4 - 3x^3 - 17x^2 - 18x - 6$ **74.** $f(x) = x^5 - 3x^4 + 8x^2 - 9x + 3$

75. $x = \dfrac{5}{2}$ **76.** No vertical asymptote   **77.** $x = 0$ **78.** $x = 3$ **79.** $y = 2x + 9$ **80.** $y = \dfrac{1}{3}$ **81.** None   **82.** $y = 0$

83.

84.

85.

86.

87. $\left(-3, \dfrac{7}{2}\right]$ **88.** $(-\infty, 2)$ **89.** $\left(-3, \dfrac{8}{9}\right) \cup (2, \infty)$

90. $(-\infty, -8) \cup (-3, \infty)$

Chapter 4 Test

1.–2. Answers will vary.   **3.** $0, 5, -4$ **4.** $\dfrac{3 \pm i\sqrt{3}}{2}$

5. x-int: $1, \pm 2$; y-int: -4
$f(x) \to \infty$ as $x \to -\infty$;
$f(x) \to -\infty$ as $x \to \infty$

6. x-int: 2; y-int: -8
$f(x) \to -\infty$ as $x \to -\infty$;
$f(x) \to \infty$ as $x \to \infty$

7. $(-6, 1)$ **8.** All real numbers except $-1, 3$

9. All real numbers

10. $40 \le x \le 50$ **11.** $x^3 + x^2 + x + 1$

12. $3x^2 i - 6x + i - 3$

13. $x^3 - 3x^2 + 8x - 16 + \dfrac{28}{x+2}$

14. k is a zero   **15.** $p\left(\dfrac{1}{2}\right) = \dfrac{5}{2}$

16. $p(-2) = -41$ **17.** $y = (x+2)(1-x)$

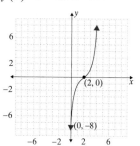

18. $y = x(x+3)(x-2)$ **19.** $-2, 1, 3$ **20.** $0, -2, \pm 2i$ **21.** 0 pos., 3 or 1 neg.   **22.** 2 or 0 pos., 1 neg.

23. $(2x-1)(x-2)(x+1)$ **24.** $(x-2)^3(x+3)$ **25.** $y = -2x^2 + 4x - 6$

26. $y = \left(x^2 - 2x + 2\right)(x-2)$

27. No

28. $f(x) = \left(x + \dfrac{2}{3}\right)\left(15x^3 - 6x + 4\right) + \dfrac{34}{3}$; $f\left(\dfrac{-2}{3}\right) = \dfrac{34}{3}$ **29.** $x = -1$ **30.** None   **31.** $y = 0$ **32.** $y = x - 2$

33.

34.

35. $x > 2$ **36.** $1 < x \le \dfrac{5}{3}$ or $x > 2$

37. No; explanations will vary.

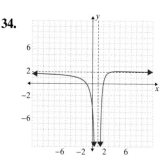

5 | Exponential and Logarithmic Functions

Section 5.1 Exponential Functions and Their Graphs

1.

3.

5.

7.

9.

11.

13.

15.

17.

19.

21.

23. $\{2\}$ **25.** $\{-2\}$ **27.** $\{-13\}$ **29.** $\{3\}$

31. $\{-2\}$ **33.** $\{-2, -1\}$ **35.** $\{7\}$ **37.** $\{3\}$

39. $\{9\}$ **41.** $\{-3\}$ **43.** $\{2\}$ **45.** $\{-1\}$

47. No solution   **49.** g   **51.** b   **53.** f

55. c   **57.** j

Section 5.2 Applications of Exponential Functions

1. $V \approx 178$ people   **3.** $C \approx \$8526.20$ **5.** $W \approx 93$ computers   **7. a.** 0.999567 **b.** 0.958 grams **c.** 0.648 grams

9. a. 3 years **b.** 9 years   **11.** 1118 rabbits   **13.** The bank offering 2.75% and monthly compounding.

15. Approximately 3.18%   **17.** \$134,392   **19. a.** 10 **b.** 7490 people **c.** The function approaches 10,000 as time goes on.

21. a. 0.965936 **b.** 0.707 kg **c.** 7.628 mg   **23. a.** \$1521.74 **b.** \$271.74   **25.** \$9459.48; \$9942.41   **27.** \$2835.71

29. \$20,000   **31.** \$7318.71   **33. a.** \$7647.95 **b.** \$7647.57 **c.** Yes; explanations will vary.

Section 5.3 Logarithmic Functions and Their Graphs

1. $4 = \log_5 625$ **3.** $3 = \log_x 27$ **5.** $3 = \log_{4.2} C$ **7.** $x = \log_4 31$ **9.** $\sqrt{3} = \log_{4x} 13$ **11.** $e^x = \log_2 11$

13. $81 = 3^4$ **15.** $4 = b^{\frac{1}{2}}$ **17.** $15 = 2^b$ **19.** $W = 5^{12}$ **21.** $2x = \pi^4$ **23.** $e^x = 2$

25.   **27.**   **29.**

31.   **33.**   **35.**

37. e   **39.** b   **41.** h   **43.** d   **45.** i   **47.** -2 **49.** 3   **51.** -1 **53.** -4 **55.** -1 **57.** 0   **59.** $-\dfrac{1}{2}$ **61.** $\{8\}$ **63.** $\left\{\dfrac{1}{8}\right\}$

65. $\{1\}$ **67.** $\{-2\}$ **69.** No solution   **71.** $\{2\}$ **73.** $\{41.96\}$ **75.** $\{19.09\}$ **77.** $\{5.60\}$ **79.** $\{0.43\}$

81. $\{4297.83\}$ **83.** $\{10,000,000,002\}$

Section 5.4 Properties and Applications of Logarithms

1. $3 + 3\log_5 x$ **3.** $2 + \ln p - 3\ln q$ **5.** $1 + \log_9 x - 3\log_9 y$ **7.** $\dfrac{3}{2}\ln x + \ln p + 5\ln q - 7$ **9.** $\log(2 + 3\log x)$

11. $1 - \dfrac{1}{2}\log(x + y)$ **13.** $\log_2\left(y^2 + z\right) - 4\log_2 x - 4$ **15.** $2\log_b x + \dfrac{1}{2}\log_b y - \log_b z$ **17.** $2 + \log_b a + b\log_b c$

19. $\log\dfrac{x}{y}$ **21.** $\log_5(x+5)$ **23.** $\log_2\left(x^{\frac{4}{3}}+3x^{\frac{1}{3}}\right)$ **25.** $\ln\left(\dfrac{3p}{q^2}\right)$ **27.** $\log\left(\dfrac{x-10}{x}\right)$ **29.** $\ln\left(\dfrac{z^2}{x^3 y^3}\right)$

31. $\log_5 4$ **33.** $\ln 45$ **35.** $\log_3 1 = 0$ **37.** $\ln 12$ **39.** $\log 11$ **41.** $\log_8\left(x^2 - y\right)$ **43.** x^2 **45.** $\dfrac{e^2 p}{x}$ **47.** $\dfrac{x^3}{y^4}$

49. x^2 **51.** 4   **53.** $12x^2$ **55.** 2.04   **57.** 0.95   **59.** 0.95   **61.** 2.45   **63.** 3.30   **65.** 0.74   **67.** 1.20   **69.** 1.86   **71.** −1

73. 3.85   **75.** 0.77   **77.** −1.76   **79.** 2   **81.** 7   **83.** 1   **85.** $4\sqrt{2} \approx 5.66$ **87.** 9.05   **89.** 2.08   **91.** 12   **93.** 1,048,576

95. 10.25   **97.** 5,011,872 times stronger   **99.** 133 decibels   **101.** 7.62; yes   **103. a.** 15.05 minutes   **b.** 7:00 p.m.

c. 112°F; no

Section 5.5 Exponential and Logarithmic Equations

1. $x \approx 0.26$ **3.** $x \approx 3.12$ **5.** $x \approx 3.89$ **7.** $x \approx -2.28$ **9.** $x \approx 8.09$ **11.** $x \approx 2.68$ **13.** $x \approx -1.12$ **15.** $x \approx 52.77$

17. $x \approx \pm 0.71$ **19.** $x = -12$ **21.** $x \approx 1.32$ **23.** $x \approx 3.27$ **25.** $x = 125$ **27.** $x = 5$ **29.** $x \approx 40.17$ **31.** $x = 35$

33. $x \approx 9.38$ **35.** $x = 4$ **37.** $x = \dfrac{1}{162}$ **39.** $x = 1$ **41.** $x \approx 100.04$ **43.** No solution   **45.** $x = 5$ **47.** $x = 8$

49. $x = \dfrac{37}{8}$ **51.** $x = 5$ **53.** No solution   **55.** $x = 6$ **57.** $x = \sqrt{2} - 1$ **59.** $x = 2, 3$ **61.** $x = 1, 2$

63. $f(x) = \log 2x^2$ **65.** $f(x) = \ln 9x^2$ **67.** $f(x) = 256^x$ **69.** $f(x) = \ln 1 = 0$ **71.** $f(x) = \ln 5^x$ **73.** $f(x) = \ln 5$

75. a. 17.36 years   **b.** 9.90 years   **77.** 4.98 hours   **79.** 4.99 years   **81.** 0.271 years (about 99 days)

Chapter 5 Review

1.

2.

3.

4.

5.

6.

7.

8.

9.

10.

11. $x = 5$ **12.** $x = -4$ **13.** $x = 2$ **14.** $x = 3$ **15.** $x = -1$ **16.** $x = -\dfrac{1}{6}$

17. $x = \dfrac{8}{5}$ **18.** $x = -1$ **19.** $x = 2$ **20.** The account at 3.95%   **21.** 8 days

22. a. $P(t) = 870\left(1.032^{t}\right)$ **b.** Approximately 68.39 minutes

23. a. 173 flies **b.** 20 flies

24. 0.58 pounds

25. $x = \log_3 8$ **26.** $\sqrt{2}\log 3a = 1$ **27.** $\log_4 4096 = 3a$ **28.** $4^x = 64$

29. $3^{-3} = \dfrac{1}{27}$ **30.** $8^3 = 2A$

31.

32.

33.

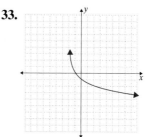

34. 2   **35.** −2   **36.** −3   **37.** −3   **38.** −2   **39.** $\dfrac{2}{3}$ **40.** $x = 125$ **41.** $x = 27$ **42.** $x = 2$ **43.** $x = \dfrac{17}{2}$ **44.** $x \approx 6.13$

45. $x \approx 155.41$ **46.** $x = 601$ **47.** $\dfrac{3}{2}\log x - \dfrac{5}{2}\log \pi - \log 2$ **48.** $\dfrac{5}{2}\ln a + \ln m + 2\ln n - 5$ **49.** $3 + 3\log_3 a$

50. $1 + \ln 2 + \ln x$ **51.** $\log_2\left(\dfrac{a^{\frac{5}{3}}}{b^{\frac{1}{3}}c}\right)$ **52.** $\ln\left(\dfrac{4}{x^2 y^7}\right)$ **53.** $\log_2(x-3)$ **54.** $\log\dfrac{a^2 b^3}{d\sqrt{c}}$ **55.** $\log_3\dfrac{x^2 - 2x}{x^2 + 4}$

56. 2.58 **57.** 6.18 **58.** -6.34 **59.** x^3 **60.** $\dfrac{x}{y^2}$ **61.** 0 **62.** 6 **63. a.** $251{,}188{,}643$ **b.** $7{,}079{,}458$ **c.** $1{,}258{,}925{,}412$

64. $\dfrac{8 - \ln 16}{5} \approx 1.05$ **65.** $\dfrac{6}{\log 321} \approx 2.39$ **66.** $\dfrac{3\ln 19}{\ln 7} + 12 \approx 16.54$ **67.** $\dfrac{\ln 5}{4 - 3\ln 5} \approx -1.94$ **68.** $\ln 8 - 2 \approx 0.08$

69. $\dfrac{\log 12}{\log 18} \approx 0.86$ **70.** 3 **71.** -2 **72.** 16 **73.** 73.5 **74.** 4 **75.** $f(x) = \ln x^3$ **76.** $f(x) = \log(8x^3)$

77. $f(x) = \log x^{10}$ **78.** $f(x) = \ln(3x^3)$ **79.** $f(x) = \log 7$ **80.** $f(x) = \ln 27$ **81.** 20.4 months (1.7 years)

82. 73.0 hours (about 3 days)

Chapter 5 Test

1.

2.
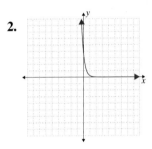

3. $x = -4$ **4.** $x = -4$ **5.** $x = \dfrac{1}{3}$

6. a. $\$18{,}726$ **b.** $\$18{,}820$ **7.** $\log_5 125 = x$

8. $x = \ln 3\pi$ **9.** $3^{-3} = \dfrac{1}{27}$ **10.** $a = 5^{10}$

11.

12.
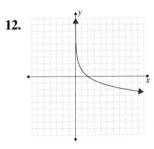

13. 2 **14.** $\dfrac{2}{3}$ **15.** $x = \dfrac{1}{\sqrt{3}}$ **16.** $x = 3^{\frac{1}{3}}$

17. $\ln 2 + 2\ln a + \ln b - \ln 3 - \ln c$

18. $1 + \log_4(3x + 5y)$ **19.** $\log\dfrac{xy}{c}$

20. $\log_3\dfrac{x^2 - 2x}{x^2 + 4}$ **21.** $x = \dfrac{1 - \log_5 7}{2}$

22. $x = \dfrac{2 + \ln 40}{5}$ **23.** $x = \dfrac{3}{\log_2 5}$ **24.** $x = \dfrac{9}{2}$ **25.** $x = \dfrac{3}{\sqrt{2}}$ **26.** $x = 16, \; x = \dfrac{1}{4}$ **27.** $x = 10$ **28.** $x = 9$

29. $I = 3.16$ watts/m²   **30.** $f^{-1}(x) = e^{\frac{x+1}{2}} - 1$

6   Trigonometric Functions

Section 6.1 Radian and Degree Measure of Angles

1. $225°$ **3.** $-67.5°$ **5.** $120°$ **7.** $150°$ **9.** $-405°$ **11.** $\dfrac{47\pi}{180}$ **13.** $\dfrac{11\pi}{15}$ **15.** $\dfrac{37\pi}{45}$ **17.** $\dfrac{8\pi}{3}$ **19.** $\dfrac{25\pi}{36}$

21. $270°$ **23.** $540°$ **25.** $-72°$ **27.** $\dfrac{\pi}{9}$ **29.** $-\dfrac{4\pi}{5}$ **31.** $\dfrac{\pi}{6}$ **33.** $\dfrac{83\pi}{90}$ **35.** $115°$ **37.** $50°$ **39.** $345°$ **41.** $295°$

43. $\dfrac{5\pi}{4}$ **45.** $\dfrac{\pi}{4}$ **47.** $\dfrac{4\pi}{3}$ **49.** $\dfrac{13\pi}{4}$

51.   **53.**   **55.**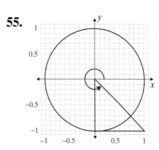

57. 4 inches   **59.** 11.78 feet   **61.** 8.64 m   **63.** 1.48 inches   **65.** 3.56 feet   **67.** 528.10 km   **69.** 43.0°

71. 611.48 km   **73.** 3004.02 km   **75.** $\dfrac{9}{2}$ **77.** $\dfrac{21}{47}$ **79.** $\dfrac{45}{4}$ **81. a.** 2000π rad/min   **b.** 87.27 ft/s

83. a. 100π rad/min   **b.** 366.52 ft/min   **85.** $\dfrac{5\pi}{12} \approx 1.31$ **87.** $29.93\ \text{cm}^2$ **89.** $15.08\ \text{m}^2$ **91.** $314.16\ \text{ft}^2$

93. $9.82\ \text{in.}^2$ **95.** 26.18 cm   **97. a.** 12π rad/s   **b.** 95.38 in./s   **99.** 38.5°

Section 6.2 Trigonometric Functions of Acute Angles

1. $\sin\theta = \cos\theta = \dfrac{\sqrt{2}}{2}$, $\tan\theta = \cot\theta = 1$, $\csc\theta = \sec\theta = \sqrt{2}$ **3.** $\sin\theta = \dfrac{1}{3}$, $\cos\theta = \dfrac{2\sqrt{2}}{3}$, $\tan\theta = \dfrac{\sqrt{2}}{4}$, $\csc\theta = 3$,

$\sec\theta = \dfrac{3\sqrt{2}}{4}$, $\cot\theta = 2\sqrt{2}$ **5.** $\sin\theta = \cos\theta = \dfrac{\sqrt{2}}{2}$, $\tan\theta = \cot\theta = 1$, $\csc\theta = \sec\theta = \sqrt{2}$ **7.** $\sin\theta = \dfrac{5}{7}$,

$\cos\theta = \dfrac{2\sqrt{6}}{7}$, $\tan\theta = \dfrac{5\sqrt{6}}{12}$, $\csc\theta = \dfrac{7}{5}$, $\sec\theta = \dfrac{7\sqrt{6}}{12}$, $\cot\theta = \dfrac{2\sqrt{6}}{5}$ **9.** $\sin\theta = \dfrac{12}{13}$, $\cos\theta = \dfrac{5}{13}$, $\tan\theta = \dfrac{12}{5}$,

$\csc\theta = \dfrac{13}{12}$, $\sec\theta = \dfrac{13}{5}$, $\cot\theta = \dfrac{5}{12}$ **11.** $\sin\theta = \dfrac{33}{65}$, $\cos\theta = \dfrac{56}{65}$, $\tan\theta = \dfrac{33}{56}$, $\csc\theta = \dfrac{65}{33}$, $\sec\theta = \dfrac{65}{56}$, $\cot\theta = \dfrac{56}{33}$

13. $\sin\theta = \dfrac{1}{2}$, $\cos\theta = \dfrac{\sqrt{3}}{2}$, $\tan\theta = \dfrac{\sqrt{3}}{3}$, $\csc\theta = 2$, $\sec\theta = \dfrac{2\sqrt{3}}{3}$, $\cot\theta = \sqrt{3}$ **15.** $\sin\dfrac{\pi}{4} = \dfrac{\sqrt{2}}{2}$, $\csc\dfrac{\pi}{4} = \sqrt{2}$

17. $\sec 60° = 2$ **19.** $\csc\dfrac{\pi}{6} = 2$ **21.** $\sec 5° \approx 1.0038$; $\tan 5° \approx 0.0875$ **23.** $\cot\dfrac{\pi}{3} = \dfrac{\sqrt{3}}{3}$ **25.** $\tan 87.2° \approx 20.4465$

27. 0.9945   **29.** 1.0355   **31.** 28.6537   **33.** 3.0777   **35.** 0.3827   **37.** 2   **39.** 38.9053°   **41.** 25.325°

43. 21.6656°   **45.** 0.1149   **47.** 0.7746   **49.** 5   **51.** True   **53.** True   **55.** False   **57.** 751.19 feet   **59.** 17.47 feet

61. 48.54 yards   **63.** 12.04 m   **65.** 314.57 feet   **67.** 13.86 feet   **69.** 6.86 feet   **71.** 20 m   **73.** 3196.80 feet

Section 6.3 Trigonometric Functions of Any Angle

1. $\sin 45° = \cos 45° = \dfrac{\sqrt{2}}{2}$, $\tan 45° = \cot 45° = 1$, $\csc 45° = \sec 45° = \sqrt{2}$ **3.** $\sin 60° = \dfrac{\sqrt{3}}{2}$, $\cos 60° = \dfrac{1}{2}$,

$\tan 60° = \sqrt{3}$, $\csc 60° = \dfrac{2\sqrt{3}}{3}$, $\sec 60° = 2$, $\cot 60° = \dfrac{\sqrt{3}}{3}$ **5.** $\sin \dfrac{5\pi}{2} = \csc \dfrac{5\pi}{2} = 1$, $\cos \dfrac{5\pi}{2} = \cot \dfrac{5\pi}{2} = 0$,

$\tan \dfrac{5\pi}{2} = \sec \dfrac{5\pi}{2} = $ undefined   **7.** $\sin 305° \approx -0.8192$, $\cos 305° \approx 0.5736$, $\tan 305° \approx -1.4281$, $\csc 305° \approx -1.2208$,

$\sec 305° \approx 1.7434$, $\cot 305° \approx -0.7002$ **9.** $\sin 6\pi = \tan 6\pi = 0$, $\cos 6\pi = \sec 6\pi = 1$, $\csc 6\pi = \cot 6\pi = $ undefined

11. $\sin \dfrac{3\pi}{2} = \csc \dfrac{3\pi}{2} = -1$, $\cos \dfrac{3\pi}{2} = \cot \dfrac{3\pi}{2} = 0$, $\tan \dfrac{3\pi}{2} = \sec \dfrac{3\pi}{2} = $ undefined   **13.** $\sin \dfrac{5\pi}{4} = -\dfrac{\sqrt{2}}{2}$, $\cos \dfrac{5\pi}{4} = -\dfrac{\sqrt{2}}{2}$,

$\csc \dfrac{5\pi}{4} = -\sqrt{2}$, $\sec \dfrac{5\pi}{4} = -\sqrt{2}$, $\tan \dfrac{5\pi}{4} = \cot \dfrac{5\pi}{4} = 1$ **15.** $\sin (-445°) \approx -0.9962$, $\cos (-445°) \approx 0.0872$,

$\tan (-445°) \approx -11.4301$, $\csc (-445°) \approx -1.0038$, $\sec (-445°) \approx 11.4737$, $\cot (-445°) \approx -0.0875$ **17.** $\theta' = \dfrac{\pi}{2}$ **19.** $\theta' = \dfrac{\pi}{4}$

21. $\theta' = 47°$ **23.** $\theta' = 12°$ **25.** $\theta' = 36°$ **27.** $\theta' = 30°$ **29.** $\theta' = 2°$ **31.** II   **33.** I   **35.** III   **37.** III   **39.** c   **41.** b

43. c   **45.** a   **47.** b   **49.** $\tan 98° = -\tan 82° \approx -7.1154$ **51.** $\cos (-60°) = \cos 60° = 0.5$ **53.** $\cos \dfrac{5\pi}{2} = \cos \dfrac{\pi}{2} = 0$

55. $\cos \dfrac{7\pi}{6} = -\cos \dfrac{\pi}{6} = \dfrac{-\sqrt{3}}{2}$ **57.** $\cos \dfrac{6\pi}{5} = -\cos \dfrac{\pi}{5} \approx -0.8090$ **59.** $\tan \dfrac{3\pi}{2} = \tan \dfrac{\pi}{2} = $ undefined

61. $\sin \dfrac{7\pi}{4} = -\sin \dfrac{\pi}{4} = \dfrac{-\sqrt{2}}{2}$ **63.** $\sin 105° = \sin 75° \approx 0.9659$ **65.** a   **67.** e   **69.** $\sec \dfrac{\pi}{2} = \csc 0 = $ undefined

71. $\cos \left(-\dfrac{3\pi}{4}\right) = \sin \dfrac{5\pi}{4} = \dfrac{-\sqrt{2}}{2}$ **73.** $\cot 313° = \tan (-223°) \approx -0.9325$ **75.** $\csc (-168°) = \sec 258° \approx -4.8097$

77. $\sec 216° = \csc (-126°) \approx -1.2361$ **79.** $\cos (-15°) = \sin 105° \approx 0.9659$ **81.** $\tan (-105°) = \cot 195° \approx 3.7321$

83. $\tan \theta \approx 4.702$, $\cot \theta \approx 0.213$ **85.** $\tan \theta \approx 3.730$, $\cot \theta \approx 0.268$ **87.** $\tan \theta \approx -0.941$, $\cot \theta \approx -1.063$

89. Prove using identities on pages 500 and 501.   **91.** $\csc \theta = \dfrac{125}{100} = \dfrac{5}{4}$, $\cot \theta = -\dfrac{3}{4}$

93. $\sec \theta = \dfrac{3}{10}$, hyp = 3, adj = 10, does not exist.   **95.** $\sec \dfrac{-13\pi}{6}$, 1.1547   **97.** $\sin \dfrac{13\pi}{6}$, $\dfrac{1}{2}$ **99.** $\csc 405°$, 1.4142

Section 6.4 Graphs of Trigonometric Functions

1.

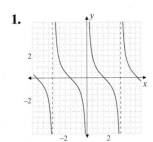

3. A = 3 inches, P = 2 seconds, $g(t) = 3 \cos \pi t$

5. A = 1.5 ft, P = $\dfrac{3}{5}$ s, $g(t) = 1.5 \cos \left(\dfrac{10}{3} \pi t\right)$ **7.** A = $\dfrac{3}{2}$, P = 2π, no phase shift

9. A = 1, P = 2π, shifted right 5 units   **11.** A = $\dfrac{1}{2}$, P = 2π, no phase shift

13. A = $\dfrac{2}{3}$, P = 2π, no phase shift   **15.** A = 3, P = 4π, no phase shift

17. $A = 1$, $P = \dfrac{2}{3}$, shifted right $\dfrac{2}{3\pi}$ units   **19.** $A = 7$, $P = 4$, shifted left $\dfrac{3}{\pi}$ units

21. $A = \dfrac{3}{4}$, $P = 2\pi$, shifted up 2 units, shifted right 3 units

23.

25.

27.

29.

31.

33.

35.

37.

39.

41.

43.

45.

47.

49.

51.

53.

55.

Section 6.5 Inverse Trigonometric Functions

1.

3.

5. $y = \dfrac{\pi}{4}$　**7.** $y = \dfrac{2\pi}{3}$　**9.** $y = -\dfrac{\pi}{6}$　**11.** $y = \pi$

13. $y = \dfrac{5\pi}{6}$　**15.** $y = \dfrac{\pi}{4}$　**17.** $y = \dfrac{\pi}{6}$　**19.** $y = \dfrac{\pi}{6}$

21. $y = \pi$　**23.** $y = 0$　**25.** $y = \dfrac{3\pi}{4}$　**27.** $y = \dfrac{5\pi}{6}$

29. Does not exist　**31.** 1.3734　**33.** 0.6747

35. Does not exist　**37.** 1.3734　**39.** 1.0472　**41.** $-\dfrac{\pi}{2}$　**43.** $-\dfrac{\pi}{6}$　**45.** $\dfrac{\pi}{4}$　**47.** 0　**49.** Does not exist

51. 0.2764　**53.** $-\dfrac{1}{2}$　**55.** 2　**57.** 2　**59.** $\dfrac{2\sqrt{3}}{3}$　**61.** $-\sqrt{3}$　**63.** $\dfrac{2\sqrt{3}}{3}$　**65.** $-\dfrac{\sqrt{3}}{3}$　**67.** 2　**69.** $\dfrac{\sqrt{3}}{2}$

71. a. 0.2898　**b.** 0.4429　**c.** 0.7956　**73.** $\dfrac{\sqrt{x^2-4}}{2}$　**75.** $\dfrac{\sqrt{3}x}{3}$　**77.** $\dfrac{4}{\sqrt{x^2+16}}$　**79.** $\theta = -14.48917361$

81. $\theta = 152.9872134$　**83.** $\theta = 1.682588837$　**85.** $\theta = 1.405647622$　**87.** $\theta = 0.7199835305$

89.

91.

Chapter 6 Review

1. $4°$ **2.** $216°$ **3.** $-315°$ **4.** $54°$ **5.** $\dfrac{7\pi}{30}$ **6.** $\dfrac{\pi}{3}$ **7.** $\dfrac{-79\pi}{180}$ **8.** $\dfrac{-5\pi}{36}$

9.

10.

11. 15.71 ft   **12.** 18.85 km   **13.** 30.72 meters

14. 10.05 inches   **15.** 1296.35 km   **16.** $\dfrac{26}{9}$ rad

17. 3 rad   **18. a.** 1200π **b.** 8.80 m/s

19. 108π ft² or 339.29 ft²

20. $\sin\theta = \dfrac{\sqrt{2}}{2}, \cos\theta = \dfrac{\sqrt{2}}{2}, \tan\theta = 1, \csc\theta = \sqrt{2}, \sec\theta = \sqrt{2}, \cot\theta = 1$

21. $\sin\theta = \dfrac{2\sqrt{5}}{5}, \cos\theta = \dfrac{\sqrt{5}}{5}, \tan\theta = 2, \csc\theta = \dfrac{\sqrt{5}}{2}, \sec\theta = \sqrt{5}, \cot\theta = \dfrac{1}{2}$ **22.** 0.9903 **23.** 4.0108 **24.** 1.0353

25. 0.2225 **26.** $36.9372°$ **27.** $15.2203°$ **28.** True   **29.** False   **30.** $10\sqrt{3}$ feet or 17.32 feet   **31.** 129.13 feet

32. $\sin 90° = 1, \cos 90° = 0, \tan 90° =$ undefined, $\csc 90° = 1, \sec 90° =$ undefined, $\cot 90° = 0$

33. $\sin(-460°) \approx -0.9848, \cos(-460°) \approx -0.1736, \tan(-460°) \approx 5.6713, \csc(-460°) \approx -1.0154,$
$\sec(-460°) \approx -5.7588, \cot(-460°) \approx 0.1763$

34. $\sin\dfrac{\pi}{4} = \cos\dfrac{\pi}{4} = \dfrac{\sqrt{2}}{2}, \tan\dfrac{\pi}{4} = \cot\dfrac{\pi}{4} = 1, \csc\dfrac{\pi}{4} = \sec\dfrac{\pi}{4} = \sqrt{2}$

35. $\sin\dfrac{7\pi}{3} = \dfrac{\sqrt{3}}{2}, \cos\dfrac{7\pi}{3} = \dfrac{1}{2}, \tan\dfrac{7\pi}{3} = \sqrt{3}, \csc\dfrac{7\pi}{3} = \dfrac{2\sqrt{3}}{3}, \sec\dfrac{7\pi}{3} = 2, \cot\dfrac{7\pi}{3} = \dfrac{\sqrt{3}}{3}$

36. $\theta' = 86°$ **37.** $\theta' = 37°$ **38.** $\theta' = \dfrac{\pi}{2}$ **39.** $\theta' = \dfrac{\pi}{4}$ **40.** I   **41.** III   **42.** $\sin 290° = \sin 70° \approx -0.9397$

43. $\tan\dfrac{4\pi}{3} = \tan\dfrac{\pi}{3} = \sqrt{3}$ **44.** $\csc 193° = \sec(-103°) \approx -4.4454$ **45.** $\sin(-42°) = \cos 132° \approx -0.6691$

46. $\cot\dfrac{3\pi}{4} = \tan\dfrac{-\pi}{4} = -1$ **47.** $\cos\dfrac{5\pi}{4} = \sin\dfrac{-3\pi}{4} = \dfrac{-\sqrt{2}}{2}$ **48.** $\theta = \dfrac{3\pi}{4}$ and $\tan\theta = -1$ **49.** $\sec\theta = \dfrac{13}{5}$

50. $A = 3, P = \dfrac{\pi}{2}$, no phase shift   **51.** $A = 6, P = 2\pi$, no phase shift   **52.** $A = -\dfrac{1}{2}, P = \dfrac{2\pi}{3}$, shifted right $\dfrac{\pi}{3}$ units

53. $A = 9, P = \pi$, shifted left π units

54.

55.

56.

57.

58.   **59.**   **60.** $y = \dfrac{\pi}{6}$ **61.** $y = \dfrac{\pi}{2}$ **62.** $y = \dfrac{-\pi}{4}$ **63.** $y = \dfrac{\pi}{3}$

64. Does not exist **65.** 0.4636 rad or

26.5651 degrees **66.** $\dfrac{\pi}{2}$ **67.** 0.9 **68.** 0.75

69. $\dfrac{-\pi}{4}$ **70.** $\dfrac{1}{2}$ **71.** $\dfrac{\sqrt{3}}{3}$ **72.** $\dfrac{\sqrt{2}}{2}$ **73.** 2

74. 1.2140171 **75.** −1.336110366 **76.** −41.81031515 **77.** 55.32339906 **78.** $\dfrac{x}{2}$

Chapter 6 Test

1. 240° **2.** $-\dfrac{73\pi}{60}$ **3.**   **4.**   **5.** 1.204 m **6.** 17.36°

7. a. 1200π rad/min
b. 125.66 in./s

8. 180π rad/min **9.** 73.55 cm²

10. $\sin\theta = \dfrac{2\sqrt{10}}{7}, \cos\theta = \dfrac{3}{7}, \tan\theta = \dfrac{2\sqrt{10}}{3}, \csc\theta = \dfrac{7\sqrt{10}}{20}, \sec\theta = \dfrac{7}{3}, \cot\theta = \dfrac{3\sqrt{10}}{20}$

11. $\tan 63° \approx 1.9626, \cot 63° \approx 0.5095$ **12.** $\sin\dfrac{3\pi}{8} \approx 0.9239; \csc\dfrac{3\pi}{8} \approx 1.0824$ **13.** 2966.02 feet **14.** 15.16 feet

15. 176.99 feet **16.** 56.93 feet **17.** $\sin(-125°) \approx -0.8192, \cos(-125°) \approx -0.5736, \tan(-125°) \approx 1.4281,$

$\csc(-125°) \approx -1.2208, \sec(-125°) \approx -1.7434, \cot(-125°) \approx 0.7002$ **18.** $\sin 5\pi = 0, \cos 5\pi = -1, \tan 5\pi = 0,$

$\csc 5\pi = $ undefined, $\sec 5\pi = -1, \cot 5\pi = $ undefined **19. a.** $\sin 179° = \sin 1°$ **b.** 0.0175 **20. a.** $\cos\dfrac{\pi}{4}$ **b.** $\dfrac{\sqrt{2}}{2}$

21. $\cot\dfrac{7\pi}{10} \approx -0.7265$ **22.** $\sec 38° \approx 1.2690$ **23.** $\sin\theta = \dfrac{-\sqrt{5}}{5}$ **24.** $\csc\theta = -\dfrac{2.6}{1} = -\dfrac{13}{5}, \cot\theta = \dfrac{5}{12}$

25. $A = 1.5, P = 1, f(t) = 1.5 \cos 2\pi t$

26.   **27.**   **28.**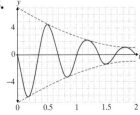

29. 0.5404 **30.** The cos function has a range of $[-1,1]$, so the expression can't be evaluated.

31. $\cos^{-1}\left(-\dfrac{1}{5.3}\right) \approx 1.7606$ **32.** $\tan^{-1}\dfrac{1}{127} \approx 0.0079$ **33.** $\dfrac{\pi}{2}$ **34.** 0.9285 **35.** $\dfrac{\sqrt{4x^2-1}}{2x}$

Trigonometric Identities and Equations

Section 7.1 Fundamental Identities and Their Uses

1. $\sec x$ **3.** $\sin t$ **5.** $\cos x - \sec x$ **7.** $\tan \alpha$ **9.** $\sec y$ **11.** $\tan^2 x$ **13.–25.** Answers will vary. **27.** $4 \tan \theta$

29. $10 \csc \theta$ **31.** $2 \cot \theta$ **33.** $12 \sin \theta$ **35.** Answers will vary.

Section 7.2 Sum and Difference Identities

1. $\dfrac{-\sqrt{6}+\sqrt{2}}{4}$ **3.** $-\sqrt{3}-2$ **5.** -1 **7.** -1 **9.** 0 **11.** 0 **13.** $\dfrac{\sqrt{6}+\sqrt{2}}{4}$ **15.** $\dfrac{\sqrt{6}+\sqrt{2}}{4}$ **17.** $\sqrt{3}+2$

19. $\dfrac{\sqrt{6}-\sqrt{2}}{4}$ **21.** $\dfrac{\sqrt{6}+\sqrt{2}}{4}$ **23.** $\sqrt{3}+2$ **25.** $\dfrac{-\sqrt{6}-\sqrt{2}}{4}$ **27.** $\dfrac{-\sqrt{6}-\sqrt{2}}{4}$ **29.** $\dfrac{\sqrt{6}-\sqrt{2}}{4}$ **31.** $\dfrac{-1}{2}$

33. $\dfrac{\sqrt{6}+\sqrt{2}}{4}$ **35.** $\dfrac{\sqrt{6}-\sqrt{2}}{4}$ **37.** $\dfrac{-\sqrt{3}}{2}$ **39.** $\dfrac{\sqrt{6}+\sqrt{2}}{4}$ **41.** $\dfrac{56}{65}$ **43.** $\dfrac{36}{85}$ **45.** $\dfrac{4\sqrt{6}-\sqrt{21}}{25}$ **47.** $\sin 45° = \dfrac{\sqrt{2}}{2}$

49. $\tan 135° = -1$ **51.** $\tan \dfrac{\pi}{4} = 1$ **53.** $\sin 150° = \dfrac{1}{2}$ **55.** $\cos 135° = \dfrac{-\sqrt{2}}{2}$ **57.** $\tan 60° = \sqrt{3}$

59.–67. Answers will vary. **69.** True **71.** False **73.** True **75.** 1 **77.** $\dfrac{2x^2 + \sqrt{1-x^2}}{\sqrt{1+4x^2}}$

79. $x\left(\sqrt{1-4x^2} - 2\sqrt{1-x^2}\right)$

81. $\sqrt{2}\sin\left(x+\dfrac{\pi}{4}\right)$ **83.** $\sqrt{2}\sin\left(2\beta - \dfrac{\pi}{4}\right)$ **85.** $13\sin(5u + 1.1760)$

  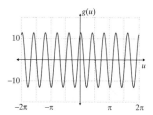

87. Answers will vary. **89.** True **91.** False **93.** Answers will vary.

Section 7.3 Product-Sum Identities

1. $\cos 2x = \dfrac{7}{25}; \sin 2x = \dfrac{24}{25}; \tan 2x = \dfrac{24}{7}$ **3.** $\cos 2x = \dfrac{1}{3}; \sin 2x = \dfrac{-2\sqrt{2}}{3}; \tan 2x = -2\sqrt{2}$

5. $\cos 2x = \dfrac{1}{2}; \sin 2x = \dfrac{\sqrt{3}}{2}; \tan 2x = \sqrt{3}$ **7.–11.** Answers will vary.

13. $\dfrac{\sin x - \sin x \cos 2x}{2}$ **15.** $\dfrac{1 - \cos 2x - \cos 4x + \cos 2x \cos 4x}{16}$

17. $\dfrac{3\sin x - 4\sin x \cos 2x + \sin x \cos 4x}{3 + 4\cos 2x + \cos 4x}$ **19.** $\dfrac{\sqrt{2 + \sqrt{2}}}{2}$ **21.** $\dfrac{\sqrt{2 + \sqrt{3}}}{2}$ **23.** $\dfrac{\sqrt{2 + \sqrt{3}}}{2}$

25. $\dfrac{\sin 6x + \sin 0}{2}$ **27.** $\dfrac{5\left(\sin 120° - \sin 90°\right)}{2}$ **29.** $\dfrac{\cos 2y - \cos 2x}{2}$ **31.** $\dfrac{\cos \dfrac{7\pi}{12} - \cos \dfrac{23\pi}{12}}{2}$ **33.** $\sin \dfrac{\pi}{2} - \sin \dfrac{\pi}{6}$

35. $2\sin 4x \cos 2x$ **37.** $-2\sin 2\beta \sin \beta$ **39.** $2\cos 75° \sin 60°$ **41.** $-2\sin \dfrac{17\pi}{24} \sin \dfrac{11\pi}{24}$

43. Answers will vary.   **45.** Answers will vary.   **47.** $\cos 2x = 2\cos^2 x - 1$; $\cos 3x = 4\cos^3 x - 3\cos x$;

$\cos 4x = 8\cos^4 x - 8\cos^2 x + 1$; $\sin 2x = 2\sin x \cos x$; $\sin 3x = \sin x \left(4\cos^2 x - 1\right)$; $\sin 4x = \sin x \left(8\cos^3 x - 4\cos x\right)$

Section 7.4 Trigonometric Equations

1. $x = \dfrac{7\pi}{6} + 2n\pi$, $x = \dfrac{11\pi}{6} + 2n\pi$ **3.** $x = \dfrac{\pi}{4} + 2n\pi$, $x = \dfrac{7\pi}{4} + 2n\pi$ **5.** $x = \dfrac{\pi}{6} + 2n\pi$, $x = \dfrac{11\pi}{6} + 2n\pi$

7. No solution   **9.** $x = \dfrac{2\pi}{3} + n\pi$ **11.** $x = n\pi$ **13.** $x = 2n\pi$ **15.** No solution   **17.** $x = \dfrac{2\pi}{3} + 2n\pi$, $x = \dfrac{4\pi}{3} + 2n\pi$

19. $x = \dfrac{5\pi}{12} + \dfrac{n\pi}{12}$ **21.** $x = \dfrac{\pi}{6}, \dfrac{5\pi}{6}$ **23.** $x = \dfrac{\pi}{4}, \dfrac{3\pi}{4}, \dfrac{5\pi}{4}, \dfrac{7\pi}{4}$ **25.** $x = 0, \pi, 2\pi$ **27.** $x = \dfrac{2\pi}{3}, \dfrac{4\pi}{3}$

29. $x = \tan^{-1}\left(-4 + \sqrt{59}\right), \tan^{-1}\left(-4 - \sqrt{59}\right)$ **31.–39.** Answers will vary.   **41.** 1.1925 and 3.8088

43. 0.8959 and 2.2457 **45.** 0.9169 and 2.2247 **47.** $0.4043, 3.7535$, and 5.4549 **49.** True   **51.** True   **53.** True

55. $x = \dfrac{2\pi}{3} + n\pi$ **57.** True   **59.** $x = \dfrac{\pi}{2}, \dfrac{7\pi}{6}, \dfrac{11\pi}{6}$ **61.** $x = \dfrac{\pi}{2}, \pi$ **63.** $x = 0, \dfrac{2\pi}{3}, \dfrac{4\pi}{3}$ **65.** $x = \dfrac{\pi}{4}, \dfrac{5\pi}{4}$

67. $x = 45°, 135°, 225°, 315°$ **69.** $x = 90°, 210°, 330°$ **71.** No solution   **73.** $x = 30°, 210°, 240°, 300°$

75. $s = -4, 3$; t has no solution.   **77.** $s = \dfrac{1 \pm \sqrt{2}}{2}$; $t = 1.7794, 4.5038$ **79.** They can have a maximum of four

solutions.   **81.** $x = 10°, 50°, 130°, 170°, 250°, 290°$ **83.** They may have as many as $2a$ solutions from $0°$ to $360°$.

85. $17.3°$

Chapter 7 Review

1. $\csc x$ **2.** $\cos^2 x$ **3.** $-\tan^2 y$ **4.** 1 **5.** $-1 - \sin \theta$ **6.** $-\sin^2 x$ **7.–12.** Answers will vary.

13. $4\sec \theta$ **14.** $8\cos \theta$ **15.** $10\cot \theta$ **16.** $6\sec \theta$ **17.** $\dfrac{\sqrt{3}}{2}$ **18.** $\dfrac{\sqrt{2} - \sqrt{6}}{4}$ **19.** $\dfrac{\sqrt{2} - \sqrt{6}}{4}$ **20.** $\dfrac{-\sqrt{6} - \sqrt{2}}{4}$

21. $\sqrt{3}$ **22.** $-2 - \sqrt{3}$ **23.** $\dfrac{13}{84}$ **24.** $\dfrac{16}{65}$ **25.** $\sin 210°; -\dfrac{1}{2}$ **26.** $\tan \dfrac{3\pi}{4}; -1$ **27.** Answers will vary.

28. Answers will vary.   **29.** $\dfrac{\sqrt{1 - x^2} - x^2}{\sqrt{1 + x^2}}$ **30.** $\dfrac{2x\left(1 - \sqrt{1 - 4x^2}\right)}{\sqrt{1 + 4x^2}}$ **31.** $2\sin\left(x + \dfrac{7\pi}{4}\right)$ **32.** $2\sin\left(4\alpha - \dfrac{\pi}{6}\right)$

33. $\cos 2x = \dfrac{-7}{25}$; $\sin 2x = \dfrac{24}{25}$; $\tan 2x = \dfrac{-24}{7}$ **34.** $\cos 2x = \dfrac{4}{5}$; $\sin 2x = \dfrac{3}{5}$; $\tan 2x = \dfrac{3}{4}$ **35.** Answers will vary.

36. Answers will vary.   **37.** $\dfrac{\sin x - \sin x \cos 4x}{8}$ **38.** $\dfrac{3\sin x - 4\cos 2x \sin x + \cos 4x \sin x}{4 + 4\cos 2x}$ **39.** $2 + \sqrt{3}$

40. $-\dfrac{1}{2}\sqrt{2+\sqrt{2}}$ **41.** $2 - \sqrt{3}$ **42.** $-\dfrac{1}{2}\sqrt{2+\sqrt{2}}$ **43.** $\dfrac{\sin 2x - \sin 2y}{2}$ **44.** $\dfrac{\cos \frac{11\pi}{2} + \cos \frac{7\pi}{2}}{2}$

45. $\dfrac{1}{2}\left(\sin 180° + \sin 150°\right)$ **46.** $\dfrac{1}{2}\left(\cos x - \cos 7x\right)$ **47.** $2\cos 4\alpha \sin \alpha$ **48.** $2\cos 120° \cos 105°$

49. $-2\sin \dfrac{17\pi}{24}\sin \dfrac{\pi}{24}$ **50.** $2\sin \dfrac{7\pi}{12}\cos \dfrac{\pi}{4}$ **51.** $x = \dfrac{\pi}{6} + 2n\pi$, $x = \dfrac{-\pi}{6} + 2n\pi$ **52.** $x = \dfrac{\pi}{6} + 2n\pi$, $x = \dfrac{5\pi}{6} + 2n\pi$, $x = n\pi$

53. $x = \dfrac{\pi}{2} + 2n\pi$, $x = \dfrac{-\pi}{2} + 2n\pi$ **54.** $x = n\pi$, $x = \dfrac{\pi}{4} + n\pi$ **55.** $x = \dfrac{\pi}{3} + 2n\pi$, $x = \dfrac{5\pi}{3} + 2n\pi$, $x = \pi + 2n\pi$

56. $x = \dfrac{7\pi}{6} + 2n\pi$, $x = \dfrac{11\pi}{6} + 2n\pi$ **57.** $x = \dfrac{\pi}{6}, \dfrac{5\pi}{6}, \dfrac{7\pi}{6}, \dfrac{11\pi}{6}$ **58.** $x = \dfrac{\pi}{2}$ **59.** True

60. $x = \dfrac{\pi}{6} + 2n\pi$, $x = \dfrac{5\pi}{6} + 2n\pi$ **61.** $x = \dfrac{\pi}{3} + 2n\pi$, $x = \dfrac{2\pi}{3} + 2n\pi$ **62.** True   **63.** $x = 0°, 180°$ **64.** $x = 60°, 300°$

Chapter 7 Test

1. $\sin^2 x - 5\csc^2 x$ **2.** -1 **3.** Answers will vary.   **4.** Answers will vary.   **5.** $13\sec\theta$ **6.** $8\sqrt{2}\cos\theta$

7. Answers will vary.   **8.** $\dfrac{\sqrt{6}-\sqrt{2}}{4}$ **9.** $\sqrt{3}-2$ **10.** $\cos 103°, -0.2250$ **11.** $\tan \dfrac{7\pi}{8}, -0.4142$

12. Answers will vary.   **13.** Answers will vary.   **14.** $\dfrac{2x + x\sqrt{1-4x^2}}{\sqrt{1+x^2}}$ **15.** $h(\beta) = 4\sin\left(4\beta + \dfrac{\pi}{6}\right)$

16. $\cos 2x = \dfrac{7}{9}$; $\sin 2x = \dfrac{-4\sqrt{2}}{9}$; $\tan 2x = \dfrac{-4\sqrt{2}}{7}$ **17.** Answers will vary.   **18.** $\dfrac{3\cos x + 4\cos x \cos 2x + \cos x \cos 4x}{8}$

19. $\dfrac{\sqrt{2-\sqrt{3}}}{2}$ **20.** $\dfrac{\sqrt{2+\sqrt{3}}}{2}$ **21.** $\dfrac{\sin 10x}{2}$ **22.** $\dfrac{\sin \frac{6\pi}{5}}{2}$ **23.** $2\sin \dfrac{x}{2}\cos \dfrac{x}{3}$ **24.** $2\cos 4x \cos x$

25. $x = \dfrac{\pi}{6} + n\pi$; $x = \dfrac{\pi}{3} + n\pi$ **26.** $x = \dfrac{\pi}{6} + 2n\pi$; $x = \dfrac{5\pi}{6} + 2n\pi$; $x = \dfrac{\pi}{2} + 2n\pi$ **27.** $x = \dfrac{n\pi}{2}$, where n is odd

28. $x = \dfrac{\pi}{3} + n\pi$; $x = \dfrac{3\pi}{4} + n\pi$ **29.** $x = \sin^{-1}\left(\dfrac{-2+\sqrt{34}}{5}\right) + 2n\pi$ **30.** $x = \cos^{-1}\left(\dfrac{-2}{3}\right) + 2n\pi$

31. $x = \tan^{-1}\left(-2+\sqrt{10}\right) + n\pi$, $x = \tan^{-1}\left(-2-\sqrt{10}\right) + n\pi$ **32.** $x = \dfrac{7\pi}{6} + 2n\pi$; $x = \dfrac{11\pi}{6} + 2n\pi$

8 Additional Topics in Trigonometry

Section 8.1 The Law of Sines and the Law of Cosines

1. 934.2 miles   **3.** 31.6 feet   **5.** 28.3 feet   **7.** 156.8 feet   **9.** 2835.6 feet   **11.** 98.1 feet   **13.** 105.5 miles

15. $C = 105°, b \approx 4.2426, c \approx 5.7956$ **17.** $C = 60°, a \approx 4.9067, c \approx 4.5221$ **19.** $A = 80°, a \approx 3.9392, b \approx 3.7588$

21. $C = 150°, b \approx 1.0154, c \approx 2.9238$ **23.** No triangle   **25.** $A \approx 20.7048°, B \approx 114.2952°, b \approx 5.1559$ **27.** No triangle

29. No triangle   **31.** $c \approx 11.54, B \approx 31.31°, C \approx 88.69°$ **33.** $C = 103.1°, a \approx 83.23, c \approx 90.42$

 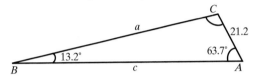

35. $b \approx 6.78, A \approx 60.90°, B \approx 19.10°$ **37.** $h \approx 0.98$. Since $0.98 < 1.5 < 2.4$, there can be two triangles.

Triangle 1: $a \approx 3.33, A \approx 115.40°, B \approx 40.60°$

Triangle 2: $a \approx 1.05, A \approx 16.60°, B \approx 139.40°$

 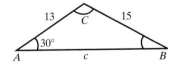

39. $C = 58°54', a \approx 2.64, b \approx 3.48$ **41.** $c \approx 24.78, C \approx 124.32°, B \approx 25.68°$ **43.** $a = \sqrt{37}, B \approx 25.2850°, C \approx 94.7150°$

 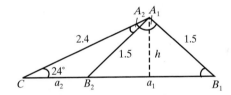

45. $b \approx 4.5985, A \approx 41.7854°, C \approx 88.2146°$ **47.** $c \approx 4.1063, A \approx 103.0643°, B \approx 46.9357°$

49. $c \approx 7.0752, A \approx 41.6113°, B \approx 68.3887°$ **51.** $A \approx 46.5675°, B \approx 104.4775°, C \approx 28.9550°$

53. $A \approx 121.8554°, B \approx 39.5712°, C \approx 18.5734°$ **55.** $A = B = C = 60°$ **57.** $A = 90°, B \approx 36.8699°, C \approx 53.1301°$

59. 65.5744 feet   **61.** 2.4×10^7 miles   **63.** 21.4413 feet   **65.** $22.6199°, 67.3801°$ **67.** $160.1188°$

69. $c \approx 8.05$, $A \approx 86.21°$, $B \approx 58.79°$ **71.** $c \approx 20.04$, $B \approx 44°15'$, $A \approx 89°38'$ **73.** $c \approx 19.65$, $A \approx 55°42'$, $B \approx 49°14'$

75. $A \approx 82.62°$, $B \approx 80.72°$, $C \approx 16.66°$ **77.** $A \approx 41.93°$, $B \approx 38.33°$, $C \approx 99.74°$ **79.** $A \approx 27.66°$, $B \approx 111.80°$, $C \approx 40.54°$

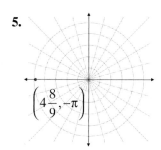

81. 46.82 **83.** 28.32 **85.** 136.67 **87.** 89.29 **89.** 85.5951 ft² **91.** 178.3882 ft² **93. a.** 61.9372 in.²

b. 584.2397 in.² **c.** 136.1041 in.²

Section 8.2 Polar Coordinates and Polar Equations

1.   **3.**   **5.**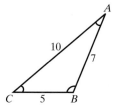

7. $(3.54, -3.54)$ **9.** $(-4.42, -4.42)$ **11.** $(-2.6, -1.5)$ **13.** $(-3, 0)$ and $(3, \pi)$

15. $\left(\sqrt{145}, -0.08\right)$ and $\left(-\sqrt{145}, 3.06\right)$ **17.** $\left(2\sqrt{21}, 1.76\right)$ and $\left(-2\sqrt{21}, -1.38\right)$ **19.** $r^2 = 25$ **21.** $r\cos\theta = 12$

23. $\sin\theta = \cos\theta$ **25.** $r\cos\theta = 16a$ **27.** $r^2 - 4ar\cos\theta = 0$ **29.** $r^2\sin^2\theta - 4r\cos\theta - 4 = 0$ **31.** $x^2 + y^2 = 5x$

33. $x^2 + y^2 = 49$ **35.** $y = \dfrac{1}{2}$ **37.** $x^4 + y^4 + 2x^2y^2 = 2xy$ **39.** $4y + 7x = 12$

41. $x^2 + y^2 = 9$ **43.** $y = -\dfrac{x}{\sqrt{3}}$ **45.** $x = 7$

47.

49.

51.

53.

55.

57.

59.

61.

63.

65.

67.

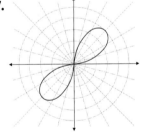

Section 8.3 Parametric Equations

1.

t	x	y
0	5	0
1	6	−1
2	7	undefined
3	8	$\sqrt{3}$
4	9	1
5	10	$\dfrac{\sqrt{5}}{3}$
6	11	$\dfrac{\sqrt{6}}{4}$

3. a. $x = 14.05t,$
$\quad y = -16t^2 + 15.61t + 7$

c. No, he won't

b.

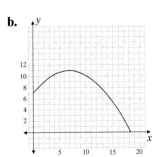

5. a. $x = 58.67t$
$y = -16t^2 + 101.61t + 10$

b.

c. 126.42 ft   **d.** 378.42 ft

e. 6.45 s   **f.** Yes

7. $y = 3x^2 + 4, x \geq 0$

9. $x = |y + 8|$

11. $y = \sqrt{\dfrac{2x}{1-x}}$

13. $x = \dfrac{4}{|y-5|}$

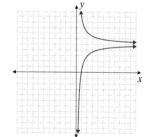

15. $y = \dfrac{\pm\sqrt{8 - 2x - x^2}}{6}$

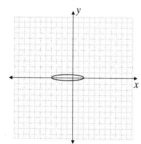

17. $y = \dfrac{3}{2}x, -2 \leq x \leq 2$

19. $y = 4 \pm 3\sqrt{1 - x^2}$

21. $x = t, y = 5t - 2$ **23.** $x = t, y = \pm 2\sqrt{1 - t^2}$ **25.** $x = t, y = t^2 + 1$ **27.** $x = t, y = \dfrac{6+t}{4}$ **29.** $x = t, y = \dfrac{t+6}{2}$

31. $x = t, y = \dfrac{1}{3t}$ **33.** $x = t, y = -2t - 12$ **35.** $x = t, y = 3t - 19$ **37.** $x = t, y = -\dfrac{3}{2}t + 6$ **39.** $x = 2 + t, y = 3t + 7$

41. $x = t + 1, y = -t^2 - 2t + 4$ **43.** $x = t - 4, y = \dfrac{t^2}{2} - 4t + 14$ **45.** $x = \cos\theta, y = \sin\theta$

47. $x = 7 + 4\cos\theta, y = -5 + 4\sin\theta$ **49.** $x = 12(\theta - \sin\theta), y = 12(1 - \cos\theta)$

Section 8.4 Trigonometric Form of Complex Numbers

1. $\sqrt{34}$

3. $\sqrt{20} = 2\sqrt{5}$

5. $\sqrt{32} = 4\sqrt{2}$

7.

9.

11.

13.

15.

17. $\sqrt{10}\left(\cos 3.46 + i\sin 3.46\right)$

19. $\sqrt{5}\left(\cos 1.11 + i\sin 1.11\right)$

21. $2\sqrt{5}\left(\cos 0.46 + i\sin 0.46\right)$

23. $2\left(\cos\left(-\dfrac{\pi}{4}\right) + i\sin\left(-\dfrac{\pi}{4}\right)\right)$ **25.** $5(\cos 0.93 + i\sin 0.93)$ **27.** $8\left(\cos\left(-\dfrac{\pi}{3}\right) + i\sin\left(-\dfrac{\pi}{3}\right)\right)$ **29.** $\dfrac{-3\sqrt{3}}{2} + \dfrac{3i}{2}$

31. $-1 - i\sqrt{3}$ **33.** $-\dfrac{5}{\sqrt{2}} + \dfrac{5i}{\sqrt{2}}$ **35.** $\dfrac{-3\sqrt{3}}{4} + \dfrac{3i}{4}$ **37.** $1.01 + 4.9i$ **39.** $16(\cos 30° + i\sin 30°) = 8\sqrt{3} + 8i$

41. $3\sqrt{6}\left(\cos\dfrac{17\pi}{12} + i\sin\dfrac{17\pi}{12}\right) = -1.9 - 7.1i$ **43.** $2\sqrt{10}\left(\cos 2.42 + i\sin 2.42\right) = \left(-3 - \sqrt{3}\right) + i\left(3\sqrt{3} - 1\right)$

45. $2(\cos 180° + i\sin 180°) = -2$ **47.** $\dfrac{10}{3}\left(\cos\dfrac{\pi}{2} + i\sin\dfrac{\pi}{2}\right) = \dfrac{10i}{3}$ **49.** $\dfrac{1}{\sqrt{2}}\left(\cos\left(-\dfrac{3\pi}{4}\right) + i\sin\left(-\dfrac{3\pi}{4}\right)\right) = -\dfrac{1}{2} - \dfrac{i}{2}$

51. $2\left(\cos\dfrac{5\pi}{12} + i\sin\dfrac{5\pi}{12}\right) = 0.52 + 1.93i$

53.

55.

57.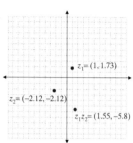

59. $32e^{\frac{\pi}{3}i}$

61. $1.04 \times 10^{13} e^{2.9i}$ **63.** $e^{2\pi i}$ **65.** $e^{\frac{\pi i}{4}}, e^{\frac{3\pi i}{4}}, e^{\frac{5\pi i}{4}}, e^{\frac{7\pi i}{4}}$ **67.** $2e^{\frac{\pi i}{12}}, 2e^{\frac{13\pi i}{12}}$

69. $4, 4e^{\frac{\pi i}{2}}, 4e^{\pi i}, 4e^{\frac{3\pi i}{2}}$ **71.** $2e^{60°i}, 2e^{240°i}$ **73.** $2\sqrt{2}e^{\frac{\pi i}{12}}, 2\sqrt{2}e^{\frac{13\pi i}{12}}$

75. $2e^{\frac{\pi i}{5}}, 2e^{\frac{3\pi i}{5}}, 2e^{\pi i}, 2e^{\frac{7\pi i}{5}}, 2e^{\frac{9\pi i}{5}}$ **77.** $5e^{\frac{\pi i}{4}}, 5e^{\frac{3\pi i}{4}}$

Section 8.5 Vectors in the Cartesian Plane

1.

3.

5.

7. $\mathbf{v} = \langle 3, -3 \rangle, \|\mathbf{v}\| = 3\sqrt{2}$ **9.** $\mathbf{v} = \langle 5, 3 \rangle, \|\mathbf{v}\| = \sqrt{34}$ **11.** $\mathbf{v} = \langle 5, -1 \rangle, \|\mathbf{v}\| = \sqrt{26}$ **13.** $\mathbf{v} = \langle -7, 7 \rangle, \|\mathbf{v}\| = 7\sqrt{2}$

15. $\mathbf{v} = \langle -4, -6 \rangle, \|\mathbf{v}\| = 2\sqrt{13}$ **17. a.** $\langle -2, 8 \rangle$ **b.** $\langle 8, -4 \rangle$ **c.** $\langle -4, 0 \rangle$ **19. a.** $\langle 1, 4 \rangle$ **b.** $\langle -11, 12 \rangle$ **c.** $\langle 6, -8 \rangle$

21. a. $\langle -5, -10 \rangle$ **b.** $\langle -8, -2 \rangle$ **c.** $\langle 6, 4 \rangle$ **23.** $-\mathbf{u} = \langle -1, -1 \rangle$, $2\mathbf{u} - \mathbf{v} = \langle -1, 5 \rangle$, $\mathbf{u} + \mathbf{v} = \langle 4, -2 \rangle$, $\|\mathbf{u}\| = \sqrt{2}$, $\|\mathbf{v}\| = 3\sqrt{2}$

25. $-\mathbf{u} = \langle 4, -4 \rangle$, $2\mathbf{u} - \mathbf{v} = \langle -12, 12 \rangle$, $\mathbf{u} + \mathbf{v} = \langle 0, 0 \rangle$, $\|\mathbf{u}\| = 4\sqrt{2}$, $\|\mathbf{v}\| = 4\sqrt{2}$ **27. a.** $\left\langle \dfrac{2}{\sqrt{5}}, -\dfrac{1}{\sqrt{5}} \right\rangle$ **b.** $\mathbf{u} = 6\mathbf{i} - 3\mathbf{j}$

29. a. $\left\langle \dfrac{-5}{\sqrt{26}}, -\dfrac{1}{\sqrt{26}} \right\rangle$ **b.** $\mathbf{u} = -5\mathbf{i} - \mathbf{j}$ **31. a.** $\left\langle \dfrac{2}{\sqrt{13}}, \dfrac{3}{\sqrt{13}} \right\rangle$ **b.** $\mathbf{u} = 2\mathbf{i} + 3\mathbf{j}$ **33.** $\|\mathbf{v}\| = 5, \theta = 30°$

35. $\|\mathbf{v}\| = 5, \theta = 36.9°$ **37.** $\langle 3\sqrt{3}, 3 \rangle$ **39.** $\langle -9\sqrt{2}, 9\sqrt{2} \rangle$ **41.** $\left\langle -\dfrac{1}{2}, \dfrac{\sqrt{3}}{2} \right\rangle$ **43.** $\left\langle \dfrac{8}{\sqrt{13}}, \dfrac{12}{\sqrt{13}} \right\rangle$ **45.** $\langle 2\sqrt{3}, 2 \rangle$

47. 38.67 mph, N 77.76° W   **49.** 1244.08 pounds

Section 8.6 The Dot Product and Its Uses

1. 17   **3.** 6   **5.** 8   **7.** 1   **9.** -7 **11.** 32   **13.** $\sqrt{13} + 2$ **15.** $\sqrt{37}$ **17.** $\sqrt{53}$ **19.** 123.7°   **21.** 14.0°

23. 161.6°   **25.** $\dfrac{\pi}{4}$ **27.** 8°, 69°, 103°   **29.** 57.1°, 60.8°, 62.1°   **31.** -62.5 **33.** $-32\sqrt{2}$ **35.** $\langle 1, 1 \rangle, \langle 5, 5 \rangle$

37. $\langle 3, 1 \rangle, \langle -6, -2 \rangle$ **39.** Neither   **41.** Orthogonal   **43.** $\text{proj}_{\mathbf{v}}\mathbf{u} = \langle 2, 1 \rangle$, $\text{perp}_{\mathbf{v}}\mathbf{u} = \langle -1, 2 \rangle$

45. $\text{proj}_{\mathbf{v}}\mathbf{u} = \left\langle \dfrac{6}{5}, \dfrac{2}{5} \right\rangle$, $\text{perp}_{\mathbf{v}}\mathbf{u} = \left\langle \dfrac{9}{5}, -\dfrac{27}{5} \right\rangle$ **47.** $\text{proj}_{\mathbf{v}}\mathbf{u} = \left\langle -\dfrac{60}{17}, -\dfrac{15}{17} \right\rangle$, $\text{perp}_{\mathbf{v}}\mathbf{u} = \left\langle \dfrac{9}{17}, -\dfrac{36}{17} \right\rangle$ **49.** 14   **51.** 3

53. 3479.3 pounds   **55.** 109.6 pounds   **57.** 579,555.5 ft-lb

Chapter 8 Review

1. Height = 22.5762 ft, angle = 15.8886°

2. $C = 105°, c \approx 5.46, a \approx 2.83$

3. $A \approx 75.40°, B \approx 47.6°, b \approx 11.45$ or
$A \approx 104.6°, B \approx 18.4°, b \approx 4.89$

4. $B = 68°40', a \approx 36.80, b \approx 35.60$

5. $A \approx 64.99°$,
$B \approx 37.01°$,
$a \approx 12.04$

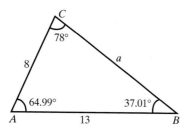

6. $a \approx 9.43$,
$B \approx 48.52°$,
$C \approx 69.48°$

7. $A \approx 22°31'$,
$C \approx 63°22'$,
$b \approx 15.62$

8. $A \approx 126.01°$,
$B \approx 12.99°$,
$C \approx 41°$

9. $A \approx 52.04°$,
$B \approx 101.99°$,
$C \approx 25.98°$

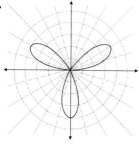

10. $(-1.725, -2.99)$ **11.** $(-6.06, -3.5)$

12. $\left(2, \dfrac{7\pi}{6}\right)$ and $\left(-2, \dfrac{\pi}{6}\right)$

13. $(15.62, 0.88)$ and $(-15.62, -2.26)$

14. $r^2 = 16a^2$ **15.** $r^2 - 9ar\cos\theta = 0$

16. $x^2 - 2x + y^2 = 0$ **17.** $x + y = 4$

18.

19.

20. $y = \dfrac{1}{1296x^2}$ $(t \neq 0)$

21. $y = |x - 7|$

22. $y = \dfrac{3}{2x} - 1$ $\left(t \neq \dfrac{1}{2}\right)$

23. $\dfrac{x^2}{16} + (y - 1)^2 = 1$

24. $x = t$, $y^2 = t^2 + 4$ **25.** $x = t$, $y = 2 - 6t$

26. $x = t,\ y = \dfrac{12}{17}t - \dfrac{100}{17}$

27. $x = 1 + \cos\theta,\ y = 1 - \sin\theta$

28. $\sqrt{29}$

29. $3\sqrt{2}$

30.

31.

32.

33.

34. $\sqrt{21}\left(\cos(-0.71) + i\sin(-0.71)\right)$

35. $\sqrt{17}\left(\cos 1.33 + i\sin 1.33\right)$

36. $2\sqrt{2} - 2\sqrt{2}i$ **37.** $\dfrac{3}{2} + \dfrac{3i\sqrt{3}}{2}$

38. $12\left(\cos\dfrac{11\pi}{6} + i\sin\dfrac{11\pi}{6}\right),\ 6\sqrt{3} - 6i$ **39.** $5\left(\cos 120° + i\sin 120°\right),\ -\dfrac{5}{2} + \dfrac{5i\sqrt{3}}{2}$ **40.** $\cos\dfrac{7\pi}{6} + i\sin\dfrac{7\pi}{6},\ -\dfrac{\sqrt{3}}{2} - \dfrac{i}{2}$

41. $24\left(\cos 315° + i\sin 315°\right),\ 12\sqrt{2} - 12i\sqrt{2}$ **42.** 64 **43.** $177{,}147 e^{120°\, i}$

44. $12e^{i\left(-\frac{\pi}{4}\right)},\ 12e^{i\left(\frac{3\pi}{4}\right)}$

45. $5e^{i\left(\frac{7\pi}{12}\right)},\ 5e^{i\left(\frac{5\pi}{4}\right)},\ 5e^{i\left(\frac{23\pi}{12}\right)}$

46. $2^{\frac{1}{8}}e^{i\left(-\frac{\pi}{16}\right)},\ 2^{\frac{1}{8}}e^{i\left(\frac{7\pi}{16}\right)},\ 2^{\frac{1}{8}}e^{i\left(\frac{15\pi}{16}\right)},\ 2^{\frac{1}{8}}e^{i\left(\frac{23\pi}{16}\right)}$

47. $2e^{i\left(\frac{\pi}{4}\right)},\ 2e^{i\left(\frac{11\pi}{12}\right)},\ 2e^{i\left(\frac{19\pi}{12}\right)}$

48. $\mathbf{v} = \langle 5,\, -5\rangle;\ \|\mathbf{v}\| = 5\sqrt{2}$

49. $\mathbf{v} = \langle -10,\, -6\rangle;\ \|\mathbf{v}\| = 2\sqrt{34}$

50.

51.

52. $-\mathbf{u} = \langle -4, -6 \rangle$; $2\mathbf{u} - \mathbf{v} = \langle 7, 14 \rangle$;

$\mathbf{u} + \mathbf{v} = \langle 5, 4 \rangle$; $\|\mathbf{u}\| = 2\sqrt{13}$; $\|\mathbf{v}\| = \sqrt{5}$

53. $-\mathbf{u} = \langle -5, -1 \rangle$; $2\mathbf{u} - \mathbf{v} = \langle 7, 1 \rangle$;

$\mathbf{u} + \mathbf{v} = \langle 8, 2 \rangle$; $\|\mathbf{u}\| = \sqrt{26}$; $\|\mathbf{v}\| = \sqrt{10}$

54. a. $\left\langle \dfrac{-4}{\sqrt{41}}, \dfrac{5}{\sqrt{41}} \right\rangle$ **b.** $-4\mathbf{i} + 5\mathbf{j}$ **55. a.** $\left\langle \dfrac{2}{\sqrt{5}}, \dfrac{1}{\sqrt{5}} \right\rangle$ **b.** $6\mathbf{i} + 3\mathbf{j}$ **56.** $\|\mathbf{v}\| = 4, \theta = 135°$ **57.** $\|\mathbf{v}\| = \sqrt{26}, \theta = -11.3°$

58. $\left(\sqrt{2}, \sqrt{6} \right)$ **59.** $\langle 3.6, -4.8 \rangle$ **60.** $\langle 45\sqrt{2}, 45\sqrt{2} \rangle$ **61.** 40.01 mph, W 49.77° N **62.** -54 **63.** $\langle -108, -270 \rangle$

64. $\sqrt{13}$ **65.** $\sqrt{10}$ **66.** 120.96° **67.** $\dfrac{5\pi}{12}$ **68.** 16 **69.** -36 **70.** $\text{proj}_{\mathbf{v}}\mathbf{u} = \left\langle \dfrac{-1}{2}, \dfrac{5}{2} \right\rangle$, $\text{perp}_{\mathbf{v}}\mathbf{u} = \left\langle \dfrac{5}{2}, \dfrac{1}{2} \right\rangle$

71. $\text{proj}_{\mathbf{v}}\mathbf{u} = \left\langle \dfrac{3}{2}, \dfrac{3}{2} \right\rangle$, $\text{perp}_{\mathbf{v}}\mathbf{u} = \left\langle \dfrac{5}{2}, \dfrac{-5}{2} \right\rangle$ **72.** 7 **73.** 31 **74.** 3449.4 lb **75.** 417,558.5 ft-lb

Chapter 8 Test

1. 6.5 feet **2.** $B \approx 20.05°, C \approx 119.95°, c \approx 20.22$ **3.** $h \approx 13.93, a < h$, so triangle cannot be constructed.

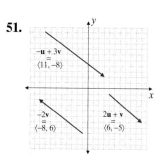

4. $A \approx 104.48°, B \approx 28.96°, C \approx 46.57°$ **5.** $A \approx 18.19°, B \approx 128.68°, C \approx 33.12°$

6. 11.62 **7.** 38.57 **8.** $\left(-3\sqrt{2}, 3\sqrt{2} \right)$ **9.** $\left(-\sqrt{5}, -1.11 \right)$ and $\left(\sqrt{5}, 2.03 \right)$ **10.** $r^2 = 49$ **11.** $y = \dfrac{1}{4}$

12.

13.

14.

15. $x = t, \ y = 8t - 3$

16. $x = t, \ y = \pm\sqrt{t - 3}$

17. $x = t, \ y = -3t + 13$

18. $x = 1 + 3\cos\theta$
$y = 2 + 3\sin\theta$

19.

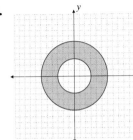

20. $2\left(\cos\dfrac{11\pi}{6}+i\sin\dfrac{11\pi}{6}\right)$ **21.** $2\sqrt{10}\left(\cos 4.12+i\sin 4.12\right)$ **22.** $3-3i\sqrt{3}$

23. $\dfrac{-5\sqrt{2}}{2}-\dfrac{-5i\sqrt{2}}{2}$ **24.** $10\left(\cos\dfrac{5\pi}{6}+i\sin\dfrac{5\pi}{6}\right)=-5\sqrt{3}+5i$

25. $\dfrac{1}{2}\left(\cos 45°+i\sin 45°\right)=\dfrac{\sqrt{2}}{4}+i\dfrac{\sqrt{2}}{4}$ **26.** $8e^{\frac{\pi}{2}i}$ **27.** $\mathbf{v}=\langle 1,-5\rangle,\ \|\mathbf{v}\|=\sqrt{26}$

28. $\mathbf{v}=\langle 2,2\rangle,\ \|\mathbf{v}\|=2\sqrt{2}$ **29. a.** $\langle 6,12\rangle$ **b.** $\langle 5,-4\rangle$ **c.** $\langle -6,0\rangle$

30. a. $\left\langle\dfrac{1}{\sqrt{5}},\dfrac{2}{\sqrt{5}}\right\rangle$ **b.** $\mathbf{u}=2\mathbf{i}+4\mathbf{j}$ **31.** $\|\mathbf{v}\|=6,\ \theta=210°$ **32.** $\langle 65.11, 54.64\rangle$ **33.** $\langle 18,6\rangle$ **34.** $3\sqrt{10}+1$

35. $33.7°$ **36.** $63.4°$ **37.** 12 **38.** $\mathrm{proj}_v\,\mathbf{u}=\left\langle\dfrac{16}{13},\dfrac{24}{13}\right\rangle,\ \mathrm{perp}_v\,\mathbf{u}=\left\langle\dfrac{-3}{13},\dfrac{2}{13}\right\rangle$

9   Conic Sections

Section 9.1 The Ellipse

1. Center: $(5,2)$; Foci: $\left(5,2\pm\sqrt{21}\right)$; Vertices: $(5,7),(5,-3)$ **3.** Center: $(-2,-5)$; Foci: $\left(-2\pm\sqrt{6},-5\right)$;

Vertices: $(1,-5),(-5,-5)$ **5.** Center: $(-3,2)$; Foci: $\left(-3\pm\sqrt{2},2\right)$; Vertices: $(-1,2),(-5,2)$ **7.** Center: $(-5,1)$;

Foci: $\left(-5,1\pm 2\sqrt{3}\right)$; Vertices: $(-5,5),(-5,-3)$ **9.** Center: $(-4,2)$; Foci: $\left(-4\pm 3\sqrt{2},2\right)$; Vertices: $\left(-4\pm 3\sqrt{3},2\right)$

11. Center: $(2,0)$; Foci: $(4,0),(0,0)$; Vertices: $\left(2\pm\sqrt{5},0\right)$ **13.** e   **15.** f   **17.** c   **19.** b

21. $\left(3\pm 2\sqrt{2},-1\right)$ **23.** $\left(3\pm\sqrt{5},4\right)$ **25.** $\left(1,4\pm\sqrt{3}\right)$

27. $\left(-1\pm\sqrt{21},-5\right)$

29. $\left(-2\pm\sqrt{7},-1\right)$

31. $\left(-1\pm\sqrt{7},2\right)$

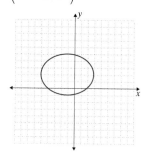

33. $\left(-5,3\pm\sqrt{15}\right)$

35. $\left(-5\pm\sqrt{5},-5\right)$

37. $\left(0,-2\pm\sqrt{3}\right)$

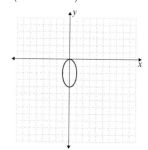

39. $\dfrac{x^2}{16}+\dfrac{y^2}{25}=1$ **41.** $(x-1)^2+\dfrac{(y-1)^2}{9}=1$ **43.** $\dfrac{(x-3)^2}{36}+\dfrac{y^2}{27}=1$ **45.** $(x+2)^2+\dfrac{(y+3)^2}{4}=1$

47. $\dfrac{(x-5)^2}{16}+\dfrac{(y-3)^2}{15}=1$ **49.** $\dfrac{(x-2)^2}{4}+\dfrac{(y+2)^2}{9}=1$ **51.** $\dfrac{(x-1)^2}{9}+\dfrac{y^2}{16}=1$

53. $e=\dfrac{\sqrt{11}}{6}$; major = 24; minor = 20   **55.** $e=\dfrac{2\sqrt{2}}{3}$; major = 12; minor = 4   **57.** $e=\dfrac{\sqrt{3}}{2}$; major = 4; minor = 2

59. $e=\dfrac{\sqrt{2}}{2}$; major = 4; minor = $2\sqrt{2}$ **61.** $e=\dfrac{\sqrt{42}}{7}$; major = 14; minor = $2\sqrt{7}$ **63.** $e\approx0.249$

65. Yes, just barely, if the boat is centered on the river.   **67.** Moving the foci closer together creates a more circular ellipse. Moving them further apart creates a more elongated ellipse.

Section 9.2 The Parabola

1. $(-1,4),\ y=2$

3. $\left(\dfrac{1}{2},4\right),\ x=\dfrac{3}{2}$

5. $(2,0),y=-2$

7. $\left(\dfrac{3}{2},0\right)$, $x=-\dfrac{3}{2}$

9. $\left(0,\dfrac{7}{4}\right)$, $y=-\dfrac{7}{4}$

11. $\left(0,-\dfrac{1}{48}\right)$, $y=\dfrac{1}{48}$

13. $\left(\dfrac{3}{2},0\right)$, $x=-\dfrac{3}{2}$

15. $(-4,0)$, $x=4$

17. $\left(0,\dfrac{1}{2}\right)$, $y=\dfrac{3}{2}$

19. $(-6,-1)$, $x=0$

21. $(-3,-3)$, $y=1$

23. $\left(\dfrac{5}{2},-3\right)$, $x=\dfrac{3}{2}$

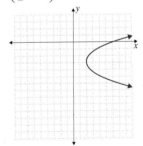

25. g   **27.** b   **29.** e   **31.** d   **33.** $(y-1)^2=-4(x+1)$ **35.** $(x-3)^2=8(y+1)$ **37.** $(y+2)^2=24(x-3)$

39. $(x+3)^2=-2(y+1)$ **41.** $(y-3)^2=10(x+4)$ **43.** $(y+1)^2=-8(x-2)$ **45.** 2 feet   **47.** 1.5 inches

Section 9.3 The Hyperbola

1. $\left(-3\pm\sqrt{13},-1\right)$

3. $\left(1,\,3\pm2\sqrt{5}\right)$

5. $\left(1\pm\sqrt{34},1\right)$

7. $\left(\pm2\sqrt{5},2\right)$

9. $\left(-2,\,2\pm\sqrt{34}\right)$

11. $(7,1),(-3,1)$

13. $\left(-7,2\pm\sqrt{113}\right)$

15. $\left(-1\pm2\sqrt{17},-7\right)$

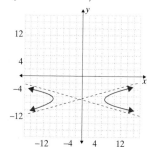

17. Center: $(-3,2)$; Foci: $\left(-3\pm\sqrt{13},2\right)$;
Vertices: $(-1,2),(-5,2)$

19. Center: $(1,-4)$; Foci: $\left(1\pm2\sqrt{3},-4\right)$;
Vertices: $\left(1\pm\sqrt{3},-4\right)$

21. Center: $(-2,1)$; Foci: $\left(-2\pm\sqrt{30},1\right)$;
Vertices: $(3,1),(-7,1)$

23. Center: $(-3,-1)$; Foci: $\left(-3\pm2\sqrt{3},-1\right)$; Vertices: $(-1,-1),(-5,-1)$ **25.** Center: $(1,0)$; Foci: $\left(1\pm\dfrac{\sqrt{5}}{2},0\right)$;

Vertices: $(2,0),(0,0)$ **27.** Center: $(8,5)$; Foci: $\left(8\pm4\sqrt{5},5\right)$; Vertices: $(12,5),(4,5)$ **29.** a   **31.** b   **33.** g   **35.** h

37. $\dfrac{x^2}{4}-\dfrac{y^2}{5}=1$ **39.** $y^2-4x^2=1$ **41.** $\dfrac{5x^2}{2}-\dfrac{5(y-4)^2}{18}=1$ **43.** $\dfrac{(x-6)^2}{9}-\dfrac{(y-5)^2}{7}=1$

45. $\dfrac{(x+4)^2}{4}-\dfrac{(y-3)^2}{16}=1$ **47.** $\dfrac{(x-1)^2}{9}-\dfrac{(y+1)^2}{4}=1$ **49.** $\dfrac{(y+4)^2}{16}-\dfrac{(x-3)^2}{25}=1$

51. $\dfrac{x^2}{\left(6\times10^7\right)^2}-\dfrac{y^2}{\left(7.2\times10^7\right)^2}=1$

Section 9.4 Rotation of Conics

1. $\left(4\sqrt{3}+3,-4+3\sqrt{3}\right)$ **3.** $\left(\dfrac{-5\sqrt{2}}{16},\dfrac{3\sqrt{2}}{16}\right)$ **5.** $(-1.2097,-13.5476)$ **7.** $(3.3485,-1.5002)$

9. Ellipse   **11.** Hyperbola   **13.** Hyperbola   **15.** Ellipse

17. Hyperbola

$\theta=\dfrac{\pi}{4},\ x'^2-y'^2=4$

19. Parabola

$\theta=\dfrac{\pi}{4},\ y'=-\sqrt{2}x'^2$

21. Ellipse

$\theta=\dfrac{\pi}{6},\ \dfrac{x'^2}{13}+\dfrac{y'^2}{25}=1$

23. Ellipse

$\theta=\dfrac{\pi}{3},\ \dfrac{x'^2}{30}+\dfrac{y'^2}{46}=1$

25.

27.

29.

31.

33. c   **35.** a   **37.** g   **39.** e

41. The objective of the rotation of axes is to eliminate the $x'y'$ term. If your final equation contains an $x'y'$ term, you know that a mistake has occurred.

43. a. Use the rotation of axes procedure to obtain the equation $4(x')^2 + 16(y')^2 - 16 = 0$. Now we know $F = -16$ and $F' = -16$. We can plug these values in $F = F'$ and obtain $-16 = -16$, which is true. **b.** Use the rotation of axes procedure to obtain the equation $4(x')^2 + 16(y')^2 - 16 = 0$. Now we know $A = 7$, $C = 13$, $A' = 4$, and $C' = 16$. We can plug these values in $A + C = A' + C'$, and obtain $7 + 13 = 4 + 16$, or $20 = 20$, which is true. **c.** Use the rotation of axes procedure to obtain the equation $4(x')^2 + 16(y')^2 - 16 = 0$. Now we know $A = 7$, $B = -6\sqrt{3}$, $C = 13$, $A' = 4$, $B' = 0$, and $C' = 16$. We can plug these values in $B^2 - 4AC = (B')^2 - 4A'C'$ and obtain $\left(-6\sqrt{3}\right)^2 - 4(7)(13) = (0)^2 - 4(4)(16)$, or $-256 = -256$, which is true.

Section 9.5 Polar Form of Conic Sections

1. c   **3.** f   **5.** b   **7.** Hyperbola, $y = \dfrac{7}{6}$ **9.** Ellipse, $x = -3$ **11.** Hyperbola, $x = \dfrac{1}{3}$ **13.** Ellipse, $x = 5$

15. Hyperbola, $x = -\dfrac{6}{5}$ **17.** Parabola, $y = \dfrac{3}{2}$ **19.** Hyperbola, $x = -\dfrac{4}{7}$ **21.** $r = \dfrac{2}{1 - \cos\theta}$

23. $r = \dfrac{3}{1 - 4\sin\theta}$ **25.** $r = \dfrac{3}{1 + \dfrac{1}{4}\cos\theta}$

27.

29.

31.

33.

35.

37.

39.

41.

43.

45.

47. Answers will vary.

Chapter 9 Review

1. Center: $(3,-1)$; Vertices: $(7,-1),(-1,-1)$; Foci: $\left(3\pm2\sqrt{3},-1\right)$

2. Center: $(-1,2)$; Vertices: $(-1,0),(-1,4)$; Foci: $\left(-1,2\pm\dfrac{2\sqrt{5}}{3}\right)$

3. $\left(-1\pm\sqrt{7},2\right)$ **4.** $\left(3\pm2\sqrt{2},-1\right)$ **5.** $\left(0,\pm3\sqrt{2}\right)$

        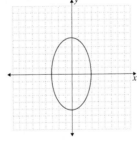

6. $\left(4,\pm\sqrt{21}\right)$ **7.** $\dfrac{(x+1)^2}{9}+\dfrac{(y-4)^2}{16}=1$ **8.** $\dfrac{(x-4)^2}{36}+\dfrac{(y-2)^2}{27}=1$

9. $\dfrac{4(x-2)^2}{9}+\dfrac{4(y+1)^2}{9}=1$ **10.** $\dfrac{(x-1)^2}{9}+\dfrac{(y+3)^2}{25}=1$ **11.** $\dfrac{(x-2)^2}{16}+\dfrac{y^2}{12}=1$

12. $\dfrac{x^2}{16}+\dfrac{(y-4)^2}{12}=1$ **13.** $\dfrac{(x+1)^2}{4}+\dfrac{(y-1)^2}{25}=1$ **14.** $(x-3)^2+\dfrac{(y+1)^2}{9}=1$

15. $30\pi a-\pi a^2$ **16.** $\dfrac{x^2}{400}+\dfrac{y^2}{100}=1$ or $\dfrac{x^2}{100}+\dfrac{y^2}{400}=1$

17. $(-6, -1), x = 0$

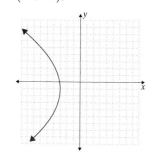

18. $\left(\dfrac{1}{2}, 4\right), x = \dfrac{3}{2}$

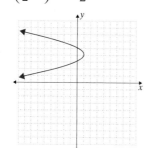

19. $(1, -1), x = -1$

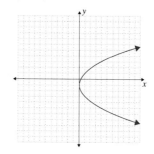

20. $(-1, 0), x = 1$

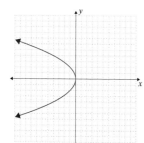

21. $\left(0, \dfrac{31}{8}\right), y = \dfrac{33}{8}$

22. $\left(\dfrac{-21}{2}, 2\right), x = \dfrac{-19}{2}$

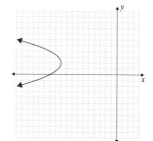

23. $(x+2)^2 = 4(y-3)$ **24.** $(x-5)^2 = 16(y+3)$ **25.** $(y+1)^2 = 2\left(x - \dfrac{5}{2}\right)$ **26.** $(x-1)^2 = -4(y+1)$

27. $(y+1)^2 = 16(x-2)$ **28.** $y^2 = 16(x+7)$ **29.** $(y+1)^2 = -16(x-5)$ **30.** $(x-2)^2 = 20(y+4)$ **31.** $\dfrac{5}{4}$ inches

32. $(2, 3), (2, -7)$

33. $\left(-3 \pm \sqrt{13}, -1\right)$

34. $\left(\pm\sqrt{2}, 0\right)$

35. $(0, \pm 13)$

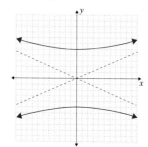

36. Center: $(-1, 2)$; Foci: $\left(-1 \pm 3\sqrt{5}, 2\right)$; Vertices: $(-7, 2), (5, 2)$

37. Center: $(0, 2)$; Foci: $\left(\pm 2\sqrt{10}, 2\right)$; Vertices: $(\pm 6, 2)$

38. Center: $(-4, 1)$; Foci: $\left(-4, 1 \pm \sqrt{5}\right)$; Vertices: $(-4, 3), (-4, -1)$

39. Center: $(3, 3)$; Foci: $\left(3, 3 \pm \sqrt{53}\right)$; Vertices: $(3, 1), (3, 5)$ **40.** $\dfrac{(x-1)^2}{9} - \dfrac{(y+1)^2}{7} = 1$

41. $\dfrac{(x+1)^2}{4} - \dfrac{(y+2)^2}{25} = 1$ **42.** $\dfrac{(y-3)^2}{9} - \dfrac{(x+1)^2}{16} = 1$

43. $\dfrac{(y-2)^2}{9}-\dfrac{(x-2)^2}{1}=1$ **44.** $\dfrac{x^2}{9}-\dfrac{y^2}{16}=1$ **45.** $\dfrac{(y-7)^2}{4}-\dfrac{(x+1)^2}{9}=1$ **46.** $\dfrac{(x-1)^2}{36}-\dfrac{(y+1)^2}{16}=1$

47. $\dfrac{(y+2)^2}{9}-\dfrac{x^2}{4}=1$ **48.** $\left(\dfrac{-\sqrt{2}}{2},\dfrac{15\sqrt{2}}{2}\right)$ **49.** $\left(11+43\sqrt{3},-11\sqrt{3}+43\right)$ **50.** $(-4.3395,-9.0299)$

51. $\left(3+3\sqrt{3},9-\sqrt{3}\right)$

52. Hyperbola

$$\theta=\dfrac{\pi}{4},\,x'^2-y'^2=12$$

53. Ellipse, $\theta=\dfrac{\pi}{3}$, $\dfrac{\left(x'-\dfrac{\sqrt{3}}{52}\right)^2}{9}+\dfrac{\left(y'-\dfrac{1}{36}\right)^2}{13}=\dfrac{5}{27{,}378}$

54. Hyperbola

$$\theta=\dfrac{\pi}{6},\,\dfrac{x'^2}{11}-\dfrac{y'^2}{17}=1$$

55. Parabola, $\theta=\dfrac{\pi}{4}$, $y'=\sqrt{2}x'^2$

56. Hyperbola, $y=\dfrac{-5}{8}$ **57.** Parabola, $y=\dfrac{7}{4}$ **58.** Ellipse, $x=\dfrac{-4}{3}$ **59.** Ellipse, $x=\dfrac{7}{2}$ **60.** $r=\dfrac{12}{1+4\sin\theta}$

61. $r=\dfrac{4}{1+\dfrac{1}{4}\cos\theta}$ **62.** $r=\dfrac{7}{1-\sin\theta}$ **63.** $r=\dfrac{3}{1+9\cos\theta}$

Chapter 9 Test

1. Center: $(2,-1)$; Vertices: $(2,8),(2,-10)$; Foci: $\left(2,-1\pm\sqrt{65}\right)$

2. Center: $(-3,1)$; Vertices: $(0,1),(-6,1)$; Foci: $\left(-3\pm\dfrac{3\sqrt{3}}{2},1\right)$

3. $\dfrac{(x-2)^2}{4}+(y-2)^2=1$ **4.** $\dfrac{(x-3)^2}{36}+\dfrac{(y-2)^2}{32}=1$ **5.** $\dfrac{x^2}{48}+\dfrac{y^2}{64}=1$ **6.** Approx. 97.98 yards apart

7. Focus: $\left(2,-\dfrac{1}{2}\right)$;

Directrix: $y=-\dfrac{3}{2}$

8. Focus: $\left(-1,-\dfrac{25}{6}\right)$;

Directrix: $y=-\dfrac{7}{6}$

9. $(x-4)^2=16(y+2)$ **10.** $(y-2)^2=-20(x+1)$

11. $8\sqrt{6}\,\text{m}\approx 19.6\text{ m}$

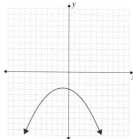

12. Center: $(-1,-2)$;
Foci: $\left(-1\pm\sqrt{145},-2\right)$;
Vertices: $(-13,-2)$,
$(11,-2)$

13. Center: $(2,-3)$;
Foci: $\left(2\pm\sqrt{10},-3\right)$;
Vertices: $(3,-3)$,
$(1,-3)$

14. Center: $(-1,2)$;
Foci: $\left(-1\pm\sqrt{29},2\right)$;
Vertices: $(-3,2),(1,2)$

15. Center: $(0,0)$;
Foci: $(\pm 5,0)$;
Vertices: $(\pm 3,0)$

16. $\dfrac{y^2}{4}-\dfrac{x^2}{12}=1$ **17.** $x^2-\dfrac{y^2}{9}=1$

18. $(3.4823,-32.2999)$

19. $\left(2\sqrt{3},0\right)$

20. Ellipse, $\theta=\dfrac{\pi}{6}$, $\dfrac{x'^2}{26}+\dfrac{y'^2}{50}=1$ **21.** Parabola, $\theta=\dfrac{\pi}{4}$, $y'=\dfrac{\sqrt{2}}{3}x'^2-2$

22. Hyperbola, $y=4$ **23.** Ellipse, $y=\dfrac{1}{2}$

24. Parabola, $x=\dfrac{-2}{3}$ **25.** $r=\dfrac{3}{1-\cos\theta}$

26. $r=\dfrac{3}{1-\dfrac{1}{5}\sin\theta}$

Systems of Equations

Section 10.1 Solving Systems by Substitution and Elimination

1. $(-5, 2)$ **3.** $(5, 3)$ **5.** $\varnothing$ **7.** $\left\{\left(\dfrac{y-3}{2}, y\right)\Big| y \in \mathbb{R}\right\}$ **9.** $(-1, 7)$ **11.** $(3, 11)$ **13.** $\left\{(x, 4x+1)\big| x \in \mathbb{R}\right\}$ **15.** $(2, 19)$

17. $(-5, 1)$ **19.** $(5, 6)$ **21.** $\left\{(-y-2, y)\big| y \in \mathbb{R}\right\}$ **23.** $(-5, 4)$ **25.** $(-1, 1)$ **27.** $(3, -5)$ **29.** $\varnothing$ **31.** $(-1, 3, 0)$

33. $(2, 2, -1)$ **35.** $\left\{\left(\dfrac{y-z+2}{3}, y, z\right)\Big| y \in \mathbb{R}, z \in \mathbb{R}\right\}$ **37.** $\varnothing$ **39.** $(1, 1, 0)$ **41.** $(9, 1, 1)$ **43.** $(3, 1, -2)$ **45.** $(4, 5, 5)$

47. $\left(\dfrac{49}{3}, \dfrac{-16}{3}, \dfrac{5}{4}\right)$ **49.** $(0, 3, 2)$ **51.** 22 pennies, 23 nickels   **53.** 25 people   **55.** Eliza is 15 years old.

57. 7 shirts and 4 pairs of shorts   **59.** 3 quarters, 11 dimes, and 28 pennies   **61.** Jim is 28 years old.

63. 3 thumb screws   **65.** Apples: \$0.78, Oranges: \$0.93, Mangos: \$1.05

Section 10.2 Matrix Notation and Gaussian Elimination

1. a. 3×2 **b.** -1 **c.** None   **3. a.** 5×2 **b.** None **c.** 10   **5. a.** 3×4 **b.** None **c.** 286   **7. a.** 3×2 **b.** 1 **c.** None

9. a. 2×5 **b.** 5 **c.** 2   **11.** $\left[\begin{array}{ccc|c} -3 & 1 & -2 & -4 \\ \frac{1}{2} & -4 & -1 & 1 \\ 0 & -3 & 3 & 1 \end{array}\right]$ **13.** $\left[\begin{array}{ccc|c} \frac{-3}{2} & -1 & 0 & -1 \\ 2 & 2 & 3 & 0 \\ 0 & -1 & 6 & 0 \end{array}\right]$ **15.** $\left[\begin{array}{ccc|c} \frac{12}{5} & \frac{1}{2} & \frac{3}{2} & \frac{1}{5} \\ 1 & 0 & 3 & 1 \\ 5 & 2 & 1 & -2 \end{array}\right]$

17. $\left[\begin{array}{ccc|c} \frac{2}{3} & \frac{-4}{3} & -2 & 0 \\ 8 & -2 & 6 & 7 \\ 3 & -2 & 0 & 0 \end{array}\right]$ **19.** $\left[\begin{array}{ccc|c} \frac{1}{2} & -14 & \frac{-1}{4} & -8 \\ \frac{1}{5} & \frac{-7}{6} & \frac{1}{4} & -3 \\ 5 & -5 & \frac{8}{3} & -5 \end{array}\right]$ **21.** $\begin{cases} x = 8 \\ y = 3 \end{cases}$ **23.** $\begin{cases} x + 3y + 6z = 16 \\ y + 2z = 9 \\ z = 4 \end{cases}$ **25.** $\begin{cases} 9y + 13z = 27 \\ 2x + 21z = 19 \\ 7x + 18y = 32 \end{cases}$

27. $\left[\begin{array}{cc|c} 2 & -5 & 3 \\ 0 & -7 & 5 \end{array}\right]$ **29.** $\left[\begin{array}{cc|c} 1 & 3 & -2 \\ 9 & -2 & 7 \end{array}\right]$ **31.** $\left[\begin{array}{cc|c} 8 & -2 & -4 \\ -6 & 2 & -14 \end{array}\right]$ **33.** $\left[\begin{array}{cc|c} 4 & 12 & -6 \\ 9 & 9 & 6 \end{array}\right]$ **35.** $\left[\begin{array}{cc|c} 4 & -1 & 5 \\ -6 & 2 & 0 \end{array}\right]$

37. $\left[\begin{array}{ccc|c} 18 & -6 & 15 & 42 \\ -7 & 19 & 2 & 3 \\ -4.5 & 5.5 & -2 & 3.5 \end{array}\right]$ **39.** $\left[\begin{array}{ccc|c} 5 & 18 & 22 & 5 \\ 32 & -9 & -27 & -23 \\ -9 & 21 & 12 & 9 \end{array}\right]$ **41.** $\left[\begin{array}{ccc|c} 0 & 1 & -9 & -3 \\ 1 & 1 & 3 & 4 \\ 0 & 0 & 0 & 0 \end{array}\right]$ **43.** $\left[\begin{array}{cc|c} -1 & 4 & -3 \\ 1 & -6 & \frac{5}{2} \end{array}\right]$

45. $\left[\begin{array}{ccc|c} 1 & 5 & -9 & 11 \\ 0 & -1 & 8 & -7 \\ 0 & -17 & 41 & 1 \end{array}\right]$ **47.** Neither   **49.** Neither   **51.** Neither   **53.** $(3, -1)$ **55.** $(1, 3)$ **57.** $(-7, 3)$

59. $\varnothing$ **61.** $(3, 2)$ **63.** $\left\{(-2y-4, y)\big| y \in \mathbb{R}\right\}$ **65.** $\varnothing$ **67.** $(4, 0, 3)$ **69.** $(15, -21, 8)$

71. $(3, -5)$ **73.** $\left\{(x, -3x-2)\big| x \in \mathbb{R}\right\}$ **75.** $(-4, 1)$ **77.** $(6, 4)$ **79.** $(-11, -5)$ **81.** $(7, 3, 3)$ **83.** $(2, 2, -1)$

85. $\left\{(1,\,y,0)\big|y\in\mathbb{R}\right\}$ **87.** $(3,-2,3)$ **89.** $(2,3,4)$ **91.** $(9,-19,7)$ **93.** $(1,-2,-1,3)$ **95.** $42,26,87$

97. Small: 10, Medium: 24, Large: 48

Section 10.3 Determinants and Cramer's Rule

1. 11   **3.** 15   **5.** $ab-x^2$ **7.** 8   **9.** -10 **11.** -39 **13.** $\{-2,3\}$ **15.** $\{-5,1\}$ **17.** $\{-5,-4\}$ **19.** $\{-6,4\}$ **21.** $\{2,5\}$

23. 3   **25.** -9 **27.** 2   **29.** -2 **31.** 159   **33.** 78   **35.** -254 **37.** 404   **39.** 4   **41.** 120   **43.** 10   **45.** x^4 **47.** x^8

49. $(76,-53)$ **51.** $\left\{(-y-2,y)\big|y\in\mathbb{R}\right\}$ **53.** $\varnothing$ **55.** $\left\{(-3z-5,-6z-10,z)\big|z\in\mathbb{R}\right\}$

57. $\left\{\left(\dfrac{-5y-z-5}{2},\dfrac{-5y+3z-19}{2},y,z\right)\Big|y\in\mathbb{R},z\in\mathbb{R}\right\}$ **59.** $\left\{(-z+8,-5z+31,-2z+37,z)\big|z\in\mathbb{R}\right\}$

61. $(1647,2071)$ **63.** $\varnothing$ **65.** $(-3,-1,0,-4)$ **67.** Candy bars: 5, Ice cream: 6

Section 10.4 The Algebra of Matrices

1. $\begin{bmatrix} 5 & -1 \\ 0 & 0 \\ 2 & 13 \end{bmatrix}$ **3.** $\begin{bmatrix} 6 & -3 \\ 18 & 30 \\ -9 & 21 \end{bmatrix}$ **5.** Not possible   **7.** $\begin{bmatrix} 14 & -14 \\ 8 & 0 \\ -4 & 14 \end{bmatrix}$ **9.** $\begin{bmatrix} -7 & 5 \\ 3 & 10 \\ -3 & -8 \end{bmatrix}$ **11.** Not possible

13. $a=3,\ b=-1,\ c=10$ **15.** $a=2,\ b=-2,\ c=-1$ **17.** Not possible   **19.** $x=10,y=5$ **21.** $x=3,y=1$

23. Not possible   **25.** $a=8,\ b=5$ **27.** $[24\ -5]$ **29.** $[35\ 18]$ **31.** Not possible   **33.** $[-30\ -3]$

35. $\begin{bmatrix} 15 & -3 & -24 \\ 25 & -5 & -40 \\ 30 & -6 & -48 \end{bmatrix}$ **37.** $[-34\ -7]$ **39.** $\begin{bmatrix} 11 & 0 \\ 0 & 11 \end{bmatrix}$ **41.** $\begin{bmatrix} 32 & -20 \\ 56 & -35 \\ -16 & 10 \end{bmatrix}$ **43.** Not possible   **45.** $\begin{bmatrix} 14 & -13 \\ -13 & 5 \end{bmatrix}$

47. $[179\ \ 76]$

Section 10.5 Inverses of Matrices

1. $\begin{bmatrix} 14 & -5 \\ 1 & 9 \end{bmatrix}\begin{bmatrix} x \\ y \end{bmatrix}=\begin{bmatrix} 7 \\ 2 \end{bmatrix}$ **3.** $\begin{bmatrix} 1 & 2 \\ 9 & -3 \end{bmatrix}\begin{bmatrix} x \\ y \end{bmatrix}=\begin{bmatrix} -6 \\ -14 \end{bmatrix}$ **5.** $\begin{bmatrix} 3 & -7 & 1 \\ 1 & -1 & 0 \\ 0 & 8 & 5 \end{bmatrix}\begin{bmatrix} x_1 \\ x_2 \\ x_3 \end{bmatrix}=\begin{bmatrix} -4 \\ 2 \\ -3 \end{bmatrix}$ **7.** $\begin{bmatrix} \dfrac{3}{5} & -\dfrac{8}{5} \\ 0 & 1 \end{bmatrix}\begin{bmatrix} x \\ y \end{bmatrix}=\begin{bmatrix} 2 \\ 2 \end{bmatrix}$

9. $\begin{bmatrix} 4 & -3 \\ 2 & -4 \end{bmatrix}\begin{bmatrix} x \\ y \end{bmatrix}=\begin{bmatrix} -9 \\ 13 \end{bmatrix}$ **11.** $\begin{bmatrix} 2 & -1 & 3 \\ -1 & 1 & 0 \\ 4 & -5 & 1 \end{bmatrix}\begin{bmatrix} x \\ y \\ z \end{bmatrix}=\begin{bmatrix} 0 \\ 17 \\ -2 \end{bmatrix}$ **13.** $\begin{bmatrix} -\dfrac{1}{20} & -\dfrac{1}{5} \\ \dfrac{1}{4} & 0 \end{bmatrix}$ **15.** $\begin{bmatrix} -5 & -4 \\ 4 & 3 \end{bmatrix}$ **17.** $\begin{bmatrix} -5 & 0 \\ 2 & 2 \end{bmatrix}$

19. Not invertible **21.** $\begin{bmatrix} 2 & 1 & -4 \\ -4 & -2 & -3 \\ -1 & -1 & -4 \end{bmatrix}$ **23.** $\begin{bmatrix} -1 & 2 & -1 \\ 0 & -1 & 1 \\ 0 & -4 & 3 \end{bmatrix}$ **25.** $\begin{bmatrix} -1 & -2 & 1 \\ -2 & 1 & -3 \\ 1 & 2 & 0 \end{bmatrix}$ **27.** $\begin{bmatrix} -2 & 1 & 1 \\ 2 & 0 & -1 \\ -1 & 0 & 1 \end{bmatrix}$

29. $\begin{bmatrix} 2 & -1 & 2 \\ 0 & 1 & -1 \\ -3 & -2 & -4 \end{bmatrix}$ **31.** No **33.** Yes **35.** No **37.** $\begin{bmatrix} \frac{-2}{11} & \frac{3}{11} \\ \frac{-1}{11} & \frac{7}{11} \end{bmatrix}$ **39.** $\begin{bmatrix} \frac{-7}{12} & \frac{-1}{3} & \frac{1}{2} \\ \frac{-5}{12} & \frac{-2}{3} & \frac{1}{2} \\ \frac{-1}{12} & \frac{-1}{3} & \frac{1}{2} \end{bmatrix}$

41. $\begin{bmatrix} -0.044 & 0.146 & -0.018 \\ -0.146 & 0.121 & -0.124 \\ -0.026 & 0.070 & -0.156 \end{bmatrix}$ **43.** $\left(-2, -\dfrac{5}{2} \right)$ **45.** $\left\{ \left(\dfrac{3y-1}{2}, y \right) \middle| y \in \mathbb{R} \right\}$ **47.** $(-2, 0)$ **49.** $(8, -19)$ **51.** $(0, 5)$
53. $(-4, 5, -1)$

Section 10.6 Partial Fraction Decomposition

1. $\dfrac{A_1}{x-3} + \dfrac{A_2}{x+2}$ **3.** $\dfrac{A_1}{x+3} + \dfrac{A_2}{x+4} + \dfrac{A_3}{(x+4)^2}$ **5.** $\dfrac{A_1}{x+3} + \dfrac{A_2}{x-2} + \dfrac{A_3}{x+2}$ **7.** d **9.** h **11.** a **13.** c

15. $\dfrac{-1}{x} + \dfrac{2}{x-2} + \dfrac{2}{x+2}$ **17.** $\dfrac{5}{72(x-2)} - \dfrac{1}{32x} - \dfrac{7}{288(x+4)} - \dfrac{x+17}{72(x^2+8)}$ **19.** $\dfrac{10}{x-4} - \dfrac{5}{x-2}$

21. $\dfrac{3}{2(x+3)} + \dfrac{3}{14(x-1)} - \dfrac{12}{7(x+6)}$ **23.** $\dfrac{1}{2(x-1)} - \dfrac{1}{2(x+1)}$ **25.** $\dfrac{1}{48(x-2)} - \dfrac{x}{80(x^2+4)} - \dfrac{3}{40(x^2+4)} - \dfrac{1}{120(x+4)}$

27. $\dfrac{5}{4(x-2)} - \dfrac{1}{4(x+2)}$ **29.** $\dfrac{1}{16(x+2)} + \dfrac{1}{16(x-2)} - \dfrac{x}{8(x^2+4)}$ **31.** $\dfrac{1}{24(x+6)} - \dfrac{1}{30(x+3)} - \dfrac{7}{40(x-2)} + \dfrac{1}{6(x-3)}$

33. $\dfrac{15}{16(x+3)} + \dfrac{1}{16(x-1)} - \dfrac{9}{4(x+3)^2}$ **35.** $\dfrac{1}{x+3} + \dfrac{1}{x-3}$ **37.** $\dfrac{1}{4(x-2)} + \dfrac{3}{4(x+6)}$

39. True

41. True

43. False

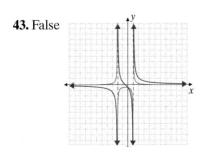

45. $\dfrac{1}{a}\left(\dfrac{1}{x} - \dfrac{1}{x+a} \right)$ **47.** $\dfrac{1}{2a}\left(\dfrac{1}{a+x} + \dfrac{1}{a-x} \right)$

49. $\dfrac{1}{a-1}\left(\dfrac{1}{x+1} - \dfrac{1}{x+a} \right)$

Section 10.7 Linear Programming

1.

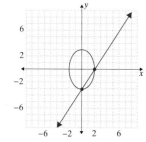

$60x + 50y \le 50,000;$

$x + 10y \le 6000;$

$x \ge 0;\ y \ge 0$

3.

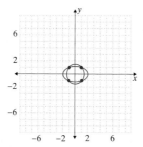

$12x + 32y \le 80;$

$2x + 3y \le 10;$

$x \ge 0;\ y \ge 0$

5. Min = 0 at $(0, 0)$; Max = 21 at $(0, 7)$ **7.** Min = 0 at $(0, 0)$; Max = 35 at $(0, 7)$ **9.** Min = 0 at $(0, 0)$;

Max = 48 at $(0, 8)$ **11.** Min = 0 at $(0, 0)$; Max = 15 at $(2.5, 0)$ **13.** Min = 0 at $(0, 0)$; Max = 50 at $(3, 4)$

15. Min = 0 at $(0, 0)$; Max = 48 at $(8, 0)$ **17.** Min = $\dfrac{172}{11}$ at $\left(\dfrac{24}{11}, \dfrac{10}{11}\right)$; Max = $\dfrac{400}{3}$ at $\left(0, \dfrac{40}{3}\right)$

19. Model X: 1000 units, Model Y: 500 units, Maximum Profit: $76,000   **21.** Ashley should buy 3 olive bundles

and 2 cranberry bundles. The total cost to make the curtains is $70.   **23.** $20,000 should be placed in

municipal bonds, $5000 should be placed in Treasury bills.

Section 10.8 Nonlinear Systems of Equations and Inequalities

1. $\{(0, -3), (2, 0)\}$

3. $\{(-1, -1), (-1, 1), (1, -1), (1, 1)\}$

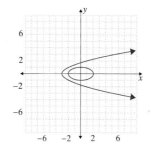

5. $\{(-1, 1), (1, 1)\}$

7. $\{(-1, 2), (1, 2)\}$

9. No solution

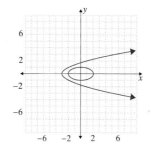

11. $\{(-3, 1), (1, 1)\}$

13. $\{(0, 0)\}$

15. $\{(\pm3, 0)\}$

17. $\{(\pm1, 0)\}$

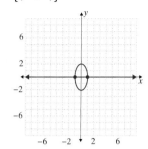

19. $\{(2, \pm3), (-2, \pm3)\}$

21. $\{(5, 3)\}$

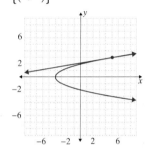

23. $\{(\pm2, 0), (0, -2)\}$

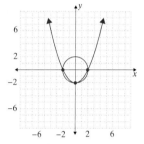

25. $\left\{\left(-i\sqrt{6}, -6\right), \left(i\sqrt{6}, -6\right), \left(-\sqrt{5}, 5\right), \left(\sqrt{5}, 5\right)\right\}$ **27.** $\{(-2, 3)\}$ **29.** $\left\{(-1, -4), (1, 4), \left(-2\sqrt{2}, -\sqrt{2}\right), \left(2\sqrt{2}, \sqrt{2}\right)\right\}$

31. $\left\{(-3i, -5), (3i, -5), (0, 4)\right\}$ **33.** $\left\{\left(-\sqrt{3}, 0\right), \left(\sqrt{3}, 0\right)\right\}$ **35.** $\left\{(-i, -6), (i, -6), \left(-\sqrt{2}, 3\right), \left(\sqrt{2}, 3\right)\right\}$

37. $\left\{(1, 1), (1, -1), \left(-\dfrac{3}{2}, \dfrac{\sqrt{14}}{2}\right), \left(-\dfrac{3}{2}, -\dfrac{\sqrt{14}}{2}\right)\right\}$ **39.** $\left\{\left(\dfrac{7}{2}, \dfrac{17}{4}\right)\right\}$ **41.** $\left\{\left(-6, -i\sqrt{15}\right), \left(-6, i\sqrt{15}\right), (2, -1), (2, 1)\right\}$

43. $\left\{(0, 1), \left(-\dfrac{100}{101}, -\dfrac{99}{101}\right)\right\}$ **45.** $\left\{\left(-\dfrac{\sqrt{42}}{6}, -\dfrac{\sqrt{66}}{6}\right), \left(-\dfrac{\sqrt{42}}{6}, \dfrac{\sqrt{66}}{6}\right), \left(\dfrac{\sqrt{42}}{6}, -\dfrac{\sqrt{66}}{6}\right), \left(\dfrac{\sqrt{42}}{6}, \dfrac{\sqrt{66}}{6}\right)\right\}$

47. $\{(-5, 0), (-2, 0), (-1, 0)\}$ **49.** $\left\{(3, 2), (2, 3), \left(\dfrac{-1 \pm i\sqrt{23}}{2}, \dfrac{-1 \pm i\sqrt{23}}{2}\right)\right\}$ **51.** $\left\{(0, -1), \left(\dfrac{\sqrt{6}}{3}, \dfrac{2\sqrt{6}}{9} - 1\right)\right\}$

53. $\left\{\left(\sqrt{2} + 1, 2\sqrt{2}\right), \left(-\sqrt{2} + 1, -2\sqrt{2}\right) \left(2\sqrt{2} + 1, \sqrt{2}\right), \left(-2\sqrt{2} + 1, -\sqrt{2}\right)\right\}$ **55.** $\left\{(4, 1), \left(1, 1 + i\sqrt{3}\right)\right\}$

57. $\{(0, 4), (0, -3), (3, -2), (3, 3)\}$ **59.** $\{(4, 0)\}$ **61.** $\left\{\left(\pm\dfrac{1}{2}, \pm\dfrac{1}{4}\right)\right\}$ **63.** b

65. b, d   **67.** a, d   **69.**

71.   **73.**   **75.**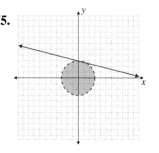

77. 9 inches by 5 inches   **79.** 60 mph and 70 mph   **81.** -12 and 7 **83.** $h = 6$ cm, $r = 3$ cm

Chapter 10 Review

1. $\varnothing$ **2.** $(4, -5, 5)$ **3.** $(3, 0, 2)$ **4.** $(2, -1)$ **5.** $\varnothing$ **6.** $(-2, 2)$ **7.** $\left\{(3 - 3y, y) \mid y \in \mathbb{R}\right\}$ **8.** $(0, -3)$ **9.** $(3, 2)$

10. $(-1, 1, 2)$ **11.** $(8, 12, 10)$ **12.** $\varnothing$ **13.** $y = \dfrac{4}{15}x^2 - x + \dfrac{11}{15}$ **14. a.** 3×3 **b.** -8 **c.** 7 **15. a.** 1×4 **b.** 8

c. None   **16.** $\begin{bmatrix} 2 & 1 & -1 & | & 3 \\ -2 & 3 & -1 & | & 2 \\ 6 & -8 & 1 & | & 3 \end{bmatrix}$ **17.** $\begin{bmatrix} 4 & 5 & -1 & | & 0 \\ 1 & 3 & 2 & | & 3 \\ 10 & -1 & -6 & | & 0 \end{bmatrix}$ **18.** $\begin{cases} 8x - 2y = 2 \\ -x + 5y = 3 \end{cases}$ **19.** $\begin{cases} 8x + 7z = 5 \\ -3y + 4z = 16 \\ 16x - 2y + z = 2 \end{cases}$

20. $\begin{cases} 3x - 7y + 6z = 9 \\ -11x + 3z = -14 \\ 8z = 2 \end{cases}$ **21.** $\begin{bmatrix} 0 & -5 & | & -11 \\ 1 & 2 & | & 3 \end{bmatrix}$ **22.** $\begin{bmatrix} 2 & 3 & | & 5 \\ 0 & 5 & | & 12 \end{bmatrix}$ **23.** $\begin{bmatrix} 1 & -4 & | & -4 \\ 1 & 7 & | & 11 \end{bmatrix}$ **24.** $\begin{bmatrix} -1 & 0 & 2 & | & -6 \\ 2 & -6 & 8 & | & 2 \\ -1 & -1 & -5 & | & 6 \end{bmatrix}$

25. $(2, -1)$ **26.** $\left(\dfrac{-50}{3}, -16\right)$ **27.** $(3, -5)$ **28.** $\varnothing$ **29.** $2x^4$ **30.** 5 **31.** 7 **32.** x^8 **33.** $9, -9$ **34.** $-15, -15$

35. $(-4, 1)$ **36.** $(2, -1, 1)$ **37.** $\varnothing$ **38.** $(4, 2, 0)$ **39.** $\begin{bmatrix} 4 & -16 & 6 \\ -5 & 8 & 12 \end{bmatrix}$ **40.** $\begin{bmatrix} 4 & 0 \\ -5 & 9 \end{bmatrix}$

41. Not possible   **42.** Not possible   **43.** $\begin{bmatrix} 9 & -23 & 3 \\ 5 & -3 & 8 \end{bmatrix}$ **44.** $\begin{bmatrix} 47 & -77 \\ 34 & 4 \end{bmatrix}$ **45.** $w = -2, x = 1, y = 3, z = -4$

46. Not possible   **47.** $x = 2, y = -3$ **48.** $x = 1, y = -1$ **49.** $[12 \quad 46]$ **50.** $\begin{bmatrix} -12 & 8 & 12 \\ -15 & 10 & 15 \\ -18 & 12 & 18 \end{bmatrix}$

51. $\begin{bmatrix} 1 & -1 & 2 \\ 2 & -3 & -1 \\ -3 & 0 & 6 \end{bmatrix} \begin{bmatrix} x_1 \\ x_2 \\ x_3 \end{bmatrix} = \begin{bmatrix} -4 \\ 1 \\ 5 \end{bmatrix}$ **52.** $\begin{bmatrix} 3 & -1 & 1 \\ 2 & 0 & -5 \\ 4 & 3 & 0 \end{bmatrix} \begin{bmatrix} x \\ y \\ z \end{bmatrix} = \begin{bmatrix} 4 \\ 1 \\ 6 \end{bmatrix}$ **53.** $\begin{bmatrix} \dfrac{3}{16} & \dfrac{1}{8} \\ -\dfrac{1}{8} & \dfrac{1}{4} \end{bmatrix}$ **54.** $\begin{bmatrix} 1 & -2 \\ -\dfrac{1}{2} & 2 \end{bmatrix}$ **55.** Not possible

56. $\begin{bmatrix} \dfrac{1}{12} & \dfrac{1}{6} & \dfrac{11}{24} \\ \dfrac{5}{12} & \dfrac{-1}{6} & \dfrac{7}{24} \\ \dfrac{1}{12} & \dfrac{1}{6} & \dfrac{-1}{24} \end{bmatrix}$ **57.** No **58.** Yes **59.** Yes **60.** No **61.** $\left(1, -\dfrac{1}{3}\right)$ **62.** $(0, -9, 7)$

63. $(-15, 20, -33), (-2, 3, -4), (-9, 11, -17)$ **64.** $\dfrac{A_1}{x} + \dfrac{A_2}{x^2} + \dfrac{A_3}{(3x-1)} + \dfrac{A_4}{(3x-1)^2}$

65. $\dfrac{A_1}{x+4} + \dfrac{A_2}{x-1}$ **66.** $\dfrac{1}{x-1} + \dfrac{1}{x+1}$ **67.** $-\dfrac{3}{2(2x-5)^2} + \dfrac{1}{2(2x-5)}$ **68.** $\dfrac{x}{(4+x^2)^2} + \dfrac{x}{(4+x^2)}$

69. $x \geq 20,\ y \geq 10,\ x \leq 40, y \leq 40,\ x+y \leq 60$ **70.** $x \geq 6,\ y \geq 6,\ x \leq 10, y \leq 8,\ x+y \geq 15$

  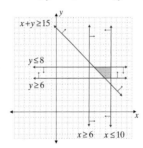

71. Min = 0 at $(0, 0)$; Max = 30 at $(5, 0)$ **72.** Min = 20 at $(0, 10)$; Max = 50 at $(10, 0)$ **73.** Min = 8 at $(0, 2)$;

Max = 24 at $\left(\dfrac{24}{7}, \dfrac{12}{7}\right)$ **74.** Min = 350 at $(5, 0)$; Max = 1028 at $(10, 4)$ **75.** 12 vases should be produced, 12

pitchers should be produced; Maximum profit: \$660 **76.** 35 bionic arms, 15 bionic legs;

Minimum cost: \$24,000

77. $\{(0, -2), (3, 1)\}$ **78.** $\left\{(i, -1), (-i, -1), \left(\dfrac{\sqrt{2}}{2}, \dfrac{1}{2}\right), \left(-\dfrac{\sqrt{2}}{2}, \dfrac{1}{2}\right)\right\}$ **79.** $\{(4, 3), (-4, 3), (-4, -3), (4, -3)\}$

80. $\{(2, 1)\}$ **81.** 9 and 16 **82.** 36 mph and 24 mph

83. None

    84.

85.   **86.**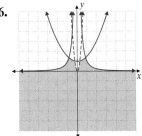

Chapter 10 Test

1. $(2,1)$ **2.** $\varnothing$ **3.** $\left(\dfrac{3}{2}, \dfrac{4}{3}, \dfrac{19}{6}\right)$ **4.** 15 **5.** $\begin{cases} 3x - y = 4 \\ 6x + 2y = 0 \end{cases}$ **6. a.** 4×2 **b.** 3 **c.** None   **7.** $\left[\begin{array}{ccc|c} 0 & 1 & -2 & 1 \\ 3 & 1 & -1 & 4 \\ 0 & -4 & 6 & -1 \end{array}\right]$

8. $\left[\begin{array}{ccc|c} 2 & -5 & -1 & 4 \\ 1 & -1 & 1 & 6 \\ -4 & -3 & 1 & 4 \end{array}\right]$ **9.** $\left[\begin{array}{cc|c} 4 & -3 & -2 \\ 1 & -2 & 3 \end{array}\right]$ **10.** $\left[\begin{array}{ccc|c} -9 & 6 & 10 & 5 \\ 5 & -10 & -8 & -7 \\ -1 & 2 & 4 & 3 \end{array}\right]$ **11.** $(0, -3)$ **12.** $(1,1,1)$ **13.** -6 **14.** 0

15. $\{\pm 4\}$ **16.** $\{5, -4\}$ **17.** $(2, -2)$ **18.** $(3, -1, 1)$ **19.** Not possible   **20.** $\{(3, -4, t)\,|\,t \in \mathbb{R}\}$ **21.** $\begin{bmatrix} -6 & 18 \end{bmatrix}$

22. $\begin{bmatrix} 0 & -1 & -5 \\ 6 & -3 & 3 \\ 8 & -6 & -6 \end{bmatrix}$ **23.** Not possible   **24.** $\begin{bmatrix} 0 & 1 \\ -1 & 3 \end{bmatrix}$ **25.** $\begin{bmatrix} \dfrac{1}{3} & \dfrac{-1}{3} \\ 1 & 0 \end{bmatrix}\begin{bmatrix} x \\ y \end{bmatrix} = \begin{bmatrix} 4 \\ -1 \end{bmatrix}$ **26.** $\begin{bmatrix} 3 & -2 & 1 \\ 1 & -1 & 0 \\ 0 & 1 & -1 \end{bmatrix}\begin{bmatrix} x \\ y \\ z \end{bmatrix} = \begin{bmatrix} 1 \\ 6 \\ 4 \end{bmatrix}$

27. $\begin{bmatrix} \dfrac{7}{11} & \dfrac{-3}{11} \\ \dfrac{-5}{11} & \dfrac{-1}{11} \end{bmatrix}$ **28.** Not possible   **29.** $\dfrac{A_1}{x+3} + \dfrac{A_2}{x+1}$ **30.** $\dfrac{A_1}{x-3} + \dfrac{B_1 x + C_1}{x^2 + 4}$ **31.** $\dfrac{1}{x+1} + \dfrac{2}{x^2 - 2x + 3}$

32. $\dfrac{5}{x-2} - \dfrac{3}{x} - \dfrac{1}{x+2}$ **33.** $900x + 1500y \le 40{,}000;$

$28x + 16y \le 1000$

34. $\left\{ (2,1), (-2,1), \left(-\sqrt{5}, 0\right), \left(\sqrt{5}, 0\right) \right\}$

35. $\left\{ (i, 3), (i, -3), (-i, 3), (-i, -3) \right\}$

36.

An Introduction to Sequences, Series, Combinatorics, and Probability

Section 11.1 Sequences and Series

1. Infinite **3.** Finite **5.** Finite **7.** Infinite **9.** Infinite **11.** $2, -1, -4, -7, -10$ **13.** $1, \dfrac{3}{2}, \dfrac{9}{5}, 2, \dfrac{15}{7}$

15. $\dfrac{2}{3}, -\dfrac{4}{9}, \dfrac{8}{27}, -\dfrac{16}{81}, \dfrac{32}{243}$ **17.** $\dfrac{1}{2}, \dfrac{4}{3}, \dfrac{9}{4}, \dfrac{16}{5}, \dfrac{25}{6}$ **19.** $-1, 6, -5, 20, -27$ **21.** $2, 2, \dfrac{18}{7}, \dfrac{16}{5}, \dfrac{50}{13}$ **23.** $2, 1, \dfrac{8}{9}, 1, \dfrac{32}{25}$

25. $10, 5, 0, -5, -10$ **27.** $\dfrac{1}{27}, \dfrac{1}{81}, \dfrac{1}{243}, \dfrac{1}{729}, \dfrac{1}{2187}$ **29.** $8, 2^{\frac{3}{2}}, 2^{\frac{3}{4}}, 2^{\frac{3}{8}}, 2^{\frac{3}{16}}$ **31.** $\dfrac{1}{3}, 1, \dfrac{9}{5}, \dfrac{8}{3}, \dfrac{25}{7}$ **33.** $-1, 2, -3, 4, -5$

35. $2, 6, 12, 20, 30$ **37.** $\sqrt{3}+1, \sqrt{6}+1, 4, 2\sqrt{3}+1, \sqrt{15}+1$ **39.** $0, -4, 0, 18, 56$ **41.** $-2, -11, -74, -515, -3602$

43. $-1, 0, -1, 0, -1$ **45.** $3, 1, 3, 1, 3$ **47.** $a_n = (-2)^n$ **49.** $a_n = \dfrac{n}{n+2}$ **51.** $a_1 = 1, \ a_n = \dfrac{1}{n} a_{n-1}, \ n \geq 2$

53. $a_n = \dfrac{4-n}{n+13}$ **55.** $a_n = 4 - 5n$ **57.** $a_1 = 1, \ a_n = n + n\left(a_{n-1}\right), \ n \geq 2$ **59.** $-3 - 12 - 27 - 48 - 75 = -165$

61. $\displaystyle\sum_{i=0}^{7} (1+3i) = 92$ **63.** $\displaystyle\sum_{i=3}^{9} i^2 = 280$ **65.** $-36 - 30 - 22 - 12 + 0 + 14 + 30 + 48 = -8$

67. $S_n = \dfrac{n}{4(n+4)}, \ S_{100} = \dfrac{25}{104}$ **69.** $S_n = 2^n - 1$, series diverges **71.** $S_n = \dfrac{n}{2(n+1)}, \ S_{49} = \dfrac{49}{100}$

73. $S_n = -\ln(n+1), \ S_{100} = -\ln 101$ **75.** $S_n = \dfrac{2n}{7(2n+7)}, \ S_{30} = \dfrac{60}{469}$ **77.** $S_n = -\ln(n+1), \ S_{65} = -\ln 66$

79. $-9, 1, -8, -7, -15$ **81.** $-17, 13, -4, 9, 5$ **83.** $2, -3, -7, -24, -79$ **85.** 987

87. $A_n = A_{n-1}(1.005) - 74; \ A_4 = 352.63; \ A_6 = 207.79;$ 9 months

Section 11.2 Arithmetic Sequences and Series

1. $a_n = 3n - 5$ **3.** $a_n = -2n + 9$ **5.** $a_n = 9n - 4$ **7.** $a_n = -6n + 9$ **9.** $a_n = 19n + 5$ **11.** $a_n = n + \dfrac{5}{2}$

13. $a_n = -\dfrac{19}{2}n + \dfrac{43}{2}$ **15.** $a_n = -2n + 1$ **17.** $a_n = -4n + 33$ **19.** $d = 2$ **21.** $d = 1$ **23.** No **25.** No **27.** 13

29. 2 **31.** 2 **33.** $d = 2.5; \ 5, 7.5, 10, 12.5, 15$ **35.** $d = 7; \ 7, 14, 21, 28, 35$ **37.** $d = 9; \ -62, -53, -44, -35, -26$

39. $d = 5$ **41.** $d = 1$ **43.** $d = -2$ **45.** 195 **47.** a_{73} **49.** 117 **51.** 26 **53.** -8 **55.** 13 **57.** 55 **59.** $14{,}350$

61. $17{,}114$ **63.** -1475 **65.** $-\dfrac{3219}{5}$ **67.** 902 **69.** -1316 **71.** 6 years **73.** 1620 pounds **75.** \$625; \$8100

Section 11.3 Geometric Sequences and Series

1. $a_n = -3(2)^{n-1}$ **3.** $a_n = 2\left(\dfrac{-1}{3}\right)^{n-1}$ **5.** $a_n = \left(\dfrac{-1}{4}\right)^{n-1}$ **7.** $a_n = \left(\dfrac{1}{7}\right)^{n-1}$ **9.** $a_n = (-3)^{n-1}$ **11.** $a_n = 3\left(\dfrac{2}{3}\right)^{n-1}$

13. $a_n = 7(-2)^{n-1}$ **15.** $a_n = \dfrac{1}{16}(2)^{n-1}$ **17.** $a_n = \dfrac{39}{68}\left(\dfrac{4}{3}\right)^{n-1}$ **19.** No   **21.** $r = \dfrac{1}{2}$ **23.** $r = 2$ **25.** $r = 7$

27. $r = 3;\ 8, 24, 72, 216, 648$ **29.** $r = 2;\ \dfrac{1}{4}, \dfrac{1}{2}, 1, 2, 4$ **31.** $r = \dfrac{1}{5};\ 62{,}500, 12{,}500, 2500, 500, 100$ **33.** $\dfrac{5}{16{,}384}$

35. $-2, 147, 483, 648$ **37.** $r = \pm 2$ **39.** $r = \pm\dfrac{1}{5}$ **41.** $\dfrac{52{,}222{,}139{,}775}{1{,}048{,}576} \approx 49{,}802.9$ **43.** $\dfrac{10{,}923}{16{,}384} \approx 0.666687$

45. $\dfrac{73{,}810}{19{,}683} \approx 3.749936$ **47.** $-109{,}200$ **49.** $-\dfrac{3}{2}$ **51.** Series diverges   **53.** $\dfrac{2{,}476{,}099}{160{,}000} \approx 15.475619$

55. Series diverges   **57.** $\dfrac{28{,}561}{152{,}064} \approx 0.187822$ **59.** $\dfrac{123}{999}$ **61.** $-\dfrac{35}{9}$ **63.** $\dfrac{989}{99}$ **65.** $\$14{,}802.44$ **67.** 1.845×10^{19}

69. Approximately 13,778 students   **71.** $S_{30} = 1.1 \times 10^{30}; r = 10$

73. Yes; explanations will vary (any example such that $r = 1$ and $d = 0$).

Section 11.4 Mathematical Induction

1. $S_{k+1} = \dfrac{1}{3k + 9}$ **3.** $S_{k+1} = \dfrac{(k+1)(k+2)(2k+3)}{4}$

5. Basic Step: $n = 1, 1 = 1$ and $\dfrac{1(1+1)}{2} = 1$; Induction Step: If $1 + 2 + 3 + \cdots + k = \dfrac{k(k+1)}{2}$, then

$$(1 + 2 + 3 + \cdots + k) + (k + 1) = \dfrac{k(k+1)}{2} + (k + 1) = \dfrac{k^2 + k + 2k + 2}{2} = \dfrac{(k+1)(k+2)}{2}$$

7. Basic Step: $n = 1, 2(1) = 2$ and $1(1+1) = 2$; Induction Step: If $2 + 4 + 6 + \cdots + 2k = k(k+1)$, then

$$(2 + 4 + 6 + \cdots + 2k) + 2(k + 1) = k^2 + k + 2k + 2 = (k+1)(k+2)$$

9. Basic Step: $n = 1, 4^{1-1} = 1$ and $\dfrac{4^1 - 1}{3} = 1$; Induction Step: If $4^0 + 4^1 + 4^2 + \cdots + 4^{k-1} = \dfrac{4^k - 1}{3}$, then

$$4^0 + 4^1 + 4^2 + \cdots + 4^{k-1} + 4^{k+1-1} = \dfrac{4^k - 1}{3} + 4^k = \dfrac{4^k - 1 + 3 \cdot 4^k}{3} = \dfrac{4 \cdot 4^k - 1}{3} = \dfrac{4^{k+1} - 1}{3}$$

11. Basic Step: $n = 1$, $\dfrac{1}{(3(1)-2)(3(1)+1)} = \dfrac{1}{4}$ and $\dfrac{1}{3(1)+1} = \dfrac{1}{4}$;

Induction Step: If $\dfrac{1}{1\cdot4} + \dfrac{1}{4\cdot7} + \dfrac{1}{7\cdot10} + \cdots + \dfrac{1}{(3k-2)(3k+1)} = \dfrac{k}{3k+1}$, then

$$\dfrac{1}{1\cdot4} + \dfrac{1}{4\cdot7} + \dfrac{1}{7\cdot10} + \cdots + \dfrac{1}{(3k-2)(3k+1)} + \dfrac{1}{(3(k+1)-2)(3(k+1)+1)}$$

$$= \left[\dfrac{1}{1\cdot4} + \dfrac{1}{4\cdot7} + \dfrac{1}{7\cdot10} + \cdots + \dfrac{1}{(3k-2)(3k+1)}\right] + \dfrac{1}{(3k+1)(3k+4)} = \dfrac{k}{3k+1} + \dfrac{1}{(3k+1)(3k+4)}$$

$$= \dfrac{3k^2 + 4k + 1}{(3k+1)(3k+4)} = \dfrac{(3k+1)(k+1)}{(3k+1)(3k+4)} = \dfrac{(k+1)}{(3(k+1)+1)}$$

13. Basic Step: $n = 1, 5(1) = 5$ and $\dfrac{5(1)(1+1)}{2} = 5$; Induction Step: If $5 + 10 + 15 + \cdots + 5k = \dfrac{5k(k+1)}{2}$, then

$$5 + 10 + 15 + \cdots + 5k + 5(k+1) = (5 + 10 + 15 + \cdots + 5k) + 5k + 5 = \dfrac{5k(k+1)}{2} + 5k + 5$$

$$= \dfrac{5k^2 + 15k + 10}{2} = \dfrac{5(k+1)(k+2)}{2} = \dfrac{5(k+1)[(k+1)+1]}{2}$$

15. Basic Step: $n = 1, 1 + \dfrac{1}{1} = 2$ and $1 + 1 = 2$; Induction Step: If $\left(1+\dfrac{1}{1}\right)\left(1+\dfrac{1}{2}\right)\left(1+\dfrac{1}{3}\right)\cdots\left(1+\dfrac{1}{k}\right) = k+1$, then

$$\left(1+\dfrac{1}{1}\right)\left(1+\dfrac{1}{2}\right)\left(1+\dfrac{1}{3}\right)\cdots\left(1+\dfrac{1}{k}\right)\left(1+\dfrac{1}{k+1}\right) = (k+1)\left(1+\dfrac{1}{k+1}\right) = k+1+\dfrac{k+1}{k+1} = (k+1)+1$$

17. Basic Step: $n = 1, 3(1) - 2 = 1$ and $\dfrac{1}{2}(3(1)-1) = 1$; Induction Step: If $1 + 4 + 7 + 10 + \cdots + (3k-2) = \dfrac{k}{2}(3k-1)$,

then $\left[1 + 4 + 7 + 10 + \ldots + (3k-2)\right] + \left[3(k+1)-2\right] = \dfrac{k}{2}(3k-1) + (3k+1)$

$$= \dfrac{k(3k-1) + 2(3k+1)}{2} = \dfrac{3k^2 + 5k + 2}{2} = \dfrac{(k+1)(3k+2)}{2} = \dfrac{k+1}{2}(3(k+1)-1)$$

19. Basic Step: $n = 2$, $3^2 = 9$ and $2(2) + 1 = 5$, so $3^2 > 2(2)+1$;

Induction Step: If $3^k > 2k + 1$, then $3^{k+1} = 3^1 \cdot 3^k > 3(2k+1) = 6k + 3 > 2k + 3 = 2k + 2 + 1 = 2(k+1)+1$

21. Basic Step: $n = 1, 1^3 = 1$ and $\dfrac{1^2(1+1)^2}{4} = 1$;

Induction Step: If $1^3 + 2^3 + 3^3 + 4^3 + \ldots + k^3 = \dfrac{k^2(k+1)^2}{4}$, then

$$(1^3 + 2^3 + 3^3 + 4^3 + \ldots + k^3) + (k+1)^3 = \dfrac{k^2(k+1)^2}{4} + (k+1)^3$$

$$= \dfrac{k^2(k+1)^2 + 4(k+1)^3}{4} = \dfrac{(k+1)^2(k+2)^2}{4} = \dfrac{(k+1)^2((k+1)+1)^2}{4}$$

23. Basic Step: $n = 1$, $a^1 = a$ so $a^1 > 1$, when $a > 1$; Induction Step: If $a^k > 1$, then $a^{k+1} = a^k \cdot a^1 > 1 \cdot a = a > 1$

25. Basic Step: $n = 1$, $1^4 = 1$ and $\dfrac{1(1+1)(2(1)+1)(3(1)^2 + 3(1) - 1)}{30} = 1$;

Induction Step: If $1^4 + 2^4 + 3^4 + \ldots + k^4 = \dfrac{k(k+1)(2k+1)(3k^2 + 3k - 1)}{30}$, then

$(1^4 + 2^4 + 3^4 + \ldots + k^4) + (k+1)^4 = \dfrac{k(k+1)(2k+1)(3k^2 + 3k - 1)}{30} + (k+1)^4$

$= \dfrac{6k^5 + 45k^4 + 130k^3 + 180k^2 + 119k + 30}{30} = \dfrac{(k+1)(k+2)(2k+3)(3k^2 + 9k + 5)}{30}$

$= \dfrac{(k+1)(k+2)(2(k+1)+1)(3(k+1)^2 + 3(k+1) - 1)}{30}$

27. Basic Step: $n \geq 2$, $\dfrac{1}{\sqrt{1}} + \dfrac{1}{\sqrt{2}} = 1 + \dfrac{\sqrt{2}}{2}$ and $1 + \dfrac{\sqrt{2}}{2} > \sqrt{2}$; Induction Step: If $\dfrac{1}{\sqrt{1}} + \dfrac{1}{\sqrt{2}} + \dfrac{1}{\sqrt{3}} + \ldots + \dfrac{1}{\sqrt{k}} > \sqrt{k}$,

then $\left[\dfrac{1}{\sqrt{1}} + \dfrac{1}{\sqrt{2}} + \dfrac{1}{\sqrt{3}} + \ldots + \dfrac{1}{\sqrt{k}}\right] + \dfrac{1}{\sqrt{k+1}} > \sqrt{k} + \dfrac{1}{\sqrt{k+1}} = \dfrac{\sqrt{k}\left(\sqrt{k+1}\right) + 1}{\sqrt{k+1}}$

$= \dfrac{\sqrt{k^2 + k} + 1}{\sqrt{k+1}} > \dfrac{\sqrt{k^2} + 1}{\sqrt{k+1}} = \dfrac{k+1}{\sqrt{k+1}} = \sqrt{k+1}$

29. Basic Step: $n = 1$, $(ab)^1 = ab$ and $a^1 b^1 = ab$; Induction Step: If $(ab)^k = a^k b^k$, then

$(ab)^{k+1} = (ab)^k \cdot (ab) = a^k b^k \cdot ab = \left(a \cdot a^k\right)\left(b \cdot b^k\right) = a^{k+1} b^{k+1}$

31. Basic Step: $n = 1$, $\ln(x_1) = \ln x_1$; Induction Step: If $\ln(x_1 \cdot x_2 \cdot x_3 \cdot \ldots \cdot x_k) = \ln x_1 + \ln x_2 + \ln x_3 + \ldots + \ln x_k$ when

$x_1 > 0, x_2 > 0, \ldots, x_n > 0$, then $\ln(x_1 \cdot x_2 \cdot x_3 \cdot \ldots \cdot x_k \cdot x_{k+1}) = \ln(x_1 \cdot x_2 \cdot x_3 \cdot \ldots \cdot x_k) + \ln(x_{k+1})$

$= (\ln x_1 + \ln x_2 + \ln x_3 + \ldots + \ln x_k) + \ln x_{k+1}$

33. Basic Step: $n = 2$, $(9^2 - 8(2) - 1) = 64$ of which 64 is a factor;

Induction Step: If $(9^k - 8k - 1) = 64p$ for some integer p, then

$(9^{k+1} - 8(k+1) - 1) = 9 \cdot 9^k - 8k - 9 = 9 \cdot 9^k - 9 \cdot 8k + 8 \cdot 8k - 9 = 9(9^k - 8k - 1) + 64k = 9(64p) + 64k = 64(9p + k)$

35. Basic Step: $n = 1$, $(1^3 - 1 + 3) = 3$, which is divisible by 3; Induction Step: If $\dfrac{k^3 - k + 3}{3} = p$

or $k^3 - k + 3 = 3p$ for some integer p, then $(k+1)^3 - (k+1) + 3 = k^3 + 3k^2 + 2k + 3 = (k^3 - k + 3) + (3k^2 + 3k)$

$= 3p + 3(k^2 + k) = 3(p + k^2 + k)$

37. Basic Step: $n = 1$, $1(1+1)(1+2) = 6$, which is divisible by 6; Induction Step: If $\dfrac{k(k+1)(k+2)}{6} = p$

or $k(k+1)(k+2) = 6p$ for some integer p, then $(k+1)(k+2)(k+3) = k^3 + 6k^2 + 11k + 6$

$= \left(k^3 + 3k^2 + 2k\right) + \left(3k^2 + 9k + 6\right) = k(k+1)(k+2) + 3(k+1)(k+2) = 6p + 3(k+1)(k+2)$.

$6p$ is clearly divisible by 6. In order for $3(k+1)(k+2)$ to be divisible by 6, it must be divisible by 2 and 3.

It is clearly divisible by 3. If k is odd, then the term $(k+1)$ must be even, making it divisible by 2.

If k is even, then the term $(k+2)$ is even, making it divisible by 2. Therefore, $3(k+1)(k+2)$ is divisible by 6.

39. $0 + 1 + 2 + 3 + \ldots + (n-1) = \dfrac{n(n-1)}{2}$; Basic Step: $n = 1$, $(1-1) = 0$ and $\dfrac{1(1-1)}{2} = 0$; Induction Step: If $0 + 1 + 2 + \ldots$

$+ (k-1) = \dfrac{k(k-1)}{2}$, then $\left[0 + 1 + 2 + \ldots + (k-1)\right] + (k+1-1) = \dfrac{k(k-1)}{2} + k = \dfrac{k^2 - k + 2k}{2} = \dfrac{k(k+1)}{2}$

$= \dfrac{(k+1)((k+1)-1)}{2}$

41. The Induction Step does not work for $n = 1$. In the case of $n = 1$, $n + 1 = 2$ and the groups formed by removing the first horse and then the last horse do not overlap.

Section 11.5 An Introduction to Combinatorics

1. Combination   **3.** Combination   **5.** $10^3 = 1000$ **7.** $9^7 = 4,782,969$ **9.** $15! \approx 1.308 \times 10^{12}$ **11.** $3! = 6$

13. $5^{10} = 9,765,625$ **15.** $36^6 = 2,176,782,336$ **17.** $26 \cdot 25 \cdot 24 \cdot 10 \cdot 9 \cdot 8 = 11,232,000$ **19.** $_{30}P_{12} \approx 4.143 \times 10^{16}$

21. $_{36}P_8 \approx 1.220 \times 10^{12}$ **23.** $_7P_6 = 5040$; $_7P_7 = 5040$ as well. (Having a child remain standing is numerically

equivalent to putting a seventh chair in the room.)   **25.** $_{26}P_3 = 15,600$ **27.** 210   **29.** 6.0823×10^{16} **31.** 6

33. 116,280   **35.** $_7C_3 = 35$ **37.** $_9C_2 = 36$ **39.** $_{75}C_5 = 17,259,390$ **41.** $\dfrac{6!}{2!3!} = 60$ **43.** $\dfrac{7!}{2!} = 2520$

45. $\dfrac{9!}{2!2!} = 90,720$ **47.** $_{10}C_4 \cdot {}_8C_4 \cdot {}_{13}C_4 = 10,510,500$ **49.** 112 cones   **51.** 96 outfits   **53.** 288 schedules

55. 120 5-letter strings   **57.** 303,600 ways   **59.** 495 pizzas   **61.** 752,538,150 groups   **63.** 420 ways

65. $x^7 - 14x^6y + 84x^5y^2 - 280x^4y^3 + 560x^3y^4 - 672x^2y^5 + 448xy^6 - 128y^7$ **67.** $x^8 - 4x^6y^3 + 6x^4y^6 - 4x^2y^9 + y^{12}$

69. $4096x^6 + 30,720x^5y^2 + 96,000x^4y^4 + 160,000x^3y^6 + 150,000x^2y^8 + 75,000xy^{10} + 15,625y^{12}$

71. $x^{15} - 5x^{12}y^2 + 10x^9y^4 - 10x^6y^6 + 5x^3y^8 - y^{10}$ **73.** -80 **75.** $8192x^{13} + 159,744x^{12} + 1,437,696x^{11}$

77. $2,008,326,144x^{14}$ **79.** $831,409,920x^4y^8$

81. $\dbinom{n}{n-k} = \dfrac{n!}{(n-k)!(n-(n-k))!} = \dfrac{n!}{(n-k)!(n-n+k)!} = \dfrac{n!}{(n-k)!k!} = \dbinom{n}{k}$

83. $2^n = (1+1)^n = \displaystyle\sum_{k=0}^{n} \dbinom{n}{k}(1)^k(1)^{n-k} = \sum_{k=0}^{n} \dbinom{n}{k} = \dbinom{n}{0} + \dbinom{n}{1} + \ldots + \dbinom{n}{n}$

Section 11.6 An Introduction to Probability

1. $\dfrac{3}{5}$ **3.** $\dfrac{9}{13}$ **5.** $\dfrac{1}{3}$ **7. a.** 0 **b.** $\dfrac{5}{8}$ **9. a.** 0 **b.** $\dfrac{3}{5}$ **11. a.** 0 **b.** 1   **13. a.** $\dfrac{1}{8}$ **b.** $\dfrac{9}{16}$ **15.** The set of all ordered

4-tuples made up of H's and T's. There are 16 such 4-tuples.   **17.** The set of all ordered pairs that have either

an H or a T in the first slot and one of the 13 hearts in the second slot. There are 26 such ordered pairs.

19. The set of all ordered triples with any of the 6 values in each slot. There are 216 such triples.

21. The set of the 38 pockets.   **23. a.** $\dfrac{2}{3}$ **b.** $\dfrac{1}{3}$ **25. a.** $\dfrac{3}{8}$ **b.** $\dfrac{1}{8}$ **c.** $\dfrac{1}{2}$ **27.** $\dfrac{3}{10}$ **29.** $\dfrac{387,420,489}{1,000,000,000} \approx 0.3874$

31. $\dfrac{3}{8}$ **33. a.** $\dfrac{11}{26}$ **b.** $\dfrac{9}{52}$ **c.** $\dfrac{2}{13}$ **35. a.** $\dfrac{1}{169}$ **b.** $\dfrac{1}{221}$ **37.** 18.75%   **39. a.** $\dfrac{1}{6}$ **b.** $\dfrac{5}{18}$ **c.** $\dfrac{1}{9}$ **41.** $\dfrac{1}{20}$

43. $\dfrac{2}{9}$ **45.** $\dfrac{3}{10}$ **47.** 27 tickets

Chapter 11 Review

1. $-3, 9, -27, 81, -243$ **2.** $-1, \sqrt[3]{2}, -\sqrt[3]{3}, \sqrt[3]{4}, -\sqrt[3]{5}$ **3.** $-3, -4, -5, -6, -7$ **4.** $1, \dfrac{1}{2}, \dfrac{2}{9}, \dfrac{3}{32}, \dfrac{24}{625}$ **5.** $a_n = 6n - 13$

6. $a_n = \dfrac{3^{n-1}}{2^n}$ **7.** $a_n = n^2 - 1$ **8.** $a_n = \dfrac{2n+1}{n+1}$ **9.** $a_1 = -2, a_n = n(a_{n-1})$ for $n \geq 2$ **10.** $a_n = n^2 + n$

11. $-3 - 5 - 7 - 9 - 11 - 13 = -48$ **12.** $-2 + 4 - 8 + 16 - 32 + 64 = 42$ **13.** $\displaystyle\sum_{i=2}^{7} i^3 = 783$

14. $-1 + 1 + 3 + 5 + 7 + 9 = 24$ **15.** $-8 - 16 - 32 - 64 - 128 = -248$ **16.** $\displaystyle\sum_{i=2}^{10} 2i^2 = 768$

17. $S_n = \dfrac{n}{2(n+2)}, S_{80} = \dfrac{20}{41}$ **18.** $S_n = \dfrac{n}{2(n+2)}, S = \dfrac{1}{2}$ **19.** $S_n = 3 - 3^{n+1}$, series diverges

20. $-2, 5, 3, 8, 11$ **21.** $-10, -12, -22, -34, -56$ **22.** $a_n = -3n + 8$ **23.** $a_n = \dfrac{5}{2}n + 9$ **24.** $a_n = -9n + 20$

25. $a_n = 3n - 1$ **26.** $a_n = \dfrac{3}{4}n - \dfrac{1}{4}$ **27.** $a_n = 9n - 14$ **28.** -146 **29.** 275   **30.** a_{61} **31.** a_{36} **32.** a_{25}

33. 8827   **34.** -7140 **35.** 66   **36.** $a_n = 2(4)^{n-1}$ **37.** $a_n = 3\left(\dfrac{1}{5}\right)^{n-1}$ **38.** $a_n = 18\left(-\dfrac{1}{3}\right)^{n-1}$ **39.** $a_n = 6(4)^{n-1}$

40. $a_n = 10(2)^{n-1}$ **41.** $a_n = 8\left(\dfrac{1}{4}\right)^{n-1}$ **42.** $r = 3$; $4, 12, 36, 108, 324$ **43.** $r = \pm\dfrac{2}{3}$; $\pm\dfrac{45}{8}, \dfrac{15}{4}, \pm\dfrac{5}{2}, \dfrac{5}{3}, \pm\dfrac{10}{9}$

44. $\pm\dfrac{1}{3}$ **45.** -2 **46.** $-\dfrac{1}{256}$ **47.** $\dfrac{381}{512}$ **48.** 40,955   **49.** -12 **50.** Series diverges   **51.** 1

52. Basic Step: $n = 1, 1^2 = 1$ and $\dfrac{1(1+1)(2(1)+1)}{6} = 1$; Induction Step: If $1 + 4 + 9 + \cdots + k^2 = \dfrac{k(k+1)(2k+1)}{6}$,

then $\left(1 + 4 + 9 + \cdots + k^2\right) + (k+1)^2 = \dfrac{k(k+1)(2k+1)}{6} + (k+1)^2$

$= \dfrac{(k+1)\left[2k^2 + k + 6(k+1)\right]}{6} = \dfrac{(k+1)\left[2k^2 + 7k + 6\right]}{6} = \dfrac{(k+1)(k+2)(2k+3)}{6} = \dfrac{(k+1)((k+1)+1)(2(k+1)+1)}{6}$

53. Basic Step: $n = 1$, $\dfrac{1}{(2(1)-1)(2(1)+1)} = \dfrac{1}{3}$ and $\dfrac{1}{2(1)+1} = \dfrac{1}{3}$; Induction Step: If

$$\dfrac{1}{1\cdot 3} + \dfrac{1}{3\cdot 5} + \dfrac{1}{5\cdot 7} + \cdots + \dfrac{1}{(2k-1)(2k+1)} = \dfrac{k}{2k+1}, \text{ then } \dfrac{1}{1\cdot 3} + \dfrac{1}{3\cdot 5} + \dfrac{1}{5\cdot 7} + \cdots + \dfrac{1}{(2k-1)(2k+1)}$$

$$+\dfrac{1}{(2(k+1)-1)(2(k+1)+1)} = \dfrac{k}{2k+1} + \dfrac{1}{(2k+1)(2k+3)} = \dfrac{2k^2+3k+1}{(2k+1)(2k+3)} = \dfrac{(2k+1)(k+1)}{(2k+1)(2(k+1)+1)} = \dfrac{(k+1)}{(2(k+1)+1)}$$

54. Basic Step: $n = 1$, $(3(1)+2) = 5$ and $\dfrac{1(3(1)+7)}{2} = 5$; Induction Step: If $5+8+11+\cdots+(3k+2) = \dfrac{k(3k+7)}{2}$,

then $5+8+11+\cdots+(3k+2)+(3(k+1)+2) = \dfrac{k(3k+7)}{2} + (3k+5)$

$$= \dfrac{3k^2+7k+6k+10}{2} = \dfrac{(k+1)(3k+10)}{2} = \dfrac{(k+1)(3(k+1)+7)}{2}$$

55. Basic Step: $n = 1$, $1(1+2) = 3$ and $\dfrac{1(1+1)(2(1)+7)}{6} = 3$;

Induction Step: If $1\cdot 3 + 2\cdot 4 + 3\cdot 5 + \cdots + k(k+2) = \dfrac{k(k+1)(2k+7)}{6}$, then $1\cdot 3 + 2\cdot 4 + 3\cdot 5 + \cdots + k(k+2)$

$$+(k+1)(k+1+2) = \dfrac{k(k+1)(2k+7)}{6} + (k+1)(k+3) = \dfrac{(k+1)[k(2k+7)+6(k+3)]}{6}$$

$$= \dfrac{(k+1)(k+2)(2k+9)}{6} = \dfrac{(k+1)((k+1)+1)(2(k+1)+7)}{6}$$

56. Basic Step: $n = 1$, $11^1 - 7^1 = 4$, which is divisible by 4;

Induction Step: If $\dfrac{11^k - 7^k}{4} = p$ or $11^k - 7^k = 4p$ for some integer p, then

$$11^{k+1} - 7^{k+1} = 11\cdot 11^k - 7\cdot 7^k = 4\cdot 11^k + 7\cdot 11^k - 7\cdot 7^k = 4\cdot 11^k + 7(11^k - 7^k) = 4\cdot 11^k + 7(4p) = 4(11^k + 7p)$$

57. Basic Step: $n = 1$, $7^1 - 1 = 6$, and $2\cdot 3 = 6$; Induction Step: If $\dfrac{7^k - 1}{3} = p$ or $7^k - 1 = 3p$ for some integer p,

then $7^{k+1} - 1 = 7\cdot 7^k - 7 + 6 = 7(7^k - 1) + 6 = 7(3p) + 6 = 21p + 6 = 3(7p+2)$

58. $8\cdot 10^3\cdot 26^3 = 140,608,000$ **59.** $8^7 = 2,097,152$ **60.** $\dfrac{8!}{3!2!} = 3360$ **61.** $_{10}P_3 = 720$ **62.** $_{21}P_5 = 2,441,880$

63. $_{12}C_4 = 495$ **64.** $-32y^5 + 80y^4 - 80y^3 + 40y^2 - 10y + 1$

65. $x^7 + 14x^6 + 84x^5 + 280x^4 + 560x^3 + 672x^2 + 448x + 128$

66. $3125x^{10} - 6250x^8y + 5000x^6y^2 - 2000x^4y^3 + 400x^2y^4 - 32y^5$ **67. a.** $\dfrac{1}{9}$ **b.** $\dfrac{2}{3}$ **68. a.** 0   **b.** $\dfrac{5}{6}$ **69. a.** $\dfrac{2}{7}$

b. $\dfrac{4}{7}$ **70. a.** $\dfrac{1}{8}$ **b.** $\dfrac{3}{4}$ **71. a.** $\dfrac{1}{10}$ **b.** $\dfrac{4}{5}$ **72. a.** $\dfrac{4}{13}$ **b.** $\dfrac{5}{26}$ **c.** $\dfrac{7}{26}$ **73.** $\dfrac{33}{108,290}$ **74.** Approximately 52.17%

Chapter 11 Test

1. $-3, -5, -7, -9, -11$ **2.** $-1, -4, -9, -16, -25$ **3.** $1, 3, 6, 10, 15$ **4.** $1, \dfrac{1}{2}, \dfrac{1}{6}, \dfrac{1}{24}, \dfrac{1}{120}$ **5.** $a_n = n^2 - 2$

6. $a_n = \dfrac{(-1)^n}{n^2}$ **7.** $-2 - 8 - 18 - 32 - 50 - 72 - 98 - 128 - 162 = -570$

8. $-2 - 9 - 20 - 35 - 54 - 77 - 104 - 135 - 170 - 209 - 252 - 299 = -1366$ **9.** $S_n = \dfrac{n}{2n+4}; S_{100} = \dfrac{25}{51}$

10. $S_n = \ln \dfrac{2}{n+2}; S_{80} = \ln \dfrac{1}{41}$ **11.** $3, 6, 9, 15, 24$ **12.** $4, -2, 0, -2, -4$ **13.** $a_n = -4n + 3$ **14.** $a_n = 8n - 12$

15. $a_n = -3 \cdot 2^{n-1}$ **16.** $a_n = \dfrac{9}{5}\left(\dfrac{2}{3}\right)^{n-1}$ **17.** Yes, $S = \dfrac{1}{2}$ **18.** No   **19.** $\displaystyle\sum_{k=1}^{6} a_k = 54$

20. Basic Step: $n = 4$, $4! = 24$, and $2^4 = 16$; so $4! > 2^4$; Induction Step: If $k! > 2^k$, then
$(k+1)! = k!(k+1) > 2^k (k+1) \geq 2^k (4+1) > 2^k \cdot 2 = 2^{k+1}$

21. Basic Step: $n = 1$, $5(1) - 3 = 2$ and $\dfrac{1}{2}(5(1) - 1) = 2$;

Induction Step: If $2 + 7 + 12 + 17 + \ldots + (5k - 3) = \dfrac{k}{2}(5k - 1)$ then

$2 + 7 + 12 + 17 + \ldots + (5k - 3) + (5(k+1) - 3) = \dfrac{k}{2}(5k - 1) + (5(k+1) - 3) = \dfrac{5k^2 - k}{2} + 5k + 5 - 3$

$= \dfrac{5k^2 - k}{2} + \dfrac{10k + 4}{2} = \dfrac{5k^2 + 9k + 4}{2} = \dfrac{(k+1)}{2}(5k + 4) = \left(\dfrac{k+1}{2}\right)(5k + 5 - 1) = \left(\dfrac{k+1}{2}\right)(5(k+1) - 1)$

22. $10,000,000$ **23.** $3,276,000$ **24.** 180 **25.** $a^6 - 18a^5b + 135a^4b^2 - 540a^3b^3 + 1215a^2b^4 - 1458ab^5 + 729b^6$

26. $8b^3 - 24b^2c + 24bc^2 - 8c^3$ **27.** $\dfrac{1}{9}$ **28. a.** $\dfrac{52}{125}$ **b.** $\dfrac{4}{5}$ **c.** $\dfrac{37}{500}$ **29. a.** $\dfrac{3}{11}$ **b.** $\dfrac{3}{44}$ **c.** $\dfrac{1}{22}$ **d.** $\dfrac{3}{55}$

Index

Symbols

2×2 matrix
  determinant 810

A

Abel, Niels 310
Absolute value 9, 653
  equations 63
  function 244
  inequalities 75
  properties of 11
  triangle inequality 11
Addition
  of complex numbers 54
  of functions 274
  of matrices 823
  of polynomials 40
  of vectors 666
Additive inverse
  of a vector 667
Adjacent leg 477
Algebra
  of matrices 830
  of polynomials 40
Algebraically complete 52
Algebraic expressions 18, 39
Almagest 552
Amplitude
  of cosine 514, 518
  of sine 514, 518
Angle 461
  complementary 500
  reference 496
Angular speed 468
Apollonius 704
Applications
  of dot products 683
  of exponential functions 404
  of inverse trigonometric
     functions 531
  of logarithmic functions 433
  of systems of equations 788
  of trigonometric functions 483
Approximating
  solutions by graphing 863
Arccosine 527

Archimedes 704
Arc length 467
Arcsine 527
Arctangent 527
Area
  of a rectangle 25
  of a trapezoid 26
  of a triangle 617
    Heron's formula 618
    sine formula 618
Argument 218, 655
Arithmetic
  progression 906
  sequence 906
    common difference 906
    general term 907
    partial sums 910, 911
  series 906
Associative property 12
Asymptote(s)
  horizontal 365, 368
  hyperbolas 733
  notation 366
  oblique 365, 368, 732
  vertical 364, 366, 367
Augmented matrices 796, 797
Axis 136
  horizontal 652
  imaginary 652
  of a parabola 228
  of symmetry 228
  polar 626
  real 652
  vertical 652

B

Back-substitution
  for matrices 797, 802
Base
  change of 433
  of exponents 19
  of exponential functions 395
  of logarithmic functions 418
Behavior 358
  of a function 240
    even exponents 241
    odd exponents 240
  of polynomials 313, 314
Binomial(s) 39
  coefficients 947
  expansion 947
  special 44

theorem 946, 947
Boundary line 179
Bounded feasible region 856
Bounds of real zeros 344
  lower 344
  upper 344

C

Cancellation properties 13
  for inequalities 72
Cardinality
  union of sets 959
Cartesian
  coordinate system 136
    plotting points 137
  plane 136
Cartesian coordinates
  converting to and from polar
    coordinates 628
Caution!
  cancelling factors 99
  composition of functions 277
  determinant notation 810
  determinant of zero 816
  domain of composite functions
    278
  extraneous solutions 65
  factoring 44
  $f(x)$ 217
  f^{-1} 289
  graphs of logarithmic functions
    419
  horizontal shifting 261
  infinity symbols 7
  intermediate value theorem 347
  inverse matrix method 839
  linear factors theorem 352
  matrix addition 823
  matrix multiplication 828
  order 5
  parentheses 136
  polynomial long division 331
  probability 956
  properties of exponents 23
  properties of logarithms 430
  properties of radicals 30
  rational inequalities 374
  rational zero theorem 340
  scientific notation 23
  sigma notation 896
  simplifying radical expressions 57
  slope 159

upper and lower bounds theorem 346

vertical line test 214

Center 189

Change of base formula 432, 433

Chebyshev, Pafnuty Lvovich 578

Chebyshev polynomials 578

Circle 189, 634

center 189

radius 189

standard form 189

unit 461, 462

Circular sector 470

Circumference 463

Closed interval 7

Closure property

of real numbers 12

Codomain 218

Coefficient 11, 39

binomial 947

leading 40, 311, 340, 359

of a polynomial 39

Cofactor

of a matrix 811

sign matrix 811

Cofunction identities 500, 553

Combination 943

formula 944

Combinatorics 939

binomial coefficient 947

binomial theorem 947

combination 943

formula 944

computing probabilities 957

factorial notation 941

multiplication principle of counting 939, 940

permutation 941

formula 942

Combining

rational expressions 99

Common

difference 906

ratio 916

Common factoring methods 42

Common logarithms 423

Commonly encountered angles 464

Commutative property 12

Complement 500

Complementary angles 500

Completing the square 84, 192, 232

Complex

conjugate(s) 55, 86

number(s) 54, 281

algebra of 54

dividing 656, 658, 660

magnitude of 653

multiplying 656, 658, 660

powers of 659

product of 656, 658, 660

quotient of 656, 658, 660

roots of 660

trigonometric form of 655

plane 652

rational expressions 101

simplifying 101

roots 356

solutions 86

Component form of a vector 668, 672

Composing functions 276

Composition

of functions and inverses 293

Compound inequalities 74

Compound interest 407, 408

application 445

continuous 411

formula 408

Compressing 264

Condensing logarithmic expressions 431

Conditional 557

Conditional equations 60

Conic sections 705

discriminant of 749

focus/directrix definition of 755–756

polar form of 758

Conjugate

pairs of zeros 356, 358

radical expression 30

roots theorem 356, 844

Consistent 781, 782

Constant 11

functions 227

nonzero 13

of proportionality 251

term 39

Constraints 854

Constructing polynomials 357

with given zeros 336

Continuous compounding interest 411, 428

Continuous function 319

Contradiction 60

Converge 899

Conversion formulas 463

Converting

between degrees and radians 463

between exponential and logarithmic forms 441

from Cartesian to polar coordinates 628

from polar to Cartesian coordinates 628

Coordinate(s)

conversion 628

polar 626, 755

system

Cartesian 136

polar 626

Cosecant 478

Cosine

amplitude of 514, 518

Cosines

law of 615

use of 610, 615

Counting

multiplication principle of 940

Counting numbers 3

Cramer, Gabriel 810

Cramer's rule 817, 818

for two-variable, two-equation systems 816

Cube root 26

D

Damped harmonic motion 520

Decibel scale 436

Decomposing functions 280

Decomposition pattern 845

Dedekind, Richard 2

Degenerate triangle 493

Degree, minute, second (DMS) notation 482

Degree(s) 461

converting to radians 463

n 311

of a polynomial 39

of terms 39

symbol 464

DeMoivre, Abraham 659

DeMoivre's Theorem 659

Dependent

system of equations 781, 782

variable 216

Descartes, René 134, 135, 704
Descartes' rule of signs 342, 343, 359
Descending order 40
Determinant(s)
  evaluating 819
  of a 2×2 matrix 810
  of an $n\times n$ matrix 812
  properties of 813
Difference
  of two cubes 44
  of two squares 44
Directed magnitudes 665
Direction 665
Directrix 706, 720, 755
Direct variation 251
Dirichlet, Lejeune 208
Discriminant 86, 87
  of a conic section 749
Disjoint 959, 961
Distance 66, 75, 641
  formula 141, 189
  on the real number line 10
Distributive property 12
Diverge 899
Division
  algorithm 327, 352
  of complex numbers 656, 658, 660
  of functions 274
  polynomial long 328, 329
    with complex numbers 331
  synthetic 332
    with complex numbers 335
Divisor 327, 329, 332
Domain 209, 218, 274, 363, 369
  implied 220
  of trigonometric functions 492–496
  restriction of 291
Domain and codomain notation 218
Dot product 677
  applications of 683
  properties of 678
  theorem 679
Double
  root 82, 86
  solution 82
Double-angle identities 576, 578, 580

E

e 411
Eccentricity 713, 755
Elementary
  exponential equations 399
  logarithmic expressions 421
    evaluating 421
  row operations 799
Elimination 784
  Gaussian 797
  Gauss-Jordan 802
Ellipse 705, 706, 707, 750, 756
  centered at origin 708
  eccentricity 713
  equations 712
  graphing 708, 710, 711
  standard form 707
Empty set ($\varnothing$) 6
Equal
  vectors 666
Equality matrix 823
Equation(s)
  absolute value 63
  conditional 60
  contradiction 60
  ellipse 712
  equivalent 61
  first-degree 61
  graph of 138
  horizontal asymptotes 368
  hyperbola 735
  identity 60
  linear 61
    in two variables 149
      point-slope form 163
      slope-intercept form 161
      standard form 149
  oblique asymptotes 368
  parabola 724
  parametric 640
    graphing 640
  polar 630
  polynomial-like 92
  quadratic 81
  radical 106
  rational 103
  second-degree 81
  simultaneous linear 781
  symmetry of 269
  system of linear 781
  vertical asymptotes 366
Equivalent equations 61

Euclid 704
Euler, Leonhard 394
Euler's Formula 656
Evaluating
  determinants 819
  inverse trigonometric functions 528, 531, 532
  logarithmic expressions 421, 424
  partial sums
    and series 899
    of arithmetic sequences 910
    of geometric sequences 918
  sums 895, 897
Even functions 267
Even/odd identities 513, 554
Event(s) 955
  independent 962
  mutually exclusive 960
    unions of 961
  probability of a union of 960
Expanding logarithmic expressions 430
Expansion along a row or column 812
Experiment 955
Explicit formula 892
Exponential
  equations
    solving 399
  functions 395
    applications 404, 445
    base 395
    compound interest 408
    continuous compounding interest 411
    e 407
    graphing 397
    natural base 412
    population growth 404
    properties 397
    radioactive decay 406
Exponential and logarithmic
  equations 418, 441
  forms
    converting between 441
Exponent(s) 19
  base 19
  fractional 47
  natural number 19
  negative integer 21
  properties of 21
  rational number 32
  zero as an 20

Expression(s)
  algebraic 18, 39
  complex rational 101
    simplifying 102
  factoring 44
  rational 98
    combining 31
    simplifying 98
Extraneous solution 65, 103, 106

F

Factorable 42
Factorial notation 941
Factoring
  by grouping 43, 46
  expressions containing fractional
      exponents 47
  greatest common factor 42
  special binomials 44
  trinomials 45, 46
Factor(s) 11
  greatest common 42
  variable 11
Feasible region 854
  bounded 856
  unbounded 856
  vertex of 856
Fermat, Pierre de 890
Fibonacci, Leonardo 900
Fibonacci sequences 900
Field properties 12
Finding
  bounds of real zeros 344
  inverse function formulas 291
  the inverse of a matrix 837
Finite
  sequence 891
  series 898
First-degree
  equations 61
  function 225
Focus 706, 720, 721, 755
FOIL method 42
Formula(s)
  change of base 432, 433
  combination 944
  compound interest 408
  continuous compounding interest
      411
  conversion 463
  distance 141, 189
  Euler's 656

Heron's 618
  interest 66
  inverse functions 291
  midpoint 143
  n^{th} term
    of a geometric sequence 916
    of an arithmetic sequence 907
  partial sum 907, 911
  permutation 942, 943
  probability 956
  quadratic 86
  rotation 743
  sector area 470
  sine (area of a triangle) 618
  special product 42
  summation 896
Frequency 514
Function(s) 212
  addition of 274
  $ax^{\frac{1}{n}}$ 243
  $\dfrac{a}{x^n}$ 242
  ax^n 240
  behavior 240
  composing 276
  continuous 319
  division of 274
  even 267
  exponential 395
    applications 445
  first-degree 225
  greatest integer 246
  inverse 289
  linear 225
  logarithmic 417, 418
    applications 433
  multiplication of 274
  multivariable 255
  notation 216
  objective, in linear programming
      856
  odd 268
  one-to-one 289
  piecewise-defined 246
  quadratic 227
  rational 363
  second-degree 227
  subtraction of 274
Fundamental theorem of algebra
      351, 358, 844

G

Galois, Évariste 310
Gauss, Carl Friedrich 310, 351,
      704, 780, 797
Gaussian elimination 797
Gauss-Jordan elimination 802
General term
  arithmetic sequence 907
  geometric sequence 916
Geometric
  formulas 25
  meaning of multiplicity 355
  progression 916
  sequence(s) 915, 916
    general term 916
    partial sums 918, 919
  series 921
Graphing
  circles 191
  ellipses 708, 710
    standard form 711
  exponential functions 397
    with a calculator 401
  factored polynomials 313
  hyperbolas 734
    standard form 733
  intervals of real numbers 73
  logarithmic functions 419
    with a calculator 424
  parabolas 722, 723
    standard form 722
  parametric equations 640
  polynomial functions 316
  quadratic functions 231
  rational functions 369, 370
  transformations of trigonometric
      functions 513
  trigonometric functions 506
Graphs
  compressing 264
  of an equation 138
  of even roots 244
  horizontal shift 260
  of odd roots 244
  reflection 263
  stretching 264
  vertical shift 261
  x-axis symmetry 270
  y-axis symmetry 267
Gravity problems 89
Greatest common factor (GCF) 42
Greatest integer function 246

Guidance systems 736

H

Half-angle identities 580
Half-closed interval 7
Half-life 406
Half-open interval 7
Half-plane 179
Hamilton, William Rowan 608
Harmonic motion
  damped 520
  simple 514
Heron's formula 618
Hertz 514
Horizontal
  asymptote(s) 365, 368
    equations 368
  axis 652
  line 152
  line test 289
  shifting 260
  velocity 641
Hyperbola 705, 706, 730, 750,
    756
  asymptotes 733
  equations 731
  graphing 733, 734
  standard form 731, 735

I

i 52
  powers of 53
Identifying
  a quadrilateral 171
  parallel and perpendicular lines
    174
Identities
  cofunction 500, 553
  double-angle 576, 578, 580
  even/odd 513, 554
  half-angle 580
  period 553
  product-to-sum 582
  Pythagorean 554
  quotient 501, 553
  reciprocal 500
  sum and difference 563
  sum-to-product 584
  trigonometric 553
  verifying 557

Identity
  matrix 835
  of an equation 60
Identity property 12
Imaginary
  axis 652
  part 652
  unit (i) 52
Implied domain 220
Inconsistent 781, 782
Independent
  events 959, 962
  variable 216
Index
  of a radical 26
  of summation 895
Induction, mathematical 928
  basic step 929
  induction step 929
  proofs 929, 931
Inequalities
  absolute value 75, 183
  graphing 185
  joined by "and" or "or" 182
  linear 71
    two variables 179–180
  nonstrict 5
  strict 5
  symbols 5
  triangle 11
Infinite
  sequence 891
  series 898
    geometric 921, 922
Infinity symbol (∞) 7
Initial point 665
Initial side 462
Inner product 677
Integers 3
  exponents 21
Intercept(s) 151
  x 151
  y 151
Interest
  compound interest formula 408
  continuous compounding
    formula 411
  formula 66
Intermediate value theorem 346,
    359
Interpreting the linear factors
    theorem 352

Intersection 8, 959
  of two sets 182
Interval(s)
  closed 7
  defined 6
  half-closed 7
  half-open 7
  notation 7
  open 7
Invariant 749
Inverse
  2×2 matrix 837
  functions 289
    formulas of 291
  matrices
    on a calculator 841
  of a matrix 834, 835
  of a relation 287
Inverse matrix method 839
Inverse property 12
Inverse trigonometric functions
    524
  applications of 531
  evaluating 528
    on a calculator 532
  using to solve trigonometric
    equations 593
Inverse variation 252
Invertible matrices 837
i of the storm 282
Irrational numbers 3
Irreducible 42, 46
Iterate 281

J

Jointly proportional 254
Joint variation 254
Jordan, Wilhelm 802

K

Kepler
  laws of planetary motion 714
Kepler, Johannes 704

L

Lambert, Johann 2
Law of cosines 615
  use of 610, 615

Law of sines 610
  use of 610
Leading
  coefficient 40, 311, 340, 359
  term
    of a polynomial 313, 314
Least common denominator
    (LCD) 62
Leibniz, Gottfried 208
Lemniscates 634
Length
  of a vector 666
Like
  radicals 31
  terms 40
Limaçons 634
Linear combination of vectors 670
Linear equation(s)
  in one variable 61
  in two variables 149
  simultaneous 781
Linear factors
  of a polynomial 328
  theorem 351
Linear function 225
Linear inequalities 71
  compound 74
  in two variables 179
Linear programming 855, 856
  method for two-variable case
    856
Linear speed 468
Linear systems
  of equations 781, 782
Line(s)
  horizontal 152
  parallel 169
  perpendicular 172
  point-slope form 163, 164
  slope-intercept form 161
  slope of 158
  standard form 149
  vertical 152
Line symmetry 633
Logarithmic
  equations 418
    solving 423, 443
  expressions 421
    condensing 431
    evaluating 421, 424
    expanding 430
  functions 417, 418
    applications 433, 445

base 418
  graphing 419, 424
Logarithms
  change of base formula 433
  common 423
  natural 423
  properties of 429, 441
LORAN 704, 736, 737
Lower bounds of zeros 344

M

Magnitude(s) 10, 665
  of a complex number 653
  directed 665
  of eccentricity 755
  of a vector 666
Mandelbrot, Benoit 281
Mathematical induction 928
  basic step 929
  induction step 929
  principle of 928
  proofs 929
Matrices
  augmented 796
  back-substitution 797
  cofactors of 811
  columns of 795
  determinant 810
  elementary row operations 799
  elements of 795
  entries 795
  Gaussian elimination 797
  invertible 837
  minors 811
  multiplication of 828
  notation 795
  on a calculator 803
  reduced row echelon form 802
  row echelon form 797
  rows of 795
  scalar multiplication 826
Matrix 795
  addition 823
  algebra 830
  equality 823
  identity 835
  inverse 834, 835
  multiplication 827, 828
  substitution 826
  subtraction 827
Maximization 233

Maximum
  of graphs 234
Midpoint formula 143
Minimization 233
Minimum
  of graphs 234
Minor of a matrix 811
Models of population growth 404
Modulus 653
Monomial 39
Multiplication
  of complex numbers 656, 658,
    660
  of functions 274
  of matrices 828
  scalar 826
Multiplication principle of
    counting 939, 940
Multiplicity
  of zeros 353, 358
Multiplying inequalities by
    negative numbers 71
Multivariable function(s) 255
Mutually exclusive 961

N

Napier, John 394, 433
Natural logarithms 423
Natural number(s) 3
  exponents 19
Negative integer exponents 21
Newton, Sir Isaac 704
Norm
  of a vector 666
Notation
  degree, minute, second (DMS)
    482
  domain and codomain 218
  factorial 941
  interval 6
  matrix 795
  radical 26
  scientific 23, 24
  set-builder 6
  summation 895
    index of summation 895
    sigma 895
n^{th}
  degree polynomial 311, 358
  iterate 281
  roots 26
Numbers

complex 54
counting 3
e 411
integer 3
irrational 3
natural 3
rational 3
real 3
whole 3
$n \times n$ matrix
determinant 812

O

Objective function
in linear programming 856
Oblique asymptote(s) 365, 732
equations 368
Oblique triangle(s) 609
Odd functions 268
One-to-one functions 289
Onto 219
Open interval 7
Opposite leg 477
Order
of a matrix 795
of transformations 265
on the real number line 5
Ordered
pair(s) 135, 136
triple 785
Orientation 643
Origin 136, 152, 626
symmetry 268
Orthogonal vectors 680
Outcomes 955
Outer product 677

P

Pair
ordered 630
Parabola 227, 705, 706, 720,
750, 756
axis 228
of symmetry 228
graphing 722, 723
standard form 721, 724
vertex of 228
Parabolic mirrors 725
Paraboloid 725

Parallel lines 169
slopes 169
Parameter(s) 640
eliminating 644
Parametric curves 640
Parametric equation(s) 640
constructing 647
graphing 640
Partial fraction decomposition 844
Partial sums 898
arithmetic sequences 910
geometric sequences 918
Pascal, Blaise 890
Pascal's triangle 954
Perfect powers 27
cube 27
square 27
Perfect square
quadratic equations 83
trinomials 46
Period 516, 518
of a function 510
Periodic 510
Periodicity 508
Period identities 553
Permutation 941
formula 942
Perpendicular vectors 680
Perpendicular lines 172
slopes 172
Phase shift 517, 518
Phi (φ) 467
pH scale 434, 435
Piecewise-defined functions 246
Pi (π) 2
Planar feasible regions 854
Plane
Cartesian 136
complex 652
in three dimensional space 786
Planetary orbits 714
Plotted
ordered pairs 136
Plotting points 137
Point(s)
initial 665
terminal 665
Point-slope form of a line 163,
165
Polar axis 626
Polar axis symmetry 633

Polar coordinates 626, 755
converting to and from Cartesian
coordinates 628
Polar coordinate system 626
Polar equation(s) 630
common 634
graphs of 634
symmetry of 633
Polar form
of conic sections 758
Pole 626
Pole symmetry 633
Polynomial-like equations 93
Polynomial(s)
addition of 41
binomials 39
Chebyshev 578
equations 311
inequalities 318
irreducible 42
leading coefficient 40
long division 328, 329, 359
with complex numbers 331
monomials 39
multiplication of 41
parabola 227
prime 42
quadratic 227
roots of 311
of a single variable 40
solutions of 311
subtraction of 41
synthetic division 332
trinomials 39
Population growth 404
Positive rational number
exponents 109
Powers
of complex numbers 659
of *i* 53
Prime
polynomials 42
Principal square roots 57
Principle
of mathematical induction 928
Probability 955
event 955
experiment 955
independent events 962
outcomes 955
sample space 955
of a union of mutually exclusive
events 961

of a union of two events 960
using combinatorics 957
when outcomes are equally likely 956
Procedure
  algebra of complex numbers 54
  completing the square 84
  determinant of an $n \times n$ matrix 812
  finding the inverse of a matrix 837
  formulas of inverse functions 291
  graphing rational functions 369
  inverse matrix method 839
  linear programming 856
  order of transformations 265
  polynomial long division 329
  solving elementary exponential equations 399
  solving polynomial inequalities
    sign-test method 320
  solving radical equations 107
  solving rational inequalities 373
    sign-test method 373
  synthetic division 334
Product
  dot 677
  inner 677
  outer 677
  scalar 677
Product-to-sum identities 582
Programming, linear 856
Projections of vectors 681, 683
Proof by mathematical induction
  basic step 929
  induction step 929
Properties
  of absolute value 11
  associative 12
  behavior of exponential functions 397
  behavior of polynomials 314
  cancellation 13
    for inequalities 72
  closure 12
  commutative 12
  of determinants 813
  distributive 12
  of dot products 678
  of exponents 21
  field 12
  geometric meaning of multiplicity 355

identity 12
inverse 12
of logarithms 429, 441
of natural logarithms 424
of radicals 29
of scalar multiplication of vectors 667
of sigma notation 896
of vector addition 667
zero-factor 13
Ptolomy 552
Pythagoras 2
Pythagorean identities 554
Pythagorean Theorem 27, 141, 465

Q

Quadrants 136
Quadratic
  equation(s) 81
    double root 86
    perfect square 83
  formula 86
  function 227
    vertex form 230
Quadratic-like equations 91
Quadrilateral 171
Quotient 327
Quotient identities 501, 553

R

rad 461
Radian measure 461
Radian(s) 461
  converting to degrees 463
Radical(s)
  equations 106
    solving 106, 107
  expression 28
    conjugate 30
  like 31
  notation 26
  properties of 29
  sign 31
  simplified form 28
Radicand 26, 28
Radioactive decay 406
Radius
  of a circle 189
Range 209, 218

Rational
  equations
    solving 103
  expressions 98
    combining 99
    complex 101
    simplifying 98
  function(s) 363
    graphing 369, 370
    inequalities 372, 375
    solving 372
  number 3
  number exponents 32
  zero theorem 340, 359
Rationalizing denominators 30
Real axis 652
Real number line
  distance on 10
Real numbers 3
  types 3
Real part 652
Real zeros 320
Reciprocal identities 500
Rectangular
  array 795
  coordinate system 136
Recursive
  formula 892
  graphs 281
Reduced row echelon form 802
Reference angle 496
Reflecting
  with respect to the axes 263
Region
  feasible 854
  of constraint 854
Relation 209
  domain 209
  inverse 287
  range 209
Remainder 327
  theorem 328
Restriction of domain 291
Richter, Charles 435
Richter scale 435
Right triangle(s)
  solving 483
Roots
  cube 26
  double 82
  of complex numbers 660
  of polynomials 311

square 27, 57
  of negative numbers 52
Roses 634
Rotation
  formulas 743
  invariants 749
  of conics
    elimination of the *xy* term 748
  relations 745
Row echelon form 797
Row operations for matrices 799
Rule of signs 342, 343, 359

S

Sample space 955
Scalar multiplication 825, 826
  of vectors, properties of 667
Scalar product 677
Scalars 665
Scientific notation 23, 24
Secant 478
Second-degree
  equations 81
  functions 227
Sector area formula 470
Sector of a circle 470
Sequence(s) 891
  arithmetic 906
    common difference 906
    general term 907
    partial sums 910
  on a calculator 901
  explicit formula 892
  explicitly defined 891
  Fibonacci 900
  finite 891
  geometric 915
    common ratio 916
    general term 916
    partial sums 918
  infinite 891
  partial sums 898
  recursive formula 892
  recursively defined 891
  sigma notation 895
    properties of 896
  summation formulas 896
  summation notation 895
    index of summation 895
    sigma 895
  terms of 891

Series 898
  arithmetic 906
  on a calculator 901
  converging 899
  diverging 899
  finite 898
  geometric 921
    sum of an infinite 922
  infinite 898
Set-builder notation 6
Set(s)
  cardinality of a union of 959
  empty 6
  intersection of 8
  union of 8
Shifting
  horizontal 260
  vertical 261
Sigma (Σ) 895
  notation 895
    properties of 896
Sign-test method
  solving polynomial inequalities
    320
  solving rational inequalities 373
Similar terms 40
Simple harmonic motion (SHM)
  514
Simple interest
  formula 66
Simplified radical form 28
Simplifying
  complex rational expressions
    101, 102
  rational expressions 98
Simultaneous linear equations 781
Sine
  amplitude of 514, 518
Sine formula
  for the area of a triangle 618
Sines
  law of 610
    use of 610
Sinusoidal 514
Size of a matrix 795
Slope-intercept form 161, 165,
  225
Slope(s) 158, 165
  of horizontal lines 160
  of parallel lines 169
  of perpendicular lines 172, 173
  of vertical lines 160

of zero 160
  undefined 159
Solution(s)
  double 82
  extraneous 65
  of equations 60
  of polynomials 311
Solution set 60
Solving
  absolute value equations 63
  compound linear inequalities 74
  equations
    by factoring 92
    with positive rational exponents
      109
  exponential equations 399
  for a variable 65
    compound inequality 74
  linear inequalities 71
    in two variables 180
    joined by "and" or "or" 182
  linear systems
    inverse matrix method 839
    using Cramer's rule 815
  logarithmic equations 423, 443
  nonlinear systems
    algebraically 865
    graphically 864
  polynomial inequalities 318
    sign-test method 320
  quadratic equations
    by completing the square 83
    by factoring 82
  quadratic-like equations 91
  radical equations 106
  rational equations 103
  rational inequalities 372
    sign-test method 373
  right triangles 483
  systems
    by elimination 784
    by substitution 782
    on a calculator 790
Special binomials
  difference of two cubes 44
  difference of two squares 44
  sum of two cubes 44
Special product formulas 42
Speed
  angular 468
  linear 468
Spirals 634
Square root method 345

Square roots 30
  of negative numbers 52
  principal 57
Standard form
  circle 189
  ellipse 707, 710
  hyperbola 731, 735
  parabola 721, 724
  system of linear equations 796
  two-variable linear equation 149, 165
Standard position of an angle 462
Stretching 264
Strict inequaltiy 5
Substitution 485
  matrix 826
  quadratic-like equations 91
  to solve systems of equations 782
  trigonometric 560
Subtend 461
Subtraction
  of complex numbers 54
  of functions 274
  of matrices 826
  of polynomials 40
Such that symbol 6
Sum
  of an infinite geometric series 922
  of two cubes 44
Sum and difference identities
  for cosine 563
  for sine 563
  for tangent 563
Summary
  of logarithmic properties 441
Summation
  formulas 896
  notation 895
    index of summation 895
    properties of 896
    sigma 895
Sum-to-product identities 584
Surface area
  of a box 25
  of a right circular cylinder 25
Surjective 219
Symbol(s)
  inequality 5
    greater than 5
    greater than or equal to 5
    less than 5
    less than or equal to 5

such that 6
Symmetry
  of equations 269
  line 633
  origin 268
  polar axis 633
  of polar equations 633
  pole 633
  x-axis 269
  y-axis 267
Synthetic division 332, 334, 359
  with complex numbers 335
System
  of two equations 485
System(s) of equations
  consistent 782
  dependent 782
  inconsistent 782
  independent 782
  linear 781
  solving
    by elimination 784
    by substitution 782
  standard form 796

T

Tangent 478
Terminal point 665
Terminal side 462
Term(s) 18, 39
  constant 39
  degree of 39
  of sequences 891
Test point 180, 320, 869
Theorem
  binomial 946
  conjugate roots 356, 844
  DeMoivre's 659
  dot product 679
  fundamental theorem of algebra 351, 358, 844
  intermediate value 346, 359
  linear factors 351
  Pythagorean 465
  rational zero 340, 359
  remainder 328
  upper and lower bounds 344, 346
Theta (θ) 461
Translating inequality phrases 77

Translation
  horizontal shifting 260
  vertical shifting 261
Triangle inequality 11
Triangle(s)
  area of 617
    Heron's formula for 618
    sine formula for 618
  oblique 609
Trigonometric equation(s)
  applying algebraic techniques to 587, 593
Trigonometric form
  of complex numbers 655
Trigonometric function(s)
  applications of 483
  domain of 492–496
  evaluating 480
    using reference angles 496
  graphing 506–509
  graphing transformations of 513
  inverse 524
  relationships between 499
  signs of 498
Trigonometric identities 553
  verifying 557
    guidelines for 557
Trigonometric substitutions 560
Trinomial 39
  factoring 45, 46
  perfect square 46
Turning point of a graph 352
Types of equations 60, 467

U

Unbounded feasible region 856
Undefined slope 159
Union(s) 8, 959
  of mutually exclusive events 961
    cardinality 959
  of sets
    cardinality 959
  of two events, probability 960
  of two sets 182
Unit circle 461, 462
Unit vector 670
Upper and lower bounds rule 359
Upper bounds of zeros 344

V

Value
  absolute 653
Variable(s) 11
  dependent 216
  factor 11
  independent 216
Variation
  constant of proportionality 251
  direct 251
  inverse 252
  joint 254
Varieties of linear systems 782
Vector(s) 665
  addition 666
    properties of 667
  component form of 668, 672
  equal 666
  length of 666
  linear combinations of 670
  magnitude of 666
  norm of 666
  operations using components
      669
  orthogonal 680
  perpendicular 680
  projections of 681, 683
  unit 670
  zero 667
Velocity
  horizontal 641
  vertical 641
Venn diagram 8
Verifying trigonometric identities
    557
  guidelines for 557
Vertex
  formula 232
  of a feasible region 856
  of a parabola 228, 232
  of a quadratic function 232
Vertex form
  of a quadratic function 230

Vertical
  asymptote(s) 364, 366, 367
    equations 366
  axis 652
  compressing 264
  line 152
  line test 214
  shifting 261
  stretching 264
  velocity 641
Viéte, François 552
Volume
  of a right trapezoidal cylinder 26
  of a sphere 26

W

Whole numbers 3
Work 684
Work-rate problems 104

X

x-axis 138, 151
  symmetry 269, 270
x-coordinate 138
x-intercept(s) 151, 315

Y

y-axis 138, 151
  symmetry 267
y-coordinate 138
y-intercept(s) 151, 165, 315

Z

Zero-factor property 13, 91, 231,
    315
Zero of multiplicity k 353
Zero(s)
  as an exponent 20
  of a polynomial 311, 328
Zero vector 667

VECTORS IN THE CARTESIAN PLANE

Magnitude of a Vector

The length of a vector is called its magnitude and is denoted $\|\mathbf{u}\|$.

$$\|\mathbf{u}\| = \sqrt{a^2 + b^2}$$

where a is the horizotal displacement and b is the vertical displacement of the vector.

Properties of Vector Addition and Scalar Multiplication

For vectors $\mathbf{u}$, $\mathbf{v}$, and $\mathbf{w}$ and scalars a and b:

Vector Addition	Scalar Multiplication
$\mathbf{u} + \mathbf{v} = \mathbf{v} + \mathbf{u}$	$a(\mathbf{u} + \mathbf{v}) = a\mathbf{u} + a\mathbf{v}$
$\mathbf{u} + (\mathbf{v} + \mathbf{w}) = (\mathbf{u} + \mathbf{v}) + \mathbf{w}$	$(a + b)\mathbf{u} = a\mathbf{u} + b\mathbf{u}$
$\mathbf{u} + \mathbf{0} = \mathbf{u}$	$(ab)\mathbf{u} = a(b\mathbf{u}) = b(a\mathbf{u})$
$\mathbf{u} + (-\mathbf{u}) = \mathbf{0}$	$1\mathbf{u} = \mathbf{u}$, $0\mathbf{u} = \mathbf{0}$, and $a\mathbf{0} = \mathbf{0}$
	$\|a\mathbf{u}\| = \|a\|\|\mathbf{u}\|$

Vector Operations Using Components

Given two vectors $\mathbf{u} = \langle u_1, u_2 \rangle$ and $\mathbf{v} = \langle v_1, v_2 \rangle$ and a scalar a,

$\mathbf{u} + \mathbf{v} = \langle u_1 + v_1, u_2 + v_2 \rangle$, $a\mathbf{u} = \langle au_1, au_2 \rangle$, and $\|\mathbf{u}\| = \sqrt{u_1^2 + u_2^2}$.

Dot Product

Given two vectors $\mathbf{u} = \langle u_1, u_2 \rangle$ and $\mathbf{v} = \langle v_1, v_2 \rangle$, the dot product $\mathbf{u} \cdot \mathbf{v}$ of the two vectors is the scalar defined by

$$\mathbf{u} \cdot \mathbf{v} = u_1 v_1 + u_2 v_2.$$

Elementary Properties of the Dot Product

Given two vectors $\mathbf{u} = \langle u_1, u_2 \rangle$ and $\mathbf{v} = \langle v_1, v_2 \rangle$ and a scalar a:

$\mathbf{u} \cdot \mathbf{v} = \mathbf{v} \cdot \mathbf{u}$ $\mathbf{0} \cdot \mathbf{u} = 0$

$\mathbf{u} \cdot (\mathbf{v} + \mathbf{w}) = \mathbf{u} \cdot \mathbf{v} + \mathbf{u} \cdot \mathbf{w}$ $a(\mathbf{u} \cdot \mathbf{v}) = (a\mathbf{u}) \cdot \mathbf{v} = \mathbf{u} \cdot (a\mathbf{v})$

$\mathbf{u} \cdot \mathbf{u} = \|\mathbf{u}\|^2$

The Dot Product Theorem

Let two nonzero vectors $\mathbf{u}$ and $\mathbf{v}$ be depicted so that their initial points coincide, and let θ be the smaller of the two angles formed by $\mathbf{u}$ and $\mathbf{v}$ (so $0 \le \theta \le \pi$). Then

$$\mathbf{u} \cdot \mathbf{v} = \|\mathbf{u}\|\|\mathbf{v}\|\cos\theta.$$

Orthogonal Vectors

Two nonzero vectors $\mathbf{u}$ and $\mathbf{v}$ are said to be orthogonal (or perpendicular) if $\mathbf{u} \cdot \mathbf{v} = 0$.

Projection of u onto v

Let $\mathbf{u}$ and $\mathbf{v}$ be nonzero vectors. The projection of $\mathbf{u}$ onto $\mathbf{v}$ is the vector

$$\text{proj}_{\mathbf{v}}\mathbf{u} = \left(\frac{\mathbf{u} \cdot \mathbf{v}}{\|\mathbf{v}\|^2} \right)\mathbf{v}.$$

ROTATION OF AXES

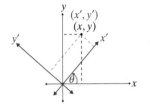

ROTATION RELATIONS

$x = x'\cos\theta - y'\sin\theta$ $x' = x\cos\theta + y\sin\theta$

$y = x'\sin\theta + y'\cos\theta$ $y' = -x\sin\theta + y\cos\theta$

ELIMINATION OF THE xy TERM

The graph of the equation $Ax^2 + Bxy + Cy^2 + Dx + Ey + F = 0$ in the xy-plane is the same as the graph of the equation $A'x'^2 + C'y'^2 + D'x' + E'y' + F' = 0$ in the $x'y'$-plane, where the angle of rotation θ between the two coordinate systems satisfies $\cot 2\theta = \dfrac{A - C}{B}$.

CLASSIFYING CONICS

Assuming the graph of the equation $Ax^2 + Bxy + Cy^2 + Dx + Ey + F = 0$ is a nondegenerate conic section, it is classified by its discriminant as follows:

Ellipse if $B^2 - 4AC < 0$

Parabola if $B^2 - 4AC = 0$

Hyperbola if $B^2 - 4AC > 0$

POLAR FORM OF CONIC SECTIONS

A conic section consists of all points P in the plane that satisfy the equation

$$\frac{D(P, F)}{D(P, L)} = e,$$

where e is a fixed positive constant. The conic is an ellipse if $0 < e < 1$, a parabola if $e = 1$, and a hyperbola if $e > 1$.

Conic Section Equation	Directrix
$r = \dfrac{ed}{1 + e\cos\theta}$	Vertical Directrix $x = d$
$r = \dfrac{ed}{1 - e\cos\theta}$	Vertical Directrix $x = -d$
$r = \dfrac{ed}{1 + e\sin\theta}$	Horizontal Directrix $y = d$
$r = \dfrac{ed}{1 - e\sin\theta}$	Horizontal Directrix $y = -d$

RADIAN AND DEGREE MEASURE

$180° = \pi \text{ rad}$

$1° = \dfrac{\pi}{180} \text{ rad}$ $\qquad \left(\dfrac{180}{\pi}\right)^{\circ} = 1 \text{ rad}$

$x° = x\left(\dfrac{\pi}{180}\right) \text{ rad} \quad x \text{ rad} = x\left(\dfrac{180}{\pi}\right)^{\circ}$

$s = \left(\dfrac{\theta}{2\pi}\right)(2\pi r) = r\theta$ $\qquad A = \left(\dfrac{\theta}{2\pi}\right)(\pi r^2) = \dfrac{r^2\theta}{2}$

$\omega = \dfrac{\theta}{t}$ $\qquad v = \dfrac{s}{t} = \dfrac{r\theta}{t} = r\omega$

TRIGONOMETRIC FUNCTIONS OF ACUTE ANGLES

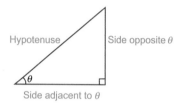

$\sin\theta = \dfrac{\text{opp}}{\text{hyp}}$ $\qquad \csc\theta = \dfrac{1}{\sin\theta} = \dfrac{\text{hyp}}{\text{opp}}$

$\cos\theta = \dfrac{\text{adj}}{\text{hyp}}$ $\qquad \sec\theta = \dfrac{1}{\cos\theta} = \dfrac{\text{hyp}}{\text{adj}}$

$\tan\theta = \dfrac{\text{opp}}{\text{adj}}$ $\qquad \cot\theta = \dfrac{1}{\tan\theta} = \dfrac{\text{adj}}{\text{opp}}$

TRIGONOMETRIC FUNCTIONS OF ANY ANGLE

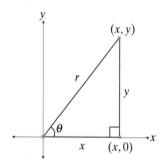

$\sin\theta = \dfrac{y}{r}$ $\qquad \csc\theta = \dfrac{r}{y} \text{ (for } y \neq 0)$

$\cos\theta = \dfrac{x}{r}$ $\qquad \sec\theta = \dfrac{r}{x} \text{ (for } x \neq 0)$

$\tan\theta = \dfrac{y}{x} \text{ (for } x \neq 0)$ $\qquad \cot\theta = \dfrac{x}{y} \text{ (for } y \neq 0)$

RECIPROCAL IDENTITIES

$\csc x = \dfrac{1}{\sin x}$ $\qquad \sec x = \dfrac{1}{\cos x}$ $\qquad \cot x = \dfrac{1}{\tan x}$

$\sin x = \dfrac{1}{\csc x}$ $\qquad \cos x = \dfrac{1}{\sec x}$ $\qquad \tan x = \dfrac{1}{\cot x}$

COFUNCTION IDENTITIES

$\cos x = \sin\left(\dfrac{\pi}{2} - x\right)$ $\qquad \sin x = \cos\left(\dfrac{\pi}{2} - x\right)$

$\csc x = \sec\left(\dfrac{\pi}{2} - x\right)$ $\qquad \sec x = \csc\left(\dfrac{\pi}{2} - x\right)$

$\cot x = \tan\left(\dfrac{\pi}{2} - x\right)$ $\qquad \tan x = \cot\left(\dfrac{\pi}{2} - x\right)$

QUOTIENT IDENTITIES

$\tan x = \dfrac{\sin x}{\cos x}$ $\qquad \cot x = \dfrac{\cos x}{\sin x}$

PERIOD IDENTITIES

$\sin(x + 2\pi) = \sin x$ $\qquad \csc(x + 2\pi) = \csc x$

$\cos(x + 2\pi) = \cos x$ $\qquad \sec(x + 2\pi) = \sec x$

$\tan(x + \pi) = \tan x$ $\qquad \cot(x + \pi) = \cot x$

EVEN/ODD IDENTITIES

$\sin(-x) = -\sin x$ $\qquad \cos(-x) = \cos x$ $\qquad \tan(-x) = -\tan x$

$\csc(-x) = -\csc x$ $\qquad \sec(-x) = \sec x$ $\qquad \cot(-x) = -\cot x$

AMPLITUDE, PERIOD, AND PHASE SHIFT

Given constants a, b (such that $b > 0$), and c, the functions $f(x) = a\sin(bx - c)$ and $g(x) = a\cos(bx - c)$ have amplitude $|a|$, period $\dfrac{2\pi}{b}$, and a phase shift of $\dfrac{c}{b}$. The left endpoint of one cycle of either function is $\dfrac{c}{b}$ and the right endpoint is $\dfrac{c}{b} + \dfrac{2\pi}{b}$.

PYTHAGOREAN IDENTITIES

$\sin^2 x + \cos^2 x = 1 \quad \tan^2 x + 1 = \sec^2 x \quad 1 + \cot^2 x = \csc^2 x$

COMMONLY ENCOUNTERED ANGLES

x	Radians	$\sin x$	$\cos x$	$\tan x$
0°	0	0	1	0
30°	$\dfrac{\pi}{6}$	$\dfrac{1}{2}$	$\dfrac{\sqrt{3}}{2}$	$\dfrac{1}{\sqrt{3}}$
45°	$\dfrac{\pi}{4}$	$\dfrac{1}{\sqrt{2}}$	$\dfrac{1}{\sqrt{2}}$	1
60°	$\dfrac{\pi}{3}$	$\dfrac{\sqrt{3}}{2}$	$\dfrac{1}{2}$	$\sqrt{3}$
90°	$\dfrac{\pi}{2}$	1	0	—
180°	π	0	−1	0
270°	$\dfrac{3\pi}{2}$	−1	0	—